D0990130

Methods in Enzymology

Volume 167

CYANOBACTERIA

METHODS IN ENZYMOLOGY

EDITORS-IN-CHIEF

John N. Abelson   Melvin I. Simon

Methods in Enzymology

Volume 167

Cyanobacteria

EDITED BY

Lester Packer

DEPARTMENT OF PHYSIOLOGY AND ANATOMY
UNIVERSITY OF CALIFORNIA, BERKELEY
BERKELEY, CALIFORNIA

Alexander N. Glazer

DEPARTMENT OF MICROBIOLOGY AND IMMUNOLOGY
UNIVERSITY OF CALIFORNIA, BERKELEY
BERKELEY, CALIFORNIA

ACADEMIC PRESS, INC.
Harcourt Brace Jovanovich, Publishers
San Diego   New York   Berkeley   Boston
London   Sydney   Tokyo   Toronto

ACADEMIC PRESS, INC.
San Diego, California 92101

United Kingdom Edition published by
ACADEMIC PRESS LIMITED
24-28 Oval Road, London NW1 7DX

LIBRARY OF CONGRESS CATALOG CARD NUMBER: 54-9110

ISBN 0-12-182068-8 (alk. paper)

PRINTED IN THE UNITED STATES OF AMERICA
88 89 90 91 9 8 7 6 5 4 3 2 1

Table of Contents

Section I. Isolation, Identification, and Culturing of Cyanobacteria

Section II. Ultrastructure, Inclusions, and Differentiation

Section III. Membranes, Pigments, Redox Reactions, and Nitrogen Fixation

Section IV. Physiology and Metabolism

Section V. General Physical Methods

Section VI. Molecular Genetics

Addendum

Contributors to Volume 167

Article numbers are in parentheses following the names of contributors.
Affiliations listed are current.

ALASTAIR AITKEN (13), *Laboratory of Protein Structure, National Institute for Medical Research, London NW7 1AA, England*

MARY MENNES ALLEN (19), *Department of Biological Sciences, Wellesley College, Wellesley, Massachusetts 02181*

HELMAR ALMON (47), *Lehrstuhl für Physiologie und Biochemie der Pflanzen, Universität Konstanz, D-7750 Konstanz, Federal Republic of Germany*

BERNARD ARRIO (77), *Bioénergétique Membranaire, C.N.R.S., Institut Biochimie, Faculté des Sciences d'Orsay, 91405 Orsay Cedex, France*

YESHAYA BAR-OR (68), *Division of Microbial and Molecular Ecology, Institute of Life Sciences, The Hebrew University of Jerusalem, 91904 Jerusalem, Israel*

JAMES A. BASSHAM (58), *Chemical Biodynamics Division, Lawrence Berkeley Laboratory, Berkeley, California 94720*

DAVID W. BECKER (29), *Department of Biology, Pomona College, Claremont, California 91711*

SHIMSHON BELKIN (40, 74, 75), *Laboratory for Environmental Applied Microbiology, The Jacob Blaustein Desert Research Institute, Ben Gurion University, Sede Boqer, Israel*

DEREK S. BENDALL (28), *Department of Biochemistry, University of Cambridge, Cambridge CB2 1QW, England*

RICHARD BLIGNY (73), *Unité Associe du Centre Nationale de la Recherche Scientifique, Centre d'Etudes Nucleaires, 38041 Grenoble Cedex, France*

PETER BÖGER (47), *Lehrstuhl für Physiologie und Biochemie der Pflanzen, Universität Konstanz, D-7750 Konstanz, Federal Republic of Germany*

GEORGE BORBELY (64, 69), *Institute of Plant Physiology, Biological Research Center, H-6701 Szeged, Hungary*

HERMANN BOTHE (52), *Botanisches Institut, Universität Köln, D-5000 Köln 41, Federal Republic of Germany*

JANE M. BOWES (28), *Department of Biochemistry, University of Cambridge, Cambridge CB2 1QW, England*

JERRY J. BRAND (29), *Department of Botany, University of Texas, Austin, Texas 78713*

M. BROUERS (70), *University of Liege, 4000 Liege, Belgium*

DONALD A. BRYANT (84), *Department of Molecular and Cell Biology, Pennsylvania State University, University Park, Pennsylvania 16802*

BOB B. BUCHANAN (43), *Division of Molecular Plant Biology, University of California, Berkeley, California 94720*

R. CAMMACK (44), *Department of Biochemistry, Division of Biomolecular Sciences, King's College, London, London W8 7AH, England*

SHULI CAO (7), *Division of Biology and Living Resources, School of Marine and Atmospheric Science, University of Miami, Miami, Florida 33149*

DONALD E. CARLSON (43), *Division of Molecular Plant Biology, University of California, Berkeley, California 94720*

N. G. CARR (65), *Department of Biological Sciences, University of Warwick, Coventry CV4 7AL, England*

RICHARD W. CASTENHOLZ (3, 5), *Department of Biology, University of Oregon, Eugene, Oregon, 97403*

B. M. CHERESKIN (30), *The Waksman Institute of Microbiology, Rutgers University, Piscataway, New Jersey 08854*

J. D. CLEMENT-METRAL (30), *Laboratoire de Technologie Enzymatique, 60202 Compiegne Cedex, France*

RAYMOND P. COX (48), *Institute of Biochemistry, Odense University, DK-5230 Odense, Denmark*

ALISON L. COZENS (87), *Medical Research Council, Laboratory of Molecular Biology, Cambridge CB2 2QH, England*

NANCY A. CRAWFORD (43), *Division of Molecular Plant Biology, University of California, Berkeley, California 94720*

T. CZERNY (39), *Biophysical Chemistry Group, Institute of Physical Chemistry, University of Vienna, A-1090 Vienna, Austria*

THIERRY DAMERVAL (88), *Unité de Physiologie Microbienne, Institut Pasteur, 75724 Paris Cedex 15, France*

ALLAN J. DARLING (42), *MRC Virology Unit, Institute of Virology, Glasgow G11 5JR, Scotland*

BERNARD DIMON (76), *Association pour la Recherche en Bioénergie Solaire, Centre d'Etudes Nucléaire de Cadarache, 13108 Saint-Paul-lez-Durance, France*

MICHEL DROUX (43), *C.N.R.S.-RPE, Rhône-Poulenc Agronchimie, Centre de la Recherche la Dargoire, 69263 Lyon, France*

JEFF ELHAI (83), *MSU-DOE Plant Research Laboratory, Michigan State University, East Lansing, Michigan 48824*

KHALIL ELMORJANI (63), *Unité de Physiologie Microbienne, Département de Biochimie et Génétique Moléculaire, Institut Pasteur, 75724 Paris Cedex 15, France*

WALTER ERBER (45), *Biophysical Chemistry Group, Institute of Physical Chemistry, University of Vienna, A-1090 Vienna, Austria*

KATHARINE G. FIELD (12), *Department of Microbiology, Oregon State University, Corvallis, Oregon 97331*

CLAUDINE FRANCHE (88), *Department of Developmental Biology, Research School of Biological Sciences, Australian National University, Canberra City, A.C.T. 2601, Australia*

DEVORAH FRIEDBERG (82), *Department of Microbiology and Molecular Ecology, Institute of Life Sciences, The Hebrew University of Jerusalem, 91904 Jerusalem, Israel*

IAN V. FRY (46), *Membrane Bioenergetics Group, Applied Science Division, Lawrence Berkeley Laboratory, Berkeley, California 94720*

DIDIER GAMET (77), *Bioénergétique Membranaire, C.N.R.S., Institut Biochimie, Faculté des Sciences d'Orsay, 91405 Orsay Cedex, France*

PIERRE GANS (76), *Service de Radioagronomie, Département de Biologie, Centre d'Etudes Nucléaires de Cadarache, 13108 Saint-Paul-lez-Durance, France*

E. GANTT (30), *Department of Botany, University of Maryland, College Park, Maryland 20742*

JANE GIBSON (60), *Section of Biochemistry, Molecular and Cell Biology, Division of Biological Sciences, Cornell University, Ithaca, New York 14853*

ALEXANDER N. GLAZER (31, 32), *Department of Microbiology and Immunology, University of California, Berkeley, Berkeley, California 94720*

SUSAN S. GOLDEN (80), *Department of Biology, Texas A&M University, College Station, Texas 77843*

D. O. HALL (53, 70), *Department of Biology, King's College London, University of London, London W8 7AH, England*

P. K. HAYES (20), *Department of Botany, University of Bristol, Bristol BS8 1UG, England*

MICHAEL HERDMAN (21, 22, 63), *Département de Biochimie et Génétique Moléculaire, Unité de Physiologie Microbienne, Institut Pasteur, 75724 Paris Cedex 15, France*

DAVID B. HICKS (38), *Department of Biochemistry, Mt. Sinai School of Medicine, New York, New York 10029*

JEAN HOUMARD (34, 89), *Unité de Physiologie Microbienne, Département de Biochimie et Génétique Moléculaire, Institut Pasteur, 75724 Paris Cedex 15, France*

BENT B. JENSEN (48), *Department of Animal Physiology and Biochemistry, National Institute of Animal Science, Forsøgsanläg Foulum, DK-8833 Ørum-Sønderlyng, Denmark*

GEORGES JOHANNIN (77), *Bioénergétique Membranaire, C.N.R.S., Institut Biochimie, Faculté des Sciences d'Orsay, 91405 Orsay Cedex, France*

BO BARKER JØRGENSEN (71), *Department of Ecology and Genetics, University of Aarhus, Ny Munkegade, DK-8000 Aarhus C, Denmark*

FRANÇOISE JOSET (79, 81), *Unité Metabolisme, Energétique LCB, Faculté des Sciences Luminy, 13288 Marseille Cedex 9, France*

UWE-JOHANNES JÜRGENS (15), *Institut für Biologie II, Mikrobiologie, Albert-Ludwigs Universität, D-7800 Freiburg, Federal Republic of Germany*

FRIEDRICH JÜTTNER (36, 67), *Max-Planck-Institut für Limnologie, D-2320 Plön, Federal Republic of Germany*

TOIVO KALLAS (86), *Division of Molecular Plant Biology, University of California, Berkeley, California 94720*

AARON KAPLAN (57), *Department of Botany, The Hebrew University of Jerusalem, 91904 Jerusalem, Israel*

SAKAE KATOH (26, 27), *Department of Botany, Faculty of Science, University of Tokyo, Bunkyo-ku, Tokyo 113, Japan*

TETZUYA KATOH (33), *Department of Botany, Faculty of Science, Kyoto University, Kita-shirakawa, Kyoto 606, Japan*

SHUZO KUMAZAWA (50), *Institute of Oceanic Research and Development, Tokai University, Shimizu 424, Japan*

DAVID J. LANE (12), *Gene-Trak Systems, Framingham, Massachusetts 01701*

WOLFGANG LUDWIG (11), *Lehrstuhl für Mikrobiologie, Technische Universität München, D-8000 München, Federal Republic of Germany*

RICHARD MALKIN (37, 86), *Division of Molecular Plant Biology, University of California, Berkeley, California 94720*

YEHOUDA MARCUS (57), *Department of Botany, The Hebrew University of Jerusalem, 91904 Jerusalem, Israel*

JEAN-BAPTISTE MARTIN (73), *Unité Associe du Centre Nationale de la Recherche Scientifique, Centre d'Etudes Nucleaires, 38041 Grenoble Cedex, France*

HIROSHI MATSUBARA (41), *Department of Biology, Faculty of Science, Osaka University, Toyonaka, Osaka 560, Japan*

J. C. MEEKS (8), *Department of Microbiology, University of California, Davis, California 95616*

ROLF J. MEHLHORN (54, 74), *Applied Science Division, Lawrence Berkeley Laboratory, Berkeley, California 94720*

AKIRA MITSUI (7, 50), *Division of Biology and Living Resources, School of Marine and Atmospheric Science, University of Miami, Miami, Florida 33149*

VÉRONIQUE MOLITOR (45, 77), *Biophysical Chemistry Group, Institute of Physical Chemistry, University of Vienna, A-1090 Vienna, Austria*

JOHN E. MORE (59), *Blood Products Laboratory, National Blood Transfusion Service, Elstree, Borehamwood, Hertfordshire WD6 3BX, England*

NORIO MURATA (23, 24), *Department of Regulation Biology, National Institute for Basic Biology, Okazaki 444, Japan*

HELGA NAES (66), *Biotechnology, Norwegian Food Research Institute, Osloveien 1, N-1430 Ås, Norway*

GABRIELE NEUER (52), *CPC/Europe, D-7100 Heilbronn, Federal Republic of Germany*

W. H. NITSCHMANN (39), *Biophysical Chemistry Group, Institute of Physical Chemistry, University of Vienna, A-1090 Vienna, Austria*

R. L. OLIVER (55), *Murray-Darling Freshwater Research Centre, Albury, NSW 2640, Australia*

GARY J. OLSEN (12), *Department of Microbiology, University of Illinois, Urbana, Illinois 61801*

TATSUO OMATA (23), *Solar Energy Research Group, The Institute of Physical and Chemical Research (RIKEN), Wakoshi, Saitama 351-01, Japan*

NORMAN R. PACE (12), *Department of Biology, Indiana University, Bloomington, Indiana 47405*

LESTER PACKER (73, 75), *Department of Physiology and Anatomy, University of California, Berkeley, Berkeley, California 94720*

ETANA PADAN (40, 61), *Division of Microbial and Molecular Ecology, Institute of Life Sciences, The Hebrew University, Jerusalem 91904, Israel*

GEORGE C. PAPAGEORGIOU (25), *National Research Center of Demokritos, Institute of Biology, Athens 153 10, Greece*

GILLES PELTIER (76), *Service de Radioagronomie, Département de Biologie, Centre d'Etudes Nucléaires de Cadarache, 13108 Saint-Paul-lez-Durance, France*

GÜNTER A. PESCHEK (39, 45, 46, 77), *Institut für Physikalische Chemie, Universität Wien, 1090 Wien, Austria*

RONALD D. PORTER (78), *Department of Molecular and Cell Biology, Pennsylvania State University, University Park, Pennsylvania 16802*

K. K. RAO (53), *Department of Biology, King's College London, University of London, London W8 7HA, England*

ROBERT H. REED (56), *Department of Biological Sciences, University of Dundee, Dundee DD1 4HN, Scotland*

LEONORA REINHOLD (57), *Department of Botany, The Hebrew University of Jerusalem, 91904 Jerusalem, Israel*

NIELS PETER REVSBECH (71), *Institute of Ecology and Genetics, University of Aarhus, Ny Munkegade, DK-8000 Aarhus C, Denmark*

ROSMARIE RIPPKA (1, 2, 22), *Unité de Physiologie Microbienne, Département de Biochimie et Génétique Moléculaire, Institut Pasteur, 75724 Paris Cedex 15, France*

MARIE-EMMANUEL RIVIÈRE (77), *Bioénergétique Membranaire, C.N.R.S., Institute Biochimie, Faculté des Sciences d'Orsay, 91405 Orsay Cedex, France*

CLAUDE ROBY (73), *Unité Associe du Centre Nationale de la Recherche Scientifique, Centre d'Etudes Nucleaires, 38041 Grenoble Cedex, France*

PETER ROWELL (42), *AFRC Research Group on Cyanobacteria and Department of Biological Sciences, University of Dundee, Dundee DD1 4HN, Scotland*

GERHARD SANDMANN (35), *Lehrstuhl für Physiologie und Biochemie der Pflanzen, Universität Konstanz, D-7750 Konstanz, Federal Republic of Germany*

NAOKI SATO (24), *Department of Botany, Faculty of Science, University of Tokyo, Hongo, Bunkyo-ku, Tokyo 113, Japan*

KAZUHIKO SATOH (27), *Department of Botany, Faculty of Science, University of Tokyo, Bunkyo-ku, Tokyo 113, Japan*

D. J. SCANLAN (65), *Department of Biological Sciences, University of Warwick, Coventry CV4 7AL, England*

AHLERT SCHMIDT (62), *Botanisches Institut, Tierärztliche Hochschule, 3000 Hannover 70, Federal Republic of Germany*

GEORGE J. SCHNEIDER (64), *Department of Biochemistry and Molecular Biology, University of Chicago, Chicago, Illinois 60637*

YOSEPHA SHAHAK (40), *Biochemistry Department, Weizmann Institute of Science, Rehovot 76100, Israel*

D. J. SHI (70), *Institute of Botany, Academia Sinica, Beijing, China*

MOSHE SHILO (68), *Division of Microbial and Molecular Ecology, Institute of Life Sciences, The Hebrew University of Jerusalem, 91904 Jerusalem, Israel*

J. M. SHIVELY (17, 18), *Department of Biological Sciences, Clemson University, Clemson, South Carolina 29634*

ARNOLD J. SMITH (59), *Department of Biochemistry, University College of Wales, Aberystwyth, Dyfed SY23 3DD, Wales*

RUSSELL L. SMITH (51), *Department of Molecular and Cell Biology, Pennsylvania State University, University Park, Pennsylvania 16802*

SUSAN SPATH (73), *Membrane Bioenergetics Group, Lawrence Berkeley Laboratory and Department of Physiology–Anatomy, University of California, Berkeley, Berkeley, California 94720*

ERKO STACKEBRANDT (11), *Institut für Allgemeine Mikrobiologie, Christian-Albrechts-Universität, D-2300 Kiel, Federal Republic of Germany*

LUCAS J. STAL (49), *Laboratory for Microbiology, University of Amsterdam, 1018 WS Amsterdam, The Netherlands*

W. T. STAM (10), *Department of Marine Biology, Biological Centre, University of Groningen, NL-9750 AA Haren, The Netherlands*

G. STANIER (COHEN-BAZIRE) (14), *Unité de Physiologie Microbienne, Institut Pasteur, 75724 Paris Cedex 15, France*

S. EDWARD STEVENS, JR. (9), *Department of Molecular and Cell Biology, The Pennsylvania State University, University Park, Pennsylvania 16802*

ALISON C. STEWART (28), *Trends in Genetics, Elsevier Publications Cambridge, Cambridge CB2 1LA, England*

WILLIAM D. P. STEWART (42), *AFRC Research Group on Cyanobacteria and Department of Biological Sciences, University of Dundee, Dundee DD1 4HN, Scotland*

B. K. STULP (10), *Department of Genetics, Biological Centre, University of Groningen, NL-9750 AA Haren, The Netherlands*

RICHARD SULLIVAN (54), *Chemistry Department, Jackson State University, Jackson, Mississippi 39217*

GYULA SURANYI (69), *Institute of Plant Physiology, Biological Research Center, H-6701 Szeged, Hungary*

F. ROBERT TABITA (51), *Center for Applied Microbiology and Department of Microbiology, The University of Texas at Austin, Austin, Texas 78712*

NICOLE TANDEAU DE MARSAC (34, 84, 89), *Unité de Physiologie Microbienne, Département de Biochimie et Génétique Moléculaire, Institut Pasteur, 75724 Paris Cedex 15, France*

MARY E. TAYLOR (28), *Department of Biochemistry, University of Cambridge, Cambridge CB2 1QW, England*

SCOTT E. TAYLOR (58), *Chemical Biodynamics Division, Lawrence Berkeley Laboratory, Berkeley, California 94720*

MARIA TRNKA (45), *Biophysical Chemistry Group, Institute of Physical Chemistry, University of Vienna, A-1090 Vienna, Austria*

MARTTI VAARA (16), *Department of Bacteriology and Immunology, University of Helsinki, SF-00280 Helsinki, Finland*

TIMO VAARA (16), *Alko Ltd., Process and Product Development, SF-00101 Helsinki, Finland*

CHASE VAN BAALEN (51), *University of Texas Marine Sciences Institute, Port Aransas, Texas 78373*

ALEXANDER VITTERBO (61), *Division of Microbial and Molecular Ecology, Life Sciences Institute, The Hebrew University of Jerusalem, 91904 Jerusalem, Israel*

KEISHIRO WADA (41), *Department of Biology, Faculty of Science, Kanazawa University, Kanazawa 920, Japan*

JOHN E. WALKER (87), *Medical Research Council, Laboratory of Molecular Biology, Cambridge CB2 2QH, England*

A. E. WALSBY (55, 72), *Department of Bot-*

any, University of Bristol, Bristol BS8 1UG, England

MARNIK WASTYN (45), *Biophysical Chemistry Group, Institute of Physical Chemistry, University of Vienna, A-1090 Vienna, Austria*

JOHN B. WATERBURY (6), *Department of Biology, Woods Hole Oceanographic Institution, Woods Hole, Massachusetts 02543*

JÜRGEN WECKESSER (15), *Institut für Biologie II, Mikrobiologie, Albert-Ludwigs-Universität, D-7800 Freiburg, Federal Republic of Germany*

JOANNE M. WILLEY (6), *Department of Biology, Woods Hole Oceanographic Institution, Woods Hole, Massachusetts 02543*

JOHN G. K. WILLIAMS (85), *Central Research and Development, E.I. Du Pont de Nemours & Co., Inc., Wilmington, Delaware 19898*

C. PETER WOLK (4, 83), *MSU-DOE Plant Research Laboratory, Michigan State University, East Lansing, Michigan 48824*

BOIHON C. YEE (43), *Division of Molecular Plant Biology, University of California, Berkeley, Berkeley, California 94720*

Preface

The cyanobacteria can be defined as "microorganisms that harbor, within a typically prokaryotic cell, a photosynthetic apparatus similar in structure and function to that located in the chloroplast of phototrophic eukaryotes."[1] The origins of cyanobacteria date back almost to the beginning of life on earth. Fossils of the progenitors of these organisms are found in strata over three billion years old.[2] The genetic, metabolic, and morphological diversity of the cyanobacteria rivals that seen among the totality of other eubacteria. As simple bacteria, the cyanobacteria would seem to be the obvious organisms of choice for the study of such fundamental processes as oxygen-evolving photosynthesis and nitrogen fixation. Indeed, in recent years this group of organisms has become the focus of intense study. This development resulted from a fundamental change in perspective and from the availability of large collections of pure cultures of cyanobacteria, collections which include representatives of each of the major groups of these organisms.

The rapid growth of this field owes much to the leadership of Roger Stanier and his associates. On September 28, 1976, Roger Stanier delivered the Emil Chr. Hansen lecture on the occasion of the centennial of the Carlsberg Laboratory in Copenhagen. He said: "Fifteen years ago, when we began work on the cyanobacteria, these organisms were still construed as algae, and their study was largely conducted, as it always had been, by phycologists, with a heavy emphasis on descriptive taxonomy and ecology. It was commonly believed that the isolation of pure cultures is exceptionally difficult, even though M. B. Allen had made a promising start on this technical problem about 1950; indeed most of the pure strains then in existence had been isolated by her. Accordingly, much of our initial work was directed to the isolation and purification of additional members of this group, an essential prerequisite for comparative experimental studies. The news that these organisms are really bacteria has spread, thus encouraging other microbiologists to take up work on cyanobacteria. More significant information about their biological properties has emerged in the past decade than in the preceding century."[1]

We hope the contributions in this volume will prove useful to investigators with little previous experience with cyanobacteria as well as to those who wish to change the emphasis of their research. This volume, with its stress on methodology, does not provide a detailed survey of the

[1] R. Y. Stanier, *Carlsberg Res. Commun.* **42,** 77 (1977).

[2] J. W. Schopf, Ed. "Earth's Earliest Biosphere. Its Origin and Evolution." Princeton University Press, Princeton, New Jersey, 1983.

literature on cyanobacteria, but extensive bibliographies are available in two recent monographs.[3,4]

We are greatly indebted for the advice we have received from many of the leaders in this field, particularly from our volume advisors, Robert Haselkorn and Germaine Stanier (Cohen-Bazire). Among those who made valuable suggestions on the organization of the volume as a whole and on the various subsections are Germaine Stanier, Rosmarie Rippka, and Michael Herdman, as well as other members of the group studying cyanobacteria at the Institut Pasteur. Their counsel on numerous occasions (while one of us was on sabbatical leave in France) gave us the courage to pursue this project. We also benefited from advice from Françoise Joset on the organization of the volume, particularly on the molecular genetics section. We also thank Peter Boger, George Papageorgiou, Elisha Tel-Or, and Gunther Peschek for comments on the sections on physiology and physical methods. Finally, we would like to acknowledge the late Nathan O. Kaplan for his enthusiasm and support in launching this volume.

LESTER PACKER
ALEXANDER N. GLAZER

[3] N. G. Carr and B. A. Whitton, Eds. "The Biology of Cyanobacteria." University of California Press, Berkeley and Los Angeles, 1982.

[4] P. Fay and C. Van Baalen, Eds. "The Cyanobacteria." Elsevier Science Publishers, Amsterdam, The Netherlands, 1987.

METHODS IN ENZYMOLOGY

EDITED BY

Sidney P. Colowick and Nathan O. Kaplan

VANDERBILT UNIVERSITY
SCHOOL OF MEDICINE
NASHVILLE, TENNESSEE

DEPARTMENT OF CHEMISTRY
UNIVERSITY OF CALIFORNIA
AT SAN DIEGO
LA JOLLA, CALIFORNIA

I. Preparation and Assay of Enzymes
II. Preparation and Assay of Enzymes
III. Preparation and Assay of Substrates
IV. Special Techniques for the Enzymologist
V. Preparation and Assay of Enzymes
VI. Preparation and Assay of Enzymes (*Continued*)
Preparation and Assay of Substrates
Special Techniques
VII. Cumulative Subject Index

METHODS IN ENZYMOLOGY

EDITORS-IN-CHIEF

Sidney P. Colowick and Nathan O. Kaplan

VOLUME VIII. Complex Carbohydrates
Edited by ELIZABETH F. NEUFELD AND VICTOR GINSBURG

VOLUME IX. Carbohydrate Metabolism
Edited by WILLIS A. WOOD

VOLUME X. Oxidation and Phosphorylation
Edited by RONALD W. ESTABROOK AND MAYNARD E. PULLMAN

VOLUME XI. Enzyme Structure
Edited by C. H. W. HIRS

VOLUME XII. Nucleic Acids (Parts A and B)
Edited by LAWRENCE GROSSMAN AND KIVIE MOLDAVE

VOLUME XIII. Citric Acid Cycle
Edited by J. M. LOWENSTEIN

VOLUME XIV. Lipids
Edited by J. M. LOWENSTEIN

VOLUME XV. Steroids and Terpenoids
Edited by RAYMOND B. CLAYTON

VOLUME XVI. Fast Reactions
Edited by KENNETH KUSTIN

VOLUME XVII. Metabolism of Amino Acids and Amines (Parts A and B)
Edited by HERBERT TABOR AND CELIA WHITE TABOR

Volume XXXII. Biomembranes (Part B)
Edited by Sidney Fleischer and Lester Packer

Volume XXXIII. Cumulative Subject Index Volumes I–XXX
Edited by Martha G. Dennis and Edward A. Dennis

Volume XXXIV. Affinity Techniques (Enzyme Purification: Part B)
Edited by William B. Jakoby and Meir Wilchek

Volume XXXV. Lipids (Part B)
Edited by John M. Lowenstein

Volume XXXVI. Hormone Action (Part A: Steroid Hormones)
Edited by Bert W. O'Malley and Joel G. Hardman

Volume XXXVII. Hormone Action (Part B: Peptide Hormones)
Edited by Bert W. O'Malley and Joel G. Hardman

Volume XXXVIII. Hormone Action (Part C: Cyclic Nucleotides)
Edited by Joel G. Hardman and Bert W. O'Malley

Volume XXXIX. Hormone Action (Part D: Isolated Cells, Tissues, and Organ Systems)
Edited by Joel G. Hardman and Bert W. O'Malley

Volume XL. Hormone Action (Part E: Nuclear Structure and Function)
Edited by Bert W. O'Malley and Joel G. Hardman

Volume XLI. Carbohydrate Metabolism (Part B)
Edited by W. A. Wood

Volume XLII. Carbohydrate Metabolism (Part C)
Edited by W. A. Wood

Volume XLIII. Antibiotics
Edited by John H. Hash

Volume XLIV. Immobilized Enzymes
Edited by Klaus Mosbach

Volume XLV. Proteolytic Enzymes (Part B)
Edited by Laszlo Lorand

VOLUME XLVI. Affinity Labeling
Edited by WILLIAM B. JAKOBY AND MEIR WILCHEK

VOLUME XLVII. Enzyme Structure (Part E)
Edited by C. H. W. HIRS AND SERGE N. TIMASHEFF

VOLUME XLVIII. Enzyme Structure (Part F)
Edited by C. H. W. HIRS AND SERGE N. TIMASHEFF

VOLUME XLIX. Enzyme Structure (Part G)
Edited by C. H. W. HIRS AND SERGE N. TIMASHEFF

VOLUME L. Complex Carbohydrates (Part C)
Edited by VICTOR GINSBURG

VOLUME LI. Purine and Pyrimidine Nucleotide Metabolism
Edited by PATRICIA A. HOFFEE AND MARY ELLEN JONES

VOLUME LII. Biomembranes (Part C: Biological Oxidations)
Edited by SIDNEY FLEISCHER AND LESTER PACKER

VOLUME LIII. Biomembranes (Part D: Biological Oxidations)
Edited by SIDNEY FLEISCHER AND LESTER PACKER

VOLUME LIV. Biomembranes (Part E: Biological Oxidations)
Edited by SIDNEY FLEISCHER AND LESTER PACKER

VOLUME LV. Biomembranes (Part F: Bioenergetics)
Edited by SIDNEY FLEISCHER AND LESTER PACKER

VOLUME LVI. Biomembranes (Part G: Bioenergetics)
Edited by SIDNEY FLEISCHER AND LESTER PACKER

VOLUME LVII. Bioluminescence and Chemiluminescence
Edited by MARLENE A. DELUCA

VOLUME LVIII. Cell Culture
Edited by WILLIAM B. JAKOBY AND IRA PASTAN

VOLUME LIX. Nucleic Acids and Protein Synthesis (Part G)
Edited by KIVIE MOLDAVE AND LAWRENCE GROSSMAN

VOLUME 85. Structural and Contractile Proteins (Part B: The Contractile Apparatus and the Cytoskeleton)
Edited by DIXIE W. FREDERIKSEN AND LEON W. CUNNINGHAM

VOLUME 86. Prostaglandins and Arachidonate Metabolites
Edited by WILLIAM E. M. LANDS AND WILLIAM L. SMITH

VOLUME 87. Enzyme Kinetics and Mechanism (Part C: Intermediates, Stereochemistry, and Rate Studies)
Edited by DANIEL L. PURICH

VOLUME 88. Biomembranes (Part I: Visual Pigments and Purple Membranes, II)
Edited by LESTER PACKER

VOLUME 89. Carbohydrate Metabolism (Part D)
Edited by WILLIS A. WOOD

VOLUME 90. Carbohydrate Metabolism (Part E)
Edited by WILLIS A. WOOD

VOLUME 91. Enzyme Structure (Part I)
Edited by C. H. W. HIRS AND SERGE N. TIMASHEFF

VOLUME 92. Immunochemical Techniques (Part E: Monoclonal Antibodies and General Immunoassay Methods)
Edited by JOHN J. LANGONE AND HELEN VAN VUNAKIS

VOLUME 93. Immunochemical Techniques (Part F: Conventional Antibodies, Fc Receptors, and Cytotoxicity)
Edited by JOHN J. LANGONE AND HELEN VAN VUNAKIS

VOLUME 94. Polyamines
Edited by HERBERT TABOR AND CELIA WHITE TABOR

VOLUME 95. Cumulative Subject Index Volumes 61–74, 76–80
Edited by EDWARD A. DENNIS AND MARTHA G. DENNIS

VOLUME 96. Biomembranes [Part J: Membrane Biogenesis: Assembly and Targeting (General Methods; Eukaryotes)]
Edited by SIDNEY FLEISCHER AND BECCA FLEISCHER

VOLUME 110. Steroids and Isoprenoids (Part A)
Edited by JOHN H. LAW AND HANS C. RILLING

VOLUME 111. Steroids and Isoprenoids (Part B)
Edited by JOHN H. LAW AND HANS C. RILLING

VOLUME 112. Drug and Enzyme Targeting (Part A)
Edited by KENNETH J. WIDDER AND RALPH GREEN

VOLUME 113. Glutamate, Glutamine, Glutathione, and Related Compounds
Edited by ALTON MEISTER

VOLUME 114. Diffraction Methods for Biological Macromolecules (Part A)
Edited by HAROLD W. WYCKOFF, C. H. W. HIRS, AND SERGE N. TIMASHEFF

VOLUME 115. Diffraction Methods for Biological Macromolecules (Part B)
Edited by HAROLD W. WYCKOFF, C. H. W. HIRS, AND SERGE N. TIMASHEFF

VOLUME 116. Immunochemical Techniques (Part H: Effectors and Mediators of Lymphoid Cell Functions)
Edited by GIOVANNI DI SABATO, JOHN J. LANGONE, AND HELEN VAN VUNAKIS

VOLUME 117. Enzyme Structure (Part J)
Edited by C. H. W. HIRS AND SERGE N. TIMASHEFF

VOLUME 118. Plant Molecular Biology
Edited by ARTHUR WEISSBACH AND HERBERT WEISSBACH

VOLUME 119. Interferons (Part C)
Edited by SIDNEY PESTKA

VOLUME 120. Cumulative Subject Index Volumes 81–94, 96–101

VOLUME 121. Immunochemical Techniques (Part I: Hybridoma Technology and Monoclonal Antibodies)
Edited by JOHN J. LANGONE AND HELEN VAN VUNAKIS

VOLUME 122. Vitamins and Coenzymes (Part G)
Edited by FRANK CHYTIL AND DONALD B. MCCORMICK

VOLUME 123. Vitamins and Coenzymes (Part H)
Edited by FRANK CHYTIL AND DONALD B. MCCORMICK

VOLUME 124. Hormone Action (Part J: Neuroendocrine Peptides)
Edited by P. MICHAEL CONN

VOLUME 125. Biomembranes (Part M: Transport in Bacteria, Mitochondria, and Chloroplasts: General Approaches and Transport Systems)
Edited by SIDNEY FLEISCHER AND BECCA FLEISCHER

VOLUME 126. Biomembranes (Part N: Transport in Bacteria, Mitochondria, and Chloroplasts: Protonmotive Force)
Edited by SIDNEY FLEISCHER AND BECCA FLEISCHER

VOLUME 127. Biomembranes (Part O: Protons and Water: Structure and Translocation)
Edited by LESTER PACKER

VOLUME 128. Plasma Lipoproteins (Part A: Preparation, Structure, and Molecular Biology)
Edited by JERE P. SEGREST AND JOHN J. ALBERS

VOLUME 129. Plasma Lipoproteins (Part B: Characterization, Cell Biology, and Metabolism)
Edited by JOHN J. ALBERS AND JERE P. SEGREST

VOLUME 130. Enzyme Structure (Part K)
Edited by C. H. W. HIRS AND SERGE N. TIMASHEFF

VOLUME 131. Enzyme Structure (Part L)
Edited by C. H. W. HIRS AND SERGE N. TIMASHEFF

VOLUME 132. Immunochemical Techniques (Part J: Phagocytosis and Cell-Mediated Cytotoxicity)
Edited by GIOVANNI DI SABATO AND JOHANNES EVERSE

VOLUME 133. Bioluminescence and Chemiluminescence (Part B)
Edited by MARLENE DELUCA AND WILLIAM D. MCELROY

VOLUME 161. Biomass (Part B: Lignin, Pectin, and Chitin)
Edited by WILLIS A. WOOD AND SCOTT T. KELLOGG

VOLUME 162. Immunochemical Techniques (Part L: Chemotaxis and Inflammation)
Edited by GIOVANNI DI SABATO

VOLUME 163. Immunochemical Techniques (Part M: Chemotaxis and Inflammation)
Edited by GIOVANNI DI SABATO

VOLUME 164. Ribosomes (in preparation)
Edited by HARRY F. NOLLER, JR. AND KIVIE MOLDAVE

VOLUME 165. Microbial Toxins: Tools for Enzymology
Edited by SIDNEY HARSHMAN

VOLUME 166. Branched-Chain Amino Acids
Edited by ROBERT HARRIS AND JOHN R. SOKATCH

VOLUME 167. Cyanobacteria
Edited by LESTER PACKER AND ALEXANDER N. GLAZER

VOLUME 168. Hormone Action (Part K: Neuroendocrine Peptides) (in preparation)
Edited by P. MICHAEL CONN

VOLUME 169. Platelets: Receptors, Adhesion, Secretion (Part A) (in preparation)
Edited by JACEK HAWIGER

VOLUME 170. Nucleosomes (in preparation)
Edited by PAUL M. WASSARMAN AND ROGER D. KORNBERG

VOLUME 171. Biomembranes (Part R: Transport Theory: Cells and Model Membranes) (in preparation)
Edited by SIDNEY FLEISCHER AND BECCA FLEISCHER

VOLUME 172. Biomembranes (Part S: Transport: Membrane Isolation and Characterization) (in preparation)
Edited by SIDNEY FLEISCHER AND BECCA FLEISCHER

Section I

Isolation, Identification, and Culturing of Cyanobacteria

[1] Isolation and Purification of Cyanobacteria

By ROSMARIE RIPPKA

Introduction

Thanks to the pioneering work of Gerloff *et al.*,[1] M. B. Allen,[2] Kratz and Myers,[3] Hughes *et al.*,[4] Van Baalen,[5] and M. M. Allen[6] in the early 1950s to late 1960s, the myth that cyanobacteria are difficult or even impossible to isolate was partly proven to be unjustified. Later disciples in the field of cyanobacterial culture work, many of whom came from the school of the late R. Y. Stanier, exploited the discoveries of these successful investigators, with the result that there are at present a good 500 axenic cyanobacteria in the various major and minor culture collections spread over different countries, the isolates originating from a wide range of habitats.

Methods for the isolation and purification of cyanobacteria have been reviewed by Castenholz[7] for thermophiles, by Walsby[8] for planktonic species, by Waterbury and Stanier[9] for organisms from marine and hypersaline environments, and by Rippka *et al.*[10] for strains of diverse origin. This chapter summarizes some of the information provided in these four review articles but is mainly an account based on personal experience. Some of the technical details described might seem unnecessary, since they are "standard knowledge" for a bacteriologist. They have been in-

[1] G. C. Gerloff, G. P. Fitzgerald, and F. Skoog, *in* "The Culturing of Algae" (F. Brunel, G. W. Prescott, and L. H. Tiffany, eds.), p. 27. Kettering Found., Yellow Springs, Ohio, 1950.

[2] M. B. Allen, *Arch. Mikrobiol.* **17,** 34 (1952).

[3] W. A. Kratz and J. Myers, *Am. J. Bot.* **42,** 282 (1955).

[4] E. O. Hughes, P. R. Gorham, and A. Zehnder, *Can. J. Microbiol.* **4,** 225 (1958).

[5] C. Van Baalen, *Bot. Mar.* **4,** 129 (1962).

[6] M. M. Allen, *J. Phycol.* **4,** 1 (1968).

[7] R. W. Castenholz, *in* "The Prokaryotes" (M. P. Starr, H. Stolp, H. G. Trüper, A. Balows, and H. G. Schlegel, eds.), Vol. 1, p. 236. Springer-Verlag, Berlin, 1981.

[8] A. E. Walsby, *in* "The Prokaryotes" (M. P. Starr, H. Stolp, H. G. Trüper, A. Balows, and H. G. Schlegel, eds.), Vol. 1, p. 224. Springer-Verlag, Berlin, 1981.

[9] J. B. Waterbury and R. Y. Stanier, *in* "The Prokaryotes" (M. P. Starr, H. Stolp, H. G. Trüper, A. Balows, and H. G. Schlegel, eds.), Vol. 1, p. 221. Springer-Verlag, Berlin, 1981.

[10] R. Rippka, J. B. Waterbury, and R. Y. Stanier, *in* "The Prokaryotes" (M. P. Starr, H. Stolp, H. G. Trüper, A. Balows, and H. G. Schlegel, eds.), Vol. 1, p. 212. Springer-Verlag, Berlin, 1981.

cluded with such precision for those who wish to isolate cyanobacteria but who may not be sufficiently familiar with these techniques to master the isolation and purification of members of this bacterial group.

Sampling and Storage of Cyanobacterial Crude Material

Sampling Techniques

Cyanobacterial populations recognized in their natural habitat[11] should be sampled with sterile instruments and placed in sterile containers in order to ensure the origin of eventual isolates. If funds permit, commercial sterile disposable scalpels, pipets, and plastic tubes are very convenient for this purpose. Only small quantities (a pea-size equivalent generally being ample) are required from habitats where macroscopic growth is visible. Wide-mouthed (2–4 cm in diameter) screw-cap tubes (10–20 ml capacity), containing a few drops of sterile water to keep the interior moist, are ideal to accommodate soil and rock scrapings (or even small chippings if the natural substrate is very hard); such samples should be stored (until further treatment in the laboratory) without additional water or media, to avoid an unnecessary increase and spreading of bacterial contaminants. The same tubes are equally useful for samples from aquatic habitats; in this case, the crude material removed should be supplemented with water from the habitat to fill the sampling tubes almost completely.[7] This will ensure that the samples are well submerged by the aqueous phase typical of their natural environment, but should also minimize the growth of aerobic bacteria during transport of the samples to the laboratory by limiting their supply of oxygen (i.e., in tubes with little air space above the liquid). This generally does not harm cyanobacteria, but increases considerably the chance of rapid purification. In addition, major changes of pH and nutrient composition resulting from bacterial metabolic activities, potentially harmful for the cyanobacterial population, will be minimized.

Sampling of endosymbiotic cyanobacteria from coralloid nodules of Cycadaceae or the stems of *Gunnera*[11] can be performed as described for soil and rock-borne cyanobacteria, but other host–cyanobacteria associations might require more special treatments. To isolate cyanobacteria from lakes and ponds in which cyanobacterial growth is not visible with the eye (or even after examination with a portable microscope) it is advisable to take larger samples: 250- to 500-ml sterile screw-cap centrifuge pots, filled almost completely with sampling water, are convenient containers for transport and allow immediate concentration (by centrifuga-

[11] R. Rippka, this volume [2].

tion) on arrival in the laboratory, the sampling volume generally being sufficient to isolate cyanobacteria present even in only low numbers.

The origin of many cyanobacteria presently in culture is poorly characterized since little more is known about their habitat than that they were derived from a soil sample, freshwater, or marine environment, which is rather restricted information (although better than source unknown, another not uncommon description). In order to improve the ecological characterization of new isolates it is advisable to make notes as thoroughly as possible (preferably in a log book) of the habitat, in addition to describing the samples with respect to color, mass appearance, etc. Precise information on the geographical location, substrate, pH, temperature, and light intensity will help to relocate the habitat for further samplings, if necessary, and is also essential for choosing the appropriate culture conditions.

Storage of Samples

Until their arrival in the laboratory the samples should be kept in very dim light or in the dark (the latter being generally more convenient, but see the cyano suitcase[12]) at room (20–25°) or lower temperature (~15°) for cyanobacteria from cold habitats (i.e., mountain lakes and rivers, alpine soil and rocks, certain marine environments). Storage at 4° or freezing should, however, be avoided since this may cause rapid lysis[7] and death.[13] Storage of thermophilic cyanobacteria at room temperature does not seem to harm their viability.[7] The chances for successful isolation and purification are best the sooner the samples are treated in the laboratory, although some more hardy cyanobacteria will survive for several weeks or even months under the storage conditions described.

Isolation and Purification

Media and Growth Conditions

Growth, preferably on solid media, is a prerequisite for the isolation and purification of any bacterium. However, attempts to isolate and purify cyanobacteria, particularly by the direct isolation method (see below), have revealed that much more needs to be learned about the nutritional requirements and growth conditions of the more fastidious members of this group: many (even those dominant in the natural habitat) consistently failed to grow on the media and under the light and temperature regimes

[12] L. Packer, S. Spath, R. Bligny, J.-B. Martin, and C. Roby, this volume [73].

[13] R. Rippka, J. Deruelles, J. B. Waterbury, M. Herdman, and R. Y. Stanier, *J. Gen. Microbiol.* **111**, 1 (1979).

employed.[14] Therefore, some of the most commonly used growth conditions are described below in the hope that this information may stimulate the interested reader to develop modifications, possibly helpful, for the isolation of those numerous cyanobacteria that so far have escaped culture.

Media. Some of the more popular media and those utilized in major cyanobacterial culture collections are compared in Tables I through V. Their salt and trace metal contents are expressed in molarity (mM and μM, respectively), which allows the reader to appreciate the differences more readily than when given in g/liter, the form in which this information is provided in most of the original descriptions. This presentation also avoids minor involuntary changes of media composition that have occurred in the literature (e.g., Refs. 4 and 15), which undoubtedly are due to errors in the reported water content of the chemicals employed. Three major trace metal solutions are listed in Table III as examples of the types of metal salts (not specified in Tables II and V) to be utilized. For technical details concerning the preparation of these media (order in which to dissolve the individual components, pH, concentrated stock solutions, etc.), the reader should consult the authentic references provided in the footnotes to Tables I through V; some of this information is also given in this volume.[16]

Tables I and II compare the composition of several freshwater media, which differ mainly with respect to their contents of combined nitrogen, phosphate, type of chelating agents, and the concentration and/or form of iron. Some media differ most dramatically in the complexity and/or concentrations of their trace metals. There is no medium that is universally suitable for the isolation of cyanobacteria, and it is best to prepare several different types if the aim is to grow all of the species in a crude sample. For the isolation and growth of N_2-fixing species the source of combined nitrogen ($NaNO_3$ or KNO_3) should be omitted and possibly be replaced by the corresponding chlorides (although the latter is not always done).

Medium BG11,[10,13] designed by M. M. Allen,[17] allows growth of a wide range of cyanobacteria from soil and freshwater habitats and is used for maintenance of the cyanobacterial collection (about 300 axenic strains) at the Pasteur Institute. It is modeled after medium No. 11,[4,15] the major changes being a 3-fold increase of $NaNO_3$ and the omission of Gaffron's minor trace elements, which are replaced by a modified A_5 solution (see Tables I and II). However, isolation attempts with medium BG11[10,13] have shown that a decrease of the $NaNO_3$ concentration (as much as 10-fold)

[14] J. Deruelles, G. Guglielmi, R. Rippka, and J. B. Waterbury, unpublished observations.

[15] A. Zehnder and P. R. Gorham, *Can. J. Microbiol.* **6,** 645 (1960).

[16] R. W. Castenholz, this volume [3].

[17] M. M. Allen and R. Y. Stanier, *J. Gen. Microbiol.* **51,** 203 (1968).

sometimes gives better results.[18] Medium ASM-1[19] has been successfully employed for the isolation of planktonic cyanobacteria by Gorham and collaborators. Medium Z8[20] is practically identical to medium No. 11,[4,15] except that citrate as a chelating agent has been omitted; it has been used by Skulberg and collaborators at the Culture Collection of Algae in the Norwegian Institute for Water Research (NIVA) for the isolation and maintenance of a rather impressive collection of planktonic cyanobacteria.

The success of ASM-I and Z8, which are sometimes superior to BG11,[10,13] could suggest that two modifications (in addition to the reduced level of nitrate) may be important: (1) the lack of citrate may be favorable, since its presence in medium KMC[3] (Table I) leads to the formation of H_2O_2 (on autoclaving), at levels toxic for growth of single cells of *Anacystis nidulans* on plates[21]; (2) the presence of Gaffron's minor trace elements (Table II) in medium Z8[20] may also be beneficial for the isolation of cyanobacteria, some of which have a growth requirement for nickel,[22] which is lacking in ASM-1 and BG11 (and in most other media, see Table II), and additional trace metal requirements may yet be discovered. Both ASM-1 and Z8 could possibly be improved by the addition of molybdenum, completely lacking in ASM-1 and present at only low concentration in medium Z8, since this element is an essential cofactor for nitrate reductase[23,24] and nitrogenase.[25] The suggested increase of molybdenum in the latter medium seems advisable since tungstate, a known competitor of molybdenum uptake, even in cyanobacteria,[26] is present at 20% of the concentration of molybdenum, and may therefore inhibit growth of some strains.

As a general rule, it is wiser to use media of rather low trace metal content (or even lacking essential trace elements) than those with excessively high concentrations, since the latter may be toxic. A known example for cyanobacteria is the concentration of copper: even in amounts as low as that in BG11 (Table II), this element is inhibitory to growth of some

[18] R. Rippka, unpublished observations.

[19] P. R. Gorham, J. McLachlan, U. T. Hammer, and W. K. Kim, *Verh.–Int. Ver. Theor. Angew. Limnol.* **15,** 796 (1964).

[20] Norwegian Institute for Water Research (NIVA), "Culture Collection of Algae: Catalogue of Strains—1985." NIVA, Oslo, Norway, 1985.

[21] C. Van Baalen, *J. Phycol.* **3,** 154 (1967).

[22] C. Van Baalen and R. O'Donnell, *J. Gen. Microbiol.* **105,** 351 (1978).

[23] E. Flores, J. L. Ramos, A. Herrero, and M. G. Guerrero, *in* "Photosynthetic Prokaryotes: Cell Differentiation and Function" (G. C. Papageorgiou and L. Packer, eds.), p. 363. Elsevier, New York, 1983.

[24] M. G. Guerrero and C. Lara, *in* "The Cyanobacteria" (P. Fay and C. Van Baalen, eds.), p. 163. Elsevier, New York, 1987.

[25] R. Haselkorn, *Annu. Rev. Microbiol.* **40,** 525 (1986).

[26] A. Kumar and H. D. Kumar, *Biochim. Biophys. Acta* **613,** 244 (1980).

TABLE I

COMPOSITION OF MEDIA EMPLOYED FOR ISOLATION AND MAINTENANCE OF CYANOBACTERIA FROM SOIL AND FRESHWATER HABITATS[a]

Component	Concentration (mM) in medium										
	No. 11[b,c]	ASM-1[d]	BG11[e]	Z8[f]	SAG1[g]	BBM[h]	AA[i]	KMC[j]	Cg-10[k]	Kn[l]	D[m]
$NaNO_3$	5.84	2.00	17.65	5.49	—	2.94	—	—	11.77	11.76	8.24
KNO_3	—	—	—	—	1.98	—	20.0	9.89	—	—	0.99
K_2HPO_4	0.22 (0.17)[c]	0.10	0.18	0.18	—	0.43	2.0[n]	5.74	0.29	4.38	—
KH_2PO_4	—	—	—	—	0.15	1.29	—	—	—	—	—
Na_2HPO_4	—	0.10	—	—	—	—	—	—	—	—	0.78
$MgSO_4$	0.30	0.20	0.30	0.10	0.08	0.30	1.0	1.01	1.02	1.01	0.41
$MgCl_2$	—	0.20	—	—	—	—	—	—	—	—	—
$CaSO_4$	—	—	—	—	—	—	—	—	—	—	0.35
$CaCl_2$	0.24	0.20	0.25	—	—	0.17	0.50	—	—	0.24	—
$Ca(NO_3)_2$	—	—	—	0.25	—	—	—	0.11	0.11	—	—
NaCl	—	—	—	—	—	0.43	4.0	—	—	—	0.14
Na_2CO_3	0.19	—	0.19	0.20	—	—	—	—	—	—	—
Na_2SiO_3	0.20	—	—	—	—	—	—	—	—	—	—
NTA	—	—	—	—	—	—	—	—	—	—	0.52
EDTA	0.003	—	—	—	—	—	0.072[o]	—	—	0.018[o]	—
Disodium EDTA	—	0.020	—	0.010[o]	0.011[o]	0.134	—	—	0.027	0.005	—
Disodium-magnesium EDTA	—	—	0.003	—	—	—	—	—	—	—	—
Citric acid	0.029	—	0.029	—	—	—	—	—	—	—	—
Trisodium citrate	—	—	—	—	—	—	—	0.561	—	0.561	—
Ferric citrate	0.018	—	—	—	—	—	—	—	—	—	—

Ferric ammonium citrate	—	—	0.030[p]	—	—	—	—	—	—	—	—
$FeCl_3$	—	0.004	—	0.010[o]	—	—	—	—	—	—	0.002
$FeSO_4$	—	—	—	—	0.013[o]	0.018	0.072[o]	—	—	0.018[o]	—
$Fe_2(SO_4)_3$	—	—	—	—	—	—	—	0.008	0.008	—	—
Glycylglycine	—	—	—	—	—	—	—	—	3.8	—	—
Soil extract	—	—	—	—	3% (v/v)	—	—	—	—	—	—
pH adjusted before autoclaving to	—	—	—	—	—	—	—	—	8.2	—	8.2

[a] For trace elements, see Table II. For preparation of soil extracts, see Ref. *g* below.

[b] From E. O. Hughes, P. R. Gorham, and A. Zehnder, *Can. J. Microbiol.* **4,** 225 (1958).

[c] From A. Zehnder and P. R. Gorham, *Can. J. Microbiol.* **6,** 645 (1960).

[d] From P. R. Gorham, J. McLachlan, U. T. Hammer, and W. K. Kim, *Verh.–Int. Ver. Theor. Angew. Limnol.* **15,** 796 (1964).

[e] From R. Rippka, J. Deruelles, J. B. Waterbury, M. Herdman, and R. Y. Stanier, *J. Gen. Microbiol.* **111,** 1 (1979).

[f] From Norwegian Institute for Water Research (NIVA), "Culture Collection of Algae: Catalogue of Strains—1985." NIVA, Oslo, Norway, 1985.

[g] From U. G. Schlösser, *Ber. Dtsch. Bot. Ges.* **95,** 181 (1982).

[h] From T. Kantz and H. C. Bold, *in* "Physiological Studies: 9. Morphological and Taxonomic Investigations of *Nostoc* and *Anabaena* in Culture, Univ. of Texas Publ. No. 6924. Univ. of Texas, Austin, Texas, 1969.

[i] From M. B. Allen and D. I. Arnon, *Plant Physiol.* **30,** 366 (1955).

[j] From W. A. Kratz and J. Myers, *Am. J. Bot.* **42,** 282 (1955).

[k] From R. W. Castenholz, *Schweiz. Z. Hydrol.* **32,** 538 (1970).

[l] From A. Neilson, R. Rippka, and R. Kunisawa, *Arch. Mikrobiol.* **76,** 139 (1971).

[m] From R. W. Castenholz, *in* "The Prokaryotes" (M. P. Starr, H. Stolp, H. G. Trüper, A. Balows, and H. G. Schlegel, eds.), Vol. 1, p. 236. Springer-Verlag, Berlin, 1981.

[n] The phosphate is autoclaved separately and added to the medium after cooling.

[o] These compounds are added as the ferric EDTA complex; for details, see appropriate references.

[p] Molarity has been calculated with respect to the iron content (28% w/w) of this compound, the molecular weight of which is unknown.

strains.[27] This generalization seems to be justified since the initial growth of single filaments might be supported by the metal impurities in the major salts of the medium; special requirements would then manifest themselves only at a later stage of growth. Media such as BBM[28] should perhaps be avoided for isolation purposes, or only be tested in conjunction with other media of lower trace metal content. Media D[7] and Cg-10[29] have mainly been used by Castenholz for the isolation of thermophilic cyanobacteria, but they could also be useful for other species, particularly since medium Cg-10, solidified with agar, often gives better results for growth on plates or slants than medium BG11.[30] Media AA,[31] KMC,[3] and Kn[32] are favorite media for experimental purposes since they often give good growth rates and high yields; however, they are not tolerated by all cyanobacteria,[18,33] possibly as a result of their relatively high phosphate concentrations.

Medium SAG1[34] is employed for the maintenance of most cyanobacteria in the Culture Collection of Algae at the University of Göttingen (SAG)[34] (and in slightly modified versions also in other algal culture collections[35,36]). The content of major salts and trace elements is relatively low compared to other media (see Tables I and II), which may explain its success for the maintenance of cyanobacteria without the latter loosing typical properties such as heterocyst and akinete formation, hormogonia production, or motility, mutations which occur quite frequently in laboratories using other media.[18] However, this success may also be partly due to the light regime employed (see below) and to the presence of soil extract in the medium (see Table I), which is generally, because of its undefined nature, disliked by bacteriologists. Medium SAG1[34] (or modifications of it) should be tested in isolation attempts, particularly when other media prove to be unsuccessful.

Tables IV and V compare three media commonly employed for marine (M) cyanobacteria, which have obligate requirements for elevated con-

[27] J. B. Waterbury, personal communication.

[28] T. Kantz and H. C. Bold, *in* "Physiological Studies: 9. Morphological and Taxonomic Investigations of *Nostoc* and *Anabaena* in Culture," Univ. of Texas Publ. No. 6924. Univ. of Texas, Austin, Texas, 1969.

[29] R. W. Castenholz, *Schweiz. Z. Hydrol.* **32,** 538 (1970).

[30] J. Deruelles, personal communication.

[31] M. B. Allen and D. I. Arnon, *Plant Physiol.* **30,** 366 (1955).

[32] A. Neilson, R. Rippka, and R. Kunisawa, *Arch. Mikrobiol.* **76,** 139 (1971).

[33] A. Neilson, unpublished observations.

[34] U. G. Schlösser, *Ber. Dtsch. Bot. Ges.* **95,** 181 (1982).

[35] R. C. Starr, *Am. J. Bot.* **47,** 67 (1960).

[36] "Culture Centre of Algae and Protozoa: List of Strains—1982" (A. Asher and D. F. Spalding, eds.). Institute of Terrestrial Ecology, National Environmental Research Council, Cambridge, England, 1982.

TABLE II

TRACE METALS OF MEDIA EMPLOYED FOR CYANOBACTERIA FROM SOIL AND FRESHWATER HABITATS[a]

Medium	Concentration (μM) in medium															Designation of trace metal solution	Ref.
	Al	B	Br	Cd	Co	Cr	Cu	I	Mn	Mo	Ni	Ti	V	W	Zn		
No. 11	0.08	4	0.08	0.04	0.04	0.008	0.04	0.04	0.8	0.04	0.04	—	0.008	0.008	0.08	Gaffron[b]	*c,d*
ASM-1	—	40	—	—	0.08	—	0.0008	—	7.0	—	—	—	—	—	3.2	A_5	*e*
BG11	—	46	—	—	0.17	—	0.32	—	9.2	1.6	—	—	—	—	0.77	A_5 + Co	*f*
Z8	0.10	5.0	0.10	0.05	0.05	0.01	0.05	0.05	1.0	0.05	0.05	—	0.01	0.01	0.10	Gaffron	*c,d*
SAG1	—	0.8	—	—	0.017	—	0.0001	—	0.045	0.02	—	—	—	—	0.017	A_5 + Co	*g*
BBM	—	185	—	—	1.70	—	6.3	—	7.3	4.9	—	—	—	—	30.7	H_5 + B	*h,i*
AA	—	46	—	—	0.17	0.19	0.32	—	9.1	1.0	0.17	0.21	0.20	0.05	0.77	A_5 + B_6	*j*
KMC	—	46	—	—	—	—	0.32	—	9.2	0.10	—	—	—	—	0.77	A_5	*i*
Cg-10	—	46	—	—	0.04	—	0.32	—	9.2	0.10	—	—	—	—	0.77	A_5 + Co	*k*
Kn	—	46	—	—	0.17	—	0.32	—	9.2	1.6	—	—	—	—	0.77	A_5 + Co	*f*
D	—	4.0	—	—	0.10	—	0.05	—	6.7	0.05	—	—	—	—	0.87	A_5 + Co	*l*

[a] Media as described in Table I.

[b] Cited in footnotes *c* and *d*. Cr and Co may be omitted.

[c] E. O. Hughes, P. R. Gorham, and A. Zehnder, *Can. J. Microbiol.* **4,** 225 (1958).

[d] A. Zehnder and P. R. Gorham, *Can. J. Microbiol.* **6,** 645 (1960).

[e] P. R. Gorham, J. McLachlan, U. T. Hammer, and W. K. Kim, *Verh.–Int. Ver. Theor. Angew. Limnol.* **15,** 796 (1964).

[f] R. Rippka, J. Deruelles, J. B. Waterbury, M. Herdman, and R. Y. Stanier, *J. Gen. Microbiol.* **111,** 1 (1979).

[g] U. G. Schlösser, *Ber. Dtsch. Bot. Ges.* **95,** 181 (1982).

[h] T. Kantz and H. C. Bold, *in* "Physiological Studies: 9. Morphological and Taxonomic Investigations of *Nostoc* and *Anabaena* in Culture, Univ. of Texas Publ. No. 6924. Univ. of Texas, Austin, Texas, 1969.

[i] W. A. Kratz and J. Myers, *Am. J. Bot.* **42,** 282 (1955).

[j] M. B. Allen and D. I. Arnon, *Plant Physiol.* **30,** 366 (1955).

[k] R. W. Castenholz, *Schweiz. Z. Hydrol.* **32,** 538 (1970).

[l] R. W. Castenholz, *in* "The Prokaryotes" (M. P. Starr, H. Stolp, H. G. Trüper, A. Balows, and H. G. Schlegel, eds.), Vol. 1, p. 236. Springer-Verlag, Berlin, 1981.

TABLE III
COMPOSITION OF THREE TRACE METAL STOCK SOLUTIONS EMPLOYED FOR CYANOBACTERIAL MEDIA

Medium	Component and concentration (g/liter)	Ref.
Gaffron	H_3BO_3, 3.1; $MnSO_4 \cdot 4H_2O$, 2.23; $ZnSO_4 \cdot 7H_2O$, 0.287; $(NH_4)_6Mo_7O_{24} \cdot 4H_2O$, 0.088; $CuSO_4 \cdot 5H_2O$, 0.125; $Co(NO_3)_2 \cdot 6H_2O$, 0.146; $Al_2(SO_4)_3K_2SO_4 \cdot 24H_2O$, 0.474; $NiSO_4(NH_4)_2SO_4 \cdot 6H_2O$, 0.198; $Cd(NO_3)_2 \cdot 4H_2O$, 0.154; $Cr(NO_3)_3 \cdot 7H_2O$, 0.037; $V_2O_4(SO_4)_3 \cdot 16H_2O$, 0.035; $Na_2WO_4 \cdot 2H_2O$, 0.033; KBr, 0.119; KI, 0.083. Volume added: 0.08–0.1 ml/liter of medium	*a,b*
$A_5 + B_6$	Solution A: H_3BO_3, 2.86; $MnSO_4 \cdot 4H_2O$, 2.03; $ZnSO_4 \cdot 7H_2O$, 0.222; MoO_3, 0.15; $CuSO_4 \cdot 5H_2O$, 0.079. Solution B (components dissolved in 0.1 *N* H_2SO_4): $Co(NO_3)_2 \cdot 6H_2O$, 0.049; $NiSO_4 \cdot 6H_2O$, 0.045; $Cr_2(SO_4)_3K_2SO_4 \cdot 24H_2O$, 0.096; NH_4VO_3, 0.023; $Na_2WO_4 \cdot 2H_2O$, 0.018; $TiO(C_2O_4K)_x \cdot yH_2O$ [this compound is prepared by precipitation (with NH_4OH) of a solution containing 0.074 g of $TiO(C_2O_4K)_2 \cdot 2H_2O$, followed by filtration and resuspension in 0.1 *N* H_2SO_4 prior to addition to solution B]. Both solutions A and B are added as 1 ml/liter of medium	*b,c*
A_5 + Co	H_3BO_3, 2.86; $MnCl_2 \cdot 4H_2O$, 1.81; $ZnSO_4 \cdot 7H_2O$, 0.22; $Na_2MoO_4 \cdot 2H_2O$, 0.39; $CuSO_4 \cdot 5H_2O$, 0.079; $Co(NO_3)_2 \cdot 6H_2O$, 0.049. Volume added: 1 ml/liter of medium	*d*

[a] E. O. Hughes, P. R. Gorham, and A. Zehnder, *Can. J. Microbiol.* **4,** 225 (1958).
[b] A. Zehnder and P. R. Gorham, *Can. J. Microbiol.* **6,** 645 (1960).
[c] M. B. Allen and D. I. Arnon, *Plant Physiol.* **30,** 366 (1955).
[d] R. Rippka, J. Deruelles, J. B. Waterbury, M. Herdman, and R. Y. Stanier, *J. Gen. Microbiol.* **111,** 1 (1979).

centrations of Na^+, Mg^{2+}, Ca^{2+}, and Cl^-.[9] The success of medium ASP-2 for non-N_2-fixing cyanobacteria[5,37] is to some extent surprising (as is the freshwater medium ASM-1, see Tables I and II), since they both lack molybdenum which is known to be essential for growth on nitrate,[23,24] the only source of combined nitrogen supplied, and can only be explained if sufficient quantities of this element are introduced as impurities with the major salts. It would seem advisable to add molybdenum in defined quantities. Media such as MN,[9] prepared with aged and filtered seawater, should always be included as isolation media for cyanobacteria from marine environments, since natural seawater contains components, absent from synthetic media, that might be essential for growth of these organisms. Many marine cyanobacteria have a requirement for vitamin B_{12},[5,9] which should therefore be routinely added to the isolation medium (2–10 μg/liter); the vitamin may subsequently be omitted if it is shown to be

[37] L. Provasoli, J. J. A. McLaughlin, and M. R. Droop, *Arch. Mikrobiol.* **25,** 392 (1957).

TABLE IV

COMPOSITION OF MEDIA EMPLOYED FOR ISOLATION AND MAINTENANCE OF MARINE (M) CYANOBACTERIA, HALOPHILES (H), AND CYANOBACTERIA FROM HIGHLY ALKALINE (A) ENVIRONMENTS[a]

	Concentration (m*M*)					
Component	ASP-2[b] (M)	ASN-III[c] (M)	MN[c] (M)	RC[d] (H)	B[e] (H)	AO[f] (A)
NaCl	308	427	—	2137	2735	17.1
KCl	8.0	6.7	—	33.5	—	—
$CaCl_2$	2.5	3.4	0.12	3.4	—	0.27
$CaSO_4$	—	—	—	—	0.35	—
K_2SO_4	—	—	—	—	—	5.7
$MgSO_4$	20.3	14.2	0.15	14.2	0.41	0.81
$MgCl_2$	—	9.8	—	49.2	—	—
K_2HPO_4	0.29	0.09	0.09	0.09	—	2.9[g]
Na_2HPO_4	—	—	—	—	0.78	—
$NaNO_3$	11.8	8.8	8.8	8.8	8.1	29.4
KNO_3	—	—	—	—	1.0	—
$NaHCO_3$	—	—	—	—	—	162[g]
Na_2CO_3	—	0.19	0.19	0.19	—	38[g]
NTA	—	—	—	—	0.52	—
Disodium EDTA	0.08[h]	0.0013	0.0013	0.0013	—	0.22
Citric acid	—	0.014	0.014	0.014	—	—
Ferric ammonium citrate	—	0.015[i]	0.015[i]	0.015[i]	—	—
$FeCl_3$	0.005[h]	—	—	—	0.0018	—
$FeSO_4$	—	—	—	—	—	0.036
Tris	8.3	—	—	—	—	—
Seawater	—	—	75% (v/v)	—	—	—
pH adjusted before autoclaving to	8.2	—	—	—	8.2	—

[a] For trace elements see Table V.

[b] C. Van Baalen, *Bot. Mar.* **4,** 129 (1962).

[c] J. B. Waterbury and R. Y. Stanier, *in* "The Prokaryotes" (M. P. Starr, H. Stolp, H. G. Trüper, A. Balows, and H. G. Schlegel, eds.), Vol. 1, p. 221. Springer-Verlag, Berlin, 1981.

[d] J. van Rijn and Y. Cohen, *J. Gen. Microbiol.* **129,** 1849 (1983).

[e] T. D. Brock, *Arch. Microbiol.* **107,** 109 (1976).

[f] S. Aiba and T. Ogawa, *J. Gen. Microbiol.* **102,** 179 (1977).

[g] These three compounds are autoclaved as a separate solution and added to the medium after cooling (for details see Ref. *f*).

[h] Disodium EDTA and $FeCl_3$ are provided in P1 solution. See L. Provasoli, J. J. A. McLaughlin, and M. R. Droop, *Arch. Mikrobiol.* **25,** 392 (1957).

[i] Molarity calculated with respect to the iron content (28%, w/w) of this compound of undetermined molecular weight.

TABLE V
TRACE METALS OF MEDIA EMPLOYED FOR MARINE (M) CYANOBACTERIA, HALOPHILES (H), AND CYANOBACTERIA FROM HIGHLY ALKALINE (A) ENVIRONMENTS[a]

Medium	Concentration (μM) in medium											Designation of trace metal solution	Ref.
	B	Co	Cr	Cu	Mn	Mo	Ni	Ti	V	W	Zn		
ASP-2 (M)	555	0.05	—	0.02	21.8	—	—	—	—	—	2.3	P1 + B	*b*
ASN-III (M)	46	0.17	—	0.32	9.2	1.6	—	—	—	—	0.77	A_5 + Co	*c*
MN (M)	46	0.17	—	0.32	9.2	1.6	—	—	—	—	0.77	A_5 + Co	*c*
RC (H)	46	0.17	—	0.32	9.2	1.6	—	—	—	—	0.77	A_5 + Co	*c*
B (H)	4.0	0.10	—	0.05	6.7	0.05	—	—	—	—	0.87	A_5 + Co	*d*
AO (A)	46	0.15	0.19	0.32	9.2	0.10	0.17	0.10	0.20	0.05	0.77	A_5 + B_6	*e*

[a] Media as described in Table IV.
[b] L. Provasoli, J. J. A. McLaughlin, and M. R. Droop, *Arch. Mikrobiol.* **25,** 392 (1957).
[c] R. Rippka, J. Deruelles, J. B. Waterbury, M. Herdman, and R. Y. Stanier, *J. Gen. Microbiol.* **111,** 1 (1979).
[d] R. W. Castenholz, *in* "The Prokaryotes" (M. P. Starr, H. Stolp, H. G. Trüper, A. Balows, and H. G. Schlegel, eds.), Vol. 1, p. 236. Springer-Verlag, Berlin, 1981.
[e] T. Ogawa and G. Terui, *J. Ferment. Technol.* **48,** 361 (1970).

nonessential for growth of the isolates. Tables IV and V also give the composition of two media (RC[38] and B[39]) employed for the isolation of halophilic (H) cyanobacteria and one (AO) so far used exclusively for the isolation and growth of *Arthrospira* (*Spirulina*) *platensis*,[40,41] a planktonic cyanobacterium from highly alkaline environments.

For the isolation and maintenance of heterocystous cyanobacteria, it is advisable to omit the source of combined nitrogen from the media and to replace their nitrate content eventually (although this is not always done; see medium $BG11_o$[13]) by an appropriate amount of a nonnitrogenous anion (the safest being chloride).

Preparation of solid media. The elegant studies of M. M. Allen[6] showed that sterilization of the mineral salts of medium BG11 together with the solidifying agent (Difco Bacto-agar) leads to production of toxic compounds. This might not necessarily be the case with other media, but it would be wise to sterilize the mineral medium and agar as separate solutions (at double strength) and mix them only after cooling to about 45°,[6] unless proof against such toxicity problems has been obtained experimentally. For media such as MN,[9] containing natural seawater, the agar has to be autoclaved separately in the appropriate volume of their freshwater content.

Freshly poured plates are often left to cool, solidify, and dry in a room or laminar flow hood sterilized by UV irradiation. Experience with medium BG11[10,13] has shown that direct UV exposure of solid medium in plastic Petri dishes produces compounds which are toxic for single cells and filaments.[18] Therefore, although sterilization of the room or hood by UV is recommended, the UV lights should be turned off while the plates are drying and stored.

Some authors found it necessary to remove toxic compounds from the agar itself by washing it prior to preparing the mineral plates.[28,42] For this, the agar was melted, solidified, cut up into small cubes (1–2 cm) and washed for several days and with several changes of sterile water. A more practical, but equally efficient,[18] way of purifying agar is described in detail by Waterburg *et al.*[43] Others included catalase[44,45] as a precaution against the potential formation of H_2O_2[21] in the autoclaved medium.

Agar is generally mildly acidic and lowers the pH to values below 7.0,

[38] J. van Rijn and Y. Cohen, *J. Gen. Microbiol.* **129,** 1849 (1983).

[39] T. D. Brock, *Arch. Microbiol.* **107,** 109 (1976).

[40] T. Ogawa and G. Terui, *J. Ferment. Technol.* **48,** 361 (1970).

[41] S. Aiba and T. Ogawa, *J. Gen. Microbiol.* **102,** 179 (1977).

[42] W. W. Carmichael and P. R. Gorham, *J. Phycol.* **10,** 238 (1974).

[43] J. B. Waterbury, S. W. Watson, F. W. Valois, and D. G. Franks, *Can. Bull. Fish. Aquat. Sci.* **214,** in press (1987).

[44] C. Van Baalen, *J. Phycol.* **1,** 19 (1965).

[45] T. Vaara, M. Vaara, and S. Niemelä, *Appl. Environ. Microbiol.* **38,** 1011 (1979).

if used for the solidification of poorly buffered media such as BG11. In such cases, organic buffers such as *N*-2-hydroxyethylpiperazine-*N*′-2-ethanesulfonic acid (HEPES)[18,46] or glycylglycine[21,29] should be tested to maintain a more favorable pH (7.0–8.0), particularly if the strains do not tolerate Tris or phosphates at concentrations which permit a good buffering capacity. However, the addition of filter-sterilized $NaHCO_3$ (10 m*M*), prior to pouring plates, often also overcomes growth problems[18]; it is not yet known whether this results from an increase of the pH or from the provision of CO_2 in a readily assimilated form ($NaHCO_3$),[47] or from both.

The freshly poured plates, while still warm, are best stacked on top of each other to prevent the formation of condensation below their covers. They should be dried at up to 37° for at least 24 hr. If not required immediately, they can be stored most conveniently at room temperature in plastic boxes (such as vegetable crispers) whose lids are opened sufficiently to allow some ventilation.

Water quality for media. The quality of the water used for preparation of cyanobacterial media is of paramount importance. The water should be deionized, and then organic compounds (extremely toxic to many strains) should be removed by double distillation or by passage through a Millipore SuperQ (or similar) purifier. If available, the use of a pyrolyzing purification system is even more highly recommended. The glassware used as culture vessels should be well washed and rinsed, taking care to eliminate traces of high-phosphate detergents.

Additions to media sterilized by membrane filtration. It is common bacteriological practice to filter-sterilize compounds that may be destroyed on autoclaving, by use of Millipore membrane or sintered glass filters (0.22–0.45 μm, maximum pore size). Although rarely done (even by bacteriologists), it seems prudent to perform this filtration with sterile instruments and autoclaved water, since these filters do not retain relatively small elements such as phages (or others) that, if present, could be potentially harmful (especially in liquid medium) to the organisms to be cultivated. The use of heat-sterilized water for this procedure is particularly advised when employing water from membrane purification systems, since the latter may become good enrichment sources for such harmful biological activities, if the filters are not changed sufficiently frequently.

Temperature, pH, and Supply of CO_2. Most cyanobacteria flourish best in neutral to alkaline environments (pH 7.0–10), and even those encountered in mildly acidic habitats (pH 5.0–6.0) such as sphagnum bogs or some hot springs are generally acid tolerant rather than acidophilic.[10] For this reason, practically all media developed for cyanobacteria are

[46] R. V. Smith and F. H. Foy, *Br. Phycol. J.* **9,** 239 (1974).
[47] A. Kaplan, Y. Marcus, and L. Reinhold, this volume [57].

alkaline. However, media of different pH values are recommended for isolation, particularly those with a pH typical of the natural habitat.

The temperature for isolation should also be varied in ranges 5 to 10° apart. Although most cyanobacteria from moderately temperate habitats will grow well at 20–30°, many do not tolerate temperatures above 20°. The routine incubation at 35°, suggested for the selective enrichment of cyanobacteria,[17] should therefore be avoided.

For growth of cyanobacteria in small quantities (i.e., on plates, slants, or in liquid cultures of relatively large surface area) the amount of CO_2 present in air is generally sufficient to fulfill the carbon requirement. However, systematic isolation studies with higher concentrations of CO_2, possibly beneficial for some cyanobacteria, have not yet been performed.

Light Regimes. The spectral range of light absorbed by cyanobacteria[48] favors the use of fluorescent light sources (i.e., cool-white, warm-white, daylight) over that of tungsten light for growth and isolation of these organisms. Unless isolation is attempted under air enriched with CO_2 (a condition leading to a higher light tolerance), the light intensities should be kept low (200–500 lux).

Cyanobacteria, particularly in the hands of bacteriologists, are normally kept under a continuous light regime rather than under a light–dark cycle, since early investigations (e.g., Allen and Arnon[31]) indicated that such a treatment is not harmful to their growth. However, it is possible that some strains are unable to grow under these conditions, and it has never been examined experimentally whether continuous light (a very unnatural regime) prevents the expression of certain cyanobacterial properties (such as synchrony of cellular division) that in the natural habitat lead to the formation of typical colonial aggregates, never or seldom observed in culture.[13] It is also possible that continuous light may favor the selection of mutants defective in regulatory mechanisms that, in the natural habitat, are under the control of light–dark cycles. Furthermore, the numerous spontaneous mutants unable to form gas vacuoles, heterocysts, hormogonia, or akinetes, or lacking motility, which are often observed in laboratory cultures,[18] may also be a consequence of a continuous light regime. It may therefore be advisable to incubate new isolates both under continuous light and under light–dark cycles. Their development should be followed much more carefully than in the past, since recent studies have shown that a number of cyanobacterial enzymes are regulated by the supply of light, with thioredoxin as the mediator.[49,50]

[48] A. N. Glazer, this volume [31], [32].

[49] P. Rowell, A. J. Darling, and W. D. P. Stewart, this volume [42].

[50] N. A. Crawford, B. C. Yee, M. Droux, D. E. Carlson, and B. B. Buchanan, this volume [43].

Tools and Isolation Methods

Isolation Tools and Their Usage. Apart from the accessory equipment (a good phase-contrast and a binocular dissection microscope, a Bunsen burner, liquid and solid media, Petri plates, some glassware, and a supply of microscope slides, coverslips, pipets, forceps, scalpels, etc.), there are only three tools essential for the isolation of cyanobacteria, their usage being largely dependent on the stage of isolation and on the size, fragility, or compactness of the colonies or filaments to be manipulated.

Classical "bacterial" loop. The first tool required is the "classical" loop used for the isolation of bacterial colonies; this can be either bought commercially or prepared from a platinum (or nichrome) wire (0.7 mm in diameter, 10 cm in length) by bending one of its extremities to form a circle of about 3–5 mm in diameter (the other end being anchored in a glass rod). For better surface contact the wire should be bent downward at a right angle 5–6 mm away from the circular part; the latter is then reoriented at a right angle opposite to the first bend to bring it back to a horizontal position, with it pointing away from the stem. This instrument is mainly employed in the initial stages of isolation (i.e., the spreading of crude material onto an agar plate), or for the streaking of single colonies for their further purification, as will be briefly described: a good streaking technique is vital for successful isolation. With the loop, heated to redness in the flame of a Bunsen burner and cooled at one edge of the agar plate, the material deposited (crude material or colony) is densely spread over the first quarter of the agar plate, passing over it several times, while trying to disentangle and homogenize aggregates of filaments and colonies. The loop is then flamed again, cooled, and used to perform several parallel streaks traversing at an angle of 90° the previously spread material (Fig. 1). This operation is then repeated twice, ensuring that the fourth set of streaks does not enter into the material on the first quarter of the plate. The principle of this streaking method is a classical dilution on solid medium; single cells (giving rise to colonies on incubation) or filaments (that can be isolated immediately as described below) will generally be located only on the streaks of the last two quarters of the plate (see Fig. 1), unless the sample contained only very low numbers of the species to be isolated.

Microspade. The second essential instrument for the isolation and purification of cyanobacteria (but equally useful for other microorganisms) is the microspade, easily constructed from a platinum (or nichrome) wire of about 0.7–1 mm thickness and 8–10 cm length. After heating to redness, approximately 5 mm of one end of the wire is flattened as much as possible, by hammer on a metal support. After cooling, the flattened part is filed to produce a rhomboidal spade with two parallel cutting sides of about 2–3 mm in length, the other extremity of the wire being then

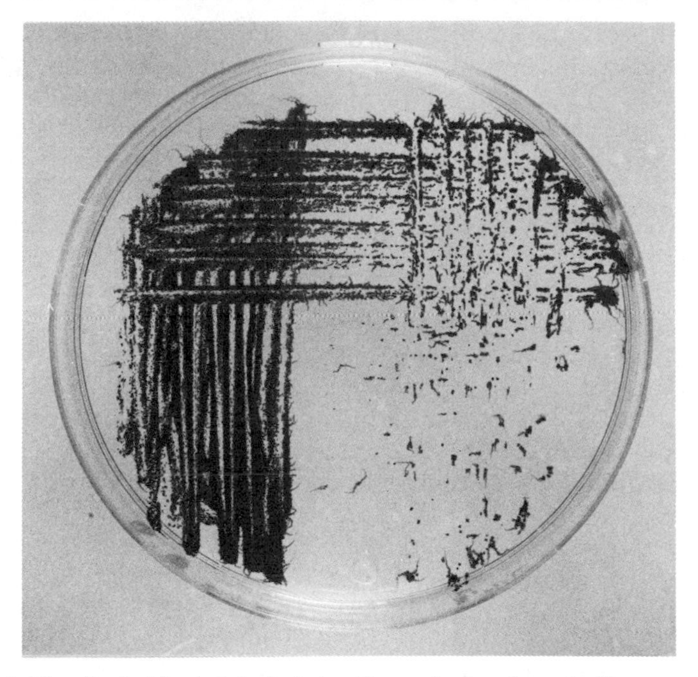

FIG. 1. The classical bacteriological streaking method performed with a nonmotile, filamentous cyanobacterium. Note the heavy growth on the first quarter of the plate; successive streaks at right angles to the first spread the material sufficiently to produce isolated colonies. See text for details.

anchored in a glass rod or any other commercially available handle. This microspade is sufficiently fine to pick or cut out with precision individual microcolonies or filaments from agar plates while observing them under a binocular dissection microscope.

Most colonies, particularly when they are dry and compact, can be picked and transferred directly, without damage to the population. More fragile organisms may also tolerate the same treatment, since the number of survivors in a colony is generally sufficient to yield at least a few new colonies. However, single filaments should never be touched directly with the microspade, or any other metallic isolation tool, the mortality rate (due to damage and drying in the course of transfer[18,51]) being too high for the few cells composing a single filament. Their chances of survival are only assured if they are transferred on their agar support with the microspade, or isolated by means of a Pasteur micropipet as described below.

To successfully remove single filaments bedded on agar the plates should be oriented under the dissection microscope in such a way that the organisms are easily accessible (avoiding steep angles for manipulation;

[51] J. W. Bowyer and V. B. D. Skerman, *J. Gen. Microbiol.* **54,** 299 (1968).

the same principle should, of course, be applied for the isolation of colonies, even of unicellular cyanobacteria). The filaments, particularly longer ones, should be located with their long axis horizontal to the observer. Two shallow incisions should be made into the agar with the cutting edges of the microspade parallel to, and on either side of, the filament; these are followed by two equally shallow incisions at both extremities of the filament. The resulting rectangle (or square), carrying the filament, can then be sliced out of its place and transferred with the broad side of the microspade. Some practice is required to master this isolation procedure since the depth and equalness of the incisions are critical for lifting the agar block out of its place without too much disturbance of the surrounding area, which might contain contaminants. For the purification of filamentous cyanobacteria that form long filaments (e.g., *Oscillatoria*) it is best to cut them into smaller fragments prior to their isolation; this will reduce the chances of carryover of bacterial contaminants possibly attached to the filaments.

Pasteur micropipet. The third essential tool is a finely drawn Pasteur pipet which is prepared as follows. A commercially available Pasteur pipet (sterilized and stored for convenience until use in a glass or metal container) is held at both ends (the capillary tube on one side, the thicker part on the other) by hand and is gently turned over the hottest part of a well-adjusted Bunsen flame so that the central part of the capillary tube is heated over a distance of about 0.5 to 1 cm. As soon as the latter area turns red and is soft enough to loose its rigidity, the Pasteur pipet is taken out of the flame while pulling immediately on both ends. The rather long (up to 20 cm) and fine (one-tenth or more of the initial diameter) capillary tube resulting from this operation is then broken (and the detached part discarded) by gentle bending or by using sterile forceps (the latter being somewhat more safe) to reduce its length to about 2–4 cm on the side adjacent to the thicker end of the original Pasteur pipet (which will serve as the isolation tool). The precise length of the capillary depends on its diameter (the smaller the latter, the shorter it should be), since too much flexibility, in addition to causing easy breakage of the micropipet, will harm the precision required to pick colonies or filaments under the dissection microscope.

The Pasteur micropipet is ideal to pick any soft microcolony of unicellular or very fine filamentous cyanobacteria, but it can also be used to pick single filaments if the latter are relatively sturdy and not too long. Before use, it is advisable to humidify the extremity of the fine tube by gently touching the surface of a sterile agar plate to fill the capillary to about 1 mm with moisture from the plate. The Pasteur micropipet is then placed (while holding it by hand) under the dissection scope well above the plate from which a colony or filament is to be isolated. While moving the pipet slowly in a horizontal direction it is lowered very gently until the capillary tube is well visible and in close proximity to the desired object. The capillary part can then be directed straight into a colony or be used (while holding it somewhat more at angle) for the capturing of a filament, both of which will enter by capillary action (unless the colonies are too compact or the filaments adhere too much to the agar surface, in which case the microspade will have to be used for their isolation). Once loaded, the content of the Pasteur micropipet is gently blown onto the surface of a new agar plate, the successful operation being immediately confirmed under the dissection microscope before proceeding any further. Needless to say, a new Pasteur micropipet will have to be prepared for each isolation attempt since the fine capillary tube would melt if passed into a Bunsen flame for re-sterilization.

Isolation Procedures

Isolation, as opposed to purification, is here defined as the separation of individual clones of cyanobacterial species from other photosynthetic organisms present in the crude material. As unicyanobacterial isolates, they are subsequently purified to the axenic state as described under Purification Methods.

Many cyanobacteria presently in axenic culture have been isolated by the so-called liquid enrichment technique, which imposes a positive selection on those members of the population most apt to proliferate in the medium and under the culture conditions (light, temperature) provided by the investigator. This method, although useful for some purposes, has the disadvantage that the cyanobacteria isolated are seldom representative of the total variety of species, and their relative number, encountered in the natural habitat. If the aim of cyanobacterial research is to study natural ecosystems (which are characterized by the totality of their indigenous population and not by the "weeds" in the test tube) it is advisable to avoid as much as possible the classical liquid enrichment technique and to establish pure cultures via direct isolation.

With both isolation methods, which are described briefly below, it is important to remember that all operations should be carried out with sterile instruments, water, media, etc. and even with sterile microscope slides and coverslips if the samples examined are subsequently to be used directly for isolation. It should also be remembered that crude material may contain bacterial or fungal spores. Isolation attempts should therefore not be performed in rooms or laminar flow hoods used for the maintenance of axenic cyanobacteria.

Isolation by Liquid Enrichment. In spite of the shortcomings mentioned above, the liquid enrichment technique is useful (if not essential) for the rapid isolation of cyanobacteria if the aim of the investigator is to select, for experimental purposes, those members of the population possessing particular physiological properties. One might, for example, wish to isolate, regardless of its generic or specific identity, a cyanobacterium that is capable of growth and nitrogen fixation at a given temperature, pH, or light regime, and/or is relatively tolerant to the ionic or trace metal composition of a selected medium or even to H_2S. For this, a soil or water sample is placed in an appropriate mineral liquid medium (in a test tube or small flask) and incubated under the light and temperature regime chosen for selection until growth is visible. In general, such liquid enrichments will contain only one or very few dominant cyanobacterial strains, and their isolation and purification is relatively easy, provided that they will grow equally well on solid media, the samples being treated from this stage on in an identical manner to those subjected to the direct isolation method (see below).

In designing liquid enrichments some common sense is required so as not to ask for the impossible, but to choose samples from appropriate habitats. For example (admittedly exaggerated), it is rather unlikely that a freshwater sample from a cold mountain river will give rise to growth of halophilic or thermophilic cyanobacteria, even though selected for under the appropriate conditions. The other restriction is the selection of cyanobacteria capable of growing on specific organic carbon or nitrogen sources: they will soon be overgrown by bacteria far more competent to utilize these compounds.

Direct Isolation. For the direct isolation technique, in contrast to liquid enrichments, the field samples should be examined under the microscope immediately on arrival in the laboratory to evaluate their content of cyanobacterial species and that of the accompanying eukaryotic and prokaryotic flora. This is important, not only to characterize the ecological niche of the cyanobacterial population at the microscopic level, but also to later reidentify those cyanobacteria that will (or will not) succeed in growing on the solid media chosen for isolation. Ideally, several aliquots of the samples should be examined, particularly from rock and soil samples, since growth might be localized and the presence of certain species could otherwise remain undetected. Depending on the consistency of the crude material, the sample for examination can be placed onto a slide [sterilized over a Bunsen flame after passage through 70% (v/v) ethanol] with a platinum loop or a Pasteur pipet, but often rather more forceful operations are necessary to extract the population so that microscopic observation is not hindered by excessive compactness or carryover of stony material. Such treatments may vary depending on the degree of aggregation: e.g., syringing, crushing between glass slides, and homogenization in a homogenizer. Liquid additions for microscopic examinations should be made with sampling water for organisms from aquatic habitats, distilled water or diluted freshwater media for rock and soil samples, unless the latter originate from marine or hypersaline environments in which case a medium of appropriate composition should be employed to avoid osmotic stress conditions.

Once the sample has been thoroughly examined and its content recorded, aliquots can be transferred directly (after carefully lifting off the coverslip, and provided all operations are carried out with sterile precautions) to solid media of appropriate choice. If the samples are rich in eukaryotic contaminants it is advisable to use plates to which 0.05–0.1 g/liter of cycloheximide (Actidione) has been added.[45,52] After streaking the deposited crude material on solid media as described above, the plates should be examined immediately under the binocular dissection microscope, since many of the larger filamentous forms or colonial aggregates of unicellular cyanobacteria may be sufficiently separated to attempt their isolation to the unicyanobacterial state without prior incubation of the plates (which increases considerably the chances of rapid isolation, since overgrowth by more dominant members is avoided). Putatively different species, as judged from their color, size, morphology, etc. under the dissection microscope, should be confirmed by cutting representatives from the plate with the microspade and examining them under the phase-contrast microscope. This additional work (although time consuming) will help the investigator greatly with the task of isolation, since one will learn to recognize with ease each of the cyanobacterial representatives in the population by their characteristic aspects under the dissection microscope. As many of the separate colonial aggregates or filaments as detectable on the crude plate should be isolated (if time and patience allow), since many of them might subsequently die or be overgrown by contaminants not recognizable under the dissection microscope. If few isolated colonies or filaments can be observed, additional plates need to be prepared from the crude material, which might be spread more successfully.

Isolated colonies of unicellular species or those composed of very fine filamentous forms should be transferred each onto an individual fresh plate on which they are immediately restreaked (after addition of a small drop of sterile medium) by the dilution technique described above. Single filaments (six to eight) of tentatively identical species can be placed, well separated from one another, onto the same plate; their location should be confirmed under the dissection microscope and be well marked on the bottom of the Petri dish.

Crude material often also contains unicellular or *Pseudanabaena* species that are easily homogenized and may, as single colonies or filaments after streaking, no longer be recognizable under the dissection microscope. If the field population is dominated by such members, the plates

[52] A. Zehnder and E. O. Hughes, *Can. J. Microbiol.* **4**, 399 (1958).

are incubated until the cells or filaments have given rise to colonies, before attempting their isolation. If, however, the crude material also contains a high percentage of larger filamentous forms (which when motile, and vigorous growers, can be very competitive) it is better to separate the latter by differential centrifugation (i.e., 1000 *g*) after suspending and mixing thoroughly an aliquot of the field sample in a small amount (~10 ml) of liquid medium. The supernatant recovered is then used to prepare serial dilutions, samples of which are either streaked directly on agar plates or used to prepare agar overlays, by mixing 2.5 ml of each dilution with 2.5 ml of a melted and cooled (to ~50°) agar solution (1%, w/v, in H_2O), that are poured onto the media plates of choice. The plates are incubated until growth, in the form of colonies, is visible under the dissection microscope.

Many filamentous cyanobacteria form very tight aggregates from which single filaments are not easily obtained on streaking of crude material as described above. Their isolation should be attempted as follows. The field material is cut by using forceps and a scalpel to yield pieces small enough to be picked and transferred by an ordinary Pasteur pipet. About a dozen of them are then placed into a test tube (~18 ml) containing 8–10 ml of medium to dilute out unicellular contaminants. After vortexing thoroughly and settling, the cyanobacterial aggregates are recuperated with a new sterile Pasteur pipet and transferred to another test tube containing fresh medium. This operation is repeated at least 10 times, at which stage the aggregates are placed individually onto the center of an agar plate with as little carryover of liquid medium as possible. After 24–48 hr of incubation single filaments or hormogonia[53] will often have moved away from the place of deposit and can then be isolated as described above. In the case of less motile or immotile species, single filaments can often be cut off with the microspade from the tips of the outgrowing material.

Water samples from very clear lakes or ponds need to be concentrated for the direct isolation procedure if cyanobacterial growth cannot be detected under the microscope. This is achieved by vacuum filtration of aliquots (e.g., 5, 10, 15 ml) through sterile Millipore filters (0.45 μm pore size), which are then placed on the agar plates. Sometimes it is convenient to first reduce the volume of the field sample by centrifugation and resuspension in an appropriately smaller amount of medium. After several days or weeks of incubation the cyanobacteria can be recognized as small colonies or diffuse growth patches on the filter; motile filaments might have moved from the filter to the surrounding agar. This method unfortu-

[53] M. Herdman and R. Rippka, this volume [22].

nately has the disadvantage that the cyanobacterial populations on the filter are heavily contaminated by bacteria (since the latter have been equally concentrated), and purification by streaking the isolated colonies directly onto fresh agar plates is generally impossible. They need to be transferred to liquid media and should be treated, after growth, by one of the purification methods described below.

Plates should be incubated (under the light and temperature regimes of choice) in transparent vegetable crispers (whose lids are slightly opened) to prevent them from drying out too rapidly. If condensation is observed under the covers of the Petri dishes the plates need to be turned upside down so that the water cannot flood the agar surface. This is particularly necessary at incubation temperatures above 25°. The plates should be examined daily under the dissecting microscope to maximize the chances of isolating filaments and colonies before they are overgrown by contaminants.

It is prudent to accompany the isolation and transfer of single colonies and filaments on plates with parallel establishments of liquid cultures for two reasons: (1) some cyanobacteria will not grow as single filaments or microcolonies on solid media,[10] and (2) if highly contaminated by bacteria, they will be completely overgrown by the contaminants on the agar surface where growth is much more localized than in liquid culture.

Purification Methods

If is often possible to obtain axenic cultures from field material by performing isolation and the subsequent purification steps on plates as described. Sometimes, however, this is not the case, particularly if the natural sample is highly contaminated by bacteria and the cyanobacterial species are immotile and incapable of self-purification by gliding away from their contaminants, or if the single filaments or microcolonies do not grow on solid media. Purification of such cyanobacteria should be attempted after the establishment of unicyanobacterial isolates in liquid culture, by one of the following methods.

Purification by Filtration. Large or colonial unicellular cyanobacteria, or filamentous forms, can be partly purified by washing procedures using Millipore membrane filters (8–10 μm pore size) for separation from their contaminants on the basis of differential cell sizes.[42,54,55] A simple but effective way of performing this method is as follows. A sample (~10 ml) of a liquid culture, grown to a density well visible by eye, is placed with a sterile disposable plastic syringe equipped with a 10-cm-long, 20-gauge

[54] S. I. Heaney and G. H. M. Jaworski, *Br. Phycol. J.* **12,** 171 (1977).

[55] M.-E. Meffert and T.-P. Chang, *Arch. Hydrobiol.* **82,** 231 (1978).

hypodermic needle in the funnel of a sterilized Millipore filtration apparatus with a sterile membrane filter (25 mm diameter, 8–10 μm pore size) in place. The sample is filtered by vacuum filtration (adjusted so that the filtrate is released only dropwise) while keeping the culture in suspension by pumping it continuously in and out of the syringe. The vacuum is released (by disconnecting the vacuum line from the Buchner flask) once the liquid level in the funnel reaches about 2–3 ml. Diluted sterile medium is then added to fill the funnel to the top. This operation is repeated 2 or 3 times. Since the filter gets clogged with bacteria and cyanobacterial cell debris, it is then exchanged aseptically with a new filter, during which time the residual culture sample is stored in the syringe (deposited on a support in such a way that the hypodermic needle is not touched). The washing operation is continued, with a total of about 500 ml of diluted medium and 6–7 changes of the filter. The washed culture is recuperated as a concentrated suspension (about 2 ml), and aliquots are streaked onto agar plates, the remainder being used to prepare a new liquid culture (in case the first purification attempt is unsuccessful and needs to be repeated).

Single filaments should be isolated and transferred to fresh plates and to liquid medium immediately. Unicellular species must be incubated on the plates until colonies are visible under the dissection microscope, unless previous experience has shown that, when axenic, they do not grow on solid media. In the latter case, the washed sample should be used for serial dilutions in sterile medium. After incubation and growth the individual dilutions need to be tested for purity (see below). It is evident that this dilution method works only if the cyanobacterial population is well superior to that of the contaminants. If this is not the case the latter need to be reduced by additional treatments such as those described in the following section.

Purification by Treatment with Bactericidal Agents or Broad-Spectrum Antibiotics. Treatment with antimicrobial agents is often the last resort to successful purification, particularly if the bacterial contaminants grow well embedded in the cyanobacterial sheath material or if the cyanobacteria do not develop as axenic clones on plates. The use of UV or γ radiation, although also successfully employed for this purpose,[1,5,56] is not advisable because of their potential mutagenic effect. The most useful bactericidal agents and antibiotics for the selective killing of bacterial contaminants are those preferentially effective on growing cells, such as those currently employed for the enrichment of bacterial auxotrophic mutants. Although several different bactericidal compounds have been

[56] M. P. Kraus, *Nature (London)* **211,** 310 (1966).

successfully employed for the purification of cyanobacteria,[42,45] only the utilization of ampicillin is described here, which may serve as a model for the use of other similarly effective agents. A cyanobacterial culture (~10 ml) from the mid-exponential phase of growth is exposed to a starvation period in the dark to deplete the cells of their endogenous carbon reserves, thereby minimizing the chances of growth and cell division at the expense of the latter during the subsequent antibiotic treatment. The medium is then supplemented with 0.02–0.1% casamino acids, 0.5% (w/v) glucose, and 1000 μg/ml ampicillin, and incubation in the dark is continued for 24–48 hr. Aliquots of the culture are removed every 6–8 hr (ensuring the survival of more delicate cyanobacteria that might not tolerate this treatment for longer incubation periods, but which may already be sufficiently liberated of viable contaminants) and are (after centrifugation and resuspension in a small amount of medium) streaked on plates. Their further treatment is identical to that described in the previous section.

Test for Purity. Presumptively pure cyanobacterial cultures are best grown in liquid medium to stationary phase. Aliquots are then examined critically under the microscope using phase-contrast objectives and oil immersion. For additional macroscopic tests small drops of the liquid culture are placed on plates prepared with the appropriate mineral medium supplemented with casamino acids (0.02–0.05%, w/v) and glucose (0.5%, w/v). More complex media should be avoided since some bacteria will not grow on them. The plates are then incubated in the dark for several days at a temperature typical for growth of the cyanobacteria to be tested, since any contaminants still present should grow at that temperature. The contaminants will then be easily detected by their superior growth, color, and other typical appearances (i.e., fungal hyphae). If the culture is heavily contaminated the entire area of the test drop will be covered with bacterial or fungal growth; a low degree of contamination will reveal itself in the form of microcolonies on the cyanobacterial lawn. Some contaminants (e.g., gliding bacteria) do not form discrete colonies on the surface of the agar. Their presence (or, it is hoped, absence) can only be confirmed by additional microscopy, after cutting small sections from the agar occupied by the cyanobacterial deposit.

Concluding Remarks

If this chapter seems overly full of advice, it is because the author learned in the course of its preparation that many unsuccessful attempts to isolate cyanobacteria could possibly have been avoided had they been performed with more imagination and respect for the organisms.

[2] Recognition and Identification of Cyanobacteria

By ROSMARIE RIPPKA

Introduction

The recognition of cyanobacteria in the natural habitat is a prerequisite for their isolation, and, in order to isolate more particular members, it helps to know their ecological distribution. Some brief comments on recognition and distribution of these organisms are presented in the first two sections of this chapter.

Following successful isolation, cyanobacteria (as other organisms) should be identified by a name, which serves as an indicator of the respective phenotypic properties and is therefore crucial for scientific communication. Unless the organisms have not been previously described, their names have to be chosen from an existing system of classification. Since cyanobacteria were first recognized more than 150 years ago, a bewildering array of genera and species has been created by botanists and ecologists. Classifications were based either on the properties observable on samples collected from the natural habitat or on those extractable from dried herbarium specimens. Furthermore, many genera and species underwent repeated taxonomic revisions, leading to a large number of synonyms that only botanical experts are capable of unraveling.

Of the various traditional taxonomic treatises published in the past, that of Geitler[1] gained the most widespread popularity among phycologists and bacteriologists. However, Geitler himself admitted that certain generic and specific assignments were purely arbitrary and not well founded.[1] This became even more evident when applying Geitler's field-based system of classification to the identification of cyanobacteria in pure culture. In an attempt to adapt this system sufficiently to be suitable for axenic cultures as well, Rippka *et al.*[2] proposed certain modifications. Given that cyanobacterial culture work was (and still is) in its infant years (although some progress has been made), these proposals were considered to be preliminary and thus subject to further changes. This is particularly so because cyanobacteria (even in the hands of bacteriologists) do

[1] L. Geitler, *in* "Rabenhorst's Kryptogamenflora von Deutschland, Österreich und der Schweiz: 14. Cyanophyceae" (R. Kolkwitz, ed.). Akademische Verlagsgesellschaft, Leipzig, Federal Republic of Germany, 1932.

[2] R. Rippka, J. Deruelles, J. B. Waterbury, M. Herdman, and R. Y. Stanier, *J. Gen. Microbiol.* **111,** 1 (1979).

not lend themselves as readily to phenotypic classification as do members of other bacterial groups: their lack of nutritional diversity excludes physiological parameters as discriminatory properties (except possibly at the species level). Therefore, the limited phenotypic characteristics of cyanobacteria (in essence restricted to morphological features) will have to be supplemented by biochemical and genetic analyses. Such studies, performed in conjunction, will determine the degree of genetic relatedness between organisms, but may also reveal biochemical properties associated with particular genotypes, which could subsequently be exploited as important taxonomic tools. The knowledge gained would lead to the establishment of a genotypic classification (indicative of evolutionary relationships) that may, however (if organisms that look alike are genetically unrelated), not be as practical for direct application to ecological studies as a phenotypic system of classification.

The major part of this chapter discusses these problems in more detail and represents a modest attempt to provide a basis for the creation of a universally acceptable *single* system of classification, which would honor, where possible, the merits of phycological tradition but would eliminate the possibilities of identical organisms being disguised, to the confusion of scientists, under different synonyms. Major culture collections carrying cyanobacterial isolates (axenic and nonaxenic) are listed in an appendix to this chapter.

Recognition of Cyanobacteria in the Natural Environment

Cyanobacteria occupy a rather wide range of illuminated niches in terrestrial, freshwater, marine, and hypersaline environments, where they often occur in such abundance that they are readily visible by eye. Tentative identification is greatly aided by their characteristic color, which may vary from green, blue–green, or olive green to various shades of red to purple, or even black; the color is mainly determined by the relative amounts of the major light-harvesting, water-soluble pigments phycocyanin and phycoerythrin,[3] minor color contributions being due to chlorophyll *a* and carotenoids.[4] Although such colors are very indicative of cyanobacteria, final proof that field specimens do belong to this group of organisms will only come from: (1) critical examination of the samples under the microscope, to ensure that they are prokaryotes; (2) a demonstration by growth requirements or O_2-evolution measurements, to prove that they are not photosynthetic bacteria (which may be similar in color

[3] A. N. Glazer, this volume [31].

[4] G. Sandmann, this volume [35].

due to carotenoids but lack phycobiliproteins and an O_2-evolving photosystem II, and therefore need an electron donor other than H_2O for growth); and (3) an analysis of their pigment content, to exclude the possibility that they may be *Prochloron*-like organisms[5] (which, like cyanobacteria, are oxygenic photosynthetic prokaryotes, but do not contain phycobiliproteins, their yellow–green color resulting uniquely from chlorophyll *a* and *b*).

The color of cyanobacteria in the natural environment, however, can be deceptive: starvation of nitrogen[3] or sulfur,[6] or conditions leading to photooxidation[7] (high light intensity together with low CO_2 concentrations), cause a reduced phycobiliprotein content, and cyanobacteria exposed to such limitations will display the yellow–green color of chlorophyll *a*, rather than the more typical colors mentioned above. Furthermore, certain cyanobacteria can be deep brown in appearance, owing either to massive akinete formation or to the production of brown sheaths[1] that mask the typical coloration of cells or filaments.

The growth pattern of cyanobacteria is variable, and the macroscopic appearance in the natural environment is generally determined by the predominant species: it may be a stratified smooth or velvetlike layer; a flocculant mass of varying compactness; a crustlike aggregate; or rather defined gelatinous or hairy colonies. Growth in homogeneous suspension is relatively infrequent, at least in densities to be visible by eye. The growth appearance of epiphytic, endosymbiotic, and endolithic cyanobacteria is determined by the host or substrate interrelationship and cannot be described in detail here.

Some Typical Habitats of Cyanobacteria

For the isolation of specific types of cyanobacteria it is important to know their ecological distribution. This information is available mainly from the taxonomic descriptions made in the past by phycologists and ecologists, who identified species not only by their morphological properties but also by their appearance in the natural environment (which by definition had to be known).

For those who have never isolated cyanobacteria a few typical habitats will be cited. Dimly lit caves and bare calcareous mountain slopes are good sources for chroococcalean cyanobacteria such as *Gloeothece* and *Gloeocapsa;* they are often accompanied by filamentous heterocystous

[5] R. A. Lewin, *in* "The Prokaryotes" (M. P. Starr, H. Stolp, H. G. Trüper, A. Balows, and H. G. Schlegel, eds.), Vol. 1, p. 257. Springer-Verlag, Berlin, 1981.

[6] A. Schmidt, this volume [62].

[7] A. Abeliovich and M. Shilo, *J. Bacteriol.* **111,** 682 (1972).

forms such as *Nostoc* and *Scytonema*.[1,8] Hot-spring environments will harbor, or often be dominated by (depending on the pH, temperature, and supply of combined nitrogen), the thermophilic cyanobacteria *Synechococcus lividus, Mastigocladus laminosus,* or *Chlorogloeopsis* sp.[9,10] During certain periods of the year, freshwater lakes may form massive water blooms composed almost exclusively of *Microcystis aeruginosa, Aphanizomenon flos-aquae,* or *Oscillatoria agardhii,* planktonic cyanobacteria that adjust their position in the water column by gas-vacuole-mediated buoyancy regulation.[11,12] The red *Trichodesmium* is a most conspicuous and well-known bloom former in tropical oceans.[13] Hypersaline alkaline lakes are often dominated by the attractive spiral and gas-vacuolate cyanobacterium *Arthrospira* (or *Spirulina*) *platensis,* which for centuries has been used as food supplement in certain parts of the world (Lake Chad in Central Africa and Mexico being the most well-known examples).[14,15] Rocks of shallow mountain rivers are often covered by *Chamaesiphon,* a budding cyanobacterium which, however, may also be found as an epiphyte on *Cladophora* or other algal species in freshwater lakes.[1,16]

Tapering cyanobacteria, traditionally assigned to *Rivularia, Calothrix,* or *Isactis,* have been readily isolated from salt marshes and rocks along the seashore.[2] Oyster shells are colonized by baeocyte-forming cyanobacteria such as *Dermocarpa, Chroococcidiopsis,* or *Pleurocapsa,*[17] whereas some of the endolithic relatives of this cyanobacterial group may inhabit limestones in intertidal regions[18,19] or even more extreme environments such as rock fissures and cracks or the structural cavities of porous rocks in hot and cold deserts.[20,21] The lovely black–green carpets covering rocks

[8] A. Zehnder, personal communication.

[9] R. W. Castenholz, *in* "The Prokaryotes" (M. P. Starr, H. Stolp, H. G. Trüper, A. Balows, and H. G. Schlegel, eds.), Vol. 1, p. 236. Springer-Verlag, Berlin, 1981.

[10] R. W. Castenholz, this volume [3].

[11] A. E. Walsby, *in* "The Prokaryotes" (M. P. Starr, H. Stolp, H. G. Trüper, A. Balows, and H. G. Schlegel, eds.), Vol. 1, p. 224. Springer-Verlag, Berlin, 1981.

[12] A. E. Walsby, this volume [72].

[13] G. E. Fogg, *in* "The Biology of Cyanobacteria" (N. G. Carr and B. A. Whitton, eds.), p. 491. Blackwell, Oxford, England, 1982.

[14] O. Ciferri, *Microbiol. Rev.* **47,** 551 (1983).

[15] R. D. Fox, *in* "Algoculture: La Spirulina, un Espoir pour le Monde de la Faim." Edisud, La Calade, Aix-en-Provence, France, 1986; co-published as R. D. Fox, "Algoculture." Laboratoire de La Roquette, St. Bauzille de Putois, France, 1984.

[16] E. Kann, *Verh.–Int. Ver. Theor. Angew. Limnol.* **16,** 646 (1966).

[17] J. B. Waterbury and R. Y. Stanier, *Microbiol. Rev.* **42,** 2 (1978).

[18] S. Golubic, *Am. Zool.* **9,** 747 (1969).

[19] T. LeCampion-Alsumard and S. Golubic, *Arch. Hydrobiol. Suppl.* **71** (*Algol. Stud.* **38/39**), 119 (1985).

[20] E. I. Friedman, *Origins Life* **10,** 223 (1980).

[21] M. Potts and E. I. Friedman, *Arch. Microbiol.* **130,** 267 (1981).

in mountainous regions irrigated by spray water or melting snow will almost certainly be composed of the rather complex true-brancher *Stigonema* (never reported to have been obtained in pure culture) and the false-brancher *Scytonema,* both of which are characterized by aerial growth, leading to the "carpet effect."[1,8] *Oscillatoria limnetica* (sensu Geitler[1]), a filamentous nonheterocystous cyanobacterium capable of anaerobic photosynthesis,[22] dominates the H_2S-rich bottom layer of the shallow mesothermal, monomictic Solar Lake during winter stratification.[23] *Microcoleus chthonoplastes* (sensu Geitler[1]) is intimately involved in the formation of intertidal mats.[24,25] And surprisingly, considering the general preference of cyanobacteria for neutral and alkaline environments, mildly acidic bog pools proved to exhibit an unimaginable array of unicellular and filamentous species, some of which have never been isolated, or of which more representatives are needed [e.g., large (10 μm wide) *Synechococcus* species, *Merismopedia, Chroococcus, Gloeothece*].[26]

Nitrogen-fixing *Nostoc* species, harbored in the outer cortex of the coralloid roots of Cycadaceae (*Encephalartos, Bowenia, Macrozamia,* etc.) or in the stem and leaf base of the angiosperm *Gunnera,* occur generally as unispecific cyanobacterial endosymbionts in such quantities, and relative purity, that their isolation and axenization preceded that of their free-living asymbiotic relatives by many years or decades.[27–29] Nitrogen-fixing endosymbionts, identified as *Anabaena azollae* (which according to the classification of Rippka *et al.*[2] would be assignable to the genus *Nostoc,* see the section on the taxonomy of cyanobacterial endosymbionts later in this article), were claimed to have been isolated and grown in pure culture from the leaf cavities of two different species of *Azolla,*[30,31] a water fern that plays an important role as biofertilizer in certain parts of the world.[32]

Excluding the more "exotic" cyanobacteria, many habitats are occu-

[22] Y. Cohen, E. Padan, and M. Shilo, *J. Bacteriol.* **123,** 855 (1975).
[23] Y. Cohen, W. E. Krumbein, and M. Shilo, *Limnol. Oceanogr.* **22,** 609 (1977).
[24] B. J. Javor and R. W. Castenholz, *Geomicrobiol. J.* **2,** 237 (1981).
[25] L. J. Stal, H. van Gemerden, and W. E. Krumbein, *FEMS Microbiol. Ecol.* **31,** 111 (1985).
[26] R. Rippka, unpublished observations.
[27] R. Harder, *Z. Bot.* **9,** 145 (1917).
[28] G. Winter, *Beitr. Biol. Pflanz.* **23,** 295 (1935).
[29] J. W. Bowyer and V. B. D. Skerman, *J. Gen. Microbiol.* **54,** 299 (1968).
[30] J. W. Newton and A. I. Herman, *Arch. Microbiol.* **120,** 161 (1979).
[31] E. Tel-Or, T. Sandovsky, D. Kobiler, H. Arad, and R. Weinberg, *in* "Photosynthetic Prokaryotes" (G. C. Papageorgiou and L. Packer, eds.), p. 303. Elsevier, New York, 1983.
[32] T. A. Lumpkin and D. L. Plucknett, *Econ. Bot.* **34,** 111 (1980).

pied by more "ordinary" members: almost any ditch or pond will harbor at least one unicellular species such as *Synechococcus* or *Synechocystis* as well as the rather cosmopolitan filamentous nonheterocystous cyanobacteria assignable to *Pseudanabaena* or to the ill-defined LPP group (*Lyngbya, Phormidium, Plectonema*).[2] Most soil samples from a garden, greenhouse, or an ordinary houseplant will reveal *Oscillatoria, Anabaena,* or *Nostoc* species among the extremely competitive members of the LPP group mentioned above; occasionally, a lucky weekend vegetable gardener–cyanobacteriologist might also discover the handsome *Cylindrospermum* growing in a patch of leeks.[33]

Identification of Cyanobacteria

Historical Background of Cyanobacterial Taxonomy

For about 150 years cyanobacteria were considered a special group of algae, the blue–green algae. As a result, their classification was developed by phycologists working under the provisions of the International Code of Botanical Nomenclature. According to this code the name of a taxon above the family level is not subject to restrictions of priority.[34] Consequently many names have been proposed for this group of microorganisms and have been used synonymously at the liberty of the authors: Myxophyceae (Wallroth, 1833) emend. Stitzenberger, Rabenhorst 1860, Phycochromophyceae Rabenhorst 1863, Cyanophyceae Sachs 1874, or Schizophyceae Cohn 1879 are the most well-known examples (see Refs. 2 and 34). The name Cyanophyceae has survived the longest and is still widely used by phycologists and botanists today. Bacteriologists believe that the name "cyanobacteria," first proposed by Stanier,[35] is more appropriate than "Cyanophyceae," in view of the prokaryotic (thus bacterial) cellular properties of these phototrophs. Given the flexibility of the Botanical Code for the naming of higher taxa, it should be relatively easy to introduce the name "cyanobacteria" under this code, if this were desired by phycologists.

The traditional characters employed for the identification of cyanobacterial genera and species were those determinable on field material: structural properties of the cells or filaments, color, shape and structure of colonial aggregates. Often the ecological niche inhabited by the organisms was also of important discriminatory value. Type materials under the

[33] M. Herdman, personal communication.

[34] P. Bourelly, *in* "Les Algues d'eau Douce: 3. Les Algues Bleues et Rouges. Les Eugléniens, Péridiniens et Cryptomonadines." Boubée, Paris, 1970.

[35] R. Y. Stanier, *in* "Bergey's Manual of Determinative Bacteriology" (R. E. Buchanan and N. E. Gibbons, eds.), p. 22. Williams & Wilkins, Baltimore, Maryland, 1974.

Botanical Code are dried herbarium specimens, descriptions, and illustrations; laboratory cultures are not yet recognized for this purpose, but an attempt to change this has recently been made.[36]

Studies on axenic cyanobacteria revealed that many characters traditionally employed to discriminate not only between species but even between genera are either not expressed in culture or vary with culture conditions. Furthermore, although phenetic and genetic entities could be recognized, it was often difficult or impossible to assign them confidently to traditional genera or species, as a result of the limited information content of the nonliving botanical type specimens.

As a remedy to these problems Stanier *et al.*[37] proposed to place the nomenclature of cyanobacteria under the rules of the International Code of Nomenclature of Bacteria, under which a living type strain, allowing modern comparative analysis, defines both species and (through type species) genus. This proposal has not yet been accepted, but hope should not be abandoned, particularly since an agreement between botanists and bacteriologists for the mutual recognition of both the Botanical and the Bacteriological Codes of Nomenclature has recently been submitted.[38] The second, more immediate, solution for the identification of axenic cyanobacteria was the proposed "working classification" of Rippka *et al.*,[2] which certainly will have to be modified as more complete data on fine structural, physiological, biochemical, and genetic characteristics become available.

Cyanobacteria: Their Place in the Hierarchy of Prokaryotes

According to the proposal by Gibbons and Murray,[39] cyanobacteria are members of the kingdom Procaryotae Murray 1968 and are included in the division Gracilicutes (bacteria with a Gram-negative cell wall); they are assignable to the class Photobacteria which, in its subclass Oxyphotobacteriae, embraces them as the order Cyanobacteriales. Their general properties are as follows: prokaryotic organisms of diverse structural properties, possessing chlorophyll *a* and phycobiliproteins; they perform oxygenic photosynthesis (using H_2O as their electron donor in the light, O_2 being a by-product of photosynthetic activity); some are also capable

[36] E. I. Friedman and L. J. Borowitzka, *Taxon* **31,** 673 (1982).

[37] R. Y. Stanier, W. R. Sistrom, T. A. Hansen, B. A. Whitton, R. W. Castenholz, N. Pfennig, V. N. Gorlenko, E. N. Kondratieva, K. E. Eimhjellen, R. Whittenbury, R. L. Gherna, and H. G. Trüper, *Int. J. Syst. Bacteriol.* **28,** 335 (1978).

[38] H. G. Trüper (Secretary for Subcommittees of the International Committee of Systematic Bacteriology [ICSB]), personal communication; proposal of Minute 7 of the Discussion Workshop "Taxonomy of Cyanobacteria" (Paris, 1985), submitted to the Executive Board of the ICSB at the 14th International Congress of Microbiology, 1986.

[39] N. E. Gibbons and R. G. E. Murray, *Int. J. Syst. Bacteriol.* **28,** 1 (1978).

of performing anoxygenic photosynthesis, using H_2S as electron donor; some are facultative photo- or chemoheterotrophs, but growth under these conditions is always much slower than under photoautotrophic conditions, and the range of carbon substrates utilized is very limited; some members have a requirement for vitamin B_{12}; many fix atmospheric nitrogen.

Classification of Cyanobacteria in Axenic Culture

In the provisional classification of Rippka *et al.*,[2] based on the results of pure culture studies performed by a number of authors,[17,40–44] cyanobacteria are subdivided into five major groups (sections), whose properties are summarized in Key 1. The genera of each section are described in Keys 2–6. It should be remembered that this classification is based entirely on the axenic representatives in the Pasteur Culture Collection of Cyanobacteria (PCC), isolated and characterized prior to 1979. For members of Section III (see Key 4) some modifications to the assignments of Rippka *et al.*[2] have been made, taking into account the recent ultrastructural studies of Guglielmi and Cohen-Bazire.[45,46] With two exceptions, specific names have not been attached formally to the genera (although some suggestions for specific assignments are proposed in this chapter) because of the difficulty of making unambiguous identifications in the complete absence of living type species. The exceptions are two recently created monotypic genera, *Chlorogloeopsis* Mitra and Pandey 1966[47] and *Gloeobacter* Rippka, Waterbury, and Cohen-Bazire 1974,[41] both of which are defined by living type species (*Chlorogloeopsis fritschii* and *Gloeobacter violaceus*). As an interim measure, Rippka *et al.*[2] designated one (or more, if necessary) reference strains that seemed reasonably representative (although not necessarily corresponding to the botanical type species, as far as could be judged) of the genera recognized by them. These reference strains are summarized in Table I. For more detailed descriptions and photographic documentation of the cyanobacterial strains examined in culture, the reader should consult the original publications.[2,17,40–46]

[40] R. Y. Stanier, R. Kunisawa, M. Mandel, and G. Cohen-Bazire, *Bacteriol. Rev.* **35,** 171 (1971).
[41] R. Rippka, J. B. Waterbury, and G. Cohen-Bazire, *Arch. Microbiol.* **100,** 419 (1974).
[42] J. B. Waterbury and R. Y. Stanier, *Arch. Microbiol.* **115,** 249 (1977).
[43] C. N. Kenyon, R. Rippka, and R. Y. Stanier, *Arch. Mikrobiol.* **83,** 216 (1972).
[44] M. Herdman, M. Janvier, J. B. Waterbury, R. Rippka, R. Y. Stanier, and M. Mandel, *J. Gen. Microbiol.* **111,** 63 (1979).
[45] G. Guglielmi and G. Cohen-Bazire, *Protistologica* **18,** 151 (1982).
[46] G. Guglielmi and G. Cohen-Bazire, *Protistologica* **18,** 167 (1982).
[47] A. K. Mitra and D. C. Pandey, *Phykos* **5,** 106 (1966).

Comments to Genera of Section I. Section I (Key 2) includes five genera (whose members divide by binary fission) of the botanical order Chroococcales and the genus *Chamaesiphon*, traditionally assigned to the order Chamaesiphonales. *Chamaesiphon* was thought to reproduce by "exospore" formation,[1] but Waterbury and Stanier[42] recognized that this mode of reproduction is analogous to that of some bacteria which divide by "budding," a process whereby small daughter cells are released successively from the apical end of a mother cell as a result of unequal binary fission. Budding being only a slight variance of the more common equal binary fission, the genus *Chamaesiphon* was included in Section I.[2]

TABLE I
DNA BASE COMPOSITIONAL SPANS AND REFERENCE STRAINS OF GENERA OR GROUPS

Section	Genus (or group)	Mean DNA base composition (mol % G + C)[a]	Reference strain[b]
Section I			
	Synechococcus		
	Type I	39–43 (5)[c]	PCC 7202[d]
	Type II	47–56 (15)[c,e]	PCC 6301[d]
	Type III	66–71 (9)[c]	PCC 6307[d]
	Gloeothece	40–43 (5)[c]	PCC 6501[d]
	Synechocystis		
	Type I	35–37 (5)[c]	PCC 6308
	Type II	42–48 (12)[c]	PCC 6714
	Gloeocapsa	40–46 (4)[c]	PCC 73106
	Gloeobacter	64 (1)[c]	PCC 7421
	Chamaesiphon	47 (2)[c]	PCC 7430
Section II			
	Dermocarpa	38–44 (6)[c]	PCC 7301
	Xenococcus	44 (2)[c]	PCC 7305
	Dermocarpella	45 (1)[c]	PCC 7326
	Myxosarcina	43–44 (2)[c]	PCC 7312
	Chroococcidiopsis	40–46 (8)[c]	PCC 7203
	Pleurocapsa group	39–47 (11)[c]	
	Type I[f]		PCC 7319
	Type II[g]		PCC 7516
Section III			
	Spirulina	53 (1)[c]	PCC 6313
	Arthrospira	44 (1)[c]	PCC 7345
	Oscillatoria	40–50 (10)[c]	PCC 7515
	Pseudanabaena	44–52 (8)[c]	PCC 7429
	LPP group A[h]	43 (1)[c]	PCC 7419
	LPP group B[i]	42–67 (19)[c]	
	"Plectonema boryanum" type[j]	46–48 (5)[c]	PCC 6306

TABLE I (*continued*)

Section	Genus (or group)	Mean DNA base composition (mol % G + C)[a]	Reference strain[b]
Section IV			
	Anabaena	38–44 (3)[c,k]	PCC 7122
	Nodularia	41–45 (2)[c,k]	PCC 73104
	Cylindrospermum	43–45 (3)[c,k]	PCC 7417
	Nostoc	39–47 (21)[c,k]	PCC 73102
	Scytonema	44 (1)[c,k]	PCC 7110
	Calothrix	40–45 (15)[c,k]	PCC 7102
	"*Tolypothrix tenuis*" type[l]	41–46 (5)[c,k]	PCC 7101
Section V			
	Chlorogloeopsis fritschii	42–43 (2)[c]	PCC 6912
	Fischerella	42–46 (8)[c,k]	PCC 7414

[a] Number in parentheses indicates the number of strains examined.

[b] Reference strains taken from R. Rippka, J. Deruelles, J. B. Waterbury, M. Herdman, and R. Y. Stanier, *J. Gen. Microbiol.* **111,** 1 (1979).

[c] M. Herdman, M. Janvier, J. B. Waterbury, R. Rippka, R. Y. Stanier, and M. Mandel, *J. Gen. Microbiol.* **111,** 63 (1979).

[d] These reference strains were proposed as types or neotypes of the following genera: PCC 7202, *Cyanobacterium stanieri;* PCC 6301, *Synechococcus elongatus;* PCC 6307, *Cyanobium gracile;* PCC 6501, *Gloeothece membranacea.* See R. Rippka and G. Cohen-Bazire, *Ann. Microbiol. (Paris)* **134B,** 21 (1983).

[e] Data for one strain (PCC 7942) taken from A. M. R. Wilmotte and W. T. Stam, *J. Gen. Microbiol.* **130,** 2737 (1984).

[f] Typical of members with symmetric baeocyte enlargement.

[g] Typical of members in which the baeocyte develops early polarity.

[h] "*Lyngbya aestuarii*" type, see comments to Section III and Key 4.

[i] This group is very heterogeneous, and designation of a reference strain is therefore meaningless.

[j] See comments to Section III and Key 4.

[k] M.-A. Lachance, *Int. J. Syst. Bacteriol.* **31,** 139 (1981).

[l] See comments to Section IV and Key 5.

Considering the structural simplicity of unicellular cyanobacteria dividing by binary fission, a surprising array of genera and species has been described on the basis of field observations, the boundaries of which, as Geitler[1] admitted, are not always easy to recognize. Furthermore, much additional confusion arose when Drouet and Daily[48] proposed a revision for the classification of chroococcalean cyanobacteria in which many of the traditional genera and species were renamed. It is for this reason that some of the rod-shaped unicellular cyanobacteria in Section I entered the

[48] F. Drouet and W. A. Daily, *Butler Univ. Bot. Stud.* **12,** 1 (1956).

Pasteur Culture Collection (see Refs. 2 and 40) under designations such as *Anacystis nidulans, Coccochloris elabens,* or *Agmenellum quadruplicatum,* which had either not been used at all, or only as older synonyms, in the Geitlerian system of classification. The revision of Drouet and Daily,[48] based entirely on the examination of herbarium specimems, has found very little acceptance even among botanists. Therefore, Rippka *et al.*[2] followed as much as possible the widely used system of classification by Geitler,[1] which is also largely in agreement with that of Bourelly,[34] a more recent taxonomic treatise. However, Geitlerian genera defined uniquely on the basis of massive slime formation (such as *Aphanothece* for rod-shaped members dividing in one plane or *Aphanocapsa,* its coccoid counterpart, dividing in more than one plane) were not recognized by Rippka *et al.,*[2] since this character did not prove to be sufficiently stable in culture.

Furthermore, botanical genera such as *Merismopedia* and *Eucapsis,* the members of which, in the natural habitat, form platelike or cubical colonies, respectively, were also not recognized in the provisional classification of Rippka *et al.,*[2] mainly because a controlled synchrony of cellular division leading to such colonial aggregates was either not observed in culture or seen only sporadically.[40] It is possible that more detailed studies on cell division (particularly under a regime of light–dark incubation[40]) of strains included in the genus *Synechocystis* (for which the number of planes of division was not specified in the original botanical description) will reveal that some of them need to be transferred to either *Merismopedia* or *Eucapsis* (see also Ref. 49). This assumption is supported by some unpublished observations[26] on a selected number of *Synechocystis* strains in the PCC: those characterized by a mean DNA base composition of 35–37 mol % G + C[44] produced platelike colonial aggregates on slide culture, whereas members of the higher GC group (42–48 mol % G + C)[44] formed only irregular colonial arrangements.

It is also difficult to decide whether planktonic coccoid cyanobacteria containing gas vacuoles (i.e., PCC 7806, 7820, and 7941) that were purified[26] after the publication of Rippka *et al.*[2] should be assigned to the traditional genus *Microcystis* (not all species of which are coccoid or contain gas vacuoles[1]) or should be more prudently identified as a gas-vacuolate species of the genus *Synechocystis*. The high frequency of spontaneous mutants observed in culture that have lost the character of gas vacuole formation, together with the rather imprecise definition of the genus *Microcystis,*[1,40] would seem to favor the latter option. However, one could also propose to restrict the ill-defined genus *Microcystis* on the basis of the properties of its most well-known species, *Microcystis aeru-*

[49] R. Rippka, this volume [1].

ginosa, which is coccoid, divides in more than one plane, and contains gas vacuoles; the PCC isolates 7806, 7820, and 7941 correspond perfectly to the description of this species. If biochemical and genetic characterization of the latter three strains (or others of similar properties) demonstrates that the generic separation from *Synechocystis* is justified, it would be possible by comparative analysis to identify those members of *Microcystis* that do not exhibit gas vacuolation and which therefore are readily (but incorrectly) identified as *Synechocystis*. Since such studies have not yet been undertaken, these latter three strains are carried in the PCC as *Synechocystis* ("*Microcystis aeruginosa*" type).

The botanical tradition[1] of separating ensheathed coccoid members of the Chroococcales into the genera *Gloeocapsa* and *Chroococcus* was abandoned by Rippka *et al.*[2] *Chroococcus* was distinguished from *Gloeocapsa* by its tightly apposed sheaths around the cells that in the postdivisional state are hemispherical; the characteristics of the genus *Gloeocapsa* are round postdivisional cells that are enclosed by sheath material which is much less tightly apposed. The factor that determines postdivisional cell shape is probably the degree of compression exerted by the more or less firm sheaths that enclose the daughter cells. Therefore, only the older genus *Gloeocapsa* was recognized for ensheathed coccoid cyanobacteria.[2]

The genus *Gloeobacter* (type species *Gloeobacter violaceus*) was created for strain PCC 7421,[41] which so far is unique among cyanobacteria: it lacks thylakoids. Its phycobilisomes are located on the internal side of the cytoplasmic membrane.[50] With respect to morphological properties visible by light microscopy it closely resembles the description of *Gloeothece coerulea*.[1,41]

The strains included in the genera *Synechococcus* and *Synechocystis* vary greatly with respect to cell size and physiological and biochemical properties. This is also reflected by the rather wide span of mean DNA base composition (39–71 mol % G + C for strains of the former genus and 35–48 mol % G + C for those of the latter).[44] For this reason, for both genera several reference strains have been designated, each of which is representative for members of a given range of GC (see Table I).

In search for a genus *Cyanobacterium* (which for obvious reasons does not exist under the Botanical Code), which is required to bring cyanobacteria as the order Cyanobacteriales legitimately under the Bacteriological Code of Nomenclature, Rippka and Cohen-Bazire[51] proposed to split the genus *Synechococcus* (sensu Geitler 1932)[1] into four generic entities: *Cyanobacterium,* for small rods (<3 μm wide) of low GC content

[50] G. Guglielmi, G. Cohen-Bazire, and D. A. Bryant, *Arch. Microbiol.* **129,** 181 (1981).
[51] R. Rippka and G. Cohen-Bazire, *Ann. Microbiol.* (Paris) **134B,** 21 (1983).

(39–41 mol %); *Cyanothece,* for larger strains (>4 μm wide) of low GC (42 mol %); *Synechococcus,* for members of variable size, but less than 4 μm wide, and a span of mean DNA base composition of 47–56 mol % G + C; and *Cyanobium,* for a rather homogenous group of small rod-shaped cyanobacteria (less than 2 μm wide) of a GC between 66 and 71 mol %. Although DNA/DNA hybridization studies[52] are largely in agreement with this proposed generic subdivision of *Synechococcus,* it would perhaps be more prudent (and certainly more practical) to continue using the classification of Rippka *et al.,*[2] described in Key 2, until more strains of each subgroup have been isolated and characterized. This is the more advisable, since members of the genus *Synechococcus* sensu Rippka and Cohen-Bazire[51] (i.e., exhibiting a GC range of 47–56 mol %) form neither tight phenotypic nor genotypic clusters and will eventually have to be subdivided further. However, it is already fairly clear that, unless further studies reveal easily recognizable phenotypic properties useful for taxonomic identification, the ultimate division of the genus *Synechococcus* (sensu Geitler 1932[1]) and possibly other chroococcalean genera will require a genotypic (rather than a phenotypic) classification.

With this goal in mind, it is suggested that the type or neotype designations proposed by Rippka and Cohen-Bazire[51] for four chroococcalean reference strains be retained: *Cyanobacterium stanieri* PCC 7202, *Synechococcus elongatus* PCC 6301, *Cyanobium gracilis* PCC 6307, and *Gloeothece membranacea* PCC 6501 (Table I, footnote *d*). They should be considered as a modest but important taxonomic foundation upon which, by comparative biochemical and genetic analyses, a more advanced classification for chroococcalean cyanobacteria can be constructed. Since Rippka and Cohen-Bazire[51] did not designate a type strain for *Cyanothece* (sensu Rippka and Cohen-Bazire[51]), strains of this genus have provisionally been retained in *Synechococcus* type I (see Table I).

Comments to Genera of Section II. Section II (see Key 3) comprises five genera and one provisional assemblage, the *Pleurocapsa* group, whose unifying character is that of reproduction by multiple fission, a process of rapid binary fission unaccompanied by growth, which results in the formation of small, spherical daughter cells. The latter were termed "endospores" by phycologists.[1] However, to avoid confusion with bacterial endospores, which share no common developmental or physiological properties with the small cyanobacterial cells resulting from multiple fission, Waterbury and Stanier[17] renamed the latter "baeocytes" (Greek: small cells). Cyanobacteria exhibiting this mode of reproduction were traditionally assigned to one order, Chamaesiphonales,[1] which was later subdivided into Chamaesiphonales (synonym Dermocarpales) and

[52] A. M. R. Wilmotte and W. T. Stam, *J. Gen. Microbiol.* **130,** 2737 (1984).

Pleurocapsales,[53] the latter order being primarily restricted to pseudofilamentous forms. The members of the genus *Chamaesiphon,* whose mode of reproduction by "exospores" (i.e., budding) was thought to be analogous to "endospore" formation (i.e., multiple fission), were also included in the order Chamaesiphonales.[1,53] For reasons outlined above, Waterbury and Stanier[42] assigned this genus to the order Chroococcales, whose members divide by binary fission (see comments to Section I). With the exclusion of *Chamaesiphon* from cyanobacteria reproducing by multiple fission, Section II corresponds most closely to the botanical order Pleurocapsales, which was the only order recognized by Waterbury and Stanier[17] for organisms of this type of reproduction.

The complexity of the developmental cycles of pleurocapsalean cyanobacteria (see Key 3) excludes precise definitions of genera (or species) on the basis of field observations alone, particularly if, as often is the case, members of several different types are growing on a common substrate.[17] Therefore, it is not surprising that many generic descriptions of this order proved to be controversial and led to several revisions (see Ref. 17). However, thanks to the elegant work of Waterbury and Stanier,[17] who characterized the ultrastructural features of 32 axenic pleurocapsalean strains (representing at least six genera), in conjunction with very detailed developmental studies, performed on slide cultures (Cooper dish method[17]), it is now possible to recognize five phycological genera unambiguously, provided that they are redefined as proposed by these authors (and as is outlined in Key 3).

Pseudofilamentous pleurocapsalean cyanobacteria are traditionally identified on the basis of colony formation and habitat (endolithic, epiphytic, or endophytic).[1] The influence of the natural substrate on growth morphology is difficult to determine in axenic culture. Therefore, Waterbury and Stanier[17] preferred to assign such cyanobacteria to a provisional assemblage: the *Pleurocapsa* group. These authors did, however, observe two major types among the 12 axenic pseudofilamentous pleurocapsalean cyanobacteria studied that could be distinguished by the presence or absence of an early polarity developed by the baeocytes on initiation of growth. Reference strains typical of each of these two subgroups are listed in Table I.

Comments to Genera of Section III. Section III is a repository for filamentous nonheterocystous cyanobacteria assignable to the botanical family Oscillatoriaceae.[1] A large number of genera have been created for members of this group that were primarily defined by such (often rather subjective) criteria as no sheath, thin, confluent, or firm sheath surrounding either individual trichomes or groups of trichomes. Apart from

[53] F. E. Fritsch, *New Phytol.* **41,** 134 (1942).

Pseudanabaena (a genus created for straight trichomes composed of barrel-shaped to ovoid cells separated from one another by deep constrictions), cellular morphological properties were used only to define species and not genera. As a result, the classical oscillatorian taxonomy is rather confusing and often arbitrary.[1] If slime and sheath production had been excluded in the past as discriminatory characters, the species described for straight trichomes overlap with respect to cell morphology to such an extent between the various traditional genera (except for *Pseudanabaena*) that identification is practically impossible. To illustrate the problems more precisely, the whole range of cellular diversity encountered among the ensemble of oscillatorian cyanobacteria, forming straight trichomes, is represented by the species of both *Oscillatoria,* for so-called "nonsheathed" members, and *Phormidium,* for "at least partly ensheathed" filamentous nonheterocystous cyanobacteria.

Little is known about the factors that determine slime or sheath production in the natural habitat, but it is likely that the amount and structural appearance of these extracellular components is influenced by nutrient limitation and/or ionic strength of the environment, periodic drying out, light intensity, or other ecological parameters. This assumption is supported by observations on some sheathed cyanobacteria in culture, which indicated that these properties may vary with age and culture conditions.[2] In addition, it should be mentioned that the unicellular cyanobacterium *Gloeothece* PCC 6909 degrades its highly characteristic sheath envelopes entirely when incubated in the dark.[26] If this is a more general property of sheath-forming cyanobacteria, descriptions based on field observations have to be taken with extreme caution.

For these various reasons, Rippka *et al.*[2] felt that the character of extracellular envelopes had in the past been overemphasized for the classification of oscillatorian cyanobacteria. Therefore, they attempted to modify the existing system of nomenclature for this group on the basis of the cellular properties of the trichomes. The approach taken by these authors was relatively conservative: they modified the definitions of two botanical genera slightly, but abstained from major redefinitions or the creation of new genera.

Rippka *et al.*[2] proposed to recognize provisionally only three genera for the axenic oscillatorian cyanobacteria represented at that time in the Pasteur Culture Collection: (1) *Spirulina,* for helically coiled members; (2) *Oscillatoria,* for filaments composed of disk-shaped cells, which are either separated from one another by very shallow constrictions or form a seemingly continuous unit in which the individual cells can only be recognized by the cross walls; (3) *Pseudanabaena,* for representatives that are either almost unicellular (1–2 cells) or form short (3–6 cells) or longer (>6 cells) trichomes and that are all composed of cylindrical to barrel-shaped cells, exhibiting polar gas vacuoles; on completed division, adjacent cells are generally separated by deep constrictions. The definition of the genus *Spirulina* by these authors corresponded entirely to that of Geitler.[1] Those of *Oscillatoria* and *Pseudanabaena* were more restrictive than the traditional generic descriptions, in which *Oscillatoria* harbors members of all cell types and *Pseudanabaena* is not restricted to forms exhibiting polar gas vacuoles. The aims of these quite arbitrary decisions were (1) the creation of a genus *Oscillatoria* of more internal structural homologies than in the past, but defined so that the inclusion of species of structural properties similar to those of the botanical type of this genus, *Oscillatoria princeps,*[1] was warranted; (2) to allow an easy distinction between certain *Phormidium* species (e.g., *P. fragile, P. mucicola, P. frigidum, P. persicinum,* etc.)[1] and *Pseudanabaena* strains that lack polar gas vacuoles. On the basis of cell morphology this distinction would be impossible. Many *Pseudanabaena* species described[54] possess polar gas vacuoles; therefore the arbitrary restrictive definition for the latter genus given by Rippka *et al.*[2] seemed to be at least partly justified.

Strains that did not comply to the definitions of *Spirulina, Oscillatoria,* or *Pseudanabaena* as described above were included by Rippka *et al.*[2] in a provisional assemblage, the "LPP group" (A and B), named after the genera *Lyngbya, Phormidium,* and *Plectonema,*[1] all three of which had been created for members of the Oscillatoriaceae that exhibit supposedly characteristic sheath properties, but whose identification has caused problems not only to bacteriologists. The prime reason for this designation was the fact that this group included a number of strains that are hosts for the cyanophage LPP-1.[55] However, this was merely a matter of convenience and equivalent to "generic assignments pending," since the names of other genera, not readily identified in culture, such as *Schizothrix* or *Microcoleus,*[1] could equally well have been chosen to typify this heterogeneous assemblage.

Recent ultrastructural studies by Guglielmi and Cohen-Bazire[45,46] showed that the strains of Section III examined could be divided into three major phenetic entities ("tribes") on the basis of cell wall structure, distribution of pores in the longitudinal peptidoglycan, and the arrangement of fimbriae[56] along the trichome (Table II). Unfortunately, this characterization is not yet complete (e.g., strains that are hosts for the LPP cyanophages were not included in these studies), and the tribe "Pseudanabaenae" (sensu Guglielmi and Cohen-Bazire[45]) is internally quite heterogeneous (Table II). Therefore, at present only one major modification to the classification of Rippka *et al.*[2] has been adopted here (see Key 4): the Genus *Spirulina* (sensu Geitler 1932[1]) was represented by two axenic members, (1) PCC 6313, which forms relatively tightly coiled heli-

[54] K. Anagnostidis and G. H. Schwabe, *Nova Hedwigia* **11,** 417 (1966).
[55] R. S. Safferman and M.-E. Morris, *J. Am. Waterworks Assoc.* **56,** 1217 (1964).
[56] T. Vaara and M. Vaara, this volume [16].

TABLE II
MAJOR DISTINGUISHING ULTRASTRUCTURAL PROPERTIES OF OSCILLATORIAN CYANOBACTERIA[a]

Property	Spirulinae[b]	Oscillatoriae[b]	Pseudanabaenae[b]
Thickness of peptidoglycan	10 nm	15–200 nm	10 nm
Outer membrane	Atypical	Typical	Atypical
Additional outer cell wall layer	Absent	Present	Present or absent
Pattern of pores in longitudinal peptidoglycan	>1 row	1 row	>1 row, or absent
Arrangement of pores around cells	Hemicircular	Circular	Circular (if present)
Disposition of fimbriae around trichome	Perpendicular	Parallel or helicoidal	Perpendicular (if present)
Organization of fimbriae	"Tufts"	"Mantle"	"Crown" (if present)

[a] Summarized from Guglielmi and Cohen-Bazire.[1-3]

[b] Strains examined and assigned[1-3] to

Spirulinae: *Spirulina* (sensu Geitler 1932[4]), PCC 6313, 7341, 7343, 7344 (the latter three strains are not axenic)

Oscillatoriae: *Oscillatoria* (sensu Rippka *et al.*[5]), PCC 6304, 6407, 7112, 8008; LPP-B strain[5] PCC 7105; *Arthrospira* (sensu Guglielmi and Cohen-Bazire[1]) PCC 7345 (formerly *Spirulina*[5])

Pseudanabaenae: *Pseudanabaena* (sensu Rippka *et al.*[5]), PCC 6406, 6802, 6901, 6903, 7367, 7402, 7403, 7429, 7955; LPP-B strains[5] PCC 7113, 7376, 7404, 7408, and 7409

References: (1) G. Guglielmi and G. Cohen-Bazire, *Protistologica* **18,** 151 (1982); (2) G. Guglielmi and G. Cohen-Bazire, *Protistologica* **18,** 167 (1982); (3) G. Guglielmi and G. Cohen-Bazire, *Protistologica* **20,** 377 (1984); (4) L. Geitler, *in* "Rabenhorst's Kryptogamenflora von Deutschland, Österreich und der Schweiz" (R. Kolkwitz, ed.), Vol. 14, p. 916. Akademische Verlagsgesellschaft, Leipzig, Federal Republic of Germany, 1932; (5) R. Rippka, J. Deruelles, J. B. Waterbury, M. Herdman, and R. Y. Stanier, *J. Gen. Microbiol.* **111,** 1 (1979).

ces composed of small cells (2 μm wide), whose cross walls are invisible by light microscopy and (2) PCC 7345, a strain of much larger cells (~10 μm wide) and more loosely coiled trichomes, with easily visible cross walls. The ultrastructural properties of these two strains are compellingly different[45,46] (Table II) and justify their assignment to two different traditional genera, *Spirulina* for the former strain and *Arthrospira* for PCC 7345. The latter genus was originally created for strains of loosely coiled trichomes and visible cross walls, but was accepted by Geitler[1] only as a subsection of the genus *Spirulina*. It should be noted that some phycologists[34] considered helical oscillatorian cyanobacteria mere variants of members assignable to the genus *Oscillatoria*. For strain PCC 7345, this point of view is justified, since its ultrastructural properties[45,46] are not distinguishable from strains assigned to *Oscillatoria* (sensu Rippka *et al.*[2]). However, until genetic evidence requires that PCC 7345 be placed into the latter genus, it should be considered a typical representative of *Arthrospira*.

The four strains of *Oscillatoria* (sensu Rippka *et al.*[2]) examined[45] proved to be very similar with respect to major ultrastructural features (Table II), although some differences, considered by Guglielmi and Cohen-Bazire[45] to be of minor importance, were observed. However, the same type of cell wall and the same pore pattern as that observed for these four strains (as well as for *Arthrospira* PCC 7345) was also shared by one strain of LPP group B,[2] PCC 7105, that forms very thin (2 μm wide) trichomes, composed of cylindrical cells, which was therefore excluded from the genus *Oscillatoria*.[2] On the basis of these results some authors might be inclined to return to the old classification system, in which cell morphology was not considered to be important for the identification of *Oscillatoria* (or other genera). However, this seems to be unwise for the following reasons. (1) There is a pronounced difference between strains of *Oscillatoria* (sensu Rippka *et al.*[2]) and PCC 7105 with respect to their intercellular spacing; when viewed by light microscopy, using bright-field illumination, the individual cells of the latter strain are interrupted by clearly visible, light-transparent gaps, which is not the case in members assigned by Rippka *et al.*[2] to *Oscillatoria*. (2) Oscillatorian trichomes composed of cells that are longer than wide and not separated from one another by deep constrictions are heterogeneous with respect to their ultrastructural properties (compare PCC 7105 and PCC 7113[45]), which is in contrast to all oscillatorian strains composed of disk-shaped cells that so far fall into a single ultrastructural group (including the heavily ensheathed strain LPP-A 7419,[57] which may correspond to *Lyngbya aestuarii* sensu Geitler[1]). (3) If the separation of *Arthrospira* from *Oscillatoria* is maintained (see above), that of PCC 7105 from the latter genus is even more highly recommended. Therefore, a modification of the definition of the genus *Oscillatoria* (sensu Rippka *et al.*[2]) on the basis of cell wall and pore properties has been postponed until genetic analyses are in favor of such a step. As an interim measure, one might want to refer to strains exhibiting ultrastructural features similar or identical to those of *Oscillatoria* (sensu Rippka *et al.*[2]), but whose trichomes are not composed of disk-shaped cells, as "*Oscillothrix*" (using PCC 7105 as reference strain), thus expressing their relatively close structural affiliation with members of *Oscillatoria* (sensu Rippka *et al.*[2]).

All *Pseudanabaena* strains (sensu Rippka *et al.*[2]) were included in the "tribe Pseudanabaenae" by Guglielmi and Cohen-Bazire[45,58] together with

[57] G. Guglielmi, personal communication.

[58] G. Guglielmi and G. Cohen-Bazire, *Protistologica* **20,** 377 (1984).

a number of strains of the LPP group B[2] (PCC 7404, 7408, 7409, 7376, and 7113). The common characters of these strains are a relatively thin peptidoglycan (~10 nm wide) and an outer cell wall that does not represent a typical triple-layered ultrastructure, and which is thus referred to as "atypical" in Table II. Except for PCC 7404, which lacks pores in the longitudinal peptidoglycan, all other strains present more than one row of pores, either aligned irregularly or parallel to each other, that form circular bands around the cells on either side of the cross walls. The total number of pores and the distinct patterns vary among the "Pseudanabaenae" strains but are very similar among nine of them (*Pseudanabaena* strains PCC 6901, 6903, 7367, 7402, 7403, 7429, and 7955, and LPP group B strains PCC 7408 and 7409).

Analyses of the electrophoretic properties of the constituents of phycobilisomes and partially purified phycobiliproteins[59] of the latter nine strains were also in support of a relatively high degree of homology among seven of them: they all synthesize four subunits of phycocyanin (α_1, α_2, β_1, and β_2) of similar molecular weights. PCC 7367 and 7403 produce only two phycocyanin subunits (α_1 and β_1); their electrophoretic properties are different in these two strains and can also be distinguished from those of all members of the former group. DNA/DNA hybridization studies[59] performed on seven of the nine structurally related strains (see above; PCC 7367 and 7955 were not included) suggest that similar ultrastructural properties are not necessarily a good taxonomic indicator of genetic relatedness, since labeled DNA of *Pseudanabaena* PCC 7403 (synthesizing only two subunits of phycocyanin) gave only 16% relative binding or less and an indeterminable $\Delta T_{m(e)}$ with each of the strains tested. The value of the electrophoretic patterns of phycobilisome constituents as a sensitive taxonomic marker (at least at the species level) was unfortunately not examined in sufficient detail, since among the strains with four subunits of phycocyanin only one (LPP-B 7409) was used as the source of labeled reference DNA. This strain was shown to be relatively closely related to LPP-B strain PCC 7408 (62% relative binding and a $\Delta T_{m(e)}$ of 7.8°) but not to any of the *Pseudanabaena* strains examined (34% relative binding or less and a $\Delta T_{m(e)}$ ranging between 14 and 17.5°). Additional crosses with other reference strains might have revealed a much higher degree of homology among some of the latter, thus demonstrating unambiguously the taxonomic usefulness of the electrophoretic patterns of phycobilisome constituents.

On the basis of these results Guglielmi and Cohen-Bazire[59] suggested a rather restricted definition of *Pseudanabaena*, by which only those seven structurally related strains synthesizing four subunits of phycocyanin

[59] G. Guglielmi and G. Cohen-Bazire, *Protistologica* **20,** 393 (1984).

could be included in this genus. As a result, four strains (PCC 6406, 6802, 7367, and 7403) previously assigned to this genus (in agreement with botanical taxonomy) became "homeless"; they were excluded from this taxon[59] because they did not comply to the new generic definition with respect to their ultrastructural characters and/or phycobilisome properties. These taxonomic changes seem to be at present as little helpful as those proposed by Rippka and Cohen-Bazire[51] for the subdivision of the genus *Synechococcus* (see comments to Section I), since one of the few oscillatoriacean genera relatively clearly defined by botanists can no longer be easily identified.

Another taxonomic proposal for a nonheterocystous filamentous cyanobacterium needs to be mentioned. It concerns strain PCC 7376, which entered the Pasteur Collection as *Phormidium fragile*,[2] although *P. persicinum* or *P. ectocarpi* would have been more appropriate botanical specific assignments.[1,26] With respect to trichome morphology, PCC 7376 resembles *Pseudanabaena* strains, but it was provisionally placed into LPP group B since it exhibited neither polar gas vacuoles nor gliding motility, and occasionally formed thin sheath material.[2] Ultrastructural studies seemed to be in favor of its placement into the genus *Pseudanabaena*,[58] but later Guglielmi and Cohen-Bazire[59] proposed to redefine the genus *Phormidium* on the so far unique properties of this strain: PCC 7376 contains hemielipsoidal phycobilisomes similar to those occurring in the red alga *Porphyridium cruentum*.[59]

The value of the studies by Guglielmi and Cohen-Bazire[45,46,58,59] is not questioned. However, the author of this chapter nevertheless abstained from major modifications (other than those concerning helically coiled trichomes) of the classification for members of Section III (in spite of its obvious inadequacies), proposed in 1979.[2] There are several reasons for this: (1) more complete genetic analyses are required for this group, particularly before suggesting major revisions, (2) the complexity of some of the taxonomic proposals by Guglielmi and Cohen-Bazire[59] are little helpful to establish a key to the identification of the oscillatorian cyanobacteria, given the present knowledge available, and (3) it seems advisable to minimize further premature changes, since they would only lead to taxonomic confusion.

However, for members included in LPP group B,[2] a minor change has been made for reasons of clarity (see Key 4). This group harbors a cluster of five strains (PCC 6306, 6402, 73110, 7410, and 7505) that seemed to be very similar with respect to a number of properties[2]: the cells which compose the trichomes are isodiametric or somewhat shorter than wide (not exceeding a width of more than 3 μm); when viewed by bright-field microscopy their interior seems "empty," implying that the thylakoids are located predominantly in the peripheral regions of the cells; end cells are not differentiated; all five strains are good photoheterotrophs using a relatively wide range of sugars; all of them are sensitive to the cyanophage LPP-1.[26,55] Three of these five strains have been shown to be closely related (at the level of nomenspecies) to eleven other strains[60,61] of similar morphological features[62] and which are equally hosts to cyanophage LPP-1.[55] The latter eleven strains (as well as PCC 6306, 6402, and 73110) are carried in the UTEX Culture Collection[63] under the numbers 426, 427, 482, 485, 487, 488, 581 (PCC 6306), 594 (PCC 73110), 595, 596, 597, 598, 790, and 1541 (PCC 6402) and were originally[64] assigned to various genera and species: *Phormidium luridum, Phormidium foveolarum, Plectonema boryanum, Plectonema notatum, Plectonema* sp., *Lyngbya* sp., etc. Collectively, these cyanobacteria (for the purpose of naming the phage, to which they are hosts) were referred to as the "LPP strains."[55]

Furthermore, Stam[61] has shown that four more strains, *Schizothrix* spp. UTEX 1817, 1818, and 1819 and PCC 6409 (UTEX 1540) (member of LPP group B, sensu Rippka *et al.*[2]), are also closely related to the former 14 strains, although LPP-1 phage sensitivity has not yet been demonstrated. In order to distinguish strains of morphological properties typical of the above genetic cluster from the rest of LPP group B (sensu Rippka *et al.*[2]) they are here referred to as LPP group B type I (*Plectonema boryanum*) (see Key 4 and Table I), a distinction that (although discussed) was not made explicitly by Rippka *et al.*[2] Since the majority of the strains shown to be hosts to the cyanophage LPP-1 were assigned by phycologists to *Plectonema,* one could propose to redefine the latter ill-defined genus on the cellular, physiological, and genetic properties characteristic of these strains, thus eliminating their identification as members of four different genera and many more species (see above)! A step in this direction has already been taken by Starr,[63] who in 1978 transferred several of these strains, previously assigned to *Lyngbya* and *Phormidium,* to the genus *Plectonema*.

Comments to Genera of Sections IV and V. The filamentous cyanobacteria assigned to Sections IV and V are distinguished from those of Section III by their potential for cellular differentiation. In the absence of combined nitrogen, certain cells in the trichome (on average about 10% of the total cellular population) develop into heterocysts, in a pattern that varies among different genera. In the course of heterocyst differentiation an ensemble of properties typical of vegetative cells is modified to ensure

[60] W. T. Stam and G. Venema, *Acta Bot. Neerl.* **26,** 327 (1977).
[61] W. T. Stam, *Arch. Hydrobiol. Suppl* **56** (*Algol. Stud.* **25**), 351 (1980).
[62] W. T. Stam and H. C. Holleman, *Acta Bot. Neerl.* **24,** 379 (1975).
[63] R. C. Starr, *J. Phycol.* **14** (Suppl.), 47 (1978).
[64] R. C. Starr, *Am. J. Bot.* **51,** 1034 (1964).

synthesis and function of the oxygen-sensitive N_2-fixing enzyme complex, nitrogenase, that is localized specifically in the mature heterocysts.[65-67] Once mature, a heterocyst can neither divide nor redifferentiate into a vegetative cell. Heterocysts can be readily distinguished from vegetative cells by their characteristic thick cell walls and their low content of phycobiliproteins, which gives them a yellow–green appearance (due to chlorophyll *a*); senescent heterocysts are completely unpigmented, empty looking, and generally become detached from the trichomes, unless the latter are enclosed by slime or sheath material.

At the point of attachment with adjacent cells, heterocysts bear light-refractile polar granules that are composed of cyanophycin,[68] the nitrogenous reserve polymer typical of cyanobacteria. An intercalary heterocyst (differentiated between two vegetative cells) has a granule at each pole; a terminal heterocyst (differentiated from the cell at the end of the trichome unit) has only one polar granule; terminal heterocysts can also be formed in an intercalary position of ensheathed trichomes, if the continuity of the latter is interrupted by dead cells. Members of Section IV produce either exclusively terminal heterocysts (see Key 5) or both terminal and intercalary heterocysts. Members of Section V may, in addition, differentiate "lateral" heterocysts. These are produced if a vegetative cell that had just undergone cell division in a plane parallel to the long axis of the trichome (typical of members of this section, see below) differentiates into a heterocyst; like terminal heterocysts, they possess only one polar granule at the site of attachment to the vegetative cell.

Many strains of Sections IV and V may also produce akinetes as cultures approach the stationary phase of growth. Akinetes are thick-walled resting cells, generally resistant to desiccation and cold, but not heat. On return to favorable growth conditions they germinate and give rise to new filaments. Their position in the filament with respect to that of heterocysts varies among genera and species (see Keys 5 and 6). More detailed information on the properties and spatial patterns of akinetes is provided by Herdman.[69]

The members of Sections IV and V are distinguished by one, very fundamental, property: those of Section IV divide always in a plane at right angles to the long axis of the trichome and are therefore uniseriate and unbranched; those of Section V, however, may as mature trichomes undergo cell divisions that are not only at right angles but also parallel to

[65] R. Haselkorn, *Annu. Rev. Plant Physiol.* **29,** 319 (1978).

[66] C. P. Wolk, *in* "The Biology of Cyanobacteria" (N. G. Carr and B. A. Whitton, eds.), p. 359. Blackwell, Oxford, England, 1982.

[67] R. L. Smith, C. van Baalen, and F. R. Tabita, this volume [51].

[68] M. M. Allen, this volume [19].

[69] M. Herdman, this volume [21].

the long axis of the trichome. Depending on the successive alternation of the planes of division and the stability of the resulting multicellular trichomes, two major types can be recognized for members of Section V in culture: (1) once division in more than one plane has been initiated, the resulting multiseriate trichome falls apart into irregular cellular aggregates, the breakpoints being the cellular intersections of the originally uniseriate trichome (*Chlorogloeopsis*); (2) the trichome does not fall apart once division in more than one plane has been initiated, and, furthermore, some cells that divided in a plane parallel to the long axis continue to do so repeatedly, producing partially multiseriate primary trichomes with uniseriate lateral branches (*Fischerella*).

Traditionally, all filamentous cyanobacteria were placed in a single order Hormogonales,[1] irrespective of whether they were capable of cellular differentiation and whether cell division could occur in more than one plane. Recognizing the fundamental difference of the latter property, Fritsch[53] proposed the assignment of filamentous cyanobacteria that divide in more than one plane into a separate order, the Stigonematales. However, the same author continued to place heterocystous and nonheterocystous cyanobacteria that divide in only one plane together in a single order, the Nostocales. Since nonheterocystous cyanobacteria have been placed in Section III (see Key 4), Section IV is only partly equivalent to the botanical order Nostocales. Section V corresponds taxonomically to the order Stigonematales.

The members of Section IV can be subdivided into two major groups on the basis of their capacity to produce hormogonia (see Key 5). Hormogonia, as defined by Rippka *et al.*,[2] are chains of cells that differ from those of their parental trichomes with respect to structure and function. Their formation is most easily observed after transfer of old cultures (i.e., stationary phase of growth) to fresh medium (or water!).[70] All hormogonia are initially devoid of heterocysts, even after transfer to media lacking combined nitrogen and, as expected,[65,66] are incapable of aerobic nitrogen fixation.[71] In general, they can be easily recognized by their characteristic cell morphology (their cells being smaller and often different in shape than those of the parental trichomes) and by their transient gliding motility. Some hormogonia also produce gas vacuoles which either are distributed throughout or located exclusively near the cross walls of the cells that compose these reproductive trichomes. Gas vacuole formation is arrested and, in the absence of combined nitrogen, heterocyst differentiation is initiated, once a hormogonium starts to develop into a mature trichome.

Three genera of Section IV, *Anabaena, Cylindrospermum,* and *Nodu-*

[70] M. Herdman and R. Rippka, this volume [22].

[71] R. Rippka and M. Herdman, unpublished observations.

laria, are reserved for heterocystous filamentous cyanobacteria never exhibiting hormogonium formation: they reproduce only by random trichome breakage and by the germination of akinetes (if produced); the resulting shorter trichomes cannot be distinguished from the parental filaments. The definitions of the genera *Cylindrospermum* and *Nodularia* correspond precisely to the phycological definitions, and strains assignable to these two genera are easily identified. The genus *Anabaena* proved to be more problematic, since there is a substantial overlap with some species of the genus *Nostoc*. Phycologists separate these two genera by the absence (*Anabaena*) or presence (*Nostoc*) of slime and sheath production that, if pronounced, lead to the formation of the characteristic spherical or hemispherical gelatinous colonies described for many *Nostoc* species in the field.[1] As mentioned above (see comments to Sections I and III), slime and sheath formation are not very useful characters for the identification of cyanobacteria in culture. However, developmental studies on axenic strains that were identified as *Nostoc* on the basis of their colony properties in the natural habitat revealed that they all produced hormogonia (as defined above). This is in contrast to species assigned to *Anabaena,* in which hormogonia production proved to be variable among the axenic strains examined. Since the developmental cycle from a hormogonium into a mature filament has been described repeatedly (sometimes in beautiful detail)[72,73] for a number of *Nostoc* species, but not for members of *Anabaena,* Rippka *et al.*[2] concluded (as did Kantz and Bold 10 years earlier)[74] that the character of hormogonia production is a much more valid criterium to distinguish between these two genera than sheath, slime, and colony appearance, which may be altered or lost in culture. The definitions of *Anabaena* and *Nostoc* were modified accordingly (see Key 5).

The results of an extensive and thorough DNA/DNA hybridization study by Lachance[75] (performed on about 65 heterocystous cyanobacteria carried in the PCC and including strains not characterized by Rippka *et al.*[2]) were generally in agreement with the proposed[2] separation of *Anabaena* and *Nostoc,* but they also revealed the problems that exist when hormogonia formation is not expressed. The 13 *Nostoc* strains examined could be subdivided into several major and minor genetic clusters.[75] One of them, composed of 8 strains (PCC 6302, 6310, 7121, 73102, 7422, 7706, 7807, and 7803), includes two (PCC 73102 and 7803) that even

[72] G. Thuret, *Ann. Sci. Nat., Bot. Biol. Veg.* **2,** 319 (1844).

[73] E. Janczewski, *Ann. Sci. Nat., Bot. Biol. Veg.* **19,** 119 (1874).

[74] T. Kantz and H. C. Bold, *in* "Phycological Studies: 9. Morphological and Taxonomic Investigations of *Nostoc* and *Anabaena* in Culture," Univ. of Texas Publ. No. 6924. Univ. of Texas, Austin, Texas, 1969.

[75] M.-A. Lachance, *Int. J. Syst. Bacteriol.* **31,** 139 (1981).

in axenic culture are identifiable as "typical" *Nostoc* species, because of their characteristic growth morphology (i.e., tightly coiling, seemingly multiseriate mature filaments). However, the same cluster also contains 3 strains that entered the PCC under the designations *Anabaena* sp. (PCC 6302 and 7422) and *Anabaena spiroides* (PCC 6310). The latter 3 strains were assigned to *Nostoc* on the basis of their developmental pattern (i.e., hormogonia formation).[2] The genetic relatedness of these three strains (53–59% relative binding and a $\Delta T_{m(e)}$ of 9–11°)[75] with PCC 73102, an "unmistakable" *Nostoc,* excludes all doubts about the validity of their generic assignment and thus the importance of hormogonia production as a discriminatory character.

On the other hand, four strains (PCC 6411, 7118, 7119, and 7120) identical with respect to morphology, physiology, and N-1 phage sensitivity,[2] were excluded from *Nostoc* and assigned to *Anabaena* (although two of them, PCC 7119 and 7120, were received as *Nostoc muscorum*), since they never produced trichomes identifiable as hormogonia.[2] The high degree of interrelatedness between these four strains was confirmed,[75] but Lachance[75] showed that the same degree of genetic similarity was also shared with two strains (PCC 6705 and 6719) that produce gas-vacuolate hormogonia and were assigned to *Nostoc*.[2] Parallel investigations[26] demonstrated that PCC 6411, 7118, and 7119 could form hormogonia, but only if care was taken to avoid carryover of old medium (by washing or using very small inocula) at the moment of transfer.[70] However, this could not be shown conclusively for PCC 7120.[26] Nevertheless, the genetic evidence[75] is such that all four strains (PCC 6411, 7118, 7119, and 7120) have to be considered nomenspecies of *Nostoc* PCC 6705 and 6719.

The conclusions that can be drawn from these results are as follows. Hormogonia, if produced, are a good discriminatory character for the generic separation of *Nostoc* from *Anabaena*. As a negative character, however, it has to be treated with caution. Strains that do not form hormogonia, either because this property has been lost (as seems to be the case in PCC 7120) or because the culture conditions leading to their production are not provided (since they may be unknown), will require genetic analyses to ensure their correct generic and specific assignments. This is even better illustrated by another extreme example: strain PCC 7121, as a consequence of prolonged cultivation in liquid medium containing combined nitrogen, has lost all properties necessary for its correct generic identification.[26,75] It is almost "unicellular" (i.e., forms very short chains of cells) and differentiates neither hormogonia nor heterocysts. According to its history,[63] this strain corresponds to the *Nostoc muscorum* isolated and characterized by Allison and collaborators[76] over 50

[76] F. E. Allison, S. R. Hoover, and H. J. Morris, *Bot. Gaz.* **98,** 433 (1937).

years ago. A replacement culture of this organism, obtained from the UTEX Culture Collection[63] (now carried under the number 7906 in the PCC), has retained the capacity to produce hormogonia and heterocysts in agreement with its assignment to *Nostoc*. The latter strain shared sufficient properties with PCC 7121 to assume their identity[26]; however, proof for this was obtained only by genetic analyses.[75,77,78]

With the transfer of PCC 6411, 7118, 7119, and 7120 to the genus *Nostoc* (see above), the genus *Anabaena* in the PCC is reduced to three strains, PCC 6309, 7122, and 7108, which form trichomes composed of cylindrical intercalary cells and conical end cells. Their akinetes are located singly, or in groups of two to three, on either side of heterocysts. This property has been lost in PCC 6309, together with the gliding motility originally typical for all three strains.[26] PCC 6309 and PCC 7122 are conspecific according to Lachance,[75] in spite of a difference of 5.6 mol % of their G + C content,[44] whereas PCC 7108 is only distantly related to them.[75] The close relatedness of PC 6309 and 7122 has been questioned by Stulp and Stam[79] on the basis of DNA/DNA hybridization studies, which included strains that in principle should have been identical to 6309 and 7122, and should therefore have confirmed the results obtained by Lachance.[75] Possibly without realizing the duplications, Stulp and Stam[79] examined two identical strains (UTEX 377[63] and CCAP 1403/4b[80]), corresponding to *Anabaena variabilis* Utrecht P40, and two strains (deposited in the UTEX collection under the number 629[63] and in the CCAP collection as 1403/2a[80]) that correspond to the well-known *Anabaena cylindrica* of Fogg (originally isolated by Chu[81,82]). Stulp's results clearly indicate that strain UTEX 377 (and therefore not surprisingly also CCAP 1403/4b) is only distantly related to UTEX 629 (and expectedly to CCAP1403/2a). Since PCC 6309 was received as UTEX 377, and PCC 7122 as UTEX 629 (admittedly via other laboratories),[2] the high degree of homology observed by Lachance[75] for the PCC strains is in conflict with the results of Stulp and Stam.[79] There is no reason to doubt the validity of the experimental data of either author: the most likely reason for the different results is that PCC 6309 corresponds neither to CCAP 1403/4b nor to UTEX 377. This seems to be highly probable, since this strain originally produced akinetes adjacent to heterocysts[26] and not, as is typical for

[77] T. Kallas, T. Coursin, and R. Rippka, *Plant Mol. Biol.* **5,** 321 (1985).

[78] T. Damerval and P. Grimont, personal communication.

[79] B. K. Stulp and W. T. Stam, *Br. Phycol. J.* **19,** 287 (1984).

[80] A. Asher and D. F. Spalding (eds.), "Culture Centre of Algae and Protozoa: List of Strains—1982." Inst. Terrestrial Ecol., National Environmental Research Council, Cambridge, England, 1982.

[81] G. E. Fogg, *J. Exp. Biol.* **19,** 78 (1942).

[82] M. B. Allen, *Arch. Mikrobiol.* **17,** 34 (1952).

Anabaena variabilis,[1,79] only away from heterocysts. Based on the existing evidence, PCC 6309 should be considered as *Anabaena cylindrica* (conspecific with PCC 7122), whose strain history is doubtful. It should also be pointed out, however, that the equivalent deposits in the Göttingen Culture Collection (SAG 1403/4b and 1403-2, i.e., Utrecht P40 and the Fogg strain, respectively) were both carried as *Anabaena cylindrica* in Pringsheim's catalog published in 1951[83] (in agreement with their identification in the PCC).

The members of the genera *Scytonema* and *Calothrix* share with *Nostoc* the ability to form hormogonia. However, the development of a hormogonium into a mature filament differs among the three genera. In *Nostoc,* the initially aheterocystous hormogonium released from the parental trichome differentiates two heterocysts, both usually in a terminal position. As growth continues and gives rise to mature trichomes (composed of larger and more rounded cells), intercalary heterocysts may also be produced. In *Scytonema* the hormogonia are less typical and could be considered borderline hormogonia. They are composed of short trichomal fragments that barely differ in cell size and cell morphology from the parental trichome and seem to be immotile (or glide very slowly?). They can be distinguished, however, from the mature trichomes by two properties (and therefore qualify as hormogonia sensu Rippka *et al.*[2]): they are only thinly ensheathed (or sheathless?) and differentiate exclusively one terminal cell into a heterocyst; this is in contrast to the mature trichomes that are heavily ensheathed and in which heterocysts are predominantly intercalary. Frequent false branching of the mature trichomes leads to aerial growth on solid media and results in a carpetlike growth appearance typical for most members of this genus.[1] Only one axenic strain of *Scytonema* (PCC 7110) has been characterized and was shown to be genetically sufficiently distant from other heterocystous cyanobacteria to justify its generic separation.[75]

Traditionally, *Scytonema* is a member of the Scytonemataceae, a family reserved for heavily ensheathed filamentous cyanobacteria that form trichomes of even width and exhibit frequent false branches.[1] The generic subdivisions of the heterocystous representatives in this family (nonheterocystous filamentous members exhibiting false branching such as *Plectonema* were also included in this family) were primarily based on characters such as thickness and multilayered structure of sheaths (*Petalonema*) or whether false branches occurred singly (*Tolypothrix*) or in pairs (*Scytonema*). The definition of *Scytonema* (Key 5) deviates from that given by Geitler,[1] since much less emphasis is placed on sheath properties and

[83] E. G. Pringsheim, *Arch. Mikrobiol.* **16,** 1 (1951).

number of false branches. Rippka *et al.*[2] observed that false branches in PCC 7110 could occur either singly or in pairs. This, together with the exclusion of sheath structure as a discriminatory character, led these authors to the conclusion that only one genus was required: *Scytonema,* as the oldest genus in this family, had priority over others.

In strains assigned to *Calothrix,* the hormogonia are generally easily recognized by their smaller cell size, active gliding motility, and, in some cases, by transient gas vacuole formation. Heavily ensheathed strains produce sheathless (or thinly sheathed?) hormogonia that are also structurally different from the mature trichomes. The newly released aheterocystous hormogonium differentiates, as in *Scytonema,* only one terminal cell into a heterocyst. However, subsequent development gives rise to a tapered trichome, a trait characteristic of *Calothrix*. The direction of the basal–apical tapering is determined by the location of the first terminal heterocyst produced. Young trichomes bear exclusively basal heterocysts; older ones (particularly when they are very long) may differentiate "terminal" heterocysts within ensheathed trichomes, if the continuity of the cellular chain is interrupted by dead cells (necridia); some mature trichomes may also produce intercalary heterocysts. Most *Calothrix* strains exhibit false branching. Hair formation at the apical end of the trichomes, typical for many *Calothrix* field specimens, is rarely observed under normal growth conditions in the laboratory. This may not be surprising, since expression of this property has been shown to be favored by lack of nutrients such as phosphate and iron.[84]

Calothrix is a genus of the botanical family Rivulariaceae.[1] Members of this genus occur solitarily, and do not form spherical or flat colonies enclosed by gelatinous sheath material, properties which define the botanical genera *Rivularia* and *Isactis,* respectively. In addition, *Calothrix* is traditionally distinguished from *Dichothrix* and *Polythrix* (two other rivularian genera) by its rather sparse formation of false branches. Since the discriminatory characters of these genera could not be ascertained (since they were variable or absent in culture), Rippka *et al.*[2] decided that all tapering cyanobacteria should be assigned provisionally to a single genus. They chose *Calothrix,* since it was defined on the basis of the structural properties of the filaments and not colony formation.

The results of DNA/DNA hybridizations[75] of a large number of PCC strains assigned to *Calothrix* are largely in support of this proposal: seven strains of marine origin (not all of which have an obligate requirement for elevated concentrations of Na^{+}, Mg^{2+}, and Ca^{2+}) that entered the Pasteur Collection as *Calothrix, Rivularia,* or *Isactis* species formed a cluster of

[84] C. Sinclair and B. A. Whitton, *Br. Phycol. J.* **12,** 297 (1977).

sufficient genetic similarity to justify their inclusion into a single genus. However, other *Calothrix* strains isolated from freshwater or soil samples were only marginally related to the former seven strains and (although several genetic clusters could be identified) exhibited considerable heterogeneity among themselves. Thus, generic subdivisions of the PCC strains assigned to *Calothrix* will eventually be required. For lack of sufficient criteria for the proposal of appropriate traditional assignments, this cannot be undertaken without major and arbitrary revisions of the classical taxonomy and has therefore been postponed.

One genetically closely related cluster of *Calothrix* (PCC 7101, 7504, 7601, 7708, 7710, and 7712, all freshwater or soil isolates) should be mentioned specifically, since it might provide a starting point for revision: this group includes strains assigned, prior to their arrival in the Pasteur Collection, to *Tolypothrix tenuis*[2] (PCC 7101) and *Fremyella diplosiphon*[85] (PCC 7601). The latter strain corresponds to UTEX 481[63] and CCAP 1429/1[80] but is most likely also identical with PCC 7710, which was obtained as *Calothrix* sp. (D-255[84] = SAG 1410-2[86]). Unless the strain history given[85] for *Fremyella diplosiphon* UTEX 481 was incorrect, and P. Strout[85] deposited two different strains in the SAG,[87] circumstantial evidence suggests that the same isolate entered the latter collection twice, once as *Calothrix* sp. 1410-2 and again later as *Fremyella diplosiphon* 1429-1a.[87] It might not, therefore, be surprising that one of the two deposits (1429-1a) is no longer listed in the latest catalog of the SAG.[86]

Contrary to PCC 7601, which has lost all properties that would allow its phenotypic identification (i.e., heterocyst differentiation, tapering, false branching), PCC 7710 exhibits these properties in agreement with its generic assignment to *Calothrix* in the Durham[84] and Göttingen[86,87] culture collections. A heterocyst forming (Het$^+$) revertant of PCC 7601[26] acquired simultaneously the capacity to form tapering ensheathed trichomes that exhibit false branching. It can no longer be readily distinguished from PCC 7710. Both PCC 7710 and 7601 (Het$^+$) share with the other four members of this genetic cluster (PCC 7101, 7504, 7708, and 7712)[75] the following characteristic traits: they all exhibit a low degree of tapering that is only sufficiently pronounced to be detected when grown in the absence of combined nitrogen; they produce relatively long hormogonia (which may break up to yield shorter fragments) with gas vacuoles distributed throughout their cells[2,26]; most of them produce sheath material and may therefore exhibit false branches[2,26]; they are good photoheterotrophs

[85] J. T. Wyatt, T. C. Martin, and J. W. Jackson, *Phycologia* **12,** 153 (1973).

[86] U. G. Schlösser, *Ber. Dtsch. Bot. Ges.* **95,** 181 (1982).

[87] W. Koch, *Arch. Mikrobiol.* **47,** 402 (1964).

and also adapt their phycobiliprotein content in response to the wavelength of light provided for growth.[75,88]

The genus *Tolypothrix,* as a member of the Scytonemataceae, was traditionally reserved for nontapering ensheathed trichomes forming single false branches.[1] Since all six strains mentioned above showed a distinct basal–apical tapering, when grown in the absence of combined nitrogen, they were placed into the genus *Calothrix* by Rippka *et al.*,[2,26] who did not take into account the previous assignment of such strains as PCC 7101 (*Tolypothrix*)[2] or PCC 7601 (*Fremyella*).[26] However, more careful inspection of the scytonematacean genera revealed that the keys for their identification[1] are misleading, since quite a number of them (*Tolypothrix, Tildenia, Scytonematopsis*) are representative of members exhibiting a certain degree of trichome polarity, and not just false branching, the classical distinguishing trait of this family. Furthermore, a low degree of tapering is also the discriminatory character of *Microchaete* (the legitimate botanical synonym of *Fremyella*[34]), which was separated from the Scytonemataceae because its members "only occasionally" produce false branches.[1] Therefore, if one would wish to exclude strain PCC 7101 and its relatives (see above) from the more "typical" *Calothrix* strains (i.e., those exhibiting a higher degree of tapering) one could propose the following revision: the genus *Tolypothrix* is redefined precisely on the structural, physiological, and genetic properties exhibited by the members of this related cluster, using PCC 7101 as reference strain. The preference for *Tolypothrix* over the other generic candidates (*Tildenia, Scytonematopsis,* or *Microchaete*) that would probably be equally suitable for this group of strains is a question of priority: the genus *Tolypothrix* antedates the others.

The DNA/DNA hybridization results of Lachance[75] for members of Section V are in full agreement with the generic assignments proposed by Rippka *et al.*[2] The two strains of *Chlorogloeopsis fritschii* (PCC 6718 and 6912) are conspecific and show a low but significant level of relatedness to the second genus, *Fischerella,* included in this section. The PCC strains assigned to the latter genus form two highly related clusters, one being composed of PCC 7115, 73103, 7414, 7520, and 7603, the second being represented by PCC 7521, 7522, and 7523. Members of both clusters are sufficiently interrelated to justify their assignment to the same genus. PCC 73103 was received as *Fischerella muscicola* (see Ref. 2), whereas most of the other strains included in the genus *Fischerella* (sensu Rippka *et al.*[2]) had previously been identified as *Mastigocladus laminosus* (see Ref. 2). The close genetic relationship of all filamentous true branchers repre-

[88] N. Tandeau de Marsac and J. Houmard, this volume [34].

sented in the Pasteur Collection (all of which are thermophilic!) are in agreement with the conclusion of Rippka *et al.*[2] who felt that only one genus was required to accommodate these strains: *Fischera* Schwabe 1837 (type species *Fischera thermalis*), later renamed *Fischerella,* since the older genus had priority over *Mastigocladus* Cohn 1863 (see Ref. 2).

Taxonomic Position of Some Cyanobacteria Not Represented in the Pasteur Culture Collection

Cyanobacterial Endosymbionts

The knowledge gained from the characterization of the *Nostoc* and *Anabaena* strains in the Pasteur Collection also allows some conclusions on the taxonomy of cyanobacterial endosymbionts that enter associations with cycads or the pteridophyte, *Azolla.* Endosymbionts of the latter eukaryotic host have always been identified as *Anabaena azollae.*[89] On the basis of published photographic documentation,[90,91] however, there is no doubt that the *Anabaena azollae* from the water fern *Azolla caroliniana* produces hormogonia and should be assigned to the genus *Nostoc* (sensu Rippka *et al.*[2]). The same is probably true for the cyanobacterial endosymbionts of other *Azolla* species, since Franche and Cohen-Bazire[92] suggested that those of *Azolla caroliniana, A. filiculoides, A. mexicana,* and *A. microphylla* are closely related (as judged from restriction maps of the *nif* region[92,93] of their genomes).

Endosymbionts from cycads have traditionally been identified mainly as *Nostoc,* but occasionally also as *Anabaena.*[89] On the basis of hormogonium formation, Rippka *et al.*[2] assigned two endosymbionts, isolated from *Macrozamia* sp. (PCC 73102) and from *Cycas* sp. (PCC 7422), to the genus *Nostoc,* even though the latter had previously been identified as *Anabaena.* DNA/DNA hybridization[75] studies confirmed this assignment, since PCC 7422 clustered genetically with seven other *Nostoc* strains, some of which are so typical that they never could be mistaken for *Anabaena* species. Furthermore, it should be mentioned that Stewart *et al.,*[89] who inspected the early descriptions of cyanobacteria entering into a symbiotic relationship with members of the cycad family, concluded that the endosymbionts are all probably assignable to *Nostoc* (sensu Rippka *et al.*[2]).

[89] W. D. P. Stewart, P. Rowell, and A. N. Rai, *in* "Nitrogen Fixation" (W. D. P. Stewart and J. R. Gallon, eds.), p. 239. Academic Press, London, 1980.

[90] G. A. Peters and B. C. Mayne, *Plant Physiol.* **53,** 813 (1974).

[91] G. A. Peters, T. B. Ray, B. C. Mayne, and R. E. Toia, Jr., *in* "Nitrogen Fixation" (W. E. Newton and W. H. Orme-Johnson, eds.), Vol. 2, p. 293. University Park Press, Baltimore, Maryland, 1980.

[92] C. Franche and G. Cohen-Bazire, *Plant Sci.* **39,** 125 (1985).

[93] C. Franche and T. Damerval, this volume [88].

Even more interesting are the results of Franche,[94] who compared restriction enzyme patterns of PCC 73102 and PCC 7422 (by using labeled *nif* probes[93]) with those obtained for the so-called *Anabaena* endosymbionts extracted from the four *Azolla* species mentioned above. Although some differences were observed, the degree of conservation seemed to be sufficient to give additional support to the postulate that *Azolla* endosymbionts are assignable to the genus *Nostoc*.

The elegant work of Enderlin and Meeks[95] has demonstrated that *Nostoc* species can be successfully isolated and purified from, and reconstituted with, their bryophyte host, *Anthoceros*. These studies were particularly important, since they permitted testing of the specificity of the host–endosymbiont relationship by reconstitution experiments with cyanobacterial strains of asymbiotic origin. Most of these results are described elsewhere in this volume,[96] but a few points concerning PCC strains, included in these studies, may be worth a brief discussion. The degree of specificity between the *Anthoceros* host tissue and the cyanobacterial endosymbiont is (within generic boundaries?) quite low, since competence was observed for *Nostoc* PCC 73102 (ATCC 29133), PCC 7107 (ATCC 29150), and PCC 6720 (ATCC 27895), the latter two of which are only distantly related to the former strain (26 and 31% relative binding, respectively, to the reference DNA of PCC 73102).[75] PCC 73102 had been isolated from its host, *Macrozamia* sp., whereas PCC 6720 and 7107 were of asymbiotic origin.[2] These three strains share certain properties, however, that might be important for competence: they grow well under heterotrophic conditions and exhibit extremely active gliding motility in the hormogonial state of development. These criteria are not met by the other PCC *Nostoc* strains that proved to be unsuccessful in establishing symbiosis,[95] since they are either good heterotrophs but produce only marginally motile hormogonia (i.e., PCC 6310 = ATCC 27986) or are obligate autotrophs in addition to forming hormogonia of poor (or zero?) gliding motility (i.e., PCC 6314 = ATCC 27904 and PCC 7413 = ATCC 29106). As mentioned before, the misidentified *Nostoc* PCC 7120 (*Anabaena* sp. ATCC 27893), equally incompetent in establishing symbiosis with *Anthoceros*,[95] does not express hormogonia development, is obligately autotrophic, and permanently immotile. No comments can be made on the competent or incompetent strains that have not been studied in the Pasteur Collection. However, the importance of actively gliding hormogonia (alone or in combination with heterotrophic potential) for the establishment of a symbiotic association with *Anthoceros* (or other hosts) should be examined.

[94] C. Franche, personal communication.
[95] C. S. Enderlin and J. C. Meeks, *Planta* **158,** 157 (1983).
[96] J. C. Meeks, this volume [8].

Recent Isolates: Dactylococcopsis and Cyanospira

Two recent reports on the isolation and characterization of cyanobacteria merit special mention, since the strains isolated are assignable to genera of which cultures were not available previously. Walsby *et al.*[97] isolated a unicellular, halophilic cyanobacterium from the hypersaline Solar Lake (Gulf of Eilat) that is rather unusual since its cells (4–8 μm wide) are extremely long (35–80 μm) and are slightly tapered at the ends. The cellular properties of this isolate suggest that it is a member of the traditional genus *Dactylococcopsis*.[1] However, since neither the habitat (i.e., hypersaline environment) nor the abundant gas vacuole formation typical of this organism fit the existing species descriptions, Walsby *et al.*[97] assigned it to the new species *Dactylococcopsis salina*. This strain would be included in Section I of the classification system described here.

The creation of the genus *Cyanospira* by Florenzano *et al.*[98] (based on the type species "*Cyanospira rippkae*") was the subject of severe criticism by some scientists. However, this work should not be misjudged in value, since these authors isolated (from a soda lake in Kenya) for the first time a number of heterocystous cyanobacteria that form helically coiled trichomes composed of vegetative cells containing gas vacuoles. Two representative strains (Mag II 702 and Mag I 504) were characterized by them, both physiologically and genetically. The creation of the new genus *Cyanospira* for these strains could have been avoided since, with respect to trichome structure, cell shape and size, gas vacuolation, position of akinetes, and, importantly, the habitat (saline lake), they correspond quite closely to *Anabaenopsis nadsonii* and/or(?) *Anabaenopsis milleri*.[1]

This confusion illustrates clearly one of the problems encountered with traditional taxonomy: one may not be able to identify the genus owing to the imprecision of the generic keys, but the description of the species may be sufficiently informative to allow unambiguous identification. In Geitler,[1] the genus *Anabaenopsis* is distinguished from other genera of nontapering, nonbranching, heterocystous filamentous cyanobacteria as follows: "heterocysts terminal or seemingly terminal; akinetes away from heterocysts." With this generic description only the botanical expert taxonomist, familiar with the numerous species of each genus, would be in the position to trace the correct generic and specific assignment for the helically coiled isolates of Florenzano *et al.*[98] The lesson from this is clear: before creating new genera, one either inspects oneself the descriptions of *all* genera and species in detail, or one consults botanical expert taxonomists for advice.

[97] A. E. Walsby, J. van Rijn, and Y. Cohen, *Proc. R. Soc. London, Ser. B* **217,** 417 (1983).
[98] G. Forenzano, C. Sili, E. Pelosi, and M. Vincenzini, *Arch. Microbiol.* **140,** 301 (1985).

It should also be stressed, however, that in the case of *Anabaenopsis* some taxonomic modifications are certainly required. This genus is poorly defined and includes species that have little in common with respect to structure and development. As demonstrated by the structural and genetic properties[2,75] of PCC 6720, identified as *Anabaenopsis circularis* prior to its entry in the Pasteur Collection, many of the traditional *Anabaenopsis* species are probably assignable to *Nostoc* (sensu Rippka *et al.*[2]). Two taxonomic strategies could therefore be envisaged: (1) redefinition and restriction of the genus *Anabaenopsis* on the basis of the *living* and well-characterized strains isolated by Florenzano *et al.*,[98] or (2) reservation of the new genus *Cyanospira* Florenzano *et al.* 1985 (as a partial synonym of *Anabaenopsis*) for helical members exhibiting properties typical of, or similar to, these new isolates. The latter approach would be equivalent to that taken by Komarek,[99] who created a new genus *Cyanothece* (type species *Cyanothece aeruginosa*) for certain species of *Synechococcus* (sensu Geitler[1]). In both cases, however, the specific epithets should be "*nadsonii*" or "*milleri*" (not "*rippkae*" or "*capsulata*"), since, in contrast to many other cyanobacterial species, the corresponding botanical descriptions allow identification with a fair amount of confidence.

Conclusions on the Present Taxonomic Status of Cyanobacteria

Those who have followed closely the development of cyanobacterial taxonomy over the last decade will have realized that this area of research did not produce the "bloom" that was hoped for with the publication of the provisional classification by Rippka *et al.*[2] in 1979. The value of some studies, however, should not be underestimated, even if their results cannot be exploited more usefully at present: it is quite obvious that the isolation and characterization of additional strains[100–103] are very worthy. Furthermore, the search for new phenotypic properties for taxonomic purposes[45,46,79,103] is of fundamental value, since together with biochemical and genetic analyses they will eventually lead to a more refined and solidly based classification of cyanobacteria than that proposed here. However, it is also evident that ultimate progress can only be made if one is willing to either create new genera and species or redefine existing ones arbitrarily on the basis of properties impossible to extract from the nonliv-

[99] J. Komarek, *Arch. Protistenkd.* **118,** 119 (1976).

[100] C. Franche and P. A. Reynaud, *Ann. Microbiol. (Paris)* **137A,** 179 (1986).

[101] T. Vaara, M. Vaara, and S. Niemelä, *Appl. Environ. Microbiol.* **38,** 1011 (1979).

[102] L. J. Stal and W. E. Krumbein, *Bot. Mar.* **28,** 351 (1985).

[103] O. M. Skulberg and R. Skulberg, *Arch. Hybrobiol. Suppl.* **71** (*Algol. Stud.* **38/39**), 157 (1985).

ing botanical type specimens. According to recent agreement[38] such decisions should not be made unilaterally, but be worked out in collaboration with botanists. Furthermore, they become sensible only once sufficient data are available as a solid background for discussions. In the meantime, it is hoped that taxonomic *suggestions* are excusable, particularly if they are dressed in quotation marks.

Key 1. Major Subgroups (Sections) for Cyanobacteria in Axenic Culture

I. Unicellular cyanobacteria; cells single or forming colonial aggregates held together by additional outer cell wall layers
- A. Reproduction by binary fission or by budding — **Section I**
- B. Reproduction by multiple fission, giving rise to small daughter cells (baeocytes), or by both multiple fission and binary fission — **Section II**

II. Filamentous cyanobacteria; cells forming a multicellular unit, the trichome
- A. Trichome always composed only of vegetative cells; reproduction only by intercalary division, followed by random trichome breakage — **Section III**
- B. In the absence of combined nitrogen the trichome is composed of both vegetative cells and heterocysts; reproduction occurs by intercalary binary fission and random trichome breakage; some members also reproduce by formation of hormogonia or by germination of akinetes
  - 1. Division in only one plane — **Section IV**
  - 2. Division in more than one plane — **Section V**

Key 2. Genera of Section I

Unicellular cyanobacteria, that occur singly or form colonial aggregates held together by additional cell wall layers (sheaths); they reproduce by binary fission or by budding.

I. Reproduction by equal binary fission
- A. Thylakoids present
  - 1. Division in only one plane, crosswise, cells rod shaped
    - a. Sheath layers absent — *Synechococcus* Nägeli 1849
    - b. Sheath layers present — *Gloeothece* Nägeli 1849
  - 2. Division in two or more planes, cells coccoid
    - a. Sheath layers absent — *Synechocystis* Sauvageau 1892
    - b. Sheath layers present — *Gloeocapsa* Kützing 1843
- B. Thylakoids absent
  - 1. Division in only one plane, crosswise, cells rod shaped
    - a. Sheath layers present — *Gloeobacter* Rippka, Waterbury, and Cohen-Bazire 1974

II. Reproduction by repeated budding at the apical end of the cells
- A. Thylakoids present
  - 1. Division in only one plane, producing small spherical daughter cells from ellipsoidal, cylindrical, club-, or pear-shaped mother cells — *Chamaesiphon* Braun and Grunow 1865 emend. Geitler 1925 (sensu Waterbury and Stanier 1977)

Key 3. Genera of Section II

Unicellular cyanobacteria that occur singly or form colonial (often pseudofilamentous) aggregates; mature cells are always enclosed by an additional fibrous outer cell wall layer; they reproduce by multiple fission only or by binary fission followed by multiple fission; the baeocytes, at the moment of their release, are either motile and lack the fibrous outer cell wall layer or possess the latter structure and are immotile.

I. Reproduction uniquely by multiple fission
  A. Vegetative growth leads to the formation of spherical cells of varying size
    1. Baeocytes motile — *Dermocarpa* Crouan and Crouan 1858 (sensu Waterbury and Stanier 1978)
    2. Baeocytes immotile — *Xenococcus* Thuret 1880 (sensu Waterbury and Stanier 1978)

II. Reproduction by both binary fission and multiple fission
  A. Growth and binary fission lead to pear-shaped structures composed of one or two basal cells and one apical cell
    1. Multiple fission of the apical cells yields motile baeocytes — *Dermocarpella* Lemmerman 1907 (sensu Waterbury and Stanier 1978)
  B. Growth and binary fission lead to cubical aggregates
    1. Multiple fission yields motile baeocytes — *Myxosarcina* Printz 1921 (sensu Waterbury and Stanier 1978)
    2. Multiple fission yields immotile baeocytes — *Chroococcidiopsis* Geitler 1933 (sensu Waterbury and Stanier 1978)
  C. Growth and binary fission lead to irregular, pseudofilamentous cell aggregates
    1. Multiple fission yields motile baeocytes — *Pleurocapsa* group
      a. Baeocytes enlarge symmetrically — Type I
      b. Baeocytes develop early polarity — Type II

Key 4. Genera of Section III

Filamentous cyanobacteria; trichomes composed uniquely of vegetative cells, even in the absence of combined nitrogen; reproduction by intercalary cell division in only one plane and random transcellular or intercellular[104] trichome breakage; hormogonia that are structur-

[104] Transcellular trichome breakage involves the sacrifice of an intercalary cell converted to a necridium (dead cell); it occurs in filamentous cyanobacteria in which transverse wall formation is restricted to ingrowth of the peptidoglycan layer, which is not accompanied by ingrowth of the outer membrane. Intercellular trichome breakage implies the separation of a trichome between two living cells; it occurs in all filamentous cyanobacteria in which transverse wall formation involves centripetal ingrowth of the peptidoglycan layer, followed by that of the outer membrane [R. Y. Stanier and G. Cohen-Bazire, *Annu. Rev. Microbiol.* **31,** 225 (1977)].

Key 4. (continued)

ally distinguishable from parental trichomes are never produced; nonsheathed (or thinly sheathed?) members generally exhibit gliding motility, which may or may not involve rotation around the long axis of the trichomes; firmly ensheathed immotile filaments may release transiently motile trichomes that lack heavy sheaths.

I. Trichome helically coiled
  A. Shape of cells not fixed; little or no constrictions between adjacent cells; sheath formation never pronounced; probably transcellular trichome breakage; motility involves rotation around the long axis of the trichomes
    1. Cross walls invisible by light microscopy — *Spirulina* Turpin 1827
    2. Cross walls visible by light microscopy — *Arthrospira* Stitzenberger 1852

II. Trichome straight
  A. Cells that compose trichome are disk shaped and separated by shallow constrictions (if any); end cells different in shape to intercalary cells; transcellular trichome breakage
    1. Trichome motile, not heavily ensheathed; motility involves rotation around the long axis of the trichome — *Oscillatoria* Vaucher 1803 (sensu Rippka *et al.* 1979)
    2. Trichome immotile and heavily ensheathed; motility restricted to sheathless (or thinly sheathed?) trichomal fragments released from the ends of the immotile ensheathed parental filaments; whether motility involves rotation around the long axis of the trichome is unknown — LPP group A ("*Lyngbya aestuarii*" type, this proposal)
  B. Cells that compose trichome are isodiametric, cylindrical, or barrel shaped; the degree of constriction between adjacent cells varies; inter- or transcellular trichome breakage; motility (if exhibited) may or may not involve rotation around the long axis of the trichome; sheath production and false branching[105] variable
    1. Trichome motile,[106] not ensheathed; cells contain polar gas vacuoles,[106] are cylindrical or barrel shaped, and are generally separated from one another by pronounced constrictions; intercellular trichome breakage; motility not involving rotation around the long axis of the trichome — *Pseudanabaena* Lauterborn 1905 (sensu Rippka *et al.* 1979)
    2. Properties not as above — LPP group B
      a. Trichome composed of isodiametric (or somewhat shorter) cells (2–3 μm wide) that are not separated by deep constrictions; probably transcellular trichome breakage; some members produce thin sheaths and may exhibit false branching[105]; motility is never pronounced and does not involve rotation around the

[105] False branching occurs as a consequence of trichome breakage of ensheathed cyanobacteria; subsequent elongation of the trichomal fragments may lead to the perforation of the sheath envelope at the breakpoints and bring them into an angular apposition to the primary trichome, thus suggesting (incorrectly) that a branch has been formed.

[106] Motility and the formation of polar gas vacuoles have been observed to be easily lost by spontaneous mutation; strains that do not exhibit these features, but comply otherwise to the description of *Pseudanabaena* as defined here, are best included in LPP group B, until their affiliation to *Pseudanabaena* has been demonstrated by additional criteria.

Key 4. (continued)

long axis of the trichome; thylakoids restricted to the peripheral regions of the cells; generally hosts to cyanophage LPP-1
LPP group B: type I ("*Plectonema boryanum*" type, this proposal)

b. Properties different to above (this assemblage harbors members that vary greatly with respect to cell morphology, degree of constrictions between cells, sheath properties, motility, and genetic complexity; it is probably representative of several different genera) LPP group B: type X

Key 5. Genera of Section IV

Filamentous cyanobacteria that divide in only one plane; in the absence of combined nitrogen, a certain percentage (~10%) of the vegetative cell population differentiates into heterocysts, the specific cellular sites of aerobic N_2 fixation; akinetes (resting cells), resistant to desiccation and cold, are produced by some members; reproduction occurs by random trichome breakage, by germination of akinetes (if produced) or, in addition, by formation of hormogonia (sensu Rippka *et al.* 1979).

I. Hormogonia not produced
  A. Trichome never exhibits basal–apical polarity
    1. Heterocysts are differentiated from terminal and intercalary cells; the position of akinetes (if produced) is variable
      a. Vegetative cells are cylindrical, spherical, or ovoid — *Anabaena* Bory de St. Vincent 1822
      b. Vegetative cells are disk shaped — *Nodularia* Mertens 1822
    2. Heterocysts are differentiated exclusively from the terminal cells at both ends of the trichome; akinetes are located adjacent to heterocysts — *Cylindrospermum* Kützing 1843

II. Hormogonia produced
  A. Trichome never exhibits basal–apical polarity
    1. Hormogonia give rise to young trichomes that differentiate terminal heterocysts at both ends of the cellular chain
      a. Mature trichomes are composed of barrel-shaped, spherical, or ovoid vegetative cells and differentiate heterocysts in both intercalary and terminal positions; they are immotile and may be enclosed by sheath material of variable consistency; some adult trichomes take on a coiled configuration; akinetes (if produced) differentiate away from the heterocysts and are often formed in chains; hormogonia are composed of cells that are generally smaller in size, different in shape (often cylindrical or isodiametric), and may contain polar or irregularly distributed gas vacuoles; they are sheathless and often exhibit active gliding motility, which (as gas vacuolation) is lost with their maturation to adult trichomes — *Nostoc* Vaucher 1803 (sensu Rippka *et al.* 1979)
    2. Hormogonia give rise to young trichomes that differentiate heterocysts from the terminal cells at only one end of the cellular chains

Key 5. (continued)

a. Mature trichomes are composed of disk-shaped, isodiametric, or cylindrical cells and differentiate heterocysts predominantly in an intercalary position; they are immotile, heavily ensheathed, and exhibit frequent false branching[105] (leading on solid media to aerial growth); akinetes not produced; hormogonia are composed of cells that are not markedly different in size or shape (although they are generally somewhat shorter) than those of the mature filaments and do not contain gas vacuoles; active gliding motility of the sheathless (or thinly sheathed?) hormogonia is never pronounced or is absent (?)
Scytonema Agardh 1824
(sensu Rippka *et al.* 1979)

B. Mature trichome exhibits basal–apical polarity

1. Hormogonia give rise to young trichomes that differentiate a single terminal cell into a heterocyst

a. Mature trichome composed of cells that are disk shaped, isodiametric, or cylindrical, and whose size decreases from the base to the apex; young trichomes bear exclusively terminal heterocysts at the basal end; mature trichomes are immotile and may contain intercalary heterocysts; differentiation of akinetes (if produced) is initiated adjacent to basal heterocysts, and they may be formed in chains; sheath production variable; false branching[104–106] frequent in ensheathed members; hormogonia are sheathless (or thinly sheathed?), generally motile, and composed of cells that are smaller in size and different in shape; some hormogonial cells contain gas vacuoles
Calothrix Agardh 1824
(sensu Rippka *et al.* 1979)

i. Degree of tapering little pronounced, even after growth in the absence of combined nitrogen; hormogonia (normally motile[107]) are composed of isodiametric cells that contain irregularly distributed gas vacuoles[107]; akinetes not produced; organisms contain phycoerythrin and adapt chromatically
Calothrix: type I
("*Tolypothrix tenuis*" type, this proposal)

ii. Properties not as above — *Calothrix:* type X

Key 6. Genera of Section V

Filamentous cyanobacteria capable of divisions in more than one plane; in the absence of combined nitrogen, heterocysts are formed that are in terminal, intercalary, or lateral positions; reproduction by random trichome breakage, germination of akinetes (if produced), or hormogonia production.

I. Hormogonia produced

A. Trichome never exhibits basal–apical polarity

1. Hormogonia give rise to young filaments that differentiate terminal or intercalary heterocysts

a. Cells of mature trichomes are spherical and divide in more than one plane; subsequent detachment of the resulting multicellular groups (derived from the individual cells that constituted the primary trichome) leads to a mode of

[107] Motility and gas vacuolation may be lost due to spontaneous mutation; if this is the case, affiliation to this group needs to be confirmed by genetic analysis.

Key 6. (continued)

growth resembling that of chroococcalean cyanobacteria (i.e., *Gloeocapsa*, see Section I); heterocysts in the multicellular aggregates are terminal or lateral (with respect to the long axis of the original trichome); rapid division of some cells in only one plane leads to the production of uniseriate, motile hormogonia, composed of small cylindrical cells; gas vacuoles not produced *Chlorogloeopsis* Mitra and Pandey 1966

2. Hormogonia give rise to young trichomes in which heterocyst differentiation occurs exclusively in an intercalary position
   a. Cells of mature trichomes are barrel shaped to spherical and may divide in more than one plane; concomitant division of some cells in only one plane leads to multiseriate primary trichomes with lateral uniseriate branches; heterocysts in the multiseriate trichomes are predominantly terminal or lateral, those of the branches are uniquely in an intercalary position; hormogonia, composed of small cylindrical cells, are produced by rapid division from the ends of the primary trichomes or from the lateral branches and are generally motile; gas vacuoles not produced; akinetes, formed in some members, occur in chains *Fischerella* Gomont 1895 (sensu Rippka *et al.* 1979)

Appendix: List of Major Collections from Which Cyanobacterial Cultures Can Be Obtained

ATCC American Type Culture Collection, 12301 Parklawn Drive, Rockville, Maryland 20852 (Dr. R. Gherna)

CCAP Culture Centre of Algae and Protozoa, Windermere, Freshwater Biological Association, The Ferry House, Ambleside, Cumbria LA 22 OLP, England (Dr. S. I. Heaney)

NIVA Culture Collection of Algae, Norwegian Institute for Water Research, P.O. Box 333—Blindern, 0314 Oslo 3, Norway (Dr. O. M. Skulberg)

PCC Pasteur Collection of Cyanobacteria, Unité de Physiologie Microbienne, Département de Biochimie et Génétique Moléculaire, Institut Pasteur, 28 rue du Dr. Roux, 75724 Paris Cedex 15, France (Mme R. Rippka)

SAG Sammlung von Algenkulturen, Pflanzenphysiologisches Institut der Universität, Nikolausberger Weg 18, D-3400 Göttingen, Federal Republic of Germany (Dr. U. G. Schlösser)

UTEX Culture Collection of Algae, Department of Botany, The University of Texas at Austin, Austin, Texas 78712 (Dr. R. C. Starr)

Additional culture collections carrying cyanobacterial isolates are listed in J. E. Staines, V. F. McGowan, and V. B. D. Skerman (eds.), "World Directory of Collections of Cultures of Microorganisms." World Data Center, University of Queensland, Brisbane, Queensland, Australia, 1986. Questions concerning this Directory should be sent to the relocated World Data Center and be addressed to: Prof. Kazuo Komagata, World Data Center on Microorganisms RIKEN, 2-1 Hirosawa, Wako, Saitama 351-01, Japan. For more specific information the reader may contact the author of this chapter at the Pasteur Collection of Cyanobacteria (PCC).

[3] Culturing Methods for Cyanobacteria

By RICHARD W. CASTENHOLZ

Introduction

An article on culturing methods for cyanobacteria should ideally lead a user through an entire procedure of finding, collecting, identifying, transporting, storing, enriching (in some cases), and, finally, isolating, culturing, and conserving the strains obtained. Since the subjects of finding, collecting, and identifying cyanobacteria are covered elsewhere in this volume ([1], [2]), this chapter begins with transportation and the measures preparatory to establishing cultures. Nevertheless, some general remarks about the earlier subjects are warranted.

For the establishment of new cultures, as much ecological information as possible about the field populations should be gathered and recorded. With this, it eventually becomes possible to relate diverse information obtained from culture strains to the survival strategies of the organism in nature, ultimately the object of all biological research unless it is directed solely to the benefit of humans.

Cyanobacteria, unlike most other prokaryotes (eubacteria or archaebacteria) are usually seen and often recognizable (even at the species level) before collection; they are present in many cases as plankton blooms or as dense turfs or mats which are composed of few other species. They are, in some cases, "macrophytes" forming branched thalloid "plants" or colonies several centimeters in length or diameter. In tropical marine shallows, they may be the most conspicuous photosynthetic organisms, and in extreme environments, such as hot springs or moist terrestrial environments of polar regions, thick spongy or gelatinous mats of seemingly "pure" cyanobacteria may predominate. Therefore, unlike many of the nonphototrophic bacteria, cyanobacteria with a direct link to the feral population can be sought, found, and cultured.

Only a miniscule percentage of cyanobacterial species have been brought into culture. This, in part, reflects the relatively few and mainly recent attempts to isolate and grow diverse cyanobacteria. However, there are real problems as well. Cyanobacteria from certain habitats and those of certain taxonomic groups seem to be extremely recalcitrant with regard to being cultured with present methods (see below).

Identification of cyanobacteria, whether of natural populations or of cultured material, is not a simple task, since at present, there are three major systems of classification: Geitler,[1] Drouet,[2] and Stanier[3] systems. The first two are based almost entirely on morphological characteristics visible with light microscopy. Geitler[1] and many other authors before and after also followed a botanical system but, being convinced that most morphological variation was genotypic rather than environmentally caused, assigned at least 400 generic and 1600 species names. In Drouet's system he reduced the species and generic names to 24 genera and 62 species, using, however, only herbarium specimens, collected material, and intuition as his base.

The system begun by Stanier in 1971,[4] and summarized by Rippka *et al.* in 1979,[3] is also one of condensation but is based on physiological, biochemical, ultrastructural, and some genetic characteristics of cultured strains. This system, with some modification, has been incorporated into the current *Bergey's Manual of Systematic Bacteriology*[5] and is meant to be tentative and ever expanding. Additional information on cyanobacterial phyletics is being supplied by information from long contiguous sequences (800–900) of 16 S rRNA nucleotides.[6]

Transportation of Sample

It is conceivable that some users of this volume may wish to establish new culture strains from natural habitats. There are some general precautions to be taken. Terrestrial mats on consolidated surfaces (rocks, wood, concrete, etc.) or unconsolidated materials (e.g., soils) can be air dried at room temperature or retained in a moist (not wet) condition in vials or plastic bags and then transported (in darkness) over a period of days to weeks with no apparent loss of viability. Several collections of this type still gave rise to plentiful live trichomes after 1 year of storage.[7] Intertidal, subtidal, or freshwater benthic mats of a gel-bound or fabriclike nature

[1] L. Geitler, "Rabenhorst's Kryptogamenflora von Deutchland, Österreich und der Schweiz: 14. Cyanophyceae." Akademische Verlag, Leipzig, Federal Republic of Germany, 1932.

[2] F. Drouet, "Revision of the Stigonemataceae with a Summary of the Classification of the Blue–Green Algae." Cramer, Vaduz, 1981.

[3] R. Rippka, J. Deruelles, J. B. Waterbury, M. Herdman, and R. Y. Stanier, *J. Gen. Microbiol.* **111,** 1 (1979).

[4] R. Y. Stanier, R. Kunisawa, M. Mandel, and G. Cohen-Bazire, *Bacteriol. Rev.* **35,** 171 (1971).

[5] R. W. Castenholz and J. B. Waterbury, *in* "Bergey's Manual of Systematic Bacteriology," Vol. 3. Williams & Wilkins, Baltimore, Maryland, 1988.

[6] S. Giovannoni, S. Turner, G. J. Olsen, S. Barns, D. J. Lane, and N. R. Pace, *J. Bacteriol.* (in press).

[7] R. W. Castenholz, unpublished observations.

may also be transported in a *moist* condition in vials, bottles, or sealed plastic bags without water being added. Deterioration is almost always slower than when such dense masses of cells and trichomes are carried in water, even when the water is from the native habitat.

Submerged aquatic cyanobacteria, marine or freshwater, may be attached to various submerged substrates, forming prostrate covers, tufts, or streamers, and, in many cases, mixed with a great diversity of filamentous macroalgae and other vegetation. It is important to transport collections of this material with a large amount of native water, e.g., a volume ratio of material to water of about 1 : 20 or less. Planktonic collections should be handled with more care; very dilute samples should be taken in vials or bottles. Decomposition and deterioration seems to be more rapid with plankton. Samples should not contain more than a few visible zooplankters and preferably none.

All types of samples should generally be kept in the dark, since light can cause severe damage when samples are outside of a normal nutrient environment. At times, however, it is beneficial to keep dilute samples in a dimly lit room at appropriate (environmental) temperatures if immediate transport is not possible. For transport of dry or moist samples, room temperature is adequate unless the moist material was taken from a cold habitat. In general, 4–15° is preferable to room temperature for aquatic samples if taken from cool waters of the same general temperature range. For some tropical marine cyanobacteria, however, even 15° represents a lethal temperature, and room temperatures (up to 35°) are preferred. Cyanobacteria from hot springs are also best kept at room temperature or as low as 14°, never at refrigerator temperatures (3–5°).[8,9]

No samples (or cultures) should ever be frozen directly unless it be at −196° (liquid N_2). Feral samples of many cyanobacteria, however, have been successfully frozen in ampules immersed in liquid N_2 and then stored for later use.[7] Freezing at about −70 to −80° (low-temperature freezer) may be satisfactory for some samples, but there are few data pertaining to cyanobacteria. The dangers of freezing at relatively high temperatures (e.g., −20°) became apparent after mailing of samples during winter to many parts of the world. Although the cargo holds of jet aircraft are heated, the time spent outside during transfers is the probable lethal period. In any case, low-temperature-sensitive cultures are air mailed with consistently better results in summer than in winter months.[7] Obviously, when samples (or cultures) have arrived at the laboratory,

[8] R. W. Castenholz, this volume [5].

[9] R. W. Castenholz, *in* "The Prokaryotes" (M. P. Starr, H. Stolp, H. G. Trüper, A. Balows, and H. G. Schlegel, eds.), Vol. 1, p. 236. Springer-Verlag, Berlin, 1981.

minimal storage times are preferable. Nevertheless, dark samples should not be stored in refrigerators unless samples were from natural temperatures of about 14° or lower.

Enrichments (Elective Cultures)

Enrichments are employed mainly for cyanobacteria in general or when desired types are not obviously present in the collection. The resultant enrichment is usually not indicative of the population makeup of the inoculum. Certainly cyanobacterial enrichment methods are not as useful as those for nonphototrophic bacteria since so many species of the latter require specific substrates that are unusable by other species. Cyanobacteria have a monotonously undifferentiated photoautotrophic metabolism with few unique characteristics for specific enrichments. There are, however, some techniques which select for cyanobacteria and some for particular groups.

1. Use an essentially inorganic medium, with a chelating compound (and sometimes an organic buffer) (see below).
2. Use more than one temperature, but note that a temperature of 30–35° definitely selects for some cyanobacteria at the expense of most eukaryotic algae; however, a significant number of cyanobacteria will not tolerate temperatures this high or will grow poorly. Obviously, to enrich for thermophilic cyanobacteria, either from thermal or nonthermal habitats, temperatures of 45° or higher should be used.[8,9] No strains have been grown in culture at temperatures above 73°.[10]
3. Take care with light intensity, particularly during enrichment or isolation procedures. Low-intensity fluorescent illumination is recommended [e.g., <500 lux (75 μW cm^{-2})] for strains in which phycoerythrin predominates. For cyanobacteria from high-light environments (e.g., hot springs) more moderate intensities are preferable [e.g., 1–2 klux (150–300 W cm^{-2})], but more than one level is safer.
4. Use media without combined nitrogen (i.e., lacking NO_3^- and NH_4^+) for aerobic N_2-fixing cyanobacteria in combination with a small quantity of inoculum (to lessen input of combined nitrogen). Nevertheless, expect a lag of 1 to several weeks.
5. Use cycloheximide (Actidione) for enrichments of cyanobacteria at temperatures below 35–40° at concentrations of 50–200 mg $liter^{-1}$.

[10] R. W. Castenholz, *Mitt. Int. Ver. Limnol.* **21,** 296 (1978).

Use filter-sterilized or nonsterilized cycloheximide. Not all eukaryotes, however, are sensitive to these levels of cycloheximide. The use of 1–10 mg liter^{-1} germanium dioxide (with or without cycloheximide) is particularly useful when diatoms are potential competitors.

Winogradsky columns are useful vehicles for the enrichment and maintenance of crude cultures of cyanobacteria. The lower portions of the columns will favor nonphotosynthetic sulfide producers and purple and green sulfur bacteria (against the glass), but the upper, illuminated parts will almost invariably develop populations of cyanobacteria. In the liquid phase on top, many eukaryotic algae may thrive, depending on the extent to which sulfide diffuses to that level. Whereas many cyanobacteria tolerate sulfide or utilize it in anoxygenic photosynthesis, few eukaryotic algae are tolerant. Winogradsky columns are well described elsewhere.[11,12] Winogradsky columns of rich sediments and water from diverse aquatic habitats have maintained populations of green, purple, and cyanobacteria for many years. A slurry of $CaSO_4$ and bits of paper (e.g., Kimwipes) should be added as substrates to the bottom of columns to be used for sediments with a low native content of sulfate.

Isolation of Cyanobacteria

Whether working with feral inocula or the results of an enrichment culture, the procedure of isolating an axenic clone of a cyanobacterium is often quite different from procedures used in traditional bacteriology.[9,13] The major differences result from the larger size and complexity of many cyanobacteria and the fact that motility, when present, is of a gliding type (with rare exceptions[14]). Most of the methods involve manual techniques working with Petri plates under a good quality dissecting microscope.

Self-Isolation of Motile Trichomes and Cells

Preparation and use of agar or Gelrite (Kelco Div., Merck & Co., San Diego, CA)[15] solidified media follows standard bacteriological procedure.

[11] S. Aaronson, "Experimental Microbial Ecology." Academic Press, London, 1970.

[12] C. B. van Niel, this series, Vol. 23, p. 3.

[13] R. Rippka, J. B. Waterbury, and R. Y. Stanier, *in* "The Prokaryotes" (M. P. Starr, H. Stolp, H. G. Trüper, A. Balows, and H. G. Schlegel, eds.), Vol. 1, p. 212. Springer-Verlag, Berlin, 1981.

[14] J. B. Waterbury, J. M. Willey, D. G. Franks, F. W. Valois, and S. W. Watson, *Science* **230,** 74 (1985).

[15] D. Shungu, M. Valiant, V. Tutlane, E. Weinberg, B. Weissberger, L. Koupal, H. Gadebusch, and E. Stapley, *Appl. Environ. Microbiol.* **46,** 840 (1983).

Concentrations of agar less than 8 g $liter^{-1}$ are not sufficient for surface gliding trichomes to "outrun" many swimming bacteria, and concentrations much over 15 g $liter^{-1}$ are often too dry for gliding. The lower agar concentrations, however (e.g., 12 g $liter^{-1}$ or less), often allow gliding trichomes to move through the medium as well as on its surface. This mode is particularly useful for outstripping contaminating bacteria.

The filamentous cyanobacteria in which normal vegatative trichomes glide include most of the nonsheathed Oscillatoriaceae (Section III)[3,5] and some of the Nostocaceae (Section IV) such as *Anabaena, Cylindrospermum,* and *Nodularia*.[3,5] The speeds vary greatly, from <30 to about 600 $\mu m\ min^{-1}$. Although uniform lighting is satisfactory, setting the major light source to the side is better. Phototaxis is usually a result of a lower frequency of directional reversals away from the light source, thus increasing the net speed away from the inoculum toward the light. Because of the diversity in gliding rates, plates must be examined with isolation tools in just a few hours after inoculation for the faster types (e.g., 240 $\mu m\ min^{-1} = 14.4\ mm\ hr^{-1}$) and in about a day for slower forms (e.g., 30 $\mu m\ min^{-1} = 1.8\ mm\ hr^{-1}$). Even growth of trichomes on or through stiff agar sometimes outstrips that of accompanying bacteria.

When whole trichomes have isolated themselves from the original inoculum and have not circled to cross an earlier trail (evident by pressing the agar surface with any instrument) the trichome should be removed by cutting out the small block of agar on which it rests. This can best be done under a dissecting microscope with a watchmaker's forceps (two parallel cuts with partially opened forceps then an upper and lower cut by closing the forceps. Transfer on an agar block to a new area or a new plate (by inversion) or to liquid medium prevents the drying that would occur if only the trichome were transferred. Since "normal" bacteriologists are not accustomed to using forceps, they should be reminded that cooling of a forceps must be hastened by a 95% alcohol dip, followed by ignition of the excess. The forceps should be kept sharpened on a ceramic or carborundum stone.

For cyanobacteria that produce hormogonia as the only motile stage, inocula from field collections or from "spent" (at least phosphate-depleted) cultures should be placed on agar medium with moderate to high phosphate. This usually results in the differentiation and release of motile, gliding hormogonia which may migrate for 1 to a few days. A few unicellular cyanobacteria exhibit gliding motility, in which case a phototactic strategy on plates may help in their isolation.[4]

Self-isolation of gliding, filamentous cyanobacteria is certainly the best way to establish clones (individual trichomes) and axenicity. Even when clonal cultures become contaminated secondarily, the same procedure repcated is by far the best, but care must be taken to avoid isolating variant or mutant clones since reisolation is, in fact, a recloning. In many cases, this possibility can be avoided, because large wisps of many trichomes often move away together from the source of inoculation and contamination. Thus, many axenic trichomes can be transferred on a larger agar block.

Agar blocks with the desired single or multiple trichomes or cells may be transferred to liquid culture or inverted and ''smeared off'' onto a new agar surface. However, the percentage of isolations that survive and grow in liquid medium is far greater than on agar.

There are many cyanobacteria that apparently will not glide (or glide poorly) on agar surfaces. This is particularly true of ''oligotrophic'' types, freshwater and marine, but particularly the latter. Liquid medium in plates over a plain, thin agar or Gelrite base will allow cloning of many filamentous forms that glide only under liquid. It is very easy to see the results under a dissecting microscope. The underwater techniques, however, do not solve the problem of contaminating heterotrophic bacteria.

Direct Manual Isolation

There are numerous large filamentous cyanobacteria that do not respond to any of the above methods even though hormogonia or other motile trichomes may eventually develop. In some cases, several days of incubation result in a complete overgrowth by heterotrophic bacteria before motility can be realized by the cyanobacteria. In marine cultures, this is often accompanied by deep pitting caused by agarotrophic bacteria.

Some of these difficult cyanobacteria can be cloned by direct manipulation of the filaments on agar plates under a dissecting microscope. The cutting off of filament tips (with fewer epiphytes) with watchmaker's forceps and the dragging of the dissected pieces over and through as much stiff agar as possible may rid the filament or trichome of most or, occasionally, all associated contaminants. Sheathed trichomes (i.e., filaments) are particularly difficult to clean. A wash of selected filaments in about 0.3% (v/v) phenol in the dark for 4–6 hr may greatly enhance the chances.[16]

The sheath structure of some cyanobacteria may be disrupted without undue mortality by the use of a common commercial disinfectant.[17] Although the product cited is no longer available, a similar product—Lysol Deodorizing Cleaner—which also contains alkyl (50% C_{14}, 40% C_{12}, 10% C_{16}) dimethylbenzyl ammonium chlorides and tetrasodium EDTA, is also effective at a 1–10% (v/v) concentration. Material should be treated for

[16] W. W. Carmichael and P. R. Gorham, *J. Phycol.* **10,** 238 (1974).
[17] R. M. Bradley and R. L. Pesano, *J. Cell Biol.* **87,** 228a (1980).

30–60 sec and then washed 2–3 times with fresh medium. Axenic cultures were commonly obtained by this treatment.[17]

The need for manual isolations of trichomes or large cells in liquid medium is also apparent for many planktonic and other forms. Several specific techniques have been described.[9,13,18,19] One method uses transparent glass depression slides (~7.6 × 5 cm) with 10 circular concavities about 1.75 mm deep. They are sterilized by swabbing in alcohol and flaming. Sterile medium is placed in each depression. The delicate trichomes (e.g., planktonic forms) are pipeted into the first depression. Thereafter, under a dissecting scope, one or a few trichomes are "pulled out" of the suspension with a glass capillary with the end pulled out to an appropriate diameter over a small flame (e.g., alcohol lamp). After blowing out the trichomes into the next depression, a new capillary is used to select a single trichome with the procedure repeated until the trichome has been washed several times. With the final removal to a flask or tube of liquid medium, the trichomal clone may also be axenic unless heterotrophic bacteria adhere to the cell surfaces. This is often the case for heterocysts in a trichome. Such trichomes may be cut in the depression slide so as to separate the heterocystous portion.

For buoyant planktonic forms treated in the above manner, culture tubes with the isolate may need to be pressurized to at least 5 bars twice weekly until the culture becomes established.[18] In this way, the single trichome is less likely to dry out at the meniscus–glass contact which recedes rapidly in test tubes owing to evaporation. For hardier trichomes, agar plates can be used for direct manual isolation. The initial inoculum can be spotted on the plate, a small block with one or more trichomes (or cells) cut out, inverted on a clean part of the plate, smeared to one side to spread apart the organisms present, then the procedure repeated several times until only the desired, "cleaned" trichome remains.

Streaking and Other Dilutions on or in Agar

Standard streaking and spreading procedures with a loop or glass rod may be used for raw inocula or for enrichment cultures. This may result in some presumptive clonal colonies of unicellular cyanobacteria, but usually not without other contaminating bacteria. Recovery and growth of unicellular cyanobacteria on agar surfaces is greatly enhanced by auto-

[18] A. E. Walsby, *in* "The Prokaryotes" (M. P. Starr, H. Stolp, H. G. Trüper, A. Balows, and H. G. Schlegel, eds.), Vol. 1, p. 224. Springer-Verlag, Berlin, 1981.

[19] R. W. Hoshaw and J. R. Rosowski, *in* "Handbook of Phycological Methods, Culture Methods and Growth Measurements" (J. R. Stein, ed.), p. 53. Cambridge University Press, Cambridge, England, 1973.

claving the agar in one-half the total H_2O to be used for the medium and the nutrient additions separately in the other half of the H_2O,[20] mixing the solutions at 45–50° just before pouring.

Shake dilution cultures (in tubes or plates) may also be used, particularly when long-term incubations may be necessary. Many unicellular cyanobacteria do not seem to recover well deep within agar, but this may be a result of CO_2 deficiency. $NaHCO_3$ may be added to satisfy the need. For example, a 1 *M* solution may be made by autoclaving 8.4 g dry $NaHCO_3$ and adding 100 ml sterile water afterward. This solution should then be sparged with pure CO_2 by sterile filtration for about 15 min to lower the pH. Then 10 ml liter^{-1} may be added to the final medium to give a concentration of 10 m*M* (0.84 g liter^{-1}), although concentrations as high as 2 g liter^{-1} are sometimes used.

Other Methods of Isolation and Purification

Gas-Vacuolate Forms. Since most planktonic cyanobacteria have a significant portion of their cell volume taken up by gas vesicles, positive buoyancy is a general feature of at least those species that are circulated within the epilimnia of lakes. Because of this property, contaminants (other types of organisms) may be separated from gas-vesiculate types by centrifugation. The procedure is well described by Walsby.[18] A general outline of the procedure is given here. Since very high centrifugation speeds result in gas vesicle collapse with applied pressures of about 1 bar or more, a mixed collection (e.g., about 10^4 trichomes ml^{-1}) is placed in tubes in a swing-out rotor and turned for the first 2 min at 100 *g* or less, after which time the buoyant cells will have risen part way in the tube. At this point, the applied pressure of 1 bar would not be exceeded at that depth until an acceleration of about 200 *g* is used. Therefore, the rotor can be increased about 100 *g* every 2 min. After 10 min all or most of the buoyant cells and trichomes will have reached the surface meniscus. The centrifuge can then be increased to its maximum safe setting for another 5 min. After a complete stop and standing for a few minutes to allow gas-vacuolate cells to float to the edge of the meniscus, merely touch the meniscus edge with the top of a flat-ended syringe needle and withdraw a very small amount and transfer to fresh medium. The centrifugation speeds and time should be varied depending on initial results. A second centrifugation series may be necessary.

Although this is an efficient way of concentrating gas-vesiculate organisms and separating them from other types, in order to establish axenic,

[20] M. M. Allen, *J. Phycol.* **4,** 1 (1968).

clonal cultures, single filament manual isolation and washing is still necessary (see Direct Manual Isolation). The tubes with isolates need to be repeatedly pressurized (to at least 5 bars) for the initial growth period of several weeks to collapse the gas vesicles.

Filter Washing. There are several relatively recent filter-washing methods that have been used to clone or to purify cyanobacteria. The generally expected result is to increase the cyanobacterial unit population relative to the numbers of heterotrophic bacteria prior to plating or some other procedure such as direct manual isolation.

Heaney and Jaworski,[21] however, have successfully isolated axenic clones of the planktonic filamentous cyanobacterium *Aphanizomenon flos-aquae* by a wash and backwash technique. They used a syringe, attached to a 25-mm-diameter Nuclepore (Pleasanton, California) Swin-Lok membrane holder. The exit was attached to the top of a vacuum flask by hoses and a T to which a side hose and clamp could be fit. Through the vertical hose and clamp vacuum pressure could be adjusted whereas sterile (filtered) air could be introduced from the side for positive pressure. An 8-μm-pore size sterile Nuclepore filter was used and first washed with about 50 ml of culture medium. Then a dilute inoculum of a few milliliters of cyanobacteria suspension was introduced (culture density about 10 μg chlorophyll *a* liter^{-1}) followed by 30–100 ml of culture medium with occasional backwashes of medium (plus cells) into the syringe. This can be achieved by adjusting the vacuum clamp and air input clamp. The point of this procedure is to wash unattached bacteria through the filter while the larger trichomes are retained. The occasional backwashes possibly resuspend a few bacteria that may have been lodged on the filter. Near the end of the procedure a small amount of medium in the syringe (hopefully with some trichomes) is transferred aseptically to culture medium. Following this, a direct manual isolation procedure could be used either with depression slides and liquid medium (see above) or on agar plates.

Filter-washing methods have been used with filters of various pore sizes.[22–24] Fitzsimons and Smith[25] used a procedure that resuspended the washed trichomes from the filter several times, replacing the filter each time. They used "Millipore" filters with either a 5- or 10-μm pore size. Filamentous bacteria have also been washed free of most small bacteria on electron microscopy grids. Carmichael and Gorham[16] have used the following method (here, somewhat simplified):

[21] S. I. Heaney and G. H. M. Jaworski, *Br. Phycol. J.* **12,** 171 (1977).
[22] M.-E. Meffert and T.-P. Chang, *Arch. Hydrobiol.* **82,** 231 (1978).
[23] M.-E. Meffert, *Arch. Hydrobiol. Suppl.* **41,** 235 (1972).
[24] A. L. Huber, *J. Phycol.* **20,** 619 (1984).
[25] A. G. Fitzsimons and R. V. Smith, *Br. Phycol. J.* **19,** 156 (1984).

1. Place culture (late exponential phase) in flasks (~30 ml) and keep in dark for 24–30 hr at appropriate temperature.
2. Add phenol to 0.3% (v/v) and keep in dark for 4–6 hr longer.
3. Add 15–20 ml of culture to 200 ml of fresh medium and gently filter using a relatively coarse sterile filter (pore size 5–10 μm).
4. Transfer filter to about 15 ml of sterile medium, swirl, and discard filter.
5. Use small volume of this medium (e.g., 0.2 ml) and incorporate in 20 ml of medium with agar still unsolidified above the gel point of about 35°.
6. Pour into a standard Petri plate and cool rapidly in freezer; incubate at growth temperature for 1–3 days (inverted).
7. Select and pick out individual cells or trichomes with a sterile, pulled Pasteur pipet. Eject into a small volume (e.g., 1 ml) of appropriate medium.
8. Observe until growth is apparent, then transfer and test for contamination.

Needless to say, heterotrophic bacteria or other contaminants that may adhere to the surface of cyanobacterial cells will not usually be removed by any of the filtration methods, although the pretreatment with phenol may eliminate some of these.

McCurdy and Hodgson[26] used a mechanical procedure that detached contaminants from cyanobacterial cells. Inocula may be dispersed by short-term mixing in an Omnimixer (Sorvall, Norwalk, CT) or brief treatment in an ultrasonic bath or with an ultrasonic probe. After dilutions, cells or trichomes are collected on membrane filters and colonies allowed to develop directly on the filter (placed on agar). Individual colonies are then removed and dispersed for 60–90 sec with 0.8 g glass beads in the Omnimixer set at 80 (arbitrary units). This is followed by surface plating to produce several axenic cyanobacterial cultures.[25]

Use of Antibiotics and Poisons. Although cyanobacteria are eubacteria, there is still a possibility of using common antibiotics to selectively eliminate contaminating heterotrophic bacteria. The need for differentiating antibiotics is great since there are many immotile cyanobacteria with sheaths or capsules that harbor heterotrophic bacteria. Whitton[27] suggested that the use of antibiotics in darkness could be useful if the cyanobacterium is an obligate phototroph. Rippka[13] has successfully used the following procedure:

1. Incubate fresh cells or trichomes in new medium under growth conditions, except in darkness for about 48 hr.

[26] H. D. McCurdy, Jr., and W. Hodgson, *Can. J. Microbiol.* **20,** 272 (1974).
[27] B. A. Whitton, *Plant Cell Physiol.* **9,** 12 (1968).

2. Then add aseptically 0.5% (w/v) glucose, 0.1% (w/v) casamino acids, and 1,000 $\mu g\ ml^{-1}$ ampicillin (all filter sterilized).
3. Continue incubation in dark for 24–48 hr.
4. Wash cells or trichomes.
5. Streak, plate, or manually isolate cells or trichomes under sterile conditions; incubate in light.

Contamination may be so low under these conditions that axenic clones can be picked off on the first set of plates.[4]

Phenol has already been mentioned as a possible discriminatory agent. However, the use of $Na_2S \cdot 9H_2O$ has proved even more effective for at least one type of planktonic cyanobacterium (*Microcystis*).[28] *Microcystis aeruginosa* is a colonial cyanobacterium. Disaggregation of the colonies into unicells occurred when colonies were placed into distilled or deionized water. When plated on agar medium many *Microcystis* survived the application of 0.05–0.1 ml of a solution containing 1.3 M Na_2S and 0.3 M Na_2SO_3 (pH 13) over the surface of the plate, while heterotrophic bacteria survived mainly on an area not reached by the sulfide/sulfite solution. Axenic cultures were established by this procedure. It is possible that this method may have much greater use, although only a limited number of cyanobacteria may be tolerant of high sulfide/sulfite or of the extremely high pH.

In conclusion, isolation and purification procedures for cyanobacteria have not been standardized, and the researcher should expect to be original and versatile. It should be emphasized, though, that single cell or trichome isolates may require many weeks or even as long as 2 months before growth becomes visible. Although growth rates of many cyanobacteria are slow (doubling times of 1 to several days, or weeks in the case of some oligotrophic forms) the reason for the great delay after the initial cloning is often the long lag in fresh (or new) medium rather than slow potential growth rate per se. Also, recall that single cell or single trichome isolates transferred to agar medium usually fail while chances of survival increase greatly with transfer to liquid medium, normally in flasks with a greater surface to volume ratio. Unexplicably, however, isolates of some types survive and initially grow better in cotton-plugged or even screw-cap tubes.[7]

Composition of Culture Medium

So many variants of diverse culture media are used today that only a relatively few basic types will be described. Needless to say, no medium is perfect for all cyanobacteria, and individual tailoring or modifications

[28] D. L. Parker, *J. Phycol.* **18,** 471 (1982).

are needed depending on the purpose intended. For example, if the need is to obtain the greatest yield of cyanobacterial cells, perhaps in the shortest time, a rich medium may be required. If the purpose is to elicit morphological, cytological, and physiological features that might be characteristic of natural populations, low nutrient medium is needed. A deficiency of combined nitrogen is generally required to elicit heterocyst differentiation and the synthesis of nitrogenase.[1] Phosphorus deficiency may be required for akinete formation in some species or for the formation of a terminal multicellular hair in tapered species of *Calothrix,* whereas the repletion of phosphate may induce differentiation and migration of hormogonia in some groups of cyanobacteria.[29] For some purposes nutrient-rich medium with a low quantity of a single nutrient may be of use in experimentation.

The lists of media should not be taken as the full range from which an investigator should choose for the needed cyanobacterium, particularly if the organism is new in culture. Various permutations or entirely new attempts should be made to establish suitable media. The attempts should be based on known habitat chemistry or specific requirements. Tables I and II are recipes to be used for batch cultures, where it is necessary for nutrients to be present in concentrations high enough to allow at least a few doublings to take place. Although proportions of various salts and nutrients may be similar in continuous culture medium, the quantities of most nutrients are greatly reduced.

One of the most important advances in medium composition in recent years is the realization that most reagent grade salts and nutrients contain levels of heavy metals or other contaminants that are particularly inhibitory to species of an oligotrophic nature (e.g., open sea types). Various methods need to be applied in order to alleviate this problem. In fact, at this point, most of the cyanobacteria in culture may be considered the "weeds" that are tolerant of most of these contaminants (see below).

Stability of pH is not always an essential aspect of culture media. An essentially unbuffered medium (e.g., D medium, Table I) is useful for most cyanobacterial stock cultures. With nitrate as the source of combined nitrogen, pH will rise from the initial level (e.g., pH 7.5, D medium) to very high levels (pH 9–10) by the end of exponential growth phase. Most cyanobacteria are not killed at this high pH, particularly since it parallels the end of exponential growth. The use of phosphates for buffering is not recommended for cyanobacterial cultures in general, since the concentrations required are inhibitory for many or most species.

[29] B. A. Whitton, *in* "The Cyanobacteria" (P. Fay and C. Van Baalen, eds.), p. 513. Elsevier, Amsterdam, 1987.

TABLE I
COMPOSITION OF FRESHWATER MEDIA FOR CYANOBACTERIA[a]

	Concentration					
Ingredient	Chu No. 10 (modified)	Gerloff *et al.*[32]	BG-11[b]	D Medium[c]	Allen and Arnon[38]	Kratz and Myers[39]
Disodium EDTA	—	—	1[e]	—	4[j]	—
Nitrilotriacetic acid (NTA)	—	—	—	100	—	—
Citric acid	3	3	6	—	—	165[k]
$NaNO_3$	—	41	1500[f]	700	—	—
KNO_3	—	—	—	100	2020	1000
$Ca(NO_3)_2 \cdot 4H_2O$	40–60	—	—	—	—	25
$K_2HPO_4 \cdot 3H_2O$	13	—	40	—	456	1000
KH_2PO_4	—	—	—	—	—	—
Na_2HPO_4	—	8	—	110	—	—
$MgSO_4 \cdot 7H_2O$	25	15	75	100	246	250
$CaSO_4 \cdot 2H_2O$	—	—	—	60	—	—
$MgCl_2 \cdot 6H_2O$	—	21	—	—	—	—
$CaCl_2 \cdot 2H_2O$	—	36	36	—	74	—
KCl	—	9	—	—	—	—
NaCl	—	—	—	8	232	—
Na_2CO_3 (H_2O)	20	20	20	—	—	20 (opt.)
Ferric ammonium citrate	—	—	6	—	—	—
Ferric citrate	3 or	3	—	—	—	—
$FeCl_3$	3	—	—	0.3[h]	—	—
$Fe_2(SO_4)_3 \cdot 6H_2O$	—	—	—	—	—	4
Micronutrients	—[d]	—[d]	1 ml[g]	0.5 ml[i]	—[d]	1 ml[g]
Vitamin mix	—[d]	—[d]	—[d]	—[d]	—[d]	—[d]

[a] Unless indicated, concentrations are in mg liter^{-1} of double-distilled or deionized water; see text for details.
[b] pH 7.4 after cooling.
[c] Prepared as a 20-fold concentrated stock, stored at 4°. Micronutrients and $FeCl_3$ included in stock. pH adjusted to 8.2 with NaOH before autoclaving. After cooling and clearing, pH is about 7.5. Several variations of this medium are described in the text.
[d] Micronutrients and vitamins optional. If used, 0.5–1.0 ml of any mixture in Table III.
[e] Disodium-magnesium EDTA is generally used.
[f] The nitrate concentration is often lowered.
[g] The medium generally uses A_5 + Co (Table III).
[h] Sometimes 2–4 times this amount is used. A stock solution of 0.29 g liter^{-1} is kept at 4°.
[i] D micro (Table III).
[j] 13% ferric-sodium EDTA.
[k] Trisodium citrate dihydrate.

When liquid medium is continuously sparged with CO_2-enriched air (1–5%) it is necessary to compensate for the decrease in pH by adding $NaHCO_3$ or N_2CO_3 to the medium (e.g., ~500 mg liter^{-1} $NaHCO_3$ when gassing with 3% CO_2 in air for Kratz and Myers Medium *C*, Table I). For axenic cultures, a variety of substituted alkyl sulfonate buffers may be used to stabilize pH (see below). Since some heterotrophic bacteria can use these as substrate, they are not useful for contaminated cultures. However, culturists are warned not to use Tris since toxic effects have been noted in numerous cases.[30]

Two tables of media are included along with a third for micronutrient and vitamin mixes. Table I includes a variety of freshwater media commonly used for cyanobacteria with the more dilute (oligotrophic) media comprising the left-hand columns, becoming progressively richer (eutrophic) toward the right. The same arrangement is followed for the marine media (Table II). Although many recipes for both freshwater and marine media list their own micronutrient (trace element) combination, only a few of these are listed since most are very similar (Table III). Almost every medium listed has been annotated in footnotes. It is noteworthy that many media, each well known by name, are essentially the same, having been derived from one initial contributor, and acquiring the names given by various uses or modifiers.

Media of Table I

None of the freshwater media are here devised with specifically designated instructions to increase the purity of the salts in order to avoid inhibition by heavy metals. The methods that pertain to Aquil medium (Table II), however, could also be applied to the preparation of any freshwater medium. Essentially any of the freshwater media of Table I could also be tried using natural or artificial seawater instead of distilled or deionized water, but various adjustments of macronutrients would also then be appropriate (see Waterbury and Stanier[31]). For example, BG-11 medium with seawater has been quite successful. In many cases, simply the addition of 20–30 g liter^{-1} NaCl will accommodate many marine cyanobacteria. Hypersaline media may also be needed in the study of halophilic species.[31]

Chu No. 10 was developed early (1942) as a general medium for various algae. It has a very low nutrient and salt content and is therefore

[30] K. Ohki, J. G. Rueter, and Y. Fujita, *Mar. Biol. (Berlin)* **91,** 9 (1986).

[31] J. B. Waterbury and R. Y. Stanier, *in* "The Prokaryotes" (M. P. Starr, H. Stolp, H. G. Trüper, A. Balows, and H. G. Schlegel, eds.), Vol. 1, p. 221. Springer-Verlag, Berlin, 1981.

acceptable for oligotrophic strains where not too many doublings of cell mass are required. The pH is usually about 6.5–7.0 without adjustment. The modification of adding citric acid and micronutrients in small quantities was not in the original recipe.[32] The medium of Gerloff *et al.*,[33,34] (more or less derived from Chu No. 10) is also of low nutrient and salt content, of higher pH, and was developed mainly for cyanobacteria. Both of these media were used along with media much richer in various nutrients to examine the growth pattern of the cyanobacterium *Scytonema* sp. Typical complex morphology (as found in field populations) developed in the low nutrient media (Chu No. 10 and Gerloff *et al.*) and not in media with high nitrate and phosphate content.[35]

BG-11[3,14] is a universal medium that evolved with only few changes from an original formulation of Hughes *et al.*[36] The media of M. M. Allen,[20] Starr,[37] and Stevens *et al.*[38] are minor modifications. BG-11 uses very high nitrate concentrations and relatively low phosphate (molar ratio 99 : 1). However, the nitrate value can be lowered considerably. BG-11_0 lacks sodium nitrate completely and is used routinely for growing N_2-fixing cyanobacteria.

D medium was developed for growing thermophilic cyanobacteria from hot springs.[9] Although it contains rather high levels of nitrate and phosphate, the molar ratio is about 12. Many modifications of this medium have also been used. D medium has proved to be a useful medium for many nonthermophilic cyanobacteria, too. Modifications include DG medium in which 0.8 g liter^{-1} glycylglycine is added as buffer (pK' 8.25). This medium, along with a medium similar to BG-11, works very well for the quantitative plating of several unicellular cyanobacteria, including *Synechococcus* AN (= *Anacystis nidulans*). However, it is essential in both cases to autoclave the nutrients in half of the water and the agar in the other half, mixing the two when the mixtures have cooled to 45–50°.[20] DGN is a variant of DG medium in which 0.2 g liter^{-1} NH_4Cl is added. For an accurate input of NH_4Cl a concentrated solution should be autoclaved

[32] H. W. Nichols, *in* "Handbook of Phycological Methods, Culture Methods and Growth Measurements" (J. R. Stein, ed.), p. 7. Cambridge University Press, Cambridge, England, 1973.

[33] G. C. Gerloff, G. P. Fitzgerald, and F. Skoog, *in* "The Culturing of Algae" (J. Brunel, G. W. Prescott, and L. H. Tiffany, eds.), p. 27. Kettering Found., Yellow Springs, Ohio, 1950.

[34] N. G. Carr, J. Komárek, and B. A. Whitton, *in* "The Biology of Blue-Green Algae" (N. G. Carr and B. A. Whitton, eds.), p. 525. Blackwell, Oxford, England, 1973.

[35] N. Jeeji-Bai, *Schweiz. Z. Hydrol.* **38,** 55 (1976).

[36] E. O. Hughes, P. R. Gorham, and Z. Zehnder, *Can. J. Microbiol.* **4,** 225 (1958).

[37] R. C. Starr, *J. Phycol. Suppl.* **23** (1987).

[38] S. E. Stevens, Jr., C. O. P. Patterson, and J. Myers, *J. Phycol.* **9,** 427 (1973).

TABLE II
COMPOSITION OF MARINE AND HYPERSALINE MEDIA FOR CYANOBACTERIA[a]

Ingredient	Grund[b]	F/2[b]	MN[c]	Ong *et al.*[d]	ASP-M[e]	Aquil[f]	Erdschreiber's[b]	Yopp *et al.*[g]
Disodium EDTA	2	10[j]	0.5	5	0.8[j]	—	10[j]	5[j]
Citric acid	—	—	3	—	—	—	—	—
$NaNO_3$	40	90	750	750	40–70	8.5	150	—
$Ca(NO_3)_2 \cdot 4H_2O$	—	—	—	—	—	—	—	1,000
$K_2HPO_4 \cdot 3H_2O$	—	—	—	—	—	—	—	—
Na_2HPO_4	4	—	—	—	—	—	40	—
$NaH_2PO_4 \cdot H_2O$	—	5–20	20	15	7–14	0.5	—	65
$MgSO_4 \cdot 7H_2O$	—	—	38	—	4,920	—	—	10,000
$MgCl_2 \cdot 6H_2O$	—	—	—	—	4,040	11,030	—	10,680
$CaCl_2 \cdot 2H_2O$	10	—	18	—	1,270	1,000–1,350	—	—
Na_2CO_3 ($\cdot H_2O$)	—	—	20	—	—	—	—	—
$NaHCO_3$	—	—	—	—	168	200	—	—
Na_2SO_4	—	—	—	—	—	4,090	—	—
NaCl	—	—	—	—	23,200	24,360	—	117,000
KCl	—	—	—	—	740	695	—	2,000
KBr	—	—	—	—	—	10	—	—
NaF	—	—	—	—	—	3	—	—
$Fe_2(SO_4)_3 \cdot 6H_2O$	0.2	—	—	—	—	—	—	—
Na_2SeO_4 (0.01 m*M* stock)	1 ml	—	—	—	—	—	—	—
$NiSO_4(NH_4)_2SO_4 \cdot 6H_2O$ (0.1 m*M* stock)	1 ml	—	—	—	—	—	—	—

Micronutrients	0.2 ml[h]	1 ml[k]	1 ml[l]	1 ml	1 ml[m]	0.5 ml[o]	—	1 ml[r]
Ferric ammonium citrate	—	—	3	—	—	—	—	—
Vitamin mix	0.5 ml[i]	1 ml[i]	—	1–2 ml	1 ml[n]	0.5–1.0 ml[p]	0.5 ml[p]	—
Natural or artificial seawater	1000 ml	1000 ml	750 ml	877 ml	—	—	950 ml	—
Distilled/deionized water	—	—	250 ml	120 ml	1000 ml	1000 ml	—	1000 ml
Soil extract	—	—	—	—	—	—	50 ml[q]	—

[a] Unless indicated, concentrations are in mg liter^{-1} of natural or artificial seawater or distilled water; see text for details.

[b] J. McLachlan, *in* "Handbook of Phycological Methods, Culture Methods and Growth Measurements" (J. R. Stein, ed.), p. 25.

[c] R. Rippka, J. Deruelles, J. B. Waterbury, M. Herdman, and R. Y. Stanier, *J. Gen. Microbiol.* **111,** 1 (1979).

[d] L. J. Ong, A. N. Glazer, and J. B. Waterbury, *Science* **224,** 80 (1984).

[e] See reference *b*; in addition to ingredients listed, 660–1320 mg liter^{-1} glycylglycine may be added as buffer for axenic cultures.

[f] See text for preparation details; 30 mg liter^{-1} H_3BO_3 and 17 mg liter^{-1} $SrCl_2 \cdot 6H_2O$ is also added.

[g] See J. B. Waterbury and R. Y. Stanier, *in* "The Prokaryotes" (M. P. Starr, H. Stolp, H. G. Trüper, A. Balows, and H. G. Schlegel, eds.), Vol. 1, p. 221. Springer-Verlag, Berlin, 1981; 500 mg liter^{-1} glycylglycine is used as buffer.

[h] A_5 + Co or D Micro generally used (Table III).

[i] Optional: mix of Ong *et al.* would generally be adequate (Table III).

[j] 13% ferric-sodium EDTA.

[k] Optional, but if used eliminate ferric EDTA of original formula.

[l] A_5 + Co (Table III).

[m] See reference *b* for suggested micronutrient addition; however, F/2 should be adequate (Table III).

[n] S-3 mix (Table III).

[o] Use PIV or F/2 (Table III).

[p] Use Ong *et al.* (Table III).

[q] See text for preparation.

[r] Sheridan and Castenholz solution (Table III).

TABLE III
COMPOSITION OF MICRONUTRIENT SOLUTIONS AND VITAMIN MIXES

Micronutrients (g liter^{-1})

Ingredient	A_5 + Co[a]	D Micro[b]	Sheridan and Castenholz[c]	PIV[d]	F/2[e]
H_2SO_4 (conc)	—	0.5 ml	—	—	—
HCl (conc)	—	—	3 ml	—	—
H_3BO_3	2.86	0.5	0.5	—	—
$MnSO_4 \cdot H_2O$	—	2.28	—	—	—
$MnCl_2 \cdot 4H_2O$	1.81	—	2.0	0.041	0.177
$ZnNO_3 \cdot 6H_2O$	—	—	0.5	—	—
$ZnSO_4 \cdot 7H_2O$	0.22	0.5	—	—	0.018
$ZnCl_2$	—	—	—	0.005	—
$CuCl_2 \cdot 2H_2O$	—	—	0.025	—	—
$CuSO_4 \cdot 5H_2O$	0.08	0.025	—	—	0.010
$Na_2MoO_4 \cdot 2H_2O$	0.39	0.025	0.025	0.004	0.007
$Co(NO_3)_2 \cdot 6H_2O$	0.049	—	0.025	—	—
$CoCl_2 \cdot 6H_2O$	—	0.045	—	0.002	0.011
$VOSO_4 \cdot 6H_2O$	—	—	0.025	—	—
$FeCl_3 \cdot 6H_2O$	—	—	—	0.097	1.90
$NiSO_4(NH_4)_2SO_4 \cdot 6H_2O$	—	0.019	—	—	—
Na_2SeO_4	—	0.004	—	—	—
Disodium EDTA	—	—	—	0.75 (add first)	4.35

Vitamins (mg ml^{-1})[f]

Ingredient	DN[g]	S-3[e]	Ong *et al.*[h]
Nicotinic acid	0.100	0.1	—
PABA	0.010	0.10	—
Biotin	0.001	0.001	0.001
Thiamin	0.200	0.5	2.0
Cyanocobalamin	0.001	0.001	0.001
Folic acid	0.001	0.002	—
myo-Inositol	0.001	5.0	—
Thymine	—	3.0	—
Calcium pantothenate	0.100	0.10	—

[a] R. Rippka, J. Deruelles, J. B. Waterbury, M. Herdman, and R. Y. Stanier, *J. Gen. Microbiol.* **111,** 1 (1979).
[b] R. W. Castenholz, *in* "The Prokaryotes" (M. P. Starr, H. Stolp, H. G. Trüper, A. Balows, and H. G. Schlegel, eds.), Vol. 1, p. 236. Springer-Verlag, Berlin, 1981.
[c] J. B. Waterbury and R. Y. Stanier, *in* "The Prokaryotes" (M. P. Starr, H. Stolp, H. G. Trüper, A. Balows, and H. G. Schlegel, eds.), Vol. 1, p. 221. Springer-Verlag, Berlin, 1981. Ni and Se have recently been added by Castenholz.
[d] R. C. Starr, *J. Phycol. Suppl.* **14** (1978).
[e] J. McLachlan, *in* "Handbook of Phycological Methods, Culture Methods and Growth Measurements" (J. R. Stein, ed.), p. 25. Cambridge University Press, Cambridge, England, 1973.
[f] Concentrations are designed for additions of usually 1 ml liter^{-1}. Vitamin mixes are generally filter sterilized and added after the medium is autoclaved. No cyanobacteria have been shown to have a complex vitamin requirement; most have none at all.
[g] D. C. Nelson, J. B. Waterbury, and H. W. Jannasch, *Arch. Microbiol.* **133,** 172 (1982).
[h] L. J. Ong, A. N. Glazer, and J. B. Waterbury, *Science* **224,** 80 (1984).

separately in tightened screw-cap tubes and added aseptically after the DG medium has cooled. Much NH_3 may be lost from media during autoclaving if the pH is high.

Other variants of D medium may use other buffers for various reasons (e.g., less toxicity, cost, and pH level required): Tricine [*N*-tris(hydroxymethyl)methylglycine, p*K*′ 8.15], EPPS [*N*-(2-hydroxyethyl)piperazine-*N*′-3-propanesulfonic acid, p*K*′ 8.0], HEPES [*N*-(2-hydroxyethyl)-piperazine-*N*′-2-ethanesulfonic acid, p*K*′ 7.5], and PIPES [piperazine-*N*,*N*′-bis(2-ethanesulfonic acid), p*K*′ 6.8]. ND is a varient of D medium which lacks combined nitrogen except in the form of the chelator nitrilotriacetic acid [NTA, or *N*,*N*-bis(carboxymethyl)glycine] which is unavailable to cyanobacteria. The sodium and potassium nitrate is withheld, and the potassium is replaced by 36 mg $liter^{-1}$ KH_2PO_4 and 70 mg $liter^{-1}$ Na_2HPO_4 (instead of 110 mg $liter^{-1}$ Na_2HPO_4). DT (D medium plus travertine) is a variant that has proved to be very successful for a number of cyanobacteria with gliding motility both for stock cultures and as medium for initial isolates. It is D medium with 100 mg $liter^{-1}$ $(NH_4)_2SO_4$, 100 mg $liter^{-1}$ $Na_2S_2O_3 \cdot 5H_2O$, 100 mg $liter^{-1}$ Na_2SO_3 (optional), and 0.5–1.0 $liter^{-1}$ hot springs travertine powder. The latter may be replaced by a 10 : 1 mixture of $CaCO_3$ and $MgCO_3$. The insoluble sediment appears to aid both the initial survival of cells or trichomes and their subsequent growth and development. A powdery sediment seems in many cases to be an essential factor for gliding cyanobacteria.

Most of the older media developed for cyanobacteria were designed for obtaining high growth yields. Of these, the media of Allen and Arnon[39] and Kratz and Myers[40] (Medium C) are given (Table I). Allen and Arnon's contains very high nitrate and moderately high phosphate (molar ratio 10 : 1). Kratz and Myers' Medium C has an inordinately high content of phosphate so that the molar nitrate to phosphate ratio is only 2.3. Cg-10 medium of Van Baalen[41] is modified Medium C in which the K_2HPO_4 is lowered to 50 mg $liter^{-1}$, glycylglycine is added to 1,000 mg $liter^{-1}$, and disodium EDTA (10 mg $liter^{-1}$) replaces sodium citrate.

Many older media did not list micronutrients as additives, except for iron. Instead, they depended on the contaminants in the macroreagents. With improvements in reagent-grade chemicals, additions are now recommended with either EDTA, NTA, or citric acid as chelators.

Although most published formulations of micronutrients do not include selenium or nickel, recent evidence indicates that they may be essential for some cyanobacteria (see Table III). It would be wise, there-

[39] M. B. Allen and D. I. Arnon, *Plant Physiol.* **30,** 366 (1955).

[40] W. A. Kratz and J. Myers, *Am. J. Bot.* **42,** 282 (1955).

[41] C. Van Baalen, *J. Phycol.* **3,** 154 (1967).

fore, to add both to all formulations. Vanadium may also be added for safety. Iron is included in micronutrient solutions only when chelator is present. Iron is more commonly included in the base medium itself (Tables I and II).

Media of Table II

The media listed in Table II, intended for marine species of cyanobacteria, again have lower nutrient levels in the left-hand columns and are richer to the right. Most employ natural seawater in the salinity range of 30–35%. This can usually be replaced, however, by various brands of artificial seawater, such as Instant Ocean (Aquarium Systems, Mentor, OH) and Marine Environments Salts (Marine Environments, San Francisco, CA).[42] The formulation of these salt mixtures is generally provided with the product and should definitely be available to the researchers using it.

Grund or modified Von Stosch medium (Table II),[42] used with natural or artificial seawater, works well as a medium for cyanobacteria from oligotrophic waters such as tropical seas. F/2 medium of Guillard and Ryther[42] is another simplified oligotrophic medium with only a few additions to natural seawater. Both Grund and F/2 media may, of course, be made without $NaNO_3$ for the enrichment or growth of N_2-fixing cyanobacteria.

Standard MN medium[3,13] or the modification by Ong *et al.*[43] are relatively rich media in which the molar ratio of nitrate to phosphate is about 100–130 : 1. In both these media more nutrient additions are made than in the two oligotrophic types, but less than full-strength seawater is used as the base. MN medium has served as a standard type for many marine strains,[3] and the modification by Ong *et al.*[43] has been used successfully for the minute *Synechococcus* isolates from oceanic and coastal waters.

Because of possible toxic substances that leach from laboratory glassware and from other sources, Waterbury[44] has suggested that autoclaving of filtered seawater be performed in polypropylene or Teflon bottles and that the major nutrient additions be autoclaved separately in distilled water in a glass container. The two should be mixed after cooling. Waterbury[44] and this author[7] have also noted improved survival of new isolates

[42] J. McLachlan, *in* "Handbook of Phycological Methods, Culture Methods and Growth Measurements" (J. R. Stein, ed.), p. 25. Cambridge University Press, Cambridge, England, 1973.

[43] L. J. Ong, A. N. Glazer, and J. B. Waterbury, *Science* **224,** 80 (1984).

[44] J. B. Waterbury, personal communication.

of marine and hot spring cyanobacteria with the addition of 1–2 mM Na_2SO_3 (see Media of Table I).

The modified Aquil medium used by Ohki *et al.*[30] for axenic cultures of the N_2-fixing, marine planktonic cyanobacteria *Trichodesmium erythraeum* and *T. thiebautii* appears to represent a large advance in the formulation of culture medium. Failed attempts to culture species of this abundant tropical marine phytoplankter have spanned more than two decades. Growth occurred also in diluted F/2 medium.[30] The important considerations appear to be the use of ultrapure reagents (where heavy metal contamination is reduced) and a high calcium content (>7.5 mM). In addition, it was shown that *Trichodesmium* was auxotrophic for cyanocobalamin (vitamin B_{12}) or for hydroxocobalamin. The previously assumed toxicity of high phosphate was at least partly due to heavy metal contamination of the phosphate reagent. The treatment of the reagent with a Chelex 100 column allowed the use of phosphate at concentrations as high as 13 μM rather than a mere 3.2 μM. Even after treatment, however, 32 μM phosphate inhibited growth. In all cases, highest grade reagents were used—for phosphate, analytical grade super pure (Merck, Darmstadt, FRG).

Aquil medium,[45] on which the slight modification of Ohki *et al.*[30] is based, requires the passing of macronutrients and of some micronutrients through a Chelex 100 column (Bio-Rad Laboratories, Richmond, CA). This includes the basic artificial salts mixture (Standard Ocean Water) which is made up as a stock solution (Table II). The sodium phosphate and sodium nitrate nutrient solutions are made up separately and also passed through the Chelex resin.[45] Sodium silicate is excluded when the medium is used for cyanobacteria, and the phosphate concentration should probably be reduced from 10 to 3.2 μM (0.5 mg liter^{-1}). The procedure for preparing the ultrapure Aquil medium is complex, and the article by Morel *et al.*[45] should be consulted.

Many modifications of the ASP marine media first developed by Provasoli *et al.* have been used.[42] All use mixtures of salts to simulate seawater of various salinities. ASP-M[42] has been included in Table II.

Erdschreiber's medium (Schreiber, 1927)[42] is a complex medium using natural seawater and soil extract with several specific nutrient additions (Table II). Soil extract has long been used as a source of chelator, iron, micronutrients, vitamins, and unknowns in the culture of algae. It is prepared in many ways. The following procedure is quite acceptable: Add

[45] F. M. M. Morel, J. G. Rueter, D. M. Anderson, and R. R. L. Guillard, *J. Phycol.* **15,** 135 (1979).

400 g of dried "garden loam" to 1 liter of water, autoclave for 30 min, and swirl while hot. Filter while hot through three layers of Whatman No. 1 filter paper (or equivalent). The liquid should be a pale or medium amber color and clear. The amount of soil used will vary with source. After addition of the reagents (dissolved in the soil extract) to the seawater, autoclaving will produce a precipitate. There are other "pasteurization" methods where this can be avoided,[42] but I have not found the precipitate to be detrimental. Many marine cyanobacteria grow well in Erdschreiber's medium, but many oligotrophic types die rapidly in this medium no matter what salinity is used.[7]

As mentioned earlier, many cyanobacteria grow well in basic freshwater media in which NaCl and, sometimes, $MgCl_2 \cdot 6H_2O$ have been added to achieve appropriate salinites. Additional amounts of these salts may be used for hypersaline cyanobacteria (Table II),[31] such as *Aphanothece* (*Synechococcus*) *halophytica*. Finally, in most cases, the pH of marine media requires little adjustment, but, in order to achieve values comparable to natural environments, the pH value should fall between 8.0 and 8.5.

Conditions for Growth and Maintenance of Cultures

The maintenance of stock cultures does not require sparging with air or other gas mixtures. Although CO_2 diffusion rates from the atmosphere may limit growth rate, particularly at high temperatures, the intention of keeping stock cultures is not to encourage high growth rates. If light intensity is also kept at a low level, a balance between this factor and CO_2 can be achieved where either is close to being the growth-limiting factor. However, for enhancement of growth rates for the purpose of experimentation or for high yields of cells, the introduction of air or N_2 with 1–5% CO_2 is generally used, together with higher but noninhibiting light intensities. If the culture is axenic, the use of an organic buffering agent may be used to prevent lowering of the pH. In other media various concentrations of $NaHCO_3$ or Na_2CO_3 may be used. Although, for most cyanobacteria, air sparging without CO_2 additions may be quite sufficient (and require less buffering), sparging with N_2 and low CO_2 encourages better growth in some cases.[7]

Nitrogen-fixing cultures of cyanobacteria (usually heterocystous forms) should always be maintained on media free of, or deficient in, combined nitrogen. Otherwise, strains have been known to lose functional heterocysts or heterocysts altogether.[13]

Light is usually provided in the laboratory by a bank of fluorescent lamps (cool white, daylight, or warm white). Since the spectral output is mainly in the visible range (400–700 nm) these afford the most efficient use of power. Naturally, normal incandescent lamps and high-intensity halogen vapor lamps also provide usable photons for photosynthesis, but a greater proportion of the output is in the far red and near-infrared region which is not available to oxygenic phototrophs. The intensity for stock culture maintenance can vary enormously, but for some an intensity lower than 500 lux is appropriate. In some cases, high-light-sensitive, phycoerythrin-containing cyanobacteria can be gradully adapted to a higher light intensity.[7]

The type of vessels used for stock or experimental cultures can vary greatly. Erlenmeyer flasks with cotton plugs have seemed the most satisfactory for stock cultures in liquid medium. Petri plates (dried at room temperature for 24–48 hr after pouring) and used in the inverted position, together with agar slants, appear to be the best for maintaining axenic cultures of most cyanobacteria. Usually 1–1.5% (w/v) agar is used. Unfortunately, there are a number of cyanobacteria (usually of Section III, Oscillatoriaceae[3,5]) that have been successfully grown only in liquid medium. It is, of course, far more difficult to maintain axenicity over many transfers in liquid medium. The same strains seem also to be those least able to be conserved in a nongrowing state (e.g., liquid N_2, lyophilization).[7]

Various types of tubes or vessels can be used for experimental purposes. Long cylinders (e.g., 5–6 cm in diameter and 30–35 cm long) attenuating at the bottom into a glass tube which is bent back up to the top of the cylinder (fused to the cylinder near the top) are easily made from tempered glass tubing by professional glassblowers.[9] Otherwise, the sparging tube can be introduced from the top with the plug or plastic cover.[13] Obviously, there is no unique demand from cyanobacteria that is much different from the demands of other microorganisms which require sparging. Various types of fermenter systems work well with many cyanobacteria as long as sufficient light entry is possible. The use of other continuous culture systems for cyanobacteria is also appropriate. There are a number of unusable strains, however, that clump tightly, form dense gel balls, or simply adhere to glass walls.

The temperature used for cyanobacterial cultures will depend on the strain and the habitat from which it was isolated. Although many cyanobacteria are dominant in very cold waters, a general rule is that the optimum temperature for growth is much higher (e.g., at 15–20° or higher). Nevertheless, there are many strains that cannot tolerate the much higher enrichment or culture temperatures (30°) used by M. M. Allen.[20] Many more species of cyanobacteria than species of eukaryotic microalgae will thrive at temperatures of 30° and above. Only at 45° and above are the thermophiles from hot springs and other warm habitats the appropriate organisms.[8] Furthermore, the growth rates of cyanobacteria in culture vary enormously, from doubling times of a few hours in the case of some unicellular types under optimal ("maximal") conditions to several days or longer for some strains from oligotrophic waters.

It is very easy to acquire one or more species of heterotrophic bacteria in axenic cultures of cyanobacteria over the course of several transfers, particularly in liquid cultures where airborne contaminants are not confined as on an agar plate. In many cases, however, the contamination is not visually obvious. Cyanobacteria, if grown photoautotrophically on essentially inorganic medium, are seldom swamped by heterotrophic bacteria even in an aged culture. Cultures should be checked periodically under phase-contrast optics. Since cell debris in older axenic clutures can often mask contaminants, the enrichment of heterotrophs with organic medium in agar is necessary. A fairly dilute addition to the "mineral" medium, such as (per liter), 0.5 g tryptone, 0.25 g yeast extract, and 0.1 g glucose, generally works well, unless unusual bacteria are involved.

Conservation of Cultures without Growth

Cultures of cyanobacteria should never be stored in a refrigerator, even for short periods, unless 4–5° represents the normal temperature of growth. However, storage at 12–18° in darkness or very dim illumination seems to retain cultures in a viable state for several weeks (with some major exceptions).[7]

For much longer periods storage in ampules submerged in liquid N_2 (−196°) is the simplest and most dependable method for most strains. Again, there are exceptions.[9] The use of dense, postexponential cultures frozen with their own medium works well, although some workers recommend 5% (v/v) dimethyl sulfoxide in the medium. Amino acid mixtures (1%), gum arabic (2–10%), glycerol (10%), and gelatin (2–10%) have also been used as cryoprotective agents.[46] Plastic ampules with screw caps are far superior to the glass type in which a slightly flawed seal in the neck may cause it to explode on removal from the liquid N_2. Direct freezing by plunging the ampules in the liquid N_2 works well. Gas-vacuolate strains must first be pressurized to collapse gas vesicles before freezing. This can be done easily in a sterile syringe by pressing the plunger tightly while the nose is held against a sterile rubber stopper. For recovery of frozen samples, a rapid thaw at room temperature to 37°, followed by rapid transfer to growth conditions, is recommended.[13] Although used for a limited

[46] M. Takano, J. I. Sude, O. Tabahira, and T. Gyozo, *Cryobiology* **10,** 440 (1973).

number of cultures, the freezing of a dense sludge of the cyanobacterium in 15% (v/v) glycerol at −70 to −80° has worked well.[5]

Lyophilization has also been successful for some cyanobacteria,[47,48] but it has not been used extensively. Lamb serum was a good suspending agent. Finally, a number of terrestrial or semiterrestrial cyanobacteria and a few others remain viable as air-dried samples when kept in darkness at room temperature, sometimes for years. It is a technique worth trying, at least for the more robust species.

[47] L. L. Corbett and D. L. Parker, *Appl. Environ. Microbiol.* **32,** 777 (1976).

[48] E. A. Leeson, J. P. Cann, and G. J. Morris, *in* "Maintenance of Microorganisms" (B. E. Kirsop and J. J. S. Snell, eds.), p. 131. Academic Press, London, 1984.

[4] Purification and Storage of Nitrogen-Fixing Filamentous Cyanobacteria

By C. PETER WOLK

Despite a history of success in the isolation of pure cultures of cyanobacteria,[1,2] there has long been a need for a simple, generally applicable, rapid technique for isolation of strains from nature and for repurification of laboratory strains that have become secondarily contaminated. There is also a need for such a technique for storing strains so that their traits will not change, as they might over the course of repetitive subculturing. This chapter addresses these two needs.

Purification

The following protocol, reminiscent of procedures described by Rippka *et al.*,[3] has proved successful for purifying a great majority of strains of N_2-fixing cyanobacteria from sewage settling ponds, from other natural waters, and from contaminated laboratory cultures. The starting point can be a single colony, a portion of a mat from an agar surface, or a culture in liquid. We normally use the medium of Allen and Arnon[4] (di-

[1] R. Rippka, J. Deruelles, J. B. Waterbury, M. Herdman, and R. Y. Stanier, *J. Gen. Microbiol.* **111,** 1 (1979).

[2] C. P. Wolk, *Bacteriol. Rev.* **37,** 32 (1973).

[3] R. Rippka, J. B. Waterbury, and R. Y. Stanier, *in* "The Prokaryotes" (M. Starr, H. Stolp, H. G. Trüper, A. Balows, and H. G. Schlegel, eds.), p. 218. Springer-Verlag, Berlin, 1981.

[4] M. B. Allen and D. I. Arnon, *Plant. Physiol.* **30,** 366 (1955).

luted 8-fold for liquid media), but have also used BG-11 and BG-11_0,[1] and normally maintain cultures at 30° and an intensity of 3500 erg cm^{-2} sec^{-1} from cool-white fluorescent lamps.

Because any adherent bacterium or fungal cell can contaminate an entire filament and all of its progeny, it is desirable both to dislodge adhering heterotrophic microorganisms and to fragment filaments so as to increase the probability of obtaining axenic clones. To this end, a few milliliters of suspension in a culture tube or small (25-ml) flash is subjected to cavitation in a sonic cleaning bath. The extent of fragmentation of filaments is monitored via a compound microscope. (Use of the sonic bath is somewhat of an art, requiring proper resonance of the system. One looks for a pattern of standing waves on the surface of the immersed suspension, for expulsion of microdroplets of liquid from that surface, or for accelerated movements of suspended particles. Cavitation with a probe is much less satisfactory, presumably because the higher energy densities involved are more deleterious to the cyanobacteria.) How short fragments can be and still remain viable varies from organism to organism, but in general the goal is to obtain many fragments with few (1–5) cells. With a Model 8845-30 ultrasonic cleaner (Cole-Parmer Inst. Co., Chicago, IL), the requisite duration of resonant cavitation varies from 0.5 to 15 min, depending on the cyanobacterium used. It is best to macerate a strongly cohering colony or mat, for example, with the tip of a pipet, before cavitating. After cavitation, it is sometimes helpful to wash the resulting fragments by centrifugation (1000 *g*, 5 min). Next, one makes a series of 10-fold dilutions of the resuspended cyanobacteria. One-tenth milliliter of each of the dilutions is spread on the surface of a Petri dish of agar-solidified medium (we normally purify the agar,[5] although this is probably often unnecessary), and the plates are set in the light.

After 3 days or, for more slowly growing organisms, as much as a week, blue–green microcolonies are observed under a dissecting microscope. With careful microscopy, one can often distinguish contaminated from presumptively axenic colonies. The best chance of obtaining axenic colonies is from the Petri dishes with the most dilute inocula that yield foci of growth, because the microcolonies on such dishes are most apt to be spatially well separated from contaminants. Because bacterial and fungal growth can spread, it is beneficial to transfer microcolonies promptly rather than to await the appearance of macroscopically visible growth of cyanobacteria.

Using an autoclaved, rounded toothpick, and while observing through the dissecting microscope, encircle the microcolony of choice with an

[5] A. C. Braun and H. N. Wood, *Proc. Natl. Acad. Sci. U.S.A.* **48,** 1776 (1962).

incision through the agar. Then transfer the tiny piece of agar bearing the microcolony to a flask or test tube of fresh liquid medium or to a fresh Petri dish. Retention of a bit of agar prevents desiccation of a filament during transfer, and also permits easy visualization that transfer has been accomplished. Very often, such microcolonies prove to be axenic. Even if not, one or two repetitions of the cycle of growth, cavitation, and plating usually result in success. Only if a residual contaminant is a bacterium that is highly motile on the agar surface has it proved prudent to return to an earlier stage in the purification, or not to persist.

Storage

There is a great range in the duration of viability of cyanobacterial cultures on agar slants or in liquid cultures. To provide reliable long-range storage, we make use of a freezer at −75°. Liquid cultures (50–200 ml) are concentrated by centrifugation at 1500 *g*, the final sedimentation being for 5 min in conical tubes (Falcon No. 2095). The supernatant fluid is removed, and dimethyl sulfoxide[6] (Pierce Chemical Co., Rockford, IL) or an autoclaved solution of 80% (v/v) glycerol[6] in water is added to the concentrated cyanobacterial sludge to a final concentration of 15% dimethyl sulfoxide or 15% glycerol (v/v; or slightly higher), e.g., 0.19 ml 80% glycerol : 0.81 ml cell pellet. The cells and the cryoprotectant are mixed thoroughly and transferred to sterile tubes (we use 2-ml Nunc cryotubes from Arthur H. Thomas, Swedesboro, NJ). The Nunc tubes are set at −75°. If the cyanobacteria settle in 80% glycerol, the tubes are left inverted at −75° until their contents have frozen.

To recover the cells, a small but visible portion is scraped from the frozen material with a sterile toothpick and either spread on the surface of agar-solidified medium or inoculated directly into liquid medium. With glycerol, recovery varies greatly from strain to strain, but at least some recovery has been obtained with 167 of 168 strains tried, including unicellular strains *Anacystis* (*Synechococcus*) R2 and PCC 6301 and *Synechocystis* PCC 6803. In preliminary experiments with a few strains, dimethyl sulfoxide has given substantially greater recoveries than were obtained by use of glycerol, and so this alternative should certainly be tried.

Acknowledgments

This work was supported by the U.S. Dept. of Energy under Contract DE-AC02-76ERO-1338.

[6] P. H. Calcott, "Freezing and Thawing Microbes." Meadowfield Press, Shildon, UK, 1978.

[5] Thermophilic Cyanobacteria: Special Problems

By RICHARD W. CASTENHOLZ

Introduction

Thermophilic cyanobacteria have been used in a number of biochemical studies.[1-3] Preparations of highly active O_2-evolving photosystem II particles from *Phormidium laminosum*[4,5] and *Synechococcus* sp.[6] have been especially successful.

Thermophilic cyanobacteria, arbitrarily defined, are those that grow well or best above 45°.[7,8] The upper temperature limit worldwide appears to be about 73–74°. In going from 45 to 74° there are progressively fewer species or strains capable of growth. Strains that will grow within this range occur usually or strictly in geothermal springs with pH levels above about 4.5. Few strains occur below pH 6.0, and culturing these has been difficult at this low a pH. Thermophilic cyanobacteria may also be present in nongeothermal environments such as desert soils and rocks, intertidal flats and sabkhas, tropical ponds and pools, cliff faces, and other habitats where temperatures often reach the 40–50° range. Many thermophiles capable of growth above 45°, nevertheless, will maintain populations in some habitats that normally are below 40°.[9] A few species produce resting spores [e.g., *Fischerella* (*Mastigocladus*) *laminosus*], but even these stages are rare in nonthermal environments. Nevertheless, it is this cyanobacterium that generally colonizes new thermal habitats first (e.g., nuclear power plant cooling ponds, Aiken, SC; geothermal streams, Mt. St. Helens, WA).

[1] D. C. Fork and K. Satoh, *Photochem. Photobiol.* **37,** 421 (1983).
[2] D. C. Fork, N. Murata, and N. Satoh, *Plant Physiol.* **63,** 524 (1979).
[3] H. Koike, K. Satoh, and S. Katoh, *Plant Cell Physiol.* **23,** 293 (1982).
[4] A. C. Stewart and D. S. Bendall, *Biochem. J.* **194,** 877 (1981).
[5] J. M. Bowes and D. S. Bendall, *in* "Photosynthetic Prokaryotes: Cell Differentiation and Function" (G. C. Papageorgiou and L. Packer, eds.), p. 163. Elsevier, Amsterdam, 1983.
[6] H. Koike and Y. Inoue, *in* "The Oxygen Evolving System of Photosynthesis" (Y. Inoue, A. R. Crofts, Govindjee, N. Murata, G. Renger, and K. Satoh, eds.), p. 257. Academic Press, London, 1983.
[7] R. W. Castenholz, *Mitt. Int. Ver. Limnol.* **21,** 296 (1978).
[8] R. W. Castenholz, *in* "The Prokaryotes" (M. P. Starr, H. Stolp, H. G. Trüper, A. Balows, and H. G. Schlegel, eds.), Vol. 1, p. 236. Springer-Verlag, Berlin, 1981.
[9] J. E. Jackson, Jr., and R. W. Castenholz, *Limnol. Oceanogr.* **20,** 305 (1975).

Collections, Culture, and Storage

The most striking feature of collected specimens of thermophilic cyanobacteria is their refractory nature with regard to decomposition. Both feral material from hot spring mats and axenic or nonaxenic cells from cultures do not deteriorate or die over many days or weeks if kept in darkness at room temperature or as low as 12–14°, at least when uncrowded in spring water or culture medium. The storage of thermophilic strains or collections in refrigerators (3–5°) is detrimental to most types. However, even crowded collections or dense cultures will retain a high percentage of viability if stored in darkness and allowed to become anaerobic. Anaerobic maintenance of axenic cultures also promotes long-term viability of some thermophilic strains[10] and facilitates survival of air shipments of cultures or collections.

The conservation of most thermophilic cyanobacteria in ampules immersed in liquid nitrogen (−196°) is a satisfactory method of conservation for periods of several years,[8] as it is for nonthermophilic forms. For thermophiles, using dense cultures (past exponential growth phase) in their own spent medium appears to work best. Several thermophilic strains, mainly Oscillatorian, rapid-gliding types, have not been successfully conserved by this or any other means.[8] Other methods (e.g., lyophilization, −80° storage) have not been tested extensively with thermophiles.

Isolation

Isolation procedures for thermophilic cyanobacteria differ little from those used for nonthermophiles. Difficulties arise, however, with manual or self-isolation methods that involve the use of agar plates. Some strains require high temperature for growth (55–72°), and since agar tends to dry out much more rapidly at 45–50° than at lower temperatures, deeper plates (20 mm deep) need to be used. However, at temperatures above 55° heavy syneresis of the agar occurs, and the medium becomes semiliquefied. Gelrite (Kelco Div., Merck & Co., San Diego, CA) is a gelling agent (produced by a species of *Pseudomonas*) that has good thermal stability and clarity, and it allows plating of thermophilic organisms at temperatures of 70° and above.[11] An overlay of agar- or Gelrite-solidified medium with cells or trichomes dispersed may be used to advantage for higher temperature. Also, shake or stab culture isolations (in tubes) are of greater use for thermophiles than for other types.

[10] L. L. Richardson and R. W. Castenholz, *Appl. Environ. Microbiol.* **53,** 2151 (1987).

[11] C. C. Lin and L. E. Casida, Jr., *Appl. Environ. Microbiol.* **47,** 427 (1984).

The cloning of high-temperature forms of unicellular cyanobacteria has been accomplished in one instance by the following procedure[12]: (1) enrichment at, e.g., 60–70°; (2) serial dilutions of resultant cultures (to <10 cells ml^{-1}); (3) gentle filtration through a sterile glass fiber filter (e.g., Whatman GF/F); (4) washing with large volume of sterile medium (e.g., 50 ml); (5) placing the *moist* filter in bottom of a culture flask containing liquid medium (e.g., D medium[13]); (6) incubation with appropriate light at 60–70° without any swirling or disturbance of the medium for 1 to several weeks, allowing putative clonal colonies to arise on the filter [sterile water may have to be added (gently) to the flasks during long-term incubations]; (7) decanting off of medium and removal of colonies from the soft, moist filter with forceps (accompanied by a small tuft of glass fibers) and inoculation into new medium, repeating after culture density has increased.

Enrichments (elective cultures) are, obviously, of greater use than for most other cyanobacteria during the isolation process. The raising of the incubation temperature to 45° or higher will eliminate nonthermophilic cyanobacteria and eukaryotes from mixed cultures or from crude inoculations. Extreme temperatures (>60°), over several generations, will often elect a single high-temperature strain, when it is present in the mixture.[7] Enrichments for other characteristic species or strains of cyanobacteria are also possible in some cases.[7]

Hot spring enrichments may be maintained for many years as modified Winogradsky columns.[14] Instead of the usual low-intensity incandescent lamp, a 75- to 150-W floodlight provides light and heat on one side of the columns, which are packed with sediment, mud, and mat material from a hot spring. The center temperature is easily raised to boiling, so the lamp distance needs to be lengthened until the center spot temperatures do not exceed 75–80°. A miniature temperature gradient will extend from the center in all directions with a concomitant zonation of cyanobacteria and other hot spring organisms.[14]

Maintenance of Cultures

There are no basic differences between maintaining thermophilic and nonthermophilic strains of cyanobacteria, merely some additional precautions to observe.[13] At temperatures of 50° or below, cotton- or latex-plugged flasks may be used as at lower temperatures, but at least 70 ml of medium is recommended in a 125-ml Erlenmeyer type. Folded caps of tissue paper (Kimwipes) or caps of other paper (Nipple caps, Asepto-thermo Indicator Co., N. Hollywood, CA) should be used over the cot-

[12] J. C. Meeks and R. W. Castenholz, *Arch. Mikrobiol.* **78,** 25 (1971).

[13] R. W. Castenholz, this volume [3].

[14] R. W. Castenholz, *Limnol. Oceanogr.* **17,** 767 (1972).

ton. Although these are porous, they reduce the evaporative loss somewhat and keep the plug free of dust. At temperatures above 50°, however, Delong flasks with Morton stainless steel closures (Bellco Glass, Inc., Vineland, NJ) are preferable since much less water is lost by evaporation. However, sterility is very difficult to maintain because of water condensing under the rim of the closure. Therefore, screw-cap flasks or tubes may be used if inorganic carbon is added as $NaHCO_3$ (usually added as a stock neutralized by sparging with CO_2) or as Na_2CO_3. Growth is generally slow under these conditions, but ideal for stock cultures.

Plates with agar or Gelrite as the solidifying agent work poorly for maintaining cultures above 50° unless taped with Parafilm or placed in refrigerator-type vegetable crispers. Condensation becomes a great problem with higher temperatures, and the inversion of plates is quite essential.

If flasks or plates are used inside incubators, the chamber must be humidified; a dishpan of water at the base is generally sufficient. Water baths with some type of screen support and overhead lighting can also be used for thermophiles, and a number of these are easier to use than incubators if a variety of temperatures is required. Also water baths of many types can easily maintain temperatures of 50–70° or above, whereas few inexpensive incubators have this capacity. It is important, however, to use water baths with automatic drip and overflow features, since even at 45° evaporation rates are considerable.

For the production of large masses of cultured cells, various types of vessels with ports or airstones for sparging with air or special gas mixtures may be used. With temperatures required for thermophiles, water loss is great. Therefore, gas needs to be streamed through a similarly heated water tower to increase humidity after entering the sterile filter system. The additional passages increase the chances of external contamination, so additional care is required. In the higher temperature range (e.g., 60–70°) the exit ports of the culture vessel may have to lead into upright condensers in order to reflux the water lost.[12]

Culture Media

High-temperature growth does not pose significant problems with regard to medium. Probably most of the media described elsewhere in this volume would be suitable.[13] A few, however, tend to have precipitation problems and are not optically clear under all conditions. Medium D, with various modifications,[8,13] was developed especially for hot spring cyanobacteria which typically grow in waters with total salt concentrations between 1000 and 2000 mg liter^{-1}. It is optically clear under all conditions and contains similarly high salt concentrations, albeit much of it in the form of salts of nitrate, phosphate, and sulfate rather than of bicarbonate and chloride as in typical hot springs. D medium may be buffered by a number of organic buffers, such as Tricine [*N*-tris(hydroxymethyl)methylglycine, p*K′* 8.15], glycylglycine (p*K′* 8.25), EPPS [*N*-(2-hydroxyethyl)piperazine-*N′*-3-propanesulfonic acid, p*K′* 8.0], HEPES [*N*-(2-hydroxyethyl)piperazine-*N′*-2-ethanesulfonic acid, p*K′* 7.5], and PIPES [piperazine-*N*,*N′*-bis(2-ethanesulfonic acid, p*K′* 6.8], and still be used at higher temperatures.

Since many hot spring waters are reducing in nature, particularly near the source, some of the strains isolated from them require reduced forms of nitrogen,[8] and in some cases the cyanobacteria are able to utilize or benefit from reduced sulfur compounds such as sulfide, sulfite, or thiosulfate. Thus, some modifications of D medium include about 3.7 m*M* NH_4Cl, 0.8 m*M* $Na_2S_2O_3 \cdot 5H_2O$, and 1.6 m*M* Na_2SO_3. The specific benefits of the sulfur compounds are unknown, but the stimulating effect of sulfite has been noted by others as well.[15]

[15] D. L. Parker, *J. Phycol.* **18,** 471 (1982).

[6] Isolation and Growth of Marine Planktonic Cyanobacteria

By JOHN B. WATERBURY and JOANNE M. WILLEY

Introduction

Cyanobacteria are widespread in marine habitats, especially in the warmer temperate and tropical regions, where they are important both as primary producers and nitrogen fixers. The overall taxonomic diversity of marine forms is comparable to that found in freshwater and terrestrial habitats. However, particular marine habitats (e.g., the intertidal and subtidal zones, coral reefs, salt marshes, and the open ocean) often contain a characteristic and restricted diversity of forms which may differ markedly from season to season and from one geographical location to another.[1,2]

[1] B. A. Whitton and M. Potts, *in* "The Biology of Cyanobacteria" (N. G. Carr and B. A. Whitton, eds.), p. 515. Univ. of California Press, Berkeley, 1982.

[2] G. E. Fogg, *in* "The Biology of Cyanobacteria" (N. G. Carr and B. A. Whitton, eds), p. 491. Univ. of California Press, Berkeley, 1982.

This is particularly evident in the open ocean where only a few genera and species have been shown to occur abundantly and to be important components of the phytoplankton community. Principal among these are marine representatives of the genera *Synechococcus, Synechocystis, Trichodesmium,* and *Richellia.* By contrast, the diversity of freshwater planktonic forms is much more extensive, encompassing over 100 species, 20 of which are capable of forming extensive water blooms.[3] This is probably due in part to the fact that freshwater planktonic habitats are diverse, ranging from oligotrophic to highly eutrophic environments. The open oceans are relatively oligotrophic, and the cyanobacteria that occur there reflect this in their growth requirements and sensitivities.

Marine representatives of the genus *Synechococcus* (sensu Rippka *et al.* 1979)[4] are small unicellular forms (0.6×1.4 μm) that are abundant within the euphotic zone of the world's temperate and tropical oceans.[5,6] Marine representatives of the genus *Synechocystis* (sensu Rippka *et al.* 1979)[4] are novel unicellular forms capable of aerobic nitrogen fixation which have been isolated from the tropical Atlantic.[7] Members of the genera *Trichodesmium* and *Richellia* are filamentous forms capable of nitrogen fixation[2] that are found in the tropical oceans. *Trichodesmium* species are free-living nonheterocystous forms, whereas *Richellia* species occur principally as intracellular symbionts in several species of diatoms.[2] The general principles of isolation and growth for cyanobacteria that are discussed elsewhere in this volume ([1], [3]) and in *The Prokaryotes*[8] are also applicable to marine planktonic forms.

Media

A variety of media has been used for the culture of marine cyanobacteria. Two in particular, one having a natural seawater base (medium MN[9]) and the other having an artificial seawater base (medium ASN-III[9]), have

[3] C. E. Gibson and R. V. Smith, *in* "The Biology of Cyanobacteria" (N. G. Carr and B. A. Whitton, eds.), p. 463. Univ. of California Press, Berkeley, 1982.

[4] R. Rippka, J. Deruelles, J. B. Waterbury, M. Herdman, and R. Y. Stanier, *J. Gen. Microbiol.* **111,** 1 (1979).

[5] H. E. Glover, *Adv. Aquat. Microbiol.* **3,** 49 (1985).

[6] J. B. Waterbury, S. W. Watson, F. W. Valois, and D. G. Franks, *Bull. Can. J. Fish. Aquat. Sci.* **214** (1987).

[7] S. W. Watson, J. B. Waterbury, F. W. Valois, and D. G. Franks, unpublished results.

[8] R. Rippka, J. B. Waterbury, and R. Y. Stanier, *in* "The Prokaryotes" (M. P. Starr, H. Stolp, H. G. Trüper, A. Balows, and H. G. Schlegel, eds.), p. 212. Springer-Verlag, Berlin, 1981.

[9] J. B. Waterbury and R. Y. Stanier, *in* "The Prokaryotes" (M. P. Starr, H. Stolp, H. G. Trüper, A. Balows, and H. G. Schlegel, eds.), p. 221. Springer-Verlag, Berlin, 1981.

been used successfully for the isolation and maintenance of pure cultures of a variety of marine cyanobacteria, mostly of coastal origin (e.g., intertidal and salt marsh isolates). Neither of these media has been successful for the isolation and maintenance of the open ocean planktonic forms, necessitating the development of two new media, designated SN[6] for the natural seawater medium and AN for the artificial seawater medium. Their composition is shown in Table I. Medium SN is used for isolation and maintenance, and medium AN is used experimentally when a defined medium is advantageous.

Preparation of Agar Plates

Solid media are prepared using Difco Bacto-agar that has been further purified using the following protocol[6]: 100 g of agar is washed by stirring with 3 liters of double-distilled water in a 4-liter beaker. After 30 min of stirring, the agar is allowed to settle, the wash water is siphoned off, and the agar is filtered onto Whatman F4 filter paper in a Büchner funnel. This procedure is repeated once more or until the filtrate is clear. The agar is then washed with 3 liters of 95% ethanol followed by a final 3-liter wash with analytical grade acetone. The agar is then dried at 50° in glass baking dishes for 2–3 days and stored in a tightly covered container. Solid media prepared with the purified agar at a final concentration of 0.6% are sufficiently stable for streaking.

To prepare 40 agar plates from 1 liter of medium, the three following solutions are prepared and autoclaved separately: (1) 750 ml of filtered seawater in a Teflon bottle, (2) 6.0 g super-clean agar in 200 ml of double-distilled water in a 2-liter glass flask, and (3) the mineral salts for 1 liter of medium (Table I) in 50 ml of double-distilled water in a 125-ml glass flask. After autoclaving, the seawater and minerals are added to the flask of agar. Vitamins[10] (Table I) and sterile sodium sulfite (2 m*M* final concentration) are added aseptically to the hot agar solution, which is then cooled to 50° before the plates are poured. It is critical that the surface of agar plates be dry prior to streaking. Following inoculation the plates are stored upside down in clear plastic vegetable crispers to minimize evaporation and contamination by fungi. Colonies appear in 2–4 weeks and are then removed from the agar surface with drawn Pasteur pipets, inoculated into liquid media, and allowed to grow between successive streakings.

Enrichment Cultures

Two media have been used for the enrichment of marine *Synechococcus*, medium SN in which the nutrients are diluted 10-fold and supple-

[10] H. C. Davis and R. R. L. Guillard, *U.S. Fish Wildl. Serv.* **58,** 293 (1958).

TABLE I
COMPOSITION OF NATURAL AND ARTIFICIAL SEAWATER MEDIA

Ingredient	Amount/liter: Medium SN, Liquid	Medium SN, Solid	Medium AN
Double-distilled water	250 ml	250 ml	1000 ml
Filtered seawater	750 ml	750 ml	—
NaCl	—	—	250 mM
$MgSO_4 \cdot 7H_2O$	—	—	30 mM
$MgCl_2 \cdot 6H_2O$	—	—	20 mM
$CaCl_2 \cdot 2H_2O$	—	—	10 mM
KCl	—	—	10 mM
EDTA (disodium salt)	15 μM	15 μM	15 μM
$NaNO_3$	9 mM	2.5 mM	9 μM
NH_4Cl	—	100 μM	—
K_2HPO_4	90 μM	22.5 μM	90 μM
$Na_2CO_3 \cdot H_2O$	100 μM	100 μM	200 μM
$NaHCO_3$	—	—	3 mM[a]
Cyano trace metals[b]	1.0 ml	0.25 ml	1.0 ml
$ZnSO_4 \cdot 7H_2O$ (0.222 g/liter)			
$MnCl_2 \cdot 4H_2O$ (1.4 g/liter)			
$CO(NO_3)_2 \cdot 6H_2O$ (0.025 g/liter)			
$Na_2MoO_4 \cdot 2H_2O$ (0.390 g/liter)			
Citric acid hydrate (6.250 g/liter)			
Ferric ammonium citrate (6.0 g/liter)			
Va vitamin mix[b,c]	0.5 ml	0.5 ml	0.5 ml
Thiamin–HCl (0.2 g/liter)			
Biotin (0.001 g/liter)			
Vitamin B_{12} (0.001 g/liter)			
Folic acid (0.002 g/liter)			
PABA (0.01 g/liter)			
Nicotinic acid (0.1 g/liter)			
Inositol (1.0 g/liter)			
Calcium pantothenate (0.2 g/liter)			
Pyridoxine–HCl (0.1 g/liter)			
Sodium sulfite	—	2 mM	—
Washed agar	—	6.0 g	—
Final pH	8.0	8.0	8.3

[a] Added after autoclaving.

[b] To prepare stock, dissolve each compound separately, then add together and bring to 1 liter.

[c] Va vitamin mix is added during enrichment and purification; in most instances purified stock cultures are maintained with only vitamin B_{12} (1 μg/liter).

TABLE II
CULTURE CONDITIONS

Cyanobacterium	Temperature (°)	Light regime	Light intensity[a] (μE m^{-2} sec^{-1})
Synechococcus	20–25	Constant or LD[b] (14/10)	10–30
Synechocystis	26–30	LD (14/10)	10–40
Trichodesmium	22–27	LD (14/10)	10–20
Richellia	20–25	Constant or LD (14/10)	10–30

[a] Light is supplied with Vitalux fluorescent lamps (Luxor Light Products, Lindhurst, NJ).

[b] LD, Light–dark cycle in hours.

mented with 100 μM ammonium chloride[6] and medium F/40 supplemented with 10 μM ammonium chloride.[11] The sole strain of *Richellia* was enriched in the diluted SN medium used for *Synechococcus*. Enrichments for *Synechocystis* have been made in medium SN diluted 10-fold from which the combined nitrogen was omitted. The sole isolate of *Trichodesmium* was enriched in medium F/2.[11,12]

Purification of Isolates

Isolates of *Synechococcus, Synechocystis,* and *Richellia* have been purified by repeated streaking on solid medium (Table I). The isolate of *Trichodesmium* was purified by putting a clump of filaments in the center of an agar plate containing the solid medium in Table I, placing the plate in a light gradient to induce phototaxis, and then cutting out agar blocks containing single filaments that had glided away from contaminating bacteria.

Stock Cultures

Axenic stock cultures are maintained in 50 ml of liquid medium in 125-ml glass flasks, *Synechococcus* and *Richellia* strains in full-strength SN medium, *Synechocystis* in SN medium without combined nitrogen, and *Trichodesmium* in medium SN diluted 4-fold. The conditions of growth used for the isolation, purification and maintenance of these cyanobacteria are shown in Table II. Stock cultures are transferred at 1- to 3-week

[11] R. R. L. Guillard, *in* "Culture of Marine Invertebrate Animals" (W. L. Smith and M. H. Chanley, eds.), p. 29. Plenum, New York, 1975.

[12] M.-R. Li, B.-Z. Bian, R. A. Lewin, L. Cheng, and Z.-Q. Pan, *Bull. Shandong Coll. Oceanogr.* **14** (1984).

intervals using heavy inocula. Representative strains of these cyanobacteria are available from the culture collection of the Bigelow Laboratory for Marine Science (Boothbay Harbor, ME 04575). The complete collection is housed at the Woods Hole Oceanographic Institution (Woods Hole, MA 02543).

Acknowledgments

The authors' work is supported by National Science Foundation Grants BSR-8607386 and DCB-8608698. Woods Hole Oceanographic Institution Contribution No. 6438.

[7] Isolation and Culture of Marine Nitrogen-Fixing Unicellular Cyanobacteria *Synechococcus*

By AKIRA MITSUI and SHULI CAO

Introduction

Nitrogen fixation by cyanobacteria contributed significantly to the nitrogen budget in the marine environment. *Trichodesmium* species (*Oscillatoria* spp. or LPP group) have been well known as marine N_2-fixing cyanobacteria.[1] More recently, however, unicellular aerobic N_2-fixing cyanobacteria of the genus *Synechococcus* have been found and isolated from the marine environment.[2-4] Since these strains can carry out photosynthetic O_2 evolution and oxygen-labile N_2 fixation in the same cell, they are good material for both ecological and biochemical studies. This chapter describes methods of sample collection, enrichment, and isolation. Culture methods for one of the fast-growing aerobic N_2-fixing marine *Synechococcus* species are also described.

[1] G. E. Fogg, *in* "Environmental Role of Nitrogen-Fixing Blue–Green Algae and Asymbiotic Bacteria" (U. Gramhall, ed.), p. 11. Swedish National Research Council, Stockholm, 1978.

[2] A. Mitsui, *Proc. Int. Ocean Dev. Conf., 5th* **1,** 29 (1978).

[3] A. Mitsui, E. J. Phlips, S. Kumazawa, K. J. Reddy, S. Ramchandran, T. Matsunaga, L. Haynes, and H. Ikemoto, *Ann. N.Y. Acad. Sci.* **413,** 514 (1983).

[4] A. Mitsui, S. Kumazawa, E. J. Phlips, K. J. Reddy, T. Matsunaga, K. Gill, B. R. Renuka, T. Kusumi, G. Reyes-Vasquez, K. Miyazawa, L. Haynes, E. Duerr, C. B. León, D. Rosner, H. Ikemoto, R. Sesco, and E. Moffat, *in* "Biotechnology and Bioprocess Engineering" (T. K. Ghose, ed.), p. 119. United India Press, New Delhi, 1985.

Sample Collection

Nitrogen-fixing cyanobacteria can be encountered in a variety of marine environments. Samples should be placed into sterilized plastic bags. Samples taken from the water column can be obtained with a discrete depth sampler, a number of which are available.[5] Sediment samples can be obtained using a grab sampler or similar apparatus. Finally, since some cyanobacteria are known to grow epiphytically, samples of macroalgae and seagrasses can be taken along with scrapings from rocks and other stationary objects.

Enrichment

In the laboratory or aboard the sampling vessel each sample is subdivided into eight subsamples, and each subsample is inoculated into sterilized test tubes or 50-ml flasks containing different enriched, combined-nitrogen-free artificial seawater media. The eight different nitrogen-free media represent a range of pH and salinity regimes (e.g., pH values of 6.5, 7.5, 8.5, and 9.5 at 30‰ salinity, and salinities of 9, 18, 27, and 36‰ at an initial pH of 8.0). The basic enriched culture medium is composed of the following in 1 liter of distilled water: $MgSO_4 \cdot 7H_2O$, 5 g; KCl, 600 mg; $CaCl_2 \cdot 2H_2O$, 370 mg; tris(hydroxymethyl)aminomethane, 50 mg; KH_2PO_4, 50 mg; disodium EDTA, 30 mg; $FeSO_4 \cdot 7H_2O$, 3.89 mg; vitamin B_{12}, 10 μg; and micronutrient solution, 10 ml. Micronutrient solution is composed of the following in 1 liter of distilled water: H_3BO_3, 3.426 g; $MnCl_2 \cdot 4H_2O$, 432 mg; $Na_2MoO_3 \cdot 2H_2O$, 130 mg; $ZnCl_2$, 32 mg; $CoCl_2 \cdot 4H_2O$, 1.2 mg; and $CuSO_4$, 0.3 mg. NaCl is added in varying amounts to obtain the desired salinities. These subsamples are placed on illuminated shelves at 100 $\mu E/m^2/sec$ at approximately 25–30° for 2–6 weeks. When growth is visually evident the subsamples are transferred to fresh, sterile, nitrogen-free media of the same pH and salinity. This process is repeated for at least 3 cycles before initiating the isolation procedures so that culture parameters become stabilized and any nitrogen-containing contaminants from the samples are minimized in the culture, thus enriching the nitrogen-fixing cyanobacteria in the subsample.

Isolation

Subsequent to the enrichment process the subsamples are examined microscopically, and those subsamples containing unicellular forms are

[5] O. T. Lind, "Handbook of Common Methods in Limnology," 2nd Ed. Kendal/Hunt, Dubuque, Iowa, 1985.

singled out for further isolation procedures. This is initiated with a serial dilution process using the same enriched medium as above. The serial dilution is followed by plating on sterile agar media (enrichment media above in 2% agar). Streaking is also employed to help isolate monoclonal colonies. After a period of growth on illuminated shelves, apparently pure colonies are picked from the surface of the agar with sterile implements and reinoculated into sterile, combined-nitrogen-free liquid medium as described above. This process is repeated until observation using phase-contrast light microscopy confirms a pure culture, which is then maintained in transfers of fresh media.

Axenic Culture and Maintenance

Scanning electron microscopic observation serves not only to confirm that a single strain has been isolated but also to detect the presence or absence of associated bacteria or fungi in the culture. However, microscopic observation is often insufficient for determining with absolute certainty whether the culture is axenic or not. Purity can be confirmed by using TYG (tryptone–yeast extract–glucose) agar and broth medium,[6] Burk's *Azotobacter* medium,[7] 2% sucrose-enriched cyanobacterial medium, and nutrient broth at varying pH values. If a contaminant is found, antibacterial or antifungal agents are utilized as part of the isolation procedure prior to physiological or biochemical studies. Cycloserine in concentrations of 1, 2, 3, 4, and 5 mg/ml and/or cycloheximide at 10, 15, 20, 25, 30, and 35 mg/liter are added to the culture media in this purification process. Classification and taxonomy of unicellular cyanobacteria has been described by Rippka *et al.*[8] and elsewhere in this volume.[9]

Growth

Batch Culture

Batch cultures of *Synechococcus* are grown in water-jacketed glass cylinders (8.5 cm in diameter and 75 cm in height). The top part of the culture cylinder is fitted with a ground glass cap. Media and inoculant are added from the top by disconnecting the ground glass cap under sterile conditions. Aeration and sampling of the culture materials are carried out

[6] T. Vaara, M. Vaara, and S. Niemela, *Appl. Environ. Microbiol.* **38,** 1011 (1979).

[7] J. Oppenheim and L. Marcus, *J. Bacteriol.* **101,** 286 (1970).

[8] R. Rippka, J. Deruelles, J. B. Waterbury, H. Herdman, and R. Y. Stanier, *J. Gen. Microbiol.* **111,** 1 (1979).

[9] R. Rippka, this volume [2].

through a port in the bottom of the cylinder fitted with a three-way valve. Air is vented through a port at the top of the ground glass cap with a sterile air filter. Illumination is provided by two 30-W fluorescent tubes (arranged parallel to the cylinder) for each culture cylinder. A handmade light reflector (aluminum foil) is added to the light fixture. In a routine experiment, 4% CO_2-enriched air is used for aeration, which is passed through a 50% H_3PO_4 solution to remove ammonia, if any, from the air and then through a sterilized cotton air filter. In addition to the bubbling through the cylinder, mixing of the culture is provided by a magnetic stirrer located at the bottom.

The basic culture medium is the same as previously described except 18 g NaCl and 2.5 g $NaHCO_3$ are added to 1 liter of culture medium. During the pH adjustment of the culture medium to 7.6, the medium is bubbled with 4% CO_2-enriched air. After autoclaving the medium, precipitates formed during the autoclaving are dissolved by bubbling the medium with CO_2 gas through a sterilized cotton air filter. The cleared medium is transferred to the sterilized culture cylinder and then bubbled with 4% CO_2-enriched air (about 200 ml/min) for pH equilibration before inoculation. The CO_2 concentration of the CO_2-enriched air is slightly adjusted in order to maintain the pH of the culture medium at 7.6.

Seed cyanobacterial cultures are grown in 1-liter flasks at pH 7.6 at room temperature (25–27°) and a light intensity of approximately 150 $\mu E/m^2/sec$. When the cultures reach midexponential growth phase, a 50-ml aliquot is inoculated into each glass cylinder containing 3.3 liters of culture medium. After the inoculation of the seed culture, the culture is bubbled with CO_2-enriched air as described above (flow rate of about 200 ml/min) and stirred by magnetic stirrer as mentioned above. The light intensity at the surface of the culture cylinders is 150 $\mu E/m^2/sec$. The temperature and pH are maintained at 30 ± 1° and 7.6 ± 0.2, respectively, throughout the culture period. As mentioned above, the pH of the culture is controlled by slight adjustment of the CO_2 concentration of the CO_2-enriched air.

At optimum culture conditions (34°, pH 7.6, 4% CO_2 in air), *Synechococcus* sp. strain Miami BG 43511 can be grown at a minimum doubling time of 14 hr in N_2-fixing conditions, when the temperature is strictly maintained. Since there is a sharp decline of growth above 34° in this strain, however, a culture temperature of 30° is recommended for ordinary use. In ammonia or nitrate culture (final concentration, 10 m*M*) a minimum doubling time of 9–10 hr can be obtained in this strain. Biomass yield in N_2-fixing optimum culture conditions at 8 days batch culture is 1.0 g dry weight per liter culture.

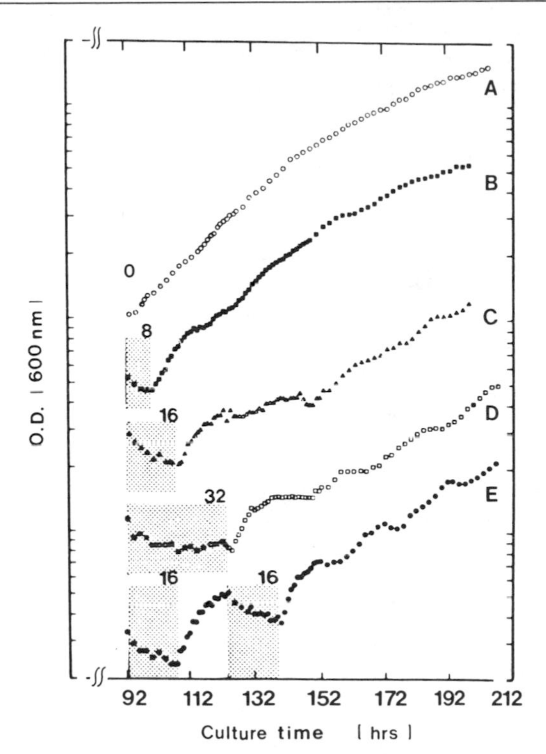

FIG. 1. Changes in turbidity of an aerobic N_2-fixing culture of marine *Synechococcus* sp. strain Miami BG 43511 during and after various dark treatments. A, B, C, and D were treated with 0, 8, 16, and 32 hr dark, respectively, and E was treated with 16 hr dark–16 hr light–16 hr dark. Numbers indicate dark treatment periods (hr). All cultures contained approximately 10^6 cells/ml (0.075 absorbance units at 600 nm) at the beginning of dark treatments (at 92 hr culture time). See text for other experimental conditions. Relative changes in absorbance are shown in a log scale.

Synchronous Culture

Under the growth conditions described above, synchronization of aerobic N_2-fixing *Synechococcus* sp. strain Miami BG 43511 can be induced by interruption during their early exponential growth phase (~10^6 cells/ml) with dark periods.[10–12] During the dark period(s), aeration is stopped, but mixing by magnetic stirrer is continued. Measurement of synchronous

[10] C. León, S. Kumazawa, and A. Mitsui, *Curr. Microbiol.* **13,** 149 (1985).

[11] A. Mitsui, S. Kumazawa, A. Takahashi, H. Ikemoto, S. Cao, and T. Arai, *Nature* (*London*) **323,** 720 (1985).

[12] A. Mitsui, S. Cao, A. Takahashi, and T. Arai, *Physiol. Plant.* **69,** 1 (1987).

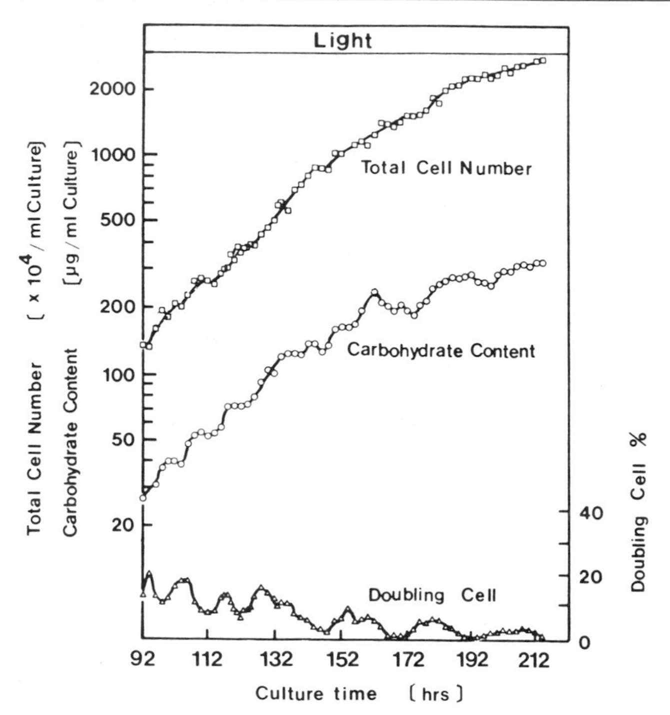

FIG. 2. Relationship among total cell number, doubling cell percentage, and carbohydrate content in a continuously illuminated batch culture of aerobic N_2-fixing marine *Synechococcus* sp. strain Miami BG 43511. See the legend to Fig. 1 and text for experimental conditions.

growth can be monitored roughly by hourly measuring the change in absorbance of the culture at 600 nm. Figure 1 shows the changes of absorbance after treatment with various dark periods. A stepwise increase is observed in well-synchronized cultures. More exact measurements, however, should be made of the doubling cells, which appear only during the segregated periods of the cell division cycle, and total cells should also be monitored. For cell counts, a 1-ml sample is taken from the culture cylinder at periodic time intervals and fixed with 50 μl of 10% Lugol's solution. Cell numbers are measured using a Petroff–Hauser bacterial cell-counting chamber. For each cell count, the total number of cells and the number of doubling (dividing) cells are counted. Doubling cells are counted as one cell.

Changes in the carbohydrate content in the culture also provide a good index of growth synchrony under nitrogen-fixing conditions[11,12] (see Figs.

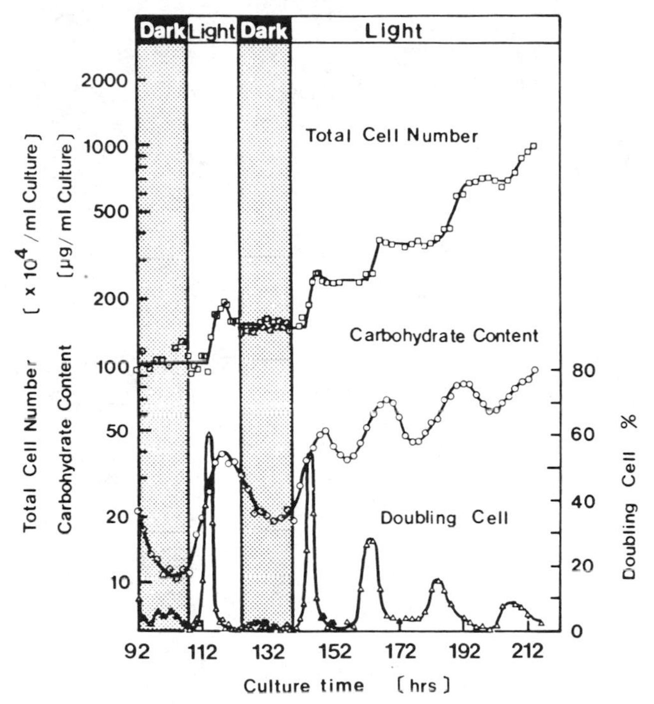

FIG. 3. Relationship among total cell number, doubling cell percentage, and carbohydrate content in a synchronous culture of aerobic N_2-fixing marine *Synechococcus* sp. strain Miami BG 43511. From Mitsui *et al.*[12]

2 and 3). Carbohydrate content of the culture is measured by a colorimetric method.[13]

To obtain well-synchronized growth in culture, insertion of proper dark and light periods is essential.[12,14] Under the culture conditions described above, a regime of 16 hr dark–16 hr light–16 hr dark yields good synchrony of *Synechococcus* sp. strain Miami BG 43511.[12,14] Longer dark periods reduce the degree of synchrony, and shorter dark periods induce irregular cell divisions.[12] Figures 2 and 3 represent the nonsynchronous (batch) culture without pretreatment and the well-synchronized culture with dark–light–dark pretreatment, respectively. After the 16 hr dark–16 hr light–16 hr dark period, cells grow synchronously during the subse-

[13] M. Dubois, K. A. Gilles, J. K. Hamilton, P. A. Rebers, and F. Smith, *Anal. Chem.* **28,** 350 (1956).

[14] T. Arai and A. Mitsui, *Prog. Photosyn. Res.* **2,** 649 (1987).

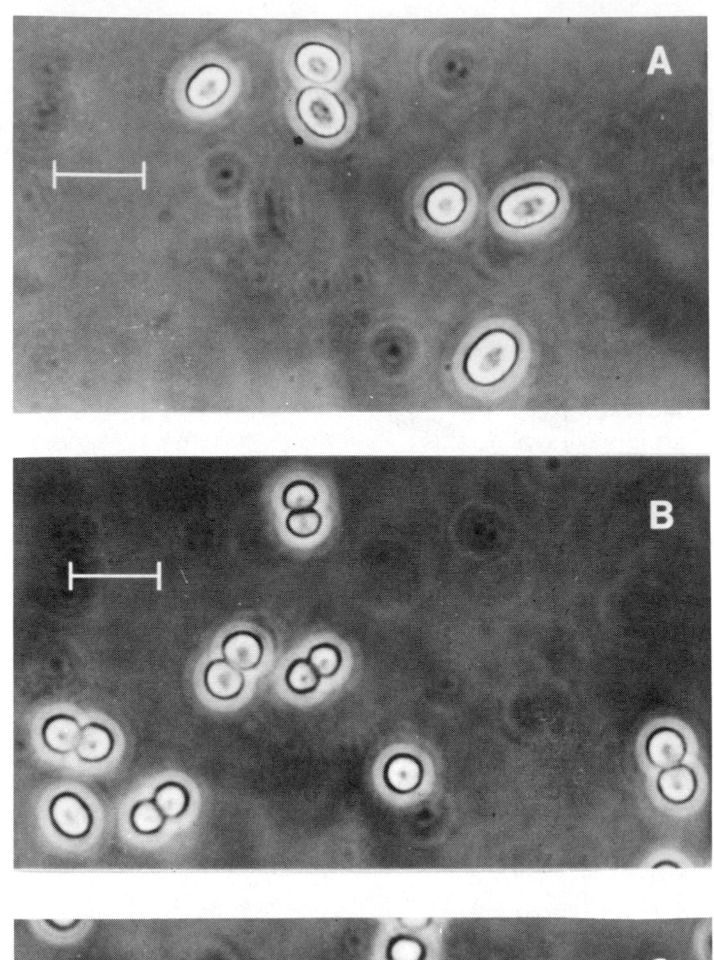

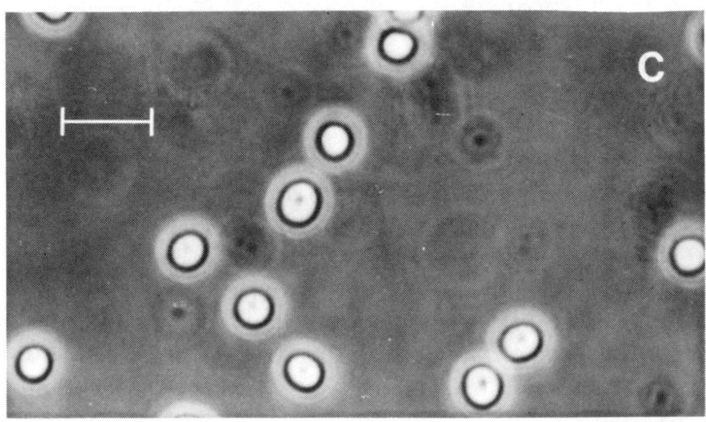

FIG. 4. Phase-contrast photomicrographs of the aerobic N_2-fixing marine unicellular cyanobacterium *Synechococcus* sp. strain Miami BG 43511 during synchronous growth. A, B, and C show cells during the photosynthetic elongation period, dividing period, and small daughter cell period, respectively, under the aerobic diazotrophic synchronous growth conditions. Bar, 10 μm.

quent 80 hr for 3–4 cell division cycles under continuous illumination (Fig. 3). As shown in Fig. 3, the distribution of doubling cell percentage is very narrow in the first cell division cycle in a well-synchronized culture. Doubling cells disappear from the culture from 8 through 18 hr until the second cell division cycle starts. In the second, third, and fourth cell division cycles, the peak doubling cell percentages decrease and the distribution of doubling cell percentage becomes wider. However, the time of occurrence of doubling cell percentage peaks during synchronous growth indicates that cell division cycles occur at approximately 20-hr intervals under the given culture conditions. The changes of carbohydrate content during well-synchronized growth also exhibit approximately 20-hr intervals for 3–4 cell division cycles during the continuous illumination period (Fig. 3).

Figure 4 represents phase-contrast photomicrographs of the culture during typical synchronous growth of *Synechococcus* sp. strain Miami BG 43511. At the start of synchronous growth of this strain, most of the cells were single and small (3.0 × 2.5 μm). During the first 5-hr period, most of the single cells became larger (3.5 × 7.0 μm) and formed septa. At 6 and 7 hr, most cells split into two daughter cells of small size (3.0 × 3.5 μm). These changes in cell size were repeated with each cycle. The cell length of *Synechococcus* sp. strain Miami BG 43522 is slightly shorter.

[8] Symbiotic Associations

By J. C. MEEKS

Introduction

Certain filamentous nitrogen-fixing cyanobacteria establish symbiotic associations with ascomycete fungi (to form lichens), with a marine diatom, and with phylogenetically diverse plants that include bryophytes, a fern, cycads (a class of gymnosperm), and an angiosperm.[1,2] Compared to free-living cultures, the cyanobacteria in association with a photosynthetic eukaryotic partner have a lower growth rate, a diminished capacity to assimilate ammonium and photosynthetically reduce CO_2, and a higher frequency of heterocysts. Thus, symbiotically associated cyanobacteria

[1] J. W. Millbank, *in* "The Biology of Nitrogen Fixation" (A. Quispel, ed.), p. 238. Elsevier, New York, 1974.

[2] G. A. Peters, R. E. Toia, Jr., H. E. Calvert, and B. H. Marsh, *Plant Soil* **90,** 17 (1986).

provide an unique physiological growth state for comparative studies on metabolic regulation and cellular differentiation. Details of experimental manipulation of the diatom and cycad associations are not given here because they have not been either cultured or reconstituted, nor have the associated cyanobacteria been isolated in quantity.

Culture of Separated Partners and Associations

Cyanobacteria. Nitrogen-fixing cyanobacteria (primarily *Nostoc* spp.) have been isolated and cultured from the various associations.[3] In the cephalodia-forming lichens and the associations with plants, the cyanobacteria are localized in colonylike masses or zones. For enrichment culture the cyanobacterial masses are manually excised and placed in liquid or on solid minimal medium, with or without combined nitrogen. The minimal media of Allen and Arnon[4] or Bg-11 of the Pasteur Culture Collection[5] have been used. Subsequent incubation conditions are as described for culture and purification of free-living cyanobacteria.[6–8]

Lichens. Because it has not been possible to culture lichens in the laboratory,[9] natural populations are the source of experimental material. Prior to use or storage, the thallus samples are rinsed with tap and distilled water, or with diluted cyanobacterial growth medium, to remove large debris.[10]

A controlled environment with a filtered air supply is essential for short- (24–72 hr[10]) or long-term (~6 months[11]) maintenance of nondried lichen samples in urban/suburban environments with high levels of air pollutants. A recommended air filtration system consists of passage through cotton wool, then a solution of 1% (w/v) sodium bicarbonate, and finally activated charcoal; the bicarbonate and charcoal are replaced monthly.[11] Other environmental provisions include tightly controlled humidification [88–96% relative humidity (RH)] and low temperature (9–20°), commonly with diurnal (16 : 8 hr, light : dark) illumination at moderate intensity (~6 W/m^2).[10,11] Excess moisture or nutrients lead to break-

[3] W. D. P. Stewart, P. Rowell, and A. N. Rai, *in* "Nitrogen Fixation" (W. D. P. Stewart and J. R. Gallon, eds.), p. 239. Academic Press, New York, 1980.

[4] M. B. Allen and D. I. Arnon, *Plant Physiol.* **30,** 366 (1955).

[5] R. Rippka, J. Deruelles, J. B. Waterbury, M. Herdman, and R. Y. Stanier, *J. Gen. Microbiol.* **111,** 1 (1979).

[6] R. Rippka, this volume [1].

[7] R. W. Castenholz, this volume [3].

[8] C. P. Wolk, this volume [4].

[9] D. H. Brown, "Lichen Physiology and Cell Biology." Plenum, New York, 1985.

[10] A. N. Rai, P. Rowell, and W. D. P. Stewart, *Planta* **152,** 544 (1981).

[11] K. A. Kershaw and J. W. Millbank, *Lichenology* **4,** 83 (1969).

down of lichen associations and promotion of heterotrophic bacterial growth[9]; acid-washed sand is preferred over vermiculite as substratum for better drainage, and the substratum is irrigated once with mineral nutrient solution, with moisture being replenished by distilled water.[11]

Lichen samples are commonly air dried, in light or dark, at room temperature and stored desiccated over silica gel at 4[12] or −20°.[13] Dried samples are rehydrated with distilled water and, after blotting to remove excess water, incubated in the light for 12 to 24 hr before use.[12,14]

Bryophytes. Laboratory cultures of bryophytes are readily established and maintained.[15] Gametophyte tissues with associated cyanobacteria of the liverwort *Blasia* and hornwort *Anthoceros* from natural populations can be maintained in growth chambers on perlite substratum, moistened with a mineral nutrient solution, and subjected to continuous illumination (~0.5 W/m^2) at 20°.[16]

Pure cultures of *Anthoceros* gametophyte tissue can be established for surface or submerged growth.[17] Four to five mature sporophytes from natural populations are treated for 0.5 to 2.0 min in 10 ml of a 0.5% (w/v) solution of sodium hypochlorite and, as a wetting agent, 0.1% (w/v) detergent followed by successive washing with a total of 20 volumes of sterile distilled water. The treated spores are spread over the surface of minimal medium solidified with 1.2% (w/v) agar in standard Petri plates; the plates are sealed with Parafilm to retain moisture. The spores germinate within 7 days, and gametophytes grow during incubation at 20° with 3.5–8 W/m^2 of illumination on a 14 : 10 hr light : dark cycle.

A basal medium for culture of *Anthoceros* was formulated by Prat[15] and contains, in g/liter: 0.2 NH_4NO_3, 0.1 each $CaCl_2$, $MgSO_4$, and KH_2PO_4, plus 1 ml/liter of a micronutrient stock solution; the pH is adjusted to 6.4 before autoclaving. The micronutrient solution contains, in g/liter: 50 EDTA (free acid), 10 H_3BO_3, 22 $ZnSO_4 \cdot 7H_2O$, 1.5 $CuSO_4 \cdot 5H_2O$, 5 $CaCl_2$, 5 $MnCl_2 \cdot 4H_2O$, 1.5 $CoCl_2 \cdot 6H_2O$, 1 $(NH_4)_6Mo_7O_{24} \cdot 4H_2O$, and 5 $FeSO_4 \cdot 7H_2O$. The EDTA is first dissolved in 750 ml distilled water and the pH adjusted to 6.5 with solid KOH; the other salts are added as listed with concurrent pH adjustment and the final volume reached with distilled water. The solution is aged to a red wine color by bubbling with membrane-filtered air for 36–48 hr, with frequent

[12] K. Huss-Danell, *Physiol. Plant.* **41,** 158 (1977).
[13] L. Hallbom and B. Bergman, *Planta* **157,** 441 (1983).
[14] D. J. Hill and D. C. Smith, *New Phytol.* **71,** 15 (1972).
[15] W. T. Doyle, *in* "Methods in Developmental Biology" (F. H. Witt and N. K. Wessells, eds.), p. 329. Crowell, New York, 1967.
[16] G. A. Rodgers and W. D. P. Stewart, *New Phytol.* **78,** 441 (1977).
[17] C. S. Enderlin and J. C. Meeks, *Planta* **158,** 157 (1983).

adjustment of pH and volume. For N_2-dependent culture of *Anthoceros–Nostoc,* the NH_4NO_3 is omitted; alternately, the NH_4NO_3 may be replaced with 2.5 m*M* NH_4Cl or 5 m*M* NO_3^- (equimolar Na and K salts) and buffered with 5 m*M* MES (*N*-morpholinoethanesulfonic acid), pH 6.4.[17]

Submerged liquid cultures of axenic *Anthoceros* are established by transfer of gametophytes to 125- or 250-ml Erlenmeyer flasks containing 50 or 100 ml of basal medium supplemented with 0.5% (w/v) glucose and 5 m*M* MES buffer, pH 6.4. The glucose and MES are made as a single stock solution at 25% and 250 m*M*, respectively, filter-sterilized (0.22-μm pore size), stored at 4°, and added to autoclaved basal medium. The liquid cultures are incubated as the surface cultures except the flasks are placed on orbital shakers (115 rpm). These culture conditions result in doubling times of about 5 and 10 days for *Anthoceros–Nostoc* with NH_4^+ and N_2, respectively, as nitrogen sources.[17] Similar N_2-dependent doubling times of *Anthoceros–Nostoc* can be obtained in the absence of glucose when liquid cultures are illuminated at 340 W/m^2 and supplemented with 5% (v/v) CO_2.[18] In all cases buffering of the medium is essential.

Azolla. The six existing species of the free-floating water fern *Azolla* normally contain a N_2-fixing cyanobacterial symbiont (*Anabaena azollae*) that is present in all stages of the *Azolla* reproductive cycle. Natural populations of the fern sporophyte can be maintained in outdoor ponds or cultured in the laboratory. To remove epiphytic microorganisms in the transfer to laboratory culture, the sporophytes are first washed extensively (20 times) with vigorous agitation in distilled water followed by surface sterilization with 0.12% (w/v) sodium hypochlorite and 0.01% (w/v) Triton X-100; the sterilizing agents are removed by washing in sterile distilled water.[19]

The conditions for optimal laboratory growth include a broad pH range between 5 and 8, a 16-hr photoperiod at a photon flux of at least 400 μmol/m^2/sec (~87 W/m^2), and 25–30°, plus an 8-hr dark period at slightly lower temperature.[20] In general, growth is best when illumination is enriched in red light (supplemental tungsten lamps). A commonly used basal medium is a 40% dilution of Hoagland's solution that contains in final concentration, as g/liter: 0.3 $CaCl_2 \cdot 2H_2O$, 0.15 KCl, 0.2 $MgSO_4 \cdot 7H_2O$, 0.03 NaCl, 0.015 Sequestrene (10–13% iron chelate), 0.05 KH_2PO_4, 0.01 $K_2HPO_4 \cdot 3H_2O$, and 2 ml/liter of a micronutrient solution; the pH is between 5.6 and 6.0 and is not adjusted.[20] The micronutrient stock solution

[18] N. A. Steinberg and J. C. Meeks, unpublished.

[19] G. A. Peters and B. C. Mayne, *Plant Physiol.* **53,** 813 (1974).

[20] G. A. Peters, R. E. Toia, Jr., W. R. Evans, D. K. Crist, B. C. Mayne, and R. E. Poole, *Plant Cell Environ.* **3,** 261 (1980).

contains, in g/liter: 2.7 H_3BO_3, 1.8 $MnCl_2 \cdot 4H_2O$, 0.22 $ZnSO_4 \cdot 7H_2O$, 0.39 $Na_2MoO_4 \cdot 2H_2O$, 0.05 $CoCl_2 \cdot 6H_2O$. The doubling time of various *Anabaena*-associated *Azolla* species under optimal conditions in 1 liter of the described medium in 2.8-liter Fernbach flasks is around 2 days.[20] Humidified surface aeration is an option, but not essential. For culture of symbiont-free *Azolla* the $CaCl_2$ and KCl are replaced with equimolar nitrate salts.

Azolla can be cured of the *Anabaena* by sequential treatment with a combination of antibiotics in a variety of regimens.[19] Although an early report indicated generation of axenic *Azolla*,[21] such isolations have not been repeated, and it appears that heterotrophic bacteria remain associated with the *Anabaena*-cured *Azolla*.[19]

Gunnera. The angiosperm *Gunnera* consists of small (New Zealand) to large (South America and Asia) tropical herbs, all with *Nostoc* spp. within a specialized cell (gland) at the base of the cotyledons and true leaves.[22] The small species are adaptable to laboratory culture and experimentation.

To eliminate epiphytic microorganisms, especially *Nostoc*, for subsequent reconstitution experiments, seeds of *Gunnera arenaria*[23] or *G. manicata*[24] are treated for 5 min in 70% ethanol then 1 h in 6% (w/v) sodium hypochlorite[23] or 1–2 min in 35% (w/v) tribasic copper sulfate.[24] The seeds are planted in pots of sterile, 3 : 1 washed sand : soil mixtures[23] or vermiculite[24] and irrigated with a standard, sterilized plant mineral nutrient solution, with or without combined nitrogen, alternating with sterile distilled water. Seedlings emerge in 3–4 weeks and are grown at 22–27°, under continuous cool-white fluorescent illumination (~4.5 W/m^2), and with greater than 60% RH.[24]

Reconstitution of Associations

Symbiotic associations have been reconstituted only with the bryophytes *Blasia*,[16] *Phaeoceros*,[25] and *Anthoceros*[16,17,25] and the angiosperm *Gunnera*.[23,24] *Nostoc* isolates from nearly all associations are capable of establishing symbioses with *Blasia*, *Anthoceros*, and *Gunnera*. The notable exceptions are isolates from three different cycads[16,17,25] (isolates from the same and other cycads are symbiotically competent) and a putative

[21] L. G. Nichell, *Am. Fern J.* **48,** 103 (1958).

[22] W. B. Silvester, *in* "Symbiotic Nitrogen Fixation in Plants" (P. S. Nutman, ed.), p. 521. Cambridge University Press, Cambridge, England, 1976.

[23] W. B. Silvester and P. J. McNamara, *New Phytol.* **77,** 135 (1976).

[24] H. T. Bonnett and W. B. Silvester, *New Phytol.* **89,** 121 (1981).

[25] J. E. Ridgway, *Ann. Missouri Bot. Gard.* **54,** 95 (1967).

symbiotic isolate from *Azolla caroliniana*.[17,24] DNA hybridization analyses imply that at least three different cyanobacteria cultured from *Azolla*, including the one from *A. caroliniana*, are not the dominant symbiotic organism.[26–28]

Bryophytes. Reconstitution of liverwort and hornwort associations occurs during surface culture in the absence of combined nitrogen when *Nostoc* spp. and gametophyte tissue are incubated together.[16,17,25] Successful reconstitution is determined by the presence of *Nostoc* colonies within the gametophytes and N_2-dependent growth of the association. Reconstitution also occurs during coculture of *Nostoc* spp. and *Anthoceros* submerged in liquid medium under *Anthoceros* growth conditions in the presence or absence of combined nitrogen.[17] After 1–2 weeks of coculture the gametophyte tissues are removed, rinsed by vigorous agitation in medium to dislodge *Nostoc* on the surface, and inoculated into fresh combined-nitrogen-free medium.

Liquid cultures of reconstituted *Anthoceros–Nostoc* are transferred to fresh combined-nitrogen-free medium every 1–3 weeks, depending on the *Nostoc* strain in association. For example, if not frequently transferred, with extensive and vigorous washings between transfers, *Anthoceros* associations with *Nostoc* strains ATCC 27895 (a free-living isolate) and ATCC 29133 (an isolate from the cycad *Macrozamia*) tend to break down because of overgrowth by *Nostoc*, while the association with *Nostoc* strain 7801, originally obtained from *Anthoceros*, has much less outgrowth and requires less frequent transfers. The outgrowth of any *Nostoc* strain is accelerated when associations are cultured in the presence of combined nitrogen. Prior to experimentation, *Nostoc* growing on the *Anthoceros* surface is eliminated by a 2- to 4-day incubation with 250 U of penicillin G/ml of medium; treated tissues are rinsed in basal medium and cultured for 2–14 days before use. Treatment with penicillin does not alter physiological activity (acetylene reduction) of *Nostoc* in symbiotic colonies, but establishment of new colonies in growing *Anthoceros* tissue is delayed for at least 3 weeks.[17] The general practice is to use *Anthoceros–Nostoc* between 3 and 12 months following reconstitution, and once penicillin treated the material is used within 3 weeks or discarded.

Gunnera. The association between *Nostoc* spp. and *Gunnera* is initiated by placing a dense suspension (single transfer loop or drop from a syringe needle) of cultured *Nostoc* on and between the cotyledons of the about 2-month-old seedlings.[23,24] Establishment of the association is determined by visual observation of infected glands in stem sections and N_2-

[26] C. Franche and G. Cohen-Bazire, *Plant Sci.* **39,** 125 (1985).

[27] S. A. Nierzwicki-Bauer and R. Haselkorn, *EMBO J.* **5,** 29 (1986).

[28] J. C. Meeks, C. M. Joseph, and R. Haselkorn, *Arch. Microbiol.* **150,** 61 (1988).

dependent growth of the mature plant. *Nostoc* in the glands at the base of the cotyledons are presumed to serve as inocula for additional infections of leaf glands and of new plants as the small *Gunnera* species vegetatively propagate through stolons.

Isolations of Cyanobacterial Symbionts in Quantity

Procedures have been devised for disruption of symbiotic tissue in lichen, *Anthoceros,* and *Azolla* associations and differential or gradient centrifugation to enrich a cyanobacterial fraction.

Lichens.[29-32] Lichen tissue (~60 mg/ml of suspension[32]), including excised cephalodia from *Peltigera aphthosa,*[31] is disrupted by 2–5 min of gentle homogenization in a mortar or all-glass homogenizer in cyanobacterial growth medium,[29] distilled water,[30] or growth medium supplemented with 10% sucrose and buffered at pH 6.8–6.9 with 10 m*M* phosphate or HEPES.[31,32] The homogenate may be strained through 0.5-mm mesh screen and homogenized a second or third time[29,30,32] before centrifugation at 300–375 *g* for 3 min[29,30] or 1000 *g* for 5 min.[31,32] The middle or upper, darkly pigmented band in the pellet fraction is collected and suspended in the homogenization solution for low-speed differential centrifugation[29,30] or sucrose gradient centrifugation.[31,32]

Successful regimes of differential centrifugation include 125 *g* starting at 30-sec duration and increasing in 10-sec intervals until 90 sec of centrifugation,[29] or two initial spins at 300 *g*, then twice at 30 *g*, and finally a repeat of 300 *g*[30]; the pellet fractions containing fungal debris are discarded. Based on chlorophyll content, the yield of cyanobacteria in the final supernatants is about 6% of the initial population; a variable amount of fungal material and heterotrophic bacteria is also present.[29,30]

Density gradient centrifugal separation is done by layering the homogenized pellet fraction, suspended in buffer plus 10% sucrose (w/v; 2–5 ml/tube), onto 80% (w/v) sucrose and centrifuging for 15 min at 3000 *g*.[31] The cyanobacteria are recovered from the 10–80% sucrose interface and washed in buffer by centrifugation at 500 *g*. A 45% yield of the *Nostoc* cells from intact lichen thallus has been reported in the combined homogenization–gradient centrifugation protocol.[32]

Anthoceros.[33,34] *Anthoceros–Nostoc* tissue is disrupted by physical

[29] E. A. Drew and D. C. Smith, *New Phytol.* **66,** 379 (1967).

[30] T. G. A. Green and D. C. Smith, *New Phytol.* **73,** 753 (1974).

[31] J. W. Millbank and K. A. Kershaw, *New Phytol.* **68,** 721 (1969).

[32] J. W. Millbank, *New Phytol.* **71,** 1 (1972).

[33] J. C. Meeks, C. S. Enderlin, C. M. Joseph, J. S. Chapman, and M. W. L. Lollar, *Planta* **164,** 406 (1985).

[34] C. M. Joseph and J. C. Meeks, *J. Bacteriol.* **169,** 2471 (1987).

and enzymatic treatment, and *Nostoc* is enriched by gradient centrifugation. The associated tissue is cultured for 2 days in the absence of glucose, then suspended 1 : 1 (generally 20 g fresh weight : 20 ml) in 5 m*M* MOPS (*N*-morpholinopropanesulfonic acid), pH 7.8, and 1% (w/v) polyvinylpyrrolidine-40 (PVP-40), and blended for 20 sec at high speed in a small container of a commercial blender. The slurry is washed 3 times in MOPS/PVP-40 by centrifugation at 1,000 *g* for 1 min. The pellet fraction is suspended in MOPS/PVP-40 and incubated at 30° for 15 min with 0.05% (w/v) digitonin. Macerase, cellulase, and β-glucuronidase are added to 1% (w/v) each and the incubation continued for 1 hr. If the tissue is grown in glucose, 0.5% (w/v), zymolyase can be used to hydrolyze excess mucilage. The digested preparation is washed 3 times as above in 5 m*M* MOPS with 2 min of vigorous vortex mixing between centrifugations.

An amount of the final pellet equivalent to not more than 5 g fresh weight of original tissue is added to a centrifuge tube containing 10 ml of each of the following solutions in a discontinuous gradient: 5% Ficoll/5% Ludox, 7% Ficoll/3% Ludox, 9% Ficoll/1% Ludox, and 11% Ficoll. Ficoll-400 is w/v and Ludox AM (colloidal silica; DuPont) is v/v; both are prepared in 5 m*M* MOPS. Just before use, 100 ml of Ludox is neutralized by mixing with 1 g of sodium Chelex 100 and 1 ml of 0.05 m*M* disodium EDTA, followed by filtration through Whatman #1 paper. The gradients are centrifuged at 10,500 *g* for 25 min. Material in the 11% Ludox pad is removed, washed 3 times in 5 m*M* MOPS as above, layered onto a second discontinuous gradient of Ludox at 12%, 16.5%, 20.4%, and 24.5% (v/v; equivalent to Ludox densitites of 0.15, 0.20, 0.25, and 0.30 g/ml, respectively) in 5 m*M* MOPS, and centrifuged at 5000 *g* for 10 min. Discrete macroscopic particles consisting of *Nostoc* in a mucilaginous matrix together with disrupted *Anthoceros* transfer cells are recovered from the 24.5% Ludox pad; the particles (colonies) are washed twice in 5 m*M* MOPS, and any *Anthoceros* tissue that is visible under a stereomicroscope is removed with forceps. The yield in this protocol is about 20% of the initial *Nostoc* colonies in associated tissue.

Azolla.[19,35,36] Physical disruption of *Azolla–Anabaena* followed by sieving or differential centrifugation has been preferred in enrichment of *Anabaena azollae*. Disruption is by variations of the "gentle roller" method,[19] the specific variation being dependent on the amount of *Azolla* tissue to be processed. In the original protocol 2–10 g (fresh weight) of *Azolla–Anabaena* are barely floated in a shallow pan with cyanobacterial growth medium, or buffer, containing 1% PVP-40. The *Azolla* fronds are

[35] T. B. Ray, G. A. Peters, R. E. Toia, Jr., and B. C. Mayne, *Plant Physiol.* **62,** 463 (1978).
[36] G. A. Peters, K. M. Busby, and D. Kaplan, unpublished.

gently squashed by multiple passes with a printer's roller or similar device until only pieces remain. The bree is filtered through 2, then 4, and, optionally, 8 layers of cheesecloth. Material retain in the cheesecloth may be disrupted a second time, filtered again, and the filtrates combined. The filtrates are harvested by centrifugation at 1000 *g* for 1 min. In the sieving protocol, the suspended pellet is passed through two layers of 100-μm nylon mesh and the filtrate harvested by centrifugation as above. Alternatively, the suspended pellet is clarified by one or two centrifugal washes at 1000 *g* for 45 sec and the material then harvested by centrifugation at 1000 *g* for 30 sec. The final pellet from either clarification consists primarily of short, heterocyst-containing *Anabaena* filaments that are active in N_2 fixation, plus simply and branched cavity hairs of *Azolla* origin. Based on phycobiliprotein content, the yield of *Anabaena* is estimated to be 90% of that present in intact *Azolla–Anabaena*.[35,36]

The basic protocol can be altered to accommodate small (<0.5 g fresh weight) or large (15–20 g fresh weight) samples by disruption with a Teflon-coated pestle in 10- or 50-ml homogenization tubes. In large-scale isolations, samples of 3–5 g (fresh weight) are separately processed and the homogenates combined.[36] The bree from homogenization is clarified as above.

Anabaena cells are separated from *Azolla* cavity hairs by first incubating crushed or homogenized preparations at about 4° for 30 min with moderate agitation (250 rpm) followed by filtration.[36] The filtration is initially through cheesecloth as above, then 53-μm nylon mesh, and successively 20- and 10-μm nylon mesh. The filtrate from the final 10-μm mesh is concentrated by centrifugation at 1300 *g* for 5 min and contains primarily short *Anabaena* filaments and detached heterocysts.[36]

Acknowledgments

The unpublished results in, and preparation of, this chapter were supported by the National Science Foundation (Grants PCM83-16384 and DCB86-08474) and the U.S. Dept. of Agriculture (Grant 83-CRCR-1-1295). I thank Dr. G. A. Peters for kind advice and sharing of unpublished information and C. S. Enderlin, C. M. Joseph, and N. A. Steinberg for helpful review.

[9] Cell Stability and Storage

By S. EDWARD STEVENS, JR.

Introduction

Cultures of cyanobacteria (blue–green algae) have normally been maintained in the laboratory in appropriate aqueous media in small flasks or dishes or on agar-solidified slants. Continuing serial transfer, in some cases for up to 40 years, has constituted the standard procedure for culture maintenance. *Synechococcus* 6301 (*Anacystis nidulans,* Tx 20) was first isolated into unialgal culture in 1947 and made axenic by W. A. Kratz in 1952. From 1952 to the present it has normally been maintained on agar slants of medium C,[1] C_s,[1] B,[1] or BG-11.[2]

We now know that the mutation frequency for single gene markers in cyanobacteria is essentially the same as that for comparable gene markers in *Escherichia coli*.[3,4] Although documentation is sparse, we may surmise that the present-day cultures of *Synechococcus* 6301 are genetically different from the original culture established in 1947. This contention is supported by the empirical observation that 6301 is less robust and not as convenient to grow as it was even 15 years ago and more directly by the spontaneous loss of the enzyme acetate thiokinase (EC 6.2.1.1, acetate–CoA ligase) rendering 6301 resistant to propionate.[5] There are also known spontaneous mutants in other species of cyanobacteria, such as the acquisition of heterocysts and nitrogen-fixing capacity in a strain of *Anabaena variabilis* (PCC 7118) known to have lost these abilities for over 20 years[6] and the spontaneous loss of the small plasmid in *Synechococcus* 73109.[7] Thus, cyanobacteria probably have the same long-term cultural instability well documented among other prokaryotes. Lastly, continuing serial cultivation is labor intensive and very time consuming.

Over a decade ago, we began a search for an alternative to continuing serial cultivation. Two procedures have proven effective. They are described below.

[1] S. E. Stevens, Jr., C. O. P. Patterson, and J. Myers, *J. Phycol.* **9,** 427 (1973).
[2] M. M. Allen, *J. Phycol.* **4,** 1 (1968).
[3] S. E. Stevens, Jr., and R. D. Porter, *Proc. Natl. Acad. Sci. U.S.A.* **77,** 6052 (1980).
[4] R. D. Porter, J. S. Buzby, A. Pilon, P. I. Fields, J. M. Dubbs, and S. E. Stevens, Jr., *Gene* **41,** 249 (1986).
[5] A. J. Smith and C. Lucas, *Biochem. J.* **124,** 23p (1971).
[6] R. Rippka, J. Deruelles, J. B. Waterbury, M. Herdman, and R. Y. Stanier, *J. Gen. Microbiol.* **111,** 1 (1979).
[7] W. F. Doolittle, personal communication.

Cryopreservation Procedures

Preparation of Cultures

Cyanobacteria are grown in the appropriate medium and under the appropriate conditions until the onset of stationary phase. The timing and culture density at the onset of stationary phase is conveniently controlled by limiting the nitrogen content of the growth medium.[8] We generally grow cells in a large test tube with a working volume of 150 ml under fluorescent illumination (540 $\mu E\ m^{-2}\ sec^{-1}$), bubbling with 1% CO_2 in air, and growth at a temperature appropriate to the species being used. If the growth temperature is above 25°, it is generally best to remove the culture from its growth bath after it has reached the onset of stationary phase and place it at room temperature overnight in a convenient place under no more than normal room illumination. This step seems to alleviate "cold shock" in those cyanobacteria that have been tested.

The total volume of the culture (generally, at a density of 5×10^7 cells/ml) is harvested under aseptic conditions by centrifugation or filtration. The collected cells are resuspended in 3–5 ml of filter-sterilized 10% polyvinylpyrrolidone (PVP).[9–11] The use of PVP, instead of the many other cryoprotectants tried,[12,13] resulted in over 90% revitalization of stored cultures. Technical grade PVP of 40,000 average molecular weight is acceptable for cryopreservation of cyanobacteria if a freshly made solution is allowed to stand under fluorescent light for 1–2 weeks before use (this allows for the decay of the trace amounts of contaminating organic peroxides usually present in PVP).

Storage in Liquid Nitrogen

Cells in 10% PVP are pipetted into plastic vials with screw caps designed to fit into the stalks or cups of commercially available liquid nitrogen freezers. The vials are rapidly frozen by plunging them into liquid N_2 in a small Dewer. We have had good results with plunge freezing and poor results with slow freezing procedures. Before storage the caps should be checked for proper closure.

[8] D. A. M. Paone and S. E. Stevens, Jr., *Plant Physiol.* **67,** 1097 (1981).
[9] M. J. Ashwood-Smith and C. Warby, *Cryobiology* **8,** 453 (1971).
[10] G. J. Morris, *Arch. Microbiol.* **107,** 57 (1976).
[11] G. J. Morris, *Arch. Microbiol.* **107,** 309 (1976).
[12] O. Holm-Hansen, *in* "Handbook of Phycological Methods" (J. R. Stein, ed.), p. 195. Cambridge University Press, Cambridge, England, 1973.
[13] L. L. Corbett and D. L. Parker, *Appl. Environ. Microbiol.* **32,** 777 (1976).

Lyophilization

Disks (1 cm diameter) of very pure absorbent paper, e.g., Schleicher & Schuell No. 740-E, are sterilized and placed in the bottom of sterilized screw-cap vacuum vials, e.g., VirTis 6516-0110, with split rubber stoppers. Cells in 10% PVP (0.1–0.2 ml) are pipetted onto the paper disks. The vials are freeze-dried (normally in 2–3 hr) with an appropriate freeze-drying apparatus equipped with quick-seal valves,e.g., VirTis 6250-2501, for the vacuum vials. It is essential to the recovery of viable cultures that the freeze-drying procedure be done in darkness. We have normally covered the apparatus and vials with several layers of black cloth and dimmed the room lights to a point just sufficient to see. After lyophilization is completed, sterile caps are screwed tightly onto the vials. If vacuum is lost in the vials, the culture will be nonviable. We have stored vials at −21° for up to a year with excellent recovery. If longer term storage (up to 5 years) is required or desirable, then a −76° freezer should be used. It is essential to protect stored cultures from any exposure to light.

Revitalization of Cultures

In our experience, the revitalization process is the step most susceptible to failure. In the absence of prior knowledge to the contrary, it is generally best to resuspend the cells stored in liquid N_2 or on lyophilized disks in a small volume of the appropriate medium (no more than 5 ml) followed by incubation at room temperature under normal room illumination. Both liquid nitrogen freezing and lyophilization result in cellular killing levels of 3–6 orders of magnitude depending on the species of cyanobacteria being used. Incubation of such cultures under higher levels of illumination often results in additional cell death due to photooxidation. Even under room illumination, however, bleaching of the culture being revitalized normally occurs, followed later by overgrowth of the survivors. If the species of cyanobacteria being revitalized is a slow growing one, it may take as much as 1–2 weeks for growth to be apparent. After initial growth has occurred, the culture may be placed under the conditions of growth normally used for that species. A few cyanobacteria, such as *Synechococcus* 7002, may be directly resuspended in medium A and revitalized at 39° under high light illumination.

[10] New Taxonomic Methods: DNA/DNA Hybridization

By W. T. STAM and B. K. STULP

Introduction

The method of DNA/DNA hybridization allows the determination of sequence similarity between genomes of different organisms without the determination of exact nucleotide sequences.[1] It is based on the feature that denatured, single-stranded DNA will renature under suitable conditions. The extent of renaturation is proportional to the complementarity of the DNA strands involved. Furthermore, the stability of the renatured hybrid molecule is proportional to the accuracy of matching between the two strands. Thus, by DNA/DNA hybridization, two parameters for genomic relatedness are obtained.

Measurement of the extent of reassociation is feasible through the use of radioactively labeled DNA. In a typical hybridization experiment a small amount of radioactively labeled DNA of a particular source, the tracer DNA, is mixed with an excess of DNA from the same or another source, the driver DNA. A hybridization between driver DNA and tracer DNA from the same source is called homologous, and the product is called a homoduplex. Hybridizations between DNA from different sources are called heterologous and the product obtained is called a heteroduplex.

Generally DNA/DNA hybridizations work well for genomic comparisons at the species level. For higher taxons the renaturation level rapidly drops down to background levels; for lower taxons the method is insufficiently sensitive. The method has become particularly popular for genetic studies of prokaryotes, owing to the rather uncomplicated nature of prokaryotic genomes and the subsequent straightforwardness of the method. Several cyanobacterial genera (*sensu* Rippka *et al.*[2]) have been studied by the use of the technique, e.g., *Synechococcus, Chroococcidiopsis, Anabaena, Nostoc, Cylindrospermum, Chlorogloeopsis, Fischerella, Calo-*

[1] For further reading, see L. Grossman and K. Moldave (eds.), this series, Vol. 65; B. D. Hames and S. J. Higgins (eds.), "Nucleic Acid Hybridization." IRL Press, Oxford, England, 1985.

[2] R. Rippka, J. Deruelles, J. B. Waterbury, M. Herdman, and R. Y. Stanier, *J. Gen. Microbiol.* **111,** 1 (1979).

thrix, and the LPP group, as well as the prochlorophycean genus *Prochloron.*[3-8] The method has been described extensively.[9]

Materials

The equipment required includes the following: adjustable pipettors (range 1–1000 μl), centrifuge (cooled, 15,000 *g*), circulating thermostatted heater, dialysis tubing, French pressure cell (optional), microfuge, mortar and pestle, multiple column block (see Hybridization Conditions), scintillation counter, scintillation vials, spectrophotometer, Teflon homogenizer, ultrasonicater, vacuum oven, water bath, water-jacketed Pasteur pipet column.[10] The following materials should be autoclaved: acid-washed sand or 0.5-mm glass beads, capillaries (5 μl), Pasteur pipets, pipet tips, siliconized glass wool.

Reagents include the following: Chelex 100 (Bio Rad) Laboratories, Richmond, CA), emulsified in TE (see below) and autoclaved, chloroform–isoamyl alcohol (CIA) (24 : 1, v/v), 100 m*M* disodium EDTA (autoclaved), 96% ethanol (EtOH), ether, radioactive deoxyribonucleotide, hydroxyapatite (see below), nick translation kit, distilled TE-saturated phenol, proteinase K (20 mg/ml stock, preincubated for 2 hr at 37°), RNase (5 mg/ml stock, pretreated 10 min at 80°), scintillation fluid (in case of ^{3}H-labeling), Sephadex G-50 medium (Pharmacia, Uppsala, Sweden), sodium acetate (3 *M* stock, autoclaved), sodium dodecyl sulfate (SDS), sodium perchlorate, sodium phosphate buffer (NaPB, and equimolar buffer of Na_2HPO_4 and NaH_2PO_4, pH 6.8, 1 *M* stock, autoclaved), sodium chloride, Tris–EDTA buffer (TE, 10 m*M* Tris, 1 m*M* disodium EDTA, pH 7.5).

Hydroxyapaptite (BioGel HTP, DNA grade, Bio-Rad) should be boiled in low molarity NaPB for 10 min and decanted and resuspended in the same buffer twice, in order to remove small HAP particles. One gram dry HAP, treated this way, will yield approximately 2.5 ml bed volume. Batches of HAP may differ in their elution characteristics, and pilot studies should be carried out to check whether the elution buffers lead to

[3] W. T. Stam, *Arch. Hydrobiol. Suppl.* **56** (Algol. Stud. 25), 351 (1980).
[4] M. A. Lachance, *Int. J. Bacteriol.* **31,** 139 (1981).
[5] B. K. Stulp and W. T. Stam, *Br. Phycol. J.* **19,** 287 (1984).
[6] A. M. R. Wilmotte and W. T. Stam, *J. Gen. Microbiol.* **130,** 2737 (1984).
[7] W. T. Stam, S. A. Boele-Bos, and B. K. Stulp, *Arch. Microbiol.* **142,** 340 (1985).
[8] B. K. Stulp and E. I. Friedmann, manuscript in preparation.
[9] R. J. Britten, D. E. Graham, and B. R. Neufeld, this series, Vol. 29, p. 263.
[10] B. K. Stulp, *J. Biochem. Biophys. Methods* **12,** 197 (1986).

separation of single-stranded and double-stranded DNA. As necessary the buffer molarities should be adjusted.

Strains

The strains used in the experiments should preferably be axenic. In case of bacterial contamination several differential centrifugations and washing steps should be carried out to remove the bacteria. The number of contaminating organisms should never exceed 3% of the total cell number. Check microscopically.

In order to carry out DNA/DNA hybridizations, the G + C content of the DNA involved should be known. If the systematic position of the strain involved is rather clear, one might rely on the G + C data of other genus members. In case these data are not available, or in case the systematic position of the strain involved is not clear, G + C determinations have to be carried out prior to hybridization experiments.

Methods

Extraction of DNA

Cyanobacterial cells are among the toughest cells to break. Chemical lysis procedures are ineffective. Mechanical disruption is necessary. Cells, collected by centrifugation and stored at −70°, are thawed and resuspended in TE buffer. Use 5 ml of TE buffer for 1 g (wet weight) of cells. Add sodium dodecyl sulfate (SDS) to 1% (w/v) final concentration and 5 *M* sodium perchlorate to a final concentration of 1 *M*. Homogenize the cells in a Teflon homogenizer. Pass the cells twice through a cold French pressure cell at 20 kpsi (138 MPa). In case a French pressure cell is not available an equally effective but more laborious method for cell disruption can be employed: Break the frozen cell mass in a −20° mortar and pestle. Gradually add acid-washed sand or glass beads (0.5 mm diameter) and continue grinding for at least 10 min. On thawing, add TE buffer, SDS, and sodium perchlorate as described above.

After cell disruption, transfer the lysate to a centrifuge tube and initiate deprotenization by phenol extraction: Add 1 volume of phenol and mix the two phases by thorough careful shaking for 20 min. Centrifuge the suspension for 10 min at room temperature at 10,000 *g* and collect the aqueous top layer. Reextract the aqueous phase with phenol. Finally extract the aqueous phase with chloroform–isoamyl alcohol (CIA). Repeat CIA extractions until the interphase between the aqueous phase and the organic phase is clear. Precipitate nucleic acids by addition of 2 volumes of −20° 96% ethanol, thorough mixing, and storage in the cold (−20°) for 1.5 hr or longer. Collect the precipitated nucleic acids by centrifugation at 15,000 *g* at 4° for 10 min. Decant, dry the pellet under reduced pressure, and dissolve in a small amount of TE buffer, preferably no more than a few milliliters.

Remove RNA by treatment with RNase (75 μg/ml final concentration) for 1 hr at 37°. Remove RNase and residual proteins by treatment with proteinase K (50 μg/ml) for 0.5 hr at 37°. Finally, extract the solution by phenol and CIA treatment as described above. Add 0.1 volume of 3 *M* sodium acetate, precipitate the solution with ethanol, and dissolve the pellet in TE buffer. Assess the concentration and purity of the DNA solution by UV spectrophotometry at 230, 260, 280, and 320 nm. DNA preparations can be considered chemically pure when the ratios $(A_{260} - A_{320})/(A_{230} - A_{320})$ and $(A_{260} - A_{320}/A_{280} - A_{320})$ both are 1.9 or more. The DNA concentration is $(A_{260} - A_{230}) \times 50$ μg/ml. For these measurements, dilutions have to be made as necessary. In case the DNA appears to be contaminated, repeated proteinase K treatments and extractions must be carried out. Store the DNA at −70°.

DNA of appropriate fragment size can be obtained by mechanical shearing as follows (Note: make sure not to use all the DNA in the shearing procedure, because unsheared DNA has to be used in labeling): Shear 0.5 ml of DNA solution in a 1.5-ml tube on ice in 2 bursts of 25 sec, with an interruption of 1 min, using a sonicator at maximum output setting. This procedure usually results in fragments of approximately 500 base pairs, but initial checks by agarose gel electrophoresis should be carried out. Treat the sheared DNA with Chelex 100 to remove bivalent cations. Transfer a small volume (~40 μl) of TE-equilibrated Chelex 100 into a glass-wool-plugged 1-ml disposable pipet tip and blow out excess TE by using a pipettor equipped with a cut-off pipet tip or a piece of tubing. Load the DNA sample onto the column; collect the eluate by gravity flow and a final blow-through.

Labeling of DNA

Of all methods to label DNA, nick-translation labeling (the enzymatic replacement of native nucleotides by radioactively labeled nucleotides) is undoubtedly the one of choice for hybridization purposes. The method is quick, reproducible, and leads to high specific activities. Customer-friendly "nick kits" are available from several biochemical companies (BRL, USA; BMB, West Germany; Amersham, UK). These kits do not usually contain the labeled nucleotides, and these have to be purchased from companies such as New England Nuclear (Boston, MA) and Amersham. For DNA hybridizations ^{3}H- or ^{32}P-labeled deoxyribonucleotide triphosphates (dNTP's) are preferable.

The researcher should carefully decide whether to use ^{3}H or ^{32}P. The obvious advantage of ^{32}P is its high β-particle energy, allowing Cerenkov counting and Geiger practice, procedures that can hardly be overappreciated in lab practice. The major drawback of ^{32}P is its short half-life of 14.3 days. ^{32}P-Labeled DNA can be used for just a few days. The β-particle energy of ^{3}H is roughly 2 orders of magnitude lower than that of ^{32}P. Counting without scintillation procedures is impossible. Monitoring by Geiger counting is also not feasible. The half-life of ^{3}H (12.35 years), on the other hand, allows more economic use of the labeled nucleotides and enables prolonged storage of labeled DNA.

The actual nick-labeling procedure should be carried out according to the specifications of the manufacturer. It is strongly suggested that the beginner make use of the nick kits mentioned previously. With more experience the researcher might be able to make up a nick kit, which might be a little more economical.

Purification of Labeled DNA

In order to obtain clean preparations of labeled DNA, the labeling mixture must be freed from unincorporated nucleotides. This is easily achieved by Sephadex chromatography. Prepare a 7.5-cm Sephadex G-50 minicolumn in a glass-wool-plugged Pasteur pipet. Wash the column with several volumes of 0.03 *M* NaPB–0.135 *M* NaCl. Load the labeling mixture onto the column and elute with 0.5-ml aliquots of the same buffer, which are collected separately. Monitor the radioactivity of the 0.5-ml eluates by counting 2 μl of each sample in scintillation fluid. Pool the fractions containing the peak radioactivity (usually fractions 2–4) in a 1.5-ml microfuge tube.

In order to remove spuriously synthesized DNA (hairpin DNA, foldback DNA), proceed as follows: Denature the labeled DNA solution in a 100° water bath for 10 min. Prevent leakage of the DNA cup by sealing it with stretch tape. Cool the mixture abruptly by plunging the cup into an ice water bath. Transfer a small amount (final bed volume 0.5 ml) of HAP to a water-jacketed, glass-wool-plugged Pasteur pipet and equilibrate the HAP with 0.03 *M* NaPB–0.135 *M* NaCl at 20°. Load the radioactive, denatured DNA onto the column, while collecting the eluate. Raise the temperature to 50° and wash with 2.5 ml 0.03 *M* NaPB–0.135 *M* NaCl. Collect the eluate in the same cup. Both eluates together represent fraction 1. Raise the temperature to 60° and carry out the following elution procedure:

Wash with 2 × 2 ml 0.03 *M* NaPB	Fractions 2 and 3
Wash with 1 ml 0.03 *M* NaPB + 100 μl 0.12 *M* NaPB	Fraction 4
Wash with 3 × 250 μl 0.12 *M* NaPB	Fractions 5, 6, and 7
Wash with 2 × 1 ml 0.3 *M* NaPB	Fractions 8 and 9

Monitor the radioactivity of the fractions by counting 5 μl of fractions 1, 5, 6, and 7; 10 μl of fraction 2; and 50 μl of fractions 3, 4, 8, and 9. Usually the bulk of radioactivity in single-stranded DNA can be found in fraction 6, and the DNA is ready for use in hybridization experiments. A specific activity of 1×10^7 dpm/μg can be achieved. All other fractions can be disposed of. Store the labeled DNA in small aliquots at −70°.

Hybridization Conditions

Adjust the stock solution of DNA to 1 mg/ml in 0.12 *M* NaPB, by either dilution or concentration after EtOH precipitation. Ten microliters of this driver DNA is used in each hybridization. Add an equal volume of radioactive tracer DNA. The final conditions of a hybridization mixture are as follows: DNA concentration, 500 μg/ml; radioactivity, 1000–10,000 dpm/μl; driver/tracer ratio, 1000 : 1.

Seal 2.5-μl aliquots of each hybridization mixture in 5-μl glass capillaries, using a torch. Denature the DNA mixtures at 100° for 10 min, followed by rapid cooling of the capillaries in ice water. Incubate the hybridization mixtures for 48 hr at the hybridization temperature (*T*). This temperature is dependent on the G + C content of the DNA involved and can be calculated accordingly:

$$T\ (\text{hybridization}) = 0.41 \times \text{GC}(\%) + 53.7 + 13.3\log[\text{Na}^+]$$

Given the concentration of DNA and the hybridization temperature, hybridization will be completed in 48 hr. Terminate the reactions by freezing the capillaries at −20° in 96% ethanol. Keep the capillaries at this temperature until HAP chromatography.

A simple tool for simultaneous handling of several HAP columns for determination of hybridization rates and thermostabilities can be obtained as follows: Construct a translucent polycarbonate plastic box of 240 × 100 × 100 mm. Make an inlet and an outlet in the sidewalls in order to connect the system to a circulating thermostatted heater. Make 12 holes in top and bottom and fit 12 Pasteur pipets in place. Use rubber rings to make the connections leakproof. Position the polycarbonate box over a tray that can hold 12 scintillation vials. In this way eluates can be collected directly from the columns into scintillation vials.

The content of a hybridization capillary is transferred onto a HAP column as follows: Carefully break off both tips of the glass capillary. Prior scratching of the tips with a diamond pen is helpful. Blow out the contents of the capillary in 0.5 ml distilled water and rinse the capillary thoroughly. Each hybridization mixture is loaded onto an 0.5 ml HAP column at 60°. Three heterologous hybridization mixtures and one homologous hybridization mixture (all in triplicate) can be treated simultaneously, using the polycarbonate box described above. The following elution procedure is performed: Collect fraction 1: it includes the hybridization mixture plus 2 washes of 500 μl 0.03 *M* NaPB. Collect fractions 2 and 3: one wash of 750 μl 0.12 *M* NaPB each. Raise the temperature to 65°. Collect fractions 4 and 5: one wash of 750 μl 0.12 *M* NaPB each. Raise the temperature to 70°. Carry out the same procedure and subsequently elute at 75, 80, 85, 90, and 95°. (Note: the elution buffer should have the appropriate temperature at each step.) The radioactivity of all samples is measured by scintillation counting.

Calculation of Hybridization Rate and Thermal Stability

The amount of radioactivity in fractions 2 and 3, mentioned above, reflects the amount of DNA that did not hybridize but remained single-stranded. The total amount of radioactivity in fractions 4–17 reflects the amount of DNA that hybridized and subsequently regained its single-stranded configuration on "melting" during the stepwise raise in temperature. The renaturation rate can be calculated:

$$\text{Renaturation rate (\%)} = \frac{\text{radioactivity in fractions 4–17}}{\text{radioactivity in fractions 2–17}} \times 100$$

The renaturation rate of homologous hybridizations will be in the range of 75–80%. The relative renaturation of a heterologous hybridization can then be calculated:

$$\text{Relative renaturation rate (\%)} = \frac{\text{renaturation rate of heteroduplex}}{\text{renaturation rate of homoduplex}} \times 100\%$$

The thermal stability of a hybrid can be established by cumulative plotting of the radioactivity eluted at each temperature against the elution temperature. A sigmoid curve results. The temperature at which 50% of the final eluted radioactivity is eluted is the thermal elution midpoint ($T_{m(e)}$) for a hybrid. The difference between the $T_{m(e)}$ values of a homoduplex and a heteroduplex yields the $\Delta T_{m(e)}$. Low $\Delta T_{m(e)}$ values indicate little mismatch in the heteroduplex, high $\Delta T_{m(e)}$ values indicate much mismatch.

Taxonomic Interpretation

DNA/DNA hybridization results, showing high relative hybridization rates and low $\Delta T_{m(e)}$ values, indicate that the cyanobacterial strains involved are genotypically closely related. There is little consensus about the exact interpretation of these values, but it is generally accepted that hybridization rates of 60–100% and $\Delta T_{m(e)}$ values of 0–4° indicate conspecificity.[11] Hybridization rates of 20–60% and $\Delta T_{m(e)}$ values of 4–10° indicate congenerity. Rates lower than 20% and $\Delta T_{m(e)}$ values higher than 10° do not permit important taxonomic conclusions. For examples of interpretations of hybridization data, one is referred to Refs. 3–8.

[11] K. H. Schleifer and E. Stackebrandt, *Annu. Rev. Microbiol.* **37,** 143 (1983).

[11] 16 S Ribosomal RNA Cataloging

By ERKO STACKEBRANDT and WOLFGANG LUDWIG

Phylogenetic studies on cyanobacteria and chloroplasts using the 16 S ribosomal RNA (rRNA) cataloging approach as a tool[1–4] have been done by two different methods: the traditional one[5] which is generally not used any more and the advanced technique[6] which is presented here in detail.

Isolation and Purification of 16 S rRNA

rRNA can be isolated either from ribosomal subunits[7] or from crude extracts. Our experience with more than 450 prokaryotic species has

[1] L. Bonen, W. F. Doolittle, and G. E. Fox, *Can. J. Biochem.* **57,** 879 (1978).
[2] E. Seewaldt and E. Stackebrandt, *Nature (London)* **295,** 618 (1982).
[3] E. Stackebrandt, *in* "Endocytobiology" (H. E. A. Schenk and W. Schwemmler, eds.), Vol. 2, p. 921. de Gruyter, Berlin, 1983.
[4] H. Reichenbach, W. Ludwig, and E. Stackebrandt, *Arch. Microbiol.* **145,** 391 (1986).
[5] T. Uchida, L. Bonen, H. W. Schaup, B. J. Lewis, L. Zablen, and C. R. Woese, *J. Mol. Evol.* **3,** 63 (1974).
[6] E. Stackebrandt, W. Ludwig, K. H. Schleifer, and H. J. Gross, *J. Mol. Evol.* **17,** 227 (1981).
[7] P. Traub, S. Mizushima, C. V. Lowry, and M. Nomura, this series, Vol. 20, p. 391.

shown that the following steps will lead in most cases to satisfactory amounts and purity of 16 S rRNA.

Cell pellets (2–4 g wet weight) are resuspended in buffer (40 m*M* Tris, 20 m*M* sodium acetate, 1 m*M* disodium EDTA, pH 7.2) at a ratio of 1–1.5 ml g^{-1} of cells and opened by passage through a precooled French pressure cell at 1.7 GPa in the presence of a few drops of sodium dodecyl sulfate (SDS, 10%). An equal volume of water-saturated, redistilled phenol is added to the lysate, which is then vortexed for 10 sec and centrifuged at 20,000 *g* for 20 min. Alternatively, cells are opened by glass beads in a Braun homogenizer (Braun, Melsungen, FRG) (2–4 g cells in 20 ml of the above-mentioned buffer with 40 g glass beads 2 times for 30 sec). Phenolization of the nucleic acid-containing upper phase is repeated until the proteinaceous interface has disappeared. The nucleic acids are then precipitated by 3 volumes of cold ethanol for at least 2 hr at −20°. Separation of rRNA species is achieved by one-dimensional polyacrylamide slab gel electrophoresis (10 V/cm^{-1}).[8] Electrophoresis is at 70 mA for 8–12 hr. Following electrophoresis the plates are separated and the rRNA visualized by placing the gel on a thin-layer plate containing a fluorescent indicator (254 nm, Merck, Darmstadt, FRG) which has been covered with plastic wrap. Under UV light (254 nm) nucleic acids appear as dark bands on the fluorescent background.

The gel region containing 16 S rRNA is removed, and the nucleic acids are isolated electrophoretically using the Biotrap apparatus (Schleicher & Schuell, Dassel, FRG) as indicated by the manufacturer. The rRNA-containing solution is subjected to two phenol extractions to remove SDS and traces of contaminants. Traces of phenol are removed by 3 volumes of ethanol, stored at −20° for 3 hr, redissolved in 1 ml distilled water, made up to 0.4 *M* by the addition of 100 μl 4 *M* NaCl, precipitated again with 3 volumes of cold ethanol, and stored at −20°. The precipitate is then dried in a Speed Vac concentrator (Savant, Hicksville, NY) and redissolved in 100 μl sterile water. The purity and the yield of 16 S rRNA are determined spectrophotometrically at 260 nm.

RNase T1 Digestion and 5′-Labeling of Oligonucleotides

Forty micrograms of 16 S rRNA is completely digested with RNase T1 (2.4 U, Calbiochem-Behring, La Jolla, CA) in 80 μl 12 m*M* Tris buffer (pH 8.0) at 37° for 3 hr, boiled for 2 min, and stored frozen. For *in vitro* labeling of 5′ termini, phosphatase-free polynucleotide kinase is used

[8] R. De Wachter and W. Fiers, *in* "Gel Electrophoresis of Nucleic Acids" (D. Rickwood and B. D. Hames, eds.), p. 77. IRL Press, Oxford, England, 1982.

(Boehringer Mannheim, FRG). Two micrograms of the digested RNA is used for enzymatic 5′-end group labeling. The reaction mixture consists of the following: 5 μl RNase T1 digest, 1 μl mercaptoethanol (0.3 *M*), 0.5 μl spermidine (0.34 *M*), 0.5 μl $MgCl_2$ (0.4 *M*), 1 μl unlabeled ATP (1 m*M*), 1 μl Tris–HCl buffer (1 *M*, pH 8.0), 10 μl of [γ-^{32}P]ATP (37,000 MBq, 37–111 TBq mM^{-1}; New England Nuclear, Boston, MA), and 1.2 μl of polynucleotide kinase (Boehringer). The mixture is incubated at 37° for 30 min, after which the reaction is stopped by boiling for 2 min and the mixture dried and redissolved in 2.4 μl of water.

Sequence Determination

The original methods of sequence determination[9–11] have been improved to facilitate sequencing of the more than 70 oligonucleotides, varying in length from 6 to 20 nucleotides, making up a 16 S rRNA catalog. These changes include the replacement of the "Homomixes"[12] by ammonium formate gradient chromatography, the use of alkaline hydrolysis rather than enzymatic cleavage for the generation of subfragments, and the introduction of additional steps to improve analysis of those oligonucleotides whose sequence could not be determined unambiguously after the secondary analysis.

Step 1: Primary Analysis

Separation of 5′-labeled oligonucleotides is in two dimensions using a combination of high-voltage electrophoresis (HVE) in the first and thin-layer chromatography (TLC) in the second dimension. For separation in the first dimension, a cellulose acetate strip (3 × 100 cm, Schleicher & Schuell) is presoaked in electrophoresis buffer (5% acetic acid, 0.05% pyridine, 5 m*M* EDTA, pH 3.5), and an area 7 cm from one end of the strip is blotted on both sides with absorbent paper to remove surplus buffer. This region of the strip is supported on two glass rods that are 2 cm apart. Using a fine glass capillary, 1 μl of the labeled rRNA fragments is then applied in a line about 1 cm long across the width of the strip where surplus buffer had been removed. On either side of the sample a marker mix containing 1% xylene cyanol (XC) and 1% orange G is applied. After the sample has been absorbed, excess buffer is removed from the rest of

[9] M. Silberklang, A. M. Gillum, and U. L. RajBhandary, *Nucleic Acids Res.* **4,** 4091 (1977).
[10] M. Silberklang, A. M. Gillum, and U. L. RajBhandary, this series, Vol. 59, p. 58.
[11] A. Diamond and B. Dudock, this series, Vol. 100, p. 431.
[12] G. G. Brownlee and S. J. Sanger, *Eur. J. Biochem.* **11,** 395 (1969).

the strip, which is then placed in a HVE tank (Savant) containing the same buffer. Varsol or a comparable solvent is used as the hydrophobic phase and cooling fluid. Electrophoresis is at 5000 V, and the XC marker is allowed to migrate 28 cm.

Labeled fragments are then transferred from the strip to the long side of a 20 × 40 cm DEAE–cellulose thin-layer glass plate (Macherey & Nagel, Düren, FRG, Cel DEAE HR-Mix-20, 0.2 mm layer). The strip is laid lengthwise on a glass rod (0.4 cm), and strips of water-soaked Whatman 3MM filter paper (5 × 45 cm) are put on either side to overlap the strip by 3 mm in the region between the blue and yellow markers. The thin-layer plate is placed onto the strip so that the line of contact between the strip and plate runs 2 cm from the long edge of the plate. The glass plate is held in place by lead weights. Oligonucleotides are transferred within a period of about 20 min. After transfer the plate is rinsed with 60 ml of ethanol to remove the HVE buffer salts.

Oligonucleotides are separated in the second dimension by ammonium formate TLC at 70°. Following prechromatography in water for about 30 min the plate is transferred to a solution containing 0.3 *M* ammonium formate, 7 *M* urea, and 1 m*M* disodium EDTA. An equal volume of a solution of 0.5 *M* ammonium formate, 7 *M* urea, and 1 m*M* disodium EDTA is added dropwise over a period of about 4–5 hr. The plate is supported on notched plastic rods, and the solution in the tank is stirred. Small oligonucleotides, running with the front, are absorbed by a filter wick made up of three sheets of 20 × 40 cm Whatman 3MM paper, fastened on top of the plate by metal binder clips. Chromatography is terminated when the orange G marker has migrated 18 cm. After removing the wick, the plate is dried, marked with radioactive ink, and the positions of the labeled oligonucleotides are located by autoradiography for about 30 min. Characteristics of the primary fingerprint, separation behavior of the oligonucleotides, and the regularity of the separation pattern have been intensively dealt with before.[6,10,11,13]

Step 2: Secondary Analysis

Each spot detected after autoradiography is scraped from the plate, and drawn by reduced pressure into the open end of a drawn-out cotton-plugged glass tube (0.4 × 3–4 cm). Nucleic acids are then eluted by 3 washes with 100 μl 1 *M* NaCl. These washings are collected by centrifugation through the glass tubes into Eppendorf test tubes with a swinging-

[13] G. G. Brownlee, *in* "Laboratory Techniques" (T. S. Work and E. Work, eds.). Elsevier/North-Holland, Amsterdam, 1972.

bucket rotor (Beckman TJ 6, 750 *g* for 2 min). Oligonucleotides are then precipitated by 3 volumes of ethanol at −20° for at least 2 hr, and the precipitates are collected by centrifugation (1000 *g* for 20 min) and dried. Controlled alkaline hydrolysis of oligonucleotides is performed in 10 μl 50 m*M* $NaHCO_3$, adjusted to pH 9.2 with 50 m*M* NaOH, at 100° for 18 min, followed by chilling in ice. Each sample is treated with 1.2 μl of 2.5 *N* HCl to remove cyclic phosphate groups, dried *in vacuo*, and redissolved in 2.4 μl of water.

A 1-μl aliquot of each sample is applied to a cellulose acetate strip for HVE separation in the first dimension. Generally, the same buffers and procedures as described for the generation of the fingerprint are involved. However, three samples are applied to one 55-cm strip, 6, 16, and 30 cm from the end. The XC marker is allowed to migrate 6 cm. Alternatively, by using 3 × 100 cm strips, 6 samples applied 6, 18, 30, 50, 62, and 74 cm from the end can be separated simultaneously. In this case the strips are immersed in a shallow tray containing the hydrophobic cooling fluid. For application the respective region of the strip is brought out of the fluid, which is removed by a stream of air, and the strip is submersed after the sample has been dried. Depending on which method is used, the three or six samples separated on a strip are transferred to one or two adjacent DEAE-cellulose plates (Macherey & Nagel, 20 × 40 cm, Polygram, Cel 300 DEAE/HR-2/15, 0.1 mm, fixed to glass plates). Transfer of the labeled material is as described for the fingerprint.

The second dimension is developed in an ammonium formate gradient. Up to 25 plates with a wick are placed in a suitable tank. The plates are supported on notched rods to allow circulation of the buffer. Excess water not used during the prechromatography is removed after 30 min. The water is immediately replaced by 1 liter of buffer (0.28 *M* ammonium formate in 7 *M* urea, 1 m*M* disodium EDTA, 1 *M* boric acid, pH 4.3, adjusted with formic acid), and an ammonium formate solution (2 → 0.5 *M* ammonium formate, as indicated above, 85 ml each) is added dropwise over a period of 3–4 hr at 70°, up to a final concentration of 0.44 *M* ammonium formate. By the time the XC marker has reached the upper edge of the plate the wicks are removed; the foils are dried, marked with radioactive ink, and the position of the 5′-labeled fragments located by autoradiography for 8–14 hr. A detailed description of the two-dimensional mobility-shift analysis including characteristics of the pattern and reading of the sequences has been presented previously.[10,14]

[14] E. Stackebrandt, W. Ludwig, and G. E. Fox, *in* "Methods in Microbiology" (G. Gottschalk, ed.), Vol. 18, p. 75. Academic Press, London, 1985.

Step 3: Tertiary Analysis

There are cases in which the sequence cannot be determined unambiguously from secondary analysis alone. Here it is necessary to do a tertiary analysis by one or a combination of several procedures.

Analysis of Long Fragments. In oligonucleotides with more than 10 bases subfragments located toward the 3′ terminus are usually insufficiently resolved. In such cases an aliquot of the oligonucleotide is separated as described in Step 2, but the ammonium formate gradient is replaced by "Homomix"[9] (50 : 75 : 100 m*M* KOH = 1 : 2 : 1, containing 1 *M* boric acid). Since fragments up to heptanucleotides are allowed to migrate into the wick (the orange G marker dye has just left the plate), the longer fragments are well resolved.

One-Dimensional Thin-Layer Chromatography. Difficulties are sometimes encountered when discriminating between adenine and cytidine in the middle or at the end of uridine-rich stretches. Such sequences can be verified by one-dimensional thin-layer chromatography of the alkaline hydrolyzate.[15] The hydrolyzate is redissolved in 100 μl distilled water, the rRNA precipitated with 3 volumes of ethanol to remove salts, and stored at $-20°$ for 2 hr. After centrifugation the precipitate is dried and redissolved in 2.4 μl of water; the sample is then applied to a DEAE thin-layer plate (20 $\times$ 30 cm, 0.2 mm) which is developed in "Homomix" as indicated above at 70° until the blue marker dye has almost reached the upper edge of the plate. Autoradiography is for 1–3 days.

Determination of 5′-Terminal Nucleotides. Determination of 5′-terminal nucleotides is necessary whenever the identity of the 5′ terminus is not obvious from the secondary analysis or when the determination of the 5′ group helps to follow the migration pattern of individual oligonucleotides in those cases where two or more fragments beginning with [^{32}P]Cp or [^{32}P]Ap comigrate in the fingerprint. The material of the 5′ terminus or of an appropriate intermediate is scraped from the thin-layer foil, and the rRNA is isolated as described above. The dried material is completely digested by 4 μl of nuclease P_1 (Boehringer, 25 ng ml^{-1} in 50 m*M* ammonium acetate, pH 4.5) at 37° for 12 hr. The labeled termini are identified by one-dimensional thin-layer chromatography on cellulose plates (Merck, 20 $\times$ 20 cm) in *tert*butanol : H_2O : concentrated HCl (14 : 3 : 3). Autoradiography is for 1 to several days.

Two-Dimensional Separation of Intermediates. Coinciding oligonucleotides or those from overlapping spots in the fingerprint may share

[15] E. Jay, R. Bambara, R. Padmanabhan, and R. Wu, *Nucleic Acids Res.* **1,** 331 (1974).

common intermediates in the secondary analysis. In order to determine the exact sequences, the fragments in question, isolated from the foil as described above, are subjected to a second controlled alkaline hydrolysis and are separated again in two dimensions using the same conditions as described for the secondary analysis. Autoradiography of these fragments is for 1–3 weeks.

Data Analysis

Each catalog is part of an ever-expanding data base, containing at present about 34,000 individual oligonucleotides from about 450 prokaryotic strains. Programs for electronic data processing have been developed.[16] Similarity coefficients (S_{AB} values) are used to compare individual catalogs, and dendrograms of relationships are generated by applying the average linkage clustering algorithm to a matrix of S_{AB} values.[17,18]

[16] J. Sobieski, K. N. Chen, J. Filiatreau, M. Pickett, and G. E. Fox, *Nucleic Acids Res.* **12,** 141 (1984).

[17] G. E. Fox, K. R. Pechman, and C. R. Woese, *Int. J. Syst. Bacteriol.* **27,** 44 (1977).

[18] M. R. Anderberg, "Cluster Analysis for Applications." Academic Press, New York, 1973.

[12] Reverse Transcriptase Sequencing of Ribosomal RNA for Phylogenetic Analysis

By DAVID J. LANE, KATHARINE G. FIELD, GARY J. OLSEN, and NORMAN R. PACE

Introduction

The cyanobacteria comprise a large group of structurally complex and ecologically important gram-negative prokaryotes whose taxonomic treatment has generated much interest and controversy over the years. By molecular taxonomic criteria (particularly 16 S rRNA oligonucleotide catalog analysis[1]), the cyanobacteria, together with chloroplasts and prochlorophytes, are a coherent, eubacterial grouping that warrants phylum status. Divisions within the phylum presently are based on morphologi-

[1] C. R. Woese, E. Stackebrandt, T. J. Macke, and G. E. Fox, *Syst. Appl. Microbiol.* **6,** 143 (1985).

cal, developmental, and physiological criteria[2] which, although pragmatically indispensible, are of uncertain evolutionary relevance. The routine application of 16 S rRNA sequences to the assessment of such lower level taxonomic assignments has been hindered by the technical difficulty of obtaining the required sequence data. We have recently described a method for more rapid determination of phylogenetically useful blocks of 16 S rRNA sequences.[3] The method has been applied to over 100 eubacterial and eukaryotic species, including about 30 cyanobacterial isolates.[4] This data base of 16 S rRNA sequences will provide a framework for evaluating the evolutionary affiliations of additional cyanobacterial isolates, and so should be of both taxonomic and determinative value to workers in the field.

Principle of the Method

The sequencing protocol is a variation of the base-specific, dideoxynucleotide-terminated, chain-elongation method.[5,6] It has been modified for the use of reverse transcriptase and ribosomal RNA templates. The method relies on the fact that certain regions of 16 S rRNA nucleotide sequence vary little or not at all between different organisms. A short oligodeoxynucleotide, complementary to such a universally conserved site, specifically anneals to its target site on the 16 S rRNA, even in the presence of other cellular RNAs. The oligonucleotide serves as a primer for the synthesis of a copy DNA strand that extends from the 3′ end of the primer. The inclusion of low levels of chain-terminating, dideoxynucleotide analogs in reverse transcription reactions results in the production of random-length copy DNAs which are specifically terminated at adenosine, guanosine, cytidine, or thymidine residues, depending on the specific nucleotide analog used. The products of chain elongation are radioactively labeled by the incorporation of [α-^{35}S]dATP during the reverse transcription. When a set of base-specifically terminated reactions is electrophoresed on a high-resolution, polyacrylamide sequencing gel, a

[2] R. Rippka, J. B. Waterbury, and R. Y. Stanier, *in* "The Prokaryotes: A Handbook on Habitats, Isolation, and Identification of Bacteria" (M. P. Starr, H. Stolp, H. G. Trüper, A. Balows, and H. G. Schlegel, eds.), p. 247. Springer-Verlag, New York, 1981.

[3] D. J. Lane, B. Pace, G. J. Olsen, D. A. Stahl, M. L. Sogin, and N. R. Pace, *Proc. Natl. Acad. Sci. U.S.A.* **82,** 6955 (1985).

[4] S. J. Giovannoni, S. Turner, G. J. Olsen, D. J. Lane, N. R. Pace, and J. B. Waterbury, *Proc. Annu. Meet. Am. Soc. Mircobiol., 86th* p. 237 (1986) (Abstr.); S. J. Giovannoni and S. Turner, personal communication (1986).

[5] F. Sanger, S. Nicklen, and A. R. Coulson, *Proc. Natl. Acad. Sci. U.S.A.* **74,** 5463 (1977).

[6] M. D. Biggin, T. J. Gibson, and G. F. Hong, *Proc. Natl. Acad. Sci. U.S.A.* **80,** 3963 (1983).

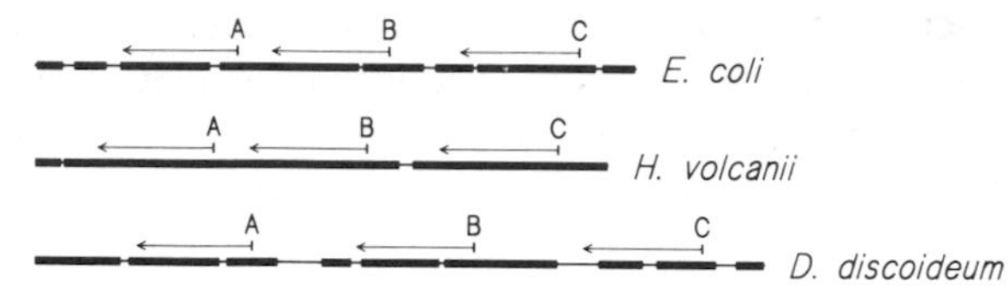

FIG. 1. Priming sites of "universal" 16 S rRNA primers. Linear representations of the small subunit rRNAs from *Escherichia coli, Halobacterium volcanii,* and *Dictyostelium discoideum* are shown. The universal primer sequences and their hybridization locations in the *E. coli* 16 S rRNA are GWATTACCGCGGCKGCTG, positions 519–536 (A); CCGT-CAATTCMTTTRAGTTT, 907–926 (B); and ACGGGCGGTGTGTRC, 1392–1406 (C). In these sequences K = G or T, M = A or C, R = A or G, and W = A or T. The solid boxes along the sequence lines are regions that display sufficient intrakingdom structural conservation to be generally useful in the inference of phylogenies. The arrow pointing to the left of each priming site indicates the approximate extent of sequence data (300 nucleotides) accessible from that primer. From Lane *et al.*[3]

pattern of bands is obtained on an autoradiogram of the gel, from which the nucleotide sequence of the copy DNA can be directly read. Three particularly useful priming sites, which provide access to the three major 16 S rRNA structural domains, collectively yield 800–1200 nucleotides of sequence from each 16 S rRNA (Fig. 1).

Materials

Reagents

STE buffer: 100 m*M* NaCl, 50 m*M* Tris–HCl, pH 7.4, 1.0 m*M* sodium ethylenediaminetetraacetic acid (NaEDTA), pH 7.4.

TE buffer: 10 m*M* Tris–HCl, pH 7.4, 1.0 m*M* NaEDTA, pH 7.4.

Phenol/IAA: Redistilled phenol–8-hydroxyquinoline–isoamyl alcohol (24 : 0.024 : 1, w/w/v) is prepared by equilibrating the organic mixture with 3 changes of STE buffer.

Phenol/chloroform/IAA: Redistilled phenol–8-hydroxyquinoline–chloroform–isoamyl alcohol (24 : 0.024 : 6 : 1, w/w/w/v) is equilibrated with STE buffer, as above.

Chloroform/IAA; Chloroform–isoamyl alcohol (24 : 1, w/v).

5× Hybridization buffer. 100 m*M* KCl, 250 m*M* Tris–HCl, pH 8.5.

5× Reverse transcription buffer: 250 m*M* Tris–HCl, pH 8.5, 50 m*M* dithiothreitol (DTT), 50 m*M* $MgCl_2$.

Oligonucleotide primers: The nucleotide sequences of the rRNA primers are given in the legend to Fig. 1; the primers are available through Boehringer Mannheim Biochemicals.

Nucleotides: 2′-Deoxyadenosine 5′-*O*-(α-thiotriphosphate) (dAT-

PαS), labeled with ^{35}S on the α-phosphate ([α-^{35}S]dATP) is purchased from New England Nuclear. The 2′-deoxynucleoside triphosphates (dNTPs), 2′,3′-dideoxynucleoside triphosphates (ddNTPs), and unlabeled dATPαS are purchased from P-L Biochemicals. Concentrated stock solutions of each are prepared, diluted to 10 m*M* (determined spectrophotometrically according to spectral information supplied by the manufacturer) in 10 m*M* Tris–HCl, pH 7.4, and stored at −70°. The 10 m*M* dATPαS stock also contains 1 m*M* DTT. Sequencing nucleotide mixtures for the reverse transcription reactions contain 10 m*M* Tris–HCl, pH 8.3, 1.0 m*M* DTT, 250 μ*M* dCTP, 250 μ*M* dGTP, 250 μ*M* dTTP, 125 μ*M* dATPαS, and either one dideoxynucleotide (30 μ*M* ddCTP, 19 μ*M* ddGTP, 30 μ*M* ddTTP, or 1.25 μ*M* ddATP) or no dideoxynucleotide (no ddNTP). The chase nucleotide mixture contains 10 m*M* Tris–HCl, pH 8.3, 5 m*M* $MgCl_2$, and 1.0 m*M* each of dGTP, dCTP, dTTP, and dATP, to which 1000 units/ml of reverse transcriptase is added just prior to use.

Reverse transcriptase: Avian myoblastosis virus enzyme (10,000–23,000 units/ml) is purchased from Seikagaku America, Inc. The enzyme and dilutions of it [1,000 units/ml in 50 m*M* Tris–HCl, pH 8.3, 2 m*M* DTT, 50% glycerol (by weight)] are stored at −20°.

Stop mix: 86% formamide, 10 m*M* NaEDTA, pH 7.4, 0.08% xylene cyanol, 0.08% bromphenol blue.

10× NNB buffer: 1.34 *M* Tris base, 0.45 *M* H_3BO_3, 25 m*M* NaEDTA.

Polyacrylamide gels: Two solutions are prepared. Top mix contains 8% polyacrylamide, 0.04% bisacrylamide, 8 *M* urea, and 0.5× NNB. Bottom mix contains 8% polyacrylamide, 0.04% bisacrylamide, 8 *M* urea, 2.5× NNB, and 10% sucrose. Both solutions are filtered (Whatman GF/F) and stored at 4°. Buffer gradient gels (0.5× NNB to 2.5× NNB) are prepared from the two solutions according to Biggin *et al.*[6] Nongradient gels are prepared from top mix. Gels are run in 0.5× NNB.

Gel fix: 10% methanol, 10% acetic acid, 2% glycerol (v/v/w) in water.

Procedures

Purification of RNA Templates

The RNA purification method is designed for use on pure cultures containing only one type of 16 S rRNA. Mixed cultures, however, can often be sufficiently purified by differential centrifugation or filtration, at the time of harvesting, for the method to work. Many different RNA purification methods have provided satisfactory RNA template preparations; the following protocol is relatively simple and has proved satisfactory for most cell types.

Start with a 0.05–0.5 g (wet weight) cell pellet from a pure (or purified) culture. Resuspend the cells in 4.5 ml of ice-cold STE buffer. Pass the suspension through a French pressure cell at 20,000 psi, collecting the lysate in a chilled, nuclease-free (baked at 200° overnight), 15-ml Corex tube. Add sodium dodecyl sulfate to 1.0% and NaEDTA to 20 m*M*. Add 5 ml of phenol/IAA, vortex vigorously for 2–3 min, chill on ice, centrifuge 5 min (10 krpm, 4°, Sorvall HB4 swinging-bucket rotor), transfer the aqueous (upper) phase to a clean 15-ml Corex tube and repeat the extraction 3 times more, substituting phenol/chloroform/IAA for phenol/IAA. Extract once with 5 ml of chloroform/IAA. Precipitate the nucleic acid from the final aqueous phase by adding one-tenth volume of 2.0 *M* sodium acetate and then 2.0 volumes of ethanol. Mix. Chill to −20° for 4 or more hours. Collect the precipitated nucleic acid by centrifugation (20 min, 12 krpm, 4°, HB4 rotor). Remove the last traces of phenol by suspending the pellet in TE buffer and reprecipitating the nucleic acid as above.

Resuspend the final pellet in a small amount of TE buffer, and determine the concentration of the nucleic acid by measuring the absorbance at 260 nm of an appropriately diluted aliquot. Assuming that a 50 μg/ml RNA solution has an A_{260} of 1, adjust the concentration of the nucleic acid solution to approximately 5 mg/ml (either by diluting with TE buffer or resuspending it in a smaller volume after reprecipitating with ethanol). Add an equal volume of 4.0 *M* NaCl, mix, and allow the high-molecular-weight RNA to precipitate overnight, on ice (this is most conveniently done in a 1.5-ml microcentrifuge tube). Collect the precipitate by centrifugation for 5–10 min at 4° in a microcentrifuge. Resuspend the pellet of high-molecular-weight RNA in a convenient volume of TE buffer. Determine the concentration of RNA spectrophotometrically, as above. Remove residual NaCl from the RNA preparation by reprecipitating once more with sodium acetate and ethanol, as above. Collect the high-molecular-weight RNA by centrifugation. Dry the pellet lightly under reduced pressure. Resuspend the RNA at 2.0 mg/ml in 10 m*M* Tris–HCl, pH 8.0, and store it at −20°. For long-term storage, the RNA is best stored as an ethanol slurry at −70°.

Reverse Transcription Reactions

For a given RNA template, a set of five sequencing reactions, four base specific (containing the four dNTPs plus one of the chain terminators, ddCTP, ddATP, ddTTP, or ddGTP) and one control reaction which contains only the four dNTPs (no ddNTP), is performed using each of the three universal primers shown in Fig. 1. The three sets of reactions are conveniently performed concurrently. Since the sequencing reactions are only 5.0 μl, common components are premixed and then aliquoted into the reaction tubes. Untreated, 0.5-ml microfuge tubes are used throughout.

For each primer, enough annealed DNA primer/RNA template "hybrid mix" is prepared for the full set of five sequencing reactions. Combine 4.0 μl of template RNA (2.0 mg/ml), 1.4 μl of 5× hybridization buffer, and 1.5 μl of primer (either A, B, or C, Fig. 1, at 0.02 mg/ml in H_2O) in a 0.5-ml microfuge tube. Mix. Heat the mixture at 65° for 2 min, slow cool to room temperature over about 10 min, and then transfer 6.0 μl of the annealed DNA primer/RNA template hybrid mix to a "premix" tube containing 30 μCi of dried [α-^{35}S]dATP and 6.0 μl of 5× reverse transcription buffer. Add 6.0 μl of reverse transcriptase (1 U/μl). Mix. Transfer 3.0-μl aliquots of the premix to each of five reaction tubes containing 2.0 μl of, respectively, ddCTP, ddATP, ddTTP, ddGTP, or no ddNTP sequencing nucleotide mixture (see Reagents).

Incubate these 5-μl reaction mixtures for 5 min at room temperature, and then for 30 min at 55°. Add 1.1 μl of chase nucleotide mixture to each tube and resume the incubation at 55° for another 15 min. Terminate the reactions by adding 6 μl of stop mix. If necessary, reactions are stored at −20° before electrophoresis. Immediately prior to electrophoresis, heat the reaction mixtures at 65° for 2 min and then load 1–2 μl of each on a sequencing gel in adjacent "wells" of a sharks tooth comb (BRL). Two 0.3-mm-thick, 40-cm-long, 18-cm-wide sequencing gels are run for each set of sequencing reactions (Reagents). The buffer gradient gels are electrophoresed at 35 W (constant power) until the bromphenol blue reaches the bottom of the gel (~2.5 hr). The nongradient gels are run at 30 W, for 1.75 times the time required for the xylene cyanol dye to reach the bottom of the gel (~7 hr). Following electrophoresis, the gels are soaked in gel fix for 1 or more hours (up to overnight), transferred to Whatman 3MM paper, dried (on a slab gel dryer), and exposed to X-ray film for 2–3 days, or as necessary.

Data Analysis

A variety of options exist for the handling and analysis of 16 S rRNA nucleotide sequence data. Many of these have been discussed at length elsewhere[7,8] and can only be mentioned here. The proper choice will

[7] G. J. Olsen, D. J. Lane, S. J. Giovannoni, and N. R. Pace, *Annu. Rev. Microbiol.* **40**, 337 (1986).

[8] G. J. Olsen, this series, Vol. 164, p. 793.

depend, of course, on the immediate experimental objective and the facilities available. Ultimately, the sequence should be made available to a laboratory where an appropriate computing facility and software exist, so that it can be incorporated into future phylogenetic analyses. As a practical matter, many investigators will find this to be the easiest and most effective first approach. The required manipulations of the sequence data are easily performed using software designed for this purpose, and, in our experience, laboratories specializing in molecular phylogeny usually are willing to provide the desired analysis in exchange for access to a new sequence. However, much information may be obtained without recourse to an elaborate computing facility.

In brief, the first task in interpreting the sequence data is the alignment of the newly derived sequence with other available 16 S rRNA sequences. Conserved features of 16 S rRNA primary and secondary structure are employed as landmarks, so that evolutionarily homologous nucleotide positions in the compared sequences are brought into one-to-one correspondence.[7,8] Following alignment, the fraction of identical nucleotides relating the new sequence to each of the others in the aligned set is determined. Three outcomes are possible. (1) The sequence is very closely related (say, 95% or greater similarity) to another sequence in the data collection. This indicates a specific affiliation between the two organisms which contributed the sequences. (2) The sequence exhibits a general affiliation (~85–95% similarity) with those derived from a previously characterized group of organisms; in this case the new organism can be identified as a member of that grouping. (3) The sequence exhibits little or no specific similarity (below ~85%) to any previously characterized sequence. This suggests that a novel organism has been characterized and that more sophisticated phylogenetic analysis is merited.

Remarks

A number of minor changes have been introduced into the rapid sequencing protocol described here. Most notably, the specificity of the primer/template annealing has been increased by (1) lowering the salt concentration in both the hybridization mix and the reverse transcription reactions, (2) lowering the primer/template molar ratio to approximately one to one, and (3) raising the reverse transcription reaction temperature to 55°. Also, the temperatures of the heating steps prior to the annealing reaction and to loading of the gel have been reduced to 65° as the less harsh conditions have proved equally effective for both steps.

[13] Protein Sequences as Taxonomic Probes of Cyanobacteria

By ALASTAIR AITKEN

Introduction

The purpose of this chapter is to outline studies carried out on primary structures of cyanobacterial proteins and methods used to compare them with the corresponding proteins in eukaryotic and eubacterial cells. This has shed some light on their evolutionary relationships and on the possible origin of chloroplasts.

Mereschowsky[1] originally proposed in 1905 that the chloroplast of eukaryotic cells originated by endosymbiosis of a cyanobacterium. Modern cytochemical and molecular methods have enabled a critical evaluation of this "endosymbiotic hypothesis." Some authors have suggested that the mitochondrion as well as the chloroplast arose from endosymbiotic associations of an aerobic bacterium and a cyanobacterium, respectively.[2] Closer association of these endosymbionts with the host cell led eventually to their development into the semiautonomous organelles as they are known today. Evidence for endosymbiotic association leading to the development of the chloroplast may be summarized briefly. Extant symbiotic associations involving cyanobacteria are common, and some may represent successive stages in the development of the chloroplast. For example, in the fungus *Geosiphon pyriforme* the symbiotic cyanobacterium is a *Nostoc* species which can be cultured separately from the host.[3] In contrast, neither the symbiont in *Cyanophora paradoxa*[4] nor that in *Glaucocystis nostochinearum*[5] can be cultured apart from the host cell. The symbiont in *C. paradoxa* has a chloroplast-type DNA[6] but has a lysozyme-sensitive bacterial type cell wall with an amino acid composition resembling a normal, cyanobacterial gram-negative peptidoglycan.[7,8]

[1] C. Mereschkowsky, *Biol. Zentralbl.* **25,** 593 (1905).
[2] J. M. Whatley, *Ann. N.Y. Acad. Sci.* **361,** 154(1981).
[3] G E. Fogg, W. D. P. Stewart, P. Fay, and A. E. Walsby, *in* "The Blue–Green Algae" (N. G. Carr and B. A. Whitton, eds.), p. 459. Blackwell, Oxford, England, 1973.
[4] W. T. Hall and G. Claus, *J. Cell Biol.* **19,** 551 (1963).
[5] W. T. Hall and G. Claus, *J. Phycol.* **3,** 37 (1967).
[6] M. Herdman and R. Y. Stanier, *FEMS Microbiol. Lett.* **1,** 7 (1977).
[7] J. M. Jaynes and L. P. Vernon, *Trends Biochem. Sci.* **7,** 22 (1982).
[8] A. Aitken and R. Y. Stanier, *J. Gen. Microbiol.* **112,** 219 (1979).

Chloroplasts themselves are semiautonomous organelles with their own prokaryotic type ribosomes and DNA.[9-11] They have 70 S type ribosomes with the appropriate prokaryotic antibiotic specificities and N-formylated methionine tRNA as the initiator of protein synthesis.

A polyphyletic origin of chloroplasts has been proposed[12,13] on the basis of the similarity in type of chlorophyll and presence of phycobiliproteins in cyanobacteria, Rhodophyta, and Cryptophyta. It was suggested that other types of oxygenic photosynthetic prokaryotes had given rise to other groups of eukaryotic algae and plants. There is certainly close similarity between cyanobacteria and red algae (or the chromatophores of the latter[14,15]). The alternative to the endosymbiotic hypothesis proposes that eukaryotic cells arose from a prokaryote by intracellular differentiation.[16,17] A certain degree of autonomy of mitochondria and chloroplasts has been suggested as essential for synthesis of proteins that cannot pass the membranes of the organelles. It has been shown, however, that plastocyanin, ferredoxin, and other proteins are encoded in nuclear DNA and synthesized on cytoplasmic ribosomes.[18] Plastocyanin is synthesized as a precursor protein of M_r 25,000, a size equivalent to the largest known precursor peptide for the chloroplast or mitochondrion.[19]

Cyanobacteria are prokaryotic organisms,[20] and like other prokaryotes they lack internal compartmentalization of their cells including the membrane-bound nucleus characteristic of eukaryotic cells. They do not undergo mitotic division. Photosynthesis and respiration is carried out on membrane systems in prokaryotes not in separate organelles. These organisms contain a peptidoglycan component characteristic of a typical bacterial cell wall.[21] In terms of DNA and ribosome structure cyanobacteria are also prokaryotic organisms.[22] In spite of their prokaryotic nature,

[9] A. Allsopp, *New Phytol.* **68,** 591 (1969).

[10] D. Boulter, R. J. Ellis, and A. Yarwood, *Biol. Rev.* **47,** 113 (1972).

[11] R. J. Ellis and M. R. Hartley, *in* "Biochemistry of Nucleic Acids," (K. Burton, ed.) MTP Int. Rev. Sci., Biochem. Ser. 1, Vol. 6. Butterworths, London, 1974.

[12] P. H. Raven, *Science* **169,** 641 (1970).

[13] L. Sagan, *J. Theor. Biol.* **14,** 225 (1967).

[14] R. Y. Stanier, *Symp. Soc. Gen. Microbiol.* **20,** 1 (1970).

[15] C. P. Wolk, *Bacteriol. Rev.* **37,** 32 (1973).

[16] R. A. Raff and H. R. Mahler, *Science* **177,** 575 (1972).

[17] T. Uzzell and C. Spolsky, *Ann. N.Y. Acad. Sci.* **361,** 481 (1981).

[18] S. Smeekins, M. DeGroot, J. van Binsbergen, and P. Weisbeek, *Nature (London)* **317,** 456 (1985).

[19] J. Bennet, *Trends Biochem. Sci.* **7,** 269p (1982).

[20] R. Y. Stanier and C. B. van Niel, *Arch. Mikrobiol.* **42,** 17 (1962).

[21] P. Echlin and I. Morris, *Biol. Rev.* **40,** 143 (1965).

[22] C. K. Leach and M. Herdman, "The Biology of Blue–Green Algae" (N. G. Carr and B. A. Whitton, eds.), p. 186. Blackwell, Oxford, England, 1973.

however, cyanobacteria have the chloroplast type of photosynthesis which uses water as an electron donor,[23] and the cyanobacteria photosynthetic apparatus bears a much stronger resemblance to that of the chloroplast.[24] Proteins common to the cyanobacterial and eukaryotic photosystems have been shown to have great similarity in primary structure. These include soluble type-*c* cytochrome *c*-552–555 or cytochrome c_6,[25–28] ferredoxin (Refs. 29 and 30 and this volume [41]–[43]), plastocyanin,[31,32] and phycobiliproteins (Ref. 33 and this volume [31]). This chapter concentrates mainly on sequence comparisons of plastocyanin and cytochrome c_6 since the other proteins are dealt with elsewhere in this volume.

Plastocyanin and Cytochrome c_6 (*c*-552–555): Their Function and Distribution

Plastocyanin is a copper-containing protein of monomer molecular weight around 11,000. The copper (1 atom/molecule protein) is in a type I ligand environment.[34] The protein has an intense blue coloration with λ_{max} 597 nm when oxidized. The molar extinction coefficient is around 4,500. Other copper proteins of this type are azurins (bacterial respiratory proteins which do not occur in photosynthetic bacteria), stellacyanin and umecyanin (which are present in nonphotosynthetic tissue in higher plants), rusticycanin, and plantacyanin. A number of other copper oxidases including laccase and ceruloplasmin contain type I copper as well as type II and type III Cu^{2+} sites.[34]

[23] W. A. Cramer, W. R. Widger, R. G. Herrmann, and A. Trebst, *Trends Biochem. Sci.* **10,** 125 (1985).

[24] J. M. Olson, *Ann. N.Y. Acad. Sci.* **361,** 8 (1981).

[25] A. Aitken, *Nature (London)* **263,** 793 (1976).

[26] A. Aitken, *Eur. J. Biochem.* **78,** 273 (1977).

[27] R. P. Ambler and R. G. Bartsch, *Nature (London)* **253,** 285 (1975).

[28] A. Aitken, *Eur. J. Biochem.* **101,** 297 (1979).

[29] M. Fukuyama, T. Hase, S. Matsumoto, T. Tsukihara, Y. Katsube, N. Tanaka, M. Kakudo, K. Wada, and H. Matsubara, *Nature (London)* **286,** 522 (1980).

[30] H. Matsubara, T. Hase, S. Wakabayashi, and K. Wada, *in* "Evolution of Protein Structure and Function" (D. S. Sigman and M. A. B. Brazier, eds.), p. 245. Academic Press, New York, 1980.

[31] A. Aitken, *Biochem. J.* **149,** 675 (1975).

[32] A. Aitken, *in* "Evolution of Protein Molecules" (H. Matsubara and T. Yamanaka, eds.), p. 251. Jpn. Sci. Soc. Press, Tokyo, 1977.

[33] A. N. Glazer, *in* "Evolution of Protein Structure and Function" (D. S. Sigman and M. A. B. Brazier, eds.), p. 221. Academic Press, New York, 1980.

[34] B. G. Malmstrom, *in* "Evolution of Protein Structure and Function" (D. S. Sigman and M. A. B. Brazier, eds.), p. 87. Academic Press, New York, 1980.

Plastocyanin occurs in higher plants,[35] in green algae,[36] and in cyanobacteria.[37] There are also reports that plastocyanin has been detected by various methods in other orders of eukaryotic algae[38] including euglenoids.[39] An immunological survey[32] of a large number of strains of cyanobacteria in the culture collection of the late R. Y. Stanier at the Pasteur Institute, Paris, indicated that the protein is present in all cyanobacteria. Previous reports by various authors had suggested that plastocyanin may have had a limited distribution among these prokaryotes. The protein has been isolated and primary structures determined from a number of higher plants,[35] the green algae *Chlorella fusca*,[40] *Scenedesmus obliquus*,[41] and *Enteromorpha prolifera*,[42] and a wide range of cyanobacteria.[31,32]

Cytochrome *c*-552–555 (cytochrome c_6) occurs in eukaryotic algae as well as cyanobacteria. It is a *c*-type cytochrome with one covalently bound heme group near the amino terminus of the protein.[43] Plastocyanin and cytochrome c_6 are localized in the photosynthetic lamellae of cyanobacteria and in the chloroplasts of photosynthetic eukaryotes. The proteins function in photosynthetic electron transfer, linking photosystem II and photosystem I at a site adjacent to each other and to membrane-bound cytochrome *f*.[23,24] The redox potentials of these two "soluble" proteins are very similar *in vitro*, but it appears to be plastocyanin that donates electrons to P700.

Primary Structure Comparisons of Plastocyanins

The complete primary structures of plastocyanin from 15 higher plants,[18,35,44] the green algae *Chlorella fusca*,[40] *Scenedesmus obliquus*,[41] and *Enteromorpha prolifera*[42] and the cyanobacterium *Anabaena variabilis*[31] are aligned in Fig. 1 along with the amino-terminal amino acid sequences of plastocyanin from two other cyanobacteria, *Plectonema boryanum*[25] and *Chroococcidiopsis* (A. Aitken, unpublished). It is seen that

[35] D. Boulter, B. G. Haslett, D. Peacock, J. A. M. Ramshaw, and M. D. Scawen, *Plant Biochem.* **13,** 1 (1977).

[36] S. Katoh, *Nature* (*London*) **186,** 533 (1960).

[37] J. J. Lightbody and D. W. Krogmann, *Biochim. Biophys. Acta* **131,** 508 (1967).

[38] P. M. Wood and D. S. Bendall, *Biochim. Biophys. Acta* **387,** 115 (1975).

[39] D. Brown, C. Bril, and W. Urback, *Plant Physiol.* **40,** 1086 (1965).

[40] J. Kelly and R. P. Ambler, *Biochem. J.* **143,** 681 (1974).

[41] J. Kelly and R. P. Ambler, unpublished observations.

[42] R. J. Simpson, R. L. Moritz, E. C. Nice, B. Grego, F. Yoshizaki, Y. Sugimura, H. C. Freeman, and M. Murata, *Eur. J. Biochem.* **157,** 497 (1986).

[43] W. A. Cramer and J. Whitmarsh, *Annu. Rev. Plant Physiol.* **28,** 133 (1977).

[44] A. G. Sykes, *Chem. Soc. Rev.* **14,** 283 (1985).

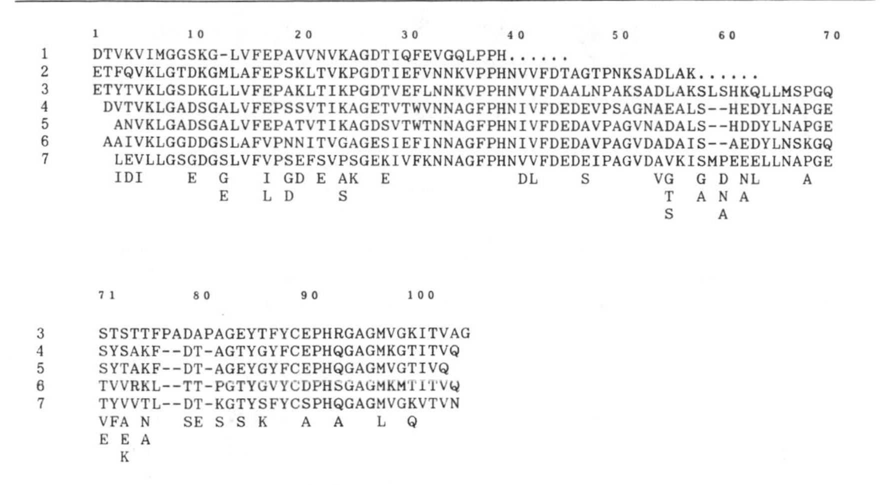

FIG. 1. Amino acid sequences of plastocyanin from cyanobacteria and chloroplasts. (1) *Plectonema boryanum*[25] CCAP 1462/2; (2) *Chroococcidiopsis* PCC 7203 (ATCC 27900) (A. Aitken, unpublished observations); (3) *Anabaena variabilis*[31] PCC 7118 (ATCC 27892); (4) *Chlorella fusca*[40]; (5) *Scenedesmus obliquus*[41]; (6) *Enteromorpha prolifera*[42]; (7) higher plants: French bean (*Phaseolus vulgaris*) plastocyanin is shown. Additional amino acids found in the other plant sequences[18,35,44] are given underneath. Sequences 1 and 3 are from filamentous and 2 from unicellular cyanobacteria; 4, 5, and 6 are from green algae. Amino acids are shown in single-letter notation. CCAP, Culture Collection of Algae and Protozoa, Cambridge, UK; PCC, Pasteur Institute Culture Collection of Cyanobacteria; and ATCC, American Type Culture Collection.

there is considerable similarity between the prokaryotic cyanobacterial sequences and the eukaryotic proteins. The best alignment of the cyanobacterial proteins is obtained by commencing their sequences two residues in advance of higher plant and *Scenedesmus* sequences (and one residue in advance of *Chlorella* and *Enteromorpha* plastocyanins). *Anabaena variabilis* plastocyanin is also one or two residues longer at the carboxy terminus and has insertions at positions 77, 78, and 81. An alternative alignment with deletions in the *Chlorella* sequence at residues 50 and 51 instead of 59 and 60 would generate three more identities with the *Anabaena* plastocyanin. This would result in two or three fewer identities between sequences of this green alga and higher plants.

The partial cyanobacterial sequences of plastocyanin also have considerable similarity to eukaryotic plastocyanins (if a deletion at position 13 is included in the *P. boryanum* sequence). With the three exceptions of parsley (R. Ambler and A. Sykes, unpublished, reported in Ref. 44),

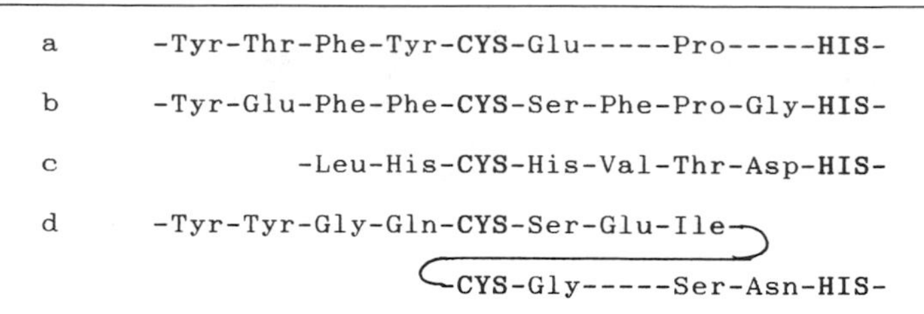

FIG. 2. Amino acid sequences around cysteine residues of type I copper proteins. (a) *Anabaena variabilis* plastocyanin;[31] (b) azurin[40] (*Pseudomonas fluoroescens*); (c) ceruloplasmin[34] (human); (d) cytochrome oxidase subunit II (bovine cardiac)[34,48]. The cysteine and histidine residues in plastocyanin and azurin function as two of the copper ligands.[44,46]

Enteromorpha, and *Scenedesmus* sequences, the highly acidic region from residues 44 to 47 present in all other eukaryotic plastocyanins is not found in *Anabaena* or *Chroococcidiopsis* plastocyanins, where the only acidic amino acid in this region is Asp-44. The cyanobacterial plastocyanins are therefore more basic than the eukaryotic proteins. The absence of this highly basic region has prompted study of electron transfer between the *A. variabilis* plastocyanin and *c*-type cytochromes.[44,45]

Some evidence was found for dimorphism at residue 84 in *Chlorella* plastocyanin.[40] This residue is usually threonine (as in most higher plant sequences), but a small proportion of glutamic acid may be present. Residue 84 is glutamate in *Anabaena* plastocyanin.

Similarity in primary structures of plastocyanin and other copper proteins around the single cysteine residue near the carboxy terminus has been noted[31,34,35,40] (Fig. 2). The degree of similarity between the region of the cyanobacterial plastocyanin and bacterial azurins is not any greater than between the eukaryotic plastocyanins and azurins. The azurins also contain copper in a similar ligand environment, and from the crystal structures of these two proteins cysteine and histidine have been shown to be two of the copper-binding ligands.[34,44,46] The weight of evidence suggests that this similarity may be caused by convergence due to functional similarity, and it is not clear support for extant azurin and plastocyanin having a common ancestor. Amino acid sequences of umecyanin,[47] stellacyanin,[47,48] cytochrome oxidase,[34,48] laccase,[34] and ceruloplasmin[34] also show similarity with plastocyanin, and it is possible that many of these copper-containing proteins share a common origin.

[45] G. D. Armstrong, S. K. Chapman, M. J. Sisley, A. G. Sykes, A. Aitken, N. Osheroff, and E. Margoliash, *Biochemistry* **25,** 6947 (1986).

[46] P. M. Colman, H. C. Freeman, J. M. Guss, M. Murata, V. A. Norris, J. A. M. Ramshaw, and M. P. Venkatappa, *Nature (London)* **272,** 319 (1978).

[47] L. Ryden and J.-O. Lundgren, *Nature (London)* **261,** 344(1976).

[48] T.-T. Wang and N. M. Young, *Biochim. Biophys. Res. Commun.* **74,** 119 (1977).

```
   1        10        20        30        40        50        60        70

1  ADIANGAKVFNSNCAQCHDMGKNVVNATKTLQKDALEKYSMNS--LEAIIN--QVTNGKNAMPAFKGRLN
2  ADIADGAKVFSANCAACHMGGGNVVMANKTLKKEALEQFGMN---SADAIM-YQVQNGKNAMPAFGGRLS
3  ADSVNGAKIFSANCASCHAGGKNLGVAQKTLKKADLEKY      SAMAIGA-QVTNGKNAMPAFKGRLK
4  ADAAAGGKVFNANCAACHASGGGQINGAKTLKKNALTA---NGKDTVEAIVA-QVTNGKGAMPAFKGRLS
5  GDVAAGASVFSANCAACHMGGRNVIVANKTLSKSDLAKYLKGFDDDAVAAVAYQVTNGKNAMPGFNGRLS
6  ----DGASIFSANCASCHMGGKNVVNAAKTLKKEDLVG----GKNAVEAIVT-QVTKGKGAMPAFGGRLG
7  ADTVSGAALFKANCAQCHVGGGNLVNRAKTLKKEALEKYNMY---SAKAIIA-QVTHGKGAMPAFGIRLK
8  GDIANGEQVFTGNCAACHSVZZZmTLELSSLWK--AKSYLANFNGDESAIV-YQVTNGKNAMPAFGGRLE
9  ADLDNGEKVFSANCAACHAGGNNAIMPDKTLKK-DVLEANSMNTIDA--IT-YQVQNGKNAMPAFGGRLV
10 ----GGADVFADNCSTCHVNGGNVISAGKVLSKTAIEEYLDGGY-TKEAIE-YQVRNGKGPMPAWEGVLS
11 IDIDNGEDIFTADCSACHAGGNNVIMPEKTLKK-DA--LADNKMVSVNAIT-YQVTNGKNAMPAFGSRLA

   71       80        90

1  VQQIEDVASYVLDKSEKGWS
2  EAQIENVAAYVLDQSSNKWAG
3  PEEIZBVAAYVLGKAEAEWK
4  DDQIQSVALYVLDKAEKGW
5  PKQIEDVAAYVVDQAEKGW
6  AEDIEAVANYVLAEAEKGW
7  AEQIENVAAYVLEQADNGWKK
8  DDEIABVASYVLSKAG
9  DEDIEDAANYVLSQSEKGW
10 EDEIVAVTDYVYTQAGGAWANV
11 ETDIEDVANFVLTBZBKGWD
```

FIG. 3. Alignment of amino acid sequences of cytochromes c_6. (1) *Chroococcidiopsis* PCC 7203 (ATCC 27900); (2) *Synechococcus* PCC 6312 (ATCC 27167); (3) *Anabaena variabilis* PCC 7118 (ATCC 27892); (4) *Plectonema boryanum* CCAP 1462/2; (5) *Spirulina maxima;* (6) *Microcystis aeruginosa;* (7) *Aphanizomenon flos-aquae;* (8) *Monochrysis lutheri* (chrysophytan alga); (9) *Porphyra tenera* (red alga); (10) *Euglena gracilis;* (11) *Alaria esculenta* (brown alga). Sequences 1 and 2 are from unicellular cyanobacteria; 3 to 7 are from a variety of filamentous forms of cyanobacteria. References to individual sequences are in Refs. 25–28, 49, and 50, except (1) *Chroococcidiopsis* (van Beeumen and Aitken, unpublished). Residue 24 (''m'') in the *Monochrysis* cytochrome is monomethyllysine. The *A. variabilis* sequence is incomplete.

Structural Comparisons of Cytochromes c_6

The cyanobacterial cytochromes[25–28,49,50] (Fig. 3) are clearly homologous to the eukaryotic algal proteins. There is close similarity between the prokaryotic and eukaryotic sequences with large variations (including insertions and deletions) in the region of residues 34–48. The cyanobacterial proteins are closer to the red algal cytochrome (57% on average) than to the other eukaryotic sequences (47% on average). This would lend support to a polyphyletic origin of chloroplasts, with repeated endosymbioses of cyanobacteria. The cyanobacterial endosymbiont that gave rise to

[49] R. E. Dickerson, *in* ''Evolution of Protein Structure and Function'' (D. S. Sigman and M. A. B. Brazier, eds.), p. 173. Academic Press, New York, 1980.

[50] E. L. Ulrich, D. W. Krogmann, and J. L. Markley, *J. Biol. Chem.* **257,** 9356 (1982).

Rhodophyta would, in this hypothesis, retain protein sequence similarity as well as many close structural features in the red algal chromatophore.[51] The close similarity between the main light-harvesting pigments (the phycobiliproteins) in these two groups (Ref. 33 and this volume [31]) is strong evidence in favor.

Origin of the Chloroplast

In addition to plastocyanin, phycobiliprotein, and cytochrome sequences, primary structures of ferredoxins (Refs. 29, 30, and 52 and this volume [41]–[43]) from cyanobacteria have been compared to the sequences of the proteins with analogous functions in eukaryotic algae and plants. The sequence comparisons indicate close homology, but the cyanobacterial sequences are not closer, or are only slightly closer, to each other than to the corresponding proteins in the eukaryotes. For example it is seen from Fig. 1 that the amino-terminal region of *A. variabilis* plastocyanin is slightly closer to the *Chlorella* plastocyanin (56% identity) than to the other filamentous cyanobacterial protein (49% identity). In contrast, *Anabaena* plastocyanin is surprisingly close to the unicellular cyanobacterium (*Chroococcidiopsis*) sequence (77% identical, whether residues 1–39 or 1–57 are compared). Regarding plastocyanin and ferredoxin, comparisons between cyanobacterial and higher plant proteins can be made, and the cyanobacterial sequences do in this case appear about 10% closer to each other than to the higher plant sequences.

Fossil evidence indicates that cyanobacteria evolved more than 3×10^9 years ago,[53] probably from a photoheterotrophic ancestor common to all types of photosynthetic bacteria. In the Gunflint Iron formations (between 1.6 and 2×10^9 years old) many representatives of extant cyanobacterial families have been identified. In more recent rocks ($\sim 0.9 \times 10^9$ years old) many species of cyanobacteria were identified at the specific or generic level. If the strains of cyanobacteria (both filamentous and unicellular forms) for which sequence information is available are as evolutionarily diverse as the fossil record suggests, then it is surprising that their amino acid sequences do not differ more. Filamentous and unicellular forms of cyanobacteria diverged around 3×10^9 years ago.[53] Cyanobacteria were even then a diverse group of organisms while eukaryotic algae diverged much more recently—on a time scale measured in hundreds of

[51] R. Y. Stanier and G. Cohen-Bazire, *Annu. Rev. Microbiol.* **31,** 225 (1977).

[52] T.-M. Chan, M. A. Hermodsen, E. L. Ulrich, and J. L. Markley, *Biochemistry* **22,** 5988 (1983).

[53] J. W. Schopf, *Origins Life* **5,** 119 (1974).

millions of years. There seems to be, however, approximately the same degree of similarity in primary structures of homologous proteins for example, the similarity between filamentous and unicellular cyanobacterial cytochrome c_6 (Fig. 1); plastocyanin (Fig. 3), and ferredoxin (this volume [41]–[43]). An approximately constant rate of evolution of a particular protein in a wide range of organisms has been postulated.[54,55] This has been shown for a number of proteins and has enabled phylogenetic trees to be constructed which correspond to paleontological evidence. Cyanobacteria, on the other hand, appear to have continued to populate similar ecological habitats and to have evolved very little in morphology and physiology since oxygenic photosynthesis was established, whereas the chloroplast has evolved and diverged in a wide range of eukaryotes over a shorter time. Cyanobacteria have continued to populate similar ecological habitats.

The rate of evolution of higher plant plastocyanin has been suggested to be approximately twice that of mitochondrial cytochrome c, that is, about 6 to 10 accepted point mutations per 10^8 years.[56] At this rate of evolution, even allowing for back-substitution, one would expect very little remaining similarity in primary structures of plastocyanin and other proteins from diverse cyanobacteria apart from a few highly invariant residues required for a specific function, e.g., binding of copper, heme attachment, and iron–sulfur binding.

Conclusions

The rates of evolution of the cyanobacterial proteins are much less than the rates of evolution of the corresponding proteins in eukaryotic algae and higher plants. These results appear to go against the hypothesis of neutral mutation in proteins.[57] The role of neutral mutation in plastocyanin and other photosynthetic proteins in cyanobacteria appears to be small, and the amount of change in these sequences may reflect functional difference.

It is also possible that the pattern of protein sequence divergence in cyanobacteria has been obscured by transfer of genetic information coding for these proteins. There is evidence that this has occurred between distantly related bacteria,[58,59] but, whatever the reason for the sequence

[54] R. E. Dickerson, *J. Mol. Evol.* **1,** 26 (1971).
[55] A. C. Wilson, S. S. Carlson, and T. J. White, *Annu. Rev. Biochem.* **46,** 573 (1977).
[56] D. Peacock and D. Boulter, *J. Mol. Biol.* **95,** 513 (1975).
[57] T. H. Jukes and M. Kimura, *J. Mol. Evol.* **21,** 90 (1984).
[58] R. P. Ambler, T. E. Meyer, and M. D. Kamen, *Nature (London)* **278,** 661 (1979).
[59] R. P. Ambler, *Syst. Zool.* **22,** 554 (1973).

comparison results, protein sequence studies have produced a considerable amount of evidence for the hypothesis of the common origin of oxygen-evolving photosynthesis. On this cautionary note it is probably wise to avoid over interpretation of protein sequence data with elaborate computer programs but to let the aligned sequences speak for themselves (Figs. 1 and 3).

Comparing only photosynthetic protein sequences may not allow a conclusion to be drawn as to whether the implied evolutionary relationships relate to the whole organism or solely to the chloroplast of eukaryotes. Schwartz and Dayhoff[60] have approached this problem by constructing a composite evolutionary tree from a wide range of *c*-type cytochromes (including c_6), ribosomal RNA, ferredoxin, and plastocyanin sequences. Chloroplast, mitochondrial, and cytoplasmic protein sequences from plants and eukaryotic algae appear on different branches of the tree. Photosynthetic proteins from cyanobacteria appear on the chloroplast branch of this tree, providing support for the endosymbiotic hypothesis. This type of study has been continued by other groups; for example, Otaka *et al.*[61] have compared bacterial and chloroplast ribosomal proteins. On a further cautionary note, it must not be forgotten that many of the proteins are synthesized in the nucleus and subsequently imported into the chloroplast.

Over the complete range of eukaryotic organisms the general similarities in cell structure, biochemistry, and function are sufficiently close to suggest that the eukaryotic cell had a monophyletic origin. Therefore, if the symbiotic hypothesis is rejected it may be difficult to avoid the alternative conclusion that a cyanobacterium (or an ancestral form) gave rise to all eukaryotic cells, some of which subsequently lost photosynthetic capability.

[60] R. M. Schwartz and M. O. Dayhoff, *Ann. N.Y. Acad. Sci.* **361,** 260 (1981).
[61] E. Otaka, T. Ooi, T. Kumazaki, and T. Itoh, *J. Mol. Evol.* **21,** 339 (1985).

Section II

Ultrastructure, Inclusions, and Differentiation

[14] Fine Structure of Cyanobacteria

By G. STANIER (COHEN-BAZIRE)

Introduction

Most of our present knowledge about the fine structure of cyanobacteria was acquired through examination via the transmission electron microscope of thin sections of fixed and resin-embedded specimens. Although the first electron micrographs obtained in the late 1950s[1] and early 1960s[2–4] were not of the best quality, they demonstrated unambiguously that cyanobacteria were prokaryotes and were endowed with characteristic features. Fixation techniques have improved over the years, and the use of fluid, unpolymerized embedding resins appears to have minimized the production of artifacts.

Cryofixation and freeze-etching techniques and the observation of freeze-etched replicas have provided additional information on the internal structure of different types of "unit membranes." First applied in structural studies of cyanobacteria by Jost in 1965,[5] this technique has been used extensively ever since,[6–8] one advantage of the method being the avoidance of fixation artifacts. Other techniques such as freeze-drying, metal shadowing, and negative staining have also been used, the former two for examination of details of outer surfaces of cells or organelles, the latter for the observation of subcellular structures such as phycobilisomes, carboxysomes, gas vesicles, fimbriae, or pili.

Methods

Preparation of Specimens for Thin Sections

Before fixation, cells actively growing in liquid medium are collected by centrifugation or concentrated on membrane filters. Alternatively,

[1] W. Niklowitz and G. Drews, *Arch. Mikrobiol.* **27,** 150 (1957).
[2] H. Ris and R. N. Singh, *J. Biophys. Biochem. Cytol.* **9,** 63 (1961).
[3] H. S. Pankratz and C. C. Bowen, *Am. J. Bot.* **50,** 387 (1963).
[4] D. C. Wildon and F. V. Mercer, *Aust. J. Biol. Sci.* **16,** 585 (1963).
[5] M. Jost, *Arch. Mikrobiol.* **50,** 211 (1965).
[6] T. H. Giddings and L. S. Staehelin, *Cytobiology* **16,** 235 (1978).
[7] T. H. Giddings and L. S. Staehelin, *Biochim. Biophys. Acta* **546,** 373 (1979).
[8] J. Golecki and G. Drews, *in* "The Biology of Cyanobacteria" (N. G. Carr and B. A. Whitton, eds.), p. 125. Blackwell, Oxford, England, 1982.

cells can be collected after growth on agar plates. The last two techniques are recommended for the preservation of fragile outer structures such as pili or to avoid the collapse of gas vesicles in gas-vacuolated strains.

Fixation. The cells are prefixed for 30 min at room temperature in a dilute solution (0.5%, w/v) of neutral glutaraldehyde in 150 m*M* Veronal acetate buffer or 100 m*M* cacodylate buffer at pH 7; for nonmarine strains, the final osmolarity of the medium should vary between 200 and 400 mOsm. Following prefixation, the cells are concentrated by centrifugation or filtration and suspended in a small volume of melted 3% agar. After hardening, the agar is cut into small blocks (1 mm wide) which are immersed for 2–3 hr at room temperature in a solution of 2–3% (w/v) glutaraldehyde in buffer. The agar blocks are then washed at least 4 times in buffer and postfixed in 1% (w/v) osmium tetroxide in 150 m*M* veronal acetate or 100 m*M* cacodylate buffer, pH 7, for 30 min at room temperature and for another 2–3 hr at 4°. After a thorough washing in distilled water, the blocks are stained for 2 hr at room temperature with 1% (w/v) aqueous uranyl acetate, washed again in distilled water, dehydrated in a graded series of ethanol solutions, washed 3 times with propylene oxide, and embedded in the low-viscosity epoxy resin of Spurr.[9] For marine strains, all buffers should be adjusted to 1100 mOsm with NaCl. Thin sections are poststained with both uranyl acetate (3% in 1% acetic acid) and lead citrate[10] before examination in the electron microscope.

Purification and Neutralization of Glutaraldehyde

Glutaraldehyde (Fischer Scientific Company, biological grade, 50% w/w) is diluted to 25% with distilled water and shaken with activated charcoal for 2 hr at 30°, filtered through Whatman paper No. 3, and shaken again with barium carbonate for 2 hr at 30°. After filtration, the solution should be at pH 7 or 7.1; it can be kept refrigerated for a few days or subdivided in small quantities kept frozen until used. High-quality glutaraldehyde solutions that do not require the treatments described above are also available.

Freeze-Etching Using the Spray-Freezing Method

Application of the spray-freezing method of cryofixation of biological specimens for freeze-etching, first described in 1971,[11] was discussed ex-

[9] A. R. Spurr, *J. Ultrastruct. Res.* **26,** 31 (1969).
[10] E. S. Reynolds, *J. Cell Biol.* **17,** 208 (1963).
[11] L. Bachmann and W. W. Schmitt, *Proc. Natl. Acad. Sci. U.S.A.* **68,** 2149 (1971).

tensively by Plattner *et al.*[12] This procedure utilizes a Balzer's spray-freezing glove box and its accessories. The cells suspended in growth medium are sprayed, without the addition of cryoprotective agents, into liquid propane at $-190°$ in a special container. The propane is allowed to evaporate, and the frozen droplets are aggregated at $-85°$ with butylbenzene and placed on a Balzer's specimen holder precooled and maintained at $-85°$. The samples are transferred to liquid N_2, placed in a freeze-etching chamber, and fractured at $-110°$ under a reduced pressure of 10^{-6} mmHg. Carbon–platinum replicas shadowed unidirectionally are cast after the desired amount of etching. The replicas are cleared of organic material by soaking overnight in sulfochromic acid and collected on 200-mesh grids after abundant washing with distilled water. The use of this technique is exemplified in Figs. 10, 11, and 12, which show, on freeze–fracture replicas of the cyanobacterium *Synechocystis* sp. PCC 6701 (Pasteur Culture Collection), the structural differences between two types of unit membranes, namely, thylakoid and cytoplasmic membranes.

Freeze-Drying Method for the Examination of Surfaces

For freeze-drying, biological specimens are fixed with 2% glutaraldehyde, washed several times with distilled water, and deposited on a thin piece of glass. After removing excess moisture, the piece of glass is immersed in liquid N_2 or liquid propane and rapidly transferred to a freeze-etching chamber. The thin superficial film of ice is then sublimated. A replica of the surface is obtained by unidirectional deposition of a carbon–platinum film. The replicas are treated as in the preceding paragraph. Fixation with glutaraldehyde is not always necessary.

Metal Shadowing of Delicate Cell Surface Structures

The metal shadowing technique was developed in 1952 by Kellemberger.[13] The principle is the following: drops of a dilute suspension of cells are deposited on an agar-containing glass Petri dish covered with a thin film of collodion. The agar has been left slightly moist so as to produce tiny wholes in the collodion film, through which the suspending medium will be absorbed by the agar. When the medium has been totally absorbed, small pieces of the collodion-covered agar are cut with a razor

[12] H. Plattner, W. W. Schmitt-Fumian, and L. Bachmann, *in* "Freeze-Etching Techniques and Applications" (E. L. Benedetti and F. Favard, eds.), p. 83. Fr. Microsc. Elect., Paris, 1973.

[13] E. Kellenberger, *Experientia* **8,** 99 (1952).

blade, and the film is floated off on distilled water and collected on a 200-mesh grid which is shadowed at the proper angle with palladium.

Negative Staining

For negative staining, glow-discharged Formvar-coated 200- or 300-mesh grids stabilized with a thin carbon film or 300-mesh grids covered with a carbon film are placed on top of a drop of medium or buffer containing the cells or the structures to be examined. After 30–60 sec, the grids are transferred on a drop of buffer containing 0.3% (w/v) glutaraldehyde. After 10 min, the grids are washed thoroughly with water or ammonium acetate (100 m*M*) before being placed on top of a drop of 2% (w/v) uranyl acetate or 1% (w/v) uranyl formate. The grids are examined immediately after removing the excess stain with the corner of a piece of filter paper. Fixation with glutaraldehyde is not always necessary.

Fine Structure of Cyanobacterial Cells Observed in Thin Sections

The great majority of cyanobacterial vegetative cells have a distinctive fine structure which is represented schematically in Fig. 1. The cell protoplast is surrounded by a multilayered cell wall similar to that of gram-negative bacteria; it consists of an outer membrane with the unit membrane structure and an electron-opaque peptidoglycan layer which varies in thickness between 5 and 10 nm but can be considerably thicker, particularly in some species of *Oscillatoria*. In *Oscillatoria* sp. PCC 6407 (Fig. 6) the peptidoglycan is approximately 25 nm thick, whereas it reaches 200–250 nm in *Oscillatoria princeps*. Many groups of cyanobacteria synthesize additional outer wall layers: diffuse slime layers, sheaths, or "decorated" extra layers, as exemplified in Figs. 5, 6, 7, 8, 9, 13, and 21.

The protoplast itself is limited by a cytoplasmic membrane of the unit membrane structure; infoldings of the cytoplasmic membrane are extremely rare and may be artifacts of fixation. The most conspicuous cytoplasmic elements are the thylakoids, parallel to one another but separated from each other by a space of about 60–70 nm. Each thylakoid is formed by two closely appressed unit membranes; they are the basic structures for the performance of oxygenic photosynthesis and contain the chlorophyll *a*–protein complexes, carotenoids, the photosynthetic reaction centers, and the electron transport system. In favorable sections (Figs. 2 and 4), alternate rows of electron-opaque discoidal structures can be distinguished in the interthylakoidal space, namely, the phycobilisomes (PBs), multimolecular complexes of phycobiliproteins, the major light-harvesting pigments of cyanobacteria. When rows of PBs are sectioned perpen-

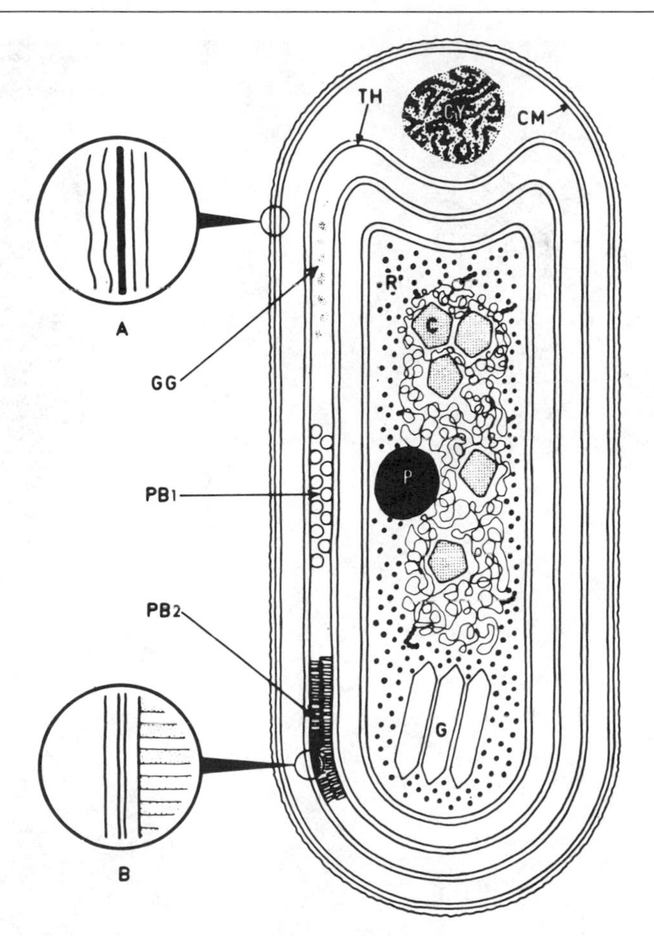

FIG. 1. Schematic diagram of a thin section of a cyanobacterial cell. CM, Cell membrane; TH, thylakoid; PB1 and PB2, face and side views of phycobilisomes attached to adjacent thylakoids; GG, glycogen granules; CY, cyanophycin granule; P, polyphosphate granule; C, carboxysome, surrounded by nucleoplasm; R, ribosomes; G, gas vesicles. (Insert A) Enlarged view of the cell envelope showing the outer membrane and peptidoglycan wall layers, and the cytoplasmic membrane. (Insert B) Enlarged view of part of a thylakoid showing the paired unit membrane with attached phycobilisomes in side view. Reproduced from R. Y. Stanier and G. Cohen-Bazire, *Annu. Rev. Microbiol.* **31,** 225 (1977), by permission of Annual Reviews, Inc.

dicularly to the thylakoid surface, they are seen in face view as hemidisks with their flat side close to the outer surface of the thylakoid membrane (Figs. 2 and 4). When the rows of PBs are cut in longitudinal section, the PBs appear as stacks of 10- to 12-nm-thick electron-opaque lines standing on the surface of the thylakoids (Fig. 3).

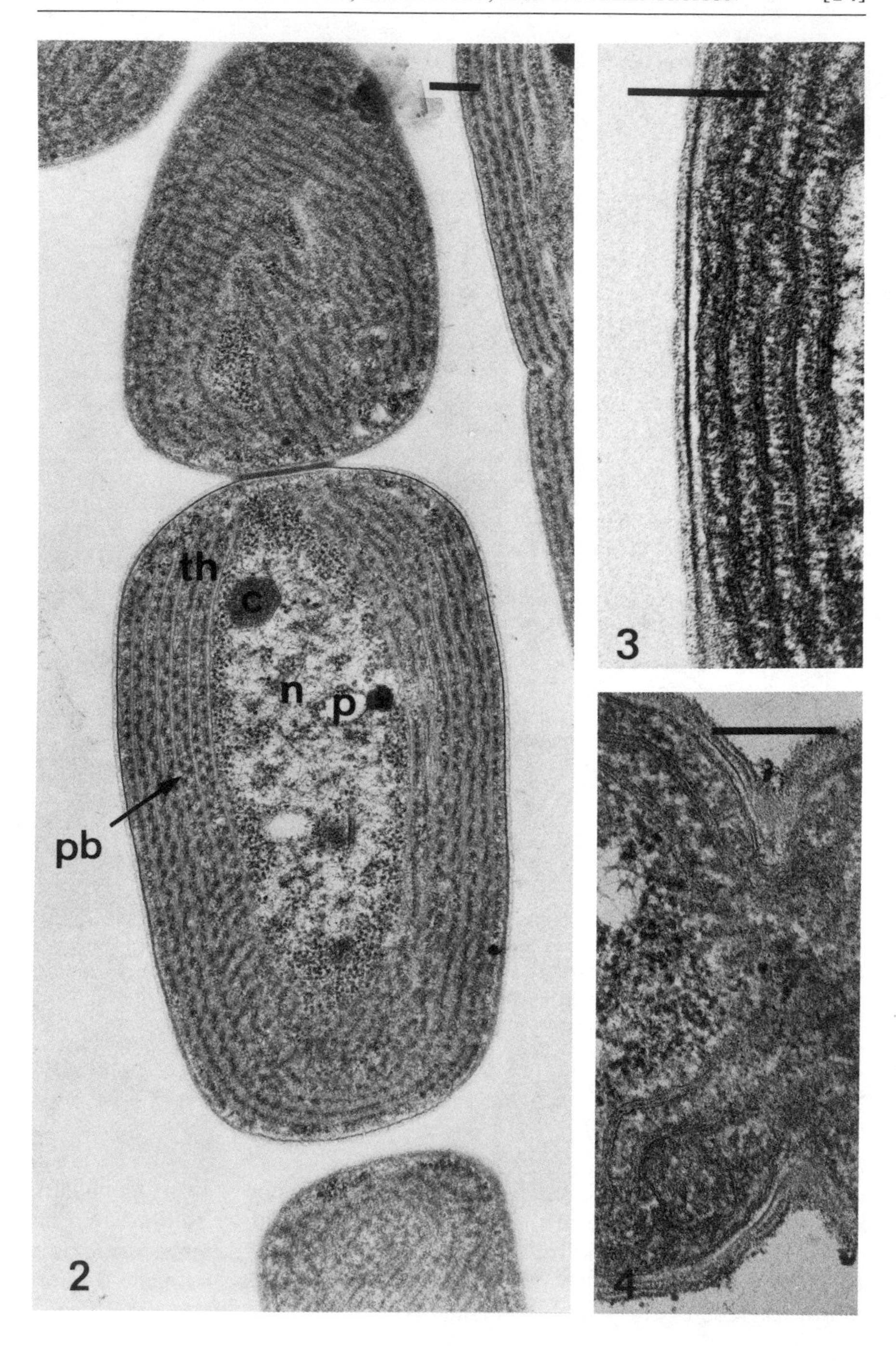
th
c
n
p
pb
2
3
4

Different thylakoid arrangements are observed among cyanobacteria. The simplest, exemplified in Fig. 2, is characteristic of most *Synechococcus* strains and of many filamentous strains. The cells contain 3–6 cortical thylakoids parallel to one another and to the outer surface of the cells, enclosing a central region, the "centroplasm," which contains the DNA fibrils, carboxysomes, polyphosphate granules, and many ribosomes. In other cyanobacteria, unicellular and filamentous, particularly in heterocystous strains, the thylakoids are convoluted and can occupy a very large portion of the cytoplasm (Fig. 5). In unicellular strains of this type (Fig. 5), the carboxysomes are often large and are usually located near the periphery of the cell, close to the cytoplasmic membrane; ribosomes and DNA fibrils are localized between the folds of the convoluted thylakoids. Arrangements of thylakoids between these two extreme types are frequent.

A cyanobacterium, encountered once so far, *Gloeobacter violaceus* (*Gloeothece coerulea*), does not contain thylakoids (Figs. 8 and 9). The only unit membrane is the cytoplasmic membrane, which has a simple topology and is devoid of cytoplasmic intrusions. This membrane plays a dual role: that of a typical cytoplasmic membrane and that of a photosynthetic membrane. This dual function has been confirmed by examination of freeze-etched replicas.[14] An electron-opaque layer, 70–80 nm thick, closely appressed on the cytoplasmic side of the cytoplasmic membrane, surrounds the cytoplasm. This layer is constituted by the phycobilisomes, which appear, when isolated, as bundles of six rods connected at the base. Typically, this strain contains only one-fourth of the amount of chorophyll *a* of a typical *Synechococcus*, it has a high phycobiliprotein content, and its phycoerythrin has an absorption spectrum of the red algal type.

Cytoplasmic Inclusions

Carboxysomes. All cyanobacterial vegetative cells contain carboxysomes, pseudocrystalline aggregates of the key enzyme of CO_2 fixation via

[14] G. Guglielmi, G. Cohen-Bazire, and D. A. Bryant, *Arch. Microbiol.* **129,** 181 (1981).

FIG. 2. Longitudinal section of *Pseudanabaena* sp. PCC 7408 showing some of the structures schematized in Fig. 1, such as thylakoids (th), phycobilisomes (pb), polyphosphate granules (p), and nucleoplasm (n). Electron micrograph by Dr. G. Guglielmi. Bar, 0.2 μm.

FIG. 3. Portion of *Synechococcus* sp. PCC 6716 showing phycobilisomes in side view. Bar, 0.2 μm.

FIG. 4. Portion of a dividing *Synechococcus* sp. PCC 6312 showing phycobilisomes in face view. Bar, 0.2 μm.

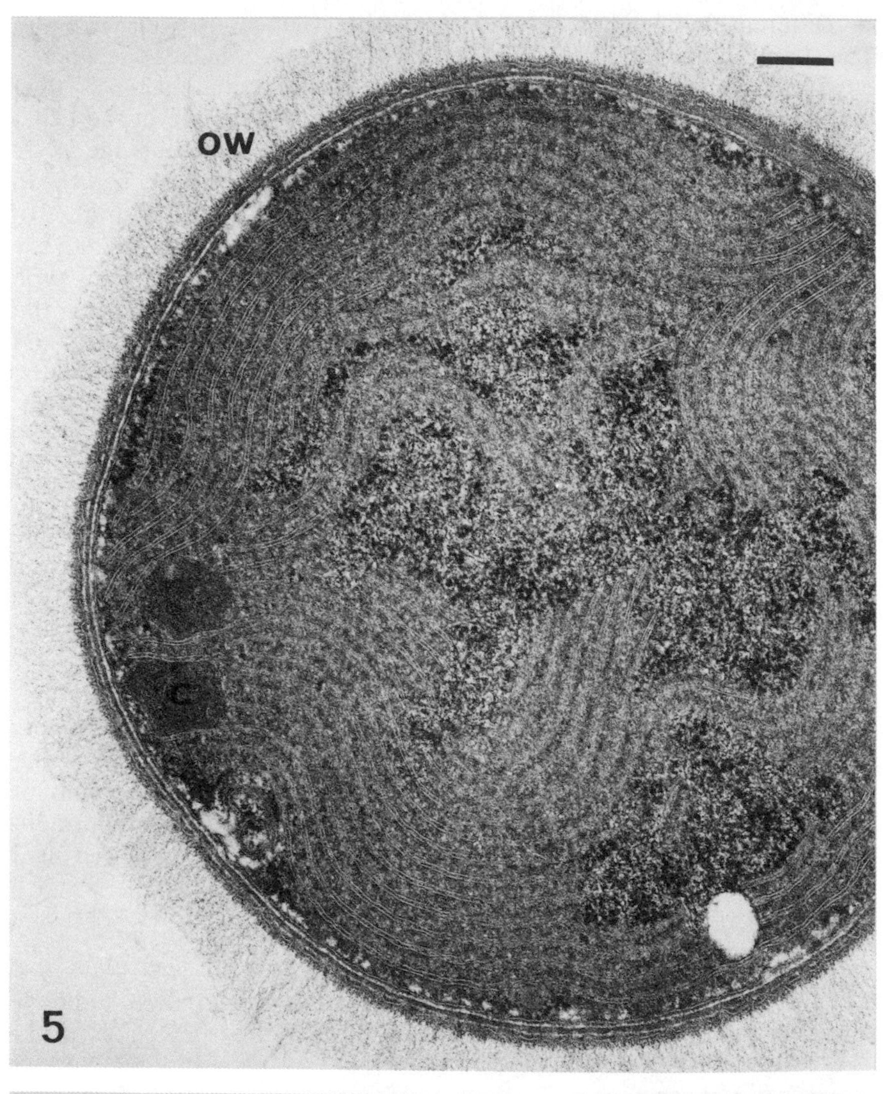
ow
c
5

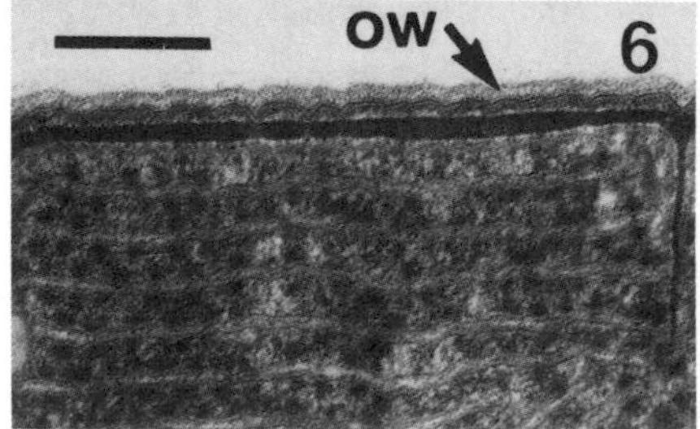
ow
6

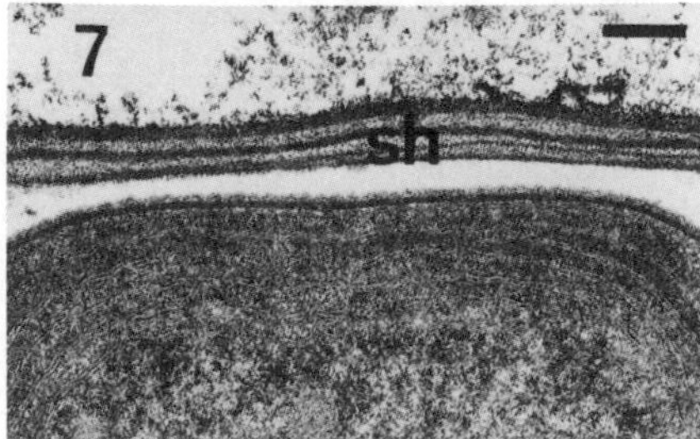
7
sh

the reductive pentose phosphate pathway, ribulose-bisphosphate carboxylase. The term "polyhedral bodies" used in the old nomenclature referred to their pseudocrystalline appearance. They are easily distinguishable in thin sections, sometimes in large numbers (5–10) (Figs. 2, 5, 9, and 13), where they appear limited by a thin (2–3 nm) electron-opaque nonunit membrane; carboxysomes are absent from fully differentiated heterocysts that do not fix CO_2.

Glycogen. Glycogen is a general carbohydrate reserve material of cyanobacteria. Its accumulation is visible in thin sections as electron-transparent irregular dots, often located between the thylakoids and prominent in nitrogen-limited photosynthesizing cells (Fig. 13). In exponentially growing cells, glycogen deposits are rarely visible; their presence can be revealed by the Thiéry[15] reaction using a silver stain specially devised to detect polysaccharides in thin sections (Fig. 15). In *Gloeobacter violaceus,* the glycogen deposits are scattered in the cytoplasm (Fig. 14).

Poly-β-hydroxybutyrate Granules. The typically prokaryotic reserve material poly-β-hydroxybutyrate is not of general occurrence in cyanobacteria. In this sections, poly-β-hydroxybutyrate granules appear as round or oval electron-transparent areas, delimited by a very thin electron-opaque nonunit membrane. Examples are shown in a thin section of *Chlorogloeopsis fritschii* (Fig. 21), a strain known to contain this reserve material. These structures can easily be mistaken, in thin sections, for ghosts of polyphosphate granules.

Cyanophycin. Cyanophycin is a nitrogenous organic reserve specific to cyanobacteria; it is composed of long polymeric molecules of multi-L-arginyl-poly(L-aspartic acid). Cells approaching the stationary phase of growth in light-limiting conditions accumulate this polymer as cyanophycin granules of irregular contour, often large enough to be resolved by

[15] J. C. Thiéry, *J. Microsc.* **6,** 987 (1967).

FIG. 5. Thin section of *Synechocystis* sp. PCC 6711. Numerous convoluted thylakoids occupy a large fraction of the cytoplasm; two large carboxysomes (c) are located at the periphery of the cell, close to the cytoplasmic membrane. ow, Outer wall layer with fine fibrils extending from it and forming a halo around the cell. Bar, 0.2 μm.

FIG. 6. Part of a cell from a filament of *Oscillatoria* sp. PCC 6407 showing the complex cell wall with a well-defined outer wall layer (ow) and a thick peptidoglycan layer. Magnification ×60,000. Photomicrograph from Dr. G. Guglielmi. Bar, 0.2 μm.

FIG. 7. Part of a cell from a filament of *Plectonema* sp. PCC 6703 surrounded by a typical multilayered sheath (sh). Bar, 0.2 μm.

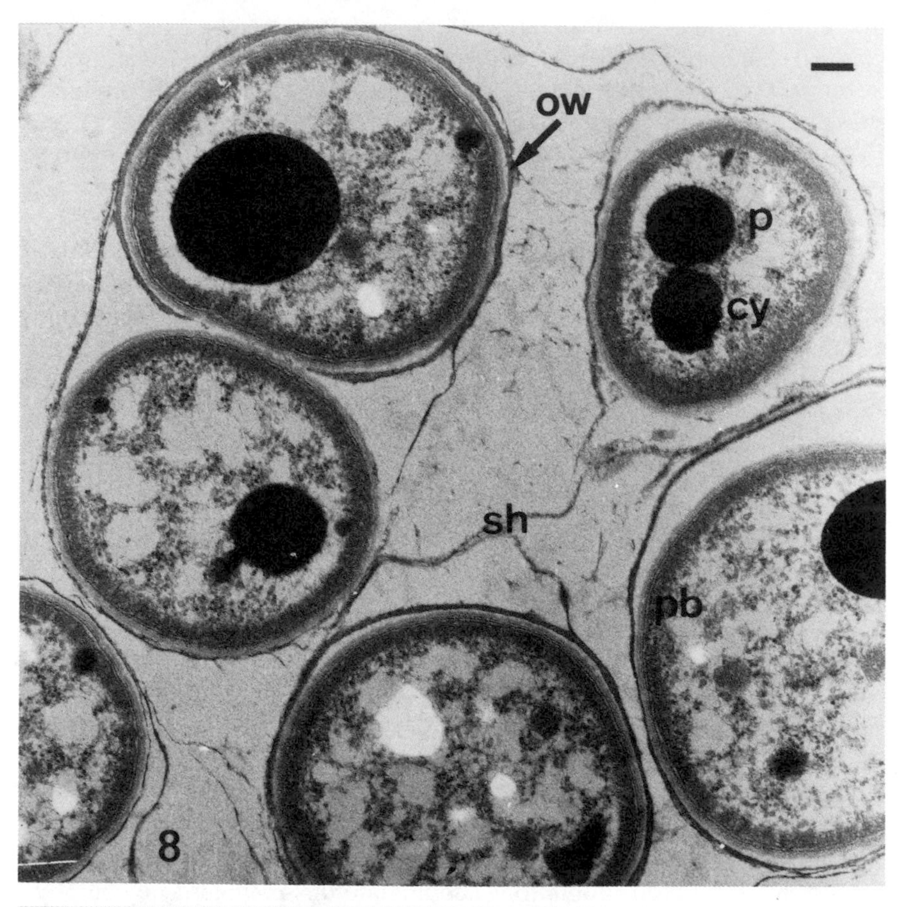
ow
p
cy
sh
pb
8

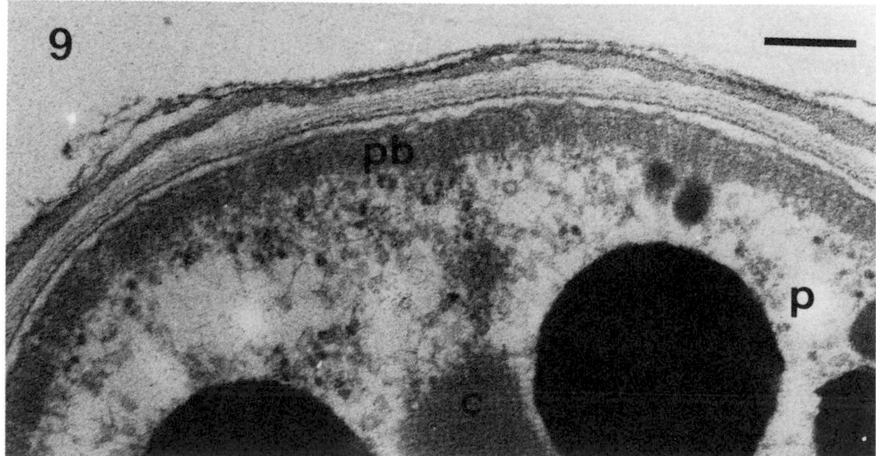
9
pb
p
c

light microscopy; they are rarely present in exponentially growing cells. Their denomination as "structured granules" depicts their appearance in the electron microscope (Fig. 19). Cyanophycin granules are particularly numerous in akinetes, the resting cells of many heterocystous cyanobacteria.

Other Inclusions. Among other inclusions are *polyphosphate granules*, which are often volatilized under the electron beam in thin sections, leaving an empty space limited by a thin electron-opaque line (Fig. 2). *Gloeobacter violaceus* contains particularly large polyphosphate granules that do not seem to be easily volatilized by the electron beam (Figs. 8 and 9). *Lipid droplets* appear as electron-opaque, small, circular objects of variable size frequently seen among the thylakoids. *Gas vacuoles*, composed of groups of gas vesicles, are found in some aquatic species of cyanobacteria and transiently during hormogonia differentiation of some filamentous heterocystous strains. Their fine structure in thin sections is characteristic (see Fig. 20): in longitudinal sections, they appear as small, empty cylinders with conical ends limited by a thin nonunit membrane; in cross section, they appear as circles. The gas vesicles of a particular species appear to have the same approximate dimensions (diameter and length), probably determined by the three-dimensional structure and packing of their major constituent protein.

Pili or Fimbriae

Pili or fimbriae cannot be visualized in thin sections. Their presence and arrangement has been observed in many unicellular and filamentous forms, mostly by using the techniques of negative staining or metal shadowing of whole cells. An example is shown in Fig. 16, and a negatively stained preparation of purified fimbriae is shown in Fig. 17.

FIG. 8. Section through part of a microcolony of *Gloeobacter violaceus* showing the overall structure of the cells and the formation of the outer wall layer (ow), which, as the cell grows, becomes detached from the cell and forms a thin sheath (sh) surrounding small groups of cells. The large, circular, very electron-opaque granules are polyphosphate granules (p). The cell at upper right contains one polyphosphate granule (p) and a granule of cyanophycin (cy) with a more irregular contour and of slightly lower electron opacity. The phycobilisomes (pb) form a 70- to 80-nm-thick layer immediately inside and adjacent to the cytoplasmic membrane. Photomicrograph from Dr. J. Waterbury. Bar, 0.2 μm.

FIG. 9. Enlarged portion of a section of a cell of *Gloeobacter violaceus* showing details of the complex cell wall and of the cortical cytoplasmic layer formed by the phycobilisomes. p, Polyphosphate granule; C, carboxysome. Photomicrograph from Dr. J. Waterbury. Bar, 0.2 μm.

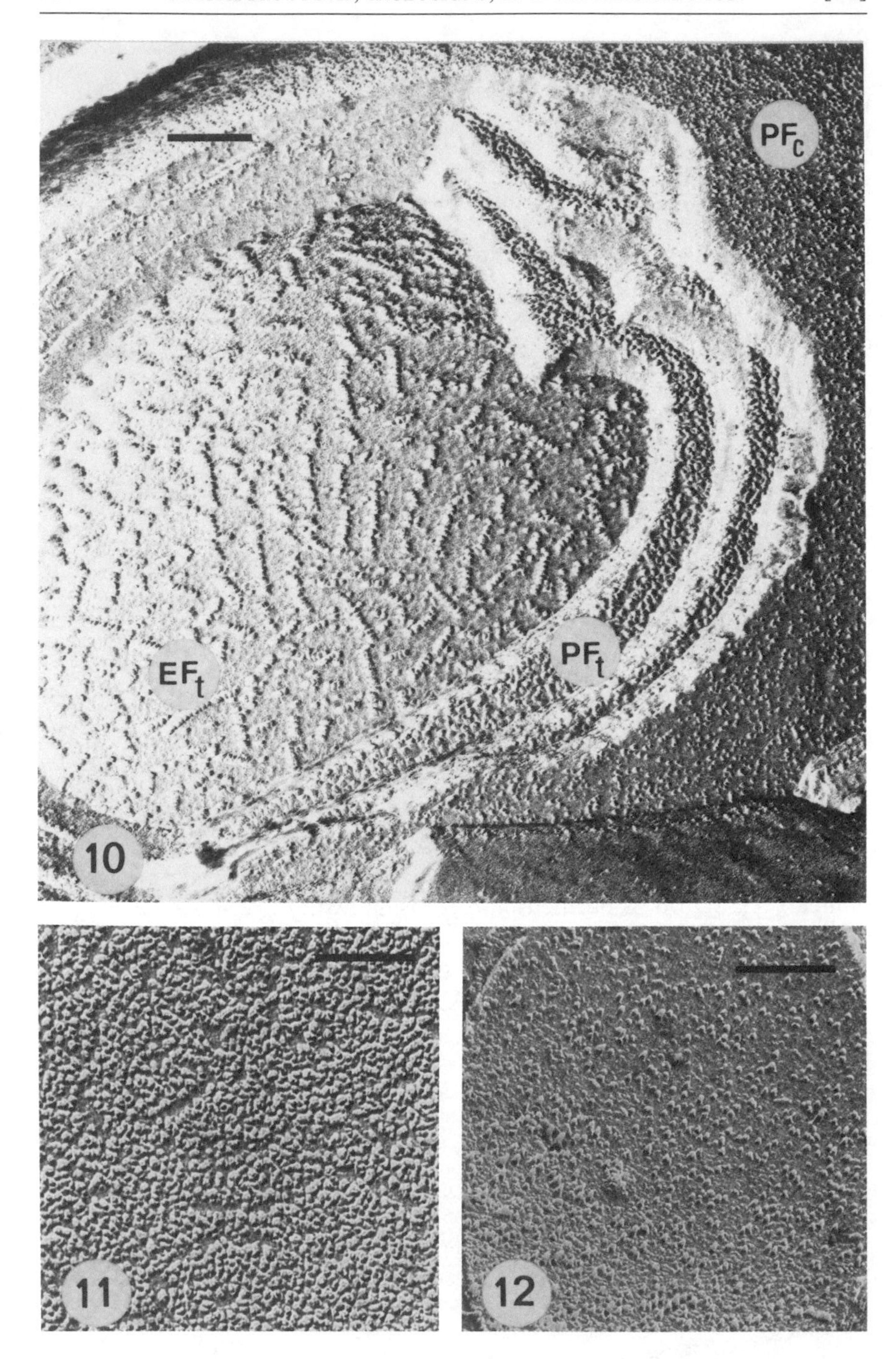
PFc
EFt
PFt
10
11
12

Pores

Observation of the peptidoglycan wall layer of many cyanobacteria, filamentous or unicellular, reveals the presence of perforations or pores.[16] A close correlation appears to exist between the repartition of the pores and that of the fimbriae. The pores are not simple perforations of the peptidoglycan but show a characteristic fine structure which is best visualized by negative staining after isolation of the peptidoglycan layer (Fig. 18). For preparation of the peptidoglycan wall layer, the cells, suspended in distilled water, are disintegrated with glass beads (0.25–0.30 mm in diameter) in a Mickle disintegrator for 1–5 min. After the elimination of the glass beads by low-speed centrifugation, the supernatant is centrifuged at 30,000 *g* for 10 min, and the pellet is treated during 1 hr with the following denaturing buffer: 250 m*M* phosphate buffer, pH 7, 8 *M* urea, 1% (w/v) sodium dodecyl sulfate, and 10 m*M* EDTA. After a 10-min centrifugation at 30,000 *g*, the pellet is thoroughly washed by centrifugation with distilled water (at least 10 times). The whitish pellet contains fragments of the peptidoglycan layer which can be observed in the electron microscope after negative staining.

In negatively stained preparations, the pores appear as circular areas of lesser electron opacity than the surrounding peptidoglycan, with a diameter of approximately 15 nm and a definite substructure which is dissolved in 50% sulfuric acid. The circular area surrounds a depression with an inner electron-transparent ring around an inner hole or depression in which the negative stain has penetrated; the inner ring of 7 nm in diameter is formed by five or six globular subunits. It is tempting to speculate that the pores are differentiated areas of the peptidoglycan through which penetrate the fimbriae and to which they are anchored. Such structures can easily be observed in the relatively thick peptidoglycan layer of

[16] G. Guglielmi and G. Cohen-Bazire, *Protistologica* **18,** 151 (1982).

FIG. 10. Photomicrograph of a freeze–fracture replica of *Synechocystis* sp. PCC 6701 which exposes the protoplasmic fracture face (PF_c) of the cytoplasmic membrane and both fracture faces of the thylakoid membranes, PF_t, the protoplasmic fracture face, and EF_t, the exoplasmic fracture face. The latter face shows aligned rows of large particles. Bar, 0.2 μm.

FIG. 11. Portion of a protoplasmic fracture face (PF_t) of a thylakoid membrane of *Synechocystis* sp. PCC 6701. The particle-free grooves among the numerous particles on this face correspond to the rows of large particles present on the EF_t face shown in Fig. 10. Bar, 0.2 μm.

FIG. 12. Exoplasmic fracture face (EF_c) of the cytoplasmic membrane of *Synechocystis* sp. PCC 6701. Bar, 0.2 μm.

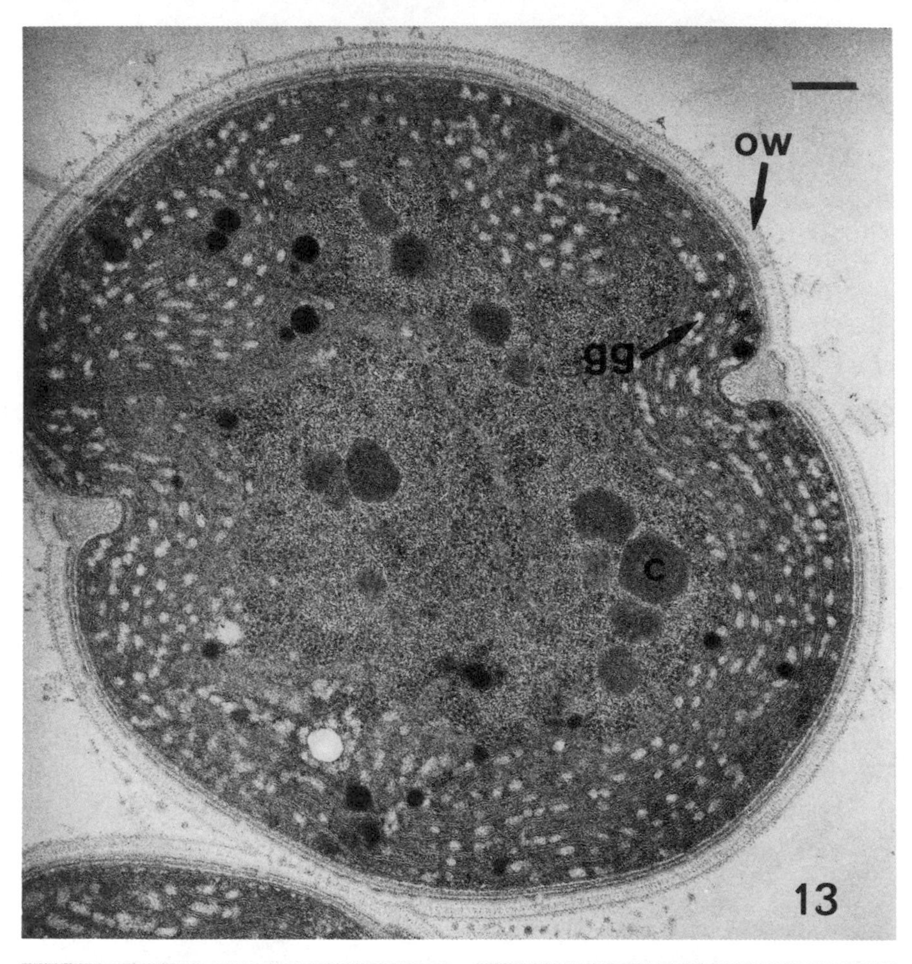
ow
gg
c
13

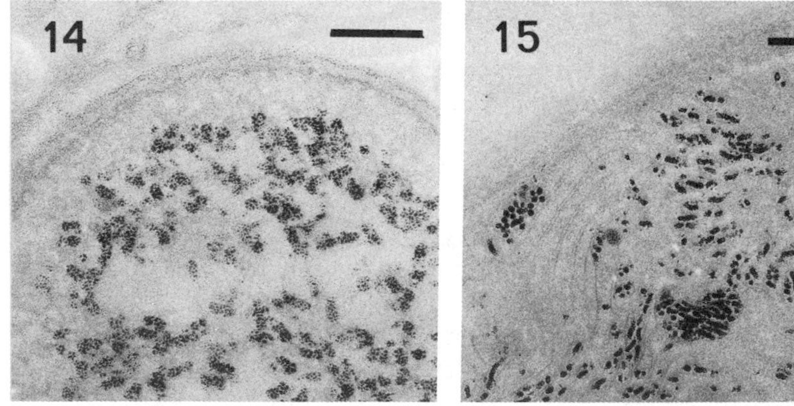
14
15

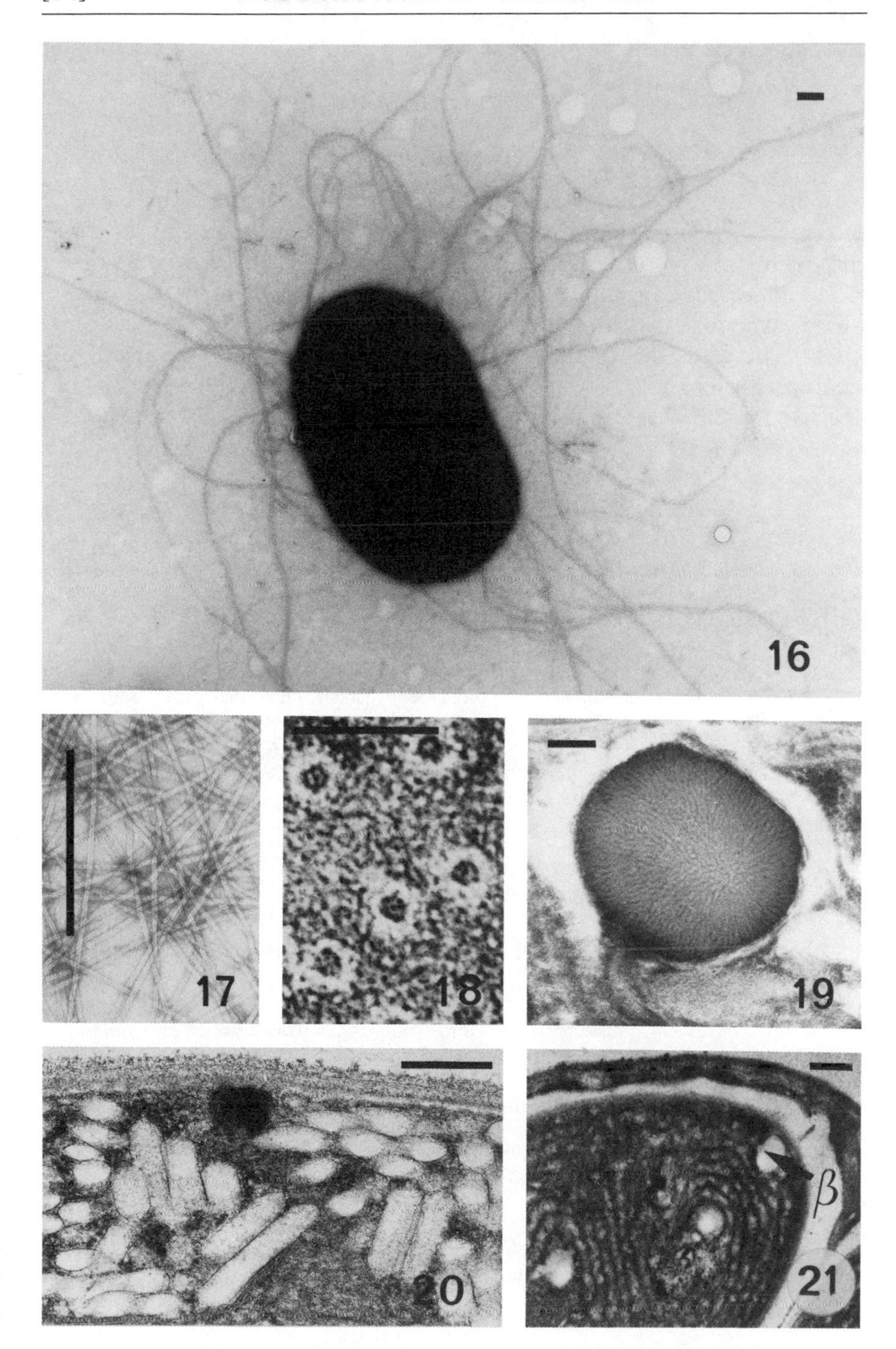
16
17
18
19
20
21
β

cyanobacteria but should be present in the peptidoglycan of all fimbriated gram-negative bacteria.

Comments

This chapter does not pretend to be an exhaustive description of all aspects of cyanobacterial cell structure and organization; for example, although pictorial representations of cell wall outer structures have been offered, we have refrained from discussing what constitutes a sheath or slime layer for lack of precise chemical or biochemical identification of such structures. Descriptions of the fine structure of differentiated cells such as hormogonia, akinetes, and heterocysts are more appropriately included in the chapters dealing with these specialized cells elsewhere in this volume ([21], [22], [51]).

FIG. 13. Thin section of a cell of *Synechocystis* sp. PCC 6702 in the process of division, showing numerous electron-transparent glycogen deposits (gg) localized between the thylakoids. Several carboxysomes (c) are scattered in the centroplasm. The outer wall layer (ow) has a well-defined, lacy fine structure. Bar, 0.2 μm.

FIG. 14. Glycogen deposits of *Gloeobacter violaceus* are scattered in the cytoplasm, not in the phycobilisomal layer. Stained with the Thiéry reagent. Photomicrograph from Dr. G. Guglielmi. Bar, 0.2 μm.

FIG. 15. Glycogen deposits in a portion of a cell of *Oscillatoria* sp. stained by the Thiéry reagent. The glycogen is localized between the thylakoids. Photomicrograph from Dr. G. Guglielmi. Bar, 0.2 μm.

FIG. 16. Negatively stained cell of *Synechococcus* sp. PCC 7942 (*Anacystis nidulans* R-2) showing the peritrichous very long pili or fimbriae. Bar, 0.2 μm.

FIG. 17. Negatively stained fimbriae purified from *Synechocystis* sp. PCC 6701. Bar, 0.2 μm.

FIG. 18. Negatively stained portion of the peptidoglycan wall layer of *Pseudanabaena* sp. PCC 7408 showing the fine structure of the pores. Bar, 0.05 μm.

FIG. 19. Cyanophycin granule in a thin section of a filament of *Lyngbya* sp. Photomicrograph from Dr. G. Guglielmi. Bar, 0.2 μm.

FIG. 20. Portion of a thin section of a cell of *Oscillatoria agardhii* showing the appearance of gas vesicles. Bar, 0.2 μm.

FIG. 21. Portion of thin section of a cell of *Chlorogloeopsis fritschii* PCC 6708 showing a well-characterized sheath and the appearance of poly-β-hydroxybutyrate granules (β). Bar, 0.2 μm.

[15] Cell Walls and External Layers

By JÜRGEN WECKESSER and UWE JOHANNES JÜRGENS

Introduction

Cells of cyanobacteria are surrounded by cell envelopes consisting of the cytoplasmic membrane, the cell wall (outer membrane plus peptidoglycan), and in many cases a sheath layer.[1] Although the presence of an outer membrane indicates a gram-negative cell wall organization, the peptidoglycan of cyanobacteria has several properties in common with that of gram-positive bacteria, such as thickness, degree of cross-linkage, and presence of covalently linked polysaccharide. Furthermore, in contrast to capsules of bacteria, sheaths of cyanobacteria are often directly visible in the light microscope, and they are stained positively with osmium tetroxide to reveal a fine structure with fibers running parallel to the cell surface. Thus, it has been suggested that cyanobacteria might have a characteristic cell envelope organization.[2] Successful isolation of the cytoplasmic membrane of *Anacystis nidulans* has been achieved recently.[3,4]

It will be shown in this chapter that procedures used with heterotrophic bacteria for the isolation of cell wall and external envelope layers, as well as their polymers, can be successfully applied to cyanobacteria. Many studies published so far were performed with the unicellular *Synechocystis* sp. PCC 6714. Thus, the following procedures refer to this cyanobacterium unless otherwise noted.

Fine Structure and Structural Models

Standard procedures, preferably using freshly harvested cells without storage, are applied for fine-structure studies on the various layers of the cyanobacterial cell envelope. These include total cell preparations stained with phosphotungstic acid, ultrathin sections of whole cells and cell walls, and freeze-etching studies.[5,6] A three-dimensional model of the regularly structured surface layer from *Synechocystis* has been suggested, based on

[1] G. Drews and J. Weckesser, *in* "The Biology of Cyanobacteria" (N. G. Carr and B. A. Whitton, eds.), p. 333. Blackwell, Oxford, England, 1982.

[2] U. J. Jürgens and J. Weckesser, *J. Bacteriol.* **168,** 568 (1986).

[3] T. Omata and N. Murata, *Plant Cell Physiol.* **24,** 1101 (1983).

[4] T. Omata and N. Murata, *Arch. Microbiol.* **139,** 113 (1984).

[5] J. R. Golecki, *Arch. Microbiol.* **114,** 35 (1977).

[6] B. Büdel and E. Rhiel, *Arch. Microbiol.* **143,** 117 (1985).

electron microscopy and Fourier reconstruction data.[7] Experimental conditions have been worked out that shift the cell surface hydrophobicity/hydrophilicity of *Phormidium* sp. strain J-1, by mechanical shearing, chloramphenicol, and proteolytic treatments after preincubation with sodium dodecyl sulfate (SDS).[8] The hydrophobicity test uses a biphasic system of growth medium and hexadecane. Hexadecane (0.5 ml) is added to 5 ml of cell suspension; the mixture is shaken and then allowed to stand for phase separation. The optical density of the aqueous phase is then compared to that of original cell suspension. The difference is expressed as percent hydrophobicity. On the basis of the surface properties and ultrastructural analysis, a surface model was proposed.

Isolation of Cell Envelope, Cell Wall, and Outer Membrane

Growth of Cells

Mass cultures of at least 5–20 g cell wet weight are usually necessary for performing quantitative analysis on cell envelopes. Many cyanobacteria can be grown photoautotrophically in BG-11 medium,[9] pH 7.5, at 25° in, e.g., a 12-liter fermentor, gassed continuously by a stream of air and carbon dioxide (250 and 2.5 liters/hr, respectively) and illuminated with white-light fluorescent lamps (5,000 lux at the fermentor surface). Harvested cells are washed once with 20 m*M* tris(hydroxymethyl)aminomethane (Tris)–HCl buffer, pH 8.0.[10] Cultures are used freshly harvested for fine-structure studies, or they may be stored at −20° for the isolation of cell envelope or sheath fractions. For the isolation of lipopolysaccharides cells are lyophilized.

Preparation of Cell Envelopes

Crude cell envelope fractions are separated from cell homogenates obtained by mechanical disruption of cells and by differential (and sucrose density) centrifugation. For mechanical disruption, cells (10–25 g wet weight, freshly harvested or stored at −20°, suspended in 20 ml Tris buffer) are mixed with glass beads (0.17–0.18 mm in diameter; cell-to-glass bead ratio, 1 : 2, v/v). The cells are broken in a Vibrogen shaker (e.g., Type Vi 2, E. Bühler, Tübingen, FRG) at full speed (4°, 15 min).

[7] B. Karlsson, T. Vaara, K. Lounatmaa, and H. Gyllenberg, *J. Bacteriol.* **156,** 1338 (1983).

[8] Y. Bar-Or, M. Kessel, and M. Shilo, *Arch. Microbiol.* **142,** 21 (1985).

[9] R. Y. Stanier, R. Kunisawa, M. Mandel, and G. Cohen-Bazire, *Bacteriol. Rev.* **35,** 171 (1971).

[10] Tris buffer is used except as noted differently.

After light microscopy examination to assess cell breakage, the glass beads are removed by filtration on a glass filter (G-1, Schott, Mainz, FRG), and unbroken cells are separated by centrifugation at 270 *g* at 4° for 10 min. The crude cell envelope fraction (orange or yellow–green in color) is sedimented from the supernatant at 12,000 *g* at 4° for 30 min and washed with Tris buffer (same centrifugation conditions) until the supernatant is colorless.

Alternatively, *Anacystis nidulans* (synonym *Synechococcus* sp. PCC 6301) cells are pretreated with lysozyme, 2 m*M* EDTA, and 600 m*M* sucrose in 30 m*M* sodium potassium phosphate buffer, pH 6.8, at 30° for 2 hr. Cells are broken by passage through a French pressure cell at 200 kg/ cm^3. The cell envelope fraction is separated from the cell homogenate by differential centrifugation (5,000 *g* for 15 min for removal of unbroken cells, then 20,000 *g* at 5° for 60 min to collect thylakoids and cell envelopes) followed by sucrose density centrifugation of the final pellet (4 ml of 30–90%, w/v, sucrose, containing 10 m*M* NaCl and 10 m*M* TES–NaOH buffer, pH 7.2; 180,000 *g* for 6 hr).[11] A similar procedure for isolation of cell envelopes of *Synechocystis* sp. PCC 6714 was performed.[4]

Preparation of Cell Walls

For isolation of cell walls, freshly harvested crude cell envelopes in Tris buffer (2–3 mg protein/ml, 10 ml) are loaded on a discontinuous sucrose gradient (10 ml of each 60, 55, 50, 45, and 40% sucrose in Tris buffer) and run in a SW 25.2 rotor (Beckman Instruments, Fullerton, CA) at 20,000 rpm for 4 hr. Cell walls are recovered from both the pellet and the band at 60% sucrose and are further purified by using the same gradient 3 times. Sucrose is removed by washing (176,000 *g* at 4° for 1 hr) the cell walls at least 5 times with ice-cold Tris buffer or, alternatively, by dialysis against Tris buffer. Cell walls, devoid of thylakoids and cytoplasmic membranes, as indicated by lack of chlorophyll *a* and phycobiliproteins,[11–13] are orange–red or yellow in color.

Gradient-purified cell walls may be further purified by treatment with 2% Triton X-100 (final concentration) in 10 m*M* *N*-2-hydroxethylpiperazine-*N*′-2-ethanesulfonic acid (HEPES) buffer, pH 7.4 (or in Tris buffer), containing 10 m*M* $MgCl_2$. Treatment is at 23°, 2 times for 30 min.[14] The Triton-insoluble fraction is collected by ultracentrifugation (176,000 *g*, 4°,

[11] N. Murata, N. Sato, T. Omata, and T. Kuwabara, *Plant Cell Physiol.* **22,** 855 (1981).
[12] C. M. Resch and J. Gibson, *J. Bacteriol.* **155,** 345 (1983).
[13] U. J. Jürgens and J. Weckesser, *J. Bacteriol.* **164,** 384 (1985).
[14] C. A. Schnaitman, *J. Bacteriol.* **108,** 553 (1971).

1 hr). The detergent is removed by washing the pellet with extraction buffer (see above) without detergent.

Purification of cell walls from *Synechococcus* sp. strains was also achieved by hydrophobic interaction chromatography of crude membrane preparations on a Phenyl–Sepharose CL-4B column preequilibrated with 750 m*M* potassium phosphate buffer, pH 7.2. A yellow-colored cell wall fraction eluted at the void volume, whereas the more hydrophobic thylakoids and phycobilisomes remained adsorbed to the column. A similar purification was achieved by aqueous phase separation of crude membrane preparations using polyethylene glycol (20,000 molecular weight). A yellow cell wall fraction partitioned into the bottom phase, while unbroken cells partitioned into the more hydrophobic lower phase.[12] Methods for the isolation of the cell wall of the gliding *Aphanothece halophytica* have been described.[15] Sonication was performed on the filamentous *Phormidium foveolarum* and *Tolypothrix tenuis* in order to remove the slimy sheaths.[16] Isolation of the cell wall from a strain of *Anabaena cylindrica* with a mucilaginous sheath requires differential centrifugation, cell breakage by sonication, and further purification by treatment with 1% SDS followed by additional centrifugation steps for removal of unbroken cells.[17]

Preparation of Outer Membrane

Outer membranes can be isolated from cell wall fractions (freshly prepared or frozen at −20°) by removing peptidoglycan by digestion with lysozyme.

Gradient-purified cell walls and Triton X-100 insoluble cell walls, respectively (see above), are incubated with hen egg white lysozyme (EC 3.2.1.17, 53,000 U/mg protein; Sigma Chemical Co., St. Louis, MO) in 20 m*M* ammonium acetate buffer, pH 6.5, at 37° for 24 hr (enzyme–substrate ratio, 1 : 25, w/w). The digested cell walls are centrifuged (176,000 *g*, 4°, 1 hr) and washed once with Tris buffer. Outer membranes are recovered from the final pellet and resuspended in a small volume of Tris buffer. The suspension is loaded on a discontinuous sucrose gradient (10 ml each of 60, 55, 50, 45, and 40% sucrose in Tris buffer) and run in a SW 25.2 rotor (Beckman Instruments, Inc., Fullerton, CA) at 20,000 rpm for 12 hr. Outer membranes, deeply orange–red or yellow-colored, are obtained from the band at 50% sucrose. Sucrose is removed by dialysis against Tris buffer, before sedimenting the outer membrane by ultracentrifugation.

[15] R. S. Simon, *J. Bacteriol.* **148,** 315 (1981).
[16] H. Höcht, H. H. Martin, and O. Kandler, *Z. Pflanzenphysiol.* **53,** 39 (1965).
[17] J. H. Dunn and C. P. Wolk, *J. Bacteriol.* **103,** 153 (1970).

Stability and Function of Cell Walls

Relative strength of cell walls of cyanobacteria can be measured by determining their resistance to disruption by sonic oscillation.[13] They are rather resistant toward detergent treatment. No change in polypeptide composition was found after a 16-hr extraction (20°) with various detergents.[11] Triton X-100 extraction (see Preparation of Cell Walls) affects neither polypeptide pattern nor fine structure of the *Synechocystis* sp. PCC 6714 outer membrane.[12] This is likely due to the addition of Mg^{2+} ions to the incubation mixture stabilizing the outer membrane. However, partial loss of lipopolysaccharide and lipids is observed under these conditions. Double-track appearance of outer membrane is retained on trichloromethane/methanol extraction[17] (see also Lipids) of *Anacystis nidulans* cells even after additional extraction with 2% SDS (37°, 30 min) prior to or after lipid extraction.[5] Extraction of cell walls with 2% SDS (37°, 30 min) removes minor proteins but not the peptidoglycan-associated major outer membrane proteins.[13] Peptidoglycan-associated outer membrane proteins are removed by treatment with SDS at temperatures above 70°. The outer membrane of *Anacystis nidulans* is solubilized by hot phenol–water extraction[5] and that of *Anabaena variabilis* by 1% Genapol X-80 treatment of lysozyme-digested cell walls.[18] The outer membrane of *Synechococcus* sp. is stripped from intact cells by incubation with 10 m*M* EDTA.[12]

Permeability measurements have been made of outer membrane proteins of *Anabaena variabilis* with porin activity reconstituted into black lipid bilayer membranes.[18] Alternatively, the liposome swelling test[19] may be used for measuring porin activity. For the latter, cell walls, outer membrane fractions, or isolated porin may be used.

Isolation of the Rigid Layer and Its Polymers

Peptidoglycan–Polysaccharide–Protein Complex: Rigid Layer

The rigid layer is considered here as peptidoglycan plus covalently bound polymers. Experimentally, it is obtained as the SDS-insoluble (100°) fraction of the cell wall.[20] In the case of *Synechocystis* sp. PCC 6714, and possibly in other cyanobacteria, the rigid layer consists of peptidoglycan with covalently linked polysaccharide and covalently linked protein.[2] For isolation of rigid layers, crude cell envelopes or cell walls

[18] R. Benz and H. Böhme, *Biochim. Biophys. Acta* **812,** 286 (1985).
[19] M. Luckey and H. Nikaido, *Proc. Natl. Acad. Sci. U.S.A.* **77,** 167 (1980).
[20] U. J. Jürgens, G. Drews, and J. Weckesser, *J. Bacteriol.* **154,** 471 (1983).

(freshly prepared or stored at −20°) are suspended in 20 ml distilled water, and the suspension is added dropwise with stirring to 200 ml of boiling 4% SDS with 0.1% (final concentration) mercaptoethanol added.[21] After boiling for 15 min and cooling to room temperature, the suspension is stirred overnight. Rigid layers are sedimented at 176,600 *g* at 20° for 1 hr. Extraction and sedimentation are repeated twice. The final sediment is washed with distilled water by resuspension and ultracentrifugation (176,000 *g* at 20° for 1 hr) until free of detergent.[22]

Peptidoglycan–Polysaccharide Complex

The presence of polysaccharide covalently bound to peptidoglycan is indicated by the finding of neutral sugars and/or nonpeptidoglycan amino sugars in rigid layer fractions. For the isolation of the peptidoglycan–polysaccharide complex, the rigid layer protein must be removed. Rigid layers (50–100 mg, freshly prepared or stored at −20°, suspended in 20 m*M* Tris–HCl buffer, pH 7.4) are incubated with pronase (from *Streptomyces griseus,* 6 U/mg protein; Sigma Chemical Co., St. Louis, MO), 5–10 mg, at 37° for 24 hr.[21] After SDS extraction 2 times and removal of the detergent by washing with distilled water (as for isolation of the rigid layer, see above), the peptidoglycan–polysaccharide complex is collected by ultracentrifugation (176,000 *g* at 4° for 1 hr) and lyophilized.

Isolated Peptidoglycan

In *Synechocystis* sp. PCC 6714 the rigid layer polysaccharide is linked to peptidoglycan via phosphodiester bridges.[2] The linking site is MurN-6-P. Thus, the polysaccharide can be removed from the rigid layer by treatment of the peptidoglycan–polysaccharide complex with hydrofluoric acid (HF) in the cold. Lyophilized peptidoglycan–polysaccharide complex (up to 30 mg of fraction dry weight) is suspended in 1 ml ice-cold 48% HF, and the suspension is kept at 0° for 48 hr.[23,24] The acid is removed by evaporation or by dialysis against ice-cold distilled water. The lyophilized residue is suspended in ice-cold distilled water and neutralized with 100 m*M* LiOH. After centrifugation at 48,000 *g* at 4° for 30 min (the supernatant contains the polysaccharide) and several washes with distilled water,

[21] V. Braun and K. Rehn, *Eur. J. Biochem.* **10,** 426 (1969).

[22] K. Hayashi, *Anal. Biochem.* **67,** 503 (1975).

[23] D. Lipkin, B. E. Phillips, and J. W. Abrell, *J. Org. Chem.* **34,** 1539 (1969).

[24] F. Fiedler, M. J. Schäffler, and E. Stackebrandt, *Arch. Microbiol.* **129,** 85 (1981).

the HF-treated peptidoglycan is collected as the final pellet. The structure of peptidoglycan from *Synechocystis* sp. PCC 6714 has been analyzed.[20]

Polysaccharide Linked to Peptidoglycan Glycopeptides

The rigid layer polysaccharide with residual glycopeptides of peptidoglycan bound is obtained by digestion of the peptidoglycan–polysaccharide complex with lysozyme followed by preparative gel filtration of the cleavage products. Freshly prepared (or frozen at −20°) peptidoglycan–polysaccharide complex (50–100 mg, in 20 m*M* ammonium acetate buffer, pH 6.5) is ultrasonicated at 4° for 10 min, adjusted to an optical density (578 nm) of 0.8–1.0, and digested with lysozyme (see Preparation of Outer Membrane) at 37° for 24 hr. Digestion is monitored by measuring the absorbance of the suspension at 578 nm and by determining the formation of reducing groups[25] during incubation. Subsequent dialysis against distilled water removes the liberated peptidoglycan glycopeptides. The polysaccharide fraction with linked peptidoglycan glycopeptides is concentrated to about 2 ml on a rotary evaporator and then chromatographed on a Sephadex G-50 (fine) column (2.6 × 90 cm; Pharmacia, Uppsala, Sweden) at a flow rate of 50 ml/hr. Fractions (~5.0 ml each) are tested for reducing sugars, amino groups, and phosphorus.[2] The glycopeptide-linked polysaccharide elutes with the void volume. The combined respective fractions are dialyzed against distilled water and lyophilized.

Isolated Polysaccharide of the Rigid Layer

The rigid layer polysaccharide is obtained from the supernatant (lyophilized) of the centrifugation at 48,000 *g* at 4° for 30 min that sediments the peptidoglycan (see Isolated Peptidoglycan) after cleavage of the peptidoglycan–polysaccharide complex by HF treatment. The supernatant is dialyzed against ice-cold distilled water and lyophilized.

Fragments of the Peptidoglycan–Polysaccharide Complex

The lyophilized peptidoglycan–polysaccharide complex (5 mg in 1 ml) is subjected to partial acid hydrolysis (1 *M* HCl, 100°, 1 hr), and the hydrolyzate is evaporated to dryness. The residue is dissolved in distilled water and adjusted to a final concentration of 50 μg/μl; 500 μg is subjected to high-voltage paper electrophoresis in pyridine–acetic acid–water (5 : 2 : 43, v/v/v), pH 5.3, at 3,000 V for 90 min. Fragments are detected

[25] G. Keleti and W. H. Lederer, "Handbook of Micromethods for the Biological Sciences." Van Nostrand–Reinhold, New York, 1974.

by staining with alkaline silver nitrate and ninhydrin, or with fluorescamine (0.05% in acetone)[2] if the separation is for preparative purposes.[20]

Isolated Protein of the Rigid Layer

The peptidoglycan–polysaccharide–protein complex (freshly prepared or stored at −20°, ~100 mg) is incubated with *N,O*-diacetylmuramidase (EC 3.2.1.17; from *Chalaropsis* sp., 500 U/mg protein)[26] in 20 m*M* ammonium acetate buffer, pH 4.8, at 37° for 24 hr. After ultracentrifugation (176,000 *g* at 4° for 1 hr), the sediment containing the rigid layer protein is extracted with 4% SDS at 100° for 15 min. Following centrifugation at 48,000 *g* at 20° for 30 min, acetone (final concentration 80%) is added to the supernatant in order to precipitate the protein. The precipitate is collected at 48,000 *g* (4°, 30 min) and lyophilized after resuspension in distilled water. The isolated protein fraction is subsequently treated with trifluoroacetic acid (TFA, 1.5 g/ml, at 20° for 30 min) in sealed tubes to remove protein-bound SDS and noncovalently bound lipids. The TFA-insoluble protein is collected by centrifugation (48,000 *g* at 4° for 30 min).

Isolation of Outer Membrane Polymers

Lipopolysaccharide

Lipopolysaccharide has been obtained from unicellular as well as filamentous cyanobacteria, using hot phenol–water extraction.[27] In some cases, the lipopolysaccharide is extracted into the phenol phase and not, as is usual, into the water phase. Only very little or even no lipopolysaccharide has been obtained from some cyanobacteria, using this method.[28]

Lyophilized cyanobacteria (10 g, suspended in 175 ml distilled water) are heated while stirring to 67°. Phenol (91%, 175 ml, preheated to 67°) is then added, and the mixture is stirred at 67° for 20 min. After cooling to 4–10° in an ice bath, centrifugation at 1,450 *g* at 4–10° for 30 min separates the upper, water phase from the lower, phenol phase, with an interphase of variable thickness in between. The water phase is removed carefully. Combined interphase phenol-insoluble material is reextracted (67°) for 5 min in 175 ml preheated distilled water. After phase separation (as above), the water phases of the two extraction steps are combined. The interphase is combined with phenol phase (however, see Isolation of the

[26] J. H. Hash and M. V. Rothlauf, *J. Biol. Chem.* **242,** 5586 (1967).

[27] O. Westphal, O. Lüderitz, and F. Bister, *Z. Naturforsch.* **7,** 148 (1952).

[28] J. Weckesser, G. Drews, and H. Mayer, *Annu. Rev. Microbiol.* **33,** 215 (1979).

Sheath, below). Phenol is removed from the water and phenol phases by dialysis against running tap water and finally against distilled water before concentration of the extracted material by rotary evaporation and centrifugation (2,500 *g* at 4° for 30 min). Additional lipopolysaccharide was found in ammonium oxalate extracts (1%, 100°, 8 hr) of *Phormidium* sp. cells that had been preextracted by hot phenol–water.[29] For purification, lyophilized water or phenol phase extracts (1–2% in distilled water) are centrifuged at 105,000 *g* at 4° for 4 hr (3 times). Lipopolysaccharide is obtained in the final sediment from either of the two phases, depending on the strain. Purity of lipopolysaccharide has been tested by isopycnic gradient centrifugation with CsCl gradients.[30]

Proteins of the Outer Membrane

The outer membrane fraction (freshly prepared or stored at −20°, 200 mg wet weight) is dissolved in 20 ml extraction buffer (2% SDS, 10% glycerol, and 0.01% mercaptoethanol in 10 m*M* Tris–HCl, pH 8.0) and heated to 70° while stirring.[31] The supernatant of a subsequent ultracentrifugation (176,000 *g* at 20° for 1 hr) contains the soluble membrane proteins and is adjusted with ice-cold acetone to a final concentration of 80% (v/v) in order to precipitate the proteins.

Alternatively, for porin activity determination of proteins (see Function and Reconstitution of Outer Membrane) the extraction mixture can be diluted with Tris buffer to a final SDS concentration below 0.2% (critical micellar concentration) before dialysis against distilled water. Acetone-precipitated or dialyzed membrane proteins are dissolved in 0.5 ml extraction buffer and chromatographed in 0.5% SDS, 20 m*M* Tris–HCl buffer, pH 8.0, on Sephadex G-200 (Pharmacia).[32]

Protein is determined quantitatively according to the method of Lowry *et al.*[33] Polypeptides can be separated by SDS–polyacrylamide gel electrophoresis (SDS–PAGE); (gradient 11–18% acrylamide, apparent molecular weight of proteins determined by comparison with standards.[34] Factors such as temperature, NaCl concentration, and 2-mercaptoethanol influence electrophoretic mobility in SDS–PAGE.[32]

[29] L. V. Mikheyskaya, R. G. Ovodova, and Y. Ovodov, *J. Bacteriol.* **130,** 1 (1977).
[30] S. Raziuddin, H. W. Siegelman, and T. G. Tornabene, *Eur. J. Biochem.* **137,** 333 (1983).
[31] T. Mizuno and M. Kageyama, *J. Biochem.* (*Tokyo*) **86,** 979 (1979).
[32] K. Nakamura and S. Mizushima, *J. Biochem.* (*Tokyo*) **80,** 1411 (1976).
[33] H. Lowry, N. J. Rosebrough, A. L. Farr, and R. J. Randall, *J. Biol. Chem.* **193,** 265 (1951).
[34] B. Lugtenberg, R. van Boxtel, C. Verhoef, and M. van Alphen, FEBS Lett. **96,** 99 (1978).

Peptidoglycan-Associated Proteins

Gradient-purified or Triton X-100-insoluble cell walls (200 mg wet weight) are suspended in 10–20 ml extraction buffer (see Proteins of the Outer Membrane) and are extracted at different temperatures (30–90°C). The rigid layer with associated proteins is sedimented at 176,600 *g* at 20° for 1 hr and washed 2 times with Tris buffer by centrifugation.[13] The sediment is suspended in 5 ml extraction buffer (the supernatant can be used for isolation of nonpeptidoglycan-associated outer membrane proteins). After boiling for 5–10 min, followed by centrifugation at 176,600 *g* at 20° for 1 hr, the peptidoglycan-associated proteins in the supernatant are precipitated by ice-cold acetone (final concentration 80%, v/v). The peptidoglycan-associated membrane proteins may be separated by SDS–PAGE (see above).

Lipids

Cell envelope, cell wall, or outer membrane fractions (~0.5 g wet weight) are suspended in a mixture of 5 ml trichloromethane and 5 ml methanol.[35] After addition of 5 ml trichloromethane and 5 ml 1% NaCl (in water), the extraction mixture is shaken vigorously. Rapid phase separation is achieved by centrifugation at 5,000 *g* at 4° for 15 min. Following removal of the trichloromethane phase, the water–methanol phase is extracted 4 times with trichloromethane (5 ml each). The combined trichloromethane phases containing the lipids are concentrated at room temperature in a stream of N_2 (250–500 μl) and kept in the dark at −80°.

Lipids are separated by two-dimensional thin-layer chromatography on silica gel plates (20 × 20 cm; Merck, Darmstadt, FRG), using in the first dimension trichloromethane–methanol–water (65 : 25 : 4, v/v/v) and in the second dimension trichloromethane–methanol–ammonia–isopropylamine (65 : 35 : 4 : 0.5, v/v/v/v).[36] Lipids are detected with the fluorescent dye primuline (0.1% in ethanol) under ultraviolet light at 350 nm.[37]

Carotenoids

Carotenoids (and/or chlorophyll) of the cell envelope, cell wall, or outer membrane fractions[13] are extracted like lipids (see above) and can be stored at −80°. Carotenoids are separated by thin-layer chromatography on silica gel plates (5 × 10 cm, aluminum foil; Merck) in the dark in the following solvent system: light petroleum (bp 40°–60°)–2-

[35] E. G. Bligh and W. J. Dyer, *Can. J. Biochem. Physiol.* **37,** 911 (1959).

[36] H. Kleinig and U. Lempert, *J. Chromatogr.* **53,** 595 (1970).

[37] N. Sato, N. Murata, Y. Miura, and N. Ueta, *Biochim. Biophys. Acta* **572,** 19 (1979).

propanol–water (100 : 11 : 0.5, v/v/v). The aluminum foil is immediately dipped into liquid N_2 in the dark after the run, to avoid oxidation and bleaching of pigments. Colored bands are then scraped off the silica gel and transferred to a small volume of the appropriate solvent (ethanol for zeaxanthin; *n*-hexane for β-carotene; acetone for echinone, myxoxanthophyll and related carotenoid glycosides, and other carotenoids.[38] Silica gel is removed by centrifugation (5,000 *g*, 4°, for 2 min). Spectra of the supernatants are recorded with a spectrophotometer. The extinction coefficients, $E^{1\%}_{1\,cm}$, at the absorption maxima are 2,160 for myxoxanthopyll and related carotenoid glycosides in acetone, 2,480 for zeaxanthin in ethanol, 2,340 for echinone and other carotenoids in acetone, and 2,592 for β-carotene in *n*-hexane.[39] Carotenoids are identified by their R_f values on thin-layer chromatography and their absorption spectra.[38]

Isolation of the Sheath

Sheaths of many cyanobacteria show considerable mechanical and physicochemical stability even when subjected to drastic methods during isolation. This greatly facilitates their isolation. Methods for separation of the sheath include cell treatment by Ultra-Turrax, ultrasonication, cell disintegration by a Vibrogen shaker followed by differential and sucrose gradient centrifugation, or hot phenol–water extraction of the cell homogenate. Criteria for success of the method are light microscopy examination of the cell homogenate (relative amount of separated sheaths should be high compared to that of remaining ensheathed cells). The purity of isolated sheath fractions can be examined by analyzing, e.g., for peptidoglycan constituents (see Amino Acids, below), for outer membrane-specific constituents such as lipopolysaccharide (see Analyses of Monomeric Compounds), or for chlorophyll *a* of internal membranes.[40] Three examples of sheath isolation procedures are given here.

Sheaths of Chroococcus and Gloeothece

Cells of *Chroococcus minutus* SAG B 41.79 (40 g wet weight) are suspended in Tris buffer and mixed with glass beads (0.25 mm in diameter, cell-to-glass bead ratio 1 : 2, v/v, to yield a viscous paste).[41] DNase

[38] H. Stransky and A. Hager, *Arch. Microbiol.* **72,** 84 (1970).

[39] T. W. Goodwin, "Chemistry and Biochemistry of Plant Pigments," Vol. 2. Academic Press, New York, 1976.

[40] G. MacKinney, *J. Biol. Chem.* **140,** 315 (1941).

[41] S. P. Adhikary, J. Weckesser, U. J. Jürgens, J. R. Golecki, and D. Borowiak, *J. Gen. Microbiol.* **132,** 2595 (1986).

(~2 mg) is added. The cells are broken in a Vibrogen shaker (Type Vi 2, Bühler) at 4° and maximum speed for 3 min up to 2 hr. A reddish brown-colored homogenate indicates a sufficient degree of cell breakage. After removal of the glass beads by filtration (see Preparation of Cell Envelopes), crude sheath fractions are separated at 750 *g* at 4° for 45 min. The supernatant is centrifuged at 12,000 *g* at 4° for 45 min, and the sediment is washed 4 times (same conditions). Discontinuous sucrose gradients (10 ml of 60%, and 5 ml of each, 55, 50, 45, and 40% sucrose, w/w, in Tris buffer) are loaded with the washed sediment and run in a swinging bucket rotor at 16,300 *g* at 4° for 4 hr. Sheaths are recovered from the 60% sucrose band and further purified once on the same gradient. The sheath material is then washed in Tris buffer 4 times at 12,100 *g* at 4° for 45 min each.

If necessary, residual cell wall contaminants are removed by treatment with lysozyme (see Preparation of Outer Membrane) followed by extraction with Triton X-100 (2%, w/v, in 10 m*M* disodium EDTA and 10 m*M* $MgCl_2$, at room temperature for 20 min) or SDS (4%, w/v, in Tris buffer, at 60° for 15 min). The purified sheath is washed 7 times with distilled water (12,000 *g* at 4° for 45 min), collected at 12,000 *g* at 4° for 45 min, and finally lyophilized.

Cells of *Gloeothece* PCC 6501 are homogenized as above. Unbroken cells are sedimented at 120 *g* at 4° for 15 min, and the supernatant is centrifuged at 3,020 *g* at 4° for 15 min, whereby a double-layered sediment with an upper, light brown-colored, slimy sheath layer is obtained. The layer is carefully removed and resuspended in a small volume of water. Centrifugation is repeated 3 times (same experimental conditions).

Sheath of Calothrix

By French Pressure Cell/Lysozyme/SDS Treatment. Lyophilized cells of *Calothrix parietina* or *C. scopulorum* (300 mg), suspended in 40 ml Tris buffer, are broken twice in a French pressure cell (12,000 psi). After sedimentation of the crude sheath fraction at 750 *g* at 4° for 45 min, contaminating cell wall is removed by treatment with lysozyme (see Preparation of Outer Membrane). The sheath (more or less colorless) is obtained in the upper layer of the sediment after centrifugation at 20,000 *g* at 4° for 30 min. The layer is carefully removed and washed several times with distilled water. Finally, the sheath material is suspended in 20 ml distilled water and added dropwise to 20 ml boiling 2% SDS. After cooling to room temperature, the sheath is obtained as the sediment of a 12,000 *g* (22°, 30 min) centrifugation. It is washed at least 7 times (same centrifugation conditions) before lyophilization.

By French Pressure Cell/Hot Phenol–Water Treatment. Cells of *Calothrix* sp. are broken with the French pressure cell as above. On treatment of the homogenate with hot phenol–water (see Lipopolysaccharide), sheath material is obtained from the phenol–water interphase. With *Gloeothece* PCC 6501, however, most of the sheath material is extracted into the water phase.

Sheath of Chlorogloeopsis

Cells of *Chlorogloeopsis* PCC 6912 are suspended in 10 m*M* Tris–HCl buffer, pH 7.8, containing 1 m*M* EDTA and 1 m*M* NaN_3. They are homogenized by shaking with glass beads (0.25 mm in diameter) at 4° for 10 min in a Braun disintegrator (Type MSK; Braun, Melsungen, FRG). Centrifugation at 160 *g* (4° for 30 min, 2 times) separates the heavy sheath from the supernatant.[42] A continuous sucrose gradient (60–40%, w/w; 600 *g* for 50 min) is used for further purification. Sucrose is removed from the sedimented sheath by dialysis against distilled water prior to lyophilization of the sheaths.

Isolation of External Cell Envelope Polysaccharides

Polysaccharides from external cell envelope layers, such as slime, capsule, or sheath, can be extracted into the water phase of hot phenol–water extracts (see Lipopolysaccharide) depending on their water solubility. They are obtained from the combined water phase supernatants (after centrifugations at 105,000 *g* at 4° C for 4 hr) treated with RNase[43] and α-amylase.[44] Enzymes are removed with hot phenol–water (see above). The polysaccharide(s) is obtained in the supernatant of a centrifugation (105,000 *g* at 4° for 4 hr) of the water phase.

Extracellular polysaccharides may be precipitated from the culture medium by the addition of two volumes of absolute ethanol, as successfully applied to *Nostoc* sp.[45] The mucilaginous sheath of the *Nostoc* strain was solubilized in hot water, with 3 successive extractions. Mucilage of *Anabaena variabilis* sheaths is similarly obtained.[17]

Cell Envelopes of Heterocysts and Spores

The work of the group of P. Wolk describes methods for structural identification of glycolipid in the envelope of heterocysts of *Anabaena cylindrica*. The study involves IR, NMR, and mass spectrometry.[46] Meth-

[42] M. Schrader, G. Drews, J. R. Golecki, and J. Weckesser, *J. Gen. Microbiol.* **128,** 267 (1982).

[43] T. Y. Lin and E. C. Gottschlich, *J. Biol. Chem.* **238,** 1928 (1963).

[44] P. Bernfeld, this series, Vol. 1, p. 149.

[45] V. B. Mehta and B. S. Vaidya, *J. Exp. Bot.* **29,** 1423 (1978).

[46] F. Lambein and C. P. Wolk, *Biochemistry* **12,** 791 (1973).

ods for structural identification of polysaccharides in envelopes of *Anabaena variabilis* and *Cylindrospermum licheniforme* are also described.[47]

Analyses of Monomeric Compounds

Only a few comments on the many quantitative procedures used for analyses of cell envelope constituents are given here (for experimental details, see elsewhere).[25] Methods suggested here can be applied to either the various cell envelope fractions or their polymers (lyophilized before analysis). Methods of investigation of lipopolysaccharides were summarized recently.[48] The metal adsorption capacity of sheaths of *Microcystis* has been measured.[49]

Neutral Sugars and Sugar Alcohols. Neutral sugars or sugar alcohols are liberated by hydrolysis in 500 mM H_2SO_4 at 100° for 4 hr. Saturated $Ba(OH)_2$ is used for neutralization. Alternatively, 0.1 N HCl at 100° for 48 hr may be used for milder liberation. Neutralization, e.g., for gas–liquid chromatographic analyses, may be done with IRA 410 $(HC)_3^-)$ ion exchanger in order to remove Cl^-. Thin-layer chromatography of neutral sugars is performed on cellulose plates (20 × 20 cm; Merck), using either ethyl acetate–pyridine–water (12 : 5 : 4, v/v/v) or *n*-butanol–pyridine–water (6 : 4 : 3, v/v/v) as the solvent system.[50] Chromatograms are stained with anilinium hydrogen phthalate[51] (hexoses stain brown with yellow fluorescence in UV light, pentoses red with reddish fluorescence) or with alkaline silver nitrate.[52]

Neutral sugars can be separated by gas–liquid chromatography as alditol acetates[53] on an ECNSS-M 2 mm × 1.5 m glass column, 3% on Gas Chrom Q, 100–200 mesh. For mass spectrometric analyses, sugar alditol acetates (reduced with $NaBD_4$)[54] are separated on a fused silica capillary column SE-54 (0.25 mm × 25 m). A column temperature gradient from 120 to 230° (in a 5°/min program) is suitable (injection temperature, 270°). Sugar alcohols are separated as trifluoroacetyl derivatives.[55] Mass spectra are taken at 70 eV in the mass range 40–400 *m/e*.

[47] L. Cardemil and C. P. Wolk, *J. Phycol.* **17,** 234 (1981).
[48] H. Mayer, R. N. Tharanathan, and J. Weckesser, *in* "Methods in Microbiology" (G. Gottschalk, ed), Vol. 18, p. 157. Academic Press, New York, 1985.
[49] Y. Amemiya and O. Nakayama, *Jpn. J. Limnol.* **45,** 187 (1984).
[50] K. V. Giri and V. N. Nigam, *J. Ind. Inst. Sci.* **36,** 49 (1954).
[51] S. M. Partridge, *Nature (London)* **164,** 443 (1949).
[52] W. E. Trevelyan, D. P. Procter, and J. S. Harrison, *Nature (London)* **166,** 444 (1950).
[53] J. S. Sawardeker, J. H. Slonegger, and A. Jeanes, *Anal. Chem.* **37,** 1602 (1967).
[54] P. E. Jansson, L. Kenne, H. Liedgren, B. Lindberg, and J. Lönngren, *Chem. Commun.* **8,** 1 (1976).
[55] J. N. Mount and H. F. Laker, *J. Chromatogr.* **226,** 191 (1981).

Amino Sugars. Amino sugars, including MurN, and GlcN-6-P from peptidoglycan, are released by hydrolysis in 4 *M* HCl at 105° for 18 hr (MurN-6P for 4 hr). Hydrochloric acid is removed by evaporation *in vacuo* or by a stream of N_2. Amino sugars may be separated by high-voltage paper electrophoresis[56] in pyridine–acetic acid–formic acid–water (2 : 20 : 3 : 180, v/v/v/v), pH 2.8, or in pyridine–acetic acid–water (5 : 2 : 43, v/v/v), pH 5.3. Electropherograms are stained with 0.1% ninhydrin (in acetone) or by alkaline silver nitrate. Separate identification and quantitative determination of all the constituents mentioned above is performed with an automatic amino acid analyzer.[2,57] Gas–liquid chromatography may also be used for separation of amino sugars.[58] MurN-6-P can be separated from peptides, disaccharides, and monomers formed by hydrolysis by cation-exchange chromatography on Dowex 50WX-8 (H^+ form, 0.9 × 20 cm column, gradient 0 to 2 *M* HCl). MurN-6-P elutes with the water fraction.[2]

2-Keto-3-deoxyoctonate and Uronic Acids. 2-Keto-3-deoxyoctonate, although considered a specific marker for lipopolysaccharide, has been found in lipopolysaccharides of some but not all cyanobacteria studied so far.[28] It is liberated by 50 m*M* H_2SO_4 at 100° for 10 min or by stronger hydrolytic conditions, depending on the strain used. Reaction with periodate–thiobarbituric acid allows a colorimetric quantitative determination.[59] For details of identification, see Mayer *et al.*[48]

Uronic acids are released by hydrolysis in 500 m*M* H_2SO_4 at 100° [neutralization with saturated aqueous $Ba(OH)_2$] or in 1 *M* HCl at 100° for 30 min (HCl removed by evaporation *in vacuo*). These uronic acids can be separated by high-voltage paper electrophoresis (see above) and stained with alkaline silver nitrate[52] or naphthoresorcine.[60] A colorimetric method for quantitative determination has been described.[61]

Amino Acids. Conditions for hydrolytic release of amino acids (including *meso*-diaminopimelic acid from peptidoglycan) are as with amino sugars (see above). D- and L-forms of amino acids including isomers of diaminopimelic acid[20] can be separated by gas-liquid chromatography as their *N*-heptafluorobutyrylisobutyl ester derivatives on a Chirasil-Val glass capillary column (0.2 mm × 25 m, temperature program: 90 to 170° C at 2°/min, then at 170° isothermal; Applied Science Europe B. V., Oud, Beijerland, Holland).[2,62]

[56] B. Kickhöfen and R. Warth, *J. Chromatogr.* **33,** 558 (1968).
[57] D. Evers, J. Weckesser, and U. J. Jürgens, *Arch. Microbiol.* **145,** 254 (1986).
[58] W. Niedermeir and M. Tomana, *Anal. Biochem.* **57,** 363 (1974).
[59] V. S. Waravdekar and L. D. Saslaw, *J. Biol. Chem.* **234,** 1945 (1959).
[60] S. M. Partridge, *Biochem. J.* **42,** 238 (1948).
[61] J. T. Galambos, *Anal. Bicohem.* **19,** 119 (1967).
[62] T. W. Larsen and R. F. Thornton, *Anal. Biochem.* **109,** 137 (1980).

Fatty Acids and Acetyl Groups. Fatty acids, liberated by hydrolysis in 4 *M* HCl at 100° for 6 hr, are esterified with Methanol–BF_3 (or -HCl) or diazomethane. They are separated as their methyl ester derivatives via gas–liquid chromatography. A Castorwax, 2.5% on Chromosorb W, 800–100 mesh, column (1.5 m × 2 mm, at 175°), an EGSS-X column (15% on Gas Chrom P, 100–200 mesh, at 165°), or a SE-30 column (3–10%) on Gas Chrom Q (100–120 mesh, at 170°) can be used for separation.[55] Capillary columns coated with CP-Sil 5 or SE-54 are suitable as well. *O*-Acetyl residues are released with 50 m*M* NaOH (23°, 4 hr). *N*-Acetyl (amide-bound) residues are calculated as the difference of total acetyl (liberated by 0.2 *M* HCl at 100° for 18 hr) and *O*-acetyl content.[63]

Organic Phosphorus. Phosphate in biological substances is determined according to Lowry *et al.*[64]

Concluding Remarks

Working on the chemistry and biology of the cell envelope of cyanobacteria, one has to consider that these prokaryotes show considerable morphologic diversity. This is to be anticipated from their taxonomic subdivision into five different sections,[65] ranging from unicellular species to those showing a complex developmental cycle. As mentioned in the introduction, the methods presented here have been applied mainly to unicellular species. Although these methods may be generally useful for studies of the cell envelope and its constituents in cyanobacteria, it must be emphasized that their applicability to other species will have to be proved in each case.

[63] I. Fromme and H. Beilharz, *Anal. Biochem.* **84,** 347 (1978).

[64] O. H. Lowry, N. R. Roberts, K. Y. Leiner, M. L. Wu, and A. L. Farr, *J. Biol. Chem.* **207,** 1 (1954).

[65] R. Rippka, J. Deruelles, J. B. Waterbury, M. Herdman, and R. Y. Stanier, *J. Gen. Microbiol.* **111,** 1 (1979).

[16] Cyanobacterial Fimbriae

By TIMO VAARA and MARTTI VAARA

Introduction

Bacterial Fimbriae

Fimbriae are stiff, tubular filaments of uniform thickness that rise from the bacterial cell surface.[1,2] They are usually 1–2 μm in length and 5–7 nm in diameter and comprise hundreds of identical subunits, fimbrillins, helically arranged along the filament. Fimbrillins are type specific, and apparent mass range from 8000 to 20,000 has been determined. In one respect, however, all fimbrillins are similar: they are particularly rich in hydrophobic amino acids, which results in increased hydrophobicity of fimbriate cells. A major part of the research has been focused on the fimbriae of pathogenic bacteria.[2,3] Some of their fimbriae act as virulence factors by mediating bacterial attachment to mammalian epithelial cells.

Cyanobacterial Fimbriae

Fimbriation in cyanobacteria was first reported by MacRae and coworkers in 1977.[4] Further studies have revealed fimbriae in more than 20 cyanobacterial strains (Table I[5–8]), including unicellular and filamentous, gliding and nongliding, symbiotic and nonsymbiotic, as well as aquatic and terrestrial strains. The best characterized cyanobacterial fimbriae belong to *Synechocystis* CB3 (ATCC 35679),[7] a unicellular planktonic strain isolated from the Gulf of Finland.[9] Its fimbriae (Figs. 1 and 2) are 1–2 μm in length and 6 nm in diameter. They attach to each other and form characteristic bundles (Fig. 1) that consist of some dozens of individual

[1] J. R. Sokatch, *in* "The Bacteria" (I. C. Gunsalus, ed.), Vol. 7, p. 229. Academic Press, New York, 1979.

[2] E. H. Beachey (ed.), "Receptors and Recognition, Vol. 6. Bacterial Adherence." Chapman & Hall, London, 1980.

[3] W. Gaastra and F. K. deGraaf, *Microbiol. Rev.* **46,** 129 (1982).

[4] T. H. MacRae, W. J. Dobson, and H. D. McCurdy, *Can. J. Microbiol.* **23,** 1096 (1977).

[5] G. Guglielmi and G. Cohen-Bazire, *Protistologica* **18,** 151 (1982).

[6] T. Vaara, *Can. J. Microbiol.* **28,** 929 (1982).

[7] T. Vaara, H. Ranta, K. Lounatmaa, and T. K. Korhonen, *FEMS Microbiol. Lett.* **21,** 329 (1984).

[8] H. Dick and W. D. P. Stewart, *Arch. Microbiol.* **124,** 107 (1980).

[9] T. Vaara, M. Vaara, and S. Niemelä, *Appl. Environ. Microbiol.* **38,** 1011 (1979).

TABLE I
FIMBRIATE CYANOBACTERIAL STRAINS

Unicellular form	Strain	Filamentous form	Strain
Synechococcus	PCC 6301[5]	*Spirulina*	PCC 6313[5]
	PCC 6312[5]	*Arthrospira*	PCC 7345[5]
	PCC 6910[6]	*Oscillatoria*	PCC 6304[5]
	PCC 7202[6]		PCC 7811[5]
	PCC 7502[6]	*Nostoc* phycobiont of *Peltigera canina*[8]	
	PCC 7942[5]	Two unidentified gliding strains[4]	
Synechocystis	CB3[6,7]		
	CLII[6]		
	PCC 6308[6]		
	PCC 6701[5,6]		
	PCC 6714[6]		
	PCC 6803[6]		
	PCC 6902[6]		
Microcystis firma	398[6]		

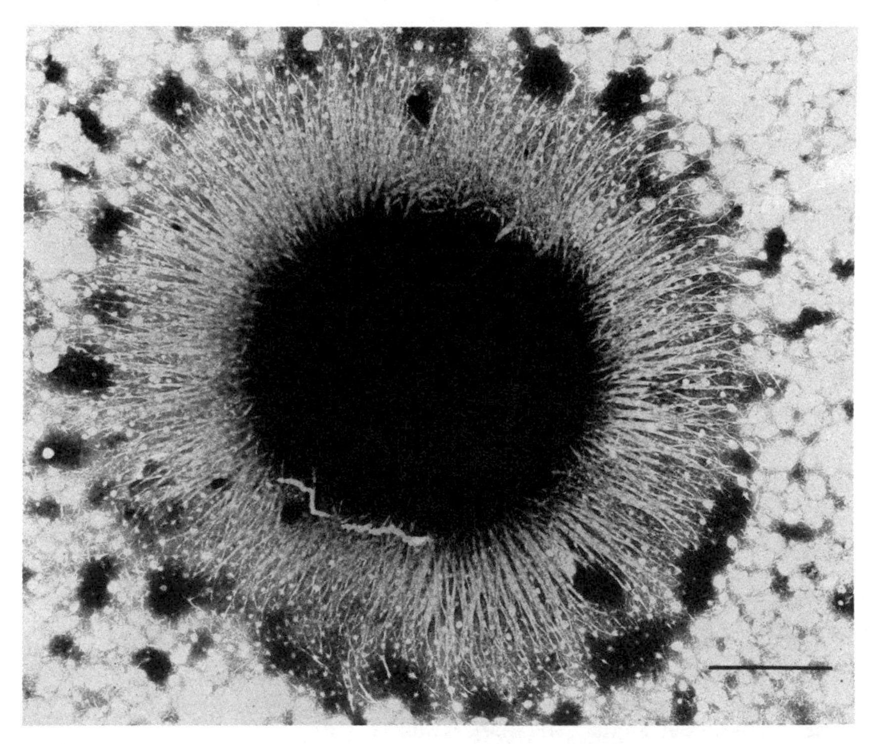

FIG. 1. *Synechocystis* CB3 cells are abundantly covered by fimbriae that tend to attach to each other. Bar, 1 μm.

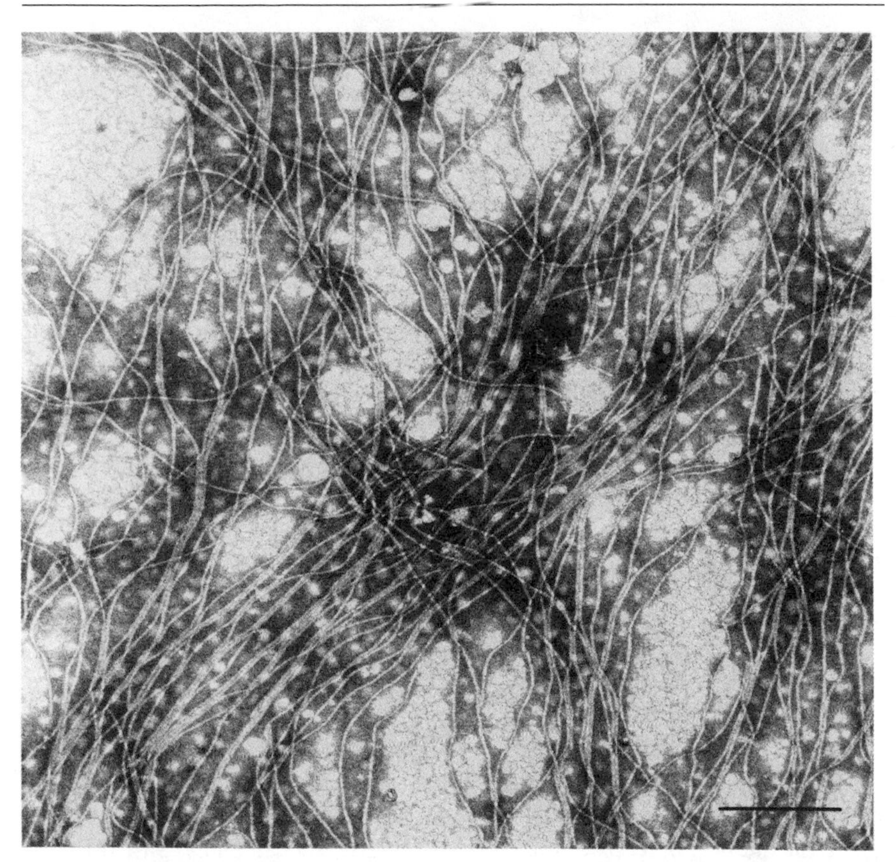

FIG. 2. Isolated *Synechocystis* CB3 fimbriae are free from contaminating cell wall material. Bar, 0.2 μm.

filaments and may be up to 60 nm in diameter.[7,10] This is in contrast to most other bacterial fimbriae, which are well separated and form a hair-like covering around the cell. Bundles of fimbriae have, however, been described for certain bacteria[11–14] including some other cyanobacteria.[15] The *Synechocystis* CB3 fimbriae consist of a single proteinaceous fimbrillin subunit with an apparent molecular weight of 21,000 and a high content (46%) of hydrophobic amino acids.[7] The following procedures have been

[10] K. Lounatmaa, T. Vaara, K. Österlund, and M. Vaara, *Can. J. Microbiol.* **26,** 204 (1980).
[11] R. Yanagava and K. Otsuki, *J. Bacteriol.* **101,** 1063 (1970).
[12] P. S. Handley and A. E. Jacob, *J. Gen. Microbiol.* **127,** 289 (1981).
[13] C. P. Novotny, J. A. Short, and P. D. Walker, *J. Med. Microbiol.* **8,** 413 (1975).
[14] P. R. Lamden, J. N. Robertson, and P. J. Watt, *J. Gen. Microbiol.* **124,** 109 (1981).
[15] T. Vaara, Ph.D. thesis. Univ. of Helsinki, Helsinki, Finland, 1984.

developed for the examination and isolation of the *Synechocystis* CB3 fimbriae. The same procedures, however, may be applicable to other cyanobacteria as well.

Cultivation and Harvesting of Cells

Materials

Synechocystis CB3 (ATCC 35679)
Medium BG-11,[16] pH 7.0

Procedure

1. *Synechocystis* CB3 is grown in ten 5-liter conical flasks each containing 1 liter of medium BG-11.[16] To avoid potential fragmentation of fimbriae, the cultures are not shaken. They are, however, gassed with sterile-filtered air that is introduced at the bottom of the culture using a glass tube firmly mounted in the cotton plug of the flask. To secure sterile air filtering, two disposable filter units (0.22 μm, 30 mm in diameter) instead of one are used.

2. The flask cultures are incubated at 26° under continuous illumination of 500–1000 lux, provided by cool-white fluorescent tubes.

3. The late-logarithmic phase of growth is usually reached within 2 weeks. At this stage, the cultures are checked for purity both by careful microscopic examination (1000×, phase contrast) and by plating on nutrient agar plates,[9] as well as analyzed for the presence of fimbriae by electron microscopic examination (see below and elsewhere in this volume [14]).

4. Uncontaminated and fimbriate cultures are cooled to 4° and harvested by centrifugation (5000 *g*, 20 min). The cell paste is suspended in 250 ml of cold medium BG-11 and washed twice in this medium.

Electron Microscopy of Fimbriae

Principle

Fimbriae are best visualized by using negative staining. This technique involves drying a small quantity of the sample on a grid for electron microscopy and staining it with an electron-dense material such as phosphotungstic acid. Regions of the specimen that the stain cannot infiltrate

[16] R. Y. Stanier, R. Kunisawa, M. Mandel, and G. Cohen-Bazire, *Bacteriol. Rev.* **35,** 171 (1971).

remain electron-transparent and are seen with high resolution against the electron-dense background.

Materials

Fimbriate cells (a sample of the cyanobacterial culture concentrated 10-fold) or isolated fimbriae (~2 mg of protein/ml)
Copper grids, coated with Formvar and stabilized with carbon[17]
Phosphotungstic acid, 1% (w/v) in water, adjusted to pH 6.5 with 1 *N* NaOH

Procedure

1. A droplet of the sample is applied on the copper grid.
2. After a few minutes, excess sample is removed by touching the droplet with a piece of filter paper. The grid should not be entirely dried.
3. A drop of sterile-filtered stain is applied on the grid.
4. After 15–60 sec, excess stain is removed as described in Step 2.
5. The grid is air-dried and examined using a transmission electron microscope. The grid should be inserted with the specimen side down, i.e., away from the electron source. The recommended magnification is 10,000× for fimbriate cells and 50,000× for fimbriae.

Comments

Phosphotungstic acid is recommended for routine staining of cyanobacterial fimbriae. Alternative stains include 1% ammonium molybdate (pH 6.8), which is less damaging but gives lower contrast, and 5% uranyl acetate (pH 4.5), which gives the highest contrast but is more difficult to use than the other two stains. To determine the optimal staining time giving the highest contrast, at least the following times should be tried: 15, 30, 45, and 60 sec.

The high electron density of negative-stained preparations makes them very sensitive to the electron beam. To limit the radiation damage of the specimen and contamination of the electron microscope, the following precautions should be taken: (1) Liquid N_2 should be used in the anticontaminators of the electron microscope. (2) Acceleration voltage should not exceed 60 kW, and, to avoid excess brightness, the beam current should be kept as low as possible. (3) Focusing should not be performed on the specimen but on an adjacent particle or hole in the supporting film.

[17] E. H. Mercer and M. S. C. Birbeck, "Electron Microscopy: A Handbook for Biologists." Blackwell, Oxford, England, 1972.

Isolation of Fimbriae

Principle

In principle, *Synechocystis* CB3 fimbriae can be isolated according to the method developed by Korhonen[18,19] for the isolation of enterobacterial fimbriae. Mechanically detached fimbriae are precipitated with ammonium sulfate and, after dialysis, solubilized with sodium deoxycholate. The concentrated supernatant is subjected to gel filtration on a Sepharose 4B column, using urea (to which fimbriae are resistant) as an eluent.

Materials

Suspension of fimbriate cells produced as described above
10 m*M* Tris–HCl buffer, pH 7.5, containing 0.05% (w/v) sodium azide as a preservative
50 m*M* Tris–HCl buffer, pH 7.5, containing 6 *M* urea
500 m*M* Tris–HCl stock buffer, pH 7.5
Homogenizer (e.g., Ato-Mix, Measuring & Scientific Equipment, Ltd., London)
Ultrafiltration equipment (e.g., Amicon ultrafiltration cell Model 52 with X50 membrane, Amicon Corp., Lexington, MA)
Sepharose 4B column (e.g., K9/15 column, Pharmacia, Uppsala, Sweden)

Procedure

1. The cell suspension (250 ml) is homogenized. The recommended time for the Ato-Mix homogenizer is 5 min at half-speed. The cells are removed by centrifugation (4°, 5,000 *g*, 20 min), and the supernatant is collected. The homogenization is repeated twice, and the supernatants are combined (750 ml).

2. Crystalline ammonium sulfate is added stepwise to the supernatant to reach a 50% saturation. After standing overnight at 4°, the precipitate is separated by centrifugation (10,000 *g*, 1 hr) and dissolved in 50 ml of 10 m*M* Tris–HCl buffer, pH 7.5.

3. To remove ammonium sulfate, the suspension is dialyzed for 48 hr at 4° against 10 m*M* Tris–HCl buffer, pH 7.5.

4. Sodium deoxycholate is added to a final concentration of 0.5% (w/v), and the suspension is dialyzed for 48 hr at 4° against 10 m*M* Tris–HCl buffer, pH 7.5, containing 0.5% sodium deoxycholate.

[18] T. K. Korhonen, E.-L. Nurmiaho, H. Ranta, and C. Svanborg Edén, *Infect. Immun.* **27,** 569 (1980).

[19] T. K. Korhonen, Ph.D. thesis. Univ. of Helsinki, Helsinki, Finland, 1980.

5. To remove deoxycholate-insoluble contaminating material, the suspension is cleared by centrifugation (10,000 *g*, 10 min).

6. The supernatant is concentrated to approximately 3 ml by ultrafiltration, after which 500 m*M* Tris–HCl buffer, pH 7.5, and crystalline urea are added to a final concentration of 50 m*M* and 6 *M*, respectively.

7. After standing for 2 hr at room temperature, the preparation is filtered through a Sepharose 4B column. Fifty millimolar Tris–HCl buffer, pH 7.5, containing 6 *M* urea is used both in equilibration of the column and as an eluent. The fimbriae are eluted in the void volume of the column.

8. After dialysis for 48 hr at 4° against distilled water containing 0.05% (w/v) sodium azide, the fimbrial preparation is stored at 4°.

Comments

The homogenization time recommended in Step 1 applies to use of the Ato-Mix only. The treatment time in other homogenizers should be carefully optimized so that no significant cell breakage takes place. The optimization work is greatly facilitated by electron microscopic examination of the homogenates. The fimbrial preparation may be checked by electron microscopy, using negative staining (Fig. 2). The purity of the preparation can be assessed by sodium dodecyl sulfate–polyacrylamide gel electrophoresis.[7,20] Further characterization of the isolated fimbriae involves amino acid analysis[7] and isoelectric focusing.[18,21]

[20] U. K. Laemmli, *Nature (London)* **227,** 680 (1970).

[21] P. G. Righetti and J. W. Drysdale, "Isoelectric Focusing." North-Holland, Amsterdam, 1976.

[17] Inclusions: Granules of Polyglucose, Polyphosphate, and Poly-β-hydroxybutyrate

By J. M. SHIVELY

The major carbon and energy or phosphate reserves of cyanobacteria are polyglucose (polyG) and polyphosphate (polyP). Poly-β-hydroxybutyrate (polyBHB) has also been reported in a few instances. Storage of these reserves by microorganisms increases their chance of survival in a changing and often stressful environment. The polymers accumulate as discrete granules in the cytoplasm of the cell and in the case of polyBHB are surrounded by a nonunit membrane. Storing the reserve materials in this fashion, i.e., as polymers in compact granules, results in a minimum change in osmotic pressure. In this chapter, methodology dealing with research on these inclusions in cyanobacteria is reported where possible. If methods are not available, procedures used for other prokaryotes are presented. The researcher will be assisted by referring to a number of available articles.[1-10]

Polyglucose Granules

The high-molecular-weight polyG polymer is composed of D-glucose linked by α-1,4-glusidic bonds with branches occurring through α-1,6-glucosidic linkages. Reported values for $\overline{Cl}$ (average chain length) vary between 9 and 26.[11-13] PolyG is deposited as a normal consequence of active photosynthesis and is mobilized during dark periods.[13-19] Under normal conditions the polymer may accumulate to 10–30% of the cell dry weight. Conditions of nutrient imbalance, e.g., nitrogen or iron limitation, result in increased deposition; amounts as high as 60% of the cell dry weight have been reported.[14-17,20-23] Heterotrophically grown cyanobacte-

[1] M. M. Allen, *Annu. Rev. Microbiol.* **38,** 1 (1984).
[2] E. A. Dawes and P. J. Senior, *Adv. Microbiol. Physiol.* **10,** 135 (1973).
[3] F. M. Harold, *Bacteriol. Rev.* **30,** 772 (1966).
[4] I. S. Kulaev and V. M. Vadabov, *Adv. Microbiol. Physiol.* **24,** 83 (1983).
[5] J. Mas, C. Pedros-Alio, and R. Guerrero, *J. Bacteriol.* **164,** 749 (1985).
[6] J. M. Merrick, *in* "The Photosynthetic Bacteria" (R. V. Clayton and W. R. Sistrom, ed.), p. 199. Plenum, New York, 1979.
[7] J. Preiss, *Annu. Rev. Microbiol.* **38,** 419 (1984).
[8] J. M. Shively, *Annu. Rev. Microbiol.* **28,** 167 (1974).
[9] A. J. Smith, *in* "The Biology of Cyanobacteria" (N. G. Carr and B. A. Whitton, eds.), p. 47. Univ. of California Press, Berkeley, 1982.
[10] C. P. Wolk, *Bacteriol. Rev.* **37,** 32 (1973).
[11] L. Chao and C. C. Bowen, *J. Bacteriol.* **105,** 331 (1971).
[12] R. D. Simon, *Biochim. Biophys. Acta* **422,** 407 (1976).
[13] M. Weber and G. Wober, *Carbohydr. Res.* **39,** 295 (1975).
[14] M. M. Allen and A. J. Smith, *Arch. Microbiol.* **69,** 114 (1969).
[15] A. Ernst and P. Boger, *J. Gen. Microbiol.* **131,** 3147 (1985).
[16] A. Ernst, H. Kirschenlohr, J. Diez, and P. Boger, *Arch. Microbiol.* **140,** 120 (1984).
[17] M. Lehmann and G. Wober, *Arch. Microbiol.* **111,** 93 (1976).
[18] A. Oren and M. Shilo, *Arch. Microbiol.* **122,** 77 (1979).
[19] L. van Liere, L. R. Mur, C. E. Gibson, and M. Herdman, *Arch. Microbiol.* **123,** 315 (1979).
[20] S. E. Stevens, Jr., D. L. Balkwill, and D. A. M. Paone, *Arch. Microbiol.* **130,** 204 (1981).
[21] C. van Eykelenburg, *Antonie van Leeuwenhoek* **46,** 113 (1980).
[22] L. B. Hardie, D. L. Balkwill, and S. E. Stevens, Jr., *Appl. Environ. Microbiol.* **45,** 1007 (1983).
[23] D. M. Sherman and L. A. Sherman, *J. Bacteriol.* **156,** 393 (1983).

ria typically have a higher polyG content.[24–26] It is well established that polyG functions as a carbon and energy reserve. Its role in buoyancy regulation in strains containing gas vesicles should also be noted.[27]

The polymer is deposited as granules between thylakoid membranes. The granules are seen in thin section as densely stained (lead citrate) spheres or rods. The spheres are commonly 25–30 nm in diameter. The rod-shaped granules of *Nostoc muscorum* are 31 × 65 nm and appear to consist of two equal but not readily dissociable parts.[11] Granules purified from this source migrated as a single component in the analytical ultracentrifuge; the *S* value was 265. The granules of *Oscillatoria rubescens* are as long as 300 nm and are composed of 7-nm disks with a central pore.[28]

PolyG synthesis in prokaryotes is accomplished by the formation of a sugar nucleotide and subsequent transfer of the glucosyl unit to a preexisting primer, as evidenced by the following reactions[7]:

$$\text{ATP} + \alpha\text{-glucose-1-P} \rightleftharpoons \text{ADP–glucose} + \text{PP}_i$$
$$\text{ADP–glucose} + \alpha\text{-1,4-glucan} \rightleftharpoons \alpha\text{-1,4-glucosylglucan} + \text{ADP}$$

The enzymes involved are glucose-1-phosphate adenyltransferase (EC 2.7.7.7.27, ADPglucose pyrophosphorylase; ATP : α-glucose-1-phosphate adenyltransferase) and starch synthase (EC 2.4.1.21, ADPglucose : 1,4-α-D-glucan 4-α-glucosyltransferase). A reaction catalyzed by α-1,4-glucan : α-1,4-glucan-6-glycosyltransferase adds α-1,6-glucosyl linkages. Potentially, α-1,4-glucans can be synthesized by other mechanisms (enzymes and nucleotide donors) including reactions catalyzed by maltodextrin phosphorylase (EC 2.4.1.1) and glycogen phosphorylase (EC 2.4.1.1). The glucose-1-phosphate adenyltransferase of *Synechococcus* 6301 is inhibited by phosphate and stimulated by 3-phosphoglycerate.[29] This is similar to the regulation in higher plants but different from that in other prokaryotes. Polyglucose utilization is not well understood in prokaryotes, but it must include an isoamylase (EC 3.2.1.68) and glycogen phosphorylase; other enzymes may also be involved.

Special techniques are available for the specific staining of PolyG in thin sections. Hardie *et al.*[22] stained polyG in thin sections of *Agmenellum quadruplicatum* by a method employing periodic acid–thiosemicarbazide–osmium tetroxide. Briefly, samples fixed with 2.5% glutaraldehyde–0.1% ruthenium red in 50 m*M* cacodylate buffer, pH 7.2, are embedded

[24] O. I. Baulina, L. A. Mineeva, S. S. Suleimanova, and M. V. Gusev, *Microbiologiya* **50,** 523 (1981).

[25] R. A. Pelroy, M. R. Kirk, and J. A. Bassham, *J. Bacteriol.* **128,** 623 (1976).

[26] B. Raboy, E. Padan, and M. Shilo, *Arch. Microbiol.* **110,** 77 (1976).

[27] J. Kromkamp, A. Kanopka, and L. R. Mur, *J. Gen. Microbiol.* **132,** 2113 (1986).

[28] M. Jost, *Arch. Microbiol.* **50,** 211 (1965).

[29] C. Levi and J. Preiss, *Plant Physiol.* **58,** 753 (1976).

and sectioned. The sections are treated with 0.1% periodic acid for 1 hr at room temperature, rinsed 2 times with distilled water, treated with 1% thiosemicarbazide for 1 hr at room temperature, and rinsed again with distilled water. Following air-drying, the sections are exposed to osmium tetroxide vapors for 4 hr at room temperature. PolyG granules stain very densely with this procedure. Westphal *et al.*[30] reported good results with *Anabaena variabilis* by using the water-compatible melamine resin Nonoplast FB 101 for embedding and staining the sections for polyG using a 1% solution of silver proteinate (pH 8.4) according to Thiéry.[31]

The total polyG content of cells can be determined by a number of methods. Pelleted cells may be resuspended, sonicated, and centrifuged (10,000 *g*, 10 min), a sample of the supernatant incubated 1.5 hr with glucoamylase (EC 3.2.1.3, glucan 1,4-α-glucosidase)/glucose oxidase (EC 1.1.3.4)/peroxidase (EC 1.11.1.7) reagent, and the optical density determined at 540 nm.[17,32] Stevens *et al.*[20] hydrolyzed the pelleted cells in trifluoroacetic acid (refluxed at 95° overnight), neutralized the solution, and assayed specifically for glucose (Glucose assay, Sigma Chemical Co., St. Louis, MO).

A number of techniques are available for the isolation of polyG from prokaryotes.[11,13,33] If one is interested in characterizing the polyG, care should be taken during the isolation procedure. Elevated temperatures, alkali, and repeated ethanol precipitations should be avoided; polyG breakdown results.[11,33] The polyG granules of *N. muscorum* were purified by sucrose density gradient centrifugation.[11] In this procedure, algal cells (40 g wet weight) are suspended in 100 ml of 0.1 *M* phosphate buffer, pH 7.4, passed through a Ribi fractionator at 2000 psi, and the extract centrifuged for 10 min at 2000 *g* to eliminate whole cells and large debris. Sodium deoxycholate is added (final concentration 1%) and the supernatant incubated at 4° for 10 hr. Large particles are removed by centrifugation at 5000 *g* for 10 min and the granules sedimented at 15,000 *g* for 10 min. The crude granules are layered on top of a 10–30% linear sucrose gradient and centrifuged for 4 hr at 23,000 *g* in a Beckman SW 25.1 rotor. Gradient fractions containing granules, as determined by electron microscopy, are pooled and characterized.

In an alternate procedure,[13] harvested, washed *Anacystis nidulans* cells are resuspended in cold distilled water and sonically ruptured. Proteins are precipitated from the resulting suspension with a cold aqueous solution of trichloroacetic acid (5% final concentration). Following cen-

[30] C. Westphal, H. Bohme, and D. Frosch, *J. Histochem. Cytochem.* **33,** 1180 (1985).

[31] J. P. Thiéry, *J. Microsc.* **6,** 987 (1967).

[32] J. J. Marshall and W. J. Whelan, *FEBS Lett.* **9,** 85 (1970).

[33] A. D. Antoine and B. S. Tepper, *Arch. Biochem. Biophys.* **134,** 207 (1969).

trifugation, the supernatant is neutralized with solid $NaHCO_3$, dialyzed, concentrated (Amicon ultrafiltration cell, PM10 Diaflo membrane), and the polyG precipitated with ethanol–potassium acetate. The precipitate is collected, dried, and resuspended in 10 m*M* phosphate buffer (pH 7). Further purification is accomplished using a DEAE–Sephadex column. Material in the void volume is collected, dialyzed, and freeze-dried.

Characterization techniques include acid and/or enzymatic hydrolysis and identification of sugar(s), λ_{max} of iodine–glucan complex, $\overline{Cl}$, yield of material resistant to α-amylase, and treatment with isoamylase to liberate 1,4-α-D-glucose linear chains which can be fractionated by gel permeation chromatography.[11–13,33,34]

Polyphosphate Granules

The majority of the polyP granules are composed of linear, high-molecular-weight molecules with greater than 500 residues/molecule.[3,35] Small linear and cyclic molecules have also been reported.[3,36] PolyP accumulates in cells when phosphate is in excess.[35,37] The accumulation is especially pronounced (as high as 60% of total cell phosphorus) when phosphate-deficient cells are transferred to a phosphate-rich medium.[35,38–41] This is commonly referred to as the "phosphate overplus" phenomenon. The polyP is lost under conditions of phosphate deficiency.[42]

The major functions of polyP in prokaryotes appear to be the storage of phosphate and the regulation of the level of orthophosphate in the cell.[3,4] It may also be a regulator of the intracellular concentrations of ATP, ADP, other nucleotides, and pyrophosphate. The polymer also represents a valuable pool of activated phosphorus that can be utilized in a variety of metabolic processes.[4,43] Mutants unable to accumulate polyP, but appearing to exist normally, question whether this is an obligate characteristic.[44,45]

[34] W. Z. Aassid and S. Abraham, this series, Vol. 3, p. 34.
[35] J. F. Grillo and J. Gibson, *J. Bacteriol.* **140,** 508 (1979).
[36] R. Niemeyer and G. Richter, *Arch. Microbiol.* **69,** 54 (1969).
[37] T. E. Jensen, *Arch. Microbiol.* **67,** 328 (1969).
[38] L. Jacobson and M. Halmann, *J. Plankton Res.* **4,** 481 (1982).
[39] N. H. Lawry and T. E. Jensen, *Arch. Microbiol.* **120,** 1 (1979).
[40] D. Livingston and B. A. Whitton, *Br. Phycol. J.* **18,** 29 (1983).
[41] L. Sicko-Goad and T. E. Jensen, *Am. J. Bot.* **63,** 183 (1976).
[42] T. E. Jensen and L. M. Sicko, *Can. J. Microbiol.* **9,** 1235 (1974).
[43] N. Okamoto, H. Tei, K. Murata, and A. Kimura, *J. Gen. Microbiol.* **132,** 1519 (1986).
[44] R. L. Harold and F. M. Harold, *J. Gen. Microbiol.* **31,** 241 (1963).
[45] S. Vaillancourt, N. Beauchemin-Newhouse, and R. J. Cedergren, *Can. J. Microbiol.* **24,** 112 (1978).

PolyP is commonly deposited in the nucleoplasmic region of the cell as non-membrane-enclosed granules with a diameter of 100–400 nm.[38,46] The appearance in the electron microscope varies from electron dense (opaque) to porus.[46] A holey or mottled image is common due to partial evaporation in the electron beam.[46] The granules were confirmed to contain phosphorus as well as potassium, calcium, and magnesium by *in situ* X-ray energy-dispersive analysis using air-dried cells which were imaged in the scanning electron microscope.[47,48] Reportedly, the granules of *Micrococcus lysodeikticus* are composed of 24% protein, 30% lipid, 27% polyphosphate, and lesser amounts of RNA and carbohydrate; however, the preparation was grossly contaminated with membranes.[49] The molecular weights of the polyP were 1000–100,000, and the metal constituents were magnesium and calcium. The granules of *Desulfovibrio gigas* were identified as magnesium tripolyphosphate and those of *Tetrahymena* consisted of hydrated mixed pyrophosphates of calcium and magnesium.[50,51]

The polymer is synthesized by the enzyme polyphosphate kinase (EC 2.7.4.1, ADP: polyphosphate phosphotransferase) according to the following reaction[3,4,35]:

$$\text{ATP} + (\text{HPO}_3)_n \rightleftharpoons \text{ADP} + (\text{HPO}_3)_{n+1}$$

The polymer may be degraded by a number of enzymes including polyphosphate kinase, adenosine monophosphate:polyphosphate phosphotransferease, polyphosphate glucokinase, polyphosphate fructokinase, polyphosphate (metaphosphate)-dependent NAD kinase, and a variety of polyphosphatases.[3,4]

The total polyP of cells can be extracted by alkaline hypochlorite.[35,52–55] The following methodology has been used by Poindexter and Fey[53]: Cells (1–25 mg dry weight) are collected by centrifugation at 4° and resuspended either in water, in medium without phosphate, or in a suitable non-phosphate-containing buffer. The centrifugation and resuspension may need to be repeated several times in order to dilute residual phosphate. Finally, the pelleted cells are resuspended in 5.0 ml of sodium hypochlorite (1.8 *M*, pH 9.8) and incubated at 30° to constant turbidity

[46] T. E. Jensen, L. Sicko-Goad, and R. P. Ayala, *Cytologia* **42,** 357 (1977).
[47] M. Baxter and T. Jensen, *Arch. Microbiol.* **126,** 213 (1980).
[48] T. E. Jensen and M. Baxter, *Microbios Lett.* **28,** 145 (1985).
[49] I. Friedberg and G. Avigad, *J. Bacteriol.* **96,** 544 (1968).
[50] N. E. Jones and L. A. Chambers, *J. Gen. Microbiol.* **89,** 67 (1975).
[51] H. Rosenberg, *Exp. Cell. Res.* **41,** 397 (1966).
[52] F. M. Harold, *J. Bacteriol.* **86,** 885 (1963).
[53] J. S. Poindexter and L. F. Fey, *J. Microbiol. Methods* **1,** 1 (1983).
[54] N. N. Rao, M. F. Roberts, and A. Torriani, *J. Bacteriol.* **162,** 242 (1985).
[55] D. H. Williamson and J. F. Wilkinson, *J. Gen. Microbiol.* **19,** 198 (1958).

(30–60 min). The precipitate is collected by centrifugation at 17,000 *g* (15 min), resuspended in 10 ml of 1.5 *M* NaCl–1 m*M* EDTA (pH 4.6), and the centrifugation repeated. The pellet is ready for analysis in most instances, but further purification may be required in some cases in order to remove traces of nucleic acid, pigments, proteins, and/or polysaccharides.[56] Rao *et al.*[54] suggest including 1 m*M* sodium fluoride to inhibit acid phosphatases. If desirable, more complex methods of purification can be used which fractionate the polyP on the basis of molecular weight.[54,56,57] The chain length of the polymer can be determined by end group titration, gel filtration, or by ^{31}P-NMR spectroscopy.[54,57] A sample can be hydrolyzed in 1 *M* HCl at 100° for 15 min and the resulting P_i quantitated by a number of methods.[58,59]

Attempts to isolate highly purified, intact polyP granules have been met with only moderate success. Jacobson and Halmann[60] presented preliminary evidence regarding the isolation of polyP granules from *Microcystis aeruginosa* using a nonaqueous gradient of carbon tetrachloride–*S*-tetrabromoethane; additional information has not been forthcoming. Friedberg and Avigad[49] resuspended *M. lysodeikticus* cells in Tris–HCl buffer (pH 7.2) containing 1.0 m*M* $MgSO_4$, treated with cells with lysozyme (100 μg/ml) and deoxyribonuclease (0.1 μg/ml) at 37°, and then pelleted the polyP granules by centrifugation at 2,000 *g* for 1 hr. As noted above, these granules were contaminated with membrane fragments. Jones and Chambers[50] centrifuged a broken cell preparation of *D. gigas* at 10,000 *g* for 15 min. The dark-colored upper portion of the pellet (cell walls, metal sulfide, and whole cells) was carefully removed using a jet of distilled water, leaving the lower, white portion (polyP granules). Two repetitions of this procedure eliminated most of the contaminating material. Rosenberg[51] resuspended *Tetrahymena*, sonicated for 2–3 min, in 9 *M* urea and centrifuged for 15 min at 18,000 rpm (MSE HS-18 centrifuge). The white pellet (granules) was removed and washed with 8 *M* urea, 10 m*M* Tris buffer (pH 9.0), ethanol, and ether.

The above-mentioned techniques might provide a useful starting point for the development of a satisfactory polyP granule isolation procedure for cyanobacteria. The addition of magnesium or calcium to buffers will aid in stabilizing the granules. Sodium fluoride can be added to preparations to inhibit phosphatases.[50] The use of chelating agents and detergents should be avoided.[49]

[56] F. M. Harold, *J. Bacteriol.* **86,** 216 (1963).
[57] P. Langen and E. Liss, *Biochem. Zeit.* **330,** 455 (1958).
[58] R. L. Dryer, A. R. Tammes, and J. I. Routh, *J. Biol. Chem.* **225,** 177 (1957).
[59] T. Onishi, R. S. Gall, and M. L. Mayer, *Anal. Biochem.* **69,** 261 (1975).
[60] L. Jacobson and M. Halmann, *J. Plankton Res.* **4,** 481 (1982).

Poly-β-hydroxybutyrate Granules

PolyBHB is a linear molecule with the following structure[2,6]:

$$\mathrm{HO{-}\underset{\displaystyle CH_3}{\underset{|}{CH}}{-}CH_2{-}\underset{\displaystyle O}{\underset{\|}{C}}}\left[\mathrm{{-}O{-}\underset{\displaystyle CH_3}{\underset{|}{CH}}{-}CH_2{-}\underset{\displaystyle O}{\underset{\|}{C}}}\right]_n\mathrm{{-}O{-}\underset{\displaystyle CH_3}{\underset{|}{CH}}{-}CH_2{-}COOH}$$

Molecular weights of the polymer extracted from various bacteria range from 1000 to 250,000.[61] Alkaline hypocholorite extraction commonly gives a lower molecular weight, indicating some breakdown as a result of this treatment. The molecule is proposed to be a compact, right-handed helix with two residues per turn.

Only a few reports have documented the occurrence of polyBHB in cyanobacteria. Campbell *et al.*[62] reported the accumulation of the polymer in *Spirulina plantensis* to be about 6% of the culture dry weight. Incorporation of acetate in the growth medium did not bring about increased levels. Other authors noted the presence of the polymer in *Chlorogloeopsis fritschii*; in this organism acetate in the medium increased polyBHB to about 10% of the dry weight.[63–65] In other prokaryotes the polymer may accumulate to greater than 50% of the cell dry weight, e.g., 96% in *Alcaligenes eutrophus* under "forced" heterotrophic conditions.[6,8,66] The accumulation takes place when carbon and energy sources are in excess and the growth limited by a nutrient deficiency.[6] The polymer functions as a reserve of carbon and/or energy.

The polyBHB is deposited in granules (100–800 nm in diameter) surrounded by a nonunit membrane approximately 3.0 nm thick.[8,64,65] The granules appear electron transparent in thin section. Campbell *et al.*[62] showed the characteristic cone-shaped morphology of the *S. platensis* granules examined by freeze-etching as originally described in *Bacillus cereus* by Dunlop and Robards.[67] The core of the granule stretches during freeze-etching. The granules of *Bacillus megaterium* were purified by Griebel *et al.*[68] and shown to consist of 98% polyBHB, 2% protein, and traces of lipid and phosphorus.

[61] D. G. Lundgren, R. Alper, C. Schnaitman, and R. H. Marchessault, *J. Bacteriol.* **89,** 245 (1965).
[62] J. Campbell III, S. E. Stevens, Jr., and D. L. Balkwill, *J. Bacteriol.* **149,** 361 (1982).
[63] N. G. Carr, *Biochim. Biophys. Acta* **120,** 308 (1966).
[64] T. E. Jensen and L. M. Sicko, *J. Bacteriol.* **106,** 683 (1971).
[65] T. E. Jensen and L. M. Sicko, *Cytologia* **38,** 381 (1971).
[66] C. Pedros-Alio, J. Mas, and R. Guerrero, *Arch. Microbiol.* **143,** 178 (1985).
[67] W. F. Dunlop and A. W. Robards, *J. Bacteriol.* **114,** 1271 (1973).
[68] R. J. Griebel, Z. Smith, and J. M. Merrick, *Biochemistry* **7,** 3676 (1968).

The pathway for synthesis is[2,6,8]

Acetyl-CoA → acetoacetyl-CoA → D-(−)-β-hydroxybutyryl-CoA → polyBHB

The polymerizing enzyme, polyBHB synthase, appears to be a part of the granule membrane. The degradation enzymes, polyBHB depolymerase, trimer hydrolyase, and dimer hydrolyase, may also be present.[2,6]

The polyBHB can be extracted from cells with alkaline hypochlorite using the same method reported for polyP. In fact, Poindexter and Fey[53] describe the determination of both polymers from the same extract. For polyBHB, the alkaline hypochlorite precipitate, after washing with NaCl–EDTA, is resuspended in 1.5 ml of water and incubated at 0° overnight. The undissolved material (polyBHB) is pelleted by centrifugation at 27,000 *g* for 20 min and then extracted with 1–5 ml of chloroform at 60°.

Isolation of polyBHB granules from cyanobacteria has not been reported. They have been isolated from *Bacillus* species by Griebel *et al.*[68] and by Ang and Nickerson[69] and Nickerson[70]; the methods seem suitable for use with cyanobacteria. In the procedure of Griebel *et al.*,[68] 20 g wet weight of *B. megaterium* cells is suspended in 60 ml of 50 m*M* Tris–HCl–16.7 m*M* $MgCl_2$ buffer (pH 8.0). Lysozyme (66 mg) and deoxyribonuclease (0.7 mg) are added, and the suspension is mixed and incubated for 30 min at room temperature. The suspension is then cooled to 0° and sonicated for 3 min. The granules are separated from the crude extract by layering the suspension on glycerol and centrifuging in a swinging-bucket rotor at 1600 *g* for 15 min at 4°. The granules accumulate at the interface. Griebel *et al.*[68] further purified the granules by differential centrifugation, a polymer two-phase system, and glycerol density gradient centrifugation. Purification of the crude granules on a linear sodium bromide gradient (1.0–1.4 g/cm^3) as described by Ang and Nickerson[69] and Nickerson[70] seems more appropriate.

The polyBHB can be identified and quantitated spectrophotometrically.[71] Characterization can include the determination of melting point, specific gravity, intrinsic viscosity, molecular weight, and solubility in a range of solvents.[2,6,61,68]

[69] B. J. Ang and K. W. Nickerson, *Appl. Environ. Microbiol.* **36,** 625 (1978).

[70] K. W. Nickerson, *Appl. Environ. Microbiol.* **43,** 1208 (1982).

[71] J. H. Law and R. A. Slepecky, *J. Bacteriol.* **82,** 33 (1964).

[18] Inclusions: Carboxysomes

By J. M. SHIVELY

Inclusions with polygonal profiles (polyhedral bodies) appear to be ubiquitous entities of the vegetative cells and akinetes of cyanobacteria.[1,2] Heterocysts, which fix nitrogen but not carbon dioxide, do not possess polyhedral bodies. From 3 to 12 (mean number) of these inclusions occur in the nucleoplasmic region of the cell.[1-6] As observed in thin section, the bodies are 100–900 nm in diameter, have a granular substructure, and are surrounded by a monolayer membrane (shell) 3–4 nm thick.

These inclusions have been isolated from *Anabaena cylindrica* and *Chlorogloeopsis fritschii* and shown to contain the enzyme ribulose-1,5-bisphosphate carboxylase (RuBisCOase) (EC 4.1.1.39), thus identifying them as carboxysomes as originally described in the autotrophic sulfur bacterium *Thiobacillus neapolitanus*.[5,7,8] In addition to the carboxysomal (particulate) RuBisCOase the cells possess a soluble form of the enzyme. These RuBisCOases appear to be identical as judged by several criteria, e.g., molecular weight of the native enzyme and of the large (LSU) and small (SSU) subunits.[9,10] For additional information on RuBisCOase, see Tabita.[11] The amount present as either the soluble or particulate form varies with growth conditions and culture age. In general, the particulate form is favored in heterotrophic growth in the dark, as well as in older cultures.[1,12] Molecular explanations for these observations are not available.

The purified carboxysomes of *C. fritschii* are composed of at least eight major polypeptides as revealed by sodium dodecyl sulfate–polyacrylamide gel electrophoresis (SDS–PAGE).[10] Cannon and Shively[9]

[1] G. A. Codd and W. J. N. Marsden, *Biol. Rev.* **59,** 389 (1984).
[2] W. D. P. Stewart and G. A. Codd, *Br. Phycol. J.* **10,** 273 (1975).
[3] J. M. Shively, F. Ball, and B. W. Kline, *J. Bacteriol.* **116,** 1405 (1973).
[4] J. M. Shively, *Annu. Rev. Microbiol.* **28,** 167 (1974).
[5] J. M. Shively, F. L. Ball, D. H. Brown, and R. E. Saunders, *Science* **182,** 584 (1973).
[6] Y. A. Holthuijzen, J. F. Van Breemen, W. N. Konings, and E. F. J. Van Bruggen, *Arch. Microbiol.* **144,** 258 (1986).
[7] G. A. Codd and W. D. P. Stewart, *Planta* **130,** 323 (1976).
[8] T. Lanaras and G. A. Codd, *Planta* **153,** 279 (1981).
[9] G. C. Cannon and J. M. Shively, *Arch. Microbiol.* **134,** 52 (1983).
[10] T. Lanaras and G. A. Codd, *Arch. Microbiol.* **130,** 213 (1981).
[11] F. R. Tabita, *Microbiol. Rev.* **52,** 155 (1988).
[12] T. Lanaras and G. A. Codd, *Planta* **154,** 284 (1982).

showed that the carboxysomes of *T. neapolitanus* consist of 12–15 polypeptides: four RuBisCOase LSU components, one RuBisCOase SSU, two shell polypeptides, and the rest unidentified. Holthuijzen *et al.*[13] recently reported 13 polypeptides in *T. neapolitanus* carboxysomes, four of which were identified as shell glycoproteins. The bodies of *Nitrobacter agilis* were shown to be composed of at least seven polypeptides.[14] The spectrophotometric scans of the SDS–polyacrylamide gels presented in all of these reports are surprisingly similar.

Additional enzymes have not been detected in the carboxysomes.[1,8,9,13,15–17] The enzymatic activities tested in various preparations include phosphoribulokinase (EC 2.7.1.19), ribose-1-phosphate isomerase (EC 5.3.1.6, phosphoriboisomerase), glyceraldehyde-3-phosphate dehydrogenase (EC 1.2.1.12, D-glyceraldehyde-3-phosphate : NAD^+ oxidoreductase), phosphoglycerate kinase (EC 2.7.2.3, ATP : 3-phospho-D-glycerate 1-phosphotransferase), fructose-bisphosphate aldolase (EC 4.1.2.13, fructose-1,6-bisphosphate D-glyceraldehyde-3-phosphate-lyase), sedoheptulose-bisphosphatase (EC 3.1.3.37, D-sedoheptulose-1,7-bisphosphate 1-phosphohydrolase), L-malate : NADH oxidoreductase, aspartate aminotransferase (EC 2.6.1.1, L-aspartate : 2-oxoglutarate aminotransferase), adenylate kinase (EC 2.7.4.3, ATP : AMP phosphotransferase), and carbonate dehydratase (EC 4.2.1.1, carbonic anhydrase). An earlier report noting the presence of many of these activities in the carboxysomes of *T. neapolitanus* was in error.[18] Recent evidence also refutes the earlier reported presence of extrachromosomal DNA in carboxysomes.[19–21]

The mechanism(s) of the formation as well as the loss of carboxysomes is unknown. The disappearance of the inclusions in *Thiobacillus intermedius* when RuBisCOase is repressed appears to be by dilution (unpublished observations). The function(s) of the carboxysome is also

[13] Y. A. Holthuijzen, J. F. Van Breemen, J. G. Kuenen, and W. N. Konings, *Arch. Microbiol.* **144,** 398 (1986).

[14] M. Biedermann and K. Westphal, *Arch. Microbiol.* **121,** 187 (1979).

[15] A. M. Hawthornthwaite, T. Lanaras, and G. A. Codd, *J. Gen. Microbiol.* **131,** 2497 (1985).

[16] T. Lanaras, A. M. Hawthornwaite, and G. A. Codd, *FEMS Microbiol. Lett.* **26,** 285 (1985).

[17] W. J. N. Marsden, T. Lanaras, and G. A. Codd, *J. Gen. Microbiol.* **130,** 2089 (1984).

[18] R. F. Beudeker and J. G. Kuenen, *FEBS Lett.* **131,** 269 (1981).

[19] K. Westphal, E. Bock, G. C. Cannon, and J. M. Shively, *J. Bacteriol.* **140,** 285 (1979).

[20] Y. A. Holthuijzen, F. J. M. Maathuis, J. G. Kuenen, R. N. H. Konings, and W. N. Konings, *FEMS Microbiol. Lett.* **35,** 193 (1986).

[21] D. Vakeria, G. A. Codd, W. J. N. Marsden, and W. D. P. Stewart, *FEMS Microbiol. Lett.* **25,** 149 (1984).

unknown. Several theories have been proposed based on a variety of experimental data, including (1) an active role in carbon dioxide fixation, providing a more favorable environment, e.g., reducing oxygen availability or concentrating carbon dioxide, and (2) an inactive role in fixation but preventing excessive turnover of the enzyme.

Determination of the amount of soluble and particulate (carboxysomal) RuBisCOase has been accomplished as follows[8]: Cells (8–13 g wet weight) are resuspended in 10 m*M* Tris–HCl buffer, pH 7.8, containing $MgCl_2$ (10 m*M*), $NaHCO_3$ (50 m*M*), disodium EDTA (1 m*M*), and 2-mercaptoethanol (12 m*M*). A buffer of low ionic strength appears to be important for carboxysome stability. All operations are at 4°. The cells are broken with a French press (110 Pa), and the whole cells and large membrane fragments are removed by centrifugation at 1500 *g* for 10 min. The supernatant is then centrifuged at 40,000 *g* for 1 hr. The resulting supernatant and pellet are assayed for soluble and carboxysomal RuBisCOase, respectively.[5,9] One should consider these data with caution; the soluble enzyme can be pelleted if trapped within and/or stuck to other cellular components. Further purification of the pelleted material should be undertaken in order to confirm its identity as carboxysomal. The soluble form of RuBisCOase can be purified from the supernatant.[5,9]

For further carboxysome purification the above pellet is resuspended in 4 volumes of buffer, and 1.0–1.5 ml is layered on Percoll–sucrose gradients. The gradients are prepared using an MSE gradient former: the low-density chamber contains 27 cm^3 of Percoll plus 3 cm^3 buffer (as above, except 50 m*M* Tris–HCl, pH 8.2), and the high-density chamber contains 12 cm^3 of Percoll plus 24 cm^3 of 2.7 *M* sucrose in buffer (8.1 m*M* Tris–HCl). The gradients are centrifuged in a 3 × 25 cm^3 swing-out rotor at 80,000 *g* (MSE Prepspin 75 centrifuge) for 45 min, fractionated from the bottom, and the carboxysomes located by RuBisCOase assay (near the center of the gradient). Alternately, the carboxysomes are purified using 10–65% linear sucrose gradients and centrifuging for 3 hr at 80,000 *g*.[8] *Note:* High sucrose may destabilize the bodies. Purification procedures used for the *T. neapolitanus* inclusions may be useful.[9,13] The carboxysomes are then ready for examination by electron microscopy and analysis by SDS–PAGE.

The carboxysomal RuBisCOase, as well as the shell, can be isolated by pelleting the carboxysomes, resuspension in 0.1 *M* Tris–HCl buffer, pH 7.8 (other components as above), sonication (MSE Soniprep 150, 9.5 mm probe, three 30-sec periods), and subjecting the material to sucrose (10–65%) density gradient centrifugation for 24 hr at 80,000 *g*.[9,10,13] RuBisCOase bands at approximately 20%, the shell at 50%, and the unbroken carboxysomes at 57% sucrose.

[19] Inclusions: Cyanophycin

By MARY MENNES ALLEN

Introduction

Cyanophycin granules,[1] although observed in cyanobacterial cells at an early date by light and electron microscopists,[2,3] were first isolated and characterized by Simon in 1971.[4] They were then shown to correspond morphologically to structured granules, which in fine structure are variable in shape and size and have a radiating pattern of substructure[5] (Fig. 1). Cyanophycin has been shown to consist of multi-L-arginyl-poly(L-aspartic acid), which has also been referred to as Arg-poly(Asp) or cyanophycin granule polypeptide (CGP), a peptide which is nonribosomally synthesized[6,7] and which consists of equimolar quantities of arginine and aspartic acid arranged as a backbone of polyaspartate with an arginine moiety attached to the β-carboxyl group of each aspartate by its α-amino group.[8] Cyanophycin ranges in molecular weight from 25,000 to 100,000.[4]

The function of cyanophycin has been shown to be nitrogen storage,[6,9,10] and its occurrence under varying environmental and growth conditions has been demonstrated.[11] Cyanophycin can accumulate in amounts up to 8–16% of cellular dry weight in stationary phase cells,[10,12] whereas a low amount of cyanophycin (0.2%) is found in nitrogen-depleted cells.[10] An enzyme that can elongate an existing cyanophycin chain by adding arginyl and aspartyl residues in the presence of ATP, $MgCl_2$, KCl, and a sulfhydryl reagent[7] has been shown to be present in crude

[1] This work was supported by grants from the National Science Foundation (PCM8004529, PCM8311035, DMB8602156), a William and Flora Hewlett Foundation Grant of Research Corporation, and Wellesley College.
[2] N. J. Lang, *Annu. Rev. Microbiol.* **22,** 15 (1968).
[3] G. W. Fuhs, *in* "The Biology of Blue–Green Algae" (N. G. Carr and B. A. Whitton, eds.), p. 117. Univ. of California Press, Berkeley, 1973.
[4] R. D. Simon, *Proc. Natl. Acad. Sci. U.S.A.* **68,** 265 (1971).
[5] N. Lang, R. D. Simon, and C. P. Wolk, *Arch. Mikrobiol.* **83,** 313 (1972).
[6] R. D. Simon, *Arch. Mikrobiol.* **92,** 115 (1973).
[7] R. D. Simon, *Biochim. Biophys. Acta* **422,** 407 (1976).
[8] R. D. Simon and P. Weathers, *Biochim. Biophys. Acta* **420,** 165 (1976).
[9] M. M. Allen and F. Hutchison, *Arch. Microbiol.* **128,** 1 (1980).
[10] M. M. Allen, F. Hutchison, and P. J. Weathers, *J. Bacteriol.* **141,** 687 (1980).
[11] M. M. Allen, *Annu. Rev. Microbiol.* **38,** 1 (1984).
[12] R. D. Simon, *J. Bacteriol.* **114,** 1213 (1973).

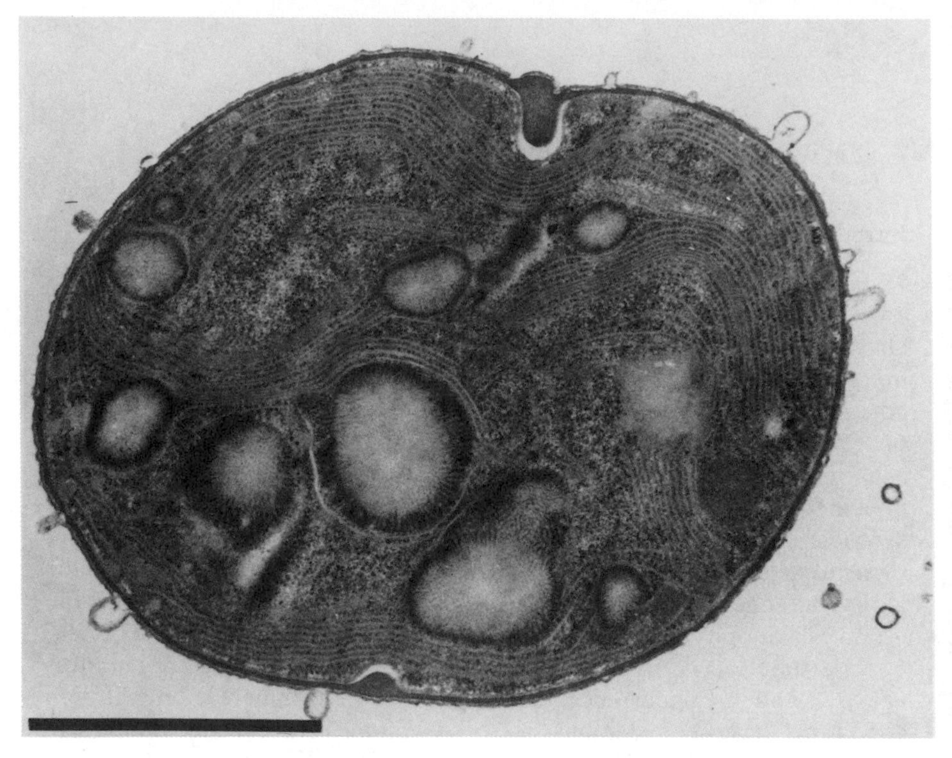

FIG. 1. Electron micrograph of a section of an osmium tetroxide-fixed *Aphanocapsa* 6308 cell grown under light-limited conditions. Cyanophycin granules are the numerous structured granules in the cell. Bar, 1 μm.

extracts of three *Anabaena* species.[7,13] Proteinases specific to cyanophycin have been detected in two *Anabaena* species[13] and in *Aphanocapsa* 6308.[14] Using *Anabaena* extracts, the major product of hydrolysis was an aspartic acid–arginine dipeptide,[13] while arginine was the major hydrolytic product using *Aphanocapsa* extracts.[14] The activity of these enzymes could account for the vast variation in cyanophycin concentration found in cells grown under different environmental conditions.

Cyanophycin is unique to, but not universally present in, cyanobacteria.[11,15] The methods discussed here are those to induce cyanophycin formation, to isolate and purify it, to analyze it quantitatively, and to

[13] M. Gupta and N. G. Carr, *J. Gen. Microbiol.* **125,** 17 (1981).
[14] M. M. Allen, R. Morris, and W. Zimmerman, *Arch. Microbiol.* **138,** 119 (1984).
[15] N. H. Lawry and R. D. Simon, *J. Phycol.* **18,** 391 (1982).

study its physiological breakdown in the non-nitrogen-fixing cyanobacterium *Aphanocapsa* (*Synechocystis*) PCC 6308 (ATCC 27150).

Induction of Cyanophycin Formation

Principle

A number of environmental conditions induce the synthesis of cyanophycin[10,11,15]; these can be used to determine if a particular strain is capable of forming the storage polymer, as well as to cause its accumulation in cells for isolation purposes. High concentrations of cyanophycin can be caused to accumulate in cells by the addition to the growth medium of a variety of nitrogen-containing compounds,[10,15] metabolic inhibitors of macromolecular synthesis,[6,10,15–18] or metals,[12] or by the limitation of nutrients such as light (Fig. 1), phosphorus, or sulfur.[10,15,19]

Procedure

Cells of *Aphanocapsa* 6308 are grown in BG-11 medium,[20] supplemented with 2.4 g/liter sodium carbonate, at 35° in 10,800 lux of cool-white fluorescent light with bubbling of 5% CO_2 in air. Volumes from 40 ml to 8 liters of cells have been used. When the cell population reaches an optical density at 750 nm of 2.0 (500 μg/ml dry weight) 5 μg/ml of chloramphenicol (2 mg/ml filter-sterilized stock) is added, and the culture is allowed to incubate for another 48 hr before cell harvesting. At this point the cyanophycin content of the cells is approximately 10% of the cellular dry weight, compared to 1–2% dry weight found in control cells grown for the same time period.[10] A typical amount of cyanophycin per cell is 1.25×10^{-5} μg.

Isolation and Purification of Cyanophycin

Principle

Cyanophycin can be isolated from broken cells on the basis of its unique size, density, and solubility properties. It is insoluble in distilled

[16] M. M. Allen, unpublished.
[17] L. O. Ingram, E. L. Thurston, and C. Van Baalen, *Arch. Mikrobiol.* **81,** 1 (1972).
[18] M. Rodriguez-Lopez, M. L. Munoz Calvo, and J. Gomez-Acebo, *J. Ultrastruct. Res.* **36,** 595 (1971).
[19] M. M. Allen and P. J. Weathers, *J. Bacteriol.* **141,** 959 (1980).
[20] M. M. Allen, *J. Phycol.* **4,** 1 (1968).

water and in nonionic detergents, but soluble in weak acids or bases.[5] The method is a modified procedure[21] of Simon.[4,12]

Procedure

Washed cells are concentrated 10 to 100 times. Aliquots are removed for dry weight measurement on washed, tared 2.4-cm Whatman GF/C glass microfiber filters which are dried for 24–48 hr at 80°. Cells are broken either in a French pressure cell at 1050 kg/cm^2 or by sonication for six 1-min treatments interrupted regularly by 30 sec of cooling on ice. After centrifugation for 15 min at 27,000 *g*, the pellet is washed with distilled water, washed twice with 2% Triton X-100, and finally washed twice with distilled water. An orange upper layer of the pellet is removed with a cotton swab. The remaining pellet, usually with a greenish tint, is resuspended in distilled water and centrifuged for 30 sec at 1000 *g*. The supernatant is spun for 15 min at 1000 *g*, and the white pellet is resuspended in distilled water and recentrifuged for 30 sec at 1000 *g*. The white, cloudy supernatant is decanted with a Pasteur pipet and centrifuged at 27,000 *g* for 15 min. Repeated short, low-speed centrifugation may be necessary to remove all green color. Further purification can be carried out by layering the suspended pellet on a 70–100% Renografin 76 (Squibb) step gradient containing 2% Triton X-100 and centrifuging for 2 hr at 23,000 *g*. The highly concentrated granule band at the bottom of the 80% gradient is collected and washed to remove the Renografin. An electron micrograph of negatively stained material collected from such a gradient is seen in Fig. 2.

Quantitative Analysis

Principle

Arginine, in a cyanophycin pellet solubilized in 0.1 *N* HCl, is measured by the modified Sakaguchi reaction of Messineo[22] as described by Simon.[12] Arginine in proteins can be assayed directly by this method without hydrolysis. Purified, lyophilized cyanophycin used as a standard shows that the assay color (absorbance at 520 nm) is proportional to the amount of unhydrolyzed protein present.[12] A correlation factor of 5 was determined between micrograms of arginine hydrochloride per milliliter of sample and micrograms of granules per milliliter of sample.[19]

[21] M. E. Dembinska and M. M. Allen, *J. Gen. Microbiol.* **134,** 295 (1988).
[22] L. Messineo, *Arch. Biochem. Biophys.* **117,** 534 (1966).

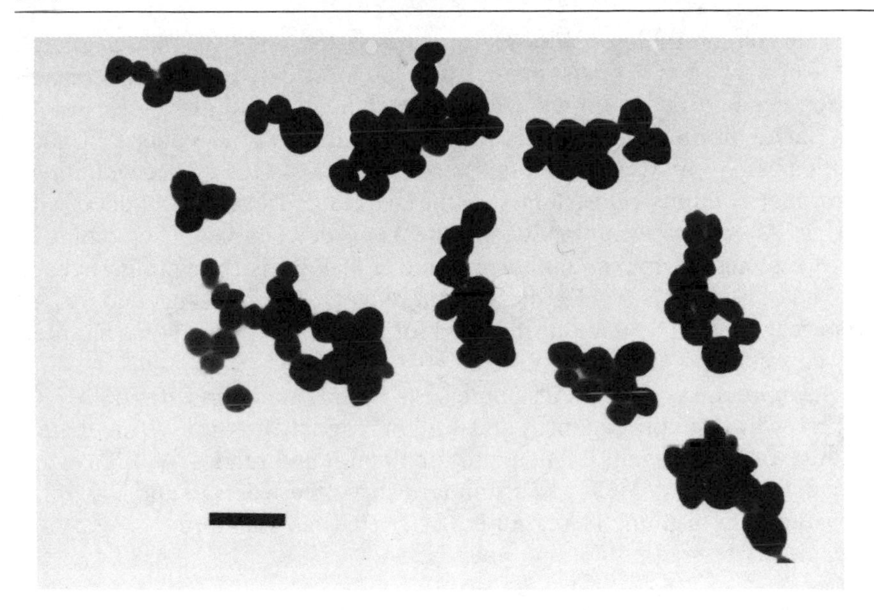

FIG. 2. Electron micrograph of uranyl acetate-stained cyanophycin granules, from *Aphanocapsa* 6308 cells grown in the presence of chloramphenicol for 48 days, purified by Renografin density gradient centrifugation. Bar, 1 μm. Courtesy of M. Dembinska and M. M. Allen.

Procedure

A pellet of cyanophycin (either purified or directly after Triton and water washes) is extracted by two successive treatments with 0.1 *N* HCl for 30 min at room temperature, followed by centrifugation at 27,000 *g* for 15 min. The supernatants are combined, and the volume is adjusted to 2 ml with 0.1 *N* HCl.

Reagents

A: 30 mg KI dissolved in 100 ml of distilled water

B: 100 ml 5 *M* KOH, 2 g potassium sodium tartrate, 100 mg 2,4-dichloro-1-naphthol, 180 ml 100% ethanol, and *x* ml of 4–6% NaOCl brought up to 50 ml with distilled water, mixed in this order; allow to stand 1 hr to stabilize color (store no longer than 1 week)

C: 2*x* ml 4–6% NaOCl brought up to 50 ml with distilled water (store no longer than 1 week)

The volume of hypochlorite to use is determined as follows: Each bottle of 4–6% NaOCl must be tested before use to determine the concentration needed for maximum color development. A dilution series of 4–6% NaOCl in distilled water is prepared, with the final volume of each dilution being 1 ml (typical volumes are 20–60 μl NaOCl). Three milliliters of reagent B minus potassium sodium tartrate and NaOCl is added with mixing. After incubation for 30 min, the A_{400} is determined. The volume of NaOCl giving the maximum absorbance at 400 nm is the volume needed for 3 ml of reagent B. x ml is the amount of NaOCl which is needed for 280 ml of reagent B. The amount required for reagent C is twice that required for reagent B.

Arginine Analysis. Mix a sample (0.1–1.0 ml) made up with 0.1 *N* HCl to 1 ml, with 1 ml of reagent A and 3 ml of reagent B. Allow to incubate 1 hr. Add 1 ml of reagent C, incubate for 10 min, and read A_{520}. Blanks are prepared with 0.1 *N* HCl, and a standard curve, between 10 and 200 μg, is prepared with arginine (1 mg/ml in 0.1 *N* HCl) as standard.

Assay of Cyanophycin Protease

Principle

Since cyanophycin accumulates under some environmental conditions and disappears under others, proteinase activity in crude extracts of cells should measure the hydrolysis of cyanophycin. Substrate is purified ^{14}C-labeled (or nonlabeled) cyanophycin prepared from cells grown in the presence of chloramphenicol. Solubilization of radioactivity as arginine or chemical analysis of arginine solubilized after incubation with cell extracts indicates proteolytic activity.[14]

Reagents

Labeled Substrate. Cyanophycin is isolated as described above from cells grown in the presence of 0.05 μCi/ml L-[U-^{14}C]arginine and 0.05 μCi/ml $NaH^{14}CO_3$. After 48 hr of incubation 5 μg/ml chloramphenicol is added, and cells are incubated for a further 48–72 hr. The amount of label incorporated into the arginine and aspartic acid moieties of cyanophycin is determined by completely hydrolyzing the labeled cyanophycin overnight at 105° *in vacuo* with 6 *N* constant boiling HCl. The free amino acids in the hydrolyzate are separated on Whatman No. 1 chromatography paper using a solvent of 88 : 12 : 0.5 liquefied phenol–distilled water–14.8 *N* NH_4OH. Spots identified by ninhydrin are cut out, and the amino acids are eluted from the paper with 20 m*M* acetate–acetic acid, pH 5.4. Aliquots are counted in a liquid scintillation counter and assayed for amino acid content so that nanomoles of amino acid could be quantitated from the disintegrations per minute of the supernatant after protease assay. Typical preparations yielded 35 ml of substrate cyanophycin with 3.5×10^3 dpm/μl.

Crude Extracts

Cells are harvested, resuspended in 20 m*M* tricine buffer, pH 8.0, and broken in a French pressure cell at 1050 kg/cm^2 followed by centrifugation at 27,000 *g* for 15 min. Protein concentration of the crude extract should be approximately 20 mg/ml.

Procedure

To 75 μl of [^{14}C]cyanophycin ($>5 \times 10^5$ dpm and $>$100 dpm/nmol arginine) in a microfuge tube is added 16 μl 1.2 *M* NaCl, 106 μl 0.8 *M* Tricine, pH 8, and 75 μl of crude extract or 20 m*M* Tricine for the blank. Incubation is at 35°. At zero time and at 20 min, 100 μl is removed and added to 100 μl of cold 10% trichloroacetic acid, followed by centrifugation for 10 min in a microfuge. Triplicate 30-μl aliquots of the supernatant are counted in a liquid scintillation counter. Increase in radioactivity in the supernatant with time indicates proteinase activity. The product(s) of hydrolysis can be qualitatively determined by paper chromatography as described above.

[20] Gas Vesicles: Chemical and Physical Properties

By P. K. HAYES

Introduction

Gas vesicles occur in a wide variety of cyanobacteria. In planktonic species they are often constitutive cell components; elsewhere their production may be restricted to differentiating hormogonia.[1] Gas vesicles provide an organism with hollow spaces which bring about a reduction in cell density, and if enough are accumulated within a cell it will float. Planktonic cyanobacteria regulate their buoyancy with gas vesicles.[2]

[1] A. E. Walsby, *in* "The Prokaryotes" (M. P. Starr, H. Stolp, H. G. Trüper, A. Balows, and H. G. Schlegel, eds.), p. 224. Springer-Verlag, Berlin, 1981.

[2] A. E. Walsby, *Bacteriol. Rev.* **36,** 1 (1972).

TABLE I
MEAN CYLINDER DIAMETER AND MEDIAN CRITICAL COLLAPSE PRESSURE FOR GAS VESICLES ISOLATED FROM DIFFERENT CYANOBACTERIA

Species[a]	Mean cylinder diameter (nm)	Median critical collapse pressure (MPa)
Dactylococcopsis salina	107	0.50
Anabaena flos-aquae	85	0.60
Aphanizomenon flos-aquae	79	0.62
Microcystis sp.	67	0.76
Oscillatoria agardhii	64	0.90
Trichodesmium thiebautii[b]	45	3.70

[a] Strain numbers are given in the legend to Fig. 1.
[b] Observations on material gathered at sea.

Gas vesicles are hollow, cylindrical structures closed at either end with a conical cap.[2] The cylinder diameter of gas vesicles varies within rather narrow limits in a given species, but marked differences in mean diameter are found when comparisons are made between some species (Table I).[3] Length varies markedly in all species because the gas vesicle grows by elongation of the central cylinder from a biconical initial.[4] Both the central cylinder and the conical caps are made up of ribs oriented perpendicular to the long axis of the cylinder. X-Ray crystallography has shown that these ribs are 4.57 nm in width and have repeats every 1.15 nm along their length.[5] The average thickness of the gas vesicle wall is 1.8 nm which means that the individual repeats along the rib have a volume of 9.46 nm.[3] A unit cell of this volume will accommodate a protein of M_r 7500. Protein is the sole structural component of isolated gas vesicles.[6,7] A protein, GVPa, of M_r 7397, very close to the value estimated from the volume of the crystallographic unit cell,[8] forms the shell of the structure.[9] Another protein, GVPc, of M_r 21,985, forms a hydrophilic surface on the outside of the structure.[10] GVPa accounts for 92% of the mass of the gas vesicle and GVPc the remaining 8%.

[3] P. K. Hayes and A. E. Walsby, *Br. Phycol. J.* **21,** 191 (1986).
[4] J. R. Waaland and D. Branton, *Science* **163,** 1339 (1969).
[5] A. E. Blaurock and A. E. Walsby, *J. Mol. Biol.* **105,** 183 (1976).
[6] A. E. Walsby and B. Buckland, *Nature (London)* **224,** 716 (1969).
[7] D. D. Jones and M. Jost, *Arch. Mikrobiol.* **70,** 43 (1970).
[8] P. K. Hayes, A. E. Walsby, and J. E. Walker, *Biochem. J.* **236,** 31 (1986).
[9] P. K. Hayes, C. M. Lazarus, A. Bees, J. E. Walker, and A. E. Walsby, *Mol. Microbiol.* (in press).
[10] A. E. Walsby and P. K. Hayes, *J. Gen. Microbiol.* (in press).

Gas vesicles are highly permeable to gases.[2,11] The outer gas vesicle surface is hydrophilic[12] and the inner surface hydrophobic.[13,14] They are rigid structures[15] which will collapse irreversibly if exposed to a pressure that exceeds a certain value known as the critical pressure (P_c). The median value of P_c (determined by the methods given elsewhere in this volume [72]) for a population of gas vesicles varies in different species[2] (Table I): it is inversely correlated with the average cylinder diameter of the gas vesicles.[3]

Isolation and Purification of Intact Gas Vesicles

The various procedures used to isolate pure gas vesicles have been described in a previous volume[16]; an updated account follows. It is essential that the gas vesicle preparation should not be exposed to pressures that approach the critical collapse pressure. For the weakest gas vesicles of some halophilic cyanobacteria this pressure may be as low as 0.2 MPa.[17] Once the gas vesicles have collapsed, buoyancy, which forms the basis of the isolation procedure, is lost.

Gas-vacuolate cells are harvested from the surface of undisturbed cultures or concentrated either by collection onto 3.0-μm pore size Nucleopore filters or by centrifugally accelerated flotation at low speed. The speed selected for the latter procedure should not produce a centrifugal acceleration exceeding 980 m sec^{-2} (100 g) for a liquid layer of 5 cm depth. Higher speeds will result in the generation of pressures in the liquid layer which may collapse the gas vesicles contained within turgid, cyanobacterial cells.

Cell lysis is achieved using one of a number of techniques. Filaments of *Anabaena, Aphanizomenon,* or *Nostoc* are suspended in 0.7 *M* sucrose; osmotic shrinkage puts the cell wall under tension, causing it to rupture. Filaments of *Oscillatoria* and *Calothrix* and cells of *Microcystis* and *Dactylococcopsis* may be broken using lysozyme (EC 3.2.1.17, 0.5 mg ml^{-1}) in a Tris–HCl buffer (pH 8.0) containing 100 m*M* EDTA. Alternatively, late exponential cultures of *Microcystis* are harvested after growth for 5 hr in the presence of benzylpenicillin (potassium salt, Sigma,

[11] A. E. Walsby, *Proc. R. Soc. London, Ser. B* **223,** 177 (1984).

[12] A. E. Walsby, *Proc. R. Soc. London, Ser. B* **178,** 301 (1971).

[13] D. L. Worcester, *Brookhaven Symp. Biol.* **27,** 37 (1975).

[14] A. E. Walsby, *in* "Relations between Structure and Function in the Prokaryotic Cell" (R. Y. Stanier, H. J. Rogers, and J. B. Ward, eds.), Symp. Soc. Gen. Microbiol. No. 28, p. 327. Cambridge Univ. Press, Cambridge, England, 1978.

[15] A. E. Walsby, *Proc. R. Soc. London, Ser. B* **216,** 355 (1982).

[16] A. E. Walsby, this series, Vol. 31, p. 678.

[17] A. E. Walsby, J. van Rijn, and Y. Cohen, *Proc. R. Soc. London, Ser. B* **217,** 417 (1983).

200–250 U liter^{-1}) and 1 mM Mg^{2+}, and the cells are ruptured, after infiltration with 1 M glycerol for 15 min, by rapid dilution with 3 volumes of 20 mM Tris–HCl (pH 7.7) and standing at 4° for 2 hr.

Shallow layers of cell lysate (2–3 cm deep) are overlaid with 0.1 M phosphate buffer (pH 7.0) to a depth of about 0.5 cm. Intact gas vesicles are collected by centrifugation up through the buffer layer at 4°. The rate of gas vesicle rise varies with both the viscosity of the fluid through which they are traveling and the centrifugal acceleration to which they are exposed. We have measured rates of rise of about 30 μm hr^{-1} g^{-1} for gas vesicles in water and of about 18 μm hr^{-1} g^{-1} for gas vesicles in the more viscous 0.6 M sucrose. After centrifugation the accumulated surface layer of gas vesicles is harvested using a syringe fitted with a narrow-gauge needle. Most impurities are removed by repeated centrifugation through phosphate buffer at 4°. Any unlysed, buoyant cells contaminating the gas vesicle preparation are removed by filtration through 0.6-μm pore size Nucleopore membranes. When large numbers of such cells are present it is advisable to remove most of them by centrifuging shallow (<0.5 cm) layers of the impure gas vesicle preparation at about 14,700 m sec^{-2} (1500 g). During this procedure the gas vesicles (which have a density of between 100 and 160 kg m^{-3}) form a well-defined layer above the cells (which have a density of ~996 kg m^{-3}). Most of the cells can be removed from below the gas vesicle layer using a syringe and needle; those remaining are removed by filtration.

Further purification of gas vesicles can often be achieved after a brief exposure to 0.2% (w/v) sodium dodecyl sulfate (SDS). The detergent disrupts contaminating thylakoid membranes which seem to anneal around gas vesicles to form buoyant globules. This treatment weakens the gas vesicles, but the weakening is reversed by dilution of the SDS.[18] The SDS concentration should be reduced to less than 0.002% (w/v) by addition of phosphate buffer prior to the final centrifugal concentration step. An SDS wash should only be included in the purification procedure after testing the stability of the gas vesicles in this detergent; for example, the gas vesicles of *Oscillatoria* and *Dactylococcopsis* are unstable in SDS, and higher concentrations will remove GVPc from gas vesicles of *Anabaena* and *Microcystis*.[10]

Measurement of Gas Vesicle Concentration[14]

Gas Volume Measurements

The volume of gas vesicle gas space in a suspension of vesicles can be determined by measuring the reduction in the volume of that suspension

[18] A. E. Walsby and R. E. Armstrong, *J. Mol. Biol.* **129,** 279 (1979).

when the vesicles are collapsed by applying pressure exceeding the maximum P_c. This is best done using the compression tube[15] described by Oliver and Walsby ([55] in this volume), but specific gravity bottles may also be used as follows. Small (~0.5 ml) specific gravity bottles are made from 0.9-mm internal diameter glass capillary tubing with a bulb at one end and a neck of about 50 mm length at the other. The volume of the bottle is calculated from the weight of water needed to fill it to a calibration mark scratched close to the open end of the neck. The bottle is filled with a gas vesicle suspension previously degassed by boiling briefly under reduced pressure. The pressure over the bottle is reduced to 0.05 MPa, and any movement in the position of the meniscus, which indicates the presence of bubbles in the bottle, is noted. If no bubbles are present the bottle and contents are weighed. The gas vesicles in the bottle are collapsed by exposure to pressure. Water is added to bring the volume back up to the calibration mark and the bottle is then reweighed. The volume of water added, and hence the volume of gas space destroyed, is calculated from the weight gain.

After measurement of gas volume by either of the above methods the contents of the bottle or compression tube, with washings, are transferred to a preweighed aluminum tray and dried to constant weight *in vacuo*. From these measurements a relationship between gas vesicle–gas space concentration and dry weight concentration is established. For subsequent samples dry weight concentration can be calculated from a measurement of gas space concentration. The relationship also holds for gas space measurements made on gas-vacuolate cell suspensions.

Optical Measurements

For isolated gas vesicles, but not for suspensions of gas-vacuolate cells, another type of measurement can be made which will allow the calculation of gas vesicle dry weight concentration. The method involves the establishment of a relationship between dry weight concentration, determined as above, and pressure-sensitive optical density (PSOD), that is, the change in light scattering which occurs when a suspension of gas vesicles is subjected to a pressure exceeding the maximum P_c. Gas vesicle suspensions are diluted to give optical density readings, at a wavelength of 500 nm, in the range 0.2–0.6 cm^{-1} (in a 1-cm path length cuvette). Optical density readings are taken before and after the collapse of the gas vesicles in the cuvette. The difference between the values is the PSOD. The relationship between PSOD and dry weight concentration remains valid only if all readings are made using the same type of spectrophotometer; different arrangements of cuvette and photocell may significantly affect PSOD readings because the light attenuation is due to scattering rather than absorption.

Gas vesicles from different species have different sizes and geometries. This means that the amount of gas space enclosed, and hence light scattered by a given amount of wall material, will vary among species. For this reason the ratios of PSOD to dry weight and of gas space to dry weight must be determined separately for each species under study.

Chemistry of Gas Vesicles

Protein is the sole constituent of isolated gas vesicles. Although the structures remain intact for years when stored in liquid suspension at 4°, they become much weaker and are not amenable to either electrophoretic or protein sequence analysis. Studies of gas vesicle chemistry should be performed only on freshly prepared material (see above).

Electrophoresis

Samples of gas vesicles are suspended in 1% SDS by boiling for 2 min. Separation of the solubilized proteins is achieved on gels with a discontinuous buffer system.[19] Proteins are made visible by staining in Coomassie blue [0.2% (w/v) in 50% methanol and 7% glacial acetic acid] and destaining in 20% methanol, 7% glacial acetic acid. Silver staining procedures,[20] which have been used with other proteins, produce little or no staining reaction with GVPs. GVPc forms a mobile band of M_r 14K to 35K in different species.[9] GVPa is thought to remain at the origin.

Protein Sequence Analysis

Sequence information for GVP in intact gas vesicles (Fig. 1) and for GVP-derived peptides[8,21,22] (see below) has been obtained by automated amino acid sequencing on either a Beckman spinning-cup sequencer or an Applied Biosystems 470B gas-phase sequencer. Amino acid phenylthiohydantoin derivatives were identified by HPLC. All quantitative amino acid analyses were performed using a Durrum D-500 amino acid analyzer after the samples had been hydrolyzed *in vacuo* at 105° in 6 *M* HCl, 0.1% (w/v) phenol.

[19] U. K. Laemmli, *Nature (London)* **277,** 680 (1970).
[20] C. R. Merril, D. Goldman, S. A. Sedman, and M. H. Ebert, *Science* **211,** 1437 (1981).
[21] J. E. Walker and A. E. Walsby, *Biochem. J.* **209,** 809 (1983).
[22] J. E. Walker, P. K. Hayes, and A. E. Walsby, *J. Gen. Microbiol.* **130,** 2709 (1984).

FIG. 1. Alignment of amino-terminal amino acid sequences for GVPa from (a) *Anabaena flos-aquae* CCAP 1403/13f, (b) *Aphanizomenon flos-aquae* CCAP 1401/1, (c) *Calothrix* sp. PCC 7601, (d) *Oscillatoria agardhii* PCC 7821, (e) *Microcystis* sp. BC (Bristol Collection) 84/1, and (f) *Dactylococcopsis salina* BC 80/4. Dashes indicate identity with the *Anabaena* sequence. The solid line indicates the part of the sequence used to construct an oligonucleotide probe. [The single-letter system of amino acids has been used. See *Eur. J. Biochem.* **138,** 4 (1984).]

Digestion of GVP with Proteolytic Enzymes[8]

Gas vesicles remain intact when incubated with trypsin (EC 3.4.21.4), chymotrypsin (EC 3.4.21.1), V8 protease from *Staphylococcus aureus* (EC 3.4.21.19, serine proteinase), proteinase K (EC 3.4.21.14), or pronase (EC 3.4.24.4). Chaotropic agents such as 6 *M* guanidine hydrochloride or 8 *M* urea also fail to disrupt the structure of intact gas vesicles. In each case the treatment results in a significant reduction in gas vesicle strength.

Limited digestion of GVP with trypsin [tosylphenylalanylchloromethane (TPCK) treated, Sigma] can be obtained after denaturation of GVP with formic acid. Freeze-dried gas vesicles are dissolved in 99% formic acid and then dialyzed at 4°, first against distilled water and then against 2 *M* urea (deionized in Amberlite MB1 and MB3 ion-exchange resins) in 50 m*M* ammonium hydrogen carbonate. The denatured protein is digested with trypsin (stock, 1 mg ml^{-1} in 1 m*M* HCl) at a substrate-to-enzyme ratio of 40 : 1 (w/w) for periods of up to 24 hr at 37°. Reaction products are initially separated on a Sephadex G-50 column (100 × 2 cm) in 50 m*M* ammonium hydrogen carbonate. Peptide-containing fractions, identified by their absorbance at 225 nm, are further fractionated by HPLC. We use a C_{18} reversed-phase column (Macherey, Nagel and Co.) eluted with a linear gradient from 0.1% (v/v) trifluoroacetic acid to 100% acetonitrile.

To digest with pepsin (EC 3.4.23.1) freeze-dried gas vesicles are dissolved in 99% formic acid and diluted with 40 volumes of an enzyme stock solution prepared in 1 m*M* HCl to give a final substrate-to-enzyme ratio of about 50 : 1 (w/w). The digestion is carried out at room temperature for up to 24 hr. Saturated ammonium hydrogen carbonate is then added to give a pH of about 7.5, and the soluble reaction products separated as above.

Succinylated (3-carboxypropionylated) proteins when digested with trypsin are cleaved only after arginine residues and not after the modified lysine residues. Succinylated GVP is prepared as follows. Formic acid-denatured GVP is dialyzed against distilled water. Guanidine hydrochloride is dissolved in the solution of denatured protein to give a final concentration of 6 *M*. A 50-fold molar excess of succinic anhydride is slowly stirred into the solution while holding the pH between 8.5 and 9.5 with small additions of 0.5 *M* NaOH. The reaction products are dialyzed first against 0.5% acetic acid, then against distilled water, and finally against 50 m*M* ammonium hydrogen carbonate. The succinylated protein can be digested with trypsin as above.

Gene Isolation

Application of standard techniques has allowed the isolation and sequencing of structural genes which code for GVPs in *Calothrix* PCC 7601,[23,24] an organism which produces gas vesicles only under special conditions, and *Anabaena*.[9] Recombinant clones carrying the genes were identified using a synthetic oligonucleotide (29-mer) made to code for a 10-amino acid section of the published protein sequence (Fig. 1).[22] Using similar techniques, we have failed to isolate the corresponding gene from *Anabaena flos-aquae* CCAP 1403/13f, an organism which unlike *Calothrix* PCC 7601 produces gas vesicles constitutively, but we have obtained a clone containing DNA from *Anabaena flos-aquae* CCAP 1403/13d (a mutant derived from CCAP 1403/13f which does not express GVP or produce gas vesicles[25]) which carries closely linked *gvp* genes encoding the GVPa and GVPc proteins.

[23] N. Tandeau de Marsac, D. Mazel, D. A. Bryant, and J. Houmard, *Nucleic Acids Res.* **13,** 7223 (1985).

[24] T. Damerval, J. Houmard, G. Guglielmi, K. Csiszar, and N. Tandeau de Marsac, *Gene* **54,** 83 (1987).

[25] A. E. Walsby, *Arch. Microbiol.* **114,** 167 (1977).

Electron Microscopy

Isolated gas vesicles can be seen in the transmission electron microscope after negative staining,[2] but for size measurements they should be shadowed.[3,26] For shadowing, gas vesicles are suspended in 33% (v/v) glycerol to a final concentration of 0.5 mg ml^{-1} and then collapsed by application of pressure. The suspension is sprayed through an atomizer (Sigma spray) onto freshly cleaved mica and dried in a vacuum desiccator. Dried specimens are shadowed with platinum/carbon from an angle of 45° and coated with carbon from above. We use an Edwards 12E6/178 shadowing unit fitted with Cressington EH5 electron guns connected to a Cressington E6602 PC power supply.

For accurate measurements of gas vesicle dimensions it is essential to have a reliable estimate of microscope magnification. The nominal magnification reading taken from the microscope may be in error by as much as 15% or more. For shadowed specimens we mix latex spheres of standard diameter (Agar Aids) with the gas vesicle suspension to act as internal size standards. Where a microscope specimen holder has positions for more than one grid a better estimate of microscope magnification can be obtained using a carbon replica of a diffraction grating.[3,26] After photographing each gas vesicle specimen a grid holding the diffraction grating replica is moved into the electron beam. The image is focused using the vertical adjustment on the specimen stage and not by the lens focus controls as any alteration in lens current will affect the final magnification. After focusing a photograph is taken. The magnification on the negative is calculated by measuring the spacing of the rulings on the diffraction grating. The grating periodicity should be calibrated as the manufacturers values may be in error by up to 3%. (We have calibrated gratings by measuring the images of an identifiable grating grid square on light and electron micrographs and measuring the distance between 100 grating rulings on the latter.)

Elastic Compressibility of Gas Vesicles

Measurements of gas vesicle elastic compressibility are relevant to the theory of critical pressures and the discussion of buoyancy at depth. They are made using a compression tube apparatus[15] (see Fig. 1 in [55] in this volume). The inner tube, volume V_i, is filled with a concentrated suspension of gas vesicles. The position of the meniscus in the capillary is noted, and its movement is measured using a vernier microscope when the outer tube is pressurized. The pressure is not allowed to reach a value at which

[26] M. Jost and D. D. Jones, *Can. J. Microbiol.* **16**, 159 (1970).

any collapse of gas vesicles occurs. The volumetric compression of the gas vesicle suspension is calculated as the area of capillary cross section multiplied by the change in meniscus position when the apparatus was placed under pressure. The tube is filled with water and the above procedure repeated.

The compression of the gas vesicles (dV_g) per 0.1 MPa is calculated from the equation

$$dV_g = dV_t - dV_w + dV_i$$

where dV_t is the total compression per 0.1 MPa for the tube filled with the gas vesicle suspension, dV_w the compression due to water, and dV_i the compression in the tube itself. The last two terms are estimated from the measurements made with the tube filled with water; dV_i is calculated equal to $-PV_i/K$, where K is the elastic modulus of glass (46 GPa) and P the applied pressure.[15] Dividing dV_g by the volume of the gas vesicles in the suspension[18] gives the elastic bulk modulus for the intact gas vesicle. The value for *Anabaena* gas vesicles is 64.5 MPa. From this value the elastic modulus of GVP can be calculated, 2.7 GPa.[15] Measurements are now needed on gas vesicles from other species.

[21] Cellular Differentiation: Akinetes

By MICHAEL HERDMAN

Although often referred to as "spores," particularly in the early literature, there is now general agreement that the word akinete (Greek: *akinetos,* motionless) is suitable to distinguish this differentiated resistant cell from the spores, cysts, endospores, and exospores of other microorganisms and from the baeocytes[1] of pleurocapsalean cyanobacteria. Akinetes are the subject of several recent reviews,[2-4] and this chapter describes only some practical aspects of their identification, production in pure culture, isolation, and properties. Although the nomenclature of cyano-

[1] M. Herdman and R. Rippka, this volume [22].

[2] J. M. Nichols and N. G. Carr, *in* "Spores" (C. H. Chambliss and J. C. Vary, eds.), Vol. 7, p. 335. Am. Soc. Microbiol., Washington, D.C., 1978.

[3] J. M. Nichols and D. G. Adams, *in* "The Biology of Cyanobacteria" (N. G. Carr and B. A. Whitton, eds.), p. 387. Blackwell, Oxford, England, 1982.

[4] M. Herdman, *in* "The Cyanobacteria: A Comprehensive Review" (P. Fay and C. Van Baalen, eds.), p. 227. Elsevier, Amsterdam, 1987.

bacteria is under discussion at present, and will certainly continue to evolve, for convenience the generic and specific names employed in the publications discussed are retained; however, many of these taxonomic assignments are clearly incorrect.

Spatial Patterns and the Recognition of Akinetes

The ability to produce akinetes is limited to certain heterocystous strains of Sections IV and V defined by Rippka *et al.*[5] In some strains, e.g., *Cylindrospermum licheniforme* and *Anabaena cylindrica,* akinetes differentiate from the cell immediately adjacent to the heterocyst; these akinetes are easily recognized in the light microscope as enlarged cells, granular in appearance (Figs. 1a, b), and often possessing a thickened outer envelope (Fig. 1b). Akinetes of *A. circinalis* (Fig. 1c) differentiate from the third vegetative cell away from the heterocyst and are similar in size to those of *C. licheniforme* and *A. cylindrica.* In other strains, e.g., *Nostoc* PCC 7524 and *Anabaena* strain CA, akinetes develop from vegetative cells distant from the heterocyst: those of *Nostoc* PCC 7524 first appear in the center of the chain of cells which separates two heterocysts (Fig. 1e), and subsequently the cells on either side successively differentiate to produce a chain of akinetes.[6] Chains of akinetes are also produced in *Anabaena* CA (see Ref. 3) and *Fischerella* (Fig. 1d); being only slightly larger than the vegetative cells, such akinetes are most easily recognized under phase-contrast optics which reveal the numerous refractile cyanophycin granules which they contain (Fig. 1e). These akinetes are often liberated from the filaments to produce a homogeneous suspension of single cells, easily mistaken for a population of unicellular cyanobacteria.

Culture Conditions Required for Akinete Formation

The formation of akinetes only by heterocystous cyanobacteria might be taken to imply that the presence of heterocysts is a prerequisite for akinete differentiation. This is true for those strains in which akinetes arise adjacent to the heterocysts (see above) and for *A. circinalis;* consequently, in order to observe akinete formation in cultures of these cyanobacteria, the organisms must be cultivated in the absence of sources of combined nitrogen which repress heterocyst differentiation. The small akinetes of *Nostoc* PCC 7524 and *Anabaena* CA, however, which first

[5] R. Rippka, J. Deruelles, J. B. Waterbury, M. Herdman, and R. Y. Stanier, *J. Gen. Microbiol.* **111,** 1 (1979).

[6] J. M. Sutherland, M. Herdman, and W. D. P. Stewart, *J. Gen. Microbiol.* **115,** 273 (1979).

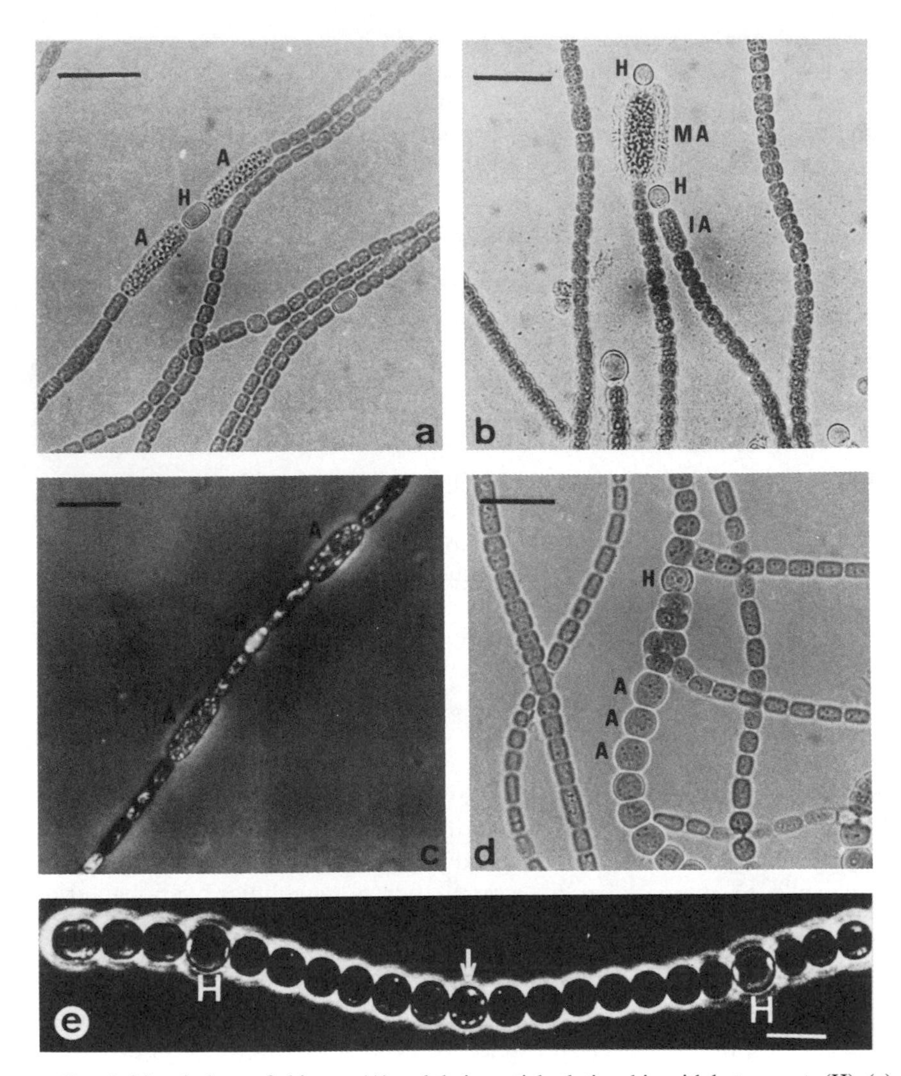

FIG. 1. Morphology of akinetes (A) and their spatial relationship with heterocysts (H). (a) Akinetes of *Anabaena* PCC 7122 (*A. cylindrica*) are elongated and usually differentiate on both sides of the intercalary heterocyst; their granular appearance is caused, as in akinetes of all cyanobacteria, by the high content of cyanophycin. (b) Immature akinetes (IA) of *Cylindrospermum* PCC 7417 are longer than the vegetative cells and give rise to a larger mature akinete (MA) surrounded by a thick envelope; since heterocysts occur only in a terminal position in the filament, the symmetry of akinete pattern observed in (a) cannot occur. (c) Akinetes of *Anabaena circinalis* normally differentiate symmetrically on either side of, but not adjacent to, the heterocyst. (d) *Fischerella* strains (PCC 7414 is shown) produce a long chain of akinetes which differentiate from the vegetative cells of the primary filament; note the lateral branches composed only of small vegetative cells. (e) The cyanophycin granules which appear first in a cell midway between two heterocysts (arrow) mark the first stage of akinete differentiation in *Nostoc* PCC 7524. (a), (b), and (d) bright field, courtesy of R. Rippka; (c) phase contrast, courtesy of P. Fay; (e) phase contrast, courtesy of J. M. Sutherland. Bar markers represent 20 μm in (a)–(d), 5 μm in (e). From M. Herdman.[4]

appear distant from the heterocyst, will differentiate even in media containing combined nitrogen, in the absence of heterocysts.

Variations in the supply of a wide range of other nutrients have been reported to induce or to inhibit akinete differentiation, but few of these have been substantiated (see Refs. 4 and 7 for review). It is clear, however, that in *Nostoc* PCC 7524,[6] *A. cylindrica,*[8,9] *A. circinalis,*[10] *Gloeotrichia echinulata,*[11] and *Aphanizomenon flos-aquae*[12] akinete differentiation in laboratory culture occurs only following the end of exponential growth when light (energy) has become limiting, and it can be prevented by maintaining steady-state growth conditions.[6,9] Since these organisms represent all of the main classes of akinete-forming cyanobacteria (with respect to the spatial pattern of akinetes and response to the presence of combined nitrogen), it is evident that stationary phase cultures are required for the production and experimental study of akinetes. Care should be taken to maintain a constant light intensity during akinete differentiation; an increase from 8000 to 11,000 lux caused akinetes to germinate immediately after their formation.[6] Typical culture conditions for akinete differentiation are as follows.

Nostoc PCC 7524.[6] Medium BG-11_0,[13] with 1 g liter^{-1} $NaHCO_3$; 34°; light intensity 8000 lux (warm-white fluorescent); gas phase 5% CO_2 in air. Exponential growth (doubling time 8.5 hr) was maintained for 30 hr; akinete differentiation commenced after 70 hr (OD_{650} of culture, 2.0) and continued for a further 60 hr, when 60% of the cells had become akinetes.

Anabaena cylindrica.[8] Medium of Allen and Arnon,[13] with 0.3 g liter^{-1} HEPES, pH 7.5; 29°; light intensity, e.g., 640 lux incandescent or 1500 lux warm-white fluorescent; gas phase 5% CO_2 in air. Exponential growth (doubling time 30 hr) occurred for 40 hr at 640 lux; akinetes first appeared at 60 hr (OD_{650} of culture, 0.6) and reached a frequency of 7% after 11 days.

Anabaena circinalis.[10] The medium of Gorham *et al.*[13] was employed, with NO_3^- and HEPES omitted; 20°; 1340 lux (daylight fluorescent) with a light–dark cycle of 14 : 10 hr; gas phase, air. Exponential growth (doubling time, 7 days) was followed by akinete production on day 9 (OD_{750}, 0.3); akinetes reached a frequency of 3% after a further 5 days.

Anabaena spp. (16 strains).[14] Medium BG-11_0; 25°; 800 lux (cool-white fluorescent) with a light–dark cycle of 12 : 12 hr; no gassing.

[7] C. P. Wolk, *Dev. Biol.* **12,** 15 (1965).
[8] J. M. Nichols, D. G. Adams, and N. G. Carr, *Arch. Microbiol.* **127,** 67 (1980).
[9] P. Fay, *J. Exp. Bot.* **20,** 100 (1969).
[10] P. Fay, J. A. Lynn, and S. C. Majer, *Br. Phycol. J.* **19,** 163 (1984).
[11] M. Wyman and P. Fay, *Br. Phycol. J.* **21,** 147 (1986).
[12] J. A. Rother and P. Fay, *Br. Phycol. J.* **14,** 59 (1979).
[13] R. Rippka, this volume [1].
[14] B. K. Stulp and W. T. Stam, *Arch. Hydrobiol. Suppl.* **63,** 35 (1982).

Gloeotrichia echinulata.[11] Culture as for *A. circinalis,* but light intensity 913 lux with a light–dark cycle of 16 : 8 hr. Doubling time 6.5 days; akinetes first appeared after 20 days (OD_{750}, 0.45). Akinete differentiation occurred 5 days earlier under green light at an intensity which gave the same growth rate as above.

The only inorganic nutrient whose effect on akinete differentiation in some strains appears to have been substantiated is phosphate. Starvation for phosphate caused akinete formation in *C. licheniforme*[15] and *A. cylindrica,*[7] even in exponentially growing cultures; this appears to be strain dependent, since phosphate stimulated akinete differentiation in *A. circinalis*[10] and akinetes were never produced in the absence of phosphate in *Nostoc* PCC 7524.[6] Cells of *C. licheniforme* produce a compound which accumulates in the medium and is capable of inducing akinete differentiation when added to young cultures, even in the presence of phosphate.[15–17] The compound was extracted from culture supernatants or whole cells with methanol[16] and was purified by relatively simple procedures[17]; analysis showed a formula of C_7H_5OSN (mol. wt. 151). The purified compound was maximally active at less than 0.3 μM. Culture filtrates of *A. circinalis* similarly increased akinete frequency by up to 50% in a concentration-dependent manner.[10] This effect is not true for all strains, however, since it was not observed in *Aphanizomenon flos-aquae,*[12] *Nostoc* PCC 7524,[6] or *A. cylindrica*.[3]

Isolation and Purification of Akinetes

Techniques used for the separation of akinetes from vegetative cells take advantage of the size of akinetes and their resistance to cold, lysozyme, or mechanical stress. Cultures of *A. cylindrica,* resuspended in distilled water at pH 8.5 (to prevent subsequent precipitation of released protein and consequent aggregation of akinetes), were passed through a French press and then centrifuged at low speed through 0.8 *M* sucrose to sediment the akinetes.[9] This technique was adapted by Simon,[18] who replaced the French press treatment by exposure to lysozyme, and akinetes of *A. variabilis* and *A. cylindrica* were successfully purified following sonication and differential centrifugation.[19–21] The resistance to cold of akinetes of *Nostoc* PCC 7524 permitted their purification: akinete viability

[15] R. W. Fisher and C. P. Wolk, *Nature (London)* **259,** 394 (1976).
[16] T. Hirosawa and C. P. Wolk, *J. Gen. Microbiol.* **114,** 423 (1979).
[17] T. Hirosawa and C. P. Wolk, *J. Gen. Microbiol.* **114,** 433 (1979).
[18] R. D. Simon, *J. Bacteriol.* **129,** 1154 (1977).
[19] R. D. Simon, *FEMS Microbiol. Lett.* **8,** 241 (1980).
[20] G. S. Björn, W. Braune, and L. O. Björn, *Physiol. Plant.* **59,** 493 (1983).
[21] Y. Yamamoto, *Plant Cell Physiol.* **16,** 749 (1975).

remained at 100% following storage at 0–4° for 7 days, during which time the fraction of vegetative cells in the samples decreased from 40 to less than 10%; akinetes were again separated from the remaining vegetative cells by differential centrifugation.[6] The latter authors also commented on the resistance of akinetes to freezing, a technique employed to produce samples of akinetes of the same organism that were 90% pure and 85% viable, following freezing at −35° for only 4 hr.[22]

The resistance of akinetes to desiccation,[6,21] which is not tolerated by vegetative cells, has apparently not yet been employed for their purification. Another potential technique which should be considered for isolation of akinetes of planktonic gas-vacuolate cyanobacteria could profit from the loss of gas vacuoles which occurs during differentiation[23]; the consequent decrease in buoyant density removes the akinetes from the water column, and it would seem that this phenomenon could be easily employed under laboratory conditions to purify akinetes by sedimentation.

Storage of Akinetes: Resistance to Extreme Conditions

Although akinetes of *Nostoc* PCC 7524 may be stored at room temperature, they slowly degrade phycocyanin[24] to meet their requirements for metabolic activity under these conditions. They retain their original properties when stored in the dark at 0–4°, 95% surviving for 15 months whereas only 10% of the vegetative cells survive after 7 days.[6] Although the akinetes tolerate short periods of freezing this is not useful for storage, since viability was reduced to 5% after only 12 hr at −35°.[22] However, freezing in liquid N_2 [with dimethyl sulfoxide (DMSO), 5% (v/v), as cryoprotectant] consistently gave good survival provided that the akinetes were thawed relatively quickly at 34°.[22] The akinetes were rapidly killed at 45° (5% survival after 10 min), only 4° higher than the maximum temperature permitting growth[6]; their resistance to desiccation (45–80% surviving in the dry state after 6 months)[6] might also be used to advantage for storage. Akinetes of *A. cylindrica* also survive desiccation, although viability steadily declines[21]: after being desiccated for 5 years in the dark at 27°, 7% of the akinetes were able to germinate; this value was increased by storage in the dark at 4° (10% germination) or at −20° (13% germination). Control (nondesiccated) akinetes germinated with an efficiency of 20%. Storage of desiccated akinetes in the light resulted in a rapid loss of viability and is not recommended.

[22] F. Chauvat and F. Joset-Espardellier, *FEMS Microbiol. Lett.* **10,** 319 (1981).

[23] J. A. Rother and P. Fay, *Proc. R. Soc. London, Ser. B* **196,** 317 (1977).

[24] J. M. Sutherland, J. Reaston, W. D. P. Stewart, and M. Herdman, *J. Gen. Microbiol.* **131,** 2855 (1985).

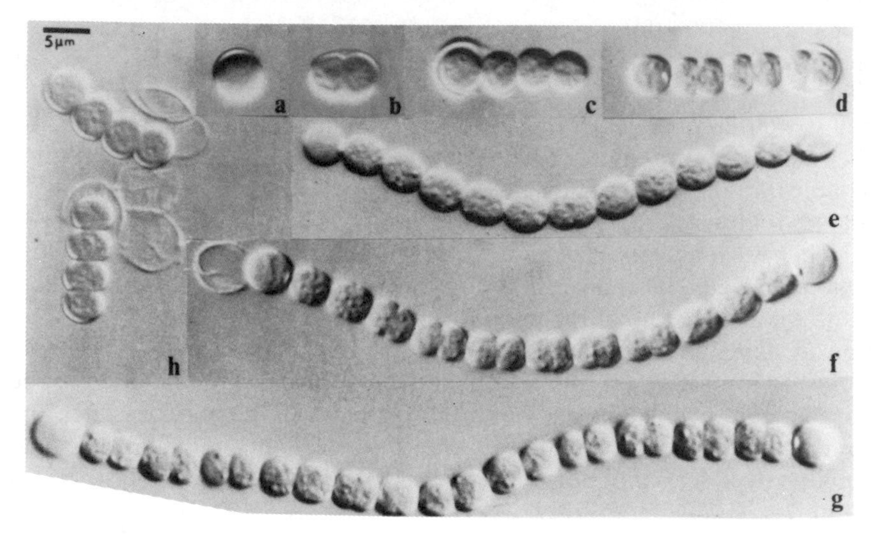

FIG. 2. Normarski differential interference contrast micrograph showing stages of akinete germination in *Nostoc* PCC 7524. The mature akinete (a) swells, and the first division often occurs inside the akinete envelope (b). After subsequent cell division the envelope ruptures but may remain attached to the developing filament (c, d, f, h); empty envelopes retain their shape (f, h). Heterocysts first differentiate from the terminal cells of the young filament (d, e), which subsequently gives rise to a mature filament (g). From Sutherland *et al.*[27]

Germination of Akinetes

The process of germination can be easily observed in the light and electron microscopes; some typical stages of this process are shown in Figs. 2 and 3. Slightly different patterns occur in strains identified as *Anabaena* spp.,[14] *A. variabilis,*[25] and *Cylindrospermum.*[26] The first cell division, depending on strain and culture conditions, would normally be expected to occur between 10 and 50 hr after the initiation of germination.

The only prerequisite for successful germination is an increase in light intensity; although this is normally achieved by diluting akinete preparations into fresh liquid medium,[14,27] the fresh medium itself is not an absolute requirement.[6] A 10-fold dilution (giving 10^7 akinetes ml^{-1}) for *Nostoc* PCC 7524 or a 50-fold dilution for a variety of *Anabaena* strains yielded cultures which were easily analyzed experimentally and permitted synchronous germination of akinetes of *Nostoc* PCC 7524[24,27] and one (strain

[25] W. Braune, *Arch. Microbiol.* **126,** 257 (1980).
[26] M. M. Miller and N. J. Lang, *Arch. Mikrobiol.* **60,** 303 (1968).
[27] J. M. Sutherland, W. D. P. Stewart, and M. Herdman, *Arch. Microbiol.* **142,** 269 (1985).

1617) *Anabaena*[14]. Culture conditions were as described above for the production of akinetes, using medium BG-11_0. The addition of $NaNO_3$ (1.5 g liter^{-1}) or $(NH_4)_2SO_4$ (0.7 g liter^{-1}) did not significantly change the rate of germination.[27] Optimum light intensity, temperature, and pH for germination of *A. cylindrica* akinetes were similar to those for growth of vegetative filaments[28] (3000 lux, 27°, pH 8); again, $NaNO_3$ had little effect on germination rate.

A slide culture technique has also been employed to investigate morphological changes during germination of akinetes of *A. variabilis*.[29] Agar slides were maintained after inoculation in humid chambers at 25° and were suitable for studies involving monochromatic light (see below). However, of 16 *Anabaena* strains examined[14] only one, strain 1617, germinated successfully in slide culture.

The rate of germination of akinetes of *A. cylindrica*[28] and *A. variabilis*[29] has been shown to depend on light intensity, and detailed studies[29] using monochromatic light revealed that the action spectrum for germination resembled that of photosynthesis, with light of 625 nm (absorbed by phycocyanin) and 670 nm (absorbed by chlorophyll) being most efficient. The requirement for photosystem II activity may be tested by adding 3′-(3,4-dichlorophenyl)-1′,1′-dimethylurea (DCMU), a specific inhibitor of this photosystem. Although the akinete envelopes of *A. variabilis* burst and akinetes were liberated in the presence of DCMU, the young germlings either did not divide or underwent only one division.[29] Germination of akinetes of *A. cylindrica* was completely inhibited by DCMU.[28] A few akinetes (4%) of *Nostoc* PCC 7524 were capable of germination under anaerobic conditions in the presence of DCMU (where only cyclic phosphorylation around photosystem I can operate[30]); the addition of air, allowing respiration also to occur, increased the efficiency of germination to 20%, and the further addition of sucrose (a utilizable carbon source for this strain) allowed all of the akinetes to germinate, although slowly. Germination did not occur in the dark in the presence of sucrose. Germination can therefore occur when photosystem II activity is inhibited, and it requires both respiration and cyclic phosphorylation; rapid germination, on the other hand, requires activity of both photosystems.[30] However, detailed analysis of akinete germination in the dark has not yet been performed with heterotrophic cyanobacteria which grow relatively rapidly under such conditions, although in 1917 Harder[31,32] had shown that

[28] Y. Yamamoto, *J. Gen. Appl. Microbiol.* **22,** 311 (1976).

[29] W. Braune, *Arch. Microbiol.* **122,** 289 (1979).

[30] F. Chauvat, B. Corre, M. Herdman, and F. Joset-Espardellier, *Arch. Microbiol.* **133,** 44 (1982).

[31] R. Harder, *Ber. Dtsch. Bot. Ges.* **35,** 58 (1917).

[32] R. Harder, *Jahrb. Wiss. Bot.* **58,** 237 (1917).

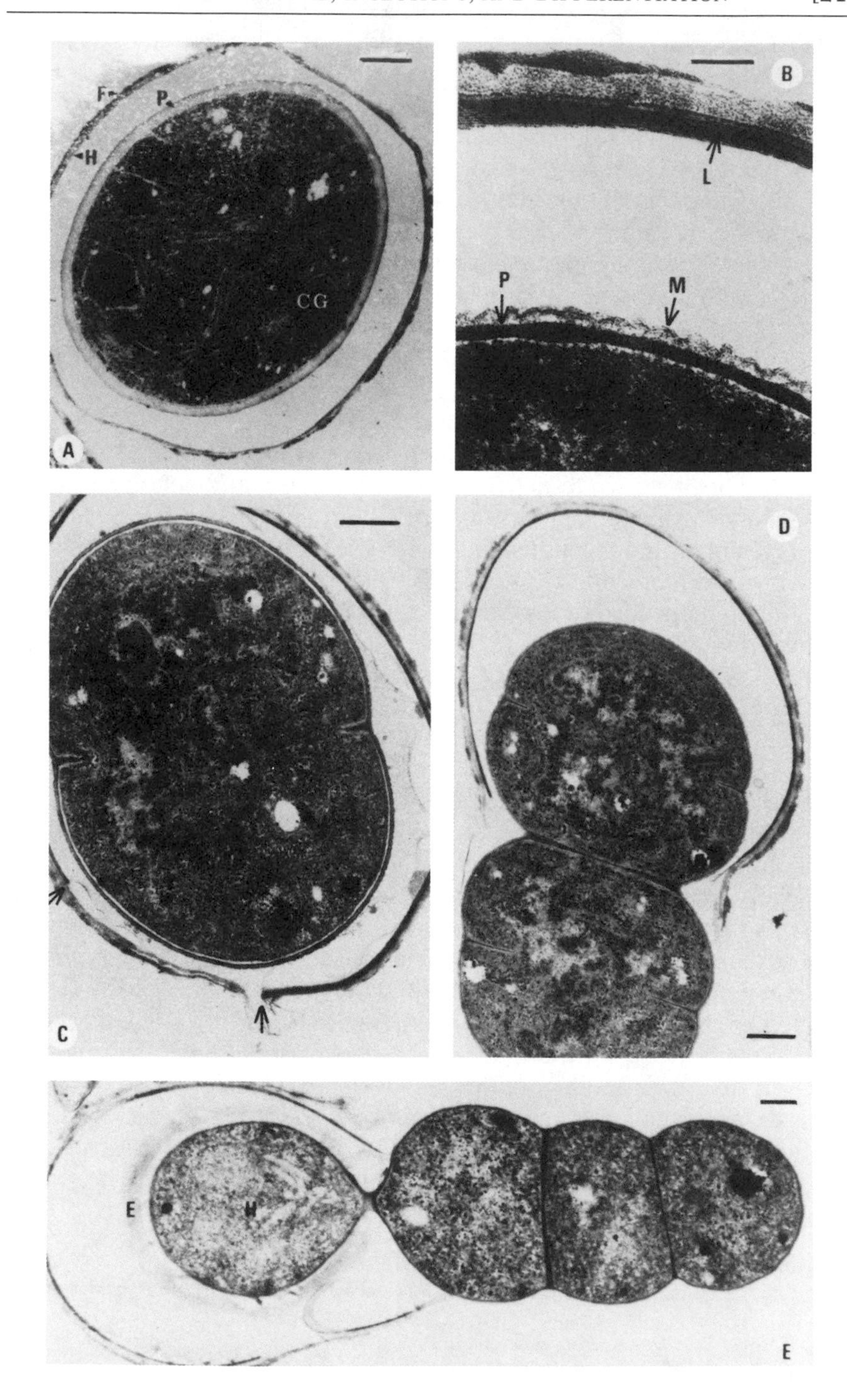
F
P
H
CG
A
B
L
P
M
C
D
E
H
E

akinetes of some strains would germinate in the dark in the presence of cane sugar. Further studies would appear necessary.

Germination of akinetes of *A. cylindrica* was not stimulated by heat shock,[28] which breaks the dormancy of bacterial and fungal spores. Among 16 strains of *Anabaena* studied, only one required a period of desiccation for akinete germination.[14]

Mutational Loss of Akinetes

Mutants of *Nostoc* PCC 7524, incapable of producing akinetes, appear to arise at relatively high frequency,[33] and many cyanobacteria which formed akinetes when first isolated can no longer do so.[34] In order to prevent such loss it would seem appropriate to maintain the strains under conditions which impose a positive selection pressure which is lethal to the mutants. No systematic study of this problem has been performed, but one can certainly suggest periodic exposure to cold and/or desiccation as a means to this end. Optimization of light intensity and temperature during growth, the imposition of light–dark cycles, and the careful use of different media[13] may be of vital importance in the maintenance of akinete-forming cyanobacteria, and these parameters require urgent study.

Concluding Remarks

Like baeocytes and hormogonia,[1] akinetes can be induced to develop relatively synchronously.[6] Under favorable conditions they can also be

[33] M. Herdman, unpublished observations.

[34] R. Rippka, personal communication.

FIG. 3. Ultrastructural changes during germination of akinetes of *Nostoc* PCC 7524. (A) Mature akinete showing the thickened peptidoglycan layer (P) of the cell wall and the extracellular envelope composed of homogeneous (H) and fibrous (F) layers. CG, Cyanophycin granule. Bar, 500 nm. (B) An akinete 2 hr after the initiation of germination. The peptidoglycan layer (P) is thinner than that of the mature akinete but has not yet reverted to the thickness of that of a vegetative cell; it is surrounded by an outer membrane (M), and a new laminated layer (L), with 4 nm between lamellae, has appeared on the inside of the envelope. Bar, 100 nm. (C) An early stage of formation of the septum during the first cell division. Note the wide break (arrows) in the laminated layer of the envelope, the gap still being covered by the outer layers. Polyhedral bodies (PB) are present, but few cyanophycin granules remain. Bar, 500 nm. (D) Division of the two daughter cells is initiated synchronously; the germling is still partly inside the akinete envelope. Bar, 500 nm. (E) A germinating akinete at the four-cell stage. A heterocyst (H) has begun to differentiate from the terminal cell of the young filament, still trapped inside the akinete envelope; the heterocyst envelope (E) is not yet completely formed. Bar, 500 nm. Figure 3A is reproduced from Sutherland *et al.*,[6] and Figs. 3B–E from Sutherland *et al.*[27]

made to germinate synchronously,[24,27] and are therefore useful for the study of the biochemical changes[24] which accompany the development of the mature filament. Their greatest potential, particularly in those strains in which they are produced at high frequency, is that they represent a unicellular stage in the life cycle of a multicellular organism: as such, they are invaluable for genetic studies since, following mutagenesis, auxotrophic mutants can be isolated as easily as in unicellular cyanobacteria.[33]

[22] Cellular Differentiation: Hormogonia and Baeocytes

By MICHAEL HERDMAN and ROSMARIE RIPPKA

Definitions

Hormogonia are produced by many, but not all, heterocystous cyanobacteria.[1] The term hormogonium (Greek: *hormos,* chain; *goneta,* generation) has been traditionally employed in the phycological literature to describe any motile filament released from an immotile, ensheathed parental trichome of both heterocystous and aheterocystous cyanobacteria. We employ here the extended definition of Rippka *et al.*[1] which, referring specifically to heterocystous cyanobacteria, describes hormogonia as "filaments, either motile or immotile, that are distinguishable from the parental trichome by cell size, cell shape, gas vacuolation or the absence of heterocysts, even when grown without a source of combined nitrogen."

Pleurocapsalean cyanobacteria reproduce by multiple fission, a large parental cell thereby liberating many small cells, termed endospores in the phycological literature. As pointed out by Waterbury and Stanier,[2] these cells differ from the classic bacterial endospore in their mode of formation, structure, and development. The use of this term is therefore confusing, and we employ here the word "baeocyte"[2] (Greek: small cell) to describe these unique structures. Since both hormogonia (as defined here) and baeocytes are produced by rapid cell division in the absence of growth (see below), they can be considered to be analogous structures, even though they are respectively filamentous and unicellular, so justifying their inclusion in a single chapter of this volume.

[1] R. Rippka, J. Deruelles, J. B. Waterbury, M. Herdman, and R. Y. Stanier, *J. Gen. Microbiol.* **111,** 1 (1979).

[2] J. B. Waterbury and R. Y. Stanier, *Microbiol. Rev.* **42,** 2 (1978).

Baeocytes

The description of baeocytes presented below is based entirely on the elegant and thorough experimental studies of Waterbury and Stanier,[2] performed on pure cultures under defined (and therefore reproducible) conditions. We do not wish to imply, however, that earlier studies (see Ref. 2 for details) are without value. The baeocytes of many strains are motile; their function (whether motile or immotile) is primarily one of dispersal since they are released, often in large numbers, from a parental colony which is generally sessile.

Media and Growth Conditions

The pleurocapsalean cyanobacteria studied were maintained in medium BG-11 (for strains of soil or freshwater origin) or media MN or ASN-III (marine strains).[2] Most were sensitive to high light and were therefore grown under intensities of 200–400 lux; temperature was maintained at 25–28° for all except one thermophilic strain, which was grown at 37°. Liquid cultures were not agitated. Developmental cycles, which often required several weeks for completion, were observed on agar in Cooper culture dishes (Falcon, No. 3009) which were originally designed for tissue culture; the plastic lid prevented dehydration and permitted periodic microscopic examination without opening the dish, thus avoiding contamination. The dishes were normally illuminated (200–300 lux) evenly from above.

Baeocyte Development

The cell cycle is similar in all strains, the small baeocyte giving rise to a much larger vegetative cell (see Ref. 3 for generic descriptions). However, the patterns of development differ: some organisms (e.g., *Dermocarpa*) reproduce exclusively by multiple fission (producing baeocytes) whereas others (e.g., *Myxosarcina*, *Pleurocapsa*) also undergo normal binary fission to produce complex aggregates of vegetative cells, not all of which necessarily undergo multiple fission to produce baeocytes. The baeocytes of *Dermocarpa* (diameter 1–2 μm) each give rise to a vegetative cell which may be up to 30 μm in diameter, representing a 3000-fold increase in cell volume; growth is accompanied by chromosome replication but not cell division. Multiple fission will subsequently give rise to as many as 1000 baeocytes which then grow and repeat the cycle. The number of baeocytes produced depends on the size of the parental cell and therefore on the stage of its life cycle at which multiple fission is initiated;

[3] R. Rippka, this volume [2].

unfortunately, nothing is known of the environmental or physiological factors which induce this process except that it always occurs rapidly (within 12–24 hr) after dilution of a stationary phase culture of vegetative cells into fresh medium. Dilution is not the only trigger, however, since baeocyte formation will occur more than once, usually asynchronously in different cells, during the growth of the culture. In some organisms, e.g., *Myxosarcina,* as few as four baeocytes are produced from a vegetative cell, the size difference therefore being much less pronounced than in *Dermocarpa.*

Nature of Multiple Fission

The vegetative cells of pleurocapsalean cyanobacteria contain many copies of the genome[2] and it is probable that multiple fission does not require *de novo* synthesis of DNA. Multiple fission is a series of rapid successive binary fissions, the "burst size" (i.e., the number of baeocytes produced from an individual cell) normally falling into the series 2^n, where n is the number of binary fissions performed. However, many burst sizes are intermediate between two successive numbers in the series, although of 281 baeocytes released from a single cell of *Dermocarpa* exactly 2^8 (256) were viable, the rest being unable to develop.[2] It may be supposed (although further studies are clearly necessary) that chromosome replication in the growing parental cell is strictly controlled, producing a number of genomes consistent with the 2^n law, and that each chromosome is segregated into a viable baeocyte at the time of fission; the fission process itself appears to be less accurately controlled and produces some (up to 10%) anucleate, and therefore nonviable, baeocytes.[2]

Baeocyte Motility

The baeocytes of many strains are motile, gliding at rates of up to 120 μm/hr at 22°; this may not seem to be extremely rapid but nevertheless represents a rate of 50–100 cell diameters per hour. To test for motility, the culture dishes are inoculated with stationary phase cultures and illuminated (50–200 lux) from the side. Motile baeocytes display a phototactic response, the direction of movement being either toward (low intensity) or away from (high intensity) the light source. Motility is dependent on the structure of the cell wall. Vegetative cells of all pleurocapsalean cyanobacteria are completely immotile and are surrounded by a thick outer fibrous cell wall layer. In many genera, this layer is synthesized during multiple fission and is consequently possessed by the baeocytes which, as a result, are also immotile. Motile baeocytes are produced only in those strains in which synthesis of this layer is suppressed during their formation. Motility is only transient, however, being lost after as little as 6 hr (although some baeocytes may remain motile for as long as 48 hr); the resynthesis of the fibrous cell wall layer appears to be responsible not only for the loss of motility but also for the subsequent ability of the baeocyte to adhere firmly to a solid substrate, thus giving rise to a new sessile colony.

Gas vacuoles,[4,4a] which permit the flotation and dispersal of the hormogonia of certain heterocystous cyanobacteria, are never produced by pleurocapsalean cyanobacteria. Consequently, except in the aquatic environment where baeocytes will be easily washed away from the parental colony, gliding motility is the only means of dispersal in these organisms.

Hormogonia

Like baeocytes, hormogonia serve for dispersal. Those of many strains are motile while others, unlike the parental filament, contain gas vacuoles. Again, they may be released from a sessile parental colony and glide, float, or be washed by water currents to a new location.

Induction of Hormogonial Differentiation

In general, hormogonia, like baeocytes, are formed abundantly after transfer of stationary phase cultures (which normally contain only vegetative trichomes) to fresh medium. Thuret[5] first showed that simple dilution of freshly harvested colonies of *Nostoc*, in a bowl of water, produced a swarm of thin, motile filaments within 2–3 days; their properties resembled those of the structures which we now term hormogonia, and they subsequently increased in diameter and gave rise to mature, immotile filaments. Dilution of a dense culture results in a sudden increase in light intensity, permitting rapid growth, which might be interpreted as suggesting that differentiation of hormogonia requires high light intensity and that their absence from stationary phase cultures is a result of light limitation. While this may be true for some strains,[6] the following results show that the control of differentiation may in fact be much more complex.

A strain (PCC 7119) originally identified as *Anabaena*[1] on the grounds that it was incapable of forming hormogonia following transfer to fresh medium, was shown[7] by DNA–DNA hybridization to be closely related to *Nostoc* strains, all of which (by our definition[1]) produce hormogonia.

[4] A. E. Walsby, this volume [72].

[4a] P. K. Hayes, this volume [20].

[5] G. Thuret, *Ann. Sci. Natl., Bot. Biol. Veg.* **2,** 319 (1844).

[6] J. R. Waaland, S. D. Waaland, and D. Branton, *J. Cell Biol.* **48,** 212 (1970).

[7] M. A. Lachance, *Int. J. Syst. Bacteriol.* **31,** 139 (1981).

Further studies[8] revealed that this organism is indeed capable of forming hormogonia (and should therefore be redesignated as *Nostoc* PCC 7119) when cells from stationary phase cultures are washed and resuspended in fresh medium; when subcultured directly, however, hormogonial differentiation is inhibited if the volume of the inoculum is more than 5% of the final volume of the new culture.[8] The organism produces, in late exponential to stationary phase, an inhibitory compound which is active even at 10- to 20-fold dilution. The formation of such a compound may be tested[8] by harvesting the culture, washing the cells, and resuspending them in fresh medium containing varying proportions (0–20%, v/v) of the supernatant of the previous culture; if an inhibitor is present, hormogonium formation will be suppressed in a dose-dependent manner. The inhibitor has not yet been identified, but it is dialyzable (and is therefore a small molecule) and is destroyed by autoclaving.[8] It follows that, in order to examine the potential of new isolates to form hormogonia, the carry-over of used medium should be strictly avoided.

Some cyanobacteria are capable of chromatic adaptation (see Refs. 9 and 10), a process whereby the cellular content of the light-harvesting phycobiliproteins changes in response to the wavelength of light supplied to the cell.[9] In such strains light quality may also play an important role in the differentiation of hormogonia, as suggested by detailed studies on *Calothrix* PCC 7601 (*Fremyella diplosiphon*).[9] Although this organism produces a low number (10–30%) of gas-vacuolate hormogonia after dilution into fresh BG-11 medium and illumination with white light, differentiation is maximal when a culture, previously grown in white or green light, is diluted and incubated under red light: 90–100% of the filaments develop into hormogonia.[8] This process occurs relatively synchronously and is normally complete 18–24 hr after the shift. Transfer from white or red to green light generally represses differentiation completely.[8] Repeated subculturing in red light gives only low frequencies, equalling at most those obtained after transfer in white light.[8]

Although dilution of the culture, with simultaneous shift from white or green to red light, is necessary for efficient production of hormogonia of *Calothrix* PCC 7601 in medium BG-11, an increase in light energy supply to the cells (which will result from the dilution if the total energy provided under the two light regimes is the same) is certainly not: differentiation occurs at high frequency when the total light energy is decreased after the shift (see below) so that the energy received per cell remains relatively

[8] R. Rippka, unpublished observations.

[9] N. Tandeau de Marsac, *Bull. Inst. Pasteur* (*Paris*) **81,** 201 (1983).

[10] N. Tandeau de Marsac and J. Houmard, this volume [34].

constant. In addition, cells grown in green light contain phycoerythrin and relatively little phycocyanin (see Refs. 9 and 10); following the shift to red light, phycoerythrin can no longer absorb the available light quanta, and this sudden lack of energy is compensated only after *de novo* synthesis of phycocyanin. Hormogonium formation precedes this event.[8]

In white light, stationary phase cultures of *Calothrix* PCC 7601 appear to produce an inhibitor of hormogonial differentiation[8]; again, carryover of old medium must therefore be avoided. The stimulatory effect of red light on hormogonium production is not necessarily restricted to strains capable of chromatic adaptation, since it has also been observed in, for example, *Nostoc muscorum* A[11] which, although capable of producing phycoerythrin,[11] does not modulate phycobiliprotein content in response to light quality. Although the physiological basis of this response is unknown, it is clear that the effect of light quality on hormogonial differentiation must be independent of the ability of the organism to adapt chromatically.

See Addendum on p. 849.

Isolation of Hormogonia

If hormogonium formation does not occur simultaneously in all of the filaments, physiological studies require that the two types of trichome be separated. This may be performed for short and gas-vacuolate hormogonia simply by permitting clumps and long filaments to settle by gravity in, e.g., a tall sterile measuring cylinder, collecting the supernatant, and concentrating by centrifugation if necessary (although the latter step will collapse the gas vacuoles). In an extension of this procedure,[12] the pellet is resuspended in growth medium, layered on a discontinuous Ficoll gradient (10–30%) and centrifuged (400 *g*, 5 min, swinging-bucket rotor). Hormogonia form a distinct band in the region of 10–15% Ficoll.

Methods for Estimation of Cell Numbers

Although the counting of cell numbers may appear to be difficult for filamentous cyanobacteria, it is feasible with an accuracy of ±5–10% by either of two methods. In the first,[13] developed for trichomes of *Oscillatoria agardhii* (which does not form hormogonia), the culture sample is

[11] N. Lazaroff, *in* "The Biology of Blue-Green Algae" (N. G. Carr and B. A. Whitton, eds.), p. 279. Blackwell, Oxford, England 1973.

[12] A. Fattom and M. Shilo, *Appl. Environ. Microbiol.* **47,** 135 (1984).

[13] L. van Liere, L. R. Mur, C. E. Gibson, and M. Herdman, *Arch. Microbiol.* **123,** 315 (1979).

placed into a Bürker–Türk (or similar) counting chamber and photographed at low magnification, sufficient fields being taken to provide a representative sample of 400–1000 trichomes. The length of each trichome is then measured on enlargements using a curvimeter (as used for measuring distances on road maps) and converted to total trichome length per milliliter of culture. Next, it is necessary to calculate the number of cells per unit length of trichome, a feat easily achieved on high-power micrographs by scoring both the number of cells in a given portion of the trichome and the length of that portion. Since the magnification of the photographs is known (by calibration with a stage micrometer), the total number of cells per milliliter can be derived by simple mathematics. However, sufficient numbers of cells must be scored to ensure that the results are representative of the whole population. This method is suitable for all filamentous cyanobacteria, although for *O. agardhii* it is necessary to collapse the gas vacuoles (in a pressure chamber as described by Walsby[14]) in order to clearly see the cross walls; collapse of gas vacuoles would be also required to visualize individual cells in gas-vacuolate hormogonia. A further step, the extraction of chlorophyll with methanol, makes the cell walls even more visible[13] and permits cell volumes to be calculated from the micrographs.

In the second method,[11a] employed for *Calothrix* PCC 7601, vegetative filaments or hormogonia (following collapse of gas vacuoles by centrifugation) are fixed by the addition of glutaraldehyde (1.25% final concentration). The number of trichomes per milliliter is then scored under low power in a counting chamber. It may be noted that the size of the chamber employed depends on the length of the filaments produced, long filaments requiring large chambers which are commercially available up to 1 ml total volume. The total number of cells in the entire trichome is then counted directly under high magnification (on a normal microscope slide), and sufficient trichomes must be counted to ensure that the results are statistically significant. Simple calculation yields the number of cells per milliliter of culture. This method, avoiding photography, is more rapid than the first but requires mental concentration and good eyesight.

Changes in Cellular Macromolecular Composition

Using the latter cell-counting technique, we[11a] demonstrated that hormogonium formation in *Calothrix* PCC 7601, induced as described above,

[14] A. E. Walsby, *Bacteriol. Rev.* **36,** 1 (1972).

is the result of rapid cell division in the absence of significant synthesis of DNA or protein. Cell numbers increased 8-fold during the 22 hr following initiation, whereas the macromolecular content of the culture changed little (Table I). Consequently, cell size (measured as protein content) decreased 7-fold. This process implies that, like the vegetative cells of baeocyte-forming cyanobacteria, those of hormogonium-forming strains must contain multiple copies of the genome. Assuming a genome size of 5.25×10^9 daltons i.e., the average value for *Calothrix*,[15] the mean DNA content of the parental cells is 35 genome equivalents and decreases to 5 in the hormogonial cells.

Properties Contributing to the Dispersal of Hormogonia

Motility. The hormogonia of many cyanobacteria are able to glide quite rapidly across a solid substrate. Relatively little is known about the mechanism of cyanobacterial motility (see Ref. 16 for a discussion). Although fimbriae were observed on the hormogonia of the symbiotic *Nostoc* within the lichen *Peltigera*,[17] and were suggested to be involved in hormogonial movement, fimbriation of the parental (immotile) trichome was not examined. Studies in our laboratory,[18] however, have shown that hormogonia of *Calothrix* PCC 7601 are fimbriated and motile, whereas the immotile parental trichomes lack fimbriae; both the differentiation of hormogonia and the synthesis of fimbriae were prevented by chloramphenicol. This observation would appear to be worthy of further study. Descriptions of cyanobacterial fimbriae and of the methods used for their examination are given elsewhere in this volume.[19,20]

Gas Vacuoles. Gas vacuoles are synthesized during the development of hormogonia of many cyanobacterial strains and permit flotation away from the parental colony. A thorough description of the properties of gas vacuoles, and of the techniques employed in their study, is given elsewhere in this volume.[4,4a]

Hydrophobicity of the Cell Surface. Cell surface hydrophobicity is known to be an important factor in the adhesion of many bacteria to solid

[15] M. Herdman, M. Janvier, R. Rippka, and R. Y. Stanier, *J. Gen. Microbiol.* **111,** 73 (1979).
[16] R. W. Castenholz, *in* "The Biology of Cyanobacteria" (N. G. Carr and B. A. Whitton, eds.), p. 413. Blackwell, Oxford, England, 1982.
[17] H. Dick and W. D. P. Stewart, *Arch. Microbiol.* **124,** 107 (1980).
[18] G. Guglielmi and R. Rippka, unpublished observations.
[19] G. Stanier (Cohen-Bazire), this volume [14].
[20] T. Vaara and M. Vaara, this volume [16].

TABLE I
MACROMOLECULAR COMPOSITION OF CELLS OF VEGETATIVE FILAMENTS AND HORMOGONIA OF *Calothrix* PCC 7601

Parameter[a]	Time after induction: 0 hr	Time after induction: 22 hr[b]
Filaments/ml ($\times 10^{-4}$)	3.30	5.92
Mean no. of cells/filament	21.5	105.3
Cells/ml ($\times 10^{-5}$)	7.08	62.3
DNA/ml (μg)	0.22	0.28
DNA/cell (fg)	3.04	0.45
Protein/ml (μg)	15.4	19.1
Protein/cell (pg)	21.8	3.1

[a] For cell counts, samples (25 ml) were centrifuged, resuspended in 1 ml of 1.25% glutaraldehyde, and stored at 4° until counting. Filaments and cells were scored (see text) using appropriate dilutions. DNA estimations each required 150 ml of culture, to which was added 15 ml of 2.75 *N* perchloric acid. The samples were left for 30 min in ice, centrifuged, resuspended in 0.5 *N* perchloric acid (2 ml), and heated (70°, 40 min). Following centrifugation, DNA was measured by the diphenylamine reaction [see I. W. Craig, C. K. Leach, and N. G. Carr, *Arch. Mikrobiol.* **65,** 218 (1969)]. For protein determinations, ice-cold trichloroacetic acid (5 ml, 50%, w/v) was added to 20 ml of culture; the samples were held in ice for 30 min, centrifuged, resuspended in 1 ml of 1 *N* NaOH, and hydrolyzed (100°, 30 min). After cooling in ice, the samples were centrifuged to remove insoluble compounds, and the supernatants were assayed by the Lowry method [D. H. Lowry, N. J. Rosenbrough, A. L. Farr, and R. J. Randall, *J. Biol. Chem.* **193,** 265 (1951)]. Hormogonium formation was induced following a shift from green to red light as described in the text, in the presence of glucose.

[b] Hormogonium formation is complete 18–22 hr after induction.

surfaces, and the elegant studies of Fattom and Shilo[12] have extended this observation to cyanobacteria. The degree of hydrophobicity may be measured by two simple and rapid methods[12]:

1. The biphasic water–hydrocarbon test: After collapsing the gas vacuoles (if present), appropriate volumes of *n*-hexadecane are added to the test cultures in Klett tubes, mixed vigorously (1 min), allowed to stand (5 min), then agitated gently and allowed to stand again. Hydrophobic filaments partition in the upper (hydrocarbon) phase whereas hydrophilic filaments remain in the aqueous phase. The phenomenon can be quantified by measuring the optical density of the aqueous phase in a Klett–Summerson colorimeter before and after treatment.

2. Adherence to Phenyl–Sepharose beads: The beads (1 ml, ~40 μmol of gel beads/ml), 45–165 μm in diameter, are added to 5 ml of culture, thoroughly mixed, and allowed to settle. The decrease in optical density of the supernatant is a measure of hydrophobicity, since hydrophobic (but not hydrophilic) organisms adhere to the beads and are therefore removed from suspension.

Mature benthic strains from a wide diversity of habitats were all found to be hydrophobic, in contrast to planktonic cyanobacteria which are hydrophilic. However, the hormogonia released by all of the benthic strains examined[12] are hydrophilic; they become hydrophobic again as soon as they commence to redifferentiate into mature organisms. These changes, therefore, coincide exactly with the various stages of the life cycle of benthic hormogonium-forming cyanobacteria and explain why the hydrophobic parental filaments occur as a mutually protective aggregate, how the hormogonia are dispersed, and finally why, as a consequence of its newly acquired hydrophobicity, the young developing vegetative filament again fixes to the substratum. As pointed out by Fattom and Shilo,[12] it is unfortunate that similar studies have not been performed on the analogous life cycles of baeocyte-forming cyanobacteria, where again the mature cells adhere to a solid surface whereas the baeocytes are dispersed.

Concluding Remarks

Baeocytes and hormogonia have great, but so far unexplored, potential for biochemical and genetic studies. Both their formation and subsequent development into a mature organism occur relatively synchronously, which should permit study of the biochemical events important in the life cycle. For example, recent results from our laboratory[21,22] have shown that the genes encoding the structural protein of the gas vacuoles of *Calothrix* PCC 7601 are transcribed *de novo* during hormogonium differentiation; changes in the cellular content of other proteins, including those involved in motility and adhesion of baeocytes and hormogonia, would be worthy of further study. Finally, unlike their parental cells, the baeocyte and the hormogonial cell contain few copies of the genome; this property should not be overlooked if one wishes to obtain mutants of these organisms since it will reduce the time required for postmutational segregation of genomes, whose high copy number in cyanobacterial vegetative cells[23–27] is probably one of the principal reasons for the difficulties encountered[28] in the isolation of auxotrophic mutants.

Acknowledgments

We thank Dr. A. H. Neilson for finding the Greek derivation of the term hormogonium.

[21] T. Damerval, J. Houmard, G. Guglielmi, K. Csiszàr, and N. Tandeau de Marsac, *Gene* **54,** 83 (1987).

[22] K. Csiszàr, J. Houmard, T. Damerval, and N. Tandeau de Marsac, *Gene* **60,** 29 (1987).

[23] C. K. Leach, J. M. Old, and N. G. Carr, *J. Gen. Microbiol.* **68,** xiv (1971).

[24] N. Mann and N. G. Carr, *J. Gen. Microbiol.* **83,** 399 (1974).

[25] J. M. Sutherland, M. Herdman, and W. D. P. Stewart, *J. Gen. Microbiol.* **115,** 273 (1979).

[26] R. D. Simon, *J. Bacteriol.* **129,** 1154 (1977).

[27] R. D. Simon, *FEMS Microbiol. Lett.* **8,** 241 (1980).

[28] M. Herdman, *in* "The Biology of Cyanobacteria" (N. G. Carr and B. A. Whitton, eds.), p. 263. Blackwell, Oxford, England, 1982.

Section III

Membranes, Pigments, Redox Reactions, and Nitrogen Fixation

[23] Isolation of Cyanobacterial Plasma Membranes

By NORIO MURATA and TATSUO OMATA

Introduction

Cyanobacteria are classified as gram-negative bacteria, in which the cell envelope is composed of outer and inner (plasma or cytoplasmic) membranes with a peptidoglycan layer between.[1] In the cytoplasm are the thylakoid membranes, which are the site of light capturing, electron transport, and ATP synthesis. The thylakoid membranes represent most of the total cellular membranes,[2] and this situation made it difficult for a long time to purify plasma membranes from the cyanobacteria. However, reliable methods to isolate plasma membranes have been developed since a pioneering study by Lockau and Pfeffer,[3,4] who partially separated plasma membranes from the filamentous cyanobacterium *Anabaena variabilis* by disruption of spheroplasts with a cell homogenizer followed by sucrose density gradient centrifugation. Subsequently developed methods[5,6] to isolate plasma membranes are essentially the same as their method,[3] although some modifications are employed such as using a French pressure cell for disruption of the cells and flotation centrifugation for isolation of plasma membranes.

Plasma membranes thus prepared can be identified by electron microscopy[5] and also by labeling the intact cells prior to disruption with a membrane-impermeant protein marker such as diazobenzene [^{35}S]sulfonate.[7] So far the method described in this chapter has been verified with *Anacystis nidulans*,[5,6] *Synechocystis* 6714,[8] and *Synechocystis* PCC 6803.[9] Subjects that have been studied using the isolated plasma membranes are phase transition of membrane lipids[10]; composition of cytochromes and quinones[11]; carrier proteins for inorganic carbon[9,12,13]; activities of cyto-

[1] M. M. Allen, *J. Bacteriol.* **96,** 842 (1968).
[2] M. M. Allen, *J. Bacteriol.* **96,** 836 (1968).
[3] W. Lockau and S. Pfeffer, *Z. Naturforsch. C* **37,** 658 (1982).
[4] W. Lockau and S. Pfeffer, *Biochim. Biophys. Acta* **733,** 124 (1983).
[5] T. Omata and N. Murata, *Plant Cell Physiol.* **24,** 1101 (1983).
[6] V. Molitor and G. A. Peschek, *FEBS Lett.* **195,** 145 (1986).
[7] V. Molitor, W. Erber, and G. A. Peschek, *FEBS Lett.* **204,** 251 (1986).
[8] T. Omata and N. Murata, *Arch. Microbiol.* **139,** 113 (1984).
[9] T. Omata and T. Ogawa, *Prog. Photosynth. Res.* **4,** 309 (1987).
[10] H. Wada, R. Hirasawa, T. Omata, and N. Murata, *Plant Cell Physiol.* **25,** 907 (1984).
[11] T. Omata and N. Murata, *Biochim. Biophys. Acta* **766,** 395 (1984).

chrome-*c* oxidase,[6,7,14] NAD(P)H dehydrogenase,[6,14] and lipid biosynthesis[15]; and carotenoid-binding proteins.[16] Variations in growth conditions such as culture age, CO_2 supply, and growth temperature do not alter effective separation and recovery of plasma and thylakoid membrane fractions.[7,12]

Isolation of Plasma Membranes[5,13]

Anacystis nidulans is photoautotrophically grown at 30° in medium C of Kratz and Myers[17] or BG-11 medium[18,19] supplemented with 20 m*M* HEPES–NaOH buffer (pH 7.0). Continuous illumination is provided by fluorescent lamps at about 120 μmol PAR[20]/m^2 sec. The culture is gassed with air containing 3% CO_2.

Approximately 3 g wet weight of *A. nidulans* cells are harvested by centrifugation at 4,000 *g* for 10 min, washed with 60 ml of 5 m*M* TES–NaOH buffer (pH 7.0), and resuspended in 30 ml of 10 m*M* TES–NaOH buffer (pH 7.0) containing 600 m*M* sucrose, 2 m*M* disodium EDTA, and 0.03% lysozyme. The cell suspension is incubated at 30° under room light for 2 hr. The lysozyme-treated cells are collected by centrifugation at 5,000 *g* for 10 min. For a thorough removal of lysozyme and EDTA,[21] cells are washed twice with 20 m*M* TES–NaOH buffer (pH 7.0) containing 600 m*M* sucrose by resuspension and centrifugation. All the abovementioned procedures are carried out at room temperature, and subsequent steps at ice temperature.

The pelleted cells are suspended in 30 ml of 20 m*M* TES–NaOH buffer (pH 7.0) containing 600 m*M* sucrose and passed through a prechilled French pressure cell at 40 MPa, which usually affords 50–70% cell breakage.[22] Immediately after the French pressure cell treatment, 0.05 ml of 10

[12] T. Omata and T. Ogawa, *Plant Cell Physiol.* **26,** 1075 (1985).

[13] T. Omata and T. Ogawa, *Plant Physiol.* **80,** 525 (1986).

[14] T. Omata and N. Murata, *Biochim. Biophys. Acta* **810,** 354 (1985).

[15] T. Omata and N. Murata, *Plant Cell Physiol.* **27,** 485 (1986).

[16] G. S. Bullerjahn and L. A. Sherman, *J. Bacteriol.* **167,** 396 (1986).

[17] W. A. Kratz and J. Myers, *Am. J. Bot.* **42,** 282 (1955).

[18] M. M. Allen, *J. Phycol.* **4,** 1 (1968).

[19] R. Y. Stanier, R. Kunisawa, M. Mandel, and G. Cohen-Bazire, *Bacteriol. Rev.* **35,** 171 (1971).

[20] PAR, Photosynthetically active radiation.

[21] It is necessary to remove EDTA from the cell suspension since EDTA inhibits DNase I.

[22] Treatment at higher pressure provides a higher proportion of cell breakage but deteriorates purity and physiological activities of the resultant plasma membrane preparation.

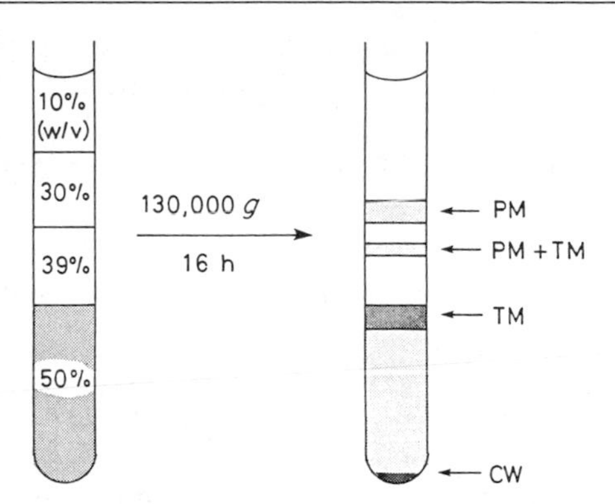

FIG. 1. Separation of plasma and thylakoid membranes from *Anacystis nidulans* by flotation centrifugation on a discontinuous sucrose density gradient. PM, Plasma membranes; TM, thylakoid membranes; CW, cell walls.

m*M* sodium acetate buffer (pH 5.6) containing 1 m*M* $MgCl_2$ and 0.1% DNase I (Sigma, DN-EP[23]) and 0.1 ml of 3% phenylmethylsulfonyl fluoride (PMSF) in methanol are added to the homogenate. After incubation for 15 min, the homogenate is centrifuged at 5,000 *g* for 10 min to remove unbroken cells and cell debris.

The red fluorescent supernatant of the centrifuged cell-free extract containing the membrane vesicles is made up to a sucrose concentration of 50% (w/v) by adding 0.74 volumes of 90% (w/v) sucrose solution.[24] A 17-ml aliquot of this suspension is placed at the bottom of a 35-ml centrifuge tube, then sequentially overlaid with 8 ml of 39%, 3 ml of 30%, and 7 ml of 10% sucrose solutions (w/v), each containing 10 m*M* TES–NaOH buffer, 5 m*M* disodium EDTA, 10 m*M* NaCl, and 1 m*M* PMSF (pH 7.0). Centrifugation at 130,000 *g* for 16 hr at 4° in a swinging-bucket rotor separates the membrane fractions. Plasma membranes (yellow) form a band in the 30% sucrose layer, and thylakoid membranes (green) band at the interface between the 39 and 50% sucrose layers. Cell walls are pelleted at the bottom. Figure 1 shows a schematic separation of the membranes by this flotation centrifugation method.

[23] Commercially available DNase I frequently contains proteinase activity. It is therefore recommended to use a minimal amount of highly purified enzyme.

[24] The 90% (w/v) sucrose solution, which is made by dissolving 90 g sucrose in 43.1 ml of 24 m*M* TES–NaOH buffer (pH 7.0) containing 12 m*M* disodium EDTA and 24 m*M* NaCl, is added dropwise to the homogenate in a 50-ml graduated cylinder until the total becomes 1.74 volumes. The cylinder is sealed with polyethylene film (Parafilm, American Can Company), and the homogenate and the sucrose solution are mixed thoroughly.

The plasma membranes are withdrawn from the gradient with a Pasteur pipet,[25] diluted 3-fold with 10 m*M* TES–NaOH buffer (pH 7.0) containing 10 m*M* NaCl, then collected by centrifugation at 300,000 *g* for 1 hr. The resultant pellet is the plasma membrane preparation without any detectable contamination by thylakoid membranes or cell walls. The yield of plasma membranes corresponds to 0.6–1.0 mg protein as determined by the Lowry method after solubilization with 0.1% sodium dodecyl sulfate (SDS).

Plasma membranes and thylakoid membranes can be prepared from *Synechocystis* PCC 6714 in the same way as above.[8] For preparation of plasma membranes from *Synechocystis* PCC 6803, which is very resistant to lysozyme treatment, the following modifications are necessary[9]: concentrations of EDTA and lysozyme during the lysozyme treatment are 5 m*M* and 0.2%, respectively, and cells are passed twice through a French pressure cell at 160 MPa to afford 50% cell breakage.

The flotation centrifugation time can be shortened to 4 hr at a centrifugal force of 300,000 *g* using an angle rotor. This method is especially suitable for studying labile activities. For details, see Ref. 14.

For an approximate estimation of the yields of plasma and thylakoid membranes from *A. nidulans,* the pelleted membranes are suspended in 10 m*M* TES–NaOH (pH 7.0), and the light absorption spectra are recorded in a wavelength range from 250 to 750 nm. The concentration of the plasma and thylakoid membranes is calculated according to the following equations:

$$[\text{Protein}]\ (\text{plasma membranes}) = 0.13 \times A_{275} \quad \text{mg ml}^{-1}$$
$$[\text{Protein}]\ (\text{thylakoid membranes}) = 0.11 \times A_{678} \quad \text{mg ml}^{-1}$$

Figure 2, which shows the absorption spectra of plasma membranes and thylakoid membranes from *A. nidulans* measured at room temperature, indicates that the plasma membrane contains a large amount of carotenoids, whereas the thylakoid membrane contains both chlorophyll *a* and carotenoids. A small amount of chlorophyll *a* is present in the plasma membrane preparation. It is not due to contamination by thylakoid membranes because the absorption peak of plasma membranes occurs at a wavelength which is shorter than that of thylakoid membranes by 5 nm.[5] The absorption spectrum of the plasma membranes depends on temperature of measurement.[5] At 0°, a peak at 390 nm becomes prominent

[25] The tip of the Pasteur pipet must be bent at a right angle so that the plasma membrane fraction can be thoroughly withdrawn with no contamination by thylakoid membranes.

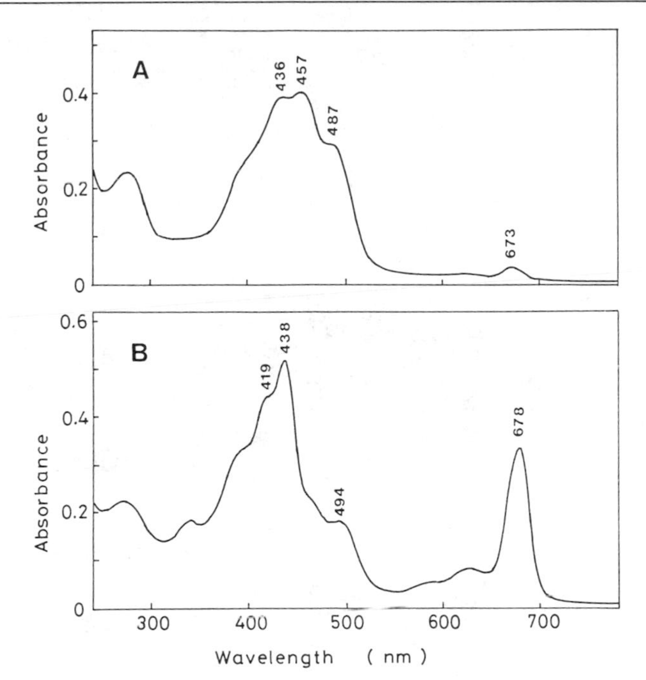

FIG. 2. Absorption spectra of plasma (A) and thylakoid (B) membranes from *Anacystis nidulans*. Temperature of measurement, 28°.

at the expense of the three peaks in the blue region.[26] This is due to a thermotropic phase transition of membrane lipids.[5]

Figure 3 compares the polypeptide composition of the plasma and thylakoid membranes from *A. nidulans* by means of SDS–polyacrylamide gel electrophoresis. Plasma membranes have a polypeptide composition distinct from that of thylakoid membranes and cell walls, and they are characterized by the occurrence of two major proteins with apparent molecular masses of 42 and 37 kDa. The 42-kDa protein is prominent in cells grown under low CO_2 (0.03%) conditions, but is hardly detectable in cells grown under high CO_2 (3%) conditions.[9,12,13] It is suggested that the protein is involved in active transport of inorganic carbon, since there is a parallelism between the content of this protein and inorganic carbon uptake activity.[13]

[26] The absorption spectrum of plasma membranes from *A. nidulans* strain R2 has a peak at 390 nm even at room temperature which is independent of the temperature of measurement.

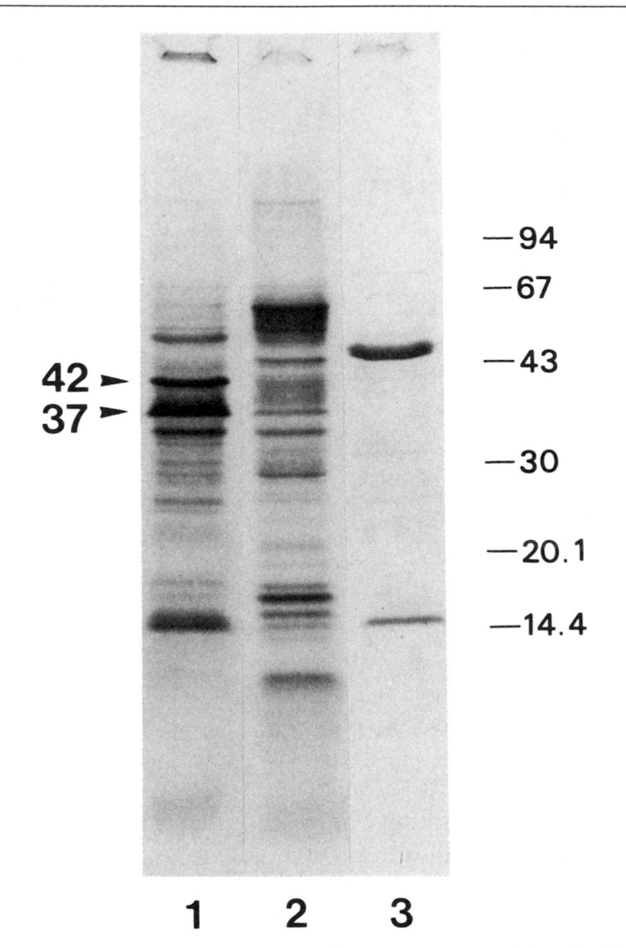

FIG. 3. Polypeptide analysis of membranes from *Anacystis nidulans* by SDS–polyacrylamide gel electrophoresis. Lane 1, plasma membranes (25 μg protein); Lane 2, thylakoid membranes (25 μg protein); Lane 3, cell walls (8 μg protein) prepared according to Murata *et al.*[27] SDS–polyacrylamide gel electrophoresis is performed in the buffer system of Laemmli.[28] Samples are solubilized in 62.5 m*M* Tris–HCl buffer (pH 6.8) containing 2% SDS, 10% glycerol, 10% mercaptoethanol, and 0.001% bromphenol blue. Plasma and thylakoid membranes are solubilized at room temperature for 30 min, whereas cell walls are solubilized at 100° for 5 min. The stacking gel contains 5% polyacrylamide, and the separating gel contains a linear concentration gradient of polyacrylamide from 8 to 15%. Electrophoresis is performed at 20°.

The 37-kDa protein migrates at apparent molecular mass of 45 kDa in SDS–polyacrylamide gels, when plasma membranes are solubilized with SDS at 100°. The heat-induced band at 45 kDa on SDS–polyacrylamide gel electrophoresis is distinct from the one due to the above-described 42-kDa protein. A corresponding heat-modifiable protein which cross-reacts with antibody against the 37-kDa protein of *A. nidulans* is found in plasma membranes from *Synechocystis* PCC 6714[16] and *Synechocystis* PCC 6803.[9] The protein from *Synechocystis* PCC 6714 is found to bind carotenoids.[16]

[27] N. Murata, N. Sato, T. Omata, and T. Kuwabara, *Plant Cell Physiol.* **22,** 855 (1981).
[28] U. K. Laemmli, *Nature* (*London*) **227,** 680 (1970).

[24] Membrane Lipids

By NAOKI SATO and NORIO MURATA

Introduction

Cyanobacterial cells contain two types of membrane, the plasma membrane and thylakoid membranes, which are distinct from each other in their composition of proteins, lipids, and pigments.[1] The composition of the fatty acids of the lipids in both types of membrane changes with growth temperature so that cyanobacterial cells adapt themselves to the environmental temperature.[2,3] Major lipid classes in cyanobacterial membranes are monogalactosyl diacylglycerol (MGDG), monoglucosyl diacylglycerol (GlcDG), digalactosyl diacylglycerol (DGDG), sulfoquinovosyl diacylglycerol (SQDG), and phosphatidylglycerol (PG).[4,5] Fatty acids of cyanobacteria are mainly unbranched chains containing 14, 16, or 18 carbon atoms and 0, 1, 2, or 3 double bonds.[4,6,7] Among them, palmitic acid and palmitoleic acid are most commonly found. The biosynthetic pathway of glycolipids in cyanobacteria is unique in that the glucolipid is a precursor to galactolipids[5,8] and stearic acid is esterified to lipids before

[1] T. Omata and N. Murata, *Plant Cell Physiol.* **24,** 1101 (1983).
[2] N. Sato and N. Murata, *Biochim. Biophys. Acta* **619,** 353 (1980).
[3] H. Wada, R, Hirasawa, T. Omata, and N. Murata, *Plant Cell Physiol.* **25,** 907 (1984).
[4] N. Sato, N. Murata, Y. Miura, and N. Ueta, *Biochim. Biophys. Acta* **572,** 19 (1979).
[5] N. Sato and N. Murata, *Biochim. Biophys. Acta* **710,** 271 (1982).
[6] C. N. Kenyon, *J. Bacteriol.* **109,** 827 (1972).
[7] C. N. Kenyon, R. Rippka, and R. Y. Stanier, *Arch. Mikrobiol.* **83,** 216 (1972).
[8] T. Omata and N. Murata, *Plant Cell Physiol.* **27,** 485 (1986).

being desaturated to oleic acid.[9,10] This chapter describes procedures in cyanobacterial lipid analysis, including extraction and fractionation of lipids, analysis of their fatty acids, determination of positional distribution of fatty acids within the lipids, and analysis of lipid molecular species. The interested reader should consult a textbook[11] for general experimental procedures in lipid analysis. Analytical data of cyanobacterial lipids and fatty acids are available.[1,2,4–9,12]

Extraction of Total Lipids

Total lipids are extracted from either intact cells or membrane preparations of cyanobacteria according to the method of Bligh and Dyer.[13] When working with lipids that contain highly unsaturated fatty acids (such as those from most filamentous cyanobacteria), excessive exposure of unsolvated lipids to air should be minimized to prevent fatty acid oxidation. The presence of an antioxidant [e.g., 2,6-di-*tert*-butyl-4-hydroxytoluene (BHT)] is recommended during long-term storage of lipids containing unsaturated fatty acids.

Procedure. One-tenth milliliter of pelleted cyanobacterial cells or membranes is suspended in 1 ml of distilled water or dilute buffer (high concentrations of sugars and other osmotics should be washed out beforehand). The suspension is transferred to a glass tube with a Teflon-lined screw-cap (e.g., Pyrex No. 8082CTF centrifuge tube). Then 3.75 ml of $CHCl_3/CH_3OH$ (1 : 2, v/v) is added and mixed by vortexing. After standing for 20 min at room temperature, 1.25 ml each of $CHCl_3$ and H_2O are added to the mixture and mixed by vortexing. The mixture is then centrifuged at 1,000 *g* for 15 min at room temperature. The clear upper phase and the intermediate fluff layer are carefully withdrawn with a Pasteur pipet and discarded. Then 2.5 ml of CH_3OH/H_2O (10 : 9) is added to the lower phase and mixed by vortexing, after which the mixture is centrifuged as above.

The lower phase is recovered and transferred to a new test tube suitable for evaporation. One-half milliliter of C_2H_5OH is added for complete evaporation of water, and the solvent is evaporated under reduced pressure using a rotary evaporator or under a stream of N_2. The total lipids are dissolved in 0.5 ml of $CHCl_3/CH_3OH$ (2 : 1, v/v). For storage times up to a few hours, the tube is covered with aluminum foil and chilled on ice to

[9] N. Sato and N. Murata, *Biochim. Biophys. Acta* **710,** 279 (1982).
[10] N. W. Lem and P. K. Stumpf, *Plant Physiol.* **74,** 134 (1984).
[11] M. Kates, "Techniques of Lipidology." North-Holland, Amsterdam, 1972.
[12] N. Murata and N. Sato, *Plant Cell Physiol.* **24,** 133 (1983).
[13] E. G. Bligh and W. J. Dyer, *Can. J. Biochem. Physiol.* **37,** 911 (1959).

minimize evaporation of the solvent. For longer storage times, BHT is added to a final concentration of 0.05%, then the lipid solution is transferred to a small glass vial with a Teflon-lined screw-cap and stored below −20°. BHT can be removed by thin-layer chromatography.

Thin-Layer Chromatographic Separation of Lipid Classes

Thin-layer chromatography (TLC) is a convenient method for separating major classes of lipids. Minor amounts of unidentified lipids may also be present, but they usually do not interfere with the analysis of the major classes of lipids, since they remain near the origin in the TLC systems described below.

Procedure. A pencil line is drawn on a precoated silica gel plate (e.g., Merck No. 5721 plate, without fluorescent dye, 5 × 20 cm or 20 × 20 cm) parallel to and approximately 2 cm from one edge. Lipid solution (10–50 μl) is applied as a streak approximately 2 cm long along the pencil line, with a microsyringe. The solvent may be removed more rapidly with an air blower. The plate is developed with either (1) $(CH_3)_2CO/C_6H_6/H_2O$ (91 : 30 : 8, by volume) or (2) $CHCl_3/CH_3OH/NH_4OH$ (28%) (13 : 7 : 1, by volume) to a height of 10–19 cm. Solvent (2) is suitable for separating MGDG, PG, SQDG, and DGDG, and Solvent (1) separates MGDG and GlcDG (Fig. 1). The plate is allowed to dry in a fume hood for about 20 min, then sprayed with 0.01% primuline in $(CH_3)_2CO/H_2O$ (4 : 1). The plate is again allowed to dry for 3 min. Lipids are detected under long-wavelength UV light (366 nm) as bluish white fluorescent bands, which are outlined with a pencil.

When the lipids are recovered from the plate, the silica gel within outlined areas is scraped into a flask and extracted in $CHCl_3/CH_3OH$ (2 : 1). The mixture is filtered through a sheet of filter paper. Water-soluble impurities may be removed by washing the filtrate with $\frac{1}{4}$ volume of distilled water. Then the solution is concentrated as described above under Extraction of Total Lipids.

Comments. Typical separation patterns are illustrated in Fig. 1. Lipid classes are easily identified by their mobility on TLC. Spray reagents are conveniently used for identification: anthrone reagent for glycolipids and Dittmer reagent for phospholipids are especially useful (see Ref. 11 for the preparation and use of spray reagents). Definitive identification can only be effected by infrared and nuclear magnetic resonance spectrometry (see a textbook[11] for a general introduction). MGDG and GlcDG can be distinguished by gas chromatographic analysis of their sugar moieties.[5]

Since the content of GlcDG is much lower than that of the other four major classes of lipids,[5,12] analysis of GlcDG requires a larger scale, e.g.,

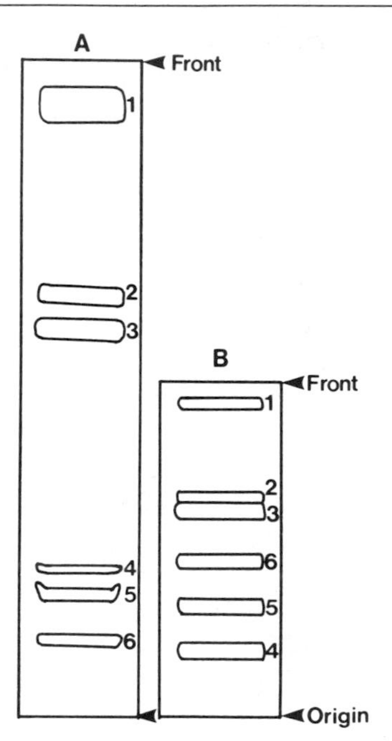

FIG. 1. Thin-layer chromatograms of total lipids from *Anabaena variabilis* strain M3. (A) A Merck 5721 plate was developed in Solvent (1) to a height of 18 cm. (B) A Merck 5721 plate was developed in Solvent (2) to a height of 10 cm. A 10-cm plate is sufficient to separate MGDG, DGDG, SQDG, and PG in Solvent (2), whereas good separation of GlcDG and MGDG requires a longer plate using Solvent (1). Identification: 1, pigments; 2, monoglucosyl diacylglycerol (GlcDG); 3, monogalactosyl diacylglycerol (MGDG); 4, digalactosyl diacylglycerol (DGDG); 5, sulfoquinovosyl diacylglycerol (SQDG); 6, phosphatidylglycerol (PG).

about 1 liter of culture in middle exponential growth (~5 μl packed cells/ml). In this case, the lipid extraction and TLC should be scaled up 10-fold. If the separation of GlcDG from MGDG is incomplete, then the GlcDG fraction should be recovered and repurified by TLC.

Gas Chromatographic Analysis of Fatty Acids in Lipids

The content and composition of fatty acids in lipids are determined by gas chromatographic analysis of the methyl esters which are obtained by methanolysis of the lipids. Methanolysis is generally performed without isolating lipids from the silica gel. Fatty acid methyl esters of cyanobacteria are easily identified in gas chromatographic analysis, though definitive identification and determination of double bond position are effected by gas chromatography–mass spectrometry (GC–MS). Consult the literature for technical details of gas chromatography.[14]

Procedure. After the detection of lipids by primuline fluorescence, the TLC plate is dried in a vacuum desiccator for 30 min. The silica gel in the appropriate area is scraped with a razor blade into a test tube (15 mm diameter by 15 cm length) with a Teflon-lined screw-cap (e.g., Pyrex TST-SCR 16-150 tube). Three milliliters of 2.5% (w/w) HCl in anhydrous CH_3OH (for this solution, HCl gas is bubbled into the CH_3OH until its weight increases by 2.5%) and 100 μl of 1 m*M* pentadecanoic acid as an internal standard in benzene are added. The tube is tightly capped and heated at 85° for 2.5 hr in an aluminum heating block or in a water bath. After the tube is cooled to room temperature, 2.5 ml of *n*-hexane is added and mixed by vortexing.

After standing for 5 min, the upper (hexane) phase is carefully transferred to a clean test tube with a narrowly tapered bottom suitable for evaporation. The lower phase is extracted twice more with additional hexane. Then 2 ml of distilled water is added to the remaining lower phase, and hexane extraction is repeated once more. After combining the hexane extracts, the solvent is removed by evaporation under reduced pressure or a stream of N_2. The resulting methyl esters are dissolved in 20 μl of *n*-hexane. The tube is covered with a piece of aluminum foil (avoid using paraffin film) and placed on ice until analysis. A 5-μl portion of the methyl ester solution is injected into a gas chromatograph equipped with a glass column (3 mm × 2 m) packed with 15% diethylene glycol succinate on Chromosorb W (acid washed). Temperatures of the column and the flame ionization detector are 180 and 260°, respectively. The flow rate of the N_2 or He gas is 30 ml/min. A chromatography data processor (e.g., Chromatopac R3A, Shimadzu Seisakusho, Kyoto) is used to record and store chromatograms and to calculate peak areas.

Comments. Figure 2 shows a typical gas chromatogram of fatty acid methyl esters of MGDG from *Anabaena variabilis*. By using appropriate calibration factors (e.g., assuming that the ratio of integrated areas is identical to the mass ratio of methyl esters of fatty acids), both fatty acid composition and the total amount of fatty acids can be calculated. Then the total amount of each lipid class can be calculated by dividing the amount of fatty acids by 2.

Positional Distribution of Fatty Acids in Lipids

The position-specific hydrolysis of lipids by lipase is used to analyze the fatty acids esterified to the *sn*-1 and *sn*-2 positions of the glycerol

[14] R. G. Ackman, this series, Vol. 14, p. 329.

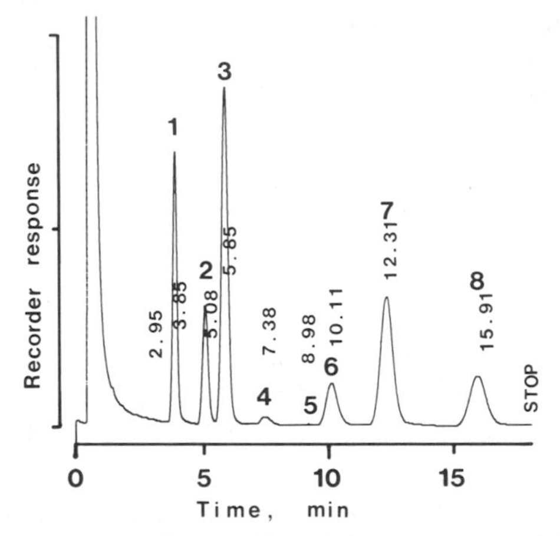

FIG. 2. Gas chromatogram of fatty acid methyl esters prepared from monogalactosyl diacylglycerol in *Anabaena variabilis* strain M3. Identification: 1, pentadecanoic acid (15 : 0, internal standard); 2, palmitic acid (16 : 0); 3, palmitoleic acid (16 : 1); 4, hexadecadienoic acid (16 : 2); 5, stearic acid (18 : 0); 6, oleic acid (18 : 1); 7, linoleic acid (18 : 2); 8, α-linolenic acid (18 : 3).

moiety. The lipase from *Rhizopus delemar,* which liberates fatty acids at the *sn*-1 position of glycolipids and phospholipids, is suitable for this purpose.[2,9,15]

Procedure. Between 0.1 and 1 mg of lipid, purified by TLC as described above for Thin-Layer Chromatographic Separation of Lipid Classes and dissolved in $CHCl_3/CH_3OH$ (2 : 1) is taken to dryness under reduced pressure or under a stream of N_2, then suspended in 0.9 ml of 50 m*M* tris(hydroxymethyl)aminomethane (Tris)–HCl buffer (pH 7.2), 0.05% Triton X-100, and mixed by vortexing. About 0.5 mg of *Rhizopus* lipase (6000 U/mg, pure grade, Seikagaku Kogyo, Tokyo), which is suspended in 0.1 ml of the same buffer, is added and mixed. The mixture is incubated at 37° for 10–60 min with gentle shaking (the reaction time should be controlled according to lipid classes as described below in Comments). Then 3 ml of C_2H_5OH is added to the mixture, and the solvent is removed using a rotary evaporator. Water should be removed completely. One-tenth milliliter of $CHCl_3/CH_3OH$ (2 : 1) is added to the residue and swirled gently in an ice-water bath (the low temperature prevents solubilization of Tris).

The supernatant is applied to a TLC plate (5 × 20 cm), which is developed in Solvent (1) (see above) for GlcDG and MGDG or in $CHCl_3$/

[15] W. Fischer, E. Heinz, and M. Zeus, *Hoppe-Seyler's Z. Physiol. Chem.* **354,** 1115 (1973).

$(CH_3)_2CO/CH_3OH/CH_3COOH/H_2O$ (10 : 4 : 2 : 3 : 1, by volume) for DGDG, SQDG, and PG. Reference compounds (purified lipids and free fatty acids) should be developed on the same plate for identification of the lysolipids and free fatty acids. Lysolipids and free fatty acids are located by primuline fluorescence as described above. Free fatty acids run just below the solvent front, whereas lysolipids appear below the original diacyl lipids. Note that Triton X-100 appears as a broad band near the solvent front. The silica gel in the fatty acid and lysolipid bands is scraped and used for fatty acid analysis as described above for Gas Chromatographic Analysis of Fatty Acids in Lipids.

Comments. The reaction time should be controlled so that the hydrolysis is just completed. Longer incubation results in partial hydrolysis of the lysolipid. Under the conditions given above, GlcDG and MGDG are hydrolyzed within 10 min, while the hydrolysis of DGDG, SQDG, and PG requires longer times (~20 min for DGDG, 30–60 min for SQDG and PG).

The fatty acid composition at the *sn*-2 position is determined by analyzing the fatty acids in the lysolipid. The fatty acid composition at the *sn*-1 position may be either determined directly from analysis of free fatty acids or estimated by comparing the fatty acid compositions of the original diacyl lipid and the lysolipid. The latter method is preferable since the free fatty acids which are located near the solvent front on the TLC plate are often contaminated.

Molecular Species of Lipids

Each class of lipids is a mixture of molecular species which contain various combinations of fatty acids at the *sn*-1 and *sn*-2 positions. Molecular species of glycolipids are separated by TLC using a $AgNO_3$-impregnated silica gel plate. The information on the positional distribution of fatty acids is necessary for the complete determination of the composition of molecular species.

Procedure. A Merck 5721 plate (10 × 20 cm) is immersed for 20 min in 5% (w/v) $AgNO_3$ in CH_3CN poured in a clean enamel dish of the type normally used for photography processing. The $AgNO_3$ solution can be stored in an amber bottle at 4° and used repeatedly. Gloves should be worn during manipulation. The plate is dried in a fume hood and activated at 105° for 30 min. The time and temperature of activation should be controlled so that the plate is not darkened.

Purified lipid (see Thin-Layer Chromatographic Separation of Lipid Classes) dissolved in $CHCl_3/CH_3OH$ (2 : 1) is applied to a $AgNO_3$-impregnated plate, and the plate is developed in Solvent (1) (see above) for GlcDG and MGDG and in $CHCl_3/CH_3OH/H_2O$ (60 : 30 : 4, by volume) for

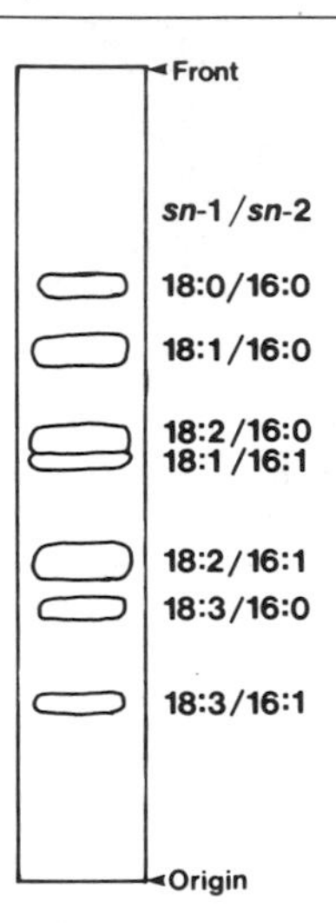

FIG. 3. Thin-layer chromatogram of molecular species of monogalactosyl diacylglycerol (MGDG) from *Anabaena variabilis*. Identification of molecular species is shown by combination of abbreviated forms of fatty acids (see the legend to Fig. 2 for names of fatty acids corresponding to specific number designations).

DGDG and SQDG to a height of 19 cm. Molecular species are detected by primuline fluorescence as described above. The use of old plates or excessively activated plates gives high background fluorescence and makes it difficult to detect lipid bands.

Each lipid molecular species is extracted from the silica gel in 4 ml of $CHCl_3/CH_3OH$ (2 : 1). Silica gel is removed by centrifugation at 1,000 *g* for 10 min. The supernatant is transferred to another tube. One milliliter of H_2O is added and mixed by vortexing. After centrifugation as above the upper layer is removed and discarded. The lower layer is washed twice with 1 ml of H_2O. These washing steps are necessary to remove $AgNO_3$ which would otherwise oxidize unsaturated fatty acids during methanolysis. The lipid molecular species are mixed with an internal standard (pentadecanoic acid) and subjected to methanolysis as described above for gas chromatography of fatty acids. Based on the positional distribution of fatty acids (see above) and fatty acid composition determined here, each molecular species of lipids is identified. The amount of molecular species is determined from the content of fatty acids, and then the composition of lipid molecular species is calculated.

Comments. A typical separation of molecular species of MGDG from *Anabaena variabilis* is shown in Fig. 3. Each molecular species is identified by its fatty acid composition. The separation is mainly effected according to the number of double bonds in the molecular species. However, 18:1/16:1 and 18:2/16:0, as well as 18:2/16:1 and 18:3/16:0, can be separated as distinct bands in close proximity.

Molecular species of DGDG and SQDG are also analyzed by this technique, but their separation is inferior to that of molecular species of GlcDG and MGDG. Molecular species of PG can be analyzed as acetyldiacylglycerol, which is prepared by digestion of PG by phospholipase C and acetylation of the resulting diacylglycerol.[9,16]

[16] M. Kito, M. Ishinaga, M. Nishihara, M. Kato, S. Sawada, and T. Hata, *Eur. J. Biochem.* **54,** 55 (1975).

[25] Rapid Permeabilization of *Anacystis nidulans* to Electrolytes

By GEORGE C. PAPAGEORGIOU

Introduction

The cell envelope of gram-negative bacteria (studied mainly in enteric bacteria)[1] consists of two hydrophobic layers, the outer membrane and the cell membrane, and a hydrophilic space, the periplasm, sandwiched in between. Peptidoglycan, an open network of polysaccharide backbones (alternating *N*-acetylglucosamine and *N*-acetylmuramic acid residues) cross-linked with oligopeptides and covalently linked to outer membrane proteins, occupies the periplasmic space next to the cell membrane. It serves as the cell exoskeleton. The cell membrane is the true selective permeability barrier. The outer membrane is only a passive molecular sieve.

In gram-negative cyanobacteria intact peptidoglycan is essential for cell envelope impermeability. Partial or total enzymic hydrolysis of peptidoglycan with lysozyme yields cells permeable to ions.[2-4] Although other means of permeabilization exist,[5,6] the enzymic method is superior when preservation of photosynthetic activity is desired. Different

[1] H. Nikaido and M. Vaara, *Microbiol. Rev.* **49,** 1 (1985).
[2] J. Biggins, *Plant Physiol.* **42,** 1442 (1967).
[3] B. Ward and J. Myers, *Plant Physiol.* **50,** 547 (1972).
[4] G. C. Papageorgiou and T. Lagoyanni, *Biochim. Biophys. Acta* **807,** 230 (1985).
[5] B. Gerhardt and A. Trebst, *Z. Naturforsch. B* **20,** 879 (1965).
[6] S. J. Robinson, C. S. DeRoo, and C. F. Yocum, *Plant Physiol.* **70,** 154 (1982).

cyanobacteria respond to lysozyme differently. Susceptibility depends on the resistance that the enzyme molecule (sphere, diameter 3.2 nm)[7] encounters as it moves through the outer membrane. An additional difficulty arises from the fact that lysozyme must get underneath the peptidoglycan layer in order to cleave glycosidic bonds.[8]

A cyanobacterium exceptionally resistant to lysozyme is the unicellular *Anacystis nidulans* (or *Synechococcus* AN).[9] It can be permeabilized to electrolytes only after long incubations with EDTA and massive amounts of lysozyme while often the derivative cells do not photoevolve O_2. In the following, we describe a rapid method that yields active ion-permeable *Anacystis* cells.

Rapid Enzymatic Permeabilization of *Anacystis*

Materials

Resuspension buffer: 50 m*M* HEPES–NaOH, pH 7.5
Assay buffer: 50 m*M* HEPES–KOH, pH 7.5, 250 m*M* KCl, 250 m*M* sorbitol
Lysozyme (EC 3.2.1.17; Sigma grade 1, from egg white; 3× recrystallized; ~40,000 U/mg protein), 90 mg/ml made up in resuspension buffer
Tetrasodium EDTA, 200 m*M*, made up in resuspension buffer
$K_3Fe(CN)_6$, 150 m*M*, in distilled H_2O.

Culture of Anacystis nidulans

Sterile *Anacystis* is grown at 30° in 3-liter Fernbach flasks.[4] Cultures (total volume 2 liters) are made in the mineral medium C of Kratz and Myers[10] and are incubated with illumination from a 40-W fluorescent lamp (Sylvania Cool White), gassed with a contaminant-free mixture of 5% CO_2 in air, and stirred at low speed. Bacteria are harvested 5–6 days after inoculation (late logarithmic phase; 4–5 mg chlorophyll *a*/liter).

Permeabilization Procedure

The bacteria are centrifuged out of the culture (6000 *g*; 5 min) and are transferred with one washing to resuspension buffer at 0.15 mg chlorophyll *a* (Chl *a*)/ml. Eight-milliliter portions (1.2 mg Chl *a*) of suspension

[7] D. C. Phillips, *Proc. Natl. Acad. Sci. U.S.A.* **57,** 484 (1967).
[8] V. Braun, K. Rehn, and H. Wolff, *Biochemistry* **9,** 5041 (1970).
[9] R. Y. Stanier and G. Cohen-Bazire, *Annu. Rev. Microbiol.* **31,** 325 (1977).
[10] W. A. Kratz and J. Myers, *Am. J. Bot.* **42,** 282 (1955).

are placed in 25-ml Erlenmeyer flasks and are equilibrated to 37° in a water bath (slow shaking). The tetrasodium EDTA and the freshly prepared lysozyme stocks are also prewarmed. At time zero, 0.05 ml of EDTA and 1 ml of lysozyme are introduced to the cell suspension to yield a mixture of the following composition: *Anacystis* equivalent to 0.132 mg Chl *a*/ml; lysozyme, 9.94 mg/ml; tetrasodium EDTA, 1.1 m*M*; and HEPES–NaOH, pH 7.5, 50 m*M*.

Progressive permeabilization is reflected by the rate of ferricyanide-supported O_2 evolution. Adequate counterion (0.2–0.3 equiv/liter of monovalent of 0.01–0.02 equiv/liter of divalent cation) should be present in the Hill reaction mixture in order to overcome coulombic repulsion of ferricyanide by the electronegative thylakoid surface. Aliquots (0.4 ml; 0.053 mg chlorophyll *a*) are transferred to an O_2 concentration cell containing 2.6 ml assay buffer and 20 μl ferricyanide solution, and the mixture is illuminated with saturating white light. We use a 1000-W lamp operated through a variable autotransformer while shielding the cells from the heat by means of a 5-cm layer of $CuSO_4$ (5%, w/v) and by circulating thermostatted H_2O around the sample. In this way, the sample can be exposed to strong photon fluxes while its temperature is under control. Three consecutive O_2 evolution traces (30 sec light–60 sec dark) are recorded. Rates are calculated by averaging the second and the third trace.

The time course of permeabilization and its dependence on the composition of the suspension medium are illustrated in Fig. 1A. Oxygen evolution starts at a slow rate, rises to a maximum as the cytoplasmic concentration of ferricyanide anion increases, and then drops off. Initial nonzero O_2 evolution here is attributed to dissolved CO_2 serving as an electron sink. Figure 1B shows that in the presence of *p*-benzoquinone, O_2 evolution starts at a fast rate, indicating unobstructed passage of the lipophilic acceptor through the cell envelope, but it falls off to a lower level as the permeabilization reaction progresses. The permeabilization reaction is fast only in the hypoosmotic HEPES–NaOH medium. When ionic or electroneutral osmotics are added to it, the kinetics are delayed and the derivative cells are less active. The reaction is stopped by taking the cells out of the reaction mixture.

Properties of *Anacystis* Permeaplasts

The permeabilized cells have the following properties: (1) Like the parent cells they are elongated, and they tolerate hypoosmotic conditions, thus evidencing a still functional exoskeleton. They have been called *permeaplasts*[3] in order to differentiate them from *spheroplasts,* cells whose modified peptidoglycan can no longer control morphology. (2)

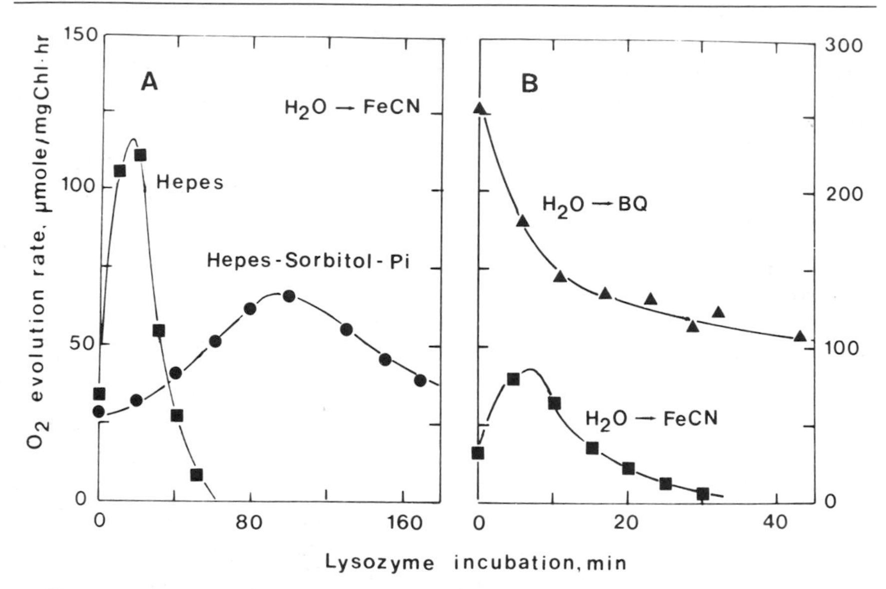

FIG. 1. Effect of lysozyme treatment on the rate of photosynthetic O_2 evolution by *Anacystis* in the presence of 1 m*M* ferricyanide (A, B) or 1 m*M* *p*-benzoquinone (B). Suspension media: 50 m*M* HEPES–NaOH, pH 7.5, and 50 m*M* HEPES–NaOH 500 m*M* sorbitol, 30 m*M* KH_2PO_4, pH 7.5.

They are capable of two-photosystem oxygenic electron transport. (3) Permeaplast thylakoids are intact and sealed. Ions pass to the lumen space only when aided by ionophores[11] or by channel-forming antibiotics.[12] (4) Ferricyanide couples at two sites of the photosynthetic electron transport: one prior to the dibromothymoquinone (DBMIB) inhibition site and one beyond it (most likely beyond the photosystem I photoreaction). Counterion compensation of the negative thylakoid charge is necessary in order to express the ferricyanide Hill reaction. (5) The shape of second derivative absorption spectra and the lack of phycocyanin-sensitized chlorophyll *a* fluorescence indicate destruction of the phycobilisome light-gathering antenna.[4] Accordingly, permeaplasts require higher photon fluxes than intact *Anacystis* cells for light saturation. In summary, the method described is capable of rapidly yielding photosynthetically active, relatively stable, ion-permeable *Anacystis* cells which contain intact and sealed thylakoids.

[11] K. Kalosaka, G. Sotiropoulou, and G. C. Papageorgiou, *Biochim. Biophys. Acta* **808,** 273 (1985).

[12] G. Sotiropoulou and G. C. Papageorgiou, *Photosynth. Res.* **10,** 445 (1986).

[26] Photosystem I and Photosystem II Preparations from Thermophilic *Synechococcus*

By SAKAE KATOH

A thermophilic cyanobacterium, *Synechococcus* sp., collected from a hot spring in Beppu, Kyushu,[1] grows maximally at 58°.[1,2] Proteins and membranes isolated from the cyanobacterium are heat stable, and, in particular, about half of the P700 photoresponse survived treatment of the thylakoid membranes at about 100° for 5 min.[3] The organism serves as an excellent material for the isolation of photosystem I (PSI) and photosystem II (PSII) preparations because photochemical activities are also strongly resistant to detergents.

Preparation of Thylakoid Membranes

For preparative purposes, cells are grown at 52–55° for 2 days in 24 liters of a modified medium of Dyer and Gafford.[4,5] The medium composition (mg/liter) is as follows: 250 KH_2PO_4, 250 K_2HPO_4, 500 $NaNO_3$, 500 KNO_3, 34 H_3BO_3, 30 disodium EDTA, 22 $(NH_4)_6Mo_7O_{24} \cdot 4H_2O$, 4 $MnCl_2 \cdot 4H_2O$, 0.66 $ZnSO_4 \cdot 7H_2O$, 0.015 $Co(NO_3)_2 \cdot 6H_2O$, 1.88 $CuSO_4 \cdot 5H_2O$, 150 $MgSO_4 \cdot 7H_2O$, 60 $CaCl_2$, and 16 $FeCl_3 \cdot 6H_2O$ (pH 7.5). The culture is continuously illuminated with a bank of tungsten lamps (10,000 lux), and an additional illumination of 10,000 lux is given to a dense culture on the second day. The medium is vigorously agitated with air containing 5% CO_2 to ensure sufficient CO_2 supply as well as homogeneous illumination and temperature distribution. The yield is 0.7–1.0 g fresh weight cells/liter.

The thylakoid membranes are prepared by the method of Biggins[6] with several modifications.[1,7] The protoplasts are passed through a French pressure cell at 300 kg/cm^2, and the homogenate is treated with DNase at 25° for 40 min. The thylakoid membranes are suspended in 50 m*M* Tris–HCl, pH 7.5, containing 0.5 *M* sucrose and 10 m*M* NaCl and stored at −30°.

[1] T. Yamaoka, K. Satoh, and S. Katoh, *Plant Cell Physiol.* **19,** 943 (1978).
[2] M. Hirano, K. Satoh, and S. Katoh, *Biochim. Biophys. Acta* **635,** 467 (1981).
[3] H. Koike, K. Satoh, and S. Katoh, *Plant Cell Physiol.* **23,** 293 (1982).
[4] D. L. Dyer and R. D. Gafford, *Science* **134,** 616 (1961).
[5] M. Hirano, K. Satoh, and S. Katoh, *Photosynth. Res.* **1,** 146 (1980).
[6] J. Biggins, *Plant Physiol.* **42,** 1442 (1963).
[7] K. Nakayama, T. Yamaoka, and S. Katoh, *Plant Cell Physiol.* **20,** 1565 (1979).

The isolated thylakoid membranes are still associated with significant amounts of allophycocyanin. In addition to detergent treatments, which will be described below, washing of the membranes with 1 *M* NaBr was recently found to be effective in largely diminishing the bound phycobiliprotein.[8] Hypotonically treated protoplasts were also shown to serve as a homogeneous right-side-out thylakoid membrane preparation for topographic studies of membrane proteins because they have leaky outer limiting membranes but intact thylakoid membranes impermeable to proteins.[8]

Preparative Gel Electrophoresis

Polyacrylamide gel electrophoresis (PAGE) is used for the preparation of PSI and PSII reaction center complexes from *Synechococcus* because the method allows good separation and resolution of chlorophyll (Chl) protein bands. In fact, the reaction center complexes of either photosystem can be isolated directly from the thylakoid membranes by a single PAGE run. To minimize contamination of comigrating extraneous proteins, however, a fraction enriched in either PSI or PSII, rather than the thylakoid membrane, is applied for electrophoresis. Slab gels of 8 mm thickness and 140 mm width are used, and 2 ml of sample suspension is applied to a gel. PAGE is carried out at 4° with a constant current of 40–60 mA to avoid excessive heating of samples. Chl proteins are extracted with 50 m*M* Tris–HCl, pH 7.5, from gel slices which are homogenized with a Teflon homogenizer. After gel fragments are spun down at 25,000 *g* for 30 min, Chl proteins are recovered from the supernatant by centrifugation at 260,000 *g* for 2 hr and suspended in Tris Buffer.

Photosystem I Reaction Center Complexes

Five PSI reaction center preparations can be simultaneously isolated from digitonin–PSI particles by PAGE in the presence of sodium dodecyl sulfate (SDS).[9,10] Thylakoid membranes containing 12 mg Chl/ml are suspended in 24 ml of 50 m*M* Tris–HCl, pH 7.5, which contains digitonin at a digitonin-to-Chl weight ratio of 50. After standing at 4° overnight with continuous stirring, 8 ml of the suspension is overlayed on 40 ml of a sucrose linear gradient (10–40%) which contains 50 m*M* Tris–HCl, pH 7.5, and 0.2% digitonin and centrifuged at 60,000 *g* for 24 hr. The fraction-

[8] I. Enami, H. Ohta, and S. Katoh, *Plant Cell Physiol.* **27,** 1395 (1986).

[9] Y. Takahashi, H. Koike, and S. Katoh, *Arch. Biochem. Biophys.* **219,** 209 (1982).

[10] Y. Takahashi and S. Katoh, *Arch. Biochem. Biophys.* **219,** 219 (1982).

ation profile shows three green bands, of which the top one often appears as a shoulder, and a precipitate. The third band from the top contains PSI particles.

The PSI fraction is dialyzed against 3 liters of 50 m*M* Tris–HCl, pH 7.5, overnight to reduce the sucrose concentration, and the particles are spun down at 300,000 *g* for 2 hr. A similar digitonin–PSI fraction is also obtained by DEAE–cellulose column chromatography.[7] The PSI particles are suspended in Tris–HCl, pH 7.5 (1 mg Chl/ml), and SDS is added to the suspension at an SDS-to-Chl weight ratio of 100. The suspension is kept at 20° for 1 hr and then applied directly for PAGE. The discontinuous buffer system of Laemmli[11] is used. The stacking gel (1 cm long) contains 0.125 *M* Tris–HCl, pH 6.8, 0.1% SDS, 4.33% acrylamide, 0.12% *N*,*N*′-methylenebisacrylamide (bisacrylamide), 0.039% ammonium persulfate, and 0.56% *N*,*N*,*N*′,*N*′-tetramethylenediamine (TEMED). The resolving gel (6 cm long) consists of 0.375 *M* Tris–HCl, pH 8.8, 0.1% SDS, 6.52% acrylamide, 0.18% bisacrylamide, 0.03% ammonium persulfate, and 0.1% TEMED. The reservoir buffer contains 25 m*M* Tris–HCl, 0.193 *M* glycine, and 0.1% SDS (pH 8.8). Typically, five sharp green bands, which are called CP1-a, CP1-b, CP1-c, CP1-d, and CP1-e in order of increasing mobility, and a fastest moving free-pigment band are resolved after 2 hr of electrophoresis. The digitonin particles often yield a split CP1-b band.[9]

CP1-a and CP1-b consist of five subunits of 62, 60, 14, 13, and 10 kDa, whereas CP1-c and CP1-d lack the 13-kDa subunit and CP1-e contains only the 62- and 60-kDa subunits. All the CP1 complexes have photoactive P700 and oxidize cytochrome *c*-553, a physiological electron donor to PSI, at high rates in the light. It is important to note that SDS strongly affects the oxidized minus reduced difference spectrum of P700 so that the differential millimolar extinction coefficient of P700 is 64 at 701 nm for the thylakoid membranes and 85 at 697 nm for CP1-a.[12] Bound electron acceptors of PSI, A_2 and P430, are present in CP1-a to CP1-d but not in CP1-e. The CP1 complexes represent different degrees of modification of the *in situ* state of the PSI reaction center complex.

CP60

A Chl protein containing only one subunit of 60 kDa (CP60) has recently been isolated from CP1-e.[13] CP1-e is incubated with 2% 2-mercaptoethanol and 2% SDS in 50 m*M* Tris–HCl, pH 7.5 (1 mg Chl/ml), at 20°

[11] U. K. Laemmli, *Nature (London)* **227,** 680 (1970).

[12] K. Sonoike and S. Katoh, *Biochim. Biophys. Acta,* in press.

[13] K. Sonoike and S. Katoh, *Arch. Biochem. Biophys.* **244,** 254 (1986).

for 1 hr and reelectrophoresed as above for 2 hr. A small Chl protein band which migrates before the CP1-e band and after the free-pigment band is CP60. The Chl protein shows a fluorescence band at 720 nm at 77 K, a characteristic of a PSI antenna, but contains no P700. CP60 is an intrinsic antenna Chl protein of the PSI reaction center core complex.

Photosystem II Reaction Center Complexes

Two purified preparations of PSII reaction center complexes have been isolated from *Synechococcus*.[14–16]

PSII Reaction Center Complex

To 10 ml of chilled thylakoid membrane suspension containing 10 mg Chl and 50 m*M* Tris–HCl, pH 7.5, is added 0.4 ml of 10% β-octylglucoside, and the suspension is kept at 0° for 20 min, followed by centrifugation at 250,000 *g* for 5 min at 0°. Large amounts of solubilized allophycocyanin remain in the supernatant. The pelleted membranes are suspended in 10 ml of the Tris buffer, and 0.8 ml of 10% octylglucoside solution is added to the suspension. After standing at 20° for 30 min, the suspension is centrifuged at 250,000 *g* for 15 min; the resultant supernatant, to which 2.5 ml of 50% glycerol is added, is subjected to PAGE using the buffer system of Davis.[17] The stacking gel (1 cm) contains 0.2% digitonin, 61.8 m*M* Tris–HCl, pH 6.7, 0.0575% TEMED, 2.5% acrylamide, 0.625% bisacrylamide, 0.0005% riboflavin, and 20% sucrose. The resolving gel (6 cm) contains 0.2% digitonin, 0.375 *M* Tris–HCl, pH 8.9, 4% acrylamide, 0.105% bisacrylamide, 0.03% TEMED, and 0.07% ammonium persulfate. Tris–glycine buffer (5 and 38.4 m*M*, respectively) is used as the reservoir buffer. Electrophoresis for 8 hr resolves four to five Chl bands which migrate after a free-pigment band containing carotenoids and biliproteins but no Chl. The reaction center complex of PSII is present in the fastest moving Chl band.

Properties of the PSII reaction center complex, together with those of other PSII preparations, are presented in Table I. The complex is competent in the primary charge separation and stabilization of the separated charges.[16,18] The subunit structure of the complex is essentially the same as that of the PSII reaction center complex isolated from spinach chloroplasts.[19]

[14] A. Yamagishi and S. Katoh, *Arch. Biochem. Biophys.* **235,** 836 (1983).
[15] A. Yamagishi and S. Katoh, *Biochim. Biophys. Acta* **767,** 118 (1984).
[16] A. Yamagishi and S. Katoh, *Biochim. Biophys. Acta* **807,** 74 (1985).
[17] B. J. Davis, *Ann. N.Y. Acad. Sci.* **121,** 404 (1964).
[18] Y. Takahashi and S. Katoh, *Biochim. Biophys. Acta* **848,** 183 (1986).
[19] K. Satoh, *Photochem. Photobiol.* **42,** 845 (1985).

TABLE I
PROPERTIES OF PSII PREPARATIONS ISOLATED FROM *Synechococcus* sp.

Preparation	Subunits (kDa)	mol/mol Q_A				Electron transport
		Chl	Mn	Ca	Plasto-quinone	
Octylglucoside	(~20 polypeptides)	48	4.0	1.0	3	$H_2O \rightarrow Q_B$
O_2-evolving complex	47, 40, 35, 31, 28, 9, 8	50	3.2–3.8	1.0	2	$H_2O \rightarrow Q_A$
Reaction center complex	47, 40, —, 31, 28, 9	46	<<1.0	<1.0	2	$Z \rightarrow Q_A$
CP2-b	47, —, —, 31, 28, 9	45	<<1.0	<<1.0	1	$P680 \rightarrow Q_A$

A PSII Reaction Center Complex Lacking the 40-kDa Subunit (CP2-b)

The PSII reaction center complex can be split into two complementary Chl protein complexes by a SDS–PAGE under mild conditions.[15] The thylakoid membranes (1 mg Chl/ml) are incubated with 0.3% lauryldimethylamine *N*-oxide in 50 m*M* Tris–HCl, pH 7.5, for 30 min at 0° and then centrifuged at 250,000 *g* for 40 min. The supernatant, after addition of glycerol to a final concentration of 10%, is subjected to PAGE. The gel system is the same as that described for preparation of the CP1 complexes except that acrylamide concentrations are 4.5 and 9.2% for the stacking and resolving gels, respectively, and 0.05% SDS is present in the reservoir buffer but not in gels. Four green bands appear after 2 hr. The second band is a Chl protein containing only the 40-kDa subunit (CP2-c), whereas the reaction center complex that lost the 40-kDa subunit (CP2-b) is present in the third band. The fastest moving band contains only free pigments.

CP2-c and CP2-b show fluorescence emission bands at 685 and 695 nm, respectively, at 77 K.[14] CP2-b is competent in the charge separation and the stabilization of separated charges,[16,18] whereas CP2-c is an intrinsic antenna Chl protein of the PSII reaction center core complex.[15]

Oxygen-Evolving PSII Preparations

For the isolation of O_2-evolving preparations, the thylakoid membranes should be suspended in a medium (medium A) which contains 1 *M* sucrose, 50 m*M* HEPES–NaOH, pH 7.5, 10 m*M* NaCl, and 5 m*M* $MgCl_2$ during preparation (except passage through a French pressure cell) and for the storage at −30° as well as for the following detergent treatments.

PAGE cannot be used for purification of O_2-evolving preparations because of its inhibitory effect on activity.

Octylglucoside Preparation[20]

The membranes are suspended in 20 ml of medium A (1 mg Chl/ml) and, after addition of 1.6 ml of 10% β-octylglucoside (see below), are kept at 20° for 1 hr. An equal volume of medium A, from which sucrose is omitted, is added to the suspension to reduce the sucrose concentration to 0.5 *M*. Then 4 ml of the suspension is layered on a cushion containing 2 ml of medium A in each of 10 centrifuge tubes. Centrifugation at 300,000 *g* for 1 hr at 0° results in separation of five bands: a top colorless band, a second blue band containing phycobiliproteins, a third dark-green band, a fourth green band, and a precipitate. Oxygen-evolving particles are present in the third band, which appears at the boundary between the 0.5 and 1 *M* sucrose layers. The fractions are combined and, after addition of 20 ml of medium A (minus sucrose), centrifuged at 300,000 *g* for 1 hr. The precipitated O_2-evolving particles are suspended in medium A.

The optimum concentration of β-octylglucoside that produces a distinct third band is usually 0.8% but may vary with membrane preparations. The detergent concentration should therefore be examined for each new membrane preparation. A too high detergent concentration results in the appearance of Chl in the second band and in contamination of PSI in the O_2-evolving preparation, whereas the third band is diminished at suboptimal concentrations.

The preparation is highly active in O_2 evolution, showing rates of 2–3 mmol O_2/mg Chl/hr at 40° and pH 6.5 with 0.4 m*M* dichloro-*p*-benzoquinone as electron acceptor. It is essential to include 1 *M* sucrose in the assay medium to obtain high rates of O_2 evolution. Addition of 0.5% digitonin stimulates O_2 evolution, but the detergent effect is considerably variable with preparations. Schatz and Witt have used a zwitterionic detergent, sulfobetaine 12, to extract a highly active O_2-evolving preparation from *Synechococcus*.[21]

Purification of Oxygen-Evolving Complex[22]

The octylglucoside preparation, which still contains about 20 polypeptides, is further purified as follows. The preparation is incubated with 0.5% sodium deoxycholate in medium A for 20 min at 25° (1 mg Chl/ml) and, to remove solubilized proteins, passed through a Sepharose CL-4B

[20] K. Satoh and S. Katoh, *Biochim. Biophys. Acta* **806,** 221 (1985).
[21] G. H. Schatz and H. T. Witt, *Photobiochem. Photobiophys.* **7,** 1 (1984).
[22] T. Ohno, K. Satoh, and S. Katoh, *FEBS Lett.* **180,** 326 (1985).

column (2.5 × 40 cm) which is equilibrated with medium A. A green fraction that elutes in the void volume is transferred to a DEAE–Toyopearl column (2.5 × 12 cm) equilibrated with medium A containing 0.1% digitonin and, in place of the HEPES buffer, 50 m*M* MES–NaOH, pH 5.5. DEAE–Toyopearl has the advantage over DEAE–cellulose that the amount of Chl remaining uneluted at the top of the column is considerably less. The charged column is washed with digitonin–medium A which additionally contains 80 m*M* NaCl until all solubilized carotenoids and phycobiliproteins are eluted. NaCl cannot be used for the elution of O_2-evolving complexes because the salt is inhibitory at concentrations above 100 m*M*. The complexes are quickly eluted with 0.75 *M* phosphate, pH 6.0, precipitated by centrifugation at 300,000 *g* for 1 hr, and finally suspended in medium A.

The purified complexes show O_2 evolution at rates between 600 and 800 μmol O_2/mg Chl/hr, with a maximum of 1300 μmol O_2/mg Chl/hr.[23] The complex, which is the PSII reaction center complex with an attached peripheral 35-kDa protein, represents a minimum functional unit for O_2 evolution. The chemical composition of the complex has recently been determined.[23]

[23] T. Ohno, K. Satoh, and S.Katoh, *Biochim. Biophys. Acta* **852,** 1 (1986).

[27] Prenylquinones

By SAKAE KATOH and KAZUHIKO SATOH

Major prenylquinones present in cyanobacteria are plastoquinone-9 and phylloquinone (vitamin K_1).[1-4] Purification, assay, and properties of prenylquinones in photosynthetic organisms including cyanobacteria have been described in detail in two previous volumes.[5,6] Extensive accounts of fractionation of quinones with column and thin-layer chromatography have previously been given.[5] A reversed-phase high-performance liquid chromatography which allows rapid and simultaneous

[1] M. D. Henninger, H. N. Bhagavan, and F. L. Crane, *Arch. Biochem. Biophys.* **110,** 69 (1965).

[2] N. G. Carr, G. Exell, V. Flynn, M. Hallaway, and S. Talukdar, *Arch. Biochem. Biophys.* **120,** 503 (1967).

[3] G. A. Peschek, *Biochem. J.* **186,** 515 (1980).

[4] M. Aoki, M. Hirano, Y. Takahashi, and S. Katoh, *Plant Cell Physiol.* **24,** 517 (1983).

[5] R. Barr and F. L. Crane, this series, Vol. 23, p. 372.

[6] R. Barr and F. L. Crane, this series, Vol. 69, p. 374.

determination of quinones and chlorophyll *a* in small amounts of samples is described here.

Assay

One milliliter of cell suspension containing 200 μg chlorophyll *a* is added to 10 ml of methanol, and the mixture is sonicated 3 times for 30 sec with cooling intervals of 5 min then left for 40 min with continuous stirring. The mixture is centrifuged at 3000 *g* for 10 min, and the precipitate is extracted successively with 10 ml of methanol and twice with 10 ml of methanol/diethyl ether (1 : 1, v/v) with sonication as above. The extracts are combined, evaporated to dryness in a rotary evaporator, and redissolved in 400 μl of methanol.

Quinones are more readily extracted from isolated thylakoid membranes or purified PSI and PSII fractions than from intact cells. A sample suspension (50–100 μl) containing 5–10 μg Chl is mixed with 5 ml of methanol, and the mixture is allowed to stand for 5 min, then centrifuged at 3000 *g* for 10 min. The precipitate is again extracted with 5 ml of the methanol/diethyl ether (1 : 1) mixture. The combined extracts are evaporated and finally dissolved in 200 μl of methanol. All extraction procedures are carried out at 0°.

A small volume (5–20 μl) of the methanol solution is applied to a Waters high-performance liquid chromatograph equipped with a Model 740 data module. A reversed-phase column, μBondapak C_{18} (3.9 × 150 mm), is used. The mobile phase is a mixture of 75% methanol and 25% 2-propanol, and the flow rate is 1 ml/min. The detector wavelength is set at 255 nm. Different combinations of columns and solvent systems have been reported in literature.[7–9]

Plastoquinone-9 and phylloquinone are eluted with widely different retention times (Table I). Each of the two quinone bands are also well separated from peaks of chlorophyll *a* and β-carotene, which have significant absorption at 255 nm. Xanthophylls are eluted before chlorophyll *a*. The chlorophyll *a* band is closely preceded by a minor band of a chlorophyll *a* isomer with identical spectral properties.

Quinones and pigments are quantified by measuring the area under the peak with the data module. Calibration curves are constructed by plotting the peak areas versus the amounts of standard samples applied. The high sensitivity and resolution of the method allow determination of quinones and chlorophyll *a* in a sample which contains only a few micrograms of chlorophyll *a*.

[7] H. K. Lichtenthaler and U. Prenzel, *Biochim. Biophys. Acta* **135,** 493 (1977).
[8] S. Okayama, *Plant Cell Physiol.* **25,** 1445 (1984).
[9] E. Lam and R. Malkin, *Biochim. Biophys. Acta* **810,** 106 (1985).

TABLE I
RETENTION TIME OF QUINONES AND PIGMENTS

Component	Retention time (min)
Chlorophyll *a* isomer	4.6
Chlorophyll *a*	5.2
Phylloquinone	7.0
β-Carotene	15.5
Pheophytin *a*	17.0
Plastoquinone	22.0

Distribution and Function

Plastoquinone and phylloquinone are associated with the thylakoid membranes in cyanobacterial cells. A *Synechococcus* PSII preparation competent in O_2 evolution and free from PSI contains three bound plastoquinone molecules per PSII reaction center but no phylloquinone.[10] The bound plastoquinone molecules function as electron acceptors, Q_A and Q_B. Plastoquinone also functions as the pool quinone connecting the PSII reaction center complexes and the cytochrome b_6/f complexes in photosynthetic electron transport. The pool which contains four to five plastoquinone molecules per PSII reaction center[10,11] is also shared by respiratory electron transport and hence serves as a common link between the two energy-transducing systems in the photosynthetic prokaryotes which lack ubiquinone.[11-13] Cytoplasmic membranes isolated from *Anacystis nidulans* contain a small amount of plastoquinone of unknown function.[14]

Most, if not all, phylloquinone molecules present in the thylakoid membranes are associated with the PSI reaction center complexes.[15,16] The PSI complexes isolated from *Synechococcus*[15] and *Anabaena*[16] contain about two phylloquinone molecules per P700 but no plastoquinone. Recent evidence suggests that phylloquinone functions as a secondary electron acceptor (A_1) of PSI.[17,18]

[10] Y. Takahashi and S. Katoh, *Biochim. Biophys. Acta* **848,** 183 (1986).
[11] M. Aoki and S. Katoh, *Plant Cell Physiol.* **24,** 1379 (1983).
[12] M. Hirano, K. Satoh, and S. Katoh, *Photosynth. Res.* **1,** 149 (1980).
[13] M. Aoki and S. Katoh, *Biochim. Biophys. Acta* **682,** 307 (1982).
[14] T. Omata and N. Murata, *Biochim. Biophys. Acta* **766,** 395 (1984).
[15] Y. Takahashi, K. Hirota, and S. Katoh, *Photosynth. Res.* **6,** 183 (1985).
[16] H.-U. Schroeder and W. Lockau, *FEBS Lett.* **199,** 23 (1986).
[17] M. C. Thurnauer and P. Gast, *Photobiochem. Photobiophys.* **9,** 29 (1985).
[18] W. E. Mansfield and M. C. W. Evans, *FEBS Lett.* **190,** 237 (1985).

[28] Oxygen-Evolving Photosystem II Particles from *Phormidium laminosum*

By DEREK S. BENDALL, JANE M. BOWES, ALISON C. STEWART, and MARY E. TAYLOR

Introduction

Photosystem II preparations that retain the capacity for photosynthetic oxygen evolution have been obtained from a number of thermophilic cyanobacteria. The first such preparation was obtained from the lysozyme-sensitive filamentous strain, *Phormidium laminosum*. Treatment of *Phormidium* thylakoids with the nonionic detergent lauryldimethylamine oxide (LDAO), which had been used to purify reaction centers from purple photosynthetic bacteria, preferentially solubilized photosystem II[1]; centrifugation then yielded a supernatant that was depleted in photosystem I and enriched in photosystem II-catalyzed O_2 evolution. A crude preparation could be recovered from this supernatant by passage through a column of Sepharose 6B to remove detergent and to allow aggregation of the particles, which were then concentrated by centrifugation. Oxygen-evolving activity of the solubilized system declined fairly rapidly unless the medium contained at least 20% glycerol, in addition to bivalent cations (Mg^{2+} or Ca^{2+}) and Cl^- or Br^-.[2]

Further purification depended on the discovery that the stability of the initial extract could be enhanced by addition of a second detergent, lauryl-β-D-maltoside (LM), which had been shown to have a similar effect on preparations of rhodopsin,[3] cytochrome oxidase,[4] and photosynthetic reaction centers isolated from *Rhodopseudomonas spheroides* with LDAO.[5] After treatment of the *Phormidium* membrane extract with LM, a purified photosystem II preparation was obtained by centrifugation through a sucrose density gradient containing LM and glycerol.[6] The particles retained a high rate of O_2 evolution and contained almost no photosystem I or cytochrome *b*/*f* complex.

[1] A. C. Stewart and D. S. Bendall, *FEBS Lett.* **107,** 308 (1979).
[2] A. C. Stewart and D. S. Bendall, *Biochem. J.* **194,** 877 (1981).
[3] P. Knudsen and W. L. Hubbell, *Membr. Biochem.* **1,** 297 (1978).
[4] P. Rosevear, T. VanAken, J. Baxter, and S. Ferguson-Miller, *Biochemistry* **19,** 4108 (1980).
[5] M. W. Kendall-Tobias and M. Seibert, *Arch. Biochem. Biophys.* **216,** 255 (1982).
[6] J. M. Bowes, A. C. Stewart, and D. S. Bendall, *Biochim. Biophys. Acta* **725,** 210 (1983).

We describe below the preparation of the crude LDAO particles and the purified LM particles. The details given are those that we currently use and differ in certain respects from those given in the original publications. Subsequently, we mention some promising results we have recently obtained with a new detergent, *N*-methyl-*N*-decanoylmaltosylamine (M1A10), in place of LM.

Isolation Procedure

Cyanobacterial Cultures. Phormidium laminosum (strain OH-1-p.Cl 1) is grown at 45° in 2-liter flasks, each containing 1.25 liters Medium D of Castenholz[7] supplemented with 0.1 g $liter^{-1}$ $NaHCO_3$, which are shaken at 100 cycles min^{-1} in a Gallenkamp orbital incubator with constant illumination from four 30-W fluorescent tubes. The incubator is gassed with air enriched with 5% CO_2 at a flow rate of 175 ml min^{-1}.

Thylakoid Membranes. Cells are harvested from 10 liters of late log-phase culture (4–5 days' growth) by centrifuging at 20° for 15 min at 4400 g_{max}. In order to prevent the loose pellet from dispersing, the supernatant is removed by suction immediately after the centrifuge comes to a halt. The filaments are washed in 400 ml of Buffer A containing 0.5 *M* sorbitol, 10 m*M* $MgCl_2$, 5 m*M* sodium phosphate buffer, pH 7.5, and 10 m*M* HEPES–NaOH, to which 12.5 m*M* disodium EDTA has been added, and resuspended in 200 ml of the same medium containing 2% (w/v) bovine serum albumin (BSA) (Sigma Fraction V). Lysozyme (0.1 g, Sigma or BDH) is predissolved in a small volume of the medium and added to the suspension, which is then briefly homogenized with a Polytron homogenizer (30 sec on speed setting 2) to break up clumps of filaments. The suspension is shaken for 1 hr at 37° in a water bath at 120 cycles min^{-1}, by which time spheroplasts (which may occur as short chains) should have formed.

The lysozyme reaction is stopped by addition of 600 ml of ice-cold Buffer A (without EDTA). Spheroplasts are harvested by centrifuging for 7 min at 3000 g_{max} and then lysed by resuspension with a Dounce homogenizer in ice-cold Buffer B containing 10 m*M* $MgCl_2$, 5 m*M* sodium phosphate buffer, and 10 m*M* HEPES–NaOH buffer, pH 7.5. DNase (BDH, ~100 μg) is added along with sufficient Buffer B to give a total volume of 800 ml. After centrifuging the suspension for 90 sec at 200 *g* to remove unbroken filaments, a pellet of thylakoid membranes is formed by further centrifugation for 15 min at 28,000 *g*, leaving a deep blue supernatant containing the bulk of the phycobilin pigments, which is discarded. The

[7] R. W. Castenholz, *Schweiz. Z. Hydrol.* **32,** 538 (1970).

membranes are washed once by resuspension with a Dounce homogenizer in at least 250 ml of Buffer B, followed by centrifugation, and are finally resuspended in 40 ml Buffer C containing 25% (v/v) glycerol, 10 mM $MgCl_2$, and 5 mM sodium phosphate and 10 mM HEPES–NaOH buffers, pH 7.5. The resuspended membranes may be stored in liquid N_2 if necessary.

Chlorophyll Assay. The chlorophyll assay needs to be accurate. Best results are obtained if a positive-displacement micropipet is used to add 20 μl of thylakoid suspension to a glass centrifuge tube containing 10 ml of freshly prepared 80% acetone while the tube is being agitated on a vortex mixer. Precipitated protein is removed by centrifuging, and the absorbance of the supernatant is read at 663 nm. The chlorophyll concentration of the suspension should be at least 1.5 mg ml^{-1}. For the procedures described here the extinction coefficient for chlorophyll *a* of 82.04 ml mg^{-1} cm^{-1} given by Mackinney[8] is used.

LDAO Extraction. To a thylakoid suspension containing 30 mg of chlorophyll *a* are added 0.3 ml 100 mM phenylmethylsulfonyl fluoride (PMSF) (freshly prepared in ethanol), Buffer C, and, finally, 0.34 ml 30% LDAO (as supplied by Onyx Chemical Co. or Calbiochem); the amount of Buffer C is sufficient to give a final volume of 30 ml. The LDAO-to-chlorophyll ratio is critical for optimum solubilization of photosystem II with minimum extraction of photosystem I. The mixture is incubated on ice in the dark for exactly 40 min and then centrifuged for 60 min at 35,000 rpm (147,000 *g*) in a Beckman SW50.1 rotor. Immediately after the rotor has come to a halt, the fluorescent bluish green supernatant is carefully removed with a Pasteur pipet, taking care to avoid dislodging the loose upper layers of the pellet. The supernatant fraction is either used to prepare the crude LDAO particles, as described in the next paragraph, or, preferably, used immediately for purification on a sucrose density gradient as described in the subsequent paragraph.

LDAO Particles. The LDAO extract is passed through a Sepharose 6B column (3.3 × 50 cm) equilibrated with Buffer C to which no detergent has been added. This treatment causes the photosystem II particles to aggregate so that they can be concentrated by centrifuging for 16 hr at 100,000 *g*. The pellet is resuspended in a minimum volume of Buffer C.

LM Particles and Sucrose Density Gradient. The volume and chlorophyll concentration of the LDAO extract are measured without delay and sufficient LM added to give a ratio of LM to chlorophyll of 40 : 1 by weight. The LM is added as a fresh solution which is prepared by dissolving 1.25 g LM in 25 ml Buffer C and then cooled on ice. The LM-treated preparation is layered in 6- to 7.5-ml portions onto linear gradients (72 ml)

[8] G. Mackinney, *J. Biol. Chem.* **140,** 315 (1941).

containing 20% glycerol, 10 m*M* $MgCl_2$, 10 m*M* HEPES–NaOH buffer, 1 m*M* PMSF, and 8–25% sucrose, plus LM at a concentration 20 times (w/w) the chlorophyll concentration of the suspension layered on the gradient. The tubes are centrifuged for 20 hr at 44,000 rpm (225,000 *g*) in a Beckman 45Ti rotor at 4°. Approximately 12 ml of the lowest green band is removed from each gradient using a syringe with a flat-tipped needle, and the preparation is concentrated to a final volume of 10–12 ml in a 50-ml Amicon ultrafiltration cell using a YM100 membrane and a pressure of 20 psi. This step takes from 5 to 9 hr and is most rapid with a fresh ultrafiltration membrane. The final preparation can be stored in liquid N_2 for several months without loss of activity. The procedure for purification of LM particles is summarized in Fig. 1.

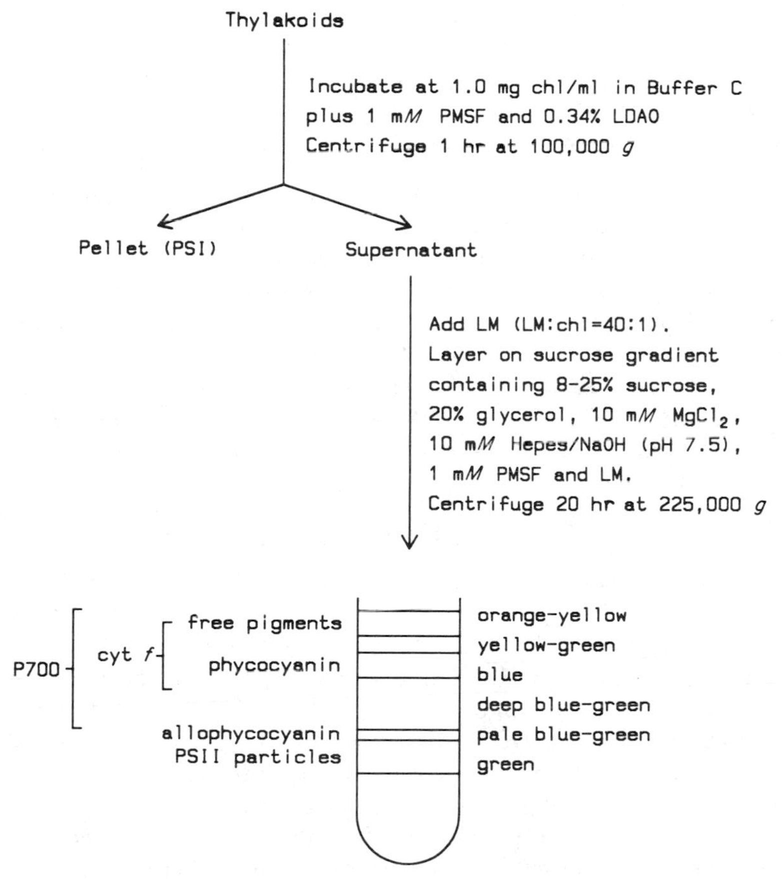

FIG. 1. Purification procedure for LM photosystem II particles from *Phormidium laminosum*.

TABLE I
PROPERTIES OF THYLAKOID MEMBRANES AND PHOTOSYSTEM II PARTICLES FROM *Phormidium laminosum*

Property	Thylakoids	LDAO particles	LM particles
Oxygen evolution[a]	300–450	1800–2500	1800–2500
P700[b]	7	1	Undetectable
Cytochrome *f*[b]	2	4	Undetectable
Cytochrome *b*-563[b]	1	4	Undetectable
Cytochrome *b*-559[b]	5	25	20
Mn[b]	16	80	80

[a] With 1 m*M* phenyl-*p*-benzoquinone as acceptor. Expressed in μmol O_2 (mg chlorophyll)$^{-1}$ hr^{-1}.
[b] Expressed in nmol (mg chlorophyll)$^{-1}$.

Properties of Photosystem II Particles

The properties of the crude LDAO particles and the purified LM particles are summarized in Table I.

Oxygen Evolution. Activity is measured at 25° in an oxygen electrode (Hansatech or Rank) with particles containing 10 μg chlorophyll suspended in 1.0 ml Buffer C. The sample is illuminated with red light (Schott RG 610 filter) from a 150-W slide projector placed as close as possible to the sample chamber, which is backed by a mirror. Maximum rates of O_2 evolution are obtained with 1 m*M* phenyl-*p*-benzoquinone (from a 100 m*M* stock solution in dimethyl sulfoxide) as a lipophilic electron acceptor. The rate should be at least 300 μmol O_2 (mg chlorophyll)$^{-1}$ hr^{-1} for thylakoids and 1800 μmol O_2 (mg chlorophyll)$^{-1}$ hr^{-1} for photosystem II preparations. There is a broad pH optimum at pH 6.5–7.5. Rates can also be measured with a high concentration (10 m*M*) of the hydrophilic acceptor ferricyanide. The optimum pH is 5.5–6.0 for this reaction, which usually gives lower rates than when phenylbenzoquinone is used. The high 3,4-dichlorophenyl-1,1-dimethylurea (DCMU)-insensitive rates sometimes observed for photosystem II preparations with ferricyanide may indicate some damage to the acceptor side, or alternatively loss of lipid, as membrane preparations are always highly sensitive. Rates of O_2 evolution are highest and most stable when both 25% glycerol and either 10 m*M* $MgCl_2$ or 10 m*M* $CaCl_2$ are included in the assay medium, although glycerol is not necessary with membrane preparations.

Photosystem I Content. Most of the photosystem I components sediment when the LDAO-treated thylakoids are centrifuged, and the remainder are separated from photosystem II on the sucrose density gradient. The concentration of P700 in LM particles should be undetectable in a reduced-minus-oxidized difference spectrum (ascorbate − ferricyanide).[6]

Cytochromes. Much of the cytochrome *b*/*f* complex is extracted by LDAO, and also cytochrome *c*-549, and these components remain with the LDAO photosystem II particles. Cytochromes *f* and *b*-563 are separated from photosystem II on the LM-containing density gradient, but cytochrome *c*-549 is distributed throughout the gradient, a significant amount remaining associated with the LM particles. Cytochrome *b*-559 is concentrated in the LM particles; a major portion tends to occur in the high-potential form (reducible by hydroquinone), but the proportion of high- to low-potential forms is variable.[2,6]

Antenna and Plastoquinone Pool Sizes. Estimates of the photosynthetic unit size of the LDAO particles have been made by measurement of the activation time for O_2 evolution and the concentration of P680[2,9] and give values in the range of 35–65 chlorophyll molecules per reaction center. Comparison of the fluorescence rise times of the LDAO and LM particles in the presence of DCMU suggests that the antenna size of the latter is about 10% smaller than that of the former.[6] Comparison of fluorescence induction curves in the presence and absence of DCMU indicates that LM particles contain 1 to 2 molecules of plastoquinone on the acceptor side (including Q_A). The LDAO particles retain a substantially larger plastoquinone pool. The initial fluorescence rise is clearly sigmoidal in both LDAO particles and thylakoid membrane fragments,[10] suggesting that photosystem II occurs as aggregates of units. With LM particles the sigmoidicity is slight or nonexistent, indicating that the particles are mainly monomeric.[6]

Manganese. LDAO particles and LM particles contain, respectively, 15 and 13 mol Mn per mol chlorophyll.[2,9] This corresponds to approximately 4 Mn atoms per reaction center.

Polypeptides. The LM particles contain a number of polypeptides in the range 7–100 kDa. They are most successfully resolved[11] by polyacrylamide gel electrophoresis at 4° in gels containing a 12–22.5% acrylamide gradient, 4 *M* urea, 0.1% lithium or sodium dodecyl sulfate, and the buffer system of Laemmli.[12] The samples are incubated with dodecyl sulfate for 30 min at 4° before being applied to the gel; heating, especially boiling, of samples with dodecyl sulfate leads to aggregation of some polypeptides.

[9] B. Ke, H. Inoue, G. T. Babcock, Z.-X. Fang, and E. Dolan, *Biochim. Biophys. Acta* **682,** 297 (1982).

[10] J. M. Bowes, P. Horton, and D. S. Bendall, *Arch. Biochem. Biophys.* **225,** 353 (1983).

[11] A. C. Stewart, U. Ljungberg, H.-E. Akerlund, and B. Anderson, *Biochim. Biophys. Acta* **808,** 353 (1985).

[12] U. K. Laemmli, *Nature (London)* **227,** 680 (1970).

Four different preparations of gradient-purified particles were run side by side in the gel illustrated in Fig. 2. The strongly staining bands with apparent molecular masses of 17 and 18.5 kDa are due to residual allophycocyanin. The 92-kDa component is presumed to be a high-molecular-mass terminal pigment (phycobilisome core–membrane linker polypeptide),[13-16] and one or more of the bands in the 30- to 35-kDa region may be colorless phycobilisome linker polypeptides. The component giving the strong band with an apparent size of 35 kDa is rapidly degraded by exposure to trypsin and cross-reacts with an antibody to the extrinsic 33-kDa polypeptide of PSII from spinach thylakoids.[11] A second component in the 30- to 35-kDa region is expected to be the Q_B (herbicide-binding) protein. Chlorophyll-binding proteins contribute to bands in the regions 39–42 and 44–50 kDa.[17] The strong band at 58.5 kDa is likely to be a contaminant, possibly a subunit of the coupling factor CF_1. The weak band at 24 kDa is also a contaminant, or possibly a proteolysis product, because it seems to be present in substoichiometric amounts and is sometimes undetectable.

Polypeptides corresponding to the 16- and 23-kDa extrinsic proteins of chloroplast photosystem II seem to be absent, and neither the LM particles nor spheroplast extracts from *Phormidium laminosum* cross-react with antibodies to the higher plant proteins.[11] Cytochrome *c*-549 runs in the region of the 17-kDa allophycocyanin band and can be detected by staining with H_2O_2 and tetramethylbenzidine.[18] In the low-molecular-mass region the polypeptide of about 9 kDa has been shown to be lost along with O_2-evolving activity when *Phormidium* particles are washed with low-glycerol media.[19] The polypeptide at about 7 kDa is probably a subunit of cytochrome *b*-559 as it has been observed to cross-react with an antibody to the spinach cytochrome.[11] Silver staining reveals further low-molecular-mass components but does not react with the 9-kDa band.

Comments on the Purification Procedure

LDAO was found to give more selective extraction of photosystem II from membrane fragments than Triton X-100 or Nonidet P-40. Zwittergent 3-12 (Calbiochem) gives good extraction and can also be used to

[13] D. J. Lundell, G. Yamanaka, and A. N. Glazer, *J. Cell Biol.* **91,** 315 (1981).
[14] M. Rusckowski and B. A. Zilinskas, *Plant Physiol.* **70,** 1055 (1982).
[15] B. A. Zilinskas, *Plant Physiol.* **70,** 1060 (1982).
[16] T. Redlinger and E. Gantt, *Plant Physiol.* **68,** 1375 (1981).
[17] A. C. Stewart, *FEBS Lett.* **114,** 67 (1980).
[18] P. E. Thomas, D. Ryan, and W. Levin, *Anal. Biochem.* **75,** 168 (1976).
[19] A. C. Stewart, M. Siczkowski, and U. Ljungberg, *FEBS Lett.* **193,** 175 (1985).

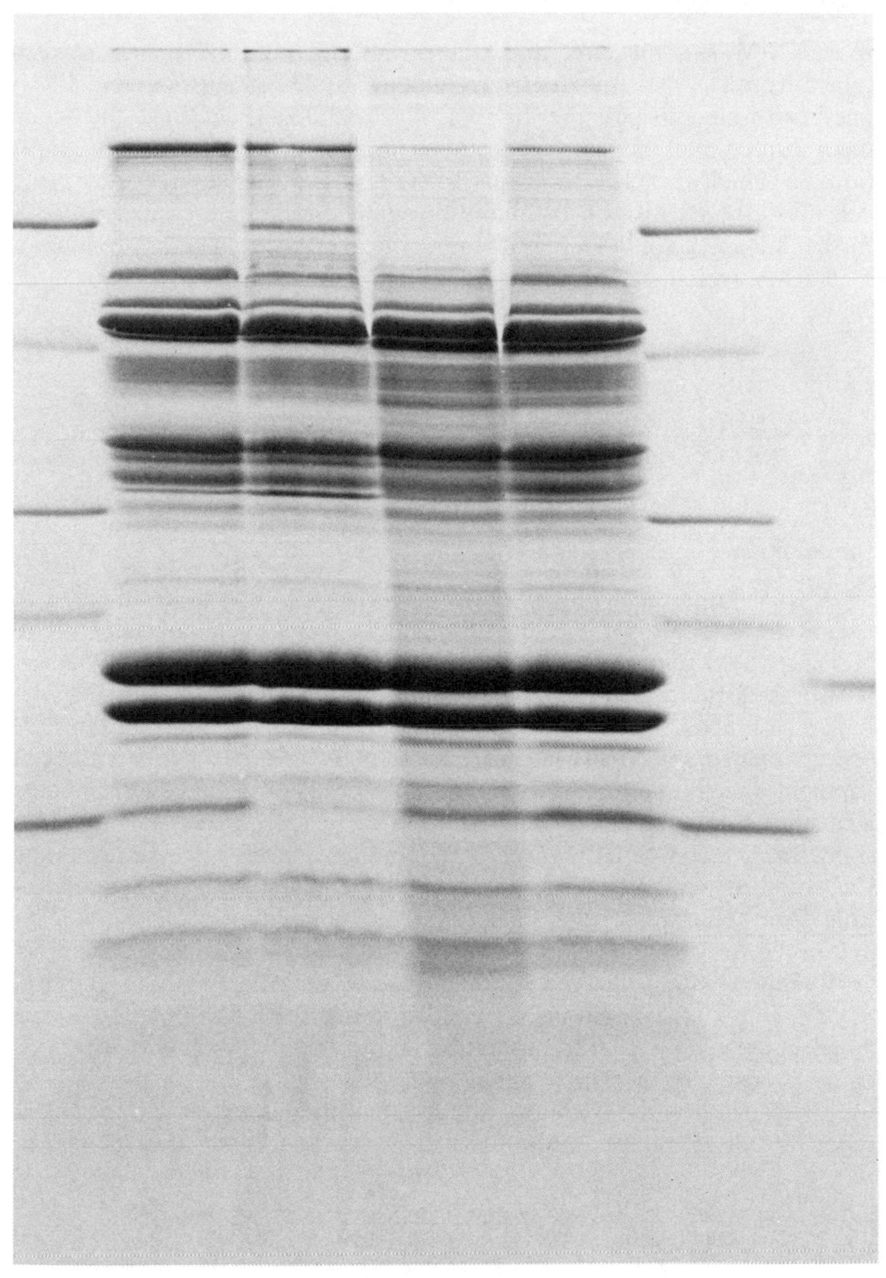

FIG. 2. Polyacrylamide gel electrophoresis of polypeptides of four different preparations of photosystem II particles from *Phormidium laminosum* stained with Coomassie blue. The two left-hand tracks correspond to LM preparations, the two right-hand tracks to particles prepared with M1A10. Molecular weight standards were bovine serum albumin (68,000), ovalbumin (43,000), carbonate dehydratase (30,000), soybean trypsin inhibitor (21,000), and cytochrome *c* (12,400). See text for details.

replace LM, but the conditions have not been optimized. M1A10 (obtained from OXYL-GmbH, Peter-Henlein-Str. 11, 8903 Bobingen, FRG) has been found to be superior to LM in stabilizing activity, and when incorporated into the sucrose gradient in place of LM it yields preparations in which the Q_A site seems to be less exposed, as judged by the sensitivity to DCMU of O_2 evolution with ferricyanide as acceptor. A further advantage of M1A10 is its high critical micelle concentration (cmc, >50 m*M*), which renders it easier to remove than LM.

[29] *In Vivo* Effect of Calcium on Photosystem II

By JERRY J. BRAND and DAVID W. BECKER

Introduction

A Ca^{2+} requirement for photosystem II (PSII) was first indicated in isolation protocols for photosynthetically active cyanobacterial membranes.[1,2] In view of the extensive homology between the photosynthetic membrane structure and function of chloroplast and cyanobacterial thylakoid membranes, it is not surprising that Ca^{2+} has been implicated in PSII of fractionated chloroplast membranes as well.[3–5] Experiments with PSII submembrane preparations from cyanobacterial cells,[6] however, indicate a site of function different from that observed in chloroplast PSII preparations. In cyanobacterial thylakoid membranes and PSII preparations the Ca^{2+} requirement lies close to the reaction center, while in PSII preparations from chloroplasts the water-oxidizing complex demonstrates a requirement for Ca^{2+}.

Cyanobacteria afford the opportunity to examine a PSII Ca^{2+} requirement *in vivo,* thus eliminating possible artifacts which may develop during membrane isolation and fractionation. Interpretable results are obtained from experiments in which intact cyanobacterial cells are stressed for Ca^{2+} because one site is clearly affected and the inhibition is total and is reversed fully by Ca^{2+} addition to the growth medium.[7] Similar experi-

[1] R. G. Piccioni and D. C. Mauzerall, *Biochim. Biophys. Acta* **504,**398 (1978).
[2] C. M.-C. Yu and J. J. Brand, *Biochim. Biophys. Acta* **591,** 483 (1980).
[3] D. F. Ghanotakis, G. T. Babcock, and C. F. Yocum, *FEBS Lett.* **167,** 127 (1984).
[4] M. Miyao and N. Murata *FEBS Lett.* **168,** 118 (1984).
[5] T.-A. Ono and Y. Inoue, *FEBS Lett.* **168,** 281 (1984).
[6] R. R. England and E. H. Evans, *Biochem. J.* **210,** 473 (1983).
[7] D. W. Becker and J. J. Brand, *Biochem. Biophys. Res. Commun.* **109,** 1134 (1982).

ments with eukaryotic algae or with higher plant cells do not yield such clear-cut results, presumably because Ca^{2+} plays important regulatory roles in the cytoplasm of all eukaryotic organisms and its removal results in irreversible disruption of cellular metabolism and subsequent cell death.

Here we report the detailed conditions which lead to Ca^{2+} deficiency in intact cyanobacterial cells, and we describe types of experiments that can be used to probe the function of Ca^{2+} in cyanobacterial thylakoid membranes using intact, Ca^{2+}-depleted cells. We also suggest how this experimental system might be used as a general tool to probe other mechanistic aspects of photosynthesis. The Ca^{2+} requirement for PSII activity in membrane preparations from cyanobacteria is not discussed, since methods for preparing photochemically active membranes and particles are described elsewhere in this volume.[8,9]

Depletion of Calcium from Cyanobacterial Cells

The methods described in this section were characterized in detail in *Anacystis nidulans* TX 20 (equivalent to *Synechococcus* 6301). With notable exceptions (indicated below) they appear to apply generally to cyanobacteria.

Liquid cultures of *A. nidulans* can be pelleted, resuspended in medium devoid of Ca^{2+}, and incubated under otherwise normal growth conditions to totally inhibit PSII activity but not cause other apparent effects on cell metabolism. Best results are obtained when cells are harvested in mid or late log phase (5–15 μg Chl/ml) from any medium which supports active growth. Subsequent manipulations are performed in acid-rinsed plasticware (polyethylene or polypropylene preferred) or glassware. Acid rinsing requires minimally a 5 min soaking in 0.2 *N* HCl, followed by exhaustive rinsing with high-purity water. Cells are pelleted and resuspended 2 times in a depletion medium (see below). For thermophilic and near-thermophilic cyanobacteria such as *A. nidulans,* it is critical that the cells remain above 20° at all times to prevent secondary chilling damage. The depletion medium of choice is a modification of cyanobacterial growth medium, Cg-10[10] (see Table I). The best grade of chemicals available should be used for preparing stock solutions, especially with regard to Na^+ and Ca^{2+} contamination. High-purity (e.g., double-distilled) water (preferably stored in rigid polyethylene) should be used for all solutions.

[8] S. Katoh, this volume [26].

[9] D. S. Bendall, J. M. Bowes, A. C. Stewart, and M. E. Taylor, this volume [28].

[10] C. Van Baalen, *J. Phycol.* **3,** 154 (1967).

TABLE I
CULTURE MEDIUM FOR PERFORMING CALCIUM DEPLETION EXPERIMENTS WITH INTACT CYANOBACTERIA

Stock solutions		Final solution[a]	
Ingredient	Amount (g/liter)	Amount of stock solution (ml/liter)	Final concentration (m*M*)
$MgSO_4 \cdot 7H_2O$	50	5.0	1.0
K_2HPO_4	10	5.0	0.3
KNO_3	100	10.0	10.0
$Ca(NO_3)_2 \cdot 4H_2O$	5	5.0[b]	0.1
Ferric EDTA	*c*	1.0	0.014 Fe^{2+}
Micronutrients	*d*	1.0	—
Glycylglycine	Add 1.5 g/liter directly to final solution		11.3

[a] For preparing the final solution the indicated volumes of stock solutions are added to approximately 850 ml of distilled water. The pH is adjusted to 8.3 with KOH, and the final volume is brought to 1.0 liter.

[b] Omit for depletion experiments.

[c] Ferric EDTA stock solution is prepared as follows: Add 1.6 g of EDTA (free acid form) to 175 ml of distilled water and adjust the pH to 5.5–7 with KOH, which will dissolve the EDTA. Add 0.8 g of $FeSO_4 \cdot 7H_2O$ and bring the final volume to 200 ml.

[d] Micronutrient stock solution is prepared by dissolving the following salts in the amounts indicated (in g/liter) in H_2O: H_3BO_3, 2.86; $MnCl_2 \cdot 4H_2O$, 1.81; $ZnSO_4 \cdot 7H_2O$, 0.22; MoO_3 (85%), 0.021; $CuSO_4 \cdot 5H_2O$, 0.79.

Calcium depletion can be demonstrated in various other media with addition of EDTA or EGTA to chelate residual Ca^{2+}. However, such strong chelators cause other effects which complicate interpretation and preclude reversibility of the inhibition.

Procedures for inhibiting PSII of cells in Ca^{2+}-deficient medium are as follows, taking into account the precautions listed above:

1. Centrifuge cells (6,000 *g* for 5 min) from normal growth medium.
2. Resuspended in depletion medium to approximately 100 μg Chl/ml and centrifuge as before.
3. Resuspended and centrifuge as in Step 2.
4. Resuspend cells in depletion medium to 5 μg Chl/ml and place in transparent culture tubes (preferably no more than 40 mm maximum diameter to ensure uniform illumination).
5. Bubble with 1% CO_2-enriched air and illuminate the cell culture.
6. Collect samples from the incubating culture at intervals to monitor photosynthetic activity with an O_2 electrode.

The Ca^{2+} depletion may be carried out conveniently in 40-ml batches of cells placed in 25 × 200 mm cylindrical glass test tubes equipped with bubblers passing through a gauze cap. Illumination is required to demonstrate Ca^{2+} depletion and accompanying loss of PSII activity, and the rate of inhibition varies greatly with light quality and intensity. Time required for loss of PSII activity due to Ca^{2+} depletion is in approximate inverse proportion to light intensity. Virtually no inhibition is seen with a photon flux density less than 50 $\mu E/m^2/sec$ of photosynthetically active radiation (PAR) in cool-white fluorescent illumination. Optimal growth temperature serves as a good temperature for performing the Ca^{2+} depletion (40° for *A. nidulans*).

The decline in photosynthetic competence which accompanies Ca^{2+} depletion may be monitored by any method which reflects PSII activity. It is convenient to monitor O_2 production in whole cells by removing samples at intervals from the culture tubes and transferring them directly to the O_2 electrode chamber. Addition of HCO_3^- to a final concentration of 50 m*M* ensures that carbon substrate is not limiting. Light intensity and quality for measuring O_2-evolution rates during the depletion process should be the same as those which afford maximum rates when assaying normal cell cultures. When aliquots of cells are collected from the culture tubes for measurement of photosynthetic activity, they should be kept in darkness or dim light and at growth temperature until a measurement is to be made. With this precaution, there is no need to rigorously remove Ca^{2+} or other ions from the collection tube or electrode chamber, because cells in the dark do not change in photosynthetic competence and thus reflect the extent of Ca^{2+} depletion. Even during the few minutes while photosynthetic activity is being measured the state of inhibition does not appreciably change.

Restoration of Photosynthetic Activity and Normal Growth to Cells Previously Inhibited by Calcium Depletion

The inhibition process is gradual and, following an initial lag, is approximately linear with time when illumination conditions are kept constant. Illumination intensity is a convenient way to regulate the rate of depletion for experimental purposes. Perhaps most useful is the fact that cells, when partially inhibited, may be transferred to darkness, where they will maintain the same level of inhibition for several hours.

During illumination the inhibition may be reversed simply by adding Ca^{2+} to the growth medium. Routinely $Ca(NO_3)_2$ is added to give a final concentration of 0.35 m*M* in order to restore activity, but 0.1 m*M* gives full restoration and 10-fold less yields partial restoration. Any Ca^{2+} salt may be used, but the nitrate, chloride, and acetate salts are especially useful since concentrated stock solutions can be prepared.

No other divalent or multivalent cation will substitute for Ca^{2+} in restoring full activity to Ca^{2+}-depleted *A. nidulans,* although Sr^{2+} affords approximately 40% restoration. The depletion medium (Table I) contains no Na^+ because in *A. nidulans* Na^+ also supports photosynthesis in apparently replacing Ca^{2+}. It acts at approximately the same concentration as does Ca^{2+}. No other cation will substitute, and in some strains of cyanobacteria (see below) Na^+ will not replace Ca^{2+}.

To demonstrate full restoration of activity to depleted cells, Ca^{2+} should be added before cells remain illuminated in a fully inhibited state for more than a few minutes. Secondary damage during illumination under Ca^{2+} stress gradually precludes full restoration. Recovery from Ca^{2+} depletion requires illumination. During recovery, as during inhibition, cells may be poised at any degree of partial restoration simply by transferring the culture to darkness.

Conditions Required to Demonstrate Calcium Depletion in Other Strains of Cyanobacteria

Other cyanobacteria with morphologies similar to *A. nidulans* can be manipulated like *A. nidulans* to demonstrate Ca^{2+} depletion. Strains which mat and those which remain as long filaments, however, may not give clear results because of difficulty in washing out trapped Ca^{2+} and in providing uniform illumination. Nevertheless, even heavily matting strains like *Oscillitoria* can be Ca^{2+}-depleted under controlled conditions by breaking up the aggregates with a hand-operated tissue grinder equipped with a Teflon pestle. The disaggregation procedure must be performed on each centrifugation pellet during the introduction of cells into the depletion medium. This generally results in shorter filaments which remain in suspension during the depletion process and permit uniform illumination.

It cannot be assumed that other strains of cyanobacteria will respond to Ca^{2+} depletion in the same way as *A. nidulans* TX 20. We have examined strains of *Anabaena, Oscillitoria, Fremyella, Phormidium,* and *Agmenellum,* in addition to *Synechococcus* strain R2 and a high-temperature strain of *Synechococcus*. Every strain examined was inhibited reversibly by depleting Ca^{2+} in Na^+-deficient medium as described above, but in some strains (e.g., *Anabaena* strain CA), Na^+ did not substitute for Ca^{2+}. In *Synechocystis* PCC 6714 Na^+ supplies the ion requirement for PSII while Ca^{2+} does not. The optimum conditions for effective depletion and restoration of photosynthetic activity may vary, even among closely related strains. For example, *A. nidulans* TX 20 (*Synechococcus* 6301) is very similar to *Synechococcus* R2, and the two strains appear morphologically identical. Yet, at a given light intensity in Ca^{2+}-deficient medium photosynthesis is inhibited more than twice as fast in strain R2 as in 6301.

Calcium Depletion as an Experimental Tool for Studying Photosynthesis and Cyanobacterial Metabolism

Several methods may be used to examine Ca^{2+} depletion effects on photosynthesis in intact cyanobacterial cells. Lipophilic artificial electron transport donors, acceptors, inhibitors, and uncouplers can be employed to examine selected portions of electron transport in intact cyanobacterial cells.[11] Combinations of these demonstrate that the site of inhibition by Ca^{2+} depletion is PSII. Variable fluorescence yield measured in intact Ca^{2+}-depleted cells remains at the F_0 level, even in the presence of 3,4-dichlorophenyl-1,1-dimethylurea (DCMU), and all ms delayed fluorescence is lost.[12] These results strongly suggest that the site of inhibition is close to the reaction center of PSII. Kinetic analysis of light saturation of photosynthesis in partially inhibited cells also suggests that inhibition is only at PSII, furthermore at a site near the reaction center.[11] Measurements of oxygen flash yields of photosynthesis in partially inhibited cell cultures lead to the conclusion that Ca^{2+} stress results in all-or-none inactivation of the active PSII centers rather than a gradual slowing of their turnover rate in saturating light.[11]

Photosynthetically active membranes are easily prepared from intact cyanobacterial cells.[2] When prepared from Ca^{2+}-depleted cells, they demonstrate inhibition at the same site as observed in the intact cells.[11] No addition of Ca^{2+} or other cation will restore PSII activity in these membranes. Thus, the site of Ca^{2+} function can be studied in considerable biochemical detail under relatively physiological conditions by depleting whole cells to afford inhibition of PSII, then isolating the membranes or preparing PSII particles for detailed analysis.

Studies of other aspects of cyanobacterial metabolism may be facilitated in whole cells by totally blocking PSII by Ca^{2+} depletion. For example, PSI is not affected, so it and cyclic photophosphorylation may be studied in whole cells with no interference from PSII, and without adding metabolic inhibitors. Also, since dark respiration is not directly affected by Ca^{2+} depletion, it can be probed in the absence of interference from PSII, yet with the quinone pool of PSII apparently unaltered.

[11] D. W. Becker and J. J. Brand, *Plant Physiol.* **79,** 552(1985).
[12] J. J. Brand, P. Mohanty, and D. C. Fork, *FEBS Lett.* **155,** 120 (1983).

[30] Photosystem II–Phycobilisome Complex Preparations

By E. GANTT, J. D. CLEMENT-METRAL, and B. M. CHERESKIN

Introduction

Energy transfer measurements have suggested a direct structural attachment of phycobilisomes and photosystem II. The hydrophilic nature of the phycobiliproteins, and ready dissociation of phycobilisomes under thylakoid membrane isolation conditions, required development of a special isolation medium. A medium consisting of sucrose–phosphate–citrate developed for *Anabaena variabilis*[1] preserved intact the phycobilisomes, energy transfer to photosystem II, and light-driven O_2 evolution. Adaptation of the medium to endocyanelles[2] and red algae[3,4] proved useful for cell-free photosynthetic activity measurements in phycobilisome-containing algae.

The main objective was to solubilize the thylakoid membrane but leave the phycobilisomes together with photosystem II and the water-splitting enzyme complex, removing photosystem I. The initial procedure developed for the unicellular red alga *Porphyridium cruentum*,[5] and since successfully modified for several cyanobacteria,[6,7] is included here.

Methods for Preparation of Photosystem II–Phycobilisome Complexes

Preparation from Porphyridium

Reagents

SPCM: 0.5 *M* sucrose, 0.5 *M* potassium phosphate, 0.3 *M* potassium citrate, 15 m*M* $MgCl_2$, pH 7.0

LDAO (lauryldimethylamine oxide) preparation: LDAO (30%), 10 ml; catalase, 10 μg/ml (peroxide scavanger)

Sucrose gradients: Prepared in 0.3 *M* potassium citrate, 0.5 *M* potassium phosphate, 15 m*M* $MgCl_2$, pH 7.0, with final sucrose in the

[1] T. Katoh and E. Gantt, *Biochim. Biophys. Acta* **546,** 383 (1979).
[2] L. P. Vernon and S. Cardon, *Plant Physiol.* **70,** 442 (1982).
[3] M. F. Dilworth and E. Gantt, *Plant Physiol.* **67,** 608 (1981).
[4] A. C. Stewart and A. W. D. Larkum, *Biochem. J.* **210,** 583 (1983).
[5] J. D. Clement-Metral and E. Gantt, *FEBS Lett.* **156,** 185 (1983).
[6] H. B. Pakrasi and L. A. Sherman, *Plant Physiol.* **74,** 742 (1984).
[7] M. Kura-Hotta, K. Satoh, and S. Katoh, *Arch. Biochem. Biophys.* **249,** 1 (1986).

following three steps: 2.0 *M* sucrose, 6 ml; 1.0 *M* sucrose, 8 ml; 0.5 *M* sucrose, 8 ml

Procedure

Step 1. Cells grown at approximately 40 μE/m^2/sec in a defined medium,[3] at 18°, with continuous agitation and aeration (5% CO_2, 95% air) are harvested in the late exponential phase of growth (~7 days) by centrifugation.

Step 2. The pelleted cells are rinsed rapidly by being suspended in distilled water (~1 g wet weight per 10 ml). The pellet of rinsed cells is then dissolved in SPCM (4 ml/g wet weight). A glass homogenizer is used to facilitate suspension of the cells in SPCM. The following steps are carried out at 4° in subdued room light or in darkness when possible.

Step 3. The cell suspension is passed through an Aminco French pressure cell at 18,000 psi (128 MPa). To the membrane fragments (~0.30 mg Chl/ml) 0.12% (v/v) lauryldimethylamine oxide (LDAO) is added to give a detergent-to-chlorophyll ratio of 3.5–4 : 1 (w/w). The detergent-to-chlorophyll ratio is more critical than the percentage of LDAO. The mixture is incubated for 30 min in the dark with very gentle stirring. The incubation mixture is cleared of cell debris and starch by centrifugation at 27,000 g_{av} for 30 min.

Step 4. Two to three milliliters of supernatant is layered per tube on discontinuous sucrose step gradients, previously prepared, and centrifuged in an angle-head rotor for 11 hr at 130,000 g_{av}.

Step 5. The photosystem II–phycobilisome (PIIP) material is collected from the 1–2 *M* sucrose interface by a syringe with a flat-tipped canula (#19), and diluted with 2 volumes of SPCM. The diluted material is centrifuged at 27,000 g_{av} for 30 min to remove any contaminating membrane fragments. The supernatant is carefully removed with a syringe, taking care not to disturb the pellet. The PIIP in the supernatant can be assayed directly, or it can be concentrated further by centrifugation for 3 hr at 270,000 g_{av}.

Assays. The density of PIIP particles is only slightly lower than that of isolated phycobilisomes. It is important not to exceed the LDAO treatment time or concentration. The PIIP preparation retains a small amount of chlorophyll (Fig. 1A), but no detectable photosystem I components (P700).[5,8] A rapid determination of the fluorescence emission (−196°), with excitation through phycoerythrin (545 nm), reveals a maximum of 688 nm (Fig. 1B). On removal of the phycobilisome components the maxi-

[8] J. D. Clement-Metral, E. Gantt, and T. Redlinger, *Arch. Biochem. Biophys.* **238,** 10 (1985).

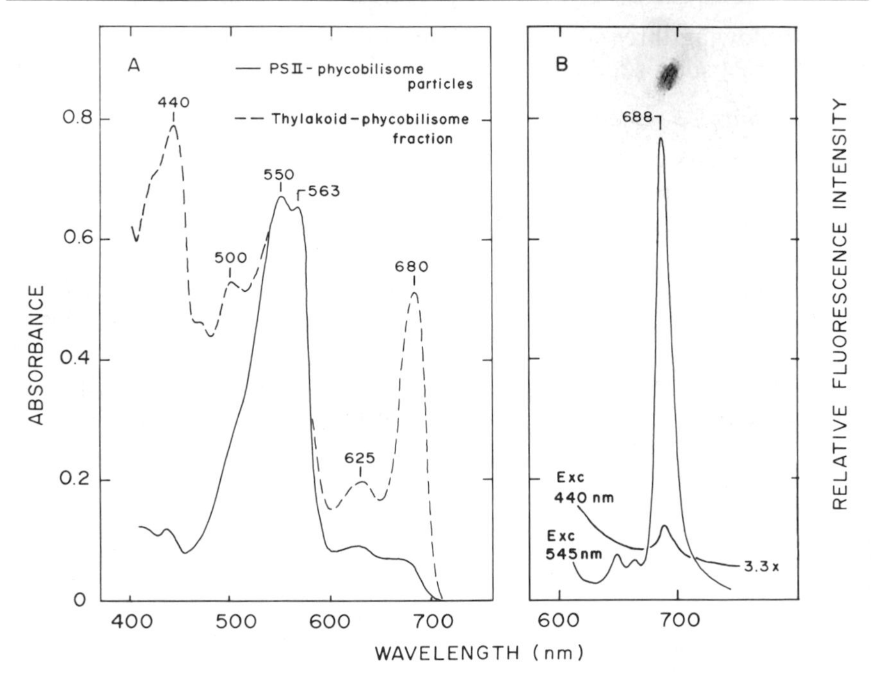

FIG. 1. (A) *Porphyridium cruentum* absorption spectra of PIIP particles and unfractionated thylakoids in SPCM medium at 20° normalized at 550 nm. (B) Fluorescence emission of PIIP particles at −196° with excitation of phycoerythrin (545 nm) and chlorophyll (440 nm).

mum becomes 698 nm,[9] a value typical of photosystem II reaction center cores. When chlorophyll is excited directly (440 nm) the PIIP preparations lack the photosystem I emission band at 715–720 nm.

Examination by electron microscopy is recommended since it is an excellent indicator of contamination especially of thylakoid membranes lacking phycobilisomes. The sample in SPCM ($A_{545\ nm}$, 1.5) is applied to carbon film-coated grids which have been freshly "ionized" by glow discharge treatment. Fixation in 0.25% glutaraldehyde for 2 min is followed by 5 rinses with distilled water and staining in 1% uranyl sulfate for 30 sec. The majority of phycobilisomes retain small putative thylakoid fragments.[8]

Photosystem II assays can be made by measuring O_2 evolution or by the reduction of 2,6-dichlorophenolindophenol (DCPIP).[10,11] Good PIIP preparations have typical O_2 evolution rates of 900–1500 μM O_2/mg Chl/

[9] B. M. Chereskin and E. Gantt, *Arch. Biochem. Biophys.* **250,** 286 (1986).

[10] B. M. Chereskin and E. Gantt, *Plant Cell Physiol.* **27,** 751 (1986).

[11] B. M. Chereskin, J. D. Clement-Metral, and E. Gantt, *Plant Physiol.* **77,** 626 (1985).

hr. Oxygen evolution is measured at 25° with a Clark-type electrode, with light from a slide projector and heat filters consisting of a 10-cm cooled water column and a Schott KG filter. The assay volume of 3.0 ml typically contains 2 m*M* FeCN, 1 m*M* dimethylbenzoquinone (DMBQ), and 0.67–1% ethanol. The electrode is calibrated by the method of Robinson and Cooper.[12] It is important to note that the solubility of O_2 in air-saturated SPCM is lower (55%) than in water.[11]

When activity is measured by dye reduction, characteristic values range from 250–450 μM DCPIP/mg Chl/hr. DCPIP measurements (40 μM DCPIP in a 3-ml sample in SPCM) are made in an Aminco DW2 spectrophotometer with side illumination from a Dolan Jenner Model 170-D illuminator. The absorbance change of 560 minus 520 nm is measured with a 3-nm slit width and at an extinction coefficient of 5.75 m*M*/cm. Complete inhibition of activity occurs with 3,4-dichlorophenyl-1,1-dimethylurea (DCMU) on the acceptor side and with NH_2OH on the donor side of photosystem II.[8,11] Oxygen-evolving activity is often inhibited 50–80% by aging, dilution, low pH, and salt washing. Bovine serum albumin (0.05%) and dithiothreitol (0.5 m*M*) can stimulate activity in all but salt-washed preparations.[10]

Preparations from Synechococcus

Several detergents have been successfully applied for PSII-enriched preparations from two cyanobacteria. Triton X-100 was used with *Anacystis nidulans* R2.[6] For *Synechococcus* sp. digitonin was found to produce PIIP preparations of greater purity than β-octylglucoside. The protocol for PIIP preparation from *Synechococcus* with digitonin, provided below, is according to Kura-Hotta *et al.*[7]

Reagents

SPC incubation medium for PIIP: 0.5 *M* sucrose, 0.5 *M* phosphate, 0.3 *M* citrate, 5 m*M* $MgCl_2$, 20 m*M* $CaCl_2$, 1 m*M* phenylmethylsulfonyl fluoride, 0.10% DNase (w/v), 1.5% bovine serum albumin, pH 6.9

Procedure

Step 1. Cells grown for 2 days at 55° are harvested. Since these cells are difficult to break, spheroplasts are produced to enhance the yield. The incubation medium consists of 0.5 *M* sucrose, 0.1 *M* phosphate (pH 7.2), and 0.15% (w/v) lysozyme. Incubation for 2 hr at 40° with gentle stirring is followed by centrifugation at 10,000 g_{av} for 10 min.

[12] J. Robinson and J. M. Cooper, *Anal. Biochem.* **33,** 390 (1970).

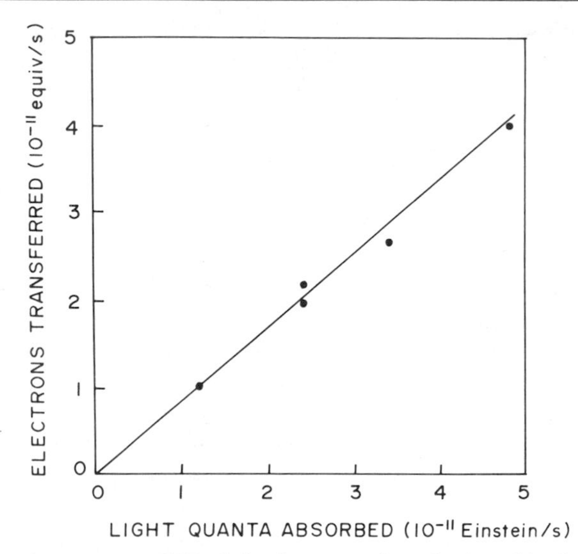

FIG. 2. *Synechococcus* sp. PIIP–digitonin preparations: ferricyanide photoreduction as a function of light quanta absorbed by phycobiliproteins.

Step 2. Spheroplasts (0.4 mg Chl/ml) are suspended in SPC incubation medium, broken in a French pressure cell at approximately 980 psi (7 MPa), and incubated for 30–60 min at 25° following addition of 0.5% digitonin with a final detergent-to-chlorophyll ratio (w/w) of 12.5.

Step 3. On removal of large membrane fragments, by centrifugation at 25,000 g_{av} for 10 min, the supernatant is layered on discontinuous sucrose step gradients in SPC medium. The sucrose gradient steps range from 0.5, 0.75, 1.0, to 1.5 *M* sucrose.

Step 4. The dark-green PIIP material is collected from the 1.0–1.5 *M* sucrose interface after centrifugation for 70 min at 300,000 g_{av}.

Assay. By their absorption spectra the phycobilisome composition characteristics of the PIIP complexes are identical to those in unfractionated thylakoids and those in isolated phycobilisomes. The energetic coupling between PSII and phycobilisomes, according to their emission at −196°, is comparable to whole cells.[7] A high quantum yield (0.85), according to measurements of ferricyanide photoreduction and quanta absorbed by phycobilisomes (Fig. 2), shows a complete coupling to PSII reaction centers.

Note. Sulfobetain 12 (0.35%) has been successfully used in isolation of PSII particles[13] and should be useful for preparing PIIP.

[13] E. Mörschel and G. H. Schatz, *Planta* **172,** 145 (1987).

[31] Phycobiliproteins

By ALEXANDER N. GLAZER

Introduction

In intact cyanobacterial cells and red algal chloroplasts, the phycobiliproteins are components of a complex assemblage, the phycobilisome.[1] The observation that upon cell breakage the products of phycobilisome dissociation are released in water-soluble form gave rise to the unfortunate impression that the phycobiliproteins represented a fairly simple family of well-behaved macromolecules. Appraisal of the true complexity of the situation required an understanding of the molecular structure of phycobilisomes. In the light of current knowledge, the definition most appropriate for purified phycobiliproteins is that they are proteins made up of $\alpha\beta$ monomers of two dissimilar polypeptide chains, α and β, each with covalently attached bilin prosthetic groups. Purified phycobiliproteins may be the monomers, $(\alpha\beta)$, or dimers, trimers, or hexamers of this building block, or equilibrium mixtures of two or more of these aggregates. For a given pure phycobiliprotein, the aggregation state depends on the organism from which the protein is isolated and the details of the purification procedure, as well as on the pH, ionic strength, composition of the solvent, protein concentration, and temperature. "Pure" phycobiliproteins are not present at significant concentrations in normal intact cells. These entities are the products of *in vitro* dissociation of phycobilisomes and of selective purification of some of the dissociation products.

During purification procedures, certain phycobilisome polypeptide components are lost through retention on chromatographic supports, or, in some cases, destroyed by proteolysis.[2] Some phycobiliproteins, such as allophycocyanin B,[3] are products of subunit scrambling during purification procedures,[4] while others, such as the $\beta^{18.3}$ polypeptide of the phycobilisome core,[5,6] which is present in only two copies per phycobilisome, may copurify undetected with one of the major phycobiliproteins such as allophycocyanin. The above points are illustrated by the following exam-

[1] A. N. Glazer, this volume [32].
[2] R. Rümbeli, T. Schirmer, W. Bode, W. Sidler, and H. Zuber, *J. Mol. Biol.* **186,** 197 (1985).
[3] A. N. Glazer and D. A. Bryant, *Arch. Microbiol.* **104,** 15 (1975).
[4] D. J. Lundell and A. N. Glazer, *J. Biol. Chem.* **256,** 12600 (1981).
[5] D. J. Lundell and A. N. Glazer, *J. Biol. Chem.* **258,** 894 (1983).
[6] R. Rümbeli, M. Wirth, F. Suter, and H. Zuber, *Biol. Chem. Hoppe-Seyler* **368,** 1 (1987).

ples. *Synechococcus* PCC 6301 C-phycocyanin, when purified to homogeneity, exists in a monomer–dimer equilibrium.[7] In intact phycobilisomes, this protein is present in the form of hexameric complexes with three different linker polypeptides of 27,000, 33,000, and 30,000, Da, respectively,[8] which dissociate during phycocyanin purification. Allophycocyanin can be purified as a trimer,[9–11] $(\alpha\beta)_3$, in equilibrium with the monomer at low protein concentration.[12,13] In the phycobilisome core this protein is a component of four different complexes[14,15] similar to those illustrated elsewhere in this volume (Fig. 4 in chapter [32]), which undergo dissociation and subunit exchange in the course of the preparation of allophycocyanin.[4]

In a few instances, as a consequence of stronger intersubunit interactions, higher order subassemblies of the phycobilisome survive even rigorous purification procedures and are isolated as homogeneous aggregates. This is the case for B- and R-phycoerythrins, which are isolated as $(\alpha\beta)_6\gamma$ species.[16,17] The γ subunit in these proteins is presumably the linker polypeptide (or a portion of one) with which the phycoerythrin interacts within the phycobilisome. Sodium dodecyl sulfate (SDS)–polyacrylamide gels of B- and R-phycoerythrin-containing phycobilisomes show several phycoerythrin-associated linker polypeptides.[18,19] It should come as no surprise, therefore, that the homogeneity with respect to aggregation state seen for B- and R-phycoerythrins is not paralleled by homogeneity of molecular species. R-Phycoerythrins contain at least two different γ subunits[20] and B-phycoerythrins probably three such subunits.[21] Since the linker polypeptides and the phycobiliproteins interact with each other with high specificity, it is to be anticipated that in many

[7] G. J. Neufeld and A. F. Riggs, *Biochim. Biophys. Acta* **181,** 234 (1969).
[8] D. J. Lundell, R. C. Williams, and A. N. Glazer, *J. Biol. Chem.* **256,** 3580 (1981).
[9] A. N. Glazer and G. Cohen-Bazire, *Proc. Natl. Acad. Sci. U.S.A.* **68,** 1398 (1971).
[10] R. MacColl, M. R. Edwards, M. H. Mulks, and D. S. Berns, *Biochem. J.* **141,** 419 (1974).
[11] R. MacColl, K. Csatorday, D. S. Berns, and E. Traeger, *Arch. Biochem. Biophys.* **208,** 42 (1981).
[12] J. R. Gysi and H. Zuber, *Biochem. J.* **181,** 577 (1979).
[13] L. J. Ong and A. N. Glazer, *Physiol. Veg.* **23,** 777 (1985).
[14] D. J. Lundell and A. N. Glazer, *J. Biol. Chem.* **258,** 902 (1983).
[15] J. C. Gingrich, D. J. Lundell, and A. N. Glazer, *J. Cell. Biochem.* **22,** 1(1983).
[16] A. N. Glazer and C. S. Hixson, *J. Biol. Chem.* **252,** 32 (1977).
[17] C. Abad-Zapatero, J. L. Fox, and M. L. Hackert, *Biochem. Biophys. Res. Commun.* **78,** 266 (1977).
[18] T. Redlinger and E. Gantt, *Plant Physiol.* **68,** 1375 (1981).
[19] M.-H. Yu, A. N. Glazer, K. G. Spencer, and J. A. West, *Plant Physiol.* **68,** 482 (1981).
[20] A. V. Klotz and A. N. Glazer, *J. Biol. Chem.* **260,** 4856 (1985).
[21] A. V. Klotz and A. N. Glazer, unpublished observations.

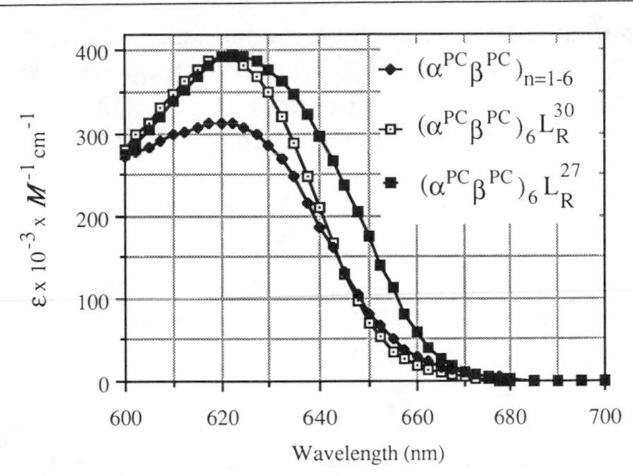

FIG. 1. Millimolar absorption spectra of *Synechococcus* PCC 6301 phycocyanin and phycocyanin–linker polypeptide complexes in 0.6 *M* sodium potassium phosphate, pH 8.0, at a phycocyanin concentration of 0.1 mg/ml.[8] Abbreviations: PC, phycocyanin; L_R^{27} and L_R^{30}, linker polypeptides of 27,000 and 30,000 Da. The extinction coefficients were calculated per $\alpha\beta$ monomer of phycocyanin.

instances complete removal of the linker polypeptides from the phycobiliproteins will require extensive purification. The linker polypeptides have a profound influence on the aggregation state, assembly properties, and spectroscopic characteristics of phycobiliproteins (e.g., Fig. 1).[5,8,15,22,23] Consequently, the presence of even small amounts of these polypeptides can alter the properties of phycobiliprotein preparations.

A polypeptide pattern obtained by SDS–polyacrylamide gel electrophoresis is an absolutely essential element in the characterization of a purified phycobiliprotein preparation. A sufficient amount of protein should be applied to the gel to ensure that components present in small amounts will be detected. A higher resolution gel electrophoretic analysis involving isoelectric focusing under denaturing conditions in the first dimension and SDS–polyacrylamide gel electrophoresis in the second dimension[24] is needed in some instances to resolve phycobiliprotein α and β subunits which may have virtually the same apparent molecular weights. Many cyanobacteria, particularly chromatic adapters,[25] produce two different phycocyanins with very similar subunit molecular weights and isoelectric points that can only be distinguished under special conditions of

[22] M.-H. Yu, A. N. Glazer, and R. C. Williams, *J. Biol. Chem.* **256,** 13130 (1981).
[23] P. Füglistaller, F. Suter, and H. Zuber, *Biol. Chem. Hoppe-Seyler* **367,** 601 (1986).
[24] P. H. O'Farrell, *J. Biol. Chem.* **250,** 4007 (1975).
[25] N. Tandeau de Marsac, *Bull. Inst. Pasteur (Paris)* **81,** 201 (1983).

gel electrophoresis.[26-28] Isoelectric focusing under denaturing conditions followed by SDS–polyacrylamide gel electrophoresis is required to resolve all of the bilin-bearing polypeptides of phycobilisomes.[15,29]

Given the above complexities in the characterization of phycobiliprotein preparations it is regretable that in numerous publications, many of recent vintage, the sole criteria used to assess the purity of the phycobiliprotein are the visible absorption spectrum and, depending on the protein, the ratio of $A_{565\text{ nm}}$, $A_{620\text{ nm}}$, or $A_{650\text{ nm}}$ to $A_{280\text{ nm}}$. Whereas the assertion of purity for a given preparation may in fact be valid, such criteria do not provide adequate objective proof.

Recent publications have appeared reviewing the amino acid sequences of the phycobiliproteins,[30,31] the structure and mode of attachment of the bilin prosthetic groups,[32-36] the aggregation behavior,[37] the spectroscopic properties of the phycobiliproteins,[37-41] as well as the crystal structure of C-phycocyanins at high resolution.[42,43] The phycobiliproteins have been widely applied as fluorescent tags in a variety of analytical and diagnostic procedures, particularly in multiparameter fluorescence-activated cell analyses.[44,45]

[26] D. A. Bryant, *Eur. J. Biochem.* **119,** 425 (1981).
[27] D. A. Bryant and G. Cohen-Bazire, *Eur. J. Biochem.* **119,** 415 (1981).
[28] G. Guglielmi and G. Cohen-Bazire, *Protistologica* **20,** 393 (1984).
[29] G. Yamanaka, D. J. Lundell, and A. N. Glazer, *J. Biol. Chem.* **257,** 4077 (1982).
[30] H. Zuber, *Photochem. Photobiol.* **42,** 821 (1985).
[31] H. Zuber, *in* "Encyclopedia of Plant Physiology: 3. Photosynthetic Membranes and Light Harvesting Systems" (L. A. Staehelin and C. J. Arntzen, eds.), Vol. 19, p. 238. Springer-Verlag, Berlin, 1986.
[32] J. C. Lagarias, A. N. Glazer, and H. Rapoport, *J. Am. Chem. Soc.* **101,** 5030 (1979).
[33] R. W. Schoenleber, D. J. Lundell, A. N. Glazer, and H. Rapoport, *J. Biol. Chem.* **259,** 5481 (1984).
[34] R. W. Schoenleber, D. J. Lundell, A. N. Glazer, and H. Rapoport, *J. Biol. Chem.* **259,** 5485 (1984).
[35] O. Nagy, J. E. Bishop, A. V. Klotz, A. N. Glazer, and H. Rapoport, *J. Biol. Chem.* **260,** 4864 (1985).
[36] J. E. Bishop, H. Rapoport, A. V. Klotz, C. F. Chan, A. N. Glazer, P. Füglistaller, and H. Zuber, *J. Am. Chem. Soc.* **109,** 875 (1987).
[37] A. N. Glazer, *in* "The Biochemistry of Plants" (M. D. Hatch and N. K. Boardman, eds.), p. 51. Academic Press, London, 1981.
[38] H. Scheer, *in* "Light Reaction Path of Photosynthesis" (F. K. Fong, ed.), p. 7. Springer-Verlag, Berlin, 1982.
[39] A. N. Glazer, *Biochim. Biophys. Acta* **768,** 29 (1984).
[40] A. N. Glazer, *Annu. Rev. Biophys. Biophys. Chem.* **14,** 47 (1985).
[41] B. A. Zilinskas and L. S. Greenwald, *Photosynth. Res.* **10,** 7 (1986).
[42] T. Schirmer, R. Huber, M. Schneider, W. Bode, M. Miller, and M. L. Hackert, *J. Mol. Biol.* **188,** 651 (1986).
[43] T. Schirmer, W. Bode, and R. Huber, *J. Mol. Biol.* **196,** 677 (1987).
[44] A. N. Glazer and L. Stryer, *Trends Biochem. Sci.* **9,** 423 (1983).
[45] M. N. Kronick, *J. Immunol. Methods* **92,** 1(1986).

Examples of Purification Procedures for the Major Cyanobacterial Phycobiliproteins

A purification procedure that works well for a phycobiliprotein from one organism may not be the method of choice for the corresponding protein from another organism. Consequently, it is strongly advised that the following procedures be applied to the phycobiliproteins of the particular organisms given as the sources. It should also be remembered that a purification procedure may in a sense be regarded as an archaeological relic; it embodies, it is hoped, the state of the art at the time it was designed. There has been extensive progress in protein purification techniques in the past decade both in the invention of new separation supports and with respect to the resolution attainable by new methods. It is intended therefore that the descriptions given below be used as guides to the parameters exploited in the preparation of phycobiliproteins rather than as methods to be followed slavishly.

Purification of Phycoerythrocyanin, C-Phycocyanin, and Allophycocyanin from Anabaena PCC 6411 and Anabaena variabilis (PCC 7118)[46]

Freshly harvested cells as well as cells which had been stored frozen at −20° for several months have been used as starting material with similar results. All buffers mentioned below contain 1 m*M* 2-mercaptoethanol and 1 m*M* sodium azide, unless otherwise specified.

Frozen cells (20 g wet weight) are suspended in 90 ml of 50 m*M* ammonium acetate buffer, pH 6.0, well dispersed in an all-glass homogenizer, and broken by passage through a French pressure cell at 14,000 psi at 4°. The suspension is centrifuged at 23,500 *g* for 30 min to remove whole cells and debris. The supernatant is decanted, and the pellet is washed once with 30 ml of buffer. The pooled supernatants are brought to 65% of saturation by the addition of solid $(NH_4)_2SO_4$ and allowed to stand overnight at 4°. The precipitate is collected by centrifugation at 27,000 *g* for 15 min, resuspended in 60 ml of 5 m*M* potassium phosphate, pH 7.0, and dialyzed to equilibrium against the same buffer at 4°. The dialyzed protein solution is centrifuged at 12,000 *g* for 10 min to remove denatured protein and aggregated membrane material and is then applied to a column (3.1 × 33 cm) of microgranular DEAE–cellulose DE-52 (Whatman), preequilibrated with the 5 m*M* potassium phosphate, pH 7.0, buffer. Chromatography is performed at room temperature (23°). The sample is washed in with 260 ml of the 5 m*M* starting buffer, and the column is developed with a linear gradient of potassium phosphate, pH

[46] D. A. Bryant, A. N. Glazer, and F. A. Eiserling, *Arch. Microbiol.* **110,** 61 (1976).

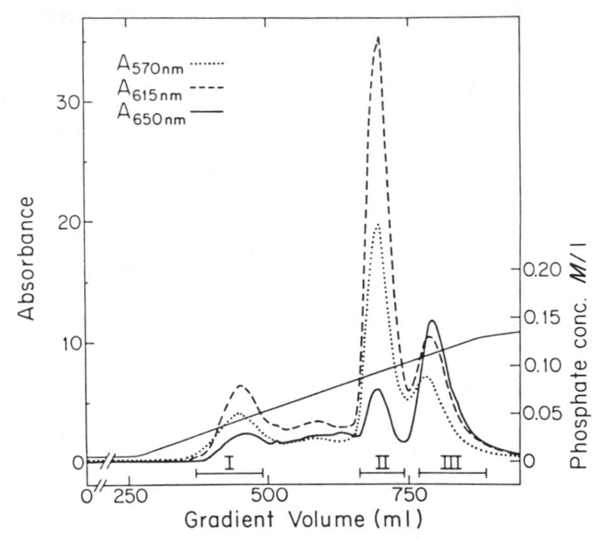

FIG. 2. Ion-exchange chromatography of *Anabaena* PCC 6411 phycobiliproteins on DEAE–cellulose at pH 7.0.[46] Fractions were pooled as indicated by the numbered bars. For experimental details, see text.

7.0 (5–200 m*M*) of total volume 1000 ml. At the conclusion of the gradient the column is washed with about 120 ml 200 m*M* potassium phosphate, pH 7.0. A typical elution profile is shown in Fig. 2.

When the fractions from this column were allowed to stand overnight at 4°, it was found that the phycocyanin had crystallized in the peak tubes where the phycocyanin concentration was about 2.5 mg/ml or greater. The phycocyanin fraction (Fraction II) could be purified further by rechromatography on DEAE–cellulose. However, analysis of washed crystals of C-phycocyanin isolated from the peak fractions showed the protein to be very pure. Consequently, an alternate means of purification is to concentrate the residual C-phycocyanin to greater than 2.5 mg/ml and to crystallize the protein by dialysis against 120 m*M* potassium phosphate, pH 7.0, at 4°.

The pooled Fraction III (Fig. 2) is brought to 65% of saturation by the addition of solid $(NH_4)_2SO_4$. The resulting precipitate is collected by centrifugation, resuspended in 25 ml of 20 m*M* sodium acetate buffer, pH 5.5, and dialyzed to equilibrium against the same buffer. The dialyzed protein is centrifuged at 14,500 *g* for 10 min, and the supernatant is applied to a column (1.9 × 30 cm) of DEAE–cellulose preequilibrated with the pH 5.5 buffer. The sample is washed in with about 25 ml of starting buffer and the column developed with a linear gradient of sodium acetate (20–250 m*M*; pH 5.5) of total volume 500 ml. A typical elution profile is shown in Fig. 3.

After 2 days at 4°, allophycocyanin crystallizes from those fractions in the region of the allophycocyanin peak (Fig. 3, Fraction II), which contains this protein at concentrations above 1.3 mg/ml. These crystals are collected by centrifugation at 14,500 *g* for 10 min and are washed and resuspended in 200 m*M* sodium acetate buffer at pH 5.5. Allophycocyanin that remains soluble is concentrated by ultrafiltration and/or precipitation with $(NH_4)_2SO_4$ and crystallized on dialysis of concentrated solutions at 4° against 200 m*M* sodium acetate, pH 5.5. Allophycocyanin crystals are collected by centrifugation, washed twice with 200 m*M* sodium acetate, pH 5.5, and stored as a suspension in this buffer at 4°.

The fraction I material from the pH 5.5 DEAE-cellulose chromatography (Fig. 3) is brought to 65% of saturation by the addition of solid $(NH_4)_2SO_4$. The resulting precipitate is collected by centrifugation, suspended in about 3.5 ml of 1 m*M* potassium phosphate–100 m*M* NaCl, pH 7.0, and dialyzed to equilibrium against the same buffer. The dialyzed protein solution is applied to a common (1.8 × 9 cm) of hydroxylapatite pre-equilibrated with 1 m*M* potassium phosphate–100 m*M* NaCl, pH 7.0. The sample is washed in with the starting buffer, and phycoerythrocyanin is eluted by washing 8 m*M* potassium phosphate–100 m*M* NaCl, pH 7.0. Under these conditions most, but not all of the phycocyanin remains on the column. (The molarity of phosphate that permits the best separation of phycoerythrocyanin from phycocyanin varies somewhat from one batch of hydroxylapatite to the next.) Fractions from this column with $A_{570\ nm}$-to-$A_{615\ nm}$ ratios above 2.14 are pooled and dialyzed to equilibrium against 1 m*M* potassium phosphate–100 m*M* NaCl, pH 7.0. Rechromatography of the phycoerythrocyanin on hydroxylapatite under identical conditions results in the collection of some pure phycoerythrocyanin fractions ($A_{570\ nm} : A_{615\ nm} = 2.60$).

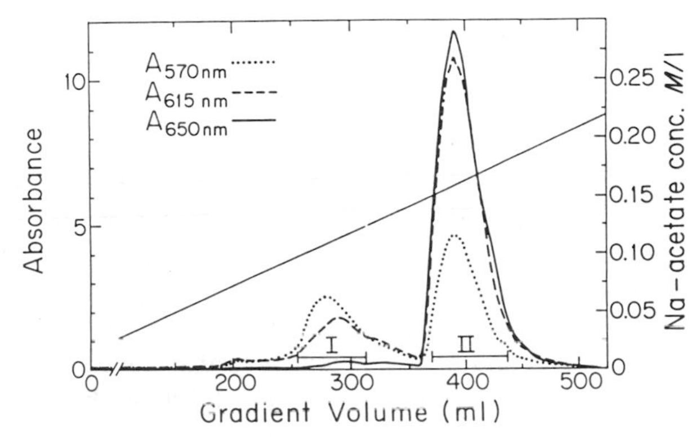

FIG. 3. Rechromatography of Fraction III (see Fig. 2) on DEAE-cellulose at pH 5.5.[46] Fractions were pooled as indicated by the numbered bars. For experimental details, see text.

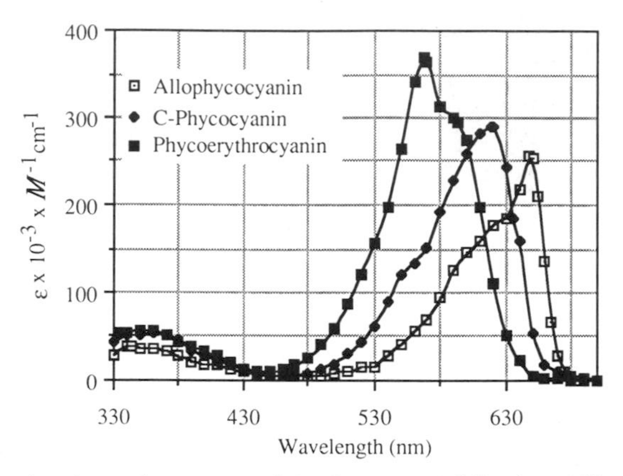

FIG. 4. Millimolar absorption spectra of *Anabaena variabilis* phycobiliproteins in 50 m*M* sodium phosphate, pH 7.0. Molar extinction coefficients were calculated per $\alpha\beta$ monomer, using a molecular weight of 36,400 for phycoerythrocyanin, 35,900 for phycocyanin, and 29,800 for allophycocyanin.

The C-phycocyanin, allophycocyanin, and phycoerythrocyanin prepared by the above procedure were pure as judged by polyacrylamide gel electrophoresis in the presence and absence of SDS and by isoelectric focusing. At protein concentrations of 0.1–1 mg/ml in 50 m*M* sodium phosphate, pH 7.0, molecular weight determinations indicated that the phycocyanin was a dimer, whereas the allophycocyanin and phycoerythrocyanin were trimers. The absorption spectra of these three proteins are shown in Fig. 4.

Purification of C-Phycoerythrin, C-Phycocyanin, and Allophycocyanin from Fremyella diplosiphon (Calothrix PCC 7601)[47]

All operations are carried out at about 4°. Cells (50 g wet weight) are suspended in 50 ml of 1 *M* sodium acetate, pH 5.0. Portions (20 ml) of the slurry are transferred to 60-ml plastic ultracentrifuge tubes and sonicated for a total of 6 min. The sonicates are pooled, stirred for 30 min, and centrifuged at 81,000 *g* for 1 hr. The procedure is repeated with the pellets, and the supernatants from both centrifugations are pooled.

The supernatant is diluted with the acetate buffer to a final $A_{280\,nm}$ of 4.5. Solid $(NH_4)_2SO_4$ is added to 35% of saturation, and the solution is allowed to stand for 90 min and then centrifuged at 16,000 *g* for 15 min. (These centrifugation conditions are employed in all subsequent steps.) The supernatant is brought to 70% of saturation with $(NH_4)_2SO_4$, left

[47] A. Bennett and L. Bogorad, *Biochemistry* **10,** 3625 (1971).

standing for 1 hr, and centrifuged. Both the phycoerythrin-rich pellet of the 0–35% $(NH_4)_2SO_4$ fraction and the phycocyanin- and allophycocyanin-rich pellet of the 35–70% fraction are resuspended in 6 ml of 100 m*M* sodium acetate, pH 5.0, and dialyzed overnight against 2 liters of the same buffer.

The dialyzed 0–35 and 35–70% $(NH_4)_2SO_4$ fractions are each passed through a column (2.5 × 40 cm) of Sephadex G-100 preequilibrated with the 100 m*M* acetate buffer. (In this buffer, the phycobiliproteins form aggregates which elute near the void volume.) After each run, the peak tubes from the early eluting biliprotein peaks are pooled. The phycoerythrin-containing solution is brought to 30% of saturation and the phycocyanin–allophycocyanin solution brought to 70% of saturation with $(NH_4)_2SO_4$, and both are allowed to stand for 1 hr prior to centrifugation. The pellets are resuspended in 5 m*M* potassium phosphate, pH 7.0, and then dialyzed overnight against 4 liters of the phosphate buffer.

The dialyzed phycoerythrin-containing solution is applied to a column (2.2 × 48 cm) of DEAE–cellulose DE-52 (Whatman), preequilibrated with the 5 m*M* phosphate buffer. After elution with 1 volume of starting buffer, the column is developed with a 550-ml linear 5–200 m*M* potassium phosphate, pH 7.0, gradient. The peak tubes from the major phycoerythrin band, eluting at about 50 m*M* potassium phosphate, are pooled, brought to 30% of saturation with $(NH_4)_2SO_4$, permitted to stand for 30 min, and centrifuged. The pellet (purified phycoerythrin) is resuspended in a small volume of 100 m*M* sodium phosphate–0.01% sodium azide, pH 7.0, and dialyzed against the same buffer. (The visible absorption spectrum of a typical cyanobacterial C-phycoerythrin is shown in Fig. 5.)

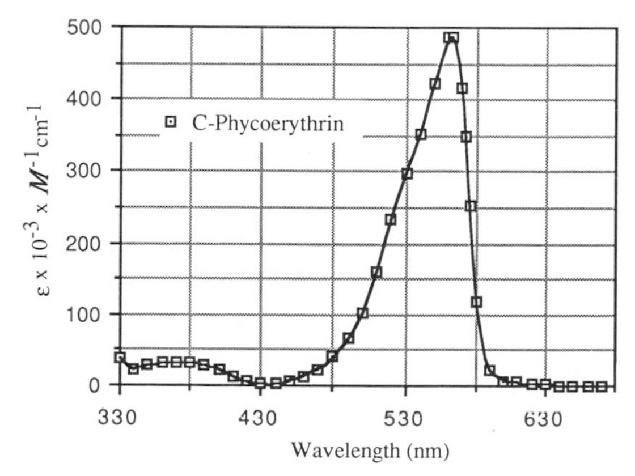

FIG. 5. Millimolar absorption spectrum (calculated per $\alpha\beta$ monomer of 38,400 Da) of *Synechocystis* PCC 6701 C-phycoerythrin in 50 m*M* ammonium acetate at pH 6.8.

The concentrated phycocyanin–allophycocyanin solution is applied to a brushite[48] column (2.5 × 25 cm), preequilibrated with 5 m*M* sodium phosphate–0.1% sodium azide, pH 7.0. One column volume of starting buffer is passed through the column, and then the column is developed with a 400-ml linear gradient of 5–100 m*M* potassium phosphate, pH 7.0. Phycocyanin is eluted from the column and allophycocyanin is retained at the top of the brushite bed. Peak tubes from the phycocyanin fractions are pooled, brought to 65% of saturation with $(NH_4)_2SO_4$, and allowed to stand for 1 hr. The portion of the brushite bed containing allophycocyanin is removed with a spatula and extracted with 500 m*M* potassium phosphate, pH 7.0. The resulting allophycocyanin solution is then brought to 75% of saturation with $(NH_4)_2SO_4$ and left to stand overnight before centrifugation. The pellets of purified phycocyanin and allophycocyanin are resuspended in small volumes of 100 m*M* potassium phosphate–0.1% sodium azide, pH 7.0, and dialyzed against the same buffer.

The purified biliproteins are stored in the dark at 4°. The purity of the phycobiliproteins is established by SDS–polyacrylamide gel electrophoresis, and the absence of cross-contamination is demonstrated spectroscopically.

Purification of C-Phycocyanin from Phormidium luridum[49]

All operations are performed at 4°. Frozen cells (300 g) are thawed with an equal weight of 50 m*M* Tris–phosphate, pH 8. The resulting suspension is homogenized in small portions (Teflon pestle) for 30 sec. The homogenate is centrifuged at 15,000 *g* for 20 min. The pellet is extracted with 300 ml of the pH 8 buffer, and the suspension is centrifuged again. The extraction is repeated and the supernatant fractions combined. The supernatant is applied to a column (5 × 10 cm) of DEAE–cellulose equilibrated with the pH 8 buffer. The column is washed with the pH 8 buffer, and the phycobiliproteins are then eluted with 100 m*M* sodium phosphate–500 m*M* NaCl, pH 7.0 (phycocyanin yield ~90%). The chromoprotein eluate is dialyzed against 10 m*M* sodium phosphate, pH 7.0, and then applied to a hydroxylapatite column (5 × 60 cm) preequilibrated with the pH 7.0 buffer. Elution is performed with 20 m*M* sodium phosphate, pH 7, and fractions are analyzed for phycocyanin (λ_{max} 617 nm) and allophycocyanin (λ_{max} 650 nm) in this and subsequent steps. C-phycocyanin (typical yield >90%) elutes ahead of allophycocyanin. The C-phycocyanin fraction is precipitated by the addition of $(NH_4)_2SO_4$ to 50%

[48] H. W. Siegelman, G. A. Wieczorek, and B. C. Turner, *Anal. Biochem.* **13,** 402 (1965).

[49] Y. Kobayashi, H. W. Siegelman, and C. H. W. Hirs, *Arch. Biochem. Biophys.* **152,** 187 (1972).

of saturation. The precipitate is dialyzed against 10 m*M* sodium phosphate, pH 7, and subjected to gel filtration on a Sephadex G-150 column (9.5 × 135 cm) in the same solvent. The C-phycocyanin fraction is adsorbed onto DEAE–cellulose and eluted as described above. C-Phycocyanin (yield >90%; 100–150 mg) is precipitated in 70%-saturated $(NH_4)_2SO_4$. Typically, only the bands corresponding to the α and β subunits are detected on analysis of the purified protein by SDS–polyacrylamide gel electrophoresis.

Purification of C-Phycocyanin from Agmenellum quadruplicatum PR-6 (Synechococcus PCC 7002)[50]

The following preparation procedure was used to prepare a C-phycocyanin later subjected to detailed study by X-ray crystallography.[42,43] Cultures are harvested with a Sharples supercentrifuge and stored at −4°. The cells are suspended in an equal volume of 200 m*M* Tris–sulfate, at pH 8 and 25°, and egg-white lysozyme is added to 1 mg/ml and disodium EDTA to 10 m*M*. This suspension is incubated at 25–35° for 1 hr. The cells are ruptured by passage through a French pressure cell, and the broken cell suspension is centrifuged at 34,000 *g* for 30 min. The supernatant is batch adsorbed to DEAE–cellulose preequilibrated with 20 m*M* Tris–sulfate, pH 8.0. The DEAE–cellulose is filtered under reduced pressure and washed repeatedly with 50 m*M* Tris–sulfate until the eluate becomes clear. The protein is eluted with several washes of 200 m*M* Tris–sulfate, pH 8. The pooled 200 m*M* Tris buffer eluates are brought to 20% of saturation with $(NH_4)_2SO_4$ allowed to stand for at least 1 hr, and centrifuged. The supernatant is brought to 55% $(NH_4)_2SO_4$ saturation, allowed to stand overnight, and then centrifuged. The resulting precipitate of C-phycocyanin is stored at 4°.

Bilin Prosthetic Groups

Four isomeric tetrapyrroles function as the visible light-harvesting chromophores of the phycobiliproteins. Their structures and modes of linkage to the polypeptide chains are shown in Fig. 6. The spectroscopic properties of these tetrapyrroles are strongly influenced by their conformation and environment within the native phycobiliproteins. Phycocyanobilin gives rise to absorption maxima above 600 nm, e.g., at approximately 620 nm in C-phycocyanin and 650 nm in trimeric allophycocyanin. Biliproteins containing phycobiliviolin (also named cryptoviolin) have ab-

[50] E. E. Gardner, S. E. Stevens, Jr., and L. J. Fox, *Biochim. Biophys. Acta* **624,** 187 (1980).

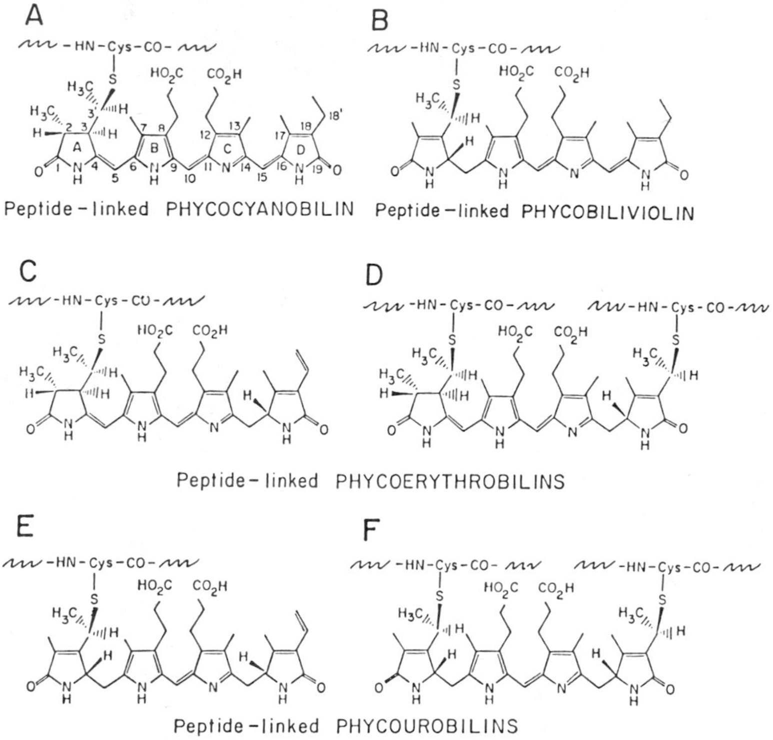

A

Peptide-linked PHYCOCYANOBILIN

B

Peptide-linked PHYCOBILIVIOLIN

C

D

Peptide-linked PHYCOERYTHROBILINS

E

F

Peptide-linked PHYCOUROBILINS

FIG. 6. Structures of peptide-linked bilins in phycobiliproteins.[32-36]

TABLE I

MILLIMOLAR EXTINCTION COEFFICIENTS OF POLYPEPTIDE-BOUND BILINS[a]

Bilin	ε (mM^{-1} cm^{-1}) at wavelength (nm)				Ref.
	495	550	590	660	
Phycourobilin	38.6	0	0	0	20
Phycoerythrobilin	18.3	53.7	8.5	0	20
Phycobiliviolin	6.8	28.4	38.6	0	36
Phycocyanobilin	1.45	6.0	16.2	35.4	51

[a] Extinction coefficients for peptide-linked phycourobilin, phycoerythrobilin, and phycobiliviolin were determined in 10 m*M* aqueous trifluoroacetic acid, those for peptide-linked phycocyanobilin were measured in 8 *M* aqueous urea, pH 1.9. Bilin absorption spectra in these two solvents are both qualitatively and quantitatively very similar.

[51] A. N. Glazer and C. S. Hixson, *J. Biol. Chem.* **250,** 5487 (1975).

TABLE II
PROPERTIES OF CERTAIN MAJOR CYANOBACTERIAL PHYCOBILIPROTEINS[a]

Protein	Aggregation state[b]	Bilin content per subunit[c]	MW ($\times 10^{-3}$)	λ_{max} (nm)[d]	λ_{em} (nm)[e]
Allophycocyanin	$(\alpha\beta)_3$	α1PCB; β1PCB	100	650	660
C-Phycocyanin	$(\alpha\beta)_n$ (n = 1–6)	α1PCB; β2PCB	36.5–220	615–620	625–645
R-Phycocyanin II[52]	$(\alpha\beta)_2$	α1PEB; β1PCB, 1PEB	72	533, 554, 615	646
Phycoerythrocyanin	$(\alpha\beta)_3$	α1PXB; β2PCB	120	568, 590(s)	625
C-Phycoerythrin[f]	$(\alpha\beta)_n$(n = 1–6)	α2PEB; β3PEB	40–240	565	575–581

[a] For original references, see Ref. 40.
[b] The molecular weights and spectroscopic properties are those of the aggregates specified in this column.
[c] Abbreviations: PCB, phycocyanobilin; PEB, phycoerythrobilin; PXB, phycobiliviolin (cryptoviolin).
[d] Visible absorption maxima; (s) denotes shoulder.
[e] Fluorescence emission maxima.
[f] Numerous strains of marine cyanobacteria *Synechococcus* spp. contain phycoerythrins with both phycoerythrobilin and phycourobilin prosthetic groups.[53,54]

sorption maxima at 568 nm, whereas those containing phycoerythrobilin have absorption maxima between 535 and 567 nm. Phycourobilin-containing biliproteins show a sharp absorption peak (or shoulder) at about 495 nm. The spectroscopic properties of the bilin prosthetic groups as well as the bilin content and spectroscopic properties of the major cyanobacterial phycobiliproteins are summarized in Tables I and II.

Acknowledgments

Work in the author's laboratory was supported by National Science Foundation Grant DMB 8518066 and National Institutes of Health Grant GM 28994.

[52] L. J. Ong and A. N. Glazer, *J. Biol. Chem.* **262,** 6323 (1987).
[53] J. B. Waterbury, S. W. Watson, F. W. Valois, and D. G. Franks, *Can. J. Fish. Aquat. Sci. Bull.* **214,** 71 (1986).
[54] L. J. Ong, A. N. Glazer, and J. B. Waterbury, *Science* **224,** 80 (1984).

[32] Phycobilisomes

By ALEXANDER N. GLAZER

Introduction

Phycobiliproteins are quantitatively the major proteins in cyanobacterial cells. They may account for up to 24% of the dry weight of the cells and well over half of the total soluble protein.[1-3] Their polypeptide chains carry covalently attached tetrapyrrole prosthetic groups (bilins) which endow the native proteins with intense colors and brilliant fluorescence. The properties of the individual phycobiliproteins are the subject of another chapter in this volume.[4]

In intact cyanobacterial cells (and red algal chloroplasts) the phycobiliproteins are assembled into particles named phycobilisomes[5] which are attached in regular arrays to the external surface of the thylakoid membranes. Phycobilisome morphology varies with the organism of origin. In all cases, however, they consist of rods made of stacked disks that radiate from a central core. These particles have molecular weights of 7×10^6 to 15×10^6, contain between 300 and 800 bilin chromophores, and absorb light over much of the visible spectrum. Energy absorbed by any of these chromophores is efficiently transferred to terminal energy acceptors in the particle via energetically favorable radiationless pathways. It is believed that the excitation energy from the terminal acceptors is funneled into the reaction centers of photosystem II.[6] The preparation and characterization of phycobilisome–photosystem II complexes is described elsewhere in this volume.[7] Numerous recent reviews should be consulted for detailed information on phycobilisome morphology,[8,9] polypeptide composition and assembly,[10-15] and energy-transfer dynamics.[13,16-19]

[1] J. Myers and W. A. Kratz, *J. Gen. Physiol.* **39,** 11 (1955).
[2] A. Bennett and L. Bogorad, *J. Cell Biol.* **58,** 419 (1973).
[3] N. Tandeau de Marsac, Doctoral thesis. Université Pierre et Marie Curie, Paris, 1978.
[4] A. N. Glazer, this volume [31].
[5] E. Gantt and S. F. Conti, *Brookhaven Symp. Biol.* **19,** 393 (1966).
[6] A. Manodori and A. Melis, *FEBS Lett.* **181,** 79 (1985).
[7] E. Gantt, J. D. Clement-Metral, and B. M. Chereskin, this volume [30].
[8] E. Gantt, *Int. Rev. Cytol.* **66,** 45 (1980).
[9] G. Cohen-Bazire and D. A. Bryant, *in* "The Biology of Cyanobacteria" (N. G. Carr and B. A. Whitton, eds.), p. 143. Univ. of California Press, Berkeley, 1982.
[10] A. N. Glazer, *Annu. Rev. Microbiol.* **36,** 173 (1982).

Preparation of Phycobilisomes

All the procedures currently in use for the preparation of phycobilisomes are variants on the procedure originally developed by Gantt and Lipschultz[20] for the isolation of these particles from the unicellular red alga *Porphyridium cruentum*. The original and the majority of the derivative procedures retain certain major features in common: a cell breakage buffer containing sodium and/or potassium phosphate buffer at a concentration ranging from 0.6 to 1.0 *M*, a pH of 7–8, and a temperature of 18–23°. Several representative procedures are described below.

Phycobilisomes from Various Cyanobacteria and Red Algae[21]

The entire procedure is carried out at room temperature (20–23°) in 0.75 *M* potassium phosphate (pH 6.8–7.0). One to two grams (wet weight) of cells is suspended in 10 ml of buffer and then broken by passage through a French pressure cell at 10,000 psi. Triton X-100 is immediately added to the broken cells to a final concentration of 2% (or more) and the mixture incubated for 20 min with stirring. Centrifugation for 30 min at 25,000 *g* is employed to remove cell debris. The supernatant is removed with a syringe from underneath a floating chlorophyll–Triton layer. Aliquots of the supernatant are then layered (2–4 ml per tube) on a step gradient containing the following sucrose molarities in the 0.75 *M* phosphate buffer: 2.0 *M* (6 ml), 1.0 *M* (4 ml), 0.5 *M* (4 ml), and 0.25 *M* (8 ml). The gradients are centrifuged for 3 hr in an angle-head rotor (Beckman 42.1) at 42,000 rpm (136,000 *g*). Phycobilisomes are recovered from the 1.0 *M* sucrose layer by suction through a flat-tipped syringe needle. The

[11] W. Wehrmeyer, *in* "Photosynthetic Prokaryotes: Cell Differentiation and Function" (G. C. Papageorgiou and L. Packer, eds.), p. 1. Elsevier, New York, 1983.

[12] A. N. Glazer, D. J. Lundell, G. Yamanaka, and R. C. Williams, *Ann. Microbiol. (Inst. Pasteur)* **134B,** 159 (1983).

[13] A. N. Glazer, *Annu. Rev. Biophys. Biophys. Chem.* **14,** 47 (1985).

[14] B. A. Zilinskas and L. S. Greenwald, *Photosynth. Res.* **10,** 7 (1986).

[15] A. N. Glazer, *in* "The Cyanobacteria" (P. Fay and C. Van Baalen, eds.), p. 69. Elsevier, Amsterdam, 1987.

[16] A. N. Glazer, S. W. Yeh, S. P. Webb, and J. H. Clark, *Science,* **227,** 419 (1985).

[17] A. N. Glazer, C. Chan, R. C. Williams, S. W. Yeh, and J. H. Clark, *Science* **230,** 1051 (1985).

[18] A. R. Holzwarth, *Photochem. Photobiol.* **43,** 707 (1986).

[19] I. Yamazaki, M. Mimuro, T. Murao, T. Yamazaki, K. Yoshihara, and Y. Fujita, *Photochem. Photobiol.* **39,** 233 (1984).

[20] E. Gantt and C. A. Lipschultz, *J. Cell Biol.* **54,** 313 (1972).

[21] E. Gantt, C. A. Lipschultz, J. Grabowski, and B. K. Zimmerman, *Plant Physiol.* **63,** 615 (1979).

phycobilisomes are either stored in the suspending medium or are diluted 3- to 4-fold with 0.75 *M* phosphate buffer and concentrated by centrifugation for 2.5 hr at 254,000 *g*.

For many species, this procedure led to the recovery of 85–90% of the total phycobiliprotein in the 1.0 *M* sucrose layer. Cell breakage of some cyanobacteria, e.g., *Synechococcus* PCC 6301 (*Anacystis nidulans*), *Synechococcus* PCC 7002 (*Agmenellum quadruplicatum*), was increased by pretreatment with lysozyme.

Preparation of Synechococcus PCC 6301 Phycobilisomes[22]

The variant procedure described here led to a greater preservation of the integrity of *Synechococcus* PCC 6301 phycobilisomes than was permitted by the first procedure.[20,21] All buffers contain 1 m*M* 2-mercaptoethanol and 1 m*M* sodium azide. All operations are carried out at room temperature unless otherwise specified. Cells are harvested by centrifugation, washed twice with 0.65 *M* NaH_2PO_4/K_2HPO_4 ($NaKPO_4$) buffer at pH 8.0, and resuspended in the same buffer at a concentration of 0.12 g wet weight/ml. The cells are broken by passage through a French pressure cell at 20,000 psi, then incubated for 30 min in the presence of 1% (v/v) Triton X-100. Cell debris is subsequently removed by centrifugation at 31,000 *g* for 30 min at 18°. Centrifugation leads to the formation of a green membrane–detergent layer at the top of an intensely blue layer. The blue supernatant is removed carefully, leaving behind the membrane–detergent layer and a trace amount of the blue layer on top of the pellet. Aliquots (1.1 ml) of the blue supernatant are layered on sucrose step gradients consisting of 1.0, 3.0, 3.0, 2.3, and 2.2 ml of 2.0, 1.0, 0.75, 0.5, and 0.25 *M* solutions of sucrose, respectively, all in 0.75 *M* $NaKPO_4$, pH 8.0, then centrifuged in a Spinco SW41 rotor at 24,000 rpm (98,000 *g*) for 13 hr at 18°. The phycobilisomes are recovered from the 0.75 *M* sucrose layer as a deep blue band, free of detectable chlorophyll, and are stable for at least 2 weeks when stored at 4° as a concentrated solution obtained directly from sucrose gradients.

When it was desired to carry out small-scale phycobilisome preparations, the following pretreatment of cells was found to be convenient.[23] Cells are harvested by centrifugation, washed once with 30 m*M* sodium phosphate, pH 6.8, and resuspended at 0.1 g wet weight/ml in a spheroplasting medium containing 350 m*M* mannitol, 30 m*M* sodium phosphate, pH 6.8, 10 m*M* disodium EDTA, and 5 mg/ml egg white lysozyme. After

[22] G. Yamanaka, A. N. Glazer, and R. C. Williams, *J. Biol. Chem.* **253,** 8303 (1978).
[23] G. Yamanaka and A. N. Glazer, *Arch. Microbiol.* **130,** 23 (1981).

3.5 hr of incubation at 37°, spheroplasts are washed twice in 500 mM mannitol, 30 mM sodium phosphate, pH 6.8, and resuspended in a lysis buffer containing 0.65 M $NaKPO_4$, pH 8.0, and 1 mM 2-mercaptoethanol. Samples are sonicated briefly to ensure good breakage, then solubilized by the addition of Triton X-100 to a final concentration of 1% (v/v). The remainder of the procedure is exactly as described above.

Preparation of Phycobilisomes of Pseudoanabaena spp.[24]

Cells are collected by centrifugation and washed once with 0.75 M potassium phosphate buffer at pH 7.0 containing a mixture of three protease inhibitors, phenylmethylsulfonyl fluoride (5×10^{-4} M), *p*-chloromercuribenzoate (5×10^{-5} M), and EDTA (1 mM). They are then resuspended in the same buffer at 1 g wet weight of cells/10 ml. The cells are broken by passage through the French pressure cell at 20,000 psi and Triton X-100 added to 2% (v/v). After 10 min, the suspension is centrifuged at 30,000 *g* for 10 min at 20°. One milliliter of the supernatant is layered on each discontinuous sucrose density gradient (0.5 M, 2 ml; 0.75 M, 3 ml; 1 M, 2 ml; 1.5 M, 2 ml), in 0.75 M potassium phosphate buffer at pH 7. Centrifugation is performed in a fixed-angle Beckman 50Ti rotor at 200,000 *g* (45,000 rpm) for 3 hr at 20°. The phycobilisomes recovered from the 0.75 M sucrose layer are dialyzed at ambient temperature against 0.75 M potassium phosphate buffer, pH 7, containing 5×10^{-4} M phenylmethylsulfonyl fluoride and 1 mM EDTA. For long-term storage, the phycobilisomes are precipitated by the addition of solid $(NH_4)_2SO_4$ (14 g/100 ml). The precipitates are pelleted by centrifugation in Eppendorf tubes and the pellets stored at −20°.

Preparation of Fremyella diplosiphon (Calothrix PCC 7601) Phycobilisomes[25,26]

This procedure was developed for the rapid isolation of *F. diplosiphon* phycobilisomes, and its general applicability is not known. All steps are performed at room temperature. Cells are collected by continuous centrifugation and suspended at 1 g wet weight/15 ml in an extraction medium containing 0.75 M potassium phosphate buffer, pH 6.8, 1% (w/v) Triton X-100. After stirring for 1 hr the extract is centrifuged for 20 min at 27,000

[24] G. Guglielmi and G. Cohen-Bazire, *Protistologica* **20,** 393 (1984).

[25] M. Rigbi, J. Rosinski, H. W. Siegelman, and J. C. Sutherland, *Proc. Natl. Acad. Sci. U.S.A.* **77,** 1961 (1980).

[26] J. Rosinski, J. F. Hainfeld, M. Rigbi, and H. W. Siegelman, *Ann. Bot.* **47,** 1 (1981).

g, and the sedimented material is discarded. Polyethylene glycol 6000 is added to the supernatant solution to 15% (w/v), the mixture is stirred for 1 hr and then centrifuged for 20 min at 27,000 *g*, and the supernatant is discarded. The sedimented material is suspended in 0.75 *M* potassium phosphate buffer, pH 6.8, 1% Triton X-100, 15% (w/v) polyethylene glycol 6000, the mixture centrifuged for 20 min at 27,000 *g*, and the supernatant discarded. The purple sediment is suspended in 0.75 *M* potassium phosphate buffer, pH 6.8, and centrifuged for 15 min at 27,000 *g*; the supernatant is saved. The purple sediment is again extracted and centrifuged, and the supernatants are combined to give a stock solution of phycobilisomes, which is stored in the dark at room temperature. For further purification,[26] the phycobilisomes are suspended in 0.7 *M* potassium phosphate buffer, pH 6.8, applied to a Sepharose CL-4B column (2.5 × 40 cm) preequilibrated with the pH 6.8 phosphate buffer, and eluted with the same buffer. The peak fractions are used for further studies.

Rapid Procedure for Isolation of Phycobilisomes from Certain Organisms[27]

This procedure can be performed in 3–4 hr and does not require ultracentrifugation. It has been employed to isolate phycobilisomes on a large scale from *Synechococcus* PCC 6301, as well as two eukaryotic organisms, *Porphyridium aerugineum* and *Cyanidium caldarium*. The procedure was not applicable to the isolation of *Porphyridium cruentum* phycobilisomes; its range of applicability to other organisms is not known.

Cells suspended in 1 *M* $NaKPO_4$ buffer, pH 7.5, are lysed by passage through a French pressure cell (15,000 psi). Triton X-100 (1%, v/v) is added to the broken cell suspension, and the mixture is incubated at room temperature for 30 min. It is then centrifuged in a Sorvall RC5B centrifuge (SS34 rotor) at 20,000 rpm for 30 min. Both the membranous material and the phycobilisomes pellet during this centrifugation. The supernatant is discarded, and the pellet is resuspended in 0.6 *M* $NaKPO_4$ buffer, pH 7.5. For homogeneous resuspension the pellet is dispersed with a ground-glass homogenizer. The suspension is incubated for 30 min after addition of Triton X-100 to 1% (v/v) and then centrifuged at 20,000 rpm for 30 min. During this centrifugation the membranes are pelleted while a large proportion of intact phycobilisomes remain in solution. The pellet is discarded and the supernatant diluted 10-fold with 1.0 *M* $NaKPO_4$, pH 7.5, and spun at 20,000 rpm for 1 hr. Intact phycobilisomes pellet under these conditions.

[27] A. Grossman and J. Brand, *Carnegie Inst. Wash. Yearbook* **82,** 116 (1983).

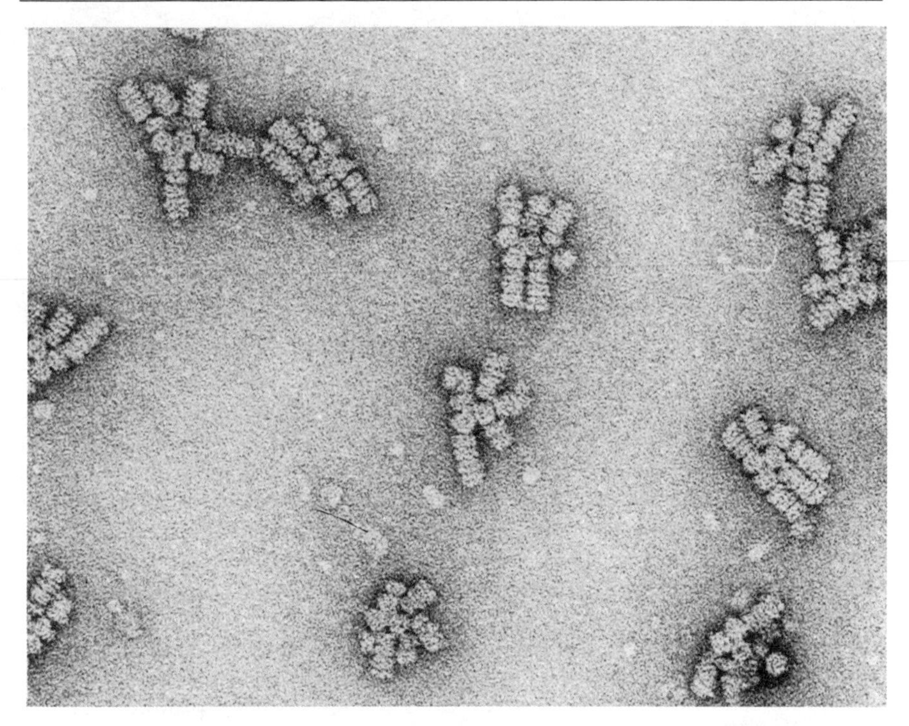

FIG. 1. Electron micrograph of glutaraldehyde-cross-linked phycobilisomes from wild-type *Synechocystis* PCC 6701 phycobilisomes. The three cylinders that make up the triangular core, seen in face view, are 110 Å in diameter. The stacked disks that make up the six rods are each 60 Å thick. Uranyl formate negative stain. Magnification: ×290,000.

Characterization of Phycobilisomes

The degree of preservation of phycobilisome structure and function can be assessed by three complementary types of analyses. Transmission electron microscopy of glutaraldehyde-cross-linked negatively stained preparations (Fig. 1) provides a measure of the preservation of the phycobilisome structure at the level of gross morphology and reveals the degree of dissociation and/or aggregation of the particles.[28] Fluorescence emission spectroscopy provides a rapid and sensitive measure of the functional integrity of a phycobilisome preparation. At room temperature, the majority of intact phycobilisomes have a fluorescence emission maximum between 670 and 680 nm independent of excitation wavelength[21] (Fig. 2).

[28] A. N. Glazer, R. C. Williams, G. Yamanaka, and H. K. Schachman, *Proc. Natl. Acad. Sci. U.S.A.* **76,** 6162 (1979).

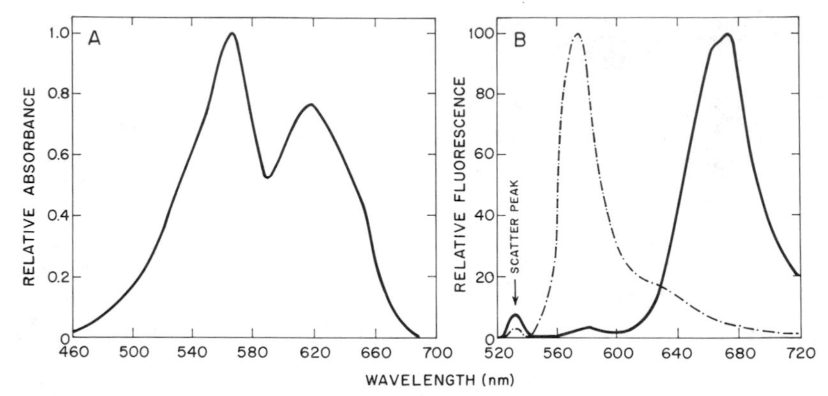

FIG. 2. Absorption and fluorescence emission spectra of intact and dissociated *Synechocystis* PCC 6701 phycobilisomes. The solid lines in A and B show the absorption and emission spectra, respectively, of phycobilisomes in 0.75 *M* sodium potassium phosphate buffer, pH 8.0. The dashed line in B shows the fluorescence emission spectrum of phycobilisomes dialyzed to equilibrium against 10 m*M* $NaKPO_4$, pH 8.0. The excitation wavelength was 530 nm.

Significant emission with maxima at shorter wavelengths is diagnostic of partial dissociation. The extent of chlorophyll *a* emission on appropriate excitation can be used to monitor possible membrane contamination of the phycobilisome preparation since pure phycobilisomes do not contain chlorophyll. Polyacrylamide gel electrophoresis in the presence of sodium dodecyl sulfate provides a means of showing that adventitious polypeptides are absent from the phycobilisome preparation. A high-molecular-weight polypeptide (75,000–120,000, depending on organismal source) is present in all phycobilisomes thus far examined. This polypeptide is believed to function in the attachment of the phycobilisome to the thylakoid membrane.[29-31] Assessment of possible proteolytic degradation is another concern. It has been observed by many investigators that this polypeptide may be subject to partial proteolysis during the phycobilisome purification procedure. The occurrence and extent of such degradation is evident from inspection of the polypeptide pattern obtained from gel electrophoresis.

[29] B. A. Zilinskas, *Plant Physiol.* **70,** 1060 (1982).

[30] E. Gantt, C. A. Lipschultz, and T. Redlinger, *in* "Molecular Biology of the Photosynthetic Apparatus" (K. E. Steinback, S. Bonitz, C. J. Arntzen, and L. Bogorad, eds.), p. 223. Cold Spring Harbor Lab., Cold Spring Harbor, New York, 1985.

[31] E. Gantt, M. Mimuro, and C. A. Lipschultz, *in* "Hungarian–USA Binational Symposium on Photosynthesis" (M. Gibbs, ed.), p. 1. Salve Regina College, Newport, Rhode Island, 1986.

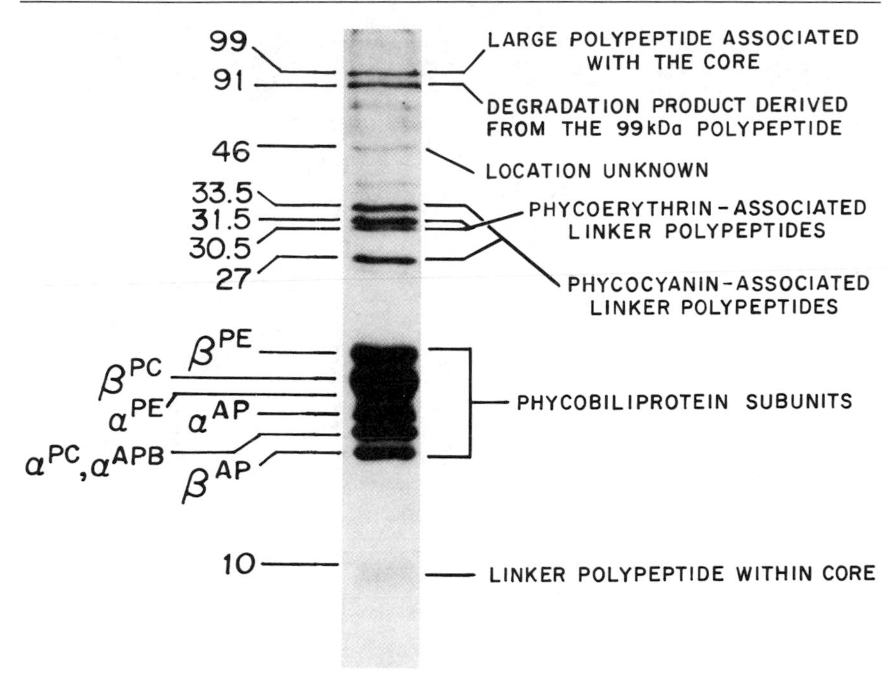

FIG. 3. Separation of the polypeptides of *Synechocystis* PCC 6701 phycobilisomes by SDS–polyacrylamide gel electrophoresis. For abbreviations, see Fig. 4.

Figures 1–3 show the application of each of these means of characterization to the phycobilisomes of *Synechocystis* PCC 6701.[32]

Polypeptide Composition and Topology in Phycobilisomes

Phycobilisomes have a complex polypeptide composition[33] (e.g., Fig. 3). Polypeptides which carry bilin chromophores are present as well as polypeptides, called linker polypeptides,[34] whose major function is in phycobilisome assembly rather than in light-energy absorption and transfer. Phycobilisomes from different organisms have distinctive polypeptide compositions. The diagrammatic representation of the *Synechocystis*

[32] R. C. Williams, J. C. Gingrich, and A. N. Glazer, *J. Cell Biol.* **85,** 558 (1980).

[33] N. Tandeau de Marsac and G. Cohen-Bazire, *Proc. Natl. Acad. Sci. U.S.A.* **74,** 1635 (1977).

[34] D. J. Lundell, R. C. Williams, and A. N. Glazer, *J. Biol. Chem.* **256,** 3580 (1981).

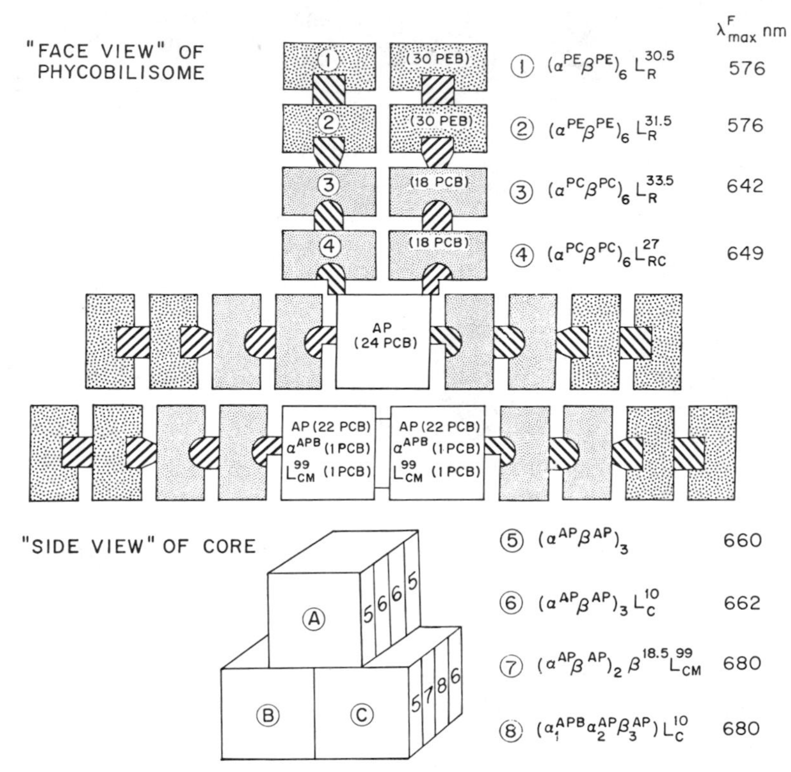

FIG. 4. Schematic representation of the *Synechocystis* 6701 phycobilisome. A rod is made up of four hexameric phycobiliprotein complexes, each of which is attached through its specific linker polypeptide to the component adjacent to it in the phycobilisome. The manner in which the core cylinders are held together is not known. The abbreviations AP, PC, and PE are used for allophycocyanin, phycocyanin, and phycoerythrin, and α^{AP} and β^{AP}, etc., for the α and β subunits of these proteins. Linker polypeptides are abbreviated L, with a superscript denoting the apparent size ($\times 10^{-3}$ D) and a subscript that specifies the location of the polypeptide: R, rod substructure; RC, rod–core junction; C, core; CM, core–membrane junction. For other details, see Ref. 13. The number of bilins present in each domain of the structure is indicated in the upper diagram. Abbreviations: PEB, phycoerythrobilin; PCB, phycocyanobilin. The fluorescence emission maximum, λ^F_{max}, is given for each of the eight subcomplexes isolated from this phycobilisome.

PCC 6701 phycobilisome shown in Fig. 4 indicates the location of the various polypeptide chains within the particle.[35]

Acknowledgments

Work in the author's laboratory was supported by National Science Foundation Grant DMB 8518066.

[35] J. C. Gingrich, D. J. Lundell, and A. N. Glazer, *J. Cell. Biochem.* **22,** 1 (1983).

[33] Phycobilisome Stability

By TETZUYA KATOH

Introduction

Phycobilisomes (PBs) require a high concentration (usually 0.75 *M*) of phosphate to be isolated in the entire and energetically coupled state.[1] This implies that the protein–protein interaction inside these particles should be significantly weaker than those in other protein assemblies. PBs are located on the outer (stroma side) surface of thylakoid membranes,[2,3] and this raised a question about the physical environment of PBs in the cytosol where PBs retain their molecular association without such a high concentration of phosphate.

Various organic and inorganic solutes in the cytosol can be extracted and separated, by appropriate techniques, and used to examine the effect of PB stability. In such studies, however, one should not ignore the concentration of soluble proteins and solutes in the free water around thylakoid lamellae. In chloroplast stroma of C_3 plants, the concentration ribulose-bisphosphate carboxylase (RuBPCase, EC 4.1.1.39) is as high as 180 mg/ml and accounts for 50% of the soluble proteins. Taking into account the hydration of protein molecules with 2.3 volumes of water,[4] the volume of free water is extremely small. Solutes which show negative affinity to the protein molecules are excluded from the peripheral layer of protein to stabilize protein assemblies,[5] so that the actual concentrations of such solutes in the free water space should be somewhat higher than the figures thus far reported. Although we have not actual figures for the concentration of soluble proteins in the cytosol of cyanobacteria, the existence of phycobiliproteins at a high level (10–20% of cell dry weight) would admittedly enhance the local concentration of solutes in the free water around the thylakoid lamellae, even if phycobiliproteins are in an aggregated form. Actually, the molecular assembly of isolated PBs is stabilized as PB concentration increases, even without the addition of other proteinaceous solutes.[6]

[1] E. Gantt, C. A. Lipschultz, J. Grabowski, and B. K. Zimmerman, *Plant Physiol.* **63,** 615 (1979).
[2] E. Gantt and S. F. Conti, *J. Cell Biol.* **29,** 423 (1966).
[3] J. C. Cosner, *J. Bacteriol.* **135,** 1137 (1978).
[4] W. Bode and T. Schirmer, *Hoppe-Seyler Biol. Chem.* **366,** 287 (1985).
[5] G. C. Na and N. Timasheff, *J. Mol. Biol.* **151,** 165 (1981).
[6] N. Kume, T. Isono, and T. Katoh, *Photobiochem. Photobiophys.* **4,** 25 (1982).

From such considerations one recognizes the importance of studying PB stability in the concentrated state. The stability of isolated PBs in a diluted solution can easily be followed by the changes in the fluorescence emission spectra over wide range of conditions. In this chapter light-scattering techniques used to check the stability of isolated PBs as well as measurements of buoyant density to estimate the extent of hydration in PBs are described.

Preparation of Phycobilisomes

Procedures for PB isolation are essentially the same as those described elsewhere in this volume [32]. Freshly harvested cells are suspended in 0.75 *M* potassium phosphate (mixture of K_2HPO_4 and KH_2PO_4 at the molar ratio of 3 : 1), disrupted by French press. The extracts are freed from cell debris, loaded on sucrose density step gradients containing 0.75 *M* potassium phosphate, and centrifuged until PBs form a colored band at the 1.0 *M* sucrose layer. Because particulate contamination, such as fragments of thylakoid membranes and mucilaginous materials of cell surface origin in the PB samples, extensively disturbs measurements of light scattering, such contaminants must be carefully removed. PBs, collected from the sucrose density step gradients, are diluted with 0.75 *M* potassium phosphate to a concentration of 1–3 mg protein/ml and filtered through a Millipore filter (pore size 0.5 μm). Filtration through larger pores depends on the amount of unfiltrable materials. Using a Diaflo membrane filter YM30 (Amicon, Lexington, MA) PBs are concentrated to 20–30 mg protein/ml and centrifuged to sediment insoluble materials. Cuvettes must be clean (see Ref. 7 for cleaning techniques).

Properties of Isolated Phycobilisomes

Light Scattering

As the sizes of PBs are 1/10 to 1/30 that of the wavelength of visible light, the intensity of light scattering should correlate directly to the particle size without correction for coherence. The absolute intensity of light scattering can be measured accurately with any device designed for this purpose, but the change in relative intensity of light scattering can be more easily measured using a fluorescence spectrophotometer. It should be equipped with a well-stabilized light source, and the wavelength of the measuring beam should be set at the minimal absorbance of PBs. For PBs containing phycoerythrin rise 450 or 740 nm light, and for those free of phycoerythrin 478 or 740 nm light is suitable. As the intensity of light scattering is in inverse proportion to the wavelength, λ^4, the light scattering at 740 nm is 1/6 or 1/7 as weak as that at 470 or 450 nm, but the absorbance of PBs at this wavelength is essentially zero so the physical cleavage of PBs can be monitored for a wider range of PB concentration. Higher sensitivities, however, are needed to detect the signal for PBs of low concentration. With PBs from *Anabaena variabilis* suspended in 0.75 *M* potassium phosphate, the intensity of light scattering is linearly proportional to the PB concentration: 5 mg/ml at 478 nm and 30 mg/ml at 740 nm.

The reduced intensity of the light scattering ($[I_{sc}/I_0]/r^2$), where I_{sc} and I_0 are the intensities of scattered and incident light and r the distance between the cuvette and the detector, is in proportion to the production of the square of molecular weight and the molar concentration of protein. The dissociation of PBs into n pieces of subparticles causes a decrease in light scattering by a factor of $(m_1^2 + m_2^2 + \ldots + m_n^2)/M^2$, where m_i and M are the molecular weights of the ith subparticle and of the entire PB complex ($M = \Sigma m_i$). When PBs split into n subfragments of equal size, the light scattering should decrease by a factor of $1/n$, and in unequal cleavage the decrease is larger than $1/n$. Thus, the physical cleavage of PBs can be directly monitored as the change in the intensity of light scattering.

Fluorescence Spectra

Unfortunately, light scattering techniques do not provide any information about the site at which PBs have been destabilized. This is achieved only by measurements of fluorescence emission spectra, although the applicable range of PB concentration is limited to optical densities lower than 0.2 at the absorbance peak. As the uncoupling of excitation energy transfer and light scattering in destabilized PBs shows practically identical kinetics over a wide range of conditions,[8] the changes in fluorescence emission reflect the physical cleavage of PBs exclusively. Gantt *et al.*[1] have specified the cleavage sites in various PBs by allowing them to dissociate in lower concentrations of phosphate. Likewise, Zilinskas and Glick[9] showed that the extent to which *Porphyridium* PBs are destabilized by various anions is in agreement with the order of chaotropicity ($SCN^- > NO_3^- >$ citrate$^{3-} > SO_4^{2-} > Cl^- > PO_4^{3-}$).[10] Recently, Mimuro and Gantt[11] observed that 2 m*M* CHAPS (3-[(3-cholamidopropyl)-dimethylammonio]-1-propane sulfonate), a zwitterionic detergent, cleaved *Nostoc* PBs into rods, consisting of a phycocyanin–phycoery-

[7] S. T. Dubin, this series, Vol. 26, p. 165.
[8] N. Kume and T. Katoh, *Plant Cell Physiol.* **23,** 803 (1982).
[9] B. A. Zilinskas and R. E. Glick, *Plant Physiol.* **68,** 447 (1981).
[10] Y. Hatefi and W. G. Hanstein, this series, Vol. 31, p. 770.
[11] M. Mimuro and E. Gantt, *Plant Cell Physiol.* **29,** in press (1988).

thrin and an allophycocyanin core, with high specificity, when incubated in 0.8 *M* phosphate.

In the case of a PB suspension having a higher optical density, changes in both the fluorescence emission and light scattering can be traced for the same sample as follows. From the PB suspension whose light scattering is being monitored, a measured volume is transferred to a cuvette filled with 0.5 *M* potassium phosphate, and the fluorescence is measured immediately. The volume to be transferred must be adjusted to give a final optical density lower than 0.2 at the wavelength of peak absorbance. The fluorescence emission spectra had no significant changes for samples in 0.5 *M* phosphate for 10 min, even when the samples were a mixture of subparticles. Figure 1 shows simultaneous measurements with *Anabaena* PBs

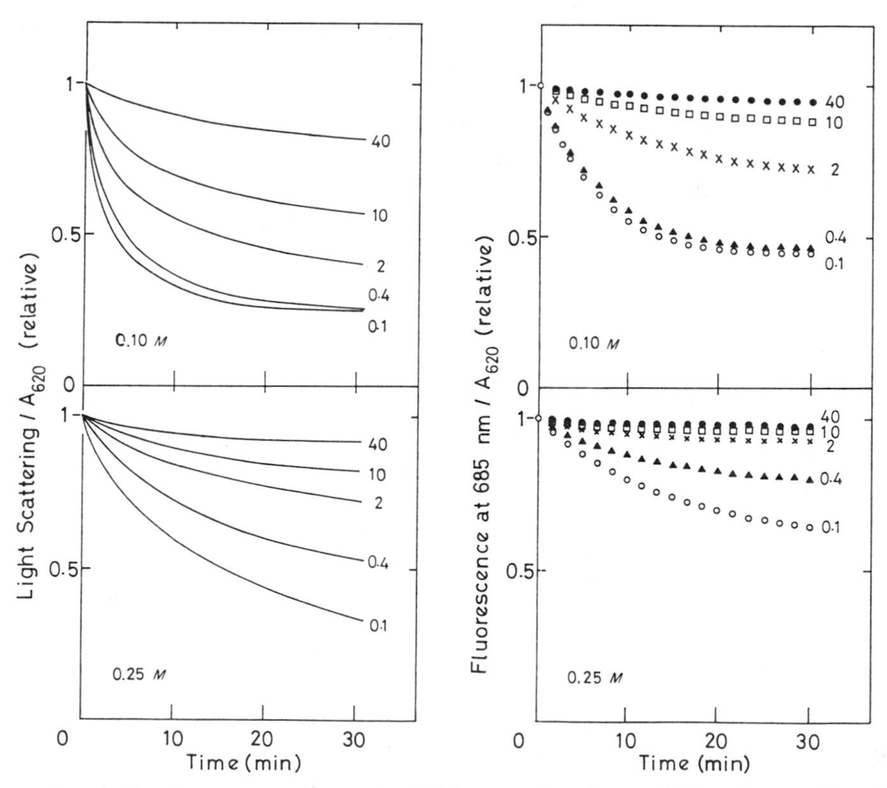

FIG. 1. Simultaneous measurements of light-scattering changes (left) and uncoupling of excitation energy transfer measured by the fluorescence yield from the terminal allophycocyanin (F_{685}, right) in *Anabaena* phycobilisomes destabilized in 0.25 *M* (top) and 0.1 *M* (bottom) phosphate and at varied phycobilisome concentrations indicated as A_{620}. The traces are normalized for the phycobilisome concentration (an A_{620} value of 1 corresponds to ~0.2 mg protein/ml).

destabilized in 0.25 and 0.1 *M* potassium phosphate, where the uncoupling of excitation transfer is measured as the decay in fluorescence (F_{685}) at intervals,[6] indicating that the cleavage was markedly repressed when the PB concentration was high. With the same arrangement, potassium salts of various organic phosphates, including β-glycerophosphate, threonine *O*-phosphate, glucose 1-phosphate, fructose 1,6-bisphosphate, and of some organic acids, such as gluconate, tartrate, succinate, and glutamate, were found to be more effective in stabilizing *Anabaena* PBs than phosphate of the same concentration. Furthermore, when the soluble proteins of *Anabaena* cells that had been collected from the upper layer of sucrose step gradients in the course of PB preparation, freed from sucrose and phosphate by ammonium sulfate precipitation, and dialyzed were added to these salts, the dissociation of PBs was found to be markedly repressed, although the addition of the soluble protein was not effective without the addition of any of these salts.[12]

Buoyant Density

The stability of protein–protein interactions in the presence of various inorganic and organic solutes is closely correlated with the extent of hydration of protein.[5] Negative interaction between protein and solute molecules (exclusion of solute from the surface of protein molecules) stabilizes the association of protein molecules, resulting in an increase in partial specific volume. To determine partial specific volume, both solution density and the weight concentration of the protein must be precisely measured. Solution density can be measured using any type of densitometer with the highest accuracy. Na and Timasheff[5] noted the increment of partial specific volume of tubulin in the presence of various concentrations of glycerol, an accelerator of tubulin assembly, using an Anton-Paar precision densitometer.

The extent of hydration of PBs can also be observed as the change in buoyant density. PBs are homogeneously dispersed in sucrose density gradients ranging in density from 1.20 (top) to 1.30 (bottom) with inclusion of the solute to be tested. Centrifuge for a period long enough that the PBs form a sharp band at the position of their buoyant density, usually at 36,000 rpm for more than 48 hr with a Hitachi RPS50T rotor. By weighing a thoroughly cleaned constriction pipet (200 μl) dried and filled with the solution at and adjacent to the position of PBs and using liquid with known density as a standard, the buoyant density of the band can be determined with an accuracy of 0.15%. In the case of a phosphate salt, the buoyant densities of *Anabaena* PBs shows a gradual decrease from 1.271

[12] T. Katoh, unpublished observations.

to 1.215 as the phosphate concentration is raised from 0.5 to 1.25 *M*, reflecting an increased hydration of PBs at higher phosphate concentration.[12] The buoyant densities of *Anabaena* PBs measured in potassium phosphate are essentially the same as those in cesium phosphate when the concentration of phosphate is equal. Because the cesium salt provides a density solution with a lower concentration of sucrose, that is, with lowered viscosity, one can obtain a sharp band of buoyant equilibrium with a shorter period of centrifugation. A disadvantage of this method is that the buoyant equilibrium cannot readily be attained when PBs have dissociated into smaller subparticles. Then the extent of hydration in the destabilized PBs has to be measured with other methods such as the precision densitometer.

[34] Complementary Chromatic Adaptation: Physiological Conditions and Action Spectra

By N. Tandeau de Marsac and J. Houmard

Introduction

Photosynthetic organisms can modulate their relative pigment content in response to changes in either light intensity or light wavelength. Generally, one observes an inverse correlation between light intensity and pigment content: the less light energy available, the more photosynthetic pigments are synthesized by the cells. The effect of light wavelength on the pigment content of the cells, termed complementary chromatic adaptation, appears to be restricted to some cyanobacteria. In this type of adaptation, changes in cell pigmentation in response to specific spectral illuminations result from modifications of the relative amounts of the red-colored phycoerythrin (PE) and the blue-colored phycocyanin (PC), with a predominance of PE in green-light-grown cells and of PC in red-light-grown cells. These phycobiliproteins (PE and PC) are the major light-harvesting pigments used to drive photosynthesis. Therefore, this mode of chromatic control allows cells to trap the available light energy with maximal efficiency.[1]

Only cyanobacteria able to synthesize PE can undergo complementary chromatic adaptation. Among the PE producers, three physiological

[1] N. Tandeau de Marsac, *Bull. Inst. Pasteur* **81,** 201 (1983).

groups can be distinguished.[1,2] Group I, no complementary chromatic adaptation, PE and PC synthesis is independent of light wavelength; Group II, "unidirectional" adaptation, only the synthesis of PE is regulated by light wavelength; Group III, "bidirectional" or complete complementary chromatic adaptation. In Group III cyanobacteria, after transfer of green-adapted cells to red light, PE synthesis is reduced or completely arrested, whereas the expression of a specific set of PC genes (PC-2) is "induced." The expression of this set of genes is turned off after transfer of red-adapted cells to green light, while PE synthesis occurs. Under both conditions a set of "constitutive" PC genes (PC-1) is expressed. Moreover, the synthesis of the linker polypeptides specifically associated with either PE or PC-2 is coordinately regulated with the corresponding phycobiliprotein.

The regulatory processes involved in complementary chromatic adaptation are controlled by a photoreceptor pigment system which presumably acts at the transcriptional level. The property of photoreversibility of the cyanobacterial photoreceptor resembles that of the phytochrome present in higher plants and algae, although its action maxima are situated at shorter wavelengths in the visible spectrum (~540 and ~640 nm for cyanobacterial photoreceptor, instead of 660 and 730 nm for phytochrome). A hypothetical model for the regulation of phycobilisomal proteins by chromatic light has been proposed[1]; however, further studies, in particular the isolation of the cyanobacterial photoreceptor, are needed to gain a deeper insight into the processes involved.

Rapid Screening for Chromatic Adaptation

When several strains or pigmentation mutants have to be analyzed, the procedure described below allows the amount of PE and PC per milliliter of culture after growth under different light conditions to be estimated with rather good accuracy. However, the values obtained for allophycocyanin (AP) are less accurate since this pigment is a minor constituent of the phycobilisome and its absorption maximum overlaps the absorption spectrum of PC.

Culture Conditions

Freshwater strains are grown in medium BG-11[3] in which the concentration of Na_2CO_3 is increased to 0.04 g/liter. Strains of marine origin are

[2] N. Tandeau de Marsac, *J. Bacteriol.* **130,** 82 (1977).

[3] R. Rippka, J. Deruelles, J. B. Waterbury, M. Herdman, and R. Y. Stanier, *J. Gen. Microbiol.* **111,** 1 (1979).

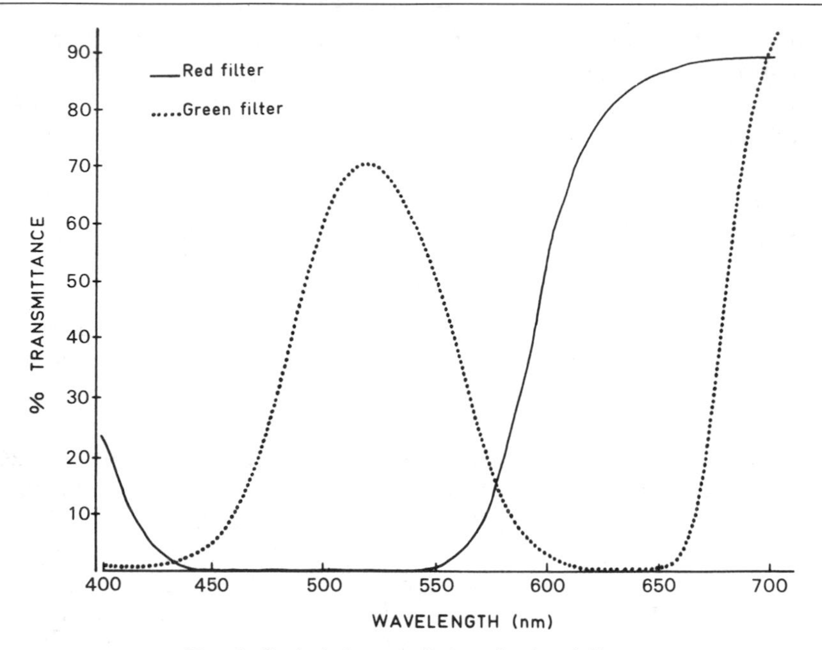

FIG. 1. Optical characteristics of colored filters.

grown either in a synthetic seawater medium such as ASN-III or in medium MN, which consists of seawater 75% (v/v) supplemented with a half-concentrated mineral solution of BG-11 medium.[3] Strains that are facultatively heterotrophic are grown in the dark in the mineral medium used for photoautotrophic growth supplemented with a filter-sterilized carbon source.[3] For rapid screening, 40 ml of liquid cultures is grown in 100-ml Erlenmeyer flasks without gassing, either in green light (~3.7 $\mu E\ m^{-2}\ sec^{-1}$) or in red light (~2.7 $\mu E\ m^{-2}\ sec^{-1}$). Measurement of the photon flux is performed with a Li-Cor quantum meter Li-185B equipped with the quantum sensor Li-190SB. The light source is a fluorescent lamp, specific chromatic illumination being provided by the interposition of plastic filters (cellulose acetate colored filters available in art supply stores). Typical transmission spectra of green and red filters routinely used are shown in Fig. 1. Pigment analyses are performed on light-grown cultures harvested after 2–3 weeks and on dark-grown cultures harvested after 6 weeks of incubation at 25°.

Absorption Spectra in Vivo

Whole cell absorbance spectra can be performed using a spectrophotometer equipped with a scattering transmission device. If this equipment is not available, an alternative method is to run spectra on cell suspensions diluted so as to obtain identical chlorophyll *a* absorption peaks [optical density (OD) ~1 between 680 and 700 nm depending on strain]. Under these conditions, interference due to light scattering will be minimized, and pigment ratios can be compared between samples grown under different chromatic illuminations. For cultures in which cells settle rapidly, it might be necessary to increase the density of the medium by adding sucrose at a final concentration of 30% (w/v).

Phycobiliprotein, Chlorophyll a, and Total Cellular Protein Determination

Determination of Phycobiliproteins. Four milliliters of cell suspension is centrifuged at 10,000 *g* for 10 min at 4°. The pellet is either stored at −20° until analysis or immediately resuspended to the same final volume in 20 m*M* sodium acetate buffer, pH 5.5.[4] Cells are broken by passage through a French pressure cell at 1,330 atm or by sonication at 4° for various times depending on the strain under study. Crude cell-free extract is precipitated with 1% (w/v) streptomycin sulfate for 30 min at 4° and centrifuged at 10,000, *g* for 10 min at 4° in order to eliminate membrane fragments containing chlorophyll. The amounts of PE, PC, and AP in the supernatant fraction are calculated from measurements of the optical densities at 565, 620, and 650 nm using Eq. (1). These equations were established by using the simultaneous equations of Bennett and Bogorad[5] and the extinction coefficients from Bryant *et al.*[6]

$$\begin{aligned} \text{PC (mg ml}^{-1}) &= \frac{\text{OD}_{620\ \text{nm}} - 0.7 \times \text{OD}_{650\ \text{nm}}}{7.38} \\ \text{AP (mg ml}^{-1}) &= \frac{\text{OD}_{650\ \text{nm}} - 0.19 \times \text{OD}_{620\ \text{nm}}}{5.65} \\ \text{PE (mg ml}^{-1}) &= \frac{\text{OD}_{565\ \text{nm}} - 2.8[\text{PC}] - 1.34[\text{AP}]}{12.7} \end{aligned} \tag{1}$$

Equations (1) can be used for all cyanobacterial strains, since interpretation of results depends on a comparison of the pigment composition (PC/AP and PE/AP ratios) of each organism studied after growth under

[4] All the buffers (sodium acetate or potassium phosphate) and sucrose solutions used either to purify phycobilisomes or to extract phycobiliproteins contain 3 m*M* sodium azide and 10 m*M* disodium EDTA.

[5] A. Bennett and L. Bogorad, *J. Cell Biol.* **58**, 419 (1973).

[6] D. A. Bryant, G. Guglielmi, N. Tandeau de Marsac, A.-M. Castets, and G. Cohen-Bazire, *Arch. Microbiol.* **123**, 113 (1979).

different chromatic illuminations, and not on the absolute values of the phycobiliprotein contents. Alternatively, the absorption spectra of the total phycobiliproteins can be compared after normalization of the OD at 650 nm. Either of the two methods allows the assignment of a given strain to a specific chromatic adaptation group.

Determination of Chlorophyll a. One-half milliliter of cell suspension is centrifuged at 10,000 *g* for 10 min at 4°, and the pellet is extracted twice with 90% (v/v) methanol for 1 hr at 4°, in dim light, followed by centrifugation at 10,000 *g* for 10 min at 4°. The chlorophyll *a* content is calculated from the absorbance of the methanolic extract at 665 nm, using Eq. (2).[7]

$$C\ (\mu g\ ml^{-1}) = OD_{665\ nm} \times 13.9 \qquad (2)$$

Protein Determination. One-half milliliter of cell suspension is precipitated with 10% (w/v) trichloroacetic acid and centrifuged at 10,000 *g* for 10 min at 4°. The pellet is resuspended in 1 *N* NaOH, boiled for 30 min, cooled, and then recentrifuged to eliminate light-scattering material. The protein content of the supernatant is determined by the method of Lowry *et al.*,[8] using bovine serum albumin as standard.

Characterization of Phycobilisomal Protein Contents

This procedure is devised to obtain precise determination of the amounts of PE, PC, and AP per cell. Moreover, when, for a given strain, the structure of the phycobilisome is known, the amount of PE and PC per phycobilisome can be calculated using AP as a reference, since the number of AP disks per core has been shown to be identical in red- and green-light-grown cells for the strains studied so far.

Culture Conditions

Cells are cultivated in 1 liter of the appropriate mineral medium (see above) gassed with air–CO_2 (99 : 1) and stirred. The red and green light fluxes are adjusted so as to obtain approximately identical generation times under both types of illumination. Cells are collected in mid log phase to avoid degradation of phycobilisomes or of free phycobiliproteins which may occur in late exponential or stationary growth phases.

[7] J. F. Talling and D. Driver, *in* "Primary Productivity Measurement, Marine and Freshwater" (M. Doty, eds., p. 142. U.S. Atomic Energy Commission, Washington, D.C., 1963.

[8] O. H. Lowry, N. J. Rosebrough, A. L. Farr, and R. J. Randall, *J. Biol. Chem.* **193,** 265 (1951).

Whole Cell Extract Analysis

Cell breakage in a low concentration buffer leads to the dissociation of the phycobilisomes into phycobiliproteins freed of their associated linker polypeptides. Cells (~1 g wet weight) are resuspended in 20 ml of 20 m*M* sodium acetate buffer, pH 5.5, and subjected to passage through a French pressure cell at 1,330 atm. For a given strain, the concentration of PE and PC can be determined with Eq. (1).

Purification of the Phycobilisomes

With the procedure described below, one can rapidly extract phycobilisomes from several samples simultaneously.[9,10] Exponentially growing cells (~1 g wet weight) are resuspended in 10 ml of 0.75 *M* potassium phosphate buffer, pH 7.0, containing protease inhibitors (10 m*M* Na_2EDTA, 500 μM PMSF, and 50 μM PCMB)[11] and subjected to passage through a French pressure cell at 1,330 atm. The cell lysate is immediately collected in 2% (v/v) Triton X-100, stirred for 10 min, and centrifuged at 20,000 *g* for 10 min at room temperature. The supernatant is loaded on top of a discontinuous sucrose gradient with the following concentrations from top to bottom: 0.5 *M* (2 ml), 0.75 *M* (3 ml), 1 *M* (2 ml), 1.5 *M* (2 ml). After centrifugation at 200,000 *g* for 3 hr at 20°, the chlorophyll membrane-containing material is carefully removed from the top of the gradient, and the fraction containing phycobilisomes (generally in the 0.75 *M* sucrose layer) is collected from the top of the gradient. To eliminate sucrose, the phycobilisome samples are diluted 4 times in 0.75 *M* potassium phosphate buffer, pH 7.0, and precipitated with $(NH_4)_2SO_4$ (40% saturation) for 15 min.

After centrifugation and careful removal of the ammonium sulfate, the phycobilisome-containing pellets can either be stored frozen or immediately resuspended in buffer. In 20 m*M* sodium acetate buffer, pH 5.5, the phycobiliproteins will be freed of their associated linker polypeptides, and the concentration of each phycobiliprotein can be determined by using Eq. (1). If the phycobilisome samples are resuspended in 0.75 *M* potassium phosphate buffer, pH 7.0, the concentrations of phycobiliproteins will be determined by using Eq. (3)[6] and the extinction coefficients given in Table I. Indeed, under these conditions, linker polypeptides remain specifically associated with the different phycobiliproteins and thus

[9] G. Guglielmi and G. Cohen-Bazire, *Protistologica* **20,** 393 (1984).

[10] G. Guglielmi, personal communication.

[11] Abbreviations: PMSF, phenylmethylsulfonyl fluoride; PCMB, *p*-chloromercuribenzoic acid; Na_2EDTA, ethylenediaminetetraacetic acid, disodium salt.

TABLE I
ESTIMATED MOLAR EXTINCTION COEFFICIENTS FOR PHYCOBILIPROTEIN–LINKER COMPLEXES[a]

Complex	ε_M^{620}	ε_M^{650}	ε_M^{560}
AP (core)[b]	160,000	309,000	~100,000
PC–27K[b]	390,000	170,000	~100,000
PC–27K–33K–30K[b]	390,000	110,000	~100,000
PE–linker[c]	0	0	~560,000

[a] ε_M values are calculated for the protomers ($\alpha\beta$).
[b] From A. N. Glazer, *Annu. Rev. Biophys. Biophys. Chem.* **14,** 47 (1985).
[c] From G. Guglielmi, personal communication.

modify their spectral properties. For example, the absorption maxima of PC-1 and PC-2 are situated at 610–615 nm and 620–630 nm, respectively.

$$
\begin{aligned}
[\mathrm{AP}] &= \frac{A_{650} - (\varepsilon_{650}^{\mathrm{PC}}/\varepsilon_{620}^{\mathrm{PC}}) \times A_{620}}{\varepsilon_{650}^{\mathrm{AP}} - (\varepsilon_{650}^{\mathrm{PC}} \times \varepsilon_{620}^{\mathrm{AP}})/\varepsilon_{620}^{\mathrm{PC}}} \\
[\mathrm{PC}] &= \frac{A_{620} - (\varepsilon_{620}^{\mathrm{AP}}/\varepsilon_{650}^{\mathrm{AP}}) \times A_{650}}{\varepsilon_{620}^{\mathrm{PC}} - (\varepsilon_{620}^{\mathrm{AP}} \times \varepsilon_{650}^{\mathrm{PC}})/\varepsilon_{650}^{\mathrm{PE}}} \qquad (3) \\
[\mathrm{PE}] &= \frac{A_{560} - [\mathrm{PC}] \times \varepsilon_{560}^{\mathrm{PC}} - [\mathrm{AP}] \times \varepsilon_{560}^{\mathrm{AP}}}{\varepsilon_{560}^{\mathrm{PE}}}
\end{aligned}
$$

Partial Purification of Components of Phycobilisomes

A more precise determination of the composition of the phycobilisomes can be obtained through hydroxyapatite (HAP) fractionation. Phycobilisomes purified from red- or green-light grown cells are resuspended in 1 m*M* potassium phosphate buffer, pH 7.0, containing 100 m*M* NaCl (~100 μg of total protein per ml). To avoid complete dissociation of the linker polypeptides from the individual phycobiliproteins, samples are rapidly loaded onto a small column of HAP (prepared in a 5-ml syringe and previously equilibrated with the same buffer). At this low phosphate concentration, all the phycobiliproteins remain bound to the HAP. Fractions containing phycobiliproteins are eluted, as rapidly as possible, by a stepwise increase of the buffer concentration. Depending on the affinity of the different phycobiliproteins for HAP, various volumes (generally 5–10 ml) of the following potassium phosphate concentrations are added: 10, 20, 30, 50 and 100 m*M* pH 7.0. Usually, the different phycobiliproteins will be eluted around 20–40 m*M* for PC, 40–60 m*M* for PE, and 50–100 m*M* for AP.

Analysis of Phycobilisomal Proteins by Electrophoresis

The protein composition of the phycobilisomes purified from red- and green-light-grown cells can be compared by performing sodium dodecyl sulfate (SDS)–polyacrylamide gel electrophoresis. Owing to their intrinsic coloration, phycobiliproteins will be visible without staining of the gel, whereas linkers will be revealed only after staining with Coomassie blue (R-250). Generally, such a gel system will separate at least one of the two subunits of PC-1 from that of PC-2 (either the α or β subunit depending on strain) and, in PC-adapting strains, the linker (L_R^{10}) located in the upper part of the rods of red-light phycobilisomes will be detectable. Scanning of the SDS gel will also permit quantification of each phycobiliprotein relative to the L_{CR}^{27} content or per phycobilisome, if one assumes that each phycobilisome contains six molecules of L_{CR}^{27}, as has been found in all the strains studied.

Since the molecular masses of the phycobiliproteins are within a very narrow range, it is easier to identify the two PC types (PC-1 and PC-2) by polyacrylamide gel electrophoresis in the presence of urea (or by isoelectrofocusing[12]) followed, if necessary, by a SDS–polyacrylamide gel in the second dimension. Scanning of the urea gels will generally allow the concentration of each phycobiliprotein subunit, including those of PC-1 and PC-2, to be determined.

SDS–Polyacrylamide Gels. Slab gel electrophoresis[13] is performed using the discontinuous buffer system of Laemmli.[14] Depending on the strain under study, either 15–20% [with a 30 : 0.1 (w/v) acrylamide-to-bisacrylamide ratio] or 10–20% [with a 30 : 0.8 (w/v) acrylamide-to-bisacrylamide ratio] linear gradient gels are used. Electrophoresis is usually performed overnight under a constant current of 15 mA (gel size, 0.15 × 15 × 20 cm). Before loading, precipitated phycobilisomes (~15 μg protein) are taken up into the sample buffer and heated for 3 min at 100°.

Polyacrylamide Gels in the Presence of Urea. Precipitated phycobilisomes (50–100 μg protein) are resuspended in 100–150 μl of running buffer [20 m*M* barbital (5,5-diethylbarbituric acid, sodium salt), 6 m*M* Tris, 6 *M* urea; the buffer is made up just before use and adjusted to pH 7.85 with HCl] made up to 8 *M* urea. Tubes (0.5 × 15 cm) are filled with a 5% gel solution [5% acrylamide–bisacrylamide (30 : 0.8, w/v), 70 m*M* Tris–HCl buffer, pH 7.85, 8 *M* urea], and polymerization is initiated by adding 0.06% (v/v) TEMED[15] and either 0.007% (w/v) ammonium persulfate or 6.6 × 10^{-7} *M* riboflavin in which case polymerization occurs

[12] P. H. O'Farrell, *J. Biol. Chem.* **250,** 4007 (1975).
[13] F. W. Studier, *J. Mol. Biol.* **79,** 237 (1973).
[14] U. K. Laemmli, *Nature (London)* **227,** 680 (1970).
[15] TEMED, *N,N,N,'N'*-Tetramethylethylenediamine.

under fluorescent or UV lamps (2–3 hr). Electrophoresis is performed for 3 hr at approximately 3.3 mA/gel at around 18°. The urea gels, which can be kept frozen, can then be laid on top of an SDS slab gel and overlayed with sample buffer.

Action Spectra for Complementary Chromatic Adaptation

Action spectra are performed at different wavelengths, to determine the efficiency of these irradiations on the triggering of phycobiliprotein synthesis. Two techniques have been described. They differ in the use of either exponentially growing cells (Method 1) or cells previously depleted of phycobiliproteins by nitrogen starvation (Method 2).

Method 1: Action Spectra of Exponentially Growing Cells

This method has been used by Haury and Bogorad[16] for the filamentous cyanobacterium *Fremyella diplosiphon,* now designated *Calothrix.*[1] One-liter cultures grown at 35 ± 2° in red light (culture free of PE) or in fluorescent light (PE-adapted culture) are transferred to the reverse light condition (12–15 hr) prior to monochromatic illumination. Duplicate 20-ml samples of cultures are transferred to test tubes which are incubated (6 hr at 36–39°) in front of a slide projector lens equipped with an interference filter. At the time of transfer to monochromatic light, the cells are actively growing (0.05–0.15 mg dry weight/ml). Culture controls are kept in the same light conditions for an extra 6-hr period and are used as a reference for phycobiliprotein production. The net pigment accumulation, per cell mass unit, resulting from the monochromatic illumination is divided by the average net pigment increase in the culture control. The ratios thus obtained for each wavelength at different photon fluxes are used to construct dose–response curves. The action spectrum is established for each phycobiliprotein (either PC or PE) by plotting wavelength versus reciprocal of the photon flux required to attain a normalized relative response value 0.555.

Method 2: Action Spectra of Phycobiliprotein-Depleted Cells

Fujita and Hattori[17] used cells depleted of their phycobiliprotein complement to show that a short exposure (3 min) to a chromatic illumination determines the subsequent resynthesis of either PE (green light) or PC (red light) in the dark. The conditions of nitrogen starvation (especially the time), however, must be predetermined for each strain so as to obtain

[16] J. F. Haury and L. Bogorad, *Plant Physiol.* **60,** 835 (1977).
[17] Y. Fujita and A. Hattori, *Plant Cell Physiol.* **1,** 293 (1960).

cells depleted of phycobiliproteins but still able to resynthesize phycobiliproteins in the dark on addition of a nitrogen source. This *de novo* synthesis occurs at the expense of the endogenous glycogen reserve which, under these conditions, provides carbon and energy sources as well as reducing power. Total cellular protein, phycobiliprotein and chlorophyll *a* derminations can be performed as described above. Cellular glycogen can be extracted, hydrolyzed, and assayed as described by Stanier *et al.*[18]

Using this method, action spectra have been determined for three different cyanobacterial strains: *Tolypothrix tenuis*,[19,20] *Fremyella diplosiphon*,[21] and *Synechocystis* PCC 6701.[22] Since it is necessary to adapt experimental conditions for each of the strains under study (especially volume of culture, as well as time and photon flux required to obtain phycobiliprotein-depleted cells) we describe below the major steps required to establish an action spectrum.

Step 1. Preillumination of Cells during Nitrogen Starvation. Exponentially growing cells are centrifuged, washed twice, and resuspended so as to obtain the original cell density ($\sim 10^7$ cells/ml) in mineral medium devoid of nitrate. The cell suspension is incubated under a high light flux (cool-white fluorescent) and bubbled with air–1% CO_2 for the period of time previously determined. At this stage, cells appear yellow–green and have lost more than 60% of their phycobiliprotein content, while chlorophyll *a* and total cellular proteins decrease by less than 10%. Simultaneously, the glycogen cell content increases to about 70% (w/w) of the total cellular dry mass. Incubation in the dark following nitrate addition results in immediate glycogen breakdown, whereas synthesis of phycobiliproteins starts after a lag of about half a generation time. Under such conditions, the time required for full restoration of the phycobiliprotein content is roughly equivalent to the period of nitrogen starvation. Resynthesis of phycobiliproteins stops when the glycogen reserve accumulated during nitrogen starvation is fully depleted.

Step 2. Determination of Dose–Response Curves and Photoreversibility. Light-bleached cells are either harvested and resuspended in growth medium containing nitrate or supplemented with nitrate. Cell suspensions are preirradiated with saturating red or green light fluxes. Aliquots are subsequently transferred to smaller flasks and flashed with complementary monochromatic light (540 or 640 nm) for various periods of time (0–

[18] R. Y. Stanier, M. Doudoroff, R. Kunisawa, and R. Contopoulou, *Proc. Natl. Acad. Sci. U.S.A.* **45,** 1246 (1959).

[19] S. Diakoff and J. Scheibe, *Plant Physiol.* **51,** 382 (1973).

[20] Y. Fujita and A. Hattori, *Plant Cell Physiol.* **3,** 209 (1962).

[21] T. C. Vogelman and J. Scheibe, *Planta* **143,** 233 (1978).

[22] N. Tandeau de Marsac, A.-M. Castets, and G. Cohen-Bazire, *J. Bacteriol.* **142,** 310 (1980).

60 min). Pigment content is determined after 15–48 hr of incubation in the dark. Experimental conditions are defined so that saturation by the second monochromatic illumination is usually reached after a 30-min exposure, half-saturation being obtained after about 10 min.

Step 3. Action Spectrum. Cells preirradiated with saturating light are subjected to flashes of monochromatic light at different wavelengths while varying photon fluxes and exposure times. Pigment content is determined after incubation in the dark to allow resynthesis of phycobiliproteins. To construct dose–response curves, phycobiliprotein ratios [PE/(PE + PC)] are plotted against the log of the photon flux. Action spectra are obtained by plotting the reciprocal of the dose required to elicit a 50% response level versus wavelength.

Alternatively, for a first qualitative analysis,[22] two aliquots of a phycobiliprotein-depleted cell suspension are preirradiated with a saturating dose of monochromatic light, one at 640 nm, the other at 540 nm. They are then exposed to nonsaturating irradiation for 10 min at wavelengths ranging from 450 to 590 nm and from 590 to 720 nm, respectively, followed by incubation in the dark after addition of nitrate. The nature and the amount of the phycobiliproteins resynthesized in the dark are compared with those present in the preirradiated control samples.

Comments

Determination of action spectra for chromatic adaptation with either Method 1 or 2 gives essentially identical results. However, when exponentially growing cells were used, Haury and Bogorad[16] showed that for each phycobiliprotein analyzed (PE or PC) there exists a second peak in the action spectrum in addition to the action maxima at 540 and 640 nm. For PC synthesis 463 nm is twice as efficient as 640 nm, whereas PE synthesis is equally efficient at 387 and 550 nm. These additional peaks were not observed by Vogelman and Scheibe[21] even though they used the same cyanobacterial strain. The observed differences might result from the degradation of a hypothetical blue photoreceptor which will occur when cells are preincubated under conditions of phycobiliprotein depletion.[16,21] On the other hand, it may be difficult to measure accurately small variations in the phycobiliprotein content of fully pigmented, exponentially growing cells.

Acknowledgments

We wish to thank Professor G. Cohen-Bazire for critical reading of the manuscript and Dr. G. Guglielmi for helpful discussions and advice.

[35] *In Vitro* Carotenoid Biosynthesis in *Aphanocapsa*

By GERHARD SANDMANN

Introduction

Membranes from cyanobacteria contain β-carotene and various xanthophylls derived from this C_{40} carotene. *Aphanocapsa* PCC 6714 is one of the few photosynthetic organisms known so far that can be used for *in vitro* studies of carotenoid biosynthesis.[1-4] As in other organisms, carotenoids are initially synthesized in *Aphanocapsa* via the general isoprenoid pathway with mevalonic acid (MVA) and prenyl pyrophosphates as intermediates. Two molecules of the resulting geranylgeranyl pyrophosphate (GGPP) are condensed to phytoene, which is the first carotene of the carotenoid pathway. This colorless carotene is subjected to a series of desaturation reactions followed by two cyclization steps yielding β-carotene and subsequently xanthophylls. For details on reactions and the enzymology of the carotenogenic pathway, see a recent review article.[5]

The *in vitro* carotenogenic system of *Aphanocapsa* has proved extremely useful, e.g., in characterizing certain herbicidal compounds that interfere with carotenoid formation and lead to general pigment bleaching.[6,7] The concern of this chapter is to describe procedures which allow the *in vitro* biosynthesis of intermediates of the carotenogenic pathway in two steps. These are the early steps from MVA to various prenyl pyrophosphates as well as the processes specific for carotenoid biosynthesis, i.e., the conversion of phytoene to various carotenes and to β-cryptoxanthin. Limited space does not permit inclusion of details on assays of single enzymes of the prenyl pyrophosphate steps and of β-carotene hydroxylase[4] in *Aphanocapsa*.

[1] I. E. Clarke, G. Sandmann, P. M. Bramley, and P. Böger, *FEBS Lett.* **140,** 203 (1982).
[2] G. Sandmann and P. M. Bramley, *Planta* **164,** 259 (1985).
[3] P. Bramley and G. Sandmann, *Phytochemistry* **24,** 2919 (1985).
[4] G. Sandmann and P. M. Bramley, *Biochim. Biophys. Acta* **843,** 73 (1985).
[5] P. M. Bramley, *Adv. Lipid Res.* **21,** 243 (1985).
[6] G. Sandmann and P. Böger, *in* "Wirkstoffe im Zellgeschehen" (P. Böger, ed.), p. 139. Universitätsverlag Konstanz, Konstanz, Federal Republic of Germany, 1985.
[7] G. Sandmann and P. Böger, *Encycl. Plant Physiol.* **19,** 595 (1986).

Assay Methods

General Remarks on Procedures and Reagents

Growth of *Aphanocapsa* (Pasteur Culture Collection, Strain 6714) is described elsewhere.[8] The cells are harvested after 4–5 days when the chlorophyll content of 1 liter has reached about 20 mg.

For the generation of phytoene from (*R*)-[2-^{14}C]MVA the *Phycomyces* mutant C5 is used in which the conversion of phytoene is blocked. This strain was obtained from the collection of the Departamento de Genética, Universidad de Sevilla, Sevilla, Spain, and was maintained and grown for 5 days, as previously described.[9] The fluffy mycelium of *Phycomyces* is harvested after 5 days. When small balls form instead, as sometimes happens, another batch must be grown and inoculated with a smaller amount of spore suspension (about 30,000 spores per liter culture medium). The mycelium is lyophilized to absolute dryness; only then does it retain carotenogenic activity, and it can be stored for several weeks in the deep freeze.

The *in vitro* synthesis of prenyl pyrophosphates from MVA as well as of colored carotenes and β-cryptoxanthin from phytoene is determined by radioactivity incorporation from the labeled substrates into the subsequent products. All reaction products are separated and purified in subsequent steps using different thin-layer chromatography (TLC) systems until constant specific radioactivity is attained. Product formation is quantitated by scraping off the bands and determining the incorporated radioactivity by liquid scintillation counting (LSC).

The plates used are silica gel 60 F_{254} and Al_2O_3 60 F_{254}, neutral, all from Merck (Darmstadt, FRG). Aluminum oxide plates are activated by heating at 120° for 30 min. Silica gel plates are impregnated with liquid paraffin by placing the plates in a 5% (v/v) petroleum ether solution until the front reaches 16 cm. Silica gel plates with a $AgNO_3$ strip are prepared by covering the plate with cardboard that is cut in such a way that a horizontal strip of 1.5 cm width at a distance of 3.5 cm away from the bottom edge can be sprayed onto the plate with a solution of 10% $AgNO_3$ dissolved in acetonitrile.

[^{14}C]Mevalonic acid was purchased from Amersham as (*R*)-[2-^{14}C]mevalonic acid lactone with a specific radioactivity of about 50 mCi/mmol or 1.9 GBq/mmol [the DL mixture is also suitable, but then twice the radioactivity has to be added to the incubations because only the (*R*) isomer reacts with the enzyme systems]. Fifty microliters of the lactone

[8] G. Sandmann, *Photosynth. Res.* **6,** 261 (1985).

[9] A. Than, P. M. Bramley, B. H. Davies, and A. F. Rees, *Phytochemistry* **11,** 3187 (1972).

dissolved in benzene (250 μCi in 2.5 ml) is dried in a stream of N_2. Then the lactone is converted to the sodium salt of MVA by addition of 500 μl of 0.01 *N* NaOH.

Marker compounds are used to identify the reaction products on the TLC plates. Farnesol (F-OH), geraniol (G-OH), and β-carotene were purchased from Sigma Chemical Co. Dimethylailyl alcohol (3-methyl-2-buten-1-ol), from Aldrich Chemical Co., is used to detect isopentenol (I-OH) on TLC plates. It runs with the same R_f value but, in contrast to the isopentenol, can be stained by iodine vapor. Phytoene and lycopene were isolated from *Phycomyces* C5 and tomato fruits (or commercial tomato paste), respectively, according to standard procedures.[10]

In Vitro Formation of Prenyl Pyrophosphates from MVA

Preparation of the Aphanocapsa Extract. A soluble extract is prepared from freeze-dried *Aphanocapsa* cells. One-half gram cells is ground with 10 ml of sand (Merck, No. 7711) in a mortar. Then 7.5 ml of 40 m*M* Tris–HCl buffer, pH 8.0, containing 0.5 m*M* dithiothreitol is added. The slury is transferred to a centrifugation tube and centrifuged at 12,000 *g* for 15 min at 4°. The supernatant is used in the following incubation immediately.

Incubation. The incubation medium contains, in 0.5 ml: 5 μmol ATP, 2 μmol $MgCl_2$, 3 μmol $MnCl_2$, 5 μmol KF, 50 μl of the (*R*)-MVA solution (containing 0.5 nmol of MVA and a total radioactivity of 10 μCi), and 100 μl *Aphanocapsa* extract (equivalent to ~2.5 mg protein) in the buffer used for the preparation of the *Aphanocapsa* extract. Incubation is carried out at 35° for 2 hr. The reaction is stopped by addition of 1 ml methanol and the mixture heated at 80–100° for 5 min.

Hydrolysis and Extraction of Prenyl Alcohols. After centrifugation at 2500 *g* for 10 min the pellet is washed with 0.5 ml Tris–HCl buffer (see above) plus 3 ml H_2O and centrifuged again. The combined supernatants are treated with 750 μl of 25% HCl at 80–100° for 10 min to hydrolyze all prenyl pyrophosphates to the corresponding alcohols. Then 750 μl of 25% NH_4OH, 750 μl of a 1% EDTA solution, and 10 μl of 1% methanol solutions each of G-OH and F-OH as well as 50 μl dimethylallyl alcohol (DMA-OH) are added.

The mixture is poured into a separatory funnel with 20 ml diethyl ether and 20 ml of a saturated NaCl solution. After collection of the organic layer, the aqueous phase is again partitioned against 20 ml diethyl ether. The combined ether extracts are dried over anhydrous Na_2SO_4. The residue is dissolved in 100 μl petroleum ether (bp 35–80°).

[10] G. Britton, this series, Vol. 18, p. 654.

Purification of Prenyl Alcohols by TLC and Radioassay

Step 1. The petrol extract is applied on the side of the reversed-phase plate that is not impregnated with paraffin. The plates are developed in methanol–H_2O (70 : 30, v/v). Before the alcohols are visualized by staining with iodine vapor, the area 0.5 cm above and below the impregnation front is scraped off and eluted with 5 ml methanol to obtain GG-OH for further purification (see Step 2). The alcohols migrate with R_f values of 0.8 for I-OH, 0.65 for G-OH, and 0.35 for F-OH. DMA-OH comigrates with I-OH and is used to identify this alcohol. Separation of I-OH and DMA-OH has shown that more than 90% of radioactivity associated with this band is found in I-OH.

Step 2. The methanol solution of the eluted GG-OH is evaporated and the residue redissolved in 100 μl petroleum ether. The petroleum ether solution is applied 2.5 cm below the $AgNO_3$ strip on the silica gel plate and chromatographed in 15% (v/v) ethyl acetate in petroleum ether (bp 100–140°). GG-OH is then adsorbed in the $AgNO_3$ zone.

Step 3. All bands with the various prenyl alcohols are scraped off and transferred to scintillation vials. After addition of 3 ml of a lipophilic scintillation fluid (e.g., Permafluor III from Packard Instruments) the radioactivity is determined by LSC.

Comments. Of all the enzymes involved in the conversion of MVA to GGPP, isopentenyl (IPP) isomerase (EC 5.3.3.2) is the most unstable. In comparison to other organisms, it is much more difficult to obtain cell extracts of *Aphanocapsa* which are active. Even in the extracts prepared as above, IPP isomerase is less active than in preparations obtained from spinach or several fungi, like *Fusarium* or *Phycomyces,* when the same procedure is employed. Apparently, inactive IPP isomerase acting as a bottleneck explains why in a previous preparation of cell homogenates from *Aphanocpasa,*[2] which forms various carotenoids from GGPP and phytoene, conversion of MVA or IPP to carotenes was not observed. Typical incorporation of radioactivity into prenyl alcohols is shown in Table I.

Phytoene Conversion to Subsequent Carotenes and β-Cryptoxanthin

Aphanocapsa membranes can be used to convert either [^{14}C]GGPP or [^{14}C]phytoene to colored carotenes. As [^{14}C]GGPP is very difficult to synthesize[11] and very unstable, only the more convenient coupled assays with simultaneous addition of cell extracts from *Phycomyces* C5 as the phytoene-generating system from (*R*)-[2-^{14}C]MVA is described here.

[11] J. Soll and G. Schultz, *Biochem. Biophys. Res. Commun.* **99,** 430 (1981).

TABLE I
INCORPORATION OF RADIOACTIVITY FROM (*R*)-[2-^{14}C]MVA INTO PRENYL PYROPHOSPHATES BY AN *Aphanocapsa* EXTRACT

Compound	Radioactivity (dpm)
Isopentenyl-PP	815
Geranyl-PP	597
Farnesyl-PP	1327
Geranylgeranyl-PP	3127

Preparation of Thylakoid Membranes from Aphanocapsa and Phycomyces C5 Cell Extracts. Thylakoid membranes are prepared from *Aphanocapsa* by lysozyme digestion of the cell wall and osmotic shock of the resulting spheroplasts in a hypertonic buffer. Spheroplasts are prepared from 500 ml of algal suspension. After centrifugation (8000 *g*, 30 min), the pellet is washed with 0.5 *M* sucrose, 10 m*M* $MgCl_2$, 2.5 m*M* potassium phosphate, and 2.5 m*M* sodium phosphate in 10 m*M* Tricine–NaOH buffer (pH 7.8). After resuspending in 12 ml of the same medium, lysozyme (50 mg) is added, and the suspension is incubated at 37° for 2 hr. The resulting spheroplasts are collected by centrifugation at 700 *g* for 5 min and resuspended in 0.4 *M* Tris–HCl buffer (pH 8.0) containing 5 m*M* dithiothreitol to reach a chlorophyll concentration of 1 mg/ml.

Cell extracts from *Phycomyces* C5 are obtained by rubbing 200 mg of freeze-dried mycelia through a sieve with a mesh size of 0.4 mm. Then 1.6 ml of the buffer mentioned above is added, and the paste is stirred with a spatula for 1 min. The supernatant after centrifugation (10,000 *g*, 15 min) is used in the coupled assay.

Incubation System. The assay medium contains the following, in 0.5 ml: 5 μmol ATP, 1 μmol NAD, 3 μmol $MnCl_2$, 2 μmol $MgCl_2$, 25 μl (*R*)-MVA solution (corresponding to 0.25 μCi or 5 nmol), 100 μl *Phycomyces* C5 extract (equivalent to 0.5–1.5 mg protein), and 300 μl *Aphanocapsa* membrane suspension (equivalent to 200 μg chlorophyll) in the 0.4 *M* Tris–HCl buffer used above. The incubation is carried out for 2 hr in the light at 35°.

Extraction of Carotenoids. The reaction is terminated by adding 3 ml methanol and marker carotenes (phytoene and lycopene, 25 μg each). All subsequent steps are carried out in dim light to avoid degradation of carotenoid pigments. The content of the reaction tubes is poured into a separatory funnel together with 50 ml of saturated NaCl solution and 15 ml diethyl ether. After partition, the upper ether layer is collected and dried with anhydrous Na_2SO_4. Then the solvent is evaporated in a stream of N_2 and the residue dissolved in 100 μl diethyl ether.

Separation of Carotenoids by TLC

Step 1. The pigment solution is quantitatively placed on a silica gel plate which is immediately developed with 15% (v/v) toluene in petroleum ether (bp 100–120°). Three bands become visible on the plate, a green one at the origin, a red one at an R_f value around 0.4, and a yellow band around 0.7. The green band (chlorophyll and xanthophylls) and the yellow band (β-carotene and phytoene) are scraped off the plate and eluted with 5 ml diethyl ether plus 1 ml methanol. The solutions are concentrated to 100 μl in a stream of N_2.

Step 2. The carotene solution is chromatographed on activated Al_2O_3 plates with 7% (v/v) toluene in petroleum ether. The β-carotene band is visible at a R_f value of about 0.4, and phytoene (R_f 0.7) is detected either under UV light by its quenching of the fluorescence of the plate or by staining with iodine vapor. When the R_f value of β-carotene exceeds 0.5, which happens on very humid days, the toluene content of the developing solvent has to be decreased.

Step 3. The chlorophyll–xanthophyll solution is rechromatographed on silica gel with toluene–ethyl acetate–methanol (75 : 20 : 5, v/v/v) as solvent. β-Cryptoxanthin runs about half the distance of the solvent front. It can be recognized as a faint yellow band at equal distance from the reddish echinenone (sometimes masked by a blackish green pigment) in the upper part of the chromatogram and from zeaxanthin, the major xanthophyll in *Aphanocapsa*. The running distance for β-cryptoxanthin relative to echinenone is 0.8 and relative to zeaxanthin, 2. If the β-cryptoxanthin band is too weak for identification, this xanthophyll has to be isolated from *Aphanocapsa* and purified from a concentrated carotenoid solution in the TLC system just mentioned. Twenty micrograms of β-cryptoxanthin should then be applied together with the extract on the silica gel plate used in this TLC step.

Step 4. β-Cryptoxanthin is eluted from the silica gel with diethyl ether–methanol (4 : 1, v/v) and is then rechromatographed on activated Al_2O_3 plates with 35% (v/v) acetone in petroleum ether.

Liquid Scintillation Counting of ^{14}C-Labeled Carotenoids. Lycopene is scraped off the TLC plate which was developed in Step 1, phytoene and β-carotene are scraped off the Al_2O_3 plate of Step 2, and β-cryptoxanthin off the Al_2O_3 plate used in Step 4 and transferred directly to LSC vials. Three milliliters of a scintillation fluid for lipophilic samples is added, and the radioactivity incorporated into the carotenoids is determined by LSC.

TABLE II
In Vitro FORMATION OF CAROTENOIDS BY *Aphanocapsa* MEMBRANES FROM [^{14}C]PHYTOENE[a]

Carotenoid	Radioactivity (dpm)
Lycopene	3975
β-Carotene	6522
β-Cryptoxanthin	8767

[a] Phytoene equivalent to 30,000 dpm was simultaneously generated by a *Phycomyces* C5 cell extract from (*R*)-[2-^{14}C]MVA in a coupled assay.

Comments. Table II shows the typical result of an experiment performed as described here. Instead of cell wall digestion with lysozyme, *Aphanocapsa* cells can be broken mechanically to obtain thylakoid membranes active in conversion of phytoene via carotene intermediates to β-carotene. The breaking conditions are crucial, however, in order to avoid inactivating carotenogenesis.[2]

The coupling of [^{14}C]phytoene generation by *Phycomyces* C5 cell extracts with *in vitro* carotenogenesis by *Aphanocapsa* membranes can be performed either simultaneously as described here or sequentially. In the latter case, the reaction is carried out as described above (Incubation System), but without *Aphanocapsa* membranes present. These membranes are added after a 2-hr incubation period, and incubation is continued for another 2 hr.[2] The advantage of the simultaneous incubations is a higher rate of both β-carotene and β-cryptoxanthin formation. This technique is also advantageous when herbicidal inhibitors interfering with carotene interconversion are assayed.[3] For time course experiments, in which the flow of radioactivity from phytoene through other carotenes is followed, however, sequential coupling has to be employed.[3] The coupling technique described here was also used to study carotenogenesis in chloroplasts of higher plants.

Acknowledgment

This work was supported by the Deutsche Forschungsgemeinschaft and by a grant from Stiftung "Umwelt und Wohnen."

[36] Carotene Oxygenase in *Microcystis*

By FRIEDRICH JÜTTNER

Introduction

Norcarotenoids are widely distributed in spent culture media of various cyanobacteria. 6-Methylhept-5-en-2-one, 6-methylhept-5-en-2-ol, and β-ionone were found in cultures of *Synechococcus* PCC 6911[1] and PCC 6301 (*Anacystis nidulans*) and *Anabaena cylindrica*.[2] The *Synechococcus* strain PCC 6301 is further characterized by its ability to excrete (*E*)-geranylacetone and β-ionone 5,6-epoxide. Quite different behavior is exhibited by members of the genus *Microcystis*, which have high capacities for the formation of β-cyclocitral[3] and 4-hydroxy-β-cyclocitral. The structures of all the aforementioned compounds indicate their genesis from carotenoids.[4] The existence of characteristic norcarotenoids for particular species points to the presence of specific carotenoid-cleaving enzymes in these organisms.

Occurrence and General Features of Carotene Oxygenase

All genuine species of the genus *Microcystis* so far studied[3] exhibit powerful carotene oxygenase activity. The enzyme system responsible is able to split oxidatively β-carotene or zeaxanthin, resulting in the formation of crocetindial, and β-cyclocitral or 4-hydroxy-β-cyclocitral, respectively. The formation of these products is strictly dioxygen dependent. Since highly efficient incorporation of 18-dioxygen has been obtained in β-cyclocitral the enzyme system has tentatively been designated β-carotene 7,8(7′,8′)-oxygenase.[5] The formal reaction can be formulated as depicted in Scheme 1.

Principles of the Assays

The axenic strain *Microcystis* PCC 7806 is a suitable organism for studies on the carotene oxygenase system. Although activity of this enzy-

[1] J. J. Henatsch and F. Jüttner, *Wat. Sci. Tech.* **15,** 259 (1983).
[2] F. Jüttner, J. Leonhardt, and S. Möhren, *J. Gen. Microbiol.* **129,** 407 (1983).
[3] F. Jüttner, *Z. Naturforsch. C* **39,** 867 (1984).
[4] C. R. Enzell, I. Wahlberg, and A. J. Aasen, *Fortschr. Chem. Org. Naturst.* **34,** 1 (1977).
[5] F. Jüttner and B. Höflacher, *Arch. Microbiol.* **141,** 337 (1985).

SCHEME 1. **1:** R = H, β-Carotene (β,β-carotene); R = OH, zeaxanthin (β,β-carotene-3,3′-diol). **2:** Crocetindial (8,8′-diapocarotene-8,8′-dial). **3:** R = H, β-cyclocitral (2,6,6-trimethylcyclohex-1-enecarboxaldehyde); R = OH, hydroxy-β-cyclocitral (2,6,6-trimethyl-4-hydroxycyclohex-1-enecarboxaldehyde).

matic system is detectable during all growth phases, use of cells at the beginning of stationary phase is recommended. The assay of the carotene oxygenase system is based on the quantitative gas chromatographic determination of the cleavage product β-cyclocitral after activation of the enzyme system by freezing and thawing. An alternative method is to analyze quantitatively crocetindial by thin-layer chromatography. The assay for β-cyclocitral monitors solely the degradation of β-carotene, while detection of crocetindial follows degradation of both β-carotene and zeaxanthin. During and after activation of the oxygenase, the entrance of dioxygen into the system must be strictly excluded to prevent the uncontrolled degradation of internal carotenoids. Gassing with an inert gas is not sufficient. Better results are obtained by additional application of an oxygen-consuming reaction mixture (glucose oxidase). The reaction is initiated by adding excess O_2. The cleavage products are subsequently quantitatively analyzed. Since very few minor products are observed, nearly stoichiometric amounts of the different products can be measured.

Assay Based on the Determination of β-Cyclocitral

The following procedure is used to analyze quantitatively the formation of β-cyclocitral.

1. Twenty milliliters of a suspension of *Microcystis* (6.5×10^7 cells/ml = 133 μg dry weight/ml = 2.0 μg chlorophyll *a*/ml) is spun down (10 min at 1200 *g*).
2. The cell pellet is resuspended in 3 ml of 20 m*M* TES–buffer, pH 7.4 [TES, *N*-tris(hydroxymethyl)methyl-2-aminoethanesulfonic acid], and the centrifuge tube is sealed with a spectum and gassed with argon for 2 min.

3. Two hundred fifty microliters of 1 *M* glucose, 30 μl catalase (130,000 U/ml), and 30 μl glucose oxidase (2,000 U/ml) are added through the septum.
4. After 8 min of incubation, the samples are frozen in a deep-freeze (−16°) for between 3 and 12 hr to activate the carotene oxygenase system.
5. Before chilling it is necessary to gas again with argon to prevent any oxygenase reaction. The active suspension prepared in this way can be used for further experiments; however, O_2 must be excluded.
6. The reaction is started by gassing with O_2 after transfer to a 100-ml round-bottomed flask in a closed loop stripping device (see elsewhere in this volume [67]). As an internal standard 1 ml of a solution of 2-decanone (10 mg 2-decanone is dissolved in 1 ml of methanol and subsequently diluted to 1000 ml with water) is added. Then β-cyclocitral is stripped for 30 min and adsorbed to a Tenax-filled cartridge.
7. β-Cyclocitral is thermally desorbed from the cartridge, transferred to a glass capillary column at 40°, and separated isothermally at 140° (see [67] in this volume). For separation, see Fig. 1.

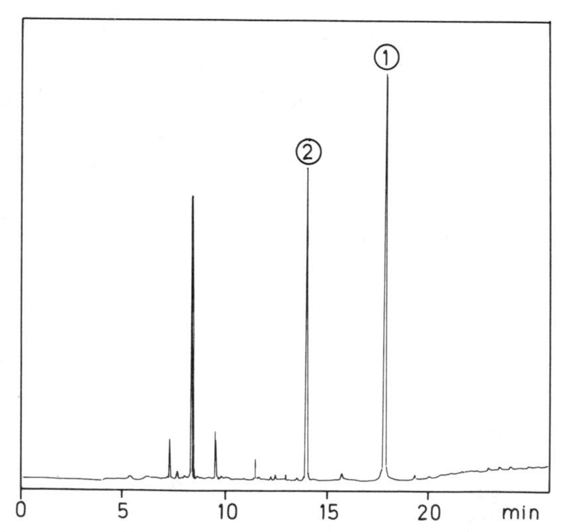

FIG. 1. Gas chromatographic separation on a 50-m glass capillary column, coated with UCON 50 HB 5100, of β-cyclocitral (1) isolated from 20 ml of a *Microcystis* suspension (2 μg chlorophyll *a*/ml) that had been activated by freezing and thawing. 2-Decanone (2) was added as internal standard. Temperature program: After the transfer, heating from 40 to 140° at a rate of 20°/min, 15 min isothermal, then heating at a rate of 50°/min up to 190° and holding for 7 min.

8. To calibrate the detector signals (peak area or height), authentic β-cyclocitral (BASF, Ludwigschafen, FRG) is added in methanolic solution (0.2 mg/ml methanol) to an equivalent amount of water and treated under the same conditions as the measured sample.

The formation of β-cyclocitral amounts to around 40 ng/μg chlorophyll *a*. Perchloric acid can be added in a final concentration of 0.6% (v/v) to terminate the reaction. β-Cyclocitral is stable under the conditions employed.

Assay Based on Crocetindial

The quantitative determination of crocetindial, the common oxidative cleavage product of β-carotene, zeaxanthin, and possibly other carotenoids, is carried out by thin-layer chromatography. The following procedure is convenient.

1. Fifty milliliters of a suspension of *Microcystis* (2.0 μg chlorophyll *a*/ml) is quickly centrifuged at high speed to obtain a very compact cell pellet.
2. The cell pellet is extracted twice with a mixture of 5 ml acetone–methanol (7 : 3, v/v) and disrupted by treating with a sonicator with a microtip for a few seconds. After two additional extractions with 3 ml of diethyl ether the solutions are combined and brought to dryness in a rotary evaporator.
3. The residue is redissolved in a small amount of chloroform–methanol (2 : 1, v/v) and applied in a streak on the unimpregnated part of a reversed-phase thin-layer plate (20 × 20 cm) that has been aged for 1–2 days.
4. The reversed-phase thin-layer plate is prepared by first dipping the whole 0.5 mm kieselguhr G layer (Kieselgur G, Merck No. 8129, Darmstadt, FRG) in a solution of antioxidant (1 g of 2,6-di-*tert*-4-methyl-phenol in 100 ml of ethanol, subsequently diluted to 1 liter by petroleum ether) and then, after drying at ambient temperature, in a solution of fat [35 g of Palmin (coconut fat), 35 g Livio (vegetable oil) in 1 liter of light petroleum], leaving the 4 cm wide lower part of the plate unimpregnated to enable the application of the carotenoid solution there in the form of a streak.[6]
5. The developing solvent is methanol–acetone–water (20 : 8 : 3, v/v/v). The sequence of the separated pigments can be seen in Fig. 2.

[6] K. Egger, *Planta* **58,** 664 (1962).

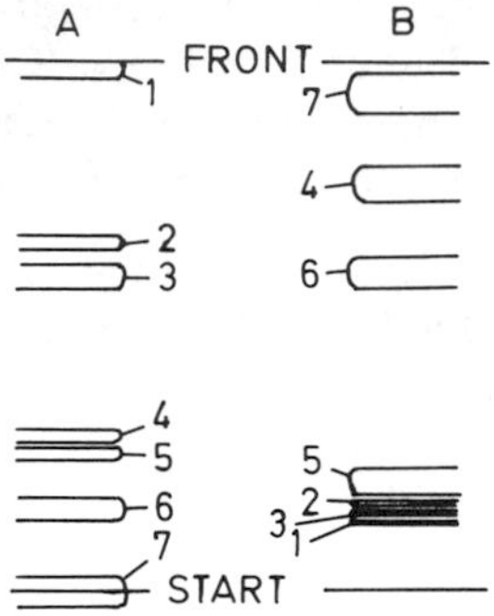

FIG. 2. Thin-layer chromatogram of the pigment extract from 50 ml of a *Microcystis* suspension activated by freezing and thawing. Trace (A) was obtained after separation on silica gel, (B) on reversed-phase plates, 1, β-Carotene (beginning of the reversed phase in B); 2, β-apo-8′-carotenal; 3, echinenone; 4, crocetindial; 5, chlorophyll *a*; 6, zeaxanthin; 7, myxoxanthophyll.

6. The crocetindial band is scraped off after separation and eluted with ethanol. After transfer to benzene, spectra are recorded from 490 to 390 nm. The calculations are performed using the specific extinction coefficient $E_{1\,cm}^{1\%}$ = 3970 at the maximum of 445 nm.[7]

An alternative way to separate crocetindial is by application to a 0.5 mm silica gel plate (Kieselgel 60 GF_{254}, Merck No. 7730) with the solvent system petroleum ether (50–70°)–2-propanol–water (100 : 10 : 0.25, v/v/v).[8] The sequence of pigments after separation is shown in Fig. 2.

Features of the Carotene Oxygenase System

The carotene oxygenase system is membrane bound and can be partially solubilized after lysozyme treatment (1 mg/ml, 30 min at 37°) and sonification. The proportion of solubilized enzyme is highly variable, depending on the treatment and growth phase of the culture. So far the partially purified preparations invariably also contained chlorophyll *a*. Though a linkage to thylakoid membranes is very likely, other carotenoid-containing membranes, such as cytoplasmic membranes[9] or cell walls[10,11]

[7] B. H. Davis, *in* "Chemistry and Biochemistry of Plant Pigments" (T. W. Goodwin, ed.), Vol. 2, p. 38. Academic Press, New York, 1976.

[8] H. Stransky and A. Hager, *Arch. Mikrobiol.* **72,** 84 (1970).

[9] G. S. Bullerjahn and L. A. Sherman, *J. Bacteriol.* **167,** 396 (1986).

[10] C. M. Resch and J. Gibson, *J. Bacteriol.* **155,** 345 (1983).

[11] U. J. Jürgens and J. Weckesser, *J. Bacteriol.* **164,** 384 (1985).

are not ruled out as reaction sites. Further purification has not been attempted because of the extreme oxygen sensitivity of the system. A close connection between carotenoid-containing structures and the enzyme is evident since application of β-carotene[5] as the substrate increases the yield of β-cyclocitral not more than 20%. β-Apo-8′-carotenal which exhibits only one β-ionone ring and which has been found in minor concentrations in activated cells, may be an intermediate in the reaction from β-carotene to crocetindial.

Cofactors do not seem to be of importance since the supernatant of activated cells can be replaced by buffer without loss of activity and since addition of supernatant from unreacted preparations to partially reacted samples did not increase the formation of β-cyclocitral. Unlike the case for carotene 15,15′-dioxygenase (EC 1.13.11.21) present in the mucosa of rats,[12] sulfhydryl compounds at a concentration of 0.1 m*M* did not have a pronounced effect on the reaction. Dithioerythritol had a slightly positive, cysteine a negative, and 2-mercaptoethanol and reduced glutathione no effect on the formation of β-cyclocitral. The most effective inhibitors were antioxidants. Complete inhibition was obtained with 0.5 m*M* *tert*-butylhydroquinone, propyl gallate, and nordihydroguaiaretic acid.

[12] D. S. Goodman and J. A. Olson, this series, Vol. 15, p. 462.

[37] Cytochrome b_6/f Complex in Cyanobacteria: Common Components of Respiratory and Photosynthetic Electron Transport Systems

By RICHARD MALKIN

The cytochrome b_6/f complex functions as a plastoquinol–plastocyanin (cytochrome *c*) reductase (EC 1.10.99.1) under conditions of photosynthetic electron transport. Recent results have also suggested the same complex functions in the cyanobacterial respiratory chain, utilizing reduced pyridine nucleotides as electron donors instead of water.[1] The cytochrome b_6/f complex from higher plants has been isolated and extensively characterized,[2,3] and, although the purification and analysis of the

[1] G. Sandmann and R. Malkin, *Arch. Biochem. Biophys.* **234,** 105 (1984).
[2] E. Hurt and G. Hauska, *Eur. J. Biochem.* **117,** 592 (1981).
[3] G. Hauska, this series, Vol. 126, p. 271.

cyanobacterial complex is not as complete,[4] the procedures used for the isolation of the higher plant cytochrome complex are generally applicable to the membranes of cyanobacteria.[3]

Purification Procedure

The procedure described for the isolation of the cytochrome b_6/f complex from cyanobacterial membranes is based on one originally developed for membranes from higher plants.[2,4] The isolation involves treatment of thylakoid membranes to remove extrinsic proteins, solubilization of the cytochrome complex with detergent, and purification of the complex by ammonium sulfate fractionation and sucrose density gradient centrifugation. Although octylglucoside was used in the original purification, other detergents such as laurylmaltoside or *N*-methyl-*N*-alkanoylglucamides (MEGA detergents), have also been used in the purification of the higher plant complex.[3] The use of these detergents for the cyanobacterial preparation has not yet been described.

Step 1. Thylakoid Membrane Preparation

Photosynthetic membranes should be isolated from whole cells of cyanobacteria by standard procedures. In general, this involves disruption of the cell, usually with the use of the French press, and removal of unbroken cells and other large membrane fragments by low-speed centrifugation. Thylakoid membranes are then isolated by differential centrifugation.

Step 2. Removal of Soluble and Extrinsic Membrane Proteins

Membranes are resuspended in a solution of 0.15 *M* NaCl to a chlorophyll concentration of 0.1–0.2 mg/ml. This suspension is centrifuged at 4° for 10 min at 13,000 *g* to pellet the thylakoid membranes. The supernatant solution, which contains soluble proteins, is discarded. The membranes are resuspended in a solution containing 2 *M* NaBr, 0.3 *M* sucrose, 5 m*M* $MgCl_2$, 20 m*M* NaCl, and 50 m*M* Tris–HCl (pH 7.8) to give a final chlorophyll concentration of 1 mg/ml. The suspension is incubated at 4° for 30 min, and, after this time, an equal volume of cold distilled water is added. The suspension is centrifuged at 4° for 20 min at 15,000 *g*. This treatment removes most of the CF_1 complex from the photosynthetic membranes, but if the final preparation of the cytochrome complex shows contamina-

[4] M. Krinner, G. Hauska, E. Hurt, and W. Lockau, *Biochim. Biophys. Acta* **681,** 110 (1982).

tion with CF_1 subunits (α and β subunits of 59 and 55 kDa on SDS–PAGE analysis) the NaBr step can be repeated an additional time. After the NaBr treatment, the pellet is resuspended in a solution containing 0.3 *M* sucrose, 5 m*M* $MgCl_2$, 20 m*M* NaCl, and 50 m*M* Tris–HCl (pH 7.8) to give a chlorophyll concentration of 0.1–0.2 mg/ml, and the membranes are reisolated by centrifugation at 4° at 13,000 *g* for 10 min. At this point, the membranes can be stored as a pellet at 4° overnight prior to detergent treatment, but it is preferable to continue the purification and complete it in a single day.

Step 3. Detergent Extraction of Cytochrome b_6/f Complex

The membrane pellet is resuspended in a solution containing 0.3 *M* sucrose, 50 m*M* Tris–HCl (pH 7.8), 5 m*M* $MgCl_2$, and 20 m*M* NaCl to give a final chlorophyll concentration of 3 mg/ml. An equal volume of the above solution, containing 60 m*M* octylglucoside, 1% (w/v) sodium cholate, and 0.8 *M* ammonium sulfate, is added. This yields final concentrations of 1.5 mg chlorophyll/ml, 30 m*M* octylglucoside, 0.5% cholate, and approximately 10% ammonium sulfate. This suspension is incubated at 4° for 30 min and then centrifuged for 1 hr at 360,000 *g*. The supernatant solution, which should be pale green in color, contains the solubilized cytochrome complex as well as contaminating pigments. It is convenient to assay for the presence of the complex at this stage by redox difference spectra. Ascorbate-reduced minus ferricyanide-oxidized spectra in the region of 540–580 nm yield a peak at 554 nm, indicative of cytochrome *f*. Usually all the cytochrome *f* present in the thylakoid membranes is extracted by this procedure; in the case of higher plant thylakoid membranes, approximately 1.5 nmol cytochrome *f*/mg chlorophyll is found using ε_{mM} = 18 at 554–540 nm.

Step 4. Ammonium Sulfate Fractionation

The detergent-solubilized supernatant solution is fractionated with ammonium sulfate. A saturated solution of ammonium sulfate adjusted to pH 7.5 is used. Ammonium sulfate is added to the supernatant solution to give 45% saturation, and the suspension is stirred for 10 min at 4°. The suspension is centrifuged at 4° at 13,000 *g* for 10 min and the pellet discarded. To the supernatant solution, ammonium sulfate is added to give 55% saturation and the incubation and centrifugation repeated as before. The supernatant solution is discarded, and the pellet is redissolved in a minimal amount of solution containing 30 m*M* Tris–succinate buffer (pH 6.5) and 0.5% sodium cholate. The cytochrome *f* concentration of this solution should be 25–50 μM by redox difference spectrophotometry. The

TABLE I
PURIFICATION OF THE CYTOCHROME b_6/f COMPLEX FROM *Anabaena variabilis*[a]

Purification step	Cytochrome b_6 (μM)	Chl/cyt b_6 (mol/mol)	Cyt b_6/protein	Activity[b]	Yield of cyt b_6 (%)
1. Extract	5.3	317	0.54	68	100
2. 300,000 *g* supernatant solution	3.7	162	3.1	50	70
3. Ammonium sulfate fraction (45–55%)	32.3	85	4.2	28	48
4. Sucrose gradient fraction	10.0	0.9	9.0	19	15

[a] Data from Krinner *et al.*[4]
[b] Activity is expressed as μmol cyt *c* reduced/nmol cyt b_6/hr.

solution is dialyzed against 1 liter of the same solution at 4° for 45 min. A longer dialysis time is to be avoided since this leads to aggregation and precipitation of the complex.

Step 5. Sucrose Gradient Centrifugation

The dialyzed solution containing the cytochrome complex is further fractionated by sucrose density gradient centrifugation. Sucrose gradients (7–30% sucrose) are prepared in a solution containing 30 m*M* Tris–succinate buffer (pH 6.5), 0.5% sodium cholate, and 30 m*M* octylglucoside. Gradients of approximately 10 ml volume can be used with a sample volume of 1.0 ml in a SW-41 swinging-bucket rotor, or gradients of approximately 35 ml and 3 ml samples can be used in a VTi50 rotor. Gradients are centrifuged at around 300,000 *g* for 16 hr in the SW-41 rotor or around 240,000 *g* for 16 hr in the VTi50 rotor. At the end of this centrifugation, the cytochrome complex is visible as a reddish brown band in the middle of the gradients. The complex is collected with a syringe and the cytochrome *f* content determined spectrophotometrically. The complex is stored in small aliquots by rapidly freezing samples to 77 K. Storage is best at 77 K, although it may be more convenient to store at −80 or −20°.

The purification scheme for the preparation of the cytochrome b_6/f complex from *Anabaena* is summarized in Table I.

Properties

Composition

The cytochrome b_6/f complex contains cytochrome b_6–cytochrome *f*–Rieske iron–sulfur center in a molar ratio of 2 : 1 : 1. Cytochrome *f* (E_m

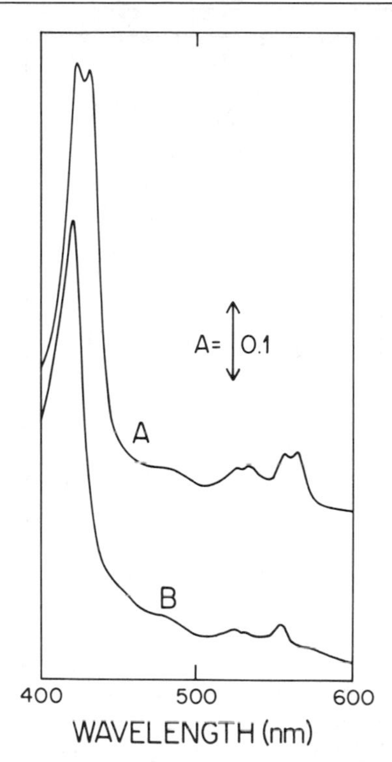

FIG. 1. Visible absorption spectra of the cytochrome b_6/f complex. (A) Spectrum recorded in the presence of sodium dithionite. (B) Spectrum recorded in the presence of sodium ascorbate. Spectra were recorded in a solution containing 30 m*M* Tris–succinate buffer (pH 6.5) and 0.4% sodium cholate. The sample contained 0.5 nmol/ml cytochrome *f*.

+350 mV) is associated with a polypeptide subunit of approximately 34 kDa. Cytochrome b_6 (E_m −40 and −170 mV at pH 6.5) is associated with a polypeptide subunit of around 23 kDa. The Rieske iron–sulfur protein (E_m +290 mV) is associated with a polypeptide subunit of 20 kDa. An additional subunit (~17 kDa), known as subunit IV, is also found in the complex, and several low-molecular-weight subunits are often found in the preparation.[5] Plastoquinone (~1 PQ/cyt *f*) is also found associated with the complex.[6]

The visible absorption spectra of the complex in the presence of dithionite (A) and ascorbate (B) are shown in Fig. 1, and oxidized-minus-reduced difference spectra in the cytochrome α-band region are shown in Fig. 2. Cytochrome *f* is analyzed by the ascorbate-reduced minus ferricy-

[5] E. Hurt and G. Hauska, *J. Bioenerg. Biomembr.* **14,** 405 (1982).
[6] R. K. Chain, *FEBS Lett.* **180,** 321 (1985).

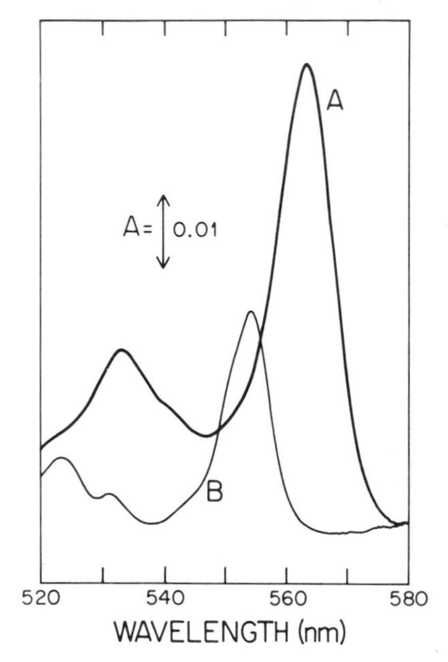

FIG. 2. Absorption difference spectra of the cytochrome b_6/f complex in the cytochrome α-band region. (A) Dithionite-reduced minus ascorbate-oxidized difference spectrum. (B) Ferricyanide-oxidized minus ascorbate-reduced difference spectrum. The sample contained 1.3 nmol/ml cytochrome *f* in a solution of 30 m*M* Tris–succinate buffer (pH 6.5) and 0.4% sodium cholate.

anide-oxidized difference spectrum (Fig. 2B) and cytochrome b_6 by the dithionite-reduced minus ascorbate-oxidized difference spectrum (Fig. 2A). The ratio of cytochrome b_6 to cytochrome *f* shown in Fig. 2 is 2 : 1. The Rieske iron–sulfur center is detected on the basis of its EPR *g* value (liquid helium temperature) at $g = 1.90$ in the reduced state. EPR signals associated with the oxidized forms of the cytochromes have also been detected at liquid helium temperatures in the $g \sim 3.0$ region; these are associated with the low-spin form of the heme groups.[7]

Enzymatic Properties

The cytochrome b_6/f complex catalyzes quinol–plastocyanin or cytochrome *c* oxidoreductase activity. The specific activity of the purified complex is in the range of 30–40 μmol acceptor reduced/nmol cytochrome *f*/hr,[2,3] although variations in activities are dependent on the specific elec-

[7] J. Bergström, L. E. Andreasson, and T. Vänngård, *FEBS Lett.* **164,** 71 (1983).

TABLE II
ELECTRON DONORS AND ACCEPTORS FOR THE CYTOCHROME b_6/f COMPLEX FROM *Anabaena variabilis*[a]

Electron donor[b] or acceptor[c]	Oxidoreductase activity (μmol/nmol cyt b_6/hr)
Electron donor	
Plastoquinol-1	16
Ubiquinol-1	15
Trimethyl-*p*-benzoquinol	0.8
Duroquinol	12
Plastoquinol-9	20
Ubiquinol-9	5
Electron acceptor	
Cytochrome *c*-553 (*A. variabilis*)	32
Plastocyanin (*A. variabilis*)	31
Cytochrome *c* (horse heart)	34
Cytochrome *c* (*Rhodopseudomonas capsulata*)	9
Cytochrome *c*-552 (*Euglena*)	1
Plastocyanin (spinach)	9

[a] Data taken from Krinner *et al.*[4]
[b] Horse heart cytochrome *c* as acceptor.
[c] Plastoquinol-9 as electron donor.

tron donors and acceptors used. In the case of the cytochrome b_6/f complex from the cyanobacterium *Anabaena,* the results of Table II indicate plastoquinol-9 is the best electron donor whereas basic electron accepting proteins (cyt *c*-553 and plastocyanin from *Anabaena,* horse heart cyt *c*) are the best electron acceptors. While plastoquinol-1, plastoquinol-2, plastoquinol-9, and duroquinol are commonly used electron donors, decylplastoquinol has recently been reported to be the most effective donor for the higher plant cytochrome complex.[8]

When incorporated into artificial liposomes, the cytochrome b_6/f complex is active in translocating protons.[9] Under conditions of electron transfer from quinol to high-potential electron acceptors, a H^+-to-e^- ratio approaching 2 is observed in the presence of the ionophore valinomycin.[9,10] The liposome-incorporated complex also shows suppressed rates of electron transfer which can be stimulated by the addition of uncoupling agents.[9,10]

[8] P. Rich, P. Heathcote, and D. A. Moss, *Biochim. Biophys. Acta* **892,** 138 (1987).
[9] E. Hurt, G. Hauska, and Y. Shahak, *FEBS Lett.* **149,** 211 (1982).
[10] I. Willms, R. Malkin, and R. K. Chain, *Arch. Biochem. Biophys.* **258,** 248 (1987).

The removal of the Rieske iron–sulfur protein and the associated plastoquinone from the complex have been reported.[6,11] In both cases, the depleted complexes are enzymatically inactive, and initially it was not possible to reconstitute the activity of the Rieske protein-depleted complex.[11] More recently, the successful reconstitution of the isolated Rieske protein into a depleted chloroplast cytochrome complex has been reported.[12] In the case of the quinone-depleted complex, restoration of activity was dependent on the addition plastoquinone (1 PQ/cyt *f*) and lipids.[6]

Inhibitors

The quinol–plastocyanin (cytochrome *c*) oxidoreductase activity of the cytochrome complex is inhibited by quinone analogs which have been found to interact with the Rieske iron–sulfur center.[13,14] The most effective inhibitor is stigmatellin, but 2,5-dibromomethylisopropylbenzoquinone (DBMIB), 2-iodo-6-isopropyl-3-methyl-2′,4,4′-trinitrodiphenyl ether (DNP-INT), and 5-(*n*-undecyl)-6-hydroxy-4,7-dioxobenzothiazole (UHDBT) are also efficient inhibitors.[2,4,15] These compounds all show inhibition in the 10^{-6}–10^{-7} *M* range. "Classical" inhibitors of the mitochondrial-type cytochrome b/c_1 complex, such as antimycin A and 2-heptyl- (or nonyl-)4-hydroxyquinoline *N*-oxide (HQNO or NQNO), have little, if any, effect on the oxidoreductase activity of the cytochrome b_6/f complex.[2,16]

Assay Methods

The most convenient electron donor for the cytochrome b_6/f complex is duroquinol, although with many systems this donor does not give the highest rate (see Table II). However, since other donors are not commercially available or are only poorly soluble in aqueous systems, duroquinol has been widely used. Most electron acceptors are naturally occurring proteins that have been isolated from photosynthetic cells, although horse heart cytochrome *c* and ferricyanide have also been used. Because plastocyanin is the physiological electron acceptor for most cytochrome b_6/f

[11] E. Hurt, G. Hauska, and R. Malkin, *FEBS Lett.* **134,** 1 (1981).

[12] Z. Adam and R. Malkin, *FEBS Lett.* **225,** 67 (1987).

[13] R. Malkin, *Biochemistry* **21,** 2945 (1982).

[14] R. Malkin, *FEBS Lett.* **208,** 317 (1986).

[15] W. Oettmeier, D. Godde, B. Kunze, and G. Hofle, *Biochim. Biophys. Acta* **807,** 216 (1985).

[16] E. Lam, *FEBS Lett.* **172,** 255 (1984).

complexes, this protein is the most suitable acceptor, and its reduction can be monitored in a dual-wavelength spectrophotometer at 590–500 nm. Rates of electron transfer are based on an $\varepsilon_{590-500\text{ nm}}$ value of 4.9 mM^{-1} cm^{-1} for plastocyanin. A *c*-type cytochrome is a more convenient electron acceptor because of its large extinction coefficient in the α-band region of its spectrum.

Preparation of Duroquinol

Duroquinol is commercially available, or it can be prepared by reduction of duroquinone by $NaBH_4$. Duroquinone is dissolved in ethanol and the tube flushed with N_2. A small amount of solid $NaBH_4$ is added and the suspension gently mixed. Complete reduction is assumed when the faint yellow color disappears. Excess $NaBH_4$ is destroyed by the addition of 5 μl of 1 *M* HCl. When stored under N_2 in a capped tube, the solution of duroquinol is stable for several hours at 4°.

Procedure

Measurements are made at 590 minus 500 nm for plastocyanin reduction with the full scale of the recorder set to 0.05 OD units. The assay mixture contains 20 m*M* Tris–HCl buffer (pH 7.0), 50 m*M* NaCl, 10 m*M* KCl, 5.8 μM plastocyanin, and cytochrome b_6/f complex containing 50 nmol cyt *f*. The reaction is started by the addition of 22 μM duroquinol with a microsyringe to a 3-ml cuvette with stirring. The noncatalyzed rate of plastocyanin reduction is measured in the absence of the cytochrome complex, and the catalyzed rate is corrected for this rate.

Comments

Activities of the cytochrome b_6/f complex of approximately 20–40 μmol acceptor reduced/nmol cyt *f*/hr have been reported for several preparations with various electron acceptors and donors.[3] These activities correspond to turnover numbers of 6–12 sec^{-1} although a value of 75 sec^{-1} has recently been reported with decylplastoquinol as the electron donor and ferricyanide as electron acceptor.[8]

[38] Preparation of Coupling Factor F_1-ATPase from Cyanobacteria

By DAVID B. HICKS

Introduction

The F_1-ATPase is the hydrophilic, catalytic sector of the proton-translocating F_1–F_0–ATPase complex of energy-transducing membranes. The F_1-ATPase has been purified from energy-transducing membranes of two different cyanobacteria[1,2] and appears to be similar to the higher plant chloroplast F_1 (CF_1) in a number of important respects, including subunit molecular weights, latency and activation of the enzyme, and ability to cross-reconstitute photophosphorylation in heterologous membranes depleted of F_1 content.[1–3] This chapter outlines procedures the author has found to be useful in the purification and characterization of the cyanobacterial ATPase. Purification of the enzyme is based largely on the work of Lien and Racker[4] and Jagendorf,[5] and readers are urged to consult these references if additional information is required.

Two distinctly different methods are effective in extracting the F_1-ATPase from the cyanobacterium *Spirulina platensis*. One method, low ionic strength treatment of membranes, works consistently and apparently solubilizes an intact, five-subunit enzyme, but involves large volumes and requires more steps to purify the enzyme to homogeneity. The alternative method, brief exposure of the membranes to chloroform, efficiently extracts ATPase activity of high specific activity in a small volume, but occasionally fails, for unknown reasons, to solubilize much ATPase activity. In addition, although not encountered in studies on the F_1 from *Spirulina platensis,* it is often observed that the ATPase purified by this procedure is deficient in its content of the δ subunit (see, e.g., Fig. 1). In general, it may be preferable to extract F_1 by low ionic strength treatment for any studies involving reconstitution of energy transducing functions, while the chloroform procedure is better suited for structural studies of F_1, particularly when only small quantities of membranes are available. Both methods are presented here.

[1] A. Binder and R. Bachofen, *FEBS Lett.* **104,** 66 (1979).
[2] D. Hicks and C. Yocum, *Arch. Biochem. Biophys.* **245,** 220 (1986).
[3] D. Hicks, N. Nelson, and C. Yocum, *Biochim. Biophys. Acta* **851,** 217 (1986).
[4] S. Lien and E. Racker, this series, Vol. 24, p. 547.
[5] A. Jagendorf, *in* "Methods in Chloroplast Molecular Biology" (M. Edelman, R. Hallick, and N.-H. Chua, eds.), p. 985. Elsevier, Amsterdam, 1982.

Assay of ATPase Activity

Reagents

Tricine–NaOH, pH 8, 1 *M*
Tricine–NaOH, pH 8.6, 1 *M*
$CaCl_2$, 0.1 *M*
$MgCl_2$, 0.1 *M*
ATP, 0.2 *M* (equivalent to Sigma, Grade I, brought to pH 6.8 with NaOH, stored in small aliquots at −20°, and used once)
Trypsin (treated with L-1-tosylamido-2-phenylethyl chloromethyl ketone), 5 mg/ml, freshly prepared in 1 m*M* H_2SO_4
Soybean trypsin inhibitor, 10 mg/ml (stored at −20°)
Methanol (reagent grade)
KH_2PO_4, 20 m*M* (P_i standard)

Activation of ATPase Activity

Like spinach CF_1-ATPase activity, detection of *Spirulina* F_1-ATPase activity is complicated by the relatively low activity catalyzed by the enzyme in the absence of specific stimulators. Substantial hydrolytic activity is observed only when *Spirulina* F_1 is trypsin-treated or when methanol is included in the reaction mixture; these treatments induce a metal specificity in the ATPase activity which resembles their effects on CF_1 activity.

Trypsin Activation (Ca^{2+}-ATPase). ATP hydrolysis is measured as released P_i spectrophotometrically as the reduced phosphomolybdate complex according to LeBel *et al.*[6] as described by Jagendorf.[5] For measurement of ATPase activity during the purification of *Spirulina* F_1, trypsin can be simply included in the reaction mixture. This treatment elicits higher Ca^{2+}-ATPase activity than is expressed by trypsin pretreatment.[7] The reaction mixture, in a final volume of 1.0 ml, contains 50 μl Tricine, pH 8.0, 100 μl $CaCl_2$, 25 μl ATP, and a volume of trypsin that yields a ratio of about 5 to 20 : 1 trypsin to sample (w/w) (optimal ratios have to be experimentally determined for different F_1 preparations). *Spirulina* F_1 is much more resistant to trypsin than is spinach CF_1. The reactions, which are carried out at 37°, are initiated by the addition of sample (2–50 μg of protein, depending on the stage of purification). The reaction is terminated at 5 min with 1.0 ml of the color reagent, which contains 2% (w/v) trichloroacetic acid (TCA), 2.4 *M* acetic acid, 0.4 *M* sodium acetate, 12 m*M* $CuSO_4 \cdot 5H_2O$, 1% (w/v) ammonium molybdate,

[6] D. LeBel, G. Poirier, and A. Beaudoin, *Anal. Biochem.* **85,** 86 (1979).
[7] D. Hicks and C. Yocum, *Arch. Biochem. Biophys.* **245,** 230 (1986).

1% (w/v) sodium sulfite, 0.4% (w/v) Elon (*p*-methylaminophenol sulfate), 1% (w/v) sodium dodecyl sulfate (SDS); the TCA in the color reagent stops the reaction. The color is allowed to develop for 5 min, and 0.1 ml of 34% sodium citrate is added to halt further color development.[8] The absorbance at 750 nm is measured against a blank (lacking sample); a P_i standard (10 μl P_i standard, or 200 nmol P_i) is run in parallel.

A qualitative procedure for the detection of F_1-ATPase activity is that described by Horak and Hill,[9] which is based on the insolubility of calcium phosphate generated during the hydrolysis of ATP. This procedure can be useful for localizing activity in fractions or on native polyacrylamide gels. For kinetic studies of *Spirulina* F_1, the enzyme must be preactivated prior to the ATPase assay by incubation at room temperature with trypsin (trypsin-to-ATPase ratio 2.5–5 : 1, w/w) in 50 m*M* Tricine; the digestion is terminated at 15 min with soybean trypsin inhibitor (3 : 1 trypsin inhibitor to trypsin, w/w). The trypsin-treated enzyme is stored on ice. Soybean trypsin inhibitor is added prior to trypsin in the controls. The protein content of samples is estimated by the method of Lowry *et al.*,[10] using bovine serum albumin as the standard; it was found that the Bradford Coomassie G binding method[11] gives substantially lower estimates of protein content relative to the Lowry method which complicates comparisons to published values of CF_1 activity.

Methanol-Stimulated Mg^{2+}-ATPase. The barely detectable Mg^{2+}-ATPase activity of purified *Spirulina* F_1 is stimulated over 50-fold by the inclusion of 25% methanol (v/v) in the reaction. The reaction contains, in a final volume of 1.0 ml, 25 μl of Tricine, pH 8.6, 25 μl $MgCl_2$, 25 μl ATP, 250 μl of methanol, and sample (2–50 μg protein). The rate of the reaction is linear for 2 min and is terminated as described above. The ease and economy of this method may make it superior to the trypsin procedure for the detection of F_1 activity.

Assay of Membranes

The assay of photosynthetic membranes of cyanobacteria is complicated by the presence of blue chromophores and also by turbidity at high protein concentrations. The assay of *Spirulina* membranes activated as described above requires relatively low amounts of membranes (equivalent to 5–10 μg Chl *a*), which generally avoids the problem of turbidity. Elimination of the interference by phycobiliproteins can be accomplished

[8] P. Lanzetta, L. Alvarez, P. Reinach, and O. Candia, *Anal. Biochem.* **100,** 95 (1979).
[9] A. Horak and R. Hill, *Plant Physiol.* **49,** 365 (1972).
[10] O. Lowry, N. Rosebrough, A. Farr, and R. Randall, *J. Biol. Chem.* **193,** 265 (1951).
[11] M. Bradford, *Anal. Biochem.* **72,** 248 (1976).

by the addition to the blank and P_i standard tubes of an equivalent amount of membranes after the addition of color reagent. Alternatively, reactions can be terminated with TCA (50 μl of 50% TCA, or 2.5% final concentration) and centrifuged in a table-top centrifuge for 10 min to pellet TCA-insoluble material, leaving a clear supernatant. P_i is then determined by addition of 0.5 ml of the supernatant to 0.5 ml of the color reagent lacking TCA.

Analysis of F_1 Preparations by Electrophoresis and Immunoblotting

The purity of F_1 preparations is often judged by the presence of only five Coomassie staining bands on SDS–polyacrylamide gels, presumably corresponding to the α, β, γ, δ, and ε subunits. When the source of the enzyme is chloroplastic or cyanobacterial, this criterion may be misleading due to possible contamination of preparations by ribulose-bisphosphate (RuBP) carboxylase (EC 4.1.1.39), whose subunits may comigrate with the β and ε subunits of the F_1. It is important, therefore, to subject F_1 preparations to native gel electrophoresis to estimate the contamination by RuBP carboxylase and to determine on SDS gels the migration of the large and small subunits of carboxylase with respect to the subunits of F_1.

This type of analysis is illustrated in Fig. 1. The F_1 from *Anabaena* PCC7120 membranes was extracted by the chloroform method. Prior to extraction, the membranes were intentionally washed inadequately (i.e., just once with 10 m*M* sodium pyrophosphate) to give a preparation substantially contaminated by RuBP carboxylase. After ion-exchange chromatography, the F_1 preparation contains two bands on 6% native gels (Fig. 1A). The faster migrating band is identified as F_1 by enzymatic staining. Following native gel electrophoresis and brief staining with Coomassie blue, the bands are cut out and subjected to SDS 10–15% polyacrylamide gradient gels (see Fig. 2B). (In native gels, F_1 often migrates as two very closely spaced bands, which are treated as a single band for the purpose of this analysis.) The large subunit of RuBP carboxylase migrates just under the β subunit in this organism, and the small subunit migrates slightly farther than the ε subunit.[12] Note that the preparation appears to be completely depleted of the δ subunit. Purified spinach CF_1, extracted by the low ionic strength method, is shown in lanes 6 and 7 for comparison.

RuBP carboxylase contamination can be largely eliminated by 3–4 washes with 10 m*M* sodium pyrophosphate, pH 7.5, as described by

[12] The genes for the F_1-ATPase from *Anabaena* 7120 have been isolated and sequenced in the laboratory of Dr. Stephanie Curtis [S. E. Curtis, *J. Bacteriol.* **169,** 80 (1987) and unpublished observations].

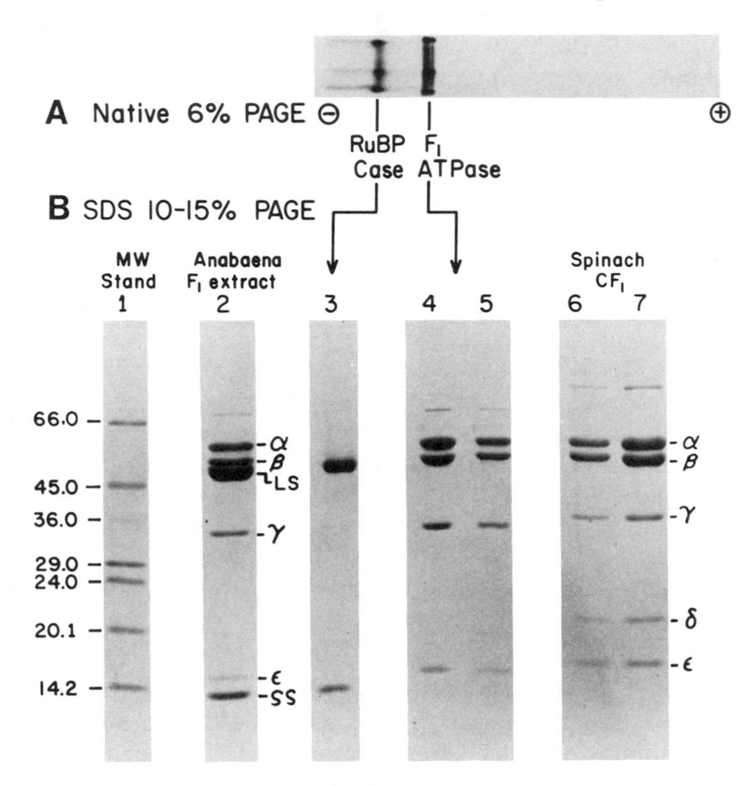

FIG. 1. Electrophoretic resolution of RuBP carboxylase and F_1-ATPase from *Anabaena* 7120. *Anabaena* F_1 was extracted by the chloroform method from membranes washed once with pyrophosphate and then partially purified by ion-exchange chromatography. The F_1 preparation was analyzed by 6% polyacrylamide native gels (A) and by SDS 10–15% polyacrylamide gradient gels (B). The Coomassie staining bands on the native gel were cut out, equilibrated with 2× sample buffer (4% SDS, 120 m*M* Tris–HCl, pH 6.8, 120 m*M* dithiothreitol, 2 m*M* phenylmethylsulfonyl fluoride, 20% glycerol) for 15 min and then placed in the appropriate well of the SDS gel and sealed in place with agarose-containing sample buffer. In lane 2, the *Anabaena* F_1 preparation was analyzed directly on the SDS gel. The protein content of samples was 25 μg of the F_1 preparation for lanes 2, 3, and 5 and 50 μg for lane 4; lanes 6 and 7 contain 10 and 25 μg, respectively, of purified spinach CF_1. Molecular weight standards ($\times 10^{-3}$) are indicated on the left margin. F_1 subunits are labeled α, β, γ, δ, and ε; carboxylase subunits are labeled LS (large subunit) and SS (small subunit).

Strotmann *et al.*[13] This method also extracts most of the phycobiliproteins, as well, and thus is especially beneficial in expediting purification of the cyanobacterial ATPase. The membranes are resuspended in the pyrophosphate solution to about 0.2 mg Chl *a*/ml and pelleted at 48,000 *g*, 1 hr. A small loss of chlorophyll-containing material that does not pellet under

[13] H. Strotmann, H. Hesse, and K. Edelmann, *Biochim. Biophys. Acta* **314,** 202 (1973).

these conditions is tolerated in order to remove large amounts of contaminating protein.

The identity of Coomassie staining bands of F_1 preparations can be determined by immunoblotting with antibodies against the subunits of ATPases from other sources as described by Rott and Nelson,[14] or by similar procedures. After electrophoresis on 10–15% gradient gels, the gel is equilibrated in transfer buffer (20% methanol, 25 m*M* Tris, 192 m*M* glycine, 0.02% SDS) for 45 min at 4°, and then transferred to nitrocellulose sheets by electrophoresis at 0.2 A constant current for 3–4 hr at 4°. The extent of transfer can be estimated by silver staining of the gel, amido black staining of the nitrocellulose sheet, and/or by visual inspection of the extent to which prestained molecular weight standards (e.g., Sigma SDS-7B) transfer to the sheet. The nitrocellulose sheets are blocked with 3% bovine serum albumin (BSA) in 20 m*M* Tris–500 m*M* NaCl, pH 7.5, washed with Tris–NaCl–0.05% Tween 20, and incubated overnight with the appropriate antibody in antibody buffer (Tris–NaCl–1% BSA–Tween 20). After washing, the sheet is incubated with goat anti-rabbit IgG conjugated to horseradish peroxidase for 1 hr, washed, and reacted with 4-chloro-1-naphthol and hydrogen peroxide. Bands appear within several minutes and increase in intensity for 10–15 min.

An example of immunoblotting is shown in Fig. 2. A single polypeptide in the ATPase from *Spirulina platensis*, *Anabaena* 7120, and spinach CF_1 reacts with anti-β from *Escherichia coli* F_1. Anti-α from spinach CF_1 is less specific, reacting strongly with the α subunit of CF_1 but showing a small amount of cross-reactivity to the β subunit. On the other hand, anti-α shows approximately equal reaction with the α and β subunits of *Anabaena* F_1. (The α and β subunits of *Spirulina* F_1 are not resolved well enough to determine the relative extent of reactivity of these subunits to the anti-α antibody.)

Isolation of Photosynthetic Membrane Vesicles from *Spirulina platensis*

Spirulina platensis is grown as described[15] and harvested in the late log or early stationary phase by centrifugation at 16,000 *g* for 20 min. The pellet can be stored in this form at −70° for at least 6–8 weeks. The pellet is gradually warmed by lukewarm water baths, dispersed with a rubber policeman in the sonication medium, and then homogenized thoroughly with a glass homogenizer. The volume of sonication medium, which contains 0.4 *M* sucrose, 20 m*M* HEPES, pH 7.5, 15 m*M* NaCl, 3 m*M* $MgCl_2$,

[14] R. Rott and N. Nelson, *J. Biol. Chem.* **256,** 9224 (1981).

[15] S. Robinson, C. Selvius DeRoo, and C. Yocum, *Plant Physiol.* **70,** 154 (1982).

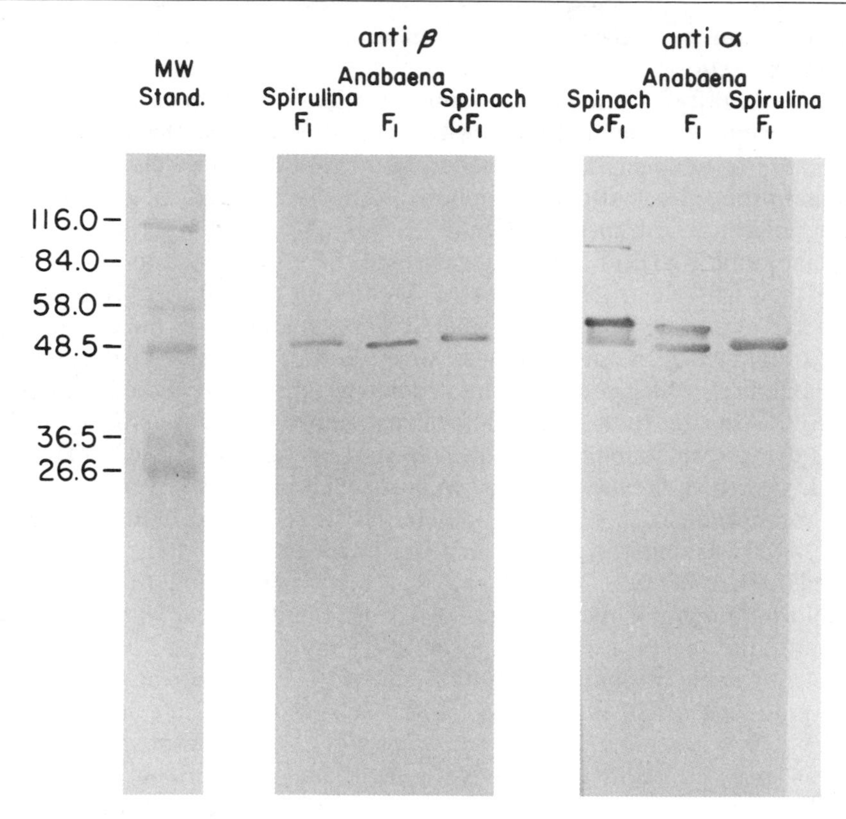

FIG. 2. Protein immunoblotting of F_1-ATPases with anti-α and anti-β antibodies. Anti-β was prepared against the β subunit of the *E. coli* F_1-ATPase, and anti-α was prepared against the α subunit of spinach CF_1. Electrotransfer and immunoblotting was carried out as described in the text. Lanes 1–3, 0.5 μg of F_1; lanes 4–6, 2.0 μg of F_1. Molecular weight standards ($\times 10^{-3}$) are indicated on the left margin.

10 m*M* $CaCl_2$, and protease inhibitors to be described in a later section, is adjusted so that the chlorophyll *a* concentration is 0.1–0.2 mg/ml. The cells are disrupted by sonic oscillation for 5 min in batches of 150–250 ml in a salt–ice bath at three-quarters full output with a Branson W-185 sonicator. The temperature is monitored occasionally to ensure that it does not rise above 20° (usually it does not go above 14–15°). Unbroken cells are removed by low-speed centrifugation (1,000 *g*, 10 min, 4°); to obtain a high yield of membranes it is advantageous to resuspend and resonicate this pellet as described above. The supernatants are pooled and centrifuged at 48,000 *g* for 1 hr at 4° to sediment the sonic vesicles.

For extraction of the F_1, the membranes are washed as described above. For energy-transducing assays, the sonic vesicles are resuspended in 0.4 *M* sucrose, 20 m*M* HEPES, pH 7.5, 15 m*M* NaCl to 0.5–1.0 mg Chl *a*/ml and stored in small aliquots at −70°. These vesicles are active in oxygen evolution and are well coupled, exhibiting a P/2*e* of 0.9 with electron transport from water to methyl viologen. The ATPase activity of the vesicles, when appropriately unmasked, is high, ranging from 13 to 16 μmol P_i/min/mg Chl.

Extraction and Purification of F_1

Early work on the *Spirulina* ATPase indicated that the enzyme is unstable at room temperature[16]; stabilization appears to require that purification be carried out at 4° in the presence of glycerol (10%). As an additional precaution, three protease inhibitors are used throughout the purification procedures at the following concentrations: phenylmethylsulfonyl fluoride (PMSF), 0.1 m*M* (1 m*M* for sonication of cells); *p*-aminobenzamidine, 1 m*M*; and L-1-tosylamido-2-phenylethyl chloromethyl ketone (TPCK), 0.05 m*M* (0.1 m*M* for sonication of cells). *p*-Aminobenzamidine is omitted during the actual extraction of the F_1 since it may prevent detachment of the F_1 from the membrane.[17]

Extraction by Low Ionic Strength Treatment

The following procedures have been carried out on membrane vesicles containing 30–200 mg Chl *a*. Pyrophosphate-washed membrane vesicles are resuspended to 0.05 mg Chl *a*/ml in 50 m*M* sucrose buffered with 2 m*M* Tricine titrated with solid Tris to pH 7.5[5,13] and stirred at 4° for 30 min. The Tris–Tricine solution is somewhat more effective than 2 m*M* Tricine–1 m*M* EDTA in extracting ATPase activity. The membranes are removed by centrifugation (48,000 *g*, 1 hr, 4°); a second extraction of the membranes does not appreciably increase the yield of ATPase and can be omitted for convenience. The light green supernatant is brought to 10% glycerol, 20 m*M* Tris–HCl, 2 m*M* EDTA, 1 m*M* ATP, 3 m*M* sodium azide prior to the addition of 100 ml of settled, equilibrated DEAE–Sephadex A-25 (or approximately 0.025 ml exchanger/ml of F_1 extract). All solutions subsequent to the extraction of F_1 contain 2 m*M* EDTA, 1 m*M* ATP, 3 m*M* sodium azide, and protease inhibitors. Greater than 90% of the ATPase activity is adsorbed to the resin after stirring for 1 hr. The resin is

[16] L. Owers-Narhi, S. Robinson, C. Selvius DeRoo, and C. Yocum, *Biochem. Biophys. Res. Commun.* **90,** 1025 (1979).

[17] G. Cox, J. Downie, D. Fayle, F. Gibson, and J. Radick, *J. Bacteriol.* **133,** 287 (1978).

gently layered on an equivalent volume of fresh exchanger. The column solutions contain 10% glycerol, 20 m*M* Tris–HCl, which are brought to pH 7.15 at room temperature and then cooled to 4°. The column is washed with several bed volumes of 0.2 *M* NaCl, and the F_1 is eluted with 0.4 *M* NaCl. Solid ammonium sulfate is added to this solution to 60% saturation and the precipitated F_1 is stored at 4°.

Purification to homogeneity requires two sucrose gradients, which are prepared essentially as described by Jagendorf.[5] Before loading on the gradient, the ammonium sulfate slurry is pelleted, redissolved, and dialyzed against 1000 volumes of 20 m*M* Tricine–NaOH, 2 m*M* EDTA, 1 m*M* ATP, 3 m*M* sodium azide, and protease inhibitors; insoluble material which sometimes results during the dialysis is removed by centrifugation. Linear 12–28% sucrose gradients having the same composition as the dialysis medium are loaded with 1 ml of the partially purified F_1 (6–10 mg/ml) and centrifuged in a VTi 50 rotor at 48,000 rpm for 4.5 hr at 4°. Depending on the equipment available, fractions can be collected from either the top or the bottom of the gradient. In the absence of specialized equipment, 30-drop fractions are collected from the bottom of the gradient as described.[5] Fractions are analyzed for protein content, for ATPase activity, and for purity, the latter by gel electrophoresis. Less active fractions at the highest sucrose densities which may contain RuBP carboxylase are discarded. A typical purification procedure is shown in Table I. For long-term storage, the enzyme can be stored as an ammonium sulfate precipitate at 4°; on a short-term basis (e.g., several days), the enzyme is stable when simply stored at 4° as long as ATP and glycerol are present.

Extraction by Chloroform Method

Pyrophosphate-washed membranes are resuspended to 0.5–1.0 mg Chl *a*/ml in 10% glycerol, 5 m*M* HEPES, pH 7.5, 5 m*M* ATP, 5 m*M* dithiothreitol (DTT), 1 m*M* EDTA, 0.1 m*M* PMSF, and 0.05 m*M* TPCK,[18] and divided into 10- to 12-ml portions in 25-ml Corex tubes. One-half volume of chloroform (reagent grade, not further purified) is added to the tube, and the suspension is mixed vigorously for 10–12 sec with a Vortex stirrer. Less effective solubilization results when larger volumes of membranes are exposed to chloroform. The phases are separated by brief (1,000 *g*, 1 min, 4°) centrifugation, and the upper aqueous phase is carefully removed (a small volume of the aqueous phase is sacrificed to ensure that the material at the interface is not collected). These operations are performed as quickly as possible. The membranes are removed by two

[18] R. Piccioni, P. Bennoun, and N.-H. Chua, *Eur. J. Biochem.* **117,** 93 (1981).

TABLE I
PURIFICATION OF *Spirulina* F_1[a]

Step	Protein (mg)	Activity (μmol P_i/min)	Specific activity (μmol P_i/min/mg)	Yield (%)
Tris–Tricine extract[b]	203.5	203.5	1.0	100
DEAE–Sephadex A-25 step gradient	20.7	118.1	5.7	58
First sucrose gradient	3.6	91.7	25.6	45
Second sucrose gradient	1.5	53.4	36.0	26

[a] From Hicks and Yocum.[2]
[b] *Spirulina* F_1 was extracted from vesicles containing 33 mg Chl *a*.

centrifugations, 10,000 *g* for 15 min followed by 73,000 *g* for 30 min, both at 4°. The resulting supernatant is either light blue or pale yellow–brown, depending on the degree to which phycobiliproteins were extracted during the membrane washes. The crude chloroform extract is dialyzed overnight against 10% glycerol, 20 m*M* Tris–HCl, and then loaded on a DEAE–Sephadex A-25 column as described above. The column is washed with 0.1 *M* NaCl, which elutes phycocyanin, and a linear gradient of NaCl from 0.1 to 0.5 *M* is used to elute the F_1. The enzymes elutes between 270 and 320 m*M* NaCl. This step is often sufficient to purify the F_1 to apparent homogeneity; however, electrophoretic analysis may indicate that a further step is required. If so, a sucrose gradient is prepared as described above and used to complete the purification.

Reconstitution of Photophosphorylation

The most sensitive assay available for the presence of an intact F_1-ATPase is the ability of the enzyme to restore photophosphorylation activity to photosynthetic membranes depleted of F_1 content. The requirements for a membrane preparation (1) which is highly depleted of F_1 content and (2) which will actively synthesize ATP on the addition of F_1 are best met by the NaBr procedure developed by Nelson and Eytan.[19]

Intact membrane vesicles, washed once to partially remove membrane-associated RuBP carboxylase, are treated with sodium bromide exactly as described,[20] except that the Chl concentration for NaBr treatment and for storage (−70°) is 0.5–1.0 mg Chl *a*/ml. The treated mem-

[19] N. Nelson and E. Eytan, *in* "Cation Flux across Biomembranes" (Y. Mukohata and L. Packer, eds.), p. 409. Academic Press, New York, 1979.
[20] N. Nelson, this series, Vol. 69, p. 301.

branes have a specific Ca^{2+}-ATPase activity of less than 0.1 μmol P_i/min/mg Chl *a*, or less than 10% of the specific activity of native membranes, and are strongly depleted in α, β, and γ subunit content according to SDS–polyacrylamide gradient gel electrophoresis. The treated vesicles catalyze negligible rates of photophosphorylation ($\leq$1.6 μmol ATP/hr/mg Chl *a*), but on addition of *Spirulina* F_1, the rate of photophosphorylation increases up to 50–60 μmol ATP/hr/mg Chl *a*.

Reconstitution of photophosphorylation is accomplished by incubating the NaBr-treated membrane vesicles (~33 μg Chl *a*), in a final volume of 0.44 ml, with *Spirulina* F_1 (between 0.1 and 5 μg F_1/μg Chl *a*) in 50 m*M* Tricine, pH 8, 25 m*M* $MgCl_2$, and 0.2 mg/ml bovine serum albumin (BSA) for 15 min on ice in the dark. One milliliter of a radioactive cocktail containing 50 m*M* Tricine, pH 8, 75 m*M* NaCl, 1.5 m*M* ADP, and 7.5 m*M* sodium phosphate (pH 8, containing 1×10^6 cpm/μmol), and 1.5 mg/ml BSA is added to the reconstituted membranes. Phenazine methosulfate is added (60 μl of a 10 m*M* solution), and the reaction is initiated by a 1-min illumination with white light (approximately 3×10^6 ergs/cm^2/sec). The final concentration of components, in 1.5 ml, is 50 m*M* Tricine, pH 8, 50 m*M* NaCl, 6.7 m*M* $MgCl_2$, 1 m*M* ADP, 5 m*M* sodium phosphate, 0.4 m*M* PMSF, 1 mg/ml BSA, and membrane vesicles equivalent to about 13 μg Chl *a*. The reaction, which is carried out in a water bath at room temperature, is terminated by the addition of 0.2 ml of 30% TCA. Unreacted phosphate is extracted as described by Avron.[21] Rates are not altered by the presence of hexokinase and glucose. [^{32}P]ATP is detected by Cerenkov counting.[22] Appropriate dark control rates are subtracted from the experimental rates.

Properties of Spirulina F_1-ATPase

The subunit M_r values of *Spirulina* F_1 are estimated at 53,500 (α), 51,500 (β), 36,000 (γ), 21,000 (δ), and 14,500 (ε),[2] according to SDS–polyacrylamide gradient gel electrophoresis. The M_r of the holoenzyme is estimated by native polyacrylamide gradient gel electrophoresis to be 320,000, whereas the M_r of spinach CF_1 is about 392,000 by this technique. The α, β, and γ subunits cross-react with antibodies against the corresponding subunits of spinach CF_1.[3] Phenazine methosulfate-dependent photophosphorylation activity in sodium bromide-treated photosynthetic membranes derived from *Spirulina platensis* and spinach chloroplasts is restored by *Spirulina* F_1.[3] Based on the specific Ca^{2+}-ATPase activity of intact membranes and purified F_1 (and the M_r of F_1), the F_1

[21] M. Avron, *Biochim. Biophys. Acta* **40,** 257 (1960).

[22] J. Gould, R. Cather, and G. Winget, *Anal. Biochem.* **50,** 540 (1972).

content of sonic membrane vesicles of *Spirulina platensis* is estimated at 1 mol F_1/830 chlorophylls, or 0.12 mol F_1/mol P700.[2] Ca^{2+}-ATPase activity of *Spirulina* F_1 increases from 4 to 20–37 μmol P_i/min/mg protein when trypsin treated and from less than 1 to about 20 μmol P_i/min/mg protein when Mg^{2+}-ATPase activity is assayed in the presence of 25% methanol.[7] A moderate stimulation of Ca^{2+}-ATPase activity is detectable when the enzyme is pretreated with dithiothreitol. The K_m for the Ca^{2+}–ATP metal–nucleotide complex is 0.42 m*M* when the trypsin-activated enzyme is assayed with free Ca^{2+} in excess; free ATP is an allosteric effector of the enzyme.[7]

Acknowledgments

This research was supported by the U.S. Department of Agriculture Competitive Research Grants Office (Grants 5901-0410-8-0103-0 and 82-CRCR-1-1127) awarded to Dr. Charles Yocum. The author is indebted to Dr. Yocum for his encouragement and counsel, to Dr. Terry Krulwich for providing time to write the manuscript, and to Dr. Stephanie Curtis for the gift of *Anabaena* 7120 cells. The author wishes to express his gratitude to Dr. Nathan Nelson for his generous gift of antibodies and for his encouragement.

[39] Respiratory Proton Extrusion and Plasma Membrane Energization

By G. A. Peschek, W. H. Nitschmann, and T. Czerny

Introduction

It is well known that living cells (microorganisms) eject protons upon energization with light (photosynthesis) or oxygen (respiration). Proton extrusion from illuminated or oxygenated cyanobacterium *Anabaena variabilis* ATCC 27892 was first observed by Mitchell and colleagues.[1] A cyanobacterial cell although prokaryotic and as such often referred to as "uncompartmentalized" nevertheless does comprise two distinct and osmotically autonomous microcompartments, namely, the cytosol and the intrathylakoid space, surrounded by two distinct and likewise bioenergetically autonomous membranes, the plasma membrane (CM) and the thylakoid membrane (ICM)[2] (Fig. 1). The enzyme(s) responsible

[1] P. Scholes, P. Mitchell, and J. Moyle, *Eur. J. Biochem.* **8**, 450 (1969).

[2] N. Murata and T. Omata, this volume [23]; G. A. Peschek, V. Molitor, M. Trnka, M. Wastyn, and W. Erber, this volume [45].

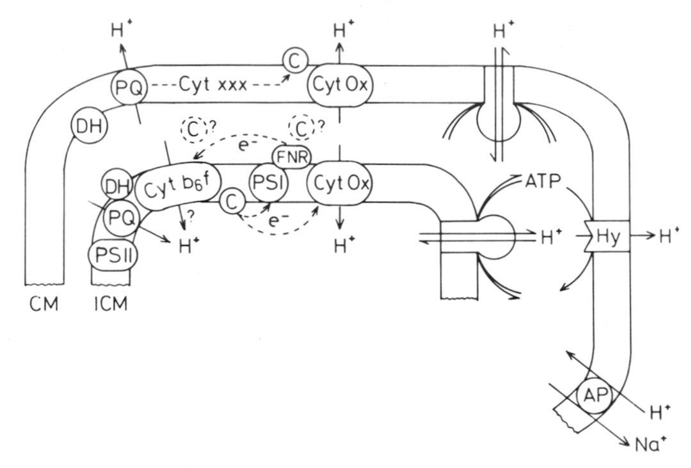

FIG. 1. Compartmentalization, polarity, and sidedness of electron transport, proton translocation, and ATPases in a cyanobacterial cell. CM, Plasma membrane; ICM, thylakoid membrane; DH, dehydrogenases; PSI and PSII, photosystems I and II; PQ, plastoquinone; Cyt, cytochrome; Cyt Ox, cytochrome oxidase; C, cytochrome(s) *c*; Cyt xxx, functionally uncharacterized cytochromes in the plasma membrane; FNR, ferredoxin: $NADP^+$ oxidoreductase; Hy, ATP hydrolase; AP, antiporter; broken lines indicate the path of electrons (e^-).

for the final step(s) of the process of proton extrusion from the cell must be situated in the plasma membrane. However, two basic mechanisms exist for proton translocation across the CM out of the energized cyanobacterial cell[3-5]: (1) Proton-translocating electron transport, and (2) proton-translocating ATPases. Among the latter, two different types of enzymes may be envisaged, (a) reversible H^+- or F_0F_1-ATPases,[4-10] and (b)

[3] Whether the initial phase of H^+ ejection from illuminated *Anabaena variabilis* ATCC 29413, as observed by P. Scherer, I. Hinrichs, and P. Böger, *Plant Physiol.* **81,** 839 (1986), could be attributed to direct light-driven H^+ pumping by a CM-bound P700 complex (similar to the bacteriorhodopsin-catalyzed process in *Halobacterium halobium*) remains to be considered.

[4] W. H. Nitschmann and G. A. Peschek, *Biochem. Biophys. Res. Commun.* **123,** 358 (1984).

[5] W. H. Nitschmann and G. A. Peschek, *Arch. Microbiol.* **141,** 330 (1985).

[6] A. Binder and R. Bachofen, *FEBS Lett.* **104,** 66 (1979).

[7] H. J. Lubberding, G. Zimmer, H. S. van Walraven, H. S. Schrickx, and R. Kraayenhof, *Eur. J. Biochem.* **137,** 95 (1983).

[8] D. B. Hicks and C. F. Yocum, *Arch. Biochem. Biophys.* **245,** 220 (1986).

[9] W. W. A. Erber, W. H. Nitschmann, R. Muchl, and G. A. Peschek, *Arch. Biochem. Biophys.* **247,** 28 (1986).

[10] G. A. Peschek, B. Hinterstoisser, M. Riedler, R. Muchl, and W. H. Nitschmann, *Arch. Biochem. Biophys.* **247,** 40 (1986).

unidirectional H^+- or E_1E_2-ATPases or ATP hydrolases.[11-13] Electrogenic proton translocation across a "H^+-impermeable" membrane, in turn, results in energization of the membrane, i.e., in the buildup of a proton-motive force (pmf) or proton electrochemical potential gradient ($\Delta\tilde{\mu}_{H+}$) across the membrane. The energy content of this gradient, in terms of millivolts, is composed of an electrical membrane potential ($\Delta\psi$) and a proton concentration gradient (ΔpH), according to Eq. (1).

$$\Delta\tilde{\mu}_{H_+} = \Delta\psi - (2.3RT/F)\Delta\text{pH} \qquad (2.3RT/F) = 60 \text{ mV at } 35° \qquad (1)$$

Thus the ultimate purpose of H^+ translocation is membrane energization. Consequently, this chapter presents (1) a description of the assay procedure for respiratory H^+ extrusion,[14] including the possible experimental discrimination between different mechanisms, and (2) a mathematical treatment of flow dialysis data from indicator distribution experiments for the separate determination of $\Delta\tilde{\mu}_{H_+}$ across plasma and thylakoid membranes, respectively[15] (cf. Fig. 1).

Proton Extrusion from Oxygen-Pulsed Cells

Proton extrusion is usually measured via the acidification of a weakly buffered suspension medium observed spectrophotometrically or spectrofluorimetrically using proper pH indicators, or by monitoring the pH of the medium with a sensitive glass electrode. While the latter method is widely used with whole cells, the former techniques which require less material are more appropriate for work on membrane vesicles.

Preparation of Cells

Cyanobacteria are grown under appropriate conditions and harvested from late logarithmic cultures by low-speed centrifugation at room temperature, washed twice with 40 m*M* HEPES/Tris buffer, pH 7.4, resus-

[11] S. Scherer, S. Stürzl, and P. Böger, *J. Bacteriol.* **158,** 609 (1984).

[12] P. Mitchell and W. H. Koppenol, *Ann. N.Y. Acad. Sci.* **402,** 584 (1982).

[13] R. M. Spanswick, *Annu. Rev. Plant Physiol.* **32,** 267 (1981).

[14] Since the plasma membrane of cyanobacteria appears to be devoid of bulk amounts of chlorophyll (and, hence, photosynthetic electron transport; cf. this volume [45]) the methodology described focuses on respiratory CM energization, i.e., on cyanobacteria (mainly *Anacystis nidulans*) rapidly shifted from dark anaerobic to aerobic conditions by means of oxygen pulses, measuring concomitant acidification of the suspension medium with a sensitive pH electrode.

[15] G. A. Peschek, T. Czerny, G. Schmetterer, and W. H. Nitschmann, *Plant Physiol.* **79,** 278 (1985).

pended in the desired assay medium (see below), and immediately used for the proton extrusion experiments.

Comments. The use of very young cultures is discouraged since respiratory capacity may not yet be fully expressed at this stage.[2,16] Treatment and storage of the cells below the lipid phase transition temperature of the plasma membrane[17,18] should be avoided as this may result in permeabilization of the cells to ions such as K^+ and H^+.

Suspension Medium

Harvested and washed cells are suspended in a buffer solution composed of 1 m*M* HEPES–Tris, 100 m*M* KCl, 1 m*M* $KHCO_3$, 10 μM valinomycin, and 50 μg/ml carbonate dehydratase (final pH 7.4). When experiments are to be conducted at different pH values the following 1 m*M* buffer solutions are used (applicable pH range given in parentheses; cf. Fig. 2): citric acid (pH 3.2–5.5), 2-(*N*-morpholino)ethanesulfonic acid (MES; pH 5.5–6.7), piperazine-*N,N'*-bis(2-ethanesulfonic acid) (PIPES; pH 6.4–7.5), *N*-2-hydroxyethylpiperazine-*N'*-2-ethanesulfonic acid (HEPES; pH 7.2–8.2), tris(hydroxymethyl)methylaminopropanesulfonic acid (TAPS; pH 7.8–9.1), 2-(*N*-cyclohexylamino)ethanesulfonic acid (CHES; pH 8.6–10.0), and 3-(cyclohexylamino)-1-propanesulfonic acid (CAPS; pH 9.7–11.0), all previously titrated to the desired pH with 1 m*M* Tris base.

Comments. In the presence of carbonate dehydratase catalyzing the fast equilibrium between (free) CO_2 and $CO_2 \cdot H_2O$ ($H_2CO_3 \rightleftharpoons H^+ + HCO_3^-$), added bicarbonate constitutes an efficient "carbonate buffer" that virtually prevents any measurable net pH change ("scalar" acidification) due to respiratory CO_2 release from the cells following the oxygen pulse; controls can be performed with saturating concentrations (50–100 μM) of a proton ionophore ("uncoupler"), such as carbonyl cyanide *m*-chlorophenylhydrazone (CCCP), leaving behind this "scalar" acidification only (see Table I).

The system K^+–valinomycin, which dissipates the transmembrane electrical potential $\Delta\psi$, is thought to promote electrogenic H^+ extrusion in, e.g., oxygen pulse experiments[19,20] by preventing the buildup of a retarding $\Delta\psi$ (inside negative) from electrogenic H^+ ejection across the

[16] V. Molitor, W. Erber, and G. A. Peschek, *FEBS Lett.* **204,** 251 (1986).

[17] T. Ono and N. Murata, *Plant Physiol.* **69,** 125 (1982).

[18] G. A. Peschek, R. Muchl, P. F. Kienzl, and G. Schmetterer, *Biochim. Biophys. Acta* **679,** 35 (1982).

[19] B. Reynafarje, A. Alexandre, P. Davies, and A. L. Lehninger, *Proc. Natl. Acad. Sci. U.S.A.* **79,** 7218 (1982).

[20] H. G. Lawford and B. A. Haddock, *Biochem. J.* **136,** 127 (1973).

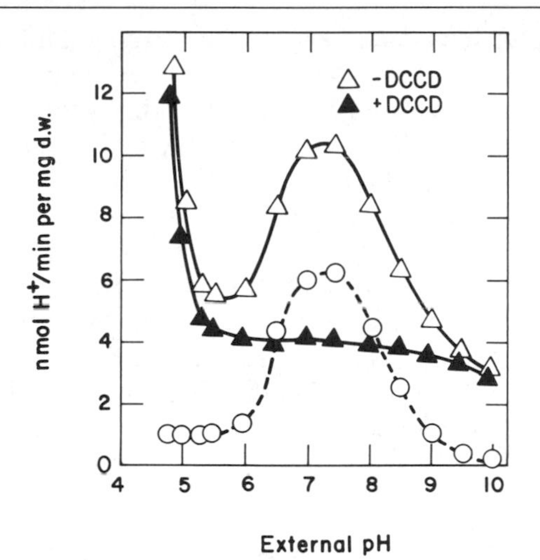

FIG. 2. pH dependence of the initial rates of H^+ extrusion from oxygen-pulsed *Anacystis nidulans* in the absence and in the presence of saturating concentrations of the ATPase inhibitor *N,N'*-dicyclohexylcarbodiimide (DCCD) (cf. Fig. 3). ○, Difference of +DCCD and −DCCD curves reflecting the activity of the reversible H^+-ATPase.

TABLE I
EFFECT OF MEDIUM COMPOSITION ON INITIAL RATES OF PROTON EXTRUSION AND OXYGEN CONSUMPTION IN AERATED SUSPENSIONS OF *Anacystis nidulans*[a]

Assay mixture	Apparent H^+ ejection		O_2 uptake		Apparent H^+/O ratio		Electrogenic H^+ ejection	
	Exp. 1	Exp. 2	Exp. 1	Exp. 2	Exp. 1	Exp. 2	Exp. 1	Exp. 2
Complete[b]	10.5	11.0	2.4	2.5	4.38	4.40	10.5	8.5
−Bicarbonate, −carbonate dehydratase	12.5	13.2	2.6	2.6	4.81	5.08	—	—
$-K^+$, −valinomycin	6.0	6.4	2.5	2.6	2.40	2.46	—	—
−Bicarbonate, −carbonate dehydratase, $-K^+$, −valinomycin	8.3	9.0	2.6	2.7	3.19	3.33	—	—
+DCCD	4.3	4.1	2.7	2.8	1.59	1.46	4.3	3.7
+CCCP	0.0	0.6	6.8	6.2	0.0	1.0	—	—
+DCCD, +CCCP	0.0	0.8	5.9	6.1	0.0	0.13	—	—

[a] Initial rates of H^+ ejection and O_2 uptake expressed as nmol H^+ or O/min/mg dry weight of cells. Results from two typical experiments are shown.

[b] The complete assay mixture contained 1 m*M* HEPES–Tris buffer, pH 7.4, 100 m*M* KCl, 1 m*M* $KHCO_3$, 10 μ*M* valinomycin, and 50 μg/ml carbonate dehydratase.

CM; alternatively, 100–150 m*M* KSCN (replacing the KCl in the suspension medium) may be used.[21] Yet, in our experiments with intact cyanobacteria, we repeatedly observed that the stimulating effect of $\Delta\psi$-dissipating agents was not as large as might have been expected (unpublished observations). This might be related to the rather low rates of H^+ extrusion from oxygen-pulsed cyanobacteria, thus permitting similarly slow rearrangement of some of the ions present to "buffer" $\Delta\psi$ changes across the CM (cf. Fig. 3; also see Table II for comparison with other bioenergetic systems). At any rate, values of H^+ extrusion rates obtained in the presence of K^+–valinomycin or KSCN are usually at least much more reproducible than those obtained in the absence of $\Delta\psi$ dissipators (unpublished observations).

Measuring Device

Cells suspended in the desired assay buffer (preferably containing ~30 mg dry weight of cells/ml, corresponding, in the case of *Anacystis nidulans*, to about 100 μl packed cells or 0.7 mg chlorophyll) are loaded into a cylindrical assay chamber which holds a total of 2.5 or 20 ml.[22] The chamber is equipped with a pH electrode (Philipps PW 2409 digital pH meter) and a Clark-type oxygen electrode (YSI oxygen monitor, Model 53). Both electrodes are connected to a Rikadenki Multi-pen Recorder. The assay chamber is kept at 35 ± 0.3° by continuously pumping water from a Thermomix (Braun, Melsungen, FRG, Model 1441) through the jacket surrounding the measuring compartment. Care is taken to maintain the cells in strict darkness throughout all preincubation and assay procedures.

Experimental Procedure

A gentle stream of water-saturated oxygen-free nitrogen (or argon) is bubbled through the cell suspension in the assay chamber for 20 min. Then the chamber is closed with a tightly fitting stopper equipped with a capillary inlet to permit addition of reagent solutions by means of a microsyringe. The oxygen electrode simply serves to monitor the oxygen concentration in the suspension which should be less than 1% air saturation for at least 20 min before the onset of the experiment. The experiment is started by rapid injection of 250 μl (1 ml) of oxygen-saturated 100 m*M* KCl solution into the anaerobic cell suspension through the capillary

[21] C. W. Jones, J. M. Brice, A. J. Downs, and J. W. Drozd, *Eur. J. Biochem.* **52,** 265 (1975).

[22] The procedure described in the following corresponds to the 2.5-ml device; parameters applicable to the 20-ml device are given in parentheses.

TABLE II
RATE OF PROTON EXTRUSION FROM OXYGEN-PULSED (CYANO)BACTERIA AND MITOCHONDRIA[a]

Organism	Proton ejection		References[h]
	−DCCD	+DCCD[b]	
Anacystis SAUG 1402-1[c], (*Synechococcus* ATCC 27144)	6.6	3.1	(1)
Nostoc PCC 8009[d]	4.1	4.2	(1)
Anabaena variabilis	11.6	3.6	(2)
ATCC 29413	1.5	0.3	(3)
Anabaena variabilis ATCC 27892	1.8	—	(4)
Paracoccus denitrificans	45–50[e]	—	(5)
Escherichia coli	15–20[f]	—	(6)
Mitochondria (rat liver)	1,300[g]	—	(7)

[a] Initial rates expressed as nmol H^+/min/mg dry weight of cells; protein content of most bacteria amounts to about 50% of the dry weight of the cells.

[b] DCCD, an inhibitor of the reversible H^+-ATPase, is added to the suspensions 20 min before the assay of proton extrusion at a concentration of 15–20 nmol/mg dry weight, which can be shown to completely eliminate all oxidative phosphorylation (cf. Fig. 3).

[c] SAUG, Sammlung von Algenkulturen der Universität Göttingen, FRG.

[d] PCC, Pasteur Culture Collection, Paris, France.

[e] Recalculated from Ref. 5.

[f] Recalculated from Ref. 6.

[g] Expressed as nmol/min/mg protein; cf. footnote *a*.

[h] (1) W. H. Nitschmann and G. A. Peschek, *J. Bacteriol.* **168,** 1205 (1986); (2) S. Scherer, S. Stürzl, and P. Böger, *J. Bacteriol.* **158,** 609 (1984); (3) W. H. Nitschmann and G. A. Peschek, *Arch. Microbiol.* **141,** 330 (1985); (4) P. Scholes, P. Mitchell, and J. Moyle, *Eur. J. Biochem.* **8,** 450 (1969); (5) H. W. van Versefeld, K. Krab, and A. H. Stouthamer, *Biochim. Biophys. Acta* **635,** 525 (1981); (6) J. M. Gould and W. A. Cramer, *J. Biol. Chem.* **252,** 5875 (1977); (7) B. Reynafarje, A. Alexandre, P. Davies, and A. L. Lehninger, *Proc. Natl. Acad. Sci. U.S.A.* **79,** 7218 (1982).

inlet, giving initial oxygen concentrations of 0.25–0.3 m*M* (0.15 m*M*) O in the suspension (cf. Table III); diffusion of air through this inlet is negligible. Final recordings are corrected for 10% (5%) dilution due to the addition of the KCl solution.

Observations

Injection of oxygen into the anaerobic cell suspension is followed by the almost immediate onset of (1) oxygen uptake, (2) H^+ ejection, and, in

TABLE III
CORRELATION OF INITIAL RATES AND EXTENTS OF PROTON EJECTION WITH OXYGEN PULSES OF DIFFERENT MAGNITUDES[a]

Initial concentration of O injected (μM)	Initial concentration of H^+ ejected (μM)	Initial rate (nmol H^+ or O/min/mg dry weight)		Apparent H^+/O ratios from	
		H^+ efflux	O_2 uptake	Initial rates	Initial concentrations
50	210	10.3	2.4	4.3	4.2
100	420	10.4	2.45	4.2	4.2
200	820	10.2	2.3	4.1	4.1
400	830	10.1	2.3	4.1	2.0
550	820	9.8	2.2	4.4	1.5

[a] For details, see Erber *et al.*[9]

the absence of ATPase inhibitors, (3) ATP synthesis as shown in Fig. 3. The following features of the traces shown in Fig. 3 deserve closer attention.

1. Immediate onset of respiratory oxygen uptake: Extrapolation of the linear trace of oxygen uptake back to zero time (oxygenation; arrow in Fig. 3) must give the amount of oxygen initially injected (e.g., 0.25 m*M* O in Fig. 3) and corresponds to the injection of 250 μl oxygen-saturated 100 m*M* KCl solution into 2.5 ml anaerobic cell suspension, assuming the 250 μl overflow to be still anaerobic.

2. Immediate onset of respiratory H^+ ejection (traces B and E in Fig. 3): The fact that in *A. nidulans* H^+ ejection does not occur with a maximum rate right from the time of oxygenation appears to reflect the composite mechanism responsible, namely, both a reversible H^+-ATPase (which has to wait for the buildup of a certain level of intracellular ATP from oxidative phosphorylation) *and* a H^+-translocating respiratory electron transport, part of which is located in the CM across which it starts to eject H^+ as soon as O_2 is available (cf. Fig. 3).

Comments. Variable lag phases between oxygen admission and the onset of *measurable* H^+ extrusion (i.e., acidification of the medium) do not necessarily indicate the absence of H^+-translocating respiratory electron transport in the plasma membrane.[11] Instead, it is more likely that any type of observed "lag phases," especially the less reproducible ones and those dependent on cell concentration, are caused by limited and varying response times of the (different!) electrodes employed for the

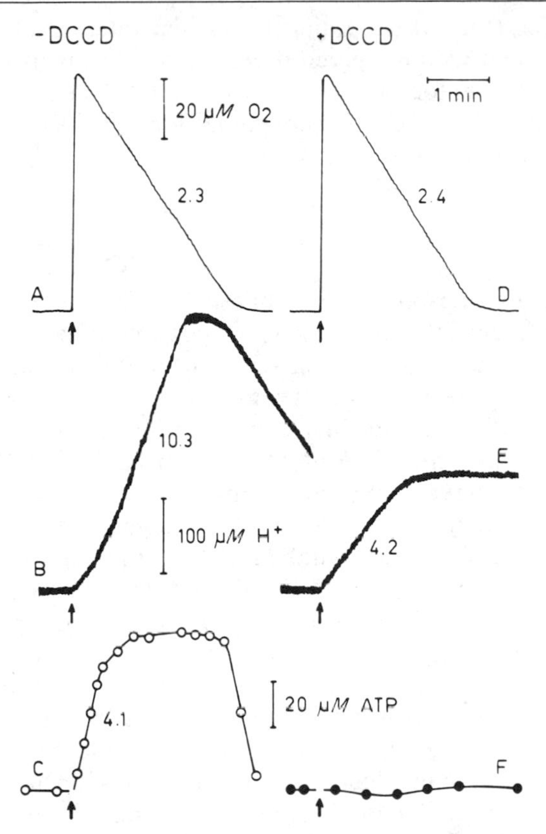

FIG. 3. Kinetics of oxygen uptake (A and D), H^+ ejection (B and E), and intracellular ATP levels (C and F) in oxygen-pulsed suspensions of *A. nidulans* in the absence (A–C) and presence (D–F) of saturating DCCD (see text). Arrows indicate the admission of oxygen. The incomplete action of DCCD on respiratory H^+ ejection indicates the presence of a H^+-translocating respiratory chain in the plasma membrane. The same result is obtained when DCCD is replaced by the F_1 inhibitor 6-chloro-4-nitrobenz-2-oxa-1,3-diazole (NBD) (see text).

measurement of, e.g., H^+ and O_2, let alone the necessarily discontinuous measurement of ATP. Electrode response time will depend on, among others, storage conditions and on the finite diffusion time of the species to be measured especially in the case of rather thick cell suspensions.

Evaluation of Data

Owing to rather low rates of H^+ ejection from oxygen-pulsed cyanobacteria (cf. Table II) it is more accurate and reliable to evaluate the recorder traces (Fig. 3) according to "initial rates" rather than "total amount of H^+ extruded per given oxygen pulse" as is usually done with bacteria[20,21,23] and mitochondria[24,25] in order to calculate H^+/O ratios.[26] Within a certain range of oxygen concentration, results from either procedure as applied to cyanobacteria appear to be reasonably consistent (Table III).

Inhibitors

Since electron transport, F_0F_1-ATPase, and E_1E_2-ATPase respond differently to different inhibitors it may be possible to discriminate experimentally between these mechanisms.[4,5,9] The following inhibitors are useful for proton extrusion measurements with cyanobacteria (stock concentration, solvent, and final concentration in parentheses): KCN, cytochrome oxidase inhibitor (0.5 *M* in 0.5 *M* HEPES–KOH buffer, pH 7.0; 1 m*M*); *N,N'*-dicyclohexylcarbodiimide (DCCD), F_0 inhibitor (10 m*M* in ethanol or dimethyl sulfoxide; 15–20 nmol/mg dry weight of cells); 6-chloro-4-nitrobenz-2-oxa-1,3-diazole (NBD), F_1 inhibitor (10 m*M* in ethanol or dimethyl sulfoxide; 15–20 nmol/mg dry weight of cells); carbonyl cyanide *m*-chlorophenylhydrazone (CCCP), proton ionophore, "uncoupler" (50 m*M* in ethanol; 50–100 μM). Inhibitor stock solutions are prepared no sooner than 3 days before use and are stored at $-20°$. Inhibitors are added to the cell suspension at neutral pH 5–20 min before the oxygen pulse. In experiments at more extreme pH values (cf. Fig. 2) the final pH is adjusted only after preincubation of the cells with the inhibitor. The effect of DCCD (or NBD) does not increase when extending the preincubation period beyond 20 min.

Comments. CCCP abolishes all electrogenic H^+ translocation (across the CM) and, hence, any net acidification of the medium due to this process. If any residual acidification in oxygen-pulsed cell suspensions is still measured in the presence of saturating concentrations of CCCP (or another H^+ ionophore) this must be attributed to respiratory CO_2 (cf. Table I) and allowed for in the calculation of true (i.e., electrogenic) H^+ ejection.

[23] H. W. Van Versefeld, K. Krab, and A. H. Stouthamer, *Biochim. Biophys. Acta* **635,** 525 (1981).

[24] P. Mitchell and J. Moyle, *Biochem. J.* **105,** 1147 (1967).

[25] K. Krab and M. Wikström, *Biochim. Biophys. Acta* **548,** 1 (1979).

[26] Since all cyanobacteria (except *Gloeobacter violaceus*) contain two types of bioenergetically competent membranes (cf. Fig. 1) H^+/O ratios formally derived from *external* acidification must not be taken as any valid measure of true "coupling" sites. Yet, they are a valuable tool in comparing relative efficiencies of respiratory H^+ ejection under different conditions.

Since the effects of DCCD and NBD, which inhibit through binding of F_0 and F_1, respectively,[27,28] depend on the number of cells present in the assay suspension their concentration is expressed in relation to cell concentrations rather than absolute units. In each series of experiments the concentration of DCCD or NBD that abolishes *all* ATP formation by oxidative phosphorylation induced by the oxygen pulse (cf. Fig. 3) should be checked. Only using this premise will the statement be valid that the H^+ extrusion measured is not effected by some H^+-ATPase. On the other hand, provided that some oxygen-induced H^+ ejection persists in the presence of saturating DCCD (or NBD) this can hardly be due to "unspecific (inhibitory) side effects" of the inhibitors.

The use of orthovanadate in H^+ extrusion experiments[11] deserves special comment. The compound Na_3VO_3 (in micromolar concentrations) has been used to inhibit ATP hydrolysis by E_1E_2-type H^+-ATPases[12,29] which covalently bind phosphate to aspartyl groups.[30] However, when *millimolar* concentrations of orthovanadate are used for long-term experiments in aqueous solution as, for example, H^+ extrusion measurements, it must be taken into account that the species VO_3^{3-} is stable under strongly acidic conditions only while, at pH 7–9 the following equilibrium occurs:

$$4\ Na_3VO_4 + 3\ H_2O \rightleftharpoons 6\ NaOH + Na_6V_4O_{13} \qquad (2)$$

Net production of OH^- from H_2O would partly neutralize the extruded protons thus simulating "inhibition of H^+ ejection." The equilibrium between ortho- and tetra- (meta−)vanadate according to Eq. (2)[31] is both concentration and pH dependent, i.e., dependent on the amount of H^+ available, including extruded H^+; therefore, control experiments without cells would not be helpful.

KCN titration of oxygen uptake, proton extrusion, and intracellular ATP level reveals the coupling of H^+ ejection to electron transport and/or ATP[5,9] (Fig. 4).

Plasma Membrane Energization

As discussed in the Introduction electrogenic proton translocation across biological membranes by some primary proton pump (electron

[27] W. Sebald, W. Machleidt, and E. Wachter, *Proc. Natl. Acad. Sci. U.S.A.* **77,** 785 (1980).

[28] D. A. Holowka and G. G. Hammes, *Biochemistry* **16,** 5538 (1977).

[29] S. G. O'Neal, D. B. Rhoads, and E. Racker, *Biochem. Biophys. Res. Commun.* **89,** 845 (1979).

[30] P. Fürst and M. Solioz, *J. Biol. Chem.* **260,** 50 (1985).

[31] "Gmelins Handbuch der Anorganischen Chemie," Vol. 48, p. 145. Verlag Chemie, Weinheim, 1967.

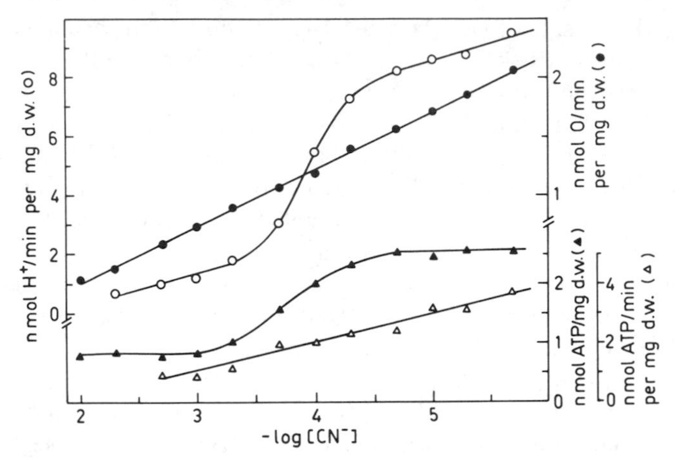

FIG. 4. Respiratory oxygen uptake (●), H^+ ejection (○), and ATP synthesis (△) in oxygen-pulsed *A. nidulans* in the presence of increasing concentrations of KCN. ▲, Steady-state levels of intracellular ATP. The biphasic inhibition pattern is typical of a composite mechanism for H^+ extrusion comprising both an ATPase (ATP dependent) and electron transport directly (ATPase independent).

transport and/or H^+-ATPase) eventually establishes a proton electrochemical potential gradient ($\Delta\tilde{\mu}_{H_+}$) or protonmotive force (pmf) across the respective membranes [cf. Eq. (1)], thereby effecting the "excited state" of the energy-transducing membrane. The degree of energization by ΔpH and/or $\Delta\psi$ [cf. Eq. (1)] can be calculated from the distribution of appropriate indicator reagents for ΔpH or $\Delta\psi$ on either side of the membrane, usually designated "in" and "out' (Δ = in *minus* out). The theoretical background for distribution methods has been amply documented elsewhere.[32,33] The reagents generally used include radioactively labeled weak acids and bases for ΔpH determination, and lipophilic anions and cations for $\Delta\psi$ determination. There are stringent criteria for the applicability of such reagents, which are generally used at concentrations around 10 μM.[34,35]

A fundamental requirement of all distribution methods is the analytical measurement of the amount of probe present in the cells (or vesicles) under given steady-state conditions, together with the known volume of the cells (or vesicles).[32,33,36] This permits calculation of c_i/c_o, the ratio of

[32] H. Rottenberg, *Bioenergetics* **7,** 61 (1975).

[33] H. Rottenberg, this series, Vol. 55, p. 547.

[34] I. R. Booth, *Microbiol. Rev.* **49,** 359 (1985).

[35] E. R. Kashket, *Annu. Rev. Microbiol.* **39,** 219 (1985).

[36] M.-E. Rivière, G. Johannin, D. Gamet, V. Molitor, G. A. Peschek, and B. Arrio, this volume [77].

probe concentrations inside and outside of the cells (or vesicles) which, in turn, is needed for the calculation of transmembrane electrical potential ($\Delta\psi$) or proton concentration (ΔpH) gradients according to established formulas.[32–35] One of the most elegant methods for the determination of c_i/c_o is flow dialysis[37] which is noninvasive and allows assessment of the time course of uptake of a given probe. (Immediate following of this uptake *in situ* is possible only with fluorescent probes whose range of applicability, however, is very limited; moreover, a comprehensive theory of "fluorescence quenching"[38,39] is still lacking.)

A particular problem for the application of distribution techniques to cyanobacteria rests in the two-compartment structure of these organisms (cf. Fig. 1). It is not clear, a priori, across which membrane, CM or ICM, the applied probe whose uptake is measured has been distributed. A mathematical treatment of flow-dialysis data will now be given that allows (1) accurate evaluation of flow-dialysis data for the calculation of c_i/c_o and (2) discrimination of individual "c_i/c_o" ratios across CM and ICM by the simultaneous use of properly "conjugated" pairs of reagents. Initial results obtained using a simplified version of evaluation procedure (2) have yielded satisfactory and consistent results for *A. nidulans*[15] and *Agmenellum quadruplicatum*.[40]

Evaluation of Flow-Dialysis Data

Figure 5a shows a typical flow-dialysis graph in the absence (i) and in the presence of cells (ii), the latter taking up the probe and characteristically modifying the monotonous dilution graph (i). After addition of the cells [phase (2)] net uptake of the probe occurs [phase (3)] until a steady state is reached [phase (4)]. The c_i/c_o ratio in phase (4) and, *cum grano salis,* from phase (4) onward (Fig. 5b), corresponds to the electrochemical potential of the probe across the cell membrane(s) under the conditions imposed. During phase (5) additional probe continues to be dialyzed out of the system (across the dialysis membrane), and, consequently, corresponding amounts of probe leave the cells (across the CM) in order to maintain c_i/c_o of phase (4) (equivalent to the "equilibrium potential"). Therefore, what we need to know is c_o and c_i characteristic of phases (4) [and (5)]. If the uptake of the probe is fast, simple calculation of c_i/c_o from M_i and M_o, together with the volume ratio v_i/v_o, is possible according to Fig. 6. This appears to hold for most of the (smaller) amines tested so far.

[37] S. Ramos, S. Schuldiner, and H. R. Kaback, this series, Vol. 55, p. 680.
[38] U. Pick and R. E. McCarthy, this series, Vol. 69, p. 538.
[39] S. Schuldiner, H. Rottenberg, and M. Avron, *Eur. J. Biochem.* **25,** 64 (1972).
[40] S. Belkin, R. J. Mehlhorn, and L. Packer, *Plant Physiol.* **84,** 25 (1987).

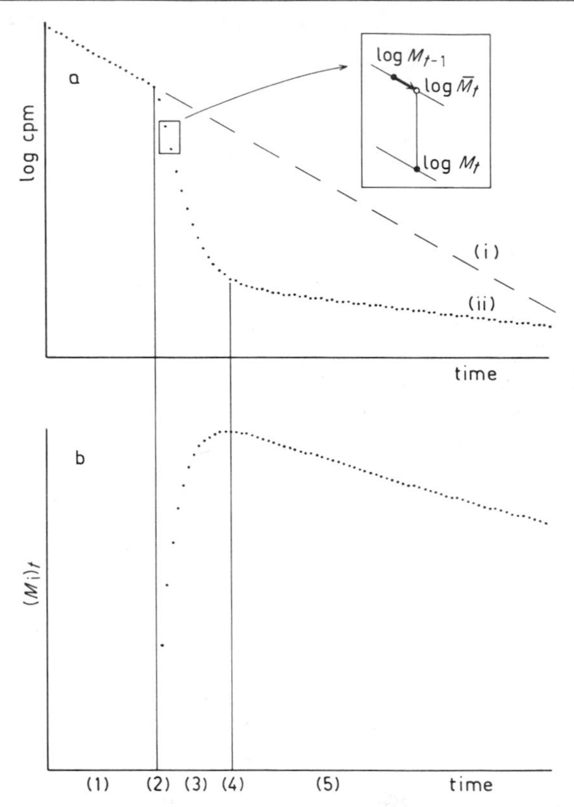

FIG. 5. (a) Typical flow-dialysis graph (time course of dilution) of a radioactively labeled indicator substance in the absence (i) and in the presence (ii) of cells which are added at time (2) and capable of taking up the probe. The inset illustrates the calculation of the differential amounts (M) of probe inside the cells. (For details, see text.) (b) Time course of the intracellular amount of probe (M_i) during a flow-dialysis experiment. The meaning of time spans/points (1) to (5) is explained in the text.

Generally, however, the uptake phase [phase (3)] takes some time (during which further probe is dialyzed out of the system), preventing such simple evaluation of the dialysis curve. This appears to be the case with tetraphenylphosphosphonium and triphenylmethylphosphonium cations, for example.

It should be emphasized that the extrapolated dilution graph [phase (1) in Fig. 5a] no longer can serve as a reference since the slope of phase (1) is steeper than that of phase (5) (because of eventual back-diffusion of probe out of the cells) and the scale is logarithmic which means that, e.g., if all probe were suddenly removed from the cells (as can be approxi-

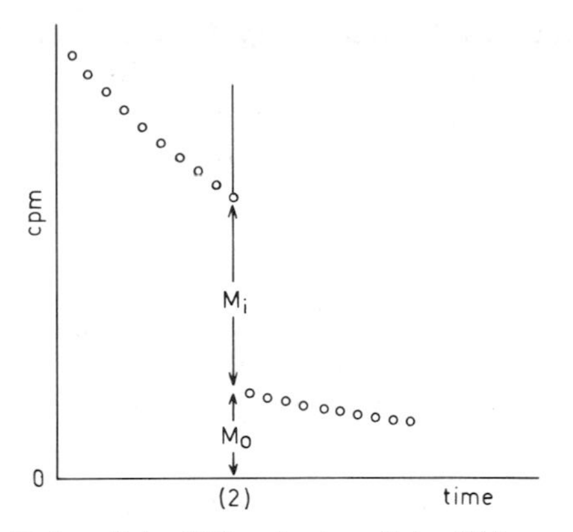

FIG. 6. Calculation of intracellular (M_i) and extracellular (M_o) amounts of probe when the latter is taken up by the cells instantaneously. Time (2), addition of the cells.

mated by the addition of an uncoupler if the probe is a weak acid) subsequent measuring points would never lie on the extrapolated curve of phase (1) (Fig. 7; also cf. Figs. 1 and 2 of Ref. 15). Similarly, back-extrapolation of phase (5) (Fig. 5a) would not be useful since, in case of slow uptake, phase (5) might not approach a straight line within a reasonable time.

For all practical purposes, therefore, calculate the total net amount of probe entering the cells until equilibrium is reached [phase (4) in Fig. 5],

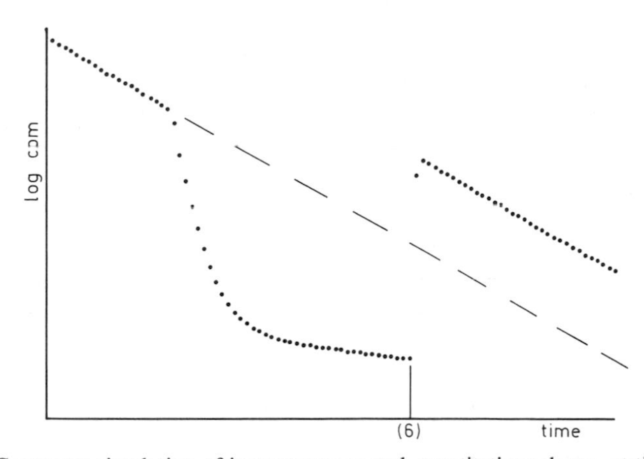

FIG. 7. Computer simulation of instantaneous and quantitative release, at time (6), of a radioactively labeled indicator substance that had previously been taken up by the cells (cf. Fig. 5a). The figure shows that the extrapolated dialysis straight line, (i) of Fig. 5a, must not be used as a reference for calculating the amount of probe taken up by the cells (see text for details).

pointwise by means of a suitable computer program as follows. With the aid of the slope k of phase (1) (Fig. 5a), from phase (2) onward each actual measuring point M_{t-1} (in the presence of cells) is used to calculate the following hypothetical "measuring" point $\bar{M}_t$ (in the absence of cells) according to Eq. (3).

$$\bar{M}_t = 10^{(\log M_{t-1} + k)} \quad (3)$$

The difference between this and the actually determined measuring point M_t (in the presence of cells; cf. Fig. 5a) equals the amount of probe that has entered the cells during the respective time interval, i.e., between t − 1 and t [Eq. (4)].

$$(\Delta M_i)_t = \bar{M}_t - M_t = 10^{(\log M_{t-1} + k)} - M_t \quad (4)$$

The total amount of probe present in the cells at time t, $(M_i)_t$, may be found, therefore, by summing all $(\Delta M_i)_t$ between $t = 0$ [or, equivalently, t = phase (2), addition of cells] and $t = t$ [Eq. (5)].

$$(M_i)_t = \sum_{a=1}^{t} (10^{(\log M_{a-1} + k)} - M_a) \quad (5)$$

As mentioned before during phase (5) this amount continuously decreases again due to maintenance of the equilibrium value of c_i/c_o.

Explanation of Subscripts and Symbols for Calculation of ΔpH and Δψ

In solving the following equations, these definitions are applicable for subscripts: T, total; o, out; i, internal; c, cytoplasm; t, thylakoid; and for symbols: v_c, ratio of cytoplasmic volume to total internal volume; v_t, ratio of thylakoid volume to total internal volume; $K = [H^+][A^-]/[AH]$; $K' = [H^+][M]/[MH^+]$; $K_1 = [H^+][DH^+]/[DH_2^{2+}]$; $K_2 = [H^+][D]/[DH^+]$; $\Delta\psi_{i-o} = (RT/F) \ln([X^-]_i/[X^-]_o) = -(RT/F) \ln ([Y^+]_i/[Y^+]_o)$; $\Delta\psi_{cm} = \Delta\psi_{c-o}$, electric potential difference across the cytoplasmic membrane; and $\Delta\psi_{tm} = \Delta\psi_{c-t}$, electric potential difference across the thylakoid membrane.

Comments. The sign of Δψ follows the definition, Δ = in minus out, taking "in" equivalent to the ATP-side, or *n*-side, of the respective membrane (cf. Fig. 1).

Calculation of ΔpH and Δψ in a Two-Membrane System[41]

In a cyanobacterial cell where CM and ICM may be simultaneously energized to different extents (cf. Fig. 1), individual bioenergetic gradi-

[41] W. H. Nitschmann, *J. Theor. Biol.* **122,** 409 (1986); B. Hinterstoisser and G. A. Peschek, *FEBS Lett.* **217,** 169 (1987).

ents across either membrane can be resolved by the use of complementary pairs of reagents, if either component of a pair (indicating a given bioenergetic gradient, i.e., ΔpH or $\Delta\psi$) is accumulated inside the cell according to an individual and independent mathematical equation. This equation will be shown for pairs of a weak acid (1) and a weak base (2), or a weak monovalent (2) and a weak divalent (3) base, in case of ΔpH determination, and for pairs of a lipophilic anion (4) and a lipophilic cation (5) in case of $\Delta\psi$ determination. In all instances, the experimentally determined parameter (e.g., from flow-dialysis experiments, see above) is the respective reagent's concentration ratio between the inside and outside of the cells, i.e., $c_i/c_o = M_i v_o / M_o v_i$. Bioenergetic gradients ΔpH, $\Delta\psi$, and $\Delta\tilde{\mu}_{H+}$ across the plasma membrane of dark aerobic and anaerobic *A. nidulans* incubated at various pH values, as measured and calculated by the procedures and formulas described in this chapter, are shown in Figs. 8A and B.[15]

Distribution ratio c_i/c_o:

(1) Weak acid (A)

$$\frac{[A]_i^T}{[A]_o^T} = \frac{[AH]_i + [A^-]_c v_c + [A^-]_t v_t}{[AH]_o + [A^-]_o} = \frac{\dfrac{1}{K} + \dfrac{v_c}{[H^+]_c} + \dfrac{v_t}{[H^+]_t}}{\dfrac{1}{K} + \dfrac{1}{[H^+]_o}} \tag{6}$$

(2) Weak monovalent base (M)

$$\frac{[M]_i^T}{[M]_o^T} = \frac{[M]_i + [MH^+]_c v_c + [MH^+]_t v_t}{[M]_o + [MH^+]_o} = \frac{K' + [H^+]_c v_c + [H^+]_t v_t}{K' + [H^+]_o} \tag{7}$$

(3) Weak divalent base (D)

$$\frac{[D]_i^T}{[D]_o^T} = \frac{[D]_i + [DH^+]_c v_c + [DH^+]_t v_t + [DH_2^{2+}]_c v_c + [DH_2^{2+}]_t v_t}{[D]_o + [DH^+]_o + [DH_2^{2+}]_o}$$

$$= \frac{K_1 K_2 + K_1[H^+]_c v_c + K_1[H^+]_t v_t + [H^+]_c^2 v_c + [H^+]_t^2 v_t}{K_1 K_2 + K_1[H^+]_o + [H^+]_o^2} \tag{8}$$

Solution for $[H^+]_c$:

$$[H^+]_c = \frac{-b - \sqrt{b^2 - 4ac}}{2a}$$

(A) if $pH_o < 7$ ($pH_t < pH_o < pH_c$), then Eqs. (6) and (7) have to be applied:

$$a = \left\{\frac{[A]_i^T}{[A]_o^T}\left(\frac{1}{K} + \frac{1}{[H^+]_o}\right) - \frac{1}{K}\right\} v_c; \qquad b = -\frac{ac}{v_c^2} - v_c^2 + v_t^2;$$

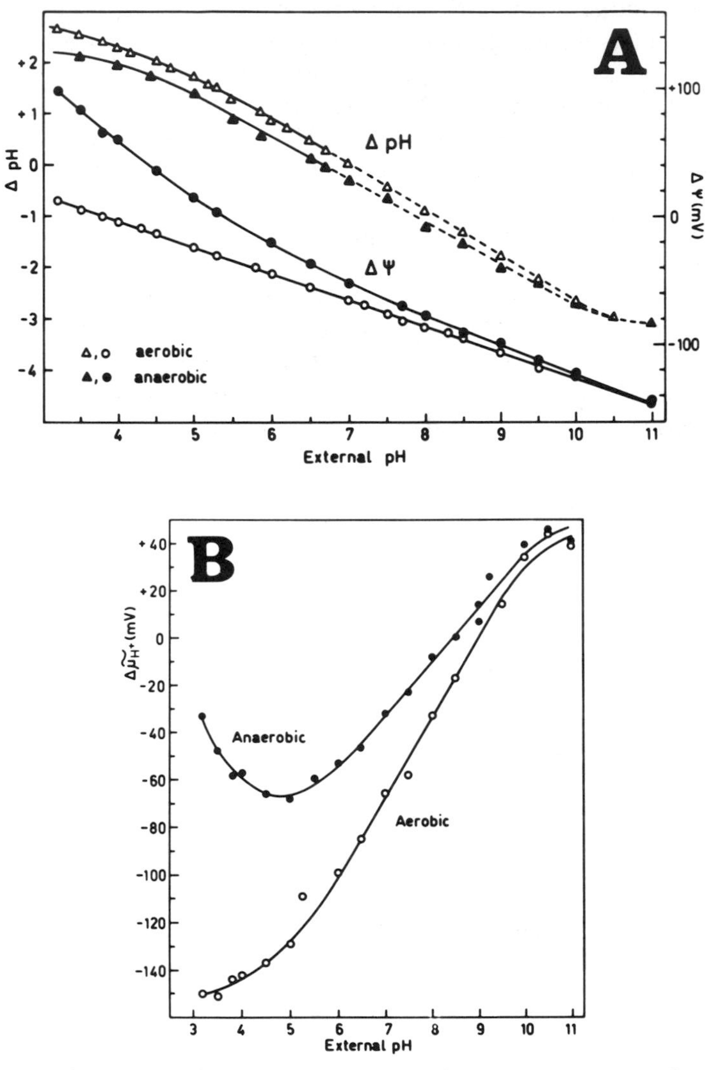

FIG. 8. pH dependence of ΔpH and $\Delta\psi$ (A), and of $\Delta\tilde{\mu}_{H_+}$ (B), across the plasma membrane of dark aerobic and anaerobic *A. nidulans*. The bioenergetic gradients are measured by the distribution method using the flow-dialysis technique as described in the text. Individual components of ΔpH and $\Delta\psi$ across the plasma membrane are resolved according to the formulas presented in the text. $\Delta\tilde{\mu}_{H^+}$ is calculated from ΔpH and $\Delta\psi$ according to Eq. (1).

$$c = \left\{ \frac{[M]_i^T}{[M]_o^T} (K' + [H^+]_o) - K' \right\} v_c$$

(B) if $pH_o > 7$ (pH_t, $pH_c < pH_o$), then Eqs. (7) and (8) have to be applied:

$$a = v_c + \frac{v_c^2}{v_t}; \qquad b = -2 \left\{ \frac{[M]_i^T}{[M]_o^T} (K' + [H^+]_o) - K' \right\} \frac{v_c}{v_t};$$

$$c = \frac{b^2 v_t}{4v_c^2} - \frac{bK_1 v_t}{2v_c} - \frac{[D]_i^T}{[D]_o^T} (K_1K_2 + K_1[H^+]_o + [H^+]_o^2) + K_1K_2$$

Substituting $[H^+]_c$, e.g., in Eq. (7) gives

$$[H^+]_t = \left\{ \frac{[M]_i^T}{[M]_o^T} (K' + [H^+]_o) - K' - [H^+]_c v_c \right\} \frac{1}{v_t}$$

(4) Lipophilic anion (X^-)

$$\frac{[X^-]_i^T}{[X^-]_o^T} = \frac{[X^-]_c v_c}{[X^-]_o} + \frac{[X^-]_t v_t}{[X^-]_o} = 10^{\frac{F\Delta\psi_{c-o}}{2.3RT}} v_c + 10^{\frac{F\Delta\psi_{t-o}}{2.3RT}} v_t \tag{9}$$

(5) Lipophilic cation (Y^+)

$$\frac{[Y^+]_i^T}{[Y^+]_o^T} = \frac{[Y^+]_c v_c}{[Y^+]_o} + \frac{[Y^+]_t v_t}{[Y^+]_o} = 10^{-\frac{F\Delta\psi_{c-o}}{2.3RT}} v_c + 10^{-\frac{F\Delta\psi_{t-o}}{2.3RT}} v_t \tag{10}$$

Solution for $\Delta\psi_{o-c}$ from Eqs. (9) and (10):

$$\Delta\psi_{c-o} = \frac{2.3RT}{F} \log \left\{ \frac{-b - \sqrt{b^2 - 4ac}}{2a} \right\} = \Delta\psi_{cm}$$

$$a = \frac{[Y^+]_i v_c}{[Y^+]_o}; \qquad b = -\frac{ac}{v_c^2} - v_c^2 + v_t^2; \qquad c = \frac{[X^-]_i v_c}{[X^-]_o}$$

Substituting $\Delta\psi_{c-o}$, e.g., in Eq. (10) gives

$$\Delta\psi_{t-o} = -\frac{2.3RT}{F} \log \left\{ \frac{[Y^+]_i^T}{[Y^+]_o^T v_t} - 10^{-\frac{F\Delta\psi_{c-o}}{2.3RT}} \frac{v_c}{v_t} \right\}$$

Finally,

$$\Delta\psi_{c-t} = \Delta\psi_{c-o} - \Delta\psi_{t-o} = \Delta\psi_{tm}$$

[40] Anoxygenic Photosynthetic Electron Transport

By SHIMSHON BELKIN, YOSEPHA SHAHAK, and ETANA PADAN

Introduction

The cyanobacteria are unique among the phototrophic prokaryotes in their ability to carry out complete plant-type, water-splitting, O_2-evolving photosynthesis utilizing both photosystems, PSI and PSII. Many cyanobacterial strains, however, can also carry out anoxygenic photosynthesis, using PSI only.[1-9] In this process, after a short induction period (2–3 hr), sulfide electrons flow freely through the electron transport chain to reduce one of the following three acceptors: (1) CO_2,[1-5] (2) protons, to produce H_2 gas (when CO_2 is absent or when CO_2 fixation is otherwise inhibited),[6-8] or (3) N_2, in a process of N_2 fixation induced by the absence of combined nitrogen.[9] Anoxygenic CO_2 photoassimilation has been shown to be widespread among cyanobacteria[3] and has been most thoroughly studied in the filamentous, nonheterocystous species *Oscillatoria limnetica*.[1-5] This chapter describes experimental conditions for assaying anoxygenic photosynthetic reactions in intact cells as well as in membrane preparations of *O. limnetica*.

Methods

Induction

Anoxygenic photosynthesis is induced in *O. limnetica* in a process that requires sulfide, light, and protein synthesis.[1-5] The cells are grown photoautotrophically under aerobic conditions to midexponential growth phase (3–4 days, 1–2 mg cell protein ml^{-1}) in artificial seawater (Turks Island salt,[10] concentrated times 2) supplemented with the constituents of

[1] Y. Cohen, B. B. Jorgensen, E. Padan, and M. Shilo, *Nature (London)* **257,** 489 (1975).
[2] Y. Cohen, E. Padan, and M. Shilo, *J. Bacteriol.* **123,** 858 (1975).
[3] S. Garlick, A. Oren, and E. Padan, *J. Bacteriol.* **129,** 623 (1977).
[4] A. Oren, E. Padan, and M. Avron, *Proc. Natl. Acad. Sci. U.S.A.* **74,** 2152 (1979).
[5] A. Oren and E. Padan, *J. Bacteriol.* **133,** 558 (1978).
[6] S. Belkin and E. Padan, *FEBS Lett.* **94,** 291 (1978).
[7] S. Belkin and E. Padan, *J. Gen. Microbiol.* **129,** 3091 (1983).
[8] S. Belkin and E. Padan, *Plant Physiol.* **72,** 825 (1983).
[9] S. Belkin, B. Arieli, and E. Padan, *Isr. J. Bot.* **31,** 199 (1982).
[10] "The Merck Index," 9th Ed., p. 1259. Merck, Rahway, New Jersey, 1976.

BG-11 medium.[11] The cultures are incubated at 35°, without shaking, at a light intensity of 25–50 $\mu E\ m^{-2}\ sec^{-1}$. After harvesting, the cells are washed, resuspended to the original cell density in an anaerobic growth medium[5] prepared in the 2-fold concentrated artificial seawater, and incubated in the presence of sulfide (3–4 m*M*). For both sulfide-dependent CO_2 fixation and H_2 evolution, 3 hr is sufficient for induction, although continuous anaerobic growth in the presence of sulfide seems to enhance the latter activity severalfold.[7] As measured by CO_2 fixation and H_2 evolution, the induction period can be shortened or altogether eliminated in the presence of sodium dithionite (10 m*M*, pH 7.0). It has been suggested that the low redox potential imposed by the dithionite exposes an acceptor for the sulfide electrons.[8] Maximal rates of N_2 fixation require approximately 24–48 hr in the presence of sulfide and in the absence of combined nitrogen.[9]

Anoxygenic Carbon Fixation

Suspensions of induced cells are prepared in 2 ml of the anaerobic growth medium lacking CO_2, buffered with 25 m*M* HEPES to the desired pH (usually 7.5), and containing 10 μM 3-(3′,4′-dichlorophenyl)-1,1-dimethylurea (DCMU). The suspension is introduced into 15-ml serum-stoppered glass vials and flushed for 15 min with N_2 or argon gas. The inlet and outlet needles are then withdrawn so that a slight positive pressure remains inside. The vials are preincubated in the dark in a shaking water bath (35°) and allowed to equilibrate for 15 min. $NaH^{14}CO_3$ is then injected (14 m*M*, 0.06 $\mu Ci\ \mu mol^{-1}$), as is $Na_2S \cdot 9H_2O$ (3 m*M*, pretitrated to pH 7.5 with concentrated HCl). The reaction is started by illumination from below (100 $\mu E\ m^{-2}\ sec^{-1}$). The bottles are opened at intervals, a sample (50 μl) is withdrawn for sulfide determination,[12] and 1 ml of the suspension is filtered under reduced pressure onto a 2.5-cm Whatman GF/C filter. The filters are washed with 40 ml of medium, acidified with a few drops of acetic acid (1.7 *M*) and left overnight in a fume hood. The radioactivity remaining on the filters may be counted with either a Nuclear Chicago gas-flow counter or a liquid scintillation counter, and the CO_2 assimilated calculated.

In a similar manner, H_2-dependent CO_2 photoassimilation (photoreduction), another PSI-dependent anoxygenic process, may be followed.[13] In this case H_2 is used as the gas phase, and sulfide is omitted.

[11] R. Rippka, J. Druelles, J. B. Waterbury, M. Herdman, and R. Y. Stanier, *J. Gen. Microbiol.* **111,** 1 (1979).

[12] H. G. Trüper and H. G. Schlegel, *Antonie van Leeuwenhoek* **30,** 225 (1964).

[13] S. Belkin and E. Padan, *Arch. Microbiol.* **116,** 109 (1978).

H_2 Evolution

The sulfide-dependent production of H_2 is monitored in the experimental system described above for anoxygenic carbon fixation, with CO_2 omitted from the experimental medium. At intervals during the incubation period, gas samples are withdrawn and immediately injected into a gas chromatograph equipped with a thermal conductivity detector and a 180 × 0.6 cm glass column of molecular sieve 5A (80 × 100 mesh). Samples for sulfide determination may also be withdrawn when required.

Total hydrogenase activity is routinely assayed in the dark without sulfide, but in the presence of 2.5 m*M* methyl viologen and 10 m*M* sodium dithionite. Assays with crude extracts (prepared by lysosyme treatment[7]) are preferable to *in vivo* experiments, since in intact cells the hydrogenase activity assayed is often only partial.

Another reduced compound which was found to serve as an anoxygenic photosynthetic electron donor is 2,3-dimercaptopropan-1-ol (BAL, British antilewisite). Aerobically, this compound is a potent inhibitor of electron transport, the inhibition site probably being the plastoquinone.[14] When added anaerobically, however, BAL serves as an electron donor for photosynthetic H_2 evolution,[15] the donation site being at or around the same location as the inhibition (see below). BAL (10 m*M*) is introduced anaerobically to the serum bottles after the N_2 flushing, with no pretitration. Owing to its volatility and toxicity, BAL should be handled in a fume hood.

Whereas H_2S supports both anoxygenic H_2 evolution and CO_2 fixation, BAL maintains only the former. It is suggested that BAL uncouples photophosphorylation and/or directly inhibits carbon fixation. An experiment of both sulfide-dependent CO_2 photoassimilation and BAL-dependent H_2 evolution is presented in Fig. 1. Figure 1 depicts the 2-hr induction period, the linear sulfide-dependent CO_2 photoassimilation which follows, the inhibition of this activity by BAL, and the BAL-dependent H_2 evolution occurring immediately upon introduction of this compound.

N_2 Fixation

The fixation of N_2 may be followed by the acetylene reduction assay.[16] Cells are preincubated anaerobically with sulfide (2 m*M*) with no combined nitrogen for 48 hr for maximal induction, and then prepared as described for carbon fixation. After the N_2 (or argon) flushing, C_2H_2 gas is

[14] Y. Shahak, G. Hind, and E. Padan, *Eur. J. Biochem.* **164,** 453 (1987).

[15] S. Belkin, Y. Siderer, Y. Shahak, B. Arieli, and E. Padan, *Biochim. Biophys. Acta* **766,** 563 (1984).

[16] W. D. P. Stewart, G. P. Fitzgerald, and R. H. Burris, *Arch. Mikrobiol.* **62,** 336 (1968).

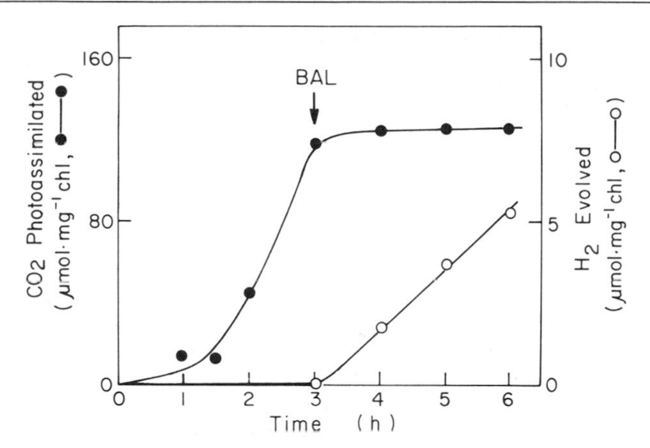

FIG. 1. Sulfide dependent CO_2 photoassimilation and BAL-dependent H_2 evolution in *Oscillatoria limnetica*. Experimental conditions as in the text, with 5 μg Chl ml^{-1}. BAL (10 m*M*) was added anaerobically after 3 hr. The CO_2 photoassimilated was determined in 1-ml suspension samples and the H_2 evolved in gas samples.

injected to a final concentration of 10–15% (v/v) in the gas phase. At the desired intervals, gas samples are withdrawn and injected into a gas chromatograph equipped with a Poropak N column and a flame ionization detector, to assay the ethylene produced.

Inhibitors of Anoxygenic Photosynthetic Electron Transport Chain

The effects of various inhibitors on the anoxygenic photosynthetic reactions have been tested. The results obtained allow the description of the anoxygenic photosynthetic electron flow in *O. limnetica* as depicted in Fig. 2. All the reactions were insensitive to the PSII inhibitor DCMU (10 μM) but were inhibited by the plastoquinone inhibitors 2,5-dibromothymoquinone (DBMIB, 25 μM) or the 2,4-dinitrophenyl ether of 2-iodo-4-nitrothymol (DNP–INT, 25 μM). Aerobic incubation of the cells in the presence of BAL (2 m*M*) is also inhibitory. The DBMIB and BAL inhibition can be bypassed by the addition of *N,N,N′,N′*-tetramethyl-*p*-phenylenediamine (TMPD, 100 μM), which channels electrons directly to plastocyanin. Under unblocked conditions, uncouplers such as carbonyl cyanide *p*-trifluoromethoxyphenylhydrazone (FCCP, 10 μM) enhance H_2 production, suggesting that electron flow from sulfide is coupled to proton pumping. All of these data point to the involvement of the plastoquinone–cytochrome b_6/f region in accepting the sulfide electrons. Since all the reactions are also inhibited by the ferredoxin inhibitor disalicylidenepropanediamine (DSPD, 1 m*M*) it is suggested that the segment of the

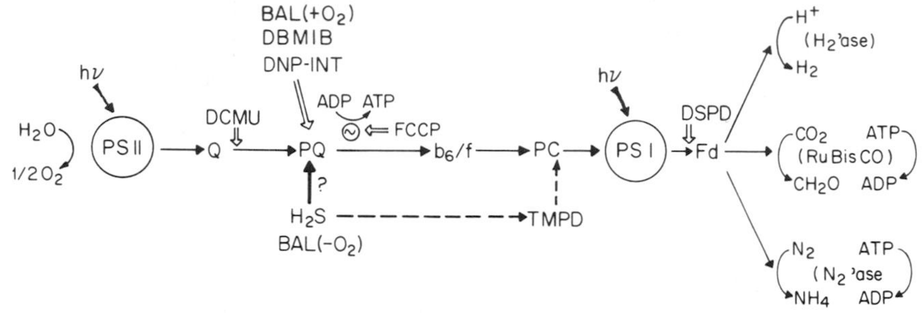

FIG. 2. Anoxygenic photosynthetic electron flow in *Oscillatoria limnetica*. Abbreviations: PSI and PSII, photosystems I and II; Q, primary electron acceptor of PSII; PQ, plastoquinone; b_6/f, cytochrome b_6/f complex, including the Rieske iron–sulfur protein; PC, plastocyanine; Fd, ferredoxin; H_2'ase, hydrogenase; RuBisCO, ribulose-bisphosphate carboxylase/oxygenase; N_2'ase, nitrogenase. Double arrows indicate inhibitor action sites, broken arrows the TMPD bypass.

electron transport chain from plastoquinone to ferredoxin is shared by all four reactions described. After ferredoxin, the electrons are channeled to CO_2, protons, or N_2 depending on the assay conditions.

Autotrophic Anoxygenic Growth

The capacity of an organism to fix CO_2 anoxygenically with sulfide electrons does not necessarily imply ability to actually grow anaerobically under these conditions. Such an ability depends on the indispensibility of O_2 for the organism's metabolism and its tolerance of low redox potentials.[17] *Oscillatoria limnetica,* in which the fatty acid metabolism is independent of O_2,[18] is as yet the only cyanobacterium in which photosynthetic sulfide-dependent anaerobic growth has been demonstrated.[5] In anaerobically growing cells the need for induction is alleviated, and electron flow rates are often higher and more stable.

Cell-Free Systems

For the molecular resolution of the anoxygenic photosynthetic electron transport system, studies of a cell-free system are essential. We have recently obtained a thylakoid preparation from *O. limnetica* which is active in anoxygenic electron flow.[19] To obtain this sulfide-oxidizing thylakoid preparation, 2.5-hr-induced cells are centrifuged, washed once,

[17] E. Padan, *Annu. Rev. Plant Physiol.* **30,** 27 (1979).

[18] A. Oren, A. Fattom, E. Padan, and A. Tietz, *Arch. Microbiol.* **141,** 138 (1985).

[19] Y. Shahak, B. Arieli, B. Binder, and E. Padan, *Arch. Biochem. Biophys.* **259,** 605 (1987).

and resuspended in a medium containing NaCl and KCl (5 mM each), $MgCl_2$ (10 mM), EDTA (0.1 mM), P_i (1 mM), sucrose (20 mM), bovine serum albumin (BSA) (0.5 mg ml^{-1}), and HEPES buffer (20 mM, pH 7.2). The cells are ruptured by passage twice through a Yeda Press (Rehovot, Israel; 1200 psi argon); the membranes are collected by centrifugation (8 min, 300 g) and resuspended in the same medium to a chlorophyll concentration of 1–1.5 mg ml^{-1}. While PSI activity is retained to a significant extent, that of PSII is lost. The membranes may be stored under liquid N_2 with no loss of activity. Glycerol (final concentration 25%, v/v) is added for storage.

The PSI activity of membranes prepared in this manner may be compared to that of similar preparations obtained from noninduced cells, by assaying for sulfide-dependent NADP photoreduction. A dual-wavelength spectrophotometer (Aminco DW-2) is used, at 340–400 nm. Side illumination is provided by a 150-W halogen lamp, and the actinic light is passed through a 4-cm water filter and a Schott RG665 filter. The photomultiplier is protected by 1.5 cm of a saturated $CuSO_4$ solution. The assay is conducted at 35° under N_2, and the reaction mixture contains (in 2.5 ml) the following: KCl (30 mM), $MgCl_2$ (3 mM), NADP (0.6 mM), *Spirulina platensis* ferredoxin (9 μM), spinach ferredoxin–NADP reductase (0.02 units, Sigma), thylakoids (8 μg Chl ml^{-1}), sodium tricine buffer (20 mM, pH 7.9), and Na_2S at the desired concentration.

As demonstrated in Table I, there is a distinct difference in the activity of the two membrane preparations. At sulfide concentrations below 0.2 mM, NADP photoreduction is observed essentially only in the thylakoids of the induced cells, in which the apparent K_m for sulfide is 30–50 μM. The reaction observed at high sulfide concentrations in noninduced thylakoids is most probably due to sulfide acting as a weak electron donor at

TABLE I
SULFIDE-DEPENDENT NADP PHOTOREDUCTION IN THYLAKOIDS PREPARED FROM INDUCED AND NONINDUCED *Oscillatoria limnetica* CELLS

	NADP photoreduction [μmol (mg Chl)$^{-1}$ hr^{-1}][a] with Na_2S (mM)			
Cell type	0.05	0.17	1.1	2.3
Noninduced	0.5	3.0	6.0	19.6
Induced	10.7	15.1	18.6	19.7

[a] For experimental conditions, see text.

the plastocyanin–P700 site. In the induced thylakoids, however, sulfide oxidation occurs mostly via the cytochrome b_6/f complex, as judged by the sensitivity to DNP–INT and 2-heptyl-2-hydroxyquinoline (HOQNO). Recently, similar results were reported by Sybesma and co-workers,[20,21] using membrane preparations from noninduced and 48-hr-induced *O. limnetica* cells. They found an apparent K_m for sulfide of 2 mM in the noninduced thylakoids and reported two values, 30 and 500 μM, for the induced system. The results from both groups may indicate that during adaptation to sulfide-dependent anoxygenic photosynthesis, changes occurring in the thylakoid membrane enable efficient utilization of sulfide electrons. Anoxygenic photosynthetic electron transport and the events leading to its induction are presently being investigated using this cell-free system.

Concluding Remarks

The cyanobacteria occupy a unique position in the evolutionary scheme of photosynthetic organisms. Among the most ancient organisms on earth, they are the only oxygenic phototrophs capable of using both sulfide and hydrogen for CO_2 photoassimilation, and their photosynthetic apparatus is intermediate between the bacterial anoxygenic system and plant-type photosynthesis. The ecological position of the cyanobacteria is also singular: they often inhabit anoxic niches and are abundant in habitats in which aerobic and anaerobic conditions regularly interchange. Unraveling the physiology and biochemistry of anoxygenic photosynthetic processes is therefore essential for our understanding of both the ecology and the evolution of this group. The experimental system described above allows the examination of all anoxygenic photosynthetic reactions, individually or combined, in *O. limnetica* as well as in other cyanobacteria.

Acknowledgments

E. Padan thanks the National Council for Research and Development, Israel, and the G.S.F., Munich, Federal Republic of Germany, for their support. The participation of B. Arieli in many stages of this work is gratefully acknowledged, as is the critical reading of the manuscript by B. Binder.

[20] C. Sybesma and L. Slooten, *Prog. Photosyn. Res.* **2,** 633 (1987).

[21] C. Sybesma, D. Schowanek, L. Slooten, and N. Walravens, *Photosyn. Res.* **9,** 149 (1986).

[41] Soluble Cytochromes and Ferredoxins

By HIROSHI MATSUBARA and KEISHIRO WADA

Soluble Cytochromes

Three soluble cytochromes of the *c* type have been isolated from various cyanobacteria. Most of the research on these cytochromes has centered on the photosynthetic cytochrome c_6[1] which functions as an electron carrier between the membrane-bound cytochrome b_6/f complex and P700 in the algal photosynthetic system. In algae, the cytochrome is functionally equivalent to plastocyanin in this electron transfer system, and the growth conditions of algae affect the ratio of these two components in the cells.[2,3] The first isolation of a cytochrome of this type was from a red alga in 1935 by Yakushiji,[4] and it was called "algal cytochrome *f*" until the membrane-bound cytochrome *f* was found in algae.[5]

Cytochrome c_6 was isolated in pure form from various cyanobacteria[6-22] and found to be soluble in water or saline solution, to have an

[1] "Enzyme Nomenclature: Recommendations (1984) of the Nomenclature Committee of the International Union of Biochemistry," p. 475. Academic Press, Orlando, Florida, 1984.

[2] G. Sandmann, *Arch. Microbiol.* **145,** 76 (1986).

[3] H. Bohner, H. Böhme, and P. Böger, *Biochim. Biophys. Acta* **592,** 103 (1980).

[4] E. Yakushiji, this series, Vol. 23, p. 364.

[5] P. M. Wood, *Eur. J. Biochem.* **72,** 605 (1977).

[6] S. Katoh, *J. Biochem. (Tokyo)* **46,** 629 (1959).

[7] W. A. Susor and D. W. Krogmann, *Biochim. Biophys. Acta* **120,** 65 (1966).

[8] R. W. Holton and J. Myers, *Biochim. Biophys. Acta* **131,** 362 (1967).

[9] R. W. Holton and J. Myers, *Biochim. Biophys. Acta* **131,** 375 (1967).

[10] T. Yamanaka, S. Takenami, and K. Okunuki, *Biochim. Biophys. Acta* **180,** 193 (1969).

[11] T. Ogawa and L. P. Vernon, *Biochim. Biophys. Acta* **226,** 88 (1971).

[12] H. L. Crespi, U. Smith, L. Gajda, T. Tisue, and R. M. Ammeraal, *Biochim. Biophys. Acta* **256,** 611 (1972).

[13] R. P. Ambler and R. G. Bartsch, *Nature (London)* **253,** 285 (1975).

[14] A. Aitken, *Eur. J. Biochem.* **78,** 273 (1977).

[15] A. A. Mutuskin, K. V. Pshenova, G. N. Vostroknutova, L. E. Makovkina, S. S. Voronkova, L. P. Nosova, and P. A. Kolesnikov, *Mikrobiologiya* **46,** 165 (1977).

[16] T. Yamanaka, Y. Fukumori, and K. Wada, *Plant Cell Physiol.* **19,** 117 (1978).

[17] H. Koike and S. Katoh, *Plant Cell Physiol.* **20,** 1157 (1979).

[18] K. K. Ho, E. L. Ulrich, D. W. Krogmann, and C. Gomez-Lojero, *Biochim. Biophys. Acta* **545,** 236 (1979).

[19] A. Aitken, *Eur. J. Biochem.* **101,** 297 (1979).

[20] M. Shin, N. Sakihama, H. Koike, and Y. Inoue, *Plant Cell Physiol.* **25,** 1575 (1984).

[21] K. K. Ho and D. W. Krogmann, *Biochim. Biophys. Acta* **766,** 310 (1984).

[22] I. Alpes, E. Stürzl, S. Schere, and P. Böger, *Z. Naturforsch. C* **39,** 623 (1984).

```
    1        10        20        30        40
(1) ADTVSGAALFKANCAQCHVGGGNLVNRAKTLKKE-A--LGKYNMY
(2) ADLAHGGQVFSANCASCHLGGRNVVNPAKTLEK--A--DLDEYGM
(3) ADIADGAKVFSANCAACHMGGGNVVMANKTLKKE-A--LEQFGMN
(4) ADAAAGGKVFNANCAACHASGGGQINGAKTLKKNALT---ANGKD
(5) GDVAAGASVFSANCAACHMGGRNVIVANKTLSKSDLAKYLKGFDD
     *   *   * **** **  *       *** *

             50        60        70        80        90
(1) SAKAIIA-QVTHGKGAMPAFGKRLKAEQIENVAAYYLEQADNGWKK
(2) ASIEAITTQVTNGKGAMPAFGAKLSADDIEGVASYALDQSGKEW
(3) SADAIM-YQVQNGKNAMPAFGGRLSEAQIENVAAYVLDQSSNKWAG
(4) TVEAIVA-QVTNGKGAMPAFKGRLSDDQIQSVALYVLDKAEKGW
(5) DAVAAVAYQVTNGKNAMPGFNGRLSPKQIEDVAAYVVDQAEKGW
            **  ** *** *   *    *  ** *        *
```

FIG. 1. Comparison of amino acid sequences of cyanobacterial cytochrome c_6. (1) *Aphanizomenon flos-aquae*,[24] (2) *Anacystis nidulans*,[25] (3) *Synechococcus* PCC 6312,[19] (4) *Plectonema boryanum*,[14] and (5) *Spirulina maxima*.[13] Several gaps denoted by – are inserted to give the most probable alignment. Asterisks below the sequences show the positions common to all five sequences. Amino acids are expressed by one-letter notation [see JCBN nomenclature and symbolism for amino acids and peptides, Recommendations 1983, cited in *Eur. J. Biochem.* **138,** 9 (1984)].

asymmetric α-band at around 553 nm, a high midpoint redox potential of about 0.34 V, and a low molecular mass, about 10,000 daltons. Its isoelectric point is not necessarily at low pH values as hitherto considered.[21] A summary of the properties of cytochrome c_6 is given in Table I. So far, 12 amino acid sequences of cytochrome c_6 are known[13,14,19,23–25]; 5 are those of cyanobacteria[13,14,19,24,25] (shown in Fig. 1). A partial sequence is also known.[26] They are similar in sequence to other algal cytochromes c_6.[24] Their evolutionary relationship is discussed in several articles.[13,14,19,23–25,27,28] The three-dimensional structure of a cytochrome c_6 (*c*-554, *Anacystis nidulans*) in the oxidized form has recently been established and found to have a chain folding similar to those of other short cytochromes.[25] One mole of heme *c* is covalently attached to the two

[23] L. T. Hunt, D. G. George, and W. C. Barker, *BioSystems* **18,** 223 (1985).

[24] J. R. Sprinkle, M. Hermodson, and D. W. Krogmann, *Photosynth. Res.* **10,** 63 (1986).

[25] M. L. Ludwig, K. A. Pattridge, T. B. Powers, R. E. Dickerson, and T. Takano, *in* "Electron Transport and Oxygen Utilization" (C. Ho, W. A. Eaton, E. Margoliash, J. P. Collman, K. Moffat, Q. H. Gibson, W. R. Scheidt, and J. S. Leigh, Jr., eds.), p. 27. Macmillan, London, 1982.

[26] S. Sanderson, B. G. Haslett, and D. Boulter, *Phytochemistry* **15,** 815 (1976).

[27] Y. Sugimura, T. Hase, H. Matsubara, and M. Shimokoriyama, *J. Biochem. (Tokyo)* **90,** 1213 (1981).

[28] A. Aitken, *Nature (London)* **263,** 793 (1976).

cysteine residues of the polypeptide chain through thioether bonds. The fifth and sixth ligands to the heme iron are the imidazole nitrogen of the histidine next to the second cysteine linking to the heme, and the sulfur of the invariant methionine at around the sixtieth position.[25,29]

Cytochrome *c*-550 was isolated from several cyanobacteria,[8–10,17,18,20,30] usually together with cytochrome c_6, and was not reducible with ascorbate, but with dithionite, indicating that it has a low midpoint redox potential (−0.26 V).[9,18] It is autoxidizable. The molecular mass is near 20,000 daltons. The function of this cytochrome is not established, but its low redox potential and the fact that the cells of the cyanobacterium containing this cytochrome released H_2S when they were disrupted may suggest its involvement in sulfur metabolism.[18] Table II shows some properties of cytochrome *c*-550 of several cyanobacteria.

Cytochrome *c*-552 was also isolated from *Anacystis nidulans*.[8,9] A basic protein was thought to correspond to mammalian cytochrome *c*, but detailed characterization was not performed. Cytochromes *c*-550 and *c*-552 also have one heme *c* per molecule.

Assay Method

Since the absorption spectra of these cytochromes *c* in their reduced forms are unique, spectra are commonly used to assay them. Cyanobacteria contain high concentrations of phycobiliproteins, chlorophyll *a*, and pteridines; therefore, the cytochrome content of the cells is difficult to determine. Reduced-minus-oxidized difference spectra may be analyzed for an indication of the cytochrome content.[31,32] Thus, a practical approach to measure the content of cyanobacterial cytochromes is to obtain absorption spectra after the removal of a large amount of phycobiliproteins by ammonium sulfate fractionation as described below. The heme content of the cytochromes is determined by measuring the alkaline pyridine ferrohemochrome spectra.[31] The spectra of *c*-type cytochrome show $\lambda_{nm} = 550 \pm 1$, $\varepsilon_{mM} = 31.18$, and $\Delta\varepsilon_{mM,\,red\text{-}ox} = 19.1$.

Purification Procedures

The first step in the purification of cytochromes is breakage of the cyanobacterial cells. Frozen cells can be used, and repeated freezing and

[29] T. Kitagawa, Y. Kyogoku, T. Iizuka, M. Ikeda-Saito, and T. Yamanaka, *J. Biochem. (Tokyo)* **78,** 719 (1975).

[30] J. Alam, J. Sprinkle, M. A. Hermodson, and D. W. Krogman, *Biochim. Biophys. Acta* **766,** 317 (1984).

[31] D. S. Bendall, H. E. Davenport, and R. L. Hill, this series, Vol. 23, p. 327.

[32] R. G. Bartsch, this series, Vol. 23, p. 344.

TABLE I
SOME PROPERTIES OF CYANOBACTERIAL CYTOCHROME c_6[a]

Organism	Absorption maxima (nm, in reduced form)			Molecular mass (daltons)	$E_{m,7}$ (V)	pI	Purity index	Yield (mg/g wet cells)	Remarks[b]
	α	β	γ						
Tolypothrix tenuis[c]	553	521	416		0.30				
Anabaena variabilis[d]	554	522	416				A_{416}/A_{275} = 5.65		
Anacystis nidulans drouet[e]	554	522.5	416.5	23,000	0.35	≤7.0	A_{554}/A_{275} = 0.84	1.5	
Synechococcus lividus[f]	552–553			9,500		4.47 (red) 5.00 (ox)	A_{552}/A_{273} = 1.50		a.a.c, heat stable
Synechococcus sp.[g]	553	522	416	9,400	0.335		A_{553}/A_{275} = 0.87		Heat stable
Synechococcus vulcanus[h]	553			10,400			A_{553}/A_{273} = 0.86		Heat stable
Synechococcus PCC 6312[i]	552							0.05	a.a.s.
Spirulina platensis Geither[j]	553	522	416	12,000–13,000		≤7.0			

Spirulina platensis[k]	553.6 (ε_{mM} = 25.3)	523	416	10,000	0.35	4.9	$A_{553.6}/A_{280}$ = 1.0	0.056	690 nm peak, a.a.c.
Plectonema boryanum[l]								0.005	a.a.s.
Aphanizomenon flos-aquae[m]	553					9.33			a.a.s.[n]

[a] Other preparations of cytochrome c_6 were also obtained from *Microcystis aeruginosa, Spirulina maxima, Anabaena variabilis, Phormidium luridum* [K. K. Ho, E. L. Ulrich, D. W. Krogmann, and C. Gomez-Lojero, *Biochim. Biophys. Acta* **545,** 236 (1979)], and several other species having basic cytochrome c_6 [K. K. Ho and D. W. Krogmann, *Biochim. Biophys. Acta* **766,** 310 (1984), including amino acid compositions].

[b] a.a.c. and a.a.s. refer to determination of amino acid composition and amino acid sequence, respectively. 690 nm peak means that a broad peak at 690 nm was observed, indicating methionine ligation to the heme iron.

[c] S. Katoh, *J. Biochem.* (*Tokyo*) **46,** 629 (1959).

[d] W. A. Susor and D. W. Krogmann, *Biochim. Biophys. Acta* **120,** 65 (1966).

[e] R. W. Holton and J. Myers, *Biochim. Biophys. Acta* **131,** 362 and 375 (1967).

[f] H. L. Crespi, U. Smith, L. Gajda, T. Tisue, and R. M. Ammeraal, *Biochim. Biophys. Acta* **256,** 611 (1972).

[g] H. Koike and S. Katoh, *Plant Cell Physiol.* **20,** 1157 (1979).

[h] M. Shin, N. Sakihama, H. Koike, and Y. Inoue, *Plant Cell Physiol.* **25,** 1575 (1984).

[i] A. Aitken, *Eur. J. Biochem.* **101,** 297 (1979).

[j] A. A. Mutuskin, K. V. Pshenova, G. N. Vostroknutova, L. E. Makovkina, S. S. Voronkova, L. P. Nosova, and P. A. Kolesnikov, *Mikrobiologiya* **46,** 165 (1977).

[k] T. Yamanaka, Y. Fukumori, and K. Wada, *Plant Cell Physiol.* **19,** 117 (1978).

[l] A. Aitken, *Eur. J. Biochem.* **78,** 273 (1977).

[m] K. K. Ho and D. W. Krogmann, *Biochim. Biophys. Acta* **766,** 310 (1984).

[n] J. R. Sprinkle, M. Hermodson, and D. W. Krogmann, *Photosynth. Res.* **10,** 63 (1986).

TABLE II
SOME PROPERTIES OF CYANOBACTERIAL CYTOCHROME *c*-550[a]

Organism	Absorption maxima (nm, in reduced form)			Molecular mass (daltons)	$E_{m,7}$ (V)	p*I*	Purity index	Remarks
	α	β	γ					
Anacystis nidulans drouet[b]	549	521.5	417.5	20,000	−0.26	≤7.0	A_{549}/A_{279} = 0.97	Autoxidizable, CO reactive, KCN, KF nonreactive
Synechococcus sp.[c]	549.5	522	417.5	19,500	$Na_2S_2O_4$[d]	≤7.0	$A_{549.5}/A_{275}$ = 0.62	Autoxidizable, heat stable
Synechococcus vulcanus[e]	550			19,000	$Na_2S_2O_4$[d]		A_{550}/A_{273} = 0.72	
Aphanizomenon flos-aquae[f,g]	550	519	414	15,500	$Na_2S_2O_4$[d]	4.4	A_{550} (red)/A_{550} (ox) = 1.37	Amino acid composition, amino-terminal sequence, no 690 nm peak[h]

[a] Other preparations of cytochrome *c*-550 were also obtained from *Anacystis nidulans* [T. Yamanaka, S. Takenami, and K. Okunuki, *Biochim. Biophys. Acta* **180,** 193 (1969)], *Spirulina maxima, Microcystis aeruginosa, Anabaena variabilis,* and *Phormidium luridum* [K. K. Ho, E. L. Ulrich, D. W. Krogmann, and C. Gomez-Lojero, *Biochim. Biophys. Acta* **545,** 236 (1979); J. Alam, J. Sprinkle, M. A. Hermodson, and D. W. Krogmann, *Biochim. Biophys. Acta* **766,** 317 (1984)].

[b] R. W. Holton and J. Myers, *Biochim. Biophys. Acta* **131,** 362 and 375 (1967).

[c] H. Koike and S. Katoh, *Plant Cell Physiol.* **20,** 1157 (1979).

[d] The cytochrome is reduced only with dithionite, and not with ascorbate.

[e] M. Shin, N. Sakihama, H. Koike, and Y. Inoue, *Plant Cell Physiol.* **25,** 1575 (1984).

[f] K. K. Ho, E. L. Ulrich, D. W. Krogmann, and C. Gomez-Logero, *Biochim. Biophys. Acta* **545,** 236 (1979).

[g] J. Alam, J. Sprinkle, M. A. Hermodson, and D. W. Krogmann, *Biochim. Biophys. Acta* **766,** 317 (1984).

[h] 690 nm peak was not observed, indicating that the ligation might not be via methionine.

thawing[12,16] can be useful. Additionally, French pressure disruption has been used.[19] Suspensions of lyophilized or fresh cells have been subjected to sonic oscillation in various aqueous solutions such as 0.4 *M* sucrose containing 10 m*M* NaCl,[7] 1 *M* acetate buffer (pH 4.7),[8] 10 m*M* potassium phosphate buffer (pH 6.9),[12] and 0.1 *M* Tris–HCl buffer (pH 7.5) containing 35% ammonium sulfate.[20] Sonication followed by acetone treatment was also used,[17] but acetone treatment alone was also effective in releasing cytochromes.[8,10,15,16] Dried cells[13,16] or fresh cells[14] were homogenized in a homogenizer. Ferredoxin is also extractable by these procedures and was separated from cytochromes by the following purification steps.

The extraction of cytochromes is carried out with dilute saline solution, and subsequent purification steps include ammonium sulfate fractionation, ion-exchange chromatography, gel filtration, hydrophobic chromatography, and high-performance liquid chromatography (HPLC). The extraction of cytochrome *c*-550 with a detergent, Tween 20, from material precipitated by ammonium sulfate (45% saturation) was reported to be useful.[16]

We describe here four examples of purification procedures adapted from the literature.[8,16,18,20] These procedures can be modified as needed. For large-scale preparations, readers should refer to the literature[18] dealing with different cyanobacterial species.

Cytochromes c-552, c-554, and c-549 from Anacystis nidulans Drouet.[8] Lyophilized cells are suspended in water (8 ml/g cells) overnight, and the suspension is centrifuged at about 20,000 *g* for 30 min. The pellet is resuspended and centrifuged as above. The combined supernatants are dialyzed against water, and the dialyzate is applied to a DEAE–cellulose column equilibrated with 5 m*M* phosphate buffer (pH 7.0). Phycobiliproteins and other pigments are also adsorbed on this column. The passed-through fraction contains cytochrome *c*-552, a basic protein. The column is washed with water until practically no absorbance at visible regions remains. After combining the washes with the passed-through fraction, the solution is lyophilized, dialyzed, and adsorbed on an Amberlite XE-64 (not on CM–cellulose) at pH 7.0. Only partial purification was achieved for this cytochrome.

The adsorbed materials on the DEAE–cellulose column are successively eluted with increasing concentrations of phosphate buffer (pH 7.0). Cytochrome *c*-554 together with phycobiliproteins is eluted at 30 m*M* and cytochrome *c*-549 at 90 m*M*. Ferredoxin remains on the column at this step and is eluted with 0.5 *M* buffer. Cytochrome *c*-554 is not successfully purified by several cation, anion, or gel filtration chromatographies. Finally, disk electrophoresis on polyacrylamide gels was applied. The detailed conditions for the electrophoresis are omitted here. The cytochrome band is eluted by electrophoresis into a dialysis bag, and the dialyzate is again chromatographed on a small DEAE–cellulose column. Cytochrome *c*-554 is eluted with 25 m*M* phosphate buffer (pH 7.0).

Cytochrome *c*-549 eluted with 90 m*M* phosphate buffer is dialyzed against water and lyophilized. The dried sample is dissolved and fractionated with ammonium sulfate (35% saturation, pH 5.5). After filtration through a Millipore filter, addition of ammonium sulfate to 56% saturation precipitates the cytochrome. It is chromatographed on a DEAE–cellulose column after dialysis. The column is developed with phosphate buffer (pH 6.0) by increasing the concentration of the buffer in a stepwise manner. Most cytochrome bands come off at higher concentrations. Further purification is again achieved on a DEAE–cellulose column with phosphate buffer (pH 7.0), and cytochrome *c*-549 elutes at 40 m*M* of the buffer.

The following two methods employ adsorption and chromatography of cytochromes on a support equilibrated with ammonium sulfate solutions at high concentrations. This procedure derived from the observation that rubredoxin and ferredoxin are retained by DEAE–cellulose in high ammonium sulfate concentrations and eluted from it by decreasing the salt concentration.[33–35] After fractionation with ammonium sulfate, the supernatant solution could be directly applied to an appropriate support column without any dilution or time-consuming dialysis which sometimes denatures proteins. This method is effective for separating proteins from nucleic acid-like materials which interfere with the purification of various proteins from cyanobacteria. Thus, the yield should be increased by adopting these processes.

Cytochrome c-554 from Spirulina platensis. This method is based on the methods previously described,[16,36] with some improvements. Fresh or frozen cells (500 g) are suspended in 1 liter of 20 m*M* Tris–HCl buffer (pH 7.5) to make a dense bree. The suspension is poured into 4 volumes of chilled acetone (−15°) with vigorous stirring followed by gentle stirring for 15 min. The acetone-treated cells are collected on filter paper in a Büchner funnel. The cells are resuspended in 1.5 liters of 20 m*M* Tris–HCl buffer (pH 7.5) and kept overnight at 5° with a gentle stirring. After centrifugation at 12,000 *g* for 40 min, the supernatant is applied to a DEAE–cellulose column equilibrated with the same buffer. The column is washed further with the same buffer, and the passed-through fractions are

[33] L. E. Mortenson, *Biochim. Biophys. Acta* **81,** 71 (1964).

[34] W. Lovenberg and W. M. Williams, *Biochemistry* **8,** 141 (1969).

[35] S. G. Mayhew and L. G. Howell, *Anal. Biochem.* **41,** 466 (1971).

[36] K. Wada, J.-Y. Cui, Y. Yao, H. Matsubara, and K. Kodo, *Biochem.* (*Life Sci. Adv.*) **6,** 53 (1987).

combined, diluted with two volumes of water, and applied to a DEAE–cellulose (OH^- form) column. The adsorbed cytochrome is eluted with 0.1 *M* Tris–HCl buffer (pH 7.5) containing 0.4 *M* NaCl.

The crude cytochrome fractions are treated with ammonium sulfate at 70% saturation, and, if necessary, centrifugation is carried out. The supernatant solution containing ammonium sulfate is directly applied to a DEAE–cellulose column equilibrated with 50 m*M* Tris–HCl buffer (pH 7.5) containing 70% saturated ammonium sulfate. Cytochrome is eluted with 0.1 *M* Tris–HCl buffer (pH 7.5) containing 0.4 *M* NaCl and dialyzed against 8 m*M* Tris–HCl buffer (pH 8.5) for 48 hr, with the outer solution changed several times. The dialyzate is applied to a DEAE–cellulose column equilibrated with the same buffer and developed with a gradient concentration of NaCl from 20 m*M* to 0.5 *M* in 8 m*M* Tris–HCl buffer (pH 8.5). Cytochrome fractions are concentrated on a DEAE–cellulose column equilibrated with ammonium sulfate solution as mentioned above and passed through a Sephadex G-100 column (2.5 × 130 cm) equilibrated with 50 m*M* Tris–HCl buffer (pH 7.5) containing 0.35 *M* NaCl. The cytochrome fractions are collected and concentrated as above.

Cytochromes c-553 and c-550 from Synechococcus vulcanus.[20] Frozen cells of *S. vulcanus* (50 g) are suspended in 0.1 *M* Tris–HCl buffer (pH 7.5) containing ammonium sulfate at 35% saturation (250 ml). The suspension is subjected to sonication for 10 min in an ice bath. After centrifugation at 8,000 *g* for 10 min, the debris is suspended in the solution (125 ml), sonicated, and recentrifuged. The supernatants were combined and treated with ammonium sulfate (66% saturation), then centrifuged to remove a blue–green precipitate. The supernatant is applied to a Toyopearl HW 65C (polyvinyl gel with hydroxyl groups) column (3 × 7 cm) equilibrated with ammonium sulfate at 66% saturation. After washing with the same solution, the column is further washed with 0.1 *M* Tris–HCl buffer (pH 7.5) containing 35% saturated ammonium sulfate to elute ferredoxin. Cytochromes *c* are eluted by decreasing the concentration of ammonium sulfate to 20% saturation. A 25-ml portion of the cytochrome solution is directly applied to a butyl-Toyopearl 650 column (2 × 30) equilibrated with 20 m*M* Tris–HCl buffer (pH 7.5) containing 0.9 *M* ammonium sulfate and developed with 20 m*M* Tris–HCl buffer (pH 7.5) containing 0.8 *M* ammonium sulfate.

Four fractions of cytochromes *c* are obtained, one of cytochrome *c*-550 and three cytochrome *c*-553. The reason why cytochrome *c*-553 separated into three fractions has been discussed.[13] The *c*-550 fractions and the major fractions of *c*-553 are brought to 1.2 *M* ammonium sulfate and separately adsorbed on a small column of Toyopearl 650 equilibrated with 20 m*M* Tris–HCl buffer (pH 7.5) containing 1.2 *M* ammonium sulfate. The cytochromes are eluted with 0.1 *M* Tris–HCl buffer (pH 7.5) containing 0.25 *M* NaCl. Purity of these preparations was very high as judged from the absorbance ratios (Tables I and II) and HPLC profiles.

The method described here is rather simple and fast, because the extraction of cytochromes and ferredoxins in high concentrations of ammonium sulfate avoids the coextraction of other proteins such as phycobiliproteins. The ammonium sulfate column is also useful.

Cytochromes c-553 and c-550 from Aphanizomenon flos-aquae.[18,30] This method was developed for large-scale preparation of cytochromes. An outline of the purification of cytochrome *c*-553 of the basic type and cytochrome *c*-550 is described here.

Cells are suspended in an equal volume of water, and a freezing–thawing process is repeated 3 times. Prolonged sonication or homogenization increases the yield of cytochromes. The broken cells are filtered through a gauze pad, and the filtrate is treated with ammonium sulfate at 45% saturation. After standing, the solution is centrifuged and the supernatant saturated with ammonium sulfate. After centrifugation, the supernatant is used for purification of ferredoxin II.

The precipitate containing ferredoxin I, cytochromes *c*-553 and *c*-550, and plastocyanin is dissolved in and dialyzed against water. The dialyzate is adjusted to 5 m*M* Tris–MES buffer (pH 7.5) and passed through a CM–cellulose column equilibrated with the same buffer. Cytochrome *c*-553 is adsorbed on the column and *c*-550 passes through. The CM–cellulose column is developed in a stepwise manner by increasing the NaCl concentration in the Tris–MES buffer from 0.1 to 0.5 *M*. The passed-through fraction is applied to a DEAE–cellulose column equilibrated with 1 m*M* Tris–HCl buffer (pH 7.2) followed by washing with the buffer containing 0.05, 0.1, and 0.2 *M* NaCl to remove several colored materials. Cytochrome *c*-550 elutes with 0.5 *M* NaCl and is dialyzed against 5 m*M* phosphate buffer (pH 7.0).

The dialyzed solution is applied to a DEAE–cellulose equilibrated with the same buffer, and the column is washed with 1 m*M* phosphate buffer (pH 7.0) containing 0.175 *M* NaCl to remove yellow pigment. The buffer containing 0.2 *M* NaCl elutes cytochrome *c*-550. The eluate is dialyzed against 1 m*M* phosphate buffer (pH 7.0) and applied to a DEAE–cellulose column which is then washed with 1 m*M* phosphate buffer (pH 6.0) containing 0.1 *M* NaCl. A linear gradient from 0.15 to 0.3 *M* NaCl in the same buffer elutes cytochrome *c*-550. After one more chromatography on DEAE–cellulose, the *c*-550 solution is dialyzed against 1 m*M* phosphate buffer (pH 7.0). The dialyzate is passed through a Sephadex G-50 column equilibrated with 1 m*M* phosphate buffer (pH 7.0). The effluent is then passed through a Sephadex G-50 column equilibrated with 0.1 *M* phosphate buffer (pH 7.0). Finally, HPLC can be used by applying a gradient of 0–80% acetonitrile in water containing 0.1% (v/v) trifluoroacetic acid. However, some denaturation occurs, and this method is probably suitable for chemical analysis only.

Ferredoxins

Excellent review articles have appeared in previous volumes of this series[37–39] which give historical background, properties of ferredoxins of various types, purification procedures, amino acid sequences, and so on. Therefore, we describe here only briefly cyanobacterial ferredoxins in terms of newly introduced purification procedures, a membrane-bound ferredoxin, a ferredoxin distinct in function isolated from heterocysts, amino acid sequences, and three-dimensional structures.

Cyanobacterial ferredoxins belong to the chloroplast type with a [2Fe–2S] cluster at the active center in a molecular mass of about 11,000 daltons. They have very acidic isoelectric points at around pH 3.3 and low midpoint redox potentials ranging from −0.34 to −0.455 V. Their absorption spectra are similar to those of higher plant and other algal ferredoxins, but the maxima at visible regions are shifted slightly to longer wavelength at around 466, 422, and 431 nm, compared with those of plant ferredoxins, except for *Microcystis aeruginosa* ferredoxin II.[40] It has a maximum at 415 nm instead of the usual maximum at around 422 nm in the oxidized form. Further, it shows a maximum at 420 nm in the reduced form never before found in [2Fe–2S] ferredoxins. This feature may be related to the unusual amino-terminal sequence described below. The millimolar extinction coefficients of ferredoxins at their maxima of around 422 nm are 10–11. Ferredoxins have multiple functions in various electron transfer systems, such as in oxygenic photosynthetic electron transfer as the terminal electron donor to $NADP^+$, in the nitrogen fixation system as the electron donor to nitrogenase, in the nitrate- and nitrite-reducing system, and in the phosphoroclastic oxidation of pyruvate.

Many ferredoxins have been isolated and characterized from various cyanobacteria: *Anabaena cylindrica,*[41] *A. flos-aquae,*[42] *A. variabi-*

[37] B. B. Buchanan and D. I. Arnon, this series, Vol. 23, p. 413.
[38] J. C. Rabinowitz, this series, Vol. 24, p. 431.
[39] K. T. Yasunobu and M. Tanaka, this series, Vol. 69, p. 228; see also addendum, p. 827.
[40] C. L. Cohn, J. Alam, and D. W. Krogmann, *Physiol. Veg.* **23,** 659 (1985).
[41] M. C. W. Evans, D. O. Hall, H. Bothe, and F. R. Whatley, *Biochem. J.* **110,** 485 (1968).
[42] P. W. Andrew, M. E. Delaney, L. J. Rogers, and A. J. Smith, *Phytochemistry* **14,** 931 (1975).

lis,[7,18,43–45] *Anacystis nidulans,*[8,42,46–48] *Aphanizomenon flos-aquae,*[18,40] *Aphanothece halophitica,*[49] *A. sacrum,*[50,51] *Chlorogloeopsis fritschii,*[52] *Mastigocladus laminosus,*[53] *Microcystis aeruginosa,*[18,40] *M. flos-aquae,*[54] *Nostoc* sp.,[43,55] *N. muscorum* (now *Anabaena* PCC 7119),[56,57] *Nostoc* strain MAC,[58–62] *N. verrucosum,*[63] *Phormidium foveolarum,*[64] *P. luridum,*[18,65] *P. persicinum,*[66] *Spirulina* sp.,[36] *S. maxima,*[18,43,44,67] *S. platensis,*[67–70] *Synechococcus* sp.,[17] *S. lividus,*[12] *S. vulcanus,*[20] and *Synechocystis* 6714.[52,71]

[43] E. Tel-Or, R. Cammack, and D. O. Hall, *FEBS Lett.* **53,** 135 (1975).
[44] T.-M. Chan and J. L. Markley, *Biochemistry* **22,** 5982 (1983).
[45] B. Schrautemeier and H. Böhme, *FEBS Lett.* **184,** 304 (1985).
[46] T. Yamanaka, S. Takenami, K. Wada, and K. Okunuki, *Biochim. Biophys. Acta* **180,** 196 (1969).
[47] R. M. Smillie, *Plant Physiol.* **40,** 1124 (1965); R. M. Smillie, *Biochem. Biophys. Res. Commun.* **20,** 621 (1965).
[48] T.-M. Chan, M. A. Hermodson, E. L. Ulrich, and J. L. Markley, *Biochemistry* **22,** 5988 (1983).
[49] T. Hase, K. Inoue, N. Hagihara, H. Matsubara, M. M. Williams, and L. J. Rogers, *J. Biochem.* (*Tokyo*) **94,** 1457 (1983).
[50] K. Wada, H. Kagamiyama, M. Shin, and H. Matsubara, *J. Biochem.* (*Tokyo*) **76,** 1217 (1974).
[51] T. Hase, K. Wada, and H. Matsubara, J. Biochem. (*Tokyo*) **78,** 605 (1975).
[52] K. J. Hutson, L. J. Rogers, B. G. Haslett, and D. Boulter, *FEMS Microbiol. Lett.* **7,** 279 (1980).
[53] T. Hase, S. Wakabayashi, H. Matsubara, K. K. Rao, D. O. Hall, H. Widmer, J. Gysi, and H. Zuber, *Phytochemistry* **17,** 1863 (1978).
[54] K. K. Rao, R. V. Smith, R. Cammack, M. C. W. Evans, D. O. Hall, and C. E. Johnson, *Biochem. J.* **129,** 1159 (1972).
[55] A. Mitsui and D. I. Arnon, *Physiol. Plant* **25,** 135 (1971).
[56] T. Hase, K. Wada, M. Ohmiya, and H. Matsubara, *J. Biochem.* (*Tokyo*) **80,** 993 (1976).
[57] K. Wada, H. Matsubara, K. Chain, and D. I. Arnon, *Plant Cell Physiol.* **22,** 275 (1981).
[58] K. G. Hutson, L. J. Rogers, B. G. Haslett, D. Boulter, and R. Cammack, *Biochem. J.* **172,** 465 (1978).
[59] G. N. Hutber, K. G. Hutson, and L. J. Rogers, *FEMS Microbiol. Lett.* **1,** 193 (1977).
[60] G. N. Hutber, A. J. Smith, and L. J. Rogers, *Phytochemistry* **20,** 383 (1981).
[61] K. G. Hutson and L. J. Rogers, *Biochem. Soc. Trans.* **3,** 377 (1975).
[62] R. Cammack, K. K. Rao, C. P. Bargeron, K. G. Hutson, P. W. Andrew, and L. J. Rogers, *Biochem. J.* **168,** 205 (1977).
[63] M. Shin, M. Sukenobu, R. Oshino, and Y. Kitazume, *Biochim. Biophys. Acta* **460,** 85 (1977).
[64] A. Mitsui and A. San Pietro, *Plant Science Lett.* **1,** 157 (1973).
[65] S. Keresztes-Nagy, F. Perini, and E. Margoliash, *J. Biol. Chem.* **244,** 981 (1969).
[66] E. Tel-Or, S. Fuchs, and M. Avron, *FEBS Lett.* **29,** 156 (1973).
[67] D. O. Hall, K. K. Rao, and R. Cammack, *Biochem. Biophys. Res. Commun.* **47,** 798 (1972).
[68] K. Wada, T. Hase, H. Tokunaga, and H. Matsubara, *FEBS Lett.* **55,** 102 (1975); H. Matsubara, K. Wada, and R. Masaki, *in* "Iron and Copper Proteins" (K. T. Yasunobu, H. F. Mower, and O. Hayaishi, eds.), p. 1. Plenum, New York, 1976.

When some cyanobacteria are grown in iron-deficient medium, ferredoxin is replaced by flavodoxin, a low-molecular-weight electron transfer flavoprotein.[47,59,60,72] Some cyanobacteria contain two distinct [2Fe–2S] ferredoxins.[40,45,51,56–63,73,74] They have similar spectroscopic characteristics, acidic natures, and molecular sizes, but their amino acid sequences and midpoint redox potentials are different. In cyanobacteria, the two ferredoxin sequences differ in 20–40 amino acids, and ferredoxin II is less acidic than ferredoxin I.

Some cyanobacteria such as *Aphanocapsa* (*Synechocystis*) 6714, *Chlorogloeopsis fritschii,* and *Nostoc* MAC can grow both autotrophically in light and heterotrophically in the dark.[75–77] Two ferredoxins were isolated from *Nostoc* MAC grown under different conditions[58,61] in a ratio of 3–5 : 1. Their midpoint redox potentials were −0.35 V for ferredoxin I and −0.455 V for ferredoxin II, which was pH dependent. Ferredoxin I was more active than ferredoxin II in the photoreduction of $NADP^+$, but ferredoxin II was more active than ferredoxin I in a phosphoroclastic reaction assay. The amino acid sequences of these two ferredoxins differ.[73] On the basis of a structural comparison of *Nostoc* ferredoxins with those of *Chlorogloeopsis fritschii* and *Synechocystis* 6714 (also capable of growing under both conditions), an explanation given for the different chemical and functional properties of *Nostoc* MAC ferredoxin II was that it had unique amino acids around the [2Fe–2S] cluster.[73] Different growth conditions both in natural blooms and in laboratory experiments affect the relative contents of the two ferredoxins.[18,40] The biological significance of the presence of two ferredoxins in one organism, however, is not clear,[40,51,57,58,74,78] and assignment of the ferredoxins to a specific metabolic function has been unsuccessful.

[69] M. Tanaka, M. Haniu, K. T. Yasunobu, K. K. Rao, and D. O. Hall, *Biochem. Biophys. Res. Commun.* **69,** 759 (1976).

[70] R. V. Smith and M. C. W. Evans, *J. Bacteriol.* **105,** 913 (1971).

[71] T. Hase, K. Inoue, H. Matsubara, M. M. Williams, and L. J. Rogers, *J. Biochem.* (*Tokyo*) **92,** 1357 (1982).

[72] E. Knight and R. W. F. Hardy, *J. Biol. Chem.* **241,** 2752 (1966).

[73] T. Hase, H. Matsubara, G. N. Hutber, and L. J. Rogers, *J. Biochem.* (*Tokyo*) **92,** 1347 (1982).

[74] T. Hase, S. Wakabayashi, K. Wada, and H. Matsubara, *J. Biochem.* (*Tokyo*) **83,** 761 (1978).

[75] P. Fay, *J. Gen. Microbiol.* **39,** 11 (1965).

[76] D. S. Hoare, L. O. Ingram, E. L. Thurston, and R. Walkup, *Arch. Mikrobiol.* **78,** 310 (1971).

[77] R. Rippka, *Arch. Mikrobiol.* **87,** 93 (1972).

[78] Y. Takahashi, T. Hase, K. Wada, and H. Matsubara, *Plant Cell Physiol.* **24,** 189 (1983).

One of the two ferredoxins isolated from heterocysts of *Anabaena variabilis* was found to be unique in functioning only as the electron donor to nitrogenase.[45] This may be related to the multiple ferredoxins in higher plants, particularly in nonphotosynthetic tissues of plants.[78–80] A thermophilic unicellular cyanobacterium, *Synechococcus* sp., isolated from a hot spring showed rapid growth and high photosynthetic activities in a high temperature range,[81] and the ferredoxin isolated from this organism was very heat stable.[17] The possibility that this ferredoxin has unique salt bridges that might contribute to the thermostability was suggested.[82]

Amino Acid Sequences, Three-Dimensional Structures, and Evolutionary Studies of Cyanobacterial Ferredoxins

So far, 14 amino acid sequences of cyanobacterial ferredoxins are known,[48,49,53,56,68,69,71,73,74,82–87] of a total of over 40 sequences of [2Fe–2S] ferredoxins. Comparison of these sequences is given in Fig. 2. Cyanobacterial ferredoxins have in general slightly longer polypeptide chain than those of eukaryotic ferredoxins. Amino-terminal sequences of *Microcystis aeruginosa* ferredoxins I and II and *Aphanizomenon flos-aquae* ferredoxins I and II have recently been determined, and one of them, *M. aeruginosa* ferredoxin II, which has an unusual absorption spectrum as mentioned above, has six residues inserted between positions 1 and 2 with three lysine residues, A-K-K-I-K-T-T-T-,[40] which has not been found before. However, all these ferredoxins are highly similar in sequence.

The three-dimensional structures of *Spirulina platensis* ferredoxin[88–91]

[79] K. Wada, H. Oh-oka, and H. Matsubara, *Physiol. Veg.* **23,** 679 (1985).
[80] K. Wada, M. Onda, and H. Matsubara, *Plant Cell Physiol.* **27,** 407 (1986).
[81] T. Yamaoka, K. Satoh, and S. Katoh, *Plant Cell Physiol.* **19,** 943 (1979).
[82] T. Hase, M. Matsubara, H. Koike, and S. Katoh, *Biochim. Biophys. Acta* **744,** 46 (1983).
[83] Y. Takashashi, T. Hase, H. Matsubara, G. N. Hutber, and L. J. Rogers, *J. Biochem. (Tokyo)* **92,** 1363 (1982).
[84] I. S. Lee, T. Hase, H. Matsubara, K. K. Ho, and D. W. Krogmann, *Biochim. Biophys. Acta* **744,** 53 (1983).
[85] M. Tanaka, M. Haniu, S. Zeitlin, K. T. Yasunobu, M. C. W. Evans, K. K. Rao, and D. O. Hall, *Biochem. Biophys. Res. Commun.* **64,** 399 (1975).
[86] M. Tanaka, M. Haniu, K. T. Yasunobu, K. K. Rao, and D. O. Hall, *Biochemistry* **14,** 5535 (1975).
[87] T. Hase, K. Wada, and H. Matsubara, *J. Biochem. (Tokyo)* **79,** 329 (1976).
[88] K. Ogawa, T. Tsukihara, H. Tahara, Y. Katsube, Y. Matsuura, N. Tanaka, M. Kadudo, K. Wada, and H. Matsubara, *J. Biochem. (Tokyo)* **81,** 529 (1977).
[89] T. Tsukihara, K. Fukuyama, H. Tahara, Y. Katsube, Y. Matsuura, N. Tanaka, M. Kadudo, K. Wada, and H. Matsubara, *J. Biochem. (Tokyo)* **84,** 1645 (1978).
[90] K. Fukuyama, T. Hase, S. Matsumoto, T. Tsukihara, Y. Katsube, N. Tanaka, M. Kakudo, K. Wada, and H. Matsubara, *Nature (London)* **286,** 522 (1980).
[91] T. Tsukihara, K. Fukuyama, M. Nakamura, Y. Katsube, N. Tanaka, M. Kakudo, K. Wada, T. Hase, and H. Matsubara, *J. Biochem. (Tokyo)* **90,** 1763 (1981).

```
             10        20        30        40        50
(1)  AT-YKVTLINEAEGINETIDCDDDTYILDAAEEAGLDLPYSCRAGACSTC
(2)  AT-YKVTLISEAEGINETIDCDDDTYILDAAEEAGLDLPYSCRAGACSTC
(3)  AS-YTVKLIT-PDG-ENSIECSDDTYILDAAEEAGLDLPYSCRAGACSTC
(4)  AS-YKVTLKT-PDG-DNVITVPDDEYILDVAEEEGLDLPYSCRAGACSTC
(5)  AT-YKVTLINEEEGINAILEVADDQTILDAGEEAGLDLPSSCRAGSCSTC
(6)  AT-YKVTLVR-PDGSETTIDVPEDEYILDVAEEQGLDLPFSCRAGACSTC
(7)  AT-YKVTLI-DAEGTTTTIDCPDDTYILDAAEEAGLDLPYSCRAGACSTC
(8)  AT-YKVTLINDAEGLNQTIEVDDDTYILDAAEEAGLDLPYSCRAGACSTC
(9)  AT-YKVTLINEAEGLNKTIEVPDDQYILDAAEEAGIDLPYSCRAGACSTC
(10) AS-YKVTLINEEMGLNETIEVPDDEYILDVAEEEGIDLPYSCRAGACSTC
(11) AT-FKVTLINEAEGTKHEIEVPDDEYILDAAEEEGYDLPFSCRAGACSTC
(12) ATVYKVTLV-DQEGTETTIDVPDDEYILDIAEDQGLDLPYSCRAGACSTC
(13) AT-YKVRLFNAAEGLDETIEVPDDEYILDAAEEAGLDLPFSCRSGSCSSC
(14) AT-FKVTLINEAEGTKHEIEVPDDEYILDAAEEEGYDLPFSCRAGACSTC
     *    * *     *         *  ***  *  * *** *** * ** *

             60        70        80        90       100
(1)  AGTITSGTI-DQSDQSFLDDDQIEAGYVLTCVAYPTSDCTIKTHQEEGLY
(2)  AGKITSGSI-DQSDQSFLDDDQIEAGYVLTCVAYPTSDCTIQTHQEEGLY
(3)  AGKITAGSV-DQSDQSFLDDDQIEAGYVLTCVAYPTSDCTIETHKEEDLY
(4)  AGKLVSGPA-PDEDQSFLDDDQIQAGYILTCVAYPTGDCVIETHKEEALY
(5)  AGKLVSGAAPNQDDQAFLDDDQLAAGWVMTCVAYPTGDCTIMTHQESEVL
(6)  AGKLLEGEV-DQSDQSFLDDDQIEKGFVLTCVAYPRSDCKILTNQEEELY
(7)  AGKLVTGTI-DQSDQSFLDDDQVEAGYVLTCVAYPTSDVTIETHKEEDLY
(8)  AGKIKSGTV-DQSDQSFLDDDQIEAGYVLTCVAYPTSDCTIETHKEEELY
(9)  AGKLISGTV-DQSDQSFLDDDQIEAGYVLTCVAYPTSDCVIETHKEEELY
(10) AGKIKEGEI-DQSDQSFLDDDQIEAGYVLTCVAYPASDCTIITHQEEELY
(11) AGKLVSGTV-DQSDQSFLDDDQIEAGYVLTCVAYPTSDVVIQTHKEEDLY
(12) AGKIVSGTV-DQSDQSFLDDDQIEKGYVLTCVAYPTSDLKIETHKEEDLY
(13) NGILKKGTV-DQSDQNFLDDDQIAAGNVLTCVAYPTSNCEIETHREDAIA
(14) AGKLVSGTV-DQSDQSFLDDDQIEAGYVLTCVAYPTSDCVIQTHKEEDLY
      *    *      ** ******   *   ******     * *  *
```

FIG. 2. Comparison of amino acid sequences of cyanobacterial ferredoxins. (1) *Spirulina platensis*,[68,69] (2) *Spirulina maxima*,[85,86] (3) *Synechocystis* 6714,[71] (4) *Aphanothece sacrum* I,[87] (5) *Aphanothece sacrum* II,[74] (6) *Synechococcus* sp.,[82] (7) *Aphanizomenon flos-aquae*,[84] (8) *Chlorogloeopsis fritschii*,[83] (9) *Mastigocladus laminosus*,[53] (10) *Aphanothece halophitica*,[49] (11) *Nostoc muscorum*,[56] (12) *Nostoc* MAC I,[73] (13) *Nostoc* MAC II,[73] and (14) *Anabaena variabilis*.[48] The gaps in the sequences, asterisks for the common amino acids, and one-letter notation of amino acids are as described in the legend to Fig. 1. Note that only one amino acid difference, at position 89, is observed between ferredoxins from *A. variabilis* (14) and *N. muscorum* (11). The latter is now reclassified as *Anabaena* PCC 7119 [R. Rippka, J. Deruelles, J. B. Waterbury, M. Herdman, and R. Y. Stanier, *J. Gen. Microbiol.* **111,** 1 (1979)].

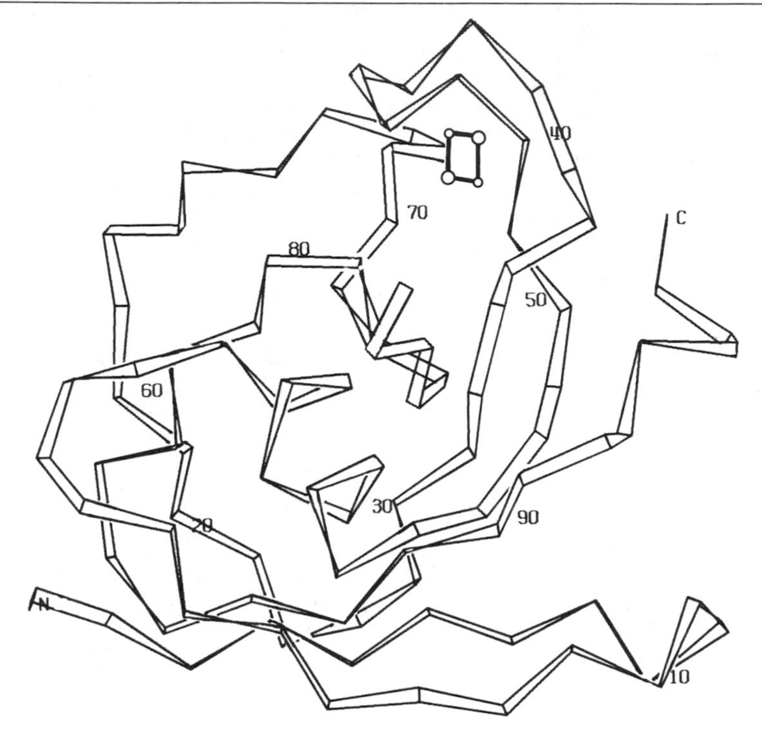

FIG. 3. Ribbon drawing of *Spirulina platensis* ferredoxin. This structure is based on previously published work,[91] courtesy of Dr. T. Tsukihara, and has been redrawn. Adjacent C_α atoms along the sequence are joined by ribbons. N and C refer to the amino and carboxy termini, respectively. The [2Fe–2S] cluster is seen at the top region of the molecule.

and *Aphanothece sacrum* ferredoxin I[92,93] (Figs. 3 and 4) showed that their main chain foldings were conserved and their iron–sulfur clusters were located near the surfaces of the molecules. The regions surrounding the clusters seem to be particularly conserved during evolution.[90] The sulfur atoms of Cys-41 and Cys-46 coordinate to one iron atom, and those of Cys-49 and Cys-79 to the other iron atom in *S. platensis* ferredoxin.[89–91] Evolutionary aspects of [2Fe–2S] ferredoxins were sometimes presented.[23,49,89,93–99]

[92] A. Kunita, M. Koshibe, Y. Nishikawa, K. Fukuyama, T. Tsukihara, Y. Katsube, Y. Matsuura, N. Tanaka, M. Kakudo, T. Hase, and H. Matsubara, *J. Biochem.* (*Tokyo*) **84,** 989 (1978).

[93] T. Tsutsui, T. Tsukihara, K. Fukuyama, Y. Katsube, T. Hase, H. Matsubara, Y. Nishikawa, and N. Tanaka, *J. Biochem.* (*Tokyo*) **94,** 299 (1983).

[94] H. Matsubara, T. Hase, S. Wakabayashi, and K. Wada, *in* "Evolution of Protein Molecules" (H. Matsubara and T. Yamanaka, eds.), p. 209. Ctr. Acad. Publ. Jpn. Sci. Soc. Press, Tokyo, 1978.

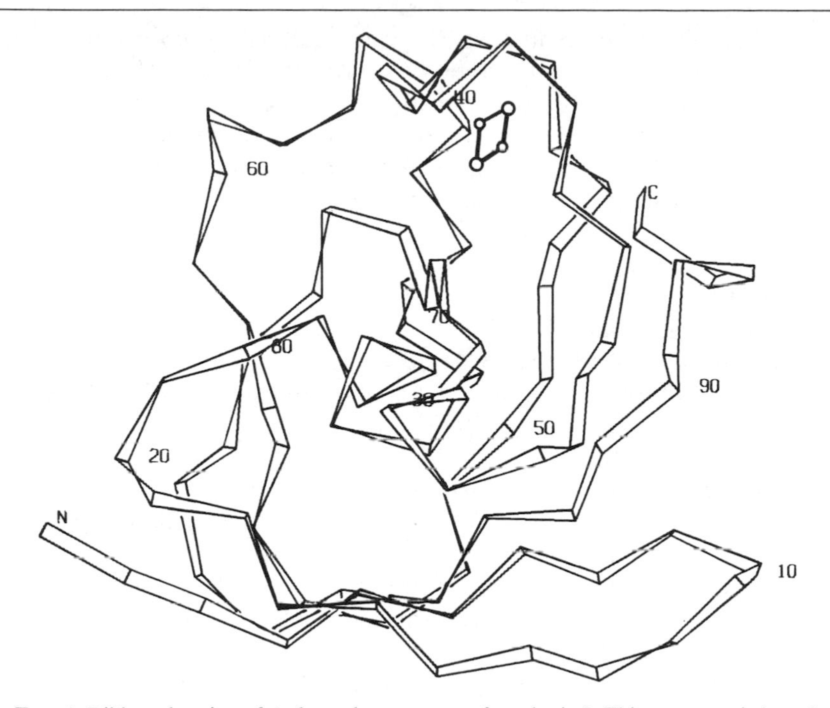

FIG. 4. Ribbon drawing of *Aphanothece sacrum* ferredoxin I. This structure is based on previously published work,[93] courtesy of Dr. T. Tsukihara, and has been redrawn. See the legend to Fig. 3 for explanation.

Assay Methods

Photochemical reduction of $NADP^+$ by isolated spinach chloroplasts is the most common procedure for the assay of ferredoxin.[37] Plant chloroplasts can be replaced by photosynthetic membrane fragments from cyanobacterial cells.[57,100,101] Ferredoxins can be assayed by their ability to

[95] H. Matsubara, T. Hase, S. Wakabayashi, and K. Wade, *in* "The Evolution of Protein Structure and Function" (D. M. Sigman and M. A. B. Brazier, eds.), p. 245. Academic Press, New York, 1980.

[96] T. Tsukihara, Y. Katsube, T. Hase, K. Wada, and H. Matsubara, *in* "Molecular Evolution, Protein Polymorphism, and the Neutral Theory" (M. Kimura, ed.), p. 299. Jpn. Sci. Soc. Press, Tokyo, and Springer-Verlag, Berlin, 1982.

[97] H. Matsubara and T. Hase, *in* "Protein and Nucleic Acids in Plant Systematics" (U. Jensen and D. E. Fairbrothers, eds.), p. 168. Springer-Verlag, Berlin, 1983.

[98] D. G. George, L. T. Hunt, L.-S. L. Yen, and W. C. Barker, *J. Mol. Evol.* **22,** 20 (1985).

[99] K. Wada, T. Hase, and H. Matsubara, *J. Biochem.* (*Tokyo*) **78,** 637 (1975).

[100] D. I. Arnon, B. D. McSwain, H. Y. Tsujimoto, and K. Wada, *Biochim. Biophys. Acta* **357,** 231 (1974).

[101] D. C. Yoch, *J. Bacteriol.* **116,** 384 (1973).

mediate $NADP^+$ photoreduction by cyanobacterial membrane fragments depleted of ferredoxin and ferredoxin–$NADP^+$ reductase. This reaction tests the ability of the photochemically reduced ferredoxin in question to reduce $NADP^+$ in the presence of ferredoxin–$NADP^+$ reductase supplied exogenously. The preparation of ferredoxin- and reductase-depleted membrane fragments is described below.

Reagents

- $NADP^+$, 2 m*M*
- Sodium ascorbate, 10 m*M*
- Dichlorophenolindophenol, 50 μM
- Photosynthetic membrane fragments (equivalent to 50–100 μg of chlorophyll *a*)
- Tricine buffer (pH 7.4), 50 m*M*
- Ferredoxin, various amounts from 0 up to 50 μg
- Ferredoxin–$NADP^+$ reductase from spinach or cyanobacteria, an appropriate amount

Procedure. The reaction is carried out aerobically at room temperature in a cuvette (3-ml capacity, 1-cm light path) containing 3 ml of the reagent mixture described above. Illumination initiates the reaction, and the absorbance increase at 340 nm due to the reduction of $NADP^+$ is recorded. The details are given in a previous volume of this series.[37]

Preparation of Photosynthetic Membrane Fragments. Membrane fragments are prepared from *Nostoc muscorum* cultured in the light on a medium of mineral salts with a gas phase consisting of N_2 and CO_2 at a ratio of 97 : 3 (v/v).[101] Two-milliliter portions of supernatant, obtained after centrifugation at 35,000 *g* of cells disrupted in a Ribi cell fractionator, in a French press, or by sonication, are layered between a 35% and a 5% sucrose solution in swing-type centrifuge tubes. The tubes are centrifuged at 25,000 *g* for 2 hr. The sucrose supernatant is discarded, and the pelleted photosynthetic membrane particles are resuspended in 20 m*M* Tricine buffer (pH 7.4). Chlorophyll *a* content is determined by the method of Arnon.[102]

Ferredoxins can also be assayed by their ability to supply electrons to nitrogenase prepared from cyanobacteria and other bacteria. Nitrogenase activity is assayed by monitoring the reduction of acetylene to ethylene by gas chromatography as previously described.[101]

Reagents

- HEPES buffer (pH 7.4), 30 m*M*
- Nitrogenase preparation

[102] D. I. Arnon, *Plant Physiol.* **24,** 1 (1949).

$MgCl_2$, 3 m*M*
ATP, 3 m*M*
Ferredoxin to be tested, about 10 μM
Sodium ascorbate, 7 m*M*
Dichlorophenolindophenol, 30 μM
Cyanobacterial membrane fragments (equivalent to 100 μg of chlorophyll *a*)
Phosphocreatine, 8.5 mg
Creatine kinase, about 10 μg

Procedure. The reaction is carried out in a Warburg vessel with a rubber stopper at 30° for 30 min under illumination. The vessel contains the reaction mixture with a total volume of 1.5 ml and a gas phase consisting of 73% argon and 27% acetylene. At an appropriate interval, one portion of the gas phase is withdrawn into a syringe through the rubber stopper and injected into a gas chromatographic apparatus.

Purification Procedures of Cyanobacterial Ferredoxins: A Generally Applicable Procedure

Step 1: Cell Breakage and Extraction of Ferredoxins. Cyanobacterial cells are broken for ferredoxin extraction as described for cytochromes in the previous section. Sometimes the homogenization is ineffective because of viscous materials enveloping the cells; therefore, dipping the cells in a hypertonic solution such as 2 *M* NaCl or kneading the cells with solid NaCl releases the ferredoxins easily. This method extracts a large amount of nucleic acid-like materials, however, and cell breakage by repeated freezing and thawing may be preferable. Acetone treatment is also effective.

Extraction of ferredoxin from the broken cells is generally accomplished by a dilute saline solution such as 0.1 *M* Tris–HCl buffer (pH 7.5). An alternative extraction method is disruption of the cells in the presence of a high concentration of ammonium sulfate[20] as described in the previous section. In this case the extract is further supplemented with ammonium sulfate at 70% saturation, and the precipitate formed is removed. Then, go to Step 4. Since ferredoxin is generally unstable below pH 6, care is required to keep the pH above 6 during extraction and purification.

Step 2: First Adsorption of Ferredoxins on DEAE–Cellulose. The extract containing ferredoxins and a large amount of phycobiliproteins is centrifuged and passed through a DEAE–cellulose (preferably a coarse grade) column equilibrated with 50 m*M* Tris–HCl buffer (pH 7.5). If the salt concentration of the extract is higher than 0.1 *M*, dilute it appropriately with water. When the volume of the extract is large and the extract contains small particles not removed by centrifugation, DEAE–cellulose, which is equilibrated with the above buffer and drained, is directly added to the extract in a beaker or a bucket and stirred for about 30 min at pH 7.5. After the DEAE–cellulose settles, the supernatant is decanted to another container, and more DEAE–cellulose is added. This process is repeated 2–3 times. The amount of wet cellulose used in this step depends on the amounts of ferredoxin and contaminants, but about 100–200 g of the wet cellulose may be enough for an extract from about 1–2 kg of wet cells. The combined DEAE–cellulose is washed in a batchwise manner with 50 m*M* Tris–HCl buffer (pH 7.5) and packed into a column. The column, with adsorbed ferredoxins, is washed with 0.1 *M* Tris–HCl buffer (pH 7.5), and ferredoxin is eluted with the same buffer containing 0.7 *M* NaCl.

Step 3: Ammonium Sulfate Fractionation. The crude ferredoxin solution is treated with ammonium sulfate at 70% saturation. The initial extract in Step 1 can be treated directly with ammonium sulfate at 70% saturation (pH 7.5). The precipitate containing phycobiliproteins is removed by centrifugation at 10,000 *g* for 10 min.

Step 4: Adsorption and Chromatography on a Column Equilibrated with Ammonium Sulfate. The supernatant solution is applied to a DEAE–cellulose column (5 × 8 cm) equilibrated with 70% saturated ammonium sulfate at pH 7.5 adjusted with powdered Tris. This method was originally applied to ferredoxin purification by Mayhew and Howell.[37] DEAE–cellulose can be replaced by Toyopearl HW-65C or Phenyl–Sepharose CL-4B.

The column is washed with the same ammonium sulfate solution, and ferredoxin is either (1) eluted with 0.1 *M* Tris–HCl buffer (pH 7.5) containing 0.7 *M* NaCl or (2) chromatographed by a linearly decreasing concentration gradient of ammonium sulfate in 0.1 *M* Tris–HCl buffer (pH 7.5) to 0% saturation. The second method is useful in removing nucleic acid-like materials and may also resolve ferredoxins not separable by conventional ion-exchange chromatography as described below. This was effective for separating several ferredoxins from a single organism.[103] An example is shown in Fig. 5.

The eluate is treated with ammonium sulfate at about 70% saturation and passed through a small DEAE–cellulose column equilibrated with 70% saturated ammonium sulfate. The concentrated ferredoxin on the column is eluted with 0.1 *M* Tris–HCl buffer (pH 7.5) containing 0.7 *M* NaCl. The eluate is dialyzed against 0.1 *M* Tris–HCl buffer (pH 7.5) containing 0.1 *M* NaCl.

Step 5: Ion-Exchange Chromatography. The dialyzed solution is applied to a DEAE–cellulose column (3 × 40 cm) equilibrated with the

[103] N. Sakihama, M. Shin, and H. Toda, *J. Biochem.* (*Tokyo*) **100,** 43 (1986).

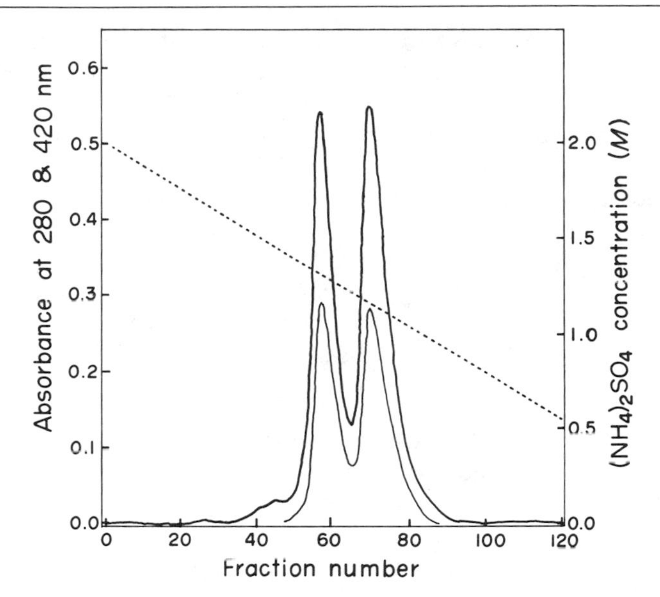

FIG. 5. Separation of ferredoxins on a Phenyl–Sepharose column equilibrated with a high concentration of ammonium sulfate. Two ferredoxins from *Alocassia macrorhiza* (a higher plant closely related to taro), which were not separated by DEAE–cellulose column chromatography, were subjected to so-called hydrophobic chromatography,[103] after adding solid ammonium sulfate to give 50% saturation. The ferredoxin solution was applied to a Phenyl–Sepharose column (2 × 22 cm) equilibrated with 50 m*M* Tris–HCl buffer (pH 7.5) containing 2 *M* ammonium sulfate. The ferredoxins adsorbed on the top of the column were developed by a linearly decreasing concentration gradient of ammonium sulfate from 2 to 0.5 *M* in 50 m*M* Tris–HCl buffer (pH 7.5) and separated into two peaks. The ferredoxins were pure as judged by several criteria and shown to have completely different amino acid sequences (K. Wada, unpublished observations).

same buffer and developed by a linear gradient of NaCl from 0.1 to 0.4 *M* (each 1 liter) in 0.1 *M* Tris–HCl buffer at a flow rate of 60–100 ml/hr. Isoforms of ferredoxin often present in cyanobacteria are usually separated at this stage, but possible minor components should not be overlooked. Ferredoxin fractions are collected, treated with ammonium sulfate at 70% saturation, and concentrated on a small DEAE–cellulose column equilibrated with ammonium sulfate solution as above. If the eluate shows an *R* value (described below) less than 0.3, DEAE–cellulose chromatography is repeated.

Step 6: Gel Filtration. The ferredoxin solution is passed through a Sephadex G-75 column (4 × 75 cm) equilibrated with 50 m*M* Tris–HCl buffer (pH 7.5) containing 0.35 *M* NaCl. Nucleic acid-like materials and other pigments are removed at this stage. Fractions showing high *R* values are collected and concentrated on a small ammonium sulfate/DEAE–cellulose column as described above. The purified ferredoxin solution is kept in an ice bath under N_2 or in a frozen state.

Comments. Polyethyleneimine is probably useful for rapid ferredoxin preparation as a specific adsorbent.[104] The initial extract may be treated with nucleases[19,58] or streptomycin sulfate[42] to remove nucleic acids. The purity of ferredoxin is expressed as an R value, namely, the ratio of A_{max} at around 420 nm to A_{max} at around 275 nm. However, the value depends on the source of the ferredoxin because the contents of aromatic amino acids vary.

Membrane-Bound Ferredoxin from Aphanizomenon flos-aquae[18,40]

A ferredoxin (II) tightly bound to the membrane of *Aphanizomenon flos-aquae* is isolated and purified by a modified method for the isolation of cytochrome *f*.[105] After repeated freezing and thawing, broken cells are passed through layers of gauze. The extract is fractionated with ammonium sulfate at 45% saturation, and the precipitate is mixed with an equal volume of 20 m*M* Tris–HCl buffer (pH 8). To the 4 liters of membrane solution is added 10 m*M* Tris–HCl buffer solution (1 liter) containing 7.5 g sodium deoxycholate and 5 g sodium cholate. The mixture is stirred for 25 hr to solubilize lipids, phycobiliproteins, and other material. The solution is mixed with an equal volume of 40% saturated ammonium sulfate. The precipitate formed is collected by decantation followed by centrifugation at 48,000 *g* for 5 min. The precipitate (~125 g) is suspended in 500 ml of 50 m*M* Tris–HCl containing 1 m*M* dithioerythritol and stirred for 12 hr.

A 4-liter bottle with a ground glass stopper is placed in a bucket, covered with ice, and chilled at −20°. Ethyl acetate (1.5 liters) is poured into the bottle and stirred in a hood. Ethanol (500 ml) chilled at −20° and NH_4OH (58% w/v, 7.5 ml) are added. The membrane suspension prepared as above is then slowly poured into the organic solvent mixture with vigorous stirring for 5 min. The bottle is stoppered and kept at −20° overnight. The mixture separates into three layers. The upper organic layer is siphoned off. The remainder is centrifuged at 48,000 *g* for 5 min at 5°, giving a green, lipid layer solidified between the yellow–brown aqueous phase and the organic phase. Careful decantation, filtration, and centrifugation separate the aqueous phase. The solution is dialyzed and applied to a DEAE–cellulose (DE-52) column (3.5 × 22 cm) equilibrated with 25 m*M* Tris–HCl buffer (pH 7.8). The column is washed with 2 liters of the same buffer containing 0.125 *M* NaCl followed by the buffer con-

[104] P. Schönheit, C. Wäscher, and R. K. Thauer, *FEBS Lett.* **89,** 219 (1978).
[105] K. K. Ho and D. W. Krogmann, *J. Biol. Chem.* **255,** 3855 (1980).

taining 0.2 *M* NaCl to elute cytochrome *f*. Ferredoxin is eluted with the same buffer containing 0.35 *M* NaCl. The ferredoxin fraction is precipitated with 4 volumes of chilled (−20°) acetone. The precipitate is dissolved in a minimum volume of 25 m*M* Tris–HCl buffer (pH 7.8) and passed through a Sephadex G-75 column (3 × 30 cm) equilibrated with the same buffer. The colored fractions are finally chromatographed and concentrated on a DE-52 column. The fractions whose ratio A_{280}/A_{420} was less than 3.0 were homogeneous on electrophoresis gels.

Aphanizomenon flos-aquae collected in natural blooms contained a large amount of soluble ferredoxin I and only very small amounts of membrane-bound ferredoxin II. The absorption spectra of these two ferredoxins were similar to those of other [2Fe–2S] ferredoxins. Their amino acid compositions and amino-terminal sequences up to 39 residues were very similar to each other. EPR spectra and redox potentials showed slight differences between them. The physiological significance of the membrane-bound ferredoxin is not clear.

Anabaena variabilis Heterocyst Ferredoxin[45,106]

Heterocysts isolated from *A. variabilis*[107] are broken by treatment in a French pressure cell (two passages at 138 MPa under H_2; 16 ml of 1 mg chlorophyll/ml) and centrifuged under H_2 at 1,500 *g* for 10 min to remove cell debris. The crude extract is centrifuged at 48,000 *g* for 30 min under H_2. The supernatant, containing nitrogenase, is centrifuged at 100,000 *g* for 1 hr. The resulting supernatant is further centrifuged at 350,000 *g* for 5 hr to precipitate nitrogenase. The light-brown supernatant is applied to a DEAE–cellulose (DE-52) column (1.5 × 7 cm) equilibrated with 50 m*M* Tris–HCl buffer (pH 8.0). The column is washed with the Tris buffer containing 0.25 *M* NaCl, and ferredoxin is eluted with the buffer containing 0.4 *M* NaCl. The reddish brown fractions are diluted with 5 volumes of cold distilled water and rechromatographed on a small DE-52 column (1 × 4 cm) equilibrated with 50 m*M* Tris–HCl buffer (pH 8.0). The ferredoxin is eluted with the buffer containing 0.4 *M* NaCl to give a pure preparation. A ferredoxin from vegetative cells grown in the presence of KNO_3 is also prepared.

The reduced ferredoxin from heterocysts transferred electrons to *A. variabilis* nitrogenase, but that from vegetative cells was inactive. Ferredoxins from both heterocysts and vegetative cells showed absorption spectra of typical chloroplast-type [2Fe–2S] ferredoxins. They were reduced by clostridial hydrogenase to an equal extent. When photoreduced

[106] B. Schrautemeier, H. Böhme, and P. Böger, *Biochim. Biophys. Acta* **807,** 147 (1985).
[107] B. Schrautemeier, H. Böhme, and P. Böger, *Arch. Microbiol.* **137,** 14 (1984).

by heterocyst thylakoids, both ferredoxins transferred electrons to nitrogenase, but the vegetative cell ferredoxin to a lesser extent. The two ferredoxins were equally active in the $NADP^+$ photoreduction system of heterocyst thylakoids. This evidence demonstrates the occurrence of a specialized ferredoxin in heterocysts that functions in the nitrogenase system. Structural differences between the two ferredoxins must be very interesting.

[42] Thioredoxin System

By PETER ROWELL, ALLAN J. DARLING, and WILLIAM D. P. STEWART

Introduction

Thioredoxin is a small (M_r usually ~12,000), ubiquitous redox protein with an active center cystine disulfide/dithiol (see Ref. 1). It may function in several oxidation–reduction reactions and have a regulatory role. For example, it can act as a protein-disulfide reductase[2] and as a hydrogen donor to ribonucleotide reductase,[3] it is required for the activity of DNA polymerase of bacteriophage T7[4] and for filamentous phage assembly,[5] and it may regulate the activities of certain enzymes. This latter function has been most thoroughly studied in the case of photosynthetic organisms.[6–8] In cyanobacteria, specifically, there is evidence for an involvement of reduced thioredoxins in the activation of the Calvin cycle enzymes fructose-1,6-bisphosphatase,[9–11] sedoheptulose-1,7-bisphosphatase,[9,10] and phosphoribulokinase[9,10]; in the activation of NADPH-

[1] A. Holmgren, *Annu. Rev. Biochem.* **54,** 237 (1985).
[2] A. Holmgren, *J. Biol. Chem.* **254,** 3672 (1979).
[3] T. C. Laurent, E. C. Moore, and P. Reichard, *J. Biol. Chem.* **239,** 3436 (1964).
[4] D. F. Mark and C. C. Richardson, *Proc. Natl. Acad. Sci. U.S.A.* **73,** 780 (1976).
[5] M. Russel and P. Model, *Proc. Natl. Acad. Sci. U.S.A.* **82,** 29 (1985).
[6] R. A. Wolosiuk and B. B. Buchanan, *Nature (London)* **266,** 565 (1977).
[7] B. B. Buchanan, *Annu. Rev. Plant Physiol.* **31,** 341 (1980).
[8] B. B. Buchanan, *in* "Thioredoxin and Glutaredoxin Systems: Structure and Function" (A. Holmgren, C. I. Branden, H. Jornvall, and B.-M. Sjoberg, eds.), p. 233. Raven, New York, 1986.
[9] N. A. Crawford, C. W. Sutton, B. C. Yee, T. C. Johnson, D. E. Carlson, and B. B. Buchanan, *Arch. Microbiol.* **139,** 124 (1984).
[10] C. W. Sutton, N. A. Crawford, B. C. Yee, D. C. Carlson, and B. B. Buchanan, *Adv. Photosynth. Res.* **3,** 633 (1984).
[11] S. M. Ip, P. Rowell, A. Aitken, and W. D. P. Stewart, *Eur. J. Biochem.* **141,** 497 (1984).

dependent malate dehydrogenase,[12] NADPH-dependent isocitrate dehydrogenase,[12] nitrite reductase,[13] and glutamine synthetase[13,14]; in the deactivation of glucose-6-phosphate dehydrogenase[15,16] and glucose dehydrogenase[17]; and as the hydrogen donor for ribonucleoside diphosphate reductase[18] and adenosine-3′-phosphate-5′-phosphosulfate and adenosine-5′-phosphosulfate sulfotransferases.[19,20]

Photosynthetically reduced ferredoxin, ferredoxin–thioredoxin reductase, and thioredoxin (the ferredoxin/thioredoxin system) provide a means whereby light can be linked to enzyme modulation, probably through the reduction of disulfides of the modulated enzymes. Such a mechanism of regulation is necessary since, in the cytoplasm of the cyanobacterial cell, enzymes of CO_2 fixation and carbohydrate degradation coexist and their activities must be adjusted to meet the varying metabolic needs of the cell. Thioredoxins have been shown to occur in a wide range of cyanobacteria (see, for example, Ref. 21). Cyanobacteria, like plant chloroplasts, have two thioredoxins, designated *f* and *m*, which selectively modulate certain enzymes,[9–11,18,22–24] and a ferredoxin–thioredoxin reductase.[25] Thioredoxin *m* is considered in this chapter.

Assay Method for Thioredoxin *m*

Principle. Cyanobacterial thioredoxin *m* may be assayed by one of several different methods (see Refs. 9, 11, 15, 18, 20, and 22). It is most convenient to use thioredoxin which is chemically reduced by dithiothreitol (DTT), rather than photoreduced thioredoxin (see chapter

[12] H. Papen, G. Neuer, M. Refraian, and H. Bothe, *Arch. Microbiol.* **134,** 73 (1983).
[13] R. Tischner and A. Schmidt, *Arch. Microbiol.* **137,** 151 (1984).
[14] H. Papen and H. Bothe, *FEMS Microbiol. Lett.* **23,** 41 (1984).
[15] J. D. Cossar, P. Rowell, and W. D. P. Stewart, *J. Gen. Microbiol.* **130,** 991 (1984).
[16] J. Udvardy, G. Borbely, A. Juhasz, and G. L. Farkas, *J. Bacteriol.* **157,** 681 (1984).
[17] A. Juhasz, V. Csizmadia, G. Borbely, J. Udvardy, and G.L. Farkas, *FEBS Lett.* **194,** 121 (1986).
[18] M. M. Whittaker and F. K. Gleason, *J. Biol. Chem.* **259,** 14088 (1984).
[19] W. Wagner, H. Follmann, and A. Schmidt, *Z. Naturforsch.* **33,** 517 (1978).
[20] A. Schmidt, *Arch. Microbiol.* **127,** 259 (1980).
[21] A. Schmidt and U. Christen, *Z. Naturforsch.* **34,** 1272 (1979).
[22] B. C. Yee, A. de la Torre, N. A. Crawford, C. Lara, D. E. Carlson, and B. B. Buchanan, *Arch. Microbiol.* **130,** 14 (1981).
[23] F. K. Gleason, *Arch. Microbiol.* **123,** 15 (1979).
[24] F. K. Gleason and A. Holmgren, *J. Biol. Chem.* **256,** 8306 (1981).
[25] M. Droux, P. Jacquot, M. Miginiac-Maslow, P. Gadal, J. C. Huet, N. A. Crawford, B. C. Yee, and B. B. Buchanan, *Arch. Biochem. Biophys.* **252,** 426 (1987).

[43]). Although the reductive activation of NADPH-dependent malate dehydrogenase (NADP–MDH) is usually used for the assay of thioredoxin *m* of plants, the activity of this enzyme in cyanobacteria is generally very low, and it is necessary to use an alternative source of the enzyme, usually spinach[18,22] or, alternatively, an alga such as *Scenedesmus obliquus*.[11] Details of the spectrophotometric NADP–MDH assay are given elsewhere[26] (see also chapter [43]) and will not be repeated here. A convenient, reliable, and sensitive procedure is the deactivation of cyanobacterial glucose-6-phosphate dehydrogenase (G6PDH),[15] and details of this are given below. The reader is referred to Refs. 9, 11, 15, 18, 20, and 22 for assay procedures employing other target enzymes.

Reagents

Tris–maleate buffer (pH 6.5), 50 m*M*
$MgCl_2$, 0.3 *M*
$NADP^+$, 22 m*M*
Glucose 6-phosphate (G6P), 150 m*M*
DTT, 75 m*M*
G6PDH preparation,[15] 50 μl; this may be purified (3 μg protein/50 μl) or partially purified (100 μg protein/50 μl), as detailed below. Partially purified G6PDH, which is rendered free of endogenous thioredoxin, is unaffected by DTT in the absence of added thioredoxin.

Partial Purification of G6PDH.[15] *Anabaena variabilis* ATCC 29413 is grown in batch culture in BG-11_0 medium.[27] Cultures are harvested by centrifugation, resuspended, and washed twice in 20 m*M* Tris–maleate buffer (pH 6.5). All subsequent operations are performed at 4°. Cells are disrupted by passage through a French pressure cell at 110 MPa, and cell debris is removed by centrifugation at 35,000 *g* for 15 min. The resulting supernatant is treated with $(NH_4)_2SO_4$ to 30% saturation for 30 min, then centrifuged at 35,000 *g* for 30 min. The resulting supernatant is treated with $(NH_4)_2SO_4$ to 80% saturation for 30 min, then centrifuged at 35,000 *g* for 30 min. The resulting pellet is resuspended in a small volume of 50 m*M* Tris–maleate buffer (pH 6.5), dialyzed extensively against the same buffer, then chromatographed on Sephadex G-75 (140 × 1.5 cm column) and eluted with 50 m*M* Tris–maleate buffer (pH 6.5).

[26] R. A. Wolosiuk, P. Schurmann, and B. B. Buchanan, this series, Vol. 69, p. 382.
[27] R. Y. Stanier, R. Kunisawa, M. Mandel, and G. Cohen-Bazire, *Bacteriol. Rev.* **35,** 171 (1971).

Assay procedure. Fifty microliters of a G6PDH preparation is incubated with 50 μl of DTT and 50 μl of a thioredoxin preparation at 30° for an appropriate time (usually 5–15 min). Then 0.55 ml of Tris–maleate buffer, 0.1 ml $MgCl_2$ solution, 0.1 ml $NADP^+$ solution, and 0.1 ml G6P solution are added, and incubation at 30° is continued. Absorbance is monitored continuously at 340 nm over a period of 5–10 min. Activity is calculated from the reaction progress curve.

Purification of Thioredoxin *m*

We have routinely used the following two methods for purification of *m*-type thioredoxins from cyanobacteria to homogeneity, as evidenced by polyacrylamide gel electrophoresis and isoelectric focusing. Immunoaffinity purification, adapted from the method of Sjoberg and Holmgren,[28] is particularly useful for the rapid purification of small quantities of thioredoxins, but it is limited by the availability of suitable antisera. The antisera which we have raised against *m*-type thioredoxins from *Anabaena cylindrica* and *A. variabilis* show cross-reactivity with only a very limited range of cyanobacterial thioredoxins tested, in the former case, and a much broader range, in the latter case (Refs. 11 and 29 and our unpublished observations).

Unless otherwise stated, all steps are carried out at 4°, and all centrifugation steps at 35,000 *g*. Cell-free extracts are prepared from 50–100 g wet weight of cells grown in batch culture, in nitrogen-free BG-11_0 medium at 26° and at a photon flux density of 100 μmol m^{-2} sec^{-1} incident at the surface of the vessel, and harvested in the late exponential phase of growth. Cells are washed and resuspended in 50 m*M* Tris–HCl buffer, pH 7.9 (buffer A), then disrupted by passage through a French pressure cell at 110 MPa (or, alternatively, by cavitation). Cell debris is removed by centrifugation for 20 min.

Conventional procedure. Starting with a cell-free extract containing 3.2 g of protein (from about 100 g wet weight of *A. variabilis* cells), 1.5 mg of thioredoxin can be obtained with a yield of about 17%. The cell-free extract is adjusted to pH 4.0 by adding 1.0 *M* acetic acid and incubated for 10 min. After centrifugation for 20 min, the resulting supernatant is adjusted to pH 7.9 by adding 2 *M* NH_4OH. Solid $(NH_4)_2SO_4$ is added to 90% saturation and, after incubating for 30 min, the suspension is centrifuged for 20 min. The resulting pellet is resuspended in a small volume of buffer

[28] B.-M. Sjoberg and A. Holmgren, *Biochim. Biophys. Acta* **315,** 176 (1973).

[29] J. D. Cossar, A. J. Darling, S. M. Ip, P. Rowell, and W. D. P. Stewart, *J. Gen. Microbiol.* **131,** 3029 (1985).

A and dialyzed extensively against the same buffer. The solution is then incubated at 75° for 3 min, rapidly cooled to 4°, and centrifuged for 20 min. The resulting supernatant is then chromatographed on DEAE–cellulose (30 × 3.5 cm column) and eluted with a gradient of 0–0.6 *M* KCl in buffer A. Active fractions are combined, dialyzed extensively against 20 m*M* sodium acetate buffer, pH 5.0, and chromatographed on CM–Sepharose (30 × 3.5 cm column) and eluted with a gradient of 20–500 m*M* sodium acetate buffer, pH 5.0. Active fractions are combined, concentrated to a small volume using an Amicon ultrafiltration cell with a YM5 membrane (M_r 5,000 cutoff), and chromatographed on Sephadex G-75 (140 × 1.5 cm column), eluting with buffer A.

We have found this procedure to be generally applicable to cyanobacterial *m*-type thioredoxins. In the case of *A. cylindrica* thioredoxin, however, we have previously used an alternative procedure from which chromatography on CM–Sepharose is omitted.[11]

Immunoaffinity purification. Antisera against purified thioredoxins are prepared as described by Ip *et al.*[11] The antiserum (containing ~30 mg protein) is coupled to cyanogen bromide-activated Sepharose 4B (Pharmacia), as directed by the manufacturer, to prepare an immunoaffinity column (14 × 1 cm). Such columns have a capacity to bind up to 100 μg of thioredoxin, and purification to homogeneity with a yield of about 90% can be achieved.

Cell-free extract (up to 40 mg protein), prepared as described above, is heated to 75° for 3 min, rapidly cooled to 4°, and centrifuged for 20 min. The supernatant is loaded onto the column which has previously been equilibrated with buffer A. The column is then washed, successively, with two column volumes of buffer A, one volume of buffer A plus 0.5 *M* NaCl and one volume of buffer A. Thioredoxin is then eluted by washing the column with 100 m*M* acetic acid (adjusted to pH 2.0 by adding HCl) to disrupt antibody–antigen complexes, and the fractions are immediately neutralized by adding Na_2CO_3. Active fractions are pooled, concentrated to a small volume, and desalted or dialyzed before use.

Properties of Thioredoxin *m*

Thioredoxin *m* consists of a single polypeptide (M_r ~11,500).[11,22,24,30] The active site dithiol, which can be reversibly oxidized to form a disulfide, is located at residues 31 and 34 in the sequence -Trp-Cys[31]-Gly-Pro-Cys[34]-Arg-. Thioredoxin *m* occurs in all cyanobacteria so far examined,

[30] F. K. Gleason, M. M. Whittaker, A. Holmgr Jornvall, *J. Biol. Chem.* **260,** 9567 (1985).

and there is extensive similarity among the amino acid sequences of cyanobacterial thioredoxin *m*, other bacterial thioredoxins, and chloroplast thioredoxin *m*.[11,30,31] Evidence has been obtained for a light-dependent reduction of thioredoxin *m* in *Anabaena cylindrica*,[32] and the protein is apparently located mainly in the centroplasm of vegetative cells of *A. cylindrica* and is absent from or present at a reduced level in heterocysts.[29] Several potential functions of cyanobacterial thioredoxin *m* have been established,[9–20] although the true physiological function(s) remains to be firmly established.

[31] F. K. Gleason, *in* "Thioredoxin and Glutaredoxin Systems: Structure and Function" (A. Holmgren, C. I. Branden, H. Jornvall, and B.-M. Sjoberg, eds.), p. 21. Raven, New York, 1986.

[32] A. J. Darling, P. Rowell, and W. D. P. Stewart, *Biochim. Biophys. Acta* **850,** 116 (1986).

[43] Ferredoxin/Thioredoxin System

By NANCY A. CRAWFORD, BOIHON C. YEE, MICHEL DROUX, DONALD E. CARLSON, and BOB B. BUCHANAN

Introduction

Light regulates enzymes of oxygenic photosynthesis via several mechanisms.[1–4] Important among these is the ferredoxin/thioredoxin system, an enzyme-mediated regulatory mechanism involving ferredoxin, ferredoxin–thioredoxin reductase (FTR), and a thioredoxin.[5] Thioredoxins are proteins, typically of 12,000 molecular weight, that are widely, if not universally, distributed in the animal, plant, and bacterial kingdoms. Thioredoxins undergo reversible reduction and oxidation through changes in thiol groups (S—S $\rightleftharpoons$ 2 SH). In the ferredoxin/thioredoxin system, a thioredoxin (Td) is reduced by photoreduced ferredoxin (Fd) via FTR, an iron–sulfur protein[6,7] [Eqs. (1) and (2)]. Thioredoxins can also be chemically reduced *in vitro* in the dark by the nonphysiological reagent

[1] B. B. Buchanan, *Annu. Rev. Plant Physiol.* **31,** 341 (1980).

[2] C. Cséke and B. B. Buchanan, *Biochim. Biophys. Acta* **853,** 43 (1986).

[3] L. E. Anderson, *in* "Photosynthesis: II. Photosynthetic Carbon Metabolism and Related Processes" (M. Gibbs and E. Latzko, eds.), Vol. 6, p. 271. Springer-Verlag, Berlin, 1979.

[4] J. Preiss, *Annu. Rev. Plant Physiol.* **33,** 431 (1982).

[5] R. A. Wolosiuk and B. B. Buchanan, *Nature* (*London*) **266,** 565 (1977).

[6] M. Droux, J.-P. Jacquot, M. Miginiac-Maslow, P. Gadal, J. C. Huet, N. A. Crawford, B. C. Yee, and B. B. Buchanan, *Arch. Biochem. Biophys.* **252,** 426 (1987).

[7] M. Droux, M. Miginiac-Maslow, J.-P. Jacquot, P. Gadal, N. A. Crawford, N. S. Kosower, and B. B. Buchanan, *Arch. Biochem. Biophys.* **256,** 372–380 (1987).

dithiothreitol (DTT), in the absence of chloroplast membranes, ferredoxin, and FTR [Eq. (3)].

$$4\ Fd_{ox} + 2\ H_2O \xrightarrow{\text{Light}} 4\ Fd_{red} + O_2 + 4\ H^+ \quad (1)$$
$$2\ Fd_{red} + Td_{ox} + 2\ H^+ \longrightarrow 2\ Fd_{ox} + Td_{red} \quad (2)$$
$$Td_{ox} + DTT_{red} \longrightarrow Td_{red} + DTT_{ox} \quad (3)$$

Two different thioredoxins, designated thioredoxin *f* and thioredoxin *m*, are a part of the ferrodoxin/thioredoxin system which has been found in different types of oxygenic photosynthetic organisms including cyanobacteria,[8] C_3,[9–11] C_4,[11] and crassulacean acid metabolism (CAM) plants.[12,13] In the reduced state, the two thioredoxins selectively activate enzymes of carbohydrate biosynthesis, including those of the reductive pentose phosphate cycle [fructose-1,6-bisphosphatase (FBPase), sedoheptulose-1,7-bisphosphatase, phosphoribulokinase, NADP–glyceraldehyde-3-phosphate dehydrogenase] and deactivate glucose-6-phosphate dehydrogenase,[1,2,14] a key enzyme of the oxidative pentose phosphate cycle, the major pathway for carbohydrate degradation in cyanobacteria. The ferredoxin/thioredoxin system also functions in chloroplasts in regulating other enzymes such as NADP–malate dehydrogenase (NADP–MDH)[1,2] and the chloroplast coupling factor (CF_1-ATPase).[15] The type of thioredoxin which interacts with each of these enzymes is shown in Table I. In cyanobacteria, the specificity of thioredoxins for some of the target enzymes may not be as rigid as their C_3 counterparts.[8,14] Cyanobacteria and certain algae appear to utilize the ferredoxin/thioredoxin system also for regulation of enzymes of sulfur[16] and nitrogen[17] assimilation. This chapter describes procedures for the isolation and assay of thioredoxins *m* and *f* and of FTR from the cyanobacterium *Nostoc muscorum*.

[8] B. C. Yee, A. De la Torre, N. A. Crawford, C. Lara, D. E. Carlson, and B. B. Buchanan, *Arch Microbiol.* **130,** 14 (1981).

[9] P. Schürmann, K. Maeda, and A. Tsugita, *Eur. J. Biochem.* **116,** 37 (1981).

[10] R. A. Wolosiuk, N. A. Crawford, B. C. Yee, and B. B. Buchanan, *J. Biol. Chem.* **254,** 1627 (1979).

[11] N. A. Crawford, B. C. Yee, S. W. Hutcheson, R. A. Wolosiuk, and B. B. Buchanan, *Arch. Biochem. Biophys.* **244,** 1 (1986).

[12] S. W. Hutcheson and B. B. Buchanan, *Plant Physiol.* **72,** 870 (1983).

[13] S. W. Hutcheson and B. B. Buchanan, *Plant Physiol.* **72,** 877 (1983).

[14] N. A. Crawford, C. W. Sutton, B. C. Yee, T. C. Johnson, D. C. Carlson, and B. B. Buchanan, *Arch. Microbiol.* **139,** 124 (1984).

[15] J. D. Mills and P. Mitchell, *FEBS Lett.* **144,** 63 (1982).

[16] J. D. Schwenn and U. J. Schrieck, *FEBS Lett.* **170,** 76 (1984).

[17] R. Tischner and A. Schmidt, *Plant Physiol.* **70,** 113 (1982).

TABLE I
TARGET ENZYMES OF THIOREDOXINS *f* AND *m*[a]

Thioredoxin type	Target enzymes
Thioredoxin *m*	NADP-malate dehydrogenase
	Chloroplast coupling factor (CF_1-ATPase)
	Glucose-6-phosphate dehydrogenase[b]
Thioredoxin *f*	Fructose-1,6-bisphosphatase
	Sedoheptulose-1,7-bisphosphatase
	Phosphoribulokinase
	NADP–glyceraldehyde-3-phosphate dehydrogenase
	NADP–malate dehydrogenase
	Chloroplast coupling factor (CF_1-ATPase)

[a] Target enzymes of sulfur and nitrogen assimilation are not included.[16,17]
[b] Inhibited by reduced thioredoxin *m*.

Thioredoxin *m*

Assay Method for Thioredoxin m

Principle. Thioredoxin *m* is assayed by measuring its capacity to promote the reductive activation of NADP–MDH.[11,18] The reducing power needed for activation may be supplied either by photoreduced ferredoxin in the presence of FTR (see below) or nonphysiologically by DTT. For convenience, we routinely use DTT as the reductant and a NADP–MDH preparation derived from corn leaves as the target enzyme. The assay is performed in two steps. The first step (activation phase) involves activation of the NADP–MDH, and the second step (catalytic or reaction phase) involves spectrophotometrically measuring the resultant activity.

Reagents

Tris–HCl buffer (pH 7.9 at 20°), 1 *M*
DTT, 100 m*M*
Corn leaf NADP–MDH, 0.15 mg/ml[19]
Thioredoxin *m* fraction
NADPH, 2.5 m*M*
Oxaloacetate, 25 m*M*

Assay Procedure. Both steps of the assay are carried out at room temperature. The activation phase of the assay is carried out in a 5-ml test

[18] R. A. Wolosiuk, P. Schürmann, and B. B. Buchanan, this series, Vol. 69, p. 382.
[19] J.-P. Jacquot, B. B. Buchanan, F. Martin, and J. Vidal, *Plant Physiol.* **68**, 300 (1981).

tube containing an appropriate amount of thioredoxin sample (usually 50 μl of the column fractions to be analyzed) plus 10 μl each of Tris–HCl buffer, DTT, and NADP–MDH. The total volume is adjusted to 0.1 ml with water. After a 5-min activation period, a 50-μl aliquot of the activation mixture is injected into a cuvette with a 1-cm light path containing the following reaction mixture: 100 μl each of Tris–HCl buffer, NADPH, and oxalacetate, plus 650 μl of water. The oxidation of NADPH is followed at 340 nm with a recording spectrophotometer.

Purification of Nostoc Thioredoxin m

Chemicals and Materials

Nostoc muscorum cells (*Anabaena,* Sp. 7119), are grown under illumination in liquid culture and N_2–CO_2 (98 : 2, v/v), as described by Arnon *et al.*,[20] and are harvested 4 days after inoculation. Frozen paste, 250 g, is used below.
Tris–HCl buffer (pH 7.9 at 20°), 1 *M* stock solution; 2 liters
2-Mercaptoethanol, 14 *M*; 60 ml
2× buffer (100 m*M* Tris–HCl, pH 7.9, plus 0.2%, v/v, 2-mercaptoethanol, freshly prepared from stock solutions
1× buffer, half-strength 2× buffer
Ammonium sulfate, crystalline, 250 g
NaCl, 25 g
Sephadex G-100 (Pharmacia Chemical Co., Piscataway, NJ), 3500 ml (swollen)
DE-52 cellulose, anion exchanger (Whatman Inc., Clifton, NJ), 400 ml (wet)
Reagent for protein determination, Bradford method (Bio-Rad Laboratories, Richmond, CA)
Dialysis tubing, Spectrapor 1, 6,000–8,000 MW cutoff, 40 mm wide (VWR, San Francisco, CA)
YM5 ultrafiltration membrane (Amicon Corp., Danforth, MA), 2.5 cm

Preparative Procedure for Nostoc Thioredoxin m. All preparative steps are carried out at 4°. Column fractions are monitored for protein by measuring the absorption at 280 nm.

Preparation of cell-free extract. Cells (250 g) are thawed overnight at 4°, homogenized in 375 ml of 2× buffer, and disrupted in a Ribi cell fractionator under N_2 with a breaking pressure of 1150 g/cm^2. Cell debris

[20] D. I. Arnon, B. D. McSwain, H. Y. Tsujimoto, and K. Wada, *Biochim. Biophys. Acta* **357,** 231 (1974).

is removed by centrifugation (40,000 *g*, 20 min), and the supernatant fraction is used below.

Ammonium sulfate fractionation. Solid ammonium sulfate (70.4 g) is added slowly and with stirring to the clarified supernatant fraction (400 ml) to make a 30% saturated solution. Centrifugation of the solution (13,000 *g*, 15 min) removes the dark green precipitate, which is discarded. The bluish red supernatant fraction (395 ml) is brought to 80% saturation by adding 140.6 g solid ammonium sulfate and is stirred for 30 min. The solution is centrifuged (13,000 *g*, 20 min), and the supernatant fraction is discarded. The bluish red pellet is resuspended in 100 ml of 2× buffer, added to dialysis tubing, and dialyzed overnight versus 13 liters of 1× buffer. In the morning, the sample is clarified by centrifugation (105,000 *g*, 2 hr).

Sephadex G-100 chromatography. The clarified bluish red sample is applied to a 5 × 150 cm Sephadex G-100 column previously equilibrated with 1× buffer. The column is developed with 1× buffer at a flow rate of 60 ml/hr, and 18-ml fractions are collected with a fraction collector. Fractions are assayed for thioredoxin and FTR (see below). If thioredoxin *f* activity is also being followed, it is important to include a minus FBPase control to monitor endogenous phosphatases (see below).

DE-52 cellulose chromatography. The fractions from the previous step showing thioredoxin activity are pooled and applied to a DE-52 cellulose column (2.2 × 33 cm) previously equilibrated with 1× buffer. The column is sequentially eluted with 250 ml of buffer, a 500-ml linear gradient of 0–0.2 M NaCl in buffer, and finally 250 ml of 0.5 *M* NaCl in buffer. Fractions (7 ml) are collected and assayed for thioredoxin *m* activity. Fractions containing thioredoxin *m* activity are pooled and concentrated by ultrafiltration in an Amicon Diaflo cell fitted with a YM5 membrane, and the protein concentration is determined by the Bradford method using reagent and instructions supplied by Bio-Rad Laboratories.

At this point in the procedure, thioredoxin *m* is not pure but is free of contaminating thioredoxin *f*, FTR, ferredoxin, and phosphatases. It may be used in studies on the ferredoxin/thioredoxin system of cyanobacteria. Thioredoxin *m* has been purified to homogeneity in other laboratories.[21,22]

Properties of Thioredoxin m

The amino acid sequence of *Anabaena* thioredoxin *m* reported by Gleason *et al.* shows 49% similarity with the *Escherichia coli* thiore-

[21] F. K. Gleason, M. M. Whittaker, A. Holmgren, and H. Jörnvall, *J. Biol. Chem.* **260,** 9567 (1985).

[22] S.-M. Ip, P. Rowell, A. Aitken, and W. D. P. Stewart, *Eur. J. Biochem.* **141,** 497 (1984).

doxin[21] and 50% similarity with spinach thioredoxin *m* (cf. Ref. 23). The active site, Trp-Cys-Gly-Pro-Cys (residues 30–34), is identical to that of all authentic thioredoxins. Thioredoxin *m* from oxygenic photosynthetic cells differs at residue 29 from *E. coli* and other aerobic bacteria sequenced to date which have Glu instead of Pro. Although *Anabaena* or *Nostoc* thioredoxin *m* cross-reacts minimally with anti-*E. coli* thioredoxin antibodies,[22,24,25] it reacts well with an anti-spinach thioredoxin *m* antibody.[24] Thus the thioredoxin *m* from cyanobacteria shares common characteristics with its bacterial and higher plant counterparts. The thioredoxin *m* gene from cyanobacteria has been cloned (Ref. 26 and E. Muller, unpublished observations).

Thioredoxin *f*

Assay Method for Thioredoxin f

Principle. The principle behind the thioredoxin *f* assay is the same as in the thioredoxin *m* assay described.[11,18] Here, any of the known target enzymes of thioredoxin *f* could be used to measure its capacity for reductive activation either with DTT or light, thylakoid membranes, and components of the ferredoxin/thioredoxin system. For convenience we routinely use DTT as the reductant and spinach chloroplast fructose-1,6-bisphosphatase (EC 3.1.3.11, FBPase) as the target enzyme. As with thioredoxin *m*, the assay is performed in two steps involving first activation of FBPase and then measurement of the resulting activity by determining the hydrolysis of fructose 1,6-bisphosphate (FBP) to fructose 6-phosphate and inorganic phosphate. The activity of FBPase can be measured either colorimetrically by analyzing the P_i released or spectrophotometrically by measuring the fructose 6-phosphate formed. Fructose 6-phosphate is determined by following the reduction of NADP in the presence of excess glucose-6-phosphate isomerase and glucose-6-phosphate isomerase and glucose-6-phosphate dehydrogenase. The latter method is useful when the sample contains phosphate buffer. In both cases, assays are carried out in the presence of a limiting concentration of Mg^{2+} because higher concentrations (greater than 1–2 m*M*) partially activate FBPase without thioredoxin. Interfering phosphate and phosphatases in crude extracts make fractionation necessary before a reliable

[23] K. Maeda, A. Tsugita, D. Dalzoppo, F. Vilbois, and P. Schürmann, *Eur. J. Biochem.* **154,** 197 (1986).
[24] T. C. Johnson, N. A. Crawford, and B. B. Buchanan, *J. Bacteriol.* **158,** 1061 (1984).
[25] F. K. Gleason and A. Holmgren, *J. Biol. Chem.* **256,** 8306 (1981).
[26] C.-J. Lim, F. K. Gleason, and J. A. Fuchs, *J. Bacteriol.* **168,** 1258 (1986).

thioredoxin assay can be performed. Even after fractionation, it is important to include a minus FBPase control in order to monitor endogenous phosphatases.

Method I: Colorimetric Assay of Thioredoxin f

Reagents

Tricine–KOH buffer (pH 7.9 at 20°), 1 *M*
$MgSO_4$, 10 m*M*
DTT, 50 m*M*
Sodium FBP, 60 m*M*
Spinach chloroplast FBPase, 1.6 mg/ml[27]
Thioredoxin *f* fraction
Trichloroacetic acid (TCA), 10% (w/v)
Mixture for P_i analysis (see below)

Assay Procedure. The reaction is carried out at room temperature in a 10-ml test tube containing 50 μl each of Tricine–KOH buffer, DTT, and $MgSO_4$; 10 μl of FBPase; the thioredoxin *f* fraction to be assayed (50 μl in the following purification protocol); and water to bring the volume to 0.45 ml. After a 10-min activation, catalysis of FBPase is initiated by the introduction of 50 μl of FBP and allowed to continue for 15 min. The reaction is stopped by the addition of 0.5 ml of TCA, and the precipitate is removed by centrifugation. A 0.5-ml aliquot of the supernatant solution is analyzed for P_i by adding 2 ml of the mixture used for P_i analysis. After 10 min, the absorbance at 660 nm is measured. In more highly purified fractions, the TCA precipitation step can be omitted because protein concentrations will be sufficiently low so as not to interfere with P_i measurement.

Reagents for P_i Analysis

H_2SO_4, 9 *N*
Ammonium molybdate, 1.65 g in 25 ml hot water
$FeSO_4 \cdot 7H_2O$, 2.5 g in 25 ml of 9 *N* H_2SO_4

Method for Preparing Mixture for P_i Analysis. $FeSO_4 \cdot 7H_2O$ (2.5 g) is dissolved in 25 ml of 9 *N* H_2SO_4 and added to 150 ml of water. Ammonium molybdate (1.65 g) is dissolved in 25 ml water by gently heating and is slowly added with stirring to the acidic $FeSO_4$ solution. This mixture can be stored for several weeks, but a standard phosphate curve should be included because the mixture changes with time.

[27] A. N. Nishizawa, B. C. Yee, and B. B. Buchanan, *in* "Methods in Chloroplast Molecular Biology" (M. Edelman, R. B. Hallick, and N.-H. Chua, eds.), p. 707. Elsevier, New York, 1982.

Method II: Spectrophotometric Assay of Thioredoxin F

Reagents

Tris–HCl buffer (pH 7.9 at 20°), 1 *M*
DTT, 100 m*M*
Spinach chloroplast FBPase, 1.6 mg/ml[27]
Thioredoxin *f* fraction
$MgSO_4$, 10 m*M*
Glucose-6-phosphate dehydrogenase (Sigma Chemical Co., St. Louis, MO)
Glucose-6-phosphate isomerase (Sigma)
NADP, 10 m*M*
Sodium FBP, 60 m*M*

Assay Procedure. To a 5-ml test tube are added 10 μl each of Tris–HCl buffer, DTT, and FBPase; 10–70 μl of thioredoxin *f*; and water to bring the volume to 0.1 ml. The mixture is incubated for 5–20 min (depending on thioredoxin *f* activity) to activate the FBPase. The resulting FBPase activity is measured by injecting a 50-μl aliquot of the activation mixture into a cuvette of 1-cm light path, containing the following reaction mixture in a total volume of 0.95 ml: 100 μl each of Tris–HCl buffer, $MgSO_4$, FBP, and NADP plus 0.75 U of glucose-6-phosphate dehydrogenase and 1.8 U of glucose-6-phosphate isomerase. NADP reduction is followed at 340 nm in a recording spectrophotometer.

Purification of Nostoc Thioredoxin f

Chemicals and Materials. Chemicals and materials needed are identical to those for thioredoxin *m* purification described above.

Preparative Procedure for Nostoc Thioredoxin f. The protocol for purifying *Nostoc* thioredoxin *f* is identical to that described for thioredoxin *m* (see above). The two thioredoxins copurify through the Sephadex G-100 step but are separated on DE-52 cellulose where *Nostoc* thioredoxin *f* is eluted at a higher salt concentration than thioredoxin *m*. A minus FBPase control must be included in the assays at each of the preparative steps to monitor interfering phosphatases. Fractions eluting from the DE-52 cellulose column which contain thioredoxin *f* activity are pooled, concentrated, and analyzed for protein concentration as described for thioredoxin *m*. Thioredoxin *f* may be purified to homogeneity using the protocol of Whittaker and Gleason.[28]

Properties of Thioredoxin f

The thioredoxin *f* from *Anabaena* is reported to have a molecular weight of 25,500[28] when analyzed by sodium dodecyl sulfate–polyacryl-

[28] M. M. Whittaker and F. K. Gleason, *J. Biol. Chem.* **259**, 14088 (1984).

amide gel electrophoresis (SDS–PAGE), a value which is more than twice as high as the molecular weight reported for the higher plant equivalent.[9,11] Furthermore, unlike the case for higher plant thioredoxin *f*, Whittaker and Gleason[28] report that *Anabaena* thioredoxin *f* is a very poor activator of endogenous FBPase and spinach NADP–malate dehydrogenase even at high concentrations (cf. Refs. 9 and 11). *Nostoc* thioredoxin *f* is reportedly effective in the activation of endogenous phosphoribulokinase, sedoheptulose-1,7-bisphosphatase, and FBPase.[14] The discrepancy of results in the capability of thioredoxin *f* to activate FBPase remains to be resolved. Ip *et al.*[22] have shown that FBPase from *Anabaena cylindrica* can be activated by a thioredoxin from the same species and found that the presence of substrate during the activation phase is necessary to see an effect of thioredoxin.[22]

Ferredoxin–Thioredoxin Reductase (FTR)

Assay Method for FTR

Principle. Like thioredoxin, FTR can be assayed by its capacity to activate any of the enzymes targeted by the ferredoxin–thioredoxin system in the presence of ferredoxin, thioredoxin, and illuminated thylakoid membranes.[6,18] As described for thioredoxin, the FTR assay is carried out in two steps, an activation phase and a catalytic phase. For convenience, we routinely monitor FTR activity using thioredoxin *f* and FBPase (both from spinach) or thioredoxin *m* (from spinach) and NADP–MDH (from corn). *Nostoc* thioredoxins could be used equally well here. As with the thioredoxin *f* assay, fractionation of crude extracts prior to assay and the inclusion of a minus FBPase control are important for monitoring FTR in FBPase-linked assays.

Method I: Colorimetric FTR Assay (FBPase as Target Enzyme)

Reagents

Tris–HCl buffer (pH 7.9 at 20°), 1 *M*
$MgSO_4$, 10 m*M*
2,6-Dichlorophenolindophenol (DPIP), 1 m*M*
Sodium ascorbate, 100 m*M*
Spinach chloroplast FBPase, 1.6 mg/ml[27]
Spinach ferredoxin, 2.5 mg/ml[29]
Spinach thioredoxin *f*, 0.5 mg/ml[11]

[29] B. B. Buchanan and D. I. Arnon, this series, Vol. 23, p. 413.

Spinach thylakoid membranes, 1 mg chlorophyll/ml[30]
Sodium FBP salt, 60 m*M*
FTR fractions, 2–10 μg if pure
TCA, 10% (w/v)
Mixture for P_i analysis (see above)

Assay Procedure. The reaction is carried out anaerobically at 20° in Warburg vessels. The main compartment contains 0.1 ml each of Tris–HCl buffer, $MgSO_4$, sodium ascorbate, DPIP, and thylakoid membranes; 0.02 ml each of thioredoxin *f*, ferredoxin, and FBPase; FTR as needed; and water to a total volume of 1.4 ml. The sidearm contains 0.1 ml FBP. After equilibration for 5 min with N_2, the vessels are incubated for an additional 5 min in 330 $\mu E/m^2/sec$ light (activation phase). The FBP is added from the sidearm to start the reaction, and the vessels are maintained in the light for the 30-min reaction phase. The reaction is stopped by adding 0.5 ml of 10% TCA after opening the vessels. The precipitate is removed by centrifugation, and 0.5 ml of the supernatant solution is analyzed for P_i as described in the assay procedure for thioredoxin *f*. It should be mentioned that, for best results, Tris buffer (Sigma 7–9 grade) should be used in the FTR assay unlike the DTT-linked FBPase assay in which Tricine buffer is preferred. We do not know the reasons for the different buffer sensitivities in the two assays.

Method II: Spectrophotometric FTR Assay (FBPase as Target Enzyme)

Reagents

Tris–HCl buffer (pH 7.9 at 20°), 1 *M*
2,6-Dichlorophenolindophenol (DPIP), 2 m*M*
Sodium ascorbate, 200 m*M*
Spinach ferredoxin, 5 mg/ml[29]
Spinach thioredoxin *f*, 0.5 mg/ml[11]
Spinach FBPase, 1.6 mg/ml[27]
Spinach thylakoid membranes, 2.4 mg/ml[30]
Catalase, 2 mg/ml (Sigma)
FTR fraction to be assayed, 2–10 μg if pure
$MgSO_4$, 10 m*M*
Sodium FBP, 60 m*M*
Glucose-6-phosphate dehydrogenase (Sigma)
Glucose-6-phosphate isomerase (Sigma)
NADP, 10 m*M*

[30] J.-P. Jacquot, M. Droux, M. Miginiac-Maslow, C. Joly, and P. Gadal, *Plant Sci. Lett.* **35,** 181 (1984).

Assay Procedure. The reaction is carried out anaerobically at 20° in white 1.5-ml Eppendorf centrifuge tubes fitted with a 11.1-mm-diameter rubber serum stopper pierced with two 20-gauge needles, one of which is connected to a N_2 source and the other used as a vent. The tubes contain 10 μl each of Tris–HCl buffer, thioredoxin *f*, FBPase, thylakoid membranes, and catalase plus 5 μl each of DPIP, sodium ascorbate, and ferredoxin. FTR and water are added as needed to make a final volume of 0.1 ml. The tubes are equilibrated with N_2 for 5 min with agitation. The vent needle is removed first and the N_2 needle second. The tubes are then illuminated without agitation for 10 min (activation phase). A 20- to 50-μl aliquot of the activation mixture is removed with a syringe and injected into a cuvette containing 0.95 ml of reaction mixture as described in the spectrophotometric assay (Method II) for thioredoxin *f* above. Change in absorbance is measured at 340 nm. As with thioredoxin *f*, this spectrophotometric assay method is useful when samples contain phosphate buffer. The reaction step of the assay is carried out in air.

Method III: Spectrophotometric FTR Assay (NADP–MDH as Target Enzyme)

Reagents

Tris–HCl buffer (pH 7.9 at 20°), 1 *M*
DPIP, 2 m*M*
Sodium ascorbate, 200 m*M*
Spinach ferredoxin, 5 mg/ml[29]
Spinach thioredoxin *m*, 0.5 mg/ml[11]
Corn NADP–MDH, 1.5 mg/ml[19]
Spinach thylakoid membranes, 2.4 mg/ml[30]
Catalase, 2 mg/ml (Sigma)
FTR fraction to be assayed, 2–10 μg if pure
NADPH, 2.5 m*M*
Oxaloacetate (OAA), 25 m*M*

Assay Procedure. The assay is carried out as described for the spectrophotometric FBPase assay except that (1) NADP–MDP and thioredoxin *m* are used in the activation mixture in place of FBPase and thioredoxin *f*, respectively, and (2) the reaction mixture contains 100 μl each of Tris–HCl buffer, NADPH, and OAA plus 650 μl of water. Here the oxidation of NADPH is followed by measuring the change in absorbance at 340 nm in a recording spectrophotometer.

Purification of Nostoc FTR

Chemicals and Materials. Chemicals, materials, and *Nostoc muscorum* cells required are identical to those described for the purification of thioredoxin *m* (see above) plus the following:

Potassium phosphate buffer, pH 7.7, 1 *M* stock solution, 100 ml
Hydroxylapatite (HTP) (Bio-Rad), 20 g (dry weight)
Mono Q, FPLC anion exchanger, and Pharmacia FPLC Unit
Reagents for SDS–PAGE[31]

Preparative Procedure for Nostoc FTR. The first three steps are identical to the procedure described above (Preparative Procedure for *Nostoc* Thioredoxin *m*). If the FBPase-linked FTR assay is used, FTR activity is calculated by subtracting a minus FBPase control from the total activity in the Sephadex G-100 column of the third step. Fractions showing FTR activity are pooled and used in the fourth through sixth steps below.

DEAE–cellulose chromatography. The FTR from the Sephadex G-100 column is applied to a 2.2 × 33 cm DE-52 cellulose column previously equilibrated with 1X buffer. After washing the column with 250 ml of buffer, the column is eluted with a 1-liter gradient of 0–270 m*M* NaCl in buffer. Fractions (6 ml) are collected, assayed for FTR, and adjusted for contaminating phosphatase activities as described above. Fractions showing FTR activity are pooled and dialyzed overnight versus 12 liters of buffer.

Hydroxylapatite chromatography. The FTR is applied to a 2.2 × 8 cm hydroxyapatite (HTP) column which was previously equilibrated with 2 liters of 1X buffer. The column is eluted sequentially with 150 ml each of 0, 50, 120, and 350 m*M* potassium phosphate, pH 7.7, in 1X buffer, and 5-ml fractions are collected. Absorbance is measured at 280 and 410 nm. The bluish red phycobilin proteins are eluted with 50 m*M* potassium phosphate and the pale yellow–brown FTR with 350 m*M* potassium phosphate. FTR may be detected at this stage by measuring absorption at 410 nm and may be assayed using a spectrophotometric assay (Methods II or III). The fractions containing FTR are pooled and concentrated to about 2 ml by ultrafiltration in an Amicon Diaflo cell fitted with a YM5 membrane. The FTR sample is dialyzed overnight versus several liters of 1X buffer. The purity of the FTR may be estimated by measuring its absorption spectrum, i.e., the ratio of its 410 nm and 278 nm absorption peaks (see below). The FTR may be pure at this point or it may require purification by the FPLC Mono Q step described below. With spinach and corn FTR preparations, a ferredoxin–Sepharose 4B chromatography step is per-

[31] U. K. Laemmli, *Nature (London)* **227,** 680 (1970).

formed at this point. We have little experience with the affinity step in the case of *Nostoc* FTR, but it would likely be effective here as well.[6]

FPLC chromatography, Mono Q. If the FTR from the HTP step is not homogeneous, it may be further purified with a Pharmacia Mono Q (1-ml anion-exchange) column attached to a FPLC apparatus. After a wash with 4 ml of buffer, the column is developed with a 20-ml linear gradient of 80–280 m*M* NaCl in buffer. Fractions (0.5 ml) are collected and absorbance is measured at 410 nm. The FTR elutes at about 195 m*M* NaCl and should be pure when analyzed by the SDS–PAGE system of Laemmli.[31]

Properties of Nostoc FTR

FTR is relatively stable to freezing but should be stored in aliquots to avoid repeated freezing and thawing. The native enzyme has a molecular weight of 28,000, and SDS–PAGE reveals two subunits migrating at 14,000 and 7,000.[6] FTR is an iron–sulfur protein (4 Fe and S^{2-} groups/mol) with absorption peaks at 278 and 410 nm and an A_{410}/A_{278} ratio of 0.34–0.40, depending on the integrity of the iron–sulfur cluster. The 14,000-Da subunit (similar subunit) is present in other FTRs examined and cross-reacts with anti-corn FTR polyclonal globulins, whereas the 7,000-Da subunit (variable subunit) is unique to the *Nostoc* preparation and fails to react with the corn antibody. Likewise, anti-*Nostoc* FTR polyclonal globulins cross-react only with the similar subunit in FTRs from corn or spinach.

[44] Electron Paramagnetic Resonance Characterization of Iron–Sulfur Proteins

By R. CAMMACK

Introduction

Iron–sulfur proteins contain clusters of iron and acid-labile sulfide atoms, coordinated to cysteine sulfurs and, occasionally, nitrogen ligands from the protein. In most cases their function is electron transport. In cyanobacteria, as in chloroplasts, the most abundant iron–sulfur clusters are those associated with photosynthetic electron transport.[1,2] They are

[1] M. C. W. Evans, *in* "Iron–Sulfur Proteins" (T. G. Spiro, ed.), p. 249. Wiley, New York, 1982.

[2] R. Malkin, *Annu. Rev. Plant. Physiol.* **33,** 455 (1982).

readily detected by low-temperature electron paramagnetic resonance (EPR) spectroscopy. The method is sufficiently sensitive to be applied to whole cyanobacterial cells. It can be used to detect the oxidation state of the iron–sulfur clusters and to investigate the effects of varying the physiological conditions such as iron deficiency[3,4] or photoinhibition.[5,6]

The relevant clusters are the [2Fe–2S] and [4Fe–4S] types, which are detectable by EPR spectroscopy only when reduced. The ferredoxins are small, soluble proteins, and all the other iron–sulfur clusters are membrane bound. The spectra of all the clusters are in the region of $g = 2.0$, but they have different line shapes and may be differentiated by selective chemical reduction, by photoreduction, and by the differing temperature dependence of the signals.

Preparation of Cyanobacterial Cells and Extracts for EPR Spectroscopy

Material. Whole cells, from cyanobacterial cultures or algal blooms harvested from the wild, may be used.[3] Better resolution of signals from membrane-bound complexes may be obtained by using washed membranes[3] or purified photosystem I (PSI) particles.[7,8] Preparations should be as concentrated as possible, to about 2–4 mg chlorophyll/ml. It is advisable to prepare all samples in duplicate or triplicate.

The outer membranes and cell walls of whole cyanobacteria present a permeability barrier to reagents such as dithionite and EDTA. It is often found that freezing and thawing renders the cells sufficiently permeable.[3] In other cases, more drastic treatments such as digestion with enzymes, e.g., lysozyme, and osmotic lysis,[9] or sonication, may be needed, depending on the type of cell wall.

Sample Tubes. Samples for low-temperature EPR spectroscopy are prepared in quartz cells of approximately 3-mm internal diameter. The signal intensity for a homogeneous cylindrical sample is proportional to the square of the internal diameter of the tube, hence wider tubes give enhanced sensitivity, the limitation being that the outer diameter of the tube must fit inside the helium cryostat of the spectrometer. Using cryo-

[3] R. Cammack, L. J. Luijk, J. J. Maguire, I. V. Fry, and L. Packer, *Biochim. Biophys. Acta* **548,** 267 (1979).

[4] G. Sandmann and R. Malkin, *Plant Physiol.* **73,** 724 (1983).

[5] C. P. Santos, R. Cammack, and D. O. Hall, *Prog. Photosynth. Res.* **4,** 63 (1987).

[6] K. Inoue, H. Sakurai, and T. Hiyama, *Plant Cell Physiol.* **27,** 961 (1986).

[7] R. Nechustai, P. Muster, A. Binder, V. Liveanu, and N. Nelson, *Proc. Natl. Acad. Sci. U.S.A.* **80,** 1179 (1983).

[8] D. J. Lundell, A. N. Glazer, A. Melis, and R. Malkin, *J. Biol. Chem.* **260,** 646 (1985).

[9] A. C. Stewart and D. S. Bendall, *Biochem. J.* **188,** 351 (1980).

stat quartz inserts constructed to close tolerances, it is possible to use tubes of up to 3.5-mm internal diameter. For quantitative work, the internal diameter of the quartz EPR tubes must be determined, for example, by weighing the quantity of water required to fill the tubes to a certain height. The volume of the sample may be estimated from knowledge of the internal diameter and the length; this is useful for estimating the concentration factor before and after centrifugation.

Techniques for filling and freezing the tubes, particularly when oxygen-sensitive reductants are used, have been described elsewhere in this series.[10] For cyanobacterial cells the principal problems are (1) EPR signals arising from contaminating inorganic ions, particularly manganese; (2) difficulty of transferring a concentrated pellet of cells into the tube; and (3) preparation of samples reduced in complete darkness.

Contamination. Inorganic particulate matter, such as precipitates from culture media, soil, and dust, gives rise to strong EPR signals, and should be washed out as much as possible by differential centrifugation. Whole cells and cell extracts contain Mn^{2+}, which will give a characteristic six-line spectrum centered around $g = 2.0$, masking the EPR signals of the reduced iron–sulfur clusters. This spectrum may be broadened out by chelation of free Mn^{2+} with EDTA.

Centrifugation. In order to obtain cyanobacterial cells at a concentration of 2 mg chlorophyll/ml or more, they are concentrated by centrifugation. It is often impossible to transfer a thick slurry of cells or membranes into the narrow tubes, particularly when the cells are associated with extracellular mucopolysaccharides. In that case a more dilute and manageable suspension can be made in a buffer such as 20 m*M* HEPES–6 m*M* EDTA, pH 7. This is injected into the quartz tubes and then concentrated by centrifugation. Centrifuge holders are machined, in pairs, from solid acrylic or other suitable plastic. They are shaped on the outside like a 50- or 100-ml centrifuge tube, to fit the rotor of a high-speed refrigerated centrifuge. Each holder is bored to take two or four quartz EPR tubes. The EPR tubes are cut down to a length of about 100 mm, depending on the internal dimensions of the rotor. They should fit snugly inside the holders, and the spaces in the holders surrounding the tubes are first filled with water. Using quartz tubes selected by destructive testing, a centrifugal force of 20,000 *g* may be applied. Normally, whole cyanobacterial cells may be sedimented at lower speeds. Most of the supernatant liquid is drawn off and the precipitate resuspended in the remaining fluid by stirring with a stainless steel wire. The sample length should be sufficient to

[10] G. Palmer, this series, Vol. 10 [94]; H. Beinert, W. H. Orme-Johnson, and G. Palmer, this series, Vol. 54 [10].

fill the sensitive region of the cavity; this will depend on the geometry of the sample holder and cavity, but is typically 15–30 mm. A longer sample is desirable if there is any likelihood of condensation of solid air at the top of the sample. This will happen if the cryostat design leaves the sample tube open to the air, and if the tube is not sealed with a vacuum-tight stopper.

Dark Conditions. Frozen samples of photosynthetic membranes are extremely photosensitive. Except where reduction conditions require illumination, samples must be kept in the dark, from the time of freezing until the measurements have been made in the spectrometer, to avoid inadvertent photoreduction of the iron–sulfur clusters of photosystem I. Laboratory lighting should be kept to a low level. Samples may be wrapped in aluminum foil or black paper, and a thick black cloth used as a cover while transferring samples. The most sensitive applications are those where the adjustment of the iron–sulfur clusters to an exact state of partial reduction is critical, such as oxidation–reduction titrations.[11,12] In these applications, the conditions required are similar to those for a photographic darkroom. It may be necessary to cover the indicator lights on the spectrometer. A faint green light is permissible.

Conditions for Reduction. In order to obtain the iron–sulfur clusters in their EPR-detectable reduced states, appropriate reducing conditions are applied to the samples before freezing. The iron–sulfur clusters are observed only in the reduced states, and they differ considerably in midpoint potential (Table I). P700 (the primary electron donor of PSI) and the iron–sulfur protein of the cytochrome b_6–f complex are reducible by ascorbate; reduction is more rapid in the presence of 20 μM dichloroindophenol, but this is not essential. Soluble ferredoxin is reduced by 5 mM dithionite within 2–5 min. Clusters A and B of PSI are also reduced by dithionite in the dark, but their reduction is greatly facilitated by illumination for 2 min before freezing. Cluster X may also be frozen in the reduced state after prolonged illumination, if the samples are continuously illuminated during freezing. Since the samples generally have a high absorbance, a high light intensity is used, and the sample is stirred during illumination. Light from a 300-W slide projector is focused on the sample with a cylindrical lens. To avoid overheating, the EPR tube is immersed in cooling water. The time required for maximum reduction is determined experimentally by freezing samples at intervals and measuring the EPR spectrum. Too long illumination may damage the iron–sulfur clusters.

[11] R. Malkin and A. Bearden, this series, Vol. 69 [21].
[12] P. L. Dutton, this series, Vol. 54 [23].

TABLE I
EPR SPECTROSCOPIC PARAMETERS OF IRON–SULFUR CLUSTERS IN CYANOBACTERIA

Fe–S protein	Cluster	Midpoint potential (mV)	*g* values	Temperature (*K*) for measurement
Ferredoxin	[2Fe–2S]	−390[c]	2.05, 1.96, 1.89	30–60
b_6–*f* complex	[2Fe–2S]	+155[d]	2.02, 1.89	30
$g = 1.92$	N.D.[a]	−275[e]	2.05, 1.92	20
Photosystem I				
Cluster A	[4Fe–4S][b]	−530[d]	2.05, 1.94, 1.86	20
Cluster B	[4Fe–4S][b]	−560[d]	2.065, 1.93, 1.88	20
Cluster A + B			2.05, 1.94, 1.92, 1.89	20
Cluster X	[4Fe–4S][b]	−705[f]	2.08, 1.88, 1.78	10

[a] Not determined.
[b] Cluster assignments are tentative.
[c] From R. Cammack, K. K. Rao, C. P. Bargeron, K. G. Hutson, P. W. Andrew, and L. J. Rogers, *Biochem. J.* **168,** 205 (1977).
[d] From Cammack and Alvarez.[15]
[e] From Cammack *et al.*[3]
[f] Potential estimated in spinach chloroplasts [S. K. Chamorovsky and R. Cammack, *Photobiochem. Photobiophys.* **4,** 195 (1982)].

EPR spectroscopy. Spectra may be measured in a conventional continuous-wave spectrometer operating at X-band frequency (9 GHz).[10] The sample is cooled to 10–70 K with a helium-flow cryostat from Oxford Instruments or Air Products. The cavity should have a port or grid for illumination of the sample at low temperature, and the cryostat should be transparent in the sample region. For photoreduction of clusters that are irreversibly photoreduced at temperatures around 10–20 K, such as clusters A and B of PSI, a relatively weak light source such as a battery flashlight or low-power laser is adequate. Where the rate of charge recombination is significant, however, it is advantageous to have an intense light source which can be focused on the sample within the cavity, such as a 1-kW slide projector with suitable lenses or a fiber-optic illuminator. To avoid raising the temperature, a heat filter, such as a Schott glass filter, Type KG1, is placed in front of the cavity. Attention should also be paid to having a good circulation of helium around the sample to dissipate heat.

EPR Spectra of Iron–Sulfur Proteins

The [2Fe–2S] ferredoxins may be observed in samples of whole cells, after treatment with EDTA and reduction with dithionite. The EPR spectrum of reduced ferredoxins, at g = 2.05, 1.96, 1.89 (Fig. 1a), may be observed at 60 K, at which temperature the iron–sulfur clusters of PSI should not interfere. Since the protein is soluble it is washed out during preparation of membranes. Better spectra (Fig. 1b) are obtained by partial purification and concentration on DEAE–cellulose. An extract from the cells in 50 mM Tris buffer, pH 8, is passed through a column of DEAE–cellulose, and the ferredoxin is eluted in a small volume with 0.8 M NaCl in the same buffer. For preparation of EPR samples the volume required is small, and the column may be made in a Pasteur pipet. Generally, cyanobacteria contain two ferredoxins,[13,14] which may be separated by high-performance liquid chromatography, but their EPR spectra show only slight differences in g values.[13]

A membrane-bound [2Fe–2S] cluster is associated with the b_6–f complex.[11] It has been described as a "Rieske-type" iron–sulfur cluster because of its similarity to that found by J. S. Rieske in Complex III of mitochondria. The cluster has an unusually high midpoint redox potential, around +300 mV in chloroplasts and photosynthetic bacteria,[11] which is attributed to having the iron coordinated to histidine nitrogenase as well as cysteine sulfur. The midpoint potential in the cyanobacterium *Phormidium laminosum* appears to be somewhat lower.[15] The [2Fe–2S] cluster is difficult to detect in whole cells, but can be observed in washed membranes of cyanobacteria after reduction by ascorbate in the dark, the most prominent feature being at g = 1.89 (Fig. 1d).

Another component, presumed to be an iron–sulfur cluster, gives rise to a narrow EPR signal at g = 1.92.[3,16,17] It is observed after reduction by dithionite at pH 6.0 (Fig. 1c); the iron–sulfur clusters of PSI are not significantly reduced by dithionite at this pH. The signal is prominent in cyanobacterial membranes but not in purified PSI preparations. It appears to be part of the axial spectrum of a reduced iron–sulfur cluster. Its function is not certain, though it has been suggested to be a component of the respiratory chain.[3]

Photosystem I

The PSI reaction center in cyanobacteria is similar to those in chloroplasts of algae and higher plants, and all of the components described here

[13] K. G. Hutson, L. J. Rogers, B. G. Haslett, D. Boulter, and R. Cammack, *Biochem. J.* **172,** 465 (1978).

[14] B. Schrautemeier and H. Bohme, *FEBS Lett.* **184,** 304 (1985).

[15] R. Cammack and V. J. Alvarez, unpublished observations (1982).

[16] J. Nugent, A. C. Stewart, and M. C. W. Evans, *Biochim. Biophys. Acta* **635,** 488 (1981).

[17] T. Hiyama, A. Murakami, S. Itoh, Y. Fujita, and H. Sakurai, *J. Biochem.* (*Tokyo*) **97,** 89 (1985).

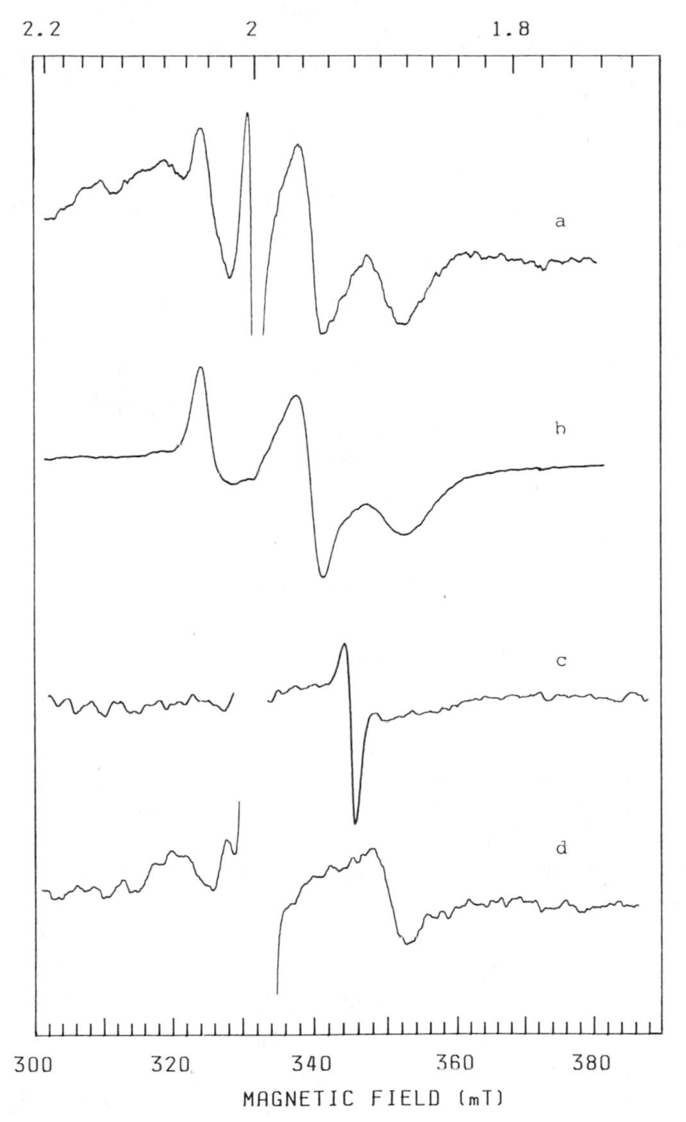

FIG. 1. EPR spectra of iron–sulfur clusters in cyanobacterial cells. These clusters are not photoreduced at low temperatures, and samples were prepared in the dark. Samples, and temperatures of measurement, were as follows: (a) ferredoxin, in cells of *Nostoc muscorum* reduced by 10 m*M* dithionite, 62 K; the narrow signal near $g = 2.0$ is a free radical, probably arising from chlorophyll; (b) *N. muscorum* ferredoxin, eluted from DEAE–cellulose, reduced by dithionite, 62 K; (c) "$g = 1.92$" signal, in *Phormidium laminosum* membranes reduced by 5 m*M* dithionite at pH 6.0, 22 K; (d) Rieske-type iron–sulfur protein, in *P. laminosum* washed membranes, reduced by 10 m*M* ascorbate, 16 K. Spectra were recorded with microwave power 10 mW, microwave frequency 9.18 GHz, modulation amplitude 1 mT, and modulation frequency 100 KHz.

have been observed in chloroplasts, except the $g = 1.92$ signal. The conditions for reduction of the iron–sulfur clusters in chloroplast PSI and examination by EPR spectroscopy have been described in this series by Malkin and Bearden.[11]

If the photosynthetic reaction centers are frozen in the resting state, with P700 reduced and the iron–sulfur clusters oxidized, illumination will result in the transfer of one electron, to either center A or B, in which it is trapped at low temperature. The intensity of the resulting EPR signal is a measure of the number of intact electron transfer chains. Detailed examination of the line shape of the EPR spectrum of chemically reduced and photoreduced preparations can be used to determine the condition of PSI, for example, whether some clusters are damaged or whether electron transfer between them is impaired.

The sequence of electron transfer (see Refs. 18 and 19) in PSI appears to be

$$\text{P700} \rightarrow \text{A}_0 \rightarrow \text{A}_1 \rightarrow \text{X} \rightarrow (\text{A,B})$$

A systematic nomenclature may be premature before the comprehensive determination of the structure and function of PSI. At present it seems likely that A_0 is a chlorophyll, A_1 may be vitamin K,[20,21] and X (also known as A_2), A (A_4), and B (A_3) are [4Fe–4S] clusters.[22]

Clusters A and B have similar midpoint redox potentials, A usually being somewhat less negative than B. The sequence of reduction of clusters A and B has not been established; possibly both of them may accept electrons from cluster X. Cluster A tends to be preferentially photoreduced at temperatures around 20 K, and its spectrum predominates (Fig. 2a), even under conditions where cluster B has a less negative midpoint potential.[23,24] Under these conditions the pathway of electron transfer is determined by kinetics rather than thermodynamic considerations.

The spectrum of reduced cluster B may be observed in various conditions. (1) In those photosynthetic membranes where it happens that cluster B has a less negative potential than cluster A, such as the cyanobacterium *P. laminosum*,[23] the spectrum of cluster B is seen after partial reduction by dithionite in the dark (Fig. 2b). (2) Treatment with glycerol

[18] A. J. Hoff, *Biophys. Struct. Mech.* **8,** 107 (1982).

[19] P. Setif and P. Mathis, *in* "Encyclopedia of Plant Physiology: Vol. 19. Photosynthetic Membranes" (L. A. Staehelin and C. J. Arntzen, eds.), p. 476. Springer-Verlag, Heidelberg, 1985.

[20] K. Brettel, P. Setif, and P. Mathis, *FEBS Lett.* **203,** 220 (1986).

[21] R. W. Mansfield and M. C. W. Evans, *FEBS Lett.* **203,** 225 (1986).

[22] E. H. Evans, D. P. E. Dickson, C. E. Johnson, J. D. Rush, and M. C. W. Evans, *Eur. J. Biochem.* **118,** 81 (1981).

[23] R. Cammack, M. D. Ryan, and A. C. Stewart, *FEBS Lett.* **107,** 422 (1979).

[24] S. K. Chamorovsky and R. Cammack, *Biochim. Biophys. Acta* **679,** 146 (1982).

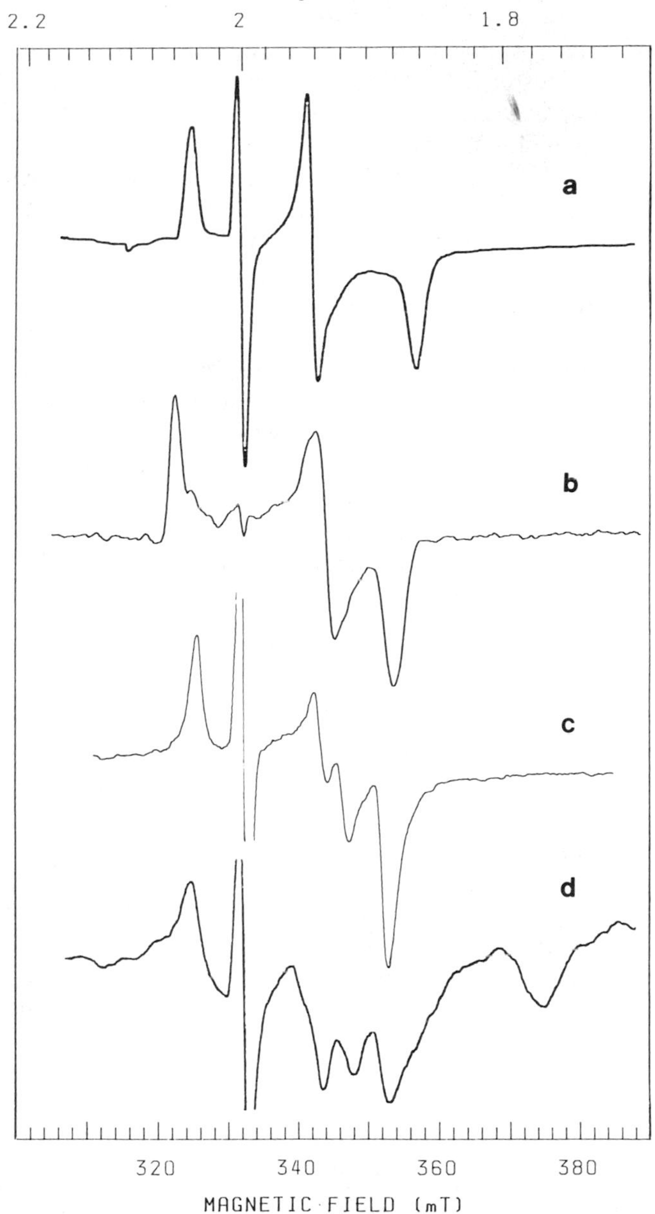

FIG. 2. EPR spectra of iron–sulfur clusters associated with photosystem I. (a) Cluster A in *Anabaena cylindrica,* sample reduced with ascorbate and illuminated at 20 K; the narrow signal near $g = 2.0$ is due to photooxidized $P700^{+}$. (b) Cluster B in *Phormidium laminosum* membranes in 30% glycerol, reduced by 5 m*M* dithionite at pH 8.0 in the dark, 22 K. (c) Cluster A + B reduced, *P. laminosum* cells reduced with 5 m*M* dithionite and illuminated for 2 min before freezing, recorded at 18 K. (d) Fully reduced reaction centers, *A. cylindrica* cells reduced with 5 m*M* dithionite and illuminated during freezing, 10 K. Other conditions of measurement were as for Fig. 1.

tends to increase the midpoint potential of cluster B relative to that of cluster A.[25] (3) Cluster B is also preferentially photoreduced in PSI preparations at temperatures around 200 K.[24]

The spectra of clusters A or B, reduced in isolation, are seen only after photoreduction of reaction centers where all the iron–sulfur clusters were previously oxidized, or under carefully controlled conditions of partial reduction. More commonly, either or both of the iron–sulfur clusters are reduced in various proportions. The resulting spectrum, measured at 20 K, is then a superimposition of the spectra of reaction centers with cluster A reduced ($X \cdot A^- \cdot B$) (Fig. 2a), cluster B reduced ($X \cdot A \cdot B^-$) (Fig. 2b), and both A and B reduced ($X \cdot A^- \cdot B^-$) (Fig. 2c). It should be emphasized that the latter spectrum is not the sum of the other two, because of the influence of spin–spin interactions between the clusters; its apparent g values (those cited are at X-band) are frequency dependent.[26] Note, for example, that the $g = 1.86$ signal of cluster A and the $g = 2.065$ signal of cluster B disappear when both clusters are reduced (Fig. 2c). The most characteristic features of the three different states of reduction are peaks at $g = 1.86$ for $X \cdot A^- \cdot B$, at $g = 2.065$ for $X \cdot A \cdot B^-$, and the intense trough at $g = 1.89$ for $X \cdot A^- \cdot B^-$. All of these features may be observed at 20–30 K. In addition, the spectrum of reduced cluster X may be observed in strongly reduced samples, at 10 K; its most prominent feature is a trough at $g = 1.78$. The spectrum of $X^- \cdot A^- \cdot B^-$ is shown in Fig. 2d.

Photoreduction of cluster X (A_1) in frozen samples by P700 may be observed in frozen samples where P700, A, and B are reduced.[27] The process is reversible, and continuous, intense illumination is required to obtain a stable $P700^+ \cdot X^-$ state.

EPR spectra of polarized triplet states are observed during illumination of reduced samples containing PSI and have been observed in concentrated PSI particles from the cyanobacterium *Mastigocladus laminosus*.[28,29] The origins of the triplet states are antenna chlorophyll or P700, depending on whether the iron–sulfur clusters, and particularly cluster X, are oxidized or reduced in the sample.[30]

[25] M. C. W. Evans and P. Heathcote, *Biochim. Biophys. Acta* **590,** 89 (1980).

[26] R. Aasa, J. Bergström, and T. Vänngård, *Biochim. Biophys. Acta* **637,** 118 (1981).

[27] A. R. McIntosh and J. R. Bolton, *Biochim. Biophys. Acta* **430,** 555 (1976).

[28] R. C. Ford and M. C. W. Evans, *Biochim. Biophys. Acta* **807,** 35 (1985).

[29] R. Nechushtai, N. Nelson, O. Gonen, and H. Levanon, *Biochim. Biophys. Acta* **807,** 35 (1985).

[30] M. B. McLean and K. Sauer, *Biochim. Biophys. Acta* **679,** 384 (1982).

[45] Characterization of Cytochrome-c Oxidase in Isolated and Purified Plasma and Thylakoid Membranes from Cyanobacteria

By GÜNTER A. PESCHEK, VÉRONIQUE MOLITOR, MARIA TRNKA, MARNIK WASTYN, and WALTER ERBER

Introduction

Recently methods have been devised for the successful separation of physiologically active plasma (CM) and thylakoid (ICM) membranes from cyanobacteria.[1,2] The results, however, with respect to the presence or absence of aa_3-type cytochrome oxidase in the CM at first were inconsistent.[2-6] In this chapter, after a brief recapitulation of membrane isolation, purification, and identification, growth conditions that permit optimum recovery of aa_3-type cytochrome oxidase activity in the CM of certain cyanobacteria are described.

Growth of the Cells

Table I summarizes the conditions of growth and harvest of the cyanobacteria so far investigated by the methods discussed in this chapter.

Isolation and Identification of the Membranes

CM and ICM are separated from crude cell-free extracts according to the method of Murata and Omata[1] and/or Peschek and colleagues[2] except that the cells are routinely harvested from late logarithmic to linear[7] or from NaCl-stressed[8] cultures. In order to show most clearly which parameters are associated with which type of membrane, both CM and ICM are withdrawn from the flotation gradient[1,2] in an intentionally cross-contami-

[1] N. Murata and T. Omata, this volume [23].

[2] V. Molitor, M. Trnka, and G. A. Peschek, *Curr. Microbiol.* **14,** 263 (1987).

[3] T. Omata and N. Murata, *Biochim. Biophys. Acta* **766,** 395 (1984).

[4] T. Omata and N. Murata, *Biochim. Biophys. Acta* **810,** 354 (1985).

[5] M. Trnka and G. A. Peschek, *Biochem. Biophys. Res. Commun.* **136,** 235 (1986).

[6] V. Molitor and G. A. Peschek, *FEBS Lett.* **195,** 145 (1986).

[7] With phototrophically growing cell suspensions, linear instead of logarithmic growth curves[8] result from increasing light limitation (self-shadowing of cells) with progressive (essentially logarithmic!) growth at rather low light intensities.

[8] V. Molitor, W. Erber, and G. A. Peschek, *FEBS Lett.* **204,** 251 (1986).

TABLE I
GROWTH AND HARVEST OF CYANOBACTERIA[a]

Species	Medium	Days of growth	Yield (μl packed cells/ml) at time of harvest	Growth temperature (°)	Illumination[b] (W/m^2)
Anacystis nidulans	D[c]	3–4	4–5	38	20–25
ATCC 27144	D plus 0.4 *M* NaCl	6–7	3–4	38	20–25
Synechocystis 6714	BG-11[d]	3	3–4	34	15–20
ATCC 27178	BG-11 plus 0.4 *M* NaCl	3–4	3–4	34	15–20
Plectonema boryanum ATCC 27894	D	7	3	35	15–20
Nostoc MAC PCC 8009	D	6–7	0.8–1.0	35	15–20[e]
Anabaena variabilis ATCC 29413	D	—[f]	4–5	35	20–25
Gloeobacter violaceus ATCC 29082	BG-11	150	4	20–23	3–5

[a] Cells are grown axenically in batch cultures [W. H. Nitschmann and G. A. Peschek, *Arch. Microbiol.* **141,** 330 (1985)], except *A. variabilis* and, occasionally, *A. nidulans* which are also grown in a turbidostat [G. A. Peschek, T. Czerny, G. Schmetterer, and W. H. Nitschmann, *Plant Physiol.* **79,** 278 (1985)]. The pH of the cultures is 7.8–8.6 throughout unless statically controlled at pH 8.2 ± 0.1 in case of turbidostat cultures. Harvesting is by centrifugation at room temperature. ATCC, American Type Culture Collection; PCC, Pasteur Culture Collection.

[b] Intensity of warm-white fluorescent light measured at the surface of the vessels using a YSI radiometer, Model 65.

[c] Medium D [W. A. Kratz and J. Myers, *Am J. Bot.* **42,** 282 (1955)] supplemented with 10 m*M* $NaHCO_3$ and 1.0 m*M* Na_2CO_3 and sparged with 1.5% (v/v) CO_2 in sterile air.

[d] Medium BG-11 [R. Rippka, J. Deruelles, J. B. Waterbury, M. Herdman, and R. Y. Stanier, *J. Gen. Microbiol.* **111,** 1 (1979)], which for *Synechocystis* 6714 is supplemented with 10 m*M* $NaHCO_3$ and 1.0 m*M* Na_2CO_3 and sparged with 1.5% (v/v) CO_2 in sterile air.

[e] Red fluorescent bulbs were used.

[f] Continuous turbidostat culture.

nated form and repeatedly purified by several subsequent steps of recentrifugation on fresh sucrose density gradients in the conventional sedimentation mode[2,8] (Fig. 1). Figure 2 shows the density profile (g/ml) of CM and ICM fractions isolated from *Anacystis nidulans,* after growth in the absence or presence of 0.4 *M* NaCl, on continuous sucrose density gradients. There is a characteristic "fine structure" of the CM band, each

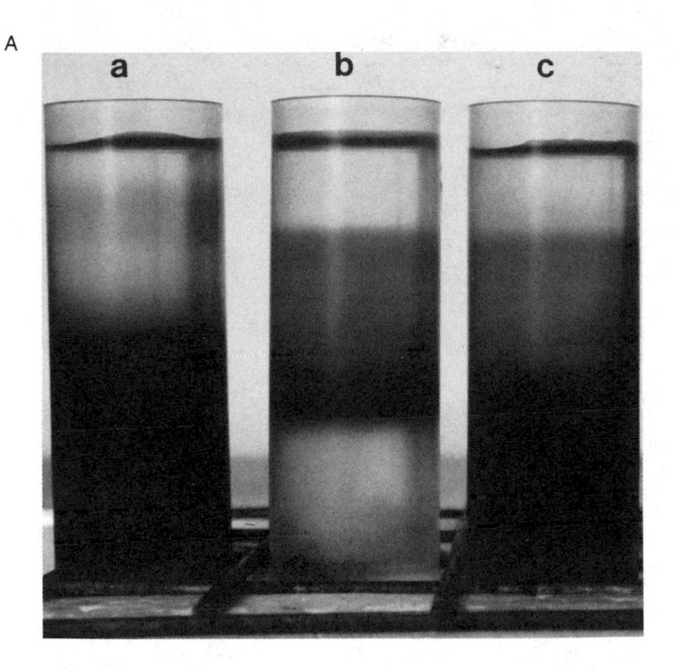

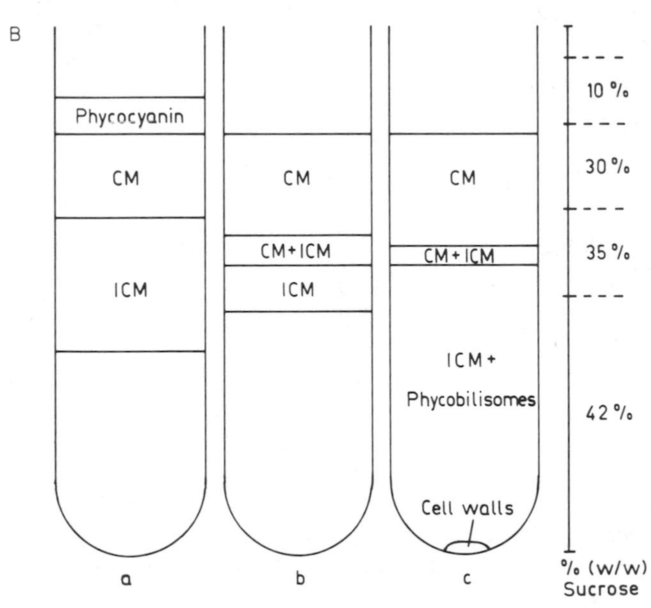

FIG. 1. Photographic (A) and schematic (B) representations of discontinuous sucrose density gradients containing isolated and separated plasma (CM) and thylakoid (ICM) membranes of *A. nidulans* after flotation centrifugation (c) and conventional recentrifugation (purification) of CM (b) and ICM (a), both withdrawn in an intentionally cross-contaminated form from (c).

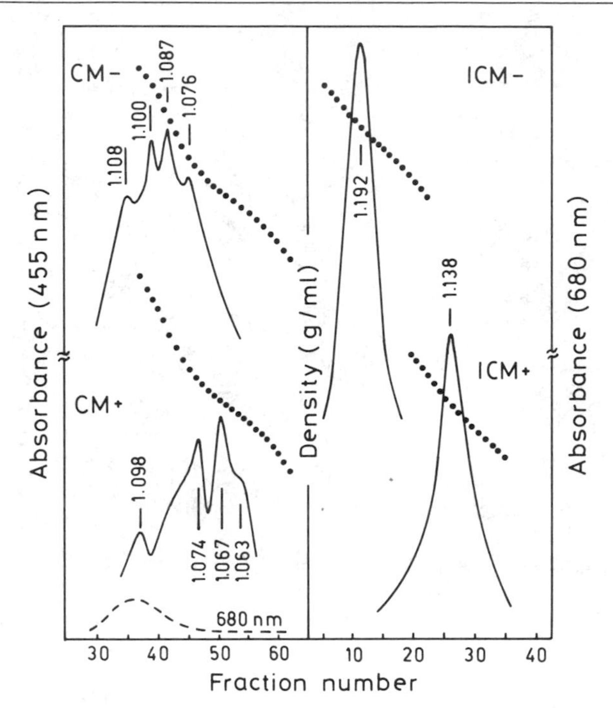

FIG. 2. Recentrifugation on continuous sucrose density gradients of prepurified CM and ICM preparations from *A. nidulans* grown in the absence (−) or presence (+) of 0.4 *M* NaCl revealing a 3- to 4-fold layered "fine structure" of the CM band from Fig. 1.

fraction of which is endowed with a distinct level of cytochrome oxidase activity (Table II). This may be one of the reasons why investigators working with only the uppermost layer of the CM band were unable to detect respiratory activity.[4]

Identification of CM can be achieved by external labeling of intact cells prior to disruption and membrane separation using an impermeant covalent protein marker, diazobenzene [^{35}S]sulfonate (DABS),[9,10] or fluorescamine which binds covalently to primary amino groups to give a fluorescent conjugate with maximum emission at 475 nm (excitation at 390 nm) but which itself is nonfluorescent and decomposes very rapidly in aqueous media to give nonfluorescent products.[11,12] Figure 3 shows the

[9] H. C. Berg, *Biochim. Biophys. Acta* **183,** 65 (1969).

[10] H. M. Tinberg, R. L. Melnick, J. Maguire, and L. Packer, *Biochim. Biophys. Acta* **345,** 118 (1974).

[11] M. Weigele, S. L. DeBernardo, J. P. Tengi, and W. Leimgruber, *J. Am. Chem. Soc.* **94,** 5927 (1972).

[12] S. Udenfriend, S. Stein, P. Böhlein, W. Dairman, W. Leimgruber, and M. Weigele, *Science* **178,** 871 (1972).

TABLE II
CYTOCHROME OXIDASE ACTIVITY ASSOCIATED WITH DISTINCT POPULATIONS (SUBFRACTIONS) OF CM VESICLES FROM *A. nidulans*[a]

Type	Density (g/ml)	Activity (nmol cyt *c*/min/mg protein)
CM−	1.076	0.0
	1.087	2.6
	1.100	16.3
	1.108	4.5
CM+	1.067	3.8
	1.074	106.3
	1.098	11.2
ICM−	1.192	4.6
ICM+	1.138	5.3

[a] See Fig. 2. Membranes were prepared from cells grown in the absence (−) and in the presence (+) of 0.4 *M* NaCl (cf. Fig. 4).

change in the specific quantities (per mg protein) of chlorophyll, ^{35}S activity, fluorescence intensity, and cytochrome-*c* oxidase activity during repeated recentrifugation ("purification") of the individual, but less and less cross-contaminated membrane fractions.[2,8] It is evident from Fig. 3 that in all cyanobacteria examined the external protein markers DABS and fluorescamine label predominantly the CM (not the ICM), that chlorophyll is almost exclusively associated with the ICM, and that the inherent proportion of cytochrome oxidase activities in CM and ICM may vary from species to species.

Cytochrome-c Oxidase Activities in CM and ICM Preparations

Table III summarizes typical cytochrome-*c* oxidase activities measured on CM-O and ICM-O preparations from several cyanobacteria grown according to Table I (without additional NaCl). As with mammalian cytochrome-*c* oxidase (EC 1.9.3.1), cyanide, carbon monoxide, azide, sulfide, salicylaldoxime, and neutral salts (at >50 mM) are all potent inhibitors of the reaction. For example, KCN causes complete inhibition at a concentration as low as 1.2 μM.[6] The different activities of cytochrome oxidase in CM and ICM isolated from *A. nidulans* at different stages of growth in the absence or presence of 0.4 *M* NaCl[8] are shown in Fig. 4.

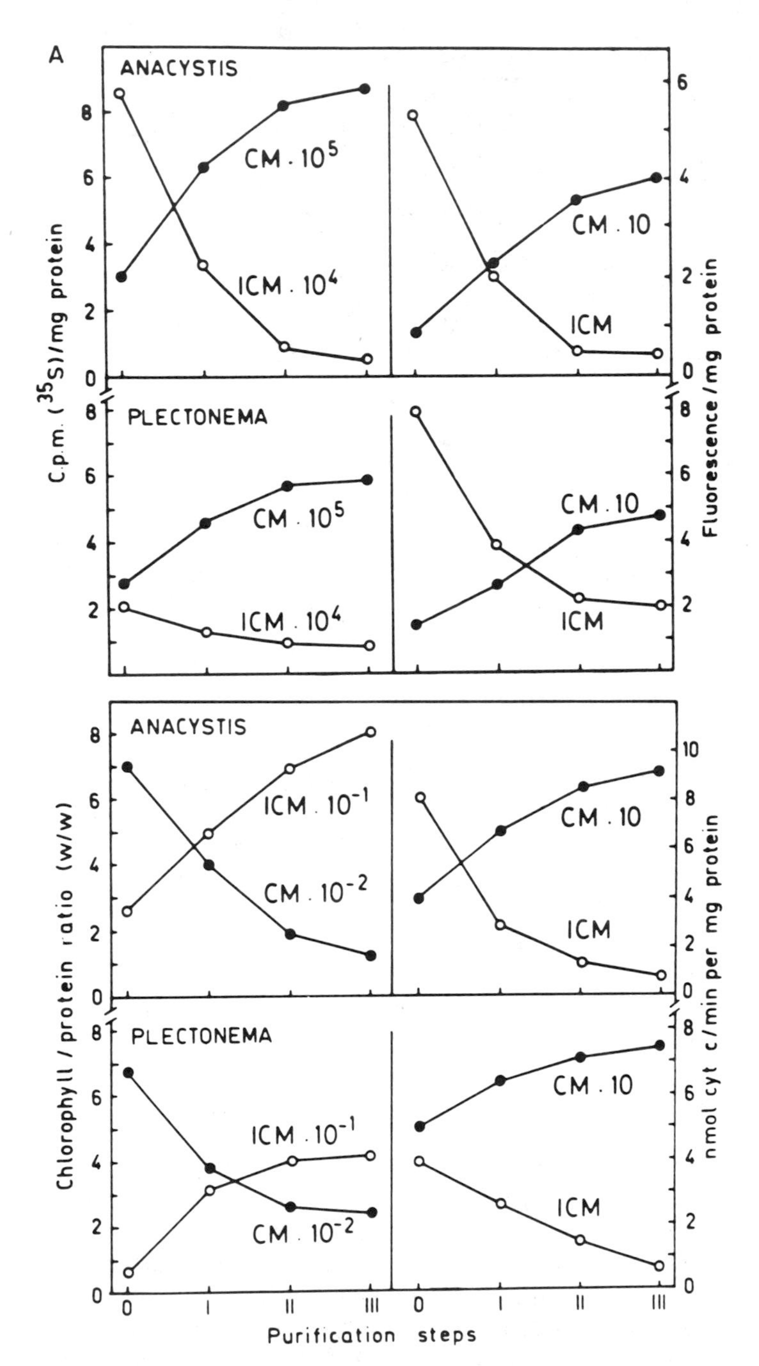

FIG. 3. Changes in diazobenzene [^{35}S]sulfonate activity, fluorescence emission, chlorophyll/protein ratio, and cytochrome oxidase activity (all per mg protein) in CM and ICM preparations from (A) *Anacystis* and *Plectonema* and (B) from *Synechocystis* and *Nostoc* during progressive purification.

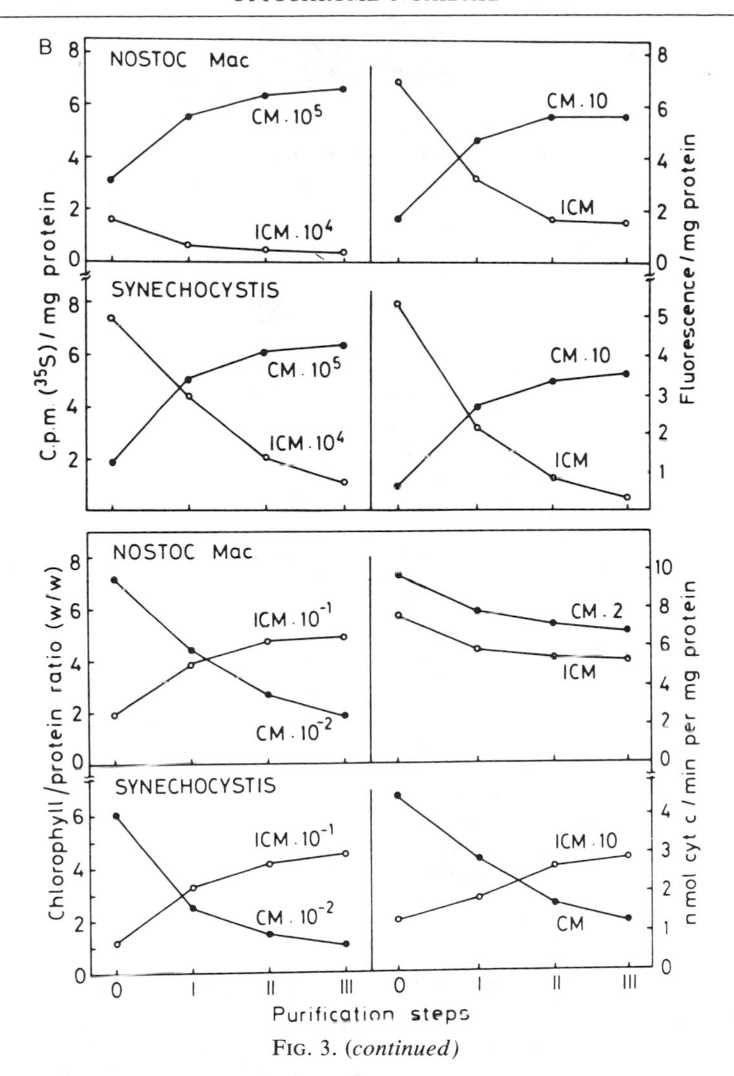

FIG. 3. *(continued)*

Identification of aa_3-Type Cytochrome Oxidase

Identification of the oxidase is carried out by sodium dodecyl sulfate–polyacrylamide gel electrophoresis (SDS–PAGE), followed by Western blotting, and cross-reaction of transferred polypeptide bands with (polyclonal) antibodies raised against subunits I and II, and the holoenzyme, of *Paracoccus denitrificans* aa_3-type cytochrome oxidase.[13] The procedure

[13] B. Ludwig and G. Schatz, *Proc. Natl. Acad. Sci. U.S.A.* **77,** 196 (1980).

TABLE III
RATES OF HORSE HEART CYTOCHROME *c* OXIDATION BY CYANOBACTERIAL CM AND ICM PREPARATIONS[a]

Species	Activity (nmol cyt *c*/min/mg protein)	
	CM	ICM
Anacystis nidulans	60	15
Synechocystis 6714	2.5	20
Plectonema boryanum	50	10
Nostoc MAC	20	7
Anabaena variabilis		
Vegetative cells	3.5	37
Isolated heterocysts	15	380
Gloeobacter violaceus	17	—

[a] Preparations were obtained by discontinuous sucrose density flotation centrifugation[1,2] from cells grown and harvested according to Table I (cf. Figs. 3A and 3B). The oxidation of ascorbate-reduced and dialyzed horse heart cytochrome *c* is measured at room temperature (21–22°) by dual-wavelength spectrophotometry.[6] Optimum reaction rates are obtained with 2–20 μM cytochrome *c* and <50 μg/ml membrane protein. Without exception, the activities shown are totally inhibitable by 1.2–30 μM KCN.

described for membranes from *A. nidulans*[2,5] can be applied to all the cyanobacteria mentioned here.

Aliquots of the membranes are subjected to SDS–PAGE according to Laemmli. Membranes are first precipitated with acetone (1 volume of membrane suspension/5 volumes of acetone) at −20° for 2 hr in the dark, pelleted in an Eppendorf centrifuge (10,000 *g*; 2 min), air-dried at room temperature, suspended in Laemmli buffer containing 5% (v/v) mercaptoethanol and 3% (w/v) SDS (protein-to-SDS ratio 1 : 8, w/w), and incubated at 95° for 3 min. After another Eppendorf centrifugation, the partly delipidated and solubilized membranes in the supernatant (about 20–30 μl each) are layered on top of the polyacrylamide gels (140 × 160 × 0.75 mm; 12.5% acrylamide) and electrophoresed in a water-cooled Bio-Rad Protean Slab Cell I operated at constant power for 2.5 hr using 8 W for the

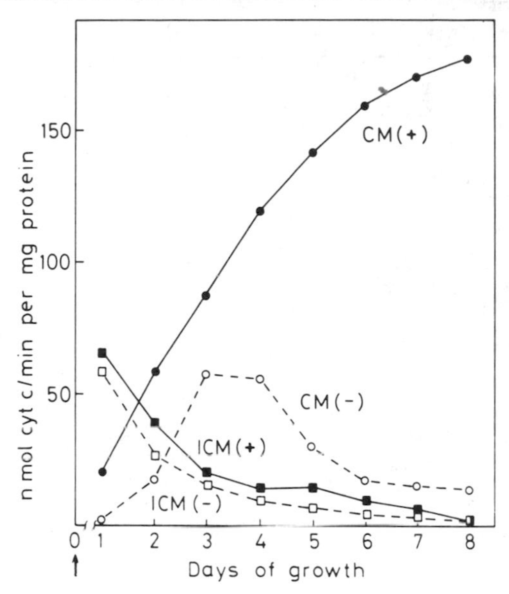

FIG. 4. Changes in the specific rate of horse heart cytochrome *c* oxidation by CM and ICM preparations isolated from *A. nidulans* grown for different time periods in either the absence (−) or presence (+) of 0.4 *M* NaCl.

stacking gel and 15 W for the separating gel. Separated polypeptide bands on the gels are stained with Coomassie brilliant blue R250 overnight, then destained for 4 hr with 40% (v/v) methanol containing 7.5% (v/v) acetic acid. Marker proteins of the Sigma MW-SDS-70L kit are used for calibration of molecular weights.

Part of the unstained SDS–polyacrylamide gel slabs containing the separated polypeptides is used for immunological cross-reaction assays which are conducted as follows. The gels are incubated with transfer buffer [25 m*M* Tris–HCl, pH 8.3; 192 m*M* glycine–20% (v/v) methanol] at room temperature for 1 hr with 3 changes of buffer. Polypeptides are electrophoretically transferred from the gels to nitrocellulose (Schleicher & Schüll, 0.45-μm pore size) in a water-cooled Bio-Rad Trans-Blot Cell using either 75 V and 0.23 A or 90 V and 0.30 A, and a transfer time of 2–2.5 hr. The nitrocellulose containing transferred polypeptides is either stained with Amido Black 10B or incubated first with 3% BSA (bovine serum albumin) in PBS (phosphate-buffered saline: 137.0 m*M* NaCl, 2.7 m*M* KCl, 1.5 m*M* KH_2PO_4, 6.4 m*M* Na_2HPO_4, pH 7.2) for 2 hr in order to saturate unspecific binding sites, and then with antibodies against subunit I or II of *Paracoccus denitrificans* cytochrome oxidase, or against the

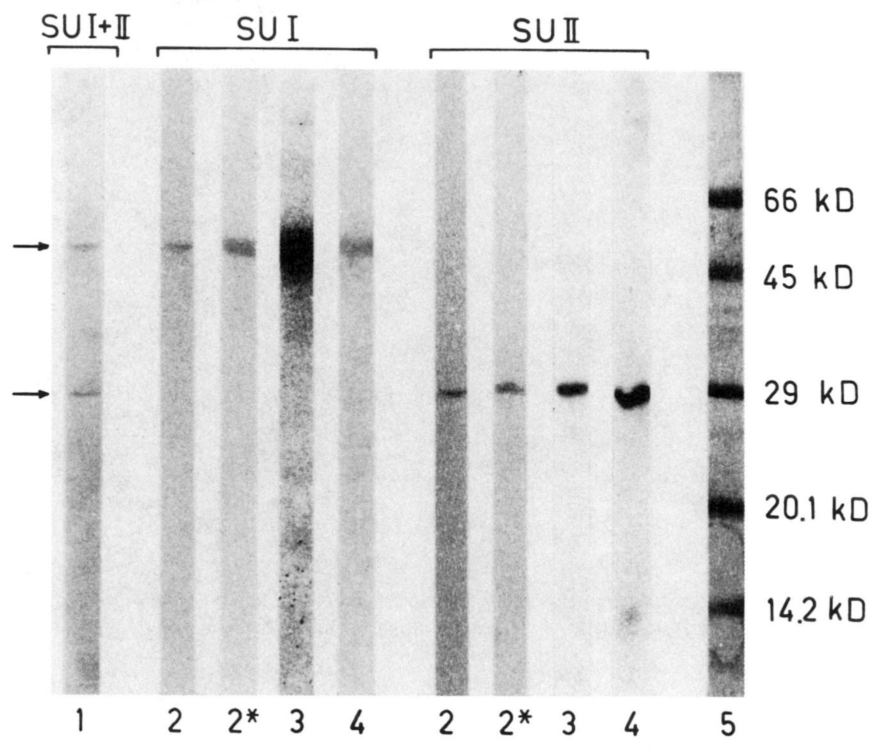

FIG. 5. Identification of aa_3-type cytochrome oxidase in CM preparations (25 μg protein/lane) from *A. nidulans*. Lanes 1, 2, 2*, with goat anti-rabbit horseradish preoxidase conjugate (GAR–HRP) as second antibody; land 2*, cells grown for 3 days in the presence of 0.4 *M* NaCl; lane 3, with ^{125}I-GAR–IgG as second antibody; lane 4, with ^{131}I-labeled protein A from *Staphylococcus aureus;* lane 5, marker proteins.

holoenzyme,[13] in appropriate dilution with 3% BSA–PBS, overnight at room temperature on a reciprocal shaker. The nitrocellulose is washed free of excess antibodies with 0.05% Tween 20 in PBS (3 times, 15 min each), then with PBS (3 times, 15 min each). Further incubation (90 min) with goat anti-rabbit horseradish peroxidase conjugate (GAR–HRP, from Sigma) in suitable dilution with 3% BSA–PBS is followed by a 90-min wash as above. Color is developed with a 4-chloro-1-naphthol-containing reagent (Bio-Rad). Alternatively, ^{125}I-labeled goat anti-rabbit IgG or ^{131}I-labeled protein A can be used as a second antibody, with detection by autoradiography. An additional control can be performed by using membranes isolated from *Escherichia coli*.[2] In neither case is a cross-reaction detected.

Figures 5 and 6 show the strictly specific and complementary cross-reaction between two of the polypeptide bands separated from *A. nidu-*

lans CM and ICM preparations, respectively, using antisera against *Paracoccus denitrificans* aa_3-type cytochrome oxidase holoenzyme, subunit I, and subunit II.[13] It is seen that the intensity of the cross-reaction (per mg protein) is higher in CM than in ICM, decreases in increasingly purified ICM preparations (Fig. 6) and increases after growth in the presence of 0.4 *M* NaCl (Fig. 5). Results obtained so far with immunoblotting experiments for the identification of aa_3-type cytochrome oxidase in isolated

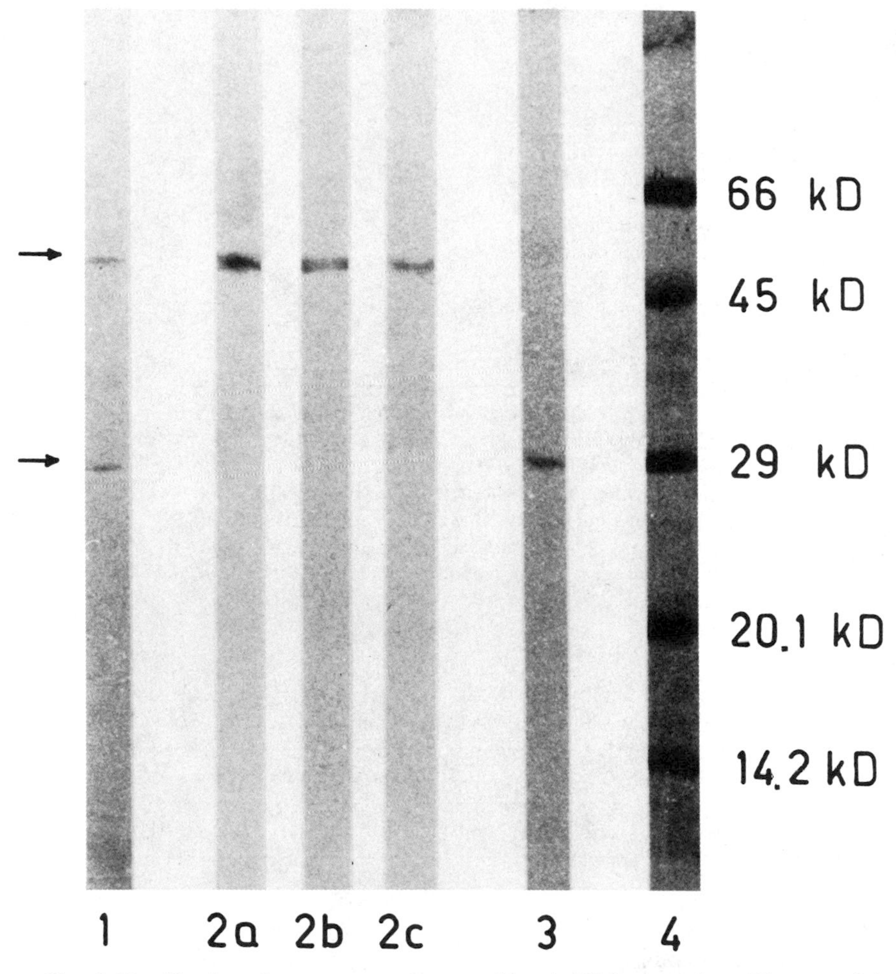

FIG. 6. Identification of aa_3-type cytochrome oxidase in ICM preparations (75 μg protein/lane) from *A. nidulans* using antibodies against the holoenzyme (lane 1), subunit I (lanes 2a, 2b, 2c), and subunit II (lane 3) of *Paracoccus denitrificans* cytochrome oxidase. The second antibody was GAR–HRP conjugate (cf. Fig. 5). ICM was applied without purification (2a) and after the first (2b) or second (2c) purification steps (cf. Fig. 3). Lane 4, marker proteins.

TABLE IV
IMMUNOBLOTTING OF CYANOBACTERIAL MEMBRANE POLYPEPTIDES USING ANTISERA AGAINST *Paracoccus denitrificans* aa_3-TYPE CYTOCHROME OXIDASE[a]

Species	Type	Relative intensity of cross-reaction (per mg protein)[b]
Anacystis nidulans	CM	+++
	ICM	++
Plectonema boryanum	CM	+++
	ICM	++
Nostoc MAC	CM	−
	ICM	−
Synechocystis 6714	CM	++
	ICM	+++
Anabaena variabilis		
Vegetative cells	CM	++
	ICM	++
Isolated heterocysts	CM	++
	ICM	++++
Gloeobacter violaceus	CM	−

[a] Organisms were grown and harvested according to Table I.
[b] Symbols: ++++, very strong and specifically complementary cross-reaction (cf. Figs. 5 and 6); ++, weak but reproducible cross-reaction possibly due to cross-contaminating membranes (cf. Figs. 6 and 3); and −, no cross-reaction.

TABLE V
APPARENT MOLECULAR WEIGHTS OF SUBUNITS I, II, AND III OF aa_3-TYPE CYTOCHROME OXIDASE FROM VARIOUS EUKARYOTIC AND PROKARYOTIC SOURCES[a]

Source	Molecular weight ($\times 10^{-3}$)			Refs.[b]
	Subunit I	Subunit II	Subunit III	
Rat liver	43	27	25.7	1, 2
Bovine heart	41.5	22.5	n.d.	2
Neurospora crassa	40	29	21	3
Saccharomyces cerevisiae	40	27.5	25	4, 5
Paracoccus denitrificans	45	28	—	6
Bacillus subtilis	57	37	21	7
PS3	56	38	22	8

TABLE V (*continued*)

Source	Molecular weight (× 10^{-3})			Refs.[b]
	Subunit I	Subunit II	Subunit III	
Nitrobacter agilis	51	31	—	9
Thiobacillus novellus	32	23	—	10
Rhodobacter sphaeroides	45	37	35	11
Pseudomonas AMI	50	30	—	12
Anacystis nidulans[a]	46–55	29–32	n.d.	13, 14
Synechocystis 6714[a]	22–26	14–17	n.d.	15
Plectonema boryanum[a]	49–50	42–43	n.d.	16
Anabaena variabilis[a]	48–49	36	n.d.	17

[a] Mitochondrial cytochrome oxidases have been reported to be composed of up to 13 polypeptides ("subunits"?) [P. Merle and B. Kadenbach, *Eur. J. Biochem.* **105,** 499 (1980)] whereas the prokaryotic counterparts never contain more than 3, and most only 2 subunits [B. Ludwig, *FEMS Microbiol. Rev.* **46,** 41 (1987)]. Data derived from SDS-PAGE and immunoblotting; other data derived from isolated enzymes. n.d., Not determined; (—) indicates "not detectable."

[b] References: (1) P. Merle and B. Kadenbach, *Eur. J. Biochem.* **105,** 499 (1980); (2) L. Höchli and C. R. Hackenbrock, *Biochemistry* **17,** 3712 (1978); (3) H. Weiss and H. J. Kolb, *Eur. J. Biochem.* **99,** 139 (1979); (4) M. S. Rubin and A. Tzagoloff, *J. Biol. Chem.* **248,** 4269 (1973); (5) M. S. Rubin and A. Tzagoloff, *J. Biol. Chem.* **248,** 4275 (1973); (6) B. Ludwig and G. Schatz, *Proc. Natl. Acad. Sci. U.S.A.* **77,** 196 (1980); (7) W. DeVrij, A. Azzi, and W. N. Konings, *Eur. J. Biochem.* **131,** 97 (1983); (8) N. Sone and Y. Yanagita, *Biochim. Biophys. Acta* **682,** 212 (1982); (9) T. Yamanaka, Y. Kamita, and Y. Fukumori, *J. Biochem. (Tokyo)* **89,** 265 (1981); (10) T. Yamanaka and K. Fujii, *Biochim. Biophys. Acta* **591,** 53 (1980); (11) R. B. Gennis, B. Ludwig, R. C. Casey, and A. Azzi, *Eur. J. Biochem.* **125,** 189 (1982); (12) Y. Fukumori, K. Nakayama, and T. Yamanaka, *J. Biochem. (Tokyo)* **98,** 493 (1985); (13) M. Trnka and G. A. Peschek, *Biochem. Biophys. Res. Commun.* **136,** 235 (1986); (14) V. Molitor, M. Trnka, and G. A. Peschek, *Curr. Microbiol.* **14,** 263 (1987); (15) M. Wastyn, A. Achatz, M. Trnka, and G. A. Peschek, *Biochem. Biophys. Res. Commun.* **149,** 102 (1987); (16) M. Wastyn and G. A. Peschek, *Proc. 5th EBEC,* Aberystwyth, Wales (1988); (17) M. Wastyn, A. Achatz, V. Molitor, and G. A. Peschek, *Biochim. Biophys. Acta,* in press.

membranes from cyanobacteria after growth according to Table I are qualitatively summarized in Table IV. Table V gives a survey of the (apparent) molecular weights of aa_3-type cytochrome oxidase subunits I–III from mitochondria, yeasts, and various bacteria including cyanobacteria.

Acknowledgments

Work in the author's laboratory is supported by grants from the Austrian Science Foundation. We are particularly indebted to Dr. Bernd Ludwig for the donation of the antisera.

[46] Electron Paramagnetic Resonance-Detectable Cu^{2+} in *Synechococcus* 6301 and 6311: aa_3-Type Cytochrome-*c* Oxidase of Cytoplasmic Membrane

By IAN V. FRY and GÜNTER A. PESCHEK

Membrane preparations from *Synechococcus* 6301 and 6311 exhibited low-temperature electron paramagnetic resonance (EPR) spectra in the $g = 2.08$ region characteristic of copper. The physical parameters of the power-saturation characteristics and the temperature-dependence profile demonstrated that the copper signals arose from a center in an environment identical to the aa_3-type cytochrome-*c* oxidases reported in mammalian, yeast, and bacterial systems. Membrane purification procedures demonstrated that the oxidase was present in the cytoplasmic membrane at 10 times the level present in the thylakoid membrane. The copper was demonstrated to be fully redox active by its reducibility with physiological electron donors.

Introduction

In addition to higher plant type photosynthesis, cyanobacteria (blue–green algae) have the capacity to carry out "mitochondrial" or dark-type respiration, and recent spectrophotometric and inhibitor studies have indicated the presence of an aa_3-type cytochrome-*c* oxidase in membrane preparations from cyanobacteria[1–7]; the presence of aa_3-type cytochrome oxidase could be verified by immunological cross-reaction with antisera against the *Paracoccus denitrificans* aa_3-type cytochrome oxidase.[8–10] Cytochrome-*c* oxidase (EC 1.9.3.1) from several sources has been shown

[1] J. C. P. Matthijs, Ph.D. thesis. Vrije Universiteit Amsterdam, Amsterdam, The Netherlands, 1984.
[2] J. C. P. Matthijs, *Adv. Photosynth. Res.* **2,** 643 (1984).
[3] J. P. Houchins and G. Hind, *Plant Physiol.* **76,** 456 (1984).
[4] G. A. Peschek, *Biochim. Biophys. Acta* **635,** 470 (1981).
[5] G. A. Peschek, *Biochem. Biophys. Res. Commun.* **98,** 72 (1981).
[6] G. A. Peschek, G. Schmetterer, G. Lauritsch, W. H. Nitschmann, P. F. Kienzl, and R. Muchl, *Arch. Microbiol.* **131,** 261 (1982).
[7] V. Molitor, W. Erber, and G. A. Peschek, *FEBS Lett.* **204,** 251 (1986).
[8] G. A. Peschek, V. Molitor, M. Trnka, W. Wastyn, and W. Erber, this volume [45].
[9] V. Molitor, M. Trnka, and G. A. Peschek, *Curr. Microbiol.* **14,** 263 (1987).
[10] M. Trnka and G. A. Peschek, *Biochem. Biophys. Res. Commun.* **136,** 235 (1986).

to contain redox active Cu^{2+},[11–14] and the growth of *Synechococcus* 6301 on Cu^{2+}-deficient medium[15] has resulted in a marked decrease in the ability of membrane preparations from this organism to oxidize exogenous *c*-type cytochromes.[16] However, another source of redox active Cu^{2+} in photosynthetic organisms is plastocyanin, and although plastocyanin has been reported to be absent from several cyanobacterial species,[17] including *Synechococcus* 6301,[18] the presence of small amounts of plastocyanin in these *Synechococcus* species could not be ruled out.

In order to distinguish between the two Cu^{2+}-containing electron transport components, as well as determine the location of the cytochrome-*c* oxidase, we have characterized the EPR-detectable Cu^{2+} signals arising from both scrambled membrane preparations (i.e., thylakoid plus cytoplasmic) and purified cytoplasmic membranes isolated from *Synechococcus* 6311 and 6301, which are believed to be independent isolates of the same species.[19] The temperature-dependence and power-saturation characteristics of the EPR-detectable Cu^{2+} signals are used to unequivocably demonstrate that the Cu^{2+} exists in an environment identical to that in typical aa_3-type cytochrome oxidases (EC 1.9.3.1) and is present in the cytoplasmic membrane. In addition, levels of EPR-detectable Cu^{2+} are correlated with the capacity to oxidize reduced cytochrome *c*. The redox activity of the EPR-detectable Cu^{2+} is demonstrated by reduction with physiological electron donors.

Materials and Methods

Synechococcus 6311 (ATCC 27145) and 6301 (strain 1402-1, Gottingen, FRG) were grown on Kratz and Myers media C and D, respectively, as described.[15] Spheroplasts were prepared according to Biggins.[20] Crude or scrambled membranes were prepared by disrupting the sphero-

[11] H. Beinert, D. E. Griffith, D. C. Warton, and R. H. Sands, *J. Biol. Chem.* **237,** 2337 (1962).

[12] B. M. Hoffmann, J. E. Roberts, M. Swanson, S. H. Speck, and E. Margoliash, *Proc. Natl. Acad. Sci. U.S.A.* **77,** 1452 (1980).

[13] A. Seelig, B. Ludwig, J. Seelig, and G. Schatz, *Biochim. Biophys. Acta* **636,** 162 (1981).

[14] J. A. Fee, M. G. Choc, K. L. Findling, R. Lorence, and T. Yoshida, *Proc. Natl. Acad. Sci. U.S.A.* **77,** 147 (1980).

[15] W. A. Kratz and J. Myers, *Am. J. Bot.* **42,** 282 (1955).

[16] P. F. Kienzl and G. A. Peschek, *Plant Physiol.* **69,** 580 (1982).

[17] G. Sandmann and P. Boger, *Biochim. Biophys. Acta* **766,** 395 (1980).

[18] P. F. Kienzl and G. A. Peschek, *FEBS Lett.* **162,** 76 (1983).

[19] R. Rippka, J. Deruelles, J. B. Waterbury, M. Herdman, and R.Y. Stanier, *J. Gen. Microbiol.* **111,** 1 (1979).

[20] J. Biggins, *Plant Physiol.* **42,** 1447 (1967).

plasts either by passage through a French press (10,000 psi) under N_2 or by ultrasonic oscillation (Branson sonifier, Model 350, microtip setting 3 for 20 min on ice). Unbroken cells were removed by centrifugation at 3,000 *g* (10 min). Membranes were sedimented at 40,000 *g* (60 min), then resuspended in 10 m*M* Tris–HCl, pH 8.0, containing 10 m*M* EDTA, and sedimented again at 40,000 *g* (60 min). The EDTA served to effectively remove any adventitious paramagnetic ions such as Mn^{2+} and Cu^{2+}. The pelleted membranes were resuspended in the same medium to approximately 50 μg chlorophyll/ml and frozen in liquid N_2. Cytoplasmic and thylakoid membranes were prepared by the method of Peschek (this volume [45]). Purified membrane preparations were stored in liquid N_2. Thawed membrane samples were centrifuged at 40,000 *g* (60 min); the pellet was resuspended in TES–HCl buffer, pH 7.0, containing 10 m*M* EDTA. Finally, the membranes were concentrated by centrifugation in quartz EPR tubes at 40,000 *g* (60 min) to give a final protein concentration of between 2 and 3 mg/ml.

Mammalian cytochrome *c* oxidase was observed in submitochondrial particles (SMPs), prepared from beef heart mitochondria as described[21] and resuspended in 10 m*M* EDTA, 10 m*M* TES, pH 7.0. Plastocyanin was observed by EPR spectroscopy in whole cells of *Gloeobacter violaceus* (I. Fry, A. Robinson, and L. Packer, unpublished observations).

Electron transport components in the membrane were oxidized with 50 m*M* potassium ferricyanide, or air, and reduced with 10 m*M* sodium dithionite or 10 m*M* ascorbate plus 25 μ*M* TMPD (NN,N′N′-tetramethyl-p-phenylenediamine), or with 30 m*M* NADPH or NADH in the presence of 0.10 m*M* horse heart cytochrome *c* and 2 m*M* KCN. In these experiments the exogenous cytochrome *c* served as an electron mediator to the cytochrome oxidase which cannot be directly reduced with NAD(P)H.[11] EPR spectra were recorded with a Varian E-109 spectrometer, fitted with an Air Products Heli-Trans liquid helium transfer system, at 20 mW power, 1 millitesla modulation amplitude, and 9.15 GHz. Cytochrome oxidase activity was determined from the KCN-sensitive oxidation of horse ferrocytochrome *c* as described previously.[6,16]

Results

Characterization of EPR-Detectable Cu^{2+}

Scrambled and purified cytoplasmic membrane preparations of *Synechococcus* 6301 and 6311 exhibited low-temperature EPR signals in the

[21] H. Low and L. Vallin, *Biochem. Biophys. Acta* **69**, 361 (1963).

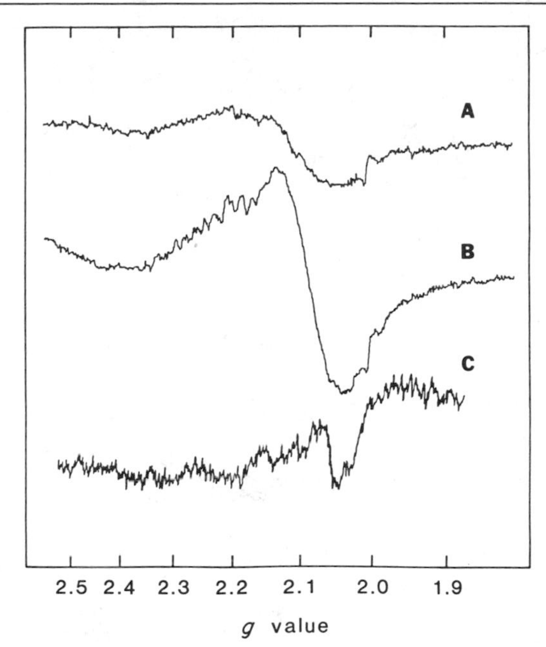

FIG. 1. Low-temperature EPR spectra of purified cytoplasmic membranes. Conditions: 20 mW power, 1 mT modulation amplitude, 9.15 GHz, 100 K. (A) Membranes prepared as indicated in the text; (B) membranes plus 2 m*M* KCN; (C) membranes plus 2 m*M* KCN, 30 m*M* NADPH, 0.1 m*M* cytochrome *c* (gain used in C is 3 times that used in A and B). Substitution of NADH for NADPH, or incubation of the membranes with 10 m*M* dithionite or 10 m*M* acorbate plus 25 μM TMPD, gave essentially the same result as in C.

$g = 2.08$ region characteristic of Cu^{2+} (Fig. 1). SMPs from beef heart and whole cells of *Gloeobacter violaceus* also exhibited Cu^{2+} EPR signals which were identical in line shape to those of the *Synechococcus* membrane preparations (data not shown; cf. Refs. 11–13 and 22–24). The temperature-dependence profiles of the respective Cu^{2+} signals, however, clearly demonstrated that they arose from different environments (Fig. 2). The temperature dependence of the Cu^{2+} signals arising from cytoplasmic and scrambled *Synechococcus* membranes, and from SMPs, showed a maximum at 100 K (Fig. 1), whereas the EPR-detectable Cu^{2+} signal arising from *Gloeobacter violaceus* plastocyanin had a temperature optimum around 10–20 K (Fig. 2).[23,24]

[22] I. V. Fry, G. A. Peschek, M. Huflejt, and L. Packer, *Biochem. Biophys. Res. Commun.* **129,** 109 (1985).

[23] G. Sandmann and P. Böger, *Plant Sci. Lett.* **17,** 417 (1980).

[24] H. Bohner, H. Merkle, P. Kroneck, and P. Böger, *Eur. J. Biochem.* **105,** 603 (1980).

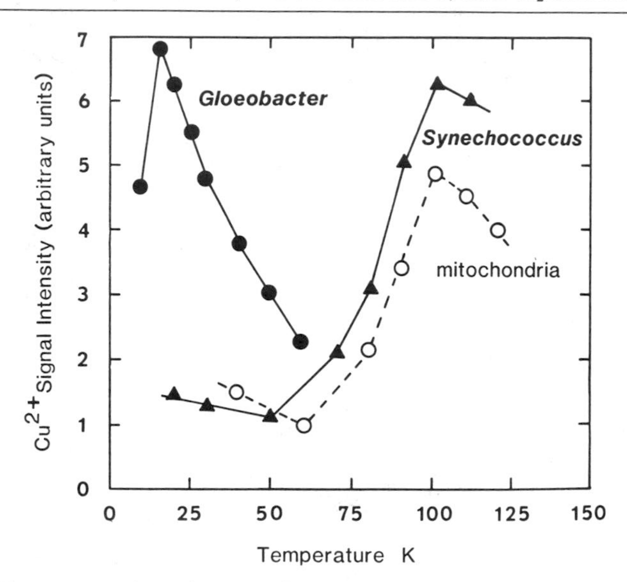

FIG. 2. Temperature-dependence profile of the $g = 2.08$ EPR-detectable Cu^{2+} signals from *Synechococcus* (scrambled or cytoplasmic) membranes, mitochondrial SMPs, and whole cells of *Gloeobacter violaceus*. Conditions as in Fig. 1.

In addition, mixed populations of Cu^{2+} could be observed in the mitochondrial SMP preparation, giving a biphasic character to the temperature profile (Fig. 3). Treatment of SMPs with repeated freeze–thaw cycles resulted in a decrease in the Cu^{2+} signal observed at higher temperatures (100 K) with a concomitant increase in the Cu^{2+} signal observed at lower temperatures (20 K) (Figs. 3A–3C). Excessive freeze–thaw cycles (Fig. 3C) resulted in an EPR-detectable Cu^{2+} signal whose temperature-dependence profile was identical to that of cupric EDTA (data not shown). The temperature-dependence profile of the Cu^{2+} signals from *Synechococcus* membranes did not show any significant population of this "adventitious" or enzyme-released Cu^{2+} present in these preparations (cf. Figs. 2 and 3).

The EPR Cu^{2+} signals from *Synechococcus* membranes and SMPs did not saturate with microwave power up to 180 mW (Fig. 4), whereas the EPR Cu^{2+} signal from *Gloeobacter* plastocyanin, observed at lower temperatures, readily saturated with microwave power (Fig. 4).[23,24] Addition of KCN resulted in an increase in the $g = 2.08$ Cu^{2+} EPR feature (Fig. 1B). Such an effect is indicative of a $Cu(II)_A$–heme *a* spin interaction, which has been observed in mammalian cytochrome *c* oxidase.[25] This

[25] G. W. Brudvig, D. F. Blair, and S. I. Chan, *J. Biol. Chem.* **259,** 11001 (1984).

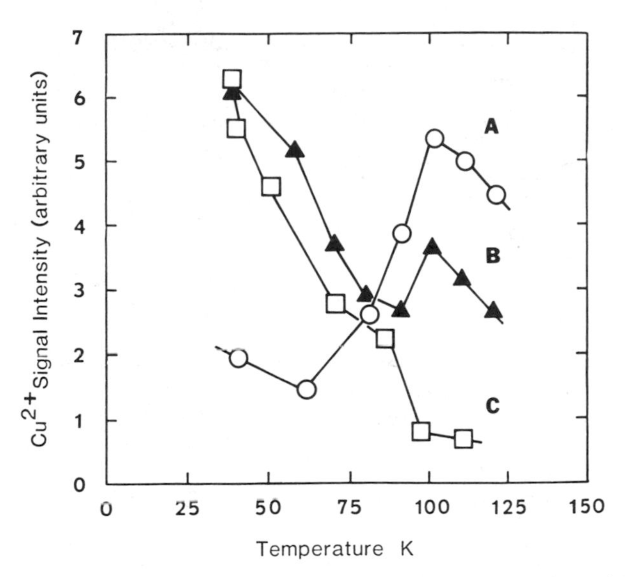

FIG. 3. Effect of repeated freeze–thaw cycles on the temperature-dependence profile of the EPR-detectable Cu^{2+} signals from mitochondrial SMPs. (A) One freeze–thaw cycle; (B) three freeze–thaw cycles; (C) five freeze–thaw cycles. Conditions as in Fig. 1.

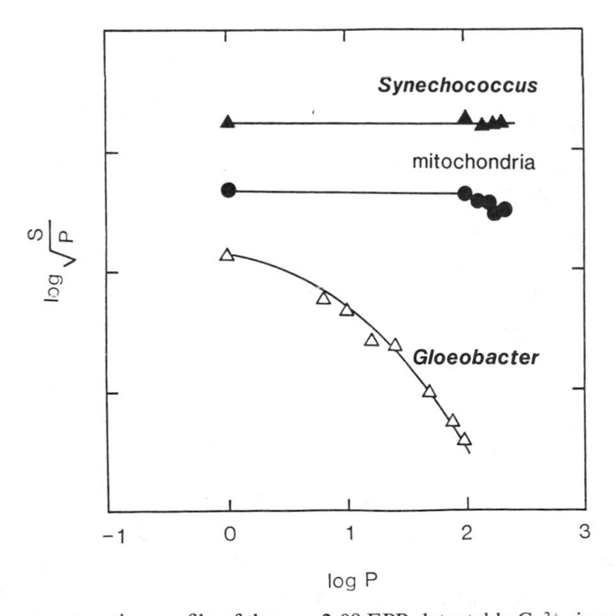

FIG. 4. Power-saturation profile of the $g = 2.08$ EPR-detectable Cu^{2+} signals arising from *Synechococcus* (scrambled or cytoplasmic) membranes, mitochondrial SMPs, and whole cells of *Gloeobacter violaceus*. Conditions as in Fig. 1, except that the spectrum of *Gloeobacter violaceus* was measured at 20 K.

TABLE I
EFFECT OF GROWTH UNDER SALINE STRESS ON COMPONENTS OF RESPIRATORY ELECTRON TRANSPORT CHAIN

Growth conditions	Whole cell endogenous respiration[a]	NADPH-dependent cytochrome *c* reduction[b]	Cytochrome-*c* oxidase activity[c]	Cu^{2+} EPR $g = 2.08$ signal[d]
0.015 *M* NaCl	2.95	9.5	7.8	0.66
0.5 *M* NaCl	20.65	25.5	89.8	3.58

[a] Expressed as μmol O_2/mg chlorophyll/hr.
[b] Expressed as μmol/mg chlorophyll/hr.
[c] Reduced cytochrome *c* oxidation, expressed as μmol cytochrome *c* oxidized/mg chlorophyll/hr.
[d] EPR-detectable Cu^{2+}, expressed as nmol Cu^{2+}/mg chlorophyll.

weak antiferromagnetic coupling between the Cu^{2+} and the iron centers is perturbed by ligand binding to the heme, resulting in the observed increase in the Cu^{2+} EPR signal.

Cytochrome Oxidase Activity and Cu^{2+} Content

Growth of *Synechococcus* 6311 under conditions of saline stress resulted in a marked increase in endogenous whole cell respiration (Table I). The increased endogenous respiration rate correlated well with both the cytochrome-*c* oxidase activity and the Cu^{2+} content of scrambled membrane preparations. However, NADPH oxidation in these preparations was not stimulated so markedly (Table I). Membranes prepared by ultrasonic oscillation did not exhibit any EPR-detectable Cu^{2+} signals in the $g = 2.08$ region (data not shown; see Ref. 22). Characteristically, cytochrome-*c* oxidase activity in sonicated or French press preparations correlated with the presence or absence of EPR-detectable Cu^{2+} (data not shown; see Ref. 22).

Physiological Electron Donors and Cu^{2+} Reduction

Cytoplasmic membranes were incubated with NADPH or NADH, which have been shown to act as electron donors to the respiratory chain and, hence, to the cytochrome-*c* oxidase of cyanobacteria,[2,26] in the presence of KCN to prevent leakage of electrons from the reduced oxidase. The membrane preparations were supplemented with horse heart cytochrome *c* to replace any soluble electron mediator(s) lost during the preparation procedure.[2] Clearly, in these experiments physiologically reduced cytochrome *c* is the immediate electron donor to the terminal oxidase;

[26] G. A. Peschek, *Subcell. Biochem.* **10,** 83 (1984).

TABLE II
REDOX REACTIONS OF EPR-DETECTABLE Cu^{2+}

Membrane type	Treatment[a]	EPR-detectable Cu^{2+} (nmol/mg protein)
Thylakoid	+KCN	0.71
Cytoplasmic	+KCN	7.01
Cytoplasmic	+KCN, +cytochrome *c*, +NADPH	0.60
Cytoplasmic	+KCN, +cytochrome *c*, +NADH	0.70
Cytoplasmic	+KCN, +dithionite	Not detectable
Cytoplasmic	+KCN, +ascorbate, +TMPD	Not detectable

[a] For reagent concentrations, see Methods and Materials.

horse heart ferrocytohrome *c* has been shown to be an excellent reductant to the membrane-bound cytochrome *c* oxidase of several cyanobacteria.[6,7,27]

The degree of reduction of the Cu^{2+} EPR signal is presented in Table II. Incubation of the cytoplasmic membrane preparation with the physiological electron donors NADH or NADPH in the presence of cytochrome *c* and KCN resulted in over 90% reduction of the Cu^{2+} EPR signal (Fig. 1C, Table II), presumably by reduction to Cu^{+}. Almost full reduction of the Cu^{2+} EPR signal demonstrates the integrity of complete electron transport pathway (except for the water-soluble cytochrome *c* lost during purification.[2] In the scrambled membrane preparations reported previously,[22] 25–60% of the total EPR-detectable Cu^{2+} was not reducible by physiological electron donors. This inaccessible Cu^{2+} was attributed to damaged and nonfunctional enzyme.[28] The levels of cytochrome-*c* oxidase damaged by the present preparation procedures (as determined by levels of physiologically reducible Cu^{2+}) were extremely low.

Discussion

The membrane preparations used in this study were washed with 10 m*M* EDTA to remove adventitious Cu^{2+}, together with other membrane-associated paramagnetic metal ions such as manganese.[29] The washing

[27] G. A. Peschek, P. F. Kienzl, and G. Schmetterer, *FEBS Lett.* **131,** 11 (1981).

[28] R. Aasa, S. P. J. Albracht, K. E. Falk, B. Lanne, and T. Vanngard, *Biochem. Biophys. Acta* **422,** 260 (1976).

[29] R. Cammack, L. J. Luijk, J. J. Maguire, I. V. Fry, and L. Packer, *Biochim. Biophys. Acta* **548,** 267 (1979).

treatment would also remove any Cu^{2+}-containing and other soluble proteins. Therefore, one of the very few possibilities of intrinsically membrane-bound Cu^{2+} in a plastocyanin-free cyanobacterium is the aa_3-type terminal oxidase,[26] which is supported by the present data. EPR spectra, temperature dependence, and power saturation of the Cu^{2+} signal closely resemble EPR spectra published for the Cu^{2+}-containing aa_3-type cytochrome oxidases from mammalian mitochondria,[8,9] yeast,[13] *Paracoccus denitrificans,*[13] and *Thermus thermophilus.*[14] The signals are, however, clearly different from those exhibited by soluble copper proteins such as plastocyanin.[17,24,30] Moreover, both EPR-detectable Cu^{2+} and cytochrome-*c* oxidase activity were absent in parallel from heavily sonicated membranes[22] in which the functional integrity of the cytochrome oxidase complex might have been disrupted and lost together with the Cu^{2+}. In addition, a concomitant increase in both the EPR-detectable Cu^{2+} and the cytochrome-*c* oxidase activity was observed in membrane preparations from *Synechococcus* sp. grown under salt stress[31] (Table I).

An aa_3-type cytochrome oxidase was recently described in terms of optical spectra, photoaction spectra, and differential reactivities toward various *c*-type cytochromes and inhibitors using membrane preparations from several axenic strains of cyanobacteria (see Ref. 24 for review). Similar findings were reported for *Plectonema boryanum*[1,2] and the heterocysts of *Anabaena* 7120.[3] In previous experiments with *Synechococcus* 6301 it was shown that the membrane-bound cytochrome aa_3 could be physiologically reduced with horse heart ferrocytochrome *c* or reduced pyridine nucleotides, or with ascorbate plus TMPD or DCPIP (2,6-dichlorophenolindophenol), and oxidized with molecular oxygen[4,6,16,26]; now the same redox behavior has been found for the tightly membrane-bound Cu^{2+} of the cytoplasmic membrane of *Synechococcus* 6311 (Table II).

In scrambled membrane preparations, full reduction of the Cu^{2+} by nucleotides did not occur, which may be due to a bleeding of electrons to other pathways (the oxidized chlorophyll radical was 90% reduced by nucleotides; data not shown) or due to population of Cu^{2+} not reducible by cytochrome *c* owing to damage of the cytochrome-*c* oxidase during membrane preparation.[28] In the purified cytoplasmic membrane preparations, however, over 90% of the EPR-detectable Cu^{2+} was physiologically active, and very little "adventitious" or enzyme-released Cu^{2+} was observed at low temperatures (20–30 K). The EPR signal intensity of the oxidized Cu^{2+} was more pronounced in the presence of KCN, owing to

[30] J. W. M. Visser, J. Amsz, and B. F. Van Gelder, *Biochim. Biophys. Acta* **333,** 279 (1974).
[31] I. V. Fry, M. Huflejt, W. W. A. Erber, G. A. Peschek, and L. Packer, *Arch. Biochem. Biophys.* **244,** 686 (1986).

perturbation of the weak $Cu(II)_A$–heme *a* magnetic interaction observed in aa_3-type cytochrome-*c* oxidases.[25] Earlier work with crude membrane fractions required Triton X-100 treatment to observe the full Cu^{2+} signal,[22] probably because of disruption of the magnetic interaction by the slight denaturing effect of the detergent at the concentration used.

Previous results from work with intact cells, spheroplasts, or isolated membranes of *Synechococcus* 6301 have indicated an aa_3-type terminal oxidase to be present in both cytoplasmic and thylakoid membranes.[26,32–35] Recent findings with a chlorophyll-free cytoplasmic membrane fraction from the same species showed it to be capable of oxidizing horse heart ferrocytochrome *c*,[7] and our present results using EPR corroborate these functional studies.

From our present findings we conclude the EPR-detectable Cu^{2+} is present in a tightly bound form in cytoplasmic membranes of *Synechococcus* 6301 and 6311, at a level which is an order of magnitude greater than in thylakoid membranes. It undergoes physiological redox reactions with respiratory electron donors and acceptors, and it is identical to the aa_3-type cytochrome-*c* oxidase (EC 1.9.3.1) which has been described in mammalian and yeast mitochondria[11–13] and in bacteria.[13,14]

[32] G. A. Peschek, G. Schmetterer, G. Lauritsch, R. Muchl, P. F. Kienzl, and W. H. Nitschmann, *in* "Photosynthetic Prokaryotes: Cell Differentiation and Function" (G. C. Papageorgiou and L. Packer, eds.), p. 147. 1983.
[33] G. A. Peschek, *J. Bacteriol.* **153,** 539 (1983).
[34] G. A. Peschek, *Plant Physiol.* **75,** 968 (1984).
[35] V. Molitor, M. Trnka, and G. A. Peschek, *Curr. Microbiol.* **14,** 263 (1987).

[47] Nitrogen and Hydrogen Metabolism: Induction and Measurement

By HELMAR ALMON and PETER BÖGER

Introduction

Nitrogen-fixing filamentous cyanobacteria have been studied intensively with respect to N_2 and H_2 gas metabolism.[1–3] This chapter summarizes some comparatively simple routine assays for the measurement of both nitrogen fixation and hydrogen gas exchange in intact cyanobacteria.

[1] W. D. P. Stewart, *Annu. Rev. Microbiol.* **34,** 497 (1980).
[2] H. Bothe, *Experientia* **38,** 59 (1982).
[3] J. P. Houchins, *Biochim. Biophys. Acta* **768,** 227 (1984).

Special attention is given to the phenomena involved in the induction of nitrogenase activity in heterocystous and nonheterocystous filamentous cyanobacteria. Many nonheterocystous cyanobacteria are also able to fix dinitrogen, when incubated under microaerobic conditions.[4] However, a sequential separation between oxygen-producing photosynthesis and oxygen-labile nitrogenase activity has to be provided either by growth in a light–dark cycle[5] or by both activities being operative in different phases of a cell cycle.[6] Additional information on nitrogenase activity in unicellular cyanobacteria is given elsewhere in this volume.[7]

Measurement of Nitrogenase Activity

Nitrogenase activity is conveniently measured using the acetylene reduction assay[8,9] [Eq. (1)]. This method requires a gas chromatograph (GC) equipped with a Porapak column, a flame-ionizing detector (details of technical equipment are described below) and acetylene (C_2H_2) as substrate. Acetylene, however, may often not be available in pure form. Contamination by methane is not critical for the nitrogenase reaction, but the methane is detected in the assay since it elutes from the GC column with a retention time shorter than that for ethylene (compare Fig. 1). Accordingly, high impurities may overlap the ethylene tracing. Some ethylene present as a contaminant in commercial acetylene usually is not critical either, provided that measurements are performed as kinetics as indicated (and not as end-point determinations), although the sensitivity of determination is decreased. Alternatively, acetylene may be synthesized in the laboratory from calcium carbide (CaC_2) by simply addition of water [Eq. (2)]. For synthesis of small amounts of acetylene this is a recommended procedure; the purity of the acetylene gas formed, however, depends on the purity of calcium carbide. It should be mentioned that the highly reactive calcium carbide should be handled with care.

$$C_2H_2 + 2\,H^+ + 2\,e^- \rightarrow C_2H_4 \tag{1}$$

$$CaC_2 + 2\,H_2O \rightarrow Ca(OH)_2 + C_2H_2\uparrow \tag{2}$$

A routine assay for determination of nitrogenase activity in the heterocystous cyanobacterium *Anabaena variabilis* (ATCC 29413) using the

[4] R. Rippka, J. Deruelles, J. B. Waterbury, M. Herdman, and R. Y. Stanier, *J. Gen. Microbiol.* **111,** 1 (1979).

[5] L. J. Stal and W. E. Krumbein, *Arch. Microbiol.* **143,** 67 (1985).

[6] C. Leon, S. Kumazawa, and A. Mitsui, *Curr. Microbiol.* **13,** 149 (1986).

[7] A. Mitsui and S. Kumazawa, this volume [50].

[8] M. J. Dilworth, *Biochim. Biophys. Acta* **127,** 285 (1966).

[9] R. Schöllhorn and R. H. Burris, *Fed. Proc., Fed. Am. Soc. Exp. Biol.* **25,** 710 (1966).

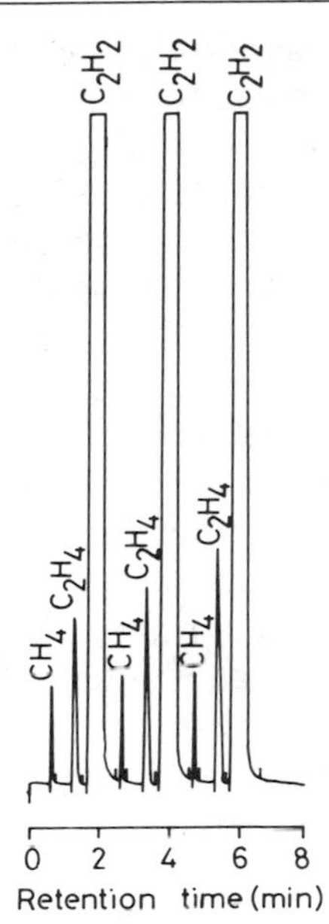

FIG. 1. Gas chromatographic analysis of three samples containing methane, ethylene, and acetylene. Samples (0.2 ml) were withdrawn from a reaction vesseL (total volume 36 ml) containing 5 ml of a cyanobacterial suspension. The gas phase of the reaction vessel consisted of air plus 12% (v/v) acetylene (slightly contaminated by methane). In 2-min intervals the samples were injected into a Carlo Erba gas chromatograph (S 2350) equipped with a flame-ionizing detector and a Hewlett Packard integrator (HP 3385). A 3-m Porapak N column (mesh 80–100, internal diameter 3 mm) and N_2 as carrier gas were used. The gas flow for the flame-ionizing detector was 30 ml H_2 min^{-1} and 300 ml N_2 min^{-1}. The column temperature was 90° and detector temperature was 150°.

acetylene reduction method may be performed as follows: 5–10 ml of cyanobacterial suspension (chlorophyll concentration 1–20 μg ml^{-1} suspended in the culture medium or suitable buffer (e.g., 20 m*M* Tris–HCl, pH 7.8) is incubated in glass vessels ~30-ml volume) sealed with rubber stoppers. After addition of acetylene to the gas phase (final concentration 10–20% C_2H_2, v/v), the vessels are incubated in the light (200 μE m^{-2} sec^{-1}) at 30° and gently shaken. Samples of 0.2 ml are withdrawn from the gas phase (using a gas-tight syringe) in 5-min intervals and injected into a gas chromatograph to determine ethylene. It is advantageous to follow the rate of ethylene formation by kinetic measurements instead of end-point determinations (Fig. 1). This is especially important when measuring dark nitrogenase activity, which is dependent on the amount of endogenous storage compounds. After depletion of these compounds nitrogenase activity ceases, resulting in nonlinear kinetics.

With respect to a conversion factor between N_2 fixation and acetylene reduction, reported figures range from 2 to 6,[10–12] whereas the theoretical value is 3 (N_2 reduction to 2 NH_3 requires six electrons; in contrast, reduction of C_2H_2 to C_2H_4 requires two electrons). Considering the theoretical factor, it has to be taken into account that N_2 fixation is always accompanied by hydrogen formation due to concomitant reduction of protons.[13] A stoichiometry of 1 mol hydrogen formed per mole dinitrogen reduced is postulated. Therefore a conversion factor of 4 appears to be appropriate, as was recently reported.[14]

Besides acetylene reduction, nitrogenase activity can be followed as hydrogen evolution, provided cyanobacteria are incubated in an argon atmosphere in the light. In the absence of reducible substrates such as N_2 or C_2H_2, nitrogenase exclusively reduces protons, resulting in enhanced hydrogen evolution from the cyanobacteria. Owing to concomitant hydrogen uptake activity catalyzed by hydrogenase, however, the rate of hydrogen formation measured is a net rate only, i.e., the total amount of hydrogen formed minus the amount of hydrogen taken up by hydrogenase. Since nitrogenase and hydrogenase activities vary, a reliable estimation of nitrogenase activity by measuring net hydrogen production is not possible.

There are other methods to determine nitrogenase activity in intact cyanobacteria; however, the majority are not applicable for routine measurements. Direct measurement of N_2 fixation has been performed using the radioisotope ^{13}N.[15] Owing to the very short half-life of ^{13}N (10 min), this method can be used only in laboratories close to a reactor producing this isotope. Promising results have been obtained using membrane-leak mass spectrometry[14]; this method, however, requires comparatively expensive and complicated equipment, so that its application for routine measurements is limited. A simple method is to measure the increase of nitrogen content of a growing cyanobacterial culture, e.g., by determining total cellular nitrogen using the Kjeldahl method (data are shown in Table I).[16] An error may be the possible excretion of ammonia by cyanobacteria,

[10] W. D. P. Stewart, G. P. Fitzgerald, and R. H. Burris, *Arch. Microbiol.* **62,** 336 (1968).

[11] R. H. Burris, *Aust. J. Plant Physiol.* **3,** 41 (1976).

[12] R. B. Peterson and R. H. Burris, *Anal. Biochem.* **73,** 404 (1976).

[13] J. Chatt, *in* "Nitrogen Fixation" (W. D. P. Stewart and J. R. Gallon, eds.), p. 1. Academic Press, London, 1980.

[14] B. B. Jensen and R. P. Cox, *Appl. Environ. Microbiol.* **45,** 1331 (1983).

[15] J. Thomas, J. C. Meeks, C. P. Wolk, P. W. Shaffer, S. M. Austin, and W.-S. Chien, *J. Bacteriol.* **129,** 1545 (1977).

[16] D. I. Arnon, B. D. McSwain, H. Y. Tsujimoto, and K. Wada, *Biochim. Biophys. Acta* **357,** 231(1974).

TABLE I
DETERMINATION OF NITROGENASE ACTIVITY IN *Anabaena variabilis* BY MEASURING THE INCREASE OF THE TOTAL NITROGEN CONTENT OF FILAMENTS[a]

Cultivation time (hr)	Nitrogen content of suspension (μmol N/ml)	Nitrogenase activity (μmol N/ml × hr)
24	2.13	—
48	10.86	0.36
53	13.08	0.44
72	20.59	0.39

so the nitrogen content of the culture medium should be controlled. Ammonia cannot be quantitatively measured in open-culture facilities allowing diffusion of ammonia into the surrounding gas phase.

Induction of Nitrogenase and Hydrogenase Activity

Heterocystous Cyanobacteria

Nitrogenase is exclusively located with heterocysts, specialized cells which differentiate from vegetative cells.[17] Nitrogenase is induced by limitation of combined nitrogen, a process counteracted by combined nitrogen (NO_3^-) or NH_4^+ when present in the culture medium. Nitrate, however, does not completely suppress nitrogenase induction and concomitant formation of heterocysts when cyanobacteria are cultivated under laboratory conditions. Even in the presence of 20 m*M* NO_3^- some heterocysts and significant nitrogenase activity are observed. In contrast, ammonia totally inhibits induction of nitrogenase activity when present in concentrations of about 1 m*M*. Cyanobacterial cultures grow rapidly when incubated under laboratory conditions at comparatively high light intensities (200 $\mu E\ m^{-2}\ sec^{-1}$). Obviously uptake of nitrate and/or reduction of nitrate to ammonia are limiting processes, and shortage of ammonia nitrogen induces induction of nitrogenase activity.

Concurrently with nitrogenase activity an uptake hydrogenase is often induced in the heterocysts. Apparently, induction of this uptake hydrogenase does not occur if there is a large pool of carbohydrate storage material (glycogen) present in the filaments or a deficiency of nickel.

[17] R. L. Smith, C. Van Baalen, and F. R. Tabita, this volume [51]; Bothe and G. Neuer, this volume [52].

Uptake hydrogenase from heterocystous cyanobacteria has been shown to be a nickel-requiring enzyme,[18,19] as are several irreversible hydrogenases from other bacteria.[20] Only very small concentrations of nickel are required in the culture medium (less than 50 μM), however, and addition of nickel to cultures is usually not necessary because other minerals of the culture medium are contaminated with trace amounts of nickel.

Nonheterocystous Filamentous Cyanobacteria

Apparently, induction of nitrogenase activity in nonheterocystous cyanobacteria requires a sequential separation between oxygen-labile nitrogen fixation and oxygen-producing photosynthesis. In addition, a microaerobic environment has to be provided to allow for enzyme activity. Since oxygenic photosynthesis cannot contribute directly to the demand of nitrogenase for reduction equivalents, large pools of carbohydrate storage material (glycogen) are necessary to support respiratory energy production. In our laboratory the phenomena involved were studied in the induction of nitrogenase activity in *Phormidium foveolarum* (SAUG 1462-1), a filamentous nonheterocystous cyanobacterium.[21] A routine procedure for induction of nitrogenase is described below, and with slight modifications this procedure will be applicable for other nonheterocystous species. Modifications will be necessary with respect to a slightly altered timetable owing to species-specific growth rates and different culture facilities.

Induction of Nitrogenase in Phormidium foveolarum. Liquid stock cultures of *Phormidium* are grown aerobically under continuous illumination (200 μE m^{-2} sec^{-1}) in a medium according to Arnon *et al.*[16] containing 20 mM NO_3^-. Cultures are inoculated with an initial chlorophyll concentration of 1 μg Chl ml^{-1} and continuously sparged with air enriched with 1.6% CO_2 (v/v). After a 2- or 3-day cultivation period, the *Phormidium* culture is transferred to a nitrate-free medium and cultivated for a further 24 hr under aerobic conditions. Owing to lack of combined nitrogen, phycobiliproteins are degraded and the color of the culture gradually turns yellowish green. Alternatively, *Phormidium* may be grown in a medium containing a limited amount of ammonia (2 mM). After depletion of ammonia, nitrogenase is formed when *Phormidium* is incubated in a microaerobic environment; heterocysts are never observed.

For the assay 10 ml (chlorophyll concentration 20 μg ml^{-1}) is trans-

[18] A. Daday and G. D. Smith, *FEMS Microbiol. Lett.* **20,** 327 (1983).
[19] H. Almon and P. Böger, *Z. Naturforsch. C* **39,** 90 (1984).
[20] R. K. Thauer, *Biol. Chem. Hoppe-Seyler* **366,** 103 (1985).
[21] H. Weisshaar and P. Böger, *Arch. Microbiol.* **136,** 270 (1983).

ferred to glass vessels (total volume 36 ml) sealed with rubber stoppers. After making these vessels anaerobic by sparging with argon, 3 ml acetylene is added, and the vessels are incubated and gently shaken in a Warburg apparatus. To allow for induction of oxygen-labile nitrogenase activity, an equilibrium between photosynthetic oxygen evolution and respiratory oxygen uptake has to be provided. This is achieved either by low light intensity (20 $\mu E\ m^{-2}\ sec^{-1}$) or by applying a higher light intensity (100 $\mu E\ m^{-2}\ sec^{-1}$) with 3,4-dichlorophenyl-1,1-dimethylurea (DCMU) (5×10^{-6} *M*) present. After 1–3 hr nitrogenase activity can be detected as ethylene formation. Induction of nitrogenase activity can also be followed as hydrogen evolution (compare Weisshaar and Böger[21]).

Hydrogen Metabolism

Hydrogen gas exchange is closely related to nitrogenase activity, since hydrogen formation is an obligatory side reaction of nitrogen fixation. As mentioned above, hydrogen produced by nitrogenase is generally taken up again by a hydrogenase, so that net hydrogen evolution by nitrogen-fixing cyanobacteria is hardly observed (at least under aerobic conditions). Apart from nitrogenase and uptake hydrogenase, the existence of a reversible hydrogenase in cyanobacteria is generally accepted[3] (for details on cyanobacterial hydrogenases, cf. Ref. 22). A variety of methods for measurement of hydrogen gas exchange in cyanobacteria have been employed. In our hands, measurements by gas chromatography gave best results with respect to reliability, sensitivity, and handling. Hydrogen is detected using a gas chromatograph equipped with a thermal conductivity detector and a molecular sieve column (for details, see the legend to Fig. 2). Gas chromatographic analysis allows for simultaneous detection of H_2, O_2, and N_2 in an identical sample, and this information often is helpful.

A routine assay for measurement of net hydrogen formation by *Anabaena variabilis* corresponds to the assay for acetylene reduction (see above). The only difference is use of an argon gas phase (instead of air) and omission of acetylene. In addition to hydrogen formation, gas chromatography can be used to measure hydrogen uptake activities, in which case assays with intact cyanobacteria should contain about 5% H_2 (v/v) in the gas phase of the reaction vessels. Uptake of hydrogen is followed by withdrawal of aliquots from the gas phase and subsequent analysis.[23,24]

[22] K. K. Rao and D. O. Hall, this volume [53].
[23] H. Weisshaar and P. Böger, *Arch. Microbiol.* **142,** 349 (1985).
[24] P.-C. Chen, H. Almon, and P. Böger, *FEMS Microbiol. Lett.* **37,** 45(1986).

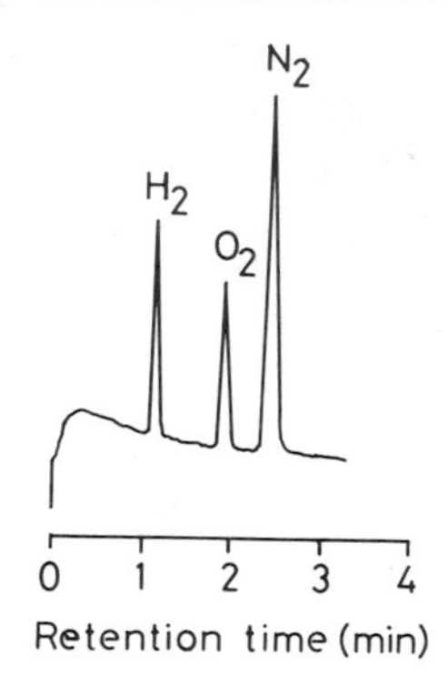

FIG. 2. Gas chromatographic analysis of a sample containing hydrogen, oxygen, and nitrogen. This figure illustrates the sensitivity of gas chromatographic determination. A clear separation between the three gases can be achieved. For determination of rates of hydrogen evolution (or hydrogen uptake), kinetic measurements (as illustrated in Fig. 1) are recommended. A sample (0.2 ml) containing 4 nmol H_2, 17 nmol O_2, and 39 nmol N_2 was injected into a Hewlett Packard gas chromatograph (Model 5880 A) equipped with a thermal conductivity detector. Gases were separated on a molecular sieve column (5 Å, length 2 m, internal diameter 3 mm) flushed with argon as carrier gas. The column temperature was 90°, injector temperature 90°, and detector temperature 200°.

Hydrogen concentrations above 10% (v/v) are not suitable because the presence of a large hydrogen signal does not allow for quantitative detection of comparatively small amounts of hydrogen which are taken up.

Hydrogen gas exchange in intact cyanobacteria can also be measured using a hydrogen electrode.[25,26] This is a rapid method allowing determination of hydrogen gas exchange rates within minutes, whereas gas chromatographic assays often require at least half an hour to give reliable results. A major problem is the diffusion of hydrogen out of the reaction measuring chamber, thus influencing especially the measurement of hydrogen uptake. Mass spectrometric techniques for hydrogen determination are described elsewhere in this volume[27] and are not discussed here.

Nitrogenase and hydrogenase(s) are involved in H_2 gas metabolism, rendering hydrogen gas exchange determinations difficult. Therefore, reactions catalyzed by nitrogenase or hydrogenase(s), respectively, must generally be distinguished. With respect to hydrogen evolution this is comparatively simple, because nitrogenase-catalyzed hydrogen evolution is an energy-requiring process and hence sensitive to uncouplers. Hydro-

[25] W. J. Sweet, J. P. Houchins, P. R. Rosen, and D. J. Arp, *Anal. Biochem.* **107,** 337 (1980).
[26] R. T. Wang, this series, Vol.69, p. 409.
[27] B. B. Jensen and R. B. Cox, this volume [48].

gen formation by a reversible hydrogenase, however, is not an ATP-consuming process. With respect to hydrogen uptake activities by intact cells, it is rather difficult, often impossible, to determine whether hydrogen taken up is consumed by an uptake hydrogenase, a reversible hydrogenase, or possibly by nitrogenase itself.[23,24]

Acknowledgments

Our research is supported by the Deutsche Forschungsgemeinschaft.

[48] Measurement of Hydrogen Exchange and Nitrogen Uptake by Mass Spectrometry

By BENT B. JENSEN and RAYMOND P. COX

Introduction

Measuring techniques allowing continuous monitoring of the parameter of interest are conceptually superior to those involving discrete sampling. Nitrogenase activity is usually measured by the acetylene reduction assay in which discrete samples of the gas phase are injected into a gas chromatograph for analysis of ethylene. This is an extremely sensitive method, but it is discontinuous and also involves the use of an artificial substrate which is inferior to direct measurements of the reduction of dinitrogen.

Gas analysis mass spectrometry involves the use of analyzers with quadrupole filters to separate ions of different mass-to-charge (m/z) ratios. A typical apparatus can separate ions up to $m/z = 100$ with a resolution of 1 mass unit. Such analyzers are relatively inexpensive (upward of $15,000 for a complete system), reliable, and compact. In combination with the use of an inlet system with a gas-permeable membrane they allow continuous monitoring in the gas or liquid phase of all gases of interest in biological investigations.[1,2] In the case of cyanobacterial photosynthesis and nitrogen fixation it is possible to make continuous measurements of H_2, N_2, and O_2.[3] In addition, CO_2 can be measured although only free CO_2 will penetrate the membrane so that measurements will also be affected by pH-dependent changes in the HCO_3^-/CO_2 equilibrium.

[1] R. Radmer and O. Ollinger, this series, Vol. 69, p. 547.

[2] H. Degn, R. P. Cox, and D. Lloyd, *Methods Biochem. Anal.* **31,** 165 (1985).

[3] B. B. Jensen and R. P. Cox, *Appl. Environ. Microbiol.* **45,** 1331 (1983).

Apparatus

A measuring system for dissolved gas analysis by membrane-inlet mass spectrometry consists of four components: (1) a quadrupole mass spectrometer, (2) a vacuum pumping system, (3) vacuum components to connect these two, and (4) a membrane-inlet system. The first three are commercially available while the last must be specially constructed.

Mass Spectrometer

Quadrupole mass spectrometers are available from a variety of manufacturers. We have used apparatus from Spectramass (Congleton, UK) or VG Gas Analysis (Winsford, UK). The latest generation of quadrupole mass spectrometers have built-in microprocessors with software allowing facilities such as automatic following of peak maxima. It is important to ensure that it is possible to obtain a hard copy of the changes in at least four m/z values with time, either via analog outputs to a recorder or through data transfer to a suitable computer. Faraday detectors are more stable than secondary electron multipliers (SEM) and have adequate sensitivity for most measurements of cyanobacterial gas exchange. An SEM is only necessary for analyses of trace components.

Pump and Vacuum Fittings

The quadrupole head and ion source of the mass spectrometer (MS) are inserted into one end of a T piece, and the inlet is attached to the other arm. The base of the T is connected to a turbomolecular pump unit [TSH 040 from Balzers (Balzers, FL)]. Turbomolecular pumps are the most satisfactory means to obtain the necessary pressure reduction in the measuring system. A nominal capacity of 40 liter/s is quite adequate.

The tightness of the vacuum fittings is important since it determines the size of the background signals due to the components of laboratory air. From this point of view joints with copper gaskets are preferable to those involving Viton O rings, particularly for the larger diameter joints.

Inlet Designs

Silicone Tubing Inlet. The simplest type of inlet system which we have employed consists of a stainless steel tube with holes covered by thin-walled silicone rubber tubing. Stainless steel catheter tubing (1.6 mm i.d., 2.0 mm o.d.) is welded closed at one end and attached to stainless steel tube of 6.0 mm o.d. at the other end. A number of small circular holes 0.4 mm in diameter are drilled near the closed end. The number of holes can be varied to obtain the optimal partial pressure of the gas to be measured inside the MS vacuum. This should be so large that the noise on the background electrical signal is negligible but not so large that the total pressure exceeds the limit for linear operation of the MS (normally 10^{-5} torr). We normally use 10 holes. These holes are covered by a length of silicone rubber tubing (Silastic #602-235 from Dow-Corning, Midland, MI). This has an internal diameter of 1.47 mm and a wall thickness of 0.25 mm. The tubing is swollen by placing it in toluene or xylene for a few minutes before sliding it onto the metal tube. This allows easy application and a tight fit once the solvent has evaporated and the tubing has shrunk again. The inlet probe is attached to the MS vacuum using 0.25-in. Ultra-Torr fittings with Viton O rings (Cajon, Macedonia, OH) and a length of flexible stainless steel vacuum tube (0.25 in. o.d.) from the same manufacturer.

Since the silicone tubing inlet can be sterilized by autoclaving, it is possible to make measurements on growing microbial suspensions in bioreactors.[4] Alcohols can penetrate a silicone rubber membrane and be detected by the mass spectrometer. This fact needs to be borne in mind if ethanol is used as a solvent for added reagents.

Teflon Membrane Inlet. An alternative type of membrane-covered inlet consists of a stainless steel tube (6.0 mm o.d., 4.0 mm i.d.) with a circular piece of sintered glass inserted into one end and the other end connected directly to the MS vacuum. The membrane is a piece of Teflon sheet (25 μm thick) which is supported by the sintered glass and held in place by a rubber O ring. Some practice is necessary to apply the membrane without rupturing the sheet.

Teflon has the advantage over silicone rubber of being about 7 times less permeable to water compared to N_2. This means that the background signal from water at $m/z = 2$ is decreased, which improves the accuracy of hydrogen measurements. Teflon is not permeable to alcohols.

Measurements

Closed Reaction Vessels

The silicone-tubing membrane inlet probe is inserted through the plexiglass cap of the reaction vessel as shown in Fig. 1. This system allows the volume to be adjusted so that the liquid surface is in the capillary tube and permits reagents to be added using a suitable microsyringe. Figure 2

[4] R. P. Cox, *in* "Mass Spectrometry in Biotechnological Process Analysis and Control" (E. Heinzle and M. Reuss, eds.), p. 63. Plenum, New York, 1987.

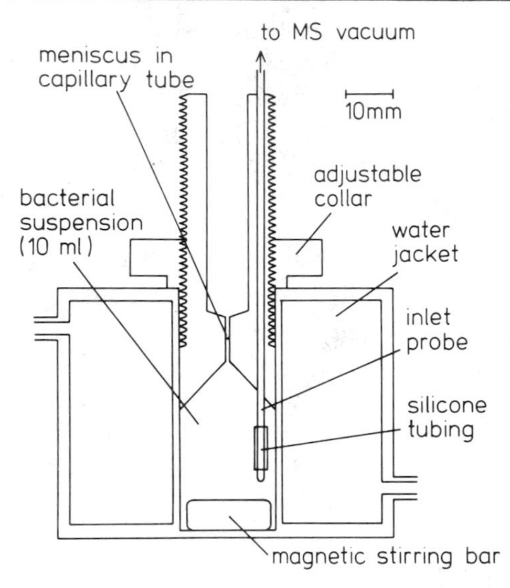

FIG. 1. Diagram of the closed reaction vessel with the membrane tubing inlet to the mass spectrometer.

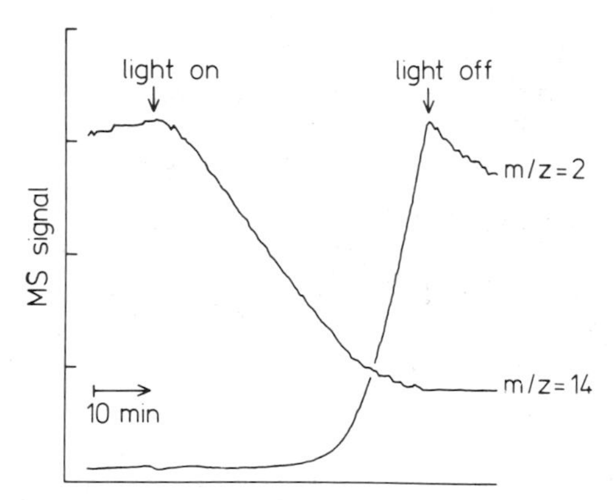

FIG. 2. Changes in the mass spectrometer signals at $m/z = 14$ (N_2) and $m/z = 2$ (H_2) in an illuminated cyanobacterial suspension. The reaction vessel contained *Anabaena variabilis* ATCC 29413 (60 μg chlorophyll/ml) resuspended in fresh growth medium supplemented with 10 m*M* HEPPSO buffer, pH 7.5 (NaOH), and 10 m*M* fructose, to which 2.5 μM 3,4-dichlorophenyl-1,1-dimethylurea (DCMU) had been added to inhibit photosynthetic oxygen evolution. At the point shown by the arrow, the suspension was illuminated from two diametrically opposed points with yellow light (Kodak Wratten 16 filter). The signal at $m/z = 32$ due to O_2 (not shown) decreased to a value indistinguishable from zero as a result of cyanobacterial respiration before the start of the experiment and remained at this value throughout the measurements. The vertical scale divisions correspond to 10^{-9} torr for $m/z = 14$ and 4×10^{-8} torr for $m/z = 2$. Temperature was 35°.

shows an example of light-induced N_2 uptake and H_2 production measured with this system. N_2 is measured at $m/z = 14$ rather than at the molecular ion ($m/z = 28$) because the latter is subject to interference by the CO^+ fragment from CO_2. As the N_2 concentration is decreased in the closed system, the rate of H_2 production increases until it exceeds the capacity of the hydrogen uptake systems. Once this point is reached the hydrogen concentration increases and inhibits N_2 reduction by nitrogenase. This process is self-reinforcing until the only reaction observed is H_2 evolution.

The H_2 uptake observed when the light is switched off is the result of loss through the membrane into the mass spectrometer vacuum. This "self-consumption" shows first-order kinetics and is typically about 10 times faster for H_2 than for N_2. It can be minimized by using an inlet with a small number of holes and a large reaction volume. Using two holes, 0.4 mm in diameter, and a 6 ml reaction volume, the half-time for self-consumption of H_2 is about 10 hr.

The same system can also be used for direct measurements of the reduction of acetylene to ethylene. The molecular ion of C_2H_4 is at $m/z = 28$ which is the same as N_2, and acetylene itself causes a background if measurements are made at $m/z = 27$. These problems can be avoided if C_2D_2 is used.[5] This can be easily made by the addition of D_2O to CaC_2, and the $C_2D_2H_2$ produced by nitrogenase can be measured at $m/z = 30$ without interference.

Open Measuring Systems

Although a closed system allows rapid measurements of initial rates, it suffers from the limitation that steady states are not attained. A system in which the cyanobacterial suspension is in contact with a gas phase is much more flexible. In this case steady states will be obtained where the measured concentration of a gas which is consumed or produced as the result of cell metabolism differs from the equilibrium value observed in the absence of the cells. The difference between these two values is proportional to the rate of exchange, and the proportionality constant is the mass transfer coefficient for transfer between the gas and the liquid phases. The theory such open-system measurements has been discussed in detail by Degn *et al.*[6]

A pseudo-open system[7,8] can be constructed by having a closed con-

[5] Y. M. Berlier and P. A. Lespinat, *Arch. Microbiol.* **125,** 67 (1980).

[6] H. Degn, J. S. Lundsgaard, L. C. Petersen, and A. Ormicki, *Methods Biochem. Anal.* **23,** 47 (1980).

[7] A. Hochman and R. H. Burris, *J. Bacteriol.* **147,** 492 (1981).

[8] B. B. Jensen and R. P. Cox, *Arch. Microbiol.* **135,** 287 (1983).

tainer with a small volume of liquid stirred by a magnetic stirring bar in the bottom in contact with a gas phase of perhaps 10 times the volume. In this case the amount of a gas such as O_2 present in the gas phase (about 1 mmol/100 ml in the case of air) can be many times greater than that dissolved in the liquid (~2.5 μmol/10 ml) and pseudo-steady states will be attained. The silicone membrane inlet probe can easily be inserted through a rubber serum stopper of such a pseudo-open system into the liquid phase. In a pseudo-open system the composition of the gas phase may be readily altered using a syringe.

In the majority of our experiments we have used an open system in which the cyanobacterial suspension is in contact with a mobile gas phase.[3] The Teflon membrane-covered inlet is inserted into the side of a stainless steel reaction vessel with a plexiglass bottom to allow for illumination.[2] The chamber has a diameter of 25 mm, and the sample (4.5 ml) is stirred by a cross-shaped stirrer on an axle connected to a motor mounted above the chamber. A downward mounted plexiglass cone on the axle prevents the formation of a vortex, which would cause variation in the mass transfer coefficient. Such an arrangement is necessary for optimal operation of an open system with a stable mass transfer coefficient, but magnetically coupled stirring as used in the pseudo-open system is adequate for all but the most critical applications as long as the formation of a vortex is avoided.

The use of a suitable gas-mixing system is essential to obtain full benefit of the possibilities of an open system with a mobile gas phase. We use the digital system described by Lundsgaard and Degn.[9]

Use of Stable Isotopes

Mass spectrometry opens the possibility of the simultaneous measurement of the uptake and production of the same compound by monitoring two isotopic forms of the same element. In studies of cyanobacterial diazotrophy it is possible to monitor the simultaneous production and uptake of hydrogen using the stable isotope deuterium. Experiments can be performed either with a gas phase containing D_2 or medium containing D_2O. With a gas phase containing D_2, experiments are in practice limited by cost to the use of a closed or pseudo-open system.

An open system can be used if the cyanobacteria are suspended in heavy water and the mobile gas phase contains hydrogen. Such an experiment is shown in Fig. 3. In the absence of N_2 there is an extensive production of D_2. This is decreased in the presence of 10% N_2 (not sufficient to saturate nitrogenase) and falls to zero in the dark. D_2 can be measured without interference at m/z = 4, but measurements of H_2 re-

[9] J. S. Lundsgaard and H. Degn, *IEEE Trans. Biomed. Eng.* **20,** 384 (1973).

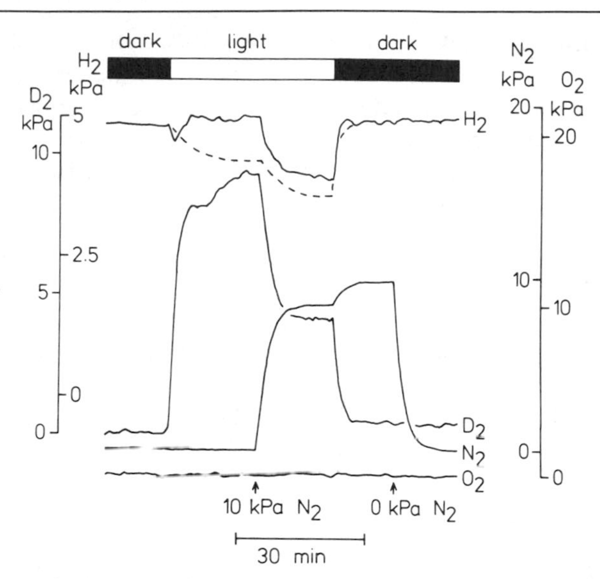

FIG. 3. Changes in the mass spectrometer signals at $m/z = 2$, $m/z = 4$ (D_2), and $m/z = 14$ (N_2) in a suspension of *Anabaena variabilis* ATCC 29413 (100 μg Chl/ml) in an open reaction vessel. The cyanobacteria were suspended in D_2O containing 50 m*M* HEPES buffer, pH 7.5 (KOH), and 2 m*M* $MgCl_2$. The gas phase was Ar with 5% H_2 and N_2 as shown. Illumination was from below with yellow light (Kodak Wratten 16 filter). The dashed line is the H_2 concentration obtained from the mass spectrometer signal at $m/z = 2$ corrected for the contribution due to D_2. Temperature was 35°.

quire correction of the signal at $m/z = 2$ to allow for the contribution due to D_2. The corrected signal shows H_2 uptake in the light at the expense of the reduction of both dinitrogen and (heavy) protons.

Ideally, experiments of this type are designed so that one isotope is vastly in excess in each form, e.g., $[D^+] >> [H^+]$ and $[D_2] << [H_2]$. This is not the case here, where D_2 accumulates to a steady-state concentration greater than H_2. Corrections could be made for this if the reuptake of D_2 is assumed to be the same as the uptake of H_2; however, the possibility of isotope effects, and nonequilibration between the intracellular gas concentration in heterocysts and that in the medium, needs to be borne in mind.

Concluding Remarks

Although the use of mass spectrometry for dissolved gas measurements in photosynthetic systems is more than 20 years old[10] and measure-

[10] G. Hoch and B. Kok, *Arch. Biochem. Biophys.* **101,** 160 (1963).

ments of dinitrogen uptake were reported in 1974,[11] the technique has not yet been widely applied. Although more expensive and less sensitive than the traditional acetylene reduction assay, the use of membrane-inlet mass spectrometry opens the possibility of new types of experimental investigations on the relationships among nitrogen fixation, hydrogen exchange, photosynthesis, and respiration in cyanobacteria.

Acknowledgments

This work was supported by the Danish Natural Sciences Research Council, the Danish Council for Scientific and Industrial Research, and Direktør Ib Henriksens Fond.

[11] H. Paschinger, *Arch. Microbiol.* **101,** 379 (1974).

[49] Nitrogen Fixation in Cyanobacterial Mats

By LUCAS J. STAL

Introduction

Intertidal sediments, hot springs, salt ponds, salt marshes, and mangrove forest sediments are often characterized by dense populations of cyanobacteria. Benthic filamentous cyanobacteria may form rigid structures, called cyanobacterial mats. Cyanobacterial mats usually contain a very high biomass, and in many environments, especially intertidal sediments, the availability of combined nitrogen alone is too low to allow development of the mat. Therefore, biological nitrogen fixation in many mats will be of paramount importance.

Many cyanobacteria are known to fix nitrogen. Not only cyanobacteria that differentiate heterocysts, but also several unicellular and filamentous cyanobacteria without heterocysts have been shown to fix nitrogen. The majority of cyanobacterial mats are built by filamentous, nonheterocystous cyanobacteria. Therefore, it can be assumed that in many mats nitrogen fixation is performed by nonheterocystous organisms. However, the physiological differences between the mechanisms of oxygen protection of nitrogenase of heterocystous and nonheterocystous cyanobacteria deserve special attention when measuring nitrogenase activity in the mat. Nitrogenase activity and oxygenic photosynthesis in heterocystous cyanobacteria are separated spatially. Therefore, under natural conditions, nitrogenase activity in heterocystous cyanobacteria is usually strictly light dependent and will show a close correlation with light intensity.

Although few data have been available up to now, it seems that, in analogy with heterocystous species, the nonheterocystous, nitrogen-fixing cyanobacteria separate oxygenic photosynthesis from nitrogen fixation temporally. Under a light–dark regime, nitrogen fixation predominantly occurs during the dark period and photosynthesis during the light period. Cyanobacterial mats, however, usually turn anaerobic during the night. Because anaerobic dark conditions are energetically not very favorable to the expensive process of nitrogen fixation, activity might be negligible in the dark. A diurnal cycling of the nitrogenase activity in mats of *Microcoleus chthonoplastes* and *Oscillatoria* sp. with activity peaks at sunrise and sunset was observed, being virtually zero at midday and during the night. These phenomena have to be taken into account when measuring nitrogenase activity in cyanobacterial mats. This is particularly important if data are extrapolated to the contribution of N_2 fixation to the total nitrogen budget.

Methods of Assay of Nitrogenase

For field measurements of nitrogenase activity only two methods are suitable. These are the incorporation of the heavy isotope ^{15}N and the reduction of acetylene to ethylene.

Acetylene Reduction Technique for Assay of Nitrogenase

The acetylene reduction technique[1–3] is convenient, inexpensive, and more sensitive than the ^{15}N method and is therefore now commonly used, both in the laboratory and in field studies. Although an indirect method, the reduction of acetylene to ethylene seems very specific for all nitrogen-fixing organisms, and no other biological process is known which can carry out this reaction. Some notes of caution have been made for using this technique in the field. Ethylene might be a substrate for methane-oxidizing bacteria,[4] and it has been shown that some bacteria may utilize acetylene.[5] These processes, however, are very slow and will interfere with nitrogenase measurements only with prolonged incubation periods.

[1] R. Schöllhorn and R. H. Burris, *Fed. Proc., Fed. Am. Soc. Exp. Biol.* **25,** 710 (1966).
[2] M. J. Dilworth, *Biochim. Biophys. Acta* **127,** 285 (1966).
[3] W. D. P. Stewart, G. P. Fitzgerald, and R. H. Burris, *Proc. Natl. Acad. Sci. U.S.A.* **58,** 2071 (1967).
[4] R. J. Flett, J. W. M. Rudd, and R. D. Hamilton, *Appl. Microbiol.* **29,** 580 (1975).
[5] T.-Y. Tam, C. I. Mayfield, and W. E. Inniss, *Curr. Microbiol.* **8,** 165 (1983).

Nevertheless, it would be wise to check for such interferences in the cyanobacterial mat to be investigated. I have not found utilization of ethylene or acetylene other than by nitrogenase-mediated reduction of acetylene in cyanobacterial mats during incubations up to 6 hr.

Acetylene and ethylene can be determined with a gas chromatograph equipped with a flame-ionization detector (FID). A 3-m-long glass or steel column with an inner diameter of 2 mm is packed with Porapak Type R, mesh 80–100. Carrier gas is N_2 at a flow rate of 30 ml min^{-1}. The flow of hydrogen and air for the FID are 20 and 200 ml min^{-1}, respectively. The oven temperature is set at 35°; the injector and detector temperatures are 70 and 90°, respectively. The gas sample to be injected may range from 100 to 500 μl, but usually 200 μl of gas sample gives satisfactory results. Under these conditions, the retention times for ethylene and acetylene are 1.9 and 2.7 min, respectively. After about 4 min, the acetylene is completely eluted from the column (with 20% C_2H_2, elution will be faster if less acetylene is used per assay). The detection limit of ethylene under these conditions is about 10^{-10} mol C_2H_4. The gas chromatograph should be calibrated using commercial or custom-made standards. For ethylene, a standard of 100 ppm and acetylene 10–20% (depending on the concentration used in the assay) is recommended. For reasons of accuracy, a commercial standard is recommended (e.g., 100 ppm C_2H_4 in helium, Scotty Specialty Gases, Troy, MI).

A considerably faster analysis of ethylene can be achieved by precipitating acetylene by adding ammoniacal silver nitrate.[6] An ammonia solution (25% w/w NH_3) is added dropwise to an aqueous solution of silver nitrate (0.5 g in 5 ml H_2O) until the precipitate just dissolves. The ammoniacal silver nitrate solution is then made up to 25 ml with H_2O. The solution should be stored in a cool, dark place. Ammoniacal silver nitrate solution is added to a final concentration of 10 mg ml^{-1} and shaken for 5 min. This precipitates acetylene quantitatively as silver carbide in about 2 min. Moreover, the ammoniacal silver nitrate solution will terminate the assay and replace, e.g., trichloroacetic acid. The gas chromatographic assay of ethylene may be accomplished in less than 1 min, and injection of the subsequent sample can be done immediately after ethylene is eluted from the column. The disadvantage of the method is that acetylene cannot be used as an internal standard, which makes quantitative nitrogenase assay more difficult.

In the nitrogenase assay, acetylene is added to a concentration of 10–20% of the gas phase. Lower concentrations may underestimate the activ-

[6] K. A. V. David, S. K. Apte, A. Banerji, and J. Thomas, *Appl. Environ. Microbiol.* **39,** 1078 (1980).

ity, especially if N_2 is present. The affinity of nitrogenase for acetylene is much higher than for N_2. Therefore, the acetylene reduction assay can be done in the presence of atmospheric concentrations of N_2 without underestimating the activity. Higher concentrations up to 50% acetylene do not affect the assay but are not necessary. Because acetylene and ethylene are very soluble in aqueous solutions, a high gas-to-liquid ratio is recommended. If the gas phase is at least 5 times the volume of the liquid, the error caused by the solubility of the gasses can be neglected. Of course, a high gas-to-liquid ratio will considerably decrease the limit of detection of ethylene production. This can be a serious problem when measuring nitrogenase in bodies of water with low biomass, but is usually not significant in the case of dense cyanobacterial mats.

In cyanobacterial mats, two different methods of nitrogenase measurement can be used. These are the bell-jar method for *in situ* measurements and the cork-borer sampling technique.

The Bell-Jar Method. The bell-jar method can be used in cyanobacterial mats with minimum disturbance of the sediments. This method is suitable for overall nitrogenase activity measurements without need of sample manipulation. Bottomless glass bottles are used with a sharp, grinded edge. We use 50-ml injection bottles with a narrow opening and the bottom cut off. These bottles are available in various sizes, but smaller bottles are not recommended. The bottles are carefully pushed into the sediment until approximately one-third of the bottle is above the sediment surface. The bottle is then sealed with a rubber stopper and secured with an aluminum seal using a seal crimper. Screw-cap bottles can also be used with rubber septa and perforated caps. It is important not to push presealed bottles into the sediment, because of the considerable increase in pressure which will disturb physicochemical gradients in the mat.

After sealing the bottle, acetylene is injected slowly, using a gas-tight syringe. The volume of acetylene should be 10–20% of the gas phase (2.5 ml C_2H_2). When acetylene is used as an internal standard (see below), it is not necessary to determine the exact volume of the gas phase. It may take up to 1 hr for the acetylene to equilibrate with the sediment.[7] Losses of acetylene by diffusion through the sediment are negligible as long as the total incubation period does not exceed about 4 hr. Sampling is most convenient using the Vacutainer system. This system, designed for blood sampling, consists of a preevacuated tube with a rubber stopper, a holder with a needle, and security valve. The security valve prevents gas from escaping when the needle is pushed into the assay bottle. To sample the gas phase, simply push the Vacutainer tube in the holder and through the

[7] K. Jones, *Limnol. Oceanogr.* **27,** 455 (1982).

security valve. The Vacutainer tubes can be kept for several weeks without serious losses of gas. Acetylene and ethylene can then be analyzed in the laboratory. Vacutainer systems are supplied by Becton Dickinson.

Cyanobacterial mats are often patchy. It is therefore recommended to incubate several replicate bottles. For each measurement, two assay bottles are necessary. One is sampled after 1 hr of incubation, assuming that equilibration of acetylene with the sediment is completed within this period. The second bottle is sampled after 2–4 hr. The difference in ethylene content in both bottles is used for the calculation of nitrogenase activity. It is strongly recommended to keep incubation times as short as possible. The number of replicates depends on the patchiness of the mat and the area of sediment enclosed by the assay bottle. The degree of patchiness of the mat may be estimated from variations in chlorophyll *a* measurements.

The nitrogenase activity, calculated from the ethylene produced, may be related to surface area or to cyanobacterial biomass (chlorophyll *a* content). For the latter, chlorophyll *a* should be extracted from the enclosed sediment after terminating the experiment (see below).

It is recommended to include several blanks in each series of measurements. Nitrogenase-independent ethylene production is checked by incubating bottles without acetylene added. The mat can be killed by 25% (w/v) trichloroacetic acid solution at the beginning of the experiment. Incubations in sediments with no cyanobacteria present can be useful to show association of nitrogenase activity with the cyanobacterial mat.

The Cork-Borer Technique. The so-called cork-borer technique allows the manipulation of samples of the mat, e.g., the addition of inhibitors or other chemicals and substrates, changing gas phases, changing light conditions, and temperature. The sampler recommended is a custom-made tube of stainless steel 1 cm in diameter. In the tube a plunger, also made of stainless steel, can be moved up and down. The tube has a sharp edge on one side to cut the mat. The plunger can be brought in fixed positions at 10 and 1.3 mm. The core is sampled by simply pushing the tube into the sediment and drawing it out. The sampler now contains a core 10 mm long, which is adjusted with the plunger to 1.3 mm (we use samples of the upper 1.3 mm of the sediment, which usually includes the cyanobacterial mat) and cut with a sharp scalpel.

The sample is then transferred (mat surface up) to a small 7-ml serum bottle (McCarthy bottle), containing 1 ml of filtered seawater from the same location. The bottle is closed with a rubber stopper and, if desired, flushed with the appropriate gas or gas mixture. Gas is introduced in the bottle through a hypodermic needle. A second needle is necessary as an outlet. To prevent overpressure in the assay bottle, first take out the gas-flushing needle and then the outlet needle. Gas flushing can be done on a manifold with several connections. After preincubation of the samples under the appropriate conditions, 1 ml of acetylene is injected using a gas-tight syringe. This results in a concentration of approximately 15%. Overpressure can be avoided by extracting 1 ml of gas with a gas-tight syringe before the addition of acetylene. In the case of anaerobic incubations, the syringes should first be flushed with the gas mixture used during the incubation. Incubation times should be kept as short as possible and should not exceed 4 hr.

The reaction is terminated by injecting 0.2 ml of 25% (w/v) trichloroacetic acid. A drop of self-vulcanizing silicone rubber or a comparable compound is used to seal any puncture holes in the rubber stopper. The gas phase is analyzed for ethylene and acetylene in the laboratory as soon as possible. Addition of 10^{-5} *M* 3-(3,4-dichlorophenyl)-1,1-dimethylurea (DCMU) (stock solution of 10^{-3} *M* in absolute ethanol) will inhibit oxygenic photosynthesis completely. Anaerobic atmospheres can be obtained by flushing with oxygen-free argon, helium, or nitrogen. At acetylene concentrations of 10–20%, nitrogen reduction will not influence acetylene reduction significantly. Oxygen concentrations can be varied by flushing with mixtures of argon, helium, or nitrogen with oxygen. Measurements can also be done in air. Care should be taken not to enrich the gas with high concentrations (3–5%) of CO_2, because this can give a shift in the pH. For short incubations, however, it is not necessary to add CO_2.

The cork-borer technique can also be applied for measuring vertical gradients in the mat, with a resolution of 1 mm. To this end, a slightly modified sampler is used. The plunger can be brought in fixed positions of 1–7 mm. The core is pushed out of the tube in steps of 1 mm and then sliced by cutting with a scalpel. The following blanks are recommended: (1) a trichloroacetic acid-killed sample, (2) filtered seawater without sediment sample, (3) no acetylene added, and (4) sediments with no cyanobacteria present.

Generating Acetylene in the Field. Acetylene is preferably transported in small tanks equipped with a regulator. If this is not available or possible, acetylene can be transported in football bladders.[8] The inflation needle of the bladder is connected to the acetylene cylinder and the collapsed bladder inflated with acetylene. Acetylene can be withdrawn from the bladder using an inflation needle cemented to the barrel of a 5-ml hypodermic syringe. The barrel is filled with water and closed with a rubber stopper. The inflation needle is inserted in the bladder, and acetylene is allowed to replace the water. Acetylene is withdrawn through the stopper with a gas-tight syringe.

[8] R. H. Burris, this series, Vol. 24, p. 415.

Acetylene can further be generated from calcium carbide by the addition of water according to the following reaction:

$$CaC_2 + 2\ H_2O \rightarrow Ca(OH)_2 + C_2H_2$$

Acetylene as an Internal Standard. It is necessary to know the total amount of ethylene produced by the microbial population enclosed in the assay bottle or bell jar. This can be calculated from the ethylene content of a known volume of gas sample and the total volume of the gas phase of the bottle or bell jar. This method, however, introduces several possible errors. It is difficult to determine the exact volume of the gas phase. Owing to over- or underpressure in the incubation bottle, it is virtually impossible to withdraw a defined volume of gas without using gas-tight syringes with a pressure lock. Moreover, there may be losses if disposable hypodermic syringes are used. Also, on injecting a gas sample into the gas chromatograph, part of the sample can sometimes escape. Finally, when field samples are fixed or when gas samples are carried in Vacutainers to be transported to the laboratory for analysis, losses of acetylene and ethylene may appear.

All these possible errors may be avoided if acetylene is used as an internal standard. The only source of error in this case is the amount of acetylene injected in the assay bottle and, of course, the calibration of the gas chromatograph. Using gas-tight syringes for the injection of acetylene, this error will be insignificant. However, no significant losses of acetylene should occur during incubation. This is generally the case if new (unpierced) butyl rubber septa are used.

For each batch of acetylene the contamination with ethylene should be analyzed as well. Total ethylene can then be calculated using Eq. (1)

$$\text{nmol } C_2H_4 \text{ (total)} = \frac{\text{ml } C_2H_2 \times \text{nmol } C_2H_4}{V_m(\text{nmol } C_2H_2 + \text{nmol } C_2H_4)} - \frac{\text{ml } C_2H_2 \times C}{V_m} \quad (1)$$

in which ml C_2H_2 is the amount of acetylene injected, nmol C_2H_2 and nmol C_2H_4 are the amounts of acetylene and ethylene, respectively, as measured in a gas sample in the gas chromatograph. V_m is the molar volume of an ideal gas and equals RT/P in which R is the molar gas constant, 8.31441 J mol^{-1} K^{-1}, T the temperature in degrees Kelvin, and P the pressure in N m^{-2} (pascals). At $T = 273.15$ K and $P = 1$ atm (101,325 N m^{-2}), V_m equals 22.4×10^{-6} ml $nmol^{-1}$. At $T = 293$ K (20° and $P = 1$ atm, V_m equals 24.0×10^{-6} ml $nmol^{-1}$. C is a correction factor to correct for the contamination of acetylene by ethylene and equals the molar ratio of C_2H_4 to C_2H_2.

Converting Acetylene Reduction to N_2 Fixed. To construct nitrogen budgets of microbial communities, it is often necessary to interpret acetylene reduced as nitrogen fixed. The conversion factor to be applied is still a matter of controversy. The equations for acetylene and dinitrogen reduction can be written as:

$$C_2H_2 + 2\,e^- + 2\,H^+ \rightarrow C_2H_4$$
$$N_2 + 6\,e^- + 6H^+ \rightarrow 2\,NH_3$$

or

$$N_2 + 8\,e^- + 8\,H^+ \rightarrow 2\,NH_3 + H_2$$

Therefore, a conversion factor of 3 or 4, respectively, would theoretically be derived, if 6 or 8 electrons are involved. Many authors have used a factor of 3, but experiments have shown that a factor 4 is more likely.[9,10] This corresponds well with the now generally accepted equation.[6] Nevertheless, in comparing $^{15}N_2$ fixation and acetylene reduction in natural environments, conversion factors may differ considerably. Therefore, without calibrating acetylene reduction with $^{15}N_2$ fixation for each environment, some reservations have to be made in converting acetylene reduction into nitrogen fixed.

$^{15}N_2$ Method for Assay of Nitrogenase

The biological reduction of acetylene to ethylene is carried out only by fully active nitrogenase, and because of the high sensitivity of the method virtually all studies on nitrogen fixation are carried out using this technique. It is, nevertheless, still an indirect method, and, particularly when data on true nitrogen fixation are wanted, measurement of the incorporation of heavy nitrogen is the obvious method.[8,9,11,12] Because the $^{15}N_2$ method is laborious, it is not suitable for routine field studies. Therefore, a calibration of the acetylene reduction technique with the $^{15}N_2$ method is usually sufficient.

Preparation of $^{15}N_2$. Commercially prepared $^{15}N_2$ can be used. To remove possible oxides of nitrogen, however, the $^{15}N_2$ is treated with a solution containing 50 g $KMnO_4$ and 25 g KOH per liter. One liter of gas is shaken with 25 ml of this solution.[8] The $^{15}N_2$ is then transferred to a reservoir and stored over a displacing fluid containing 20% (w/v) Na_2SO_4 in 5% (v/v) H_2SO_4. This displacing fluid will absorb any NH_3 in the gas.

[9] M. Potts, W. E. Krumbein, and J. Metzger, *in* "Environmental Biogeochemistry and Geomicrobiology" (W. E. Krumbein, ed.), Vol. 3, p. 753. Ann Arbor Science, Ann Arbor, Michigan, 1978.

[10] B. B. Jensen and R. P. Cox, *Appl. Environ. Microbiol.* **45,** 1331 (1983).

[11] W. D. P. Stewart, *Ann. Bot. N.S.* **31,** 385 (1967).

[12] R. H. Burris and P. W. Wilson, this series, Vol. 4, p. 355.

$^{15}N_2$ can be prepared from $^{15}NH_4Cl$ or $(^{15}NH_4)_2SO_4$ by passage over heated copper oxide.[12] A simple method for generating ^{15}N-enriched N_2 is by treating $(^{15}NH_4)_2SO_4$ with an alkaline solution of hypobromite[9]:

$$2\ ^{15}NH_3 + 3\ NaBrO\ (LiBrO) \rightarrow 3\ NaBr\ (LiBr) + ^{15}N_2 + 3\ H_2O$$

Two milliliters of $(^{15}NH_4)_2SO_4$ (27 g liter^{-1}) is injected into a 5-ml Vacutainer, using a gas-tight syringe. Then 1 ml of a solution containing 40 ml NaOH (480 g liter^{-1}), 50 ml KI (1.8 g liter^{-1}), and 10 ml Br_2 is injected, and the Vacutainer is shaken vigorously. All solutions should be gassed and stored under argon, to minimize contamination with $^{14}N_2$. Generally, gas purity will be about 80 atom % ^{15}N, but this should be checked by analyzing a subsample of 100 μl in a mass spectrometer. The alkaline solution will ensure absorption of nitrogen oxides which may be formed during the reaction. The disadvantage of this method is the contamination with $^{14}N_2$, which introduces an error in the estimation. This can be minimized by flushing the Vacutainers with argon and then evacuating them again.

Incubation under $^{15}N_2$. Samples from the cyanobacterial mat are taken using the cork-borer technique, as described for the acetylene reduction. Samples of 100 mm^3, containing the upper 1.3 mm of the cyanobacterial mat, are placed in 7-ml serum bottles, to which 1 ml of filtered seawater from the same location is added. The bottles are closed with rubber stoppers and flushed with argon (or argon/oxygen mixtures, if desired), to replace any $^{14}N_2$. Flushing can be done using two hypodermic needles that are introduced through the rubber stopper. One is connected to the gas tank, and the other is used as an outlet. Several samples can be flushed at the same time by connecting them to a manifold. To regulate the gas flow, a needle valve is strongly recommended. To prevent overpressure, the outlet needle should be removed after the bottle is disconnected from the manifold.

After flushing has been completed, 1 ml of gas is withdrawn and 1 ml of $^{15}N_2$ is injected. Argon-flushed, gas-tight syringes should be used throughout. With this method a pN_2 of only 0.2 is reached. This might not, however, saturate nitrogenase and could give rise to too low fixation rates.[8,9,11] Experiments should be carried out in order to determine whether higher pN_2 supports higher rates of nitrogen fixation. In that case a conversion factor can be applied. Otherwise, the procedure of injecting $^{15}N_2$ should be repeated 2–3 times. It is not recommended to evacuate the bottle before adding $^{15}N_2$ in order to reach higher pN_2. Incubation should be restricted to 1–2 hr.

The assay is terminated by injection of 0.2 ml of 5 *N* H_2SO_4. The rubber stopper is treated with self-vulcanizing silicone rubber or comparable sealant, to seal any puncture holes. Before the bottles are opened, 100

μl of gas is withdrawn using an argon-flushed, gas-tight syringe. This gas sample is analyzed in the mass spectrometer to determine atom % ^{15}N to which the samples are exposed. If more gas sample is to be withdrawn, at least the same volume of liquid should be injected before, to produce the appropriate overpressure. This replacing fluid can be freshly boiled water or the inactivating 5 *N* H_2SO_4.

Determination of Atom % ^{15}N in Samples. The samples are subjected to Kjeldahl digestion, distillation, and conversion to ammonia, and the total nitrogen is determined by Nesslerization.[12] Ammonia is converted to N_2 with alkaline hypobromite using an evacuated Rittenberg tube[13] and atom % excess ^{15}N is determined with the mass spectrometer. A much faster and more convenient method is the combination of a CHN elemental analyzer and mass spectrometer.[14] The samples are dried, and 1 mg dry weight of sample is pyrolyzed at 1060° under a flow of helium (30 ml min^{-1}) under the addition of oxygen. The gasses are separated gas chromatographically, using a column packed with Poropak Q, and N_2 is determined using a thermal conductivity detector (TCD). The gas outlet of the TCD is connected to the slit separator of the mass spectrometer. As soon as N_2 elutes from the column, as many mass scans as possible are run. The masses 28, 29, and 30 are measured in all samples at a scanning speed of 100 masses sec^{-1}. The nitrogen fixed can then be calculated using Eq. (2).

$$N_{fixed} = \frac{\text{atom \% excess } ^{15}N_{sample} \times \text{total } N_{sample}}{\text{atom \% excess } ^{15}N_{gas}} \qquad (2)$$

Chlorophyll Measurements in Cyanobacterial Mats

Nitrogenase activity in cyanobacterial mats can be related to surface area or to cyanobacterial biomass. Cyanobacterial biomass can be determined satisfactorily as the chlorophyll *a* content of a sample. Specific activities of nitrogenase, expressed as C_2H_4 produced per milligram of chlorophyll *a*, may provide important information. We have found very high specific nitrogenase activities in freshly colonized sediment, almost exclusively consisting of nitrogen-fixing *Oscillatoria* sp., although absolute activity (expressed as surface area) was quite low. Much higher absolute activities were found in established mats of *Microcoleus chthonoplastes* with low numbers of *Oscillatoria* sp., but here the specific activity was relatively low. From such calculations evidence was obtained that only *Oscillatoria* sp. and not *M. chthonoplastes* fixed nitrogen.

[13] A. San Pietro, this series, Vol. 4, p. 473.

[14] J. Metzger, *Fresenius Z. Anal. Chem.* **292,** 44 (1978).

Chlorophyll *a* measurements in microbial mats are often hampered by the presence of interfering compounds such as chlorophyll degradation products, pheophytin, and bacteriochlorophylls. Satisfactory measurements of chlorophyll *a* can be obtained by a two-phase solvent extraction method.[15] Samples to which some $MgHCO_3$ is added are homogenized in methanol by grinding. Samples are extracted twice for 2 hr in the dark at room temperature. The extracts are centrifuged and pooled. The volume of methanol depends on the chlorophyll *a* content of the sample. We routinely use 5–10 ml of methanol for a 100-mm^3 sample, containing the top 1.3 mm of the mat. To 10 ml of the methanol extract, 3 ml of 0.05% solution of NaCl (w/v) and 13 ml *n*-hexane is added, and the mixture is vigorously shaken. The phases are allowed to separate. When necessary, a short, low-speed centrifugation may aid separation. The hexane hyperphase contains 90% of the chlorophyll *a* and pheophytin *a*. The hyperphase is divided into two parts, and one part is acidified with 100 μl 5 *N* HCl in 5 ml of hexane. Following vigorous shaking of the acidified hexane and drying with anhydrous Na_2SO_4, the extinction at 660 nm is measured in the neutral and acidified extracts. Chlorophyll *a* and pheophytin *a* are calculated according to Eqs. (3) and (4), where E_n and E_a are the absorptions measured at 660 nm in the neutral and acidified hexane extracts, respectively.

$$\text{Chlorophyll } a \text{ (g/liter hexane)} = 0.0110(E_n - E_a) \quad (3)$$
$$\text{Pheophytin } a \text{ (g/liter hexane)} = 0.0107(4.14E_a - E_n) \quad (4)$$

[15] L. J. Stal, H. van Gemerden, and W. E. Krumbein, *J. Microbiol. Methods* **2,** 295 (1984).

[50] Nitrogen Fixation by Synchronously Growing Unicellular Aerobic Nitrogen-Fixing Cyanobacteria

By AKIRA MITSUI and SHUZO KUMAZAWA

Introduction

Unicellular aerobic nitrogen-fixing cyanobacteria are unique in that both oxygenic photosynthesis and anoxygenic nitrogen fixation are carried out in a single cell. Nitrogenase in nitrogen-fixing cyanobacteria must be protected from both internal photosynthetically produced oxygen and external atmospheric oxygen diffusing into the cells.[1,2] In order to eluci-

[1] R. Y. Stanier, and G. Cohen-Bazire, *Annu. Rev. Microbiol.* **31,** 225 (1977).
[2] W. D. P. Stewart, *Annu. Rev. Microbiol.* **34,** 497 (1980).

date the mechanism by which these apparently incompatible reactions are carried out within a single cell, cultures grown under several different conditions are examined. These include batch cultures under continuous illumination[3] or light–dark cycles[4] and synchronous cultures under continuous illumination[5–7] or light–dark cycles.[5] It has been shown with *Gloeothece* (*Gloeocapsa*) sp. that the changes in activities of photosynthesis and nitrogen fixation are inversely correlated during the batch growth with continuous illumination.[3] Under the batch-culture conditions with light–dark cycles,[4] activities of photosynthesis and nitrogen fixation are shown to be temporally separated into light and dark periods of growth, respectively. Synchronous culture,[8,9] which is ideally suited for the observation of temporally phased events within a cell cycle, can also be applied.

Recently, aerobic nitrogen-fixing marine *Synechococcus* spp. strains Miami BG 43511 and 43522 were grown under synchronized culture conditions.[5–7] In these strains, phases of photosynthesis and nitrogen fixation are temporally separated within the cell cycle. During synchronous growth, the phase of nitrogen fixation is observed in the light under continuous illumination[5,7] or in the dark under a 12 hr light–12 hr dark (12L–12D) regime.[5] The former condition is suitable for studying the relationships between the activities of photosynthesis and nitrogen fixation and the latter condition the relationships between those of respiration and nitrogen fixation. Thus, materials prepared under synchronous growth conditions can ideally be used for analyzing the mechanism by which nitrogen fixation in unicellular aerobic nitrogen-fixing cyanobacteria is regulated. The method which we have been using for the study of nitrogen fixation by unicellular cyanobacteria is described below.

Induction of Synchronous Growth

Culture conditions are described elsewhere in this volume (Mitsui and Cao [7]). Synchronization of aerobic nitrogen-fixing *Synechococcus* sp. Miami BG 43511 was induced by interruption during their early exponen-

[3] J. R. Gallon, T. A. LaRue, and W. G. W. Kurz, *Can. J. Microbiol.* **20,** 1633 (1974).

[4] P. M. Mullineaux, J. R. Gallon, and A. E. Chaplin, *FEMS Microbiol. Lett.* **10,** 245 (1981).

[5] A. Mitsui, S. Kumazawa, A. Takahashi, H. Ikemoto, S. Cao, and T. Arai, *Nature* (*London*) **323,** 720 (1986).

[6] C. León, S. Kumazawa, and A. Mitsui, *Curr. Microbiol.* **13,** 149 (1986).

[7] A. Mitsui, S. Cao, A. Takahashi, and T. Arai, *Physiol. Plant.* **69,** 1 (1987).

[8] H. Lorenzen and G. S. Venkataraman, *in* "Methods in Cell Physiology" (D. M. Prescott, ed.), p. 373. Academic Press, New York, 1972.

[9] N. I. Bishop and H. Senger, this series, Vol. 23, p. 53.

tial growth (~10^6 cells/ml) by dark periods (see chapter [7] by Mitsui and Cao, this volume). During the dark periods, aeration was stopped but mixing by magnetic stirrer was continued. After the dark induction period, cells grew synchronously during the subsequent light and dark periods as indicated in Fig. 1.

Assay of Cellular Activities

Experiments with batch cultures show that growth changes from an exponential phase to a light-limited phase when the cell density increases to around 10^7 cells/ml. Thus, under the conditions described above (see chapter [7] by Mitsui and Cao, this volume), synchronized growth can be optimally observed when the density of the cells is in the range of about 10^6–10^7 cells/ml.

Growth. For cell counts, a 1-ml sample is taken from the culture cylinder at periodic intervals and fixed with 50 μl of 10% Lugol's solution. Cell numbers are measured using a Petroff–Hausser bacterial cell counting chamber.[10] For each sample, the total number of cells and the number of doubling (dividing) cells are counted. The doubling cell is counted as one cell. Dry weight of the culture sample is measured after the cells are collected on Whatman 934AH glass microfiber filters, washed twice with distilled water and dried at 90° for 24 hr.

Photosynthesis. Changes in the capability for photosynthetic oxygen evolution can be measured by a Clark-type oxygen electrode with samples taken directly from the culture cylinder. At periodic time intervals, a sample (2 ml) is transferred from the culture cylinder to the reaction vessel without any further treatment. The reaction vessel is similar to the one described by Wang.[11] The culture sample is transferred using a 2.5-ml Glaspak syringe (Becton Dickinson) fitted with a long needle (22-gauge, ~12 cm in length, Hamilton). Then the culture sample is bubbled with 4% CO_2-enriched air in the dark for 2–3 min at 30° to establish equilibrium. Bubbling is carried out with the aid of a long needle as described above. After the bubbling, air bubbles remaining in the reaction vessel are removed using a syringe with a long needle. Then the reaction vessel is closed with a glass stopper (see Ref. 11), and rates of respiration and photosynthetic oxygen evolution are measured in the dark and the light (2,000 $\mu E/m^2/sec$), respectively.

[10] R. R. L. Guillard, *in* "Handbook of Phycological Methods: Culture Methods and Growth Measurements" (J. R. Stein, ed.), p. 289. Cambridge Univ. Press, Cambridge, England, 1973.

[11] R. T. Wang, this series, Vol. 69, p. 409.

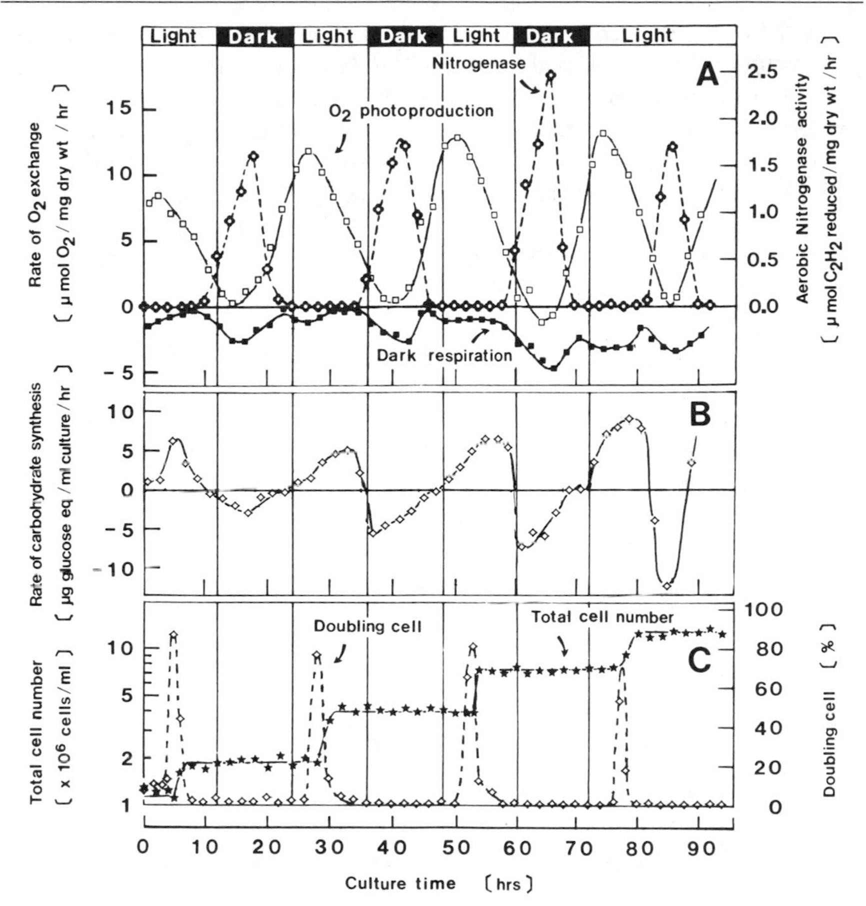

FIG. 1. Changes (A) in photosynthetic oxygen evolution, respiratory oxygen uptake, and acetylene reduction capabilities, (B) in the rate of carbohydrate synthesis, and (C) in total cell number and occurrence of doubling cells during synchronous growth of *Synechococcus* sp. Miami BG 43511 under diurnal light–dark cycles. After Mitsui *et al.*[5] After the dark induction period, synchronized growth was monitored during the light and dark periods as indicated. Continuous illumination was initiated at hour 72. Aliquots of culture samples were taken every 2 hr and assayed for oxygen exchange and acetylene reduction capabilities, dry weight, carbohydrate content, and cell number. Carbohydrate content was measured by the phenol–sulfuric acid method [M. Dubois, K. A. Gilles, J. K. Hamilton, P. A. Rebers, and F. Smith, *Anal. Chem.* **28,** 350 (1956)]. The rate of carbohydrate synthesis was calculated from the change in carbohydrate content (per ml culture basis) measured every 2 hr. See text for other details.

As an alternative measurement of the capacity for photosynthetic oxygen evolution, $NaHCO_3$ and HEPES–NaOH buffer (pH 7.6) are added to the culture sample to give final concentrations of 5 and 20 m*M*, respectively, and then the bubbling is carried out with argon for 2–3 min before the reaction vessel is closed with a glass stopper. As shown in Figs. 1A and 2A, oscillation of net photosynthetic oxygen evolution was observed under both conditions, that is, with 12L–12D cycles (Fig. 1A) and with continuous illumination (Fig. 2A). As expected from the oscillation of

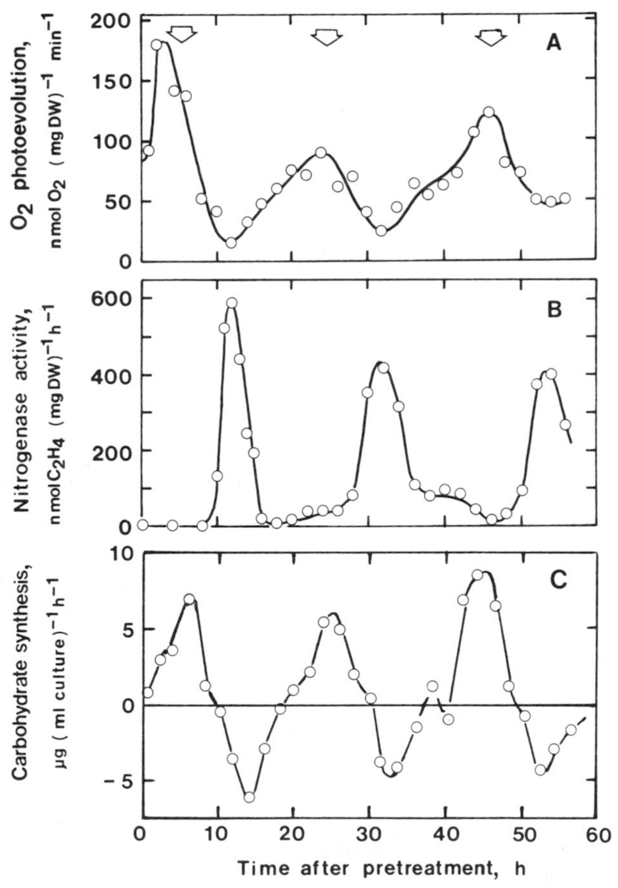

FIG. 2. Changes (A) in photosynthetic oxygen evolution capability, (B) in acetylene reduction capability, and (C) in the rate of carbohydrate synthesis during synchronous growth of *Synechococcus* sp. Miami BG 43511 under continuous illumination. From Mitsui *et al.*[7] Growth synchrony was induced by 16 hr dark–16 hr light–16 hr dark pretreatment. See text and the legend to Fig. 1 for further details.

photosynthetic oxygen evolution capability, carbohydrate content in the culture samples showed distinct phases of net accumulation and net consumption (Figs. 1B and 2C).

It must be mentioned here, however, that whole cells of Strain Miami BG 43511 are quite impermeable to the agents (such as diaminodurene, oxidized form of methyl viologen, and ferricyanide) used for the measurement of photosynthetic electron transport studies. Thus, in order to analyze the mechanism by which photosynthetic oxygen evolution oscillates, the cells have to be converted to permeaplasts or spheroplasts.

Nitrogen Fixation. The activity of nitrogenase can be measured by the acetylene reduction method without concentration of the culture material. In routine experiments, 5-ml aliquots are taken from the culture cylinder at periodic intervals and transferred to 25-ml Fernbach flasks. The flasks are then sealed with rubber stoppers, and 2 ml of acetylene gas is injected into each of the flasks. Incubation is carried out in a shaking bath at light intensity of 150 $\mu E/m^2/sec$ at 30° for 30 min. The reaction is terminated by injecting 1 ml of 20% trichloroacetic acid. Formation of ethylene is measured by gas chromatography as described by Burris.[12] As shown in Figs. 1A and 2B, nitrogenase activity appears during a segregated period in the cell cycle. The appearance of nitrogenase activity coincides consistently with the phase of declining photosynthetic oxygen evolution, and the maximum nitrogenase activity is observed when the oxygen evolution capability is at a minimum (Figs. 1 and 2).

Nitrogen Fixation and Hydrogen Metabolism. Since nitrogenase activity appears in a segregated manner within the cell cycle, correlation between the changes in activities of nitrogenase and hydrogenase can be examined with synchronously growing cultures. Changes in nitrogenase activity can be measured as described above, and hydrogenase activity by the amperometric hydrogen electrode. The method for the measurement of hydrogen has been described previously in this series by Wang[11] and Hanus *et al.*[13] A combination of oxygen and hydrogen electrodes is useful for the measurement of H_2 oxidation (oxyhydrogen reaction).[14,15]

Comments

The observations shown in Figs. 1 and 2 indicate that temporal separation of the phases of photosynthesis and nitrogen fixation occurs within the cell cycle as a mechanism by which unicellular aerobic nitrogen-fixing

[12] R. H. Burris, this series, Vol. 24, p. 415.

[13] F. J. Hanus, K. R. Carter, and H. J. Evans, this series, Vol. 69, p. 731.

[14] S. Kumazawa, T. Ogawa, Y. Inoue, and A. Mitsui, *Plant Cell Physiol.* **26,** 1485 (1986).

[15] S. Kumazawa and A. Mitsui, *Appl. Environ. Microbiol.* **50,** 287 (1986).

cyanobacteria avoid the adverse effects on nitrogenase activity of photosynthetically produced oxygen. This implies that kinetic changes in the activities related to nitrogen fixation can ideally be measured by using synchronous cultures. Synchronous cultures, such as those described here, will also provide quantities of uniform cell material useful for further investigations to determine the biochemical, physiological, cytological, and genetic regulatory mechanisms involved in nitrogen fixation and photosynthesis.

[51] Isolation of Metabolically Active Heterocysts from Cyanobacteria

By RUSSELL L. SMITH, CHASE VAN BAALEN,† and F. ROBERT TABITA

Introduction

Several species of filamentous cyanobacteria, when grown under conditions favorable for nitrogenase synthesis, differentiate specialized cells known as heterocysts.[1-3] Of the various biochemical and enzymological changes expressed during heterocyst induction, loss of the oxygen-evolving photosystem I components, biosynthesis of a complex cell envelope, and derepression of nitrogenase synthesis are the primary features which allow the heterocysts to serve as the primary site for dinitrogen reduction. As opposed to the unicellular cyanobacteria, which seem to temporally separate oxygenic photosynthesis from nitrogen fixation,[4,5] the differentiation of vegetative cells into heterocysts permits the simultaneous operation of both photosynthesis and nitrogen fixation by spatially separating the two processes. The heterocyst is thus able to take advantage of the tremendous reducing power available to the organism under phototrophic conditions.

Not only does the heterocyst present a novel system to study the physiology and enzymology of biological nitrogen fixation, but it has also

† Deceased.

[1] R. Haselkorn, *Annu. Rev. Plant Physiol.* **29,** 319 (1978).

[2] W. D. P. Stewart, *Annu. Rev. Microbiol.* **34,** 497 (1980).

[3] C. P. Wolk, *in* "The Biology of Cyanobacteria" (N. G. Carr and B. A. Whitton, eds.), p. 359. Blackwell, Berkeley, California, 1982.

[4] N. Grobelaar, T. C. Huang, H. Y. Lin, and T. J. Chow, *FEMS Microbiol. Lett.* **37,** 173 (1986).

[5] A. Mitsui, S. Kumazawa, A. Takahashi, H. Ikemoto, S. Cao, and T. Arai, *Nature (London)* **323,** 720 (1986).

recently been shown to exhibit a unique type of genomic rearrangement for the control of gene expression.[6,7] During heterocyst differentiation, a portion of the vegetative cell chromosome is excised, allowing the uninterrupted transcription from the *nif* H promoter of the entire *nif* HDK operon (i.e., the structural genes of the iron–molybdenum protein and iron protein of nitrogenase). Such rearrangements are usually ascribed to eukaryotic systems, so its occurrence in a differentiated prokaryotic cell may provide a model system to study genomic rearrangement as a mechanism for gene control and cellular differentiation.

The procedures described here for the isolation of metabolically competent heterocysts are based on the methods described by Fay[8] and further refined by Kumar *et al.*,[9] who used a fragile vegetative cell wall mutant of *Anabaena* sp. Strain CA. With few exceptions,[10–13] most active heterocyst preparations have been dependent on the addition to the final assay of materials not unlike those employed with *in vitro* assays using cell extracts.[14–19] For example, nitrogenase activity of heterocyst preparations (acetylene reduction) has most often been measured in the presence of either an ATP-generating system or artificial electron donors, or both. Without these additions, such preparations had a very low or short-lived activity, calling into question the metabolic competency of the heterocysts. The addition of exogenous compounds is not required to achieve high and sustained metabolic activities when the procedures described here or those described earlier are employed.[10–13] Exogenous thiols, ATP-generating systems, additional reductants or cofactors, and various osmoregulators are not necessary to obtain highly active preparations. Thus, the isolated heterocysts described in this chapter rely only on those processes or factors endogenous to the cell.

[6] R. Haselkorn, J. W. Golden, P. J. Lammers, and M. E. Mulligan, *Trends Genet.* **2,** 255 (1986).

[7] J. W. Golden, S. J. Robinson, and R. Haselkorn, *Nature (London)* **314,** 419 (1985).

[8] P. Fay, this series, Vol. 69, p. 801.

[9] A. Kumar, F. R. Tabita, and C. Van Baalen, *Arch. Microbiol.* **133,** 103 (1982).

[10] B. B. Jensen, R. P. Cox, and R. H. Burris, *Arch. Microbiol.* **145,** 241(1986).

[11] A. Kumar, F. R. Tabita, and C. Van Baalen, *J. Bacteriol.* **155,** 493 (1983).

[12] R. B. Peterson and C. P. Wolk, *Proc. Natl. Acad. Sci. U.S.A.* **75,** 6271 (1978).

[13] R. L. Smith, D. Kumar, Z. Xiankong, F. R. Tabita, and C. Van Baalen, *J. Bacteriol.* **162,** 565 (1985).

[14] G. Eisbrenner and H. Bothe, *Arch. Microbiol.* **123,** 37 (1979).

[15] T. H. Giddings and C. P. Wolk, *FEMS Microbiol. Lett.* **10,** 299 (1981).

[16] W. Lockau, R. B. Peterson, C. P. Wolk, and R. H. Burris, *Biochim. Biophys. Acta* **502,** 298 (1978).

[17] R. B. Peterson and R. H. Burris, *Arch. Microbiol.* **116,** 125 (1978).

[18] L. S. Privalle and R. H. Burris, *J. Bacteriol.* **157,** 350 (1984).

[19] W. D. P. Stewart, A. Haystead, and H. W. Pearson, *Nature (London)* **224,** 226 (1969).

Methods

Growth of Cultures

The use of physiologically compromised cultures defeats, from the beginning, any attempt to isolate competent heterocysts. Therefore only cultures which maintain high diazotrophic growth rates or require other rigidly defined parameters (as in the case of argon-treated cultures described below) should be used. Cultures of *Anabaena* sp. Strain CA (ATCC 33047), a marine filamentous cyanobacterium,[20] are routinely grown in Pyrex culture tubes containing 20–30 ml of medium ASP-2[21] (NaCl content at 5 g/liter) bubbled with 1% CO_2 in air at 39°. The culture is illuminated with four F36T12/D/HO fluorescent lamps on either side, 12 cm from the center of the growth tubes, at an average intensity of 300 μE/m^2/sec (see Ref. 22 for a further description of these growth conditions).

It is preferable to work with rapidly growing cultures which are in the early to mid log phase of growth.[23,24] In general, cultures of *Anabaena* sp. CA from which heterocysts are to be isolated do not exceed 0.10–0.12 mg dry wt/ml (~1 μg Chl *a*/ml). Older or more dense cultures tend to give erratic and lower overall recovery of nitrogenase activity when compared to the activity of the original intact filaments.[24]

Heterocyst Derepression

The problems associated with older cultures may be related to differences in the age of the heterocysts when they are isolated from the filaments. Therefore, one of the most important manipulations is to treat the filaments such that the synthesis of heterocysts occurs in a synchronous or nearly synchronous fashion.[9] Initially, cultures are grown to a density of 0.14–0.16 mg dry wt/ml in a medium where heterocyst formation and nitrogenase synthesis have been completely suppressed (in the case of *Anabaena* CA, in the presence of 10 m*M* NH_4NO_3[25]). Filaments are then transferred to a medium free of fixed nitrogen, using a 1% inoculum. With a generation time of about 4.0 hr, full nitrogenase expression and heterocyst formation is complete within about 10–12 hr (with little or no interruption in the growth rate[25]) following transfer to diazotrophic conditions. If the cultures are harvested for the isolation of heterocysts 18–20 hr

[20] G. Stacey, C. Van Baalen, and F. R. Tabita, *Arch. Microbiol.* **114,** 197 (1977).

[21] C. Van Baalen, *Bot. Mar.* **4,** 129 (1962).

[22] J. Myers, *in* "The Culturing of Algae" (J. Brunel, G. W. Prescott, and L. H. Tiffany, eds.), p. 45. Kettering Found., Yellow Springs, Ohio, 1950.

[23] C. P. Wolk, personal communication (1986).

[24] R. L. Smith, unpublished observations (1985).

[25] P. J. Bottomley, J. F. Grillo, C. Van Baalen, and F. R. Tabita, *J. Bacteriol.* **140,** 938 (1979).

following transfer, then the age of any two mature heterocysts in filaments differs by no more than one to two generations. This technique yields preparations of heterocysts which, presumably, all have similar (high) metabolic capacity.

Heterocyst Isolation

Following synchronized heterocyst formation, cell suspensions of 20–30 ml, at a density of 0.10–0.12 mg dry wt/ml, are washed twice in medium ASP-2, with the total concentration of KCl raised to 30 m*M* and NaCl to 370 m*M* (assay medium). The addition of NaCl to the assay medium is probably more closely related to the fact that this is a marine cyanobacterium rather than to any osmotic requirement.[9,10] Mannitol used in place of the added NaCl does not support the activity of heterocysts isolated from this strain. The requirement for increased levels of KCl is as yet, unexplained. After washing, the cells are resuspended in 5 ml assay medium with lysozyme added to 1.0 mg/ml. This assay medium should be thoroughly sparged with either 1% CO_2–99% N_2 or 1% CO_2–99% Ar (depending on the desired experiment). The cells are then transferred by gas-tight syringe to a stoppered glass tube containing the samc gas mixture, and the tube is subsequently placed into a shaking water bath set at 39°. Although anaerobic or microaerobic conditions are desired, this does not seem to be critical at this point in the preparation, since heterocysts isolated from cells incubated in the light (thus exposed to photosynthetically generated oxygen) and those incubated in the dark during lysozyme treatment show almost identical levels of nitrogenase activity.[24]

After a 30-min incubation in the presence of lysozyme, the cell suspension is subjected to sonic disruption. Sonication is carried out under a stream of Ar or N_2, simply by immersing an appropriate sparging device to the bottom of thc lysozyme solution in the tube in which sonication is to be performed. We routinely use a Heat Systems Model W-10 ultrasonicator equipped with a standard 4.5-inch titanium microprobe, set at full output, for the 5.0-ml suspensions. The small volume, low density cell suspensions lend themselves to rapid (of the order of 15–20 sec) sonication with almost complete disruption of the vegetative cells.[26] Alterna-

[26] The use of small culture volumes (20–30 ml) is a matter of convenience, since these samples are easily manipulated during the isolation procedure. These cells are highly active, and the analytical techniques employed are more than sufficiently sensitive to measure the activity of these samples, thus precluding the need for large cultures for most purposes. However, scale-up to larger culture volumes may be accomplished quite easily provided that the cells are rapidly dividing and are in early to mid logarithmic growth phase. Indeed, using a larger, more powerful sonicator probe, we have isolated hetero cysts from 1 liter of culture (incubation in 250 ml assay medium plus lysozyme) with no loss in the overall activity of the final preparation.

tively, we employ a Branson Model 350 Sonifier with microprobe attachment; three sonication intervals of 10 sec at setting No. 1 are used (5 sec between each sonication interval).[27] Perhaps because of the rapidity with which the vegetative cells are destroyed, structural damage to the heterocysts is minimized. The incubation time in the presence of lysozyme can be reduced somewhat (to 15–20 min); however, a significant increase in the sonication time is required, therefore increasing the possibility of damage to the heterocysts.

Following sonication, separation of the heterocysts from vegetative cell debris is accomplished by slow, differential centrifugation, followed by resuspension of the sonicated preparations in freshly gassed assay medium. Two rounds of centrifugation with rotor speeds not exceeding about 200 *g* are usually sufficient to obtain heterocyst preparations free from particulate vegetative cell material. After centrifugation, microscopic observation of the heterocyst preparation should indicate that the heterocysts possess intact polar bodies and that they have retained their blue–green color. If it is noted that the polar bodies become dislocated or the heterocysts become bleached or yellowed, it is likely that sonication was too harsh; invariably an inactive heterocyst preparation is the end result. On the other hand, if it is found that there is incomplete breakage of vegetative cells, it is possible to resonicate the suspension. However, it is advisable to decrease, by at least half, the output power of the sonicator probe, since the cultures are considerably more dilute at this stage.

It is at this point that the heterocysts may be subjected to any of several assays to determine metabolic competency. Routinely, the final heterocyst pellet is resuspended in 2.0 ml of the assay medium under a suitable anaerobic atmosphere with 10% C_2H_2 added to determine the acetylene-reducing capacity of the preparation. It was found that ethylene production becomes maximal 20–30 min immediately following the isolation procedure.[13] Therefore, if any assay is to be attempted which is dependent on short-term observations or very rapid changes in activities (e.g., those concerned with fast, photomediated events), it is advisable to allow the heterocysts to preincubate in the light for a few minutes prior to the actual measurement.

Heterocysts with High, Endogenous Metabolic Activity

For cultures grown in either 1% CO_2 in air or in 1% CO_2–99% N_2, the highest rates of acetylene reduction are achieved when isolated heterocysts are incubated in the light under an atmosphere of 10% C_2H_2–90% H_2. Rates upward of 3.5–4.5 μmol ethylene produced/mg dry wt of het-

[27] L. A. Li and F. R. Tabita, unpublished observations (1986).

erocysts/hr are routinely achieved. This rate represents about 50–60% of the rate of acetylene reduction of the whole filaments when taken on a per heterocyst basis. These rates are usually linear for at least 3–4 hr. The rate of acetylene reduction under 10% C_2H_2–90% Ar is 4–5 times less than the above activity, and is linear for about 2 hr. The increase in activity under a hydrogen atmosphere[9,12,14,18] is presumably due to the function of an uptake hydrogenase within the heterocysts, with hydrogen acting as the reductant for nitrogenase.

Heterocysts can be prepared which contain a high endogenous activity in the absence of hydrogen.[11] This is accomplished by preincubation of the cultures from which the heterocysts are to be isolated under an atmosphere of 1% CO_2–99% Ar for a sufficiently long time period to deplete internal nitrogen stores. For *Anabaena* sp. Strain CA, 16–18 hr is sufficient. Visually, a yellowing of the cultures within 3–5 hr of treatment with Ar indicates the utilization and breakdown of the phycobiliproteins and signals the onset of nitrogen starvation. Microscopically, the heterocysts retain their blue–green color, however. The combination of nitrogen-limiting conditions and continuous availability of photoassimilable CO_2 leads to a condition wherein heterocysts acquire enhanced levels of compound(s) capable of serving as readily available sources of reductant for nitrogenase. Preincubation of isolated heterocysts under 100% Ar does not impart high acetylene-reduction capacity. Using this procedure for the generation of internal reductant, rates of acetylene reduction are as high under 10% C_2H_2–90% Ar as they are under 10% C_2H_2–90% H_2, and the activity remains linear for over 4 hr. However, the simple buildup of vast amounts of reductant cannot be the only process which increases heterocyst activity in these preparations since the levels of reductant needed (based on overall ethylene production) are about 4 times higher than the actual observed buildup of excess carbohydrate in heterocysts obtained from argon-treated filaments. Clearly, some other mechanism functions to supply the needed reductant for such high and sustained metabolic activity.

Concluding Remarks

Using these procedures, we have been able to manipulate heterocysts for a wide variety of experimental procedures.[9,11,13] While the methods described here were developed for the marine cyanobacterium *Anabaena* sp. Strain CA, appropriate modifications for other species (based on their general physiology and growth requirements) should be feasible if attention is paid to the more important physical manipulations outlined here. The isolation of highly active heterocysts from other species of filamentous cyanobacteria would be highly desirable.

[52] Electron Donation to Nitrogenase in Heterocysts

By HERMANN BOTHE and GABRIELE NEUER

Introduction

Nitrogenase in heterocystous cyanobacteria is confined to specialized cells, heterocysts, under aerobic growth conditions. When the filaments grow in the absence of oxygen or under reduced oxygen tensions, both heterocysts and vegetative cells perform N_2 fixation.[1,2] Cyanobacteria synthesize a [2Fe–2S] ferredoxin and a flavodoxin (the latter only under iron deficiency), and both function as electron carriers in C_2H_2 reduction of cell-free nitrogenase with comparable efficiencies and independently of each other.[3] Heterocysts have recently been reported to contain a special type of ferredoxin[4] which awaits chemical characterization.

The reduction of ferredoxin and the generation of reducing equivalents for N_2 fixation in heterocysts is more complicated. These cells do not restrict themselves to a single pathway.[5–8] This may be due to the fact that heterocysts being unable to perform photosynthetic CO_2 fixation are supplied with various carbon compounds from vegetative cells (Fig. 1). Among these, monosaccharides are degraded mainly via the hexose monophosphate shunt to glyceraldehyde 3-phosphate, which is further metabolized by glycolysis and the incomplete tricarboxylic acid cycle.[9] The resulting reductants (NADH, NADPH, pyruvate) as well as H_2 are used as electron donors for the reduction of ferredoxin in N_2 fixation (Fig. 2). The reactions between these electron donors and the electron carriers (ferredoxin, flavodoxin) are enzyme mediated. Some of the electron donors (pyruvate, NADPH) can reduce ferredoxin in the dark. In contrast, the reduction of ferredoxin by H_2 or NADH is strictly dependent on photosystem I. With all the different electron donors, C_2H_2 reduction is stimulated by light, because the ATP required for N_2 fixation can effec-

[1] R. Haselkorn, *Annu. Rev. Plant Physiol.* **29,** 319 (1978).

[2] H. Bothe, *in* "The Biology of Cyanobacteria" (N. G. Carr and B. A. Whitton, eds.), p. 87. Blackwell, Berkeley, California, 1982.

[3] H. Bothe, *Ber. Dtsch. Bot. Ges.* **83,** 421 (1970).

[4] B. Schrautemeier and H. Böhme, *FEBS Lett.* **184,** 304 (1985).

[5] S. K. Apte, P. Rowell, and W. D. P. Stewart, *Proc. R. Soc. London, Ser. B* **200,** 1 (1978).

[6] W. Lockau, R. B. Peterson, C. P. Wolk, and R. H. Burris, *Biochim. Biophys. Acta* **502,** 298 (1978).

[7] J. P. Houchins and G. Hind, *Biochim. Biophys. Acta* **682,** 86 (1982).

[8] G. Neuer and H. Bothe, *Arch. Microbiol.* **143,** 185 (1985).

[9] G. Neuer and H. Bothe, *FEBS Lett.* **158,** 79 (1983).

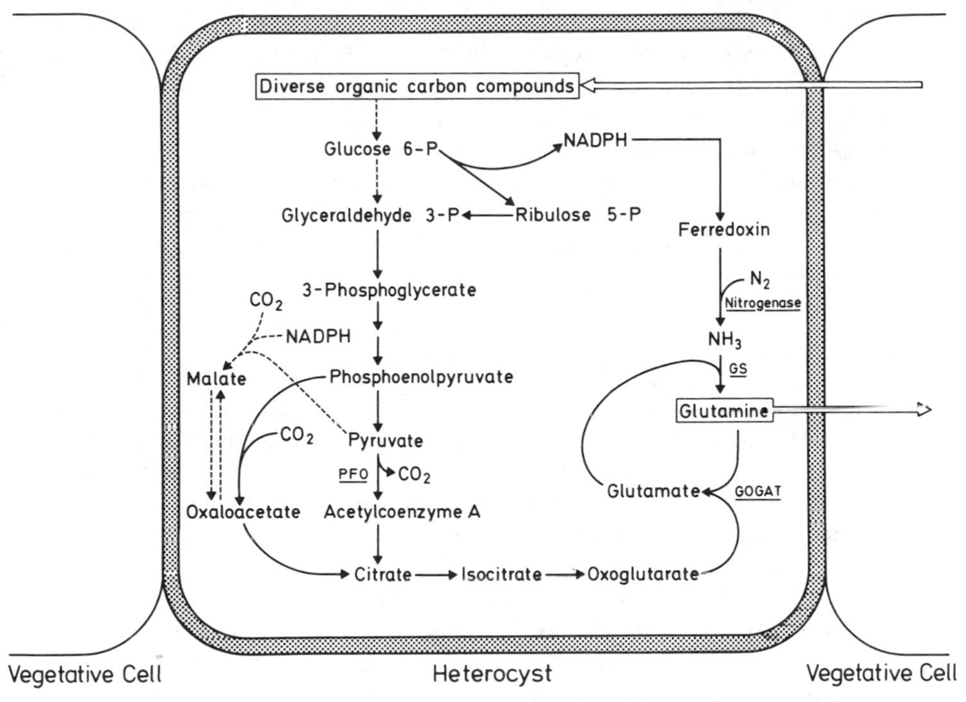

FIG. 1. Carbohydrate catabolism in heterocysts and the exchange of compounds between heterocysts and vegetative cells.

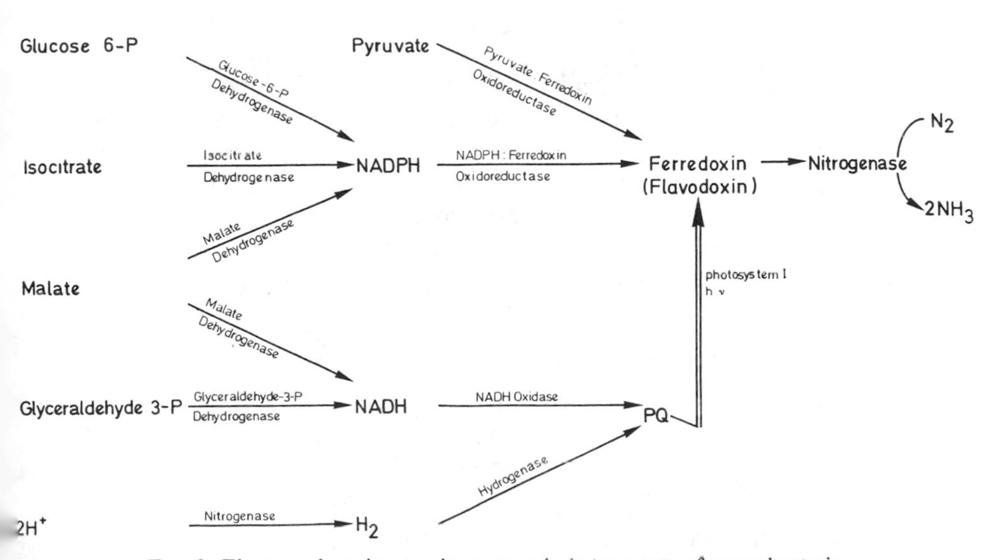

FIG. 2. Electron donation to nitrogenase in heterocysts of cyanobacteria.

tively be supplied by cyclic photophosphorylation but by respiration only to a limited extent in heterocysts. The flow of electrons from the different electron donors to nitrogenase requires regulation which is not yet understood. The membrane potential or the energy charge may or may not be involved in the generation of reductants for N_2 fixation.[10–12] C_2H_2 reduction by heterocysts is stimulated by several simple organic compounds (erythrose 4-phosphate, glycolate, glyoxylate, fructose[8,13,14]), and the generation of reductants from these sources to nitrogenase has not yet been elucidated.

Heterocyst Isolation and Preparation of Extracts

The method to isolate heterocysts was critically evaluated in this series.[15] Generally, heterocysts are prepared by prolonged incubation of the filaments with lysozyme at elevated temperature, followed by passage through a French or Yeda press at very low pressure and centrifugation at low speed. However, complete lysis of the cells by lysozyme requires an incubation of 1 to several hours[15] and is, therefore, critical. Several enzymes previously tested did not retain activities over such long periods. Therefore, we routinely do not incubate with lysozyme. The following procedure for isolating heterocysts and for preparing extracts from these cells can be recommended for *Anabaena cylindrica*.[16] The filaments are grown as described[17] and harvested at mid logarithmic growth phase. They are concentrated by rapid self-sedimentation and suspended in one-tenth of the original volume with buffer consisting of 30 m*M* HEPES/KOH, pH 7.6, 1 m*M* $MgCl_2$, 3 m*M* dithiothreitol. The cyanobacteria are passed twice through a chilled French pressure cell at 1600 $lb/in.^2$ (1 $lb/in.^2 \approx 7$ kPa) and centrifuged (1000 g, 5 min). The supernatant is used as the "extract from vegetative cells." The pellet is suspended in a minimal volume of buffer and is usually assayed as a heterocyst preparation directly. This heterocyst preparation contains less than 2% vegetative cells and no or very small amounts of debris. If not, the preparation is once more passed through the French press and treated as above. "Ex-

[10] M. J. Hawkesford, R. J. Reed, P. Rowell, and W. D. P. Stewart, *Eur. J. Biochem.* **115,** 519 (1981).

[11] L. S. Privalle and R. H. Burris, *J. Bacteriol.* **154,** 351 (1983).

[12] A. Ernst, H. Böhme, and P. Böger, *Biochim. Biophys. Acta* **723,** 83 (1983).

[13] T. Murai and T. Katoh, *Plant Cell Physiol.* **16,** 789 (1975).

[14] L. S. Privalle and R. H. Burris, *J. Bacteriol.* **157,** 350 (1984).

[15] P. Fay, this series, Vol. 69, p. 801.

[16] G. Neuer and H. Bothe, *Biochim. Biophys. Acta* **716,** 358 (1982).

[17] G. Eisbrenner, P. Roos, and H. Bothe, *J. Gen. Microbiol.* **125,** 383 (1981).

tracts from heterocysts" from *A. cylindrica* are obtained by breaking the heterocysts in the French press at 20,000 lb/in.[2] and by centrifuging (45,000 *g*, 20–60 min) to remove debris.

All procedures must be performed in ice and under argon when nitrogenase and pyruvate: ferredoxin oxidoreductase (pyruvate synthase) are to be assayed. The activity levels of glucose-6-phosphate dehydrogenase and isocitrate dehydrogenase are indicative of the purity of these preparations from *A. cylindrica:* Glucose-6-phosphate dehydrogenase activity is 6–8 times higher in heterocysts than in vegetative cells, whereas isocitrate dehydrogenase is almost inactive in extracts from vegetative cells. With other cyanobacteria (*Anabaena* 7119, *Anabaena* 7120, or *Anabaena variabilis*), the pressure in the French press is raised slightly (~2000 lb/in.[2]) to obtain optimal heterocyst preparations.[17]

Standard Assay Procedure

Electron donation to nitrogenase in the heterocyst preparation is routinely measured by the C_2H_2 reduction method using a gas chromatograph equipped with a flame-ionization detector and a Porapak R column (80–100 mesh, 3 ft × $\frac{1}{8}$ in., N_2 as the carrier gas[18]). The reaction can also be followed with an H_2 electrode when H_2 is the electron donor.[19] A Clark-type electrode has also been used for measuring the disappearance of N_2O catalyzed by nitrogenase. The experiment failed with heterocyst preparations, however, but not with intact *Klebsiella pneumoniae*.[20] The standard assay for the electron donor-dependent C_2H_2 reduction by heterocysts is performed in small flasks (serum bottles, Fernbach flasks, preferentially 5–7 ml) covered with Suba Seals. The vessels contain in a final volume of 2 ml[8]: creatine phosphokinase, 0.3 mg, heterocysts with 0.4–0.6 mg protein, and, in μmol, the following: HEPES/KOH buffer, pH 7.6, 100; $MgCl_2$, 10; ATP, 15; dithiothreitol, 5; creatine phosphate, 25. The assay has to be supplemented with one of the different electron donors (5 m*M* each if not indicated otherwise): NADPH; glucose 6-phosphate plus $NADP^+$ (0.05 m*M*); isocitrate plus $NADP^+$ (0.05 m*M*); H_2 (10 m*M*); $Na_2S_2O_4$ (10 m*M*); pyruvate (10–20 m*M*) plus coenzyme A (0.25 m*M*); malate plus NAD^+ (0.05 m*M*); glyoxylate; glycolate; or erythrose (the latter not with *A. cylindrica* but with *Anabaena* 7120 and *A. variabilis*). The reaction is started by injecting C_2H_2 through the Suba Seals [10% (v/v) final concentration], and the experiment is performed in a shaking water bath at 28 and 35,000 lux for 15–16 min.

[18] H. Bothe, J. Tennigkeit, G. Eisbrenner, and M. G. Yates, *Planta* **133,** 237 (1977).
[19] H. Papen, T. Kentemich, T. Schmülling, and H. Bothe, *Biochimie* **68,** 121 (1986).
[20] W. Zimmer, G. Danneberg, and H. Bothe, *Curr. Microbiol.* **12,** 341 (1985).

With H_2 as the electron donor, rates are 10–20% of those of intact filaments. The addition of ATP and the ATP-generating system creatine phosphate and creatine phosphokinase enhances the activity to a variable extent. Claims have been forwarded, however, that these compounds are not required.[21] The heterocysts contain the enzymes NADPH : ferredoxin oxidoreductase, glucose-6-phosphate dehydrogenase, malate dehydrogenase, pyruvate synthase (pyruvate : ferredoxin oxidoreductase), isocitrate dehydrogenase, and glyceraldehyde-3-phosphate dehydrogenase[8,22] in such quantities to meet the requirements for electron donation to nitrogenase (see Fig. 2).

Assay Conditions for Pyruvate : Ferredoxin Oxidoreductase (Pyruvate Synthase)

Pyruvate is one of the electron donors to nitrogenase in heterocysts and is converted in the presence of coenzyme A and oxidized ferredoxin to acetyl-CoA, CO_2, and reduced ferredoxin. The occurrence of this enzyme in heterocysts was debated, and the tests seem to be critical. The enzyme can be assayed either by the decarboxylation of $^{14}CO_2$ from carboxyl-labeled pyruvate, by following the reduction of ferredoxin or methyl viologen spectrophotometrically, by the formation of acetyl-CoA from pyruvate and coenzyme A which can be coupled to the synthesis of citrate from oxaloacetate in the presence of citrate synthase, by pyruvate-dependent C_2H_2 reduction, or by the reverse reaction, the synthesis of [^{14}C]pyruvate from acetyl-CoA, $^{14}CO_2$, and reduced ferredoxin or methyl viologen.[8,16] Strict anaerobic conditions are mandatory for assaying pyruvate : ferredoxin oxidoreductase (pyruvate synthase). The activities of the enzyme do not appear to be controlled by adenosine phosphates,[8,16] in contrast to earlier findings.[23] However, the concentration of coenzyme A is critical in all assays, particularly in the pyruvate-dependent C_2H_2 reduction by heterocyst preparations.[8]

The assay of [^{14}C]pyruvate formed from $^{14}CO_2$, acetyl-CoA and reduced ferredoxin (or methyl viologen) is indicative of the occurrence of the enzyme because pyruvate is not synthesized by the pyruvate dehydrogenase complex or pyruvate decarboxylase. This so-called reverse reaction is typically performed in a final volume of 3 ml containing, in μmol the following[16]: Tris–HCl buffer, pH 7.6, 500; $MgCl_2$, 20; dithiothreitol, 30; coenzyme A, 0.5; acetyl phosphate, 50; thiamin pyrophosphate, 2;

[21] H. Böhme and H. Almon, *Biochim. Biophys. Acta* **722,** 401 (1983).
[22] H. Papen, G. Neuer, A. Sauer, and H. Bothe, *FEMS Microbiol. Lett.* **36,** 201 (1986).
[23] C. K. Leach and N. G. Carr, *Biochim. Biophys. Acta* **245,** 165 (1971).

deazaflavin, 0.5; methyl viologen, 2; $NaH^{14}CO_3$, 40 (0.1 Ci/mol); phosphate acetyltransferase (Boehringer), 20 U; extract from heterocysts, 0.5–1.0 mg protein. Methyl viologen effectively substitutes for ferredoxin. Deazaflavin (potassium salt of 7,8-bisnor-5-deazalumiflavin-3-propanesulfonic acid) is the photoreductant of methyl viologen and can be substituted by broken spinach chloroplasts.[24] After incubating the reaction mixture at 28° and 35,000 lux for 2 hr, the radioactive pyruvate formed is precipitated as phenylhydrazone and counted by scintillation spectrometry.[25] Rates of this reverse reaction are normally 50-fold lower than those of the forward reaction.

The easiest way to measure the enzyme activity is to follow the reduction of methyl viologen spectrophotometrically in an anaerobic cuvette sealed with rubber stoppers containing in a volume of 1 ml, in μmol, the following: HEPES/KOH buffer, pH 7.6, 100; $MgCl_2$, 6; dithiothreitol, 7; thiamin pyrophosphate, 2.5; methyl viologen, 5 (or ferredoxin, 80 nmol); coenzyme A, 0.25. After repeatedly evacuating and flushing with argon, extract from heterocysts (0.15–0.35 mg protein) is injected into the cuvette. The reaction is started by adding pyruvate.[20] The reduction of methyl viologen is followed at 578 nm ($\varepsilon_{578\,nm} = 9{,}400\ M^{-1}\ cm^{-1}$ [26]). Alternatively, reduction of ferredoxin can be recorded over the spectral range 350–600 nm.[8]

[24] H. Bothe, B. Falkenberg, and U. Nolteernsting, *Arch. Microbiol.* **96,** 291 (1974).

[25] G. Neuer, Ph.D. thesis. University of Cologne, Cologne, Federal Republic of Germany, 1982.

[26] H. P. Blaschkowski, G. Neuer, M. Ludwig-Ferstl, and J. Knappe, *Eur. J. Biochem.* **123,** 563 (1982).

[53] Hydrogenases: Isolation and Assay

By K. K. Rao and D. O. Hall

Types and Distribution of Hydrogenases

In cyanobacteria three different types of enzymes are involved in H_2 metabolism.[1–3] (1) An *uptake hydrogenase* mainly catalyzes the consumption of H_2. This enzyme is bound to the heterocyst membranes of the nitrogen fixers and has also been found in the membranes of the non-

[1] G. R. Lambert and G. D. Smith, *Biol. Rev.* **56,** 589 (1981).

[2] H. Bothe, *Experientia Suppl.* **43,** 65 (1982).

[3] J. P. Houchins, *Biochim. Biophys. Acta* **768,** 227 (1984).

nitrogen-fixing cyanobacterium *Anacystis nidulans*. Only electron acceptors with a positive midpoint redox potential actively support H_2 uptake by this hydrogenase. This enzyme catalyzes H_2 consumption associated with both photosynthetic and respiratory electron transport, although it is uncertain as to whether the same enzyme is involved in both pathways or whether two separate enzyme activities occur in the membranes. (2) A *soluble hydrogenase* preferentially catalyzes H_2 evolution with artificial electron donors such as reduced methyl viologen. Since this enzyme also catalyzes H_2 uptake, albeit with very low activity, it is often referred to as reversible hydrogenase. Reversible hydrogenase activity has been reported in the cytoplasm of both heterocysts and vegetative cells of cyanobacteria—again, it is not definitely established as to whether the same type of "reversible" hydrogenase occurs in both types of cells. This enzyme couples very poorly with biological redox mediators such as ferredoxin and flavodoxin. The physiological role of the enzyme and the *in vivo* electron donor to the enzyme, if any, are not known. (3) *Nitrogenase-associated hydrogenase* activity occurs in heterocysts and is responsible for the irreversible evolution of hydrogen. It should be borne in mind that cyanobacterial species vary markedly in their hydrogen metabolism activities and in the composition of the three types of hydrogenase, which will depend on the species, strains, age of the culture, and growth conditions.

Problems Encountered in the Isolation of Hydrogenases

The two major problems associated with the isolation and purification of hydrogenases are their separation from phycobilins and the instability of the isolated hydrogenase activity. Some of the phycobiliproteins cosediment with the hydrogenase during ammonium sulfate fractionation and comigrate with the enzyme in ion-exchange and/or gel-permeation chromatography. A series of exploratory experiments are therefore required to select the best conditions for purification of a particular hydrogenase. The isolated enzyme is extremely unstable in oxygen; in fact, the enzyme activity is not very stable even when stored in liquid N_2. There are also practical difficulties in separating cyanobacterial filaments into heterocysts, pure membranes, and soluble fractions when attempting to localize the hydrogenase activities in the cells.

Isolation of "Soluble" Hydrogenases from Cyanobacteria

Because the enzyme is oxygen-sensitive, the buffers used in all isolation procedures are flushed with nitrogen or argon gas, and all steps are carried out at 0–5°.

Hydrogenase from the Nonheterocystous Cyanobacterium Spirulina maxima (Procedure of Llama et al.[4]*).* Fresh cells of *S. maxima* grown in natural populations in Lake Texcoco, Mexico, and supplied by Sosa Texcoco are stored in liquid N_2. Wet cells (50 g) are thawed, suspended in 250 ml buffer (20 m*M* Tris–HCl, pH 8.0, containing 10 m*M* mercaptoethanol), and stirred for an hour to lyse the cells. The lysate is centrifuged for 20 min at 40,000 *g*. The sediment is discarded, and the supernatant is filtered through cheesecloth (muslin) and the filtrate centrifuged for 1 hr at 100,000 *g*. The blue supernatant is decanted and applied to a DEAE–cellulose (Whatman DE-52) column, 2.5 × 30 cm, equilibrated with buffer. The column is washed with 0.2 *M* KCl in buffer until no more blue material is eluted. Then 0.3 *M* KCl in buffer is passed through the column, and 10-ml fractions of the eluate are collected and assayed for H_2-evolution activity.

Fractions with hydrogenase activity are pooled, and solid ammonium sulfate is added to 65% saturation. After stirring for 30 min the mixture is centrifuged for 30 min at 40,000 *g*. The sediment is resuspended in a minimum volume of buffer and applied to a 2.2 × 90 cm column of Sephacryl S-200 (Pharmacia Biochemicals) equilibrated with 50 m*M* KCl in buffer. The active eluate fractions are pooled and, if required, are concentrated by ultrafiltration in a Diaflo apparatus fitted with an XM50 membrane (Amicon, Lexington, MA). The results of a typical preparation are given in Table I, where the activity is expressed in units of micromoles H_2 evolved per hour with reduced methyl viologen as electron donor. Further purification of the enzyme from contaminating proteins (as judged by polyacrylamide gel electrophoresis) can be achieved by rechromatography of the Sephacryl S-200 eluate on a DE-52 column equilibrated with 50 m*M* KCl, washing with 0.2 *M* KCl, and eluting with a linear KCl gradient (0.2–0.3 *M*), all in buffer.

The hydrogenase in the S-200 fraction has a molecular weight of 56,000 (± 2,000) as determined by gel filtration. The purified enzyme was able to evolve H_2 only with reduced methyl or benzyl viologens; no activity could be observed with reduced *Spirulina* ferredoxin, NADH, NADPH, or FMNH. The presence of ATP and Mg^{2+} (at 5 m*M*) did not show any activation effect on H_2 evolution, suggesting that the enzyme was free of nitrogenase activity. Carbon monoxide reversibly inactivated hydrogen evolution by the enzyme.

Partial Purification of Two Hydrogenases from Oscillatoria limnetica (Procedure of Belkin et al.[5]*).* Fresh cells of the facultatively anoxygenic phototrophic, nonheterocystous cyanobacterium *O. limnetica* grown in

[4] M. J. Llama, J. L. Serra, K. K. Rao, and D. O. Hall, *FEBS Lett.* **98,** 342 (1979).

[5] S. Belkin, K. K. Rao, and D. O. Hall, *Biochem. Int.* **3,** 301 (1981).

TABLE I
PURIFICATION OF HYDROGENASE FROM *Spirulina maxima*[a]

Step	Total protein (mg)	Total activity (units)[b]	Specific activity (units/mg protein)	Purification (-fold)	Yield (%)
Lysate	12,800	12,860	1.0	1.0	100
40,000 *g* supernatant	1,207	6,265	5.2	5.2	48.7
100,000 *g* supernatant	800	6,027	7.5	7.5	46.8
DE-52 eluate	107	3,750	35.0	35.0	29.0
Sephacryl S-200	6.6	1,453	112.4	112.0	11.3

[a] From Llama *et al.*[4]
[b] The activity units are μmol of H_2 evolved/hr.

natural populations in Solar Lake, Israel, are stored at −20° until needed. Frozen cells (50 g) are thawed, suspended in buffer (20 m*M* Tris–50 m*M* EDTA, pH 7.8), and stirred for 10 min. The suspension is sonicated for 3 min with a Dawe soniprobe (Dawe Instruments Ltd., Concord Road, London) at 4 A and the sonicate centrifuged at 35,000 *g* for 30 min. The sediment is resuspended in the buffer, sonicated, and spun as before. The combined supernatants are centrifuged at 77,000 *g* for 1 hr and the sediment discarded. To the supernatant ammonium sulfate is added to 75% saturation and the mixture centrifuged at 35,000 *g* for 30 min. The sediment is resuspended in buffer, dialyzed, and then adsorbed on top of a DEAE–Sephacel (Pharmacia) column. Elution is carried out with a linear gradient of NaCl (0.1–1 *M*) in buffer. The eluted fractions contain two hydrogenase activities eluting at 0.18 and 0.4 *M* NaCl, respectively. The two fractions are purified separately by rechromatography on DEAE–Sephacel followed by gel filtration on Sephacryl S-200. The relative activities of the two hydrogenases are shown in Table II.

Both hydrogenases evolved hydrogen using reduced methyl viologen as electron donor; the apparent K_m values were 0.023 and 0.26 m*M* for forms I and II, respectively. Unlike *S. maxima* hydrogenase, the enzymes isolated from *O. limnetica* were able to utilize *O. limnetica* ferredoxin, reduced with sodium dithionite, for hydrogen formation in *in vitro* assays. Both hydrogenases catalyzed H_2 uptake with methylene blue or methyl viologen as electron acceptor, methylene blue (E_0 = 11 mV) being much more readily reduced than methyl viologen (E_0 = −440 mV). Hydrogenase I was able to couple with photosynthetically reduced ferredoxin and evolve H_2, albeit with low activity. Both forms of hydrogenase were unstable in the presence of O_2.

TABLE II
PARTIAL PURIFICATION OF *Oscillatoria limnetica* HYDROGENASES[a]

Step	Total protein (mg)	Total activity (units)[b]	Specific activity (units/mg protein)	Purification (-fold)	Recovery (%)
Crude sonicate	5,190	4,418	0.85	1	100
35,000 *g* supernatant	1,875	3,225	1.72	2.0	73
77,000 *g* supernatant	1,540	3,004	1.95	2.3	68
75% $(NH_4)_2SO_4$	311	928	2.98	3.5	21
First DEAE–Sephacel					
Form I	41	353	8.7	10.2	8.0
Form II	43	530	10.5	12.4	12
Sccond DEAE–Sephacel					
Form I	15	191	13.0	15.3	4.3
Form II	19	295	15.6	18.4	6.7
Sephacryl S-200					
Form I	6	106	17.6	20.7	2.3
Form II	6	158	25.8	30.3	3.6

[a] From Belkin *et al.*[5]

[b] The activity units are μmol of H_2 evolved/hr.

Hydrogenase from the Thermophilic Cyanobacterium Mastigocladus laminosus (Procedure of Rieder and Hall[6]*)*. Cells of *M. laminosus* are collected from hot springs (55–60°) near Reykjavik, Iceland. The cells are suspended in buffer (50 m*M* Tris–HCl, pH 8.0), and filaments are disrupted using a Polytron blender (Kinematica GmbH, Lucerne, Switzerland). The homogenate is sonicated using a Dawe Soniprobe for 5 min and the sonicatc centrifuged at 40,000 *g* for 1 hr. Ammonium sulfate is added to the supernatant to 35% saturation, and after centrifugation at 40,000 *g* for 20 min the pellet containing chlorophyll material is discarded. More ammonium sulfate is added to the supernatant to 55% saturation and the mixture centrifuged again. The resulting pellet is suspended in a small volume of buffer containing 0.1 *M* NaCl and 20 m*M* $MgCl_2$. This fraction contains 54% of the hydrogenase activity originally present in the sonicate and has a specific activity of 0.72 U/mg protein. Further purification to a specific activity of 10 U/mg protein is achieved by gel filtration on Sephadex G-100. Gel filtration and ion-exchange chromatography techniques fail to separate completely the phycobiliproteins from the hydrogenase activity present in the 35–55% ammonium sulfate cut.

[6] R. Rieder and D. O. Hall, *Biotechnol. Lett.* **3,** 379 (1981).

Assay of Hydrogenase Activity

The following characteristics of the hydrogenases found in cyanobacteria help distinguish between the different types of hydrogen metabolism carried out by the respective enzymes.[7,8] Nitrogenase-mediated H_2 evolution is inhibited by low concentrations (<1 m*M*) of reduced methyl viologen and by C_2H_2, but this enzyme activity is insensitive to CO. Hydrogen production with reduced methyl viologen catalyzed by the soluble (reversible) hydrogenase is insensitive to C_2H_2 but is inhibited by CO and by O_2. Hydrogen consumption by the uptake hydrogenase (membrane-bound) is sensitive to CO but is unaffected by atmospheric levels of O_2. Thus, in whole filaments of cyanobacteria, H_2 evolution mediated by nitrogenase can be specifically assayed by incubating the filaments in light, in an argon–carbon monoxide (4%) atmosphere, and measuring the H_2 production at various time intervals. The uptake hydrogenase activity can be measured by the oxyhydrogen (Knallgas) reaction in a gas phase containing 20% O_2. Reversible hydrogenase activity can be selectively assayed (in the presence of nitrogenase) as H_2 production from reduced methyl viologen (5 m*M*), if necessary in the presence of C_2H_2.

Hydrogenase Activity of Whole Cells of Cyanobacteria (See Also Chapter [70], This Volume). Nitrogenase-mediated H_2 production from whole cells of cyanobacteria can be assayed by incubating a suspension of cells in culture medium in glass vials (7–10 ml capacity) fitted with Suba Seal rubber stoppers (Gallenkamp, London) in an argon plus 4% CO atmosphere. The incubates are shaken continuously in a thermostatted (25–30°) water bath and illuminated with white incandescent light at 250 μmol photons m^{-2} sec^{-1}. Hydrogen in the gas phase is monitored by withdrawing aliquots at intervals and assaying in a gas chromatograph fitted with a molecular sieve column and thermal conductivity detector.

Reversible hydrogenase activity of the cells is usually measured as H_2 evolution from reduced methyl viologen. Since methyl viologen does not readily permeate the cell membrane, the filaments are either gently sonicated in buffer (20 m*M* phosphate, pH 7, or 50 m*M* Tris, pH 8) or subjected to a freeze (liquid N_2)–thaw (H_2O at 50°) cycle to permeabilize the cell walls.[9] Cells thus modified are incubated in stoppered glass vials in a N_2 atmosphere at 30° with 5 m*M* methyl viologen and 10 m*M* sodium dithionite. The H_2 evolved is measured at intervals using a gas chromatograph.

Evolution of H_2 from the cyanobacterial extract or whole cells can

[7] A. Daday, G. R. Lambert, and G. D. Smith, *Biochem. J.* **177,** 139 (1979).
[8] J. P. Houchins and R. H. Burris, *J. Bacteriol.* **146,** 209 (1981).
[9] G. D. Smith, A. Muallem, and D. O. Hall, *Photobiochem. Photobiophys.* **4,** 307 (1982).

also be followed amperometrically in a Clark-type electrode.[8,10,11] H_2 consumption catalyzed by the uptake hydrogenase can be measured by the oxyhydrogen (Knallgas) reaction as H_2 uptake in a hydrogen electrode at atmospheric levels of O_2.

Assay of Isolated Hydrogenases. H_2 evolution from reduced methyl viologen is determined either amperometrically using a hydrogen electrode or by the use of a gas chromatograph.[12,13] The latter method is routinely employed in the authors' laboratory. Fifty microliters of 100 m*M* methyl viologen and 1.8–1.85 ml 20 m*M* phosphate buffer, pH 7.0 (to make a final volume of 2 ml), are taken in a series of glass vials, closed with Suba Seal rubber stoppers and connected to a gas manifold via syringe needles. A vial containing 200 mg sodium dithionite and another with 2 ml phosphate buffer are also attached to the manifold. An additional syringe needle is inserted through the center of the Suba Seal stopper of each vial for nitrogen circulation. A continuous stream of oxygen-free N_2 is passed through the manifold into the vials for about 10 min. (For better removal of oxygen the manifold can be connected to a vacuum pump and a N_2 tank and the contents of the vials can be alternately exhausted and flushed with N_2.) Aliquots (10–50 μl) of the hydrogenase sample are then injected into the vials containing methyl viologen. Five milliliters of 50 m*M* Tris–HCl, pH 8.5, is injected into the vial containing sodium dithionite to give a 200 m*M* solution. After removing the syringe needles, all the vials except that containing sodium dithionite are transferred to a water bath maintained at 30°, and the bath is shaken vigorously. After incubation in the bath for 10- min 100-μl sodium dithionite solution is injected into each vial containing hydrogenases, at 30-sec intervals. The contents of the vials should turn deep blue. One hundred microliters of pure H_2 from a cylinder is then injected into the vial with buffer alone.

The H_2 evolved in each vial is determined by withdrawing, at intervals, aliquots from the gas phase with a gas-tight syringe, injecting into a gas chromatograph (GC) column, and noting the peak height on a recorder. The instrument is calibrated by withdrawing 10- to 100-μl aliquots of the H_2–N_2 mixture from the vial containing buffer only, injecting into the GC column and constructing a standard curve of peak height versus H_2 concentration. The GC in use in the authors' laboratory for measuring H_2 (Taylor Servomex, Crowborough, Sussex, UK) is fitted with a Poropak Q column and a thermal conductivity detector with N_2 as carrier gas.

[10] R. Wang, F. P. Healey, and J. Myers, *Plant Physiol.* **48,** 108 (1971).
[11] S. Kumazawa and A. Mitsui, *Appl. Environ. Microbiol.* **50,** 287 (1985).
[12] K. K. Rao, L. Rosa, and D. O. Hall, *Biochem. Biophys. Res. Commun.* **68,** 21 (1976).
[13] L. Packer, this series, Vol. 69, p. 625.

The H_2 peak in this device appears on the recorder about 13 sec after sample injection, and hence gas samples can be withdrawn from the vials and assayed at 30-sec intervals. If the response time of the GC is longer the addition of the dithionite to the vials should be spaced at longer intervals so that consecutive measurements can be made. Alternatively, the H_2 evolution can be stopped completely (say, after a 15-min incubation with dithionite) by the addition of 20% trichloroacetic acid to the vials, and the gas phase can then be assayed at leisure. Other gas chromatographs with microprocessor controls fitted with integrators and printers are available which directly print out the amount of H_2 injected into the GC.

The protein concentration of the hydrogenase sample can be determined by standard techniques (Lowry or Biuret) and the specific activity of the preparation expressed as micromoles H_2 evolved per milligram protein per hour (or minute). For comparative purposes the activity can also be reported as micromoles H_2 evolved (or consumed) in 1 hr per gram of wet cells or per milligram of chlorophyll.

Spectrophotometric Assay for H_2 Uptake. The reduction of dyes by H_2, catalyzed by hydrogenases, can be followed by measuring absorbance changes in a standard spectrophotometer.[14] The electron acceptors usually used and their extinction coefficients are as follows: methyl viologen, 12,000 M^{-1} cm^{-1} at 600 nm; benzyl viologen, 8,100 M^{-1} cm^{-1} at 555 nm; and methylene blue, 7,000 M^{-1} cm^{-1} at 601 nm. The reaction mixture (3 ml) of 20 mM potassium phosphate buffer, pH 7.0, and the electron acceptor (1 mM) are added to the main compartment of an anaerobic cuvette (e.g., No. 26, Starna Ltd., London) and the hydrogenase in the side arm. After repeated degassing and flushing of the cuvette with H_2 at atmospheric pressure, the cuvette is placed in the sample cavity of a thermostatted (30°) spectrophotometer. The reaction is started by tipping the hydrogenase into the main compartment and mixing thoroughly. The reduction of the acceptor with time is followed by recording the absorbance change. Since the enzyme has to be reduced by H_2 before it becomes active there may be an initial lag period of 5–10 min before constant rates of dye reduction are observed.

Reduction of methyl viologen by hydrogenase in the presence of H_2 is the classic method for detecting and locating hydrogenase activity in a mixture of proteins.[15] The sample is subjected to analytical disk gel electrophoresis, and the gel rods are stained with Coomassie blue and destained electrophoretically to detect the protein bands. The gel is then

[14] M. W. W. Adams and D. O. Hall, *Arch. Biochem. Biophys.* **195,** 288 (1979).
[15] B. A. C. Ackrell, R. N. Asato, and H. F. Mower, *J. Bacteriol.* **92,** 828 (1966).

transferred to a test tube containing 2.5% (w/v) methyl viologen in 20 m*M* phosphate buffer, pH 7.0, and an equal volume of 2,3,5-triphenyltetrazolium chloride [2.5% (w/v)] is added. The tube is closed with a Suba Seal stopper and incubated under H_2 for 1 hr or more. The methyl viologen becomes reduced by the hydrogenase, and reduced methyl viologen reacts with the tetrazolium salt to give a bright red band.[14]

Hydrogenases from Other Cyanobacteria. Two distinct hydrogenases, "uptake" and "reversible," have been partially purified and characterized from *Anabaena* sp. Strain 7120 by Houchins and Burris.[16] Kerfin and Boger[17] have reported partial purification of a soluble hydrogenase from *Nostoc muscorum*. Cell-free preparations catalyzing both H_2 uptake and H_2 evolution activities have been obtained from the nonheterocystous non-nitrogen-fixing cyanobacterium *Anacystis nidulans*.[18]

Acknowledgments

We are indebted to J. L. Serra, M. J. Llama, R. Rieder, S. Belkin, and M. W. W. Adams for some of the work described in this article, and to the European Economic Community (DGXII), Brussels, for grant support.

[16] J. P. Houchins and R. H. Burris, *J. Bacteriol.* **146,** 215 (1981).
[17] W. Kerfin and P. Boger, *Symp. Tech. Mikrobiol., 4th,* p. 313 (1979).
[18] G. A. Peschek, *Biochim. Biophys. Acta* **548,** 187 (1979).

[54] Measurement of Bioenergetics Phenomena in Cyanobacteria with Magnetic Resonance Techniques

By ROLF J. MEHLHORN and RICHARD SULLIVAN

Introduction

Electron spin (ESR) and nuclear magnetic (NMR) resonance methods can yield information about intracellular concentrations of organic and inorganic solutes without cell disruption. ESR provides substantially greater sensitivity than NMR, but relies on probes, which may adversely affect structure or function. NMR detects molecules and ions that are present naturally, or employs isotopes of natural molecules that do not perturb biological structure significantly. Thus, each method has its advantages. For organic solutes being observed with carbon-13 NMR, sensitivity is greatly enhanced with enrichment of this isotope. Enrichment provides the added advantage that pulse–chase strategies can be used to monitor carbon turnover processes. The ESR method has recently been expanded to include measurements with multiple probes that are sensitive to pH; this has raised the possibility of determining the pH values of several subcellular compartments and, in principle, the volumes of these compartments in unbroken cells.

Transmembrane electrochemical potentials are integral to energy-conversion processes in biological organisms as articulated in Mitchell's chemiosmotic hypothesis.[1] For many years, the measurement of these potentials relied on extracellular probes, whose intracellular concentrations could not be measured directly, but, rather, required making assumptions about the distribution of the probes in intracellular environments.[2] Of particular concern was the problem of probe binding to membranes, since many probes that were utilized were either hydrophobic or amphiphilic. The introduction of ESR probes, in conjunction which membrane-impermeable paramagnetic complexes, allowed direct determinations of the distribution of probes to be made.[3-8] Spectral resolution of probe molecules that were tumbling freely in aqueous solution from those bound to membranes or macromolecules made it possible to perform accurate and unambiguous calculations of pH gradients and electrical potentials, free of assumptions about binding effects. This chapter deals with the application of these magnetic resonance methods to the problem of cyanobacterial bioenergetics, particularly as it pertains to mechanisms of salt tolerance.

Methods

Confirmation of Probe Purity

The measurement of volumes, pH gradients, and electrical potentials can be distorted considerably if probes are impure. We have found that commercial nitroxide preparations are not always reliable and now routinely confirm that volume probes are inert toward electrochemical poten-

[1] P. Mitchell, "Chemiosmotic Coupling in Oxidative and Photosynthetic Phosphorylation." Glynn Research, Bodmin, Cornwall, England, 1966.

[2] H. Rottenberg, this series, Vol. 55, p. 547.

[3] A. T. Quintanilha and R. J. Mehlhorn, *FEBS Lett.* **91,** 104 (1978).

[4] R. J. Mehlhorn and I. Probst, this series, Vol. 88, p. 334.

[5] R. J. Mehlhorn, P. Candau, and L. Packer, this series, Vol. 88, p. 751.

[6] R. J. Mehlhorn and L. Packer, *Ann. N.Y. Acad. Sci.* **414,** 180 (1983).

[7] R. J. Mehlhorn, L. Packer, R. Macey, A. T. Balaban, and I. Dragutan, this series, Vol. 127, p. 738.

[8] P. Candau, R. J. Mehlhorn, and L. Packer, *in* "Photosynthetic Procaryotes: Cell Differentiation and Function" (L. Packer and G. Papageorgiou, eds.), p. 91. Elsevier, New York, 1983.

tials and that pH- or electrical potential-responsive probes behave as expected in liposome model systems. Liposomes are prepared in buffer of defined pH and then treated with either acid or base to establish pH gradients across their membranes. Asolectin, a heterogeneous lipid extract of soybeans, readily forms liposomes and has proven adequate for these purposes.[9] pH measurements with an electrode before and after imposition of the pH gradient establishes the magnitude of the gradient that should be observed. Then the volumes and pH gradients are measured with ESR probes.

As described in previous publications, the ESR method consists of adding a membrane-permeable spin probe with or without a membrane-impermeable paramagnetic ion complex to cells or vesicles, and measuring ESR signal intensities.[4,5] The relative line height before and after addition of the paramagnetic ion complex represents the internal fraction of the spin probe. Probes synthesized in our laboratory, or previously established to be reliable, are used as standards, and their behavior in the liposomes is checked against the behavior of the new (or suspect) spin probes. The purity of electrical potential probes is also checked in liposomes that have been treated with buffers to maintain pH gradients. For electrical potential measurements, we exploit the fact that liposomes prepared in impermeable buffers are selectively permeable to protons and establish electrical potentials[10] whose magnitudes can be calculated from the magnitude of the imposed pH gradient ($\Delta\psi = -\Delta pH$).

Confirmation That Paramagnetic Quenching Agents Do Not Enter Cells or Impair Photosynthetic Functions

Penetration into cells of paramagnetic quenching agents like MnEDTA,[11] NiTEPA (Ni-tetraethylene pentamine),[5] or the dissociated free paramagnetic ions, can be detected in terms of an apparent volume loss as measured from line heights of spin probes like 2,2,6,6-tetramethylpiperidone-*N*-oxyl (Tempone). Tempone is well suited for measuring volumes because it has intrinsically narrow linewidths, that can readily be discerned from lines that are broadened by collisions with paramagnetic quenching agents. However, its narrow linewidths also render Tempone highly sensitive to other line-broadening effects, like viscosity, which is known to be higher inside of cells than in bulk water.[12] To ensure that Tempone volume measurements are not distorted by intracellular line-broadening phenomena, it is important to establish, in control experi-

[9] V. K. Miyamoto and W. Stoekenius, *J. Membr. Biol.* **4,** 252 (1971).
[10] D. S. Cafiso and W. L. Hubell, *Biophys. J.* **44,** 49 (1983).
[11] T. L. Lomax and R. J. Mehlhorn, *Biochim. Biophys. Acta* **821,** 106 (1985).
[12] A. D. Keith and W. Snipes, *Science* **183,** 666 (1974).

ments, that another nitroxide with a different intrinsic linewidth gives the same values of cell volume. We have found a proxyl nitroxide (containing a five-membered saturated ring) with a single alcohol group on the ring to be useful for checking volumes inferred from Tempone. The five-membered ring nitroxides are advantageous for biological studies because of their resistance to chemical reduction.

Since cells are known to reduce nitroxides to their nonparamagnetic hydroxylamines, either through respiratory/photosynthetic electron transport activity or via thiol-mediated reduction,[13] slow accumulation of the quenching agents in cells cannot readily be discerned from chemical probe reduction. Therefore, experiments to check impermeability to the quenchers should be performed by incubating cells with the paramagnetic complexes prior to adding the volume probe. Such studies are also potentially valuable for checking membrane integrity, since increased leakiness to the paramagnetic complexes, as detected with volume measurements, represents a sensitive measurement of compromised membrane permeability barriers.

To assess the functional integrity of cyanobacterial cells in the presence of quenching agent, rates of photosynthetic oxygen evolution and respiratory oxygen consumption, as measured by polarographic methods, can be used. Light attenuation by the colored quenching agents must be taken into account, e.g., by checking the effect of inserting a light filter containing the quencher solution between the assayed cell suspension and the light source.

Ion Permeabilities as Inferred from ESR Volume Measurements

ESR volume measurements also provide a useful and rapid assay system for assessing whether or not nonparamagnetic solutes can enter cells.[14,15] Impositions of hypertonic concentrations of an impermeable solute is reflected as a decrease in volume. The extent of volume compression expected in the absence of membrane damage can be calculated from the assumption that cyanobacterial cells behave as perfect osmometers, an assumption that has been validated for a large variety of cells and subcellular membranes.[11,16–18] If the cells reswell, this reflects a permeabil-

[13] R. J. Mehlhorn and L. Packer, this series, Vol. 105, p. 215.

[14] E. Blumwald, R. J. Mehlhorn, and L. Packer, *Proc. Natl. Acad. Sci. U.S.A.* **80,** 2599 (1983).

[15] E. Blumwald, R. J. Mehlhorn, and L. Packer, *Plant Physiol.* **73,** 377 (1983).

[16] B. A. Melandri, R. J. Mehlhorn, and L. Packer, *Arch. Biochem. Biophys.* **235,** 95 (1984).

[17] M. C. Ball, R. J. Mehlhorn, N. Terry, and L. Packer, *Plant Physiol.* **78,** 1 (1985).

[18] R. J. Mehlhorn, B. Schobert, L. Packer, and J. Lanyi, *Biochim. Biophys. Acta* **809,** 66 (1985).

ity of the membrane to the solute, and the rate of reswelling can be analyzed, in principle, to yield quantitative permeability information. This approach was employed to show that the membranes of *Synechococcus* 6311 were highly permeable to NaCl when the cells were treated with hypertonic solutions of this salt.[15] Cells treated with sucrose underwent plasmolysis and deplasmolysis, consistent with impermeability to this solute. The approach can readily be extended to other salts; indeed, we have performed preliminary experiments that showed a high permeability to KCl and a significantly lower permeability of *Synechococcus* 6311 to $NaHCO_3$. These preliminary experiments also showed that the membranes were much more sensitive to disruption by extracellular KCl than by comparable concentrations of NaCl.

Determination of Turgor Pressure

The accurate measurement of cell volumes with the ESR method lends itself to the analysis of excess osmotic pressure within cells with cell walls. The differential osmotic pressure between cell interior and exterior is of interest in analyzing responses of cyanobacteria to growth in saline media. This parameter should be used to determine, quantitatively, the total concentration of osmotically active solutes within cells. To determine the intracellular osmotic pressure, ESR volumes are measured as a function of increasing osmotic strength, using impermeable solutes. Assuming that the cells exhibit ideal behavior, a plot of ESR volumes as a function of inverse osmotic strength should yield a linear relationship, intersecting the origin, for high solute concentrations, but should be independent of solute concentration when cell volumes can no longer increase, due to the restraining force of the cell wall. This is illustrated in Fig. 1, which plots cell volumes as a function of increasing concentration of the quencher salt of MnEDTA. A reasonably accurate determination of the osmotic strength of the cytoplasm is made by determining the break in the plot, assuming that the two portions of the plot conform to the linear relationship stated above. By relating the results to the osmotic strength of the growth medium, the differential osmotic pressure is readily calculated. The data depicted in Fig. 1 suggest that cells grown at the high salinity exhibit a higher incremental cytoplasmic pressure than do cells grown in low salt media.

Assessment of Membrane Structure in Terms of Resistance to Salt-Induced Disruption

Previously it was observed that cells of two salt-tolerant cyanobacterial strains were permeable to NaCl[15] (unpublished observations) at salt

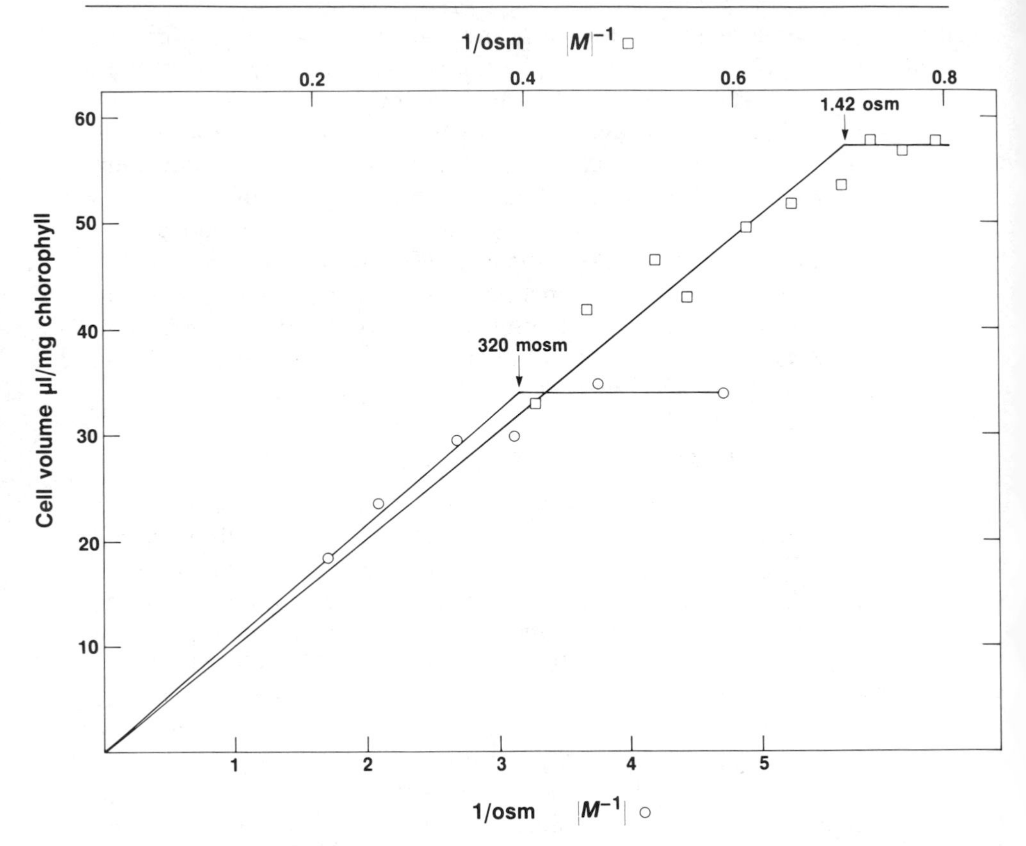

FIG. 1. Volumes of *Synechococcus* 6311 as a function of the inverse osmotic strength of the paramagnetic complex ditetramethylammonium manganese EDTA. The breaks in the plots, indicated by downward arrows, represent the osmotic strengths of the cytoplasms. Circles designate results from cells grown in buffer at an osmotic strength of about 80 mOs*m*, whereas squares denote cells grown in 0.5 *M* NaCl plus 80 mOs*m* buffer.

concentrations considerably higher than could be tolerated for growth (e.g., concentrations exceeding 1 *M* NaCl for *Synechococcus* 6311). At higher concentrations of salt, however, a loss of ESR volumes was observed that could not be reversed by subsequently suspending the cells in hypotonic solutions of either ionic or nonionic solutes. Treatment of the cells with hypertonic salt solutions in the presence of quenching agents showed that ESR volumes were substantially smaller than when cells were exposed to salt solutions in the absence of the quenching agent. This suggested that high-salt concentrations induced a transient membrane

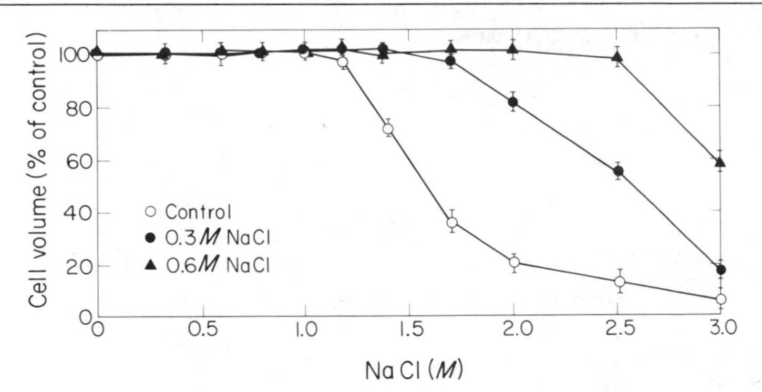

FIG. 2. Volumes of *Synechococcus* 6311 as a function of NaCl concentration for cells grown in low- and high-salt conditions. From Ref. 15.

disruption, which allowed quencher to enter, followed by a resealing of the membranes. The onset of salt-induced irreversible ESR volume loss was seen to depend on growth conditions of the cells. *Synechococcus* 6311 grown in high-salt conditions was considerably more resistant to volume loss than those grown in low-salt solutions (Fig. 2). Thus the threshold concentration of imposed salt that leads to a loss of ESR volumes appears to be a good index of membrane stability and correlates strongly with the growth salinity of the cells that we have studied.

Analysis of Organic Solute Turnover Using NMR Pulse–Chase Studies

The utilization of organic solutes, like sugars or betaines, for the maintenance of osmotic tonicity when cells are subjected to hypersaline environments can be exploited to study the metabolism of major organic solutes noninvasively by carbon-13 NMR.[19] Intracellular concentrations of sugars can reach several hundred millimolar under saline growth conditions. By enrichment with carbon-13, such concentrations of sugars can be detected with NMR methods, using only a few minutes of signal acquisition (Fig. 3). The strategy is to grow cells in the light with carbon-13 bicarbonate, causing it to be fixed into low- or high-molecular-weight molecules depending on the ionic strength of the growth medium. High-molecular-weight glycogen gives NMR features distinct from those of low-molecular-weight carbohydrates, particularly at about 101 ppm.[19] Previously this distinct spectral characteristic was used to analyze the interconversion of low- and high-molecular-weight carbohydrates under

[19] E. Tel-Or, S. Spath, L. Packer, and R. J. Mehlhorn, *Plant Physiol.* **82,** 646 (1986).

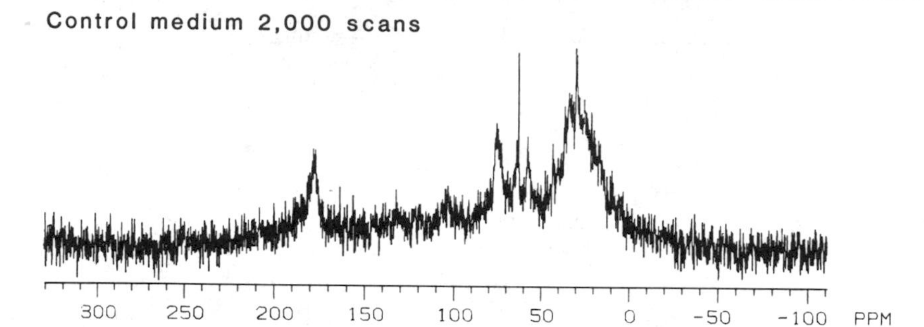

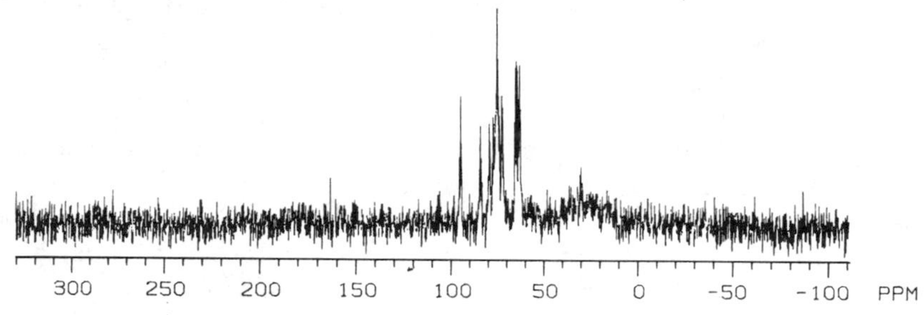

FIG. 3. ^{13}C-NMR spectra of *Synechococcus* 6311 grown in 10% carbon-13 bicarbonate under low and high salinity.

salt shock, both in the light and in the dark.[19] In combination with ESR volume measurements, NMR analysis of intracellular organic solutes can be used to obtain concentrations of solutes without disrupting cells.

In a typical experiment, cultures for carbon-13 NMR analysis are grown in 250-ml Erlenmeyer flasks containing 100 ml of buffered medium and 10 m*M* sodium bicarbonate enriched with 30% carbon-13, at 32–34°. (A 30% enrichment in carbon-13 provides high signal-to-noise ratios without undue distortion from carbon–carbon spin interactions.) To promote glycogen accumulation, cells are grown under limiting nitrate concentrations.[19] The harvested cells are concentrated by centrifugation (5000 *g* for 10 min) and resuspended in a final volume of 3.6 ml in the growth medium. Four-tenths milliliter of D_2O is added to provide a lock signal. The concentrated cells are placed in a 1-cm-diameter NMR tube containing two 1-mm-diameter capillaries with degassed methyl iodide as a line height standard. The Fourier transform, proton-decoupled ^{13}C-NMR spectra are obtained using a line broadening of 20 Hz and 1-sec intervals between pulses. NMR data can be obtained either within a few hours of harvesting cells or, if necessary, with frozen cells.

pH Gradient Determinations

The use of spin-labeled weak acids and amines for measuring pH gradients relies on the same principle as the volume measurement and exploits the free permeability of these probes in their uncharged forms as well as their impermeabilities as charged species to infer pH gradients. The method as it applies to single compartment cells and vesicles has been described previously.[4-6] The application of these probes to multicompartment cyanobacteria, having aqueous domains with different pH values, fully exploits the availability of both amine and acid probes to quantitate the pH values of two separate compartments, provided that the volumes of these compartments can be estimated by some independent technique, e.g., electron microscopy.

Consider the example of a cyanobacterium, like *Agmenellum quadruplicatum,* which contains a cytoplasmic and a thylakoid compartment. The thylakoid compartment is more acidic than that cytoplasm by at least one pH unit. Moreover, the fractional volume of the thylakoid compartment is probably less than 0.1. Taking these two factors together, we can confidently neglect the accumulation of weak acid probes in the thylakoids and treat the unquenched signal of spin-labeled weak acids as though it arose exclusively from the cytoplasm. Hence an analysis of the distribution of spin-labeled weak acids in such cyanobacteria yields the pH gradient across the plasmalemma. With a separate measurement of the extracellular pH, e.g., with a pH electrode, the cytoplasmic pH can be immediately calculated. However, since amines accumulate in acidic compartments they will be more concentrated in the thylakoids than in the cytoplasm. An analysis of the distribution of a spin-labeled amine will yield an intracellular concentration greater than expected for the cytoplasm, as calculated from the previously inferred cytoplasmic pH. The excess amine must reside in the thylakoids, and it can be quantitated if the thylakoid volume is known. This approach was used to estimate pH values of both cytoplasmic and thylakoid compartments in *A. quadruplicatum* both in the dark and in the light.[20]

In principle, spin-labeled diamines or dicarboxylates can be used to gain additional information about subcellular compartments. Whereas the distribution of simple amines is a direct function of pH gradient, a diamine will distribute as a quadratic function of pH gradient. Thus, a diamine should accumulate to a much greater extent in thylakoids of *Agmenellum quadruplicatum* than the spin-labeled monoamine that was used previously.[20] A mathematical analysis of ESR distribution data of four nitroxides (weak acid, volume probe, monoamine, and diamine) can yield the

[20] S. Belkin, R. J. Mehlhorn, and L. Packer, *Plant Physiol.* **84,** 25 (1987).

thylakoid volume directly and would obviate the need for a separate determination of this parameter. Spin-labeled diamines have been synthesized, and work is in progress to implement the research strategies we have just outlined.

Acknowledgment

We thank Dr. Ian Fry and Ms. Jana Hrabeta for providing cells. Supported by the Department of Energy (Grant DE-AT03-80-ER10637) and NSF R11-8405345.

Section IV

Physiology and Metabolism

[55] Buoyancy and Suspension of Planktonic Cyanobacteria

By R. L. OLIVER and A. E. WALSBY

Cyanobacteria are unique among phytoplankton in their ability to control their depth in lakes by buoyancy regulation. They derive their buoyancy from gas vesicles, which if present in sufficient quantity will offset the excess buoyant density of other cell constituents. Cyanobacteria may regulate buoyancy by one, or more, of three mechanisms: gas vesicle collapse; regulation of gas vesicle formation; or changes in other components, especially carbohydrate. This chapter describes the techniques developed to elucidate the mechanisms of buoyancy regulation in laboratory cultures and to study the phenomena of vertical migration or stratification by cyanobacteria in lakes.[1]

Proportion of Cells Floating and Sinking

The first measurement to be made in any investigation of buoyancy is the proportion of cells, filaments, or colonies that float or sink: almost none will have exactly the same density as water. For filaments or small colonies introduce a drop of the cyanobacterial suspension into the 0.1 or 0.2 mm gap between the coverslip and platform of a hemocytometer.[2] Let stand for 10 min and then use a microscope to count the number of cyanobacteria that have settled down onto the square grid ruled on the platform. Next, adjust the focus to count the number in the same area floating under the coverslip.

For very small cells that may take an hour or more to come to rest, draw up a drop of suspension into a Microslide (a flattened, thin-walled glass capillary tube). Place the filled Microslide flat on a microscope slide and seal the ends by heating fragments of embedding wax with a soldering iron. Place the slide flat in the bucket of a swing-out centrifuge. Centrifuge for 2 min at about 2000 m sec^{-2} (200 *g*): buoyant cells will be forced to the upper, and negatively buoyant cells to the lower, Microslide surfaces where they can be separately counted as before. The pressure generated by centrifugation (0.4 kPa) is insufficient to cause gas vesicle collapse and will not affect buoyancy.[3] For large colonies the proportion that are float-

[1] A. E. Walsby and C. S. Reynolds, *in* "The Physiological Ecology of Phytoplankton" (I. Morris, ed.), p. 371. Blackwell, Oxford, England, 1981.

[2] A. E. Walsby and M. J. Booker, *Br. Phycol. J.* **15,** 311 (1980).

[3] R. H. Thomas and A. E. Walsby, *J. Gen. Microbiol.* **131,** 799 (1985).

ing or sinking can be determined by settling the suspension in a sedimentation chamber.

Buoyant Density of Cells

The density of negatively buoyant cells, filaments, or colonies can be determined by isopycnic banding on continuous gradients of a silica sol, Percoll[4] (Pharmacia). Gradients can be formed by high-speed centrifugation: their final shape depends on (1) the medium with which Percoll is mixed, (2) initial density of the mixture, (3) centrifugal acceleration and time of run, and (4) rotor geometry and tube size.[5] A 15-ml tube containing 5 ml Percoll added to 5 ml of double-strength ASMI medium, when centrifuged in a 24° fixed-angle rotor at 25,500 *g* for 2 hr at a temperature of 5°,[4] will form a smooth gradient ranging in density from 1000–1200 kg m^{-3} with osmolality between about 30 and 50 mOsm kg^{-1} H_2O. A sample of a cyanobacterial suspension is layered on top of the preformed gradient and carefully centrifuged in a swing-out rotor at 1000–3000 m sec^{-2} (100–300 *g*) for 20 min. With gas-vacuolate organisms care must be taken not to generate sufficient pressure in the centrifuge to collapse gas vesicles; an acceleration of 1000–1500 m sec^{-2} (100–150 *g*) should not be exceeded. The actual critical acceleration can be calculated from the expression for centrifugal acceleration,[6] $a = p/h\rho$ where h is depth below the surface of the liquid, ρ the liquid density, and p the incipient collapse pressure of gas vesicles in the organisms.[6] For many gas-vacuolate cyanobacteria p exceeds 100 kPa. When h is 8 cm and ρ is 1200 kg m^{-3} the maximum permissible acceleration is 1042 m sec^{-2} (106 *g*).

Samples may be removed from the visible bands of cyanobacteria using a microsyringe, and the density measured by sinking droplets from the syringe in a calibrated kerosene–carbon tetrachloride gradient[7]; the isopycnic position of the droplets is determined from the cylinder graduations. The organic column, with a density from 996 to 1340 kg m^{-3}, is prepared by layering ten 100-ml mixtures of CCl_4 and kerosene (25 ml up to 70 ml CCl_4 in 5-ml intervals) in a 1000-ml graduated cylinder. The mixtures are introduced gently, in order of increasing density, into the bottom through a narrow-bore glass tube connected to a funnel by plastic tubing which can be clamped to adjust the flow rate. Before removing the glass tube the cylinder is given a single stir to assist in mixing the boundaries.[7] After 24 hr, diffusion will have resulted in the formation of a linear

[4] R. L. Oliver, A. J. Kinnear, and G. G. Ganf, *Limnol. Oceanogr.* **26,** 285 (1981).

[5] Pharmacia Fine Chemicals, "Percoll: Methodology and Applications." Pharmacia Fine Chemicals, Piscataway, New Jersey.

[6] A. E. Walsby, this series, Vol. 31, p. 678.

[7] D. A. Wolff, this series, Vol. 10, p. 85.

gradient that can be calibrated using drops of standard sucrose solutions, the densities of which are determined by refractive index measurement and reference to the table of Weast.[8]

The density of *buoyant* gas-vacuolate cyanobacteria has been determined by suspending cells in varying mixtures of medium and isolated gas vesicles[6] and observing whether the cells sink or float, using a hemocytometer.[2] Droplets of the mixture that results in cells being neutrally buoyant (50% floating and sinking) are applied to an organic column prepared in a 500-ml measuring cylinder using five CCl_4–kerosene mixtures ranging from 76 to 68% kerosene, which has a density range of around 950–1000 kg m^{-3} but is not linear over its entire length. The column is calibrated with drops of NH_3 solutions, whose densities are determined by specific gravity bottle. NH_3 is volatile and must not be allowed to evaporate. Large droplets of the gas vesicle suspension should be used in the column, and readings should be made rapidly because the organic solvents gradually pervade the droplet, causing gas vesicle collapse. The densities of filaments from two buoyant cultures of *Anabaena flos-aquae* measured using this method were 996.0 and 996.8 kg m^{-3}, respectively, compared with 998.2 kg m^{-3} for water at 20°.

Weight of Cellular Constituents

To determine the cause of density changes in organisms, information is needed on the mass and volume of each of the major cell constituents before and after the change. From this can be calculated the weight of each component under water.[9,10] The sum of these "ballast" weights divided by the volume of the cells is then equivalent to the excess density of the organism in water. Carbohydrate, protein, lipid, and phosphorus may account for 90% or more of cell weight.[3,9,10,11] Protein is determined by using Folin–Ciocalteau reagent.[12] The carbohydrate content can be determined on 2–4 ml of culture (~100 μg dry weight ml^{-1}) using anthrone reagent.[12] The total phosphorus concentration of cells is determined by the ammonium molybdate method after acid digestion.[13] The density of cell constituents can be determined by stepwise or gradient centrifugation in mixtures of appropriate liquids that do not dissolve them. For example,

[8] R. C. Weast, "CRC Handbook of Chemistry and Physics," 60th Ed. CRC Press, Boca Raton, Florida, 1980.

[9] R. L. Oliver and A. E. Walsby, *Limnol. Oceanogr.* **29,** 879 (1984).

[10] H. C. Utkilen, R. L. Oliver, and A. E. Walsby, *Arch. Hydrobiol.* **102,** 319 (1985).

[11] R. L. Oliver, R. H. Thomas, C. S. Reynolds, and A. E. Walsby, *Proc. R. Soc. London, Ser. B* **223,** 511 (1985).

[12] D. Herbert, P. J. Phipps, and R. E. Strange, *Methods Microbiol.* **5B,** 209 (1971).

[13] G. R. Bartlett, *J. Biol. Chem.* **234,** 466 (1959).

cyanophycin granules isolated from *Anabaena flos-aquae* were neutrally buoyant in a mixture of carbon tetrachloride–paraffin that was found by specific gravity measurement to have a density of 1394 kg m^{-3} (unpublished observations). Glycolipids isolated and purified by thin-layer chromatography from the same organism were isopycnic in a solution of NaCl of density 1050 kg m^{-3}.[3]

Gas Vesicle Volume

To estimate the buoyant lift provided by gas vesicles the absolute gas vacuole space is measured with a capillary compression tube[14] (Fig. 1). The tube has a water jacket to minimize temperature fluctuations. The precision bore capillary tube (30 nl mm^{-1}) and attached reservoir (2–4 ml) are filled completely with the gas-vacuolate suspension and then closed at the reservoir end with a nylon stopper through which passes a smooth, tight-fitting piston. The stopper is firmly attached to a metal plug with O rings that make a seal against an outer glass pressure tube. The screw-driven piston can be used to adjust the level of the meniscus in the capillary tube. The other end of the pressure tube is connected to a gas pressure system with a pressure gauge. The compression tube is mounted on the stage of a Vernier microscope with a scale reading to 0.05 mm and allowed to equilibrate thermally, ensuring that liquid remains in the capillary by using the screw adjustment. Once the meniscus position has stabilized, and a reading has been taken, pressure is slowly applied to 1.4 MPa to collapse the gas vesicles. The pressure is then slowly released and the new meniscus position is recorded. Its change from the first reading is used to calculate the gas vesicle volume. The sample volume in the capillary tube is determined from the difference in its weights filled with sample and empty. Gas vesicle volumes can be measured with a precision better than ±5%.[9,10]

Measurement of Cell and Sheath Volume

Calculation of the excess density of a cell requires measurement of cell volume, including the volume of mucilage sheath which may influence sinking rate. A sufficient volume of culture is concentrated by centrifugation to approximately 2 ml and mixed with a suitable marker compound which is excluded either from the cells or from the cells plus sheath.[9] [^{14}C]Sucrose (1 μCi ml^{-1} in 5 m*M* sucrose) penetrates the sheath but not the cells of *Anabaena*[15] and other cyanobacteria.[16] One milliliter of the [^{14}C]sucrose solution is added to the algal concentrate.

[14] A. E. Walsby, *Proc. R. Soc. London, Ser. B* **216,** 355 (1982).
[15] A. E. Walsby, *Proc. R. Soc. London, Ser. B* **208,** 73 (1980).
[16] A. E. Walsby, *Proc. R. Soc. London, Ser. B* **178,** 301 (1971).

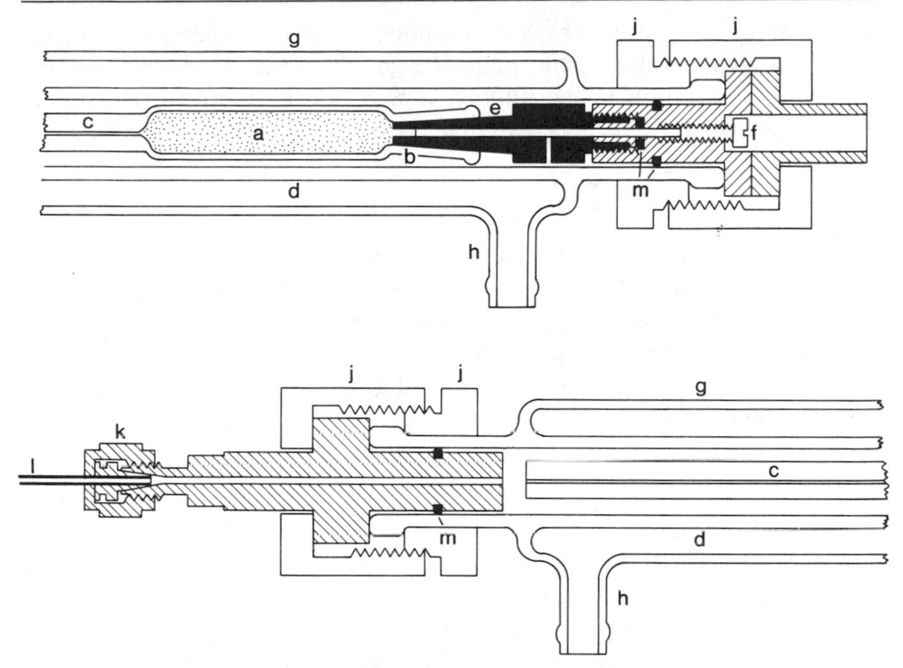

FIG. 1. Compression tube for measuring volume of gas vesicle gas space: a, tube containing suspension; b, socket; c, capillary; d, thick-walled glass pressure tube; e, conical stopper; f, screw that drives piston through stopper; g, glass water jacket; h, connection to water bath; j, screw couplings; k, pressure connector; l, Teflon tubing; m, O rings. Modified from Walsby.[14]

The volume of cell plus sheath is determined using blue Percoll prepared as follows.[9] Add 1 ml of a 1% (w/v) solution of cold-water dye (Dylon Int. Ltd., Riviera Blue) to 5 ml of Percoll and 2 ml of water. Dialyze overnight: the dye stays bound to Percoll in the sac. Adjust the volume to 12 ml to give a 42% Percoll mixture. Dilute this stock 1 : 4 with the appropriate medium and mix 1 ml with the algal concentrate. The blue Percoll is assayed by spectrophotometry at 596 nm (E_{max}).

Mix the algal concentrate with the appropriate exclusion-volume marker and place in a modified hematocrit centrifuge tube (a graduated tube 40–50 mm long and of 3–4 mm i.d. attached below a reservoir of volume ~5.5 ml). Weigh the tube plus sample, pellet the cells at 650 *g* in a swing-out centrifuge rotor, remove the supernatant, and make up the sample to a fixed volume (e.g., 5 ml = V_s). Reweigh the hematocrit tube to determine the supernatant volume removed (*a*).

Note the height of the pellet before removing it to a volumetric flask. Make up to a similar volume (V_p) before recentrifuging to sediment the cells. The pellet volume (p) is determined from the weight of water required to fill the tube to the pellet height. From a comparison of the marker concentration in the diluted supernatant (C_s) and in the diluted interstitial fluid (C_p) the cell volume (x) can be calculated[9,17] using the expression

$$x = p - \{[aC_p(V_p - p)]/[V_sC_s - aC_p]\}$$

We have shown that [^{14}C]sucrose penetrates the sheath and provides an estimate of cell volume, whereas blue Percoll does not penetrate the sheath and provides a measure of cell plus sheath volume.[9] The densities of cells measured by centrifugation on Percoll gradients correspond with densities calculated from the Percoll-excluded cell volume and combined ballast weights.[10]

Sinking Velocity on Sucrose Gradients

The sinking velocity of algal cells can be measured on linear sucrose gradients (between 0.1 *M* sucrose at the bottom and 0.01 *M* sucrose at the top) formed in 100-ml measuring cylinders.[18] Layer the sample over the gradient in a room temperature water bath to prevent convection. Allow the turbid algal band to sink about a third of the distance down the column, and then withdraw the liquid in a series of 5-, 10-, or 20-ml layers using a syringe needle held in contact with the surface meniscus. Preserve the samples with Lugol's iodine and determine the number of cells in each layer by counting with an inverted microscope. The distance that the cells would have sedimented in a column of water is given by the expressions[18] $y' = u_0t$ and

$$u_0t = \frac{(\rho - \rho')}{\rho' b\alpha}\left\{\left[1 + \frac{\beta(\rho - \rho')}{\rho'\alpha}\right]\ln\left[\frac{\rho - \rho'(1 + \alpha a)}{\rho - \rho'(1 + \alpha a) - \rho'\alpha by}\right] - \beta by\right\}$$

where ρ is the cell density (kg m^{-3}), ρ' the density of water, α the coefficient of specific gravity change per concentration (molar^{-1}), β the coefficient of relative viscosity change (η sucrose/η_0 where η_0 is viscosity of water) per concentration (molar^{-1}), a the sucrose concentration at the top of the gradient (molar), b the sucrose concentration gradient (molar^{-1}), y the measured depth (m), u_0 the sinking velocity in water, and t the sinking time.

The midpoint (y) of each depth sample removed from the sucrose gradient column at the end of the experiment is corrected (y') using the above equation, and the mean distance settled in time t is then calculated

[17] A. E. Walsby and A. Xypolyta, *Br. Phycol. J.* **12,** 215 (1977).

[18] M. J. Booker and A. E. Walsby, *Br. Phycol. J.* **14,** 141 (1979).

as $z = \Sigma(ny')/\Sigma n$ where n is the number of filaments counted in each depth sample and Σn the total number in all depth samples. The mean sinking velocity in water is then $\bar{u}_0 = z/t$. The gradient column may be used to measure floating velocities by introducing samples (made denser with sucrose or salt than the bottom layer) at the base of the column.[19] Nongradient columns are always unstable and are suitable only for measuring the sinking velocities of large, rapidly moving colonies.[1]

Theory of Sinking Velocity

The sinking velocity of a spherical cell is given by the Stokes equation, $u = \frac{2}{9}gr^2(\rho - \rho')/\eta\phi$, where g is the acceleration (due to gravity or centrifugation), r the radius of the sphere, and η the coefficient of viscosity. ϕ is the coefficient of form resistance, equal to u_s/u where u is the observed velocity and u_s the velocity of the sphere of identical volume and density. Exact solutions for ϕ are given by McNown and Malaika[20] for all classes of ellipsoids falling in various attitudes. These solutions may be applied to shapes that approximate to ellipsoids, with only small errors.[21] The Stokes equation applies in reverse for floating cells; when $\rho < \rho'$, u_s is negative and equals the floating velocity.

Concluding Remarks: Internal Checks on Measurements

If accurate measurements of cell size and density have been made by the methods given above, the sinking (or floating) velocity calculated by substitution in the Stokes equation should agree with the velocity measured by the sucrose gradient (e.g., see Fig. 3 of Davey and Walsby[21]). If the mass and density of the major cell constituents have been correctly measured the sum of their ballast mass divided by the cell volume should be equal to the excess density determined by the Percoll gradient (as e.g., results in Table 1 of Utkilen *et al.*[10]). Finally, changes in buoyancy, as occur in response to light intensity, should be accounted for by the changes in the ballast mass of each of the major constituents. In this way the physiological mechanisms of buoyancy changes can be elucidated.[3,9,10,11]

[19] A. E. Walsby, J. van Rijn, and Y. Cohen, *Proc. R. Soc. London, Ser. B* **217,** 417–447 (1983).

[20] J. S. McNown and J. Malaika, *Trans. Am. Geophys. Union* **31,** 74 (1950).

[21] M. C. Davey and A. E. Walsby, *Br. Phycol. J.* **20,** 243 (1985).

[56] Osmotic Adjustment: Organic Solutes

By ROBERT H. REED

Introduction

Cyanobacteria survive and grow in environments with dissimilar salinity regimes, ranging from freshwater through brackish and marine conditions to hypersaline habitats approaching salt saturation. Several recent studies have shown that long-term osmotic balance in salt-stressed cyanobacteria may be achieved by the synthesis of specific low-molecular-weight organic solutes. This was first demonstrated for the heteroside 2-O-α-D-glucopyranosylglycerol (glucosylglycerol, lilioside) in the marine unicell *Synechococcus* RRIMP N 100.[1] Subsequent research has identified other organic osmolytes in salt-stressed cyanobacteria: these include the disaccharides sucrose and trehalose and the quaternary ammonium compounds glycine betaine (trimethylglycine) and glutamate betaine (trimethylglutamate).[2] These metabolites share the common features of (1) a high solubility in water and (2) a lack of toxicity when assayed *in vitro* at physiologically relevant concentrations against a range of enzymes. In general, a single metabolite is accumulated, although certain strains may synthesize more than one organic solute, particularly in response to increased temperature.[3] This chapter outlines some of the experimental procedures used to identify and quantify these solutes.

Natural Abundance ^{13}C-Nuclear Magnetic Resonance Spectroscopy

^{13}C-NMR spectroscopy relies on the "resonance" of individual ^{13}C nuclei in a magnetic field between (1) a low-energy state (magnetic moment parallel to the applied field) and (2) a high-energy state (magnetic moment antiparallel to the applied field). Each ^{13}C nucleus gives a characteristic resonance peak, revealing the functional group to which the resonance belongs and enabling individual organic molecules to be identified by their ^{13}C "fingerprint." Natural abundance ^{13}C-NMR spectra of salt-stressed cyanobacteria are dominated by resonances from the low-molec-

[1] L. J. Borowitzka, S. Demmerle, M. A. Mackay, and R. S. Norton, *Science* **210,** 650 (1980).

[2] R. H. Reed, L. J. Borowitzka, M. A. Mackay, J. A. Chudek, R. Foster, S. R. C. Warr, D. J. Moore, and W. D. P. Stewart, *FEMS Microbiol. Rev.* **39,** 57 (1986).

[3] S. R. C. Warr, R. H. Reed, and W. D. P. Stewart, *New Phytol.* **100,** 285 (1985).

ular-weight organic osmolytes which are responsible, in part at least, for the generation of intracellular osmotic pressure. The technique has been used in recent years to screen a wide range of cyanobacteria for organic solutes, demonstrating that halotolerance may be linked to the class of accumulated organic solute (disaccharide, heteroside, or betaine).[2]

Natural abundance ^{13}C-NMR spectroscopy is noninvasive and nondestructive: analysis of intact cells can be used to determine unequivocally all of the principal low-molecular-weight organic compounds which are in solution intracellularly in osmotically significant amounts. Additionally, a comparison of the ^{13}C-NMR spectra of a sample containing intact cells and of an aqueous extract from a similar biomass will show whether all of the osmolyte is freely mobile within the cells.[1] Additional information on the physical state of the cytoplasm of cyanobacteria, including the rotational motion and mobility of organic osmotica, can be obtained from ^{13}C-NMR spin–lattice relaxation time measurements of intact cells, as shown for *Synechococcus* RRIMP N 100.[4]

The natural abundance level of ^{13}C is approximately 1.1%, the remaining ^{12}C fraction being NMR silent. Additionally, ^{13}C is less sensitive to NMR analysis than other nuclei (e.g., ^{1}H, ^{31}P). These factors impose a requirement for (1) a high magnetic field strength, (2) long accumulation times, and/or (3) large amounts of biological material. Resolution can be improved in most cases by using homogeneous extracts, rather than intact cells, enabling samples containing more than 10 mg organic osmolytes to be analyzed in less than 12 hr on most commercially available lower-field spectrometers. Figure 1 shows a typical spectrum of an extract from salt-stressed *Synechocystis* PCC 6714, using 10 liters of exponentially growing cells centrifuged and extracted in 80% (v/v) ethanol : water (24 hr), dried down, and redissolved in 1 ml D_2O (obtained with a Brüker WP60 FT spectrometer operating at 15.08 MHz).[5] The spectrum contains 8 resonances, corresponding to those of authentic glucosylglycerol,[1] and confirms the osmotic significance of this metabolite in this strain. Higher-field NMR spectrometers would allow considerable savings in assay time for such samples, together with a further potential increase in sensitivity.

Recent studies have shown that ^{13}C-NMR spectroscopy can be used to investigate the biosynthesis and turnover of organic osmolytes in cyanobacteria, using ^{13}C-enriched bicarbonate as a source of inorganic carbon for photosynthesis. This enables the changes in cell osmolytes to be monitored *in vivo*, without recourse to the time-consuming extraction and

[4] R. S. Norton, M. A. Mackay, and L. J. Borowitzka, *Biochem. J.* **202,** 699 (1982).

[5] S. R. C. Warr, R. H. Reed, J. A. Chudek, R. Foster, and W. D. P. Stewart, *Planta* **163,** 424 (1985).

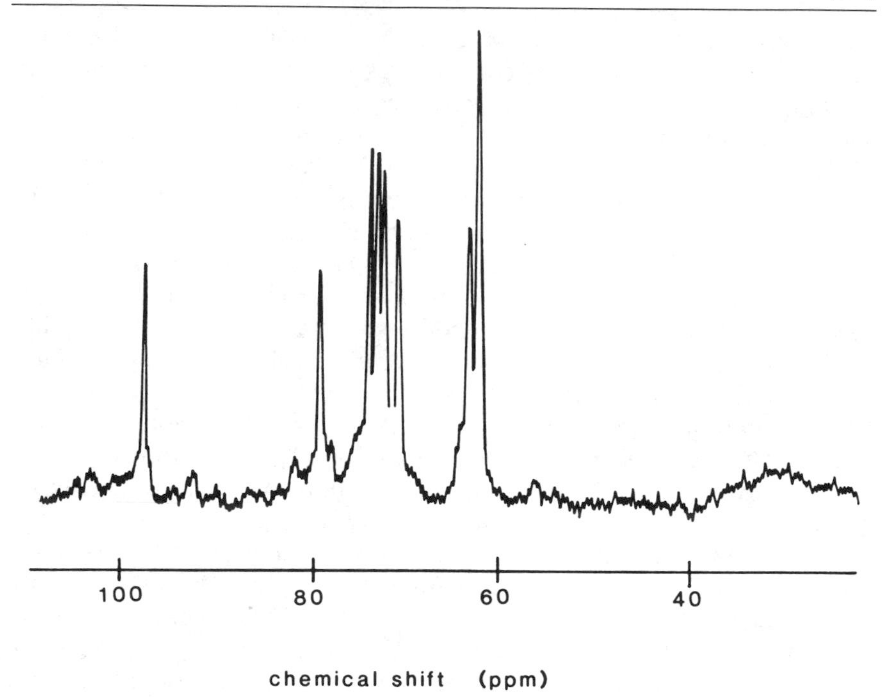

FIG. 1. Natural abundance ^{13}C-NMR spectrum of an extract of the glucosylglycerol-accumulating euryhaline unicell *Synechocystis* PCC 6714.

separation procedures required for radiocarbon tracer experiments. Further details are contained in Refs. 6 and 7.

Low-Molecular-Weight Carbohydrates

The ^{13}C-NMR procedures described above can be used to quantify individual carbohydrates in cyanobacterial extracts. However, limitations on sample size and spectrometer access time may prevent analysis of large numbers of samples by this method. An alternative procedure for routine analytical work involves the separation of trimethylsilyl (TMS) ethers of carbohydrates by gas–liquid chromatography (GLC).

Extraction of cyanobacterial samples (each containing carbohydrate at 0.05–2.00 mg) is carried out by incubation in boiling 80% (v/v) ethanol : water (containing 0.2–1.0 mg arabitol as an internal standard) for 5 min, followed by reextraction of the residual material in 80% (v/v) ethanol : water at 25° for 18 hr. This procedure extracts over 99% of the

[6] M. A. Mackay and R. S. Norton, *J. Gen. Microbiol.* **133,** 1535 (1987).

[7] E. Tel-Or, S. Spath, L. Packer, and R. J. Mehlhorn, *Plant Physiol.* **82,** 646 (1986).

low-molecular-weight carbohydrates from cyanobacterial cells. The ethanolic extracts are then pooled, evaporated to dryness at 40°, and stored in a vacuum desiccator for 24–48 hr. Samples are then dissolved in 0.2–1.0 ml of a suitable organic solvent, e.g., pyridine[8] or dimethyl sulfoxide.[9] TMS ethers are produced by the sequential addition of 2 volumes of hexamethyldisilazane and 1 volume of trimethylchlorosilane (0.1 ml and 0.05 ml, respectively). Samples are shaken for at least 90 sec and left for over 12 hr at room temperature to ensure that all free hydroxyl groups are converted to TMS ethers. Since all solvents containing free hydroxyl groups will react with the reagents used to produce the TMS derivatives, it is essential to exclude both ethanol and water from the later stages of sample preparation.

While pyridine has been widely used as a reaction solvent for the analysis of low-molecular-weight carbohydrates from plant tissues,[8] dimethyl sulfoxide has several advantages since (1) low-molecular-weight carbohydrates dissolve more rapidly in dimethyl sulfoxide prior to derivatization and (2) an upper phase of hexamethyldisiloxane is formed after reaction. TMS–carbohydrates show a high partition coefficient for this phase which provides (3) a useful additional selective procedure for the removal of interfering substances and (4) a method of increasing the concentration of TMS derivatives in a smaller reaction volume. Furthermore, the hexamethyldisiloxane phase gives a rapidly eluted and reduced solvent peak compared with pyridine. This removes any requirement to dry down samples and redissolve in a more suitable solvent prior to analysis.

Samples can be analyzed by conventional GLC techniques, with flame-ionization detection, using a 2 m × 6 mm (i.d.) column containing 2% methyl phenyl silicone gum (SE 52, Pye Unicam, Cambridge, UK) with diatomite as solid support, or by capillary GLC using a 10 m × 0.53 mm silica heliflex column coated at 0.25 μm with RSL 150 polydimethylsiloxane (Alltech Assoc., Carnforth, UK). A temperature program is required to achieve optimum separation: with a temperature change from 140 to 280° at 20° min^{-1}, holding the initial and final temperatures for 2 min and with suitable carrier gas flow rates (30 and 3 ml min^{-1}, respectively), analysis is complete in under 10 min. Representative retention times and response factors (relative to the internal standard, arabitol) are shown in Table I, as obtained using a Varian 3700 GLC (Varian Instruments, Walnut Creek, CA). We have used this procedure to screen a wide range of cyanobacteria from freshwater, brackish, and marine habitats,[10] showing

[8] P. M. Holligan and E. A. Drew, *New Phytol.* **70,** 271 (1971).

[9] D. J. Moore, R. H. Reed, and W. D. P. Stewart, *J. Gen. Microbiol.* **131,** 1267 (1985).

[10] R. H. Reed, D. L. Richardson, S. R. C. Warr, and W. D. P. Stewart, *J. Gen. Microbiol.* **130,** 1 (1984).

TABLE I
GLC CHARACTERISTICS OF CARBOHYDRATES FROM CYANOBACTERIA

Stationary phase and characteristic	Carbohydrate		
	Glucosylglycerol	Sucrose	Trehalose
SE 52			
Retention time	1.47	1.77	1.84
Response factor	0.75	0.60	0.60
RSL 150			
Retention time	2.18	2.88	3.06
Response factor	0.77	0.64	0.64

that the least halotolerant forms accumulate disaccharides in response to salt stress while the heteroside glucosylglycerol is accumulated by the more halotolerant strains, irrespective of their source of isolation.[11]

Direct extraction of freeze-dried samples in dimethyl sulfoxide (at 100° for 18 hr) has been proposed as a means of reducing assay time and sample preparation procedure.[12] We have used this experimental procedure to investigate the synthesis of sucrose as a secondary osmolyte in unicellular glucosylglycerol-accumulating cyanobacteria grown at high temperature in hyposaline media,[3] showing that this method is a viable alternative to ethanolic extraction.

Quaternary Ammonium Compounds

A range of analytical procedures can be used to identify and quantify quaternary ammonium compounds. However, many of the methods are time-consuming, nonspecific, and/or insensitive, relying on spectrophotometric analysis of periodide/reineckate salts. Alternative procedures, including thin-layer chromatography,[13] pyrolysis–gas chromatography,[14] and high-performance liquid chromatography,[15] often require several additional ion-exchange purification steps per sample prior to analysis, increasing costs and assay times. We have developed a procedure using

[11] D. L. Richardson, R. H. Reed, and W. D. P. Stewart, *FEMS Microbiol. Lett.* **18,** 99 (1983).

[12] R. H. Reed and I. R. Davison, *Br. Phycol. J.* **19,** 381 (1984).

[13] G. Blunden, S. M. Gordon, W. F. H. McLean, and M. D. Guiry, *Bot. Mar.* **25,** 563 (1982).

[14] W. D. Hitz and A. D. Hanson, *Phytochemistry* **19,** 2371 (1980).

[15] R. D. Guy, P. G. Warne, and D. M. Reid, *Physiol. Plant.* **61,** 195 (1984).

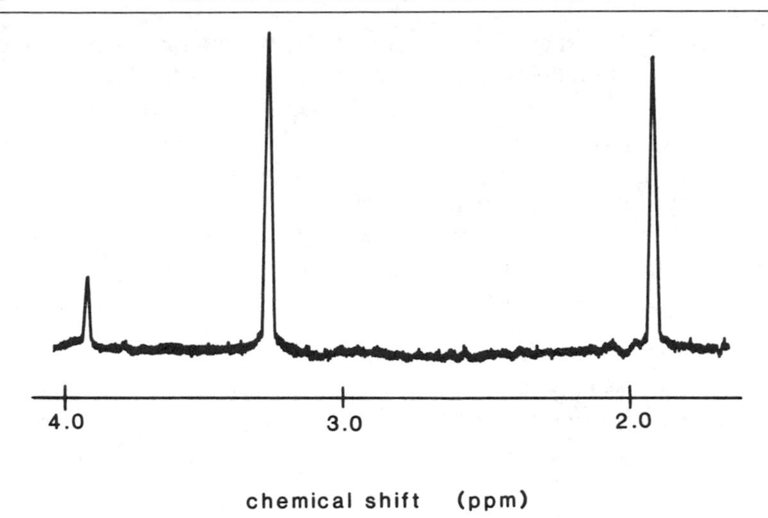

FIG. 2. ^{1}H-NMR spectrum of an extract of *Synechococcus* PCC 7418 with added sodium acetate (resonance 1.91 ppm) as an internal standard.

high-resolution continuous wave ^{1}H-NMR spectroscopy to identify and quantify methylated osmolytes in algae and cyanobacteria.[16] Owing to the chemical shift equivalence of the protons in all three of the methyl groups, the dominant feature of ^{1}H-NMR spectra of betaines is a singlet resonance of relative intensity 9. This singlet provides a sensitive and distinctive probe for quantitative purposes, by comparing the peak size with that of a known amount of an internal standard with dissimilar ^{1}H absorption peaks.

Cyanobacterial samples (containing 0.5–5.0 mg betaine) are extracted in 80% (v/v) ethanol : water, as described above, with 2–10 mg sodium acetate as internal standard. Samples are then dried, redissolved in 0.5 ml D_2O and analyzed by ^{1}H-NMR spectroscopy without further purification. If glycine betaine is present, the major 9-proton resonance will be observed at a chemical shift (δ_H) of 3.27, with an additional ($C\underline{H}_2$) resonance at 3.88, as shown in Fig. 2 for an extract of the halotolerant unicell *Synechococcus* PCC 7418 (*Aphanothece halophytica*), obtained using a Brüker HX90 continuous wave spectrometer operating at 90 MHz. Similar spectra have been obtained for other halotolerant cyanobacteria, including *Dactylococcopsis salina*, *Synechocystis* DUN 52,[17] and *Oscillato-*

[16] J. A. Chudek, R. Foster, D. J. Moore, and R. H. Reed, *Br. Phycol. J.* **22,** 169 (1987).
[17] R. H. Reed, J. A. Chudek, R. Foster, and W. D. P. Stewart, *Arch. Microbiol.* **138,** 333 (1984).

ria limnetica,[18] showing glycine betaine to be the major organic solute in these strains. Glutamate betaine, which has been reported for two isolates of *Calothrix*,[19] may be quantified using the same procedure. Fourier transform ^{1}H-NMR spectroscopy offers a further increase in sensitivity: samples containing 0.01–0.50 mg glycine betaine can be analyzed in approximately 60 min.[16] However, this degree of sensitivity should not be required for those strains which contain glycine betaine or glutamate betaine in osmotically significant amounts.

Acknowledgments

Research supported by the Royal Society, Natural Environment Research Council UK, and Science and Engineering Research Council UK. RHR is a Royal Society Research Fellow.

[18] R. H. Reed, A. Oren, and J. A. Chudek, unpublished observations (1986).
[19] M. A. Mackay, R. S. Norton, and L. J. Borowitzka, *J. Gen. Microbiol.* **130,** 2177 (1984).

[57] Inorganic Carbon Uptake by Cyanobacteria

By Aaron Kaplan, Yehouda Marcus, and Leonora Reinhold

The ability to concentrate inorganic carbon (C_i) within their cells enables cyanobacteria to compensate for the discrepancy between the $K_m(CO_2)$ of ribulose-bisphosphate carboxylase (EC 4 : 1 : 1 : 39, Rubisco) (200 μM) and the concentration of dissolved CO_2 in equilibrium with air (10 μM). This ability to accumulate C_i internally is light dependent[1] and develops as a function of the concentration of CO_2 experienced by the cells during growth.[2] The rate of C_i uptake as well as the extent of accumulation are far greater in cells adapted to the air level of CO_2 then in cells grown at elevated CO_2 concentration, with the result that the apparent photosynthetic affinity for CO_2 shown by the former cells is considerably higher.[2] The extent of C_i accumulation is far greater than would be predicted on the basis of passive penetration of C_i species along their electrochemical potential gradient with subsequent interspecies equilibration according to intracellular pH.

Two major questions have been addressed in the study of the CO_2 concentrating mechanism: the nature of the C_i concentrating system and

[1] A. Kaplan, D. Zenvirth, Y. Marcus, T. Omata, and T. Ogawa, *Plant Physiol.* **84,** 210 (1987).
[2] A. Kaplan, M. R. Badger, and J. A. Berry, *Planta* **149,** 219 (1980).

the mechanisms involved in the adaptation to varying levels of CO_2. We shall briefly review the major experimental approaches taken to investigate the former.

Filtering Centrifugation Technique

The filtering centrifugation technique is the one most commonly used to determine the accumulation of acid-stable carbon (photosynthetic products) and acid-labile carbon (C_i) over short periods of exposure to CO_2 or HCO_3^-. A microcentrifuge such as Beckman microfuge B, 400-μl centrifuge tubes, silicone fluid of the proper specific gravity and viscosity, glass mixing rods, syringes, and a suitable light source are required.[2] In a typical experiment 200 μl of a cell suspension is introduced as the top layer in a centrifuge tube containing (from bottom to top) 20 μl of killing medium (0.1 *N* NaOH) and 60 μl of the silicone fluid. The cells are allowed to adjust to the experimental conditions, after which the desired $^{14}C_i$ concentration is injected into the cell suspension layer which is then mixed. Exposure to $^{14}C_i$ is terminated after the desired interval by centrifugation into the killing medium. The tip of the centrifuge tube is frozen in liquid N_2 and cut into 0.1 *N* NaOH. Samples of the latter are then taken, and some of them are acidified. The radioactivity in both alkaline and acidified samples is measured in order to assess the C_i and acid-stable carbon. The ^{14}C in the extracellular space of the pellet is allowed for by incubating parallel samples in tritiated H_2O and a relatively nonpenetrating solute such as [^{14}C]sorbitol. The intracellular space is estimated as tritium-permeable space minus sorbitol-permeable space. Time course experiments are required since the cells may take up sorbitol to some extent.

When it is desired to estimate the rates of photosynthesis and of C_i accumulation during any given interval, in particular immediately after the supply of C_i, a major difficulty is the assessment of the specific activity of the C_i within the cells. The amount of ^{14}C detected in the acid-stable or acid-labile carbon represents the integration with time of the rates of photosynthesis and accumulation of C_i, respectively. However, the specific activity of the C_i within the cells is changing with length of exposure until a steady state is reached, which may take 60–90 sec depending on the concentration of C_i provided. This introduces error into the estimation of the rates of photosynthesis and of the internal C_i concentration. There are two major possible experimental approaches to control or evaluate the specific activity of the C_i within the cells. One is the "zero trans" type of experiment where the cells are allowed to utilize the C_i in the medium in an O_2 electrode chamber prior to transfer to the microfuge tubes.[2] The concentration of C_i when the O_2 compensation point has been reached is very low both in the medium and within the cells, particularly when low-CO_2-adapted cells are used. A known concentration and specific activity of $^{14}C_i$ is supplied, and the concentration of C_i within the cells can then be followed with time. It is assumed that the transport system does not discriminate against $^{14}C_i$.

In a second type of experiment (the disequilibrium experiment) the cell suspension is bubbled vigorously with a known concentration of CO_2. The C_i concentration in the medium under these conditions can be calculated from the equation $pH = pK + \log[HCO_3^-]/CO_2]$. (In a modification of this method a dilute suspension of cells is exposed to a known, usually high concentration of C_i.) A trace of $^{14}C_i$ of known radioactivity is injected and the specific activity of the $^{14}C_i$ in the medium calculated.[3,4] It is necessary to evaluate the concentration of unlabeled C_i within the cells. This can be accomplished by first performing the "zero trans" type of experiment for the same extracellular C_i concentration until the steady-state internal concentration has been reached. Since total internal C_i is constant with time, the rise in intracellular $^{14}C_i$ following the injection of the trace gives a measure of the changing internal specific activity. The photosynthetic rate is calculated from the rate of accumulation of acid-stable ^{14}C and the specific activity, assuming rapid mixing of the intracellular C_i pool. The internal CO_2 concentration can be calculated from the above equation assuming that the various C_i species are at equilibrium at the estimated internal pH. This may not be the case, however, and, further, gradients of CO_2 concentration may exist within the cells.[5] Another source of possible error is the estimated pK value within the cells which might be somewhat lower than in the medium (6.1 instead of 6.4) due to the higher intracellular ionic strength.

In the case of unicellular cyanobacteria where the centrifugation time required for separating the cells from the medium may be longer than a few seconds, airfuge or alternative methods should be considered.

Gas-Exchange Systems

The rate of CO_2 uptake can be assessed from the difference in CO_2 concentration between the air arriving and that leaving a chamber containing the cell suspension. Such a system is composed of a temperature-controlled transparent chamber, a gas mixing system, flow meters, a

[3] G. S. Epsie, K. A. Gehl, G. W. Owttrim, and B. Colman, *Adv. Photosynth. Res.* **3,** 457 (1984).

[4] G. S. Espie and B. Colman, *Plant Physiol.* **80,** 863 (1986).

[5] L. Reinhold, M. Zviman, and A. Kaplan, *Proc. Int. Congr. Photosynth., 7th,* in press.

strong light source, and a CO_2-detecting device such as an infrared gas analyzer[6,7] or a mass spectrometer.[8] The geometry of the chamber and the volume of the gas between the chamber and the detector strongly affect the response time of the system. Exchange of CO_2 between the gas phase and the C_i in the buffered medium may lead to a considerable error in the estimated rate of CO_2 uptake, particularly following a change in the experimental conditions.[1] Rapid equilibrium among the different C_i species in the medium may be achieved by adding carbonate dehydratase.

While the data from the gas-exchange system are immediately available, unlike those obtained by the filtering centrifugation technique, the method does not enable direct differentiation between uptake of CO_2 into the acid-stable pool and that into the acid-labile (C_i) pool. Mutants defective in their ability to fix CO_2[1] as well as inhibitors[7] have been used to allow estimation of CO_2 taken up into the intracellular C_i pool. The latter has also been estimated from the amount of CO_2 released to the medium on darkening.[7] The initial rate of CO_2 release on darkening was also used to evaluate the steady-state efflux of CO_2.[8] This procedure is dubious, however, since the intracellular CO_2 concentration may rise on darkening due to a rapid fall in the intracellular pH and in the rate of photosynthetic CO_2 consumption.

Use of heavy isotopes of carbon and oxygen in a system connected to a mass spectrometer[8] may enable determination of the rate of interconversion between the different C_i species during transport. This may be of particular importance in experiments aimed at understanding the interaction between the various C_i species during transport[9] and the possible role of carbonate dehydratase.[10]

Species of C_i Taken Up

The nature of the species of C_i taken up has attracted attention because of its importance from the point of view of the mechanism of C_i uptake. The dependence of the apparent photosynthetic affinity for extracellular C_i on external pH has been interpreted as indicating that HCO_3^- as well as CO_2 is taken up from the medium.[11] However, the pH may

[6] T. Ogawa, A. Miyano, and Y. Inoue, *Biochim. Biophys. Acta* **808,** 77 (1985).

[7] T. Ogawa, T. Omata, A. Miyano, and Y. Inoue, *in* "Inorganic Carbon Uptake by Aquatic Photosynthetic Organisms" (W. J. Lucas and J. A. Berry, eds.), p. 287. Waverly, Baltimore, Maryland, 1985.

[8] M. R. Badger, M. Basset, and H. N. Comins, *Plant Physiol.* **77,** 465 (1985).

[9] M. Volokita, D. Zenvirth, A. Kaplan, and L. Reinhold, *Plant Physiol.* **76,** 599 (1984).

[10] K. Aizawa and S. Miyachi, *FEMS Microbiol. Rev.* **39,** 215 (1986).

[11] A. G. Miller, *in* "Inorganic Carbon Uptake by Aquatic Photosynthetic Organisms" (W. J. Lucas and J. A. Berry, eds.), p. 17. Waverly, Baltimore, Maryland, 1985.

affect the photosynthetic V_{max} via an alteration in cytoplasmic pH leading to an apparent but not a real effect on the affinity. Uptake of HCO_3^- has also been invoked where the rate of photosynthesis observed was higher than that expected on the basis of the uncatalyzed rate of formation of CO_2 from HCO_3^- in the medium.[12]

These methods, however, provide only qualitative indications. Quantitative evidence allowing analysis of the kinetic parameters for uptake of both CO_2 and HCO_3^- can be obtained from experiments performed under non-steady-state conditions. In a typical experiment the cells are exposed to radioactive CO_2 or HCO_3^- for very short time intervals (up to 5 sec) to minimize the interconversion between the C_i species. (The rate of formation of CO_2 from HCO_3^- and vice versa can be calculated from the rate constants of the appropriate reactions.[11] At alkaline pH the contribution of external HCO_3^- to CO_2 uptake due to conversion during the experiment is relatively small. On the other hand, the rapid conversion of CO_2 to HCO_3^- may lead to overestimation of the rate of HCO_3^- uptake. At acidic pH the rate of CO_2 uptake may be overestimated due to formation of CO_2 from HCO_3^-.) The stock of radioactive C_i is brought to pH 8.5–9.0 to minimize the amount of CO_2 supplied with HCO_3^-. When CO_2 is supplied, the stock is adjusted to pH 5.0–5.2 in the presence of the desired C_i concentration. The stock is allowed to stand at room temperature for at least 60 sec, after which aliquots are injected into the microfuge tube containing the cells. Another sample is injected into NaOH to measure the activity of the CO_2 supplied. In experiments of this type evidence has been obtained that in *Anabaena* both CO_2 and HCO_3^- are taken up via a system(s) which exhibits saturable kinetics, but the different K_m and V_{max} values.[9]

Carbonate dehydrotase activity in the periplasmic space has been reported in *Chlamydomonas* but not as yet in cyanobacteria. Possible extracellular carbonate dehydrotase activity must be tested for because it greatly complicates distinction between CO_2 and HCO_3^- uptake.

Permeability to CO_2 (P_{CO_2})

It is difficult to accommodate the observed accumulation of CO_2 at the carboxylation site (to levels well above that in the medium) with the accepted view that this lipophilic molecule permeates biological membranes easily. Measurements of CO_2 fluxes[8,13] have been used to assess P_{CO_2} in cyanobacteria, yielding very low values (10^{-5} to 10^{-4} cm sec^{-1}). One major difficulty is assessment of the exact CO_2 concentration at the

[12] A. G. Miller and B. Colman, *Plant Physiol.* **65,** 397 (1980).

[13] Y. Marcus, R. Schwarz, D. Friedberg, and A. Kaplan, *Plant Physiol.* **82,** 610 (1986).

source of the efflux since it is not clear whether the different c_i species are at equilibrium throughout the cell. It is possible that CO_2 might be formed primarily in the carboxysomes.[5] It would therefore be desirable to develop and apply a different experimental approach to investigate P_{CO_2}.

Role of Na^+

Sodium ions are required for HCO_3^- uptake but only to a lesser extent for CO_2 uptake.[14] This finding has been used to differentiate between CO_2 and HCO_3^- uptake.[15] While the exact role of Na^+ in C_i transport is yet to be established, it is clear that the ionic composition in the medium will influence the results and should be carefully controlled.

[14] A. Kaplan, *in* "Inorganic Carbon Uptake by Aquatic Photosynthetic Organisms" (W. J. Lucas and J. A. Berry, eds.), p. 325. Waverly, Baltimore, Maryland, 1985.
[15] A. G. Miller and D. T. Canvin, *FEBS Lett.* **187,** 29 (1985).

[58] Carbon Metabolism: Studies with Radioactive Tracers *in Vivo*

By JAMES A. BASSHAM and SCOTT E. TAYLOR

Introduction

Photosynthetic carbon metabolism in cyanobacteria may be studied by exposing photosynthesizing cells to $^{14}CO_2$, subsequently killing the cells, and then separating and identifying the labeled metabolites by a suitable chromatographic technique. Early indications[1] and subsequent detailed studies[2] showing that photosynthetic carbon dioxide incorporation in cyanobacteria occurs via the reductive pentose phosphate cycle (Calvin cycle) were obtained by these methods. For study of dark metabolism in these organisms, the cells may be first labeled by photosynthesis with $^{14}CO_2$. Alternatively, ^{14}C-labeled compounds such as glucose may be employed as substrates in such studies. The use of ^{32}P-labeled phosphate, in light and dark, provides additional information about phosphorylated metabolites and levels of ATP and ADP.[3,4]

[1] L. Norris, R. E. Norris, and M. Calvin, *J. Exp. Bot.* **16,** 64 (1955).
[2] R. A. Pelroy and J. A. Bassham, *Arch. Mikrobiol.* **86,** 25 (1972).
[3] R. A. Pelroy and J. A. Bassham, *J. Bacteriol.* **115,** 937 (1973).
[4] R. A. Pelroy, G. A. Levine, and J. A. Bassham, *J. Bacteriol.* **123,** 633 (1976).

Certain types of kinetic tracer studies are complex and require specialized equipment. While these requirements are described in this chapter, simpler alternatives are presented for those interested in less complex or less quantitative investigations. The information obtained from tracer studies *in vivo* can be used not only to map metabolic pathways but also to study metabolic regulation, as has been the case with light–dark regulation in green algae.[5] A complete understanding of carbon metabolism requires other kinds of studies such as enzymology and gene expression which are not covered in this chapter but are found in other chapters in this volume and in others. In general, techniques used to measure carbon metabolism in other organisms can be applied to cyanobacteria.[6]

The present discussion is devoted mostly to the use of labeled substrates with whole cells and to alternative methods for analysis of the resulting labeled metabolites. The classic method of analysis by two-dimensional paper chromatography is still the method of choice for some workers because of its capability of separating a broad range of metabolites such as sugar phosphates, sugars, amino acids, and carboxylic acids. This method is time-consuming and somewhat tedious, however, and it requires equipment lacking in many laboratories. Variations involving thin-layer chromatography and electrophoresis have been used with success and provide useful alternatives to paper chromatography. As an alternative to radioactive tracer studies, analysis of some metabolites by ^{13}C-NMR spectroscopy is briefly described.

In Vivo Metabolic Studies with Radioisotopes

Cell Culture

A number of different species of cyanobacteria have been used successfully in tracer metabolic studies,[1–4] and it is likely that any photosynthetically active, single cell species would serve. Multicellular species, such as filamentous cyanobacteria, also can be used, provided they do not present sampling problems. In general it is desirable to harvest the cells by centrifugation at room temperature, typically 20 min at 20,000 *g*, followed by resuspension in fresh growth medium. Provided other environmental conditions are kept the same as during growth (temperature, light intensity, CO_2 concentration, etc.), the metabolic pattern identified should be representative of that of the growth conditions.

The subsequent analysis of metabolites by chromatography can be

[5] J. A. Bassham, *Science* **172,** 526 (1971).

[6] A. San Pietro (ed.), this series, Vol. 24, p. 261.

adversely affected by high salt concentrations in the media, so that experiments with marine algae present great difficulty. Desalting methods can be attempted, but often at the cost of loss of quantitative results. It may be possible to grow the cells with a more dilute medium than that prescribed in the literature without serious alteration in cell metabolism. For some studies, a deliberate change in the suspending medium may be made. For example, the metabolic effects of short-term nitrogen deprivation might be investigated.

After centrifugation, the cells are resuspended in the experimental medium at 1% packed cell volume/suspension volume. Such a concentration usually ensures that, when the cells are later killed in 80% methanol, an aliquot sample of sufficiently small volume can be used for analysis and will contain enough labeled material for quantitative measurement of the labeling of each metabolite, provided other requirements described below are met.

Preillumination and Gas Mixtures

Since it is generally desirable that the experiment be carried out under conditions permitting steady-state metabolism, the cells should be placed immediately in the experimental vessel and be allowed to photosynthesize with unlabeled CO_2 for 30 min to 1 hr. If the experiment is to be carried out with air levels of CO_2, the cell suspension is provided with a bubbler and air stream during this period. Studies with gas mixtures other than air require some system for bubbling the prescribed gas through the cells both during preillumination and during the exposure to $^{14}CO_2$. While appropriate gas mixtures can be supplied during preillumination by gas tanks and flow metering devices, maintenance of $^{14}CO_2$ at constant specific radioactivity and CO_2 pressure for more than 1 or 2 min may require a fairly complicated gas-handling system, such as that described below.

Sources of $^{14}CO_2$

For the simplest, short-term experiments, a solution of radioactive bicarbonate, $H^{14}CO_3^-$, may be injected into a closed, illuminated vessel for a few seconds to 1 min, and the contents of the flask can then be poured into a killing solution (ethanol) in a fume hood. Such an experiment might serve, for example, to test for the presence of carboxylases in a mutant or wild type.

For quantitative, kinetic experiments it is necessary to maintain the concentration and specific radioactivity of the carbon dioxide and to control the physiological conditions, either keeping them constant or varying one element (such as light intensity) in a predetermined way. A gas-handling system is required, and usually this will be a closed, gas recirculating system, employing a gas pump, since conservation of $^{14}CO_2$ is necessary for both environmental and economic reasons. Provision must be made for either storing any unused radioactive gas in a closed vessel or absorbing it in alkali at the end of the experiment. If a sufficient quantity is stored in a closed vessel that can be reopened to the system later, it may be possible to conduct several experiments before the concentration or specific radioactivity falls below acceptable levels.

Gas circulation and control require the development of a closed gas recirculating system which meets various requirements. For circulation, a small vibrating diaphragm gas pump capable of circulation rates of several liters per minute may be used. Considerable gas-impervious tubing, such as glass or stainless steel, is required to connect the pump with various inlet and outlet valves, including those leading to the cell vessel, and others for the inlet of gases such as air and radioactive carbon dioxide, to gas storage reservoirs, and to the sensors of the gas monitoring instruments. This tubing should be kept small enough (of the order of 3 mm i.d.) to keep the overall volume of the minimum system from becoming large compared to the gas storage vessels. In developing a gas-handling system one may have to spend some weeks or months adjusting flow rates and pressures with various valves, flow meters, etc., before achieving satisfactory operation. Once developed, however, such a "steady-state" apparatus offers many possibilities for productive studies of metabolism.

The gas should be circulating in the closed gas recirculating system before the end of the preillumination period, separated from the vessel containing the cells only by a four-way stopcock or equivalent valve system. The volume of gas in the tubing leading from these valves to the cell vessel and back should be small compared to the gas flow rate so that once the valves are switched to admit radioactive gas to cells, no more than a few seconds are required to replace ordinary carbon dioxide with labeled gas. If instantaneous change is required, labeled bicarbonate solution of the same specific radioactivity can be injected directly into the vessel at the moment the valves are switched.

A gas cylinder equipped with suitable reducing valves and flow valves may be used for addition of $^{14}CO_2$ to the system. Such a radioactive storage tank should be enclosed in a safety chamber vented through a CO_2-absorbing material (i.e., a commercial preparation of soda lime designed for this purpose). Any radioactive gas-handling equipment should be approved and monitored by the health safety officers of the institution. For experiments with lesser requirements, $^{14}CO_2$ can be generated on a vacuum line (again, this should be an enclosure approved for this purpose by safety officers!) by addition of acid to $Ba^{14}CO_3$ and trapping in a loop closed by a four-way stopcock (so the radioactive gas can be flushed out when needed) and equipped with quick-fit joints for attaching to the gas-recirculating system.

Light Intensity and Vessel Shape

If a 1% suspension is used, it may be necessary to use a thin vessel and rather strong illumination in order that the cells are exposed to light at an intensity comparable to that of the growth conditions. For experiments at relatively high light intensities, a flat illumination vessel with plastic walls and a space about 6 mm wide and 50–100 ml in total volume has been found useful.[2,7,8] With illumination from both sides, an individual cell is never shaded by more than 3 mm of suspension, and with rapid stirring from the gas bubbling all cells are exposed to an average high light intensity. If strong illumination is used, provision for temperature control, such as a transparent water jacket, is required.

Sampling

Sampling requirements vary with the objective. For a relatively slow metabolic conversion, such as the conversion of an applied labeled metabolite to an end product in a secondary biosynthetic pathway, samples probably could be withdrawn by opening a stopcock or by removing the sample with a syringe through a rubber stopper and placing in methanol (final concentration after addition of sample = 80% v/v). For quantitative, kinetic studies of the fast reactions of carbon metabolism in photosynthesis, sampling and quenching (killing) must be accomplished within about 1–2 sec, since longer times permit unacceptable amounts of enzymatic conversions during non-steady-state conditions.

One solution to this problem is to equip the vessel with an outlet valve operated by a push-button-controlled solenoid.[7] The stainless steel conical valve, seated on the end of a Teflon outlet tube, is lifted by a wire attached to the tip of the cone and running down through the outlet tube to a lever attached to the solenoid. When activated, the solenoid operates against a compressed spring to raise the valve for a period of about 0.2 sec. As a further refinement, N_2 is injected just at the base of the valve to flush the outlet tube. Sample size is determined by weighing the vial (plus methanol) before and after sampling.

[7] J. A. Bassham and M. Kirk, *Biochim. Biophys. Acta* **43,** 447 (1960).

[8] J. A. Bassham and M. Kirk, *Biochim. Biophys. Acta* **90,** 553 562 (1964).

Other Illumination Vessel Refinements

For certain types of experiments, the illumination vessel can be fitted with sensing devices, including a pH electrode and a device for monitoring density.[7,8] The outputs from these sensors can be employed to control the automatic addition of acid or base, or of medium, to maintain pH, density, or both. For example, kinetic studies of the effects of addition of ammonium ion, as ammonium chloride, required the automatic addition of dilute ammonium hydroxide, since uptake of NH_4^+ caused the green algae used to acidify the medium. Automatic density control is not likely to be needed in most cases since experiments usually will not last long enough for significant density change due to growth to occur. To ensure rapid gas–liquid equilibration, particularly at low CO_2 pressures such as in air, the tip of the bubbler may have to be made of sintered glass to provide fine bubbles.

Gas Monitoring

For experiments requiring addition of $^{14}CO_2$ as a gas and employing a closed gas recirculating system, instruments to monitor levels of CO_2 and ^{14}C are a necessity. Carbon dioxide can be measured with an instrument which makes use of the absorption of infrared radiation by CO_2. It should be noted that these instruments are somewhat nonlinear in response and require a calibration curve. Moreover, they are much less sensitive to $^{14}CO_2$ than to $^{12}CO_2$, so that total carbon dioxide level can be underestimated when an appreciable fraction is $^{14}CO_2$. The level of ^{14}C can be monitored with an ionization chamber and a vibrating reed electrometer. For studies of photosynthesis and respiration, it is also useful to monitor O_2 with an oxygen meter. Solid-state devices are much easier to employ than the older paramagnetism-measuring instruments. A record of the signals from these gas-monitoring instruments made with a multichannel recorder allows the specific radioactivity, rates of photosynthesis and respiration, and composition of gas mixture to be calculated during and after the experiments.

Specific Radioactivity

For quantitative, kinetic tracer experiments, it is usually necessary to start with a high level of radioactivity. The best chromatographic separation is achieved when the quantity of material applied to the chromatogram is small. For the procedure described below, the amount of material applied to the paper chromatogram is derived from about 2 μg of wet-packed cells. Only submicrogram amounts of a given intermediate compound may be present. Pure $^{14}CO_2$ would be roughly 60 $\mu Ci/\mu mol$, and it is useful to employ from 10 to 50 $\mu Ci/\mu mol$. This gas should be circulating in the closed gas recirculating system before the end of the preillumination period, separated from the vessel containing the cells by a four-way stopcock or equivalent valve system.

Other Labeled Substrates

Other labeled substrates such as ^{14}C-labeled glucose (either with uniform labeling or with the label in a specific position), ^{32}P-labeled inorganic phosphate or other phosphorylated compounds, or even tritium (^3H)-labeled compounds may be added as solutions injected into the cell vessel through a soft plug with a needle. In the case of ^{14}C-labeled compounds, usually unlabeled carbon dioxide is supplied. High specific radioactivities are required, as discussed above. For ^{32}P-labeled compounds, the radioactivity emissions are much stronger than for ^{14}C, so there is no difficulty in obtaining enough radioactivity, but special precautions must be taken to avoid undue exposure of workers and laboratory contamination. Detailed plans should be discussed with health safety officers.

Dual tracer experiments, employing $^{14}CO_2$ and $[^{32}P]P_i$, have been useful in studies of carbon metabolism during photosynthesis and respiration in both green algae[9] and in cyanobacteria.[2-4] The specific radioactivities of the two labeled substrates should be chosen to give comparable film exposures when radioautographs are prepared.

Exposure to $^{14}CO_2$

At the conclusion of preillumination to achieve steady-state photosynthesis, the four-way stopcock or valve system is turned, shutting off the flow of unlabeled gas to the vessel and starting the flow of gas containing $^{14}CO_2$. Samples are taken at predetermined intervals. If initial rates of labeling are of interest, samples will be taken every 5 sec or so at first, then less frequently up to 10 min. By that time, the reductive pentose phosphate cycle intermediates will be "saturated" with radiocarbon in an actively photosynthesizing culture. The levels of radioactivity in these compounds "at saturation," when later determined following analysis and radioactivity counting, may be used to calculate the pool sizes, that is, the concentration of the intermediates in the cells, provided the specific radioactivity of the administered carbon dioxide is known.[10]

For cell suspensions of 1%, 1-ml samples may be taken directly into 4

[9] T. A. Pedersen, M. Kirk, and J. A. Bassham, *Physiol. Plant.* **19,** 219 (1966).

[10] J. A. Bassham and G. H. Krause, *Biochim. Biophys. Acta* **189,** 207 (1969).

ml of ethanol at room temperature in 10-ml vials, preweighed with ethanol and with screw caps. The caps are removed briefly for sample taking, then are replaced and the vials reweighed to obtain precise sample sizes.

Changes in Physiological Conditions

When enough time (10 min or more) has passed for metabolites to become fully labeled or "saturated," a physiological condition may be suddenly changed, and samples of the cells may be taken and killed for analysis to study the most immediate metabolic effects of the change. Some changes that have been studied include light to dark, changes in CO_2 level, O_2 level, and addition of NH_4^+. Since transient changes in the primary photosynthetic pathways and some secondary pathways (for example, synthesis of certain amino acids such as alanine and aspartate) occur within seconds, samples should be taken as rapidly after the environmental change as possible. For example, in one study of light–dark changes in three species of cyanobacteria,[2] labeled 6-phosphogluconate, not detected in the light, rose to its maximum value within 1 min after the light was turned off.

Separation and Identification of Labeled Metabolites

Two-Dimensional Paper Chromatography

Two-dimensional paper chromatography, which played a key role in the discovery of the photosynthetic reductive pentose phosphate cycle (Calvin cycle) in green algae and higher plants,[11] was used, with modifications,[12] to examine light–dark regulation and other types of regulation in green algae and carbon metabolism in several species of cyanobacteria.[2–4] In addition to the products of photosynthetic and respiratory metabolism with $^{14}CO_2$, labeled metabolites formed in *Aphanacapsa* 6714 from ^{14}C-labeled glucose and from ^{32}P-labeled phosphate during dark aerobic and anaerobic metabolism were separated by this method. It should be noted that the specific radioactivity of the uniformly labeled glucose used as substrate was 20 μCi/μmol. Since this amounts to only about 3.3 μCi/g atom of carbon, it is close to the lower limit of specific radioactivity that can be used and still give enough label in individual compounds on the paper chromatograms to permit their location by radioautography in a reasonable time on the developed paper chromatograms.

[11] N. Erdmann and U. Schiewer, *Biochem. Physiol. Pflanzen* **180,** 515 (1985).

[12] B. Feige, H. Gimmler, W. D. Jeschke, and W. Simonis, *J. Chromatogr.* **41,** 80 (1969).

As mentioned earlier, in a typical experiment, 1-ml samples of a 1% cell suspension are killed in 4 ml of methanol. A one-fifth aliquot of this algae–water–methanol mixture is applied slowly with an air stream to the origin of Whatman No. 1 chromatographic paper (47 × 56 cm). The chromatograms are developed in two directions, by descending chromatography, in a vapor-tight cabinet. The first solvent, used in the long direction of the paper, is made up of 840 ml liquefied phenol (about 88% pure phenol and 12% water), 160 ml water, 10 ml glacial acetic acid, and 1 ml 1.0 *M* ethylenediaminetetraacetic acid (EDTA) adjusted to pH 4.0. The second solvent is made up just before use by mixing equal volumes of *n*-butanol–water (370 : 25, v/v) and of propionic acid–water (180 : 220, v/v). For a trough with two papers, about 100 ml of chromatographic solvent is needed. For movement of the solvents to the bottom of the paper, the time for development in each direction can range from about 16 to 24 hr, depending on the number of papers, the dimensions of the cabinet, room temperature, etc. Since higher resolution of the phosphorylated compounds may be desired, duplicate sets of chromatograms can be prepared, and one set developed as long as 48 hr in each dimension to obtain greater migration of these compounds, even though this results in other labeled metabolites being lost from the paper. Constant room temperature is required for good results. After development in the first direction, the papers must be dried before running in the second direction. A map of a typical chromatogram is provided in Fig. 1.

Two-Dimensional Thin-Layer Chromatography

Radiolabeled intermediates (both ^{14}C and ^{32}P) can also be separated on cellulose-coated plates. To improve separation of ^{14}C-labeled compounds, one-dimensional ion-exchange paper chromatography is performed as a first step.[11] Thin strips (2 × 30 mm) of Amberlite SB-2 paper (Serva Co.) are activated with 10 *M* formic acid and then washed with distilled H_2O. A 10-μl aliquot is applied to one end of the paper strip, and the neutral fraction is eluted with 100 μl of 10 *M* formic acid. These two fractions are then chromatographed separately.

Cellulose plates are prepared by application of a 0.25-mm-thick cellulose–water slurry prepared with Cellulose MN 300 (Machery, Nagel and Co.). A 5- to 20-μl sample is applied as a small spot to a corner of the plate, 2.5 cm from each edge. For ^{32}P-labeled compounds, the chromatogram is developed in the first dimension using a solvent system of isobutyric acid–NH_3–H_2O–EDTA (1000 : 50 : 550 : 0.5, v/v/v/w). The chamber is preequilibrated for 5 hr, and the chromatogram then is run for 8 hr.

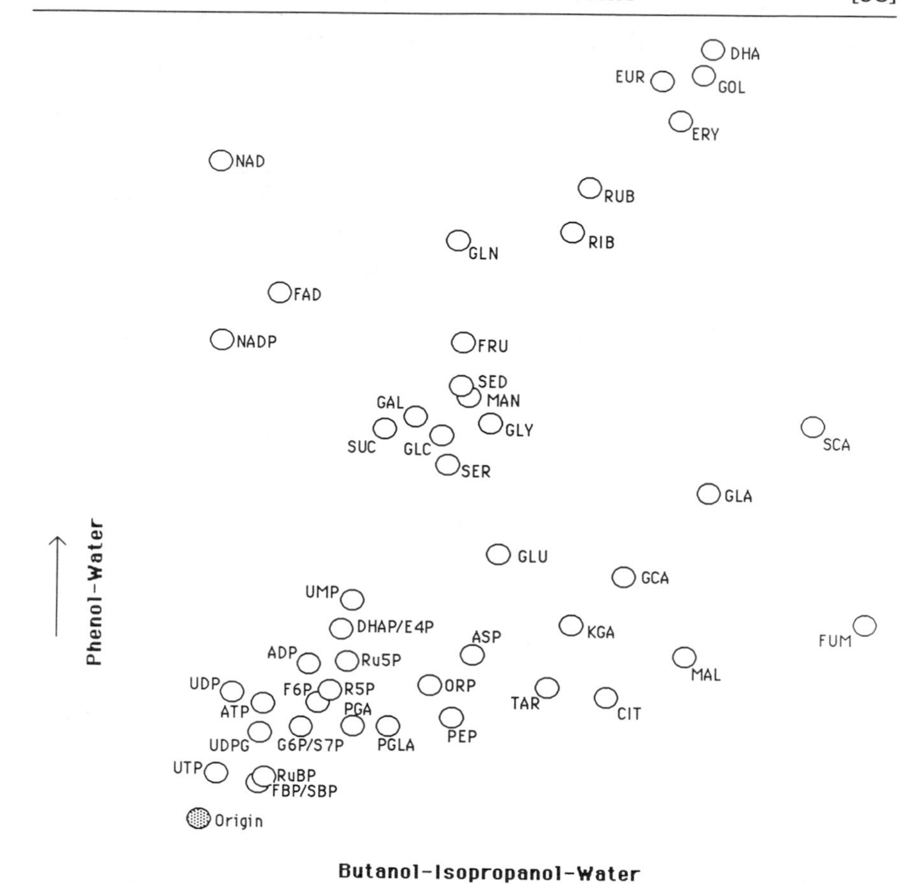

FIG. 1. Map of a typical two-dimensional paper chromatogram of cyanobacterial metabolites. The actual distances traveled may vary from chromatogram to chromatogram, but the relative positions remain constant. Abbreviations: ASP, aspartic acid; CIT, citric acid; DHA, dihydroxyacetone; DHAP, dihydroxyacetone phosphate; ERY, erythrose; EUR, erythrulose; E4P, erythrose 4-phosphate; FAD, flavin adenine dinucleotide; FBP, fructose 1,6-bisphosphate; FRU, fructose; FUM, fumaric acid; F6P, fructose 6-phosphate; GAL, galactose; GCA, glyceric acid; GLA, glycolic acid; GLC, glucose; GLN, glutamine; GLU, glutamic acid; GLY, glycine; GOL, glycerol; G6P, glucose 6-phosphate; KGA, α-ketoglutaric acid; MAL, malic acid; MAN, mannose; ORP, orthophosphate; PEP, phosphoenol pyruvate; PGA, 3-phosphoglyceric acid; PGLA, phosphoglycolic acid; RIB, ribose; RUB, ribulose; RuBP, ribulose 1,5-bisphosphate; R5P, ribose 5-phosphate; Ru5P, ribulose 5-phosphate; SBP, sedoheptulose 1,7-bisphosphate; SCA, succinic acid; SED, sedoheptulose; SER, serine; SUC, sucrose; S7P, sedoheptulose 7-phosphate; TAR, tartaric acid; UDPG, UDP–glucose.

After drying, the TLC is developed twice in the second direction, first for 5 hr (after a 3-hr preequilibration) with *n*-butanol–*n*-propanol–*n*-propionic acid–H_2O (400 : 175 : 285 : 373) and then for 5 hr (after a 3-hr preequilibration) with *n*-butanol–acetic acid–H_2O (5 : 1 : 4). For ^{14}C-labeled compounds, the chromatogram is developed in the first dimension with isobutyric acid–*n*-butanol–2-propanol–*n*-propanol–NH_3–EDTA (210 : 6 : 6 : 29 : 760 : 1, v/v/v/v/v/w). The chamber is preequilibrated for 5 hr, and the chromatogram is run for 8 hr. The plate is developed twice in the second dimension as described above for the ^{32}P system.[12]

Thin-Layer Electrophoresis/Chromatography

Labeled intermediates may also be separated by a combination of electrophoresis and chromatography.[13] This technique has been found to be superior to two-dimensional TLC for the separation of polar intermediates.[11]

Cellulose thin-layer plates are prepared and spotted as above. The sample spot is covered with a small piece of Parafilm and the plate sprayed with the electorphoresis buffer (pyridine–acetice acid–H_2O, 1 : 3.5 : 95.5, pH 4.0) until shiny. The Parafilm is removed, and electrophoresis is performed at 0° for 1 hr at 900 V on a horizontal electrophoresis apparatus. Whatman No. 1 paper wicks are used to connect the plate to the buffer trays which contain the electrophoresis buffer. The starting point is on the cathode side. After completion of the electrophoresis run, the plates are dried under cool air and developed in the second dimension using a system of 1-butanol–formic acid–H_2O (6 : 1 : 2). The chromatography is stopped when the front travels 14 cm; the plate is then dried and rechromatographed with the same solvent system until the front reaches 17 cm.[13]

Autoradiography

Location of the labeled compounds on the chromatograms is accomplished by placing the papers in contact with a sheet of medical X-ray film (single emulsion) for 3–7 days (if possible at −70°), after which the film is developed. If ^{32}P-labeled inorganic phosphate was used as a substrate, it is necessary to avoid fogging the film. This is accomplished by a preliminary autoradiography for 30 min to reveal the center of the inorganic phosphate spot. A somewhat larger area is cut from the paper, which is then replaced on a fresh film for the longer exposure.

[13] P. Schurmann, *J. Chromatogr.* **39,** 507 (1969).

Identification

A preliminary indication of the identity of a labeled compound can be made from its position. The next step in identification is to obtain an authentic sample of the suspected compound, chromatograph a few micrograms of it, and then convert it to a colored spot on the paper by spraying it with a chemical reagent. Some of the most useful ones are the ninhydrin spray for amino acids,[14] the Haynes–Isherwood spray for phosphates,[15] and silver sprays for free sugars.[16] Other tests may be obtained from standard work on color reagents.[17]

Once the procedure has been tested, the appropriate amount of authentic compound is mixed with enough of the eluted labeled compound to expose an X-ray film in a reasonable time, the mixture is chromatographed and an autoradiograph is prepared, after which the paper is treated to give the colored spot. Coincidence of the spot not only with respect to its location but also the fine details of the spot shape is good evidence for identity. Further chromatographic evidence can be obtained if the compound can be transformed chemically, rechromatographed to a new location, and again sprayed to give a colored spot. For example, a sugar phosphate could be first identified as a phosphate and later, after hydrolysis of a fresh sample, as a sugar. In dual tracer experiments, using ^{14}C and ^{32}P, the first of these two steps (identification as a phosphate) is already accomplished by the tracer. If two films are placed in contact with the paper, the first will register both isotopes, while the second will be exposed only by the more energetic β particles from the ^{32}P.

Counting

Radiolabeled incorporation can be determined by slicing the chromatographic spots into small pieces, then placing the pieces in a 20-ml scintillation vial, adding 5 ml of H_2O, and shaking for 1–2 hr. About 15 ml of a high-quality, all-purpose scintillation cocktail is added, and the vial is shaken to form a one-phase system (gel). Samples can then be counted by liquid scintillation counting.

Characterization of Carbohydrates by ^{13}C-NMR Spectroscopy

The most abundant carbohydrate components can be determined using proton-decoupled ^{13}C NMR. *In vivo* levels of sucrose, glucosylgly-

[14] R. Consden, A. H. Gordon, and A. J. P. Martin, *Biochem. J.* **38,** 224 (1944).

[15] C. S. Hanes and F. A. Isherwood, *Nature* (*London*) **164,** 1107 (1949).

[16] J. W. H. Lugg and B. T. Overell, *Aust. J. Sci. Res.* **1,** 98 (1948).

[17] R. J. Block, E. L. Durrum, and G. Zweig, "A Manual of Paper Chromatography and Paper Electrophoresis." Academic Press, New York, 1955.

cerate, glucosylglyceral, and free glutamate have all been measured using NMR.[18,19] To perform this technique, cells are grown in enriched (20–30%) $^{13}CO_2$ environments. Cells are concentrated by centrifugation (5000 *g* for 10 min at 25°)[18] and resuspended in growth media to a final concentration of 0.25 g dry weight/liter.[18] The solution is made 10% D_2O to provide a lock signal.[19] The sample is placed in an NMR tube of 1 cm diameter, containing two 1-mm-diameter tubes containing either degassed methyl iodide[19] or tetramethylsilane[18] as line standards. Fourier transformed, proton-decoupled pulsed ^{13}C-NMR spectra are then obtained, using a line broadening of 20 Hz and 1-sec intervals between pulses.[19] Volume measurements can be obtained by incubating the packed cells with 1 m*M* Tempone and 90 m*M* MnEDTA to determine the ESR signal of the intercellular probe as described previously.[20]

Acknowledgments

The work described herein was supported by the Office of Energy Research, Office of Basic Energy Sciences, Biological Energy Research Division, of the U.S. Department of Energy under Contract No. DE-AC03-76SF00098.

[18] V. Kollman, J. L. Hanners, R. E. London, E. D. Adame, and T. E. Walker, *Carbohydr. Res.* **73,** 193 (1979).
[19] E. Tel-Or, S. Spath, L. Packer, and R. J. Mehlhorn, *Plant Physiol.* **82,** 646 (1986).
[20] R. J. Mehlhorn, P. Candau, and L. Packer, this series, Vol. 88, p. 751.

[59] Transport of Carbohydrates into Cyanobacteria

By Arnold J. Smith and John E. More

Introduction

To characterize a transport system mediating the uptake of a carbohydrate or, for that matter, of any molecule or ion by a microorganism, the vectorial process should be isolated as effectively as possible from the further metabolism of the molecule or ion once it is within the cell. This separation of sequential biochemical events can be achieved to a greater or lesser extent by different experimental strategies.

Biochemical Isolation. This strategy is based on the availability of a mutant of the organism which lacks the first enzyme involved in the intracellular metabolism of the substrate transported. Such strains are thus able to transport the physiological substrate into the cell but are unable to metabolize it to a significant extent.

Temporal Isolation. This approach is based on the assumption that when organisms are incubated with the transport substrate for periods of time no longer than 1 min, the compound enters the cell in such small amounts that there is insufficient time for significant metabolism and conversion to other metabolites and cell constituents.

Chemical Isolation. This is achieved using an analog of the compound transported which, though a substrate for the uptake process, is not metabolized intracellularly and, therefore, accumulates unchanged within the cell.

Of these three approaches, biochemical isolation is closest to the ideal because it ensures the complete isolation of the transport process even with the physiological substrate. The other two strategies are both subject to specific and not insignificant limitations: Using the physiological substrate in conjunction with the wild-type strain, the uptake process can never be isolated completely from further metabolism, while with analogs of the physiological substrate uncertainties arise because the substrate for the transport process is a nonphysiological compound. When suitable mutants are not available the investigation is, however, limited to a combination of the other two strategies. The physiological substrate can be used in transport studies with wild-type organisms only in short-term incubations; these are suitable for establishing the kinetics of transport in terms of the initial rate of uptake and its dependence on substrate concentration as well as the demonstration of saturation and inhibition effects. The use of nonmetabolizable analogs where they are available complements this approach and remains the only option when wild-type organisms have to be incubated with a transport substrate for more than 1 min or so.

Carbohydrate transport in cyanobacteria has not been investigated by techniques involving biochemical isolation because suitable mutant strains are not available. Fortunately, the vectorial aspect of the assimilation of several carbohydrates by cyanobacteria has proved amenable to investigation by the technique of temporal isolation.[1] This approach together with the analog strategy has been restricted to studies of the transport of D-glucose[1-3] because appropriate derivatives of other carbohydrates are not available even in unlabeled form. Analogs of D-glucose suitable for screening as transport substrates are marketed by the Sigma Chemical Co.; they include 1-deoxy-D-glucose (1-DOG), 2-deoxy-D-glucose (2-DOG), 6-deoxy-D-glucose (6-DOG), α-methyl-D-glucoside (α-MG), and 3-*O*-methyl-D-glucose (3-OMG) and of these 2-DOG, α-MG,

[1] J. E. More and A. J. Smith, unpublished observations (1981).

[2] A. A. D. Beauclerk and A. J. Smith, *Eur. J. Biochem.* **82,** 187 (1978).

[3] B. Raboy and E. Padan, *J. Biol. Chem.* **253,** 3287 (1978).

and 3-OMG can be obtained specifically labeled with ^{14}C from Amersham International.

Principle of Assay

Whichever strategy is employed, the experimental approach is essentially the same. Freshly harvested organisms are incubated with a preparation of the selected transport substrate consisting of appropriate proportions of the ^{14}C-labeled compound and unlabeled substrate as carrier, so that there is sufficient isotope present to detect uptake and enough substrate in total to give the required concentration. Samples are removed at set times, the organisms separated rapidly from the suspending medium by filtration under vacuum through microporous membrane filters, and the amount of isotope taken up determined. Two different methods are available for measuring the amount of isotope on filters. The simplest approach uses a thin end-window, gas-flow Geiger detector. Modern equipment of this type can measure amounts of ^{14}C with efficiencies up to 30%. Scintillation spectrometry can also be used to determine the isotope content of organisms on filters, but, because the cells contain high levels of light-absorbing pigments, the extent of quenching has to be monitored rigorously and appropriate corrections applied.

Growth of Cyanobacteria and Suspension Preparation

Anabaena strain CCAP 1403/13, which is capable of growth on D-fructose in the light in the presence of 50 μM 3-(3,4-dichlorophenyl)-1,1-dimethylurea (DCMU) or photoautotrophically, is grown in medium BG-11.[4] For rapid photoautotrophic growth this medium is supplemented with Na_2CO_3 (0.4 g/liter) and gassed continuously with a mixture of air and CO_2 (95 : 5 by volume).[5] The organisms are harvested by centrifugation at 10,000 *g* at 10° for 15 min, washed twice, and resuspended in sterile medium to a density of approximately 0.5 mg dry wt/ml. The washed cell suspension is incubated at 35° in the light in a glass tube and gassed with compressed air.

Transport Assay

The kinetics of this process in terms of initial rates are established by the following procedure. Portions of the washed cell suspension are

[4] R. Rippka, J. Deruelles, J. B. Waterbury, M. Herdman, and R. Y. Stanier, *J. Gen. Microbiol.* **111,** 1 (1979).
[5] A. J. Smith, J. London, and R. Y. Stanier, *J. Bacteriol.* **94,** 972 (1967).

mixed with 1-ml volumes of medium containing 0.03–2.0 μmol of D-[U-^{14}C]fructose of specific activity 0.2 μCi/μmol in conical glass centrifuge tubes and incubated in the light for 0.5–1 min. The contents of each tube are dispersed rapidly, together with washings, in 10 ml of medium in a filter holder containing a microporous membrane filter of average pore size 0.2 μm. The medium is drawn through the filter under vacuum, and the organisms retained on the filter are washed twice with 10-ml volumes of medium. Unlabeled transport substrate is not included in the wash medium as it is likely to exchange rapidly with the radioactive compound in the cell.[2] The filters are allowed to dry before being glued to aluminum planchets for the measurement of isotope incorporation using a thin end-window, gas-flow Geiger detector. All of these operations are conducted at room temperature (~18°).

The initial rate of uptake is dependent on substrate concentration at low concentrations but reaches a maximum at higher concentrations (Fig. 1), which implicates a carrier-mediated uptake process. The kinetics of growth substrate uptake by other heterotrophic (versatile) cyanobacteria are essentially the same.[2,3] Using these data, substrate uptake can be characterized in terms of maximum initial velocity of substrate uptake (V_{max}) and substrate affinity (K_m) in a manner analogous to enzyme-catalyzed reactions that exhibit Michaelis–Menten kinetics (Table I). Conclu-

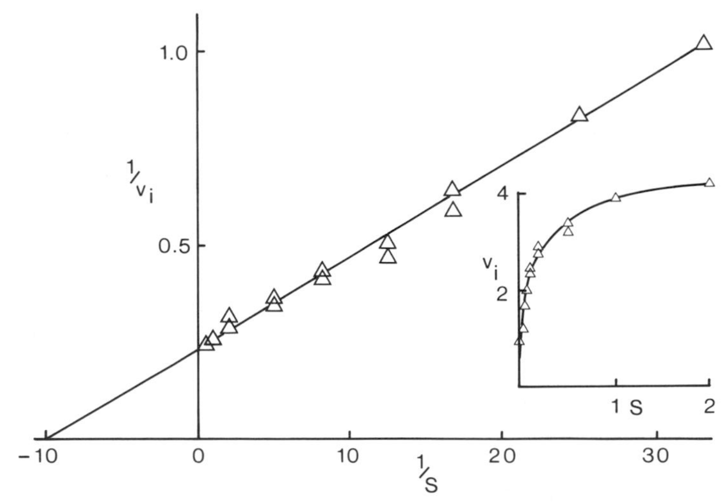

FIG. 1. Dependence of initial rates of D-[U-^{14}C]fructose uptake by *Anabaena* CCAP 1403/13 on substrate concentration. Suspensions of photoautotrophically grown organisms (0.58 mg dry wt/ml) were incubated in the light (10 klux) with D-[U-^{14}C]fructose (0.2 μCi/μmol) at 35° for 1 min. Initial velocity (v_i) is expressed as nmol/min/mg dry wt and substrate concentration (S) in mM.[1]

TABLE I
TRANSPORT OF CARBOHYDRATE GROWTH SUBSTRATES AND THEIR ANALOGS BY VERSATILE[a] CYANOBACTERIA: KINETIC PARAMETERS

Organism	Growth[b] conditions	Substrate	Initial rate of uptake, V_{max}[c] (nmol/min/mg dry wt)	Substrate affinity, K_m[c] (mM)	Inhibitor constant, K_i[c] (mM)	Ref.
Chlorogloeopsis ATCC 27193[d]	PA	D-Glucose	10.0	1.25	—	1
Nostoc Strain MAC	PA	D-Glucose	18.2	0.44	0.64 (3-OMG)	1
		3-OMG	7.2	0.77	0.50 (D-glucose)	
Plectonema boryanum	PH	D-Glucose	3.4	0.27	—	3
		α-MG	3.1	0.12	0.29 (D-glucose)	
Synechocystis ATCC 27178	PA	D-Glucose	17.2	0.43	0.72 (3-OMG)	2
		3-OMG	11.1	0.42	0.33 (D-glucose)	
Anabaena ATCC 29413	PA	D-Fructose	14.2	0.14	—	1
Anabaena CCAP 1403/13	PA	D-Fructose	4.3	0.10	—	1
Plectonema PCC 73110	PA	Sucrose	2.0	1.25	—	1

[a] Versatile cyanobacteria can grow with an organic compound as primary carbon source, in contrast to specialist strains which are restricted to CO_2.

[b] PA, Photoautotrophic; PH, photoheterotrophic.

[c] The kinetic parameters K_m and V_{max} were obtained by fitting initial velocities to the Michaelis–Menten equation.

[d] ATCC, American Type Culture Collection; CCAP, Cambridge Collection of Algae and Protozoa; PCC, Pasteur Culture Collection of Cyanobacteria.

sions about the control of carrier synthesis can be drawn from the effect of growth conditions on the magnitude of V_{max}.

With strains that can grow on D-glucose the effect of various analogs on the uptake of the physiological substrate can also be examined[1,2]; D-[U^{14}C]glucose uptake by *Synechocystis* (*Aphanocapsa*) ATCC 27178 and *Nostoc* Strain MAC is inhibited competitively by 3-OMG and 6-DOG, which suggests that these analogs are also substrates for the same transport system. This conclusion is justified only when it can be demonstrated that the preparations of analogs are free from contamination with D-glucose. Complementary experiments demonstrating the uptake of labeled 3-OMG by organisms and the competitive inhibition of analog uptake by D-glucose[1,2] (Table I) confirms that such analogs are substrates for the D-glucose transport system. In contrast, a different analog, α-MG, is a substrate for the D-glucose transport system in *Plectonema boryanum*.[3] Although D-glucose is assimilated by many cyanobacteria,[5] high rates of incorporation that are inhibited by specific analogs are found only with versatile strains capable of heterotrophic growth on this substrate.[2] Other cyanobacteria that do not grow on D-glucose appear to be the natural counterparts of D-fructose-resistant, transport-negative mutants that have been derived from strains capable of growth on D-glucose.[6]

Analog Screening and Applications

For analogs to be of use in transport studies, particularly those involving extended incubations, it is necessary to demonstrate not only that they are substrates for the vectorial event but also that they are not metabolized within the cell. This requirement is satisfied if all of the labeled analog taken up by organisms can be extracted with hot water and, in addition, if analog is the only radioactive compound in the extract. Using an experimental system based on that described above for the measurement of initial rates, the maximum amount of analog taken up by organisms can be determined from the kinetics of accumulation as distinct from initial rates of uptake. In conjunction with precise measurements of cell water this gives the intracellular concentration of the analog. The method for the measurement of cell water that employs radioactive permeant and impermeant substrates[7] is the most convenient as it avoids the need to measure the isotope content of samples containing heavily pigmented, light-absorbing cell material.

The accumulation within the cell of analog at concentrations appreciably greater than those outside, particularly at low extracellular substrate concentrations,[2,3] is characteristic of a process in which the energy metabolism of the cell drives analog accumulation against the concentration gradient. The features of this process can be probed by analyzing the effect on analog uptake of shifts from light to dark, aerobic to anaerobic conditions, and between different pH values. The energy coupling process can also be probed with compounds known to interfere with the generation and conservation of energy including ionophores, uncouplers, and inhibitors of proton-translocating ATPases.

[6] E. Flores and G. Schmetterer, *J. Bacteriol.* **166,** 693 (1986).

[7] D. B. Kell, S. J. Ferguson, and P. John, *Biochim. Biophys. Acta* **502,** 111 (1978).

[60] Ammonium Transport

By JANE GIBSON

Introduction

Ammonia is used as a nitrogen source by all cyanobacteria, and it is taken up preferentially even if another potentially usable compound such as nitrate, nitrite, or dinitrogen is also present. Alternative nitrogen sources are all reduced to ammonia intracellularly before incorporation into amino acids, a process which involves first amidation of glutamate to give glutamine through the glutamine synthetase (glutamate–ammonia ligase, EC 6.3.1.2) reaction. Ammonia thus is the central inorganic form of nitrogen in cyanobacterial metabolism. As a weak base, whose uncharged form is a gas, free diffusion of NH_3 has been assumed to supply growth needs at about the pK_a of the NH_3/NH_4^+ system (close to 9.5 under physiological ionic and temperature conditions) or above, but a transport system must be postulated to explain the ability of cells to maintain the observed intracellular pools of about 1 mM in the pH range from 7.0 to about 8.0.

Actual measurements of ammonia permeability have rarely been published, and it has been widely assumed that the permeabilities of NH_3 and of small organic amines such as CH_3NH_2 or $C_2H_5NH_2$ are very similar. Recent measurements[1] have given values of the permeability coefficient, P, in *Synechococcus* of 5–7, 85, and 110 μm sec^{-1} for the three amines, indicating that NH_3 is about 10 times less permeable than the organic amines in this prokaryote. Even so, a continuous loss of NH_3 from the cell, whose internal pH when illuminated is about 7.6,[2] must be occurring, since the internal concentration of the unprotonated NH_3 under these conditions would be approximately 10 μM. Yet the concentration of total ammonia in the medium of growing cells using nitrate is below the detection limit of the chemical assay of about 2 μM[3] and was estimated from lyophilized medium to be no more than one-tenth of this.

Because cyanobacterial cells maintain a large intracellular pool of ammonia, and because there is no convenient and easily available radioactive isotope of nitrogen, most investigations of ammonia transport in cyanobacteria and other prokaryotic and eukaryotic organisms have used

[1] R. J. Ritchie and J. Gibson, *J. Membr. Biol.* **95,** 131 (1987).

[2] J. Gibson, *Arch. Microbiol.* **130,** 175 (1981).

[3] S. Boussiba, C. M. Resch, and J. Gibson, *Arch Microbiol.* **138,** 287 (1984).

indirect methods (see Refs. 4 and 5). A lower limiting value for NH_4^+ transport can be obtained by measuring the net rate at which it is removed from the medium by metabolizing cells; since the total intracellular ammonia pool remains remarkably constant over a pH range from 6.8 to over 8,[3] this uptake rate reflects more closely the rate of glutamine synthetase-catalyzed amidation than the rate at which NH_4^+ is transported into the cell. A second approach makes use of ^{14}C-labeled methylammonium since this cation, unlike ethylammonium, is also accumulated against a concentration gradient. This accumulation can greatly exceed any difference to be anticipated from response to a pH gradient between the cell and its surroundings. Some advantages and uncertainties of each approach become obvious in what follows.

Direct Measurement of NH_4^+ Uptake

Maximum uptake rates are observed in suspensions incubated in saturating light and CO_2 concentrations. In *Synechococcus* and in the limited number of filamentous cyanobacteria studied,[2] uptake of NH_4^+ is as rapid in the presence of 5–10 m*M* $NaNO_3$ as in its absence. The choice of suspending medium will depend on the assay procedure to be used but should include 10–20 m*M* HCO_3^-. Consistent uptake rates are obtained after a 10-min equilibration. After adding NH_4Cl to 0.05–0.2 m*M*, samples of the suspension are freed of cells by filtration through membrane filters at fixed time intervals. Appropriate sample timing and cell densities will vary with the organism and assay conditions. Uptake rates of 30–40 nmol min^{-1} mg^{-1} protein at 25° are common with illuminated suspensions of *Synechococcus*. Residual ammonia in the filtrate may be determined either chemically or with a NH_3 electrode; since sample volumes of at least 50 ml after alkalinization are needed for the Orion electrode, we have preferred chemical determination. The modification of an indophenol-yielding reaction[6] given below requires handling suspension samples of less than 1 ml, is reproducible, and is sensitive down to 2–3 nmol.

Reagents

Sodium citrate solution: 20% trisodium citrate in 0.1 *N* NaOH
Nitroferricyanide solution: 5 mg ml^{-1} $[Fe(NO)(CN)_6] \cdot 2H_2O$
Phenol solution: 10% (w/v) in 95% ethanol

[4] D. Kleiner, *Biochim. Biophys. Acta* **639,** 41 (1981).
[5] D. Kleiner, *FEMS Microbiol. Rev.* **32,** 87 (1985).
[6] L. Solorzano, *Limnol. Oceanogr.* **14,** 799 (1969).

Hypochlorite solution: household bleach

NH_4Cl standard: 0.1 m*M* prepared in freshly distilled water and kept in a glass-stoppered bottle.

Procedure

Mixtures of sodium citrate and hypochlorite solutions (3 : 1, v/v) and of phenol plus nitroferricyanide solutions (1 : 1, v/v) are made freshly each day. Samples or standards (0.7 ml) containing 20–100 nmol NH_3 are mixed with 50 μl phenol–nitroferricyanide, followed by 70 μl sodium citrate–hypochlorite solutions. Tubes are incubated at 37° for 1 hr and absorbance read at 640 nm. The blue color formed is stable for at least 24 hr but increases very slowly during that period after the first hour; this can be compensated for if standards are run at the same time.

Interferences

The pH in the NH_4^+ assay mix must be between 10 and 11; acidic or strongly buffered solutions should be brought to pH 7–8 before use. Tris, Good (HEPES, MOPS, etc.), and Triethanolamine buffers interfere strongly with color development, but phosphate or carbonate buffers are inert. $CH_3NH_3^+$ yields about one-tenth of the color produced by NH_4^+. The same assay procedure can sometimes be employed for neutralized extracts in order to determine intracellular pools; *Synechococcus* extracts contain factors which reduce the assay color yield by about 10%, so that accurate determinations require internal standards. With some other organisms interference may be severe, and it is necessary to separate NH_3 from the extract using microdiffusion[7] before determination. Intracellular volumes needed to calculate pool concentrations are conveniently measured using permeant and impermeant solutes and centrifugation through silicone oil barriers.[8]

Indirect Assay: $^{14}CH_3NH_3^+$ Uptake

Both unicellular and filamentous cyanobacteria growth with NO_3^- transport $CH_3NH_3^+$ against a concentration gradient.[9–11] Uptake of the analog is strongly inhibited by NH_4^+, and accumulated analog can be

[7] E. J. Conway, "Microdiffusion Analysis and Volumetric Error." Crosby, Lockwood, London, 1939.

[8] E. Padan and S. Schuldiner, this series, Vol. 125, p. 340.

[9] A. M. Rai, P. Rowell, and W. D. P. Stewart, *Arch. Microbiol.* **137,** 241 (1984).

[10] S. Boussiba, W. Dilling, and J. Gibson, *J. Bacteriol.* **160,** 204 (1984).

[11] N. W. Kerby, P. Rowell, and W. D. P. Stewart, *Arch. Microbiol.* **143,** 353 (1986).

driven out of poisoned cells by exchange diffusion, indicating that the two ions can be carried by the same transport system. Transport kinetics are readily measured using a standard filtration assay[10]; uptake is initiated by addition of $^{14}CH_3NH_3Cl$ to aliquots of suspensions containing up to 0.3 mg cell protein ml^{-1}, and samples of 0.1 ml withdrawn for rapid filtration through Nuclepore polycarbonate membrane filters. As these bind very little label nonspecifically, washing the samples can generally be omitted, but controls should be done with cells uncoupled with a protonophore. Uptake of the analog is slower than that of NH_4^+, and maximum rates of 10–15 nmol min^{-1} mg $protein^{-1}$ are common, with an apparent K_m of 7–15 μM. Uptake curves are generally markedly biphasic, and the second, slower phase is accompanied by accumulation of the γ-methyl analog of glutamine.[10,11] The maximum intracellular concentration of unchanged amine is about 0.7 mM and may be 100-fold higher than that in the external concentration.

The extent of metabolic conversion can be ascertained by chromatography of neutralized $HClO_4$ extracts on cellulose (MN 300) thin-layered plates, using 2-propanol–formic acid–water [40 : 20 : 10 (v/v)][10] or on silica gel using methanol–acetic acid–water [60 : 2 : 38 (v/v)].[12] Depending on the length of time for which the cells were incubated with $^{14}CH_3NH_3^+$, up to 10 mM γ-glutamylmethylamide was accumulated. Production of this metabolite is completely inhibited by preincubation of suspensions with 50 μM methionine sulfoximine.[10,11] Thus, transport can be clearly separated from metabolism[13]; incubation with methionine sulfoximine should be carried out only for the minimum time needed to inactivate glutamine synthetase completely, as endogenous NH_4^+ production can interfere with $^{14}CH_3NH_3^+$ transport after prolonged treatment with inhibitor.

In *Synechococcus,* net uptake of $^{14}CH_3NH_3^+$, but not of NH_4^+, requires at least 5 mM Na^+ in the suspending medium, and optimal uptake is obtained in 10–20 mM Na^+. Transport is energy dependent and is reduced by over 90% in darkness or by incubation for 5–10 min with 20 μM CCCP, 25 μM monensin, 1 μM gramicidin, or 20 μM DCCD. Addition of KCl (100 mM) strongly inhibits $CH_3NH_3^+$ accumulation. These findings suggest that $\Delta\psi$ is the driving force for transport. The system is sensitive to sulfhydryl reagents, and incubation with 0.25 mM N-ethylmaleimide (NEM) for 1 min resulted in 80% reduction of uptake[10]; neither $CH_3NH_3^+$ nor NH_4^+ protected against NEM inactivation.

Differences between the uptake of NH_4^+ and of $CH_3NH_3^+$ thus include the following: (1) the concentrative uptake of $CH_3NH_3^+$, but not of NH_4^+,

[12] E. M. Barnes, P. Zimniak, and A. Jayakumar, *J. Bacteriol.* **156,** 752 (1982).
[13] S. Boussiba and J. Gibson, *FEBS. Lett.* **180,** 13 (1985).

is repressed by growth with NH_4^+ as nitrogen source; (2) Na^+ (5–20 m*M*) is required for transport of $CH_3NH_3^+$ but not of NH_4^+; (3) $CH_3NH_3^+$ transport is driven by $\Delta\psi$; (4) NH_4^+ uptake is much less sensitive to perturbation of $\Delta\psi$ by manipulating extracellular K^+. It is not yet clear whether these differences reflect the presence of multiple transport systems.

Acknowledgments

Unpublished investigations from this laboratory were supported in part by Grant DMB 84 15628 from the National Science Foundation.

[61] Cation Transport in Cyanobacteria

By ETANA PADAN and ALEXANDER VITTERBO

Introduction

Although not often studied with respect to active transport, cyanobacteria appear to have several ion transport systems which catalyze the movement of cations and anions across the cytoplasmic membrane. This chapter considers the systems responsible for cation transport in cyanobacteria, excluding proton[1,2] and NH_4^+ [3] transport, which are reviewed elsewhere in this volume.

Cells use both primary and secondary transport systems for ion translocation across their membranes. Primary systems couple chemical energy directly to the performance of electrochemical work resulting in an ion gradient. In the reversed mode osmotic energy can be coupled to chemical bond formation. Among the most widely recognized primary pumps are the respiratory and photosynthetic H^+-translocating ATPases and the H^+-translocating respiratory and photosynthetic chains. The photosynthetic H^+-ATPase (F_0F_1-ATPase type[2]) and the photosynthetic electron transport-linked proton pump of cyanobacteria[1,2] exist on the thylakoid membrane where they maintain an electrochemical gradient of protons and are responsible for photosynthetic phosphorylation.[4] It is not clear yet whether both types of pumps or only the H^+-ATPase exist on the

[1] Review on proton transport of cyanobacteria, this volume.
[2] D. B. Hicks, this volume [38].
[3] J. Gibson, this volume [60].
[4] E. Padan and S. Schuldiner, *J. Biol. Chem.* **253,** 3281 (1978).

cytoplasmic membrane.[5–10] Similarly, it is still debatable whether primary respiratory pumps exist on the two membranes.[5,8–11] Different species may differ in this respect.[5–7] In any event, an electrochemical gradient of protons ($\Delta\tilde{\mu}_{H^+}$) is formed across the cyanobacterial cytoplasmic membrane which is comprised of both ΔpH and $\Delta\psi$[4–6] as opposed to the $\Delta\tilde{\mu}_{H^+}$ maintained at the thylakoid membrane which is composed mainly of ΔpH.[4] The electrochemical gradient is expressed as

$$\frac{\Delta\tilde{\mu}_{H^+}}{F}\,(\mathrm{mv}) = \Delta\psi - \frac{2.3RT}{F}\,\Delta\mathrm{pH} \tag{1}$$

where R is the gas constant, T absolute temperature, F Faraday, and ΔpH the difference of pH inside versus outside the membrane, $\mathrm{pH_i} - \mathrm{pH_o}$. There is evidence for the existence of other primary cation pumps at the cytoplasmic membrane. Potassium[7,12,13] and calcium[14] ion transport systems are the best documented of these and are discussed below.

The $\Delta\tilde{\mu}_{H^+}$ generated at the cytoplasmic membrane can be utilized for osmotic work. This is accomplished by secondary transport systems called antiporters or symporters which respectively either exchange or symport H^+ with the heterologous ion. Thus downhill proton movement into the cell will drive either uphill uptake or export of the respective ion. The best characterized secondary ion transport system in cyanobacteria is the Na^+/H^+ antiporter.[6,12,15] This system maintains a Na^+ gradient (directed inward) across the cytoplasmic membrane which in turn can drive transport systems catalyzing Na^+–solute translocation.

In bacteria as well as in eukaryotic cells, demonstration of the existence of a particular ion-transport system, and determination of its apparent kinetics and energy source, can be achieved in intact cells. However, for further detailed study of energy coupling and mode and mechanism of action, the experimental preparations of choice are isolated membrane vesicles and/or proteoliposomes containing the purified carrier protein. These *in vitro* preparations, devoid of cytoplasm and many nonrelevant

[5] B. Raboy and E. Padan, *J. Biol. Chem.* **253,** 3287 (1978).
[6] H. Paschinger, *Arch. Microbiol.* **113,** 285 (1977).
[7] R. H. Reed, P. Rowell, and N. D. P. Stewart, *Eur. J. Biochem.* **116,** 323 (1981).
[8] P. Scholes, P. Mitchell, and J. Moyle, *Eur. J. Biochem.* **8,** 450 (1969).
[9] G. A. Peschek, *Plant Physiol.* **75,** 968 (1984).
[10] S. Scherer, E. Sturzl, and P. Böger, *J. Bacteriol.* **158,** 609 (1984).
[11] W. Lockau and S. Pfeffer, *Z. Naturforsch. C* **37,** 658 (1982).
[12] M. A. Dewar and J. Barber, *Planta* **113,** 143 (1973).
[13] R. H. Reed and A. E. Walsby, *Arch. Microbiol.* **143,** 290 (1985).
[14] W. Lockau and S. Pfeffer, *Biochim. Biophys. Acta* **733,** 124 (1983).
[15] E. Blumwald, J. M. Wolosin, and L. Packer, *Biochem. Biophys. Res. Commun.* **122,** 452 (1984).

components, allow the investigator to approach directly the structure and function of the transport system.

Recently, the isolation of a cytoplasmic membrane fraction from a number of cyanobacteria[11,16–18] has been claimed. Only in the case of *Anabaena variabilis*, however, was transport activity (ATP-dependent Ca^{2+} transport) shown to occur in this fraction.[11,14] Clearly, efforts must now be focused on the isolation and biochemical characterization of cyanobacterial cytoplasmic membranes from different strains.

Determination of Active Transport *in Vivo*

Measurement of intracellular ion concentration, as dependent on extracellular concentration and on cellular metabolic state, is the routine experiment for identifying an active transport system *in vivo*. Although different species may vary with respect to small details, the protocol found satisfactory for the study of transport in *Plectonema boryanum*,[5] *Anabaena variabilis*,[7] and *Synechococcus*[15] may serve as guiding examples.

For calculation of intracellular ion concentration it is assumed that the intracellular ion is free in solution. The intracellular osmotic volume can be determined by any of the techniques already published and thoroughly discussed.[7,12,19] An innovative approach for measuring intracellular volume has been recently developed in cyanobacteria.[20] This method is based on spin-labeled, metabolically inert, small molecules that rapidly equilibrate across the cytoplasmic membrane. The procedure for volume measurement consists of obtaining an ESR spectrum of the probe in the presence of cells and a second spectrum after differential quenching of the external probe with impermeable spin quenchers. From the nonquenched probe concentration the internal volume is directly calculated. Utilizing this technique, intracellular osmotic volumes of 70–80 and 30–40 μl/mg chlorophyll were estimated for *Synechococcus* PCC 6311[20] and for *Agmenellum quadruplicatum*,[21] respectively.

An electrochemical gradient of an ion across the cytoplasmic membrane indicates active transport of that ion which must be coupled to energy supply.

[16] T. Omata and N. Murata, *Plant Cell Physiol.* **24,** 1101 (1983).
[17] W. H. Nitschmann, G. Schmetterer, R. Muchl, and G. A. Peschek, *Biochim. Biophys. Acta* **682,** 293 (1982).
[18] N. Murata and T. Omata, this volume [23].
[19] E. Padan and S. Schuldiner, this series, Vol. 125, p. 337.
[20] R. J. Mehlhorn, P. Candau, and L. Packer, this series, Vol. 88, p. 751.
[21] S. Belkin, R. J. Mehlhorn, and L. Packer, *Plant Physiol.* **84,** 25 (1987).

Determination of Energy Source *in Vivo*

Cyanobacteria are either obligate or facultative photoautotrophs. Although the former grow only photoautotrophically they have some respiratory activity. The latter have both photosynthetic and respiratory growth capacities.[22] Whereas all cyanobacteria possess the plant-type oxygenic photosynthetic system, some also have the bacterial-type anoxygenic system, utilizing sulfide as an electron donor.[23] There are few indications that glycolytic anaerobic metabolism occurs in cyanobacteria.[24]

Dependence of a transport system on photosynthetic or respiratory metabolism is determined by the use of the respective electron transport inhibitors as well as by consideration of the effect of light and O_2.[5] The results show that the dominant metabolism in cyanobacteria is the photosynthetic one. It is also apparent that cyclic electron flow can efficiently maintain transport activity.[5]

Both photosynthetic and respiratory metabolism yield several similar forms of energy transmitters which serve most endergonic reactions of the cell. When grown photosynthetically, cells generate ATP and $\Delta\tilde{\mu}_{H^+}$ at the thylakoid membrane. These are also produced at the cytoplasmic membrane if it possesses both photosynthetic electron transport and the thylakoid type H^+-ATPase. If, on the other hand, the membrane bears only H^+-ATPase, $\Delta\tilde{\mu}_{H^+}$ will be generated by reversion of its activity with ATP being supplied by the thylakoids. The same alternatives for energy flow may exist for respiratory metabolism. The route of energy supply can be more complicated; ATP and $\Delta\tilde{\mu}_{H^+}$ may support reactions that produce other gradients or energy-rich molecules which in turn can serve as the energy source for transport.

To distinguish between primary and secondary coupling of energy to transport in intact cells, there is a need to determine whether a particular transport system is coupled to chemical or electrochemical energy. It is possible to measure the magnitude of the candidate driving force, such as $\Delta\tilde{\mu}_{H^+}$, under steady-state conditions[19,25–27] and to calculate whether it could account for the concentration gradient of the ion, assuming a certain stoichiometry of ion per H^+.[25] A more direct approach is to isolate the different forms of energy from each other by creating intracellular condi-

[22] G. A. Peschek, W. H. Nitschmann, and T. Czerny, this volume [39].

[23] S. Belkin, Y. Shahak, and E. Padan, this volume [40].

[24] A. Oren and M. Shilo, *Arch. Microbiol.* **122,** 77 (1979).

[25] H. Rottenberg, this series, Vol. 125, p. 3.

[26] E. Padan, D. Zilberstein, and S. Schuldiner, *Biochim. Biophys. Acta* **650,** 151 (1984).

[27] See review on measurement of $\Delta\tilde{\mu}_{H^+}$ in cyanobacteria, this volume [39].

tions under which either chemical energy alone or electrochemical energy alone is available for transport.

Cyanobacterial F_0F_1-ATPases, like those in mitochondria, chloroplasts, and other bacteria, are sensitive to *N,N'*-dicyclohexylcarbodiimide (DCCD).[2,4-6] In photoautotrophically grown cyanobacteria that have electorn transport at the cytoplasmic membrane, addition of DCCD will deprive cells of ATP but will leave $\Delta\tilde{\mu}_{H^+}$ across the membranes intact. This can be measured as described.[26,27] In such cells, demonstration of light-dependent active transport, which is sensitive to uncouplers and photosynthetic electron transport inhibitors, will suggest a secondary energy coupling mechanism via $\Delta\tilde{\mu}_{H^+}$. In fact, existence of light-dependent and uncoupler-sensitive glucose transport in DCCD-treated cells was taken as an indication for the presence of photosynthetic electron transport at the cytoplasmic membrane of *Plectonema boryanum*.[5] If, on the other hand, DCCD inhibits transport (but does not interfere with the $\Delta\tilde{\mu}_{H^+}$ across the cytoplasmic membrane), it may be implied that the system is coupled directly to chemical energy. Although ATP is a reasonable candidate for the energy donor in such a case, rigorous identification of the immediate energy source for transport requires direct *in vitro* studies.

It is important to note that in species which contain only H^+-ATPase or other ATP-dependent electrogenic primary cation pumps at the cytoplasmic membrane DCCD will abolish the generation of $\Delta\tilde{\mu}_{H^+}$ across the membrane as well as inhibit ATP synthesis. In this type of cyanobacteria, therefore, addition of DCCD does not allow the identification of the energy source for transport. Since substrate level phosphorylation is absent or very low in cyanobacteria, experimental conditions for providing cells with ATP in the absence of $\Delta\tilde{\mu}_{H^+}$ are not known.

An independent way to determine the role of $\Delta\tilde{\mu}_{H^+}$ is to impose it artificially in energy-depleted cells poisoned by DCCD and electron transport inhibitors. ΔpH, alkaline inside, is created by equilibration of the cells at alkaline pH followed by a rapid shift to acidic extracellular pH. $\Delta\psi$ is imposed by creating a diffusion potential of K^+ in the presence of valinomycin.[28] Several cyanobacteria are sensitive to ionophores[4,5]; others must be pretreated with EDTA to expose their cytoplasmic membrane to the drug.[19,29] If the artificial ΔpH and $\Delta\psi$ drive ion transport which is sensitive to uncouplers, $\Delta\tilde{\mu}_{H^+}$ can be implicated as the driving force.

[28] R. Muchl and G. A. Peschek, *FEBS Lett.* **164,** 116 (1983).

[29] W. W. A. Erber, W. H. Nitschmann, R. Muchl, and G. A. Peschek, *Arch. Biochem. Biophys.* **247,** 28 (1986).

K^+ Transport

Cyanobacterial cells concentrate K^+ from the medium.[6,7,12,30,31] A typical example is given in Fig. 1. *Plectonema boryanum* cells[4] that were grown at 5–10 mOs*M* and extracellular K^+ concentration ($[K^+]_{out}$) of 10–150 μM concentrate K^+ to 160–240 m*M* and maintain a concentration gradient of up to 10^4, implying the existence of an active transport system(s) for K^+. Below $[K^+]_{out}$ of 10 μM, the growth capacity of *P. boryanum* is impaired (Fig. 1).

Kinetic parameters of K^+ transport were calculated in *Anabaena variabilis* from tracer ($^{42}K^+$) exchange rates, yielding similar flux rates of efflux and influx in the range of 14.4–15.2 nmol m^{-2} sec^1.[7] Comparison of K^+ transport to that of Rb^+ suggested that there are at least two transport systems for K^+, one with low affinity (4.5 m*M*) and little discrimination and one with higher affinity (40 μM) that discriminates against Rb^+.[31,32]

Additional kinetic parameters can be determined from measurement of net uptake of K^+. For this purpose it is essential to deplete the cells of the ion, a goal unattained in many trials.[7,12] It was suggested that the membrane is highly impermeable to K^+ and that the K^+ transporters are unidirectional.[7,12] Among the many manipulations used to deplete bacterial cells of K^+,[30] we found that *N*-ethylmaleimide (NEM) treatment, previously used for *Escherichia coli*[31,33,34] is very efficient for *P. boryanum* (Fig. 2A). Best results were obtained with 10 μmol NEM/mg cell protein. The rate of leakage from the treated cells exhibited first-order kinetics with respect to intracellular K^+ (not shown), implying a single intracellular compartment of free K^+. However, there was always about 5–10 m*M* K^+ (less than 5% of the total $[K^+]_{in}$) bound in the cells which could only be released by disruption of the cells by lysozyme treatment, detergents, or boiling. Therefore, we assumed that this fraction is osmotically inactive and subtracted it from the total intracellular K^+ for calculation of $[K^+]_{in}$.

NEM treatment did not induce morphological changes in the cells and caused neither ATP nor phycocyanin leakage, leading to the conclusion that the effect of NEM on cell permeability may be confined to K^+ as is the case in *E. coli*.[31,33,34] However, the ion specificity of the leakage has not yet been determined in the cyanobacterium. It will be interesting to

[30] E. M. Alison and A. E. Walsby, *J. Exp. Bot.* **32,** 241 (1981).
[31] M. O. Walderhaug, D. C. Dosch, and W. Epstein, *in* "Ion Transport in Bacteria" (B. P. Rosen and S. Silver, eds.). Academic Press, New York, 1986.
[32] R. H. Reed, P. Rowell, and W. D. P. Stewart, *FEMS Microbiol. Lett.* **11,** 233 (1981).
[33] J. Meury, S. Lebail, and A. Kepes, *Eur. J. Biochem.* **113,** 33 (1980).
[34] E. P. Bakker and W. E. Mangerich, *FEBS Lett.* **140,** 177 (1982).

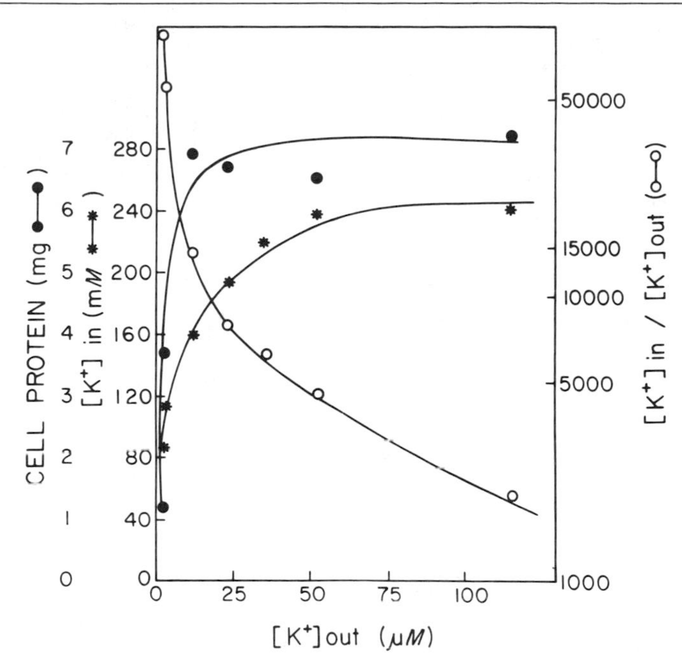

FIG. 1. Active concentration of K^+ by *Plectonema boryanum*. Cells of *P. boryanum* were grown photoautotrophically in 100 ml mineral medium (Padan and Schuldiner[4]) containing various K^+ concentrations ($[K^+]_{out}$). Phosphate concentration and osmolarity were kept constant by addition of Na_2HPO_4. After 5 days of growth, intracellular K^+ ($[K^+]_{in}$) and $[K^+]_{out}$ were measured by atomic absorption. For determination of $[K^+]_{in}$, cells (1 mg cell protein) were washed, collected on Millipore filters (3 μm pore size), and boiled in 10 ml H_2O for 10 min. Aliquots for atomic absorption measurement were taken from the supernatant obtained by centrifugation at 25,000 *g* for 10 min. Protein concentration was determined by the method of Lowry *et al.*[3] [O. H. Lowry, N. J. Rosebrough, A. L. Farr, and R. J. Randall, *J. Biol. Chem.* **193,** 265 (1951)]. The protein content of the freshly inoculated culture was 0.4 mg. An intracellular osmotic volume of 5 μl/mg cell protein[5] served to calculate $[K^+]_{in}$.

test whether the mechanism underlying NEM treatment is similar to that of *E. coli*.[35]

As has been observed in *E. coli*, NEM treatment is reversible with respect to $[K^+]_{in}$ by the addition of mercaptoethanol (ME) (Fig. 2B). The mechanism of this activation is unknown.[31,33,34] In the presence of light, K^+ is taken up by the NEM–ME-treated *Plectonema* cells to the initial values. This uptake is completely light dependent and insensitive to 3-(3,4-dichlorophenyl)-1,1-dimethylurea (DCMU). Since DCMU-poisoned cells do not evolve O_2, we conclude that it is cyclic electron flow which provides the metabolic energy for transport.

[35] J. Meury and A. Kepes, *EMBO J.* **1,** 339 (1982).

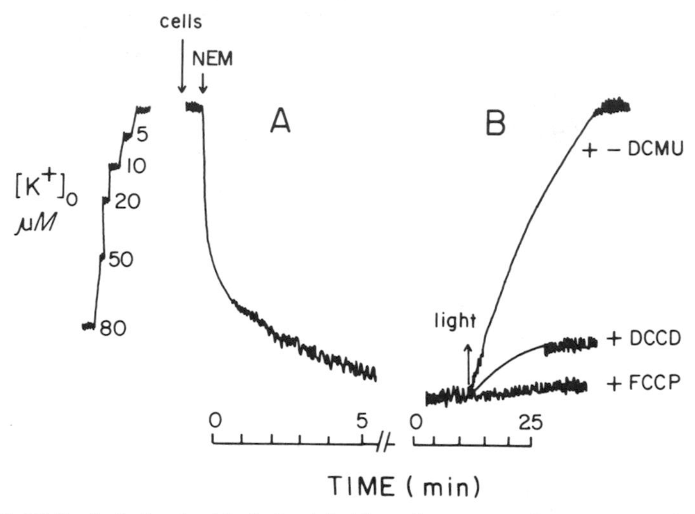

FIG. 2. $[K^+]_{in}$ depletion by *N*-ethylmaleimide and recovery after mercaptoethanol treatment. (A) *Plectonema* cells were washed and resuspended to 1 mg protein in 20 ml buffer lacking added K^+ (contamination was below 1 μM) but containing 0.1 mM $MgSO_4 \cdot 7H_2O$, 0.17 mM $NaHPO_4$, 17.6 mM $NaNO_3$, 4 mM $NaHCO_3$, 0.08 mM $CaCL_2 \cdot 2H_2O$, and 5 mM Tricine. The final pH was adjusted to 7.5 with NaOH. The cell suspension was exposed to a K^+ electrode (Orion Model 93-19) and extracellular K^+ recorded (Goerz, Wien, FRG, RE511). The concentration of NEM was 0.5 mM, and $[K^+]_{in}$, measured by atomic adsorption, was 220 mM at time 0 and 10 mM at 25 min. (B) Cells at 40 min of NEM treatment were harvested and resuspended in 20 ml of the buffer lacking K^+ as in A. After addition of 4 mM ME, the cells were incubated for 5 min, harvested, and resuspended as above. At 0 time 100 μM K^+ was added and K^+ uptake monitored as in A. Light intensity was 5×10^2 erg cm^{-2} sec^{-1}. Where indicated 5 μM DCMU, 40 μM FCCP, or 200 μM DCCD was added at time 0.

The K^+ transport in the NEM–ME-treated cells showed marked dependence on intracellular ATP ($[ATP]_{in}$) (Fig. 2B and Fig. 3). DCCD decreased, in parallel, both $[ATP]_{in}$ and K^+ transport. This dependence of K^+ transport system on $[ATP]_{in}$ can be explained by direct coupling of the transport system to ATP or by secondary coupling to $\Delta\tilde{\mu}_{H^+}$ if the ATP is needed to maintain the $\Delta\tilde{\mu}_{H^+}$ via the H^+-ATPase at the cytoplasmic membrane. A requirement for both driving forces in the transport could also account for the results.

Measurement of $\Delta\tilde{\mu}_{H^+}$ in *A. variabilis* showed that it cannot account for the large K^+ gradient maintained by these cells.[7] We have similarly calculated the $\Delta\psi$ across the cytoplasmic membrane of the NEM–ME-treated *Plectonema* cells from the distribution of K^+ in the presence of valinomycin (Fig. 3A) or from the distribution of tetraphenylphosphonium[26] (TPP^+, not shown). In both cases a $\Delta\psi$ of about 134 mV (negative inside) was found.

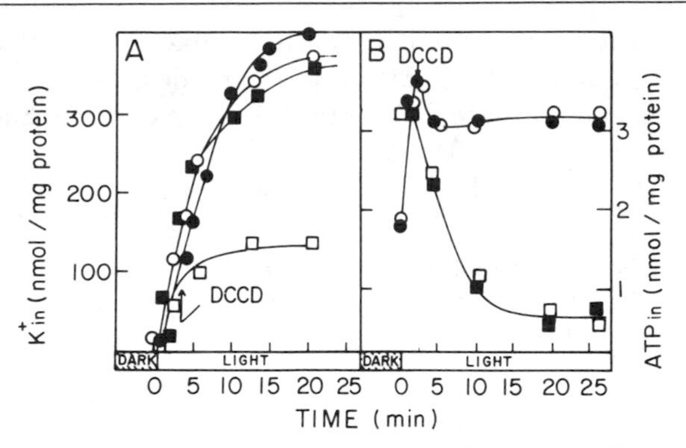

FIG. 3. K^+ transport in NEM–ME-treated cells. *Plectonema* cells were treated with NEM and ME and incubated as described in Fig. 2B. (○) Control, (●) 20 μM valinomycin added at time 0; 200 μM DCCD was added at 2.5 min to reaction mixtures containing (■) or lacking (□) valinomycin. At different time intervals $[K]_{in}$ (A) and $[ATP]_{in}$ (B) were determined. Values of $[K^+]_{in}$ determined by atomic absorption as in Fig. 1 or from the change in $[K^+]_{out}$ monitored by a K^+ electrode as in Fig. 2 gave equal results. $[ATP]_{in}$ was measured as in Nitschmann *et al.*[7]

In corroboration with the previous suggestion[5] that $\Delta\tilde{\mu}_{H^+}$ is maintained at the cytoplasmic membrane by photosynthetic electron transport, DCCD had no effect on the distribution of either TPP^+ or K^+ in the presence of valinomycin (Fig. 3A). Nevertheless, DCCD blocked ATP production (Fig. 3B) and K^+ transport in the absence of valinomycin (Fig. 3A). It is thus suggested that either ATP has a regulatory role or it is an immediate driving force for K^+ transport observed in NEM–ME-treated *Plectonema* cells. The latter conclusion has been reached for the driving force of K^+ transport in both *A. variabilis*[7] and *Synechocystis* PCC 6714.[35] Unfortunately, as yet it is not possible to deplete cyanobacterial cells of $\Delta\tilde{\mu}_{H^+}$ without effecting $[ATP]_{in}$, a condition necessary for further study of the role, if any, of $\Delta\tilde{\mu}_{H^+}$ in this transport system.

Utilizing the NEM–ME-treated *Plectonema* cells we determined the kinetic parameters of the transport system and found a K_m of 100 μM and a V_{max} of 114 nmol (mg protein)$^{-1}$ min^{-1}. This transport system accounts for the growth capacity in low-K^+ medium around 100 μM $[K^+]_o$ (Fig. 1), which is characteristic for many cyanobacteria. Involvement of K^+ transport in regulation of turgor pressure of cyanobacteria has also been demonstrated.[13,30,31,36–38] In the gas-vacuolate cyanobacterium *Anabaena flos-*

[36] R. H. Reed and W. D. P. Stewart, *Biochim. Biophys. Acta* **812,** 155 (1985).

[37] D. M. Miller, J. H. Jones, J. H. Jopp, D. R. Tindell, and W. E. Schmid, *Arch. Microbiol.* **111,** 145 (1976).

[38] R. H. Reed, D. L. Richardson, and W. D. P. Stewart, *Biochim. Biophys. Acta* **814,** 347 (1985).

aquae, rise in turgor pressure is supported by light-stimulated K^+ uptake, which causes collapse of the gas vesicles and destroys buoyancy. It has therefore been suggested that K^+ transport is involved in buoyancy regulation of planktonic cyanobacteria.[13,30]

The studied cyanobacterial K^+ transport system is reminiscent of the *kdp*-coded K^+-ATPase of *E. coli* and other bacteria.[31] As observed in enteric bacteria, K^+ transport in *A. variabilis* and *Synechocystis* PCC 6714 behaves as a turgor-sensitive process, increasing dramatically when turgor is reduced.[31,36]

Na^+ Transport

The existence of a concentration gradient of Na^+ directed inward, which is dependent on photosynthetic or respiratory metabolism, has been demonstrated in cyanobacteria.[6,7,12,15,38] These results indicate the existence of an active transport system extruding Na^+ from cyanobacterial cells.

Monitoring $^{22}Na^+$ fluxes, Paschinger[6] has presented evidence for a Na^+/H^+ antiporter in *A. nidulans*. He found that cells poisoned with DCCD no longer actively pump out Na^+ but instead take up Na^+ passively in exchange for intracellular H^+. It was suggested that the H^+-ATPase alone maintains a $\Delta\tilde{\mu}_{H^+}$ across the cytoplasmic membrane of *Anacystis*, which in turn drives the Na^+/H^+ antiporter.

Utilizing acridine orange as a probe for ΔpH, Blumwald *et al.*[15] showed that ΔpH (acid inside) is formed in response to artificially imposed $\Delta\tilde{\mu}_{H^+}$ (directed outward) across the cytoplasmic membrane. This activity was specific to Na^+, sensitive to protonophores, inhibited at low extracellular pH, and greatly enhanced in salt-adapted cells. A stoichiometry of 1 H^+ to 1 Na^+ was calculated from the ΔpH and the Na^+ concentration ratio.

Cyanobacteria prefer alkaline extracellular pH for growth.[39] A requirement for Na^+ in the alkaline growth of these organisms has been demonstrated recently.[39] It was also shown that the Na^+ requirement is for intracellular pH regulation. Thus, involvement of a Na^+/H^+ antiporter has been suggested in pH_i homeostasis of cyanobacteria. In this respect cyanobacteria appear very similar to both neutrophilic and alkalophilic bacteria.[19,26] Interestingly, a Na^+ requirement for cell division even at neutral pH_o has been implicated in *Synechococcus leopoliensis*.[39] This effect can be explained by perturbation of pH_i in the absence of Na^+ and pH_i being linked to cell division machinery. Recent results in bacteria

[39] A. G. Miller, D. H. Turpin, and D. T. Danvin, *J. Bacteriol.* **159,** 100 (1984).

show that alkaline shift in pH_i induces the SOS system in *E. coli* and affects cell division.[40]

Involvement of Na^+ in cyanobacterial bicarbonate transport has also been suggested.[41] Operation of an active bicarbonate transport with a high flux rate will flush the cytoplasm with OH^- as CO_2 is consumed by ribulase-bisphosphate carboxylase. Operation of a most important Na^+ cycle can be envisaged in these organisms. Na^+ may enter the cells in symport with bicarbonate and be extruded by the Na^+/H^+ antiporter, maintaining the cytoplasmic pH constant. As yet, however, there is only indirect evidence for a Na^+/HCO_3^- symport transporter. Without such a system a high leak for Na^+ must be predicted in order to maintain a sodium cycle.

Ca^{2+} Transport

The intracellular Ca^{2+} concentration of cyanobacteria is low, suggesting the existence of an export mechanism for this ion.[14] The so-called cytoplasmic membrane fraction[18] from *Anabaena variabilis* was shown to carry a Ca^{2+}-stimulated, Mg^{2+}-dependent ($Ca^{2+}Mg^{2+}$-) ATPase.[11,14] Ca^{2+} transport into this vesicular fraction was investigated using either $^{42}Ca^{2+}$ or the membrane-impermeant optical indicator of Ca^{2+} murexide.[11] Uptake of $^{42}Ca^{2+}$ was assayed at 22° in a reaction mixture containing 0.1–0.3 mg protein/ml, 50 m*M* Tricine–NaOH (pH 8), 50 m*M* KCl, 5 m*M* $MgCl_2$, 50 m*M* sucrose, and about 0.1 μCi/nmol $^{42}Ca^{2+}$. Following 10 min of preincubation, ATP (0.8 m*M*) was added, and $^{42}Ca^{2+}$ accumulated in the vesicles was determined by filtering 0.05- to 0.15-ml aliquots mixed with 2 ml ice-cold assay buffer (lacking Ca^{2+} and ATP) on a cellulose nitrate filter (0.2 μm pore size, Sartorius, Gottingen). The filters were washed with 3.5 ml of the buffer and the radioactivity determined. When murexide was used to measure Ca^{2+}, concentration absorption changes were monitored in a temperature-controlled Aminco DW-2 spectrophotometer (540 minus 507 nm, optical bandpass 4 nm).

It was found that the "cytoplasmic membrane" vesicles of *A. variabilis* take up Ca^{2+} on addition of ATP. Uncouplers do not affect this reaction, but the Ca^{2+} ionophore A23187 rapidly reverses the uptake. Resistance of calcium accumulation in the membrane vesicles to a variety of ionophores and DCCD, and its sensitivity to orthovanadate, strongly suggest it to be primary active transport not requiring a chemiosmotic force.

The cyanobacterial Ca^{2+}-ATPase is similar to a bacterial one found in *Streptococcus faecalis*.[42] In the latter case it has been suggested that in

[40] E. Padan and S. Schuldiner, *J. Membr. Biol.* **95,** 189 (1987).
[41] See A. Kaplan, Y. Marcus, and L. Reinhold, this volume [57].
[42] D. L. Heefner and F. M. Harold, *Proc. Natl. Acad. Sci. U.S.A.* **79,** 2798 (1982).

addition to Ca^{2+} extrusion the Ca^{2+}-ATPase serves to energize the cytoplasmic membrane which is devoid of electron transport.

Concluding Remarks

In addition to K^+, Na^+, and Ca^{2+} transport, only NH_4^+ transport has been extensively studied in cyanobacteria.[3] Nevertheless, judging from the many cations and anions comprising the growth medium of cyanobacteria and the remarkable capacity of cyanobacteria to grow at very low concentrations (micromolar range) of these ions, it is tempting to suggest that many more ion transport systems are yet to be discovered. Recently, molecular genetics has been extensively applied to study ion transport in bacteria, and several genes coding for transport systems have been cloned. This calls for utilizing DNA sequence similarity and recombinant DNA techniques to search for and study cyanobacterial transport systems.

Acknowledgments

We thank Dr. A. Kaplan, Dr. B. Binder, and Dr. I. R. Booth for critical reading of the manuscript. This work was supported by a grant from the United States–Israel Binational Foundation given to E.P.

[62] Sulfur Metabolism in Cyanobacteria

By AHLERT SCHMIDT

Introduction

Living systems are dependent on reduced sulfur in the amino acids cysteine and methionine for protein formation and coenzyme biosynthesis of glutathione, coenzyme A, iron–sulfur proteins (ferredoxins), or biotin. Organisms capable of synthesizing these compounds from inorganic sulfate are termed assimilatory sulfate reducers, because they reduce sulfate for further incorporation (assimilation).[1] Organisms using sulfate as a terminal electron acceptor under anaerobic conditions are designated dissimilatory sulfate reducers because the reduced sulfur is not assimilated but dissimilated as sulfide. Cyanobacteria (and other oxygenic organisms) normally reduce sulfate for protein and coenzyme synthesis and thus are

[1] H. D. Peck, Jr., *Bacteriol. Rev.* **26,** 67 (1962).

not dependent on reduced sulfur compounds, in contrast to animals and humans who need methionine and/or cysteine as essential amino acids.

Photosynthetic organisms including cyanobacteria need sulfur compounds at the oxidized level for sulfolipid biosynthesis. Sulfolipids are diacyl lipids containing the sulfonic acid sulfoquinovose (6-deoxy-6-sulfo-D-glucose[2]), predominantly found in photosynthetic membranes. Certain cyanobacteria are capable of using hydrogen sulfide as an electron donor for photosystem I, eliminating the need for oxygen production. Light-dependent growth under anaerobic conditions is termed anoxygenic photosynthesis.[3–5] We address here pathways for sulfate assimilation into cysteine and further transfer reactions to methionine, glutathione, and other sulfur-containing molecules necessary for sulfur metabolism and its regulation in cyanobacteria.

Sulfur Sources for Growth of Cyanobacteria

Cyanobacteria are normally grown using 0.3 m*M* sulfate concentrations[6–8]; this has to be compared with sulfate available in freshwater lakes (~0.1 m*M*), eutrophic lakes (~0.2 m*M*), and seawater (~27 m*M*).[8] Obviously, algae growing in seawater are adapted to sulfate concentrations available normally within a cell (~0.6 m*M* in *Anacystis nidulans*[9]), whereas in freshwater systems the sulfate concentration could be limited (see above). Therefore, cyanobacteria of marine origin should be cultured using a high-sulfate concentration (about 30 m*M*[7]), whereas for freshwater cyanobacteria a sulfate concentration of 0.3 m*M* is sufficient for normal growth of batch cultures.

We have determined that 1 μmol/liter sulfate leads to an optical density at 680 nm of 0.125 with *Synechococcus* 6301[10]; such material is in a sulfur-deficient state with a reduced C-phycocyanin content. *Synechococcus* 6301 achieved good growth at 0.3 m*M* concentrations with a limited spectrum of sulfur compounds: sulfate, thiosulfate, thioacetic acid, mercaptoacetic acid, thioacetamide, L-cysteine, L-cystine, and reduced glu-

[2] A. A. Benson, *Adv. Lipid. Res.* **1,** 387 (1963).
[3] Y. Cohen, B. B. Jørgensen, E. Padan, and M. Shilo, *Nature* (*London*) **257,** 489 (1975).
[4] E. Padan, *Adv. Microbiol. Ecol.* **3,** 1 (1979).
[5] E. Padan, *Annu. Rev. Plant Physiol.* **30,** 27 (1979).
[6] R. Y. Stanier, R. Kunisawa, G. Mandel, G. Cohen-Bazire, *Bacteriol. Rev.* **35,** 7 (1971).
[7] R. Rippka, J. Deruelles, J. B. Waterbury, M. Herdman, and R. Y. Stanier, *J. Gen. Microbiol.* **111,** 1 (1979).
[8] N. G. Carr, "CRC Handbook in Nutrition and Food," p. 283. CRC Press, Boca Raton, Florida, 1978.
[9] H. C. Utkilen, M. Heldal, and G. Knutsen, *Physiol. Plant.* **38,** 217 (1976).
[10] A. Schmidt, I. Erdle, and H.-P. Köst, *Z. Naturforsch.* C **37,** 870 (1982).

tathione. Best growth rates were achieved in our research using reduced glutathione,[10] probably due to the nutritional aspects with reduced sulfur and nitrogen. Methionine could not sustain growth in this cyanobacterium. Most sulfur sources analyzed did not inhibit growth of *Synechococcus* 6301; however, we noticed growth inhibition with aminomethanesulfonic acid, aminoiminomethanesulfinic acid, cysteamine, methional, *O*-methyl-L-cysteine, dithioerythritol (DTE), *O*-ethyl-L-cysteine, sulfanilic acid, and *S*-methylmethionine. L-Cysteine at concentrations of 1 m*M* completely blocks growth of cyanobacteria and green algae, whereas DL-ethionine, thioacetic acid, and 2,3-dimercaptopropanol allow growth of *Synechococcus* 6301; however they inhibit growth of the green alga *Chlorella fusca* which might be useful for discrimination in enrichment cultures. We detected good growth on the sulfonic acids taurine and ethane sulfonate with *Anabaena variabilis* and *Plectonema* 73110[11] and limited growth with *Anabaena cylindrica* 7122. Other strains including *Synechococcus* 6301, *Anabaena variabilis* 7120, *Nostoc* 6310, *Spirulina* 6313, and *Calothrix* 7101 could not metabolize sulfonic acids.[11]

Uptake of Sulfur Sources

Sulfate Uptake. Sulfate uptake catalyzed by a sulfate permease is an energy-dependent process.[12] Uptake rates are determined using ^{35}S-labeled sulfate and removing excess sulfate either by centrifugation of the cells or washing on membrane filters; the rates can be calculated from the specific activity of sulfate added. Sulfate uptake in cyanobacteria is pH dependent with optimal rates between pH 8 and 9.[9,11–15] The K_m data for sulfate uptake in *Anacystis nidulans* were determined as 0.75 μM[9]; for *Spirulina platensis* K_m data of 0.1 and 0.45 μM[15] were obtained. Sulfate uptake is competitively inhibited by sulfite and thiosulfate[9]; however, the structural analog selenate and chromate are inhibitory as well.[14,15] Sulfate uptake rates are higher after about 3 hr of starvation, demonstrating regulation of the sulfate permease as in other photosynthetic organisms.[16]

Uptake of Sulfate Esters. *Plectonema* 73110 can utilize sulfate esters for growth[17] inducing arylsulfatase activity. Evidence for localization at the cell periphery is presented, suggesting hydrolysis of sulfate esters

[11] S. Biedlingmaier and A. Schmidt, *Z. Naturforsch. C* **42,** 891 (1987).
[12] R. Jeanjean and E. Broda, *Arch. Microbiol.* **114,** 19 (1977).
[13] G. Prakash and H. D. Kumar, *Arch. Mikrobiol.* **77,** 196 (1971).
[14] H. D. Kumar and G. Prakash, *Ann. Bot.* **35,** 697 (1971).
[15] V. K. N. Menon and A. K. Varma, *FEMS Microbiol. Lett.* **13,** 141 (1982).
[16] A. Schmidt, *Prog. Bot.* **48,** 1330 (1986).
[17] S. Müller, A. Schmidt, *Z. Naturforsch. C* **41,** 820 (1986).

outside the cell (probably in the periplasmic space) and transport of sulfate afterward.[17]

Sulfonate Uptake. Uptake of taurine and ethane sulfonate, which has been analyzed in *Anabaena variabilis*,[11] has a pH optimum around 6.5 (for sulfate permease, about 9). Furthermore, sulfonate uptake is not discriminated by sulfate but by other sulfonic acids, as shown also for the green alga *Chlorella*.[18] Techniques for sulfonate uptake are similar to sulfate uptake measurements as discussed above, using ^{35}S-labeled sulfonic acids.

Hydrogen Sulfide Uptake. A variety of cyanobacteria can use sulfide as an electron donor for photosystem I under conditions of anoxygenic photosynthesis.[4,5] Transport systems for sulfide, however, have not been characterized so far. This is discussed below under anoxygenic photosynthesis.

Cysteine, Methionine, and Glutathione Uptake. Carriers for transport of organic thiol compounds have to be present since *Synechococcus* 6301 is able to use cysteine and glutathione as the only sulfur source for growth.[10] Reports suggest that in contrast to *Synechococcus* 6301 methionine can be used as the sole sulfur source for growth in *Anabaena variabilis* and *Anacystis nidulans*.[13] None of these uptake systems has been characterized in detail.

Metabolism of Oxidized Sulfur Compounds

Sulfate Activation and Degradation. Sulfate has to be activated for further metabolic transformations.[19] This is achieved by the two coupled enzymes ATP-sulfurylase (sulfate adenylyltransferase, EC 2.7.7.4) and APS kinase (adenylylsulfate kinase, EC 2.7.1.25) forming adenosine 5′-phosphosulfate (APS) and 3′-phosphoadenosine 5′-phosphosulfate (PAPS). These enzymes can be measured using radioactive sulfate and analyzing the sulfonucleotides formed by electrophoretic separation,[20] and the same technique can be used for enzymes degrading sulfonucleotides.[21] The ATP-sulfurylase forming APS can be directly measured as phosphate liberated in the presence of molybdate and a pyrophosphatase. This method is not advisable in crude extracts due to inhibition of other ATPases by molybdate. A better technique is the back-reaction

[18] S. Biedlingmaier and A. Schmidt, *Biochim. Biophys. Acta* **861,** 95 (1986).

[19] A. Schmidt, *in* "On the Origin of Chloroplasts" (J. A. Schiff, ed.), p. 179. Elsevier/North-Holland, Amsterdam, 1982.

[20] S. K. Sawhney and D. J. D. Nicholas, *Planta* **132,** 189 (1976).

[21] S. K. Sawhney and D. J. D. Nicholas, *Plant Sci. Lett.* **6,** 103 (1976).

forming ATP from APS and pyrophosphate and measuring ATP formed with the luciferin–luciferase assay.[22,23]

ATP-sulfurylase forming APS is present in *Anabaena cylindrica*,[23] *Spirulina platensis*,[24] and *Synechococcus* 6301.[25] The ATP-sulfurylase from *Spirulina* and in a partially purified fraction from *Anabaena cylindrica* have pH optima around 8, are inhibited by AMP, ADP, and phosphate, and thus are correlated to the energy level of the cell.[23,24] Other characteristics of cyanobacterial ATP-sulfurylases have not been determined. An APS kinase catalyzing the ATP-dependent formation of PAPS from APS has been demonstrated in *Synechococcus* 6301[25] and *Anabaena cylindrica*.[23] The *Anabaena* APS kinase was measured using a coupled assay of phosphoenolpyruvate generating pyruvate from ADP formed following pyruvate formation by NADH oxidation coupled to lactate dehydrogenase. Such a method should, however, be used with care since cyanobacterial extracts contain ATPases and pyridine nucleotide oxidases which require proper controls. No biochemical data are available for APS kinases from cyanobacteria.

Sulfonucleotides can be degraded by phosphatases and hydrolases yielding either APS from PAPS, 5′-AMP form APS, or 3′,5′-ADP from PAPS. These degradative activities are found in *Anabaena cylindrica*; inhibition of APS hydrolases was possible using fluoride.[21] Our experiments support this, using either 5 m*M* fluoride or 0.5 *M* sodium sulfate concentrations for inhibition of sulfatase activities.

Metabolism of Sulfate Esters. Plectonema induces an arylsulfatase during growth with *p*-nitrophenyl sulfate in the absence of sulfate. Enzyme activity can be followed by formation of *p*-nitrophenol from *p*-nitrophenyl sulfate (2 m*M*) in 50 m*M* Tris–HCl buffer, pH 8.0. This arylsulfatase has been partially purified and can be classified as a type I arylsulfatase.[17] No other reports of cyanobacterial arylsulfatases are available.

Sulfolipid Formation. Sulfolipids [1,2-diacyl-3-(6-sulfo-α-D-quinovosyl)-L-glycerol] are integral parts of photosynthetic membranes from cyanobacteria to higher plants. They contain the sulfonic acid sulfoquinovose.[2] There is little information on its biosynthesis[26] with regard to

[22] G. J. E. Balhary and D. J. D. Nicholas, *Anal. Biochem.* **40,** 1 (1971).
[23] D. Schmutz and C. Brunold, *Anal. Biochem.* **121,** 151 (1982); see also S. K. Sawhney and D. J. D. Nicholas, *Biochem. J.* **164,** 161 (1977).
[24] V. K. N. Menon and A. K. Varma, *Experientia* **35,** 854 (1978).
[25] A. Schmidt, *FEMS Microbiol. Lett.* **1,** 137 (1977).
[26] J. L. Harwood, *Biochem. Plants* **4,** 301 (1980).

sulfoquinovose formation; limited data are available for cyanobacteria.[27] Sulfolipids are extracted using normal procedures for lipid separation. The sulfolipid can be specifically located, since it is the only ^{35}S-labeled lipid if [^{35}S]sulfate is used, as shown for *Anabaena variabilis*.[28] Its biosynthesis is related to either transfer of activated sulfate from PAPS or oxidation from cysteine to cysteic acid with further metabolism to 3-sulfolactate.[25] Sulfolipids have been reported to be present in cyanobacteria but absent from photosynthetic bacteria.[27] They were reported, for instance, in *Prochloron*,[29] *Anabaena variabilis*,[30] *Rivularia atra*,[31] *Anacystis nidulans*, *Oscillatoria rubescens*, *Spirulina platensis*,[32] *Microcystis aeruginosa*, *Anabaena cylindrica*, *Nostoc calcicola*, and *Oscillatoria chalybea*.[31,33]

Sulfotransferases for Further Reduction. The activated sulfate from either APS or PAPS has to be transferred to suitable acceptors for further reduction, yielding free sulfite either directly or indirectly by isotopic exchange reactions. If ^{35}S-labeled sulfonucleotides are used, free sulfite can be separated from sulfate and sulfonucleotides by distillation using carrier sulfite. This procedure allows rapid analysis of APS- and PAPS-dependent sulfotransferases; the sensitivity of the test is determined by the specific activity of labeled sulfonucleotides used.

Sulfotransferases specific for APS or PAPS can be detected in cyanobacteria[25] and phototrophic bacteria.[34] APS-dependent sulfotransferases have been detected in all higher plants studied,[35] green algae,[36] the cyanobacteria *Plectonema* 73110,[25] *Chroococcidiopsis* 7203, and *Synechococcus* 6312,[37] and certain phototrophic bacteria,[34] whereas PAPS-dependent sulfotransferases are found in the cyanobacteria *Synechococcus* 6301, *Synechocystis* 6714, *Cyanophora paradoxa*,[38] and *Spirulina platensis*.

[27] R. Y. Stanier and G. Cohen-Bazire, *Annu. Rev. Microbiol.* **31,** 225 (1977).
[28] T. H. Giddings, Jr., P. Wolk, and A. Shomer-Ilan, *J. Bacteriol.* **146,** 1067 (1981).
[29] G. J. Perry, G. T. Gillan, and R. B. Johns, *J. Phycol.* **14,** 369 (1978).
[30] N. Sato and N. Nurata, *Biochim. Biophys. Acta* **710,** 271 (1982).
[31] J. L. Harwood and R. G. Nicholls, *Biochem. Soc. Trans.* **7,** 440 (1979).
[32] M. Piorreck, K.-H. Baasch, and P. Pohl, *Phytochemistry* **23,** 207 (1984).
[33] H. D. Zepke, E. Heinz, M. Radunz, M. Linscheid, and R. Pesch, *Arch. Microbiol.* **119,** 157 (1978).
[34] J. F. Imhoff, *Arch. Microbiol.* **132,** 197 (1982).
[35] A. Schmidt, *Plant Sci. Lett.* **5,** 407 (1975).
[36] A. Schmidt, *Encyclopedia Plant Physiol.* **6,** 481 (1979).
[37] A. Schmidt, *in* "Biology of Inorganic Nitrogen and Sulfur," p. 327. Springer-Verlag, Berlin, 1981.
[38] A. Schmidt and U. Christen, *Z. Naturforsch. C* **34,** 222 (1979).

Eubacteria such as *Escherichia coli* have sulfotransferases specific for PAPS.[19]

Sulfotransferases from bacteria and cyanobacteria are activated by thioredoxins,[37,39-43] which are present in cyanobacteria.[41] pH optima of either APS- or PAPS-specific sulfotransferases are about 8.5, and the enzymes are inhibited by either AMP or 3′,5′-ADP.[43] A possible relationship between sulfonucleotide specificity and the evolution of chloroplasts has been discussed.[19]

Reduction to the Level of Sulfide

Reduction of sulfite to sulfide is a 6-electron step catalyzed by a sulfite reductase using as the electron donor either NADPH in bacteria or ferredoxin in photosynthetic organisms. Sulfite reductases analyzed to date contain a specific heme derivative called siroheme as a prosthetic group.[44-47] Cyanobacteria can be grown on sulfate as discussed above, and sulfide emission was observed for *Synechococcus lividus* Y 52 from sulfate and thiosulfate in the light[48] and was inhibited by darkness or 3,4-dichlorophenyl-1,1-dimethylurea (DCMU).[49] This suggests the presence of a sulfite reductase and a coupling of reducing power via ferredoxin as in higher plants and green algae[36]; however, sulfite reductases have not been isolated from cyanobacteria. Metabolic studies of taurine utilization in intact *Anabaena variabilis* cells suggests possible reduction of a sulfonate group to a thiol group.[11] No evidence for a trithionate pathway[50] is available to date, although a thiosulfate reductase has been demonstrated in *Plectonema*.[17]

[39] A. Schmidt and U. Christen, *Planta* **140,** 239 (1978).
[40] A. Schmidt, *Arch. Microbiol.* **127,** 259 (1980).
[41] A. Schmidt and U. Christen, *Z. Naturforsch. C* **34,** 1272 (1979).
[42] W. Wagner, H. Follmann, and A. Schmidt, *Z. Naturforsch. C* **33,** 517 (1978).
[43] A. Schmidt, *in* "Thioredoxins: Structure and Functions," Colloq. Int. CNRS/NASA Ed., p. 119. Centre Natl. Rech. Sci., Paris, 1983.
[44] H. D. Peck, Jr., S. Tedro, and M. D. Kamen, *Proc. Natl. Acad. Sci. U.S.A.* **71,** 2404 (1974).
[45] M. J. Murphey, L. M. Siegel, H. Kamen, and D. Rosenthal, *J. Biol. Chem.* **248,** 2801 (1973).
[46] M. Murphey and L. M. Siegel, *J. Biol. Chem.* **248,** 6911 (1973).
[47] J. M. Vega, R. H. Garrett, and L. M. Siegel, *J. Biol. Chem.* **250,** 7980 (1975).
[48] R. P. Sheridan and R. W. Castenholz, *Nature* (*London*) **217,** 1064 (1968).
[49] R. P. Sheridan, *J. Phycol.* **9,** 437 (1973).
[50] J.-H. Kim and J. M. Akagi, *J. Bacteriol.* **163,** 472 (1985).

Synthesis and Metabolism of Sulfur Amino Acids

Synthesis of L-Cysteine. One report of two distinct cysteine synthases catalyzing the formation of L-cysteine from *O*-acetyl-L-serine has appeared for *Synechococcus* 6301.[51] Formation of cysteine was analyzed using the acidic ninhydrin method of Gaitonde.[52] It should, however, be pointed out that this method is not specific for cysteine, since it will detect all compounds being hydrolyzed to cysteine including glutathione.[53]

Cysteine synthases I and II from *Synechococcus* 6301 have molecular weights of about 56,000, K_m values for *O*-acetyl-L-serine of 0.8 m*M* (I) and 0.6 m*M* (II), and K_m values for sulfide of 0.8 m*M* (I) and 0.9 m*M* (II). Both cysteine synthases catalyze an isotopic exchange reaction between free sulfide and cysteine, probably due to pyridoxal phosphate chemistry. Formation of *O*-acetylserine from acetyl-CoA and serine has not yet been reported for cyanobacteria. Dual cysteine synthases are found in phototrophic bacteria[54] and eubacteria,[55] one catalyzing the formation of *S*-sulfocysteine from thiosulfate and *O*-acetylserine, whereas the other used only sulfide. These have not been determined in cyanobacteria.

Degradation of D- and L-Cysteine. Cyanobacteria can grow on D- and L-cysteine or glutathione as the only source of sulfur.[10] Since sulfolipid formation is required for growth, oxidation of sulfur compounds in either free or bound form is necessary, suggesting that cysteine desulfhydrase activity is present. No systematic studies have been reported for L-cysteine desulfhydrase activity or D-cysteine desulfhydrase (cystathionine δ-lyase, EC 4.4.1.1), activity in cyanobacteria.

Formation of Glutathione. The tripeptide glutathione (GSH) has been detected in *Nostoc muscorum* (2.5 μmol/g[56]), *Anabaena variabilis*,[28] and *Synechhococcus* 6301 (between 9.6 and 15.7 nmol/10^{10} cells[77]). Because *Synechococcus* can use glutathione as the only sulfur source for growth (discussed above), formation and degradation of GSH is possible. GSH has been determined either electrophoretically[28] or using the recycling glutathione reductase assay coupled to the reduction of 5,5′-dithiobis(2-nitrobenzoic acid) (DTNB; Ellman's reagent),[57] which we have found

[51] W. Diessner and A. Schmidt, *Z. Pflanzenphysiol.* **102,** 57 (1981).
[52] M. K. Gaitonde, *J. Biochem.* (*Tokyo*) **104,** 627 (1967).
[53] K. F. Krauss, Thesis. Univ. of Munich, 1984.
[54] G. Hensel and H. G. Trüper, *Arch. Microbiol.* **109,** 101 (1976).
[55] T. Nakamura, Y. Kon, and Y. Eguchi, *J. Bacteriol.* **156,** 656 (1983).
[56] R. C. Fahey, W. C. Brown, W. B. Adams, and M. B. Worsham, *J. Bacteriol.* **133,** 1126 (1978).
[57] F. Tietze, *Anal. Biochem.* **27,** 502 (1969).

useful. Glutathione reductase has been purified from *Anabaena* 7119.[58] This cyanobacterial glutathione reductase has a molecular weight of 104,000 and a pH optimum around 9.0. A K_m of 210 μM was determined for GSSG and of 9.4 μM for NADPH.[58] The role of glutathione reductase and GSH is discussed as a detoxification mechanism for H_2O_2, although no direct function for glutathione has been demonstrated in cyanobacteria and no studies for its synthesis or degradation are available. Glutathione reductase activity was detected in *Anabaena* 73110 (see above), *Spirulina maxima*,[59] and *Synechococcus* 6301.

Methionine Synthesis and Metabolism. For methionine synthesis cysteine is conjugated with an activated homoserine by the cystathionine γ-synthase to form cystathionine.[60] Cystathionine is cleaved by a β-cystathionase to yield pyruvate, ammonia, and homocysteine, which is further methylated to methionine. Higher plants use *O*-phosphohomoserine for cystathionine formation.[60] In *Anabaena flos-aquae* no activity was detected with the phosphorylated homoserine; however, activity was found using acetylated or succinylated homoserine, as in *Eschericia coli*.[60]

A homoserine succinyltransferase (EC 2.3.1.46) used for activation of homoserine by succinly-CoA has been studied in *Anacystis nidulans*.[61] This enzyme has a pH optimum around 7.5 and is completely inhibited by cystathionine or homocysteine and partly inhibited by methionine.[61] Cystathionine γ-synthases have not been purified from cyanobacteria. β-Cystathionase activity needed for cystathionine degradation should be present; however, no detailed study is available. Methylation of homocysteine to methionine has not been studied in detail in cyanobacteria, but methionine formation from radiolabeled sulfate was easily detected in *Anabaena variabilis*.[28]

Formation of [Fe–S] Clusters. Iron–sulfur centers of ferredoxins obviously require the insertion of sulfide derived from L-cysteine in *Escherichia coli*.[62] Two enzymes possibly involved have been detected in cyanobacteria: (1) A thiosulfate reductase (thiosulfate sulfotransferase, rhodanese) is detected in *Synechococcus* 6301 and *Plectonema* 73110[17] that catalyzes the transfer of the sulfide sulfur from thiosulfate to suitable acceptors including cyanide. The concentration of this enzyme is high under sulfur starvation conditions. (2) A β-mercaptopyruvate sulfurtransferase (EC 2.8.1.2) is constitutive in *Synechococcus* 6301 as in other

[58] A. Serrano, J. Rivas, and M. Losada, *J. Bacteriol.* **158,** 317 (1984).
[59] R. N. Ondarza, J. L. Lendon, and M. Ondarza, *J. Mol. Evol.* **19,** 371 (1983).
[60] A. H. Datko, J. Giovanelli, and S. H. Mudd, *J. Biol. Chem.* **249,** 1139 (1974).
[61] S. F. Delaney, A. Dickson, and N. G. Carr, *J. Gen. Microbiol.* **79,** 89 (1973).
[62] R. H. White, *Biochemistry* **21,** 4271 (1982).

photosynthetic organisms,[63] transferring sulfide to suitable acceptors including DTE to form pyruvate and sulfide (or thiosulfate using sulfite). β-Mercaptopyruvate can be formed from cysteine either by transamination reactions or by an amino acid oxidase.[63] Both reactions could lead to emission of sulfide.

Emission of Volatile Sulfur Compounds. Higher plants emit sulfide when fed L-cysteine[64] and methylmercaptane when fed L-methionine.[65] Emission of sulfide by cyanobacteria was also reported by Sheridan,[48,49] using *Synechococcus lividus* strain Y 52 with either sulfate or thiosulfate. This sulfide emission was light dependent and detected only in the absence of CO_2. Another possible source of sulfide formation is anaerobic fermentation using zero-valent sulfur as the electron acceptor, leading to the liberation of sulfide. This has been shown for *Oscillatoria limnetica*.[66]

Further possibilities for emission of sulfur compounds were reported as well as formation of dimethyl sulfide in cyanobacteria, probably due to degradation of either dimethyl-β-propiothetin or *S*-methylmethionine.[67] Presence of dimethylsulfonium compounds was reported for the cyanobacterium *Microcoleus lyngbyacus*.[68]

Formation of Elemental Sulfur. Elemental sulfur can be detected outside of *Oscillatoria limnetica* cells due to anoxygenic photosynthesis,[3] which will be discussed in detail below. Zero-valent sulfur was detected in *Oscillatoria limnetica* 3, *Oscillatoria salina*, *Aphanoteca halophytica*, and *Phormidium*.[69]

Anoxygenic Photosynthesis

Bacterial photosynthesis with only one photosystem is dependent on an electron donor for photosynthesis, since the water-splitting photosystem II is not available. Cyanobacteria have both photosystems typical for oxygenic photosynthesis of plants and algae. Surprisingly some cyanobacteria are capable of using sulfide as the electron donor for growth under anaerobic conditions. Such a situation is termed anoxygenic photosynthesis[1–3] since it allows normal light-dependent growth without production of O_2.

[63] A. Schmidt, *Z. Naturforsch. C* **39,** 916 (1986).
[64] H. Rennenberg, *Annu. Rev. Plant Physiol.* **35,** 121 (1984).
[65] A. Schmidt, H. Rennenberg, L. G. Wilson, and P. Filner, *Phytochemistry* **24,** 1181 (1985).
[66] A. Oren and M. Shilo, *Arch. Microbiol.* **122,** 77 (1979).
[67] G. A. Maw, *in* "The Chemistry of the Sulfonium Group," p. 703. Wiley, London, 1981.
[68] R. H. White, *J. Mar. Res.* **40,** 529 (1982).
[69] S. Garlik, A. Oren, and E. Padan, *J. Bacteriol.* **129,** 623 (1977).

One should differentiate between anoxygenic photosynthesis and anaerobic CO_2 fixation. This can be shown for *Anacystis nidulans*, which is capable of CO_2 fixation with thiosulfate or cysteine as the electron donor under light-limited conditions. However no growth could be observed using thiosulfate and blocking photosystem II with DCMU.[70] Similar results were obtained for the same alga using reduced nitrogen and sulfur compounds and hydrogen for anaerobic CO_2 photoreduction.[71] Thus, electron transfer from sulfide to photosystem I coupled to CO_2 fixation (anoxygenic photosynthesis) should not be confused with growth under anoxygenic conditions, since several algae and cyanobacteria have oxygen-dependent reactions for either fatty acid or sterol synthesis[4] which allow no growth under anaerobic conditions even though the requirements for anoxygenic CO_2 fixation with sulfide or other electron donors for photosynthesis are met.

Anoxygenic photosynthesis was detected in *Oscillatoria limnetica*[3] with optimal sulfide concentrations around 4 m*M*, which completely inactivates photosystem II.[72,73] Anoxygenic photosynthesis in *Oscillatoria limnetica* is independent of photosystem II. Electrons are fed into photosystem I after the DCMU block of the electron transport chain[72] but before DBMIB-inhibitory site,[70,71] suggesting that plastoquinone is involved. Plastoquinone itself, however, is not the first electron acceptor to form sulfide, since the change to anoxygenic photosynthesis requires an induction period of about 2 hr[72,73] and protein synthesis is necessary for adaption to sulfide as an electron donor.[73] As pointed out above, sulfide is toxic to photosystem II, causing inhibition in the range of about 0.1 m*M* in cyanobacteria,[72,73] whereas after induction optimal conditions for CO_2 fixation are around 4 m*M*[3] in *Oscillatoria limnetica*.

Since anoxygenic photosynthesis is operative only with photosystem I, the quantum yield for CO_2 fixation is high compared to oxygenic photosynthesis with no enhancement for CO_2 fixation. The quantum yield for 1 mol of CO_2[74] is highest around 700 nm and clearly different than quantum yield measurements of oxygenic photosynthesis with optimal CO_2 fixation rates around 620 nm due to preferential coupling of phycocyanin to photosystem II. Furthermore, anoxygenic photosynthesis is also operative under monochromatic light of about 700 nm (conditions without activation of photosystem II).[72-74] A stoichiometry of 2 mol of sulfide oxidized to elemental sulfur for 1 mol of CO_2 was found for *Oscillatoria limnetica*

[70] H. Utkilen, *J. Gen. Microbiol.* **95,** 177 (1976).
[71] G. A. Peschek, *Arch. Microbiol.* **119,** 313 (1979).
[72] Y. Cohen, E. Padan, and M. Shilo, *J. Bacteriol.* **123,** 855 (1975).
[73] A. Oren and E. Padan, *J. Bacteriol.* **133,** 558 (1978).
[74] A. Oren, E. Padan, and M. Avron, *Proc. Natl. Acad. Sci. U.S.A.* **74,** 2152 (1977).

with elemental sulfur deposited outside the cells.[3] Elemental sulfur thus formed[3] may be used as an electron acceptor under anaerobic dark conditions,[66] allowing a light–dark cycle of anoxygenic growth and respiration.

Occurrence of anoxygenic photosynthesis was detected in several cyanobacterial strains as summarized by Garlik *et al.*[69] It is of interest to note that anoxygenic photosynthesis has not been found in heterocystous cyanobacteria.[69] The impact of anaerobic photoautotrophic metabolism on growth of cyanobacteria in different habitats is discussed by Padan.[4]

General Remarks on the Regulation of Sulfur Metabolism in Cyanobacteria

Sulfur limitation in cyanobacteria leads to an increase in sulfate uptake capacity as discussed above. Furthermore, sulfur limitation induces degradation of C-phycocyanin[10] and phycoerythrin.[75] Sulfur limitation affects nitrogen metabolism, causing nitrite excretion into the growth medium at least in *Synechococcus* 6301.[75] Sulfate and nitrate assimilation must be coordinated because of the fixed ratio of reduced sulfur and nitrogen in proteins as discussed for plant cells.[16,76,77] Thus, nitrogen and sulfur metabolism are interrelated, although the specific regulatory signal(s) is not known. A specialized situation for sulfur metabolism was detected in heterocysts. Normal assimilatory sulfate reduction seems to be restricted to vegetative cells, whereas heterocysts are restricted to sulfur metabolism at the reduced level only,[28] suggesting that heterocysts must be provided with reduced sulfur compounds. Such a specialization is similar to differences in sulfur metabolism in C_3 and C_4 plants.[10]

Obviously, sulfide is oxidized to elemental sulfur only when used for anoxygenic photosynthesis.[3] However, oxidation of thiosulfate to sulfate is observed for *Anacystis nidulans*,[70] thus demonstrating that oxidation of reduced sulfur is possible in cyanobacteria. Metallothioneins for the detoxification of heavy metals have also been detected in cyanobacteria.[78]

Acknowledgments

This work was supported by the Deutsche Forschungsgemeinschaft.

[75] A. Schmidt, unpublished observations.
[76] T. Reuveny and P. Filner, *J. Biol. Chem.* **252,** 1858 (1977).
[77] T. Reuveny, *Proc. Natl. Acad. Sci. U.S.A.* **74,** 619 (1977).
[78] R. E. Olafson, K. Abel, and R. G. Sim, *Biochem. Biophys. Res. Commun.* **89,** 36 (1979).

[63] Cyclic Nucleotides

By MICHAEL HERDMAN and KHALIL ELMORJANI

Introduction

Adenosine 3′,5′-cyclic monophosphate (cAMP) plays a major role in the transcription of those bacterial genes which are subject to regulation by catabolite repression (see Ref. 1 for review), whereas guanosine 3′,5′-cyclic monophosphate (cGMP), although present in some prokaryotes, has been more thoroughly studied in eukaryotic cells where it acts as a second messenger involved in processes under hormonal control.[2] The activities of many biosynthetic pathways in cyanobacteria are not controlled at the genetic level by repression or induction.[3] However, many strains show complex developmental cycles which involve the differention of heterocysts,[4] akinetes,[5] hormogonia,[6] or baeocytes.[6] These events must be controlled, and it is surprising that the roles of cyclic nucleotides in morphogenesis have been little studied.

This chapter describes methods for the extraction and measurement of cyclic nucleotides, their typical intracellular concentrations, and the estimation of the activities of adenylate cyclase (EC 4.6.1.1) and cAMP phosphodiesterase (EC 3.1.4.17). Some details of the tetra- and pentaphosphates of guanosine (ppGpp and pppGpp), although not cyclic nucleotides, are also included. Since the taxonomy of cyanobacteria is somewhat confused at present, we have retained generic and specific names as published in the articles described, even though we do not necessarily agree with them.

Extraction of Nucleotides

Harvesting of Cells. Previous reports[7–9] on the intracellular concentration of cAMP in cyanobacteria have employed centrifugation to harvest

[1] A. Ullmann and A. Danchin, *Adv. Cyclic Nucleotide Res.* **15,** 1 (1983).
[2] N. D. Goldberg and M. K. Haddox, *Annu. Rev. Biochem.* **46,** 823 (1977).
[3] N. G. Carr, *in* "The Biology of Blue–Green Algae" (N. G. Carr and B. A. Whitton, eds.), p. 39. Blackwell, Oxford, England, 1973.
[4] R. L. Smith, C. Van Baalen, and F. R. Tabita, this volume [51].
[5] M. Herdman, this volume [21].
[6] M. Herdman and R. Rippka, this volume [22].
[7] E. E. Hood, S. Armour, J. D. Ownby, A. K. Handa, and R. A. Bressan, *Biochim. Biophys. Acta* **588,** 193 (1979).
[8] D. A. Francko and R. G. Wetzel, *Physiol. Plant.* **49,** 65 (1980).
[9] D. A. Francko and R. G. Wetzel, *J. Phycol.* **17,** 129 (1981).

the cells. While this may be necessary when large volumes of culture are to be treated, the intracellular level of at least cGMP in *Synechocystis* PCC 6803 increases approximately 2-fold during centrifugation,[10] probably as a result of the sudden change of temperature and illumination. If accurate values are required (as in kinetic studies), therefore, centrifugation should be avoided. A more appropriate method involves filtration: samples (up to 20 ml at OD_{650} 2.0, equivalent to 4 mg total protein) can be filtered rapidly (<15 sec) (Millipore, 0.45 μm pore size, 4.7 cm diameter) under reduced pressure.[10] The filter can then be quickly transferred to a vessel containing acid (for extraction of the nucleotides, see below), which also arrests the metabolism of the cells and therefore prevents any further change in the concentration of the nucleotides.

Extraction Procedure. Francko[11] reviewed the evidence which suggests that extracts of eukaryotic algae and higher plants contain compounds which interfere with the estimation of cAMP. Extraction methods employed for cyanobacteria have usually utilized one or more purification steps to eliminate such compounds, which appear to be present in at least *Anabaena variabilis*.[7] A typical procedure[7] involves the addition of perchloric acid (0.4 *M* final concentration) to harvested cells which are then disrupted by sonication; the acid-soluble extract is neutralized with KOH, and cAMP is purified by chromatography on Norit A-activated charcoal followed by (after air-drying and redissolving in distilled water) Bio-Rad AG50 (100–200 mesh, H^+ form), eluting with water. The first 2 ml is discarded and the next 5 ml collected, air-dried, and analyzed. A similar method[8] involves extraction with $HClO_4$ (0.5 *M*) and elution from Norit A; samples are redissolved in 50 m*M* Tris–HCl, pH 6.8, and further purified by chromatography on neutral alumina and Dowex 50.

We[10] have routinely extracted cGMP, following rapid filtration (see above) of *Synechocystis* PCC 6803, by placing the filter in 2 ml of trichloroacetic acid (TCA; 5%, w/v) in ice for 30 min. The TCA is removed, the filter washed with TCA (2 ml), and the two acid extracts combined. Following centrifugation (Eppendorf) for 2–5 min, to remove cells and precipitated protein, the TCA is eliminated by 5 successive washings with equal volumes of diethyl ether. The final aqueous phase (pH 6.5–7.0) may be evaporated to dryness overnight at 60° or lyophilized; the residue is then dissolved in buffer (250 μl) appropriate for radioimmunoassay. Treatment of the filtered cells with HCl (0.1–0.2 *M*, 10–15 min, 90°) avoids the tedious ether extractions employed for the removal of TCA,

[10] K. Elmorjani, Ph.D. thesis. Univ. of Orsay, Orsay, France, 1986.

[11] D. A. Francko, *Adv. Cyclic Nucleotide Res.* **15,** 97 (1983).

since the acid extract may be lyophilized and assayed directly[12]; however, this method appears to extract less cGMP (70–80%) than does TCA.[12]

Identification of Cyclic Nucleotides by Chromatography

It is clear that studies of cyclic nucleotide levels in organisms which have not been examined previously should incorporate strict controls to demonstrate that the extracts do in fact contain the cyclic nucleotides which appear to be detected in subsequent tests, even though these tests are thought to be specific. Such controls have been performed by sequential column chromatography as described above.[7,8] The true identity of both cAMP and cGMP in *Synechocystis* PCC 6803 was shown[13] by labeling the cells overnight with ortho[^{32}P]phosphate, carrier free (3.7 MBq ml^{-1}), in medium BG-11[14] with the concentration of K_2HPO_4 decreased to 10^{-4} *M*. Typically, cultures (10 ml) are centrifuged, the cells treated with TCA (1 ml, 5%, w/v) as above, and the acid removed with ether. To the extract is added imidazole (1.1 M)–HCl (0.55 *M*), pH 7.0 (1/10 volume), and the sample is poured directly into a disposable plastic column containing 1 g of neutral alumina (Bio-Rad, AG7, 100–200 mesh).[15] The effluent (~0.2 ml) is collected; cyclic nucleotides are eluted with 4 ml of imidazole (100 m*M*)–HCl (50 m*M*), pH 7.0, with collection of 1-ml fractions which are subjected to TLC on Polygram CEL 300 PEI plates (without fluorescence indicator) (Macherey & Nagel, Dürem, FRG) with 1 *M* ammonium acetate–95% ethanol (35 : 65), pH 9, as solvent. Following autoradiography, ^{32}P-labeled spots comigrating with authentic standards are identified. In this way, cGMP (R_f 0.27), eluting in the first 1.2 ml, and cAMP (R_f 0.41), eluting in the next 3 ml, were selectively eluted,[15] avoiding the use of the sequential column chromatography steps previously employed (see above); all other labeled compounds remain bound to the column. If the alumina chromatography step is omitted, cAMP and cGMP are not easily resolved by TLC from the much larger quantities of other phosphorylated compounds in the extract.

Other solvents employed for TLC (following purification of cAMP by column chromatography) include the following[7]: (1) methanol–ethyl acetate–NH_4OH–1-butanol (3 : 4 : 4 : 7, v/v); (2) methanol–1 *M* ammonium acetate (7 : 3, v/v); (3) 2-propanol–NH_4OH–H_2O (7 : 1 : 2, v/v); (4) 0.1 *M* LiCl. To date, these have not been used to separate cGMP. As a final control of the authenticity of the compounds, the disappearance of the

[12] C. Seailles and M. Herdman, unpublished observations.
[13] M. Herdman unpublished observations.
[14] R. Rippka, this volume [1].
[15] M.-L. Lacombe, personal communication.

presumptive cyclic nucleotide following treatment with phosphodiesterase should be examined; the rate of degradation should be identical to that of known standards.

Estimation of Cyclic Nucleotides

Although cAMP and cGMP are easily identified by TLC as described above, quantitative estimation is most easily performed by specific radioimmunoassay. This employs antiserum to the cyclic nucleotide, radioactively labeled nucleotide (^{125}I or ^{3}H), and the extracts to be tested. The unlabeled (extracted) nucleotide interferes competitively with the labeled standard for binding to the antiserum, and its concentration can therefore be easily determined by reference to standard curves. Radioimmunoassay kits are available from several suppliers of radiochemicals and appear to give reproducible results. However, TCA must be rigorously removed from the extracts (see above). Another method employed[8] for the estimation of cAMP utilizes cAMP-dependent protein kinase,[16] measuring the consumption of ATP (the phosphate donor) by the firefly luciferin–luciferase system.

Typical Concentrations of Cyclic Nucleotides in Cyanobacteria

Cyclic AMP. Intracellular concentrations of cAMP during exponential growth of *A. variabilis* ranged from 1.0 to 2.7 pmol (mg protein)$^{-1}$ and increased 4-fold during nitrogen starvation[7]; extracellular levels were approximately 9 times higher. Cells of *Anabaena flos-aquae, Synechococcus leopoliensis* (*Anacystis nidulans*), and *Microcystis aeruginosa* contained, respectively, 90, 90, and 340 pmol cAMP (g fresh weight)$^{-1}$,[8] equivalent to approximately 0.2, 0.2, and 0.7 pmol (mg protein)$^{-1}$. The extracellular concentrations of cAMP in late exponential growth showed wide variability (480, 9, and 43 pmol liter^{-1}, respectively); since culture densities were not given, these values cannot be compared to the intracellular concentrations. In *Synechocystis* PCC 6803[10] total cAMP was [as pmol (mg protein)$^{-1}$] 13–24 in exponential growth, 11 in stationary phase, and, contrary to the results in Ref. 7, did not change significantly during nitrogen starvation. For comparison, reported values for total cAMP in cultures of *Escherichia coli* range from 270 to 4700 pmol (mg protein)$^{-1}$ (see Ref. 1).

Cyclic GMP. Cyclic GMP has been observed at varying concentrations in several cyanobacterial strains. Autotrophically growing cells (at

[16] A. K. Handa and R. A. Bressan, *Plant Physiol.* **59,** 490 (1978).

34°, 3000 lux, in a Gallenkamp illuminated orbital incubator gassed with CO_2 at 6 liters hr^{-1}) of *Synechocystis* PCC 6803 in exponential and stationary phase contained cGMP in the range 4.6–6.0 and 3.5 pmol (mg protein)$^{-1}$, respectively.[10] The intracellular content of cGMP increased to 9.5–9.8 pmol (mg protein)$^{-1}$ in cells growing under mixotrophic (light plus glucose) and photoheterotrophic [light plus glucose plus 3′(3,4-dichlorophenyl)-1′,1′-dimethylurea (DCMU), an inhibitor of photosystem II activity] conditions and decreased to 3 pmol (mg protein)$^{-1}$ in cells growing heterotrophically in the dark.[10] Following transfer to medium lacking combined nitrogen, intracellular levels of cGMP increased to approximately 20 pmol (mg protein)$^{-1}$, in contrast to those of cAMP which remained constant[10] (see above). We[12] have also routinely found intracellular concentrations of cGMP in the range of [as pmol (mg protein)$^{-1}$] 17–35 in *Synechocystis* PCC 6308, 4 in *Plectonema* PCC 73110, and 1–3.5 in *Nostoc* MAC (*Nostoc* PCC 8009), all growing at 27°, 2000 lux, gas phase 1% CO_2 in air, with shaking. It should be noted, however, that under these less optimal growth conditions the intracellular levels of cGMP in *Synechocystis* PCC 6803 increased, implying that direct comparison of different strains is possible only under optimal culture conditions, probably different for each strain. In comparison, low intracellular levels of cGMP have been reported in other bacteria, for example [all values as pmol (mg protein)$^{-1}$], *E. coli* 0.02–0.2[17]; *Bacillus megaterium*[18] 0.02 (exponential growth), 0.001 (stationary phase), 0.0007 (spores); *Caulobacter crescentus*[19] 0.67 (exponential growth with glucose), 0.16 (stationary phase). It is apparent, therefore, that cyanobacteria contain high levels of this cyclic nucleotide in comparison to other bacteria; study of its function(s) is urgently required.

Adenylate Cyclase and cAMP Phosphodiesterase

The two enzymes involved in the synthesis and degradation of cAMP have been measured in cyanobacteria.

Adenylate Cyclase Activity. Adenylate cyclase was detected in *A. nidulans*[20] by the method of Salomon *et al.*[21] For detection of the enzyme activity, cells are broken in 50 m*M* Tris–HCl buffer, pH 7.4, containing sorbitol (0.5 *M*), Ficoll (2.5%, w/v), dithiothreitol (2 m*M*), and EDTA (1

[17] W. R. Cook, V. F. Kalb, A. A. Peace, and R. W. Bernlohr, *J. Bacteriol.* **141,** 1450 (1980).
[18] B. Setlow and P. Setlow, *J. Bacteriol.* **136,** 433 (1978).
[19] N. Kurn, L. Shapiro, and N. Agapian, *J. Bacteriol.* **131,** 951 (1977).
[20] R. Hintermann and R. W. Parish, *Planta* **146,** 459 (1979).
[21] Y. Salomon, C. Londos, and M. Rodbell, *Anal. Biochem.* **58,** 541 (1974).

m*M*), and the supernatant of a low-speed centrifugation is assayed. The incubation medium (pH 7.8) contains, in a final volume of 100 μl, the following: 0.1 μmol cAMP, 0.05 μmol ATP, 1 μmol $MgCl_2$, 1 μmol creatine phosphate, 2–3 U creatine kinase, 2 μmol dithiothreitol, 3.7 μmol Tris–HCl, [α-^{32}P]ATP (7.5×10^{11} Bq $mmol^{-1}$) 7.6×10^4 Bq, and protein 20–30 μg. Samples are incubated for 10 min at room temperature, and the reaction is stopped by adding 100 μl of 2% sodium dodecyl sulfate (SDS), 40 m*M* ATP, and 1 m*M* cAMP in 25 m*M* Tris–HCl buffer, pH 7.5. [^{3}H]cAMP (10^{12} Bq $mmol^{-1}$, 5×10^4 cpm) is added as an internal standard to monitor recovery. Following purification by column chromatography on Dowex resin (AG50-WX8, 200–400 mesh) and neutral alumina, the [^{32}P]cAMP formed is detected by liquid scintillation counting. Activities of 0.13–0.22 pmol cAMP min^{-1} (mg protein)$^{-1}$ were recorded.

cAMP Phosphodiesterase Activity.[22] Cells of *A. variabilis* in the late exponential phase of growth are resuspended in ice-cold 40 m*M* Tris–HCl, pH 7.6, containing 10 m*M* cysteine and 2 m*M* $MgSO_4$, broken by sonication, centrifuged (3000 *g*, 5 min), and the supernatant used for assay. The reaction essentially follows the method of Fischer and Amrhein.[23] The reaction mixture (400 μl) contains the following: crude extract (100 μl, 50–150 μg protein), buffer as above (200 μl), and 100 μl of [5,6-^{3}H]cAMP (80 nmol, 138 Bq $nmol^{-1}$). Following incubation at 32° for 90 min, the reaction is stopped by boiling (1 min), and, after cooling, the precipitated proteins are removed by centrifugation. The supernatant is mixed with 200 μg (200 μl) of *Crotalus adamanteus* venom to hydrolyze 5′-AMP to adenosine. Following incubation (20 min at 32°), the sample is boiled, centrifuged, and the nucleotides removed from the supernatant with an ethanolic slurry of Dowex 1-X8, 200–400 mesh. Adenosine is determined by liquid scintillation counting.

Activities varied dramatically during growth, being maximal [2.5 nmol 5′-AMP formed min^{-1} (mg protein)$^{-1}$] at the end of exponential growth and virtually undetectable in the early exponential or stationary phases of growth.[22] Phosphodiesterase activity decreased 3-fold in cells transferred to medium lacking combined nitrogen, consistent with the increased (3- to 4-fold) intracellular levels of cAMP observed[7] in the same organism under these conditions. Much lower activities were reported in *Anabaena cylindrica*[24]; for comparison, *E. coli* shows at least 10-fold greater activity[25] than *A. variabilis*.

[22] J. D. Ownby and F. E. Kuenzi, *FEMS Microbiol. Lett.* **15,** 243 (1982).
[23] U. Fischer and N. Amrhein, *Biochim. Biophys. Acta* **341,** 412 (1974).
[24] N. Amrhein, *Z. Pflanzenphysiol.* **72,** 249 (1974).
[25] L. D. Nielsen, D. Monard, and H. V. Rickenberg, *J. Bacteriol.* **116,** 857 (1973).

Other Potential Regulatory Nucleotides

The magic spot nucleotides guanosine tetraphosphate (ppGpp) and guanosine pentaphosphate (pppGpp), long known to be involved in the regulation of stable RNA synthesis in *E. coli,* have been found in some[26–28] but not all[10,29] cyanobacteria.

Extraction and Identification. To ensure efficient incorporation of label, the concentration of phosphate in the medium is normally decreased to about 5×10^{-5} *M*, and the medium is buffered with either Tris (0.4 g $liter^{-1}$)[27] or HEPES (0.3 g $liter^{-1}$)[29] to maintain a pH of 7.5–7.8. Cells are labeled as described above with ortho[^{32}P]phosphate for at least one generation time. Samples (0.2 ml) are added to 0.1 ml of ice-cold 98% formic acid, incubated in ice for 20 min, and centrifuged. Samples (10 μl) of the supernatant are chromatographed on PEI cellulose TLC plates (G-1440, Schleicher & Schüll, Dassel, FRG) with 1.5 *M* KH_2PO_4, pH 3.4. R_f values for nucleotides in this system are given by Cashel[30]; the identity of the compounds should, of course, be checked by cochromatography with authentic samples. Relative contents of the nucleotides can be estimated by cutting the spots (identified by autoradiography) from the plates and subjecting them to liquid scintillation counting.

Conclusions and Perspectives

Although the presence of regulatory nucleotides is now well established in cyanobacteria, their functions are largely unknown. The guanosine tetra- and pentaphosphates, which control the stringent response in, e.g., *E. coli* following nutritional step-down, have been found in *A. nidulans*,[26,29] *Nostoc* MAC,[29] and *A. cylindrica*[28] but not in *Aphanocapsa* (*Synechocystis*) PCC 6308[29] or *Synechocystis* PCC 6803.[10] Where present, these nucleotides have been suggested to be involved in the control of the synthesis of stable RNA. Their increased intracellular levels following light step-down of *A. nidulans*[26] and nitrogen starvation of *A. cylindrica* (CCAP 1403/2a)[28] were associated with a decrease in the rate of RNA synthesis; in contrast, these nucleotides were not produced in *Synechocystis* PCC 6803 following deprivation of nitrogen, and RNA synthesis continued at an unchanged rate.[10] Cells of *A. cylindrica* CCAP 1403/2a subjected to light step-down, however, showed a rapid decrease of RNA

[26] N. Mann, N. G. Carr, and J. E. M. Midgley, *Biochim. Biophys. Acta* **402,** 41 (1975).
[27] R. J. Smith, *FEMS Microbiol. Lett.* **1,** 129 (1977).
[28] J. Akinyanju and R. J. Smith, *FEBS Lett.* **107,** 173 (1979).
[29] D. G. Adams, D. O. Phillips, J. N. Nichols, and N. G. Carr, *FEBS Lett.* **81,** 48 (1977).
[30] M. Cashel, *J. Biol. Chem.* **244,** 3133 (1969).

synthesis without producing these magic spot nucleotides,[29] and, in *Anabaenopsis circularis,* the inhibition of glutamine synthetase activity by DL-methionine sulfoximine (1 m*M*), which caused rapid cessation of protein synthesis, was not accompanied by an increase in ppGpp[31] as would have been expected from results obtained with certain other bacteria. This strain, however, produced ppGpp following transfer to the dark. It appears, therefore, that these guanosine phosphates may not be universally involved in the control of macromolecular synthesis in cyanobacteria, and further studies would seem appropriate.

The 4-fold increase in intracellular cAMP levels during nitrogen starvation of *A. variabilis*[7] might suggest involvement of this cyclic nucleotide in the process of heterocyst differentiation; the constant levels observed in *Synechocystis* PCC 6803,[10] a unicellular nonheterocystous strain, under the same conditions, are not inconsistent with this hypothesis. However, detailed studies are required to investigate the physiological basis of the control of cAMP concentration and its potential involvement in metabolic regulation in cyanobacteria. Finally, the high levels of cGMP observed in cyanobacterial cells, far greater than those observed in any other group of organisms,[10,12] suggest that this cyclic nucleotide may play an important role in control processes. For example, the enzymatic degradation of phycocyanin *in vivo* in nitrogen-starved cells of *Synechocystis* PCC 6803 is associated with increased intracellular concentrations of cGMP; in the presence of glucose, however, levels of cGMP rapidly decline and, concomitantly, phycocyanin degradation is arrested.[10,32] This inhibition is tightly correlated with cGMP concentration. The glucose effect observed in *Synechocystis* PCC 6803 is therefore the first clear example of the probable involvement of a cyclic nucleotide in metabolic control in a cyanobacterium and would appear to be the system of choice for detailed studies of control processes in these organisms.

[31] A. C. Rogerson, P. Rowell, and W. D. P. Stewart, *FEMS Microbiol. Lett.* **3,** 299 (1978).
[32] K. Elmorjani and M. Herdman, *J. Gen. Microbiol.* **133,** 1685 (1987).

[64] Cyanobacterial DNA-Dependent RNA Polymerase

By GEORGE BORBELY and GEORGE J. SCHNEIDER

Introduction

Cyanobacterial gene expression requires the transfer of genetic information from DNA to RNA molecules which is mediated by DNA-dependent RNA polymerase (EC 2.7.7.6). The RNA polymerases purified from cyanobacterial sources are similar to bacterial RNA polymerases in the following respects: divalent metal ion requirement for enzymatic activity, rifamycin sensitivity, α-amanitin insensitivity, and requirement for a DNA template. The enzyme catalyzes the initiation, elongation, and termination of polyribonucleotide chains using ribonucleoside triphosphates as substrates. Since RNA polymerases of cyanobacterial origin have a number of properties which are common to other prokaryotic RNA polymerases, purification steps and strategies established for *Escherichia coli* and/or other enteric bacteria are useful guides in purification of the enzyme.[1,2] The role and functions of cyanobacterial RNA polymerases and the mechanism by which the enzyme functions require a great deal of further investigation. Here we describe the purification of *Anabaena* sp. PCC 7120 DNA-dependent RNA polymerase.[3]

Assay Method

The RNA polymerase assay mixture contains, in a final volume of 35 μl, the following: 50 mM Tris–HCl, pH 8.0 at 25°, 10 mM $MgCl_2$, 2.5 mM 2-mercaptoethanol, 0.5 mM each of UTP, GTP, and CTP, 0.125 mM [8-^{3}H]ATP (0.5–1.0 μCi of tritiated ATP diluted to a specific radioactivity of 0.06 μCi/nmol), and 0.1 mg/ml chicken erythrocyte DNA. DNA synthesis is initiated by adding enzyme extract (5–10 μg enzyme protein) to the assay mixture. After 10 min at 37° the reaction mixture is spotted onto Whatman GF/C filters which are washed batchwise 3 times for 5 min in 5% (w/v) trichloroacetic acid. After drying under a heat lamp, the filters are counted in toluene base scintillation fluid by standard procedures.

[1] R. R. Burgess, *in* "RNA Polymerase" (R. Losick and M. Chamberlin, eds.), p. 69. Cold Spring Harbor Lab., Cold Spring Harbor, New York, 1976.

[2] M. J. Chamberlin, *in* "RNA Polymerase" (R. Losick and M. Chamberlin, eds.), p. 20. Cold Spring Harbor Lab., Cold Spring Harbor, New York, 1976.

[3] G. J. Schneider, N. E. Tumer, C. Richaud, G. Borbely, and R. Haselkorn, *J. Biol. Chem.* **262,** 14633 (1987).

One unit of RNA polymerase activity is defined as 1 nmol of [^{3}H]ATP incorporated in 10 min at 37°. Specific activity is defined as units/mg protein. In the early phase of enzyme purification, the use of volume activity units (nmoles of radioactive substrate incorporated by 1 ml enzyme fraction in 10 min at 37°) more conveniently follows the RNA polymerase activity. In addition to measurement of enzymatic activity by incorporation of radioactive substrates into acid-precipitable material, the purified cyanobacterial RNA polymerase should be analyzed for subunit structure by sodium dodecyl sulfate (SDS)–polyacrylamide gel electrophoresis.[4]

Template DNAs such as calf thymus or chicken erythrocyte DNA, cyanophage AS-1 or N-1 DNA (obtained by standard purification steps and phenol extraction), and the dAT copolymer are prepared in 10 m*M* Tris–HCl, pH 8.0, 10 m*M* NaCl, 0.1 m*M* EDTA solution, and usually 20, 5, or 2 μg are added per assay, respectively. The assay mixture is prepared by mixing the ingredients just prior to use from more concentrated solutions containing the salts, buffer, unlabeled nucleoside triphosphates, DNA, labeled substrate, and RNase-free double-distilled H_2O.

Comments on the Assay

Beside [^{3}H]ATP, ^{14}C-labeled ATP or UTP (0.1–1 mCi/mmol) can be used as alternative substrates for the measurement of RNA polymerase activity (theoretically, any one of the nucleoside triphosphates may be used as the label). Control experiments using UTP as a labeled substrate and employing rifampicin (10 μg/ml) are recommended to ensure that the incorporation is the result of DNA-dependent RNA polymerase activity and not another contaminating and copurified enzyme(s) in the extract. For further discussion and review, see Burgess,[5] Burgess and Jendrisak,[6] and Chamberlin and Ryan.[7]

Purification Procedure

The procedure described is for the isolation of the RNA polymerase holoenzyme and core polymerase from *Anabaena* sp. PCC 7120 (ATCC 27893). Cell disruption is accomplished by French pressure cell treatment. After ultracentrifugation of the lysate, RNA polymerase is further

[4] U. K. Laemmli, *Nature (London)* **227,** 680 (1970).

[5] R. R. Burgess, *J. Biol. Chem.* **244,** 6160 (1969).

[6] R. R. Burgess and J. J. Jendrisak, *Biochemistry* **14,** 4634 (1975).

[7] M. Chamberlin and T. Ryan, *in* "The Enzymes" (P. D. Boyer, ed.), 3rd Ed., Vol. 15, p. 87. Academic Press, New York, 1982.

purified by Polymin P (polyethyleneimine, Sigma) precipitation. Final purification involves gel filtration chromatography (BioGel A-1.5 m, Bio-Rad) and binding to Heparin–Sepharose (Pharmacia). The holoenzyme and core RNA polymerase are separated on a BioRex 70 (Bio-Rad) column. The σ factor is purified by SDS–gel electrophoresis.[8]

Buffers. All buffers and solutions are prepared from double-distilled water and the highest grade chemicals available.

Buffer A: 20 m*M* Tris–HCl, pH 8.0, 10 m*M* $MgCl_2$, 50 m*M* KCl, 0.1 m*M* EDTA, 0.1 m*M* dithiothreitol, 1.0 m*M* phenylmethylsulfonyl fluoride, 10% (v/v) glycerol

Buffer G: 20 m*M* Tris–HCl, pH 8.0, 10 m*M* $MgCl_2$, 50 m*M* KCl, 0.1 m*M* EDTA, 0.1 m*M* dithiothreitol, 1.0 m*M* phenylmethylsulfonyl fluoride

Polymin P: 10% solution of polyethyleneimine titrated with 12 *N* HCl to pH 8.0 and stored at 4°

Storage Buffer: Buffer A solution with 50% (v/v) glycerol

Saturated ammonium sulfate: Enzyme grade ammonium sulfate is used to prepare a saturated solution at 25° in water and the pH then adjusted to 7.5–8.0 with KOH

Growth of Cyanobacterial Cells. Anabaena sp. filaments are grown in Kratz and Myers' medium[9] with combined nitrogen under axenic conditions to mid exponential phase of growth. Illuminated 10- to 15-liter carboys are stirred with magnetic stirrers and aerated with sterile air containing 1% CO_2. Yields are 1.5–2.0 g (wet weight) cells per liter. (Growth of *Anabaena* sp. in Allen medium[10] or BG-11[11] works equally well for RNA polymerase purification.)

Purification Steps

The cells are harvested with a Sorvall continuous flow system (Szent-Györgyi–Blum apparatus) in an SS-34 rotor at 4° and washed in Buffer G. The packed cell pellet is suspended in 2 volumes of Buffer A and then passed through a cold French pressure cell twice at 20,000 psi. The cell lysate is ultracentrifuged for 90 min at 120,000 *g*.

The RNA polymerase is precipitated from this supernatant by slow addition with stirring in 10% polyethyleneimine solution to a final concentration of 0.05%. After centrifugation (12,000 *g* for 10 min at 4°C in a

[8] D. A. Hager and R. R. Burgess, *Anal. Biochem.* **109,** 76 (1980).

[9] W. A. Kratz and J. Myers, *Am. J. Bot.* **42,** 282 (1955).

[10] M. M. Allen, *J. Phycol.* **4,** 1 (1968).

[11] R. Rippka, J. Deruelles, J. B. Waterbury, M. Herdman, and R. Y. Stanier, *J. Gen. Microbiol.* **111,** 1 (1979).

Sorvall SS-34 rotor), the pellet is extracted with 25 ml of Buffer A containing 400 m*M* KCl by resuspension with a Teflon homogenizer.

After centrifugation (as above) the washed pellet is dissolved in 10 ml Buffer A containing 2.0 *M* KCl and the polyethyleneimine removed by 3 ammonium sulfate precipitations; each time the pellet is resuspended in 2.0 *M* KCl–Buffer A and 2.3 volumes of saturated ammonium sulfate solution added. The final pellet is resuspended in 5 ml Buffer A containing 400 m*M* KCl and dialyzed against the same buffer.

Gel Filtration on BioGel A-1.5 m. The dialyzed extract is loaded on a 2.8 × 83 cm column of BioGel A-1.5 m equilibrated with Buffer A containing 400 m*M* KCl and then eluted at a flow rate of 15 ml/hr with the same buffer. The active enzyme fractions are pooled and precipitated by adding solid ammonium sulfate to 60% saturation. The precipitate, after centrifugation, is resuspended and dialyzed against Buffer A containing 200 m*M* KCl.

Heparin–Sepharose Chromatography. The dialyzed protein fraction from the previous step is applied to a 1.5 × 7 cm Heparin–Sepharose column equilibrated with Buffer A containing 200 m*M* KCl. After washing the column with 2 volumes of Buffer A containing 200 m*M* KCl, the RNA polymerase is step-eluted with Buffer A containing 300 m*M* KCl. The pure enzyme is precipitated with 60% saturated ammonium sulfate as above and resuspended and dialyzed against storage buffer (containing 50 m*M* KCl and 50% glycerol) for storage at −20°.

BioRex 70 Chromatography for Core and Holoenzyme. Three milligrams of RNA polymerase from the Heparin–Sepharose column is separated into core- and holoenzyme using a 1 × 6 cm freshly prepared BioRex 70 column equilibrated with Buffer A containing 50 m*M* KCl. The column is washed with 2 column volumes of this buffer, and the core enzyme is step-eluted with Buffer A containing 200 m*M* KCl. The holoenzyme elutes from the BioRex 70 column with 500 m*M* KCl in Buffer A.

The purification of *Anabaena* sp. RNA polymerase from 16 g wet weight of cells is summarized in Table I. A total of 2.7 mg of enzyme protein is obtained; the overall yield of 46% is based on the amount of β subunit analyzed by Coomassie blue-stained SDS–polyacrylamide gels.[7] Protein samples from each step of the purification are compared in Fig. 1.

Purification of Anabaena sp. RNA Polymerase σ Factor

Since it has not been possible to obtain the RNA polymerase σ subunit chromatographically, the σ subunit is purified from 1.25 mg of BioRex 70 holoenzyme pool separated into component subunits on SDS–polyacrylamide gels. The σ subunit is isolated and renatured as described.[8] The

TABLE I
PURIFICATION OF *Anabaena* RNA POLYMERASE

Purification step	Volume (ml)	Total protein (mg)[a]	Total units[b]	Specific activity (U/mg)	Yield (%)[c]
High-speed supernatant	78	733	854	1.2	100
PEI eluate	6.2	42	1061	25	95
BioGel pool	3.65	6.1	483	79	66
Heparin–Sepharose pool	1.08	2.7	422	156	46

[a] Measured by Bio-Rad protein assay with bovine serum albumin as standard.
[b] One unit equals incorporation of 1 nmol [^{3}H]ATP in 10 min at 37°.
[c] Calculated by quanitation of β subunit on Coomassie-stained SDS gels.

electrophoretically purified σ factor fully reconstitutes the activity of holoenzyme when added to core in a run-off transcription assay using a DNA fragment containing a T4 bacteriophage early promoter as a template[12]; the transcripts are analyzed on a 5% polyacrylamide sequencing gel (Fig. 2).

Comments on the Purification Steps. The Polymin P precipitation may be performed on a French pressure cell-treated crude extract and afterward extracted with high-salt-containing Buffer A.[7] As an additional step DEAE–cellulose (DE-52, Whatman) chromatography can be helpful in the purification of RNA polymerase; the enzyme activity is step eluted with Buffer A containing 250 m*M* KCl.

Properties of Cyanobacterial RNA Polymerases. Thus far, DNA-dependent RNA polymerase has been purified and partially characterized from *Anacystis nidulans* (*Synechococcus* sp.),[13–17,20] *Fremyella diplosiphon*,[18] *Anabaena cylindrica*,[19] and *Anabaena* sp. PCC 7120.[3] RNA

[12] M. Chamberlin, R. Kingston, M. Gilman, J. Wiggs, and A. DeVera, this series, Vol. 101, p. 540.
[13] K. van der Helm and W. Zillig, *Hoppe-Seyler's Z. Physiol. Chem.* **348,** 302 (1967).
[14] K. van der Helm and W. Zillig, *FEBS Lett.* **3,** 76 (1969).
[15] F. Herzfeld and W. Zillig, *Eur. J. Biochem.* **24,** 242 (1971).
[16] F. Herzfeld and N. Rath, *Biochim. Biophys. Acta* **274,** 431 (1974).
[17] F. Herzfeld and M. Kiper, *Eur. J. Biochem.* **62,** 189 (1976).
[18] S. S. Miller and L. Bogorad, *Plant Physiol.* **62,** 995 (1978).
[19] G. Borbely and N. G. Carr, unpublished observations.
[20] M. Kumano, N. Tomioka, K. Shinozaki, and M. Sugiura, *Mol. Gen. Genet.* **202,** 173 (1986).

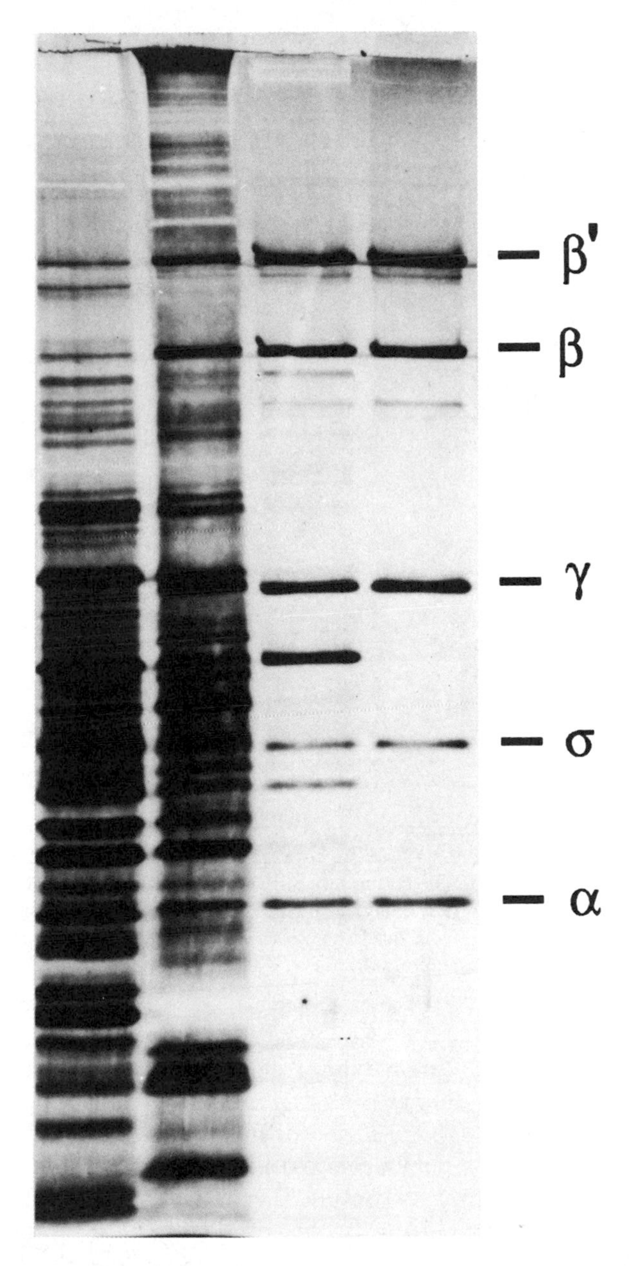

FIG. 1. Silver-stained gel of *Anabaena* RNA polymerase purification fractions. Samples from each step of the purification were separated on a 12.5% SDS–polyacrylamide gel and stained with silver. Lanes: 1, 9.4 μg high-speed supernatant; 2, 6.7 μg PEI eluate; 3, 1.7 μg BioGel A-1.5 m pool; 4, 1.25 μg Heparin–Sepharose pool. Subunit assignments are at right.

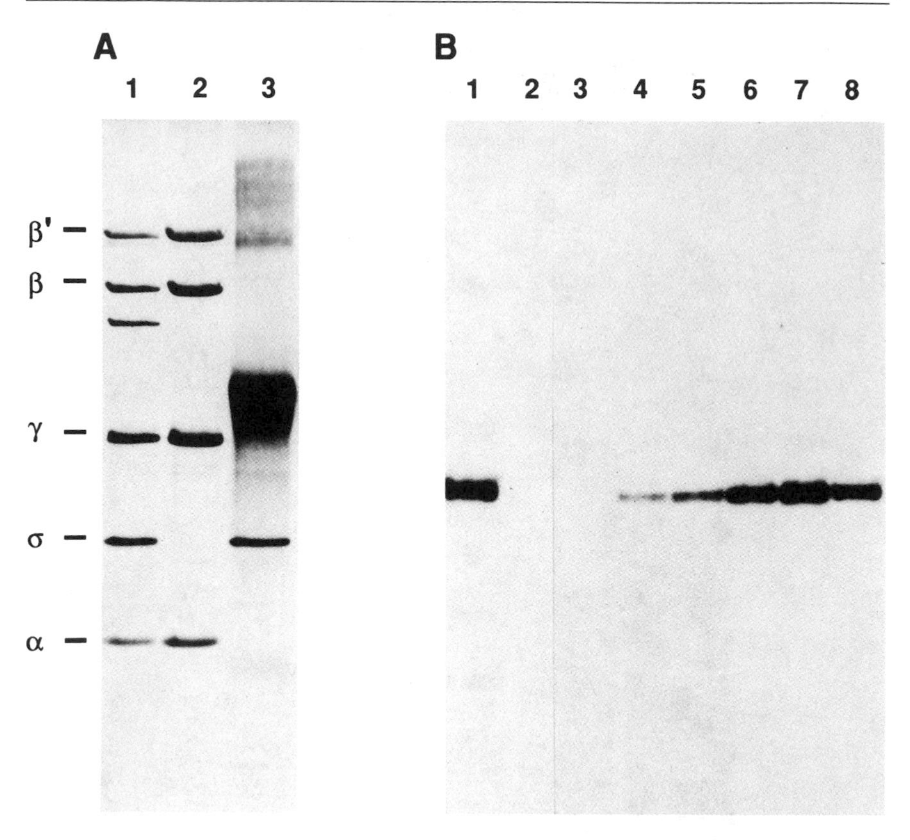

FIG. 2. Sigma subunit identification and holoenzyme reconstitution. (A) Silver-stained gel of (lane 1) 2.6 μg holoenzyme, (2) 4.0 μg core enzyme separated on BioRex 70, (3) 0.2 μg gel-purified σ subunit with bovine serum albumin. (B) Run-off transcription assays using a bacteriophage T4 early promoter, analyzed on a 5% acrylamide sequencing gel. Lanes: 1, 1.3 μg holoenzyme; 2, 0.2 μg core enzyme; 3, 0.2 μg σ subunit; 4–8, 0.2 μg core enzyme plus (4) 0.01 μg σ, (5) 0.02 μg σ, (6) 0.05 μg σ, (7) 0.1 μg σ, or (8) 0.2 μg σ.

polymerases of different origin exhibit slight differences for divalent metal ion requirement. The *Anacystis nidulans* and *Anabaena* sp. PCC 7120 enzymes require Mg^{2+} (10–20 m*M*) for maximal activity; Mn^{2+} can partially replace Mg^{2+} for enzyme activity. *Anabaena cylindrica* enzyme requires Mn^{2+} (2–4 m*M*) for optimal activity, and Mg^{2+} shows only a limited ability to replace Mn^{2+}. In the case of the *Fremyella diplosiphon* enzyme, Mg^{2+} (30 m*M*) or Mn^{2+} (5 m*M*) works equally well for maximal activity, but the enzyme shows an extreme instability in the presence of salts.

TABLE II
SUBUNIT MOLECULAR WEIGHTS OF VARIOUS CYANOBACTERIAL RNA POLYMERASES

Source	Subunit[a]						Ref.
	β'	β	γ	σ	α	x^b	
Anabaena sp. PCC 7120	171	124	66	52	41		3
Anabaena cylindrica	185	125	47		30	93	19
Anacystis nidulans	190	145	72		38		16
Fremyella diplosiphon	161	134	72		41	91	18

[a] Subunits are given by their sizes on SDS gels in kilodaltons.
[b] Unrelated to *Anabaena* sp. PCC 7120 subunits.

Anabaena 7120 RNA polymerase has, at 66 kDa, an additional subunit as part of the core enzyme. This subunit copurifies in a 1:1:1:2 ratio with the other core subunits β', β, and α (named by analogy to the *E. coli* RNA polymerase subunits). It has been proposed that this subunit be named the γ subunit.[3] On the basis of published copurification and stoichiometric data, the 47-kDa subunit from *Anabaena cylindrica*[19] and the 72-kDa subunits from *Anacystis nidulans*[16] and *Fremyella diplosiphon*[18] have been assigned as the corresponding γ subunits. The 93-kDa *Anabaena cylindrica* and the 91-kDa *Fremyella diplosiphon* subunits, although close to the apparent size of *E. coli* σ when analyzed by SDS–polyacrylamide gel electrophoresis, do not have a counterpart in the *Anabaena* sp. PCC 7120 enzyme. Table II summarizes the apparent molecular sizes of the subunits of cyanobacterial RNA polymerases, as established by SDS–polyacrylamide gel electrophoresis.

[65] Extracellular Proteins

By D. J. SCANLAN and N. G. CARR

Introduction

Cyanobacteria have been widely reported to excrete peptides and proteins into their culture media. In spite of their apparent quantitative importance little is known regarding the composition, and less about the function of extracellular proteins and peptides, some aspects of which are discussed below. It is, of course, established that some of the potent cyanobacterial toxins are peptides, but there is no clear evidence that these are exported, as distinct from being products of cell lysis. It is necessary to define the criteria[1] by which proteins are considered extracellular. Extracellular proteins are those which are substantially and selectively enriched in cell-free media and, for the purposes of this chapter, are those greater than 10,000 molecular weight.

Electrophoretic separations of cytoplasmic proteins of all cyanobacteria are characterized by the presence of major bands attributable to the subunits of the light-harvesting biliproteins. Since these biliproteins often account for approximately 30% of total soluble protein, and since these proteins are stable, they form a readily detectable feature of any cytoplasmic protein analysis. The absence of biliprotein subunits in an analysis of putative extracellular material is prima facie evidence that one is dealing with truly extracellular molecules, rather than the products of lysis of a proportion of the cyanobacterial culture. It should be noted that the absence of the phycocyanin and phycoerythrin proteins in a culture supernatant does not exclude leakage of proteins from the cyanobacterial periplasm. It is known that extrinsic proteins exist on the inner[2] (cytoplasmic) and outer membranes,[3] but we have no substantial information to the extent to which proteins exist in the intermembrane area. Along with the absence of biliproteins, true extracellular protein profiles on sodium dodecyl sulfate–polyacrylamide gel electrophoresis (SDS–PAGE) should be expected to be substantially different from those of cytoplasmic proteins.

Preparation of an Extracellular Protein Fraction

All work should be conducted with exponentially growing cultures, and it should be noted that the rate of secretion of proteins into the supernatant changes during exponential growth.[4] Maximum rates were obtained at the end of the exponential phase (Fig. 1).

Batch Cultures of 100 ml

Organisms are centrifuged at 10,000 rpm for 10 min at 20° (low-temperature harvesting leads to lysis) and the supernatant passed through a 0.2-μm Millex sterilizing filter. Total protein precipitation is obtained by addition of ammonium sulfate (47.2 g/100 ml, yielding a 70% saturated solution), which is ground to a fine powder and added to the supernatant at 4° over 10–15 min with stirring, and is allowed to continue for a further

[1] A. P. Pugsley and M. Schwartz, *FEMS Microbiol. Rev.* **32,** 3 (1985).
[2] T. Omata and N. Murata, *Arch. Microbiol.* **139,** 113 (1984).
[3] D. J. Scanlan and N. G. Carr, manuscript in preparation.
[4] D. J. Scanlan and N. G. Carr, manuscript in preparation.

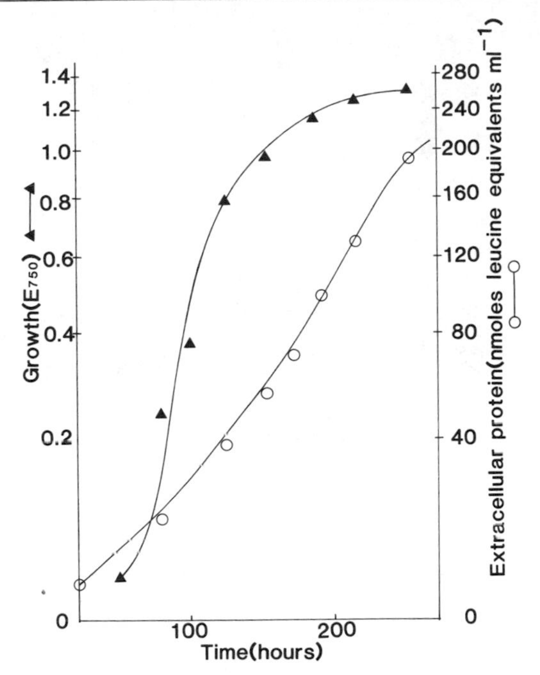

FIG. 1. Relationship of extracellular protein production to culture growth of *Synechococcus* PCC 7942.

30 min before spinning in transparent tubes at 10,000 rpm for 40 min at 4°. The supernatant is poured off, the tubes allowed to drain, and the pellet (which can be unstable, and is not always visible) resuspended in 0.5 *M* Tris buffer, pH 6.8. The sample is then subjected to SDS–PAGE.

Batch Culture of 4 Liters

Organisms are centrifuged at 10,000 rpm for 10 min at 20°, and protein in the supernatant is then fractioned by ultrafiltration through an Amicon hollow-fiber cartridge filter (Type H1P01-8), giving a mean selection of 10,000 molecular weight and above. The supernatant is then passed through a 0.2-μm Millex filter and precipitated with ammonium sulfate as above.

Qualitative Examination

Much of what we know of the functions of individual extracellular proteins has been derived from specific biological assays. Some of those employed are listed below.

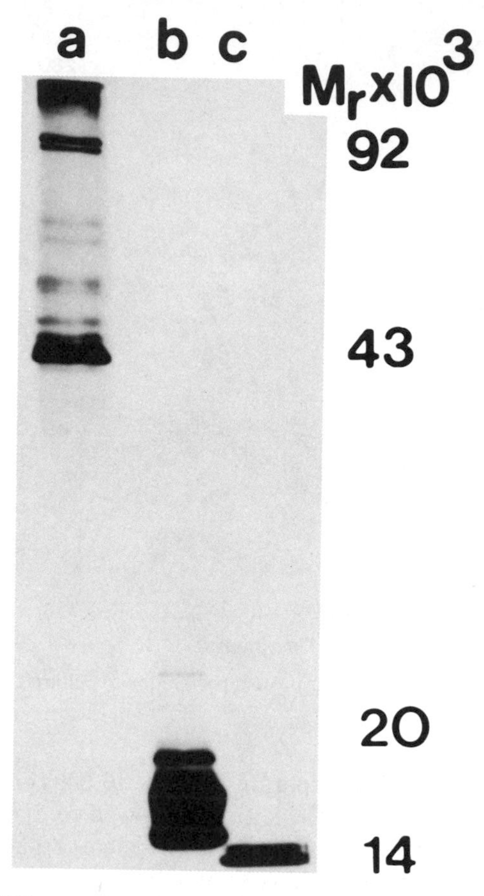

FIG. 2. SDS–PAGE separation of extracellular proteins after silver straining. (a) Extracellular protein from *Nostoc* PCC 8009; (b) partially purified phycocyanin from *Synechococcus* 7942; (c) extracellular proteins from *Synechococcus* 7942. Note the absence of biliproteins in tracks a and c.

SDS–PAGE. The overall size distribution of extracellular proteins in concentrated supernatants may be characterized electrophoretically.[4] Routinely, 10–30% exponential gradient SDS gels are run at 4°, 20 mA, overnight, and followed by silver staining (sensitivity 0.38 ng/mm^2 BSA) or Coomassie staining (sensitivity 0.2–0.5 μg of a single protein) for samples derived from 100-ml or 4-liter cultures, respectively (Fig. 2).

Antimicrobial Activity. In a study of antimicrobial products from filamentous cyanobacteria[5] it was found that a *Nostoc* sp. secreted a protein

[5] E. Flores and C. P. Wolk, *Arch. Microbiol.* **145,** 215 (1986).

which inhibited the antibiotic action, and a second *Nostoc* sp. produced a bacteriocin of low molecular weight (i.e., a proteinaceous antibiotic that is active against bacterial strains clearly related to the bacterium producing the substance).

Enzyme Activities. DNase,[6] phosphatases,[7] and protease[8] are among the enzymes which have been extracellularly detected from cyanobacteria.

Quantitative Measurement

The value of a standard quantitative method is a combination of sensitivity and the extent to which it may be applied to all proteins. Thus, in principle, a method which measures peptide bond frequency is more use than, say, a method based on the amount of a particular amino acid. To these considerations should be added the extent to which the method is open to influence by other chemicals and the convenience of the method, especially if it is only intermittently employed. The following methods have been employed, the descriptions of many of which are published in this series.

Biuret.[9] The biuret procedure, which detects peptide bonds, requires in excess of 100 μg of protein, which effectively excludes it from the kinetic analysis of cyanobacterial protein secretion.

Microbiuret.[10] The microbiuret method is based on the measurement of the UV absorption of the complex formed between protein and copper in strongly alkaline solution. All proteins may be measured. In the presence of DNA the range of the assay is 0.15–3 mg, whereas in the absence of DNA as little as 0.075 mg protein can be assayed.

Ninhydrin.[11] The ninhydrin spectrophotometric method is based on the reactivity of amino groups and is thus more sensitive to proteins which have greater than average basic amino acid composition.

Lowry.[12] The Lowry reaction is similar to the biuret assay in that there is a reaction of protein with copper ion. While this method is sensitive (minimum 1 μg required), it is open to interference by other chemicals.

[6] C. P. Wolk and J. Kraus, *Arch. Microbiol.* **131,** 302 (1982).

[7] F. P. Healey, *J. Phycol.* **9,** 383 (1973).

[8] C. P. Wolk, *in* "Nitrogen Fixation" (W. E. Newton and W. H. Orme-Johnson, eds.), Vol. 2, p. 279. University Park Press, Baltimore, Maryland, 1980.

[9] E. Layne, this series, Vol. 3, p. 450.

[10] R. F. Itzhaki and D. M. Gill, *Anal. Biochem.* **9,** 401 (1964).

[11] J. R. Spies, this series, Vol. 3, p. 468.

[12] E. Layne, this series, Vol. 3, p. 448.

Bio-Rad.[13] Bio-Rad Laboratories markets a reagent employing the method described by Bradford.[14] The assay is based on the observation that Coomassie blue undergoes a shift in absorption maxima after binding to protein. Using the Bio-Rad microassay procedure, 1–20 μg of protein can be detected. Eight-tenths milliliter of protein sample or standard bovine serum albumin solution is added to 0.2 ml of dye reagent concentrate, gently mixed, and the OD_{595} read against a reagent blank after 20 min. Full details of the compatability of the Bio-Rad protein assay with metabolites and other chemicals may be obtained from Bio-Rad laboratories.[13] Reference 13 also includes a comparison of the reactivity of 23 different proteins with several protein assay procedures. Wide ranges of divergence between methods are reported, in some cases 2-fold differences being observed when identical proteins were assayed by different procedures.

Fluorescamine.[15] The fluorescamine method is the most sensitive, requiring only 10–100 ng of protein, and is readily applicable to measuring the kinetics of protein release during cyanobacterial growth. The method is based on the reaction of fluorescamine with primary amines[16] to yield a fluorescent product. An account of the reaction under various conditions and an evaluation of the earlier literature is given by Castell *et al.*,[15] from which the method below has been derived. The standard for this assay may be a protein or an amino acid.

Materials

0.2 *M* sodium borate buffer, pH 8.0 (3.09 g boric acid in 250 ml H_2O, adjusted to pH 8.0 with 2 *M* NaOH)
Fluorescamine (3.4 μg/ml, in acetone dried over anhydrous $CaCl_2$)

Procedure. Prepare six tubes containing between 0 and 10 nmol of standard (0–40 μl of 0.25 m*M* standard such as leucine). Make up the volume to 150 μl with water. Add 0.9 ml of borate buffer to each tube. While mixing the sample on a vortex mixer, add 50 μl of fluorescamine reagent. The reaction is complete in 2–3 sec. Add a further volume of buffer to each tube to bring the final volume to that appropriate for the fluorimeter, typically, for the Perkin-Elmer 3000, a further 1.0 ml. Measure the fluorescence as soon as possible using an excitation wavelength

[13] "Bio-Rad Protein Assay Instruction Manual." Bio-Rad Laboratories, Richmond, California, 1981.
[14] M. Bradford, *Anal. Biochem.* **72,** 248 (1976).
[15] J. V. Castell, M. Cervera, and R. Marco, *Anal. Biochem.* **99,** 379 (1979).
[16] S. Udenfriend, S. Stein, P. Böhlen, W. Dairman, W. Leimgruber, and M. Weigele, *Science* **178,** 871 (1972).

of 390 nm and emission of 475 nm. Because of variations in the fluorimeter and the instability of fluorescamine, it is necessary to make up fresh fluorescamine every 2–3 days and to include a standard curve each time the assay is performed. The correlation coefficient of the line should be better than 0.98.

[66] Geosmin Production

By HELGA NAES

Geosmin (*trans*-1,10-dimethyl-*trans*-2-decalol) is an earthy smelling substance produced by actinomycetes and cyanobacteria. The method widely used for geosmin determination is the closed-loop stripping analysis (CLSA) system developed by Grob for analysis of trace organics in water.[1] This method was later modified by Borén *et al.* to the open-loop stripping system (OLSA),[2] which offers more flexibility in the stripping conditions and lower the contamination level. In this chapter, the OLSA method for the determination of geosmin in cyanobacterial cultures will be described.

A 1-liter stripping bottle is filled with distilled water to the level shown in Fig. 1. Internal standards and 5–10 ml of culture are added to the water sample. The internal standards used are 1-chlorooctane, 1-chlorodecane, 1-chlorododecane, 1-chlorotetradecane, and 1-chlorohexadecane dissolved in acetone. Their final concentration in the stripping bottle are 100 ng/liter. Thereafter the sample is purged with N_2 for 1.5 hr in a water bath at 60°. Even when using the purest quality of N_2, additional purification with activated carbon is recommended. The gas flow rate is 1 liter/min. Volatile compounds stripped from the culture are adsorbed to a 1.5 mg activated carbon filter kept at 80° in an oven. The filter and filter holder are manufactured by Bender & Hobein (Zürich, Switzerland).

After stripping, the filter is transferred to a conical vial and extracted by placing 10 μl dichloromethane on top of the filter disk. The solvent is forced 10–15 times back and forth through the filter by a syringe. At last when the drop is just below the carbon layer the syringe is removed and the system shaken 2 times to transfer the extract to the bottom of the vial (Fig. 2). This procedure is repeated 3 times. The filter is then cleaned with and stored in dichloromethane.

[1] K. Grob, *J. Chromatogr.* **84,** 255 (1973).

[2] H. Borén, A. Grimvall, and R. Sävenhed, *J. Chromatogr.* **252,** 139 (1982).

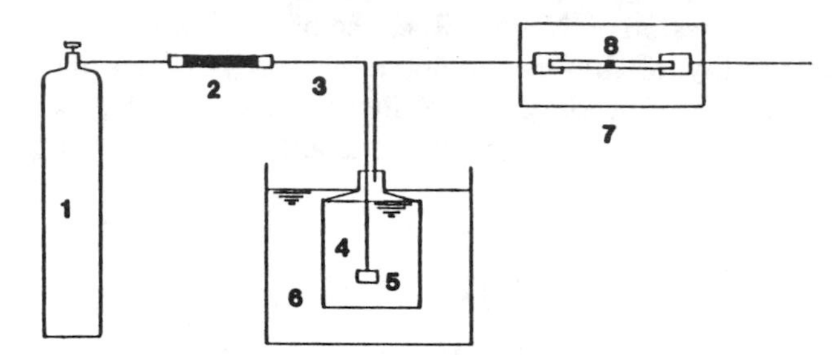

FIG. 1. Open-loop stripping system: 1, nitrogen gas; 2, gas cleaning filter (activated carbon); 3, PTFE tubing; 4, stripping bottle; 5, sintered glass filter; 6, water bath; 7, oven; 8, analytical filter (1.5 mg activated carbon).

The extract is analyzed by capillary gas chromatography on a 25-m WCOT BP 5 capillary column in a gas chromatograph equipped with a flame-ionization detector. The conditions for the gas chromatographic analysis are as follows: carrier gas, helium; injector and detector temperature; 280°; column temperature, 3 min at 35°, increased from 35 to 90° at 30°/min, increased from 90 to 230° at 5°/min. Quantitative determination of geosmin is done relative to 1-chlorododecane by an integrator.

The selection of appropriate equipment materials is of primary importance in avoiding contamination. The materials recommended are glassware, polytetrafluoroethylene (PTFE), and stainless steel. All the materials must be thoroughly washed with appropriate solvents before building the system. Between each run the sintered glass filter must be cleaned with chromic acid. The stripping conditions described above are found to be optimal for determining the total amount of geosmin in a cyanobacterial culture. Experiments that lead to this conclusion are described below.

Nothing is known about the function of geosmin in cyanobacteria. Is geosmin an intracellular compound, released only after death and decay, or is it actively excreted into the medium? These questions are important to have in mind for the choice of stripping conditions. Because of these uncertainties, studies on the effect of different cell disruption methods on

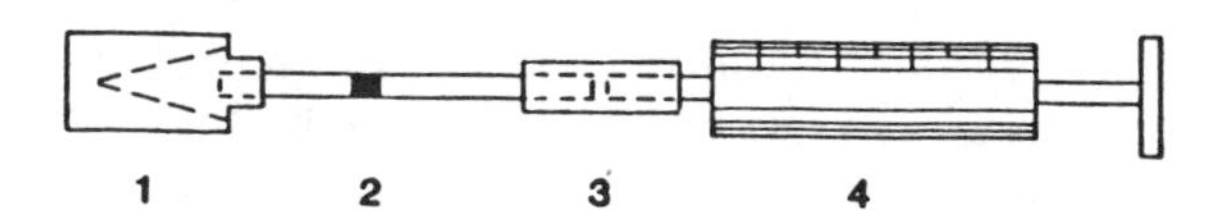

FIG. 2. Equipment used for extraction: 1, conical vial; 2, analytical filter (1.5 mg activated carbon); 3, PTFE tubing; 4, syringe.

TABLE I

EFFECT OF DIFFERENT DISRUPTION METHODS COMBINED WITH VARYING STRIPPING CONDITIONS ON THE AMOUNT OF GEOSMIN ISOLATED FROM *Oscillatoria brevis* CULTURES[a]

Sample	Disruption method	Stripping conditions	ng geosmin/ μg dry weight
A1	Untreated	60°	0.47
A2	Lysozyme, sonication	60°	0.49
A3	Lysozyme, SDS, sonication	60°	0.51
B1	Untreated	20°, NaCl	0.02
B2	Lysozyme, sonication	20°, NaCl	0.55
B3	Untreated	60°	0.43

[a] The letters A and B denote different batches.

cyanobacteria and variations in stripping conditions on the recovery of geosmin from the organisms were performed.

Experiments are outlined with batch cultures of *Oscillatoria brevis* cultivated without aeration at low light intensity (5 μE m^{-2} sec^{-1}) and at room temperature (20°). The cultures to be subjected to different disruption methods are centrifuged and washed with 0.1 *M* Tris–EDTA buffer, pH 8, and the supernatant discarded. Lysozyme is dissolved in the buffer to give a final concentration of 5 mg/ml. This solution (0.5 ml) is added to the pellet, incubated at 37° for 1 hr, and then sonicated 3 times for 1 min, and then the suspension is added to the stripping bottle. Another cell fraction is additionally treated with sodium dodecyl sulfate (0.5 ml 1% SDS in 0.1 *N* NaOH) for 1 hr before sonication.

TABLE II

EFFECT OF SALT ON THE RECOVERY OF STANDARD GEOSMIN RELATIVE TO THE INTERNAL STANDARD (C_{12}) AT LOW STRIPPING TEMPERATURE

Geosmin/C_{12}	Stripping conditions
0.18	20°
0.95	20°, NaCl
0.90	60°

TABLE III
AMOUNT OF GEOSMIN IN THE CULTURE (CELLS PLUS MEDIUM) COMPARED TO THE FILTRATE (CULTURE MINUS CELLS)

Sample	Stripping conditions	ng geosmin/μg dry weight
Culture 1	60°	0.41
Filtrate 1	60°	0.18

Stripping at 20–25° is widely used to isolate volatile compounds from cyanobacteria.[3-5] The results in Table I, however, show that this method is not adequate to obtain the total amount of geosmin from the culture. The measured concentration of geosmin in cells disrupted with lysozyme, SDS, and sonication is higher than that of untreated cells stripped at 20°. Stripping fully disrupted cells at 20° gave the same results as stripping untreated cells at 60°. This indicates that stripping at 60° is sufficient to make the cells leaky and thereby permits the total amount of geosmin to be determined.

Sodium chloride is added to the stripping bottle (150 g/l) when stripping is performed at 20°, to obtain equal recovery of geosmin relative to the internal standard at this temperature. Without salt, the recovery of geosmin is much less than the internal standard, and the two stripping conditions (20 and 60°) are then not comparable (Table II).

Because geosmin is present both in the culture medium and in the cells (Table III), stripping of the whole culture is required in order to determine the total concentration of the compound. In conclusion the results so far indicate that the OLSA method at 60° is adequate for determining the total amount of geosmin in cyanobacterial cultures.

[3] G. Izaguirre, C. J. Hwang, S. W. Krasner, and M. J. McGuire, *Wat. Sci. Tech.* **15,** 211 (1983).

[4] S. W. Krasner, C. J. Hwang, and M. J. McGuire, *Wat. Sci. Tech.* **15,** 127 (1983).

[5] S. Möhren and F. Jüttner, *Wat. Sci. Tech.* **15,** 221 (1983).

[67] Quantitative Trace Analysis of Volatile Organic Compounds

By FRIEDRICH JÜTTNER

Introduction

Despite the fact that volatile organic compounds (VOC) are widely distributed in cyanobacteria, the present knowledge of them can at best be regarded as in its infancy. The lack of microanalytical techniques[1] and axenic cyanobacterial strains[2] prevented any substantial progress in this field for a long time. Although geosmin and 2-methylisoborneol, both highly odoriferous compounds, were described for axenic cyanobacteria as early as 1976,[3] other compounds such as volatile thio compounds,[4] mesityl oxide,[5] norcarotenoids,[5,6,7] alkenes, unsaturated aldehydes, ketones, and alcohols were detected only recently.[8] It is therefore not surprising that investigations of the physiology and biochemistry of these compounds in cyanobacteria[5,7] are still only in their initial stages.

The present availability of numerous axenic cultures of cyanobacteria enables a clear differentiation between metabolites excreted by cyanobacteria and contaminating heterotrophic bacteria. However, the distinction between genuine cyanobactrial products on the one hand and contaminants and analytical artifacts on the other is still the subject of some confusion. The elucidation of the structures of compounds present in growth media does not necessarily indicate the origin of the compounds. Decisive information allowing discrimination between biogenic products, contaminants, and artifacts can be obtained from studies of the dynamics of these compounds. A prerequisite for such studies is a micromethod enabling quantitative analyses of the VOC. In this chapter, a procedure is described that can be used for structural elucidations, as well as for physiological and biochemical studies of the VOC which for quantitative analyses are needed.

[1] K. Grob, *J. Chromatogr.* **84,** 255 (1973).
[2] R. Y. Stanier, R. Kunisawa, R. Mandel, and G. Cohen-Bazire, *Bacteriol. Rev.* **35, 171 (1971).**
[3] J. L. Tabachek and M. Yurkowski, *J. Fish. Res. Board Can.* **33,** 25 (1976).
[4] F. Jüttner, *Z. Naturforsch. C* **39,** 867 (1984).
[5] F. Jüttner, J. Leonhardt, and S. Möhren, *J. Gen. Microbiol.* **129,** 407 (1983).
[6] J. J. Henatsch and F. Jüttner, *Wat. Sci. Tech.* **15,** 259 (1983).
[7] F. Jüttner and B. Höflacher, *Arch. Microbiol.* **141,** 337 (1985).
[8] S. Möhren and F. Jüttner, *Wat. Sci. Tech.* **15,** 221 (1983).

Quantitative Determinations of VOC in Small-Scale Cultures

Closed culture systems which restrict gas exchange are preferred for quantitative studies on VOC. This applies both to fermenters and shaking cultures. Shaking cultures are best performed in 300-ml Erlenmeyer flasks with ground glass joints and are inoculated with 100 ml of cyanobacterial suspension.[5] Flasks are sealed by hollow threaded stoppers with ground glass ends, at whose outer ends are positioned PTFE-protected septa, through which samples can easily be added or removed without opening the system. The pH of the culture and the supply of CO_2 can be maintained for an experimental period of 2–3 days by the addition of 20–30 m*M* $NaHCO_3$. It should be stressed that organic materials should be avoided whenever possible because autoclaving and flaming release numerous compounds from the organic matter.

In general, the supply of different samples with identical light and temperature conditions is easier to achieve than the withdrawal of representative aliquots. The most critical step is the opening of vessels, which leads to an exchange of air with a subsequent new distribution of the VOC between the liquid and the gas phase. Optimum results are obtained when different sample units are treated under identical conditions and one sample is removed and worked up per unit of time.

Quantitative Analysis of VOC in Fermenter Cultures

When fermenter cultures are used for experiments, highly volatile substances (e.g., isopropylthiol) can be trapped at the exit port of the aeration gas. The VOC cartridge described below can easily be fastened to the exit of the fermenter with its ground glass joint. When aliquots are removed from the cyanobacterial suspension for analysis of less volatile compounds, the samples should be as small as possible (10–20 ml of a 2-liter fermenter) to keep the changes of the proportion of the gas to liquid phase in the vessel as small as possible. To reduce any loss of VOC by the aeration gas stream, the latter should be reduced to low flow rates in combination with increased CO_2 concentrations.

Performance of Quantitative Analyses of VOC

The stripping procedure is an ideal means for extracting VOC from culture media (with and without biomass). It was developed in its original form for water samples by Grob.[1] Using this procedure the VOC were stripped from the liquid to be analyzed by a gas stream and sorbed on activated charcoal. The sorbed compounds were subsequently eluted from the charcoal by application of a suitable solvent, and part of the eluate was injected into a gas chromatograph for separation. We have improved this procedure using Tenax TA as a sorbant and thermal desorption for the transfer of compounds to the gas chromatograph.[9] In this way all the VOC which have been stripped from a sample are used for a single chromatogram, avoiding dilution by the solvent. In addition, data we have obtained indicate that the recoveries of polar VOC (alcohols, aldehydes, ketones) are substantially increased after addition of nearly saturating concentrations of sodium chloride to the solution to be stripped. The recovery of lipophilic substances, however, is not significantly affected by salt treatment. With a stripping time of 30–60 min, depending on the volatility of the VOC, between 1 and 40% of the amount present in the sample is made available for gas chromatography. The amounts of VOC thus obtained are sufficient to provide good mass spectra by combined gas chromatography–mass spectrometry. It should be pointed out, however, that the addition of salt in saturating concentrations activates lipoxygenases and carotene oxygenases and thus leads to the liberation of corresponding degradation products.

Construction of the VOC Cartridge

To trap a wide range of different VOC, a cartridge with the design shown in Fig. 1 should be constructed. Its entrance has a ground glass joint with which it can easily be connected to an exit port of a fermenter or introduced into the circuit of a stripping system. A metal Luer lock is soldered to the other side of the glass barrel. VOC cartridges are distributed by SGI Laborbedarf (D-7909 Dornstadt, FRG). The VOC cartridge is filled with 150 mg of Tenax TA, 40–60 mesh, held on each side by quartz wool.

Concentration of VOC by Stripping

The VOC are extracted by a stripping system as outlined in Fig. 2. Water is added to 1–50 ml of cyanobacterial suspension to make up a volume of 20–50 ml in a 100-ml round-bottomed flask equipped with a ground glass joint, 20% (w/w) NaCl and a drop of antifoaming agent are added, and the temperature is adjusted to 22°. Then a head with a gas input tube and a gas exit tube is placed on the flask and connected to the stripping system. Use of a fritted disk in the head is not recommended because heavy foam formation occasionally occurs. The VOC cartridge with the Luer lock and ground glass joint is then attached to the stripping system, which is constructed out of flexible copper tubes. To avoid any condensation of water before the gas stream reaches the Tenax bed, two

[9] F. Jüttner and K. Wurster, *J. Chromatogr.* **175,** 178 (1979).

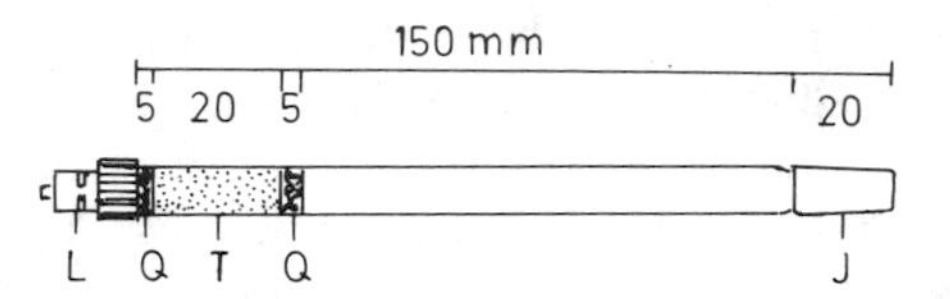

FIG. 1. Construction of the VOC cartridge which consists of a glass barrel (7 mm i.d., 10 mm o.d., length 150 mm) with a ground glass joint (J) and a soldered Luer lock (L). The Tenax bed (T) is held in place by quartz wool (Q).

40-W incandescent lamps are positioned at the entrance of the VOC cartridge and adjacent to the Tenax bed, or a thermostatted (40°) copper tube is coiled around the cartridge. The air in the closed stripping system is circulated with a stainless steel pump (Metal Bellows Corp., Model MB 21) with a flow rate of 1.2 liter/min. After 30–60 min stripping time the VOC cartridge is removed and stored in a gas-tight glass vessel. Loss on storage has been observed only for highly volatile compounds. Even after storage for 3 years, the VOC pattern of a natural cyanobacterial bloom that was identical to the original could be obtained.

Transfer of Sorbed VOC to a Gas Chromatograph

To enable the transfer of compounds sorbed to Tenax to a gas chromatograph, the loaded VOC cartridge is introduced into a movable heat-

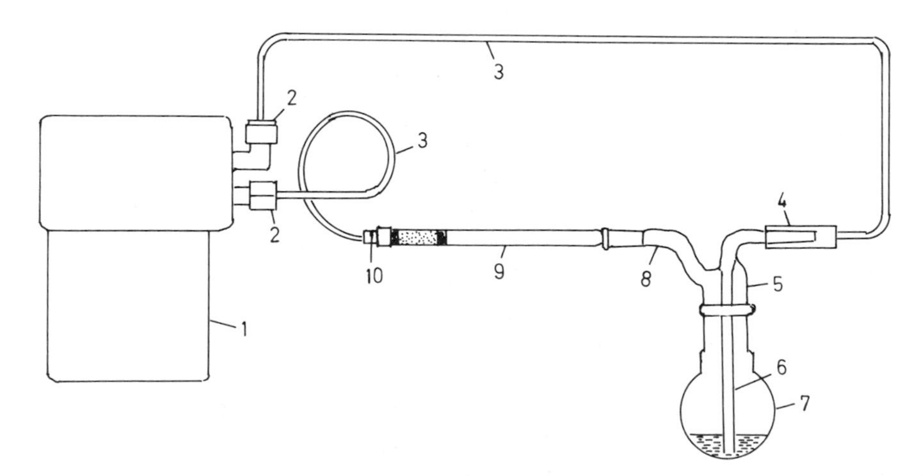

FIG. 2. Closed-loop stripping system for the separation of VOC from aqueous samples. 1, Stainless steel pump; 2, union; 3, copper line (3 mm i.d.); 4, female metal joint; 5, head; 6, gas input tube; 7, 100-ml round-bottomed flask; 8, gas exit tube; 9, VOC cartridge; 10, Luer lock.

ing block (Fig. 3) (40-W capacity) and connected via its ground glass joint to a hydrogen or helium gas supply with a flow rate of 50 or 31 ml/min, respectively. At the Luer lock a side-port needle (0.2–0.3 mm i.d., 0.5 mm o.d.) is fixed. After lowering the heating block the needle is introduced through the septum into the injection port. The internal helium carrier gas stream (1.2 ml/min) is switched off and replaced by the externally supplied stream which passes through the VOC cartridge, and which has the same initial pressure. The oven of the gas chromatograph must be cooled at the beginning of the transfer. When a UCON 50 HB 5100-coated glass capillary column is used, an initial temperature of 0° is sufficient. This can easily be obtained by two tins each filled with 200 ml of liquid N_2 placed in the oven. An OV-101-coated column needs a much lower initial temperature of about −60° to obtain sharp peaks of low-boiling substances. For the analysis of high-boiling compounds, such as geosmin, however, ambient temperature suffices. To desorb the VOC from the Tenax, the VOC cartridge is heated by the heating block to 280° within 3 min. For complete

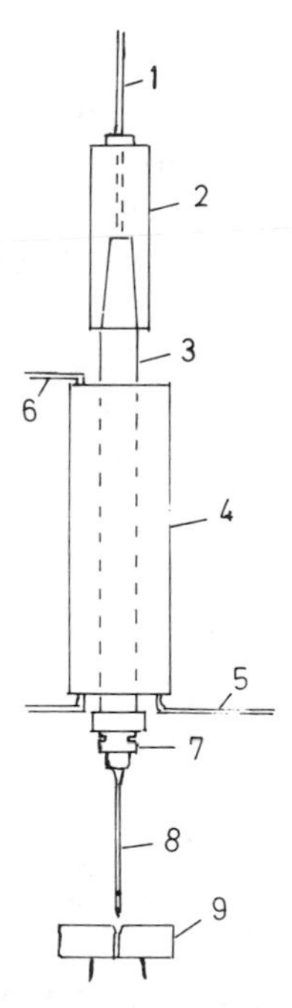

FIG. 3. Movable heating device for desorption and transfer of VOC to gas chromatographs. 1, Hydrogen or helium line; 2, female metal joint; 3, VOC cartridge; 4, heating block; 5, temperature control; 6, heating wire; 7, Luer lock; 8, side-port needle; 9, injection port of the gas chromatograph.

desorption this temperature is maintained for further 2 min. When the transfer has been completed, the needle is removed from the injection port by lifting the heating system. The internal carrier gas stream is turned on again and the separation of the VOC obtained by a temperature program. After cooling the VOC cartridge in a gas-tight vessel it can be reused directly.

Gas Chromatography of Volatile Organic Compounds

The separation of VOC is conducted on a 50-m glass capillary column (0.3 mm i.d.) coated with UCON 50 HB 5100. A suitable temperature program is 5°/min with a hold at 180° for 15 min. As an example, a gas chromatogram of the VOC isolated from a 100-ml culture of *Synechococcus* PCC 6301 (*Anacystis nidulans*) is presented in Fig. 4. If only one compound is present in higher concentrations, e.g., mesityl oxide in *Anabaena cylindrica* or geosmin in *Fischerella muscicola,* a 20-m column is sufficient for characterization and is recommended to shorten the time required for analysis. The amount of a compound is calculated from the peak area and may be obtained by using one of several commercially available integrators or simply from the peak height. Calibration is performed by diluting a substance with methanol in such a way that addition of about 50 μl of this methanolic solution to a water sample, which is then treated in the same manner as the measuring sample, gives an identical detector signal. Methanol is used as a solvent because it is only weakly retained on Tenax. Depending on the concentration and the property of a compound, either linear or nonlinear recoveries may be obtained from

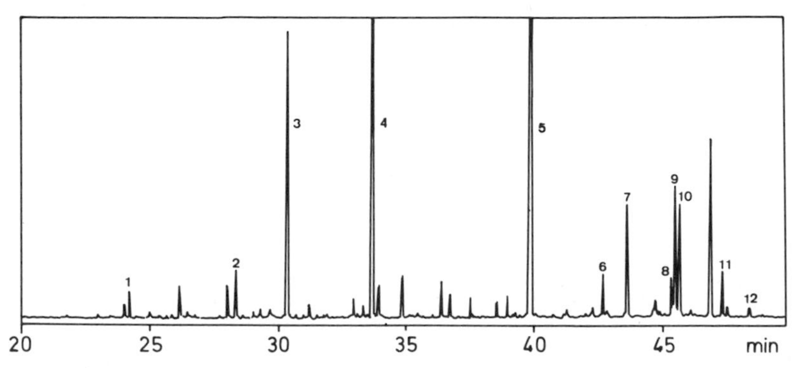

FIG. 4. Gas chromatogram (total ion current) of the VOC isolated from a 100-ml culture of *Synechococcus* PCC 6301 (*Anacystis nidulans*). The conditions of separation are the same as stated in Table I. 1, 2-Methyl-2-pentenal; 2, dimethylallylic alcohol; 3, 6-methyl-5-hepten-2-one; 4, 6-methyl-5-hepten-2-ol; 5, pentadecane; 6, hexadecane; 7, geraniol; 8, (*E*)-geranyl acetone; 9, heptadecane; 10, heptadecene; 11, β-ionone; 12, β-ionone 5,6-epoxide.

dilution series. A calibration curve in the measuring range must be determined for each compound.

The qualitative determination of VOC requires combined gas chromatography–mass spectrometry or, when the compounds have already been

TABLE I
ELUTION TIMES OF CYANOBACTERIAL VOC[a]

Compound	Elution time (min)	Compound	Elution time (min)
Dimethyl sulfide	8.90	6-Methyl-5-hepten-2-ol	33.66
Isopropyl thiol	9.78	*Benzaldehyde*	33.92
Acetone	10.22	Citronellal	35.32
Butanone	14.12	1-Octanol	36.71
Benzene	15.92	Diisopropyl trisulfide	37.43
3-Pentanone	17.60	*Acetophenone*	37.72
1-Penten-3-one	18.55	(*E*)-2-Octen-1-ol	37.74
1,3-Octadiene	19.22	2-Methylisoborneol	38.43
Toluene	20.23	β-Cyclocitral	38.99
Dimethyl disulfide	20.62	2,6,6-Trimethylcyclohex-2-ene-1,4-dione	39.22
(*Z*)-2-Pentenal	21.55	Pentadecane	39.83
Hexanal	21.80	1-Pentadecene	40.60
(*E*)-2-Pent-enal	22.40	Dimethyl tetrasulfide	41.15
Mesityl oxide	23.07	Citronellol	42.05
1-Penten-3-ol	23.40	Tetrahydrogeranyl acetone	42.14
(*E,Z*)-Octa-1,3,5-triene	23.63	Nerol	42.61
2-Menthyl-2-pentenal	24.17	Hexadecane	42.69
p-Xylene	24.22	Geraniol	43.62
m-Xylene	24.44	Germacrene D	43.74
Methylisopropyl disulfide	25.36	γ-Cadinene	45.00
o-Xylene	25.73	Geosmin	45.03
1-Pentanol	26.73	(*E*)-Geranyl acetone	45.32
Dimethylallylic alcohol	28.32	Heptadecane	45.50
3-Octanone	28.57	(*Z*)-5-Heptadecene	46.01
Diisopropyl disulfide	29.05	1-Heptadecene	46.33
1-Hexanol	30.24	*p*-Cresol	46.94
6-Methyl-5-hepten-2-one	30.33	β-Ionone	47.37
Dimethyl trisulfide	31.08	β-Ionone 5,6-epoxide	48.50
3-Octanol	32.30	Dihydroactinidiolide	60.08
1-Octen-3-ol	33.43		

[a] VOC were separated on a 40-m glass capillary column (0.3 mm i.d.) with an initial temperature of 0° for 4 min and a subsequent temperature program of 5°/min up to 200° and a hold for 20 min at that temperature. Helium gas was used as carrier. Frequently encountered contaminants and degradation products of the Tenax sorbant (benzaldehyde and acetophenone) are included (italics).

identified, the superposition of peaks by addition of the authentic substance to the sample. To facilitate the identification of compounds known to be released by cyanobacteria, the elution time of important VOC are listed in Table I. Some contaminants which are often present in samples and which can be used as internal standards are also included.

Production of Background-Free Water and Salt

Tap water and distilled water which is produced from deionized water are often especially rich in VOC which may disturb the pattern of the chromatogram. Spring water or groundwater is usually of much better quality. The elimination of VOC from water can also be achieved using UV irradiation after addition of H_2O_2 and/or stripping for several hours with repeated change of the VOC cartridge. Sodium chloride is fired for 24 hr at 280° for purification.

Application of the Method for Sensitive Molecules

The technique described above is suitable for a large number of compounds. The trapping of very-low-boiling substances (methyl mercaptan), however, requires cooling the Tenax. Oxygen-sensitive compounds (thiols) cannot be stripped successfully in the closed system without oxidation. In this case stripping by a stream of inert gas is the method of choice. The method presented is inexpensive, effective, and extremely sensitive, allowing the analysis of nanogram amounts of volatile organic compounds in aqueous solution as well as in any material that can be dissolved or suspended in water.

[68] Cyanobacterial Flocculants

By YESHAYA BAR-OR and MOSHE SHILO

Introduction

Benthic cyanobacteria are attached to various submerged surfaces in aquatic habitats by means of their hydrophobic cell surface properties.[1] Fixed to their substrata, they cannot escape unfavorable environmental conditions, as opposed to planktonic species which can position themselves along light and nutrient gradients in the water column.[2] Survival and proliferation of the benthic cyanobacteria thus depend on a complex

[1] A. Fattom and M. Shilo, *Appl. Environ. Microbiol.* **47,** 135 (1984).
[2] A. E. Walsby, *Bacteriol. Rev.* **36,** 1 (1972).

set of adaptations to the varying conditions prevailing at their sites of attachment. These include metabolic flexibility,[3,4] modulation of cell surface hydrophobicity to enable temporary detachment,[5–7] and production of extracellular flocculants, capable of flocculating and sedimenting suspended clay particles,[8] thus allowing light to reach the benthic interface even in turbid waters. In this chapter we describe the production, purification, and properties of two benthic cyanobacterial flocculants and methods for their assay.

Preparation and Properties of Flocculants

After screening many benthic cyanobacteria, high levels of flocculant were discovered in cell extracts and culture supernatants of *Phormidium* J-1 (ATCC 39161) and *Anabaenopsis circularis* [Pasteur Culture Collection (PCC) 6720]. Both strains were grown in BG-11 medium[9] at 35°, with shaking, under cool-white fluorescent lamps (incident light intensity 2.5×10^3 ergs/cm^2 sec).

Phormidum J-1 Flocculants.[8] Four-week old cultures are centrifuged (10,000 *g*, 10 min) and the supernatant decanted and rotary-evaporated at 45° to 1/10 of the original volume. The concentrate is treated with pronase (Sigma, 100 μg/ml) for 1 hr at 37° and dialyzed overnight in the cold. Two volumes of cold ethanol is added to the dialyzate, and the precipitate is separated by centrifugation (10,000 *g*, 15 min) and dried under reduced pressure. Cellular flocculant is obtained by suspending the cell pellet in 20 m*M* trishydroxymethylaminomethane (Tris), 10 m*M* $MgSO_4$ buffer (pH 7.0), and disintegrating it with glass beads (0.1 mm diameter, 0.5 g/ml) in a Braun cell disintegrator for 2 min. The homogenate is filtered through Whatman GF/C filter paper and then treated with pronase and precipitated as above. Flocculant could be extracted from the external cell wall layers of *Phormidium* J-1 by treating the cells successively with sodium dodecyl sulfate (SDS), pronase, and lysozyme.[7,10]

[3] S. Garlick, A. Oren, and E. Padan, *J. Bacteriol.* **128,** 623 (1977).

[4] E. Padan, *Adv. Microbial. Ecol.* **3,** 1 (1979).

[5] M. Shilo, *Philos. Trans. R. Soc. London Ser. B* **297,** 565 (1982).

[6] A. Fattom and M. Shilo, *FEMS Microbiol. Ecol.* **31,** 3 (1985).

[7] Y. Bar-Or, M. Kessel, and M. Shilo, *Arch. Microbiol.* **142,** 21 (1985).

[8] A. Fattom and M. Shilo, *Arch. Microbiol.* **139,** 421 (1984).

[9] G. Stanier, this volume [14]; see also R. Y. Stanier, R. Kunisawa, M. Mandel, and G. Cohen-Bazire, *Bacteriol. Rev.* **35,** 171 (1971).

[10] The washed cells are suspended in 0.25% SDS at 37° for 15 min, centrifuged, and resuspended in fresh BG-11 medium containing 200 μg/ml pronase for 15 min at 37°. After centrifugation, they are suspended in 10 m*M* tricine–NaOH, 10 m*M* K_2HPO_4, 10 m*M* NaH_2PO_4, 10 m*M* $MgSO_4$, 0.5 *M* mannitol, pH 7.8, containing 0.1% lysozyme (Sigma). Incubation is at 35° for 100 min.

The protein content of the extracellular preparation is high (32%) and can be lowered without loss of flocculating activity by treatment of the aqueous flocculant solution with hot phenol.[11] The crude dry flocculant is "wetted" with a minimum amount of ethanol, dissolved in water (2 mg/ml), and warmed to 70°. An equal amount of 90% aqueous phenol at 70° is then added. The mixture is stirred at 70° for 15 min, centrifuged (12,000 *g*, 15 min), and the lower phenolic phase is separated and extracted with an equal amount of water at 70°. The centrifuged water phases are combined, dialyzed, and lyophilized. The yield is approximately 50% of the crude preparation on a weight basis, containing 70–80% of the initial flocculating activity. Protein content is 9%. Further purification is achieved by gel filtration chromatography on Sepharose CL-4B (Pharmacia) (2.5 × 60 cm). The running buffer is 0.1 *M* NaCl, 2 m*M* Tris, pH 8.5. Two peaks appear in the eluate, the first at the void volume, corresponding to an apparent molecular weight of 1,200,000 or higher and containing about 45% of the initial flocculating activity and 12% of the dry weight loaded. The second peak contains a lower molecular weight material, lacking any activity.

A rapid alternative way to obtain a crude flocculant preparation is the addition of cetyltrimethylammonium bromide (CTAB) (Eastman, final concentration 0.2%) to the centrifuged culture medium. The coarse precipitate formed is stirred in $CaCl_2$-saturated ethanol for several hours and then dissolved in 50 m*M* Na_2SO_4, reprecipitated with 2 volumes cold ethanol, washed with ethanol, and dried. The main characteristics of the *Phormidium* J-1 extracellular flocculant are summarized in Table I.

Anabaenopsis circularis Flocculant. The centrifuged culture supernatant is treated with CTAB (final concentraiton 0.2%) overnight at 35° with stirring. The fine precipitate formed is separated by centrifugation or collected on cotton wool and incubated in 0.15 *M* NaCl at 35° overnight for complete dissolution. The biopolymer is reprecipitated with 2 volumes of cold ethanol, washed with ethanol, and dried. Its main characteristics are summarized in Table I.

Determination of Bioflocculant Concentration and Activity

Production and excretion of bioflocculant into culture supernatants can be checked qualitatively by adding a 10- to 100-μl sample to 0.5 ml of a bentonite suspension (Fisher, 0.6 mg/ml in 2 m*M* $MgSO_4$). After mixing, the suspension is allowed to stand for 5 min and then observed visually for flocculation of the bentonite particles. Quantitative measurement of

[11] O. Westphal, O. Luderitz, and F. Bister, *Z. Naturforsch. B* **7,** 148 (1952).

TABLE I
CHARACTERISTICS OF EXTRACELLULAR CYANOBACTERIAL FLOCCULANTS

Cyanobacterium	MW[a]	Composition (%)				
		Neutral sugars[b]	Uronic acids[c]	Keto acids[d]	*Lipids*[e]	Sulfate[f]
Phormidium J-1	≥1,200,000	40	32	—	12	1.5
Anabaenopsis circularis	≥1,200,000	28	—	11	0	—

[a] Derived from gel permeation chromatography.
[b] Determined with the anthrone reagent [W. E. Trevelyan and J. S. Harrison, *Biochem. J.* **50,** 298 (1952)].
[c] According to J. T. Galambos [*Anal. Biochem.* **19,** 119 (1967)].
[d] According to H. Katsuki, T. Yoshida, C. Tanegashima, and S. Tanaka [*Anal. Biochem.* **43,** 349 (1971)].
[e] Preparations were extracted with acidic hexane and the residue reweighed.
[f] Determined with benzidine according to T. Terho and K. Hartiala [*Anal. Biochem.* **41,** 471 (1971)].

bioflocculant concentrations in culture supernatants or cell extracts is done by either one of the following methods.

Bentonite Sedimentation Test.[8] To a series of Klett tubes are added 0.1- to 2.5-ml volumes of the sample, 0.5 ml of a stock bentonite suspension (6 mg/ml), and 0.5 ml of Tris–Mg buffer (20 m*M* Tris, 20 m*M* $MgSO_4$). The volume in each tube is made to 5 ml with water; the tubes are stirred thoroughly and then allowed to stand at room temperature. The turbidity of the assay mixture is measured at short time intervals in a Klett–Summerson colorimeter (Filter 54). The time required for a 50% decrease in initial turbidity is inversely proportional to the flocculant concentrations (Fig. 1). Such decrease after a 5-min incubation period is expressed as 5 flocculation units. Figure 1 shows that at flocculant concentrations of up to 25 μg/ml there is a linear relationship with the reciprocal of the sedimentation time. The inclusion of divalent cations, such as Mg^{2+} or Ca^{2+}, in the test buffer is essential for flocculation. A concentration of 2 m*M* $MgSO_4$ satisfies this requirement.

Alcian Blue Binding Test. The Alcian blue binding test is a slight modification of the method proposed by Ramus[12] for algal anionic polysaccharides. Alcian blue 8NGX (Ingrain blue 1, C.I. 74240, obtained from MCB, Cincinnati, OH) is dissolved in 0.5 *M* acetic acid to a concentration of 1 mg/ml. After adding 4.25 ml of 0.5 *M* acetic acid to a 0.5-ml sample,

[12] J. Ramus, *J. Phycol.* **13,** 345 (1977).

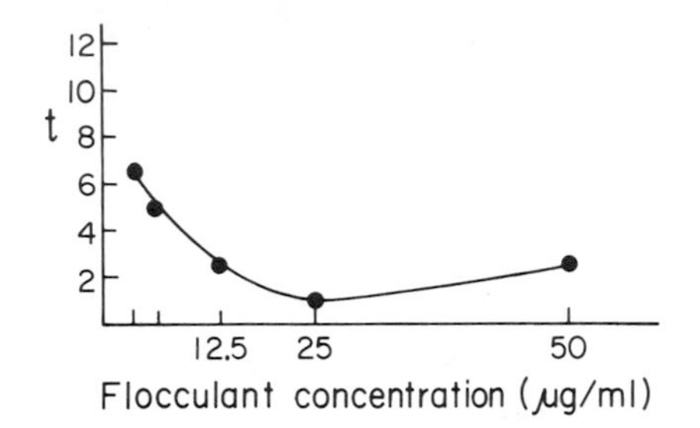

FIG. 1. Bentonite sedimentation time as a function of flocculant concentration. Aliquots of an extracellular *Phormidium* J-1 flocculant solution were added to a standard bentonite suspension (600 μg/ml in 2 m*M* $MgSO_4$), and the time (*t*, minutes) required for a 50% decrease in initial turbidity was recorded.

the solution is mixed. Then 0.25 ml of the Alcian blue solution is added, and the solution mixed again. After standing overnight the solution is mixed, centrifuged (24,000 *g*, 20 min), and the optical density of the supernatant read at 610 nm. The same is done with a control containing BG-11 medium instead of the sample, with 0.5 *M* acetic acid serving as a blank. The difference between the control and the sample is proportional to the bioflocculant concentration. Figure 2 shows that a linear relationship exists between flocculant concentrations and the difference in absorbance in the range of 0–100 μg/ml.

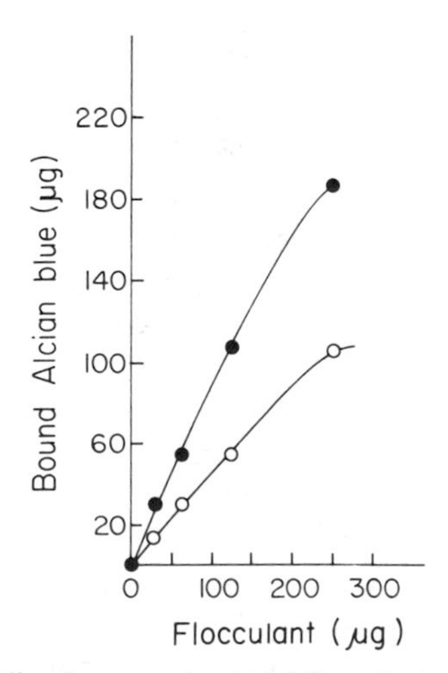

FIG. 2. Alcian blue binding by cyanobacterial flocculants. The experiments were conducted as described in the text, using preparations of extracellular flocculants from (●) *Phormidium* J-1 and (○) *A. circularis*.

TABLE II
PRODUCTION OF FLOCCULANT AND COFLOCCULATION OF CYANOBACTERIA WITH BENTONITE

Cyanobacterium	Flocculant production	Coflocculation
Phormidium J-1	+	+
Anabaenopsis circularis	+	+
Calothrix desertica	+/−	−
Anacystis nidulans	−	−
Plectonema boryanum	−	−

This method is much less sensitive to variations in temperature and salt concentration than the bentonite sedimentation test, and is more reproducible. Also, it can be used to compare the charge density of different bioflocculant preparations by measuring the amount of dye removed by constant flocculant concentrations. The disadvantages of the method include the following: (1) a minimum incubation period of 6 hr is an absolute requirement before readings can be taken; (2) any other polyanions present in the sample, not having flocculating activity, might interfere with the test; (3) flocculation activity is not measured directly. The Alcian blue binding test is therefore a suitable method either for purified flocculant preparations or as a semiquantitative method for crude culture supernatants.

Coflocculation of Cyanobacterial Cells with Clay Particles

Several cyanobacterial species are capable of binding clay particles onto the cell surface and of coflocculating with them.[13] Correlation was found between production of flocculant and the capacity to coflocculate with clay particles (Table II). Our results showed that while young (7-day) cells of *Phormidium* J-1 coflocculated with bentonite particles, older (14-day) cells were not active in this respect.

Flocculants are not the only known exopolymers produced by benthic cyanobacteria. We have found *Phormidium* J-1 and *A. circularis* to produce amphiphilic substances, termed by us emulcyans,[6] which coat the cell surface and make it progressively more hydrophilic, thus allowing detachment and dispersal of adherent cells. Coflocculation of young cells with clay is apparently due to surface-bound flocculants, which in older cells are probably masked by emulcyan. Attachment to, and detachment

[13] Y. Bar-Or and M. Shilo, *FEMS Microbiol. Ecol.* **53,** 169 (1988).

from, the benthos may therefore be a complex process involving not only shifts in cell surface hydrophobicity but in surface-bound flocculants as well.

Conclusions

The discovery of two biochemically different cyanobacterial flocculants indicates the evolution of diverging pathways for production of functionally similar exopolymers. A still larger variation may be expected in other cyanobacteria, and this could be important for potential utilization in a broad range of industrial applications, such as wastewater treatment, clarification of solar ponds, and sedimentations of colloids in the chemical industry.

[69] Cyanobacterial Heat-Shock Proteins and Stress Responses

By GEORGE BORBELY and GYULA SURANYI

Introduction

Growing cyanobacteria are strongly influenced by their nutritional, chemical, and physical environments. Of those factors, temperature plays a critical role: cyanobacteria exposed to temperatures higher than those for normal growth respond by altering patterns of growth and protein synthesis. An exposure of cyanobacterial cells to elevated temperature induces the synthesis of a new set of highly conserved proteins, the heat-shock proteins or HSPs.[1] The phenomenon of heat-shock response has been observed in a variety of organisms—prokaryotic and eukaryotic—and is believed to be universal.[2-5] Some of the proteins are induced by various other stresses, including starvation and exposure to toxic chemicals.

One of the most emphasized aspects of light-dependent cyanobacterial gene expression is its pleiotropic nature; hence, a specific manipulation of

[1] G. Borbely, G. Suranyi, A. Korcz, and Z. Palfi, *J. Bacteriol.* **161,** 1125 (1985).
[2] M. J. Schlesinger, M. Ashburner, and A. Tissières (eds.), "Heat Shock from Bacteria to Man." Cold Spring Harbor Lab., Cold Spring Harbor, New York, 1982.
[3] F. C. Neidhardt, R. A. Van Bogelen, and V. Vaughn, *Annu. Rev. Genet.* **18,** 295 (1984).
[4] S. Lindquist, *Annu. Rev. Biochem.* **55,** 1151 (1986).
[5] M. J. Schlesinger, *J. Cell Biol.* **103,** 321 (1986).

cyanobacterial genes or regulons could simplify attempts to understand the regulation of gene expression in these organisms.[6,7] Accordingly, studies on heat shock and other stress responses are of interest not only in their own right but from the viewpoint of light-dependent and regulated cyanobacterial gene expression as well.

This chapter concentrates primarily on techniques involved in studies of the heat-shock phenomenon in cyanobacteria. For convenience of discussion only the heat shock of *Synechococcus* Strain PCC 6301 is described, but these methods have direct applicability in studies directed toward the stress responses of other cyanobacterial strains as well. The techniques are simple and easy to execute. The specific pattern of protein synthesis can be monitored readily by use of two-dimensional polyacrylamide gel electrophoresis. Although the procedures described below were used for *Synechococcus* (a unicellular cyanobacterium) stress responses, these techniques work equally well for other strains, including filamentous species (*Anabaena* PCC 7120, *Nostoc* MAC, etc.).[8]

Growth

Synechococcus PCC 6301 (*Anacystis nidulans* ATCC 27144) is grown in the liquid medium of Allen.[9] Jacketed culture vessels (20 ml) are thermostatically maintained at 39° by circulating water; illumination is with warm-white fluorescent light (360 W m^{-2}). Agitation and carbon dioxide are provided by continuous bubbling of sterile 5% (v/v) CO_2 in air through cultures. Cell growth can be monitored either by spectrophotometry at 800 nm or by measurement of the chlorophyll *a* content, using 90% acetone extracts,[10] or both. Since the heat-shock or other stress treatment can be harmful for the cells, regular assessment of cell viability is recommended by determination of colony forming units per milliliter. Samples (0.1 ml) are serially diluted aseptically into sterile Allen medium at room temperature and plated on this medium solidified with agar (1.5%, w/v); incubation is under light for 1–1.5 weeks at 28°. In addition, the oxygen-evolving capacity of cells can be measured by a Clark-type oxygen electrode.

There are several reasons to grow cells under conditions of near maximal growth rate. The major one is that slight changes in the pattern of protein synthesis and growth are easier to recognize. It is especially im-

[6] W. F. Doolittle, *Adv. Microb. Physiol.* **20,** 1 (1979).
[7] G. Suranyi, A. Korcz, Z. Palfi, and G. Borbely, *J. Bacteriol.* **169,** 632 (1987).
[8] E. R. Leadbetter, G. Suranyi, and G. Borbely, unpublished observations.
[9] M. M. Allen, *J. Phycol.* **4,** 1 (1968).
[10] T. Kallas and R. W. Castenholtz, *J. Bacteriol.* **149,** 229 (1982).

portant to provide an adequate supply of carbon dioxide, since in cultures kept on shakers, and heat-shocked, a carbon starvation may easily occur at higher temperature. In general, cells are heat-shocked during exponential growth.

Heat-Shock Conditions

Expontentially growing cultures (0.4–0.7 A_{800} units) are heat-shocked in 10- to 20-ml glass vessels by transfer to elevated temperature in a thermostat-controlled circulating-type water bath. If for the experimental purposes a larger amount of cells is needed, sterilized glass culture vessels (e.g., 2-liter Roux bottles equipped with an inlet/outlet for gassing and with an intake for liquid) are preincubated at the higher temperature in a transparent, aquariumlike thermostatted bath. Cultures of appropriate cell densities are transferred from normal growth conditions (39°) to the prewarmed vessels through a 1.5-m-long glass coil, prewarmed and incubated at the higher temperature, using a high capacity peristaltic pump and sterilized transparent Tygon tubing. Under the described conditions, the culture attains the required elevated temperature level within 60–90 sec, as measured by an electronic digital thermometer (LED Thermometer, Fisher), equipped with semisolid probe.

Characterization of *Synechococcus* sp. Heat-Shock Response

A first step in the description of the heat-shock response of cyanobacteria is the analysis of RNA and protein synthesis. Since the mere accumulation of radioactive precursors under these conditions may not be a true measure of the macromolecular syntheses (RNA and protein) a pulse-labeling type of analysis is recommended.

Solutions and Materials

Trichloroacetic acid (TCA): 12.5 and 5% (w/v)
Sodium hydroxide: 0.2 and 1.0 *M*
Bleaching solution: 10% (w/v) TCA solution containing 1.5% hydrogen peroxide (prepared freshly before use)
Ethanol
Acetone
Glycerol, 75% (v/v)
Radioactive precursors: [2-^{14}C]uracil (50 mCi/mmol); [5,6-^{3}H]uracil (20–30 Ci/mmol); uniformly ^{14}C-labeled protein hydrolyzate (50 mCi/mg atom carbon); sodium [^{14}C]bicarbonate (50 mCi/mmol); L-[^{35}S]methionine (800 Ci/mmol)
Whatman filter disks, No. 540 or GF/C

Effect of Elevated Temperature on RNA Synthesis of Synechococcus sp.

To measure the rate of total RNA synthesis, *Synechococcus* cultures are labeled for several generations (usually overnight) with [^{14}C]uracil (0.1 μCi/ml). The rate of total RNA synthesis is determined by pulsing 0.5-ml samples throughout the desired time period, using an overlapping time regimen, with [^{3}H]uracil (30 μCi/ml) for 10 min in prewarmed thin-walled glass test tubes, capped with metal caps, and positioned to receive identical illumination. Labeling is stopped by withdrawing a 0.2-ml sample of each culture, adding this to an equal volume of ice-cold solution containing 0.4 mg bovine serum albumin and immediately mixing with 0.8 ml of cold 12.5% TCA solution. After 30 min on ice the precipitates are collected by centrifugation in an Eppendorf-type centrifuge and washed twice with cold 5% TCA solution and ethanol. The pellet, dried briefly in a vacuum desiccator, is then hydrolyzed in 0.5 ml 0.2 *M* sodium hydroxide overnight at room temperature. After this treatment the samples are clarified by centrifugation, and aliquots of the supernatant are taken to measure their radioactivity by standard scintillation counting in water-miscible counting solution.

The rate of total RNA synthesis is given as normalized values of [^{3}H]uracil incorporated into alkali-soluble material per 10 min of radioactive pulses. The ratio of incorporation of [^{3}H]uracil to that of [^{14}C]uracil is defined as a unit at the beginning of the experiment, depending on experimental purposes, and the rate of RNA synthesis is calculated. After an initial oscillation the rate of total RNA synthesis reaches a new but lower steady-state level under heat-shock conditions. These alterations in the rate of total RNA synthesis are in inverse correlation with the dramatic changes in guanosine tetraphosphate (a key signal molecule in prokaryotes) pool size under light and (even more apparent) in the dark.[7]

Effect of Elevated Temperature on Protein Synthesis

For the analysis of alterations in rates of total protein synthesis during heat shock, 0.5-ml samples of *Synechococcus* cultures are pulse-labeled for 10 min with L-[^{35}S]methionine (100 μCi/ml) during the heat shock using an overlapping time regimen. Then 0.2-ml aliquots of treated or control cultures are pipetted into an equal volume of a solution containing 0.4 mg bovine serum albumin and immediately mixed with 0.8 ml of 12.5% TCA solution. The samples are kept on ice for 30 min and then treated for 10 min in a boiling water bath. The precipitates are collected by centrifugation and washed twice with 5% TCA solution and ethanol. The pellet is briefly dried in a vacuum desiccator and dissolved in 0.2 *M* sodium hydroxide solution. The radioactive precursor incorporation is determined by standard scintillation counting. The rate of protein synthesis in heat-shocked (47°) *Synechococcus* cells decreases immediately after heat treatment and reaches a new but lower rate as before.[1]

Effect of Heat Shock on the Protein Synthesis Pattern in Synechococcus Strain PCC 6301

For the analysis of changes in the pattern of protein synthesis of heat-shocked *Synechococcus,* cells are cultured in jacketed vessels and shifted to elevated temperature (47°) as described above. One-milliliter samples are removed from the culture at appropriate intervals, pulse-labeled (in capped, thin-walled glass test tubes positioned in a water bath under identical conditions) with 10 μCi of uniformly ^{14}C-labeled protein hydrolyzate for 0.5–1.0 hr (a longer time period is recommended for the pulse under conditions of energy deprivation). The radioactive samples are collected by centrifugation and processed for one- or two-dimensional polyacrylamide gel electrophoresis.[11,12] For two-dimensional analysis the samples are resuspended in O'Farrell's[12] lysis buffer (50–100 μl); otherwise, the sample buffer described by Laemmli[11] is used. For one-dimensional separation a 10–18% linear sodium dodecyl sulfate (SDS)–polyacrylamide gel supported with a glycerol gradient (from 1 to 10%) is used. Usually, proteins from 1 ml of labeled culture are loaded onto the gels, or samples with equal radioactivity are processed for SDS–polyacrylamide electrophoresis, depending on the experimental needs.

[Note: Since the high pigment content of the cyanobacterial samples may cause quenching in scintillation counting, radioactivity of the samples is assayed, using filter paper disks (Whatman No. 540 or GF/C, 24 mm), after bleaching. Aliquots (5 μl) of cell lysates are spotted on filter paper disks, and, after they are dry, the disks are treated with cold 10% TCA solution (10–15 ml/disk) for 10 min in a beaker. Then the filter disks are bleached by boiling for 3–5 min in 10% TCA solution containing 1.5% hydrogen peroxide. The bleached disks are washed twice with 5% TCA solution, then twice with ethanol and acetone. The filters are dried under a heat lamp prior to counting.]

After electrophoresis, the gels are processed for staining and/or fluorography.[13] The gels, after drying on a slab dryer (Bio-Rad) are exposed to preflushed X-ray film at −70°. The relative rate of HSP synthesis may be determined from the developed films by the method of Suissa,[14] on the

[11] U. K. Laemmli, *Nature (London)* **227,** 680 (1970).
[12] P. H. O'Farrell, *J. Biol. Chem.* **250,** 4007 (1975).
[13] R. A. Laskey and A. D. Mills, *Eur. J. Biochem.* **56,** 335 (1975).
[14] M. Suissa, *Anal. Biochem.* **133,** 511 (1983).

basis of spectrophotometric quantitation of silver grains isolated from bands. Briefly, the desired band and an appropriate blank area (same size from the area not exposed to radioactivity) are excised from the films and further cut into smaller pieces. The pieces of film are submerged in 0.5 ml of 1.0 *M* sodium hydroxide solution and incubated at room temperature until all silver grains are eluted; occasional vortex mixing is helpful. Then an equal volume of 75% glycerol is added, vortexed carefully to avoid trapping air bubbles, and the absorbance (500 nm) determined in a spectrophotometer.

To establish the optimal conditions for the expression of heat-shock regulons in cyanobacterial strains, SDS–polyacrylamide gel electrophoresis seems to be a valuable tool. Sodium [^{14}C]bicarbonate (15 μCi/ml) or [^{35}S]methionine (15 μCi/ml) may be used to label the stress proteins. The protein pattern using sodium [^{14}C]bicarbonate is identical to that obtained by using radioactive protein hydrolyzate. In extracts labeled with [^{35}S]methionine, however, several bands that are identified with ^{14}C-labeled precursors are missing, thus indicating proteins with either no or a low number of sulfur-containing amino acids. Therefore, care should be taken when [^{35}S]methionine is used as label.[1]

The main feature of the cyanobacterial heat-shock response is the rapid but transient activation of a small number of specific genes previously either silent or active at a very low level (Fig. 1). In that sense it seems to share many properties in common with the heat-shock response of other prokaryotic systems, both eubacteria and archaebacteria, and eukaryotic cells of fungi, plants, and animals which have been studied thus far.[1–5,7,15]

In *Synechococcus* cultures simultaneously light deprived (dark) and heat shocked (47°) there is a synthesis of (1) the main heat-shock proteins, (2) several transiently synthesized dark-specific proteins, and (3) polypeptides synthesized in the cells under light, dark, and heat shock in dark conditions. Under normal growth conditions, however, several of the transiently synthesized dark-specific proteins are abolished by an elevated temperature (47°) treatment in the dark and the main heat-shock proteins are synthesized even in the dark. This phenomenon might be of aid in study of light-dependent cyanobacterial gene expression.[7]

In *Synechococcus* sp., the synthesis of HSPs is accompanied by accumulation of adenylated nucleotides ("alarmons"). In heat-shocked cells, under light conditions, there is an increase in the pool sizes of diadenosine tetraphosphate, adenosine–guanosine tetraphosphate, diadenosine triphosphate, and adenosine–guanosine pentaphosphate (at 30–60 min after

[15] R. Mannar Mannan, M. Krishnan, and A. Gnanam, *Plant Cell Physiol.* **27,** 377 (1986).

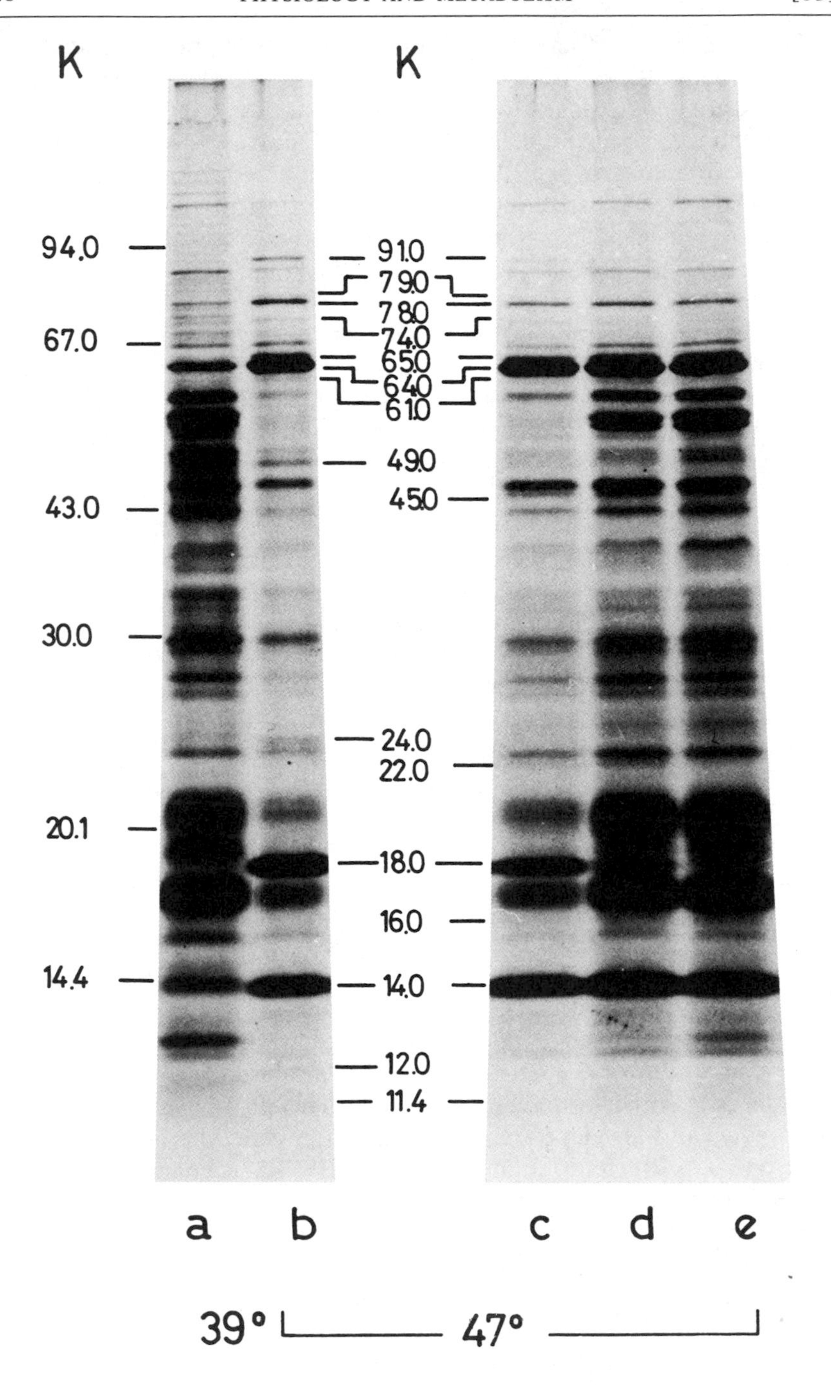
K
K
94.0
67.0
43.0
30.0
20.1
14.4
91.0
79.0
78.0
74.0
65.0
64.0
61.0
49.0
45.0
24.0
22.0
18.0
16.0
14.0
12.0
11.4
a
b
c
d
e
39°
47°

FIG. 1. Effect of shift to elevated temperature on polypeptide synthesis in *Synechococcus* PCC 6301 cells. An exponential-phase culture was shifted from normal (39°) to elevated temperature (47°). Portions (1 ml) were removed before (zero time, a) or after temperature shift and were pulse-labeled (30 min) with ^{14}C-labeled protein hydrolyzate (10 μCi) at different times during a 2-hr experimental period (b, 0–30 min; c, 30–60 min; d, 60–90 min; e, 90–120 min). The samples (equal volumes, 1 ml) were analyzed for labeled proteins on a 10–18% linear gradient SDS–polyacrylamide gel and processed for fluorography. Columns K mark positions of molecular weight markers ($M_r \times 10^{-3}$), as well as of polypeptides affected by temperature shift (HSPs). Reprinted, with permission, from Borbely *et al.*[7]

the stress). In darkened and heat-shocked cells expansion of the diadenosine triphosphate pool is more dramatic than the increases in pool size of the other adenylated nucleotides. Toxic heavy metal ions like Cd^{2+}, Zn^{2+}, Cu^{2+}, and Hg^{2+} provoke the accumulation of adenylated nucleotides as well.[16]

[16] Z. Palfi, G. Suranyi, and G. Borbely, unpublished observations.

[70] Immobilization Methods for Cyanobacteria in Solid Matrices

By M. BROUERS, D. J. SHI, and D. O. HALL

Introduction

There is a serious interest in light energy conversion technologies aimed toward the development of "product-oriented" photobiological systems for the production of fuels (H_2 and H_2O_2), fertilizers (NH_3), and other specific biomolecules (ATP, $NADPH_2$, carbohydrates, pharmaceuticals, etc.). Because of their wide distribution and metabolic adaptability, cyanobacteria are most suitable for the production of NH_3 and H_2.[1] They can grow in sunlight with water, air, and salts as the major substrates and can produce H_2 and reduce atmospheric N_2 to NH_3 via hydrogenase and ATP-dependent nitrogenase activity.

The stability of these photosynthetic systems was found to improve on entrapment (immobilization) of the cells in solid matrices. Additional advantages of using immobilized cells are ease of separation of products from the catalyst, reuse of the catalyst, and ease of manipulation in bioreactors. Entrapment in porous gels and (increasingly) polymer foams

[1] D. O. Hall, D. A. Affolter, M. Brouers, D. J. Shi, L. W. Yang, and K. K. Rao, *Ann. Proc. Phytochem. Soc. Eur.* **26**, 161 (1985).

is a popular method for immobilization of photosynthetic materials since these supports are relatively inert (especially the foams), stable during autoclaving, and are reasonably transparent to light.

A general review of entrapment techniques for chloroplasts, cyanobacteria, and hydrogenases was published recently.[2] This chapter deals with selected methods for both entrapment of filamentous heterocystous cyanobacteria and characterization of some biological activities of the immobilized cells.

Strains of Cyanobacteria and Growth Conditions

Various strains of heterocystous cyanobacteria were used. Strains of *Nostoc muscorum* and *Mastigocladus laminosus* were obtained from the Cambridge Culture Collection, Storey's Way, Cambridge (now located at Culture Centre of Algae and Protozoa, Fresh Water Biological Assn., Ambleside, Cumbria 4A22 OLP) and *Phormidium laminosum* (Strain OH-1-P) from Dr. R. W. Castenholz, University of Oregon. *Mastigocladus laminosus* and *N. muscorum* were grown at 28° in Allen and Arnon and Kratz and Myers media, respectively,[3] under cool-white fluorescent light at a light intensity of 60 μmol photons m^{-2} sec^{-1}. *Phormidium laminosum* was grown at 45° in medium D of Castenholz.[4] *Anabaena azollae,* a presumptive isolate from *Azolla filiculoides* capable of growth in independent culture, obtained from Dr. E. Tel-Or of The Hebrew University, Rehovot, Israel,[5] was grown in BG-11 medium[6] without nitrate at 28° under cool-white fluorescent light at a light intensity of 75 μmol photons m^{-2} sec^{-1}. All cells were grown in a 5% CO_2–air mixture in 250-ml Erlenmeyer flasks kept agitated on a rotary shaker at 125–140 rpm. They were used for immobilization at the end of the exponential growth phase.

Immobilization of Cyanobacteria

Methods for Entrapment of Cyanobacteria in Alginate Gels

Alginate is a polysaccharide isolated from marine brown algae (e.g., *Laminaria hyperborea*) where its function is skeletal. Chemically it is a

[2] P. E. Gisby, K. K. Rao, and D. O. Hall, this series, Vol. 135, p. 440.

[3] G. D. Smith, A. Muallem, and D. O. Hall, *Photobiochem. Photobiophys.* **4,** 307 (1982).

[4] R. W. Castenholz, *Schweiz. Z. Hydrol.* **32,** 538 (1970).

[5] E. Tel-Or, O. T. Sandovsky, D. Kobiler, C. Arad, and R. Weinberg, *in* "Photosynthetic Prokaryotes—Cell Differentiation and Function" (G. Papageorgiou and L. Packer, eds.), p. 303. Elsevier, New York, 1983.

[6] R. Y. Stanier, R. Kunisawa, M. Mandel, and G. Cohen-Bazire, *Bacteriol Rev.* **35,** 171 (1971).

linear copolymer of two uronic acids, D-mannuronic (M) and L-guluronic acid (G), linked together by β-1,4 and α-1,4 glycosidic bonds, respectively. The two monomers are arranged in homopolymeric blocks, M-blocks and G-blocks, and in copolymeric regularly alternating MG-blocks. Alginate forms strong gels with divalent cations like Ca^{2+}; the content of G-blocks is the main structural feature contributing to gel strength and stability of the gel. Alginate has been used in the immobilization of many cell types, including cyanobacteria.[7-9]

Immobilization of cyanobacteria in sterile alginate beads is carried out as follows: 5.5 g sodium alginate (Protonal 10/60, supplied by Protan A/S, Drammen 3001, Norway) is dissolved in 150 ml growth medium without phosphate at 80°. The solution is autoclaved and after cooling is mixed with 50 ml of a sterile concentrated suspension of cyanobacteria (obtained by centrifugation at 2,500 *g* for 5 min of 250 ml of culture and resuspension in 50 ml of growth medium without phosphate). The sodium alginate–cyanobacterial mixture is then added dropwise from a separating funnel to a 0.1 *M* $CaCl_2$ solution at room temperature. The alginate beads formed by cross-linking via calcium ions are kept for 2 hr in the Ca^{2+} solution with gentle mixing for hardening. They are then harvested, washed in growth medium, and resuspended in fresh growth medium. The growth and wash media are free of phosphate to prevent the formation of calcium phosphate which would result in disruption of the beads.

Methods for Entrapment of Cyanobacteria in Preformed Polyvinyl and Polyurethane Foams

Immobilization of cyanobacterial cells is carried out in polyurethane foam (Codes 3300A, 4300A, and 74165A) and in hydrophilic polyvinyl foam (codes D, Jan. 84, and PR22/60) supplied by Caligen Foam Ltd. (Accrington BB5 2BS, UK). The foams are cut into 5-mm cubes and washed 5 times in distilled water for a few days. One hundred pieces of foam are added to each 250-ml flask containing 140 ml growth medium. The flask and contents are then autoclaved at 120°. After cooling the liquid medium is discarded and replaced by 140 ml sterile fresh growth medium before inoculation with the cyanobacteria. After a few days, cell growth is observed in the internal pores of the foam where the cells adhered (Fig. 1). When the culture is kept for a prolonged time the growth media are renewed aseptically every 2 weeks.

[7] M. Kierstan and C. Burke, *Biotechnol. Bioeng.* **19,** 387 (1977).

[8] P. E. Gisby and D. O. Hall, *Nature (London)* **287,** 251 (1980).

[9] S. C. Musgrave, N. W. Kerby, G. A. Codd, and W. D. P. Stewart, *Biotechnol. Lett.* **4,** 647 (1982).

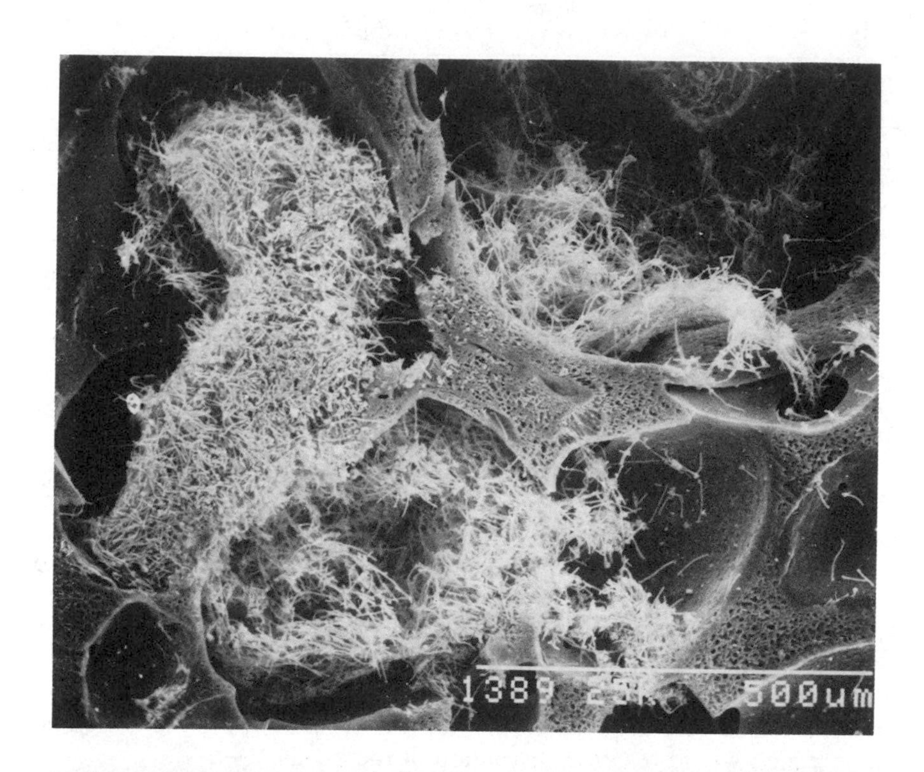
1389
500um

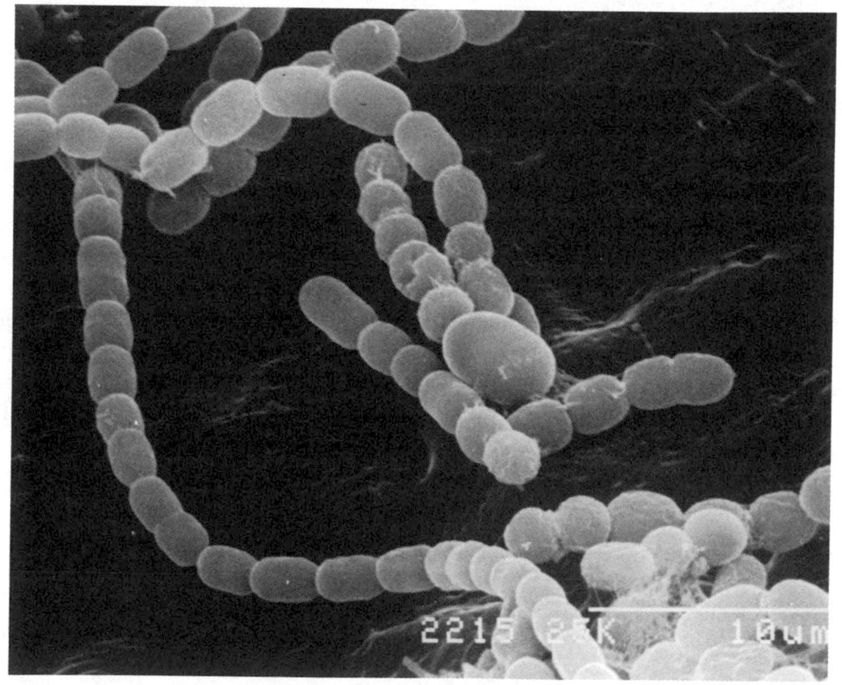
10um

Methods for Entrapment of Cyanobacteria in a Urethane Prepolymer

Polyurethane foam can be obtained by hydrolytic condensation of a liquid viscous prepolymer formed by the reaction between an isocyanate (TDI: toluene 2,4-diisocyanate; MDI: diphenylmethane diisocyanate) and a hydroxyl group (polyether or polyester diol). The foam structure is due to production of CO_2 during hydrolysis according to the reaction

$$2\ R{-}N{=}C{=}O + H_2O \rightarrow R{-}NH{-}CO{-}NH{-}R + CO_2$$

Direct immobilization of the biocatalyst can be obtained by mixing an aqueous suspension of cells with the prepolymer. In order to avoid toxicity effects during the polymerization process, the temperature must be controlled and prepolymers with low TDI content must be used. Urethane prepolymers have been used for the entrapment of bacteria, enzymes, chloroplasts, and algae.[10–14]

We have routinely used the commercially available TDI-based hydrophilic urethane prepolymer Hypol FHP 2002 with low free TDI content (W. R. Grace Ltd., Northdale House, London NW10 7UH, UK). A typical procedure is as follows: 100 ml cyanobacterial suspension is centrifuged at 2,500 *g* for 5 min, and the pellet is resuspended in 15 ml sterile growth medium. For immobilization 12 ml of cell suspension is added to 10 g Hypol prepolymer FHP 2002. The mixture is maintained in an ice bath and mixed with a spatula for 1 min; polymerization is then continued at room temperature without mixing. The reaction requires about 15 min, after which a spongy blue–green mass forms in the vial. After polymerization the foam is cut into small pieces and washed several times in fresh growth medium in order to remove toxic by-products of the polymerization reaction. Foam pieces are then suspended in fresh nutrient medium for continued growth or direct use in bioreactors. It is advisable to renew the growth medium every day during the first 10 days following immobilization to ensure washout of any toxic polymerization products.

[10] S. Fukui, K. Sonomoto, N. Itoh, and A. Tanaka, *Biochimie* **62,** 381 (1980).

[11] M. F. Coquempot, B. Thomasset, J. N. Barbotin, G. Gellf, and D. Thomas, *Eur. J. Appl. Microbiol. Biotechnol.* **11,** 193 (1981).

[12] M. Brouers, F. Collard, J. Jeanfils, and R. Loudeche, "Photochemical, Photoelectrochemical and Photobiological Processes" (D. O. Hall, W. Palz and D. Pirrwitz, eds.), Vol. 2, p. 171. Reidel, Dordrecht, The Netherlands, 1983.

[13] C. Baillez, Ph.D. thesis. Universite Pierre et Marie Curie, Paris, 1984.

[14] C. Thepenier and C. Gudin, *Biomass* **7,** 225 (1985).

FIG. 1. (top) *Anabaena azollae* growing in polyvinyl foam (PR22/60). (bottom) *Anabaena azollae* filament on a polyvinyl foam surface. Mucilage film normally covers the cells and foam surface but is only seen with low-temperature scanning electron microscopy.

Characterization of Biological Activities and State of Immobilized Cyanobacteria

The chief criterion we have established for immobilization techniques in our laboratory is maximum retention of biological activity. Retained activity can be expressed in terms of the photosynthetic electron transport rate in the presence of an artificial electron acceptor, as the rate of product formation (e.g., NH_3, H_2, etc.), or as retention and stabilization of enzyme activity (e.g., nitrogenase). The state of pigment–protein complexes can be analyzed by 77 K fluorescence spectroscopy. The state of the immobilized cells can be observed by light and electron microscopy.

PSI (photosystem I) and PSII (photosystem II) activities are measured using a Clark-type oxygen electrode at 25° under saturating light from a quartz iodine projector lamp with an orange filter. In the electrode chamber a metal grid is positioned to keep the foam pieces above the stirring bar, thus ensuring even liquid circulation and O_2 evolution. PSII activity is measured as DCPIP (2,6-dichlorophenolindophenol) light-mediated O_2 evolution in the presence of NH_4Cl as an uncoupler. The 4-ml reaction mixture (growth medium with small pieces of polymer with immobilized cells) contains approximately 100 μg Chl *a*, 20 μl 1 *M* NH_4Cl, and 20 μl 0.1 *M* DCPIP. PSI activity is then measured as O_2 consumption by photoreduction of methyl viologen as the electron acceptor (10 μl, 10 m*M*) in the presence of 3,4-dichlorophenyl-1,1-dimethylurea (DCMU) as the PSII inhibitor (10 μl, 10 m*M*), 2,3,5,6-tetramethylphenylenediamine (TMPD) as the electron mediator (10 μl, 60 m*M*) and NaN_3 as the inhibitor of catalase activity (20 μl, 0.8 *M*).

Fluorescence spectra (77 K) are obtained as described.[15] Excitation wavelengths are 436 nm (excitation of chlorophyll *a*) or 578 nm (excitation of phycobilins). Chlorophyll estimation is made at 665 nm in methanol; however, complete extraction requires repeated homogenization.

Nitrogen fixation activity is determined as acetylene reduction. Ethylene formation is followed with a gas chromatograph equipped with a flame-ionization detector and a Poropak S column operated at 45° with N_2 as the carrier gas. Sealed vials with cells in growth medium are incubated with a gas phase of 15% C_2H_2 in argon; they are shaken continuously at 25° in white incandescent light at intensity of 250 μmol photons m^{-2} sec^{-1}.

Hydrogen production by nitrogenase is measured with a gas chromatograph fitted a Poropak Q 80–100 mesh column heated to 55° and a microkatharometer equipped with a thermal conductivity detector. The gas phase is 4% CO in argon.[16] Hydrogen production by hydrogenase is

[15] C. Sironval, M. Brouers, J. M. Michel, and Y. Kuiper, *Photosynthetica* **2,** 268 (1968).

[16] J. R. Benemann and N. M. Bucke, *Biotechnol. Bioeng.* **19,** 387 (1977).

measured in the presence of nitrogenase as H_2 formation from reduced methyl viologen (MV).[17] This experiment is done in 7-ml glass vials containing 2 ml of cells. After flushing with nitrogen, a MV solution containing 288 m*M* sodium dithionite is injected into the vial and flushing continued for a while. The reaction vials are incubated in the dark with shaking, and hydrogen is detected by gas chromatography as above.

Ammonia production in the vials is measured colorimetrically at various time intervals by the method of Solorzano.[18] Production by immobilized cells in a continuous-flow, packed-bed photobioreactor is measured with an ammonia electrode using the experimental device shown in Fig. 2. The reactor consists of a thermostatted column (internal diameter 2 cm, length 25 cm) with 50 ml liquid medium, packed with about 200 polymer foam pieces containing immobilized cyanobacteria (1–4 mg Chl *a*). Nutrient medium [with or without addition of the inhibitor of glutamine synthetase activity L-methionine DL-sulfoximine (MSX)] is added to the top of the column and collected at the bottom using peristaltic pumps with a dilution rate of 0.4 hr^{-1}. The outflow effluent is directed to a thermostatted (28°) mixing cell (5-ml volume) for continuous addition (10%, v/v) of a 1 *M* NaOH–0.1 *M* NaEDTA solution in order to liberate NH_3. The resulting solution is pumped to an ammonia electrode (Kent Industrial Measurements, Stonehouse, Gloucestershire GL10 3TA; Model 8002-8) fitted with a flow device. Continuous measurement of ammonia is thus monitored. Calibration is obtained by addition of standard NH_4^+ solutions. The bioreactor is operated at 28° under white fluorescent light at 100 μmol photons m^{-2} sec^{-1}.

Heterocysts are observed with a Standard 14 Microscope (Zeiss, FRG) under phase-contrast illumination and high-power magnification (×400). Material is freshly extracted with a homogenizer and then fixed in glutaraldehyde (2.5% final concentration). Heterocyst frequency is estimated using a hand tally counter and standard counting chamber. Triplicate counts of at least 1000 cells are done on each sample. Heterocyst frequency is expressed as a percentage of the total number of cells.

Perhaps the best way to observe the situation of immobilized cells *in situ* is low-temperature scanning electron microscopy.[19,20] The foams or beads are cut to expose the cells, rolled on a piece of tissue paper, and allowed to drain. Traces of remaining water at the surface are evaporated under a gentle stream of warm dichlorofluoromethane, during which time

[17] A. Daday, G. R. Lambert, and G. D. Smith, *Biochem. J.* **177,** 139 (1979).

[18] L. Solorzano, *Limnol. Oceanogr.* **14,** 799 (1969).

[19] M. J. C. Rhodes, R. J. Robins, R. J. Turner, and J. I. Smith, *Can. J. Bot.* **63,** 2357 (1985).

[20] R. J. Robins, D. O. Hall, D. J. Shi, R. J. Turner, and M. J. C. Rhodes, *FEMS Microbiol. Lett.* **34,** 155 (1986).

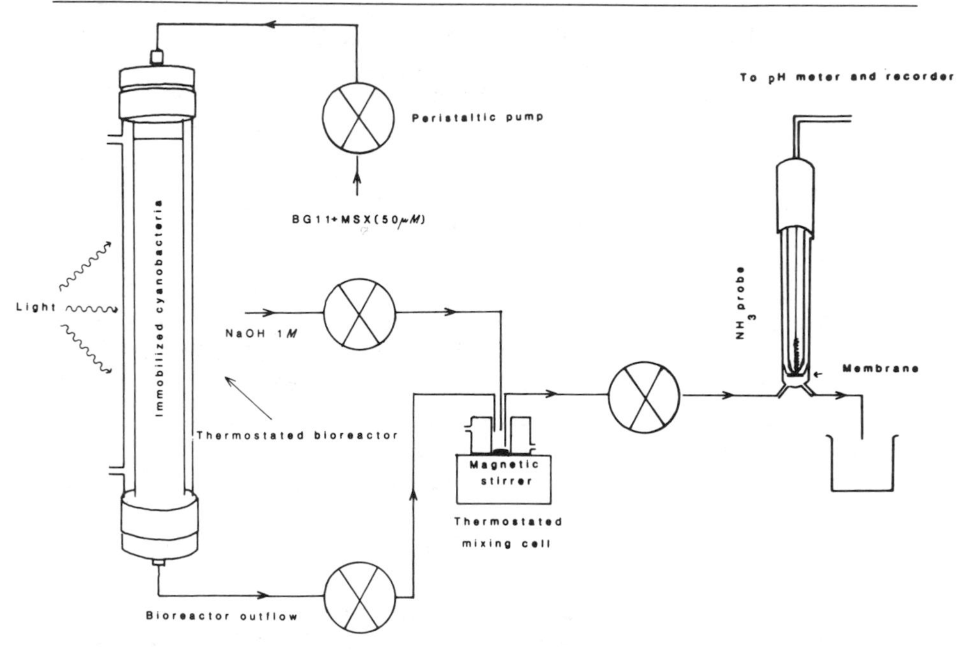

FIG. 2. Experimental device for continuous measurement of ammonia production (details in the text).

(15–30 sec) the specimen is constantly viewed under a dissecting microscope. The samples are frozen in slushed liquid N_2 under argon gas using an EM Scope Sputter Cryo SP 2000 apparatus.[19] Material prepared in this way is partially freeze-dried when water is sublimed from the surface of the specimen by heating at a power input to maintain a constant temperature of −80°. Finally, the specimens are fixed to specimen stubs and earthed with silver dag.[20] Except where it is intended to sublime water from the sample in the microscope, a gold coating of approximately 20 nm is applied using the EM Scope Sputter Cryo. All microscopy is performed using a Phillips PSEM 50 1B at an accelerating voltage of 20 kV for coated material and 15 kV for uncoated samples. Images are recorded on Ilford FP4 black and white film at 32 lines/sec.

Section V

General Physical Methods

[71] Microsensors

By BO BARKER JØRGENSEN and NIELS PETER REVSBECH

Introduction

Physiological studies of cyanobacteria have been conducted mostly on organisms cultured in homogeneous suspension. While this may be a satisfactory environment for planktonic cyanobacteria, many cyanobacteria in nature grow in aggregates or attached to sediment particles. The environment of attached cyanobacteria is characterized by steep gradients, both physical and chemical, which cannot be simulated in homogeneous liquid culture. The microbial metabolism within the stagnant microenvironments formed by the substratum and its attached microorganisms creates steep concentration gradients of the chemical species involved. These chemical gradients can be studied by the use of microelectrodes as described below, and metabolic rates can be calculated from these gradients.

Light also exhibits pronounced gradients in both intensity and spectral composition within cyanobacterial communities. We describe a fiber optic probe which can be used to monitor light intensity and spectral composition in microscale.

Microelectrodes

Four types of microelectrodes, which are useful for chemical analysis in dense populations of cyanobacteria, are shown in Fig. 1. Figures 1A and 1B show two different types of oxygen microelectrodes, Fig. 1C is a sulfide microelectrode, and Fig. 1D is a pH microelectrode.

Construction of Microelectrodes

In the following, construction of the Clark-type oxygen microelectrode illustrated in Fig. 1B is described in detail. Construction of the cathode-type oxygen microelectrode shown in Fig. 1A can easily be deduced from this description, as the techniques are similar to the ones used for the cathode of the Clark-type microelectrode. The application of a membrane on this type of electrode is described elsewhere.[1] Construction of the sulfide microelectrode is also described, whereas the construction

[1] N. P. Revsbech, *in* "Polarographic Oxygen Sensors: Aquatic and Physiological Applications" (E. Gnaiger and H. Forstner, eds.), p. 265. Springer-Verlag, Heidelberg, 1983.

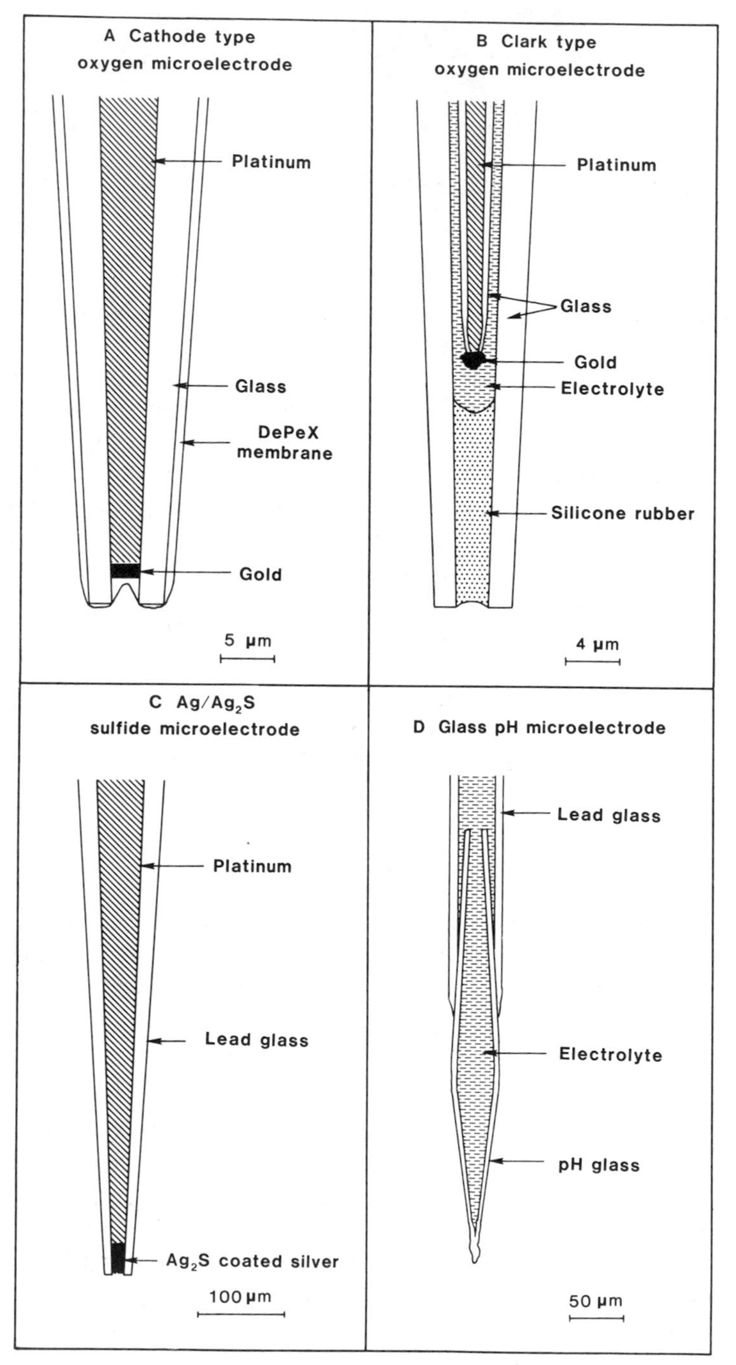

FIG. 1. Tips of microelectrodes useful for oxygen, sulfide, and pH analyses in dense microbial communities. From Revsbech and Jørgensen.[3]

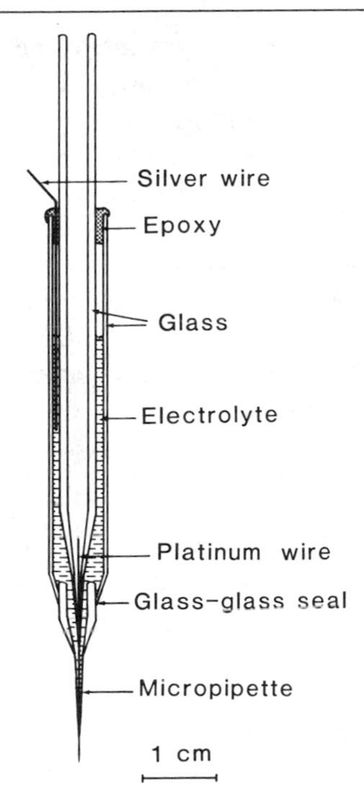

FIG. 2. Clark-type oxygen microelectrode. The glass–glass seal in the outer casing is not described, as the electrodes are now made without this complicating seal. From N. P. Revsbech and D. M. Ward, *Appl. Environ. Microbiol.* **45,** 755 (1983).

of pH microelectrodes is well described by other authors.[2,3] pH microelectrodes can be made with tips smaller than 1 μm.

It is usually advisable to make one's own microelectrodes, as they can then be constructed according to specific needs and as their lifetime is limited. Both oxygen and pH microelectrodes are, however, commercially available. A Clark-type oxygen microelectrode is available from Micro-sense, P.O. Box 361, Ramat Gan 52103, Israel; pH microelectrodes and a cathode-type oxygen microelectrode are available from Diamond Electro-Tech, P.O. Box 2387, Ann Arbor, MI 48106.

Construction of Clark-Type Oxygen Microelectrode. The Clark-type oxygen microelectrode is composed of a cathode immersed in a KCl electrolyte (Fig. 2). The electrolyte is contained in an outer glass casing

[2] R. C. Thomas, "Ion-Sensitive Intracellular Microelectrodes: How to Make and Use Them." Academic Press, London, 1978.

[3] N. P. Revsbech and B. B. Jørgensen, *Adv. Microb. Ecol.* **9,** 293 (1986).

equipped with a silicone rubber membrane at the tip. Oxygen can diffuse through the silicone rubber and reach the gold tip of the cathode, where it is reduced and produces a current in the measuring circuit (see below).

The cathode is made of platinum (99.99% purity, annealed, slightly hard-drawn condition, 0.1-mm thermocouple wire, Engelhard Industries, Surrey, UK). The platinum wire is cut into 4-cm pieces, straightened by hand, and etched in saturated KCN solution while 2–7 V AC is applied. A graphite rod (e.g., from a pencil or a conventional 1.5-V battery) constitutes the other electrode immersed in the KCN thus completing the circuit. One centimeter of the wire is initially immersed in the solution while 7 V AC is applied. A rapid movement of the distant 1 cm in and out of the solution during the initial 3 sec will result in a gentle taper of the final tip which facilitates the subsequent manufacturing steps. When the wire is etched so thin that the distant approximately 2 mm has disappeared, the voltage is reduced to 2 V, and the etching is continued for an additional 3 sec. The etching should result in a smooth tip, typically being 3 μm wide 0.1 mm from the tip, 6 μm wide 0.5 mm from the tip, and 20 μm wide 3 mm from the tip. The electrodes will work better the thinner they are, but extremely thin platinum wires are difficult to coat with glass. Some batches of platinum wire produce a very irregular tip when etched and thus cannot be used. The etched platinum wires are washed in water, in concentrated sulfuric acid, in water again, and in ethanol. The tips are very delicate, so it is advisable to hold the thick end of the platinum wire with forceps while immersing the etched, thin part in liquids.

After washing, the platinum wires are dropped, tip downward, into prewashed and predrawn soda lime glass capillaries. Not all soda lime glasses have sufficient insulating properties to be used (GW Glas, Glaswerk Wertheim, D-6980 Wertheim, FRG, and melting glass 8510, Jenaer Glaswerk Schott & Gen., D-6500 Mainz, FRG, both have the desired properties[4]). The capillaries should be thin-walled. A ratio of inner diameter to outer diameter of 0.8 or more is preferable. The platinum wires are coated with glass as shown in Fig. 3. The glass layer is often so thin that it is invisible even when the electrode is examined under the microscope at 400× magnification. Diffuse light often facilitates observation of the glass coating. Too hot a heating loop may result in bubbles in the glass coating rendering the cathodes useless, whereas too cold a heating loop often leads to broken platinum wires. The wires are completely coated with glass if the coating was successful, and the glass at the

[4] H. Baumgärtl and D. W. Lübbers, *in* "Polarographic Oxygen Sensors: Aquatic and Physiological Applications" (E. Gnaiger and H. Forstner, eds.), p. 37. Springer-Verlag, Heidelberg, 1983.

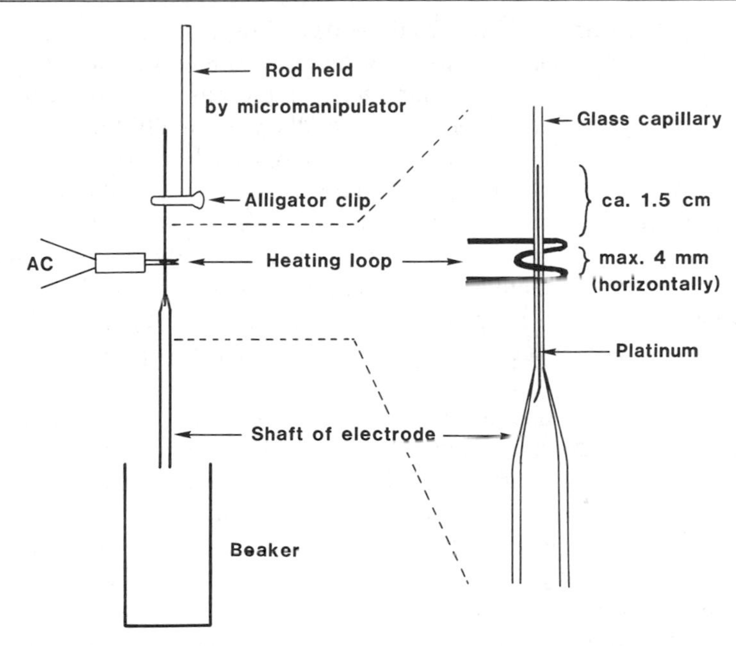

FIG. 3. Construction of the cathode of a Clark-type oxygen microelectrode. The procedure shown is the coating of the platinum core with glass. From Revsbech and Jørgensen.[3]

tip must therefore be removed. We grind the tips on a rotating brass wheel covered by a *thin* layer of diamond paste with a grain size of 0.25 μm. Grinding paste can be removed from the ground tip by washing in ethanol.

The area of exposed platinum (i.e., the actual cathode) should be very small after the grinding, but it may be increased by gold plating in 5% $KAu(CN)_2$ while 4.5 V DC (cathode negative) is applied. A large resistor (about 1 MΩ) should be inserted into the circuit to prevent deposition of excessive amounts of gold. The plating may be performed under the microscope at 100× magnification while the gold-plating solution is contained in a capillary. The plating should be continued until a small coating of gold can be seen at 100× magnification. The tip of the electrode should be allowed to soak in water for several hours after the gold plating. Long-term storage of the finished cathodes is best done in desiccant-containing test tubes with the electrode shafts fastened in rubber stoppers.

The outer casing of the microelectrodes is also made from soda lime glass. Thin-walled Pasteur pipets with an outer diameter of 7 mm are suitable if the cathodes are made from 5-mm glass tubes. The washed pipets are heated in a soft flame about 2 cm in front of the shoulder and drawn to capillaries. The capillaries should be approximatcly 0.6 mm in diameter at a distance of 1 cm from the new shoulder, i.e., around 3 cm in front of the large shoulder. This approximately 0.6 mm capillary is pulled to a very thin capillary in the same heating loop as used for coating the platinum with glass (Fig. 3). The final capillary is usually too thin walled, but the relative wall thickness may be increased by shrinking the glass where the future tip is going to be shaped. To do this, about 0.02 g of tape is attached to the capillary tip, and the capillary is positioned in the heating loop, shaft upward. Heat is applied carefully until the internal diameter within the capillary is reduced from about 60 to about 30 μm. The weight of the tape keeps the capillary straight during this procedure. The casing is then suspended in the capillary just as in the arrangement shown in Fig. 3.

A small heating loop made from 0.1-mm platinum wire is used to shape the tip. The white-hot heating loop is first brought close to the thick-walled part of the capillary so that it is pulled to an approximately 1-mm-long tube with a minimum internal diameter of about 10 μm. A gradual taper from 30 to 10 μm over a length of about 0.5 mm is preferable. The faintly red-hot heating loop is now brought close to the thinnest section of the capillary until the glass pulls apart. This last step should produce a cone 200–500 μm long, starting with an inner diameter of 10 μm and narrowing to a diameter of less than 1 μm. The tip is now broken off under the microscope, using a glass rod held by a micromanipulator. An opening equivalent to the diameter of the cathodes should be obtained (2–6 μm).

A membrane is applied by touching the tip with uncured silicone rubber (Silastic, Medical Adhesive Type A, Dow Corning, Midland, MI 48640) which enters by capillary suction. A 20-μm-deep membrane is usually preferable; however, membranes as thin as 10 μm may be applied if especially fast electrodes are needed, and slightly thicker (and more durable) membranes may be applied if a fast response is not essential.

The cathode is now pushed into the outer casing until the distance between cathode tip and the silicone membrane is 5–10 μm. It is essential that the casing be as narrow as possible for the specific cathode to be inserted. Too much space between cathode and casing over the distant approximately 300 μm results in slow response and high currents for zero oxygen. A partial seal of fast-curing epoxy (Fig. 2) is applied between cathode and casing, and the assembled sensor can be removed from the microscope after about 15 min. The silicone and epoxy should be allowed to cure in room air for at least 24 hr, after which the sensor can be filled with electrolyte (1 *M* KCl saturated with thymol). If the sensors are not going to be used within the next few days, they should be stored dry as described for the cathodes. The filling with electrolyte is best accomplished by first adding a few drops of methanol followed by evacuation until the methanol boils. The sensors are thereby filled to the very tip, and the methanol can afterward be replaced with electrolyte. The filling is done with a 1-ml plastic syringe with the tip drawn over a flame into a long, flexible capillary. Finally, a chlorinated[2] silver wire (Fig. 2) is inserted, and the epoxy seal is completed. The electrodes should be left for several hours before a test of performance.

Construction of Sulfide Microelectrode. The sulfide microelectrode is made from a tapered platinum wire coated with lead glass (e.g., 28% PbO Bleiglas, No. 8095, Schott Glaswerke, Mainz, FRG), using the same procedure as described above for the cathode of the oxygen microelectrode. The tip is ground to a suitable diameter (larger than 30 μm) with a somewhat coarser abrasive than used for the oxygen microelectrode. The platinum is then etched in saturated KCN as described above until an approximately 40-μm-deep recess is formed. This recess is filled with silver by electroplating at 0.2 V DC (platinum negative) in a solution containing 50 mM $AgNO_3$ and 0.3 M KCN. The Ag_2S layer is formed during the calibration of the electrode in relatively high concentrations of sulfide.

Measuring Equipment and Calibration

A measuring circuit for an oxygen microelectrode is shown in Fig. 4. The negative polarization of the cathode (which in the Clark-type electrode is in the same unit as the anode) relative to an anode results in a

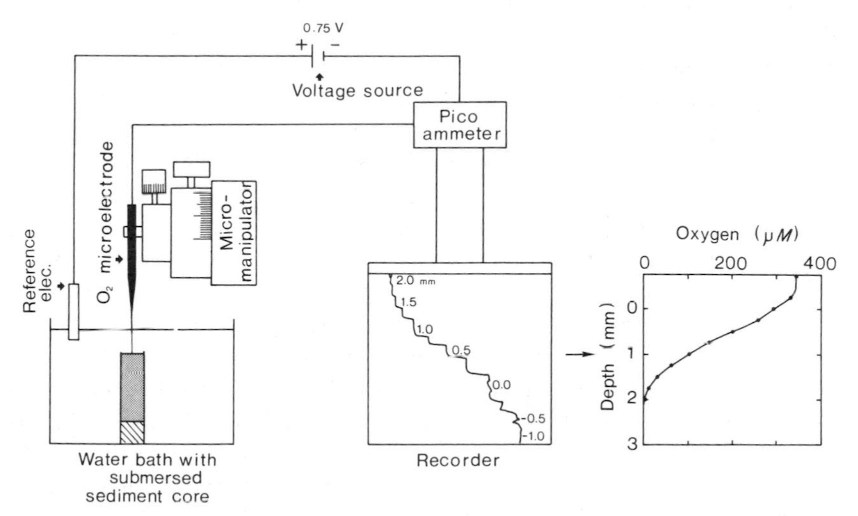

FIG. 4. Measuring circuit used for oxygen microelectrodes. The recorder output shows actual data from the measurement of an oxygen profile in a dark-incubated sediment. From Revsbech and Jørgensen.[3]

current in the circuit which is directly proportional to the oxygen concentration in the medium. The current is measured with a picoammeter (e.g., Keithley 480, Keithley Instruments, Cleveland, OH 44139) connected to a strip-chart recorder, and the actual oxygen concentrations can be calculated from the tracing produced. The microelectrodes are introduced into the medium using a micromanipulator (e.g., Mertzheuser, Steindorf/Wetzlar, FRG). Micromanipulators with stepper motors (e.g., Fairlight Scientific & Industrial Equipment, Postbus 4055, 3006 AB Rotterdam, The Netherlands) should be used if computerized data collection[3] is planned. Measuring circuits for sulfide and pH microelectrodes differ from that shown in Fig. 4 in that a high-resistance ($>10^{13}\ \Omega$) voltmeter is used instead of the voltage supply plus picoammeter.

It is essential that all signal-conducting parts of the microelectrode circuits are efficiently shielded. All shielded wires should have a graphite layer below the copper shield (''low-noise'' quality). Using such efficient shielding, it is possible to measure oxygen with the Clark-type oxygen microelectrode with a noise level approaching 0.1 pA, even when people are moving in the same room. No external Faraday cage enclosing the experimental setup is necessary.

The oxygen microelectrodes are usually calibrated in water equilibrated with N_2, air, and pure O_2. The current is typically 10–30 pA for N_2 and 70–400 pA for air. The calibration curve should be linear. The response time may be checked by quickly moving the electrode tip from air into N_2-equilibrated water. The 90% response time is often about 1 sec, but response times down to 0.2 sec can be obtained. The current from the electrode should ideally be independent of medium turbulence or diffusional characteristics. The effect of stirring is therefore checked by measuring the difference in the signal between stagnant and stirred water at constant O_2 concentration. This difference is typically 1–3% and should always be less than 5% of the current.

The stability of the response of the cathode-type oxygen microelectrode is inferior to that of the Clark-type electrode, but the electrode is easier to construct. Cathode-type oxygen microelectrodes will work satisfactorily only in some media, one of the requirements being that Ca^{2+} and Mg^{2+} constitute only a minor fraction of the salt content.

The sulfide microelectrode senses only S^{2-}, and the potential is thus dependent on both the total concentration of dissolved sulfide and the pH. Calibration should thus be performed at various concentrations of dissolved sulfide and at various pH values. The response is nonlinear at concentrations of dissolved sulfide below around 100 μM.

The pH microelectrode is calibrated in standard buffers as normal, commercial pH electrodes. Interference of sodium may be a problem by

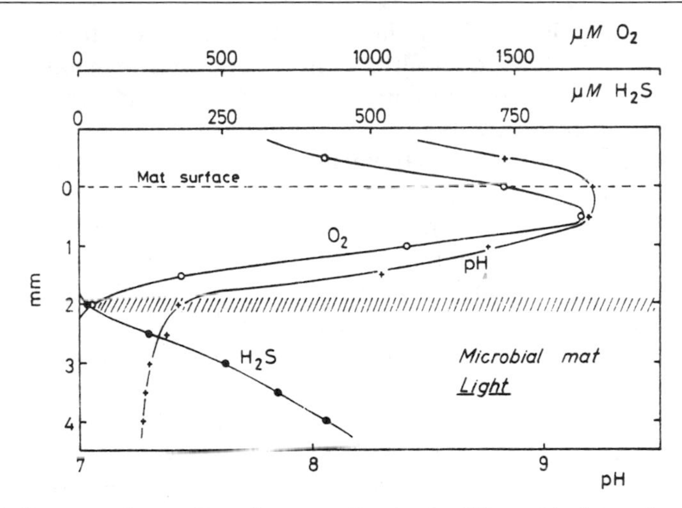

FIG. 5. Concentration profiles of oxygen, dissolved sulfide, and hydrogen ion (pH) in an illuminated cyanobacterial mat from Solar Lake, Sinai. From B. B. Jørgensen, *in* "Microbial Geochemistry" (W. E. Krumbein, ed.), p. 91. Blackwell, London, 1983.

excessively hydrated electrodes, and this source of error should consequently be checked.

Examples of Data Obtained with Microelectrodes

A few illustrative examples of the use of microelectrodes in communities dominated by cyanobacteria are described below. Further references to results obtained by use of microelectrodes are listed in the review of Revsbech and Jørgensen.[3]

Concentration Profiles. The concentration profiles of O_2, H^+ (pH), and dissolved H_2S in an illuminated mat dominated by *Microcoleus chthonoplastes* are shown in Fig. 5. The concentration profiles shown are typical for such dense photosynthetic communities, which normally produce very high oxygen concentrations and high pH values during illumination. The overlap zone between sulfide and oxygen is very narrow, and chemoautotrophic bacteria here catalyze the oxidation of sulfide with a very high efficiency.[5] The cyanobacteria may also utilize sulfide as an electron donor and can shift between oxygenic and anoxygenic photosynthesis. This has been demonstrated in intact cyanobacterial mats using oxygen, sulfide, and pH microelectrodes.[6]

[5] B. B. Jørgensen and N. P. Revsbech, *Appl. Environ. Microbiol.* **45,** 1261 (1983).

[6] B. B. Jørgensen, Y. Cohen, and N. P. Revsbech, *Appl. Environ. Microbiol.* **51,** 408 (1986).

The gradient above the mat in Fig. 5 was due to a diffusive boundary layer which exists around all surfaces in aquatic environments. The diffusive boundary layers around communities of cyanobacteria not only impede the export of oxygen but may also lower the availability of inorganic carbon and nutrients. The microelectrode techniques allow a very direct observation of these layers.[7]

Photosynthetic Rates. The dynamics of the oxygen pool inside cyanobacterial mats can be studied using oxygen microelectrodes. Figure 6 shows an example of the changes in oxygen concentration at various depths within a hot spring microbial mat dominated by the cyanobacterium *Synechococcus lividus* during a light–dark cycle. Steady-state oxygen concentrations were not obtained during the 3-min light period, but illumination for a somewhat longer time period would have accomplished this. It can be shown[8] that the initial rate of decrease in oxygen concentration after turning off the light is identical to the former rate of gross photosynthesis at that particular depth if the oxygen concentration was at a steady state before the darkening. The spatial resolution of the determination is about 0.1 mm if the calculation is based on the rate of decrease 1 sec after darkening.[9,10]

The microelectrode method for measuring oxygenic photosynthesis is very rapid, and rates can often be measured with good accuracy and reproducibility. The lowest specific rates of photosynthesis which can be detected are often about 0.5 nmol cm^{-3} sec^{-1}, but lower rates can be measured if dark incubations of more than 1 sec can be used, i.e., if maximum spatial resolution is not essential. Physical instability of the substratum is often a problem, as it results in unstable readings which may be useless for determinations of photosynthetic rates. Variations in turbulence in the water above the substratum and hence variations in the thickness of the diffusive boundary layer often result in such unstable oxygen concentrations near the substrate–water interface.

The very rapid determination of photosynthetic rates by the microelectrode method may facilitate otherwise very time-consuming studies. Determination of an *in situ* action spectrum of a layer of cyanobacteria in a microbial mat may thus be accomplished within less than 2 hr at 10 nm resolution. Figure 7 shows an example of an action spectrum of a predominantly *Microcoleus chthonoplastes* cyanobacterial mat. The correct determination of action spectra within microbial mats is also dependent on

[7] B. B. Jørgensen and N. P. Revsbech, *Limnol. Oceanogr.* **30,** 11 (1985).
[8] N. P. Revsbech, B. B. Jørgensen, and O. Brix, *Limnol. Oceanogr.* **26,** 717 (1981).
[9] N. P. Revsbech and B. B. Jørgensen, *Limnol. Oceanogr.* **28,** 749 (1983).
[10] N. P. Revsbech, unpublished observations, (1986).

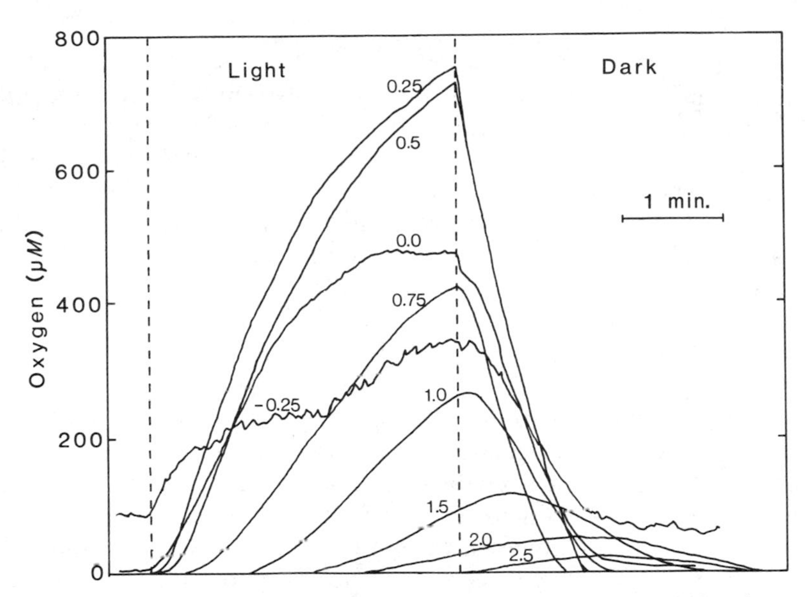

FIG. 6. Oxygen concentrations at various depths (in millimeters) within a hot spring cyanobacterial mat which was subjected to 3 min light–5 min dark cycles. from N. P. Revsbech and D. M. Ward, *Appl. Environ. Microbiol.* **48,** 270 (1984).

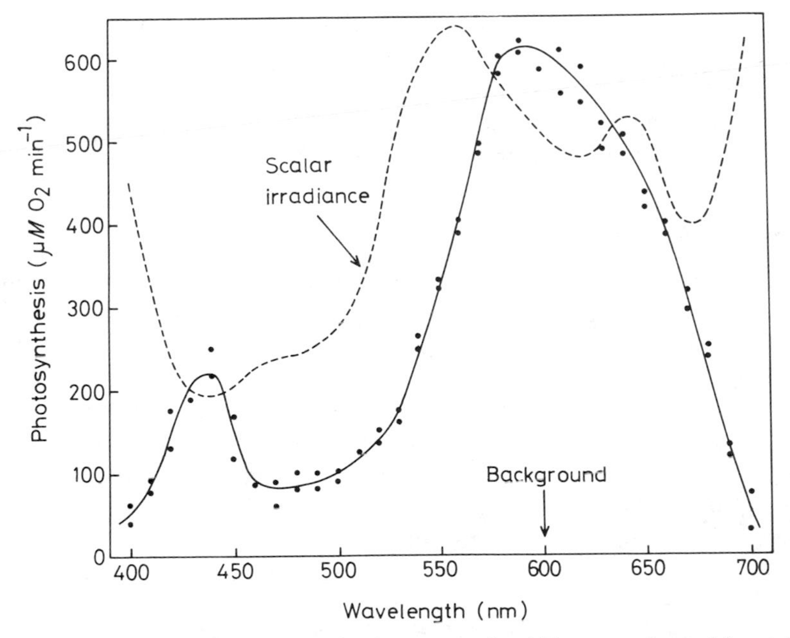

FIG. 7. Action spectrum of oxygenic photosynthesis within a cyanobacterial mat dominated by *Microcoleus chthonoplastes*. Also shown is the relative scalar irradiance as a function of wavelength. From Jørgensen *et al.*[11]

the use of micro light sensors. The relative spectral space irradiance within the analyzed cyanobacterial layer is also shown in Fig. 7.[11]

Respiratory Rates. Rates of respiration can be calculated from the diffusion fluxes of oxygen from or to individual layers of the microbial mats, subtracting possible concurrent rates of photosynthesis measured as described above. The diffusion flux can be calculated from Fick's first law of diffusion:

$$J = -D_s\Phi \frac{\partial C}{\partial x}$$

where J is the diffusion flux, D_s the apparent diffusion coefficient in the substrate, Φ the porosity, and $\partial C/\partial x$ the slope of the concentration profile at depth x. Methods have been developed which allow estimation of $D_s\Phi$ in thin microbial films.[12] Figure 8 shows profiles of oxygen and photosynthesis at various light intensities in a cultured biofilm[12] containing the cyanobacterium *Pseudoanabaena* sp. as the only phototroph. The oxygen gradients in the diffusive boundary layer just above the biofilm and in the deepest layer that was analyzed allow reasonably accurate calculations of the diffusion flux out of the analyzed layer and hence of the rate of oxygen consumption within this layer. Figure 9 shows the calculated rates of photosynthesis and respiration as a function of light intensity. The rate of respiration seemed only marginally affected by the variations in light, photosynthesis, and oxygen concentration. Depth profiles of respiration can be obtained by inserting the measured data for oxygen, diffusion coefficient, and photosynthetic rate into computer models.[10]

Fiber Optic Microprobe

Gradients of spectral light distribution in dense communities of cyanobacteria can be analyzed at a spatial resolution which matches that of the oxygen microelectrodes used in photosynthesis measurements. A fiber optic microprobe[13] is described here which is both inexpensive and simple to build and use. It consists of an 80-μm optical fiber with a rounded sensing tip of 20–30 μm diameter. When used with a single photodiode detector the probe has a sensitivity of 0.01 μE m^{-2} sec^{-1} and a spectral range of 300–1100 nm.

Microprobes are constructed from single-stranded optical fibers of 80–250 μm diameter (e.g., Corning or American Optical). Graded index fibers

[11] B. B. Jørgensen, Y. Cohen, and D. J. Des Marais, *Appl. Environ. Microbiol.* **53,** 879 (1987).

[12] J. Jensen and N. P. Revsbech, unpublished observations (1986).

[13] B. B. Jørgensen and D. J. Des Marais, *Limnol. Oceanogr.* **31,** 1374 (1986).

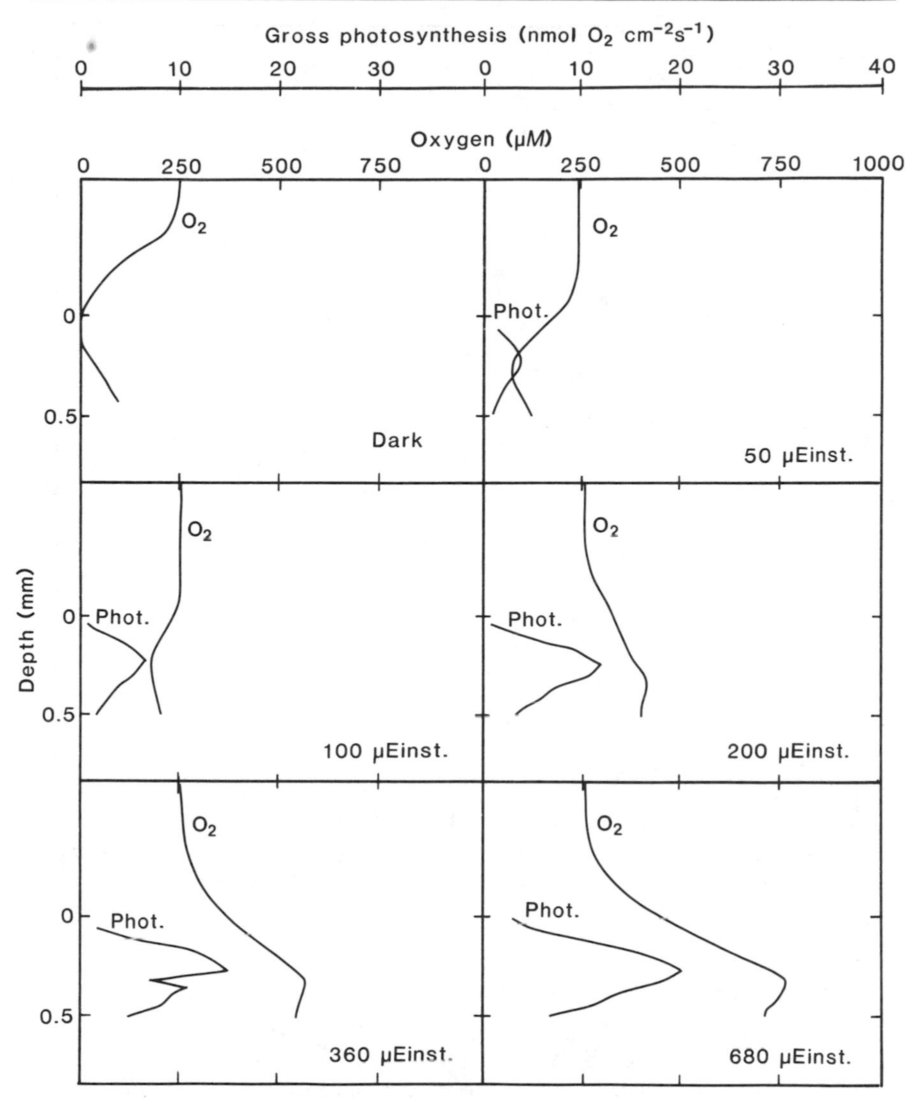

FIG. 8. Profiles of oxygen and oxygenic photosynthesis at 6 different light intensities (in $\mu E\ m^{-2}\ sec^{-1}$) within a cultured biofilm containing *Pseudoanabaena* sp. as the only phototroph. The biofilm was growing on a permeable membrane, and the temperature was 24°.

with silica cores and cladding are used without the polymer jackets. Quartz fibers (e.g., Schott) may be used for UV detection. We have used 80-μm fibers obtained from the multistranded light guide of a halogen lamp. Coated fibers can be submerged in dichloromethane for 1 min and

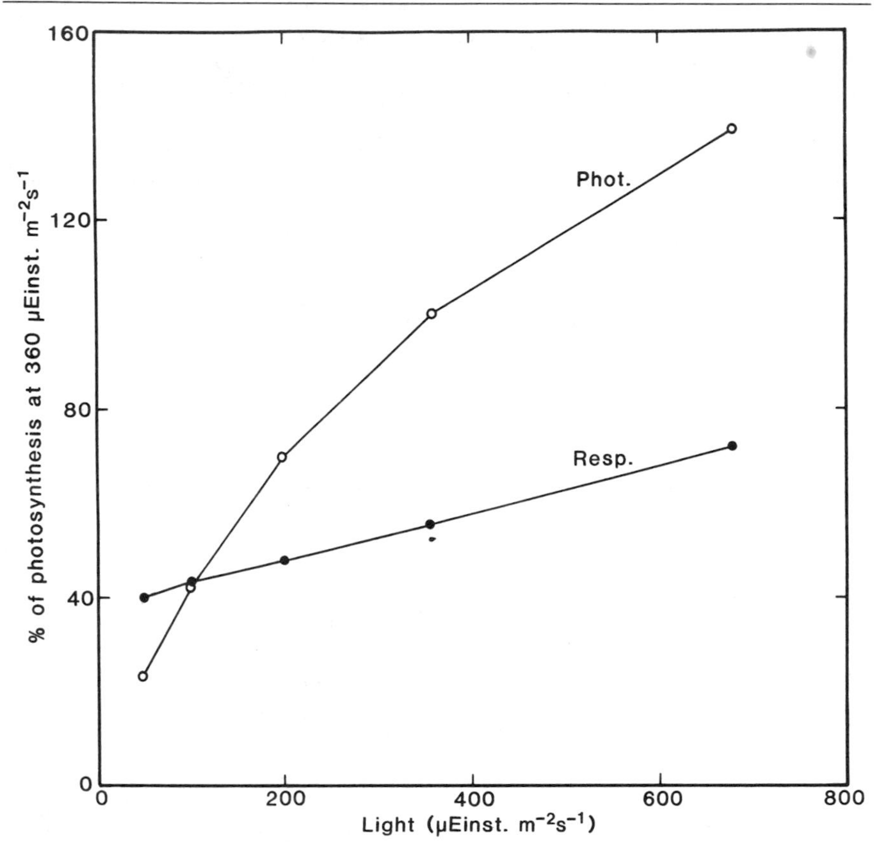

FIG. 9. Rates of photosynthesis and respiration integrated over all depths of the *Pseudoanabaena* biofilm. The rates were calculated from the data shown in Fig. 8.

wiped with clean tissue to remove the jacket. The fiber may be used with a flat sensing tip or, better, with a tapered sensing tip.

A fiber can be tapered by the use of a heating loop made of 0.1-mm platinum wire as described for the microelectrodes. The fiber is suspended vertically and stretched with a 1-g weight in the lower end. The red-hot loop is then carefully advanced toward the fiber until a fine constriction is drawn. The fiber is broken at the constriction, and the fine tip is rounded by approaching the hot loop again. The fiber is then broken about 20 cm from the tip by making a fine score with the edge of Carborundum paper and pulling the fiber apart. A flat, smooth surface perpendicular to the fiber axis is required for the optical coupling to the detector.

The fiber is mounted in a shaft as shown in Fig. 10A. The fiber is inserted in a 10-cm-long 1/16 inch stainless steel tubing with 0.01 inch internal diameter (e.g., Alltech, precut for HPLC). The flat fiber end is aligned flush with one end of the steel tubing. The fiber is then sealed to the other end of the tubing with epoxy which is painted black. An outer glass casing is made from 5-mm-o.d. sodium glass tubing pulled to a capillary at one end. The tapered fiber is mounted in the capillary with 5–10 mm protruding. The glass tubing is then sealed at both ends with epoxy.

A hybrid, low-noise photodiode/amplifier is a simple and sensitive detector solution (e.g., EG&G Electro Optics, TCN-1000-93). An aluminum holder with an O-ring attachment as shown in Fig 10B can be sealed to the detector. Care should be taken that no stray light enters and that the optical geometry ensures that all light falls on the active area of the detector. The output signal from the detector is 0–500 mV and is read on a millivoltmeter. A good integrating meter with <0.1 mV sensitivity is required. The response of the detector described here with a 20–30 μm fiber tip diameter is about 10 mV at 1 μE m^{-2} sec^{-1}. With a typical noise level of 0.05–0.1 mV, the detectability of white light is about 0.01 μE m^{-2} sec^{-1}.

The optical fiber can also be coupled to a spectroradiometer in order to obtain a spectral analysis of light on the detector side. An important consideration is here the very low photon flux coming through the optical fiber. Even with a good scanning spectroradiometer, integration times of many minutes may be required to analyze one complete spectrum of scattered light. The ideal solution seems to be an array-type detector built into a good spectrograph in which parallel detection of many wavelengths occurs simultaneously over 500–1000 sensitive elements. Such optical multichannel analyzers with special intensified detectors are available with high sensitivity (1 digital count per <10 photons per element) but at high cost.

The directional sensitivity of the fibers is important for the quantitative discrimination between collimated light and scattered light. We use optical fibers with 25°–30° acceptance half-angle (Fig. 11). The acceptance angle increased and became less sharply defined by tapering and rounding the fiber tip. Fiber tips can be made with good optical symmetry as shown in Fig. 11, but each individual fiber probe should be checked by fixing the fiber tip in a collimated light beam and varying the light-to-fiber angle over +90° to −90°.

In order to measure spectral light gradients with a single photodiode it is necessary to illuminate the samples with monochromatic light which can be scanned over the spectrum. We use a halogen lamp with a flexible

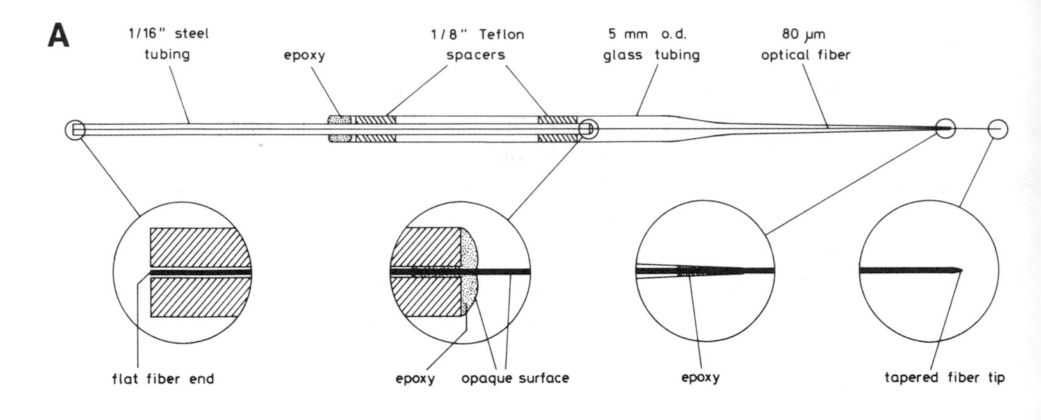

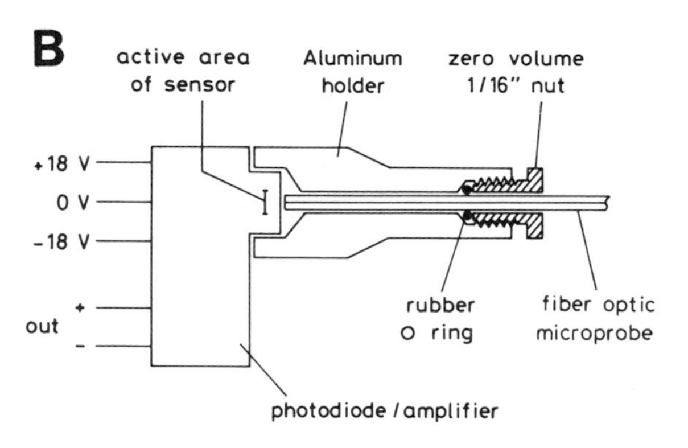

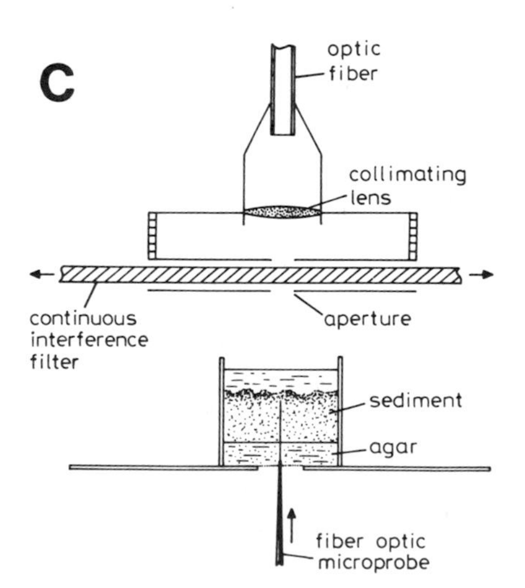

FIG. 10. Fiber optic microprobe. (A) Construction of a probe with single-stranded optical fiber. (B) Simple photodetector with adapter for microprobe. (C) Experimental setup for measurement of spectral light gradients. From Jørgensen and Des Marais.[13]

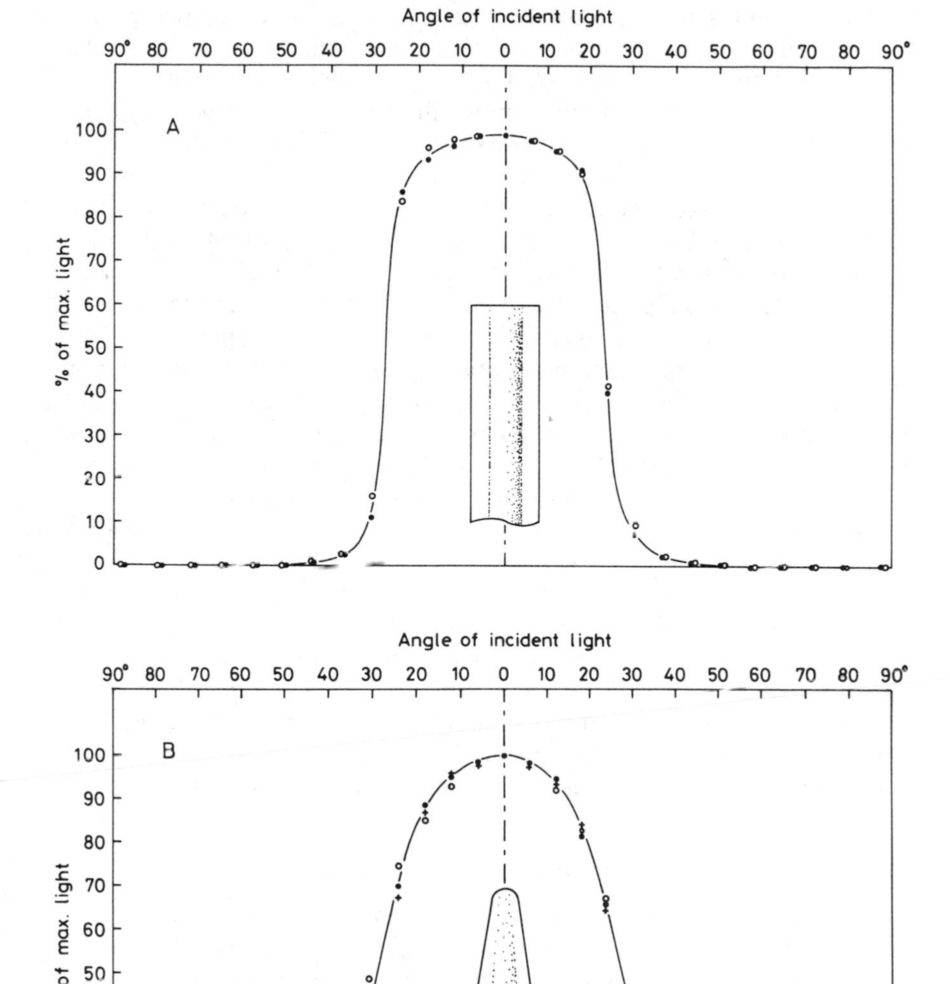

FIG. 11. Acceptance angles in air of fiber optic microprobes with (A) a flat fiber tip or (B) a tapered and rounded fiber tip. The numerical apertures were 0.53 and 0.65, respectively. (A) The fiber was rotated 90° around its axis between the first (○) and second (●) set of measurements of white light. (B) Measurements were done in monochromatic light at 400 nm (○), and 700 nm (●), and 1000 nm (+). From Jørgensen and Des Marais.[13]

light guide and a terminal condensing lens as a source of white light. Monochromatic light is provided by a continuous interference filter sliding between two apertures (Fig. 10C). With a 400–700 nm interference filter (Schott) a 12–14 nm half-bandwidth may be obtained. For higher spectral resolution, better and more expensive monochromators with reflecting gratings are available (e.g., Oriel).

Measurements of spectral radiance gradients in cyanobacterial mats or other dense cyanobacterial communities can be made of either the collimated or the scattered light as shown in Fig. 10C. A small core is cut out of the sample and is then plugged in the bottom by pressing it down over a solidified agar plate. The agar plug prevents the overlying water from leaking out. The core is then placed on a plastic plate over a 10-mm hole through which the fiber can penetrate into the core. The fiber optic microprobe is attached to a micromanipulator. The position of the sensing tip may be difficult to determine but can be observed under a dissecting microscope when it breaks through the cyanobacterial layer.

With the sensing tip diameter of 20–30 μm of a 80-μm optical fiber, it is realistic to measure spectral gradients at 50- to 100-μm depth increments in a cyanobacterial community. It is a surprising observation, however, that the vibration of many laboratory buildings sets the limit of spatial resolution. These vibrations may be damped in a similar way as, e.g., for microbalances by placing the equipment on a large, heavy concrete plate resting on special rubber supports or on foam rubber. Another problem is the friction between sample and fiber which may cause the sample to move slightly unless special precautions are taken.

Spectral Gradients of Radiance and Scalar Irradiance

Most of the compact cyanobacterial communities have such high light attenuation coefficients that a resolution of 0.1 mm is indeed required. One example is shown in Fig. 12 from a cyanobacterial mat dominated by *Microcoleus chthonoplastes*. A freshly collected mat core was illuminated on the surface, and the spectra of down-welling radiance, L, were measured from the surface to a depth of 0.9 mm. A high concentration of Chl *a* was evident from the absorption peak at about 670 nm and, less pronounced, at 430 nm. The blue absorption band fused with depth into a broad absorption maximum between 430 and 500 nm, due to carotenoid pigments. A less pronounced absorption peak due to phycocyanin was present at 620 nm. Violet (400 nm) and far-red (700 nm) light penetrated deepest into the mat. Light attenuation coefficients, K_L, varied throughout the visible spectrum from 3 to 11 mm^{-1}, but values of up to 34 mm^{-1} have been found in other cyanobacterial mats. Thus, already at a depth of

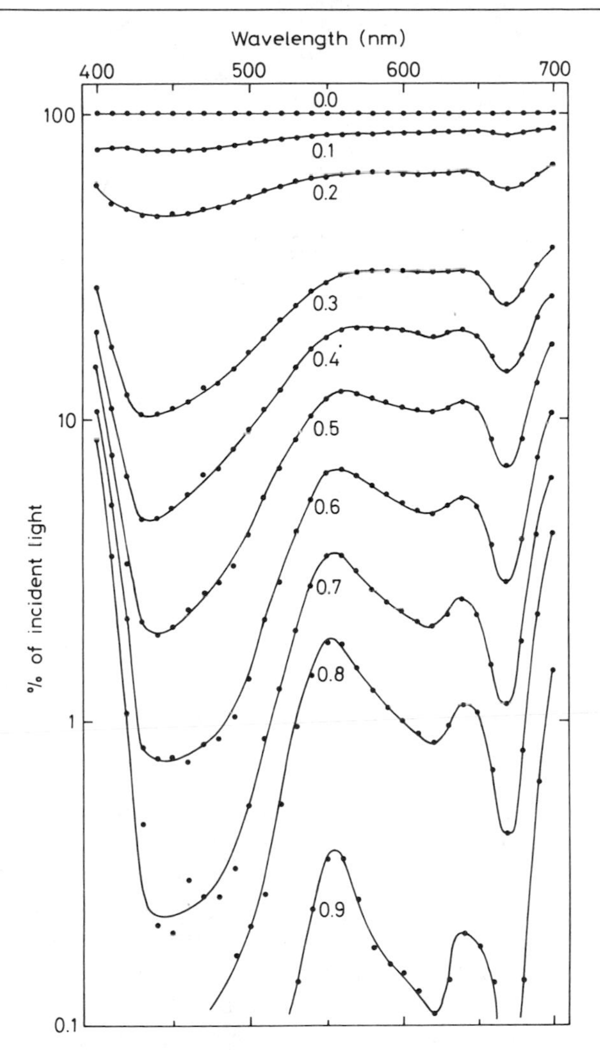

FIG. 12. Spectral distribution of down-welling radiance, L, in a *Microcoleus chthonoplastes*-dominated cyanobacterial mat. Radiance values were calculated as percentage of the surface value for each depth and wavelength. From Jørgensen *et al.*[11]

0.8 mm the down-welling radiance had been reduced to about 1% of the surface radiance. Microelectrode measurements showed that this was the deepest layer of detectable oxygenic photosynthesis.

The scalar irradiance is the integral of the radiance distribution at a point over all directions around the point. The scalar irradiance is there-

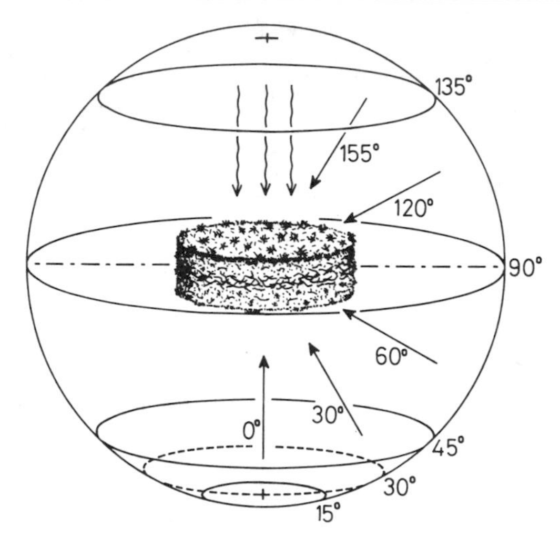

FIG. 13. Measuring angles of fiber optic microprobes (straight arrows) and spherical zones used to calculate scalar irradiance, E_0 in a cyanobacterial mat. The mat was vertically illuminated by a collimated, monochromatic light beam. From Jørgensen and Des Marais.[15]

fore a more relevant parameter for the individual cyanobacterial cells because photosynthesis is regulated by the total incoming radiant flux.[14] In strongly scattering media such as a cyanobacterial mat or coastal sediment it is therefore important to analyze light coming from all directions. Owing to the well-defined acceptance cone of the optical fibers such as a spherical integration can be done by measuring gradients of spectral radiance in a few representative directions as shown in Fig. 13.[15] The spectra measured for each of the five light-fiber angles are representative of an angular interval corresponding roughly to the acceptance angle of the fiber. Owing to the vertical illumination of the mat surface, the light field in the mat will approach axial symmetry around the collimated beam. The total relative radiant flux from different angular intervals can therefore be calculated by multiplying the measured relative radiance from that angle with an appropriate weighing factor. The weighing factor applied is the surface area of the spherical zone covered by that angular interval, when the radius of the sphere is one. When the total relative radiant fluxes from the five zones are added, all solid angles of the sphere have been covered. The sum of the weighing factors is consequently 4π, equal to the surface area of a sphere with unit radius.

Figure 14 shows the radiance spectra from the different angles at 0.2

[14] T. C. Vogelmann and L. O. Björn, *Physiol. Plant.* **60,** 361 (1984).

[15] B. B. Jørgensen and D. J. Des Marais, *Limnol. Oceanogr.* **33,** 99 (1988).

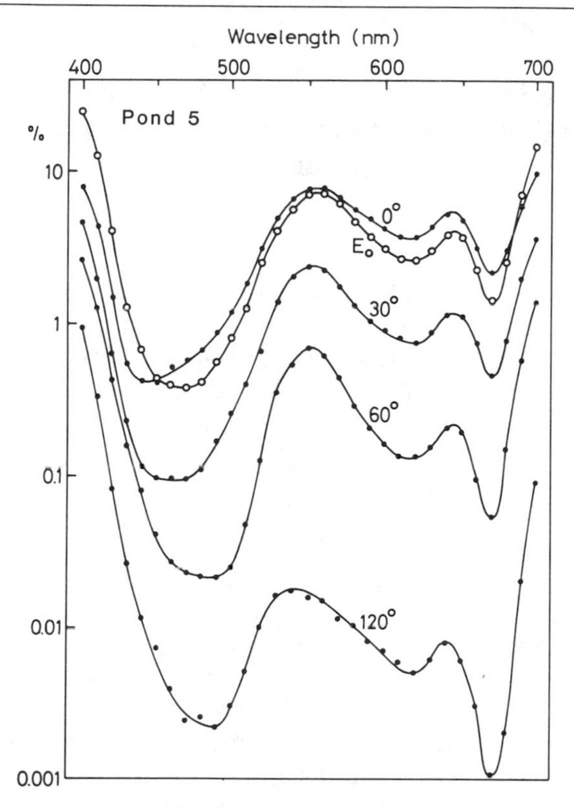

FIG. 14. Spectral radiance measured at different angles and calculated relative to the surface (0°) radiance. The data set is used to calculate the scalar irradiance, E_0, which is expressed relative to the downward space irradiance at the surface. From Jørgensen and Des Marais.[15]

mm depth below the surface of a dense cyanobacterial mat. There was 100- to 1000-fold lower radiance of backscattered light than of downwelling radiance from the collimated beam; forward scattered light (60°) was intermediate. Yet, the scattered light contributed strongly to the total space irradiance spectrum due to the much larger weighting factor.

Measurements of scalar irradiance, E_0, in cyanobacterial mats have shown that E_0 at the mat surface may reach almost 200% of the incident irradiance, E_d, at wavelengths with minimum absorbance. The spherically integrated space irradiance thus differs strongly from the incident irradiance, with respect to both total intensity and spectral composition. Since this applies even to thin films of cyanobacteria, it is important to take it into account when analyzing, e.g., light saturation and action spectra for oxygen evolution (Fig. 7).

[72] Determination of Turgor Pressure and Other Cell–Solute Relations by Using Gas Vesicles as Pressure Probes

By A. E. WALSBY

The gas vesicle is a rigid, hollow structure. It will collapse irreversibly at a precise pressure, known as its critical pressure (p_c). This property enables gas vesicles to be used in the measurement of cell turgor pressure[1] and in the determination of other properties of cells that depend on turgor pressure.[2]

Determination of Gas Vesicle Critical Pressure (p_c)

The critical pressure is the minimum pressure difference, between the outside of the gas vesicle and gas pressure inside, that causes the structure to collapse. Because the structure is very permeable to gases[3,4] the gas inside will always be in equilibrium with that in the surrounding solution. In a suspension that has been equilibrated with air the pressure of the gas inside will balance the atmospheric pressure, and the critical pressure will be equal to the pressure applied above atmospheric pressure that just causes gas vesicles to collapse.[1,2]

The critical pressures of suspensions of isolated gas vesicles can be determined by the pressure nephelometer (Fig. 1). The gas vesicle suspension, preequilibrated with air, is placed in the thick-walled glass tube and the pressure-tight coupling is attached. The nephelometer is first zeroed against a 'blank' tube of water and then set to full-scale reading with the sample tube in place. The sample tube must not be knocked or jolted. The outlet valve on the gas-handling system is closed and the inlet valve operated to allow the pressure to rise, in steps, to 1.5 MPa or more if required. The light-scattering reading falls as gas vesicles are collapsed. The reading is left to stabilize for a few seconds at each step before proceeding to the next. When no further decrease occurs all the gas vesicles will have been collapsed. On releasing the pressure there should be no change in turbidity reading (gas vesicle collapse is irreversible).

[1] A. E. Walsby, *Proc. R. Soc. London, Ser. B* **178,** 301 (1971).
[2] A. E. Walsby, *Proc. R. Soc. London, Ser. B* **208,** 73 (1980).
[3] A. E. Walsby, *Proc. R. Soc. London, Ser. B* **223,** 117 (1984).
[4] A. E. Walsby, *Proc. R. Soc. London, Ser. B* **226,** 345 (1985).

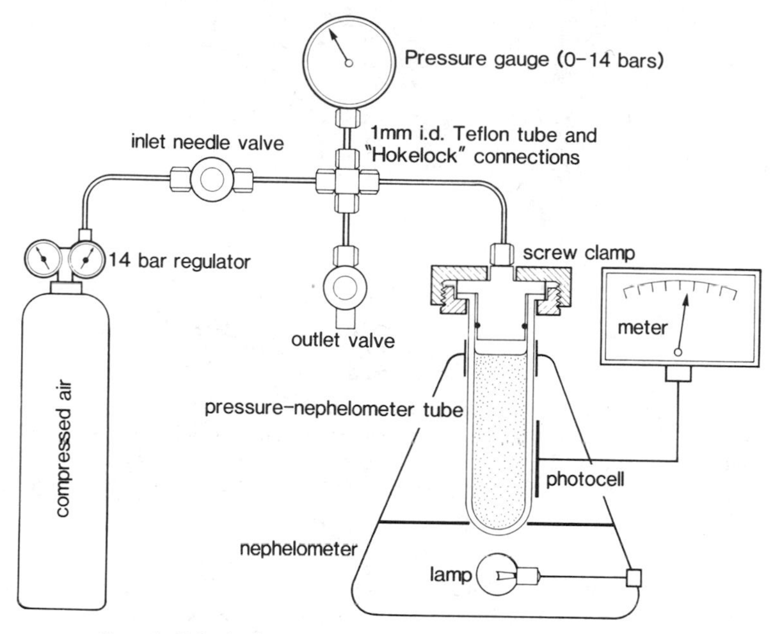

FIG. 1. Principal components of the pressure nephelometer.

The percentage of gas vesicles collapsed at any pressure is $[(a - r)/(a - b)] \times 100$ where r is the reading at that pressure, a the initial reading, and b the final reading with all gas vesicles collapsed. A plot of percent gas vesicle collapse versus pressure applied is shown in Fig. 2.

Determination of Cell Turgor Pressure

The turgor pressure of a cyanobacterium is the hydrostatic pressure (P) that develops to balance the difference between the internal osmotic pressure (π_i) of the cytoplasmic fluids and the external osmotic pressure (π_e) of the suspending solution; hence

$$P = \pi_i - \pi_e \tag{1}$$

Direct determination of the turgor pressure can be made using gas vesicles in the following way.

The critical pressure distribution of the gas vesicles is first made by pressure nephelometry but using a suspension of cells (or filaments) freshly suspended in culture medium containing 0.5 M sucrose prepared by placing 3 ml of 1.0 M sucrose in the nephelometer tube, layering over it 3 ml of the cyanobacterial suspension, closing the gas-tight connection,

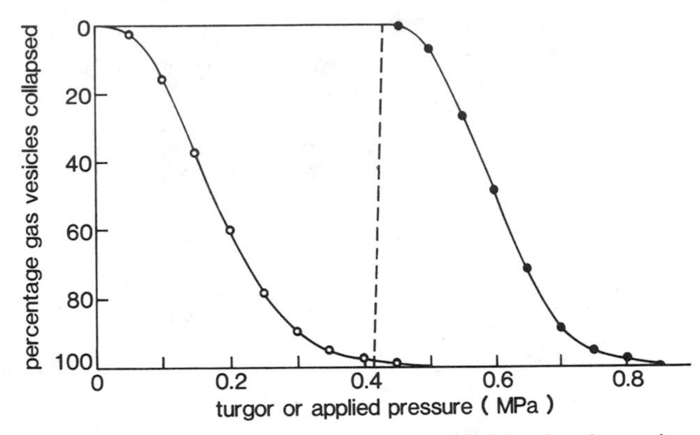

FIG. 2. Collapse of gas vesicles by pressure, indicated by the change in turbidity: gas vesicles in cells of *Anabaena flos-aquae* suspended in culture medium (○) and in medium containing 0.5 *M* sucrose (●). The difference between these two curves, indicated by the dashed line, gives a measure of turgor pressure. The results obtained with isolated gas vesicles are similar to those with cells in 0.5 *M* sucrose (●).

and then rocking the tube to and fro for about 40 sec to mix the contents. After the initial reading has stabilized the pressure should be applied in steps of 0.05 MPa (0.5 bar). The procedure should then be repeated with the cyanobacterial suspension without added sucrose. This gives a curve of *apparent critical pressure* (p_a) with values less than that of the first curve (see Fig. 2) because gas vesicles already are subjected to the cell turgor pressure. A measure of this turgor pressure is given by

$$P = \bar{p}_c - \bar{p}_a \tag{2}$$

where p_a and p_c are the median values, when 50% of the gas vesicles are collapsed.

The critical pressure distribution curve for sucrose is similar to that obtained for isolated gas vesicles[1] because π_e for the sucrose solution exceeds π_i and the turgor pressure falls to zero. Subtracting the curve for the cells in medium alone gives a plot of the turgor pressure for cells with different proportions of gas vesicles remaining (dashed line in Fig. 2). If the cells all have the same turgor pressure this line will be nearly vertical. (However, cells must shrink slightly and lose some turgor as gas vesicles are collapsed.[2]) When turgor pressure varies between cells the line will slope, having a higher value at the top: the midpoint indicates the median value and the top and bottom indicate the maximum and minimum.

Turgor pressures of different cyanobacteria vary, usually within the range 0.2–0.5 MPa (2–5 bar) for freshwater species[1,5]; halophilic cyano-

[5] M. T. Dinsdale and A. E. Walsby, *J. Exp. Bot.* **23,** 561 (1972).

bacteria have lower values.[6] Turgor pressure may rise in high irradiance and cause spontaneous collapse of the weaker gas vesicles; this is the basis of one mechanism of buoyancy regulation in planktonic cyanobacteria.[5,7]

Volumetric Elastic Modulus (E)

The cell wall of a cyanobacterium limits osmotic swelling but it does stretch measurably in response to turgor changes. The relationship of the cell volume change (ΔV) to the pressure change (ΔP) is described by the bulk or volumetric elastic modulus (E) of the cell:

$$E = V_r \Delta P / \Delta V \tag{3}$$

where V_r is a reference volume (see Fig. 3); this and the pressure–volume relationship are determined in the following way.

A series of sucrose solutions is prepared, in culture medium, that will generate osmotic pressures between 0 and 1.2 MPa in steps of 0.05 MPa or 0.1 MPa. The apparent median critical pressure and turgor pressure of the cells suspended in each solution is measured as above, and these values of $\bar{p}_a$ and P are plotted against π_e: the decrease in P is always less than the increase in π_e, indicating that the cells have shrunk by losing water as P falls (see Fig. 3a). If the initial turgor pressure P falls to P' when the external osmotic pressure rises from its initial value of π_e (~0 MPa) to π'_e, then the relative change of cell water volume from its initial value of V_w is calculated from the relationship

$$V'_w / V_w = (P + \pi_e)/(P' + \pi'_e) \tag{4}$$

The total volume of the cells in the sample, V_t, is equal to $V_w + V_s + V_g$, where V_s is the corresponding volume of cell solids (see Refs. 2 and 7) and V_g the volume of the gas vesicle gas space (see chapter [55] in this volume). The value of V_t for turgid cells can be found from measurements of packed cell volume minus interstitial volume[2,7]; cell volume determination by Coulter counter[8] gives V'_t, the value for partially turgid cells in electrolyte solution (of known π'_e); and the absolute volumes of V_w and V'_w can then be calculated. Finally, the total cell volume V'_t can be calculated for cells in each sucrose solution, and this can be plotted against the turgor pressure of the cells in the corresponding solutions, as shown in Fig. 3b. The slope of this graph gives $\Delta V/\Delta P$ and the intercept gives V_r.

[6] A. E. Walsby, J. Van Rijn, and Y. Cohen, *Proc. R. Soc. London, Ser. B* **217,** 417 (1983).
[7] R. L. Oliver and A. E. Walsby, *Limnol. Oceanogr.* **29,** 879 (1984).
[8] R. H. Reed and A. E. Walsby, *Arch. Microbiol.* **143,** 290 (1985).

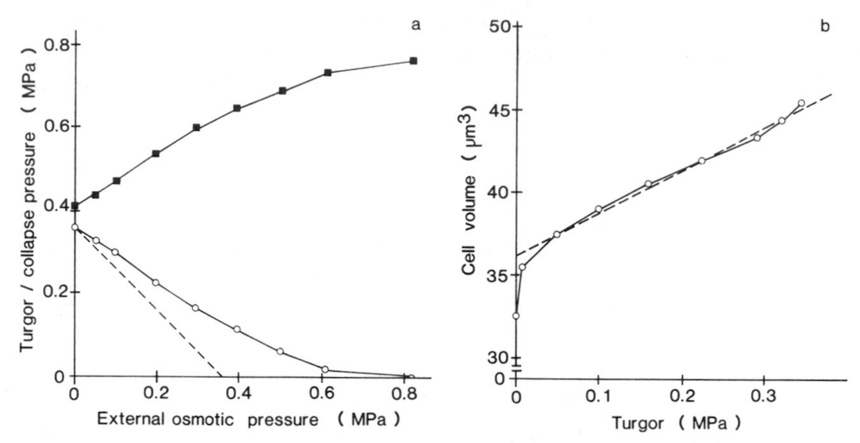

FIG. 3. (a) Effects of osmotic pressure on cell turgor (○) and median critical collapse pressure (■), obtained by incubating *Microcystis* sp. in sucrose solutions, up to 335 mOsmol kg^{-1} (with an osmotic pressure of 0.82 MPa). The dashed line shows the change in turgor pressure expected if no change in cell volume had occurred. (b) Relationship between total cell volume (V_t) and turgor pressure (P), calculated from the data shown in a. From Reed and Walsby.[8]

The elastic modulus has been calculated by this method for cells of *Anabaena flos-aquae*[2] (E = 1.06 MPa) and *Microcystis aeruginosa*[8] (E = 1.33 MPa). The values indicate that the cells change volume by about 9.4 and 7.5%, respectively, for a 0.1 MPa (1 bar) change in turgor pressure.

Permeability (k) and Hydraulic Conductivity (L_p)

A brief mention is made here of two other aspects of cell–solute and cell–water relations that can be investigated by using gas vesicles to measure rates of turgor pressure change. Full details are given by Walsby.[2]

Permeability to Solutes. When *Anabaena* filaments are suspended in 0.15 *M* sucrose solution the turgor pressure is permanently reduced, indicating that sucrose does not subsequently enter the cells. With other solutes, such as mannitol or erythritol, the turgor shows an exponential recovery as solute diffuses passively into the cell. From the half-time for recovery, T, measured as the half-time for decline to the original apparent critical pressure, the permeability coefficient of the cells, k, can be calculated:

$$k = (V \ln 2)/TA \tag{5}$$

where V is the cell volume and A the cell surface area in contact with the external solution. Turgor changes may be caused by other factors, such as active uptake of K^+,[8] and this should be investigated.

Hydraulic Conductivity. If the osmotic equilibrium of a cell is disturbed, e.g., by a change in π_i, it must be reestablished by the uptake or loss of water to restore a balance in both Eqs. (1) and (3). The rate of water loss to the external solution under a given driving pressure difference is given by the hydraulic conductivity:

$$L_p = (V \ln 2)/[T'A(E + \pi_i)] \quad (6)$$

An estimate of L_p can be made by measuring the rate of gas vesicle collapse after a sudden and sustained increment of pressure, with a chart recorder connected to the output of the nephelometer photocell. The initial instantaneous collapse of gas vesicles causes a reduction in cell volume and hence in turgor pressure [Eq. (3)]. Water enters the cell to restore the osmotic balance [Eq. (1)] and as it does so the turgor rises and more gas vesicles collapse. The factor T' in Eq. (6) is a derivative of the half-time for turgor recovery that is itself indicated by the time course of the second phase of gas vesicle collapse. In *A. flos-aquae* this method gives $L_p = 1.4 \times 10^{-12}$ m sec^{-1} Pa^{-1}. The method is immune from the problems associated with unstirred layers around the cell that plague other measurements of L_p.

Gas Permeability of Heterocysts

The combination of two properties of gas vesicles, their precise critical pressure distribution and their extreme permeability to gas, permits them to be used in the determination of rates of gas diffusion into heterocysts. The following summarizes the method which is described in detail by Walsby.[10,11]

A small volume (~1 μl) of the cyanobacterial culture is placed on a bacterial counting slide and the coverslip secured over it so as to leave a free gas pathway between the suspension and the air. Gas-vacuolate heterocysts that are (1) very close to the gas–water interface and (2) remote from it near the center of the slide are observed, photographed, and their positions noted under a microscope. The slide is removed to a pressure chamber where the gas pressure is raised to a pressure p' (that is approximately equal to $6p_a$) at an exponential rate coefficient λ [i.e., $p = p'(1 -$

[9] N. Tandeau de Marsac, D. Mazel, D. A. Bryant, and J. Houmard, *Nucleic Acids Res.* **13**, 7223 (1985).

[10] A. E. Walsby, *Proc. R. Soc. London, Ser. B* **223**, 177 (1984).

[11] A. E. Walsby, *Proc. R. Soc. London, Ser. B* **226**, 345 (1985).

$e^{-\lambda t})]$ determined by recording from a pressure transducer. Returning the slide to the microscope, the heterocysts are inspected for changes in gas vacuole content. The heterocysts remote from the gas–water interface will have lost its gas vesicles (because $p' \gg p_a$) but if the rate of pressure rise, λ, has been slow enough gas vacuoles will remain in the heterocysts close to the interface because during the pressure rise gas will have diffused through the heterocyst surface and into the gas vesicles, so that the net pressure difference across the gas vesicle wall will not have exceeded the apparent critical pressure, p_a. The procedure is repeated, taking up the gas pressure at increasing rates, until it is found that the heterocysts at the interface have lost half of their gas vacuoles. From this it is deduced that the difference between the gas pressure outside the heterocyst (p_o) and inside the gas vesicles within the heterocyst (p_i) reached a maximum (P_m) that is equal to p_a. This maximum can be shown to be equal to

$$P_m = p'\chi \exp[-\chi/(\chi - 1)] \tag{7}$$

where $\chi = \alpha/\lambda$. The term α is the rate coefficient for gas diffusion into the heterocyst. Since α is the only unknown in Eq. (7) its value can be determined, by numerical methods. The value obtained for heterocysts of *Anabaena flos-aquae* is $\alpha = 0.3\ \text{sec}^{-1}$ using N_2 and O_2. The permeability coefficients are relevant to the mechanism of nitrogen fixation and oxygen protection in heterocysts.[11]

Modifications of this method can be used to measure the much more rapid diffusion rates into vegetative cells. Quantitative assessments of percentage gas vacuole changes in single cells can be made by video image analysis systems (unpublished observations).

Concluding Remarks

The methods described above can, of course, be used only with those cyanobacteria that happen to produce gas vesicles. If the capacity for gas vesicle production can be transferred to other species by recombinant DNA technology[9] these methods may have universal application.

[73] Energetic Status of Cyanobacteria: ^{31}P-NMR Spectra

By LESTER PACKER, SUSAN SPATH, RICHARD BLIGNY, JEAN-BAPTISTE MARTIN, and CLAUDE ROBY

Phosphorus-31 nuclear magnetic resonance spectroscopy allows the determination of levels of ATP, ADP, P_i, and H^+ in living cells under physiological conditions.[1,2] Procedures are described below for obtaining good quality ^{31}P-NMR spectra for the unicellular, freshwater cyanobacterium *Synechococcus* 6311.

Methods

Strain and Culture Conditions. Synechococcus 6311 (ATCC Strain 27145, PCC Strain 6311), obtained from the American Type Culture Collection, is maintained on medium BG-11[3] solidified with 1.5% agar. Batch cultures are grown at 30–32° in medium C of Kratz and Myers,[4] bubbled with 1% CO_2 mixed with air, stirred, and continuously illuminated with cool-white fluorescent light, intensity 4000–5000 lux.

Transport of Cultures. When it is necessary to transport cultures to the NMR site, the culture flasks (1.5 liters) are placed directly on top of a battery-powered fluorescent lamp (Camping Gaz Lumotube) in a styrofoam-lined carrying box (similar to an insulated ice chest commonly used for camping). The light intensity at the surface of the lamp in contact with the bottom of the culture vessels is 3500–4000 lux. The temperature in the box remains at 25–30° for up to 5 hr. The cultures are occasionally aerated during transport by swirling the flasks.

Preparation of Cells for NMR. Aliquots of cultures are centrifuged at low speed (100 *g* for 5 min) to pellet inorganic precipitates and aggregated cell debris. This material contains inorganic phosphate and is discarded. The resulting supernatant is centrifuged at 3000 *g* to pellet cells. (Centrifugation at higher accelerations appeared to result in less physiologically active cells). The pellets are resuspended in buffered (50 m*M* HEPES, pH 8.3) fresh culture medium without phosphate or trace elements. Cells are washed a minimum of 3 times to ensure the removal of external phosphate from the sample. To chelate Mn^{2+} and reduce broadening of resonance

[1] J. K. M. Roberts and O. Jardetzky, *Biochim. Biophys. Acta* **639,** 53 (1981).

[2] J. K. M. Roberts, *in* "Nuclear Magnetic Resonance: Modern Methods of Plant Analysis" (H. F. Linskens and J. F. Jackson, eds.), Vol. 2. Springer-Verlag, New York, 1986.

[3] M. M. Allen, *J. Phycol.* **4,** (1968).

[4] W. A. Kratz and J. Myers, *Am. J. Bot.* **42,** 282 (1955).

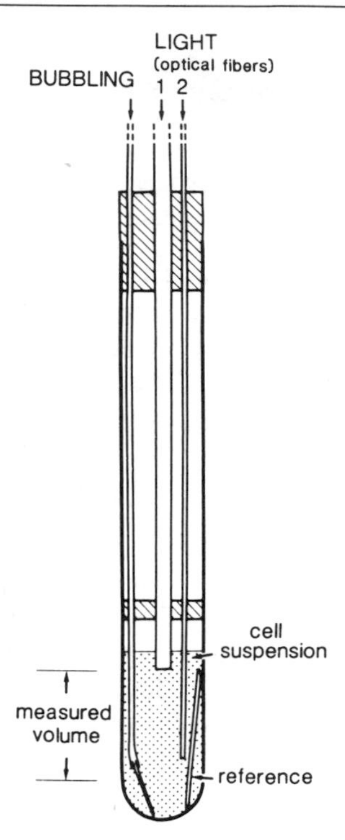

FIG. 1. Apparatus for illuminating and aerating cell suspensions in 25-mm NMR tubes.

peaks, EDTA (2 m*M*) is added to the final cell suspensions just prior to NMR analysis.[5] Sodium bicarbonate (10 m*M*) is also routinely included in samples.

Acceptable spectra were obtained with as little as 0.3 g wet weight of cell pellet. The spectrum shown later in Fig. 2 was obtained with 4 g wet weight of cell pellet (corresponding to ~1.2 mg/ml chlorophyll).

NMR Spectroscopy. The NMR spectrum shown here was recorded with a Bruker WM 200 spectrometer operating in the pulsed Fourier transform mode at 81 MHz. The repetition times were 0.4 sec. It was verified that spectra obtained with longer repetition times (2 sec) were not significantly different. Routinely, D_2O (5%, v/v) is included in the samples to serve as a field lock. Field homogeneity is adjusted on the proton signal. The temperature of the sample cavity is 25 ± 2°. Methylene diphosphonate in a sealed capillary serves as a reference; its NMR signal occurs at 16.4 ppm (see Fig. 2).

Illumination and Aeration of Samples in the NMR Cavity. Cold-white

[5] T. Kallas and F. W. Dahlquist, *Biochemistry* **20**, 5900 (1981).

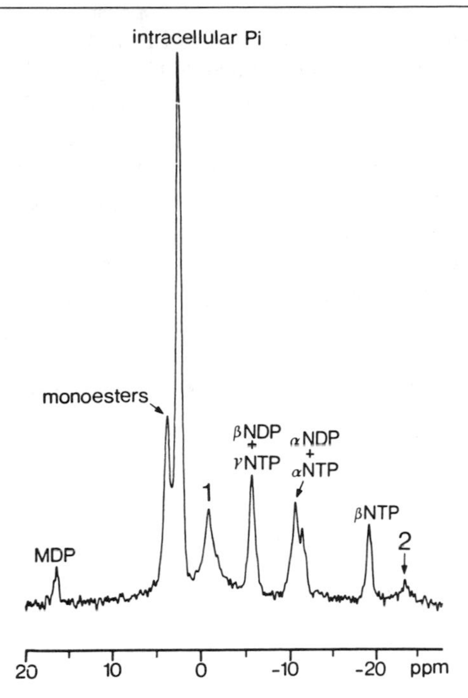

FIG. 2. ^{31}P-NMR spectrum (81 MHz) for *Synechococcus* 6311 cells. The spectrum was obtained from 4 g wet weight of cells, bubbled with O_2 at a flow rate of 20 ml/min, and illuminated as described in the text.

light sources and optical fibers are used to illuminate cell suspensions as needed during spectral acquisition. The NMR tubes (25 mm) are equipped with three aeration capillaries, one optical cable (5000 glass fibers, total diameter, 4 mm) connected to a 250-W light generator (Fort Lux 250, France), and one optical cable divided into six PMMA fibers (1 mm diameter), connected to a 150-W light generator (Fort GLI 150, France) (Fig. 1). Samples are aerated during spectral acquisition through capillary tubes connected to a peristaltic pump (Fig. 1).

^{31}P-NMR Spectra

A ^{31}P-NMR spectrum (9600 scans, 64 min) for *Synechococcus* 6311, illuminated and aerated, shows a pattern of peaks typical of plant material (Fig. 2).[1,2,6,7] Resonances from the β-phosphoryl group of NTP occur at

[6] L. P. Packer, S. Spath, J.-B. Martin, C. Roby, and R. Bligny, *Arch. Biochem. Biophys.* **256,** 354 (1987).

[7] J. B. Martin, R. Bligny, F. Rebeille, R. Douce, J. J. Leguay, Y. Mathew, and J. Guern, *Plant Physiol.* **70,** 1156 (1982).

−19 ppm, the α-phosphoryl groups of NTP and NDP at about −8 ppm, and the terminal phosphoryl groups of NTP and NDP at about −5 ppm. The peak labeled 1, at −1 ppm, corresponds to the unidentified peak "D" observed by Kallas and Dahlquist,[5] and may be a phosphoenolpyruvate signal.[8] The prominent peak at −2 ppm is assigned to intracellular inorganic phosphate at pH 7.4–7.5. Signals from monoesters, e.g., glucose 6-phosphate, occur at 4.3 ppm. Peak 2, at −23 ppm, occurs in the region for polyphosphate.[5] The signal at 16.4 ppm is from the methylene diphosphonate standard. Cells prepared as described above can also be used for ^{23}Na-NMR studies.[6]

[8] J. Sianoudis, A. C. Kusel, T. Nasujokat, J. Offerman, A. Mayer, L. H. Grimme, and D. Leibfrits, *Eur. Biophys. J.* **13,** 89 (1985).

[74] Electron Spin Resonance Oximetry

By SHIMSHON BELKIN, ROLF J. MEHLHORN, and LESTER PACKER

Introduction

Nitroxides, the relatively stable free radicals, are used as spin probes in electron spin resonance (ESR) spectroscopy for a variety of chemical and biological purposes. They have been shown, for instance, to be efficient tools for the determination of cell or organelle volumes, as well as of bioenergetic parameters such as pH and potential gradients.[1–4] In these and other assays involving nitroxide spin probes, the parameter monitored is usually the line height of the molecule's ESR signal (H_m, Fig. 1B). The lines of the spectrum may, however, be characterized by an additional parameter, their width; this is the horizontal distance, in units of magnetic flux density (Gauss or Tessla), between the line's peak and its trough (W_m, Fig. 1C). This parameter is sensitive to various effectors, among them the viscosity of the medium and the presence of other paramagnetic species; the latter cause line broadening due to collision-dependent spin exchange, the degree of broadening being directly proportional to the concentration of the paramagnetic agent.[5] Since molecular oxygen is paramagnetic, measurements of line width should lead to its quantita-

[1] R. J. Mehlhorn, P. Candau, and L. Packer, this series, Vol. 88, p. 751.

[2] R. J. Mehlhorn and L. Packer, *Ann. N.Y. Acad. Sci.* **414,** 180 (1983).

[3] S. Belkin and L. Packer, this volume [75].

[4] S. Belkin, R. J. Mehlhorn, and L. Packer, *Plant Physiol.* **84,** 25 (1987).

[5] A. D. Keith, W. Snipes, R. J. Mehlhorn, and T. Gunter, *Biophys. J.* **19,** 205 (1977).

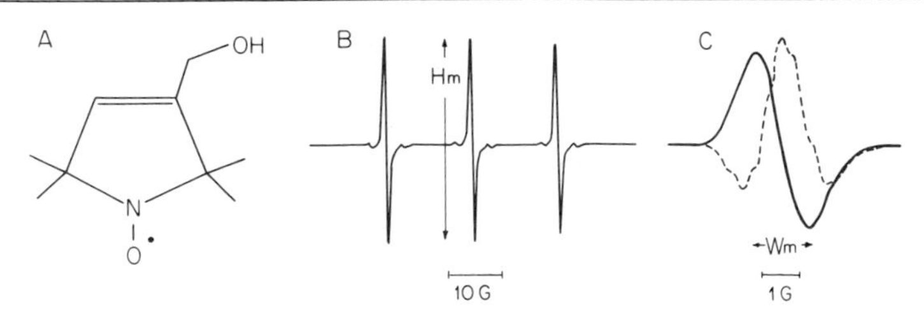

FIG. 1. The spin probe 2,2,5,5-tetramethyl-3-methanolpyrroline-*N*-oxyl (PCAOL). (A) Chemical structure. (B) ESR spectrum (field center 3396 G, field sweep 100 G, modulation amplitude 1 G, microwave power 10 W). (C) Midfield line of the spectrum, expanded scale; the solid line represents the first derivative, the dashed line the second derivative. H_m and W_m represent the height and width, respectively, of the midfield line.

tion. This chapter describes this methodology, recently termed ESR oximetry,[6] as applied to total and intracellular measurements of dissolved O_2 in a cyanobacteria system. The organism used in the work described here is the unicellular, marine species *Agmenellum quadruplicatum*.

Two main approaches have been developed in ESR oximetry. In the first, it is not the width of the main line itself which is being followed, but rather the height of the superhyperfine lines, which tend to "flatten out" on signal broadening. This phenomenon, and several variations, have been applied to O_2 measurements with a high degree of accuracy and sensitivity.[6-10] The superhyperfine lines, however, often totally disappear in the presence of relatively low O_2 concentrations. For high O_2 levels, for instance, above air saturation (~0.25 m*M*), this approach is therefore of limited value. In such cases, the range of the method can be multiplied manyfold by using the complete line width,[11] as described below.

Choice of Spin Probe

Several factors must be considered when choosing the appropriate spin probe molecule for the desired application. (1) Permeability: the probe should be freely permeable across biological membranes, especially

[6] W. Froncisz, C.-S. Lai, and J. S. Hyde, *Proc. Natl. Acad. Sci. U.S.A.* **82,** 411 (1985).
[7] T. Sarna, A. Duleba, W. Korytowski, and H. Swartz, *Arch. Biochem. Biophys.* **200,** 140 (1980).
[8] C.-S. Lai, L. E. Lopwood, J. S. Hyde, and S. Lukiewicz, *Proc. Natl. Acad. Sci. U.S.A.* **79,** 1166 (1982).
[9] J. S. Hyde and W. K. Subczynski, *J. Magn. Reson.* **56,** 125 (1984).
[10] P. D. Morse and H. M. Swartz, *Magn. Reson. Med.* **2,** 114 (1985).
[11] S, Belkin, R. J. Mehlhorn, and L. Packer, *Arch. Biochem. Biophys.* **252,** 487 (1987).

if intracellular O_2 concentrations are to be measured. The size and charge of the side group attached to a nitroxide ring would therefore be of importance. Small, uncharged molecules are preferable. (2) Binding: binding of the probe to the cell membrane would alter its line shape and width and should be avoided. (3) Reducibility: spin probes often tend to lose their signal in biological environments due to reduction by several processes, including respiratory and photosynthetic electron transport. In some cases, irreversible probe destruction may also occur. The probe to be selected has to exhibit stability in the specific experimental system. As a general rule,[12] five-membered ring probes, containing the basic structures pyrroline or pyrrolidine, are more stable than six-membered ring nitroxides (piperidine- or hydropyridine-based). Uncharged molecules are usually advantageous in this regard as well. (4) Intrinsic line width: the amount of broadening measured is proportional to the O_2 concentration and not to the original signal width, and under identical conditions will be the same for any probe. Nevertheless, the narrower the original O_2-free line, the more evident and measurable will any broadening be. A choice of a probe with a narrow line width is thus preferable, although it is of relatively minor importance when compared to points 1–3 above. In the examples given here, the use of the nitroxide 2,2,5,5-tetra-methyl-3-methanopyrroline-N-oxyl (PCAOL,[13] Fig. 1A) is demonstrated.

Probe Concentration

Oxygen-induced line broadening should theoretically be independent of probe concentration. At high probe densities, however, self-quenching, owing to interactions among the probe molecules, may lead to an increase in line width. A self-quenching curve may be easily obtained by examining line width as a function of concentration, and a concentration should be chosen below the point at which the width starts to increase. Normally, a range of 0.5–1 m*M* is within safe limits. In some cases, probe concentrations inside the cells may be much higher than in the medium, due, for instance, to pH or potential gradients. In these cases, self-quenching may occur intracellularly even though the overall probe level is low. Unless specifically desired, this is another good reason why charged (or ionizable) probes should be avoided.

[12] S. Belkin, R. J. Mehlhorn, K. Hideg, O. Hankovszky, and L. Packer, *Arch. Biochem. Biophys.* **256,** 232 (1987).

[13] R. H. Hammerstedt, A. D. Keith, P. C. Bolz, and P. W. Todd, *Arch. Biochem. Biophys.* **194,** 565 (1979).

Viscosity

The more viscous the solution, the broader the lines will be, and less sensitive to the effects of oxygen. Biological media of any composition are not viscous enough to cause any problems in this respect, but the situation may be different in the intracellular environment. When internal O_2 concentrations are assayed, this factor has to be checked. A good indication of viscosity may be obtained by determining the height ratio of two peaks in the spectrum, the high-field and midfield lines.[14] This ratio decreases with increasing viscosity and can be calibrated by comparison to a standard concentration series of a compound such as glycerol or sucrose. In *A. quadruplicatum,* for example, intracellular viscosity corresponds to that of approximately 50% glycerol, which is around 0.05 poise. The effect of this viscosity on line width is negligible when compared to that of O_2.

Calibration

In order to obtain a calibration factor correlating line width with O_2 concentrations, spectra should be taken with probe solutions equilibrated with different amounts of O_2 in N_2. For this purpose, the sample should be introduced into the spectrometer cavity in thin-walled gas-permeable tubing (e.g., 0.005 inch wall thickness, 0.032 inch internal diameter, Zeus Industrial Products, Raritan, NJ). The desired gas mixture (1–2 liter min^{-1}) is then flushed over the sample for 2–3 min, and the width of the midfield line (W_m) is measured for each O_2 concentration. The line width is the absence of O_2 is then subtracted from the other values and the degree of broadening calculated. In the case of PCAOL (0.5 m*M*), in the saline medium of *A. quadruplicatum* (18 g $liter^{-1}$ NaCl), at 25°, the calibration factor was calculated to be 3.94 m*M* O_2 $Gauss^{-1}$. Data relating O_2 solubility to temperature, salinity, and atmospheric pressure may be obtained from many references (see, e.g., Ref. 15).

O_2 Measurements: Total

For actual measurement of dissolved O_2, a thick cell suspension (1–2 mg Chl ml^{-1}) is recommended. It should be introduced into the cavity in thin (50–100 μl) glass capillaries, sealed at one end. Sample volume will

[14] R. J. Mehlhorn and A. D. Keith, *in* "Membrane Molecular Biology" (C. F. Fox and A. D. Keith, eds.), p. 192. Sinauer, Stamford, Connecticut, 1972.

[15] M. L. Hitchman, "Measurement of Dissolved Oxygen." Wiley, New York, 1978.

depend on the size of the capillary: for a 75-μl capillary, 50 μl of the preparation (including the spin probe at the desired concentration) should suffice. Illumination, when required, can be supplied via the grid of the cavity. When illuminating, care should be taken to minimize sample heating. This may be achieved by including water filters, by rapid air flow through the cavity, or by using a thermostated unit.

The width of a single line (preferably the midfield line) may be followed by one of two ways, employing either the first or the second derivative of the ESR spectrum.

Use of the First Derivative. The "regular" ESR spectrum, as presented in Fig. 1B, is actually the first derivative of the original resonance spectrum. In the approach, the line to be measured is traced before, during, and after illumination. Line width may then be measured for each trace and actual O_2 concentrations calculated as above. In Fig. 2A, evolu-

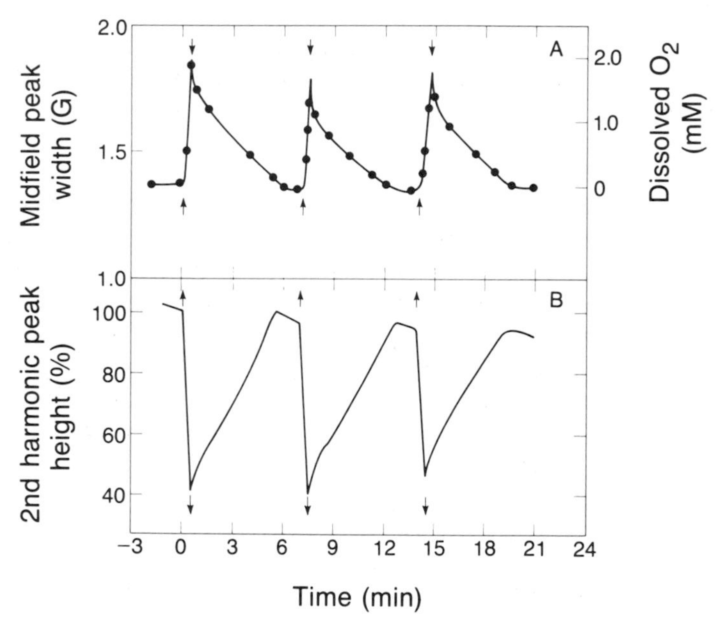

FIG. 2. O_2 evolution and uptake total. *Agmenellum quadruplicatum* cells (1.4 mg Chl ml^{-1}), with 0.5 m*M* PCAOL, were illuminated 3 times (0.5 min light–6.5 min dark intervals). (A) First derivative method: the midfield line was repeatedly traced, its width measured for each trace, and the data plotted as a function of time. (B) Second derivative method: the peak of the second derivative was continuously followed, 100% being its height at time zero. Upward point arrows indicate light on, downward pointing arrows, light off.

tion and uptake of dissolved O_2 by *A. quadruplicatum* are plotted as a function of time during a series of 30-sec illuminations, both in line-width units (Gauss) and O_2 concentrations.

Use of the Second Derivative. Some ESR spectrometers are equipped with the option of plotting the second derivative ("second harmonic"). As can be surmized from Fig. 1C, when the width of the first derivative line increases, the slope of the line between its peak and trough decreases. The second derivative's peak represents this slope, and it is correspondingly reduced in height. It is possible, therefore, rather than to repeatedly trace and measure the first derivative line, to "lock" the magnetic field onto the peak of the second derivative and follow the changes in its height with time. An example is presented in Fig. 2B, in which the cells were subjected to treatment identical to that in Fig. 2A. It can be seen that the results obtained from both methods mirror each other. It should be noted, however, that although the second derivative method is more convenient for continuous measurements, and is more sensitive to fast transients in O_2 concentrations, care should be taken in quantitative interpretation of the results obtained. By its nature, the second derivative is not linearly related to the first; thus, the higher the O_2 concentration, the lower the sensitivity will be. Above 2 m*M* O_2, the first derivative method is more reliable.

O_2 Measurement: Internal

One of the unique advantages of ESR oximetry is the possibility of isolating the fraction emanating from the intracellular space from the total signal observed. This is achieved by selectively quenching the ESR signal of all molecules not enclosed by the cell membrane. For this purpose a strongly paramagnetic molecule is used, which is impermeable to the cell. At the correct concentration, such a compound (usually a transition metal ion complex) will completely broaden the signals of all the nitroxide molecules outside the cells and only the internal signal will be "visible." The width of the midfield line of the latter may be used to measure intracellular O_2 levels, in a manner identical to that of the total O_2 concentration. As noted earlier, this assay should be calibrated separately. In Fig. 3, an example of such a measurement is presented, with the aid of 100 m*M* $Na_2MnEDTA$ as the impermeable broadening agent. Both the first and the second derivative methods are demonstrated. In the latter (Fig. 3A), a fast transient in intracellular O_2 is observed, which is not expressed in the first derivative method owing to the time gaps between the separate points.

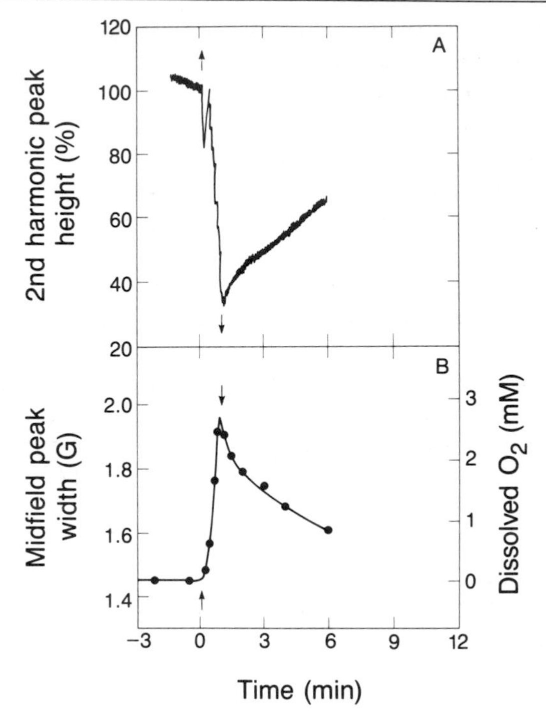

FIG. 3. O_2 evolution and uptake: internal. The same cell suspension as in Fig. 2, with 2 mM PCAOL and 100 mM $Na_2MnEDTA$, was used. Illumination was for 1 min. Upward arrows indicate light on, downward arrows, light off. (A) Second derivative. (B) First derivative.

Concluding Remarks

The application of ESR oximetry for cyanobacterial research presents unique advantages unshared by other methods. The use of a concentrated cell sample in a very small enclosed volume allows the rapid generation of extremely high O_2 concentrations (5 mM and above[11]). Due to the linearity of the method, these can be accurately measured in a concentration range unobtained by a regular oxygen electrode. The combination of high O_2 levels and externally provided illumination can be usd for fast generation of photooxidative conditions, as well as for an immediate study of their effects on the cyanobacterial system.[11] Finally, the method allows for direct observations of the changes in intracellular O_2 levels immediately following illumination, or in the course of the photosynthetic process.

Acknowledgments

This research was supported by the National Institutes of Health (Grant AG-04818) and by the Department of Energy (Grant DE-AT0380-ER10637). S. Belkin was a recipient of the Dr. Chaim Weizmann Postdoctoral Fellowship for Scientific Research.

[75] Determination of pH Gradients in Intact Cyanobacteria by Electron Spin Resonance Spectroscopy

By SHIMSHON BELKIN and LESTER PACKER

Introduction

The most accepted methodology for the measurement of internal pH in cells, organelles, or membrane vesicles involves the determination of the distribution of weak acids or bases across the membrane. The theory on which this approach is based has been discussed often in the past,[1-4] and the reader is referred to these reviews for a comprehensive discussion of its underlying principles. The main points may be summarized as follows: (1) The neutral species of the acid or the base (usually an amine) is freely permeable through the membrane, while the ion is not. (2) At equilibrium, the concentrations of the neutral species on both sides of the membrane will be equal. The concentration of the ionized species, on each side, will be determined by the pH. (3) If a weak acid is used, and pH_{in} is higher than pH_{out}, the neutral molecule, upon penetrating, will lose its proton, become negatively charged, and will be "trapped" inside. Uncharged molecules will thus continue to diffuse inward until equilibrium is reached, leading to an overall C_{in} higher than C_{out}. (4) When pH_{in} is lower than pH_{out}, an amine is used, to the same effect: the neutral amine will diffuse inside, to be protonated and trapped, leading again to $C_{in} > C_{out}$. (5) From the experimental determination of C_{in}/C_{out}, and the knowledge of pH_{out}, pH_{in} may be calculated.

Since the introduction of this approach, our understanding of pH gradients in cells, organelles, and various membrane preparations has greatly increased. This was mainly accomplished by the use of two types of weak acid or base probes, radioactively labeled or fluorescent. Spin-labeled

[1] H. Rottenberg, this series, Vol. 55, p. 547.
[2] E. Paden, D. Zilberstein, and S. Schuldiner, *Biochim. Biophys. Acta* **650,** 151 (1982).
[3] I. R. Booth, *Microbiol. Rev.* **49,** 359 (1985).
[4] E. R. Kashket, *Annu. Rev. Microbiol.* **39,** 219 (1985).

probes—nitroxides—have also been employed for the same purpose,[5–8] but were put to a much more limited use.

Nitroxides[9] are relatively stable free radicals, which can be designed to be of a varied chemical nature, for instance, that of a weak acid or an amine. They are easily detected by room-temperature electron spin resonance (ESR) spectroscopy, and unlike radiolabeled probes their distribution data are available immediately on the introduction of the sample into the spectrometer. In addition, the method allows for fast and reliable measurements of the water volume of the cell, under the same conditions employed for the determination of internal pH.

The ESR method has been shown to be effective and accurate in the determination of internal pH in various systems[5–8]; it has, however, very rarely been applied to intact cells. One of the main obstacles was that nitroxide spin probes are susceptible to signal loss in biological environments, either by reduction to the hydroxylamine or by irreversible destruction. The problem, however, may be satisfactorily circumnavigated if relatively stable probes are chosen (see below). This chapter describes the use of such probes for the determination of pH gradients in intact cyanobacterial cells.

Notation

The following notation is used:

P_{in}	% of probe molecules inside the cells
P_{out}	% of probe molecules outside the cells ($100\% - P_{in}$)
C_{in}	intracellular probe concentration
C_{out}	extracellular probe concentration
V_{in}	intracellular sample volume (%)
V_{out}	extracellular sample volume ($100\% - V_{in}$)
V_{cyt}	fraction of total cell volume occupied by the cytoplasm
V_{thy}	fraction of total cell volume occupied by the intrathylakoid space ($1 - V_{cyt}$)
pH_{out}	external pH
pH_{in}	internal pH
pH_{cyt}	cytoplasmic pH

[5] R. J. Mehlhorn and I. Probst, this series, Vol. 88, p. 334.

[6] R. J. Mehlhorn, P. Candau, and L. Packer, this series, Vol. 88, p. 751.

[7] R. J. Mehlhorn and L. Packer, *Ann. N.Y. Acad. Sci.* **414,** 180 (1983).

[8] B. A. Melandri, R. J. Mehlhorn, and L. Packer, *Arch. Biochem. Biophys.* **235,** 97 (1984).

[9] L. J. Berliner (ed.), "Spin Labeling: Theory and Applications," Vol. 2. Academic Press, New York, 1979.

pH_{thy}	intrathylakoid pH
pH_{av}	internal pH, assuming equal distribution of the probe in the total cell volume
ΔpH	$pH_{in} - pH_{out}$
$\Delta pH_{cyt-out}$	$pH_{cyt} - pH_{out}$
$\Delta pH_{cyt-thy}$	$pH_{cyt} - pH_{thy}$

Choice of Spin Probe

The choice of a spin probe for determination of pH gradients in a particular experimental system should be dictated by several factors, the most important of which is probe stability. As a general rule,[10] five-membered ring nitroxides are much less susceptible to signal loss than six-membered ring nitroxides. Thus, probes based on pyrroline or the pyrrolidine ring structures are preferable to those containing piperidine or hydropyridine rings, often used in the past.

Within these groups, differences between individual probes are often minor, as long as two more characteristics are selected for: lack of binding to cellular constituents and free permeability across the membrane. The degree of binding may be estimated from changes in the ESR spectrum created by the presence of the cells[5]; probes that showed indications of binding should be avoided. Permeability can be measured by following with time the intensity of the intracellular signal (determined as described below). If an equilibrium is quickly reached, i.e., if the probe is freely permeable, the observed signal should not increase with time. Unless a significantly long side chain is attached to the nitroxide,[7] equilibrium is immediate.

A set of three probes answering to these specifications is required for a complete pH assay: a neutral, acidic, and an amine nitroxide (see Fig. 1). The use of these three probes is described below.

Measurement and Calculation of C_{in}/C_{out}

The ESR assay is preferably conducted with a dense cell suspension (0.5–1.5 mg Chl ml^{-1}), containing the probe at the required concentration. The sample is introduced into the spectrometer cavity in a thin glass capillary (50–100 μl), sealed at the lower end, which is inserted in a regular-size ESR tube.

Probe concentrations should be determined experimentally. For the

[10] S. Belkin, R. J. Mehlhorn, K. Hideg, O. Hankovszky, and L. Packer, *Arch. Biochem. Biophys.* **256,** 232 (1987).

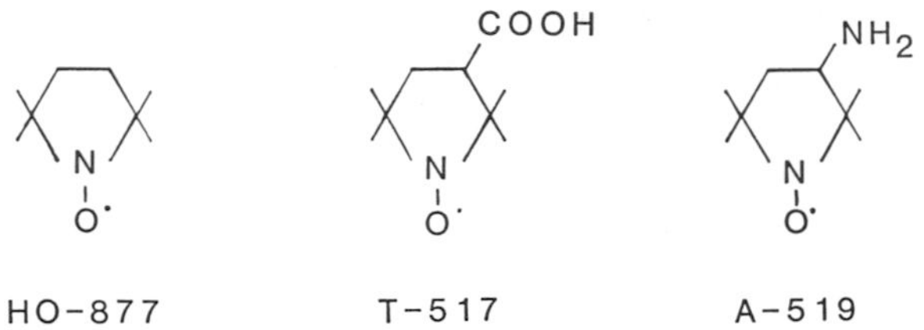

COOH NH2 N O HO-877 T-517 A-519

FIG. 1. Structures of the three nitroxide spin probes: HO-877, 2,2,5,5-tetramethylpyrrolidin-1-oxyl; T-517, 2,2,5,5-tetramethylpyrrolidin-1-oxyl-3-carboxylic acid; A-519, 3-amino-2,2,5,5-tetramethylpyrrolidin-1-oxyl. HO-877 was a generous gift from K. Hideg and O. Hankovsky (University of Pecs, Hungary); the other probes were obtained from Molecular Probes, Inc. (Junction City, OR).

gradient probes, the acid or the amine, the concentration should be low enough to prevent collapse of the pH gradients, but high enough to be detectable even when internal concentrations are low. A range of 20–100 μM is usually applicable.[11] Neutral probes for volume determination may be used at higher concentrations, 0.5–1 mM.

For the determination of the magnitude of the pH gradient, the ESR assay is basically conducted to yield a single experimental datum, P_{in}: *the fraction of the total probe signal in the preparation which emanates from the intracellular space*. For that purpose, the spectrum is traced twice: in the absence and in the presence of an impermeable paramagnetic broadening agent. At the correct concentration, such a compound (usually a transition metal complex) will totally quench the signal of the free nitroxide molecules in the medium[12] but, being impermeable, will not affect the probe inside the cells. In its presence, therefore, only the internal signal will be "visible." By measuring the height of a single line of the spectrum (preferably the midfield or the low-field line), with the broadening agent present, and calculating its ratio to the height of the same peak in the absence of this compound, the intracelluar fraction (P_{in}) of the total probe molecules is obtained.

If a neutral probe is used, P_{in} represents the total water volume of the cells in the preparation ($P_{in} = V_{in}$). If, on the other hand, a concentration gradient is established, the concentration ratio c_{in}/C_{out} should be calculated. Since $C_{in} = P_{in}/V_{in}$ and $C_{out} = P_{out}/P_{out}$, then

$$C_{in}/C_{out} = P_{in}/P_{out} \times V_{out}/V_{in} \tag{1}$$

[11] S. Belkin, R. J. Mehlhorn, and L. Packer, *Plant Physiol.* **84,** 25 (1987).

[12] A. D. Keith, W. Snipes, R. J. Mehlhorn, and T. Gunter, *Biophys. J.* **19,** 205 (1977).

Using C_{in}/C_{out} to Calculate pH_{in}

The simplest case for calculating pH_{in} is that in which the cell is comprised of a single compartment. Only one such cyanobacterium is known, *Gloeobacter violaceus*.[13] For the acid probe T-517 in this case,

$$\Delta pH = \log C_{in}/C_{out} \tag{2}$$

This relationship holds as long as $pH_{out} > pK + 1$.[1] Since the pK of T-517 is low (<4.0), for any pH_{out} higher than 5.0 no further calculations are necessary.

The distribution of the amine probe in a single-compartment organism should be reciprocal to that of the acid probe, and the pH calculated as in Eq. (2), with an opposite sign:

$$\Delta pH = -\log C_{in}/C_{out} \tag{3}$$

Equation (3) is applicable when $pH_{out} < pK - 1$. Since, however, the pK of A-519 is relatively low (7.55), it should be taken into account.[1] The internal pH, when calculated according to this probe is, therefore,

$$pH_{in} = -\log[C_{in}/C_{out}(10^{-pK} + 10^{-pH_{out}}, - 10^{-pK}] \tag{4}$$

In the example given in Table I, pH_{in} was calculated for *G. violaceus* at a pH_{out} of 7.1. The two probes were used independently, with a difference of only 0.04 pH units in the final results.

The situation is more complex in most cyanobacteria, in which the space enveloped by the thylakoid membrane is separate from the cytoplasm.[14,15] To calculate pH values for both intracellular compartments, their fractional volumes (V_{cyt} and V_{thy}) must be known.[15,16] Commonly used values are 0.93 and 0.07 for V_{cyt} and V_{thy}, respectively, as calculated for *Anacystis nidulans*,[17] though a V_{cyt} of 0.9 has also been used.[15]

Owing to the expected low pH of the thylakoid space, and its relatively small volume, it is assumed that practically all of the acidic T-517 probe will be found in the cytoplasmic compartment. C_{in}/C_{out} has to be corrected accordingly, and the pH gradient between the medium and the cytoplasm is

$$\Delta pH_{cyt-out} = \log(C_{in}/C_{out} \times 1/V_{cyt}) \tag{5}$$

[13] R. Rippka, J. B. Waterbury, and G. Cohen-Bazire, *Arch. Microbiol.* **100,** 419 (1974).

[14] G. A. Peschek, *Subcell. Biochem.* **10,** 85 (1984).

[15] G. A. Peschek, T. Czerny, G. Schmetterer, and W. H. Nitschmann, *Plant Physiol.* **79,** 278 (1985).

[16] H. W. Heldt, K. Werdan, M. Milovancev, and G. Geller, *Bichim. Biophys. Acta* **314,** 224 (1973).

[17] M. M. Allen, *J. Bacteriol.* **86,** 836 (1968).

TABLE I
DETERMINATION OF pH_{in} IN *Gloeobacter violaceus*[a]

Probe	Peak height (arbitrary units)		P_{in}	C_{in}/C_{out}	ΔpH	pH_{in}
	−Q	+Q				
HO-877 (200 μM)	2100	10.8	0.51	1.0	—	—
T-517 (100 μM)	720	4.5	0.63	1.23	0.09	7.19
A-519 (100 μM)	724	3.0	0.41	0.81	0.13	7.23

[a] Cell preparations (77μg Chl ml^{-1}), containing the appropriate probe, were scanned in the absence and presence of 95 mM of the broadening agent nickel tetraphenylenepentaamine [NiTEPA (Q)]. Power was 10 mW, modulation amplitude 1 gauss, time constant 0.064 sec, and the scan speed 10 cm min^{-1}. Peak heights of the midfield lines were measured, and the values were corrected for reduction and normalized to a single gain setting (10^4). Calculations were as described in the text. pH_{out} was 7.10.

The amine probe, on the other hand, will tend to accumulate in the thylakoid volume. Because of the large V_{cyt}, however, the amount of the probe present in the cytoplasm may not be ignored. To solve that,[15,16] an average pH (pH_{av}) is calculated using Eq. (4), as if the probe were equally distributed in the total cell volume. The relationship between pH_{cyt}, pH_{av}, and pH_{thy} is

$$10^{-pH_{av}} = 10^{-pH_{thy}} \times V^{thy} + 10^{-pH_{cyt}} \times V_{cyt} \tag{6}$$

and, therefore,

$$pH_{thy} = -\log[(10^{-pH_{av}} - 10^{-pH_{cyt}} \times V_{cyt})/V_{thy}] \tag{7}$$

An example of the measurement of peak heights and the subsequent calculations, as performed with the cyanobacterim *Agmenellum quadruplicatum,* is presented in Table II.

Effects of Oxygen and Light

In the experimental system described above, the samples are incubated in the dark. Since a dense cell suspension is recommended, O_2 levels in the preparation drop very rapidly to zero; this can be verified by

TABLE II
DETERMINATION OF INTERNAL pH VALUES OF *Agmenellum quadruplicatum*[a]

Probe	Peak height (arbitrary units) −Q	+Q	P_{in}	C_{in}/C_{out}	ΔpH (cyt − out)	pH_{cyt}	pH_{av}	pH_{thy}
HO-877 (1 m*M*)	7560	190	2.51	1.0	—	—	—	—
T-517 (100 μ*M*)	820	24	2.93	1.17	0.10	7.30	—	—
A-519 (20 μ*M*)	85	5.3	6.23	2.59	—	—	6.68	5.64

[a] Experimental conditions as in Table I, with an *A. quadruplicatum* suspension containing 0.82 mg Chl ml^{-1}. pH_{out} was 7.20.

measuring line width.[18,19] In order to carry out aerobic pH determinations, the samples are placed not in glass capillaries as before, but in thin-walled gas-permeable tubing.[18,19] The tubing is placed in a bottomless regular-size ESR tube, which is inserted into the spectrometer cavity. A stream of air (1–2 liter min^{-1}) is continuously passed through the tube and over the sample, and the assay conducted as above. Owing to the presence of O_2, the spectrum lines will be slightly broader and hence somewhat shorter.

The effect of light on ΔpH may also be followed. Gas-permeable tubing is again used; in this case a stream of O_2-free N_2 is passed over the sample, to avoid line broadening by accumulation of photosynthetically evolved O_2. Care should be taken, when illuminating, to avoid sample heating by the absorbed light. This is achieved by including a water filter the in the light path, by rapid air flow around the sample tube, or by the use of a thermostatted unit.

To measure probe distribution in the light, the spectrum is first traced in the dark. The sample is then illuminated through the cavity's grid, and changes in peak height are monitored. This is carried out either by consecutive traces of the measured line or, more conveniently, by "locking" the magnetic field onto the peak. Line height is then followed with time, that is, before, during, and after illumination. Light-induced changes in probe distribution, as obtained for both *G. violaceus* and *A. quadruplicatum*, are presented in Fig. 2. For each probe the change in peak height is

[18] S. Belkin, R. J. Mehlhorn, and L. Packer, *Arch. Biochem. Biophys.* **252,** 487 (1987).
[19] S. Belkin, R. J. Mehlhorn, and L. Packer, this volume [74].

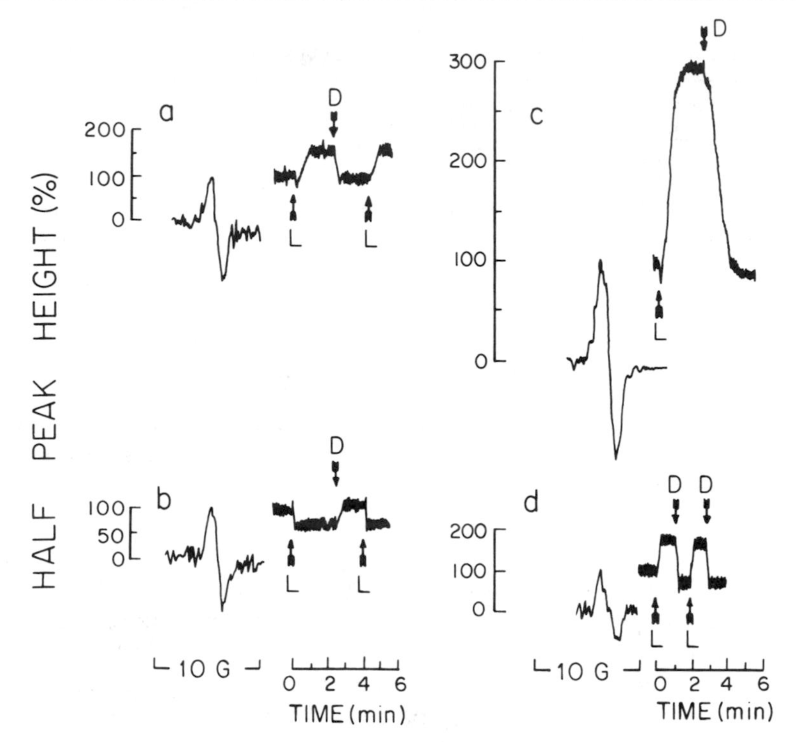

FIG. 2. Light-dependent pH changes in *Gloeobacter violaceus* and *Agmenellum quadruplicatum*. Cell suspensions (pH_{out} 7.5), containing the appropriate probe and NiTEPA (95 m*M*) as the broadening agent, were incubated in the spectrometer cavity in gas-permeable tubing under a flow of N_2 gas. For each probe, the midfield line was traced and the magnetic field then "locked" onto the peak of the line. Line height was followed with time during illumination, as marked by arrows: L, light on; D, light off. (a, b) *Gloeobacter violaceous* (77 μg Chl ml^{-1}). (c, d) *Agmenellum quadruplicatum* (0.82 mg Chl ml^{-1}). Probes were as follows: a, c, T-517, 100 μM; b, d, A-519, 100 and 20 μM, respectively. In all cases the power was 10 mW and the modulation amplitude 1 gauss. The scan range was 100 gauss, centered at 3390 gauss. Time course scans were conducted at 40 cm hr^{-1}, with the field set at zero and locked on the peak.

measured, and a new C_{in}/C_{out} ratio is calculated and incorporated into Eqs. (2)–(7) to yield internal pH values in the light.

In *G. violaceus* the light-dependent changes in the two probes mirror each other: the internal concentration of the acidic probe (T-517) increases, while that of the amine probe (A-519) decreases. Both phenomena point at an increase in pH_{in} of about 0.2 units. In *A. quadruplicatum,* the internal levels of both probes increase, indicating an increase in pH_{cyt} and a decrease in pH_{thy} of about 0.4 units each.

Correcting for Probe Reduction

In some cases, probe reduction (or destruction) may occur to some extent, even when relatively stable probes are used. In these instances, changes in line height are followed over several minutes, and the value at time zero (probe addition) is calculated. This value is then used in all subsequent calculations.

Concluding Remarks

In the past, the use of nitroxide spin probes for the determination of internal pH has been restricted mostly to various membrane preparations. Thanks to the introduction of more stable probes, we believe the methodology describe herein will serve to enhance our knowledge regarding this parameter in intact cyanobacteria. The advantages of this technique over other assays are numerous:

1. The results are immediately available; the internal ESR signal can be visualized as soon as the sample is introduced into the cavity of the spectrometer.
2. Cell volumes are also immediately apparent, and actual probe concentrations can therefore be calculated with no lag period.
3. Cell volumes are determined under conditions identical to those employed for assaying the distribution of the charged probes; the effects of osmotic changes in cell volume, or of "bound" water inaccessible to the probe molecules, are therefore canceled out.
4. The precision is high, and even minor differences in probe concentrations inside and outside the cell can be reproducibly discerned. Relatively small pH gradients can therefore be reliably determined.
5. For the same reason, one can use spin probes even at "the wrong side" of their p*K* without significant loss of sensitivity.
6. Since very small internal probe concentrations can be detected, negative probe accumulation can be measured, up to a certain limit, as easily as a positive one.
7. Probe binding may still be a problem in correct interpretation of the experimental results, as in other methods; the easily demonstrated reciprocity of oppositely charged probes, however, may serve as a convincing control.

Acknowledgments

Supported by the National Institutes of Health (Grant AG-04818) and by the U.S. Department of Energy (Grant DE-AT03-80-ER 10637). S. Belkin was a recipient of the Dr. Chaim Weizmann Postdoctoral Fellowship for Scientific Research.

[76] Mass Spectrometric Measurement of Photosynthetic and Respiratory Oxygen Exchange

By BERNARD DIMON, PIERRE GANS, and GILLES PELTIER

Introduction

When photosynthetic organisms are illuminated, net oxygen exchange generally results from the superposition of evolution and uptake processes. In order to determine the levels of each unidirectional oxygen flux, oxygen isotopes can be used. In this respect, the stable isotope ^{18}O is the most useful, and, therefore, mass spectrometry is the technique of choice to measure the isotopic content.

The method based on the analysis of dissolved gases admitted to the mass spectrometer via a membrane inlet system[1] has been widely used to measure oxygen exchange in green algae,[2-4] cyanobacteria,[5] isolated cells,[6,7] and chloroplasts[7,8] and has been described elsewhere in this series.[9] The short response time of the membrane inlet makes this technique particularly suited to the study of rapid induction phenomena. The alternative method we present here is based on the continuous analysis of the gas phase in contact with the cellular suspension. It presents the advantage that oxygen exchange rates can be determined under steady-state conditions (i.e., at constant concentrations of CO_2 and O_2).

Principle of the Method

When placed in the presence of ^{18}O-enriched oxygen, an illuminated photosynthesizing organism takes up all the isotopic species present without significant discrimination. At the same time, it produces oxygen with the natural isotopic content (^{16}O = 99.8%, ^{18}O = 0.2%) from the photolysis of intracellular water. In our experiments the ^{18}O content of the oxy-

[1] G. Hoch and B. Kok, *Arch. Biochem. Biophys.* **101,** 160 (1963).
[2] R. Gerster, B. Dimon, and A. Peybernes, *Proc. Int. Congr. Photosynth., 3rd,* p. 1589 (1974).
[3] R. Radmer and B. Kok, *Plant Physiol.* **58,** 336 (1976).
[4] G. Peltier and P. Thibault, *Plant Physiol.* **79,** 225 (1985).
[5] G. Hoch, O. Owens, and B. Kok, *Arch. Biochem. Biophys.* **101,** 171 (1963).
[6] P. Behrens, T. Marsho, and R. Radmer, *Plant Physiol.* **70,** 179 (1982).
[7] R. Furbank, M. Badger, and C. Osmond, *Plant Physiol.* **70,** 927 (1982).
[8] H. Egneus, U. Heber, U. Matthiesen, and M. Kirk, *Biochim. Biophys. Acta* **408,** 252 (1975).
[9] R. Radmer and O. Ollinger, this series, Vol. 69, p. 547.

gen added to the atmosphere was higher than 99%. Under these conditions, the contribution of the $^{16}O^{18}O$ species can be neglected and total oxygen uptake (U_O) can be deduced from the only variation of the heavy species ($^{18}O_2$) in the following manner:

$$U_O = -(1 + {}^{16}O_2/{}^{18}O_2)\ d{}^{18}O_2/dt$$

The "true" evolution of oxygen (E_O) is then

$$E_O = d{}^{16}O_2/dt - ({}^{16}O_2/{}^{18}O_2)\ d{}^{18}O_2/dt$$

To correct for variations in gas concentration arising from the mass spectrometer gas consumption and from eventual instabilities caused by the inlet system, argon contained in air is taken as an internal reference. The intensities of the three mass peaks $m/e = 32$ ($^{16}O_2$), $m/e = 36$ ($^{18}O_2$), and $m/e = 40$ (Ar) are continuously recorded. For this purpose, the ionic current corresponding to each mass peak is integrated during 1024 msec every 20 sec. The result of the integration is stored in the computer memory and recorded on a floppy disk at the end of the experiment. After calibration the results are expressed as cumulated moles of oxygen exchanged (consumed and produced) as a function of time. The net O_2 evolution is calculated as the difference between E_O and U_O.

Description of the Experimental Device

The experimental device contains the following components:

(1) A *mass spectrometer* based on a 80° magnetic sector of 14 cm radius (Type MM 14-80, V.G. Instruments); this machine, monitored by an Apple II microcomputer, is equipped with three collectors permitting the simultaneous measurement of the ion species 32, 36, and 40. The multicollection associated with a differential pumping system allows a precision better than 1/1000 in the measurement of isotopic ratios. (2) A *variable leak thermovalve* (UDV 135, Balzers) is used to continuously introduce gases in the ion source. A thermovalve controller (RVG 050, Balzers), acting on the valve aperture, maintains the total ionic current constant. (3) A *closed gas circuit* as shown in Fig. 1 is used; particular attention has been focused on the quality of the circuit components to limit leaks to a minimum level. The gases are bubbled through the suspension contained in the reaction vessel at a rate of about 30 liter hr^{-1}. The CO_2 concentration in the gas phase can be changed by injecting a bicarbonate solution in the liquid phase with a hypodermic needle introduced through a septum located at the bottom of the reaction vessel. The pH and the temperature are regulated during the experiment at ±0.02 pH and ±0.5°, respectively. A saturating light intensity of about 1000 $\mu E\ m^{-2}$

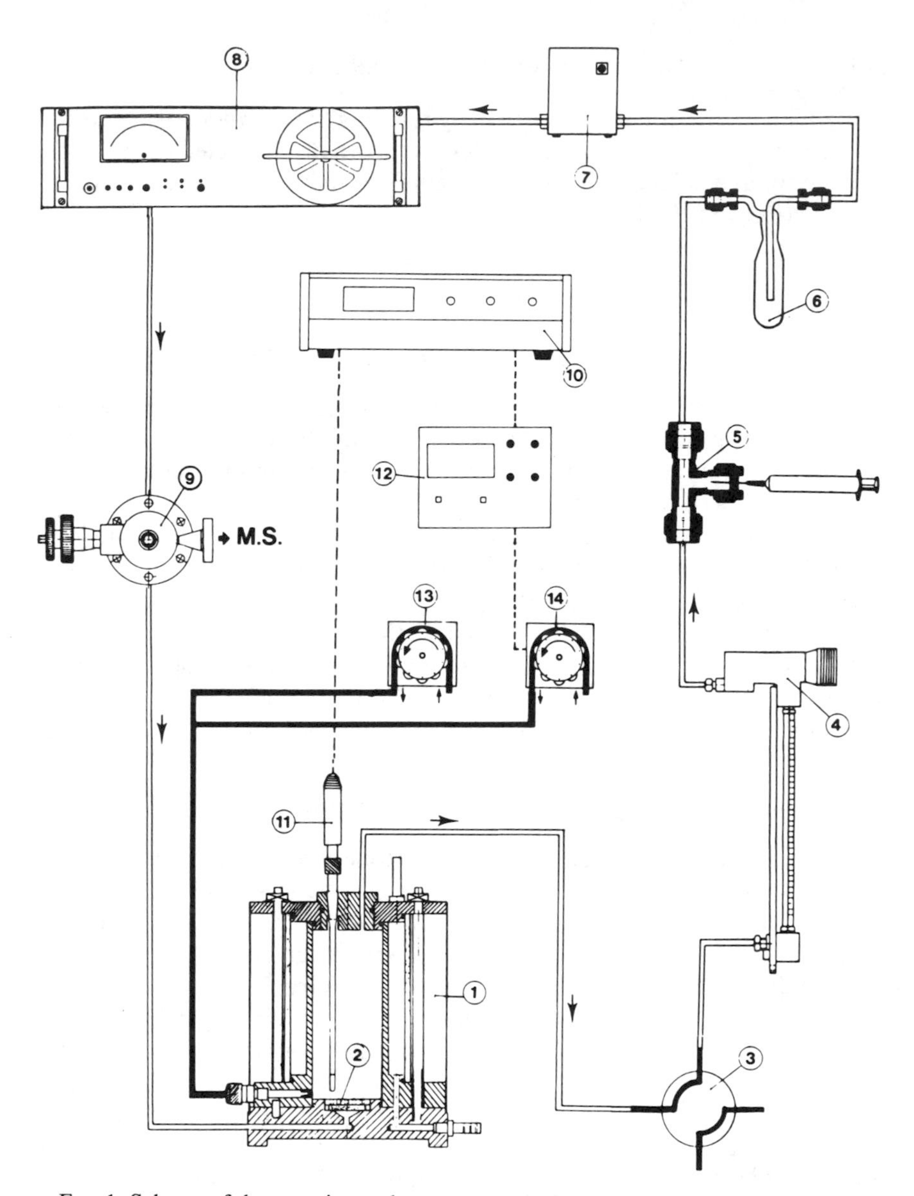

FIG. 1. Scheme of the experimental system: (1) plexiglass reaction vessel (inner volume 100 ml) thermoregulated by a water circulation jacket, (2) fritted steel disk for bubbling the gas phase through the suspension, (3) four-port valve (Valco), (4) rotameter fitted with a fine metering value (A.S.M.), (5) gas-injector equipped with a septum, (6) water trap cooled at $-15°$ in a cryogenic bath, (7) membrane pump (Metal Bellows, 41E) ensuring circulation and quick homogenization of the gas phase, (8) infrared CO_2 analyzer (Maihak, UNOR 6-N), (9) variable leak valve (Balzers, UDV 135) for the introduction of gases to the ion source of the mass spectrometer, (10) pH meter (Radiometer, PHM 82), (11) pH electrode, (12) pH regulator monitoring the peristaltic pump, (13) and (14) peristaltic pumps (Gilson, Minipuls 2) for acid (H_2SO_4, 0.1 *N*) and bicarbonate (20 m*M*) injections.

sec^{-1} (400–700 nm) is supplied with a projector (Oriel) equipped with a 1000-W quartz–halogen lamp. A water filter is used to reduce unwanted infrared radiations.

One Application of the Method

The method was applied to study the effect of CO_2 concentration on oxygen-exchange rates in the cyanobacterium *Anabaena* 7120. For this purpose, the following procedure was used. Cells were harvested by centrifugation (10 min at 5000 *g*) during exponential growth and resuspended in 70 ml of a phosphate buffer (pH 6.0). The total chlorophyll content was about 1 mg. The suspension was introduced into the reaction vessel, and the circuit was sparged with air for about 5 min. After completion of the sparging, the circuit was closed and 3 ml of $^{18}O_2$ (99.1 ^{18}O atom %) was injected into the gas phase (total volume of 135 ml) with a gas syringe. After switching on the light, the CO_2 initially present in the circuit was consumed by the photosynthetic activity of the suspension until the CO_2 compensation point was reached. At this time, the values of the mass peaks 32, 36, and 40 were recorded as a function of time.

Oxygen exchange was followed at the CO_2 compensation point for at least 15 min, the minimum period to obtain sufficient accuracy of measurement of the exchange rates from the slope of the integral curves (standard deviation of ~0.01 μmol min^{-1} mg^{-1} Chl for a chlorophyll content of ~1 mg). Bicarbonate is then injected at a constant rate. When the bicarbonate injection rate is not sufficient to saturate the photosynthetic activity, the CO_2 concentration in the gas phase reaches a constant value. Under such conditions the oxygen-exchange activity can be measured at a constant CO_2 level. At the same time, the relative increase of the oxygen concentration is less than 2% because of the high ratio between the volume of the gas phase and the quantity of oxygen evolved from the cells.

The data reported in Fig. 2 are relative to one typical experiment performed on *Anabaena* 7120. The CO_2 concentration measured in the gas phase is shown in Fig. 2A. Each point shown in Fig. 2B represents the cumulative amounts of oxygen exchanged ("true" photosynthesis, net photosynthesis, and uptake) from the beginning of the experiment. At the CO_2 compensation point (about 12 μl $liter^{-1}$ CO_2), oxygen evolution was strictly compensated by uptake and no net O_2 production was observed.

When bicarbonate was injected (at 20 min), the CO_2 concentration increased and reached a constant level (~95 μl $liter^{-1}$). "True" photosynthesis was stimulated (from 1.14 to 1.40 μmol min^{-1} mg^{-1} Chl) while oxygen uptake was slightly inhibited (from 1.14 to 1.04 μmol min^{-1} mg^{-1} Chl). By increasing the bicarbonate injection rate (at 40 min), the CO_2

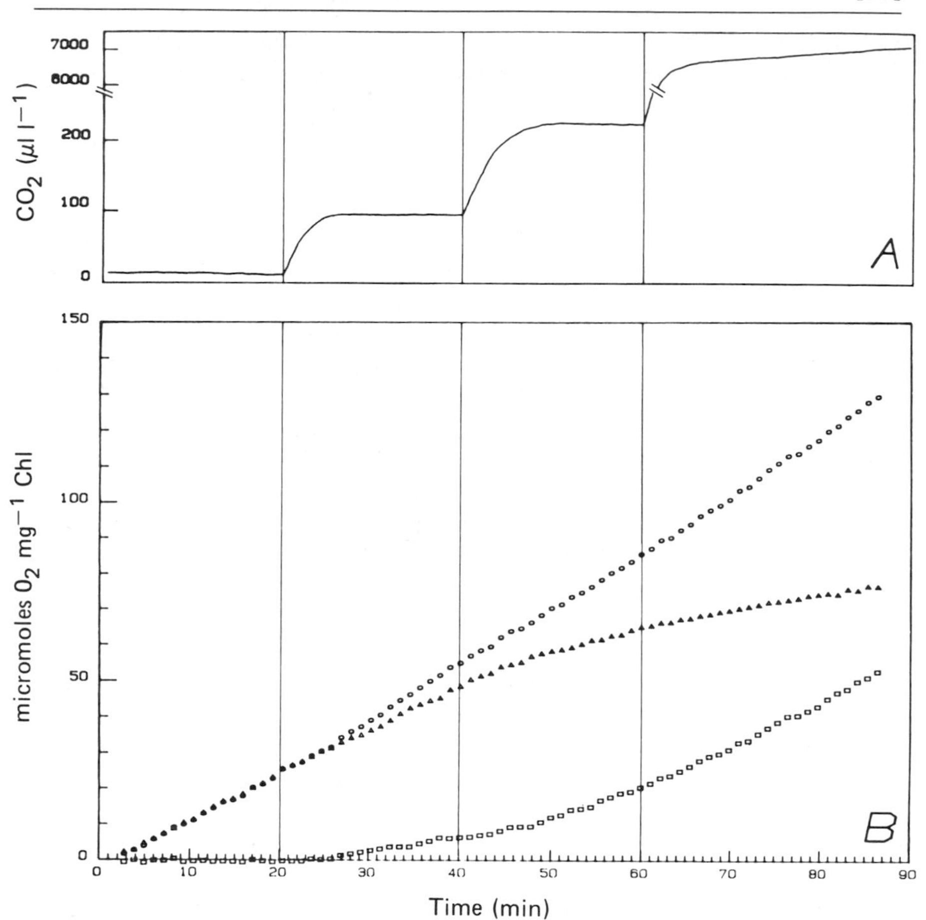

FIG. 2. Oxygen exchange exhibited in the light (1000 μE m^{-2} sec^{-1}) by *Anabaena* 7120 in the presence of various CO_2 concentrations. (A) CO_2 concentration in the gas phase. (B) ○, "True" photosynthesis; △, oxygen uptake; □, net photosynthesis. During the experiment the temperature was 25 ± 0.5°, and the pH value was regulated at 6.00 ± 0.02.

concentration reached an higher constant level (about 220 μl liter^{-1} CO_2). "True" photosynthesis remained practically unchanged whereas oxygen uptake was dramatically inhibited (from 1.04 to 0.60 μmol min^{-1} mg^{-1} Chl). After 60 min, the bicarbonate injection rate applied was higher than the maximum CO_2 assimilation ability of the cells, thus explaining the continuous increase of the CO_2 concentration. Finally, when the CO_2 concentration is higher than 100 μl liter^{-1} we observe a complete competition between O_2 and CO_2 for utilization of the reducing power. Under

CO_2-limiting conditions, O_2 photoreduction is unable to sustain the electron flow at its maximum rate.

In contrast to the oxygen evolution process which is due to only the activity of photosystem II, several oxygen uptake phenomena might be involved in the light [Mehler reactions, oxygenase activity of ribulose-bisphosphate carboxylase (Rubisco), respiratory oxidations]. In most photosynthetic organisms, and particularly in cyanobacteria, the nature and the relative proportion of the different O_2 uptake mechanisms implicated remain to be established. This problem has been partially solved in the green alga *Chlamydomonas* where the use of steady-state measurements allowed determination of the fraction of O_2 uptake due to photorespiration.[10] This emphasizes the fact that, under some circumstances, measurements obtained under steady-state conditions can be of a great help.

Acknowledgments

The authors thank P. Carrier and F. Sarrey for their helpful technical assistance.

[10] G. Peltier and P. Thibault, *Plant Physiol.* **77,** 281 (1985).

[77] Laser Light Scattering Techniques for Determining Size and Surface Charges of Membrane Vesicles from Cyanobacteria

By MARIE-EMMANUEL RIVIÈRE, GEORGES JOHANNIN, DIDIER GAMET, VÉRONIQUE MOLITOR, GÜNTER A. PESCHEK, and BERNARD ARRIO

Introduction

Laser light scattering techniques allow accurate determination not only of the diffusion coefficient and, hence, the volume of membrane vesicles but also of intact cells. The surface potential is also obtained by these techniques which have the great advantage of being noninvasive.[1,2] Both internal volume and surface charges of vesicles and cells are important parameters for the evaluation of transmembrane chemiosmotic gradients.[3] The internal volume can be determined by widely different methods

[1] M. W. Steer, *NATO ASI Ser.* **59,** 43 (1983).

[2] J. H. Kaplan and E. E. Uzgiris, *J. Immunol.* **117,** 115 (1976).

[3] H. Rottenberg, this series, Vol. 55, p. 547.

such as radioactive tracer techniques which result in exclusion volume,[3] electron paramagnetic resonance spectrometry of spin probes enclosed in the cells or vesicles,[4,5] and planimetry on electron micrographs.[6,7] For surface charge determination, ESR, spectrophotometric, and fluorometric probes give information on local charge densities. The aim of this chapter is to illustrate the application of quasi-elastic laser light scattering (QELS) and laser Doppler velocimetry (LDV) to the determination of volumes and surface charges of plasma (CM) and thylakoid (intracellular, ICM) membrane vesicles prepared from normal and salt-stressed cyanobacteria according to a recently established method.[8,9]

Methodology and Evaluation of Data

The theory and instrumentation of QELS and LDV, as well as details regarding data analysis, have been extensively reviewed in this series,[10–12] so only a few relevant statements and equations are briefly summarized here. When a beam of monochromatic light is scattered by a suspension of identical particles the scattered light spectrum has a Lorentzian shape. This is due to the Doppler frequency shift provoked by the random motion of the particles in suspension. The phenomenological equations that describe the power spectrum of the photocurrent may be briefly summarized as follows:

$$P \simeq \frac{\mathbf{K}^2 D_t}{\omega^2 + (\mathbf{K}^2 D_t)^2} \tag{1}$$

$$P \simeq \frac{\mathbf{K}^2 D_t}{[\omega - \mathbf{K}V\cos(\theta/2)]^2 + (\mathbf{K}^2 D_t)^2} \tag{2}$$

Equations (1) and (2) are valid in the absence and in the presence of an external electric field, respectively, where $\mathbf{K}$ is the scattering vector $[\mathbf{K} = (4\pi n/\lambda)\sin\theta/2]$, λ the laser wavelength *in vacuo*, n the refractive index of the medium, θ the scattering angle, D_t the translational diffusion coefficient, $\omega = 2\pi\nu$ the frequency of the fluctuations of the photocurrent

[4] E. Blumwald, R. J. Mehlhorn, and L. Packer, *Proc. Natl. Acad. Sci. U.S.A.* **80,** 2599 (1983).

[5] D. S. Cafiso, W. L. Hubbell, and A. Quintanilha, this series, Vol. 88, p. 682.

[6] M. M. Allen, *J. Bacteriol.* **96,** 836 (1968).

[7] W. H. Nitschmann and G. A. Peschek, *J. Bacteriol.* **168,** 1205 (1986).

[8] N. Murata and G. A. Peschek, this series.

[9] V. Molitor, W. Erber, and G. A. Peschek, *FEBS Lett.* **204,** 251 (1986).

[10] S. B. Dubin, this series, Vol. 26, p. 119.

[11] V. A. Bloomfield and T. K. Lim, this series, Vol. 48, p. 415.

[12] B. Ehrenberg, this series, Vol. 127, p. 678.

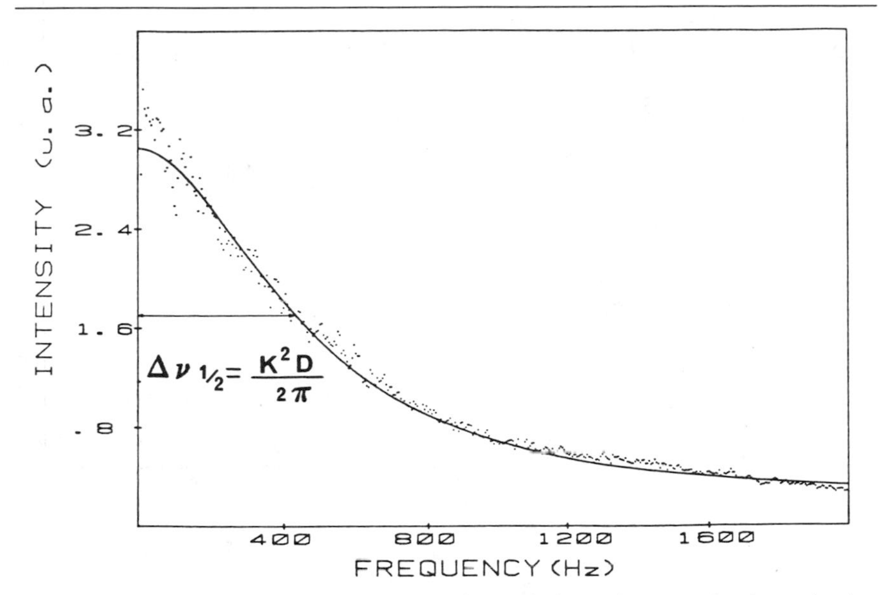

FIG. 1. Power spectrum of the photocurrent in a typical experiment on size determination by QELS of ICM vesicles obtained from normally grown cells. Buffer was 10 m*M* HEPES, pH 7.2; observation angle 90°, temperature 25°, laser wavelength 5140 Å. Experimental points are shown as dots; —, fitted Lorentzian curve.

arising from the scattered light falling on the phototube, and V the velocity of the particles in the electric field E. These two situations are illustrated in Figs. 1 and 2a.

Equation (1) corresponds to a Lorentzian spectrum (Fig. 1) whose half-width at half-height is proportional to the translational diffusion coefficient of the scattering particles according to Eq. (3):

$$\Delta\nu_{1/2} = \frac{\Delta\omega_{1/2}}{2\pi} = \frac{\mathbf{K}^2 D_t}{2\pi} \tag{3}$$

In the presence of a uniform electric field, the particles move with a constant velocity V, and, in this case, the Doppler effect is revealed by an overall shift of the photocurrent spectrum (Fig. 2a):

$$\Delta\nu = \frac{\mathbf{K}V\cos(\theta/2)}{2\pi} \tag{4}$$

The velocity V is related to the electrophoretic mobility μ and to the electric field strength E:

$$V = E\mu \tag{5}$$

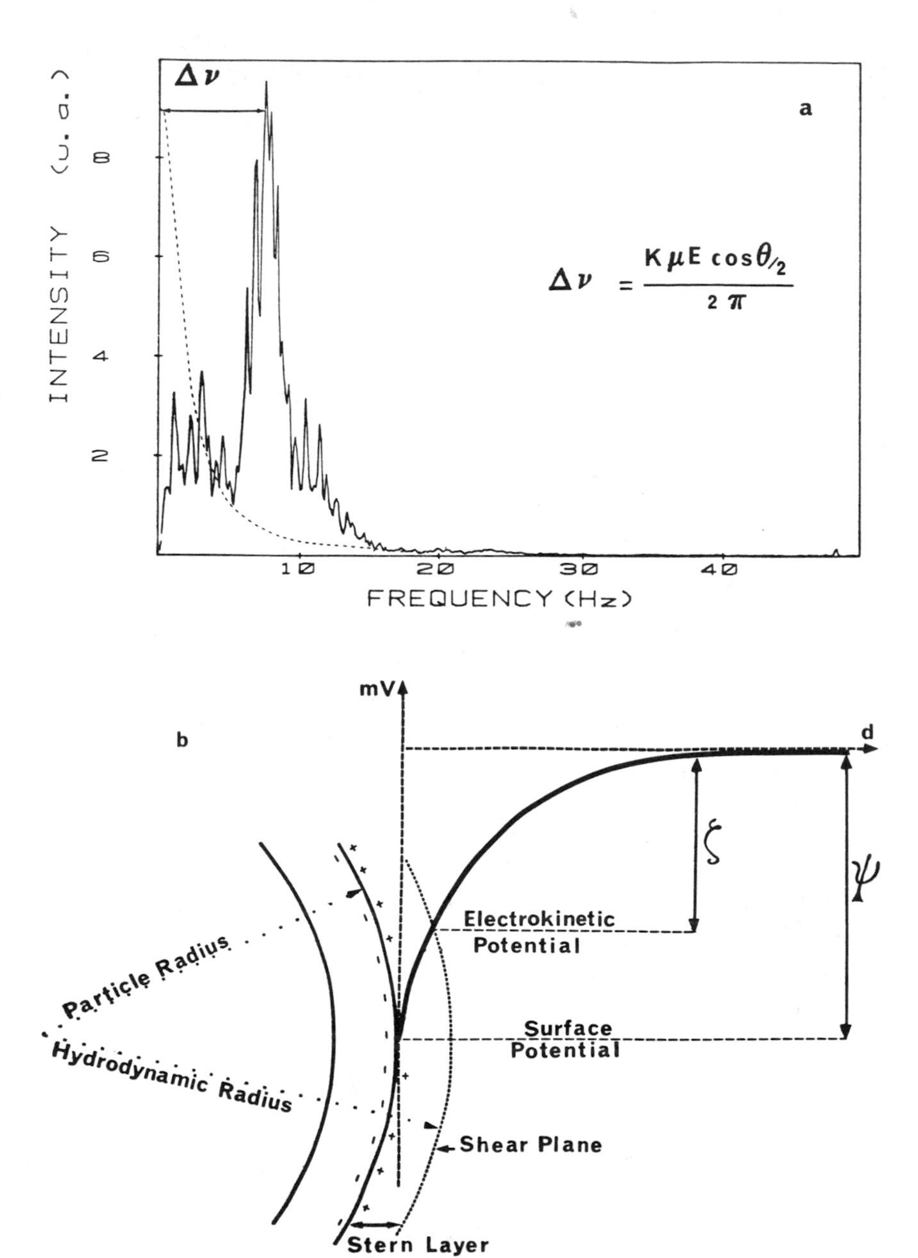

FIG. 2. (a) Power spectrum of the photocurrent in a LDV experiment on charge density determination of ICM vesicles obtained from normally grown cells. Buffer was 10 mM HEPES, pH 7.2; observation angle 5°, temperature 25°, laser wavelength 5140 Å, electric field strength 30 V/cm. Note the different scattering angle (5°) compared to Fig. 1. The dotted line represents the broadening of the laser line in the absence of electric field (Lorentzian spectrum of the vesicles observed at 5°). (b) Schematic representation of a membrane defining the terminology used to describe its electrokinetic properties.

Thus, the relation between the Doppler shift $\Delta\nu$ and the electrophoretic mobility μ is

$$\Delta\nu = \frac{\mathbf{K}\mu E \cos(\theta/2)}{2\pi} \tag{6}$$

Finally, the translational diffusion coefficient D_t allows calculations of the hydrodynamic radius R for spherical particles through the Stokes equation

$$D_t = \frac{kT}{6\pi\eta R} \tag{7}$$

where k is the Boltzmann constant, T the absolute temperature, and η the viscosity of the medium.

QELS and LDV Apparatus

The optical device for LDV and QELS experiments is similar to that described previously.[13-15] Measurement of the translational diffusion coefficient D_t is carried out in a thermostatted fluorescence cuvette. Two lasers are used for this work: He–Ne and Ar (Models 124B and 165 from Spectra Physics). The scattered light is detected by a 2234B Radiotechnique photomultiplier equipped with an Oltronix high-voltage power supply. The power spectrum of the photocurrent is displayed on a Rockland spectrum analyzer (Model 512). Data are transferred and analyzed on a DEC computer (Model MINC), by means of a program already described.[16]

In order to check the validity of the results obtained by these methods, all the tests described in Ref. 14 are performed. The dependence of $\Delta\nu_{1/2}$ and $\Delta\nu$ on laser wavelength, scattering angle, temperature, and viscosity of the medium are in agreement with theoretical equations. A linear relationship exists between the Doppler frequency shift and the electric field strength (from 10 to 100 V/cm). Usually, the LDV experiments were recorded at 20 V/cm. For QELS experiments, controls with Dow latex spheres (R = 86.5 nm) may be done before the measurement on the vesicles, usually giving an accuracy of 3%.

[13] B. A. Smith and B. R. Ware, *Anal. Clin. Chem.* **2,** 29 (1978).

[14] R. Mohan, R. Steiner, and R. Kaufmann, *Anal. Biochem.* **70,** 506 (1976).

[15] B. Arrio, G. Johannin, A. Carrette, J. Chevallier, and D. Brethes, *Arch. Biochem. Biophys.* **228,** 220 (1984).

[16] D. Brethes, D. Dulon, G. Johannin, B. Arrio, T. Gulik-Krzywicki, and J. Chevallier, *Arch. Biochem. Biophys.* **246,** 355 (1986).

Evaluation of Electrokinetic Parameters

A general expression of the relationship between the electrophoretic mobility and the electrokinetic potential (the zeta potential) is given by Henry's equation[17]:

$$\mu = \frac{2\varepsilon_0\varepsilon_r}{3\eta}\zeta f(\kappa R) \tag{8}$$

where ε_0 and ε_r are the vacuum permittivity and the dielectric constant, respectively, η the viscosity of the medium, ζ the zeta potential, and $f(\kappa R)$ Henry's function in which R is the radius of the particle and κ the inverse of the "electrical double layer" thickness defined by the Debye–Hückel theory. κ is a function of the ionic composition of the medium according to Eq. (9):

$$\kappa = \left(\frac{2e^2NI}{1000\varepsilon_0\varepsilon_r kT}\right)^{1/2} \tag{9}$$

where e is the electron charge, N Avogadro's number, k Boltzmann's constant, T absolute temperature, and I ionic strength of the solution. The values of Henry's function, $f(\kappa R)$, are determined from the curve $f(\kappa R)$ versus $\log(\kappa R)$.[18]

The relationship between the electrophoretic mobility and the electric charge of the particles is based on the "electrophoretic" theory developed by Henry and the "electrical double layer" theory of Gouy and Chapman. The use of these theories requires some assumptions and approximations which give a relation between charge density and electrophoretic mobility for spherical particles. Since it is not the purpose of this chapter to discuss these theories in detail, interested readers are referred to the many excellent reviews which deal with these points.[19–24] In this study the surface charge density is calculated from an approximate equation for spherical particles[25]:

$$\sigma = \frac{Q}{4\pi R^2} = \varepsilon_0\varepsilon_r\zeta\,\frac{1 + \kappa R}{R} \tag{10}$$

[17] D. C. Henry, *Proc. R. Soc. London, Ser. A* **133,** 106 (1931).

[18] C. Tanford, "Physical Chemistry of Macromolecules." Wiley, New York, 1961.

[19] R. J. Hunter, *in* "Zeta Potential in Colloid Science: Principles and Applications" (R. H. Ottewill and R. L. Rowell, eds.). Academic Press, New York, 1981.

[20] P. H. Wiersema, A. L. Loeb, and J. T. G. Overbeek, *J. Colloid Interface Sci.* **22,** 78 (1966).

[21] R. W. O'Brien and L. R. White, *J. Chem. Soc. Faraday Trans. 2* **74,** 1607 (1978).

[22] S. Levine, M. Levine, K. A. Sharp, and D. E. Brooks, *Biophys. J.* **42,** 127 (1983).

[23] D. A. Haydon, *Biochim. Biophys. Acta* **50,** 450 (1961).

[24] M. Mille and G. Vanderkooi, *J. Colloid Interface Sci.* **61,** 455 (1977).

[25] J. T. G. Overbeek and J. Lijklema, *Electrophoresis* **1,** 1 (1959).

where Q is the electric charge of the particle. The combination of Eqs. (8) and (10) gives

$$\sigma = \frac{3}{2}\eta\mu\frac{1 + \kappa R}{Rf(\kappa R)} \tag{11}$$

Equation (11) is valid only for low values of zeta potential (<25 mV).[25] In similar studies many authors used the Gouy equation, which describes the relationship between the surface charge density and the surface potential (ψ_0). For charged particles in solution containing symmetrical electrolytes at concentration C and valency Z, the equation is

$$\sigma = (8CN\varepsilon_0\varepsilon_r kT)^{1/2}\sinh\left(\frac{Ze\psi_0}{2kT}\right) \tag{12}$$

Equation (12) is determined for a planar surface and provides a good approximation when the "electrical double layer" thickness ($1/\kappa$) is assumed to be small in comparison to the particle radius. For unsymmetrical electrolytes a more general form of this equation must be used:

$$\sigma = \left((2\varepsilon_0\varepsilon_r NkT)\sum_i C_i\left[\exp\left(-\frac{Z_i e\psi_0}{kT}\right) - 1\right]\right)^{1/2} \tag{13}$$

TABLE I
HYDRODYNAMIC RADIUS (R) AND ELECTRICAL SURFACE CHARGE DENSITY (σ) OF VESICLES[a]

Membrane vesicle	R (μm)	Charge density ($e^-/\mu m^2 \times 10^{-4}$)
CM (−)	0.200 ± 0.013	2.0
CM (+)	0.08 ± 0.02	2.0
ICM (−)	0.08 ± 0.01	1.5
ICM (+)	0.08 ± 0.01	1.2

[a] Isolated and purified plasma (CM) and thylakoid (ICM) membrane vesicles from *Anacystis nidulans* were measured after growth in the absence (−) and presence (+) of 0.4 M NaCl by QELS and LDV. Membrane vesicles were prepared according to Ref. 8; cells were grown as described in Ref. 9. Values of R are from up to five independent preparations; values of σ are from only one representative preparation. (Large variations are found with different preparations, however, without a change of internal consistency, meaning that values varied as CM > ICM.)

We point out that electrophoretic mobility measurements lead to the determination of the electrokinetic potential which reflects the surface charge at the shear plane of the particle. Although a usual simplification is to consider that the zeta potential matches the surface potential ψ_0, one should keep in mind that they can be quite different (see Fig. 2b).

Plasma and Thylakoid Membrane Vesicles from Cyanobacteria: A Case Study

Membrane vesicles are prepared according to Ref. 8 from *Anacystis nidulans* cells grown as described in Ref. 9 for freeze–fracture electron

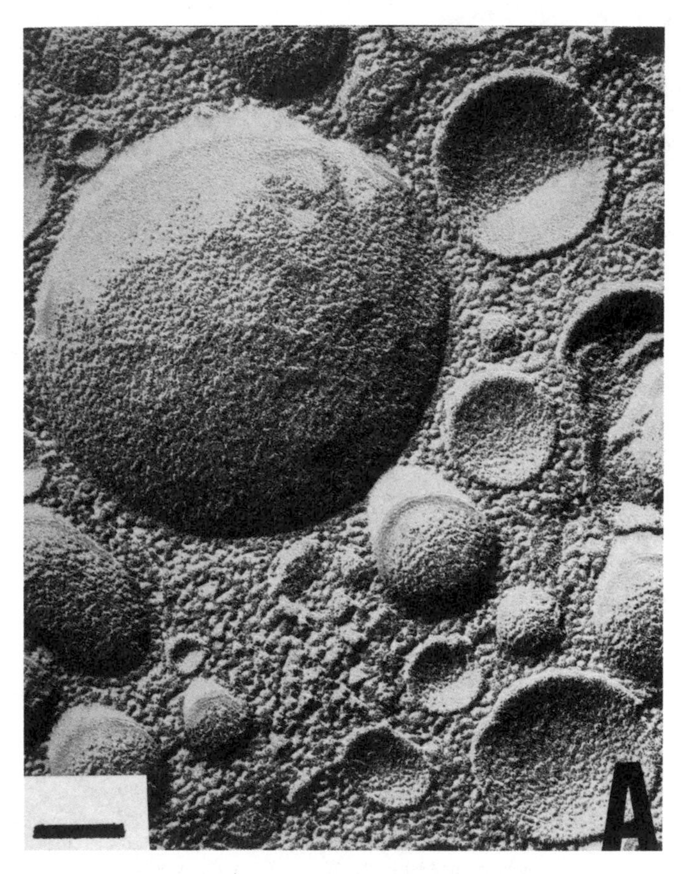

FIG. 3. Freeze–fractured plasma (A) and thylakoid (B) membrane vesicles of *Anacystis nidulans* from mid to late logarithmic cultures. Glutaraldehyde fixation and glycerination of the vesicles were performed according to Armond and Staehelin.[26] Bar, 100 nm.

FIG. 3. (*continued*)

microscopy, QELS, and LDV experiments. Figure 3[26] shows the appearance of plasma (A) and thylakoid (B) vesicles on freeze–fracture electron micrographs. QELS and LDV data obtained on these vesicles, and on similar vesicles from cells grown in the presence of 0.4 *M* NaCl,[9] are given in Table I. The main observation is the variation in the CM vesicle radius compared to that of the ICM vesicle after the salt adaptation. This decrease in the radius of the CM vesicle may be related to an increase in lipid content as revealed by electron microscopy.

Surface charge densities are not affected by the salt adaptation, but this parameter is very sensitive to the external medium as shown by Eqs. (12) and (13). Further studies on the interaction between these mem-

[26] P. A. Armond and L. A. Staehelin, *Proc. Natl. Acad. Sci. U.S.A.* **76,** 1901 (1979).

branes and ions and on the biochemical modifications of the membranes during the salt adaptation might reveal hidden differences. We have already observed that an increase in the number of negative charges on a membrane leads to a lower variation of charge density than the value predicted by theory.[16] This may be a compensation for the increase of electrostatic energy due to ionic interactions and structural modifications.

It should also be mentioned that the different methods available for determining the size and the surface charges of particles in suspension might lead to some discrepancies in the results. These discrepancies might be due in part to the different physical principles involved but, certainly, they deserve careful examination. We would also like to emphasize the fact that, of all these different methods, QELS is the only one that provides immediate results without any modification of the biological sample.

Acknowledgments

We thank Dr. U. B. Sleytr (Department of Ultrastructure, University of Agriculture, Vienna) for help in preparing the electron micrographs.

Section VI

Molecular Genetics

[78] DNA Transformation

By RONALD D. PORTER

Introduction

The full power of genetics in research with any organism involves a need to be able to manipulate the genetic information by introducing new or altered DNA nucleotide sequences into a viable cell. A conjugation system involving triparental matings with *Escherichia coli* is now available for some cyanobacteria, and that system is described in detail elsewhere in this volume [83]. The most common gene transfer mode used with these organisms, however, is DNA-mediated transformation. To date, DNA-mediated transformation has been clearly demonstrated only in those unicellular cyanobacteria that have been classified as *Synechococcus* or *Synechocystis*.[1] It may only be a matter of time, however, before transformation systems are described for some of the more morphologically and developmentally complex cyanobacteria.

A primary key to the functioning of a transformation system in any microbe is the development of cellular competence for DNA uptake. Depending on how it is developed, competence is either physiological or artificial in nature. Physiological competence describes the situation when at least some of the cells in the population possess a natural ability to productively internalize exogenous DNA without any special treatment to the cells before their exposure to the donor DNA. Artificial competence, on the other hand, is when DNA internalization by cells occurs as a result of a treatment regimen that would not be part of a normal growth cycle. The most common example of artificial competence is the $CaCl_2$ treatment protocols that are used for plasmid transformation of *Escherichia coli*. The vast majority of the cyanobacterial transformation studies that have been reported involve physiological, or natural, competence for DNA uptake. We therefore deal most extensively with methods used in physiological transformation systems, and we discuss separately artificial competence later.

Transformation is often classified as being either chromosomal or plasmid, and both types have been described in cyanobacteria. Chromosomal transformation involves the recombination of internalized donor chromosomal DNA with homologous chromosomal DNA in the recipient cell.

[1] R. D. Porter, *CRC Crit. Rev. Microbiol.* **13,** 111 (1986).

This type of transformation can be done using donor DNA from mutant strains or using cloned genomic DNA that has been subjected to recombinant DNA manipulations *in vitro*. The *in vitro* introduction of drug-resistance genes into cloned cyanobacterial genomic DNA fragments has allowed the use of powerful drug selections following transformation to obtain insertion and deletion mutants in cyanobacterial genomes. Plasmid transformation normally involves the introduction of a replicating plasmid DNA molecule into the recipient cell. The ability to introduce replicating plasmids into cyanobacteria has allowed the development of plasmid cloning vehicles for cyanobacteria that can be used to produce cyanobacterial partial diploids, or merodiploids, as well as to introduce foreign genes. When the cyanobacteria are transformed with nonreplicating plasmids containing homologous chromosomal DNA, one of the possible outcomes is that the entire nonreplicating plasmid may be added to the recipient cell chromosome by homologous recombination.

This chapter deals primarily with the practical aspects of transformation in cyanobacteria as it is practiced for those members of the group where it has already been described. This information, it is hoped, will be of use to those individuals attempting to document and utilize transformation in other cyanobacteria. A recent review of transformation in cyanobacteria is available for those readers desiring a more detailed discussion of mechanistic considerations.[1]

Preparation of Cyanobacterial DNA

Although the methods for the preparation of either plasmid or chromosomal DNA from cyanobacteria are in large part very similar to those used for many other bacteria, the successful production of a high yield of high-quality DNA is not a trivial concern. Many cyanobacteria produce large quantities of extracellular polysaccharide which may complicate DNA purification. The use of the classical Marmur procedure for chromosomal DNA purification[2] often suffers from problems with attaining clean lysis of the cells. The alkaline–sodium dodecyl sulfate (SDS) procedure of Birnboim and Doly[3] often works reasonably well for the purification of the smaller plasmids from *Agmenellum quadruplicatum* PR-6, but clean lysis can still be a problem at times. The chance of success can be increased for both of these procedures by increasing the time for which the cells are incubated with lysozyme and by carrying out that incubation at 37° rather than at room temperature.

[2] J. Marmur, *J. Mol. Biol.* **3,** 208 (1961).

[3] H. C. Birnboim and J. Doly, *Nucleic Acids Res.* **7,** 1513 (1979).

Large-Scale Preparation of Total Cyanobacterial DNA

The following procedure for the extraction of total DNA from large-scale cultures of cyanobacteria has been assembled from the combined efforts of a number of people, and it has gained widespread acceptance within the cyanobacterial research group at Penn State. The procedure works well for the preparation of total DNA from *A. quadruplicatum* PR-6 and has been used in the preparation of DNA from a variety of other cyanobacteria. With minor modifications, this procedure should allow the purification of DNA from most cyanobacteria. The reader may also want to consult several published procedures that have been successfully used for obtaining genomic and/or plasmid DNA from *A. nidulans* R2[4–6] or *Synechocystis* 6803.[7]

1. Harvest cells grown to late exponential phase by centrifugation. Cells grown into stationary phase may be somewhat refractory to lysis. With *A. quadruplicatum* PR-6, a cell concentration of 1–2 × 10^8 cells/ml is appropriate.

2. Resuspend the cell pellet in 0.05 volumes of lysis solution. Lysis solution consists of 10% sucrose in 50 m*M* Tris–HCl, pH 8, 100 m*M* disodium EDTA. Disperse the cells thoroughly by some combination of vortexing and pipetting. With some cyanobacteria, washing the cell pellet with 50% of the original culture volume of TES (10 m*M* Tris, 50 m*M* NaCl, 50 m*M* EDTA, pH 7.2) before treating the cells with lysis solution improves DNA yield.

3. The cells may be stored at −80° at this point in the procedure. Repeated cycles of freezing at −80° and thawing at 37° will probably increase the DNA yield.

4. Centrifuge the cell suspension for 20 min at 7000 rpm in a Sorvall SS-34 or equivalent rotor. Decant the supernatant to waste and again resuspend the cell pellet in lysis solution. The volume of lysis solution should be about 80% of that used in the first resuspension. Dispersion should again be thorough, and a tissue homogenizer may be used to help with the dispersion of filamentous organisms.

5. Transfer the cell suspension to a sterile Erlenmeyer flask and add lysozyme to a final concentration of 10 mg/ml. Swirl the flask until the lysozyme is completely dissolved and then incubate at 37° for 30 min.

6. Use a 10% (w/v) solution of sarkosyl in water to bring the final

[4] S. V. Shestakov and N. T. Khyen, *Mol. Gen. Genet.* **107,** 372 (1970).

[5] C. A. M. J. J. van den Hondel, W. Keegstra, W. E. Borrias, and G. A. van Arkel, *Plasmid* **2,** 323 (1979).

[6] F. Chauvat, C. Astier, F. Vedel, and F. Joset-Espardellier, *Mol. Gen. Genet.* **191,** 39 (1983).

[7] G. Grigorieva and S. Shestakov, *FEMS Microbiol. Lett.* **13,** 367 (1982).

detergent concentration to 1%. Mix thoroughly by gentle swirling and then incubate at 37° for 15 min.

7. Measure the volume of the solution and add 1 g of solid CsCl for every milliliter of solution. Continue the incubation at 37° with gentle mixing until the CsCl is completely dissolved. Ethidium bromide (EtBr) is not used for this initial density gradient centrifugation step.

8. Load the mixture into polyallomer ultracentrifuge tubes and centrifuge for 40 hr at 44,000 rpm at 15° in a Beckman 60Ti rotor.

9. Bottom-collect the gradient, taking 0.5- to 1.0-ml fractions. The fractions containing DNA will generally be noticeably viscous, but such fractions can be determined by running 2-μl samples of the fractions on agarose gels that are subsequently stained with EtBr.

10. The pooled DNA-containing fractions from the first gradient are then mixed with an equal volume of a 50% (w/w) solution of CsCl in TE (10 mM Tris, 1 mM EDTA, pH 8), and EtBr is added to a final concentration of 200 μg/ml using a stock solution of 5 mg/ml in water.

11. Load the mixture into polyallomer ultracentrifuge tubes and spin for 40 hr at 44,000 rpm at 15° in a Beckman 70.1Ti rotor. With many cyanobacteria, there will be a bottom band of supercoiled plasmid DNA and an upper band with chromosomal and relaxed plasmid DNA. The DNA band(s) can be seen for purposes of collection by illuminating the gradient with a long-wavelength UV lamp.

12. Extract the EtBr from the collected DNA fractions with 2-propanol that has been equilibrated with NaCl-saturated TE buffer. Dialyze the aqueous material against TE for several hours with at least 3 changes of buffer. The DNA obtained can then be analyzed for concentration, purity, and composition by absorption spectroscopy and agarose gel electrophoresis.

Preparation of Total DNA from Small-Scale Cyanobacterial Cultures

Situations often arise where it is desirable to make small amounts of total DNA from a number of cyanobacterial isolates while avoiding the CsCl centrifugations used with the large-scale protocol. The following variation can be used with 10-ml cultures of cyanobacteria grown to $1-2 \times 10^8$ cells/ml.

1. Resuspend the pelleted cells in 0.5 ml of lysis solution as above, except the lysis solution should contain 25% sucrose.

2. After one or more freeze–thaw steps in a 1.5-ml microcentrifuge tube, add 5 mg of lysozyme, mix thoroughly by vortexing, and incubate for about 30 min at 37°.

3. Add SDS to 1% and proteinase K to 100 μg/ml. Incubate at 50° for at least 2 hr. Overnight incubation is preferred.

4. Carry out two or three phenol–chloroform and then two chloroform extractions as described by Maniatis *et al.*[8]

5. Add 1/3 volume of 10.5 *M* ammonium acetate and 2 volumes of 2-propanol to the aqueous phase from the final chloroform extraction. Place the sample at −20° for at least 15 min and then centrifuge for 20 min at 4° in a microfuge.

6. Wash the pellet twice with 70% (v/v) ethanol, centrifuging for 5 min after each wash.

7. Dry the pellet under reduced pressure in a desiccator and then resuspend in 0.3 ml of TE. With *A. quadruplicatum* PR-6, this procedure should yield between 10 and 50 μg of total DNA.

Preparation of Plasmid DNA from Small-Scale Cyanobacterial Cultures

The procedure that we typically use for the preparation of plasmid DNA from small cultures of *A. quadruplicatum* PR-6 is very similar to that described for *E. coli* by Birnboim and Doly.[3] The following adjustments improve the reproducibility and yield from that procedure.

1. A 10- to 20-ml culture of the cyanobacterium is grown to saturation, and the cells are collected by centrifugation and washed with 0.5 volumes of TES. The cell pellet is then subjected to one or two cycles of freezing at −20° and thawing at 37°.

2. The pellet is resuspended in 0.5 ml of Birnboim and Doly Solution I containing 5 mg/ml, rather than 2 mg/ml, of lysozyme. This suspension is then incubated at 37° for at least 30 min.

3. With appropriate adjustments in volume, the published procedure is followed until after the RNase treatment step. At that point, proteinase K is added to a concentration of 100 μg/ml, and the material is incubated at 50° for at least 2 hr.

4. The plasmid DNA is then recovered by ethanol precipitation, desiccation, and resuspension of the pellet in 0.3–0.5 ml of TE. With *A. quadruplicatum* PR-6, approximately 30–50 μg of plasmid DNA is obtained.

Preparation of Plasmid DNA from Large-Scale Cyanobacterial Cultures

Large-scale preparations of cyanobacterial plasmid DNA may be obtained either by collecting the plasmid DNA band from the large-scale

[8] T. Maniatis, E. F. Fritsch, and J. Sambrook, "Molecular Cloning." Cold Spring Harbor Lab., Cold Spring Harbor, New York, 1982.

total DNA purification procedure or by scaling up the small-scale plasmid DNA procedure (see above). In the latter case, CsCl–EtBr centrifugation may be added as a final purification step.

Standard Transformation Protocol

This section contains a generic transformation protocol that can be used for a variety of naturally competent cyanobacteria. The reader is also referred to two publications that directly address the question of optimizing transformation in *A. nidulans* R2.[6,9]

Growth and Transformation

Growth. A culture of the recipient strain to be transformed should be diluted back and grown to approximately 2–4 $\times$ 10^7 cells/ml under optimal growth conditions in a suitable medium. Medium A with nitrate[10] is routinely used for *A. quadruplicatum* PR-6, while BG-11 medium[11] is frequently used for *A. nidulans* R2. Any medium supporting the optimal growth of the particular cyanobacterium being transformed may be used for this step.

Washing and Concentration. Cyanobacterial transformation protocols often call for harvesting the cells by centrifugation and resuspending them in approximately 0.1 volumes of fresh media to yield a final cell concentration of approximately 2–5 $\times$ 10^8 cells/ml. A wash step is generally included, and the wash may be done with either fresh media or a salts–buffer solution. The addition of 0.1 m*M* EDTA to the wash solution may be useful in reducing the action of nucleases present in the culture on the transforming DNA. This step is generally not employed in *A. quadruplicatum* PR-6 transformation experiments.

Addition of Donor DNA. The desired donor DNA is added to the recipient cells at a volume-to-volume ratio of no more than 1–9. As saturating DNA concentrations with either plasmid or chromosomal DNA can vary considerably from system to system, the actual amount of donor DNA to be used is largely dependent on the particular type of experiment being performed and the particular cyanobacterium being used. The donor DNA is generally stored in TE buffer (10 m*M* Tris, 1 m*M* EDTA, pH 8) prior to its addition. Other DNA storage buffers may be used, but the final concentration of chelating agents in the transformation mix should be no more than 0.1 m*M* for EDTA or 0.2 m*M* for citrate.

[9] S. S. Golden and L. A. Sherman, *J. Bacteriol.* **158,** 36 (1984).

[10] S. E. Stevens, Jr., C. O. P. Patterson, and J. Myers, *J. Phycol.* **9,** 427 (1973).

[11] M. M. Allen, *J. Phycol.* **4,** 1 (1968).

Incubation for DNA Uptake. (1) Timing: Quantitative transformation frequency experiments require that DNA uptake be terminated by the addition of DNase as continuing competence can allow DNA uptake and the production of transformants to occur even after the cells have been plated on solid media. Uptake periods of 15–60 min are typically allowed for experiments involving transformation frequency determinations. Experiments are conducted where the sole objective is to obtain transformants for further study, and in these experiments the uptake termination step can be omitted with the end result that more transformants will generally be obtained. Uptake termination is generally achieved by adding DNase I to a final concentration of 10 μg/ml under conditions where the final Mg^{2+} concentration is at least 10 m*M*.

(2) Light versus dark incubation: Published reports involving *A. nidulans* R2 transformation have come to differing conclusions regarding whether DNA uptake and processing occurs most efficiently in the light or in the dark.[6,9] Reports indicating superior transformation of *A. nidulans* R2 by maintaining the cultures in the dark during uptake have, however, often involved prolonged periods of DNA uptake. It would appear, on the other hand, that light versus dark incubation may have much less, if any, effect during shorter periods of DNA uptake. In the development of new systems, where long periods of DNA uptake are allowed in the hope of obtaining any transformants at all, dark incubation during uptake may be of some benefit. Transformation cultures of *A. quadruplicatum* PR-6 are incubated under normal light conditions during DNA uptake. Although a short period of dark incubation does not significantly reduce chromosomal transformation efficiency with this organism,[12] prolonged periods of dark incubation have resulted in reduced levels of genomic transformation.[13]

Expression and Selection

Many of the selections used in cyanobacterial transformation involve drug-resistance traits that are phenotypically recessive. This may be due to the time required for the expression of a gene product that carries out the destruction or alteration of the selective agent. It may also be due to a need for the cell to dilute out cellular components that are sensitive to the selective agent before the potential transformants are challenged with that agent. This time for expression generally varies from 12 to 48 hr. Although the transformation culture may be permitted to grow in liquid culture while expression of the new genetic information and/or dilution of

[12] S. E. Stevens, Jr., and R. D. Porter, *Proc. Natl. Acad. Sci. U.S.A.* **77,** 6052 (1980).

[13] E. Essich, S. E. Stevens, Jr., and R. D. Porter, unpublished observations.

sensitive cellular components occurs, platings on solid media containing the selective agent after expression in liquid media do not allow accurate determinations of frequency or the separation of siblings arising from a single transformation event. Most experiments therefore utilize platings on nonselective media shortly after DNA uptake is terminated. These plates are incubated under the appropriate growth conditions for the requisite period of time, and then challenged with the desired selective agent by one of the following three methods.

Aerosol Spray. A spraying device is loaded with a solution of the selective agent. Aerosol devices that are commercially available for use with throat sprays work well for this purpose. The sprayer is thoroughly cleaned with ethanol and dried before it is loaded with a sterile solution of the selective agent. The plate to be sprayed is uncovered and placed on its edge in a fumehood. The sprayer is placed about 8 inches from the plate and three or four bursts of the selective agent are sprayed on the plate. Care must be taken to ensure a uniform coverage of the surface while not permitting the accumulation of large drops of liquid on the surface of the plate. This method works very well when the minimum inhibitory concentration (MIC) of the agent for the resistant strain is at least 10-fold greater than the MIC for the sensitive strain and when the selective agent is highly soluble in aqueous solution.

Underlaying. Underlaying, a commonly used method, involves lifting up the edge of the agar and placing an appropriate volume of an aqueous solution of the selective agent under the agar. The quality of the results obtained may be affected by the placement of the selective agent, and an attempt should be made to place it as close to the middle of the plate as possible. A variation on this procedure can be carried out by placing a thin layer of agar containing the selective agent into a fresh Petri dish and transferring the entire slab of agar from the original plate onto the base layer of drug-containing agar. This variation has the advantage of providing a more uniform distribution of the selective agent throughout the media.

Overlaying. We have found that the drug overlay method gives the most consistent and reproducible results for the selection of transformants with *A. quadruplicatum* PR-6. The cells are initially plated on a nonselective plate of 100 mm diameter in 2.5 ml of top agar that consists of 0.8% agar in distilled water. After the plates have been incubated under growth conditions for a suitable length of time, another layer of top agar (0.6% in distilled water this time) containing the required amount of drug for the entire plate is poured on top of the layer of agar that contains the cells. After continued incubation, drug-resistant colonies readily grow up within the agar layer. Although we initially had some concern as to the availability of CO_2 within the agar, we have found that CO_2 limitation does not seem to be a significant problem.

Transformation without Selection

Although the typical transformation experiment involves a direct selection for the desired class of transformant, experiments can be done in situations where there is no direct selection available. This brute force approach is seldom undertaken unless there is a reasonable method for screening the colonies for the desired transformant. If the transformant is expected to produce colonies with altered pigmentation or morphology, visual examination of the colonies obtained will constitute a satisfactory screening procedure. If the desired transformant will possess a new growth factor requirement, it can be detected by gridding and replica plating colonies that have initially been plated on appropriately supplemented media. The feasibility of screening by replica plating is largely dependent on the expected transformation frequency, and we have used this method successfully in *A. quadruplicatum* PR-6 transformations where the desired transformants occur at a level between 10^{-2} and 10^{-3}.[14]

It is also possible to use DNA–DNA hybridization techniques to screen cyanobacterial colonies when the desired transformant is expected to contain a piece of DNA that is not present in the parental cells. In the case of *A. quadruplicatum* PR-6, the colonies can be patched or replica plated onto nitrocellulose membranes, although care must be taken to ensure that the membranes are properly wetted prior to use. It is also important to incubate the plates in such a way that the incident light is passing through the membrane. Although this results in some reduction in light intensity, growth is often poor when the side of the membrane containing the colonies is directly illuminated. Colony hybridization with a radiolabeled DNA probe can be carried out by a standard procedure developed for *E. coli*[15] which has been used to detect the presence of the *E. coli lac* operon in *A. quadruplicatum* PR-6 transformants.[16]

The sensitivity and reproducibility of probing cyanobacterial colonies on nitrocellulose filters can be improved by the following pretreatment of the nitrocellulose membranes prior to the alkaline lysis step. The membrane is placed in a baking dish on Whatman #3 filter paper saturated with TE buffer, pH 8. The membrane is then sprayed with a 5 mg/ml solution of lysozyme in TE. The baking dish is then sealed with cellophane and

[14] R. D. Porter, J. S. Buzby, A. Pilon, P. I. Fields, J. M. Dubbs, and S. E. Stevens, Jr., *Gene* **41,** 249 (1986).

[15] D. Hanahan and M. Meselson, this series, Vol. 100, p. 333.

[16] J. S. Buzby, R. D. Porter, and S. E. Stevens, Jr., *Science* **230,** 805 (1985).

incubated at 42° for 15 min. The spraying and 42° incubations are repeated until the total incubation time at 42° reaches 60 min. The membrane is then transferred to filter paper that has been saturated with a 10% sarkosyl solution. An open container of chloroform is placed in the baking dish, which is then sealed and incubated at 42° for 60 min. The filters are then blotted on dry Whatman #3 filter paper and carried through the standard alkaline lysis procedure.

It is also possible to detect transformants expressing a protein not present in the parental cells. We have carried patches of *A. quadruplicatum* PR-6 on solid media through a procedure that has been described for antibody screening in *E. coli*[17] and detected the presence of a foreign protein in the cyanobacterium by the use of a chromogenic substrate for that protein.[16] Although we have not used this procedure for an actual antibody-screening procedure, it should be possible to detect foreign proteins in cyanobacterial transformants by immunological techniques.

Transformation of Artificially Competent Cyanobacteria

Although most of the cyanobacterial transformation experiments described in the literature involve naturally occuring competence, transformation has been achieved with artificial competence regimens. An artificial competence scheme involving $CaCl_2$ treatment has been described for the genomic transformation of *Synechocystis* 6308 with DNA–RNA complexes,[18] and *A. nidulans* has undergone plasmid transformation after a lysozyme–EDTA treatment (or permeaplast) protocol.[19]

[17] D. Anderson, L. Shapiro, and A. M. Skalka, this series, Vol. 68, p. 428.
[18] C. I. Devilly and J. A. Houghton, *J. Gen. Microbiol.* **98,** 277 (1977).
[19] H. Daniell, G. Sarojini, and B. A. McFadden, *Proc. Natl. Acad. Sci. U.S.A.* **83,** 2546 (1986).

[79] Transformation in *Synechocystis* PCC 6714 and 6803: Preparation of Chromosomal DNA

By FRANÇOISE JOSET

The protocol described in the preceding chapter[1] applies to the two *Synechocystis* strains PCC 6714 and 6803. One of the characteristics of these strains, however, is a particular resistance toward the usual lysing

[1] R. D. Porter, this volume [78].

and cell-disruption procedures. The following modifications of classic protocols, resulting from our experience over several years[2] (F. Chauvat, personal communication; C.-C. Zhang and F. Joset, unpublished observations), are suggested.

1. Lysis is facilitated if the cells are grown at fairly low temperatures (25–30° instead of their optimal 34°). Harvesting at mid exponential phase is also important.

2. These cells appear to be very refractory to lysozyme. Incubation in lysis buffer, containing sucrose or glucose, as described in Ref. 1, increases the efficiency of the enzyme. A pretreatment with penicillin, ampicillin, or any antibiotic which interferes with cell wall synthesis also enhances the action of lysozyme. Concentrations of 200 U/ml penicillin (specillin G, free of any additive) or 200 μg/ml ampicillin are appropriate for suspensions containing 10^8 cells/ml. The antibiotic concentration should produce bacteriostasis (measured as OD) but no cell lysis. Since these drugs act most efficiently on dividing cells, the suspension should be treated when cell growth is still fast ($\sim 10^8$ cells/ml is a maximum). Treatment is usually performed overnight (2–3 generation equivalents). The cells can be harvested by centrifugation in the morning. A slight leaking of pigments, yielding a brownish supernatant, is common and should not present any problems.

3. Incubation at 60–65° for about 20 min, before the lysozyme treatment, greatly increases the final yield. This probably results from two effects: an inactivation of exo- and endonucleases and a fragilization of the cell envelope.

4. The lysozyme treatment, under conditions described in Ref. 1, should last for at least 30 min and up to 90 min at 37°. This long incubation justifies the pretreatment at 60°.

5. Sodium dodecyl sulfate (SDS) (1–2% final concentration) may be preferred to sarkosyl. Very high quality SDS should be used (British Drug House may be recommended). Incubation should be for at least 60 min at 37°, with gentle stirring.

6. Direct purification in CsCl, as described in Ref. 1, can then be performed, and yields DNA fragments of fairly high molecular weight [$\gg$30 kilobase pairs (kbp)]. However, comparable results in yield, size, and homogeneity of fragments can be obtained by either of the following procedures: (a) Add sodium perchlorate to 0.5 *M*, final concentration. Incubate at 4° with gentle stirring for 10 min. (b) Add 50–100 μg/ml proteinase K and incubate for 90 min at 37 or 50°, with gentle agitation. This treatment can be performed simultaneously with the lysozyme step.

[2] C. Astier and F. Espardellier, *C.R. Acad. Sci. Paris* **283,** 795 (1976).

This saves time and decreases the risk of degradation by nucleases. The DNA can then be isolated on a CsCl gradient or deproteinized and purified according to classic procedures.

Yields of 1 mg of DNA can be expected from 2 liters of culture having approximately 10^8 cells/ml.

[80] Mutagenesis of Cyanobacteria by Classical and Gene-Transfer-Based Methods

By SUSAN S. GOLDEN

Introduction

One of the most powerful approaches to the study of cellular phenomena is the analysis of mutants which are altered in particular functions. The utility of cyanobacterial mutants is enhanced by the recent advances in genetic transfer techniques for many species,[1-3] which afford the opportunity of identifying genes by complementation. This chapter outlines methods which have been used successfully to obtain mutants in a variety of cyanobacteria, with a description of the system in which each method was developed. In many cases the protocol can be modified to apply to other cyanobacterial species. Classical methods are described, including use of chemical mutagens and UV irradiation, which are applicable to both unicellular and filamentous species. Transposon mutagenesis has been demonstrated only in a transformable strain, but it should also be possible in filamentous strains which can receive plasmids by conjugation from *Escherichia coli*. Techniques for both site-directed and random mutagenesis are also described which, at this time, have been shown to be useful only in transformable unicellular strains. These methods are based on recombination between the chromosome and added DNA which possesses sequence similarity to the chromosomal DNA. This chapter also discusses some procedures that are necessary when working with filamentous species to separate mutagenized from wild-type cells on the same filament. The following protocols for mutagenesis and mutant enrichment should provide a starting point for developing the appropriate selection and screening methods to obtain a desired cyanobacterial mutant.

[1] R. D. Porter, this volume [78].
[2] F. Joset, this volume [79].
[3] J. Elhai and C. P. Wolk, this volume [83].

Basic Reagents, Supplies, and Techniques for Mutagenesis of Cyanobacteria

Liquid and Solid Media

Basic media and culture conditions for diverse cyanobacterial species have been established and are described elsewhere in this volume.[4] Several methods for preparing agar plates to optimize cyanobacterial plating efficiencies have been described. These include the following: (1) preparing twice-concentrated solutions of medium and agar (in water) which are autoclaved separately and mixed when cooled to pouring temperature,[5] (2) adding sterile sodium thiosulfate to the mixture to a final concentration of 1 m*M* before pouring,[6] (3) using agar that has been purified by the method of Braun and Wood,[7] and (4) preparing plates with only 1% agar.[8] All of these precautions are probably of some benefit, and some can be combined. Separate sterilization of agar and mineral salts is strongly recommended.

At least some cyanobacterial strains do not survive quantitatively when plated directly on medium containing a selective agent to which the cells should be resistant.[6] The following technique for underlaying the selective agent improves plating efficiency.

Procedure

1. Spread 100- to 150-μl aliquots of each culture sample on 100-mm plates containing 40 ml of basal medium solidified with 1.5% agar.
2. Incubate the plates under standard illuminated growth conditions for 4–6 hr prior to the addition of the required selective agent.
3. Add the appropriate selective agent by lifting the agar slab with an ethanol-flamed spatula and dispensing 400 μl of a 100× concentrated stock underneath. Distribute the solution by rotating the spatula under the agar and removing it gently to reseat the agar slab.
4. Continue incubation of the plates until colonies form (4–7 days).

Another method of plating cells and adding selective agents is to use soft agar overlays.[5] Inocula can be added to 2 ml portions of 0.7–1.0% agar melted and cooled to 48°, mixed, and poured onto the surface of the

[4] R. Castenholz, this volume [3].
[5] M. M. Allen, *J. Phycol.* **4,** 1 (1968).
[6] S. S. Golden and L. A. Sherman, *J. Bacteriol.* **158,** 36 (1984).
[7] A. C. Braun and H. N. Wood, *Proc. Natl. Acad. Sci. U.S.A.* **48,** 1776 (1962).
[8] T. C. Currier, J. F. Haury, and C. P. Wolk, *J. Bacteriol.* **129,** 1556 (1977).

agar. A second layer containing a selective agent can be poured over the first at a later time.

Heterocystous strains which are otherwise able to grow in the absence of combined nitrogen may plate poorly on nitrogen-free media after sonication[9] (see Special Considerations for Mutagenesis of Filamentous Cyanobacteria). This may be because cells deplete their supplies of stored nitrogen before filaments have reached a critical length necessary for heterocyst differentiation.[10] Therefore the following protocols should include a combined nitrogen source in the basal medium to avoid starvation for nitrogen.

Centrifugation to Harvest Cells

Conditions for harvesting and washing cyanobacterial cultures vary somewhat with different strains. Cells can usually be pelleted by centrifugation at 1000–4000 *g* for 5–10 min at 10–20°. Some strains are cold sensitive and should not be subjected to refrigeration during the harvesting.[11] Filamentous strains may not form a tight pellet at these or higher centrifugal forces. Use of conical tubes and a swinging-bucket rotor improves the pellet compaction.

Classical Methods of Mutagenesis

Chemical Mutagenesis

Classical methods of mutagenesis, such as treatment with chemical mutagens, have been used successfully for both unicellular[12] and filamentous[8] cyanobacterial species. In both types of strains, chemical mutagens provide a means of producing a variety of lesions at random loci. A recent report by Chapman and Meeks[9] describes a careful study of parameters that affect mutagenesis frequency induced by *N*-methyl-*N′*-nitro-*N*-nitrosoguanidine (MNNG) in *Anabaena* ATCC 29413. Their measure of mutagenesis was the frequency of obtaining colonies that are resistant to the analog 5′-fluorocytosine, for which there is a positive selection. The central findings of this study are that concentration and exposure time are not important variables in themselves, but that a combination of the two

[9] J. S. Chapman and J. C. Meeks, *J. Gen. Microbiol.* **133,** 111 (1987).

[10] M. Wilcox, G. J. Mitchison, and R. J. Smith, *in* "Microbiology—1975" (D. Schlessinger, ed.), p. 453. Am. Soc. Microbiol., Washington, D.C., 1975.

[11] R. Rippka, J. Deruelles, J. B. Waterbury, M. Herdman, and R. Y. Stanier, *J. Gen. Microbiol.* **111,** 1 (1979).

[12] L. A. Sherman and J. Cunningham, *Plant Sci. Lett.* **8,** 319 (1977).

which achieves 99% lethality produces the highest frequency of mutants. The pH of the incubation was found to strongly influence mutation frequency, with an optimum at pH 6.0. Another important factor was the duration of the expression period after mutagenesis to allow segregation of mutant and wild-type chromosomes. The following procedure is adapted from the findings of Chapman and Meeks, and is similar to the procedure used by others.[9,12] Mutants should be obtained at a frequency of $1-4 \times 10^{-4}$ in *Anabaena* ATCC 29413,[9] as compared to a spontaneous rate of $1-2 \times 10^{-7}$.

Reagents for MNNG Mutagenesis

MNNG stock solution: 10 mg/ml in glass-distilled water. Filter-sterilize and freeze 2.0-ml aliquots. Do not refreeze. Note that MNNG is a hazardous compound; follow the supplier's handling instructions carefully.

Sterile 10 m*M* citrate buffer, pH 6.0

Procedure

1. Grow 50 ml of cyanobacterial culture to 10^6–10^7 cells/ml in liquid basal medium under standard conditions.
2. If working with a filamentous strain, fragment filaments to an average length of 2 cells (see Special Considerations for Mutagenesis of Filamentous Strains).
3. Wash the fragmented culture by centrifugation twice with basal medium and resuspend at 1×10^6–2×10^7 cells/ml in 20 ml of 10 m*M* citrate buffer, pH 6.0.
4. Treat with MNNG at 1.0 mg/ml for 15 min (or at a lower concentration for a period of time which results in 99% killing) in the light at room temperature.
5. Remove the mutagen by three centrifugal washes and resuspend cells in 50 ml basal medium.
6. Incubate under standard growth conditions for 6 days. Note that some mutants grow better at room temperature than at 30° (J. C. Meeks, personal communication).
7. Repeat fragmentation of filaments (if appropriate).
8. If the desired mutant is expected to have impaired growth, include steps for mutant enrichment (see below).
9. Plate survivors on solid medium containing appropriate nutritional supplements or selective agents to screen or select for desired mutant phenotypes. Concentrate cells to approximately 1×10^8 cells/ml and plate 200 μl/plate on 4–10 plates prior to adding a selective agent. To screen for a nonselectable phenotype, plate

dilutions of mutagenized cells on a permissive master plate. This may contain nutritional supplements for auxotrophs or may be incubated at a permissive temperature for conditional mutants.

10. Incubate plates under standard growth conditions until colonies form. Replica plate if necessary for mutant screening.

Similar frequencies of mutation can be achieved by treating 20 ml of cell suspension with 5–20 μl of diethyl sulfate (DES) for 30 min.[9] A major drawback of both MNNG and DES is the tendency of both compounds to produce multiple clustered lesions.[13] Nitrous acid is more likely to produce single site mutations, but this agent induces mutations at a much lower frequency than the other compounds. A 5-min treatment of cells with 50 m*M* nitrous acid resulted in a mutation frequency of 1.5×10^{-5}; increasing the exposure time reduced viability without increasing the mutation frequency.[9]

Mutagenesis by Irradiation with Ultraviolet Light

Ultraviolet (UV) radiation has been used successfully to increase the incidence of mutation 25- to 1000-fold in the unicellular species *Synechocystis* PCC 6714[14] (*Aphanocapsa* 6714) and *Synechocystis* PCC 6803[15] (*Aphanocapsa* 6803), as measured by the frequency of colonies resistant to *p*-fluorophenylalanine or fructose, respectively. Thiel and Leone[16] obtained mutants of the filamentous species *Anabaena variabilis* ATCC 29413 following exposure to germicidal UV light which caused greater than 99% killing. However, Thiel and E. Levine report that this strain, as well as the other filamentous strains *Anabaena* PCC 7118, *Anabaena* PCC 7120, and *Anabaena* sp. M-13 (University of Tokyo) are all quite resistant to UV (T. Thiel, personal communication). Wolk and colleagues[17] used UV irradiation to obtain mutants of *Anabaena* PCC 7120 which are phenotypically nitrogen-fixation defective. All of these researchers report that the cyanobacteria that have been tested have active photoreactivation systems. For this reason an important requirement for obtaining mutants is a period of growth under nonphotoreactivating conditions. In strains which can not grow heterotrophically in the dark, this condition can be met by growth in yellow light.[17] The following procedure was

[13] N. Guerola, J. L. Ingraham, and E. Cerda-Olmedo, *Nature* (*London*), *New Biol.* **230,** 122 (1971).

[14] C. Astier, F. Joset-Espardellier, and I. Meyer, *Arch. Microbiol.* **120,** 93 (1979).

[15] E. Flores and G. Schmetterer, *J. Bacteriol.* **166,** 693 (1986).

[16] T. Thiel and M. Leone, *J. Bacteriol.* **168,** 769 (1986).

[17] C. P. Wolk, E. Flores, G. Schmetterer, A. Herrero, and J. Elhai, *Proc. Int. Symp. Nitrogen Fixation, 6th,* p. 491 (1985).

compiled from the conditions used in the reports summarized above and from additional suggestions of some of these researchers.

Supplies for UV Mutagenesis

30-W germicidal UV light source, prewarmed for 30 min
Yellow lights for cyanobacterial growth (e.g., General Electric Bug-lites)

Procedure

1. If working with a filamentous strain, fragment the filaments to 1–3 cell lengths (see Special Considerations for Mutagenesis of Filamentous Cyanobacteria).
2. Suspend cyanobacterial cells in basal medium at $1–2 \times 10^7$ cells/ml. Remove 1 ml as a control sample for steps 4–5.
3. Irradiate 5 ml of cells in a 100-mm petri dish at a distance of 45 cm from a 30-W germicidal UV light. Remove 1-ml portions at 2-min intervals.
4. Incubate aliquots for at least 36 hr under yellow lights at the same intensity used for growth in white light. Facultative chemoheterotrophs can be incubated in the dark with an appropriate carbon source.
5. Plate serial dilutions of mutagenized cells on appropriate basal or nutrient-enriched solid medium to assess survival.
6. Plate the remainder of each aliquot by spreading or by incorporating in a soft agar overlay.
7. If a selective agent is needed, incubate plates under white light for at least 24 hr before adding the agent, either in a soft agar overlay or underneath the agar.
8. Continue incubating under normal growth conditions until colonies form.

Gene Inactivation by Insertion of Selectable Heterologous Sequences

Mutagenesis by Transposition of Tn901

Cyanobacterial transposons have not yet been described, but a cyanobacterial gene can be tagged and cloned by transposition of an *E. coli* element. Tandeau de Marsac *et al.*[18] identified a gene involved in methionine biosynthesis by transposition of transposon Tn*901* (encoding

[18] N. Tandeau de Marsac, W. E. Borrias, C. J. Kuhlemeier, A. M. Castets, G. A. van Arkel, and C. A. M. J. J. van den Hondel, *Gene* **20,** 111 (1982).

ampicillin resistance) into the chromosome of *Synechococcus* PCC 7942 (*Anacystis nidulans* R2). They introduced the transposon into the cyanobacterium by transformation with the plasmid pCH1,[19] which replicates in this organism and carries Tn*901*. All of the transformed cells were resistant to the antibiotic ampicillin, encoded by the transposon. The mutant was detected by screening the total population of ampicillin-resistant transformants for methionine auxotrophs which were induced by insertion of the transposon into a structural gene. The inactivated methionine gene was isolated by cloning DNA from the mutant in *E. coli* and selecting ampicillin resistance encoded by the transposon. In turn the wild-type gene was identified from a library of *Synechococcus* PCC 7942 DNA by sequence similarity to the segment flanking the Tn*901* element.

Although Tn*901* transposition occurs at a low frequency, this technique may be applicable in filamentous strains that can receive conjugatable shuttle vectors from *E. coli* but that have not yet been shown to recombine foreign DNA into the chromosome. Discovery of conditions that induce transposition at a higher frequency or identification of elements that readily transpose in cyanobacterial strains would enhance the usefulness of this procedure. Dr. W. E. Borrias reports that similar experiments using Tn*5* (kanamycin resistance) showed instability of the insertion sequences at the ends of the element during replication in the cyanobacterium and resulted in an immobile kanamycin-resistance trait. Mutagenesis by transposition of elements from nonreplicating "suicide" vectors to the chromosome has not been reported for any of the conjugatable or transformable strains. However, transposition from a nonreplicating *E. coli* plasmid to an endogenous plasmid in *Synechococcus* PCC 7942 has been reported for two different transposons.[19,20]

Procedure. Plasmid pCH1 is an autonomously replicating plasmid for *Synechococcus* PCC 7942 which carries Tn*901* and confers resistance to 1 μg/ml ampicillin to the cyanobacterium.[19] Other shuttle vectors for transformation or conjugation between *E. coli* and a cyanobacterium should be suitable for carrying the transposon into the cell.

1. Introduce the transposon-containing plasmid into the cell by transformation, selecting for the transposon antibiotic-resistance marker and/or markers on the replicating vector.
2. Grow transformants to 10^8 cells/ml and culture for at least 10 generations in liquid medium in the presence of 1 μg/ml ampicillin (for

[19] C. A. M. J. J. van den Hondel, S. Verbeek, A. van der Ende, P. J. Weisbeek, W. E. Borrias, and G. A. van Arkel, *Proc. Natl. Acad. Sci. U.S.A.* **77**, 1570 (1980).

[20] L. A. Sherman and P. van de Putte, *J. Bacteriol.* **150**, 410 (1982).

Tn*901*) and any nutritional supplement required by the expected mutant.

3. Wash cells by two sequential centrifugation steps, resuspending in growth medium at each step. Resuspend the final cell pellet at 5×10^6 cells/ml in growth medium and continue with a 50-ml sample.
4. If working with a filamentous strain, disrupt filaments as described below (section on Special Considerations for Mutagenesis of Filamentous Strains).
5. If the desired mutant should have impaired growth under normal, unsupplemented conditions, include steps for mutant enrichment (see below).
6. Harvest the cells and resuspend in 5 ml medium.
7. Plate cells onto solid medium with appropriate nutritional supplements (for an expected auxotroph). Plate 100 μl of undiluted cells, as well as 10^{-1}, 10^{-2}, and 10^{-3} dilutions to ensure obtaining a master plate with many well-separated colonies.
8. Incubate plates approximately 1 week, until colonies are visible.
9. Replica plate onto solid media with or without nutritional supplements, or otherwise under conditions that allow detection of mutant colonies.

Recombinational Methods of Mutagenesis

Some methods of mutagenesis are uniquely applicable to the transformable unicellular cyanobacterial strains,[2] typified by *Synechococcus* PCC 7942 (*Anacystis nidulans* R2), *Synechococcus* PCC 7002 (*Agmenellum quadruplicatum* PR-6), and *Synechocystis* PCC 6803 (*Aphanacapsa*). Successful mutagenesis of each of these strains by integration of foreign DNA into the chromosome has been reported in the literature.[21-23] It is likely that these methods will be more widely applicable as protocols for genetic transfer in other strains are developed or improved. The following paragraphs provide strategies rather than protocols for recombinational mutagenesis. The requirements for use of these procedures are (1) a transformable host which can incorporate nonreplicating DNA, (2) purified DNA from the host, and (3) an antibiotic-resistance gene "cassette" which is appropriate for selection in the host. Transformation protocols vary from strain to strain and are discussed elsewhere in this volume.[1,2] A

[21] S. S. Golden, J. Brusslan, and R. Haselkorn, *EMBO J.* **5,** 2789 (1986).
[22] J. S. Buzby, R. D. Porter, and S. E. Stevens, *Science* **230,** 805 (1985).
[23] J. G. K. Williams, this volume [85].

recent chapter in this series deals specifically with these methods and describes techniques for the analysis of nucleic acids from the resulting mutant transformants.[24]

Ectopic Mutagenesis

Ectopic mutagenesis was used by Buzby *et al.*[22] to produce mutations at random sites in the chromosome of *Synechococcus* PCC 7002. The basis of this technique is the ligation of random fragments of chromosomal DNA from the host to a DNA fragment bearing an antibiotic-resistance gene. The ligation is performed in such a manner that the ligated molecules remain linear and the DNA lacks sequences for replication in the cyanobacterial host. The entire ligation mix is used to transform the host to antibiotic resistance. Transformation occurs by an unknown mechanism which results in the integration of the marker into the chromosome, presumably at the locus of the linked host DNA. These authors used random fragments of *Synechococcus* PCC 7002 DNA, partially digested with *Sau*3A, ligated to a *Bam*HI/*Pvu*II fragment carrying an ampicillin-resistance gene. The compatability of *Sau*3A and *Bam*HI sticky ends allowed ligation to form linear ligated molecules, which could not circularize by the remaining *Sau*3A and *Pvu*II ends. (Circular molecules bearing a single region of sequence similarity to the chromosome may integrate by an apparent single crossover which results in duplication of the chromosomal locus.[24]) The ectopic mutagenesis method is useful when a particular phenotype is desired which requires mutation at an unknown locus. The resulting antibiotic-resistant transformants can be screened for that phenotype, and the mutated locus is marked by the presence of heterologous DNA. The interrupted gene, and the corresponding wild-type gene, can then be cloned by the same strategy used to isolate a *met* gene by Tn*901* insertion[18] (see Mutagenesis by Transposition of Tn*901*).

Inactivation of Specific Genes

Another insertional mutagenesis method which is applicable in the transformable strains is the inactivation of a known gene in the chromosome by replacement with an interrupted allele. This procedure differs from that of ectopic mutagenesis in that the former method uses random site insertion to obtain a desired phenotype whereas the following method targets a particular locus to assess the phenotype which results from gene

[24] S. S. Golden, J. Brusslan, and R. Haselkorn, this series, Vol. 153, p. 215.

inactivation. This method was used recently to show that all three of the *psbA* genes in *Synechococcus* PCC 7942 are functional.[21]

A cyanobacterial gene cloned in an *E. coli* plasmid is digested with a restriction enzyme that cuts within the open reading frame of the gene to be inactivated and does not otherwise cut the plasmid. An antibiotic-resistance gene, hereafter referred to as the inactivation cassette, is ligated to the ends to recircularize the plasmid. The desired plasmid is selected in *E. coli* by the antibiotic-resistance markers on the vector and from the inactivation cassette. Alternatively, the cyanobacterial gene can be interrupted by transposon insertion while it is maintained on an *E. coli* plasmid.[25] A preparation of the recombinant plasmid is then used to transform the cyanobacterial host to antibiotic resistance encoded by the inactivation cassette. The resulting transformants will have replaced the wild-type gene in the chromosome with the inactivated allele, and the (nonreplicating) plasmid vector sequences will be lost from the cell. Southern analysis is necessary to determine the new restriction pattern at that locus and thereby confirm inactivation of the gene. If a marker on the plasmid vector is selected, the transformants will be the result of a single crossover event which integrates the plasmid and causes a duplication at the site of chromosomal insertion. In *Synechococcus* PCC 7942, the double-crossover or gene conversion event which replaces the chromosomal gene occurs at a sufficiently higher frequency than the single-crossover event that plasmid integration is not observed unless specifically selected. Inactivation of presumed essential genes has been shown to result in the apparent selection of cells carrying a mixed population of chromosomes, so that the selectable marker and uninterrupted gene are both present.[21]

Site-Directed Mutagenesis

Another method made possible by recombination of cloned genes with the chromosome is site-directed mutagenesis, causing a single nucleotide mutation in a gene of interest. Site-directed mutagenesis of genes cloned in *E. coli*, by repair of a single strand hybridized to a mismatched oligomer, has been described.[26] When this method is used to alter a cloned cyanobacterial gene, the mutant gene can be returned to the cyanobacterial host by DNA-mediated transformation. Strategies for detecting integration of the mutated DNA into the chromosome at its native locus are (1) direct selection for an expected phenotype,[27] (2) restoration of some

[25] G. B. Ruvkun and F. M. Ausubel, *Nature (London)* **289,** 85 (1981).
[26] T. A. Kunkel, *Proc. Natl. Acad. Sci. U.S.A.* **82,** 488 (1985).
[27] S. S. Golden and R. Haselkorn, *Science* **229,** 1104 (1985).

level of function (pseudorevertants) in a host which has had that gene previously inactivated,[23] (3) screening for loss of an antibiotic-resistance marker in a gene-inactivated strain,[23] and (4) selection for a linked marker. In the last case, an antibiotic-resistance gene is inserted within a cloned (cyanobacterial DNA) fragment as in the gene inactivation scheme. In this case the marker insertion is made at a restriction site outside of the coding region of the gene. Cyanobacterial DNA flanking the heterologous marker again directs it to the appropriate locus in the chromosome. Evidence that this strategy is feasible is provided by the insertion of Tn*5* downstream of a herbicide-resistance allele of the *Synechococcus* PCC 7942 *psbAI* gene.[21] Ninety-six percent of the kanamycin-resistant transformants carried the linked, single nucleotide mutation conferring herbicide resistance. As is the case in the gene inactivation procedure, heterologous DNA outside of the borders of the cloned cyanobacterial DNA is lost during the recombination event unless the vector is specifically selected.

Special Considerations for Mutagenesis of Filamentous Cyanobacteria

Mutagenesis of filamentous strains poses a special problem in that the colony forming unit is a filament, whereas the target of mutagenesis is a chromosome in a single cell. This means that the resulting colony following mutagenesis can be a mixture of mutant and wild-type cells. If undisrupted filaments are used for mutagenesis and plating, the phenotype of a single mutant cell in a filament would be lost at the level of screening colonies. Physical disruption of filaments to 1–3 cell lengths prior to mutagenesis, and again after mutagenized cells have been allowed to recover and divide, minimizes the incidence of mixed colonies.[8,28]

Cavitation in a sonic cleaning bath, or by a sonic cell disruptor equipped with a microprobe tip, is useful for breaking filaments. Because there is considerable variability between individual sonicators, and because tube dimensions and volume will affect cavitation, it is not practical to provide a specific protocol here. Treatment times in the literature vary from less than 15 sec to at least 15 min. With either type of apparatus there are two parameters, filament length and cell survival, which should be assessed during treatment to establish a protocol. Microscopic examination of aliquots is necessary to determine the number of cells per filament and whether the cell suspension is being disrupted uniformly. It should be possible to achieve an average length of 1–3 cells and still maintain close to 100% plating efficiency.[9,28]

[28] C. P. Wolk and E. Wojciuch, *Arch. Mikrobiol.* **91,** 91 (1973).

Nonuniformity of filament length is a potential problem when using a sonicator microprobe tip which is placed into the cell suspension, from which most of the energy is directed downward such that filaments at the meniscus receive less exposure. To use this type of apparatus, the probe can be sterilized by immersing in 70% ethanol and rinsing in sterile distilled water. To maximize exposure, the depth of the probe in the cell suspension should be the minimum penetration which does not cause splattering. Care should be taken to avoid heating during the sonication. Delivering the energy in measured pulses will help to control and reproduce the exposure. Gentle mixing at intervals during treatment will improve uniformity. Using a sonic cleaning bath requires fewer precautions. The sample, in a sterile Erlenmeyer flask, can be placed directly in the bath. Intermittent swirling of the culture will improve uniform disruption.

The following simple procedure for estimating cell survival was developed by Dr. D. Parker. Because it does not depend on counting colony forming units, survival can be determined independent of filament length. The assay is also appropriate for estimating survival from treatments other than sonication.

Estimated Survival Curve for Sonicated Samples

1. Make serial 2-fold dilutions of the unsonicated culture in basal medium, up to a 128-fold dilution. Spot 10 μl of each dilution and of the undiluted culture onto a basal medium–1% agar plate.
2. Remove two 10-μl samples from the culture after different times of sonication. Spot one 10-μl sample of each time point and a 4-fold dilution of the other onto the agar plate.
3. Incubate under normal growth conditions for 2.5–3 days until growth is evident.
4. Score plates as early as possible. Compare the appearance of each spot of sonicated cells with the dilutions of unsonicated cells. A sonicated spot that is equivalent to the 2-fold dilution of unsonicated cells represents approximately 50% survival, one equivalent to the 4-fold dilution represents 25% survival, etc.

Mutant Enrichment Methods

Antibiotic Selection of Growth-Impaired Mutants

The antibiotics penicillin, ampicillin, and cycloserine can be used to kill wild-type cells selectively in a mutagenized cyanobacterial culture as with chemoheterotrophic bacteria. It is necessary to carry out this enrichment procedure under conditions in which the desired mutants have impaired growth and are less susceptible to the drug. For auxotrophs or nitrogen fixation-deficient mutants, this is achieved by a starvation period in which the needed nutrient is deleted from the medium for a number of generations. Temperature-sensitive mutants can be treated following an incubation at the restrictive temperature to halt cell division. The following procedure is recommended as an addition to the classical and transposon-mediated mutagenesis methods.

Procedure

1. Mutagenize cells and continue with the chosen protocol through a postmutagenesis growth period to allow segregation of mutant and wild-type chromosomes. Fragment filaments if working with a filamentous strain (see Special Considerations for Mutagenesis of Filamentous Cyanobacteria).
2. Transfer the culture to restrictive conditions for 24 hr as a preenrichment step. For auxotrophs, wash cells twice with basal medium to remove nutritional supplements and carry out this incubation under normal growth conditions in basal medium. For temperature-sensitive mutants, incubate at the restrictive temperature under standard illumination.
3. Add 150 μg/ml ampicillin or cycloserine[18] or 200 U/ml penicillin G[8] and incubate under illuminated, restrictive conditions for 16–24 hr or at least two generations.
4. Remove the antibiotic by two centrifugal washes. Resuspend cells for plating; see mutagenesis protocol for recommended plating densities.

Metronidazole Enrichment of Photosynthetically Impaired Mutants

Guikema and Sherman[29] have developed a protocol for the isolation of temperature-sensitive photosynthesis mutants of *Synechococcus cedrorum* (UTEX 1191, IU 1191) using the redox-active drug metronidazole (2-methyl-5-nitroimidazole-1-ethanol). Approximately one-half of the temperature-sensitive mutants (30° permissive/40° restrictive) examined were impaired in photosynthetic function, with abnormalities throughout the photosynthetic electron transport chain. Metronidazole is an effective electron acceptor of photosystem I and is specifically toxic to photosynthesis-proficient cells. Incubation of *S. cedrorum* cells in the presence of the drug caused an 80% reduction of viable cells in the dark, and killing was enhanced five orders of magnitude by illumination. Inhibitors of elec-

[29] J. A. Guikema and L. A. Sherman, *J. Bioenerg. Biomembr.* **12,** 277 (1980).

tron transport such as DCMU block this enhancement of toxicity. Because maximum toxicity is electron transport dependent, metronidazole enrichment is primarily useful for *conditional* photosynthesis mutants in facultative photoheterotrophs and chemoheterotrophs as well as obligate photoautotrophs.

Reagents for Metronidazole Enrichment

Metronidazole (Sigma) stock solution: prepare a fresh 50 m*M* solution in glass-distilled water and filter sterilize before use

Procedure

1. Mutagenize cells as previously described and incubate under permissive conditions to allow segregation of mutant chromosomes.
2. Transfer cells to restrictive conditions for one doubling to allow cessation of electron transport in photosynthetically impaired mutants.
3. Add metronidazole to 1 m*M* and incubate 6 hr under illuminated restrictive conditions.
4. Remove the drug by two centrifugal washes and resuspend in basal medium at a concentration which, assuming 10^{-5} survival, will yield 1–2 × 10^3 cells/ml.
5. Plate 100-μl aliquots and incubate in permissive illuminated conditions for approximately 1 week, until colonies form. Replica plate to screen for a photosynthesis-impaired phenotype.

Acknowledgments

This chapter is a synthesis of methods which were worked out in the laboratories of the researchers who are referenced here. I gratefully acknowledge their contributions to this chapter. I especially thank Drs. Gerard van Arkel, Mies Borrias, Jack Meeks, Dorothy Parker, Terry Thiel, and Peter Wolk, who provided me with protocols and advice.

[81] Selection for Mutants in Energetic Pathways: Enrichment Procedures

By FRANÇOISE JOSET

Introduction

As is clear from many chapters in this volume, variations in the physiological capacities of different species of cyanobacteria, even those taxonomically close, is a reality faced by workers studying these organisms. This variation has proved true also of the responses of cyanobacteria to either mutagenesis or mutant selection procedures. This chapter aims at emphasizing this point through one example, the search for mutants impaired in the functioning of the energetic pathways.

This example was chosen as being fairly specific to cyanobacteria, because the constituents of the two energetic pathways in these cells, photosynthesis and respiration, must be coded by the same chromosome (except for possible plasmidic genes), may coexist on the thylakoid membranes, and may even share some common electron carriers.[1] The physiological relationships between these two pathways are still not clear. In particular, the role of respiration, its relative importance in the cells, and the cellular localization of its constituents, are incompletely understood. Recent information suggests that these characteristics may vary from strain to strain[1] (G. A. Peschek, personal communication). The physiological equilibrium between these electron transport chains, and the capacities of the cells to adapt to varying conditions, influence their responses to the means of selecting mutants in these processes. Several considerations should be kept in mind when starting such a program.

1. At least to our present knowledge most species cannot grow on reduced carbon substrates, by using their respiratory capacity.[2] This implies that mutants deficient in the unique energetic pathway of these obligate phototrophs can be found only as conditionals. Mutants isolated from *Synechococcus cedrorum*,[3,4] for example, are thermosensitive.

[1] G. A. Peschek, *in* "The Cyanobacteria: A Comprehensive Review" (P. Fay and C. Van Baalen, eds.), p. 119. Elsevier, Amsterdam, 1987.

[2] R. Rippka, J. Deruelles, J. B. Waterbury, M. Herdman, and R. Y. Stanier, *J. Gen. Microbiol.* **111,** 1 (1979).

[3] L. A. Sherman and J. Cunningham, *Plant Sci. Lett.* **8,** 319 (1977).

[4] S. S. Golden, this volume [80].

Other conditions can be exploited such as cryosensitivity and light sensitivity. Nonconditional mutants can be obtained only from facultative phototrophic species, such as certain *Anabaena* strains, or the unicellular species *Synechococcus* PR-6,[5] *Synechocystis* PCC 6714, and *Synechocystis PCC* 6803.[6,7] Mutations affecting constituents of the respiratory electron transfer chain, though operationally possible in all strains since they all seem to contain this function, can be phenotypically recognized only through impairment of the function. Thus, until the role of the respiratory chain is known in obligate phototrophs, mutants can be obtained directly only in the facultative species. Introduction of a mutated gene, recognized by the presence of a transposon marker, for instance, is obviously always possible. The procedures for this approach are described elsewhere in this volume.[4]

2. Electron-transfer chains are organized in membrane systems, whose fine structure is of utmost importance to ensure the efficiency of function. A modification (or deletion) of one of the constituents, beside preventing the realization of its own function, can disrupt a larger region of the structure and thereby impair other steps of electron transfer. It has been our experience that most mutants show a pleiotropic phenotype[7] (unpublished observations). The determination of the initial impairment may be difficult, particularly when analyzing spontaneous mutants, which often correspond to an alteration, not a loss, of a protein.

3. Possibly because of cell viability, or the conditions for selection (see Mutant Recovery Procedures, below), a large proportion of the mutants resulting from spontaneous events or classical mutagenesis (i.e., not from the insertion or loss of DNA) are only partially deficient. This renders more difficult the discrimination on functional criteria of the intrinsically mutated step among the pleiotropic modifications. Preparation of mutants through molecular engineering, either *in vitro* or *in vivo*,[4] should help to overcome this problem through the production of clones totally deficient for a protein and, therefore, its function.

Generally, a phenotype of deficiency in an energetic pathway, whether in a facultative or an obligate phototrophic species, will be detrimental compared to the wild type. Owing to the low frequencies of mutations (even after mutagenic treatments), an enrichment procedure is necessary to select for mutants. As pointed out in the previous chapter,[4] enrichment for mutants through selective killing of wild-type cells, adapted from clas-

[5] D. H. Lambert and S. E. Stevens, *J. Bacteriol.* **165,** 654 (1986).

[6] C. Astier, F. Joset-Espardellier, and I. Meyer, *Arch. Microbiol.* **120,** 93 (1979).

[7] C. Astier, K. Elmorjani, I. Meyer, F. Joset, and M. Herdman, *J. Bacteriol.* **158,** 659 (1984).

sical procedures,[8] applies to energy mutants. In principle, any agent or conditions that will kill wild-type cells while preserving the mutant phenotype may be used. The important points to consider are the definition of the phenotype, the efficiency of the selective treatment, and the optimal conditions for recovery and maintenance of the mutants.

Definition of the Phenotype

In gross terms, a deficiency can be regarded simply as photosynthetic or respiratory, which directly implies the permissive and nonpermissive growth conditions that will be applicable for the enrichment. A deficiency can also be specified more precisely, for instance, as in a PSII function, giving more power to the selective treatment. Table I indicates the basic rationale for the choice of enrichment conditions, which can be adapted to specific cases. For instance, mutants defective in phycobilins would grow slowly in either white or red light (absorbed by chlorophyll *a*), due to a limitation in the number of photons reaching PSII. The wild-type cells will also grow slowly under red light, but will show a faster rate of division under white light. Counterselection could thus be achieved, even in obligate photorophs, under the latter conditions.[9]

A priori, a more specific determination of the required phenotype, such as impairment of a precise electron carrier, is possible. The deficiency is recognizable by its rescue only by the addition of an artificial electron donor feeding downstream from the mutated step. Because most electron donors do not enter the cells, however, this approach is often impractical.

Choice of Enrichment Agents

Penicillin

Cyanobacteria, as gram-negative bacteria, are highly sensitive to penicillin and related antibiotics. These drugs have been successfully used to obtain photosynthetic mutants in *Synechococcus cedrorum*[3] and *Anabaena variabilis* UTEX 1444,[10] and respiratory mutants in *Synechocystis* PCC 6803.[11] The concentrations of antibiotics used in these studies ranged from 200 to 5000 U/ml, for durations equivalent to 1–3 generations. Survival of wild-type cells (not specified in the case of *Anabaena variabilis*

[8] D. B. Davis, *J. Am. Chem. Soc.* **70,** 4267 (1948).
[9] J. Myers, J. R. Graham, and R. T. Wang, *Arch. Microbiol.* **124,** 143 (1980).
[10] P. W. Shaffer, W. Lockau, and C. P. Wolk, *Arch. Microbiol.* **117,** 215 (1978).
[11] R. Jeanjan and F. Joset, manuscript in preparation.

TABLE I
POSSIBLE PERMISSIVE AND NONPERMISSIVE GROWTH CONDITIONS FOR MUTANTS OF THE ENERGETIC PATHWAYS

Mutation	Type of autotrophy	Growth conditions: Light (Temperature) Low	Light (Temperature) High	Light, DCMU, reduced C	Dark, reduced C
Photosynthesis	Facultative	−	−	+/−[a]	+
	Obligate (thermosensitive)	+	−	+/−[a]	−
PSII	Facultative	−	−	+	+
	Obligate (thermosensitive)	+	−	+	−
Respiration	Facultative	+	+	+/−[b]	−

[a] Whether block is upstream or downstream of PSI.
[b] Whether block is in light or dark reactions.

and *Synechococcus cedrorum*) decreased to 1–0.1% with *Synechocystis* PCC 6714[7] and 6803 (unpublished) treated with 200–500 U/ml.

For autotrophic growth conditions, it is advisable to treat the culture when it is growing exponentially at a high rate, that is, at a low initial cell density (1–5 × 10^7/ml). This implies the manipulation of fairly large volumes, since the total number of cells treated should be large enough to ensure with good probability the presence of more than one mutant of the desired phenotype in the population, taking into account the possible range of mutation frequencies. Assuring 10 mutants (i.e., 10^8 cells for an expected frequency of 10^{-7}) can be a safe and convenient estimate. It may be preferable to increase the concentration of the antibiotic, and decrease the duration of treatment (maximum 3 generation equivalents), rather than the reverse. Decreases in viability of 10^{-2} to 10^{-4} may be expected after one treatment. The suspension should then be washed free of the antibiotic before further use.

If necessary, several successive similar cycles may be applied. In such cases, the surviving population should be allowed to undergo several generations under nonselective conditions, so as to increase the absolute number of cells and to increase the probability of bringing into an appropriate physiological state the wild-type cells which had resisted the first treatment. Nonselective conditions are those under which both wild type and presumptive mutants can divide, to the extent possible, at similar growth rates. For instance, during the selection of photosynthetic mutants from a facultative phototroph, the optimal nonselective growth condition is chemoheterotrophy rather than photoheterotrophy or mixotrophy. Otherwise, the remaining wild-type cells, capable of using the light, would divide faster than the mutants, dependent solely on their respiratory capacities. The enrichment previously obtained would be compromised. For the same reason, if looking for spontaneous mutants, it may be fruitful to grow the suspension under nonselective conditions for a few generations so as not to lose preexisting mutants.

Redox Agents

Enrichment agents more specifically active on the energetic pathways may be of interest. The principle and a procedure for this type of work are presented elsewhere in this volume,[4] as first used in cyanobacteria with *S. cedrorum*.[12] It is important to emphasize that the response to such redox agents, unlike to penicillin, depends on the species. The reason for this probably lies in the regulation and relative roles of the electron-transfer pathways and their interaction[1] in each species. These enrichment agents act through competition with normal carriers (and probably toxicity consequent to their reduction). It is reasonable to assume that the relative sensitivities of different species to a given agent reflect specificities of interaction between these pathways.

The principle of the enrichment implies that electrons will not reach the competitor in the mutants because of a block caused by the mutation. It should be remembered, though, that this is the case only for mutations affecting transfer steps upstream of the level of competition of the drug. It is not true for mutations downstream of this point, which means that the probability of losing those mutants is high. The potential enrichment agent must be appropriate in its redox character and toxicity, and in its site of action in relation to the mutational block.

Metronidazole. Metronidazole, known to take up electrons from ferredoxin, downstream from PSI,[13] becomes toxic when reduced by a functioning photosynthetic electron-transfer chain. Metronidazole's efficient selective killing of wild-type *S. cedorum* under phototrophic conditions[12] is in agreement with this mode of action. However, with similar dosages of the drug, a different result is obtained with the facultative phototrophs *Syncechocystis* PCC 6714 and 6803.[7] The sensitivities of these two species are at least 10 times higher under heterotrophic than phototrophic conditions. The slight reduction of the killing effect when glucose (the growth

[12] J. A. Guikema and L. A. Sherman, *J. Bioenerg. Biomembr.* **12,** 277 (1980).
[13] G. W. Schmidt, K. S. Malkin, and H. H. Chua, *Proc. Natl. Acad. Sci. U.S.A.* **74,** 610 (1977).

substrate) is omitted in the dark indicates the existence of an important endogenous heterotrophic metabolism that feeds into an electron-transfer chain, which can reduce metronidazole. This result could also be due to a metronidazole-sensitive enzymatic step specific to glucose catabolism, particularly active in cells in the dark. Whatever its site of action, metronidazole could clearly be used in these species as an agent to enrich for respiratory mutants, but not photosynthetic ones.

Using different experimental conditions, however, S. V. Shestakov (personal communication) succeeded in using metronidazole in *Synechocystis* 6803 to enrich for photosynthetic-deficient mutants. The protocol was as follows: after mutagenesis by MNNG, the cells were washed and incubated on plates containing 1% glucose for 4 days. Metronidazole was then added, at 10 μg/ml, underneath the agar, as described by Porter.[14] Surviving clones checked for metronidazole sensitivity in photoautotrophy were recovered.

Sodium p-Mercurihydroxybenzoate. Sodium *p*-mercurihydroxybenzoate (PMB) is described as a competitor to metronidazole,[15] and due to its Hg constituent as a complexing agent for sulfhydryl groups. It can thus be expected to have either or both an effect on electron-transfer chains and a widespread inhibitory effect on proteins with SH-containing active sites, as is the case for several electron carriers.[16] Its level(s) of action on the photosynthetic chain, while not defined precisely, is probably on a step upstream of PSI. It has proved an efficient enrichment agent for photosynthetic mutants in the two *Synechocystis* species mentioned above, in which it acts as metronidazole does in *S. cedrorum*.[7,17] The general procedure is similar to that described in the preceding section on metronidazole enrichment. In these species, the restrictive, selective conditions are light and CO_2, at standard temperature, the permissive ones are glucose (1%) in the dark. Optimal conditions for PMB treatment have been determined to be 200 μM for 8 hr and 500 μM for 6 hr, for *Synechocystis* PCC 6803 and 6714, respectively, at a cell concentration of 3–4 $\times$ 10^7/ml, and under 3500 lux at 34°.[7] Variations in cell concentrations resulted in important differences in apparent sensitivity to the drug, probably as a result of differences in mean incident light intensity, and thus photosynthetic activity, per cell.

[14] R. D. Porter, this volume [78].

[15] R. M. Hochster and J. H. Quastel, "Metabolic Inhibitors." Academic Press, New York, 1963.

[16] K. K. Ho and D. W. Krogman, *in* "The Biology of Cyanobacteria" (N. G. Carr and B. A. Whiton, eds.), p. 191. Blackwell, Oxford, England, 1982.

[17] K. Elmorjani, J.-C. Thomas, and P. Sebban, *Arch. Microbiol.* **146,** 186 (1986).

Other Enrichment Agents

The previous descriptions of enrichment conditions, based on experimentally tested procedures, should be considered as a basis for the development of other, maybe more appropriate, selective means.

Specific Inhibitors of Electron Transfer. Electron-transfer inhibitors acting at specific sites, such as DCMU or atrazine for photosynthetic electron transfer, can also be used, provided they have a bactericidal and not only a bacteriostatic action on the wild-type cells. DCMU, at 10 μM, kills 99.9% of a population, under photoautotrophic conditions, in approximately one generation time (unpublished observations). No inhibitor totally specific to the respiration chain is presently known.

Tetracycline. An inhibitor of protein synthesis in prokaryotes, tetracycline utilizes an active transport system, possibly coupled to proton exchange,[18] to enter the cells. This energy dependence can easily be checked by the appearance of a shift in the absorption peak of the antibiotic during its transport, resulting probably from a chemical, or charge, modification of the drug. The light dependence of this transport confers an apparent insensitivity to tetracycline to mutants impaired in the building up of energy under specific conditions. Successfully used in *Rhodopseudomonas sphaeroides,*[18] this procedure could also be used with cyanobacteria. A higher toxic effect has been observed in cells of *Synechocystis* 6714 treated with 1 mg/ml tetracycline in the light than in the dark. The efficiency of killing was equivalent to that of penicillin at 500 U/ml (survival of 10^{-2}–10^{-4} after 8 hr, at 34°) in *Synechocystis* 6714.[7]

Mutant Recovery Procedures

It should be noted that mutants may show abnormal sensitivities to otherwise normal environmental factors, as a consequence of their disturbed metabolism. Increased light or O_2 sensitivity is frequent in photosynthetic mutants. Care must be taken in setting the growth conditions, during the enrichment treatment as well as during the screening, and for their maintenance. We have observed enhancement of secondary lethality by the addition of too large concentrations of glucose in certain mutants (S. Bédu, personal communication).

The normal permissive growth conditions for photosynthesis-deficient mutants, in facultative phototrophs, should be chemoheterotrophy, that is, the addition of an appropriate reduced carbon substrate and the omission of light. However, though no difficulties are usually encountered with liquid cultures, growth of clones on plates has never been achieved

[18] J. Weckesser and J. A. Magnuson, *J. Bacteriol.* **138,** 678 (1979).

under this regime. Washing the agar with activated charcoal, as frequently used with other bacteria, was not successful for *Synechocystis* 6714 and 6803 (M. Herdman and F. Joset, unpublished observations). Growth with 100% efficiency of plating is obtained, with these strains at least, only under a very weak light intensity (50–100 lux). Omission of the reduced substrate (glucose, in this case) under similar conditions prevented all growth. The estimated generation times were characteristic of chemoheterotrophic metabolism.

For reasons stated above, the direct selection of a precise phenotype is usually not possible. Screening of individual clones recovered after the enrichment procedure cannot be avoided. Utilization of side effects of the expected deficiency can sometimes facilitate screening. Taking advantage of the dependence on respiratory metabolism of the capacity to tolerate high salt concentration,[19] it has been possible to screen, by replica plating on media containing 0.4 *M* NaCl, for mutants impaired in this energetic pathway.[11]

Conclusion

Any piece of information available concerning direct or secondary expected effects of a mutation should be considered in designing selection procedures. Use can be at the level of selection (direct, as for resistant mutants, or indirect, by replica plating), or at the level of enrichment, equivalent to an indirect selection. The enriching power, through the efficiency of killing of the wild-type cells and the accuracy of selection of a precise mutant phenotype, will largely depend on the amount of knowledge available on the related physiology of the metabolic trait considered and of the organism studied.

[19] I. V. Fry, M. Huflejt, W. W. A. Erber, G. A. Peschek, and L. Packer, *Arch. Biochem. Biophys.* **244,** 686 (1986).

[82] Use of Reporter Genes in Cyanobacteria

By DEVORAH FRIEDBERG

Introduction

Reporter genes fused to various regulatory sequences have proved useful in analyzing gene expression in many cell systems (for reviews, see Refs. 1 and 2). They have been used to detect and monitor signals which regulate gene expression. The technique involves the use of a gene trimmed of its promoter and, when desired, of its translating initiation signals. The 5′ end of this truncated gene is then fused to the transcriptional and/or translational signals of interest.

The product of the reporter gene should be readily quantitatable by simple enzymatic assays, and its synthesis should allow scoring or selection of cells expressing the gene. Regulatory elements can be studied by isolating clones expressing the gene, and the level of transcription is determined by assaying its product.

Recent developments in gene cloning techniques in cyanobacteria[3,4] permit the use of the *cat*,[5] *lacZ*,[6,7] and *lux*[8,9] genes, previously applied in other cell systems.[1,10,11] Their respective gene products, chloramphenicol acetyltransferase (EC 2.3.1.28, CmActase), β-galactosidase (EC 3.2.1.23), and luciferase can be assayed in these organisms.[5,7,8,12] Using

[1] P. J. Bassford, J. Beckwith, M. Berman, E. Brickman, M. Casadaban, L. Guarente, I. Saint-Girons, A. Sarthy, M. Schwartz, H. A. Shuman, and T. Silhavy, *in* "The Operon" (J. H. Miller and W. S. Reznikoff, eds.), p. 245. Cold Spring Harbor Lab., Cold Spring Harbor, New York, 1978.

[2] M. Rosenberg, M. Brawner, J. Gorman, and M. Reff, *in* "Genetic Engineering: Principles and Methods (J. K. Setlow and a Hollaender, eds.), Vol. 8, p. 151. Plenum, New York, 1985.

[3] C. J. Kuhlemeier and G. A. van Arkel, this series, in press.

[4] S. Golden, J. Brusslan, and R. Heselkorn, this series, in press.

[5] D. Friedberg and L. Seijffers, *Mol. Gen. Genet.* **203,** 505 (1986).

[6] G. Gasparich, J. S. Buzby, D. A. Bryant, R. D. Porter, and S. E. Stevens, Jr., *in* "Progress in Photosynthetic Research" (J. Biggins, ed.), Vol. 4, p. 761. Nijhoff, Dordrecht, The Netherlands, 1987.

[7] J. S. Buzby, R. A. Porter, and S. E. Stevens, Jr., *Science* **230,** 805 (1985).

[8] G. Schmetter, P. Wolk, and J. Elhai, *J. Bacteriol.* **167,** 4111 (1986).

[9] E. P. Wolk, E. Flores, G. Schmetterer, A. Herrero, and J. Elhai, *Proc. Int. Symp. Photosynth. Prokaryotes, 5th,* p. 122.

[10] D. S. Goldfarb, R. L. Rodriguez, and D. H. Doi, *Proc. Natl. Acad. Sci. U.S.A.* **79,** 5886 (1982).

[11] J. Engelbrecht, M. Simon, and M. Silverman, *Science* **227,** 1345 (1985).

[12] D. Friedberg and J. Seijffers, manuscript in preparation.

reporter genes, regulated gene expression was demonstrated in cyanobacteria. Thus *cat*, fused with a DNA fragment carrying λ phage operator promoter regions O_LP_L, O_RP_R, and the *cI*857 gene coding for a temperature-sensitive repressor protein, is expressed in *Anacystis* in response to temperature shifts.[5] The induced promoter activity is correlated with the finding of a protein compatible in its cross-reactivity and size with the λ repressor.[5] Vectors carrying these signals might thus be applicable for regulated gene expression in *Anacystis*. A further example is the *cpc* promoter of *Synechococcus* PCC 7002 which responds in *LacZ* fusions to light intensity and nitrogen availability.[6] A reduced promoter activity in response to elevated light intensity and to nitrogen depletion is correlated with *cpc* mRNA levels.[6] The activity of each of the P_L and *cpc* promoters in the heterologous host is compatible with that in the homologous host.[5,6] Monitoring of these activities gave rise to the suggestion that cyanobacteria and *Escherichia coli* are capable of recognizing efficiently each others' promoters.[5,6] This might indeed be tenable in view of close fit of some cyanobacterial promoters[13] with the *E. coli* consensus promoter.[14]

The chapter addresses the utilization of reporter genes and vectors employed in the assessment and detection of transcription regulatory sequences in cyanobacteria.

Principles

Two strategies have been employed for gene cloning in cyanobacteria. One uses shuttle plasmid vectors that carry both a cyanobacterial and an *E. coli* replicon,[3] each of which functions in its indigenous host. Depending on the sequence of the cloned gene, the selection applied, and the cyanobacterial host, the cloned gene can be maintained on the autonomous plasmid or can be integrated into the cyanobacterial chromosome. The other strategy exploits the efficient recombination system of *Anacystis nidulans* R2[15] and *Aphanocapsa* PCC 6803[3] for a directed integration of the cloned gene into the chromosome.[4] By the latter strategy a DNA vector that lacks a cyanobacterial replicon, but carries similar chromosomal sequences, is applicable.[4] According to the imposed selective conditions, the cloned gene, either alone or together with the vector sequences, integrates into the chromosome.[15] Both strategies have also

[13] D. A. Bryant, J. M. Dubbs, P. I. Fields, R. D. Porter, and R. de Lorimier, *FEMS Microbiol. Lett.* **29,** 343 (1985).

[14] M. Rosenberg and D. A. Court, *Annu. Rev. Genet.* **13,** 1319 (1979).

[15] J. G. K. Williams and A. A. Szalay, *Gene* **24,** 37 (1983).

been employed for cloning transcriptional regulatory sequences and reporter genes.[5,6,8,12,16]

A typical vector suitable for studying promoter activity by means of a reporter gene as depicted in Fig. 1 contains the following elements: (1) a cyanobacterial and an *E. coli* replicon; (2) a selectable marker other than the reporter gene; (3) a DNA fragment carrying the transcriptional regulatory sequences of interest; (4) a unique restriction site downstream the promoter sequences, affording the insertion of a reporter gene. The *cat* gene fused to a DNA fragment which contains transcriptional regulatory signals of bacteriophage λ is used for monitoring the activity of these signals in a cyanobacterium background.[5]

In addition to the manipulations with the shuttle vector, the *cat* gene also has been used to detect promotor activity of cyanobacterial DNA fragments.[12,16] A known *E. coli/Bacillus subtilis* promoter analysis plasmid[10] into which cyanobacterial DNA fragments are inserted can be applied for this purpose. Using *E. coli* as an heterologous host, cyanobacterial constitutive—but not inducible—promoters can be detected. In another approach, this system may be applied with the cyanobacterium as initial host following integrative transformation.[12] However, since the DNA fragments that show promoter activity are relatively short,[16] and the frequency of integrative transformation is low for short inserts,[17] the initial screening in *Anacystis* might be less efficient than in *E. coli.*

When measuring promoter activity in cyanobacteria and *E. coli,* one should consider the differences in copy number of the gene fusions in the two hosts. In the case of the shuttle vector, its copy number in *E. coli* is usually higher than in *Anacystis*. Whereas merely one copy of the gene fusion is expected in the cyanobacterial chromosome when an integrative vector is applied,[15] in *E. coli* the copy number will be much higher and equivalent to that of the vector.

Choice of a particular selection following transformation should be guided by the conditions which afford transcription. Thus *cat* which confers chloramphenicol resistance (Cm^R) may serve as a selectable marker when constitutive promoters are studied or inducible conditions are sought. In *Anacystis,* however, even nonpermissive conditions might lead to Cm^R phenotypes, brought about by very low levels of CmActase.[5] Therefore, once a Cm^R phenotype is selected, the strength of the fused promoter should be addressed by the expressed CmActase activity.

[16] D. Friedberg and D. Goldfarb, *Proc. FEBS Meet., 15th,* p. 112 (1983).

[17] K. S. Kolowsky, J. G. K. Williams, and A. A. Szalay, *Gene* **27,** 289 (1984).

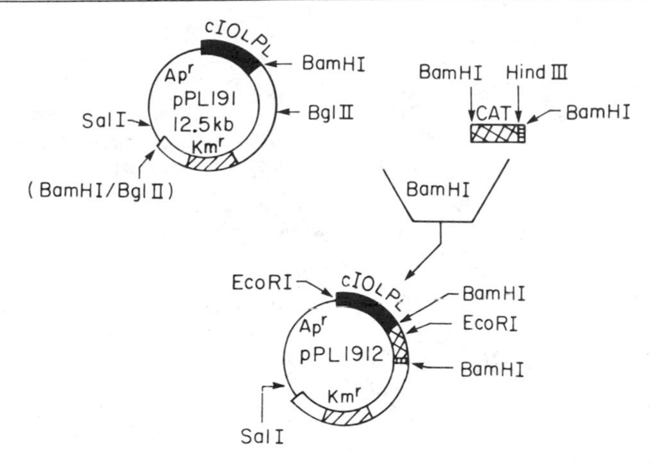

FIG. 1. Schematic presentation of a typical *Anacystis/E. coli* shuttle vector carrying *cat* as a reporter gene and the λ phage transcription regulatory sequences to be analyzed in *Anacystis*. The following notation is used: ▭, cyanobacterial 8-kb plasmid sequences; —, pBR322 sequences; ■, λ phage sequences; ▨, KM^R gene; ▩, promoterless *cat* gene; ⊟, vector sequences.[25]

For maintenance of the autonomous or the integrative vector, a selection utilizing the vector trait, rather than the *cat* gene, is preferable. The *nptI* (derived from Tn*903*[18]), which encodes for neomycin phosphotransferase and confers kanamycin resistance (Km^R), is an effective marker in *Anacystis*.[5]

The utilization of the *lacZ* as a scorable trait in cyanobacteria is complicated by the fact that the color of the colonies obscures that of the indicator, an inconvenience that can be partly circumvented in *Anacystis*[12] and in *Synechococcus* 7002.[7] Nonetheless the encoded β-galactosidase activity can be easily measured in extracts[12] or in permealized cells,[7] thus allowing the study of transcriptional regulatory signals fused to *lacZ*.

The cotranscribed *luxA* and *luxB* genes from *Vibrio harveyi* and *Vibrio fischeri* have been cloned in conjugative shuttle vectors and mobilized into *Anabaena* spp.[8,9] The extreme sensitivity of the assay that measures the encoded luciferase in the cyanobacterium makes these genes good candidates for studying gene expression in this organism. Indeed, the enhanced expression of *lux* in fusions with the strong *rbc* promoter, as compared with that of *nif/lux*,[8,9] confirms its applicability as reporter gene.

[18] A. Oka, H. Sugisaki, and M. Takanami, *J. Mol. Biol.* **147,** 217 (1981).

Methods

Strains and Growth Conditions

Anacystis nidulans R2K is a derivative of the transformable strain R2[19] deleted of its indigenous 8-kb plasmid, retaining the 50-kb plasmid.[5] Cells are grown in BG-11 medium[20] at 34° and a light intensity of 30 $\mu E\ m^{-2}\ sec^{-1}$. Additional growth details are essentially as described elsewhere.[3] When desired, antibiotics are added at the following concentrations: chloramphenicol (Cm) 5 $\mu g\ ml^{-1}$ and kanamycin (Km) 25 $\mu g\ ml^{-1}$. *Escherichia coli* DR100 is a *recA hsdR* derivative of C600[21] used in *cat* experiments; *E. coli* MC1061[22] is a *hsdR lac* used in *lacZ* experiments with *Anacystis*.

DNA Methods

Restriction digests, ligations, transformation of *E. coli* and plasmid isolation from *E. coli* are performed according to standard procedures.[23] Methods for plasmid minipreps from *Anacystis* are as described elsewhere.[3]

Chromosomal DNA from *Anacystis* is purified as follows: 400–1000 ml cell culture of OD 0.800 at 720 nm is centrifuged at 10,000 rpm for 10 min at 4°, washed in 40 ml H_2O and suspended in 8 ml of 1× SSC (8.76 g NaCl and 4.41 g sodium citrate in 1 liter of H_2O, adjusted to pH 7.0) containing 20 m*M* EDTA. One-half milliliter of freshly made lysozyme in H_2O is added to a final concentration of 1 $mg\ ml^{-1}$, and the mixture is incubated for 60 min at 37° with occasional swirling. Lysis is achieved by the addition of 800 μl of 10% sarkosyl and reincubation for 15 min at 37°. After transferring 8.3 ml of the lysate to a suitable ultracentrifuge tube, the volume is made up to 25 ml with 25.0 g $CsCl_2$ and 1× SSC/EDTA, to which 1.0 ml of ethidium-bromide (10 $mg\ ml^{-1}$) is added. The DNA band is isolated after 20 hr of centrifugation at 45,000 rpm, 18°, in an ultracentrifuge using a VTi50 rotor. The extraction of the ethidium bromide by isoamyl alcohol and subsequent dialysis and precipitation of the DNA follows standard methods.[23]

[19] S. V. Shestakov and N. T. Khyen, *Mol. Gen. Genet.* **107,** 372 (1970).

[20] R. Rippka, J. Deruelles, J. B. Waterbury, M. Herdman, and R. Y. Stanier, *J. Gen. Microbiol.* **111,** 1 (1979).

[21] A. Laban and A. Cohen, *Mol. Gen. Genet.* **184,** 200 (1981).

[22] S. K. Shapira, J. Chou, F. V. Richard, and M. J. Casabadan, *Gene* **25,** 71 (1983).

[23] T. Maniatis, E. F. Fritsch, and J. Sambrook, "Molecular Cloning: A Laboratory Manual." Cold Spring Harbor Lab., Cold Spring Harbor, New York, 1982.

Application of *cat*

cat DNA

The promoterless Tn*9* *cat* gene (~780 bp) can be isolated from known *E. coli* promoter analysis vectors[24] or from derivatives thereof.[25] It can be excised as a DNA fragment flanked by the resriction sites *Hin*dIII, *Bam*HI or *Sal*I. This gene derivative contains the complete coding sequence and an *E. coli* ribosome-binding site. An upstream promoter will give rise to Cm^R transformants which express CmActase activity. The sequence of the gene is known.[26]

Regulated cat Expression in Anacystis

Selection of cat fusions in E. coli. A chimeric gene[25] composed of the operator promoter regions O_LP_L, O_RP_R, and the temperature-sensitive repressor gene *cI*857 of bacteriophage λ is carried by the *Anacystis/E. coli* shuttle vector. A unique *Bam*HI site is present downstream to the promoter region (Fig. 1). A *Bam*HI DNA fragment carrying *cat* is ligated with a *Bam*HI digest of the vector carrying the transcriptional regulatory signals of interest. The ligation mixture is used to transform *E. coli*. Km^R transformants are selected at 30° and screened for Cm^R at 30° and 38°. Some will be Cm^R at 38° and Cm^S at 30°. This is the expected phenotype when *cat* transcription is driven by P_L (or P_R), which is regulated by the cI temperature-sensitive repressor protein.[27] Plasmid DNA isolated from these phenotypes is digested with *Bam*HI in order to confirm the cloning of *cat*. The orientation of the inserted gene is verified by restriction with *Eco*RI, which shows that in these phenotypes *cat* is oriented with its 5′ end toward the operator promoter sequences. The plasmid carrying the λ cIO_LP_L/cat fusions is designated pPL1912 (Fig. 1). Transformants carrying pPL1912 are grown at 30° and analyzed for *cat* expression upon a temperature shift to 42°, as described below for *Anacystis*.

Selection of cat Fusions in Anacystis. pPL1912 is used to transform *Anacystis nidulans* R2K as follows: A culture of 0.5–1 × 10^8 cells ml^{-1} is concentrated by centrifugation (10,000 rpm, 10 min, 15°) to 5 × 10^8 ml^{-1} in BG-11. Aliquots of 0.6 ml are incubated with 1 μg plasmid DNA under gentle mixing at 30° in the dark for 18 hr. Thereafter, 0.1 ml of the original transformation mixture and of its dilutions (10^{-1} and 10^{-2}) are spread on

[24] T. J. Close and R. L. Rodriguez, *Gene* **20,** 305 (1982).

[25] A. Honigman, J. Mahajna, S. Altuvia, S. Koby, D. Teft, H. Locker, H. Hayman, C. Kronman, and A. B. Oppenheim, *Gene* **36,** 131 (1986).

[26] N. K. Alton and D. Vapnek, *Nature* (*London*) **282,** 864 (1979).

[27] M. Rosenberg, Y. Ho, and A. Shatzman, this series, Vol. 101, p. 123.

plates containing 30 ml BG-11–agar. After 24 hr of incubation at 34° (selection at 28° is inefficient) under light, 0.5 ml of Km is added underneath the medium to a final concentration of 25 μg ml^{-1}. Single colonies, which appear within 5–10 days, are picked onto two BG-11–agar plates containing Cm and incubated at 34° and 38°, respectively. Unlike *E. coli*, all Km^R transformants of *Anacystis* R2K are also Cm^R at 34° due to residual CmActase activity.[5] Km^R/Cm^R transformants are purified and grown in 2 ml of BG-11 in the presence of Km at 34°. Plasmid is purified from 100-ml cultures by the miniprep method[3] and analyzed by restriction enzymes, as described above for *E. coli*. In addition, the plasmid integrity is verified by Southern blot analysis[23] of plasmid DNA restricted with suitable enzymes and hybridized with a probe of pPL1912.[5]

cat Expression and Preparation of Cell-Free Extracts. Samples from *Anacystis* cultures grown at 34° to a cell density of 1 × 10^8 ml^{-1} are transferred to 42° (or 39°) and incubated under standard light conditions. Cell samples (1–3 ml) removed at different times are centrifuged for 2 min in an Eppendorf microfuge at 12,000 rpm, washed once, and resuspended in 50 m*M* Tris, pH 7.8. The cells are disrupted at 0° by two 30-sec bursts by means of a microtip-equipped ultrasonic cell disrupter (MSC) at maximal setting. The cell debris is removed by a 5-min centrifugation at 4°. The supernatant is used for spectrophotometric assays of CmActase activity and for electrophoresis (Fig. 2). At least a 20-fold increase in CmActase activity is observed on a temperature shift to 42° within 24 hr.[5] Heat induction does not occur at 37°. Extracts prepared from cells grown at 34 and 39° are electrophoresed on sodium dodecyl sulfate (SDS)–polyacrylamide gels. The presumptive heat-induced CmActase protein (24 kDa) is detected in 39° samples in the Coomassie blue-stained gel.

Spectrophotometric Assay of CmActase. A spectrophotometric assay quantitates the rate of Cm acetylation[28] and is applied to cell extracts of *Anacystis*. The reaction is carried out in a cuvette (1 cm path) using a recording spectrophotometer equipped with a temperature-controlled chamber at 37°. The reaction mixture is freshly made, as follows: 4 mg of 5,5′-dithiobis(2-nitrobenzoic acid) (DTNB) dissolved in 1 *M* Tris, pH 7.8, to which 9.8 ml H_2O and 0.2 ml of 5 m*M* acetyl-CoA are sequentially added. To 0.5 ml of the reaction mixture 5–20 μl of cell-free extract, or a 10^{-1} dilution thereof, containing 0.5–10 μg protein, is added. The reaction is started by the addition of 10 μl of 5 m*M* Cm. The net change in the absorption at 412 nm min^{-1} is divided by 13.6 to give the results in terms of micromoles of Cm acetylated per minute. This value is normalized to the protein concentration.

[28] W. V. Shaw, this series, Vol. 43, p. 737.

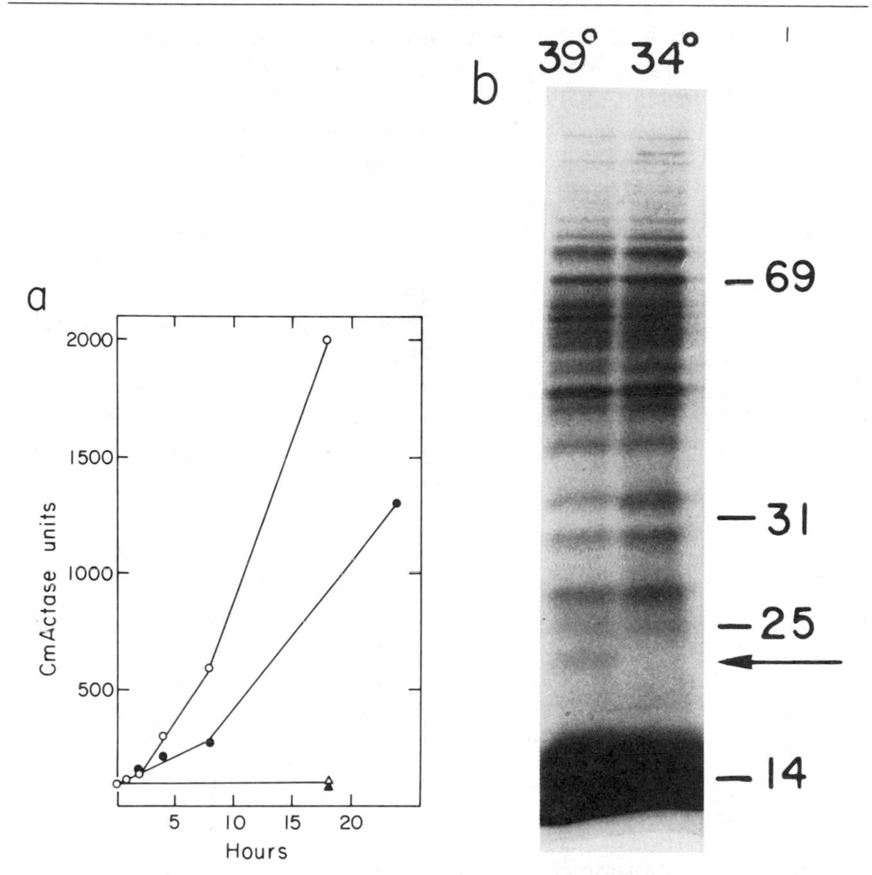

FIG. 2. Heat induction of CmActase in *Anacystis* carrying cIO_LP_L/cat. (a) Late-phase exponentially growing cells at 34° are transferred to 37° (▲), 39° (●), and 42° (○) under light or to 39° (△) in the dark. CmActase activity is determined as described in the text. One unit of CmActase is equal to 1 nmol Cm acetylated min^{-1} (mg protein)$^{-1}$. (b) Coomassie blue-stained SDS–15% polyacrylamide gel analysis of cell-free extracts (60 μg) obtained from cells grown at 34° and from cells further incubated at 39° for 24 hr. The arrow denotes the presumptive CmActase protein.

Detection of Anacystis DNA Fragments Which Promote cat Expression

The plasmid pGR71, designed for promoter analysis in *E. coli* and *B. subtilis*,[10] can also be applied for detection of *Anacystis* promoters in *E. coli*.[16] This vector contains the promoterless *cat* gene, a *Hin*dIII site at its 5′ end, a Km^R gene and, a ColE1 replicon.[10] A dephosphorylated *Hin*dIII digest of pGR71 DNA is ligated with a *Hin*dIII digest of *Anacystis nidulans* PCC 6311[20] DNA. This gene library is used to transform *E. coli*.[16]

Transformants are selected in the presence of Km and thereafter screened for their level of resistance to Cm (or directly selected as resistant to 40 μg ml^{-1} Cm). Restriction analysis with *Hin*dIII and *Eco*RI of plasmids isolated from clones which reveal significant resistance to Cm confirms the insertion of DNA fragments into the *Hin*dIII site. These clones are assayed for their CmActase activity. The level of the enzymatic activity measured in clones resistant to 100 μg ml^{-1} Cm varies from 200 to 8000 nmol Cm acetylated min^{-1} (mg protein)$^{-1}$.[12] Plasmid DNA isolated from various Cm^R clones is used to transform *E. coli* DR103, which produces minicells.[29] This system allows the detection of *de novo* synthesis of plasmid-encoded proteins. Purified minicells are labeled by [^{35}S]methionine,[30] and the cell-free extracts are electrophoresed. The presumptive CmActase protein (24 kDa) is displayed on autoradiograms and in Coomassie blue-stained gels (Fig. 3). The apparent intensity of the 24-kDa protein is compatible with the level of CmActase activity.[12,16]

A typical plasmid DNA carrying an *Anacystis* 6311 DNA fragment with promoter activity can be used to transform *Anacystis* R2[12] (such transformation is possible because of the close sequence similarity of the two strains[12]). Km^R clones verified to be also Cm^R are expected to contain the intact plasmid in their genome following the integrative transformation.[15] The activity of CmActase measured in a typical clone is 30% less than that found in *E. coli*,[12] conceivably due to the lower copy number of the *cat* fusions in the cyanobacterium host.

Application of *lacZ* Fusions

Truncated *lacZ,* trimmed of its promoter, its translation initiation site, and its early codons, is carried by a number of plasmid vectors.[22] The sequence of *lacZ* is known.[31] DNA fragments to be analyzed are inserted into a correct translational reading frame of the *lacZ* for construction of *lacZ* fusions. A translational fusion composed of a DNA fragment carrying the *trp* operator promoter region, its attenuator, as well as the first codons of *trpE* fused to *lacZ*,[32] when cloned into an Anacystis shuttle vector, is found to give low levels of β-galactosidase activity, as compared with the respective *E. coli* clones,[12] a phenomenon partly explained by the presence of attenuation signals. Nevertheless, this low activity is

[29] H. I. Adler, W. D. Fisher, A. Cohen, and H. Hardigree, *Proc. Natl. Acad. Sci. U.S.A.* **57,** 321 (1966).

[30] D. Friedberg and J. Seijffers, *Gene* **22,** 267 (1983).

[31] A. Kalnins, K. Otto, and B. Muller-Hill, *EMBO J.* **2,** 593 (1983).

[32] J. Kopelowitz, R. Schoulaker-Schwarz, A. Lebanon, and H. Engelberg-Kulka, *Mol. Gen. Genet.* **196,** 541 (1985).

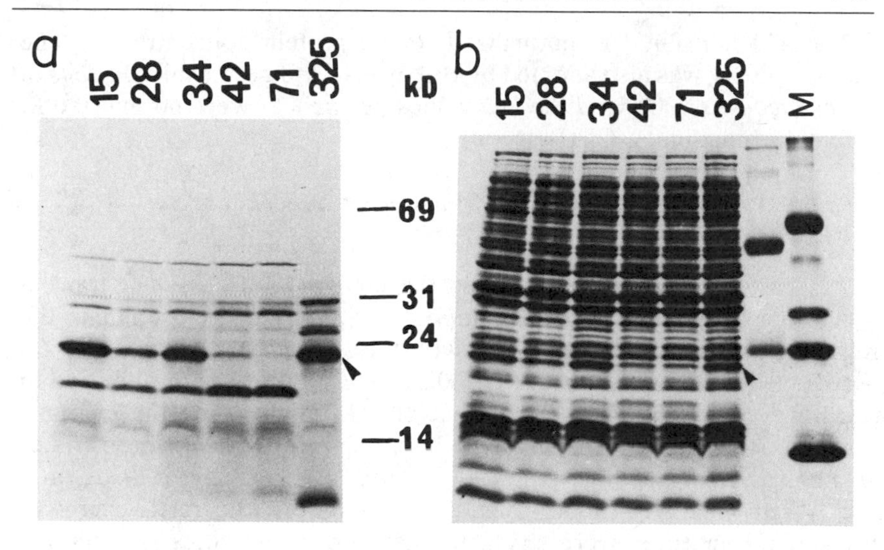

FIG. 3. *cat* expression driven by cyanobacterial sequences in *E. coli*. (a) Autoradiogram of SDS–15% polyacrylamide gel analysis of ^{35}S-pulse-labeled proteins synthesized in minicells. The numbers 15, 28, 34, and 42 designate different pGR71[10] derivatives carrying *Anacystis* DNA fragments; 71 is pGR71 (carrying the promoterless *cat*), and 325 is pBR325 (carrying the native *cat*). (b) Coomassie blue-stained gel of total cell proteins (60 μg). The arrow denotes the CmActase protein. Lane M was run with molecular size markers.

sufficiently significant for applying methods that measure *lacZ* expression in *Anacystis*.

β-Galactosidose Assays

In Plates. *Anacystis* colonies composed of cells carrying the *trp/lacZ* fusion are streaked to form heavy patches (3 × 3 mm) on BG-11/Km–agar plates containing 40 μg ml^{-1} of the indicator 5-bromo-4-chloro-3-indolyl-β-D-galactoside (X-Gal). Within 5 days the color of the growth areas turns from green to turquoise, surrounded by blue halos, indicating the expression of *lacZ*. No such color change occurs in controls carrying vectors lacking the *lacZ* fusion.[12] A different procedure which employs chloroform vapor for discerning the color is used with *Synechococcus* 7002 carrying native *lacZ* or fusions thereof.[7]

In Cell-Free Extracts. The standard procedure is based on cleavage of *o*-nitrophenyl-β-galactopyranoside (ONPG)[33] and is applied to *Anacystis* extracts, prepared as above, but using β-galactosidase buffer.[33] The net

[33] J. H. Miller (ed.), "Experiments in Molecular Genetics." Cold Spring Harbor Lab., Cold Spring Harbor, New York, 1972.

OD at 420 nm min^{-1} is normalized to the protein concentration. The standard assay was also adopted by Buzby *et al.*[7] to permeabilized cells of *Synechococcus* 7002, and the OD values per minute were normalized to colony forming units (CFU).

Regulated Expression of cpc/lacZ Fusions in Synechococcus PCC 7002[6]

Gasparich *et al.*[6] have described the construction of a *lacZ* translational fusion of a 1.2-kb *Eco*RI/*Bam*HI DNA fragment, containing the regulatory region of the *cpc* gene, with the *lacZ* gene. This gene fusion was cloned into a *Synechococcus* 7002/*E. coli* shuttle vector and used to transform *Synechococcus*. The effect of light intensity on the *cpc* promoter activity was investigated by comparing the β-galactosidase values of cultures grown at high (282 $\mu E\ m^{-2}\ sec^{-1}$) versus low (119 $\mu E\ m^{-2}\ sec^{-1}$) light intensity (under these light conditions the relative growth rates of the cultures were similar). The ratio of activities for cultures grown at low versus high light was 1.75 as opposed to 1.0 for control strains, implying that, under reduced light, the *cpc* promoter is more active. Nitrogen starvation also affected the activity of this promotor in that a significant 50% drop in β-galactosidase activity occurred during the first 3 hr of starvation, suggesting reduced activity of the *cpc* promoter under these circumstances.

Application of *lux*[8,9]

The promoterless *luxA* and *luxB* genes from *V. fischeri* and *V. harveyi* trimmed from plasmids pJE205[34] and pBB128,[35] respectively, were cloned by Schmetterer *et al.*[8,9] into *Anabaena* spp. conjugative shuttle vectors[8] or a modification thereof.[9] The latter was designed to insert promoters to be analyzed in a polylinker present between the *lux* and a *cpc* transcriptional termination region.[9] The restriction of the *lux* genes in *Anabaena* PCC 7120 is prevented by mobilizing the vector from *E. coli* bearing a methylase that modifies *Ava*II sites.[36]

Luciferase Assay

Luciferase catalyzes the oxidation of an aldehyde and of a reduced flavin mononucleotide ($FMNH_2$). The resultant light emission is quanti-

[34] J. Engebrecht, K. Nealson, and M. Silverman, *Cell* **32,** 773 (1983).

[35] R. Belas, A. Mileham, D. Cohn, M. Hilmen, M. Simon, and M. Silverman, *Science* **218,** 791 (1982).

[36] J. Elhai, personal communication.

tated. Cells containing luciferase produce light on addition of the aldehyde. The standard assay,[37] as modified by Elhai *et al.*[8,36] is applied to *Anabaena* spp: mid to late log-phase cultures are diluted 1 : 100 to 1 : 1000 in BG-11 (1×10^5–1×10^3 cells ml^{-1}), subsequent to fragmentation of the filaments. The reaction is performed in a scintillation vial containing 2 ml of the suspension. Preincubation in the dark should be avoided. At time zero 200 μl of the aldehyde emulsion is added. The latter is composed of thoroughly sonicated 0.1% (v/v) *n*-decanal in 20 mg ml^{-1} bovine serum albumin and 50 m*M* phosphate buffer, pH 7.8. Thc light is quantitated as described.[8] As luciferase from *V. fischeri* is temperature sensitive, growth of the cells in question as well as the assays are to be carried out below 34°. Colonies producing light can be visually identified on plates, and single cells can be discerned under the microscope, in the presence of *n*-decanal.[8]

Acknowledgments

The author is indebted to R. D. Porter and D. A. Bryant for data on *lacZ* fusions in *Synechococcus* PCC 7002 and to J. Elhai for data on the application of the *lux* genes in *Anabaena* spp. This research was supported by grants from the National Council for Research and Development, Israel, from the Gesellschaft für Strahlen und Umweltforschung, München, Federal Republic of Germany, and from the Basic Research Foundation of the Israel Academy of Science and Humanities.

[37] J. W. Hastings, T. O. Baldwin, and M. Z. Nicoli, this series, Vol. 57, p. 135.

[83] Conjugal Transfer of DNA to Cyanobacteria

By JEFF ELHAI and C. PETER WOLK

Introduction

Conjugation appears to bc a gcneral means to introduce DNA from *Escherichia coli* into cyanobacteria,[1] using the broad host range conjugal apparatus of an IncP plasmid, such as RP4.[2] RP4, originally isolated from *Pseudomonas,* has been shown to mediate the transfer of DNA into a

[1] C. P. Wolk, A. Vonshak, P. Kehoe, and J. Elhai, *Proc. Natl. Acad. Sci. U.S.A.* **81,** 1561 (1984).
[2] C. M. Thomas and C. A. Smith, *Annu. Rev. Microbiol.* **41,** 77 (1987).

wide range of gram-negative bacteria, including such distantly related organisms as myxobacteria,[3] thiobacilli,[4] and unicellular and filamentous cyanobacteria.[1] To obtain stable conjugal transfer it is necessary that (1) conjugal contact be made, (2) transferred DNA escape restriction or degradation, and (3) the DNA replicate autonomously or integrate into one of the replicons of the recipient. The first requirement is probably met in the great majority of gram-negative eubacteria, cyanobacteria included. The task for the experimenter is to find conditions that permit the last two requirements to be met as well.

Even with transformable unicellular cyanobacteria,[5] conjugation may be the preferred route of DNA transfer in certain cases. DNA taken up by unicellular cyanobacteria appears to be randomly cut early during the transformation process, so that the efficiency of transfer of a segment of nonhomologous DNA decreases exponentially with its length.[5,6] When a foreign plasmid containing cloned chromosomal DNA is transformed into such a cyanobacterium, gene replacement (suggesting double recombination) is favored over merodiploid formation (suggesting single recombination) by as much as two orders of magnitude, explicable if few plasmids enter the cell intact (single recombination of the chromosome with linearized plasmid DNA would result in a broken chromosome). Conjugation proceeds by a wholly different mechanism,[7] and transfer of such plasmids by conjugation leads to a great excess of merodiploid formation over gene replacement.[8] Thus, conjugation may be advantageous if a large segment of foreign DNA is to be transferred or if merodiploids are sought (e.g., if the intent is to direct a plasmid to a specific site in the chromosome without interfering with gene function).

Principle of Conjugal Transfer of Nonconjugal Plasmids

Conjugal plasmids are all quite large (RP4 is 57 kb), because many genes are required to encode the conjugal apparatus.[2,9] Some smaller plasmids have evolved to take advantage of the conjugal opportunities provided by a coresident conjugal plasmid, and the availability of these smaller plasmids makes isolation and cloning much easier. There are two

[3] A. M. Breton, S. Jaoua, and J. Guespin-Michel, *J. Bacteriol.* **161,** 523 (1985).

[4] C. F. Kulpa, M. T. Roskey, and M. T. Travis, *J. Bacteriol.* **156,** 434 (1983).

[5] R. D. Porter, *CRC Crit. Rev. Microbiol.* **13,** 111 (1987).

[6] K. S. Kolowsky, J. G. K. Williams, and A. A. Szalay, *Gene* **27,** 289 (1984).

[7] N. Willetts and B. Wilkins, *Microbiol. Rev.* **48,** 24 (1984).

[8] E. Van Haute, H. Joos, M. Maes, G. Warren, M. Van Montagu, and J. Schell, *EMBO J.* **2,** 411 (1983).

[9] N. Willetts and R. Skurray, *Annu. Rev. Genet.* **14,** 41 (1980).

requirements that must be satisfied for a plasmid to be mobilized by a conjugal plasmid.[7] First, the plasmid must contain a stretch of DNA called the *bom* site (for basis of mobility; also called the *oriT* region). Second, the plasmid must encode or be provided with a DNA-nicking protein (encoded by a gene termed *mob*) that specifically recognizes the *bom* site.[7,10]

ColE1 is one such molecular passenger: RP4 can transfer the plasmid wherever RP4 can transfer itself. pBR322[11] and other widely used derivatives of pMB1, a ColE1-like plasmid, retain the *bom* site but not the *mob* gene of its natural parent. These derivatives can be transferred only when the nicking protein is provided in trans, by a helper plasmid.[10] The donor may therefore contain several plasmids to facilitate the transfer of DNA to cyanobacteria: (1) the *conjugal* plasmid, e.g., RP4; (2) the *helper* plasmid(s), e.g., a plasmid encoding the ColE1-specific nicking protein or one carrying a gene for a methylase to circumvent restriction by the recipient; and (3) the *cargo* plasmid, the plasmid you really want to transfer. Table I provides some examples of each category.

Conjugal Plasmids. We have used RP4 and derivatives of it almost exclusively, particularly pRL443, an RP4 derivative that has lost kanamycin resistance, but other IncP plasmids are also effective. R702 and R751 have also been employed.[12]

Helper Plasmids. pDS4101 (a gift of David Sherratt, University of Glasgow, Glasgow, Scotland) and pGJ28[8] are plasmids that encode nicking proteins recognizing the *bom* site of ColE1-like plasmids and that are compatible with ColE1 derivatives. We now routinely use an alternative helper plasmid, pRL528 (J. Elhai and C. P. Wolk, unpublished), a mob^+ derivative of ColK that contains the gene encoding the *Ava*I methylase from *Anabaena* PCC 7118 (a gift of K. Lunnen, New England Biolabs, Beverly, MA) and the gene encoding the *Eco*47II methylase (a gift of A. Janulaitis, ESP Fermentas, Vilnius, U.S.S.R.). The latter methylase, which modifies GGNCC sites, protects against restriction by *Ava*II (which recognizes $\mathrm{GG}^{\mathrm{A}}_{\mathrm{T}}\mathrm{CC}$). Isoschizomers of *Ava*I and *Ava*II are found in a wide variety of cyanobacteria.[13] Other helper plasmids have been constructed that contain the gene for one but not the other of the methylases (see Table I).

Cargo Plasmids. Cargo plasmids carry the DNA intended for transfer and fall into two classes: those that can replicate in the target cyanobacte-

[10] J. Finnegan and D. Sherratt, *Mol. Gen. Genet.* **185,** 344 (1982).

[11] P. Balbás, X. Soberón, E. Merino, M. Zurita, H. Lomeli, F. Valle, N. Flores, and F. Bolivar, *Gene* **50,** 3 (1986).

[12] T. Thiel and C. P. Wolk, this series, Vol. 153, p. 232.

[13] N. Tandeau de Marsac and J. Houmard, *in* "The Cyanobacteria" (P. Fay and C. van Baalen, eds.), p. 251. Elsevier, Amsterdam, 1987.

TABLE I
PLASMIDS USEFUL IN CONJUGATION INTO CYANOBACTERIA

Plasmid or Vector	Drug resistance	Comment
Conjugal plasmids		
RP4	Ap Km Tc	IncP
pRL443	Ap Tc	Spontaneous mutant of RP4
R702	Km	IncP
R751	Tp	IncP
Helper plasmids		
pDS4101	Ap	Based on ColK, carries *mob*
pGJ28	Km	Based on ColD, carries *mob*
pRL449	Cm	Based on pACYC184, carries gene for *Eco*47II methylase
pRL542	Cm	Based on ColK, carries *mob*
pRL518	Cm	Based on ColK, carries *mob* and gene for *Eco*47II methylase
pRL530	Cm	Based on ColK, carries *mob* and gene for *Ava*I methylase
pRL528	Cm	Based on ColK, carries *mob* and genes for *Eco*47II and *Ava*I methylases
Cargo vectors[a]		
pRL6	Cm Nm	Many *Anabaena* and *Nostoc*[b,c]
16C12AK	Ap Km	*Anabaena* ATCC 29413[d]
pRL25C	Nm	Many *Anabaena* and *Nostoc*; cosmid vector[e]
pJCF62	Cm Nm	*Fremyella diplosiphon*[f]
pSG111M	Ap Cm	*Synechococcus* R2[c]

[a] Mobilizable shuttle vectors for *Agmenellum quadruplicatum* PR-6 were described by S. Buzby, R. D. Porter, and S. E. Stevens, Jr., *J. Bacteriol.* **154,** 1446 (1983). Conjugation into this strain, however, has yet not been demonstrated. Comment indicates known range.

[b] E. Flores and C. P. Wolk, *J. Bacteriol.* **162,** 1339 (1985).

[c] C. P. Wolk, A. Vonshak, P. Kehoe, and J. Elhai, *Proc. Natl. Acad. Sci. U.S.A.* **81,** 1561 (1984).

[d] C. P. Wolk, unpublished observations.

[e] C. P. Wolk, Y. Cai, L. Cardemil, E. Flores, B. Hohn, M. Murry, G. Schmetterer, B. Schrautemeier, and R. Wilson, *J. Bacteriol.* **170,** 1239 (1988).

[f] J. Cobley, "Molecular Biology of Cyanobacteria Workshop: Abstracts," SPM-4. St. Louis, Missouri, 1987.

rium (called shuttle vectors) and those that cannot (suicide vectors). In either case, if the plasmid is based on a ColE1-like replicon, it is essential that the *bom* site be retained. This means that pBR322 and many of its derivatives[11] can serve as cargo plasmids but pBR328[11] cannot, and neither can pUC7 and beyond in the series of vectors constructed by Messing and co-workers.[14] Of the many shuttle vectors described for unicellular cyanobacteria, several are based on pACYC184 (e.g., pUC303,[15] constructed for *Synechococcus* R2) or pACYC177 (e.g., pLF8,[16] constructed for *Synechocystis* 6803) and would be poorly mobilized,[17] and others are based on derivatives of pBR322 that have lost the *bom* site (e.g., pSG111,[18] constructed for *Synechococcus* R2). A derivative of pSG111 called pSG111M was constructed that contains the *bom* site and is mobilizable.[1] There now exist shuttle vectors for several strains of *Anabaena* and *Nostoc* (e.g., PCC 7120, ATCC 29413) and for *Fremyella* that can be transferred by conjugation. A conjugally introduced suicide vector might be the vector of choice for transposon mutagenesis,[19] gene replacement, or integration of the vector into the cyanobacterial chromosome.

Methods

This protocol describes how to do triparental matings. The three parents are (1) *E. coli* bearing the conjugal plasmid, (2) *E. coli* bearing the cargo plasmid plus helper(s), and (3) the target cyanobacterium. The three parents are brought together on a filter resting on solid cyanobacterial medium. After a day on nonselective medium, the filter is transferred to selective medium. It is sometimes necessary to use different dilutions of cyanobacteria in the matings, for at high concentration nonmated cyanobacteria may grow despite selection against them. Two types of matings are described: spot mating, useful in testing a new strain or a new plasmid, and plate mating, useful in finding rare exconjugants. For details and variations, see Thiel and Wolk.[12]

Preparation of E. coli

Growth of Cultures. Grow each *E. coli* strain (e.g., J53[RP4], termed here the conjugal strain, and HB101[cargo;helpers], termed here the cargo strain) in L broth plus the appropriate antibiotics. We keep frozen transformable HB101[pRL528] on hand so that it is easy to construct a

[14] C. Yanisch-Perron, J. Vieira, and J. Messing, *Gene* **33,** 103 (1985).

[15] C. J. Kuhlemeier, A. A. M. Thomas, A. van der Ende, R. W. van Leen, W. E. Borrias, C. A. M. J. J. van den Hondel, and G. A. van Arkel, *Plasmid* **10,** 156 (1983).

[16] F. Chauvat, L. de Vries, A. van der Ende, and G. A. van Arkel, *Mol. Gen. Genet.* **204,** 185 (1986).

[17] A. C. Y. Chang and S. N. Cohen, *J. Bacteriol.* **134,** 1141 (1978).

[18] S. S. Golden and L. A. Sherman, *J. Bacteriol.* **155,** 966 (1983).

[19] G. S. Bullerjahn and L. A. Sherman, *in* "Abstracts of the American Society of Microbiology," p. 157. Am. Soc. Microbiol., Washington, D. C., 1985.

desired cargo strain. Most strains of *E. coli* restrict DNA carrying 5-methylcytosine,[20] so that if a helper plasmid (such as pRL528) that methylates cytosine residues is used, care must be taken in choosing a host strain. We have generally harvested *E. coli* in late exponential phase, but overnight cultures appear to work about as well.[12] In our laboratory we use 50 μg/ml ampicillin (Ap), 50 μg/ml kanamycin (Km), 15 μg/ml tetracycline (Tc), 25 μg/ml chloramphenicol (Cm), and 25 μg/ml streptomycin (Sm) for *E. coli*. For plate matings cargo strains (10 ml per plate of each) and a conjugal strain (volume equal to sum of cargo volumes) are needed. For spot matings cargo strains (anything greater than 0.75 ml each) and a conjugal strain (0.75 ml per cargo strain) are needed. There is no need to worry if the cargo plasmid and the helper share drug resistances. pGJ28 and the helper plasmids based on ColK are extremely stable without selection.

Harvesting. For spot matings, sediment (a microfuge may be used) 0.75 ml of each strain (0.75 ml × number of cargo strains for the conjugal strain) and resuspend in the same volume of L broth without antibiotics. For plate matings, spin down the entire volume and resuspend in 10–25 ml of L broth without antibiotics. Avoid vortexing the conjugal strain, for fear of shearing conjugal pili.

Mixing. Mix equal volumes of each cargo strain with the conjugal strain (0.75 ml + 0.75 ml for spot matings; 10 ml + 10 ml per plate for plate matings) and sediment (in a microfuge or large centrifuge). Resuspend in a convenient minimal volume (e.g., 60 μl for spot matings, 200 μl per plate for plate matings). Allow the mixture stand while preparing the cyanobacteria.

Preparation of Cyanobacteria

Growth of Cultures. The growth phase of the cyanobacteria does not seem to matter much.[12] If it is important that each colony appearing after conjugation represents a distinct conjugal event, then filamentous strains should be fragmented to an average of less than two cells per filament.[21] However, fragmentation is not necessary for efficient conjugation.

Harvesting. Sediment culture (1.5 ml of a mid log culture is enough for spot matings, 40 ml for several plate matings). Resuspend in a minimal volume that is easily pipettable.

Dilution. Make serial dilutions, 1 : 10 (final volume 100 μl or less), up

[20] E. A. Raleigh, N. E. Murray, H. Revel, R. M. Blumenthal, D. Westaway, A. D. Reith, P. W. Rigby, J. Elhai, and D. Hanahan, *Nucleic Acids Res.* **16,** 1563 (1988).

[21] C. P. Wolk and E. Wojciuch, *Planta* **97,** 126 (1971).

to 1 : 100000 in order to obtain single colonies. Measure colony-forming units and/or measure chlorophyll (OD_{665} in methanol × 13.43 = μg chlorophyll/ml[22]; very roughly, $OD_{665} \times 4.5 \times 10^7$ = cells/ml).

Conjugation

Application of Filter. Lay a sterile filter[12] on the mating plate containing cyanobacterial medium without antibiotic. On theoretical grounds we originally added L broth to the medium to 5%, but in fact this supplementation appears to have little or no effect on conjugal efficiency.[12] If, as is often the case, the nitrocellulose filters to be used contain detergent, it is essential that they be washed thoroughly before use. It is more convenient to use filters without detergent. We generally use unwashed Nucleopore REC-85 filters, which come prestcrilized, and Millipore HATF filters.

Mixing of Parents. For plate matings, mix 200 μl of each *E. coli* suspension (conjugal strain plus cargo strain) with 10 μl of undiluted cyanobacteria and, separately, with 10 μl of 1 : 10 diluted cyanobacteria. A surprisingly effective way to spread the mixtures evenly is to slowly pipet the 210 μl at one end of the filter while tipping the Petri plate to let the mixture drain. For spot matings, apply a 5-μl spot of each dilution of cyanobacteria on an empty sterile Petri dish. To each spot apply 5 μl of one *E. coli* mixture, mixing thoroughly. Take 2 μl from each mixture and apply it to the filter. Work out a grid with cyanobacterial dilutions in one dimension and *E. coli* strains in the other.

Incubation of Filter without Selection. Incubate the filters for about 24 hr under conditions of optimal cyanobacterial growth. This period allows time for the expression of drug resistance.

Transfer of Filter to Selective Medium. The proper concentration of antibiotic to kill spots or lawns of the recipient must be determined for each strain. The minimal inhibitory concentration of antibiotic in liquid may differ substantially from that on a plate.

Making Colonies Axenic. Any colonies obtained will most likely be contaminated with *E. coli*. The easiest way to rid an exconjugant of *E. coli* is to grow it in liquid. By the time that the culture is green, the *E. coli* are often totally gone, presumably starved to death. It is also possible to streak out a presumptive exconjugant, picking resulting colonies under a dissecting microscope so as to avoid obvious bacterial microcolonies and checking at each streaking for contamination by subculturing colonies to L plates.

[22] G. Mackinney, *J. Biol. Chem.* **140,** 315 (1941).

Expected Results

Despite selection, you may see growth at the higher concentrations of cyanobacteria in control matings (in which one or more of the three kinds of plasmids is omitted). This will normally yellow and die with time, but in the highest concentration(s), growth in the control matings may never disappear. The presence of a colony (as opposed to green patches of growth) is not definitive evidence of stable conjugative transfer. Nonreplicating, non integrating plasmids can produce colonies at a low frequency. However, these cannot be made axenic: apparently, the presence of *E. coli* is required to maintain these false positives. It is therefore essential to rid a colony of *E. coli*.

Under the best conditions, conjugal efficiencies approaching one drug-resistant colony per cell can be achieved with *Anabaena* PCC 7120. However, shuttle vectors carrying sites for *Anabaena* PCC 7120 restriction enzymes (*Ava*I and *Ava*II, at least) are transferred less efficiently or not detectably. Preliminary results suggest that transfer efficiency is diminished by about one order of magnitude per unmodified *Ava*I site and by a lesser amount per unmodified *Ava*II site (J. Elhai and C. P. Wolk, unpublished). Restriction has long been known to lower the efficiency of conjugation, at least between strains of *E. coli*,[23] so this result is not surprising.

Modification of DNA in *E. coli* by coresidency with pRL528, a plasmid that encodes a methylase that modifies *Ava*I and *Ava*II sites, dramatically increases the efficiency of conjugation into *Anabaena* PCC7120. For example, the transfer efficiency of a shuttle vector carrying two *Ava*I sites and three *Ava*II sites is increased approximately 1000-fold by the presence of pRL528 (J. Elhai and P. C. Wolk, unpublished). Available clones of genes for other methylases sharing site specificity with known cyanobacterial restriction enzymes[24–26] might increase the efficiency of conjugation into other strains of cyanobacteria as well.

Acknowledgments

This work was supported by National Science Foundation Grants PCM-8402500 and DCB-8702368 and by the U.S. Department of Energy under Contract DE-AC02-76ERO-1338.

[23] W. Arber and M. L. Morse, *Genetics* **51,** 137 (1964).

[24] E. Raleigh and G. Wilson, *Proc. Natl. Acad. Sci. U.S.A.* **83,** 9070 (1986).

[25] U. Guenthert, L. Reiners, and R. Lauster, *Gene* **41,** 261 (1986).

[26] K. Mise, K. Nakajima, N. Terakado, and M. Ishidate, Jr., *Gene* **44,** 165 (1986).

[84] Isolation of Genes Encoding Components of Photosynthetic Apparatus

By DONALD A. BRYANT and NICOLE TANDEAU DE MARSAC

Introduction

In recent years cyanobacteria have become an increasingly popular model system for investigations into the structural and functional aspects of oxygenic photosynthesis. The cyanobacterial photosynthetic apparatus consists of five multiprotein complexes: the phycobilisome; the photosystem II reaction center and oxygen evolution complex; plastoquinol–plastocyanin reductase (EC 1.10.99.1); the photosystem I reaction center complex; and ATP synthase (coupling factor).[1] Each of these complexes, with the exception of the phycobilisome, is the structural and functional equivalent of its higher plant analog.[1] The role of the phycobilisome in cyanobacteria is assumed in higher plants by chlorophyll *a*/*b*-binding proteins known as light-harvesting complex II (LHC II) for photosystem II and light-harvesting complex I (LHC I) for photosystem I.[2] In addition to these membrane-associated complexes, many enzymes (e.g., ribulose-1,5-bisphosphate carboxylase, EC 4.1.1.39[3]) and water-soluble electron-transport proteins (e.g., ferredoxin,[4] plastocyanin,[5] and ferredoxin–NADP$^+$ reductase, EC 1.18.1.2[6]) are functionally and structurally similar to their higher plant analogs.

The cyanobacterial photosynthetic apparatus differs from that of eukaryotic algae and higher plants in one significant aspect: amenability to genetic manipulation. A number of cyanobacteria can be transformed by exogenously added DNA[7] or can be the recipients of plasmid DNAs mobilized for transfer from *Escherichia coli*.[8] This ability to manipulate the genes encoding components of the photosynthetic apparatus via recombinant DNA techniques clearly differentiates the cyanobacteria from

[1] D. A. Bryant, *Can. Bull. Fish. Aquat. Sci.* **214,** 423 (1987).
[2] J. P. Thornber, *Encycl. Plant Physiol.* **19,** 98 (1986).
[3] H. M. Miziorko and G. H. Lorimer, *Annu. Rev. Biochem.* **52,** 507 (1983).
[4] H. Matsubara and T. Hase, *in* "Proteins and Nucleic Acids in Plant Systematics" (U. Jensen and D. E. Fairbrothers, eds.), p. 168. Springer-Verlag, Berlin, 1983.
[5] J. A. M. Ramshaw, *Encyl. Plant Physiol.* **14A,** 229 (1982).
[6] Y. Yao, T. Tamura, K. Wada, H. Matsubara, and K. Kodo, *J. Biochem.* (*Tokyo*) **95,** 1513 (1984).
[7] R. D. Porter, this volume [78].
[8] J. Elhai and C. P. Wolk, this volume [83].

higher plants and eukaryotic algae. Although some purple bacteria can be genetically manipulated,[9] their photosynthetic apparatus differs from that found in oxygenic organisms in many respects.[10,11]

The facility for analyzing gene structure and function would be meaningless unless one could isolate the genes of interest. Four methods for the isolation of cyanobacterial genes encoding components of the photosynthetic apparatus are described in this chapter. The first and probably most widely applied method involves heterologous hybridization. In higher plants genes encoding components of the photosynthetic apparatus are frequently encoded by the chloroplast DNA. The complete nucleotide sequences of two chloroplast genomes have been determined, and the genes and potential genes which these genomes encode have been extensively analyzed.[12,13] Molecular biologists studying higher plants have also isolated genes or cDNA clones for many nuclear-encoded proteins which are components of the photosynthetic apparatus.[14–17] The availability of characterized genes or cDNA clones for higher plant proteins which play a role in photosynthesis provides a ready source of materials for the isolation and subsequent manipulation of cyanobacterial genes. The close homology of the cyanobacterial photosynthetic apparatus to that of higher plants makes many of these genes readily isolatable by heterologous hybridization.

Structural homology of the components of the photosynthetic apparatuses of cyanobacteria and higher plants can also be demonstrated by immunological methods. For example, the α, β, and γ subunits of the ATP synthase complex are closely related to their plant homologs as shown by immunodecoration,[18] and this similarity has been confirmed by nucleotide sequence analysis of the genes encoding the ATP synthase

[9] D. C. Youvan, S. Ismail, and E. J. Bylina, *Gene* **38,** 19 (1985).

[10] D. C. Youvan and B. L. Marrs, *Cell* **39,** 1 (1984).

[11] G. Drews, *Microbiol. Rev.* **49,** 59 (1985).

[12] K. Ohyama, H. Fukuzawa, T. Kohchi, H. Shirai, T. Sano, S. Sano, K. Umesono, Y. Shiki, M. Takeuchi, Z. Chang, S. Aota, H. Inokuchi, and H. Ozeki, *Nature (London)* **322,** 572 (1986).

[13] K. Shinozaki, M. Ohme, M. Tanaka, T. Wakasugi, N. Hayashida, T. Matsubayashi, N. Zaita, J. Chunwongse, T. Obokata, K. Yamaguchi-Shinozaki, C. Ohto, K. Torazawa, B. Y. Meng, M. Sugita, H. Deno, T. Kamogashira, K. Yamada, J. Kusuda, F. Takaiwa, A. Kato, N. Tohdoh, H. Shimada, and M. Sugiura, *EMBO J.* **5,** 2043 (1986).

[14] G. Coruzzi, R. Broglie, A. Cashmore, and N.-H. Chua, *J. Biol. Chem.* **258,** 1399 (1983).

[15] S. Smeekens, J. van Binsbergen, and P. Weisbeek, *Nucleic. Acids Res.* **13,** 3179 (1985).

[16] S. Smeekens, M. de Groot, J. van Binsbergen, and P. Weisbeek, *Nature (London)* **317,** 456 (1985).

[17] J. Tittigen, J. Hermans, J. Steppuhn, T. Jansen, C. Jansson, B. Andersson, R. Nechushtai, N. Nelson, and R. G. Herrmann, *Mol. Gen. Genet.* **204,** 258 (1986).

[18] D. B. Hicks, N. Nelson, and C. F. Yocum, *Biochim. Biophys. Acta* **851,** 217 (1986).

complexes of *Synechococcus* sp. PCC 6301[19] and *Anabaena* sp. PCC 7120.[20] Similar results have been obtained for components of photosystem II, photosystem I, the plastoquinol–plastocyanin reductase, and many enzymatic and soluble electron-transport proteins.[21–23] Hence, antibodies directed against either cyanobacterial or higher plant proteins can provide a second reagent for the isolation of cyanobacterial genes.

A third method for the identification and isolation of cyanobacterial genes relies on the use of synthetic oligonucleotide probes whose sequences are derived from the amino acid sequences of proteins. In a few instances comparative analyses of amino acid sequences for plant and cyanobacterial proteins (e.g., plastocyanin,[5] ferredoxin,[4] and ferredoxin–$NADP^+$ reductase[6]) may allow the identification of suitable target sequences for the design of oligonucleotide probes for the cyanobacterial gene. Finally, cyanobacteria allow the possibility for isolation of genes by complementation of isolated and characterized mutations. This is an important capability, since this method is not available for eukaryotic systems and since this method allows the identification of genes, and potentially gene products, which do not produce structural components of complexes and hence would not be detected in typical purified preparations of the complexes.

The basic mechanics of gene isolation from cyanobacteria are similar, if not identical, to the processes applied to other prokaryotes and eukaryotes. A number of excellent how-to books and general reference works describing methods in molecular genetics and recombinant DNA technology, including several volumes in this series,[24–26] have been published.[27–30] The discussion in this chapter is limited to comparison of the

[19] A. L. Cozens and J. E. Walker, *J. Mol. Biol.* **194,** 359 (1987).

[20] S. E. Curtis, *J. Bacteriol.* **169,** 80 (1987).

[21] W. F. J. Vermaas, J. G. K. Williams, A. W. Rutherford, P. Mathis, and C. J. Arntzen, *Proc. Natl. Acad. Sci. U.S.A.* **83,** 9474 (1986).

[22] R. Nechushtai, P. Muster, A. Binder, V. Liveanu, and N. Nelson, *Proc. Natl. Acad. Sci. U.S.A.* **80,** 1179 (1983).

[23] S. M. van der Vies, D. Bradley, and A. A. Gatenby, *EMBO J.* **5,** 2439 (1986).

[24] This series, Vol. 68.

[25] This series, Vols. 100 and 101.

[26] This series, Vols. 152–155.

[27] T. Maniatis, E. F. Fritsch, and J. Sambrook, "Molecular Cloning: A Laboratory Manual." Cold Spring Harbor Lab., Cold Spring Harbor, New York, 1982.

[28] B. Perbal, "A Practical Guide to Molecular Cloning." Wiley (Interscience), New York, 1984.

[29] T. J. Silhavy, M. L. Berman, and L. W. Enquist, "Experiments with Gene Fusions." Cold Spring Harbor Lab., Cold Spring Harbor, New York, 1984.

[30] L. G. Davis, M. D. Dibner, and J. F. Battey, "Basic Methods in Molecular Biology." Elsevier, New York, 1986.

four basic methods described above for specific gene isolation, includes some specific guidelines for the performance of low-stringency heterologous hybridization, and provides examples from the literature of application of the various methods to the isolation of specific cyanobacterial genes.

Isolation of Genes by Hybridization with Cloned, Heterologous DNA Fragments

A large number of genes encoding components of the cyanobacterial photosynthetic apparatus have been cloned through the use of heterologous hybridization probes. In most but not all cases, these probes have been derived from genes originally cloned from the chloroplast genomes of higher plants. Examples of genes which have been cloned in this manner include the following: *psaA* and *psaB*[31,32]; *psbA*[33–36]; *psbB*[36,37]; *psbC*[36,38]; *psbD*[36,38]; *psbE* and *psbF*[36]; *petA, petB,* and *petD*[39]; *rbcL* and *rbcS*[40–43]; *petF*[44,45]; and *atpA, atpB, atpE, atpH,* and *atpI.*[19,20,32] Low-stringency hybridization with heterologous cloned DNA fragments has also been widely employed to isolate members of the phycobiliprotein/linker polypeptide multigene families.[46–53]

[31] A. Cantrell and D. A. Bryant, *Plant Mol. Biol.* **9,** 453 (1987).

[32] D. H. Lambert, D. A. Bryant, V. L. Stirewalt, J. M. Dubbs, S. E. Stevens, Jr., and R. D. Porter, *J. Bacteriol.* **164,** 659 (1985).

[33] S. E. Curtis and R. Haselkorn, *Plant Mol. Biol.* **3,** 249 (1984).

[34] B. Mulligan, N. Schultes, L. Chen, and L. Bogorad, *Proc. Natl. Acad. Sci. U.S.A.* **81,** 2693 (1984).

[35] S. S. Golden, J. Brusslan, and R. Haselkorn, *EMBO J.* **5,** 2789 (1986).

[36] J. C. Gingrich and D. A. Bryant, unpublished observations.

[37] W. F. J. Vermaas, J. G. K. Williams, and C. J. Arntzen, *Plant Mol. Biol.* **8,** 317 (1987).

[38] J. G. K. Williams and D. Chisholm, *Prog. Photosynth. Res.* **4,** 89 (1987).

[39] T. Kallas, S. Spiller, and R. Malkin, *Prog. Photosynth. Res.* **4,** 801 (1987).

[40] S. E. Curtis and R. Haselkorn, *Proc. Natl. Acad. Sci. U.S.A.* **80,** 1835 (1983).

[41] K. Shinozaki, C. Yamada, N. Takakata, and M. Sugiura, *Proc. Natl. Acad. Sci. U.S.A.* **80,** 4050 (1983).

[42] S. A. Nierzwicki-Bauer, S. E. Curtis, and R. Haselkorn, *Proc. Natl. Acad. Sci. U.S.A.* **81,** 5961 (1984).

[43] K. Shinozaki and M. Sugiura, *Nucleic Acids Res.* **11,** 6957 (1983).

[44] J. van der Plas, R. P. de Groot, P. J. Weisbeek, and G. A. van Arkel, *Nucleic Acids Res.* **14,** 7903 (1986).

[45] J. van de Plas, R. P. de Groot, M. R. Woortman, P. J. Weisbeek, and G. A. van Arkel, *Nucleic Acids Res.* **14,** 7804 (1986).

[46] P. B. Conley, P. G. Lemaux, and A. R. Grossman, *Science* **230,** 550 (1985).

[47] P. B. Conley, P. G. Lemaux, T. L. Lomax, and A. R. Grossman, *Proc. Natl. Acad. Sci. U.S.A.* **83,** 3924 (1986).

[48] J. Houmard, D. Mazel, C. Moguet, D. A. Bryant, and N. Tandeau de Marsac, *Mol. Gen. Genet.* **205,** 404 (1986).

As a general rule, low-stringency heterologous hybridization is quite likely to be successful if the protein products of a previously cloned gene and of the gene of interest are closely related antigenically. This is not an absolute rule, however, since hybridization of phycocyanin gene sequences to those encoding both phycoerythrin and allophycocyanin has been reported despite the fact that antisera to these proteins are not cross-reactive.[50,52] Hybridization between the *psbA* and *psbD* genes also occurs at very low stringency even though the gene products are antigenically distinct.[53,54]

The successful application of low-stringency heterologous hybridization is dependent on several factors that are in practice rather simple to control. First, the hybridization probe should be an isolated, purified DNA fragment that is largely, preferably entirely, gene-internal. The best results are obtained when the gene-internal fragment corresponds to a portion of the encoded protein which is conserved in a functional and structural sense, although this information may not often be readily available. Stringent washes of blots should be avoided; low-ionic-strength washes at the hybridization temperature should not be performed. The conditions described below have been developed empirically in the laboratory by one of us (D.A.B.) and have been successfully employed in identifying and isolating sequences which share as little as 45–55% nucleotide sequence similarity.[50,52,55] Since the procedures involved are specific adaptations of widely established methods, only the general procedure and a few specifics are discussed.

Probe Preparation

As described above, the best results are obtained when gene-internal fragments corresponding to highly conserved regions of the protein are employed as probes. Restriction endonuclease digestion and preparative agarose or polyacrylamide gel electrophoretic separation of DNA fragments are carried out according to standard procedures.[27–30] We prefer to recover all DNA fragments by electroelution into hydroxyapatite. The DNA is recovered from the hydroxyapatite by washing with 2.0 *M* potas-

[49] D. Mazel, G. Guglielmi, J. Houmard, W. Sidler, D. A. Bryant, and N. Tandeau de Marsac, *Nucleic Acids Res.* **14,** 8279 (1986).

[50] J. M. Dubbs and D. A. Bryant, *Prog. Photosynth. Res.* **4,** 765 (1987).

[51] W. R. Belknap and R. Haselkorn, *EMBO J.* **6,** 871 (1987).

[52] P. G. Lemaux and A. R. Grossman, *EMBO J.* **4,** 1911 (1985).

[53] D. A. Bryant, unpublished observations.

[54] P. J. Nixon, T. A. Dyer, J. Barber, and C. N. Hunter, *FEBS Lett.* **209,** 83 (1986).

[55] R. C. Murphy, D. A. Bryant, R. D. Porter, and N. Tandeau de Marsac, *J. Bacteriol.* **169,** 2739 (1987).

sium phosphate, 10 mM EDTA, pH 8.0. The DNA is desalted by spin-desalting[27] with 10 mM ammonium acetate buffer, pH 6.8, and concentrated by lyophilization. We routinely achieve very high recoveries of DNA by this method. The DNA may be radiolabeled by the nick-translation[27–30] or random primers method.[56,57] We use 200–400 ng of nick-translated DNA as the probe for a hybridization volume of 20 ml. The DNA is denatured by treatment with 1/5 volume of 1.0 M NaOH for 3–5 min; this is followed by neutralization with 1/5 volume of 1.0 M HCl and rapid addition to the hybridization solution (see below).

Preparation of Southern Blots

DNA fragments are transferred to nitrocellulose or nylon filters by the method of Southern.[58] We have routinely used the BA85 nitrocellulose (0.45 μm pore size) of Schleicher & Schuell (Keene, NH). The blots are prehybridized for 4 to 24 hr at 65° in a sealed plastic bag containing 20 ml of the following solution (for a 15 × 20 cm blot):

6× SET (1X SET = 150 mM NaCl, 1 mM EDTA, 30 mM Tris–HCl, pH 8.0)
5× Denhardt's solution (1X Denhardt's solution = 0.02% Ficoll, 0.02% polyvinylpyrrolidone, 0.02% bovine serum albumin)
0.5% sodium dodecyl sulfate

In general prehybridization should be longer when oligonucleotide probes are being employed (see below).

Hybridization Conditions

The base-denatured and neutralized probe DNA is added directly to the prehybridization solution in the plastic bag which is then resealed. No carrier DNA, such as sonicated calf thymus or salmon sperm DNA, or RNA is added. The hybridization temperature employed depends on the expected degree of similarity of the probe and the target sequence. The useful range extends from about 55 to 85°. Empirically, we have determined that hybridizations at 55° allow hybridization of sequences which share as little as 45% or greater nucleotide sequence similarity.[50] Hybridizations at 60° allow hybridization of sequences with about 55% or greater nucleotide sequence similarity.[55] Hybridizations at 70° allow hybridization of sequences with about 75–80% or greater nucleotide sequence similarity. These empirically determined guidelines have been confirmed by

[56] A. Feinberg and B. Vogelstein, *Anal. Biochem.* **132,** 6 (1983).
[57] A. Feinberg and B. Vogelstein, *Anal. Biochem.* **137,** 266 (1983).
[58] E. Southern, *J. Mol. Biol.* **98,** 503 (1975).

nucleotide sequence analyses of the fragments cloned using the conditions stated.[50,55]

Wash Conditions

For hybridizations between highly similar sequences, washes (4 × 20 min) with 20 ml of 2× SET (see above) at room temperature are employed. For hybridizations between weakly similar sequences, or for all oligonucleotide probes (see below), washes (4 × 20 min) with 20 ml of 6× SET at room temperature are employed.

Results

Interpreting results of low-stringency hybridizations can be tricky and sometimes requires a bit of "faith." A commonly observed phenomenon is nonspecific hybridization to standard marker fragments (e.g., restriction fragments of phage λ). This is not unexpected because these marker fragments are present at more than 1000-fold higher concentration than a specific fragment in a typical genomic digest. We have found that, if specific hybridization to a genomic fragment occurs, this fragment can be cloned from a suitable library in phage λ even if some hybridization to the λ marker fragments was originally detected when checking genomic Southern blots. If no hybridization is observed, one can lower the temperature of hybridization until hybridization occurs. Practically, no meaningful results are obtained below about 55° under the ionic strength conditions described (6× SET). If too many hybridization signals result, the hybridization temperature can be increased. We routinely move upward or downward in 5° increments and generally use 60° as the starting temperature unless we anticipate a very high degree of nucleotide sequence similarity, in which case we use 65–70° to start.

Reuse of Blots

To strip probes from nitrocellulose blots, the filter is placed in 100 ml of distilled water in a plastic dish and treated at full power in a microwave oven (650–700 W) for 3 min. In general, one treatment is sufficient to remove weakly similar sequences or oligonucleotides; three treatments are generally sufficient for highly similar sequences. The number of treatments should be kept to the minimum necessary to remove the probe DNA from the blot, since some bound DNA will also be removed with each successive treatment. The blot should be prehybridized after stripping as described above. We have successfully reused nitrocellulose blots up to 6 times using these methods.

Isolation of Genes by Immunological Screening of Expression Libraries

Immunological screening of expression libraries has been extensively discussed in other volumes of this series[59–62] and will not be reviewed in detail here. Immunological screening, of course, requires an antiserum capable of recognizing the product of the gene of interest. This antiserum can be directed against the actual product of the gene of interest or can be directed against an antigenically related analog. The method is also dependent on adequate expression of the protein product of the gene of interest in *E. coli*. Since many cyanobacterial genes may be only weakly expressed (if at all) from their own promoters in *E. coli*, the screening of expression libraries in plasmids (in the pUC vectors[63]) or in phage λ vectors (λgt10 or λgt11[64]) greatly improves the chances of success when this method is selected. Immunological screening has been successfully employed in the isolation of clones encoding the *cpcB* gene of *Cyanophora paradoxa*,[65] the 94-kDa linker phycobiliprotein of *Nostoc* sp. PCC 8009,[66] and the 35-kDa extrinsic protein of the oxygen evolution complex of PSII from *Synechocystis* sp. PCC 6803.[67] In the latter case the antiserum was a polyclonal antiserum raised against the 33-kDa extrinsic protein of the oxygen evolution complex of spinach. This antiserum is specifically cross-reactive with a cyanobacterial protein of approximately 35 kDa; a partial nucleotide sequence of the cloned gene indicates that the cyanobacterial gene product is approximately 55% identical in sequence to the spinach protein. Hence, it may not be necessary to purify the protein product of the gene of interest if a closely related homolog can be purified and used to produce an antiserum.

Synthetic Oligonucleotide Probes in the Isolation of Genes

Synthetic oligonucleotides have become an increasingly popular and commonplace weapon in the technical arsenal of the molecular biologist. The rapid advances that have allowed the routine synthesis of DNA fragments 10–150 base pairs in length have recently been reviewed.[68,69] Cer-

[59] D. Anderson, L. Shapiro, A. M. Shalka, this series, Vol. 68, p. 428.
[60] L. Clarke, R. Hitzeman, and J. Carbon, this series, Vol. 68, p. 436.
[61] H. A. Erlich, S. N. Cohen, and H. O. McDevitt, this series, Vol. 68, p. 443.
[62] D. A. Kaplan, L. Greenfield, and R. J. Collier, this eries, Vol. 100, p. 342.
[63] J. Vieira and J. Messing, *Gene* **19,** 259 (1982).
[64] R. Young and R. Davis, *Proc. Natl. Acad. Sci. U.S.A.* **80,** 1194 (1983).
[65] P. G. Lemaux and A. Grossman, *Proc. Natl. Acad. Sci. U.S.A.* **81,** 4100 (1984).
[66] B. A. Zilinskas and D. A. Howell, *Prog. Photosynth. Res.* **2,** 161 (1987).
[67] J. B. Philbrick and B. A. Zilinskas, personal communication.
[68] K. Itakura, J. J. Rossi, and R. B. Wallace, *Annu. Rev. Biochem.* **53,** 323 (1984).
[69] M. H. Caruthers, *Science* **230,** 281 (1985).

tainly one of the most important applications of synthetic DNA is the production of DNA probes which are used to isolate specific DNA sequences from libraries of DNA fragments. If a portion of the amino acid sequence of a protein has been determined or can be predicted by homology, then this amino acid sequence can be "reverse-translated" to deduce those DNA sequences which could encode that particular amino acid sequence. From considerations of DNA–DNA hybridization specificity and the genomic complexity of cyanobacteria, oligonucleotides corresponding to the equivalent of 5 to 6 amino acid codons (i.e., greater than 14 base pairs) are required as hybridization probes. Because of the degeneracy of the genetic code, only in extremely rare cases will this process yield a unique DNA sequence.

Three strategies can be employed to deal with this problem. In the first strategy, if only a small number of potential sequences exist, all potential sequences can be synthesized and tested by hybridization individually. This strategy was employed in the cloning of the *cpcA* gene of *Synechococcus* sp. PCC 7002[70] and the *apcB* gene of *Cyanophora paradoxa*.[71] Alternatively, all potential oligonucleotide sequences are synthesized as a mixture, and the mixture of sequences is radiolabeled and used as the hybridization probe. Mixtures of oligonucleotides have been used to clone the *cpcB* gene of *Synechococcus* sp. PCC 7002,[72] the *petF* genes encoding ferredoxins in *Anabaena* sp. PCC 7120[73] and *Synechococcus* sp. PCC 7942,[74] and the *cpeA* gene encoding that α subunit of phycoerythrin in *Calothrix* sp. PCC 7601.[49] The use of oligonucleotide mixtures, and of oligonucleotides in general, is substantially improved if two independent mixtures are employed to minimize the possibility of false-positive hybridization signals. When mixtures are employed, the hybridization conditions are varied (usually the temperature is varied) so that mismatched sequences do not hybridize. Practically speaking, this is monitored by the specificity and strength of the signals obtained from a series of hybridization experiments. Approximate temperatures for hybridizations employing short oligonucleotides can be determined as described by Hanahan and Meselson.[75]

[70] R. De Lorimier, D. A. Bryant, R. D. Porter, W.-Y. Liu, E. Jay, and S. E. Stevens, Jr., *Proc. Natl. Acad. Sci. U.S.A.* **81,** 7946 (1984).

[71] D. A. Bryant, R. de Lorimier, D. H. Lambert, J. M. Dubbs, V. L. Stirewalt, S. E. Stevens, Jr., R. D. Porter, J. Tam, and E. Jay, *Proc. Natl. Acad. Sci. U.S.A.* **81,** 3242 (1985).

[72] T. J. Pilot and J. L. Fox, *Proc. Natl. Acad. Sci. U.S.A.* **81,** 6983 (1984).

[73] J. Alam, R. A. Whitaker, D. W. Krogmann, and S. E. Curtis, *J. Bacteriol.* **168,** 1265 (1986).

[74] M. E. Reith, D. E. Laudenbach, and N. A. Straus, *J. Bacteriol.* **168,** 1319 (1986).

[75] D. Hanahan, and M. Meselson, this series, Vol. 100, p. 333.

A major disadvantage of the mixtures method is that, for highly degenerate target sequences, the correct oligonucleotide sequence can represent a very small percentage of the total oligonucleotide mixture. This situation arises because the nucleotide derivatives employed for DNA synthesis do not have identical reaction rates in the condensation step that adds the nucleotide to the growing chain. Hence, any mixture will be biased in composition toward those bases which react fastest in this reaction. A third strategy which avoids this problem has been used to clone the *gvpA* gene of *Calothrix* PCC 7601[76] and the *psbE* and *psbF* genes of *Synechocystis* sp. PCC 6803[77] and *Cyanophora paradoxa*.[78] In this method unique-sequence oligonucleotide probes 30–50 bp in length are synthesized. Codon usage data, if available, is used to predict the most likely base at degenerate positions. Potential purine–purine mismatches are generally avoided, unless severe codon usage bias is known to exist, and generally G–T mismatches are selected rather than A–C mismatches since mismatches containing guanine are more stable than most other combinations.[79] A recently introduced minor variation is the use of long synthetic probes containing deoxyinosine at ambiguous codon positions.[80] This base analog is relatively inert and neither destabilizes nor stabilizes the formation of a DNA duplex at a site of mismatching. Careful selection of target sequences should allow probe sequences to be predicted which will be approximately 80% correct. Hybridization experiments are then conducted under various conditions in order to determine those conditions which allow highly specific hybridization to occur. These experiments are similar to hybridization experiments at low stringency carried out with nick-translated, heterologous probes (see above).

In Vivo Complementation of Mutations Affecting Photosynthetic Capabilities

A highly promising approach that has been little examined at present is the *in vivo* complementation of photosynthetic mutants by the introduction of cloned DNA fragments into the mutant on a suitable shuttle vector plasmid. This procedure involves the selection of mutations of the photosynthetic apparatus for which the mutants could be maintained by virtue

[76] N. Tandeau de Marsac, D. Mazel, D. A. Bryant, and J. Houmard, *Nucleic Acids Res.* **13,** 7223 (1985).

[77] H. Pakrasi, personal communication.

[78] A. Cantrell and D. A. Bryant, *Prog. Photosynth. Res.* **4,** 659 (1987).

[79] F. Adoul-ela, D. Koh, and I. Tinoco, Jr. *Nucleic Acids Res.* **13,** 4811 (1985).

[80] Y. Takahashi, K. Kato, Y. Hayashizaki, T. Wakabayashi, E. Ohtsuka, S. Matsuki, M. Ikehara, and K. Matsubara, *Proc. Natl. Acad. Sci. U.S.A.* **82,** 1931 (1985).

of growth under permissive conditions, e.g., either a permissive temperature for temperature-sensitive mutations or under photoheterotrophic or heterotrophic conditions for strains capable of these metabolic modes. Shotgun cloned fragments, isolated from genomic DNA from the wild-type strain and cloned in a suitable shuttle cloning vector, would then be transformed[7] or conjugatively introduced[8] into the mutant. After some suitable time period under nonselective conditions to allow expression of the markers on the introduced DNA, the growth conditions would be shifted to the nonpermissive, selective conditions. In general coselection for a vector-borne drug-resistance marker would be employed with a shift to a nonpermissive growth condition. Complementing DNA fragments would then be identified by their ability to confer growth under conditions otherwise incompatible with growth of the mutant. Complementation would be confirmed by isolation of the recombinant plasmid and demonstration that the introduction of this plasmid into the mutant was responsible for the phenotypic difference observed, as evidenced by a greatly increased frequency of phenotypic conversion in the transformants. This general strategy has been employed in cloning DNA fragments conferring resistance to a variety of herbicides in *Synechococcus* sp. PCC 7002[81] and has also been employed in cloning DNA fragments complementing photosynthetically impaired mutants of *Synechocystis* sp. PCC 6803.[82] A distinct advantage of this method is that it can be used to isolate DNA fragments complementing mutations in components which have not been previously isolated, characterized, nor perhaps even suspected to exist.

In rare instances genes encoding polypeptides that are peripheral to the photosynthetic process may be cloned by complementation of well-characterized mutations in *E. coli*. This strategy has been employed in cloning the phosphoenolpyruvate carboxylase (*ppc*) gene of *Synechococcus* sp. PCC 6301,[83] the thioredoxin gene of *Anabaena* sp. PCC 7119,[84] and the *glnA* gene of *Anabaena* sp. PCC 7120.[85]

[81] J. S. Buzby, R. O. Mumma, D. A. Bryant, J. Gingrich, R. H. Hamilton, R. D. Porter, C. A. Mullin, and S. E. Stevens, Jr., *Prog. Photosynth. Res.* **4,** 757 (1987).

[82] V. A. Dzelzkalns and L. Bogorad, *Prog. Photosynth. Res.* **4,** 841 (1987).

[83] T. Kodaki, F. Katagiri, M. Asano, K. Izui, and H. Katsuki, *J. Biochem.* (*Tokyo*) **97,** 533 (1985).

[84] C.-J. Lim, F. K. Gleason, and J. A. Fuchs, *J. Bacteriol.* **168,** 1258 (1986).

[85] R. Fisher, R. Tuli, and R. Haselkorn, *Proc. Natl. Acad. Sci. U.S.A.* **78,** 3393 (1981).

[85] Construction of Specific Mutations in Photosystem II Photosynthetic Reaction Center by Genetic Engineering Methods in *Synechocystis* 6803

By JOHN G. K. WILLIAMS

Introduction

A new genetic technique has been developed in *Synechocystis* 6803 for the molecular analysis of electron transport in the photosystem II reaction center (PSII). This methodology permits specific PSII genes to be deleted from the cyanobacterial genome. The deleted genes can then be replaced with copies modified by site-directed mutagenesis. In this way, specific amino acid changes can be engineered in the polypeptides that bind the pigments and electron carriers in the PSII protein complex. No comparable methodology for the molecular analysis of PSII is available in higher plants. This experimental system depends on two important characteristics of *Synechocystis* 6803 found in few other cyanobacteria: a naturally occurring genetic transformation system[1] and the ability to grow photoheterotrophically on glucose, which is necessary for the propagation of PSII mutants that are incapable of photosynthesis.[2]

PSII is functionally and structurally similar in the chloroplasts of higher plants and in cyanobacteria.[3] This has allowed chloroplast genes to be used as hybridization probes to clone the corresponding genes from *Synechocystis* 6803 that encode polypeptides of the PSII "core complex" (Table I). The core is isolated as a subcomplex of PSII comprising five polypeptides that bind all of the electron carriers except for the water-oxidizing catalyst.[4] The PSII core is intrinsic to the thylakoid membrane,

[1] G. Grigorieva and S. Shestakov, *FEMS Microbiol. Lett.* **13,** 367 (1982).

[2] R. Rippka, *Arch Mikrobiol.* **87,** 93 (1972). In photoheterotrophic metabolism, both glucose and light are needed for growth when PSII is inactivated. The action spectrum for photoheterotrophic growth of *Synechocystis* 6803 is the absorption spectrum for chlorophyll plus a minor contribution from phycobilin [W. F. J. Vermaas and T. Ogawa, unpublished observations (1986)]. This indicates that photosystem I (PSI) is needed for photoheterotrophic growth, probably to generate ATP through cyclic electron transport. In this regard, we have found that *Synechocystis* 6803 grows very slowly on glucose in the dark, with a doubling time of 84 hr. Therefore, PSI is not absolutely required for growth, but it is important for achieving practical growth rates. We believe that PSI is important, even in the presence of glucose, and that it is therefore unavailable to modification by the methods described here for PSII.

[3] H. B. Pakrasi, H. C. Reithman, and L. A. Sherman, *Proc. Natl. Acad. Sci. U.S.A.* **82,** 6903 (1985).

[4] C. J. Arntzen and H. B. Pakrasi, *Encycl. Plant Physiol., New Ser.* **19,** 457 (1986).

TABLE I
NINE GENES ENCODE THE FIVE PROTEINS OF THE PSII CORE COMPLEX IN *Synechocystis* 6803[a]

Protein	Gene	Length of gene (bp)	Sequence similarity (%) with spinach chloroplast gene	Ref.
D1	*psbAI*	1079	72	*b*
D1	*psbAII*	NA[i]	NA	*c*
D1	*psbAIII*	NA	NA	*d*
D2	*psbDI*	1056	74	*e*
D2	*psbDII*	1056	73	*e*
CP-43	*psbC*	NA	NA	*f*
CP-47	*psbB*	1521	68	*g*
Cyt *b*-559	*psbE*	243	78	*h*
Cyt *b*-559	*psbF*	132	72	*h*

[a] All of the genes listed have been cloned. Sequence similarities with the corresponding spinach genes are calculated from nucleotide sequence data. Cytochrome *b*-559 consists of two polypeptides which are encoded by the *psbE* and *psbF* genes. All of the other proteins comprise a single polypeptide. There are three copies of the *psbA* gene and two copies of the *psbD* gene.
[b] H. Oziewaks, J. Williams, and L. McIntosh, unpublished observations (1986).
[c] R. Debus and L. McIntosh, unpublished observations (1986).
[d] J. Metz, P. Nixon, H. B. Pakrasi, and B. A. Diner, manuscript in preparation (1988).
[e] J. Williams and D. A. Chisholm, *Prog. Photosynth. Res.*, in press (1987).
[f] D. Chisholm and J. Williams, *Plant Mol. Biol.* **10,** 293 (1988).
[g] W. F. J. Vermaas, J. G. K. Williams, and C. J. Arntzen, *Plant Mol. Biol.* **8,** 317 (1987).
[h] H. B. Pakrasi, J. G. K. Williams, and C. J. Arntzen, *EMBO J.* **7,** 325 (1988).
[i] Data not available (NA).

while the water-oxidizing catalyst is bound near the membrane surface where it is protected by three "extrinsic" polypeptides.[5] Deletion from *Synechocystis* 6803 of the genes for the PSII core polypeptides D2, CP-43, CP-47, or cytochrome *b*-559 in all cases produced mutants that specifically lack PSII function and depend on glucose for growth.[6,7] This chapter describes a procedure for deleting PSII genes from *Synechocystis* 6803 to create a PSII mutant; replacement of the deleted genes to restore photosynthetic function; and some of the properties of the genetic transformation system in this cyanobacterium.

[5] X.-S. Tang and K. Satoh, *FEBS Lett.* **201,** 221 (1986).
[6] W. F. J. Vermaas, J. G. K. Williams, A. W. Rutherford, P. Mathis, and C. J. Arntzen, *Proc. Natl. Acad. Sci. U.S.A.* **83,** 9474 (1986).
[7] This chapter, and also footnote *h* in Table I.

Deletion and Replacement of the PSII Genes *psbDI* and *psbC*

To demonstrate that PSII mutations can be constructed in *Synechocystis* 6803, the overlapping genes *psbDI* and *psbC* were deleted from the organism using the genetic transformation system as outlined in Fig. 1. This method, known as gene replacement, depends on homologous recombination to replace a section of the recipient chromosome with donor DNA that contains a selectable marker.[8] Gene replacement has been described in two other cyanobacteria with naturally occurring transformation systems.[9-11]

Construction of Donor DNA for Deletion of PSII Genes

The first step in the procedure was to delete the PSII genes from a recombinant plasmid using standard construction methods with restriction enzymes. A 1536-bp segment of DNA was deleted from the plasmid pKW1218 and replaced with a 1.2-kb kanamycin-resistance marker. The deleted DNA encodes 856 bp of *psbDI* plus 680 bp of *psbC* (Fig. 1A). The new plasmid, pKW1224, was used to delete the specified segments of *psbDI* and *psbC* from the cyanobacterial genome by genetic transformation.

Transformation of Synechocystis 6803

To prepare a culture of *Synechocystis* 6803 for transformation, cells were scooped up from a fresh stock plate (see Materials and Methods) and suspended in BG-11 medium at a density of 0.5×10^8 cells ml^{-1} (A_{730} = 0.25 corresponds to 1×10^8 cells ml^{-1}). The culture was grown to a density of 2 to 5×10^8 cells ml^{-1}, and the cells were harvested by centrifugation at 4,500 *g* for 6 min at room temperature. The cell pellet was suspended in fresh BG-11 medium at a density of 1×10^9 cells ml^{-1} and used immediately for transformation. One hundred microliters of the cell suspension was mixed with 2 μl of pKW1224 DNA (the DNA was in 10 m*M* Tris–HCl, pH 7.5, 0.1 m*M* EDTA). The final concentration of DNA was 10 μg ml^{-1}. The mixture of cells and DNA was incubated for 4 hr in a sterile test tube under standard temperature and light conditions. The mixture was spread onto membrane filters[12] resting on BG-11–agar plates.

[8] R. J. Rothstein, this series, Vol. 101, p. 202.

[9] J. G. K. Williams and A. A. Szalay, *Gene* **24,** 37 (1983).

[10] K. S. Kolowsky, J. G. K. Williams, and A. A. Szalay, *Gene* **27,** 289 (1984).

[11] D. A. Bryant, personal communication (1986).

[12] Filters were Rec-85 membranes, 85 mm diameter, presterilized, Stock Number 145318, from Nuclepore Corp., 7035 Commerce Circle, Pleasanton, CA 94566-3294. Nitrocellulose membranes from Millipore or Schleicher & Schuell have also been used successfully.

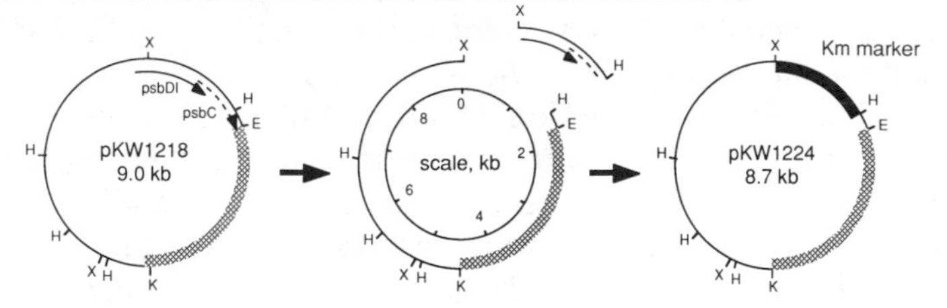

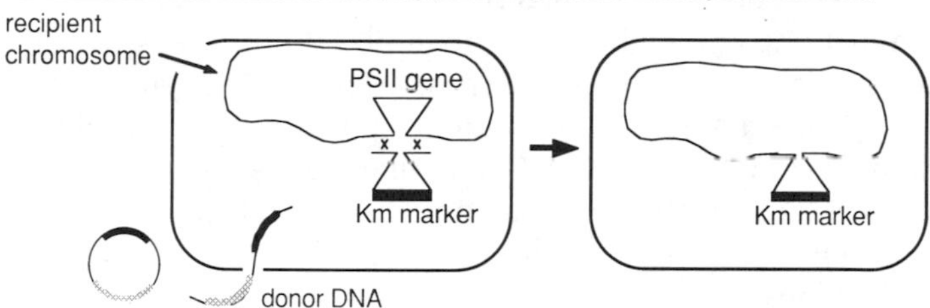

FIG. 1. Scheme for deletion of the *psbDI* gene and part of the *psbC* gene. (A) The plasmid pKW1218 contains the entire *psbDI* gene (1.05 kb total length) plus 899 bp of the *psbC* gene (1.4 kb total length). The two genes overlap for 50 bp and are translated in different reading frames. pKW1224 was constructed by deleting the indicated restriction fragment and replacing it with the 1.2-kb kanamycin-resistance gene (Km) from Tn*903* [J. Vieira and J. Messing, *Gene* **19,** 259 (1982)]. Restriction enzyme recognition sites are as follows: E, *Eco*RI; H, *Hinc*II; K, *Kpn*I; X, *Xba*I. The thin line is a continguous fragment of *Synechocystis* 6803 DNA; thick cross-hatched line, the plasmid vector pUC18 which encodes ampicillin resistance; thick solid line, the kanamycin-resistance marker. (B) The deletion constructed in pKW1224 is crossed into the *Synechocystis* 6803 genome by homologous recombination, through a process called gene replacement. Although there is no evidence either way, the donor DNA is shown in the diagram to be cleaved during DNA uptake, as occurs in other naturally transformable prokaryotes [R. D. Porter, *CRC Crit. Rev. Microbiol.* **13,** 111 (1986)].

After incubating the plates under standard conditions for 18 hr to allow for expression of kanamycin resistance in transformed cells, the filters were transferred to plates containing BG-11 medium supplemented with 5 m*M* glucose and kanamycin at 5 μg ml^{-1}. After 4 days of incubation, colonies of transformed cells could be seen. Ten thousand colonies were obtained per 10^8 cells (0.1 ml) plated. No colonies appeared in a control experiment from which pKW1224 DNA was omitted.

Importance of Segregation in Mutant Isolation

In these transformants, the *psbDI–psbC* genes should have been replaced with the kanamycin-resistance marker (see Fig. 1). Loss of the *psbDI* gene might have no effect on photosynthesis, since the second copy, *psbDII,* should still be present in the organism (see Table I); however, loss of the single-copy *psbC* gene was expected to destroy PSII function. Contrary to expectation, all of the kanamycin-resistant colonies tested were still capable of photosynthetic growth in the absence of glucose. Analysis of DNA from these transformants by Southern hybridization suggested that they contained intact copies of the *psbDI–psbC* genes, as well as copies that had been replaced with the kanamycin-resistance marker (date not shown). *Synechocystis* 6803 contains six to eight chromosomal DNA copies.[13] If integration of the kanamycin-resistance gene occurs in just one of the chromosome copies during transformation, then it would be necessary to grow the transformants through several generations to segregate homozygous mutants. In the presence of glucose, cells that retain PSII activity grow 1.5–2.0 times faster than PSII mutants; to facilitate segregation, the medium should be supplemented with atrazine or diuron to inhibit PSII and eliminate the advantage to cells that retain intact PSII genes.

Indeed, a failure to segregate was the reason that no mutants were found initially. A mutant defective in photosynthesis was obtained only after three serial streak-purifications of a single colony on plates that contained kanamycin (10 μg ml^{-1}), glucose (5 m*M*), and atrazine (10 μM). A period of 1 month was required to segregate a homozygous mutant. Southern hybridization analysis of this mutant showed complete substitution of the *psbDI–psbC* genes by the kanamycin-resistance marker (Fig. 2).

Restoration of Photosynthesis by Transformation

To prove that absence of photosynthetic function in the deletion mutant was due only to loss of the *psbDI–psbC* genes, and not to other genetic lesions, the deletion mutant was transformed with pKW1218 DNA which contains the segments of *psbDI* and *psbC* missing from the mutant (Fig. 1A). After incubation with DNA, the cells were plated on BG-11 medium supplemented with glucose and were incubated for 18 hr to allow for expression of PSII function in transformants. The filters were transferred to unsupplemented BG-11 medium to select for transformants capable of photosynthetic growth. Colonies were obtained at a frequency of 10^{-4} per cell. Restoration of photosynthesis by addition of the *psbDI* and *psbC* genes indicates that the only lesion in the mutant is an absence of these PSII genes. No photosynthetic colonies were obtained from

[13] B. Haskell and L. A. Sherman, personal communication (1986).

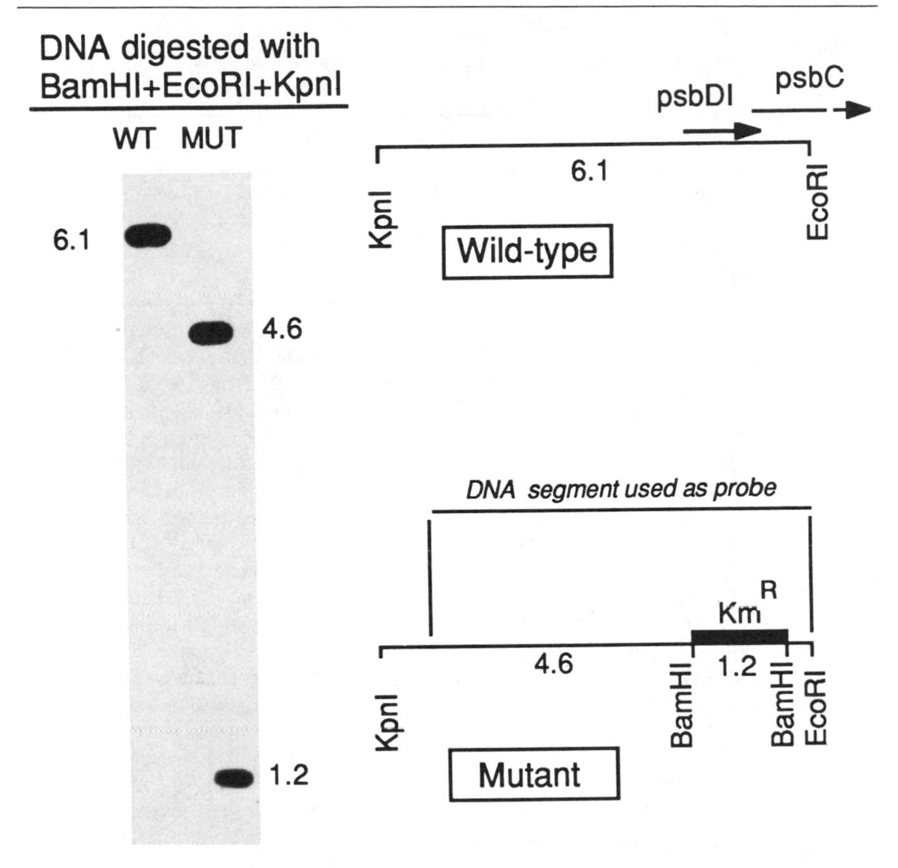

FIG. 2. Southern hybridization analysis of the *psbDI–psbC* deletion mutant. One microgram of chromosomal DNA (WT, from wild-type cells; MUT, from the deletion mutant) was digested with *Bam*HI plus *Eco*RI plus *Kpn*I. The digested DNA was fractionated on an agarose gel, transferred to a Gene Screen Plus membrane (New England Nuclear), and probed with radioactivity labeled pKW1224 DNA, as indicated. The hybridization pattern shows that the 6.1-kb *Kpn*I–EcoRI fragment in wild-type cells (top) has been completely replaced with the deletion structure of pKW1224 (bottom).

among 10^8 cells plated when DNA was omitted, which shows that the deletion mutation is stable.

As measured by oxygen uptake in the presence of methyl viologen,[14] PSII is inactive in the *psbDI–psbC* deletion mutant, while PSI activity is 60% of the wild-type level on a chlorophyll basis. Electron transport was fully restored to the revertant by replacement of the *psbDI–psbC* genes (Table II). This experiment demonstrates that it is possible to delete PSII

[14] G. Hauska, *Encycl. Plant Physiol., New Ser.* **5,** 255 (1977).

TABLE II
PHOTOSYNTHETIC ELECTRON TRANSPORT IN THE DELETION MUTANT AND THE REVERTANT[a]

Conditions	Light-dependent uptake, μmol O_2 (mg Chl)$^{-1}$ hr^{-1}		
	Wild type	Deletion mutant	Revertant
PSII plus PSI	54	0	55
PSI only	83	52	91

[a] Light-induced oxygen uptake (the Mehler reaction) was measured in whole cells with a Clark-type oxygen electrode. Cells were suspended in BG-11 medium at a chlorophyll concentration of 0.5 mg ml^{-1}. A 2-ml sample of the cell suspension was illuminated with a saturating intensity of blue light (produced by filtering through a solution of $CuSO_4$). To measure electron transport through both PSII and PSI, cells (2 ml) were mixed with 67 μl of methyl viologen (100 m*M* in water). To measure PSI activity alone, cells (2 ml) were mixed with 1 μl of diuron (20 m*M* in methanol; to block PSII), 100 μl of sodium ascorbate (100 m*M* in water; to donate electrons to PSI), 2 μl of durohydroquinone (50 m*M* in ethanol; to mediate electron transfer from ascorbate outside of the cells to PSI inside the cells), and 67 μl of methyl viologen (100 m*M* in water; to catalyze electron transfer from PSI to oxygen). To calculate chlorophyll concentration, the optical density of a cell suspension was measured, at 678, 720, and 750 nm: μg chlorophyll ml^{-1} = 14.96-$(A_{678} - A_{750}) - 0.616(A_{720} - A_{750})$. The standard method of measuring chlorophyll, involving extraction of cells or chloroplasts with 80% acetone, is not reliable with *Synechocystis* 6803, since extraction of chlorophyll is often incomplete.

genes from *Synechocystis* 6803 and to grow the resulting PSII mutants on glucose. The deleted genes can be replaced to restore photosynthetic function. This methodology provides a way by which PSII genes may be deleted and replaced later with copies modified by site-directed mutagenesis.[15,16]

Characteristics of Genetic Transformations in *Synechocystis* 6803

Three different donor DNA molecules were prepared to characterize transformation in *Synechocystis* 6803. Each donor contains a different fragment of *Synechocystis* 6803 DNA that is cloned in the *E. coli* vector pUC9 and that is interrupted by a kanamycin-resistance marker (Fig. 3).

[15] W. F. J. Vermaas, J. G. K. Williams, and C. J. Arntzen, *Z. Naturforsch. C* **42,** 128 (1987).

[16] R. J. Debus, B. A. Barry, G. T. Babcock, and L. McIntosh, *Proc. Natl. Acad. Sci. U.S.A.* **85,** 427 (1988).

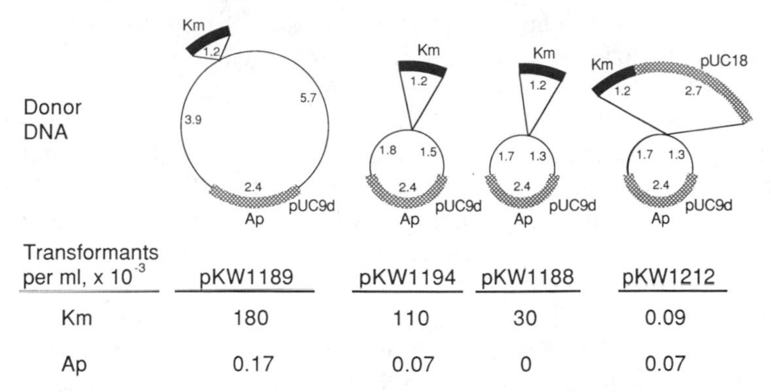

Transformants per ml, x 10^{-3}	pKW1189	pKW1194	pKW1188	pKW1212
Km	180	110	30	0.09
Ap	0.17	0.07	0	0.07

FIG. 3. Transformation efficiency depends on the nature of the donor DNA. Four different donor DNA constructs were used to transform *Synechocystis* 6803 as described in the section "Deletion and Replacement of the PSII Genes *psbDI* and *psbC*." Each donor contains a different fragment of *Synechocystis* 6803 DNA, except for pKW1212 which has the same fragment as pKW1188. The thin lines represent *Synechocystis* 6803 DNA which is interrupted by a kanamycin-resistance marker where indicated. Thick cross-hatched lines represent the 2.7-kb *E. coli* plasmid pUC9 [J. Vieira and J. Messing, *Gene* **19,** 259 (1982)] or a derivative, "pUC9d," identical to pUC9 except for a deletion of 0.3 kb [J. G. K. Williams, unpublished observations (1984)]. Lengths of selected DNA segments are indicated in kilobases. The transformation mixtures (0.5 ml) contained 0.5 μg of donor DNA and were incubated for 2.5 hr before plating on filters. Ampicillin (Ap) was included in the selection medium at 0.3 μg/ml, kanamycin at 5 μg/ml.

The three fragments were selected from a gene library for their ability to mediate insertion of the kanamycin-resistance marker. Kanamycin-resistant transformants are generated at different frequencies by the different donors (Fig. 3). The most efficient is pKW1189, distinguished by long stretches of cyanobacterial DNA on both sides of the kanamycin-resistance marker. This suggests that homologous pairing between donor and recipient DNA is facilitated by long pieces of cyanobacterial DNA in the donor. However, other factors also influence efficiency, since two smaller donors of the same size, pKW1194 and pKW1188, differ in efficiency by a factor of 4. Other important factors could involve specificity of DNA uptake[17] or differences in recombination frequency at different sites in the recipient chromosome. Integration of the kanamycin-resistance marker by gene replacement was confirmed by Southern hybridization analysis (data not shown).

The low yield of ampicillin-resistant transformants which would be generated by integration of the pUC9 portion of donor DNA is puzzling, since ampicillin-resistant transformants were readily obtained in analo-

[17] D. B. Danner, H. O. Smith, and S. A. Narang, *Proc. Natl. Acad. Sci. U.S.A.* **79,** 2393 (1982).

gous experiments with another cyanobacterium, *Anacystis nidulans* R2.[10] Southern hybridization analysis of DNA from an ampicillin-resistant *Synechocystis* 6803 transformant confirmed that pUC9 DNA, which encodes ampicillin resistance, had integrated into the cyanobacterial chromosome. The donor DNA in this case was pKW1188 from which the kanamycin-resistance marker had been deleted *in vitro*. The integrated pUC9 DNA plus flanking cyanobacterial sequences (the 1.7- and 1.3-kb fragments in pKW1188) were excised by cleavage with a restriction enzyme and subsequently ligated into circular form. This DNA was successfully returned to *E. coli* to give the original donor DNA construct. These data show that the plasmid's *E. coli* replication function and ampicillin-resistance gene were unaffected by passage through *Synechocystis* 6803. The plasmid recovered from *Synechocystis* 6803 was amplified in *E. coli* and was used to transform the cyanobacterium again; however, it was still unable to transform the cyanobacterium efficiently (data not shown). Therefore, once integrated into the *Synechocystis* 6803 genome, pUC9 DNA is stable. The block to integration must occur during DNA uptake or assimilation into the recipient chromosome. Inhibition of transformation could be due to a nucleotide sequence encoded in the pUC9 DNA. In support of this notion, it was found that linkage of pUC9 DNA directly to the kanamycin-resistance marker blocked the generation of kanamycin-resistant transformants (see pKW1212, Fig. 3).

The effects of DNA concentration and incubation time on transformation are shown in Fig. 4. The yield of kanamycin-resistant transformants changed only 2-fold when the time of incubation of cells with DNA was varied from 1 to 24 hr. In contrast, a 3-log increase in DNA concentration produced a 4-log increase in transformant yield. This suggests that there may be cooperation between donor molecules to enhance transformation.[18] In practice, a convenient incubation time is between 2 and 6 hr, and the donor DNA should be used at the highest practical concentration if it is important to maximize the yield of transformants.

Materials and Methods

Special Strain of Synechocystis 6803 for Making PSII Mutants

Our original isolate of *Synechocystis* 6803 was from Dr. C. P. Wolk, who had obtained it from the American Type Culture Collection. As

[18] If donor DNA were clipped into smaller segments during transport as occurs in naturally transformable bacteria [M. L. Pifer and H. O. Smith, *Proc. Natl. Acad. Sci. U.S.A.* **82,** 3731 (1985)], then overlapping fragments could be pieced together to regenerate an intact donor molecule capable of integration; overlapping fragments could only be produced if two or more donor molecules were transported into a single cell.

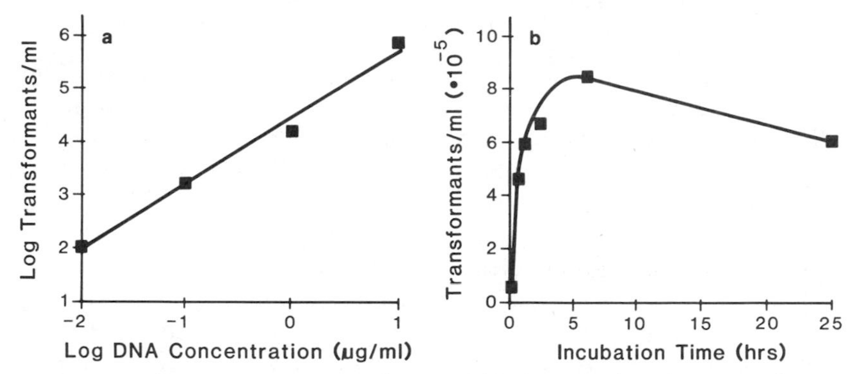

FIG. 4. (a) Effect of DNA concentration on yield of kanamycin-resistant transformants. Cells (0.2 ml) were mixed with 3 μl of pKW1188 DNA (Fig. 3) at the final concentrations indicated. After incubation for 6 hr, the samples were diluted in BG-11 medium, plated on filters, and transferred 20 hr later to the selection plates. The yield increases exponentially with the 1.8th power of the DNA concentration. (b) Effect of incubation time on yield of kanamycin-resistant transformants. Cells (2 ml) were mixed with 30 μl of pKW1188 DNA at a final concentration of 10 μg ml^{-1}. At the indicated times, samples were withdrawn, plated on filters, and transferred 20 hr later to medium containing kanamycin.

reported,[2] this strain grew photoheterotrophically on 5 m*M* glucose when PSII was inhibited by the herbicides atrazine or diuron. However, colonies would not grow on agar medium containing more than 0.2 m*M* glucose when herbicide was omitted. In this strain of *Synechocystis* 6803, glucose is toxic when PSII is functional. Moreover, kanamycin-resistant transformants could not be recovered under photoheterotrophic growth conditions (PSII blocked by herbicides), which indicated that this strain would be useless for making PSII mutants. A spontaneous mutant of *Synechocystis* 6803 was isolated under photoheterotrophic conditions which were nonpermissive for the original isolate; 20 m*M* glucose plus 5 μM atrazine. The new strain is much more tolerant toward glucose (it can grow on 10 m*M* glucose in the absence of herbicides), and kanamycin-resistant transformants can be recovered under photoheterotrophic conditions.[19] We now work exclusively with the glucose-tolerant strain of *Synechocystis* 6803.

[19] Fructose is toxic to the original strain of *Synechococcus* 6803 obtained from Dr. C. P. Wolk [E. Flores and G. Schmetterer, *J. Bacteriol.* **166,** 693 (1986)]. The new strain, which was selected for glucose tolerance, was found also to be fructose resistant. Fructose is not utilized for growth by either the original strain or the new strain. An impairment in sugar transport has been ruled out as an explanation for sugar tolerance in the new strain, since it transports glucose normally [G. Schmetterer and J. G. K. Williams, unpublished observations (1986)].

Medium and Growth Conditions

All microbiological manipulations are done in a laminar flow hood (Baker Company, Sanford, ME). *Synechocystis* 6803 is grown on BG-11 medium.[20] To prepare agar medium, a solution of 2× concentrated BG-11 salts plus 0.6% sodium thiosulfate and 20 m*M* TES–KOH, pH 8.2, is autoclaved, cooled to 55°, and supplemented as required (Table III), and then this solution is mixed with an equal volume of separately autoclaved 3% Difco Bacto agar, also cooled to 55°. Inclusion of 10 m*M* TES buffer and 0.3% thiosulfate[21] in agar medium greatly improves the growth of colonies; these supplements are unnecessary in liquid medium. Growth conditions are 30°, 28 μE m^{-2} sec^{-1} of an equal mix of warm-white and Gro-light fluorescent light, and the relative humidity is maintained at 70% to prevent drying of the agar medium in petri plates.

Liquid cultures are perfused with air at 50–150 ml min^{-1} per 100 ml of culture medium. The air is prehumidified by bubbling through a solution of 1% $CuSO_4$ and then sterilized by passage through a filter (Gelman #12123). The $CuSO_4$ prevents growth of contaminating organisms in the flask of water. A dry flask trap is placed downstream of the $CuSO_4$ solution to prevent contamination of the system with the toxic salt. Fish aquarium gang valves are used to distribute the humidified air to culture flasks, each individually equipped with an in-line bacterial filter (Gelman #4210). Standard Erlenmeyer flasks were modified in a glass shop by permanently sealing the top of the flask with a melted-in glass plug. A bubbling tube (1.0 mm i.d., × 7.5 mm o.d.) is sealed into the top plug; the tube extends from 30 mm above the sealed top to 5 mm above the flask bottom. A side-arm is fused to the flask at a 45° angle and is capped with a metal test tube closure (Bellco Glass). This arrangement effectively prevents contamination of the cyanobacterial cultures. Under these conditions, the doubling time of a liquid culture is 8–12 hr. Glucose (5 m*M*) in liquid cultures extends growth to higher culture densities where light is limiting, and it is added routinely to medium when growing cultures for the isolation of DNA.

Culture Maintenance

Cultures are stored in two ways. First, the cells are maintained by streaking on plates every 2 weeks. For streaking to a fresh plate, 20–40 single colonies are picked up together on a microbiological loop from a

[20] R. Rippka, J. Deruelles, J. B. Waterbury, M. Herdman, and R. Y. Stanier, *J. Gen. Microbiol.* **111,** 1 (1979).

[21] H. C. Utkilen, *J. Gen. Microbiol* **95,** 177 (1976).

TABLE III
ANTIBIOTIC-RESISTANCE GENES USEFUL IN *Synechocystis* 6803 AND SUPPLEMENTS FOR GROWTH MEDIUM

Supplement[a]	Concentration in medium	Source of antibiotic-resistance gene[b]
Kanamycin sulfate	5 μg/ml	(1)
Chloramphenicol	7 μg/ml	(2)
Spectinomycin dihydrochloride	5 μg/ml	(3)
Atrazine	10 μM	
Diuron	10 μM	
Glucose	5 mM	

[a] The antibiotics are from the Sigma Chemical Co., atrazine from Serva Fine Biochemicals, diuron from the Du Pont Co.

[b] References: (1) J. Vieira and J. Messing, *Gene* **19,** 259 (1982); (2) A. C. Y. Chang and S. N. Cohen, *J. Bacteriol.* **134,** 1141 (1978); (3) P. Prentki and H. M. Krisch, *Gene* **29,** 303 (1984).

region of closely spaced colonies on the old plate. By picking more than one colony, we avoid the possibility of isolating a variant that has lost the ability to be transformed with DNA[22]; by picking from a region of separated colonies, we minimize the chances of picking up contaminating microorganisms. Second, cultures are stored in 15% glycerol at −80°. Cells from 100 ml of liquid culture (A_{730} = 3 to 6) are harvested by centrifugation at 4,500 *g* for 6 min at room temperature. The cell pellet is resuspended in the residual growth medium present in the pellet; the volume of the dense suspension is measured, and then it is mixed with 0.24 volumes of autoclaved 80% glycerol at room temperature and placed in a freezer at −80° (this general method for storing cyanobacteria is from Dr. C. P. Wolk, personal communication). For recovery, the cells are thawed and either spread onto plates or innoculated into liquid medium. In our experience, growth of the thawed cells occurs with a success rate of about 90%.

DNA Isolation

To prepare DNA, cells from 300 ml of a thick culture (A_{730} = 5 to 10) are harvested by centrifugation at 4,500 *g* for 6 min. The wet weight of the cell pellet ranges from 1 to 2 g. For each gram of cells, add 2 ml of saturated sodium iodide (4 g of NaI plus 2 ml of water yields 3 ml of

[22] Transformation-defective variants of another cyanobacterium, *Synechococcus* R2, have been isolated fortuitously by picking single colonies in the course of culture maintenance [J. G. K. Williams and K. S. Kolowsky, unpublished observations (1983)].

saturated NaI solution; heat to dissolve[23]). Suspend the cells by vortexing, incubate at 37° for 20 min, add water to a final volume of 40 ml, and centrifuge at 10,000 *g* for 10 min to remove the NaI. Suspend the cell pellet by vortexing in 10 ml of 50 m*M* Tris–HCl, pH 8.5, 50 m*M* NaCl, 5 m*M* NaEDTA, and add 1.5 ml of lysozyme (50 mg ml^{-1} in water). Incubate at 37° for 45 min, add 1 ml of 10% (w/v) *N*-lauroylsarcosine, and incubate at 37° for 20 min. The green lysate should be viscous and clear. Reduce the viscocity by expelling the lysate forcefully through a 10-ml polystyrene pipet (Falcon) several times. To a tared flask add 3 ml of ethidium bromide (10 mg ml^{-1}), the lysate, and water if necessary to a final weight of 23 g. Add 21 g of CsCl and centrifuge at 45,000 rpm for 20 hr in a Beckman 50VTi rotor at 15°. Collect the DNA (chromosome upper band, plasmid lower band), dialyze against 4 liters of water for 90 min, extract once with an equal volume of phenol,[24] extract once with 5 volumes of chloroform, and precipitate the DNA with ethanol.[24] Wash the DNA pellet with 70% ethanol, dry under reduced pressure in a Speed Vac (Savant Co.), and dissolve the pellet in 100 μl of 10 m*M* Tris–HCl, pH 7.5, 0.1 m*M* EDTA. DNA prepared in this way ranges in size from 20 to 30 kb. To prepare high-molecular-weight DNA, extract the clear green lysate with an equal volume of phenol for 1 hr by gentle inversion of the sample in a test tube. A rotating wheel is convenient for this procedure. Centrifuge at 10,000 *g* for 10 min. Extract the top aqueous phase with an equal volume of chloroform for 1 hr, centrifuge, and precipitate the DNA from the aqueous phase with ethanol.

Acknowledgments

Initial characterization of the transformation system was done in the laboratory of Dr. L. McIntosh at Michigan State University, East Lansing, Michigan, where the author was supported by a grant from the McKnight Foundation. Application of the system to PSII was done with colleagues W. F. J. Vermaas and H. B. Pakrasi at the Du Pont Co., under Dr. C. J. Arntzen. The author acknowledges the excellent technical assistance of D. A. Chisholm.

[23] *Synechocystis* 6803 is notoriously resistant to digestion with lysozyme. Pretreatment of cells with 4% sodium dodecyl sulfate (SDS) at 75° for 15 min, or with saturated NaI as described above, allows lysozyme to digest the cell wall. These procedures are thought to remove a protective coating from the cell surface. NaI is preferred over SDS because it is easier to wash away from the cells.

[24] T. Maniatis, E. F. Fritsch, and J. Sambrook, "Molecular Cloning: A Laboratory Manual," p. 353. Cold Spring Harbor Lab., Cold Spring Harbor, New York, 1982.

[86] Isolation and Characterization of Genes for Cytochrome b_6/f Complex

By TOIVO KALLAS and RICHARD MALKIN

Introduction

The proton-translocating thylakoid membrane cytochrome b_6/f complex catalyzes electron transfer between the two photosystems (PS) and is required for cyclic electron flow around PSI. In cyanobacteria as in plant chloroplasts, the complex consists of four essential polypeptides: cytochromes *f* and b_6, the Rieske Fe–S protein, and a subunit IV.[1] In the cyanobacterium *Anabaena* these have apparent molecular weights of 31,000, 22,500, 22,000 and 16,000, respectively.[2] Cloned genes can now be introduced by transformation[3] or conjugation[4] into several strains of cyanobacteria enabling, among other applications, a molecular genetic analysis of the structure and function of thylakoid membrane protein complexes. An additional requirement for this approach is the availability of cloned genes for these proteins and the determination of their nucleotide and derived amino acid sequences. Here we describe procedures used for the isolation and preliminary characterization of genes encoding the cytochrome b_6/f complex from the filamentous cyanobacterium *Nostoc* PCC 7906 (*sensu* Rippka *et al.*[5]). We have chosen this organism because it can grow facultatively as a heterotroph, differentiate heterocysts, and because of the availability of a unicellular mutant, *Nostoc* PCC 7121[6] into which DNA can be transferred by conjugation.[7]

Strategy for Isolation of Genes by Heterologous Hybridization

Heterologous hybridization (the use of a gene from one source as a probe for isolating a gene from another source) was the approach first

[1] R. Malkin, this volume [37].
[2] M. Krinner, G. Hauska, E. Hurt, and W. Lockau, *Biochim. Biophys. Acta.* **681,** 110 (1982).
[3] C. J. Kuhlemeier, W. E. Borrias, C. A. M. J. J. van den Hondel, and G. A. van Arkel, *Mol. Gen. Genet.* **184,** 249 (1981).
[4] C. P. Wolk, A. Vonshak, P. Kehoe, and J. Elhai, *Proc. Natl. Acad. Sci. U.S.A.* **81,** 1561 (1984).
[5] R. Rippka, J. Deruelles, J. B. Waterbury, M. Herdman, and R. Y. Stanier, *J. Gen. Microbiol.* **111,** 1 (1979).
[6] T. Kallas, T. Coursin, and R. Rippka, *Plant Mol. Biol.* **5,** 321 (1985).
[7] T. Kallas, unpublished data (1988).

used to isolate cyanobacterial genes[8] and perhaps remains the most widely used one. Its success ultimately depends on the level of nucleotide sequence similarity between the probe and the target DNA. In this regard, the well-characterized chloroplast genomes, which, from all available evidence, share a common ancestry and conserved sequences with cyanobacterial genes, provide an excellent source of probes for the isolation of cyanobacterial genes. Three of the polypeptides of the b_6/f complex, namely Cyt$_f$, Cyt b_6, and subunit IV (SU IV), are encoded by chloroplast genes.[9]

The material requirements for gene cloning by heterologous hybridization are as follows: (1) purified cyanobacterial genomic DNA, (2) a library of genomic DNA fragments maintained as separate clones in a suitable vector, and (3) a previously characterized gene or DNA fragment for use as a hybridization probe. These materials are used as follows: (1) Radioactively labeled probe DNA is hybridized to blots of electrophoretically separated genomic DNA restriction fragments (Southern[10] analysis) in order to determine the size and number of hybridizing fragments and the appropriate hybridization conditions. (2) Probes are hybridized, under the determined conditions, to the gene library. (3) Positively hybridizing clones are identified and purified. (4) DNA is extracted from positive clones and initially characterized by restriction site mapping. (5) Positive identification and further characterization of the cloned genes is accomplished by means of nucleotide sequence analysis and/or expression of the genes and identification of protein products. Below we describe the preparation of these materials and each step in greater detail.

Construction of Cyanobacterial DNA Libraries in Bacteriophage λ Vectors

Purification of Cyanobacterial DNA

A general procedure, based on the method of Schwinghamer,[11] for purifying high-molecular-weight DNA from cyanobacteria has been described previously.[6] Here we describe a simplified procedure for obtaining DNA from readily lysed strains such as *Nostoc* PCC 7906 and 7121. Sterile solutions are used throughout. Cultures are grown with stirring at a light intensity of approximately 2000–4000 lux (cool-white fluorescent) in

[8] B. J. Mazur, D. Rice, and R. Haselkorn, *Proc. Natl. Acad. Sci. U.S.A.* **77,** 186 (1980).

[9] J. Alt, P. Westhoff, B. B. Sears, N. Nelson, E. Hurt, G. Hauska, and R. G. Herrmann, *EMBO J.* **2,** 979 (1983).

[10] E. Southern, this series, Vol. 68, p. 152.

[11] E. A. Schwinghamer, *FEMS Microbiol. Lett.* **7,** 157 (1980).

2-liter flasks containing 1 liter BG-11 medium[5] supplemented with Na_2CO_3 (0.2 g/liter) and gassed with 1% CO_2 in air. Dense, stationary phase cultures are harvested by centrifugation at about 6000 *g*, resuspended in 200 ml 50 m*M* Tris–HCl, pH 8.5, 50 m*M* EDTA, 0.1% sodium sarkosyl, and incubated at 30–37° for 1 hr with gentle shaking. Spheroplasts and cell debris are pelleted at 6000 *g*, the supernatant is collected, and to it are added 5 *M* NaCl and then polyethylene glycol (MW 6000–8000) crystals to concentrations of 0.54 *M* and 10%, respectively. These are allowed to dissolve during incubation for a period ranging from about 3 hr to overnight with gentle shaking at 4°, and the solution is then centrifuged at 6000 *g* for 15 min at 4°.

The supernatant is discarded and the greenish DNA-containing pellet resuspended in 10 m*M* Tris–HCl, pH 7.5, 1 m*M* EDTA to a volume of 16 ml (~16 g). This solution is transferred to a capped 40-ml polypropylene centrifuge tube, 15.7 g CsCl and 1.6 ml ethidium bromide (10 mg/ml) are added, and the contents gently mixed until the CsCl dissolves (the mixture may remain slightly turbid and flocculent). The suspension is then centrifuged approximately 26,000 *g* for 15 min at 4°, causing protein and other materials to either form a pellet or aggregate at the surface. The clear, reddish DNA solution is transferred with minimal shearing to Beckman "Quick-Seal" polyallomer tubes by means of a funnel and tubing attached to the neck of the tube. Tubes are filled to the top with buffer–CsCl–ethidium bromide solution, sealed, and centrifuged to equilibrium for a minimum of 24 hr at 15° and 45,000 rpm in a 70.1 Ti rotor (g_{max} ~190,000). DNA bands are visualized under long-wavelength ultraviolet light and the upper band containing linear DNA collected with an 18-gauge needle. The linear DNA fractions from two tubes are pooled and subjected to a second ethidium bromide–CsCl centrifugation; the linear DNA is again collected and then extracted with NaCl-saturated 2-propanol[12] to remove ethidium bromide. Two volumes of H_2O and then twice the final volume of ethanol are added, mixed, and the DNA precipitated by centrifugation at around 17,000 *g* for 30 min at 4°. The pellet is washed with 70% ethanol, dried, and resuspended in 1–2 ml 10 m*M* Tris–HCl, pH 7.5, 0.1 m*M* EDTA, and the DNA concentration is determined by diluting a small portion approximately 1/200 in H_2O and measuring the OD at 260 nm. The DNA yield from the sarkosyl supernatant is about 200–500 μg and contains both chromosomal and plasmid DNA because some of the latter invariably becomes nicked and therefore fractionates with linear DNA.

[12] R. W. Davis, D. Botstein, and J. R. Roth, "Advanced Bacterial Genetics." Cold Spring Harbor Lab., Cold Spring Harbor, New York, 1980.

Preparation of Cyanobacterial DNA for Cloning

DNA fragments of 15–20 kilobase (kb) size suitable for cloning into the *Bam*HI site of the λ EMBL3 vector (described below) are obtained by partial *Sau*3AI digestion and sucrose gradient fractionation of genomic DNA as described in Maniatis *et al.*[13] with minor modifications. After appropriate test digests and analysis by gel electrophoresis, 100 μg cyanobacterial DNA is partially cleaved with *Sau*3AI to obtain a random population of DNA fragments spanning the desired size range. The DNA fragments are precipitated with ethanol,[14] resuspended in 20 m*M* Tris–HCl, pH 8, 10 m*M* EDTA, 1 *M* NaCl, and 50-μg samples in 0.5 ml are loaded onto 10–40% sucrose gradients (made up in the same buffer) in Beckman SW41 tubes and centrifuged for 18 hr at 18° and 26,000 rpm (g_{max} ~110,000). A 50-μl capillary with attached polyethylene tubing is placed at the bottom of each gradient tube, gentle suction is applied, and 0.5-ml fraction are collected by gravity. Portions of these fractions are analyzed by electrophoresis on an agarose gel; fractions containing 15- to 20-kb fragments are pooled, diluted with 2 volumes H_2O, 2 volumes ethanol are added, and the DNA is precipitated, washed in 70% ethanol, and resuspended in a small volume of 10 m*M* Tris–HCl, pH 7.5, 0.1 m*M* EDTA to a concentration of approximately 500 μg/ml.

Preparation of Bacteriophage λ EMBL3 DNA for Cloning

DNA fragments of up to but not larger than about 22 kb, which have ends digested by the endonuclease *Sau*3AI, can be ligated to *Bam*HI digested arms of the λ replacement vector EMBL3[16] and form viable phage. This precludes the cloning of noncontiguous segments of genomic DNA from a population of 15- to 20-kb fragments. Recombinant phage show a Spi^- phenotype and can therefore be selected by their ability to grow on *Escherichia coli* harboring a phage P2 lysogen.[17] As a host we use *E. coli* LE392[13] for routine growth and screening of EMBL3 and recombi-

[13] T. Maniatis, E. F. Fritsch, and J. Sambrook, "Molecular Cloning: A Laboratory Manual." Cold Spring Harbor Lab., Cold Spring Harbor, New York, 1982.

[14] Unless stated otherwise, purified DNA is precipitated by addition of 3 *M* sodium acetate (0.1 volume) and ethanol (2.5 volumes); mixed; stored at −80° for 1 hr, at −20° overnight, or on ice for 10 min (the latter appears to be equally effective[15]); and pelleted by centrifugation at a minimum of 15,000 *g* for 30 min at 4°. The pellet is washed in 70% ethanol (care must be taken to avoid losing pellets at this stage), dried in air or under reduced pressure, and resuspended in the desired buffer.

[15] Bethesda Research Laboratories, *Focus* **7** (1985).

[16] A.-M. Frischauf, H. Lehrach, A. Poustka, and N. Murray, *J. Mol. Biol.* **170,** 827 (1983).

[17] J. Karn, S. Brenner, and L. Barnett, this series, Vol. 101, p. 3.

nants and a P2 lysogen of *E. coli* LE392 (EM18[18]) for detecting recombinants. For DNA preparations, growth on *E. coli* K802 (su^- strain[19]) appears to give a better lysis and yield. Phage are grown and DNA extracted as described in Maniatis *et al.*[13] with some modifications. Dialysis of DNA after phenol–chloroform extraction is replaced by precipitation with ethanol (2 volumes) or 2-propanol (0.6 volume) followed by a 70% ethanol wash and resuspension of the DNA in 10 m*M* Tris–HCl, pH 7.5, 0.1 m*M* EDTA. The yield from a 750 ml culture is usually 6–700 μg. For vector preparation, EMBL3 DNA (10–20 μg) is doubly digested with 4-fold excesses of *Bam*HI and *Eco*RI in the medium and high-salt buffers, respectively, of Maniatis *et al.*,[13] then extracted once with an equal volume of phenol–chloroform–isoamyl alcohol (25 : 24 : 1), extracted once with chloroform–isoamylalcohol (24 : 1), precipitated with ethanol, and resuspended in 10 m*M* Tris–HCl, pH 7.5, 0.1 m*M* EDTA to a concentration of 500 μg/ml. *Eco*RI cleaves several times within the nonessential "stuffer" fragment thereby minimizing the possibility of its religating to the arm fragments to form wild-type EMBL3.

Ligation of Cyanobacterial DNA to λ EMBL3 Vector

Equimolar amounts of cyanobacterial insert DNA (assuming an average size of 20 kb) and *Bam*HI/*Eco*RI digested EMBL3 DNA (total concentration ~200 μg/ml) are ligated in a 5- to 10-μl volume containing 50 m*M* Tris–HCl, pH 8, 10 m*M* $MgCl_2$, 1 m*M* ATP, 50 μg/ml bovine serum albumin, 10 m*M* dithiothreitol, and 1 U T4 DNA ligase for 1–4 hr at 29°. The DNAs are heated 5 min at 65° prior to the addition of 10× concentrated ligation buffer. Ligation can be monitored by electrophoresis of 0.5-μl samples, taken before and at times after the addition of T4 ligase, on an agarose gel. The ligated recombinant DNA is heated 10 min at 70° and stored at 4° until needed for packaging. Packaging efficiency may be improved by ethanol precipitation, 70% ethanol wash, and resuspension of the DNA to approximately 200–500 μg/ml in H_2O prior to packaging.[20]

Preparation of Extracts and Packaging of DNA into Phage Capsids

Packaging extracts are prepared from *E. coli* strains BHB2688 and BHB2690[21] according to the method of Sternberg *et al.*[22] as described in

[18] E. D. Muller, J. Chory, and S. Kaplan, *J. Bacteriol.* **161,** 469 (1985).
[19] W. B. Wood, *J. Mol. Biol.* **16,** 118 (1966).
[20] U. Wienand, personal communication (1983).
[21] B. Hohn, this series, Vol. 68, p. 299.
[22] N. Sternberg, D. Tiemeier, and L. Enquist, *Gene* **1,** 255 (1977).

protocol II of Maniatis *et al.*[13] with some modifications.[20] Aliquots are frozen in liquid N_2, stored at $-80°$, and can be used for at least 1 year. Extracts prepared by this procedure package wild-type EMBL3 and hybrid cyanobacteria–EMBL3 DNAs at efficiencies of $1-2 \times 10^8$ and $1-5 \times 10^5$ plaque-forming units (pfu)/μg DNA, respectively. Packaging extracts and buffers are thawed on ice and added to the DNA to be packaged (~1 μg in 4 μl), on ice, in the following order: 14 μl sonication buffer (20 m*M* Tris–HCl, pH 8, 1 m*M* EDTA, 3 m*M* $MgCl_2$, 5 m*M* 2-mercaptoethanol), 2 μl packaging buffer (6 m*M* Tris–HCl, pH 8, 50 m*M* spermidine, 50 m*M* putrescine, 20 m*M* $MgCl_2$, 30 m*M* ATP, 30 m*M* 2-mercaptoethanol), 10 μl sonicated extract (SE from BHB2690), and 10 μl freeze–thaw lysate (FTL from BHB2688). For maximum efficiency, the optimal ratio of SE to FTL needs to be determined for each batch of extracts. The mixture is incubated 10 min on ice, 30 min at 29°, and then diluted with 0.5 ml SM (50 m*M* Tris–HCl, pH 7.5, 0.1 *M* NaCl, 8 m*M* $MgSO_4$, 0.1% gelatin). The titer of viable phage is determined by plating: dilutions in SM (0.1 ml) are mixed with 0.1 ml host bacteria [grown overnight in LB medium[13] (10 g/liter Bacto-tryptone, 5 g/liter yeast extract, 5 g/liter NaCl, pH 7.4) containing 0.2% maltose, resuspended in 10 m*M* $MgSO_4$ to an OD_{600} of ~2, equal to ~200 Klett units, and stored at 4° for periods of several weeks], incubated 15 min at 37°, 3.5 ml 0.7% top agar in λ medium (LB lacking yeast extract) added (at 47–50°), and the mixture poured onto λ medium 1.2% agar plates. Plaques form after 8–12 hr of incubation at 37°. Prolonged incubation should be avoided as it may result in overgrowth of plaques with resistant bacteria.

Amplification and Storage of the Library

Viable phage are amplified as a plate lysate stock.[13] Host bacteria are infected as described above and approximately 10^5 pfu are plated onto a single 90-mm Petri dish. After 8–12 hr at 37°, SM buffer is pipetted onto the agar surface; the phage diffuse into the liquid, are treated with chloroform, and then stored at 4° in SM with 0.3% chloroform. The titer of bacteriophage appears to remain relatively stable over a period of several years. Optimal long-term storage is probably in 7% dimethyl sulfoxide or 15% glycerol at $-80°$.[13]

Characteristics of the Nostoc DNA Libraries

Our *Nostoc* PCC 7906 and 7121 genomic DNA libraries, constructed as described above, are comprised of approximately 1.2 and 1.8×10^5 pfu, respectively. Of these about 66% form plaques on a P2 lysogen

TABLE I
CHARACTERISTICS OF THE *Nostoc* DNA LIBRARIES

Source of DNA	μg DNA packaged	10^5 pfu per DNA packaged: Total	10^5 pfu per DNA packaged: Recombinant[a]	10^3 recombinants among which a unique sequence occurs at $P = 99.9\%$	Titer 10^{10} pfu/ml after amplification in *E. coli* host strain: LE392	Titer 10^{10} pfu/ml after amplification in *E. coli* host strain: LE393(P2)
Nostoc PCC 7906	0.4	1.2	0.8	3.5	7.4	5.2
Nostoc PCC 7121	0.4	1.8	1.2	3.5	3.8	2.5

[a] The number of recombinants is calculated from the percentage of phage able to form plaques on *E. coli* LE392(P2).

and are therefore recombinants (Table I). The equation of Clarke and Carbon[23]:

$$N = \ln(1 - P)/\ln(1 - f)$$

is used to calculate the number, N, of recombinant clones among which a unique DNA sequence should occur at a probability, P. The variable f represents the average DNA insert size, expressed as a fraction of the genome size. For *Nostoc,* the genome size is around 7500 kb,[24] and we used an insert size of 15 kb to arrive at the value of N shown in Table I. Consequently only about 5000 pfu (plated on *E. coli* LE392) need be screened in order to find (at $P = 99.9\%$) a unique gene sequence. This is a very manageable number and thus we chose not to ampify our libraries on a P2 lysogen.

Screening and Isolation of Positive Clones

Preparation of Probe DNA

Purified gene fragments are used as hybridization probes. This minimizes nonspecific hybridization to vector sequences (which can occur at low hybridization stringency) and maximizes the effective concentration of probe, thereby helping to drive the reaction toward duplex formation. The best probes are generally small (e.g., 0.2–0.4 kb), gene-internal frag-

[23] L. Clarke and J. Carbon, this series, Vol. 68, p. 396.
[24] M. Herdman, M. Janvier, R. Rippka, and R. Y. Stanier, *J. Gen. Microbiol.* **111,** 73 (1979).

TABLE II
DNA FRAGMENTS USED AS PROBES

Gene	Protein encoded	Isolated fragment size (kb)	Description	Source of plasmid
Pea *petA*	Cyt *f*	0.4	Internal fragment near the amino terminus	A. Barkan
Spinach *petB*	Cyt b_6	0.3	Internal fragment near the amino terminus	H. Bohnert
Pea *petD*	SU IV	0.4	Most of the coding region and some 5′ noncoding sequence	A. Barkan
Maize *psbB*	51 kDa PSII	0.3	Internal fragment near the amino terminus	C. Rock, A. Barkan

ments of regions likely to be highly conserved. The probe fragments used to isolate the b_6/f genes are described in Table II. The *psbB* gene encoding a 51-kDa polypeptide of PSII was of interest because in chloroplasts it can be cotranscribed with the *petB* (Cyt b_6) and *D* (SU IV) genes.[25]

Probe fragments are cleaved from their vectors (10–50 μg DNA) by digestion with appropriate restriction endonucleases and separated by electrophoresis on 0.8% low-melting-temperature agarose gels (containing 0.25 μg/ml ethidium bromide) in 1× TBE buffer (89 m*M* Tris base, 89 m*M* boric acid, 2.5 m*M* disodium EDTA, pH ~8.3). We use a commercial minigel apparatus which can accommodate a 20-ml gel volume and a 200-μl sample. DNA bands are visualized with a minimum exposure to long-wavelength ultraviolet light, cut from the low-melt gel, and purified by passage over Schleicher & Schuell ELUTIP-d minicolums[26]: 10–12 volumes low-salt buffer (LSB: 0.2 *M* NaCl, 20 m*M* Tris–HCl, pH 7.5, 1 m*M* EDTA) are added to an excised DNA band, heated 30 min at 70°, and then cooled to 37°. Two milliliters of high-salt buffer (HSB: 1.0 *M* NaCl, 20 m*M* Tris–HCl, pH 7.5, 1 m*M* EDTA) and then 10 ml LSB are forced through an ELUTIP-d column at 37°. The DNA solution (15–20 ml maximum) is slowly passed through the column (~1 drop/sec) followed by a 3-ml LSB wash. DNA fragments are eluted with 0.4 ml HSB into a microfuge tube, precipitated by addition of 1 ml ethanol, washed with 70% ethanol, and resuspended in a small volume (10–25 μl) of 10 m*M* Tris–HCl, pH 7.5, 0.1 m*M* EDTA. Recovery is usually 50% or better. An

[25] W. Heinemeyer, J. Alt, and R. G. Herrmann, *Curr. Genet.* **8,** 207 (1984).

[26] Schleicher & Schuell, "Tips," #206. Schleicher & Schuell, Keene, New Hampshire, 1984.

alternative though somewhat more tedious method, involving extraction with hexadecyltrimethylammonium bromide (CTAB)–butanol,[27] has worked equally well. DNA fragments purified by either procedure can be labeled to high specific activities by nick translation.

Labeling of Probe DNA Fragments

A commercial kit (N5000, Amersham) is used to label purified DNA fragments by nick translation[28] to specific activities of 2–5 × 10^8 cpm/μg DNA. The kit provides ingredients for around 100 reactions (200 ng DNA each), and aliquots can be stored at −80° for at least 2 years. As label, we use [α-^{32}P]dCTP at 3000 Ci/mmol (Amersham or ICN) diluted as needed with cold dCTP. Thus the label can be used for at least 1 month to generate probes of high specific activity. DNA (200 ng), nucleotide buffer mix, enzyme mix (DNase I and DNA polymerase I), and label are mixed (25 μl volume), incubated at 15° for 2–3 hr, and the reaction terminated by addition of 2.5 μl 0.2 *M* EDTA, 1% sodium dodecyl sulfate. Labeled DNA is separated from free nucleotides by centrifugation (2–5 min at ~2000 *g* on a clinical swing-out rotor) through 1 ml of Bio-Rad BioGel P-6DG matrix (a polyacrylamide resin which will not crush on centrifugation) which has been equilibrated with 10 m*M* Tris–HCl, pH 7.5, 1 m*M* EDTA, and packed into a 1-ml syringe held in a polypropylene tube. A decapped microfuge tube collects the labeled DNA effluent. The column is washed with 40 μl of buffer, and the combined effluents (~70 μl) are stored at −20°. The incorporation of labeled nucleotide into purified DNA fragments is consistently about 50%.

Preparation of Genomic DNA Blots

In heterologous hybridizations the sequence similarity of probe to target DNA, and therefore the intensity of the hybridization signal above background, may be very low. In hybridizations to genomic DNA, the amount of probe will be in excess over the target DNA which represents but a minute fraction of the genome. Therefore the hybridization signal is maximized by loading a maximum amount of target DNA on the gel. In our experiments with b_6/f genes, 5–10 μg of *Nostoc* DNA digested to completion with a 2- to 5-fold excess of endonuclease are loaded per lane (0.45 cm wide) of an 0.7% agarose gel (containing 0.5 μg/ml ethidium bromide) in 1× TBE buffer (described above) and separated on a horizontal submarine gel apparatus by electrophoresis at approximately 6 V/cm

[27] J. Langridge, P. Langridge, and P. L. Berquist, *Anal. Biochem.* **103,** 264 (1980).
[28] P. W. J. Rigby, M. Dieckman, C. Rhodes, and P. Berg, *J. Mol. Biol.* **113,** 237 (1977).

for 3 hr or at approximately 1.5 V/cm overnight. Figure 1 (left) shows a stained gel. Little banding detail is visible because of the large quantity of DNA loaded.

DNA fragments are transferred from the gel to nylon (Amersham Hybond-N) or nitrocellulose (Schleicher & Schuell BA85) membranes by the unidirectional capillary transfer method of Southern.[10] Nylon membranes bind more DNA and can be more easily rehybridized[29] than nitrocellulose although background hybridization may be higher. Our conditions for Southern transfer are the following: DNA in the gel (~150 cm^2) is partially depurinated (thereby improving the transfer of large fragments) by a 10-min treatment in 400 ml 0.25 *N* HCl, then washed with H_2O, denatured by two successive 15-min treatments in 400 ml 1.5 *M* NaCl, 0.5 *M* NaOH, washed with H_2O, neutralized by two 30-min treatments in 400 ml of 1.5 *M* NaCl, 0.5 *M* Tris–HCl, pH 7.2, 1 m*M* EDTA, and placed directly onto a unidirectional blotting stack.[10,13] A membrane prewetted in H_2O and then 2× SSC (1× SSC: 0.15 *M* NaCl, 15 m*M* trisodium citrate) is placed on top of the gel and the rest of the blotting stack installed.[10,13] The blotting buffer is 20× SSC. After overnight transfer, the stack is disassembled and the membrane marked to correspond to the gel slots, washed for 15 min in 2× SSC, blotted dry, and baked for 2–3 hr at 70–75° (we do not bake under reduced pressure). DNA can be covalently linked to nylon membrane by a 5-min ultraviolet light irradiation (e.g., on a standard gel transilluminator). Membranes are stored at room temperature between sheets of filter paper.

Preparation of Bacteriophage λ Plaques for Screening

Typically we screen 5 to 6 90-mm petri dishes each with 1–2000 phage plaques from our λEMBL3/*Nostoc* libraries grown on *E. coli* LE392 as described above. DNA from phage plaques is transferred to filters (Schleicher & Schuell BA85) by the method of Benton and Davis[30] essentially as described by Maniatis *et al.*[13] A dry, labeled, filter disk is placed face down for 2–5 min on the top agarose surface (agarose usually does not adhere to the filter) of each dish, and its position is marked with four asymmetric needle holes. Filters are removed with forceps and placed successively, face up, for 5 min at a time on filter papers soaked in denaturing solution (1.5 *M* NaCl, 0.5 *M* NaOH), then neutralizing solution (3 *M* NaCl, 0.5 *M* Tris–HCl, pH 7.5), and finally 2× SSPE (20× SSPE: 0.2 *M* NaH_2PO_4, 3 *M* NaCl, 0.2 *M* disodium EDTA). The filters are blotted

[29] Amersham, "Membrane Transfer and Detection Methods." Amersham, Arlington Heights, Illinois, 1985.

[30] W. D. Benton and R. W. Davis, *Science* **196,** 180 (1977).

dry, placed between sheets of filter paper, baked at 70° for 2–3 hr, and stored at room temperature. The petri dishes bearing phage plaques are stored at 4° until needed for replication of positive phage.

Hybridization to Genomic DNA Blots

The parameters, T_m (the melting temperature of a DNA duplex) and ΔT (the stringency of the hybridization expressed as degrees Celsius below the T_m) are of paramount importance for heterologous hybridization experiments. The T_m for a perfectly matched DNA duplex can be estimated from the "consensus" expression of Beltz *et al.*[31]:

$$T_m = 81.5 + 0.41(\mathrm{G} + \mathrm{C}) + 16.6 \log[\mathrm{Na}^+] - 0.63(\%\ \text{formamide}) - \frac{300 + 2000[\mathrm{Na}^+]}{d}$$

where G + C is the percent guanosine plus cytosine, $[Na^+]$ the molarity of Na^+ or equivalent monovalent cation, and *d* the length of the hybridizing duplex in nucleotides. As a gross generalization, the T_m decreases by 1° for every 1 ± 0.3% mismatch.[31,32] Data from numerous sequence comparisons, including our own for cytochrome b_6/f genes,[33] show that chloroplast and cyanobacterial genes generally share 70% or better nucleotide sequence similarity. Therefore a heterologous plastid/cyanobacterial DNA duplex might melt as much as 30–40° below the T_m for a homologous duplex, and hybridizations must be performed at a stringency low enough to permit duplex formation but not so low as to permit nonspecific hybridization or result in unreasonably slow reaction kinetics. For isolation of the *Nostoc* b_6/f and PSII genes described here, a ΔT of 40–45° below the T_m of *Nostoc* DNA met these requirements. In principle, any combination of formamide, Na^+, and temperature (within certain limits[31]) could be used to attain desired stringency. The proper stringency, however, is critical as illustrated by our failure to hybridize the *petA* (cyt *f*) probe to *Nostoc* DNA in initial experiments at ΔT approximately 30° below T_m (these are standard conditions which work well for *nif* gene hybridizations[6]).

Hybridizations are performed by the procedure of Wahl *et al.*[34] with modifications. Genomic blot filters are rehydrated in 2× SSC (described above); then preprehybridized in 20% formamide (BRL specialty reagent

[31] G. A. Beltz, K. A. Jacobs, T. H. Eickbush, P. T. Cherbas, and F. C. Kafatos, this series, Vol. 100, p. 266.
[32] R. J. Britten, D. E. Graham, and B. R. Neufeld, this series, Vol. 29, p. 363.
[33] T. Kallas, S. Spiller, and R. Malkin, *Prog. Photosynth. Res.* **4,** 801 (1987).
[34] G. M. Wahl, M. Stern, and G. R. Stark, *Proc. Natl. Acad. Sci. U.S.A.* **76,** 3683 (1979).

used without further treatment), 5× SSC, 50 mM sodium phosphate, pH 6.8, at 42° for 30–60 min in a sealed plastic bag; then prehybridized in the same bag with 20% formamide, 5× SSC, 50 mM sodium phosphate, pH 6.8, 5× Denhardt's solution[35] (100× Denhardt's: 2% bovine serum albumin, 2% polyvinylpyrrolidone, 2% Ficoll—MW 400,000), 0.01% sodium pyrophosphate, 0.1% sodium dodecyl sulfate, and 250 μg/ml sonicated, denatured salmon testes DNA at 42° for 3 hr to overnight. After removal of prehybridization solution, blots are hybridized in 20% formamide, 5× SSC, 20 mM sodium phosphate, pH 6.8, 1× Denhardt's solution, 0.005% sodium pyrophosphate, 0.1% sodium dodecyl sulfate, 125 μg/ml sonicated, denatured salmon DNA, and 10% dextran sulfate (this latter reagent accelereates the rate of hybridization[34] which is lowered in the presence of formamide and low-stringency conditions[31]). ^{32}P-Labeled probe DNA is denatured by heating 10 min at 100°, quenched on ice, added to 2 ml hybridization buffer at 70°, mixed, then added to the blot and the rest of the hybridization buffer in the plastic bag. The bag is sealed, the probe solution mixed well, air bubbles are chased away from the filter, and hybridization proceeds at 42° usually for 48–72 hr with gentle shaking. All solutions are added at approximately 0.2 ml/cm^2 of filter, and the final probe concentration is 5 ng/ml. The T_m of a 400-bp *Nostoc* DNA duplex (G + C ~43%) under these conditions is approximately 81°, and therefore incubation at 42° gives a ΔT of about 40° below T_m.

After hybridization, the blots are washed at room temperature, first in 2 changes (~100 ml) of 0.5× SSC, 0.01% sodium pyrophosphate, 0.05% sodium sarkosyl for 5–10 min and then in 4–5 changes (~400 ml/100 cm^2 filter) of 0.05× SSC, 0.01% sodium pyrophosphate, 0.05% sodium sarkosyl about 1 hr per time. Washes at 25° in this solution give a ΔT of about 40° below the T_m. The membrane filters are blotted dry and placed, still damp, onto plastic-wrapped filter paper which has been marked with spots of ^{35}S radioactive ink. The ensemble is wrapped in plastic film and exposed (usually overnight) with an intensifying screen to X-ray film at −80°. If autoradiograms show a high background or evidence of nonspecific hybridization, blots can be rewashed at higher stringency and reexposed to film. Figure 1 shows the results of hybridizations (under the above conditions) with the probes described in Table II to replicate, *Nostoc* genomic DNA blots. Each lane shows one or a few relatively intense bands, suggesting that the hybridizations were specific and the conditions appropriate for screening *Nostoc* gene libraries.

[35] D. T. Denhardt, *Biochem. Biophys. Res. Commun.* **23,** 641 (1966).

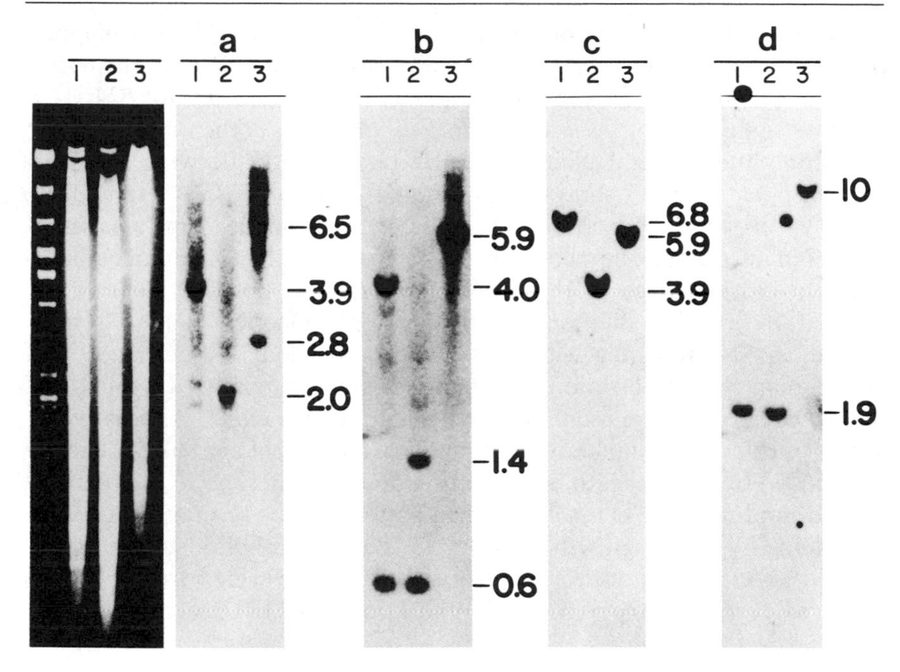

FIG. 1. (Left) Stained gel of electrophoretically separated *Hin*dIII (1), *Hin*dIII/*Xba*I (2), and *Xba*I (3) digests of *Nostoc* PCC 7906 DNA, approximately 8 μg/lane. The extreme left lane shows restriction fragments of λ DNA used as size markers. (Right) Autoradiograms of (a) Cyt *f* (*petA*), (b) Cyt b_6 (*petB*), (c) SU IV (*petD*), and (d) 51 kDa PSII (*psbB*) probe hybridizations to replicate (nitrocellulose) blots of the stained gel. For rehybridization of the two original blots, old probe was removed by submerging rehydrated membrane in 0.05% sodium sarkosyl at 100°, cooling slowly to 55°, and then incubating for approximately 1 hr. Fragment sizes are expressed in kilobases (kb).

Hybridization to Recombinant Bacteriophage Plaques

Filter disks for plaque hybridization are rehumidified in 2× SSC, prewashed (to remove agarose and bacterial debris) in 50 m*M* Tris–HCl, pH 8, 1 *M* NaCl, 1 m*M* EDTA, 0.1% sodium dodecyl sulfate[13] at 42° for 30–60 min, and then preprehybridized, prehybridized, hybridized, and washed in the solutions described above for genomic blot hybridizations. Five to ten filter disks can be incubated per 20 ml of solution in a 90-mm petri dish. The dishes are gently shaken on a rotary shaker and during overnight or longer incubations are sealed in plastic bags to prevent evaporation. After washing, the filter disks are mounted, damp, on plastic-wrapped filter paper (marked with ^{35}S ink for orientation), wrapped in plastic film, and exposed to X-ray film as described above for genomic blots. Positive hybridization spots on the autoradiograms are mapped back to the filter disks and then to the original petri plates from which positive plaques are picked with a Pasteur pipet into 0.5 ml SM. If a positive signal cannot be assigned to a single plaque, then a larger agar plug containing several plaques can be picked with the wide end of a Pasteur pipet. Figure 2 shows an autoradiogram of *petB* (Cyt b_6) and *petD* (SU IV) probe hybridizations to replicate blots made from a lawn of recombinant phage. The two probes hybridized to the same clones, indicating a close linkage of these genes in *Nostoc*. Positive plaques are purified by making dilutions of resuspended plaques ($\sim 10^6$–10^7 pfu/plaque) in SM, replating with host bacteria to about 100 pfu/plate, and rescreening by hybridization as described above. Usually two rounds of plaque purification are required to obtain an axenic clone. Alternatively, if there are a large number of putatively positive plaques, these can be rescreened by transfer into SM, spotting dilutions (in, e.g., 1-μl volumes) of these and negative controls in a grid pattern onto a lawn of host bacteria,[31] allowing phage growth, transfer to filters, and hybridization as described above. Plate stocks are made and DNA purified from positive phage as described above for the preparation of EMBL3 vector.

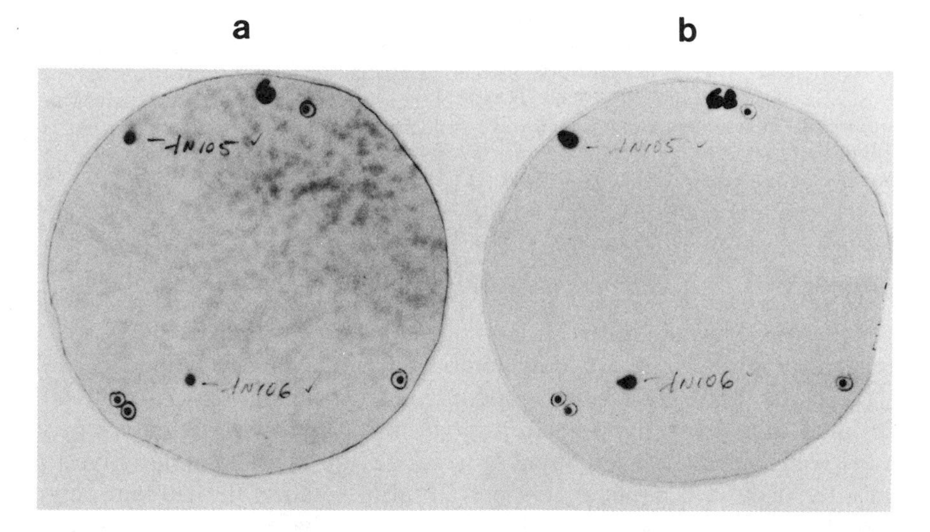

FIG. 2. Autoradiograms showing hybridization of (a) Cyt b_6 (*petB*) and (b) SU IV (*petD*) probes to replicate blot filters made from a petri dish carrying approximately 1000 recombinant *Nostoc*/EMBL3 phage plaques (grown on *E. coli* LE392). The encircled spots are orientation marks. The intense spots labeled λN105 and λN106 represent positive hybridization signals.

Characterization of Positively Hybridizing Bacteriophage Clones

Because of potential artifacts in heterologous hybridization experiments, the identities of isolated genes must be independently established. Available approaches include complementation of defined mutations,[23] nucleotide sequence analysis and comparison with known sequences,[36] or association of the cloned gene with an identifiable protein product. The latter includes *in vivo* expression of the cloned gene in plasmid[37] or phage[38] vectors, *in vitro* coupled transcription–translation,[39] and hybridization selection.[40] Most of these procedures have been described in previous volumes. Because of the known conservation of cyanobacterial and chloroplast sequences and the detailed characterization of several chloroplast genomes,[41,42] nucleotide sequencing is often the method of choice for the characterization of cyanobacterial genes isolated by hybridization with plastid gene probes.

When starting from phage clones that contain large inserts, preliminary characterization usually involves the construction of a simple restriction map by digestion of phage DNA with combinations of restriction endonucleases, electrophoretic separation of the fragments, transfer to filters, and hybridization with probe DNA as described for genomic blots. Comparisons of genomic and cloned hybridization fragment sizes and maps constructed for different probes can provide information about gene organization and can also indicate whether cloned fragments accurately represent the order of genes and restriction fragments as they occur in the genome. Fragments of interest are purified from low-melting agarose as described above and, for our purposes, subcloned into phage M13 vectors[36] for sequencing by the Sanger[43] dideoxy method. The identities of the *Nostoc* genes for Cyt *f*, Cyt b_6, and SU IV of the b_6/f complex and the 51-kDa polypeptide of PSII have been established by comparison of their sequences with those from the chloroplast genes.[33]

Recently, we have used the techniques described here to locate the

[36] J. Messing, this series, Vol. 101, p. 20.

[37] B. P. Nichols and C. Yanofsky, this series, Vol. 101, p. 155.

[38] R. A. Young and R. W. Davis, *Proc. Natl. Acad. Sci. U.S.A.* **80,** 1194 (1983).

[39] H.-Z. Chen and G. Zubay, this series, Vol. 101, p. 674.

[40] H. Bunemann and P. Westhoff, this series, Vol. 100, p. 400.

[41] K. Ohyama, H. Fukuzawa, T. Kohchi, H. Shirai, T. Sano, S. Sano, K. Umesono, Y. Shiki, M. Takeuchi, Z. Chang, S. Aota, H. Inokuchi, and H. Ozeki, *Nature* (*London*) **322,** 572 (1986).

[42] K. Shinozaki, M. Ohme, M. Tanaka, T. Wakasugi, N. Hayashida, T. Matsubayashi, N. Zaita, J. Chunwongse, J. Obokata, K. Yamaguchi-Shinizaki, C. Ohto, K. Torazawa, B. Y. Meng, H. Deno, T. Kamogashira, K. Yamada, J. Kusuda, F. Taikawa, A. Kato, N. Tohdoh, H. Shimada, and M. Sugiura, *EMBO J.* **5,** 2043 (1986).

[43] F. Sanger, S. Nicklen, and A. R. Coulson, *Proc. Natl. Acad. Sci. U.S.A.* **74,** 5463 (1977).

remaining gene for the *Nostoc* cytochrome b_6/f complex (*petC*, encoding the Rieske protein) closely linked to the gene (*petA*) for cytochrome *f*.[44] A fragment of a cDNA copy of the spinach nuclear gene (kindly provided by Prof. R. Herrmann[45]) was used as the probe in these experiments.

Acknowledgments

We wish to thank Therese Coursin for assistance in constructing the phage libraries, Alice Barkan and Hans Bohnert for chloroplast gene probes, Reinhold Herrmann for the spinach Rieske cDNA clone, Eric Muller for strain EM18, and Robert Fischer for valuable information and suggestions. This work was supported by the National Institutes of Health.

[44] T. Kallas, S. Spiller, and R. Malkin, *Proc. Natl. Acad. Sci. U.S.A.* (in press).

[45] J. Steppuhn, C. Rother, J. Hermans, T. Janson, J. Salnikow, G. Hauska, and R. G. Herrmann, *Mol. Gen. Genet.* **210,** 171 (1987).

[87] Genes for ATP Synthase in *Synechococcus* 6301

By ALISON L. COZENS and JOHN E. WALKER

Introduction

The genes for the eight subunits of *Escherichia coli* ATP synthase are in a single transcriptional unit, the *unc* or *atp* operon.[1,2] They are arranged in two subclusters, corresponding to the three subunits of the E_0 segment and the five subunits of the F_1 sector of the enzyme complex,[2] and are preceded by a ninth gene, *uncL*, of obscure function.[3-6] Operons for the five F_1 subunits only are found in the *Rhodospirillaceae, Rhodopseudomonas blastica*,[7] and *Rhodospirillum rubrum*,[8,9] the genes being arranged in the same order as their *E. coli* homologs (Fig. 1). However they differ from *E. coli* in that neither a homolog of *uncI* nor genes for F_0 components are associated with the F_1 loci. Two separate gene clusters for six out of

[1] F. Gibson, *Proc. R. Soc. London, Ser. B* **215,** 1 (1982).

[2] J. E. Walker, M. Saraste, and N. J. Gay, *Biochim. Biophys. Acta* **768,** 164 (1984).

[3] N. J. Gay and J. E. Walker, *Nucleic Acids Res.* **9,** 3919 (1981).

[4] K. von Meyenburg, B. B. Jørgenson, J. Nielsen, and F. G. Hansen, *Mol. Gen. Genet.* **188,** 240 (1982).

[5] W. S. Brusilow, A. C. G. Porter, and R. D. Simoni, *J. Bacteriol.* **155,** 1265 (1983).

[6] N. J. Gay, *J. Bacteriol.* **158,** 820 (1984).

[7] V. L. J. Tybulewicz, G. Falk, and J. E. Walker, *J. Mol. Biol.* **179,** 185 (1984).

[8] G. Falk, A. Hampe, and J. E. Walker, *Biochem. J.* **228,** 391 (1985).

[9] G. Falk and J. E. Walker, *Biochem. J.* **229,** 663 (1985).

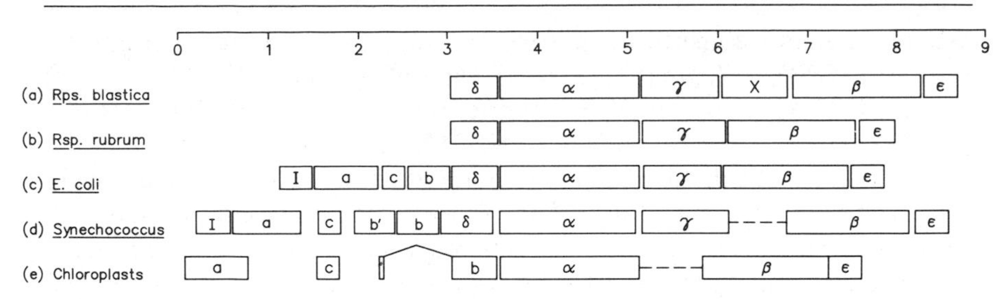

FIG. 1. Organization of genes for ATP synthase in eubacteria and chloroplasts. The symbol X represents a gene of unknown function in the midst of the F_1 cluster of *Rps. blastica*; *b'* is a duplicated and diverged form of *b* in *Synechococcus* 6301.

the nine ATP synthase subunits of the chloroplast enzyme are found in chloroplast DNA.[10] Again gene orders are related to that of the *E. coli unc* operon.

As described here, we have extended this study of ATP synthase genes to the obligate phototrophic unicellular cyanobacterium *Synechococcus* 6301. From its genome we have cloned and determined sequences around regions that hybridize to the α and β genes for ATP synthase from *Rps. blastica,*[7] and the *c* gene from pea chloroplast DNA.[11] The DNA sequence has been determined by primed synthesis in the presence of dideoxy chain terminators[12,13] of DNA fragments generated by sonication[14] or by restriction enzyme digestion. The methodologies employed in these experiments are described below.

Cloning the Genes

Preparation of Genomic DNA

Synechococcus Strain PCC 6301 was kindly provided by Professor A. E. Walsby as an axenic stationary phase culture. It was grown in

[10] A. L. Cozens, J. E. Walker, A. L. Phillips, A. K. Huttly, and J. C. Gray, *EMBO J.* **5,** 217 (1986).

[11] J. C. Gray, A. L. Phillips, C. J. Howe, D. L. Willey, A. K. Huttly, A. Doherty, C. M. Bowman, and T. A. Dyer, *in* "Biosynthesis of the Photosynthetic Apparatus" (J. P. Thornber, A. Staehel and R. Hallick, eds.), UCLA Symp., p. 295. University of California, Los Angeles, 1984.

[12] F. Sanger, S. Nicklen, and A. R. Coulson, *Proc. Natl. Acad. Sci. U.S.A.* **74,** 5463 (1977).

[13] M. D. Biggin, T. G. Gibson, and G. F. Hong, *Proc. Natl. Acad. Sci. U.S.A.* **80,** 3963 (1983).

[14] P. L. Deininger, *Anal. Biochem.* **129,** 216 (1983).

BG-11 medium[15] constituted as follows: Na_2SiO_3, 43.5 mg; iron citrate, 6.0 mg; K_2HPO_4, 37.1 mg; $MgSO_4$, 75 mg; $CaCl_2$, 18 mg; Na_2CO_3, 20 mg; sodium EDTA, 1 mg; $NaNO_3$, 1.5 g; citric acid, 6 mg; H_3BO_4, 2.9 mg; $MnCl_2$, 1.8 mg; $ZnSO_4$, 0.2 mg; Na_2MbO_7, 0.4 mg; $CaSO_4$, 79 μg; $Co(NO_3)_2$, 49 mg; distilled H_2O, 1 liter. Stock solutions of media were sterilized by filtration. Cultures (500 ml) were grown in 2-liter flasks at 33° in an illuminated incubator with slow shaking (~100 rpm). Cells were harvested by centrifugation after 5 days when the midpoint of exponential growth had been reached. They were suspended in buffer (1 g wet cells/10 ml) containing 0.15 *M* $CaCl_2$, 0.1 *M* EDTA, 50 m*M* Tris–HCl, pH 8.0, and transferred to a capped polypropylene centrifuge tube (50 ml). The suspension was incubated first at 37° for 30 min with lysozyme (1 mg/ml) and then at 60° for 10 min with 1% sodium dodecyl sulfate (SDS). DNA was recovered from the lysed cells as described by Marmur.[16]

DNA Hybridization

The DNA was digested with restriction endonucleases as recommended in the New England Biolabs catalog (Beverley, MA). The resulting fragments were fractionated on a 0.8% agarose gel. DNA was transferred from the gel to nitrocellulose filters as described by Southern.[17] After transfer, the nitrocellulose filters were incubated at 65° for 30 min with yeast RNA (50 μg/ml) dissolved in a solution (pH 6.2) containing 0.7 *M* NaCl, 0.2 *M* Na_2HPO_4, 7 m*M* EDTA, 0.1% bovine serum albumin, 0.1% Ficoll, 0.1% polyvinylpyrrolidone, and 0.5% *N*-laurylsarcosine. Then the filters were hybridized at the same temperature for 15–20 hr in the presence of radioactive "prime-cut" probes[18] dissolved in the same solution as for prehybridization except containing in addition 10% dextran sulfate. Then the filters were washed 4 times in a solution containing 0.3 *M* NaCl, 30 m*M* trisodium citrate at 65° for 20 min. Autoradiographs were exposed at −70° with fluorescent screens for 36 hr. Single-stranded M13 prime-cut probes were labeled with [α-^{32}P]dATP. These M13 clones contained nucleotides 7096–7302 of the gene for the α subunit of ATP synthase in *Rps. blastica,* nucleotides 9989–10,201 of the gene for the β subunit from the same bacterium,[7] and a *Taq*I fragment (kindly provided by Dr. J. C. Gray) containing the entire gene for subunit III of ATP synthase from pea (*Pisum sativum*) chloroplasts.[11]

[15] R. Y. Stanier and G. Cohen-Bazire, *Annu. Rev. Microbiol.* **31,** 225 (1977).

[16] J. Marmur, *J. Mol. Biol.* **3,** 208 (1961).

[17] E. M. Southern, *J. Mol. Biol.* **98,** 503 (1975).

[18] P. J. Farrell, P. L. Deininger, A. Bankier, and B. G. Barrell, *Proc. Natl. Acad. Sci. U.S.A.* **80,** 1656 (1983).

Preparation and Screening of Libraries of Genomic DNA

High-molecular-weight DNA prepared from cells of *Synechococcus* 6301 was partially digested with restriction endonuclease *Sau*3A and fractionated by electrophoresis on low-melting-point agarose.[19] Fragments with sizes in the range 15–20 kb were recovered by phenol extraction and ethanol precipitation and then were ligated with the vector λ2001[20] which previously had been cut with the restriction endonucleases *Bam*HI and *Eco*RI. The recombinant DNA was packaged *in vitro*[21,22] and used to infect *E. coli* Q359.[23] Phage libraries were stored over chloroform at 4° in λdil medium (20 m*M* $MgSO_4$, 0.2 *M* NaCl, 0.1% gelatin, 10 m*M* Tris–HCl, pH 7.4). Each library contained over 10,000 individual recombinants. Libraries of 2,500 plaque-forming units (pfu) on *E. coli* Q358 were screened according to the plaque hybridization method of Benton and Davis.[24] The probes and conditions used were as described above. Recombinant phage were grown in *E. coli* Q358 (50-ml cultures) and DNA prepared from them according to Maniatis *et al.*[25]

Subcloning for DNA Sequencing

A 5.6-kb *Bam*HI fragment hybridizing to the probes for the genes for α and *c* (III) subunits and a 4.7-kb *Bam*HI–*Sac*I fragment hybridizing to the probe for the β subunit were subcloned into the vectors pUC8[26] and pUC12[27] respectively. Subsequently a 4-kb *Pst*I fragment and 1.7-kb *Kpn*I fragment overlapping the ends of the 5.6-kb *Bam*HI fragment were identified by hybridization and subcloned into M13*mp*8.[28] The amplified fragments were prepared by restriction endonuclease digestion followed by electrophoresis in low-melting agarose. Before sequencing by the random strategy, fragments were sonicated and subcloned into M13*mp*8 which had previously been cut with *Sma*I,[14] except for the 1.5-kb *Kpn*I fragment which was digested separately with *Alu*I and *Rsa*I and the restriction fragments cloned into M13*mp*8 that had been predigested with *Sma*I.

[19] L. Wieslander, *Anal. Biochem.* **98,** 305 (1979).

[20] J. Karn, S. Brenner, and L. Barnett, this series, Vol. 101, p. 3.

[21] B. Hohn and K. Murray, *Proc. Natl. Acad. Sci. U.S.A.* **74,** 3259 (1977).

[22] J. Collins and B. Hohn, *Proc. Natl. Acad. Sci. U.S.A.* **75,** 4242 (1978).

[23] J. Karn, S. Brenner, L. Barnett, and G. Cesarini, *Proc. Natl. Acad. Sci. U.S.A.* **77,** 5172 (1980).

[24] W. D. Benton and R. W. Davis, *Science* **196,** 180 (1977).

[25] T. Maniatis, E. F. Fritsch, and J. Sambrook, "Molecular Cloning: A Laboratory Manual." Cold Spring Harbor Lab., Cold Spring Harbor, New York, 1982.

[26] J. Vieira and J. Messing, *Gene* **19,** 259 (1982).

[27] J. Messing, this series, Vol. 101, p. 20.

[28] J. Messing and J. Vieira, *Gene* **19,** 269 (1982).

DNA Sequencing

DNA sequences were determined using the modified[13] dideoxynucleotide chain termination method.[12] Three strategies were employed at various stages.

Random Sequencing of Sonicated Fragments. Random sequencing of sonicated fragments was the method of choice for fragments greater than 1.5 kb in length.

Random Sequencing of Restriction Fragments. Random sequencing of restriction fragments was used only for small DNA fragments that proved difficult to self-ligate and sonicate. The two restriction enzymes *Rsa*I and *Alu*I were employed. They both recognize specific 4-bp sequences and cut, leaving a blunt end that can be cloned into the *Sma*I site of M13*mp*8. Use of each of these separately gave an overlapping set of restriction fragments that were used to construct a consensus sequence. Unfortunately, this method depends on the distribution of restriction sites and so is not entirely random. Any gaps in the sequence were determined by the use of directed methods.

Directed Sequencing Using Specific Sequencing Primers. Directed sequencing using primers was useful as a rapid method for sequencing small fragments of DNA as well as for completion of sequences generated by the random strategies. Fragments of DNA cloned in M13*mp*8 were subjected to successive rounds of DNA sequencing, the new sequence data produced at each step being used to design synthetic oligonucleotide primers 17 bases in length for the next round. These were made with the aid of an Applied Biosystems 380B oligonucleotide synthesizer. The limiting factor is the speed of production of synthetic oligonucleotides.

Sequencing Reactions and Electrophoresis

Hybridization of a synthetic 17-mer oligonucleotide,[29] complementary to a region of DNA adjacent to the cloning site, primes the template DNA for DNA synthesis by the Klenow fragment of DNA polymerase. Annealing of the template and all sequencing reactions were carried out in Falcon U-bottomed microtiter plates (Becton Dickinson, Oxnard, CA). These plates have wells arranged in an 8×12 matrix and can accomodate 24 sets of four sequencing reactions. It was convenient, however, to carry out 20 sets of four reactions, i.e., 20 clones, per plate, as the gels employed for analysis of the reaction products could hold only 40 tracks, the products of 10 clones, each. After each addition of reagents the plates were centrifuged in an IEC Centra 3 centrifuge in order to mix the solutions. All

[29] M. L. Duckworth, M. J. Gait, P. Goelet, G. F. Hong, M. Singh, and R. C. Titmas, *Nucleic Acids Res.* **9,** 1691 (1981).

samples and reagents were dispensed onto the sides of the wells of the plate using a Hamilton 100-μl repetitive dispenser which was set to deliver 2 μl; hence, for convenience, all volumes of reagents were adjusted to 2 μl. The method is described below for a single clone.

Primer Annealing. Template DNA (2 μl) was dispensed into each of four wells of the microtitre plate. A primer–TM mix was made up containing LMB2 sequencing primer (0.2 pmol/μl; 1 μl) TM buffer (1 μl; TM buffer: 50 m*M* Tris–HCl, pH 8.0, 25 m*M* $MgCl_2$) and sterile distilled water (6 μl). Portions, 2 μl, were added to each well. The plate was covered with Saranwrap and incubated in an oven at 55° for 45 min. It was then centrifuged briefly to recover the condensation formed on the sides of the wells during annealing.

Sequencing Reactions. The following solutions were used for the sequencing reactions:

Stock solutions	NTP mixes, volume dispensed (μl)			
	T	C	G	A
0.5 m*M* dTTP	25	500	500	500
0.5 m*M* dCTP	500	25	500	500
0.5 m*M* dGTP	500	500	25	500
10 m*M* ddTTP	50	—	—	—
10 m*M* ddCTP	—	8	—	—
10 m*M* ddGTP	—	—	16	—
10 m*M* ddATP	—	—	—	2
TE buffer	1000	1000	1000	1000

The pH of each of these solutions was adjusted to pH 7.8 with 1 *M* Tris.

To the first ("T") well, 2 μl "T" mix was added, to the second 2 μl "C" mix, to the third 2 μl "G" mix, and to the fourth, 2 μl "A" mix. An enzyme–radioactive label cocktail was made up containing the Klenow fragment of DNA polymerase (5 U/μl; 0.4 μl), [α-^{35}S]dATP (8 μCi/μl, 410 Ci/mmol; 0.5 μl), 0.1 *M* dithiothreitol (1.0 μl), and 10 m*M* Tris–HCl, pH 8.0 (6.1 μl). Portions, 2 μl, were dispensed into each of the four wells to start the sequencing reaction. This was allowed to proceed at room temperature for 20 min. A chase solution was made up containing 0.5 m*M* dTTP, 0.5 m*M* dCTP, 0.5 m*M* dGTP, and 0.5 m*M* dATP. Portions (2 μl) were added to each reaction and the plates again allowed to stand at room temperature for 20 min. Electrophoresis of the reaction mixtures was carried out according to Biggin *et al.*[13] as described in detail earlier.[30]

[30] V. L. J. Tybulewicz, G. Falk, and J. E. Walker, this series, Vol. 125, p. 230.

Long Runs. In order to extend short regions of the sequence to produce an overlap or to determine the sequence on both strands of the DNA, it was sometimes necessary to read more than 300 nucleotides from a single M13 clone. In this case the sequencing reactions were carried out as usual, but the electrophoresis was performed at 37 W for 20 min, then at 30–33 W for 7–8 hr. The gels measured 20 cm × 50 cm × 4 mm as before, and contained 6% acrylamide, 7 *M* urea, and 0.5× TBE buffer. They were run in 1× TBE buffer as usual. Under these conditions, DNA fragments of 250 nucleotides or less, corresponding to the known sequence, migrated off the bottom of the gel, whereas fragments up to 400–450 nucleotides from the cloning site were fractionated; thus the sequence up to this point could be deduced.

Compressions. The biggest problem in the interpretation of DNA sequencing gels is band compression in G–C-rich areas of the DNA. These are areas on the gel where the band spacing is compressed, such that bands coalesce and so may run together or even appear in inverted order. Compressions occur when two or more consecutive G–C residues are present with a complementary sequence within about 3–15 nucleotides. In the newly synthesized DNA chains these sequences may form hairpin loops which are stable under normal fractionation conditions when they occur at the 3′ end of the chains. This causes these chains to migrate as if they were shorter.

Unless multiple overlapping potential base-paired structures are present, the sequence of the compressed region can often be deduced from the sequence of the complementary strand. Although a compression would be evident on both DNA strands, it would occur on opposite sides of the center of the base pairing on the two DNA strands, enabling the sequence of one side of the complementary sequences to be determined successfully on one strand, and of the other side from the other strand. The gel distortion due to compressions can be very slight and may pass unnoticed if only one strand is sequenced. For this reason it is essential that all parts of the sequence be determined at least once on each DNA strand.

Severe compressions were resolved by substituting inosine triphosphate (dITP) for dGTP in the sequencing reactions so replacing G–C base pairs (containing three hydrogen bonds) with weaker I–C base pairs (only one hydrogen bond).[31] This usually abolished the compression entirely, but tended to produce other artifact bands, giving a high background with some clones. The NTP mixes were the same as before except that 0.5 m*M* dGTP was replaced with 2.0 m*M* dITP in all four mixes, and the ddGTP in the "G" mix was reduced from 16 μl to 2 μl.

[31] D. R. Mills and F. R. Kramer, *Proc. Natl. Acad. Sci. U.S.A.* **76,** 2232 (1979).

Sequencing with Specific Primers. Sequencing with specific synthetic primers was carried out in exactly the same way as described above. After synthesis, oligonucleotides were purified on reversed-phase C_{18} SepPak columns (Water Associates) according to the manufacturer's instructions. The concentration of each oligonucleotide primer was calculated by measurement of its absorbance at 260 nm. The primer was diluted in water to a concentration of 0.2 pmol/μl, and 1 μl of the solution/clone was used in the sequencing reactions.

Compilation and Analysis of Data

The random sequencing strategy depends on good computing facilities to compile the data. The programs described earlier in Vol. 30 of this series were run on a Digital Equipment Corporation VAX/VMS computer.

Results

The probes for the α and c subunits were found to hybridize with a 5.6-kb *Bam*HI fragment in *Synechococcus* 6301 DNA. The probe for the β subunit did not hybridize with this fragment but with a 7.0-kb fragment in the same digest, suggesting that the α and c genes were in one locus and the β gene in a second, separate locus (Fig. 2). This was confirmed by the isolation of λ phage from the *Synechococcus* 6301 genomic library. Two

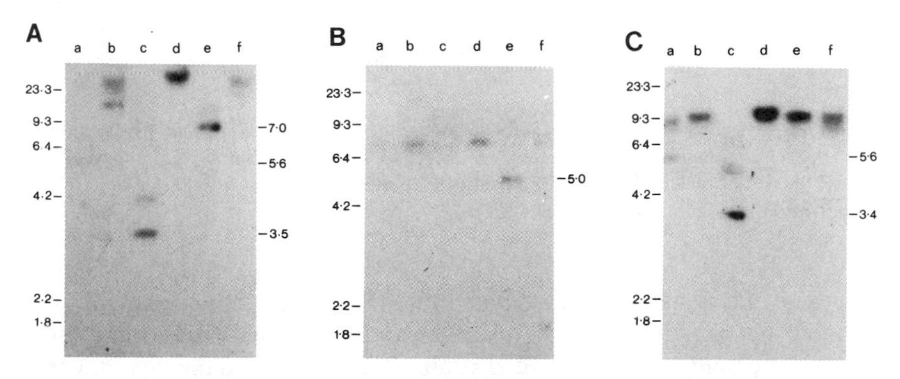

FIG. 2. Hybridization of *Synechococcus* 6301 genomic DNA with DNA probes containing genes for ATP synthase subunits. The probes employed corresponded to (1) nucleotides 7096–7302 of the gene for the α subunit in *Rps. blastica,* (2) nucleotides 9989–10,201 of the gene for the β subunit in the same organism,[7] and (3) a *Taq*I fragment including the entire gene for subunit c (III) from pea chloroplast.[11] The DNA was digested with (a) *Bam*HI, (b) *Sac*I, (c) *Pst*I, (d) *Hin*dIII, (e) *Eco*RI, or (f) *Xho*I. Marker and fragment sizes are in kilobases (kb).

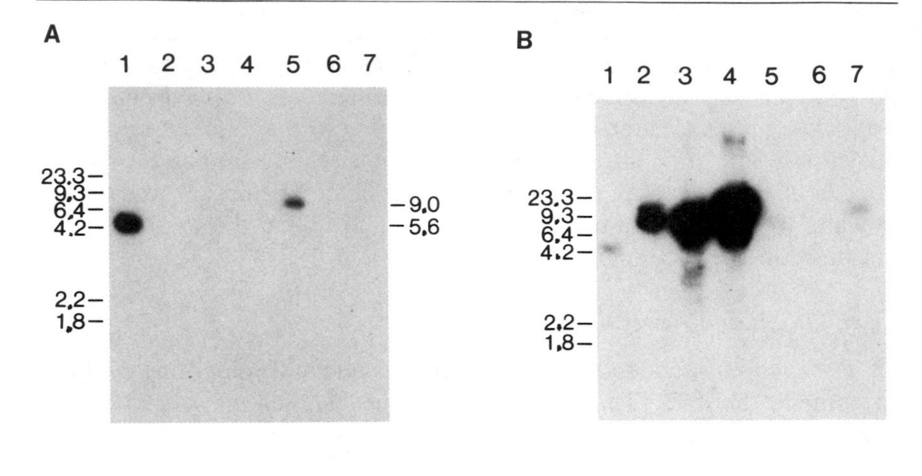

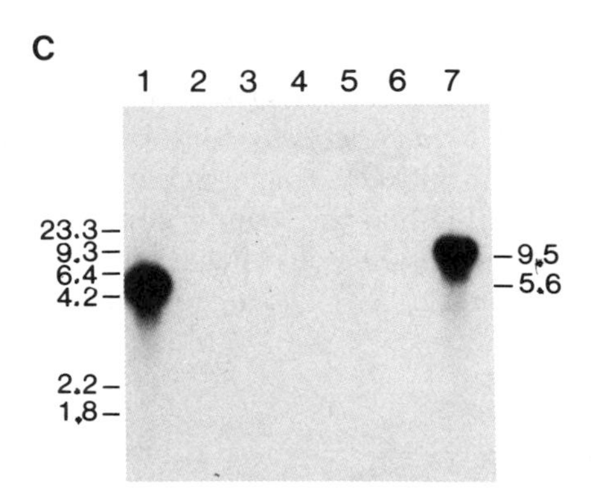

FIG. 3. Hybridization of recombinant λ phage containing *Synechococcus* 6301 genomic DNA to DNA probes containing genes encoding ATP synthase subunits. Columns 1–7 are *Bam*HI digests of phage λS1–λS7. The probes employed are the same as those used in the hybridization to genomic DNA digests (Fig. 2).

classes of the phage were isolated: those containing the 5.6-kb *Bam*HI fragment hybridizing to α and c probes and those containing the 8.0-kb *Bam*HI fragment hybridizing to β probe (Fig. 3). This indicates that the two loci must be separated by at least 10–15 kb of DNA.

The DNA sequences of the hybridizing regions and flanking sequences have been presented elsewhere.[32] These experiments have shown that the genes for the *Synechococcus* 6301 ATP synthase are indeed grouped in

[32] A. L. Cozens and J. E. Walker, *J. Mol. Biol.* **194,** 359–383 (1987).

two separate loci. One contains the genes for the subunits *a–c–b′–b–δ–α–γ* (*b′* is a duplicated and diverged form of *b*) and the second, the genes for the β and ε subunits. The gene for the *a* subunit is preceded by a gene for an analog of the *E. coli uncI* protein. The seven genes in this locus and the two genes, β–ε, in the second are probably operons for which the names *atp1* and *atp2* are proposed. The gene for the γ subunit is followed in the *Synechococcus* genome by a gene for a [2Fe : 2S] ferredoxin.[33]

NOTE ADDED IN PROOF. We have recently reported that the order of genes for the F_0 portion of ATP synthase in *Rsp. rubrum* is I-a-c-b′-b [G. Falk and J. E. Walker, *Biochem. J.* **254,** 109–122 (1988)]. In the cyanobacterium *Synechococcus* 6716, we found the same gene order as in strain 6301, except that the gene for the γ subunit is at the separate locus (H. S. van Walraven and J. E. Walker, unpublished results). In the thermophilic bacterium, PS3, the gene order for ATP synthase subunits is as in *E. coli* (Ohta *et al., Bichim. Biophys. Acta* **933,** 141–155).

[33] A. L. Cozens and J. E. Walker, *Biochem. J.,* **252,** 563–569 (1988).

[88] Tests on *nif* Probes and DNA Hybridizations

By CLAUDINE FRANCHE and THIERRY DAMERVAL

Introduction

The molecular genetics of the organization and regulation of *nif* (nitrogen fixation) genes in cyanobacteria has been initiated in a filamentous heterocystous cyanobacterium, *Anabaena* sp. PCC 7120.[1] Using DNA hybridization techniques with *Klebsiella pneumoniae nif* probes and DNA from vegetative cells, four genes involved in nitrogen fixation in *Anabaena* PCC 7120 have been identified[1]: *nifH*, *nifD*, and *nifK*, encoding the polypeptide components of the nitrogenase complex, and *nifS*, a gene involved in *Klebsiella pneumoniae* in the maturation of nitrogenase. Whereas in the enterobacteria the *nif* structural genes are adjacent and cotranscribed, *nifK* is separated from *nif*DH by 11 kilobases of DNA in *Anabaena* vegetative cells.[2] Using *Anabaena* PCC 7120 *nif* genes as hybridization probes, the *nif* structural gene arrangement has been studied in several other cyanobacteria including unicellular strains of Section I[3,4,5]

[1] B. J. Mazur, D. Rice, and R. Haselkorn, *Proc. Natl. Acad. Sci. U.S.A.* **77,** 186 (1980).
[2] D. Rice, B. J. Mazur, and R. Haselkorn, *J. Biol. Chem.* **257,** 13157 (1982).
[3] T. Kallas, M. C. Rebiere, R. Rippka, and N. Tandeau de Marsac, *J. Bacteriol.* **155,** 427 (1983).
[4] T. Kallas, T. Coursin, and R. Rippka, *Plant Mol. Biol.* **5,** 321 (1985).
[5] B. Saville, N. Straus, and J. R. Coleman, *Plant Physiol.* **85,** 26 (1987).

and filamentous strains of Sections III,[4-6] IV,[3-5,7,8] and V.[5] *nif*H, *nifD*, and *nifK* appear clustered in all nonheterocystous cyanobacteria examined, and in a strain of *Fischerella* sp. (Section V) whereas there is a separation of *nifK* from *nifD* and *nifH* in DNA of vegetative cells of all heterocystous cyanobacteria of Section IV.

DNA hybridization methods using *nif* probes are not only useful in cyanobacteria to isolate *nif* genes and to study the *nif* gene organization and structure, but can also provide information for the regulation of *nif* gene transcription *in vivo*.[9,10] *nif* probes can also be used (1) to demonstrate taxonomic relatedness among cyanobacteria[4] or to distinguish different strains among cyanobacteria which are phenotypically similar,[11,12] (2) to investigate the presence of *nif* genes in non-nitrogen-fixing cyanobacteria,[4,5] and (3) to study at a genetic level the process of differenciation of vegetative cell in heterocyst.[13]

In this chapter, the cyanobacterial *nif* probes available, the techniques of DNA extraction, and the procedures for DNA hybridization are presented. Limited details of standard procedures are given, and major emphasis is placed on modifications which have been developed with cyanobacteria.

nif Probes

The *Anabaena* PCC 7120 *nif* genes which can be used as probes are described in Table I. *NifH** is a second copy of *nifH*; the function and location of this gene in *Anabaena* PCC 7120 remain unknown.[14] The gene *xisA* is required in *Anabaena* PCC 7120 for excision of the 11-kb DNA region during heterocyst differentiation.[13,15]

Preparation of Probes

Plasmids used as a source of probes are purified by means of cesium chloride–ethidium bromide density gradient centrifugation.[16] DNA frag-

[6] S. R. Barnum and S. M. Gendel, *FEMS Microbiol. Lett.* **29,** 339 (1985).
[7] R. Hirschberg, S. M. Samson, K. E. Kimmel, K. A. Page, J. J. Collins, J. A. Myers, and L. R. Yarbrough, *J. Biotech.* **2,** 23 (1985).
[8] A. Herrero and C. P. Wolk, *J. Biol. Chem.* **261,** 7748 (1986).
[9] R. Haselkorn, D. Rice, S. E. Curtis, and S. J. Robinson, *Ann. Microbiol. (Paris)* **134,** 181 (1983).
[10] S. A. Nierzwicki-Bauer and R. Haselkorn, *EMBO J.* **5,** 29 (1986).
[11] C. Franche and G. Cohen-Bazire, *Plant Sci.* **39,** 125 (1985).
[12] C. Franche and G. Cohen-Bazire, *Symbiosis* **3,** 159 (1987).
[13] J. W. Golden, S. J. Robinson, and R. Haselkorn, *Nature (London)* **314,** 419 (1985).
[14] S. J. Robinson and R. Haselkorn, *Proc. Int. Symp. Photosynth. Prokaryotes, 5th,* p. 112.
[15] P. J. Lammers, J. W. Golden, and R. Haselkorn, *Cell* **44,** 905 (1986).
[16] T. Maniatis, E. F. Fritsch, and J. Sambrook, "Molecular Cloning: A Laboratory Manual." Cold Spring Harbor Lab., Cold Spring Harbor, New York, 1982.

TABLE I
Anabaena PCC 7120 *nif* GENES

Gene	Gene product	Ref.
nifH	Nitrogenase reductase	2, 25[a]
*nifH**	—	2, 10
nifD	α subunit of nitrogenase	2, 26[b]
nifK	β subunit of nitrogenase	2, 27[c]
nifS	Polypeptide involved in nitrogenase maturation	2
xisA	Excisase	45

[a] M. Mevarech, D. Rice, and R. Haselkorn, *Proc. Natl. Acad. Sci. U.S.A.* **77,** 6476 (1980).
[b] B. J. Mazur and C. F. Chui, *Proc. Natl. Acad. Sci. U.S.A.* **79,** 6782 (1982).
[c] P. J. Lammers and R. Haselkorn, *Proc. Natl. Acad. Sci. U.S.A.* **80,** 4723 (1983).

ments carrying the *nif* genes are separated from the vector following electrophoresis of the restricted plasmids in agarose gels (preferably low-melting-point agarose), and fragments are recovered by electroelution.[16] If the entire plasmids are used as hybridization probes, control experiments should be done to verify that there is no hybridization of the vector to the genomic DNAs.

Probe DNAs are labeled with [α-^{32}P]dCTP to specific activities of 10^8 cpm/μg DNA by nick-translation.[16] In experiments requiring a high specific activity (2×10^8 cpm/μg DNA), DNA fragments can be labeled with both [α-^{32}P]dCTP and [α-^{32}P]dATP. This modification was found necessary when *nifK* was used as a probe since the hybridization bands observed in the cyanobacterial DNA digests were weak.[4,11,12] Labeled DNA is separated from unincorporated nucleotides as described in Maniatis *et al.*[16] Before hybridization the probe is denatured either by heat treatment (10 min at 95°) or by adding NaOH to 0.1 *N* and incubating 5 min at 37°.

Preparation of Genomic DNA

Vegetative Cell DNA

Depending on the strain studied, on the size of DNA fragments required, and on the volume of cyanobacterial culture used, several methods of DNA extraction have been developed. These methods include modifications[1,3,8,17] of the Marmur technique,[18] the modified osmotic

[17] N. Tomioka, K. Shinosaki, and M. Sugiura, *Mol. Gen. Genet.* **184,** 359 (1981).
[18] J. Marmur, *J. Mol. Biol.* **3,** 208 (1961).

shock–lysozyme treatment[3,4] of Schwinghamer,[19] and procedures[20] based on the method of van den Hondel *et al.*[21]

We have found that the following procedure is successful with most of the unicellular and filamentous cyanobacteria we have tested: 5 g of cells is harvested from a culture in midexponential phase in BG-11 medium,[22] washed first in 20 ml of TE buffer,[16] and then washed in 20 ml of TE buffer containing 0.1% sarkosyl. The pellet is resuspended in 20 ml of sucrose buffer (25%, w/v, sucrose, 50 m*M* Tris–HCl, 1 m*M* EDTA, pH 8) and kept for 2 hr at 4°. The suspension is then treated with 5 mg/ml lysozyme at 37° for 1 hr in the presence of 100 m*M* EDTA (pH 8). Proteinase K (to 100 μg/ml) and sodium dodecyl sulfate (SDS) (to 2.5%) are added, and the suspension is incubated at 37° until clear (2 hr to overnight). The viscous lysate is extracted twice with phenol–chloroform (v/v) and once with chloroform–isoamyl alcohol (24 : 1, v/v). The aqueous phase is then dialyzed against TE buffer and treated with RNase A at a final concentration of 100 μg/ml for 3 hr at 37°. The lysate is then reextracted and dialyzed as previously described. A subsequent DNA purification by cesium chloride gradient centrifugation is sometimes necessary to allow complete DNA digestion by restriction endonucleases. When the yield of DNA extracted is not high enough to allow this procedure, the DNA solution is incubated for 10 min at 37° in the presence of 5 m*M* spermidine before digestion.

Heterocyst DNA

Filamentous heterocystous cyanobacteria are grown under nitrogen-fixing conditions in 10 liters of BG-11_0 medium.[22] Vegetative cells are eliminated as described in Section I. The extraction of heterocyst DNA is achieved following the procedure developed by Golden *et al.*[13]: purified heterocysts are resuspended in 50 m*M* Tris–HCl, 50 m*M* EDTA, pH 8, treated with 0.5% sarkosyl and 0.5% Triton X-100, and then vortexed in the presence of 1 volume of 0.45-mm glass beads. After several phenol extractions, heterocyst DNA is purified in cesium chloride–ethidium bromide gradient.

Restriction Enzyme Cleavage

Two micrograms of genomic DNA are digested with restriction endonucleases at 5 U/μg of DNA. DNA digests are routinely elec-

[19] E. A. Schwinghamer, *FEMS Microbiol. Lett.* **7,** 157 (1980).

[20] G. R. Lambert and N. G. Carr, *Biochim. Biophys. Acta* **781,** 45 (1984).

[21] C. A. van den Hondel, W. Keegstra, W. E. Borrias, and G. A. van Arkel, *Plasmid* **2,** 323 (1979).

[22] R. Rippka, J. Deruelles, J. B. Waterbury, M. Herdman, and R. Y. Stanier, *J. Gen. Microbiol.* **111,** 1 (1979).

trophoresed in horizontal 0.7% agarose gels for 16 hr at 40 V using Tris–borate buffer.[16] DNA fragments of small size (0.2–1 kb) are separated on 1.5% agarose gels to allow better resolution of the bands.[16] Molecular weights of DNA fragments are determined by their mobilities relative to *Hin*dIII–*Eco*RI fragments of λ DNA.

Preparation of Southern Blots

To ensure equal transfer efficiencies of large and small DNA molecules, gels are first treated with 0.2 *M* HCl for 15 min or exposed to a UV light treatment.[23] Gels treated sequentially in standard denaturing and neutralizing solutions[16] are then Southern blotted onto Millipore filters with 20× SSC.[16] After overnight transfer, filters are washed briefly in 2× SSC and dried under reduced pressure for 2 hr at 80°. Though the solid support of choice for most hybridization experiments is nitrocellulose, the use of nylon membrane[23] which is of high strength is recommended for experiments involving multiple use of blots.

DNA Blot Hybridization

Factors Affecting DNA Blot Hybridization

As reviewed by Meinkoth and Wahl,[23] the melting temperature (T_m) of nucleic acid hybrids obtained with probes greater than 50 nucleotides follows the equation

$$T_m = 81.5° + 16.6 \log(M) + 0.41(\%G + C) - 500/n - 0.61(\% \text{ formamide})$$

where M is the ionic strength of the hybridization solution and n the length of the shortest chain in the duplex. For hybrids longer than 150 bp, it has been established that the T_m decreases by 1° with every 1% of base pairs which are mismatched.[24] From these data, it appears that the hybridization stringency must be empirically determined by either adjusting the formamide concentration and/or changing the hybridization temperature. An alternative approach is to conduct the hybridization at low stringency and to wash the filters at increasing stringencies, analyzing the results after each wash.

Hybridization Procedure

Routinely, when the relatedness of the probe and the genomic DNA is unknown, hybridizations are done in the following conditions: either in

[23] J. Meinkoth and G. Wahl, *Anal. Biochem.* **138,** 267 (1984).

[24] T. I. Bonner, D. T. Brenner, B. R. Neufeld, and R. J. Britten, *J. Mol. Biol.* **81,** 123 (1973).

aqueous hybridization buffer at 50, 55, 60 or 65°, or at 42° in presence of formamide at a final concentration of 10, 30, or 50% (v/v). Hybridization in the presence of formamide is preferred because it solves the evaporation problem encountered with aqueous buffer at temperatures up to 60°. It should be noted, however, that the rate of hybridization is 2 times slower in the presence of 50% formamide than in an aqueous solution.[16]

Filters are prehybridized at the desired temperature and concentration of formamide in a 6× SSC–1× Denhardt's solution[16] containing 15 μg denatured herring sperm DNA solution per milliliter. The denatured probe is then added to the hybridization mix (routinely 1×10^6 cpm/filter) and hybridized to the filter for 20 hr. Filters are then washed 2 times for 10 min in 1× SSC, 0.1% SDS at room temperature and 2 times for 15 min in 0.1× SSC, 0.1% SDS at 42°. To reduce nonspecific hybridization and give optimal signal-to-noise ratios, an additional high-stringency wash in 0.1× SSC, 0.1% SDS at 50–65° for 5–20 min may be necessary. After the final wash, filters are air dried and autoradiographed. The rate of hybridization can be increased by the addition of dextran sulfate (final concentration of 10%) to the hybridization buffer. Nevertheless, it should be noted that the use of this polymer may lead to increased background.

Up to now, hybridizations of the *Anabaena* PCC 7120 *nifD* and *nifH* probes to DNA from all the nitrogen-fixing unicellular and filamentous cyanobacteria tested have been performed in stringent conditions (65° in aqueous buffer).[3,4,11,12] Fragments hybridizing to *nifK* probe[4] and *nifS* probe (unpublished observations) are observed under less stringent conditions (58–60° in aqueous buffer).

[89] Cyanobacterial Genetic Tools: Current Status

By JEAN HOUMARD and NICOLE TANDEAU DE MARSAC

In recent years, cyanobacteria have become increasingly popular, and the development of cyanobacterial genetics has arisen with the new possibilities offered by molecular genetics.[1] The following tables are a compilation of the information concerning the genetic tools available for various cyanobacteria.

The literature dealing with cyanobacterial taxonomy is somewhat confused by different names and numbers appended to the same strain. Table

[1] N. Tandeau de Marsac and J. Houmard, *in* "The Cyanobacteria" (P. Fay and C. Van Baalen, eds.), p. 251. Elsevier, Amsterdam, 1987.

I lists alternative strain designations and numbers in the different culture collections for the cyanobacteria mentioned in the other tables. Whenever possible, we have referred to specific strains by their number in the Pasteur Culture Collection (PCC).

Nearly all cyanobacteria harbor extrachromosomal DNA elements, but, unfortunately, no function has yet been shown to be encoded by any of these plasmids. Table II presents the plasmid content of nearly a hundred cyanobacterial strains in which both the number and the size of these plasmids vary greatly. On the other hand, it is important to avoid DNA restriction when introducing DNA into heterologous strains. Most of the strains examined so far do contain sequence-specific endonucleases. They are listed in Table III together with their specific recognition sequences.

Since none of the endogenous plasmids contain selectable genetic markers, a large number of cloning vectors has been developed for the strains which are able to receive DNA either by transformation or by conjugation. The available cyanobacterial cloning vectors are listed in Table IV.

Finally, Table V is a compilation of cyanobacterial genes which were cloned up to November 1, 1987. Since several multigene families have been recognized in different cyanobacteria, it appears that the nomenclature of cyanobacterial genes will rapidly become confusing. Consequently, we propose the following rules: Gene designations proposed for *Escherichia coli*[2] and/or *Bacillus subtilis*[3] must be used whenever possible. For functions specifically related to photosynthesis, gene designations employed for photosynthetic bacteria or plants will be used. The designations already used for bacterial genes must be avoided if gene products either have not been identified or are functionally different from the ones which were previously published. Genes involved in the formation of multimolecular complexes (structure, assembly, and/or regulation) or which are part of a given metabolic pathway can be designated by difference capital letters appended to a unique three-lowercase-letter root. The following are given as examples:

apcA to *apcZ:* genes related to allophycocyanin
atpA to *atpZ:* genes related to the ATPase complex
cpcA to *cpcZ:* genes related to phycocyanin
cpeA to *cpeZ:* genes related to phycoerythrin
metA to *metZ:* genes related to the methionine biosynthetic pathway
nifA to *nifZ:* genes involved in nitrogen fixation

[2] B. J. Bachmann, *Microbiol. Rev.* **47,** 180 (1983).
[3] P. J. Piggot and J. A. Hoch, *Microbiol. Rev.* **49,** 158 (1985).

psaA to *psaZ:* genes related to photosystem I
psbA to *psbZ:* genes related to photosystem II

Arabic numerals following a gene designation are used to identify the various copies within a multigene family (*psbA1, psbA2, psbA3, . . .*). Allele numbers of a given locus must be indicated by a hyphen preceding arabic numerals (*atpA-1, cpcB2-1, psbD1-3, . . .*).

Acknowledgments

We are deeply grateful to Prof. A. de Waard for compilation of the cyanobacterial restriction enzymes presented in Table III. Our thanks also go to colleagues who provided us with information in advance of publication and to Dr. A. Pugsley for critical reading of the manuscript.

TABLE I
ALTERNATE STRAIN DESIGNATIONS AND NUMBERS IN THE CULTURE COLLECTIONS[a]

Designation in Tables II to V	Alternate designation	Number in culture collection[b]				
		PCC	ATCC	CCAP	UTEX	Other
Section I						
Genus *Gloeobacter*						
Gloeobacter sp.	*Gloeobacter violaceus*	7421	29082			
Genus *Gloeothece*						
Gloethece sp.	*Gloeocapsa* sp.	6909	27152	1430/3	795	
Genus *Synechococcus*						
Synechococcus sp.	*Anacystis nidulans*, *Synechococcus leopoliensis*	6301	27144	1405/1	625 1550 0100	TX20
	Coccochloris peniocystis, *Cyanobium* sp.	6307			1548	
	Anacystis sp.	6311	27145		1549	
	Anacystis sp.	6312	27167			
	Synechococcus lividus	6715	27149			
	Synechococcus cedrorum	6908	27146		1191	
	Agmenellum quadruplicatum PR-6	7002	27264			
	Coccochloris elabens 17-AR	7003	27265			
	Synechococcus cedrorum, *Cyanobacterium stanieri*	7202	29140	1479/2a and /2b		
	Synechococcus sp.	7335	29403			
	Agmenellum quadruplicatum BG-1	73109	29404			
	Aphanothece halophytica, *Cyanothece halophytica*	7418	29534			
	Microcystis sp.	7425	29141			
	Synechococcus sp.	7502	29172			

(continued)

TABLE I (*continued*)

Designation in Tables II to V	Alternate designation	Number in culture collection[b]				
		PCC	ATCC	CCAP	UTEX	Other
	Anacystis nidulans R-2	7942				
	Anacystis nidulans 602	7943				
	Anacystis nidulans				2196	
Genus *Synechocystis*						
Synechocystis sp.	*Aphanocapsa* HA	6701	27170			
	Aphanocapsa sp.	6711	27175			
	Aphanocapsa sp.	6714	27178			
	Aphanocapsa sp.	6803	27184			
	Aphanocapsa sp.	6808	27170			
	Eucapsis sp.	6906	27266			
	Microcystis aeruginosa	7005	27153	1450/1		
Section II						
Genus *Chroococcidiopsis*						
Chroococcidiopsis sp.	*Chlorogloea* sp.	6712	27176	1411/2		
Section III						
Genus *Pseudanabaena*						
Pseudanabaena sp.	*Synechococcus* sp.	6802	27183			
	Synechococcus sp.	6901	27263			
	LPP	7409	29541			
LPP group						
Microcoleus sp.	*Microcoleus* sp.				2220	
Phormidium sp.	*Phormidium faveolarum*			1462/1	427	
Plectonema sp.	*Plectonema boryanum*	6306	27894	1463/1	581	M-9.2.6
					1542	
	Plectonema sp.	6402	27902		1541	
	Plectonema boryanum	73110	29407		594	M-9.2.1
	Phormidium luridum var. *olivacea*	7602		1462/2	426	
	Plectonema calothricoides				598	M-9.3.1

	Plectonema notatum, *Plectonema boryanum*				482	
	Plectonema boryanum				596	M-9.2.3
					597	M-9.2.4
					790	M-9.2.5
	Plectonema					FS180
	Plectonema boryanum, *Lyngbya* sp.			1446/2 and /3	487 and 488	
	Phormidium sp., *Plectonema boryanum*			1462/4	485	
Section IV						
Genus *Anabaena*						
Anabaena sp.	*Anabaena subcylindrica*	6309	29211	1403/4b	377	
	Anabaena sp., strain W	7108	29208			
	Anabaena cylindrica	7122	27899	1403/2a	629	
			29142			
	Anabaena doliolum	7936				
	Anabaena variabilis UW, *Anabaena flos-aquae*	7937	29413	1403/13a	1444	IUCC1444
	Anabaena cylindrica, *Anabaena inequalis*	7938	29414	1446/1C	381	
	Anabaena catenula			1403/1		
	Anabaena "*oscillarioides*"			1403/9		
	Anabaena oscillarioides			1403/11		
	Anabaena sp., strain H			1403/12		
	Anabaena flos-aquae			1403/13f		
	Anabaena inequalis			1446/1a		
	Anabaena sp., strain TAi					
Genus *Calothrix*						
Calothrix sp.	*Tolypothrix tenuis*	7101	27914	1482/3a and /3b		

(*continued*)

TABLE I (*continued*)

Designation in Tables II to V	Alternate designation	Number in culture collection[b]				
		PCC	ATCC	CCAP	UTEX	Other
	Nodularia sphaerocarpa	7103	27905	1466/1	583	
	Fremyella diplosiphon	7601		1429/1	481	
	Calothrix scopulorum			1410/5		
Genus *Cylindrospermum*						
Cylindrospermum sp.	*Cylindrospermum* "*licheniforme*"		29412		1828	
	Cylindrospermum licheniforme				2014	
Genus *Nostoc*						
Nostoc sp.	*Anabaena* sp.	6302	27897		1551	
	Anabaena spiroides	6310	27896		1552	
	Anabaena sp.	6411	27898		1597	
	Nostoc sp., strain B	6705	29131			
	Nostoc muscorum	6719	29105			UC142
	Anabaenopsis circularis	6720	27895	1402/1		
	Nostoc sp.	7107	29150			
	Anabaena variabilis, *Cylindrospermum* sp.	7118	27892			
	Nostoc muscorum UW, *Anabaena* sp.	7119	29151			
	Nostoc muscorum ISU, *Anabaena* sp.	7120	27893			

	Nostoc sp.	73102	29133			
	Nostoc sp., strain H	7413	29106			
	Nostoc sp.	7422	29132			
	Nostoc sp., strain C	7524	29411			
	Nostoc muscorum	7906		1453/12	486	
	Nostoc sp. MAC	8009				
	Nostoc linckia					
	Nostoc muscorum					M-131-G
	Nostoc ellipsosporum			1453/17		B1453-17
Section V						
Genus *Chlorogloeopsis*						
Chlorogloeopsis sp.	*Chlorogloeopsis fritschii*	6912		1411/16		
Genus *Fischerella*						
Fischerella sp.	*Mastigocladus* sp.	7115	27929			
	Fischerella musicola	73103	29114	1427/1	1301	
	Mastigocladus laminosus	7414	29161	1447/1		

[a] Differences in structure and development have led to the recognition of five large subgroups or sections among the cyanobacteria [R. Rippka, J. Deruelles, J. B. Waterbury, M. Herdman, and R. Y. Stanier, *J. Gen. Microbiol.* **111,** 1 (1979)].

[b] PCC, Pasteur Culture Collection, 75724 Paris Cedex 15, France [R. Rippka and G. Cohen-Bazire, *Ann. Inst. Pasteur/Microbiol.* **134B,** 21 (1983); G. Guglielmi and G. Cohen-Bazire, *Protistologica* **20,** 377 (1984)]; ATCC, American Type Culture Collection, Rockville, MD 20852; CCAP, Culture Collection of Algae and Protozoa, Ambleside LA22 OLP, UK; UTEX, Culture Collection of Algae at the University of Texas, Austin, TX 78712 [R. Starr, *J. Phycol. Suppl.* **14,** 47 (1978)].

TABLE II
CYANOBACTERIAL PLASMIDS

Organism	DNA base ratio (mol % GC)[a]	Number of plasmids	Size of plasmids[b] (kb)	Ref.[f]
Section I				
Synechococcus sp. PCC 6301	55.1	2	7.0, 55	1
		2	7.9 (pUH1), 48.8 (pUH2)	2
		2	8.5, 45.5	3
		2	8.0, 48.7	4
		2	7.9[c], 45.5	5, 6
Synechococcus sp. PCC 6307	67.7	2	7.9 (pUH3), 49.3 (pUH4)	2
Synechococcus sp. PCC 6311	54.8	2	8.0 (pDF3), 50	7, 8
		2	7.9, 49.3	9
Synechococcus sp. PCC 6312	50.2	1	23.7 (pUH9)	2
Synechococcus sp. PCC 6908	56.0	2	7.9 (pUH5), 48.4 (pUH6)	2
		2	7.6, 45.2	5
Synechococcus sp. PCC 7002	49.1	6	4.0, 5.0 (pAQEI), 9.7, 15.4, 30, 36.9, 112	10, 11
		5–7	5.2, 9.3, 15.8, 38.8, ≥74.6	6
		5	4.5, 15.4, 30, 36.9, 112	12
Synechococcus sp. PCC 7003	49.3	5	2.8, 4.3, 5.2, 49.3, ≥74.6	6
Synechococcus sp. PCC 7335	47.4	2	32, 35	12
Synechococcus sp. PCC 7336		3	5.4, 32, 70	12
Synechococcus sp. PCC 73109	49.0	4–5	8.8[c], 15.8, ≥74.6	6
Synechococcus sp. PCC 7425	48.6	2	8.1 (pUH7), 35.8 (pUH8)	2
Synechococcus sp. PCC 7942	55.0	2	7.9 (pUH24)[d], 49.3 (pUH25)	13
		2	8.0, 48.7	4
		2	8.0 (pANS), 48.7 (pANL)	14, 15
		3	7.9, 50.8, 1000–1500	12

Synechococcus sp. PCC 7943		2	7.9, 49.3	9
Synechococcus sp. UTEX 2196		2	8.0, 48.7	4
Synechocystis sp. PCC 6701	35.8	8	2.0–40	6
		8	5.0 (pSCY1), 7.5 (pSCY2), 13.5 (pSCY3), 15 (pSCY4) + 4 other bands	16
		10	4.1, 6.4, 8.4, 14.9, 18.1, 22.7 + 4 others 35–60	12
Synechocystis sp. PCC 6711	37.0	3	2.1, 16.8, 27	12
Synechocystis sp. PCC 6714	47.9	1	44.8	6
Synechocystis sp. PCC 6803	47.5	2	2.3 (pUG1), 5.2 (pUG2)	17
		4	2.5, 5.2[c], 50, 100	12, 18
		3	2.1 (pSS2), 4.5, 90	19
Synechocystis sp. PCC 6808	36.0	5	3.8, 22, 26, 50, 60	12
Synechocystis sp. PCC 7005	45.4	1	3.0 (pUH10)	2
Section III				
Oscillatoria limnetica		1	7.6	7
Phormidium sp. UTEX 427	45.7	3	1.3, 14.9, ~17.9	20
Plectonema sp. PCC 6306	45.8–48.8	1	14	1
		3	1.3, 14.9, ~17.9	20
Plectonema sp. PCC 6402	46.3–47.4	3	4.5, 14.9, ≥45.8	20
Plectonema sp. PCC 73110	45.4–49.9	3	1.3, 14.9, ~17.9	20
Plectonema sp. PCC 7602	46.0	2	<3.1, 13.4	1
		3	1.3, 14.9, ~17.9	20
Plectonema sp. FS180		3	1.3 (pMP1), 14.9 (pMP2), ~17.9 (pMP3)	20
Plectonema sp. UTEX 482, 485, 487, 488	45.4–46.0	3	1.3, 14.9, ~17.9	20
Plectonema sp. UTEX 596, 597, 790	45.6–46.1	3	1.3, 14.9, ~17.9	20
Plectonema sp. UTEX 598	45.4	2	14.9, ~17.9	20
Pseudanabaena sp. PCC 6802	45.9	4	2.5, 37, 40, 100	12
Pseudanabaena sp. PCC 7409	43.6	6	2.7, 4.2, 6.6, 10, 130, 180	12

(*continued*)

TABLE II (*continued*)

Organism	DNA base ratio (mol % GC)[a]	Number of plasmids	Size of plasmids[b] (kb)	Ref.[f]
Section IV				
Anabaena sp. PCC 7936		3	8.5, 38.8, 54	1
Anabaena sp. PCC 7937		2	37.3, 47.8	1
Anabaena sp. PCC 7938		5	20.9, 35.8, 60, 79, 110	1
Anabaena sp. 74S08		3	37.3, 44.8, >150	21
Anabaena sp. 74S12		1	>150	21
Anabaena sp. 74S18		3	40.3, 47.8, >150	21
Anabaena sp. 74S19		2	37.3, 44.8	21
Anabaena sp. 74S23		1	56.7	21
Anabaena sp. 74S24		4	3.0, 10.4, 26.9, 32.8	21
Anabaena sp. 74S25		1	47.8	21
Anabaena sp. 74S26		3	22.4, 35.8, 38.8	21
Anabaena sp. 77S15		3	37.3, 44.8, >150	21
Anabaena sp. 77S19		3	20.9, 37.3, 47.8	21
Anabaena sp. 79S01		2	37.3, 47.8	21
Anabaena sp. 79S02		1	22.4	21
Anabaena sp. 79S03		3	25.4, 29.9, 38.8	21
Calothrix sp. PCC 7601	42.7	4	35.8, 47.8, 60, 72	1
		>4	15.5 (pFdB), 15.9 (pFdB′), 18.5 (pFdA) + others	22
		>10	9.1, 15, 19.1, 90 + others ≥200 (A)[e]	12
		>4	11.9, 15, 16.7, 19.7 (B)[e]	12
Nostoc sp. PCC 6705	43.4	3	3.9 (pGL2), 20.9, 60	3, 23
Nostoc sp. PCC 7118	41–43	2	5.4 (pGL1), 20.9	3, 23
Nostoc sp. PCC 7120	42.5	4	4.9, 49.3, 63, 110	1
Nostoc sp. PCC 7524	39.0	3	6.1 (pDU1), 11.8 (pDU2), 37.3 (pDU3)	24, 25

Nostoc sp. PCC 8009	3	8.4 (pDC1), 30, 60	3, 23, 26
	6	7.9, 30, 38, 40, 225, ~400	12
Nostoc sp. 74S04	3	11.9, 22.4, >150	21
Nostoc sp. 74S06	3	3.7, 17.9, 25.4	21
Nostoc sp. 74S09	1	17.9	21
Nostoc sp. 74S51	4	3.0, 5.2, 28.4, 34.3	21
Nostoc sp. 74S53	2	53.7, >150	21
Nostoc sp. 74S54	1	22.4	21
Nostoc sp. 74S56	3	34.3, 41.8, 89.6	21
Nostoc sp. 74S60	4	2.2, 5.2, 17.9, 37.3	21
Nostoc sp. 74S66	1	16.4	21
Nostoc sp. 77S17	1	>150	21
Nostoc sp. 79S04	5	6.7, 12.7, 22.4, 28.4, 38.8	21
Nostoc sp. 79S05	4	5.2, 10.4, 19.4, 37.3	21

Strains that have been found not to contain plasmids are the following:

Cylindrospermum sp. UTEX 942 and UTEX 1828, *Calothrix* sp. PCC 7103 and PCC 7101, *Fischerella* sp. PCC 73103 and PCC 7115 [R. D. Simon, *J. Bacteriol.* **136,** 414 (1978)]; *Synechococcus* sp. PCC 6715 [C. A. M. J. J. van den Hondel, W. Keegstra, W. E. Borrias, and G. A. van Arkel, *Plasmid* **2,** 323 (1979)]; *Synechococcus* sp. PCC 6307 and PCC 7502, *Gloeobacter* sp. PCC 7421, *Gloeothece* sp. PCC 6909, *Chroococcidiopsis* sp. PCC 6712, *Microcystis aeruginosa* NRC-1, *Gloeocapsa alpicola* W, *Pseudanabaena* sp. PCC 6901, *Chlorogloea fritschii, Anabaena flos-aquae* (CCAP 1403/13d; 1403/13f; 1403/13g), and some natural isolates (4 strains) [R. H. Lau, C. Sapienza, and W. F. Doolittle, *Mol. Gen. Genet.* **178,** 203 (1980)]; *Gloeocapsa alpicola* (A. Smith), *Anabaena variabilis* (CCAP 1403/12), *Anabaena flos-aquae* (UTEX 1555), *Nostoc* sp. PCC 73102 and PCC 7422 [G. R. Lambert and N. G. Carr, *Arch. Microbiol.* **133,** 122 (1982)]; *Chroococcus* S24 and N41 [M. Potts, R. Ocampo-Friedman, M. A. Bowman, and B. Tözun, *Arch. Microbiol.* **135,** 81 (1983)].

[a] Base composition expressed in mol % G + C, taken from R. Rippka, J. Deruelles, J. B. Waterbury, M. Herdman, and R. Y. Stanier, *J. Gen. Microbiol.* **111,** 1 (1979); R. Y. Stanier, R. Kunisawa, M. Mandel, and G. Cohen-Bazire, *Bacteriol. Rev.* **35,** 171 (1971); W. T. Stam, *Arch Hydrobiol./Suppl.* **56** (*Algol-Stud.* **25**), 351 (1980); M. A. Lachance, *Int. J. Syst. Bacteriol.* **31,** 139 (1981); and A. M. R. Wilmotte and W. T. Stam, *J. Gen. Microbiol.* **130,** 2737 (1984).

(continued)

Footnotes to TABLE II (*continued*)

[b] Published plasmid designations are indicated in parentheses. Plasmid sizes were estimated either by gel electrophoresis and/or from contour length electron micrographs. Values published in megadaltons were converted to kilobases by dividing by the factor 0.67.

[c] Plasmids which can be lost spontaneously in clonally derived strains.

[d] The complete nucleotide sequence of pUH24 has been determined [P. J. Weisbeek, R. Teerstra, M. van Dijk, G. Bloemheuvel, D. de Boer, J. van der Plas, W. E. Borrias, and G. A. van Arkel, *Abstr. Int. Symp. Photosynth. Prokaryotes, 5th,* p. 328 (1985)].

[e] For explanations concerning the patterns A and B see M.-C. Rebière, A.-M. Castets, J. Houmard, and N. Tandeau de Marsac, *FEMS Microbiol. Lett.* **37,** 269 (1986).

[f] References to Table II: (1) R. D. Simon, *J. Bacteriol.* **136,** 414 (1978); (2) C. A. M. J. J. van den Hondel, W. Keegstra, W. E. Borrias, and G. A. van Arkel, *Plasmid* **2,** 323 (1979); (3) G. R. Lambert and N. G. Carr, *Arch. Microbiol.* **133,** 122 (1982); (4) K. S. Engwall and S. M. Gendel, *FEMS Microbiol. Lett.* **26,** 337 (1985); (5) R. H. Lau and W. F. Doolittle, *J. Bacteriol.* **137,** 648 (1979); (6) R. H. Lau, C. Sapienza, and W. F. Doolittle, *Mol. Gen. Genet.* **178,** 203 (1980); (7) D. Friedberg and J. Seijffers, *FEBS Lett.* **107,** 165 (1979); (8) D. Friedberg and J. Seijffers, *Gene* **22,** 267 (1983); (9) C. A. M. J. J. van den Hondel and G. A. van Arkel, *Antonie van Leeuwenhoek* **46,** 228 (1980); (10) T. M. Roberts and K. E. Koths, *Cell* **9,** 551 (1976); (11) J. S. Buzby, R. D. Porter, and S. E. Stevens, Jr., *J. Bacteriol.* **154,** 1446 (1983); (12) M.-C. Rebière, A.-M. Castets, J. Houmard, and N. Tandeau de Marsac, *FEMS Microbiol. Lett.* **37,** 269 (1986); (13) C. A. M. J. J. van den Hondel, S. Verbeek, A. van der Ende, P. J. Weisbeek, W. E. Borrias, and G. A. van Arkel, *Proc. Natl. Acad. Sci. U.S.A.* **77,** 1570 (1980); (14) S. Gendel, N. Straus, D. Pulleyblank, and J. Williams, *J. Bacteriol.* **156,** 148 (1983); (15) D. E. Laudenbach, N. A. Straus, S. Gendel, and J. P. Williams, *Mol. Gen. Genet.* **192,** 402 (1983); (16) L. K. Anderson and F. A. Eiserling, *FEMS Microbiol. Lett.* **29,** 193 (1985); (17) F. Chauvat, L. de Vries, A. van der Ende, and G. A. van Arkel, *Mol. Gen. Genet.* **204,** 185 (1986); (18) A.-M. Castets, J. Houmard, and N. Tandeau de Marsac, *FEMS Microbiol. Lett.* **37,** 277 (1986); (19) S. Shestakov, I. Elanskaya, and M. Bibikova, *Abstr. Int. Symp. Photosynth. Prokaryotes, 5th,* p. 109 (1985); (20) M. Potts, *FEMS Microbiol. Lett.* **24,** 351 (1984); (21) C. Franche and P. A. Reynaud, *Ann. Inst. Pasteur/Microbiol.* **137A,** 179 (1986); (22) L. Bogorad, S. M. Gendel, J. H. Haury, and K.-P. Koller, *in* "Photosynthetic Prokaryotes: Cell Differentiation and Function" (G. C. Papageorgiou and L. Packer, eds.), p. 119. Elsevier, New York, 1983; (23) G. R. Lambert, J. G. Scott, and N. G. Carr, *FEMS Microbiol. Lett.* **21,** 225 (1984); (24) J. Reaston, C. A. M. J. J. van den Hondel, A. van der Ende, G. A. van Arkel, W. D. P. Stewart, and M. Herdman, *FEMS Microbiol. Lett.* **9,** 185 (1980); (25) J. Reaston, C. A. M. J. J. van den Hondel, G. A. van Arkel, and W. D. P. Stewart, *Plasmid* **7,** 101 (1982); (26) G. Lambert and N. Carr, *Plasmid* **10,** 196 (1983).

TABLE III
SEQUENCE-SPECIFIC ENDONUCLEASES FROM CYANOBACTERIA

Organism	Endo-nuclease	Isoschizomer	Recognition sequence[a]	Ref.[b]
Section I				
Gloeothece sp. PCC 6909	*Gsp*I	*Pvu*II	CAGCTG	1
Synechococcus sp. PCC 7002	*Aqu*I	*Ava*I	C ↓ PyCGPuG	2
Synechococcus sp. PCC 7003	*Cel*I	*Bam*HI	GGATCC	1
	*Cel*II	*Esp*I	GCTNAGC	
Synechococcus sp. PCC 7202	*Sce*I	*Fnu*DII	CGCG	1
Synechococcus sp. PCC 7418	*Aha*I	*Cau*II	CC$^{C}_{G}$GG	3
	*Aha*II	*Acy*I	GPu ↓ CGPyC	
	*Aha*III	*Dra*I	TTT ↓ AAA	4
Synechococcus sp. PCC 7942	*Ani*I		n.d.	5
Synechocystis sp. PCC 6701	*Sec*I		C ↓ CNNGG	6
	*Sec*II	*Msp*I	CCGG	
	*Sec*III	*Mst*II	CCTNAGG	
Synechocystis sp. PCC 6711	*Sci*I	*Bst*EII	GGTNACC	1
	*Sci*II	*Pvu*II	CAGCTG	
Synechocystis sp. PCC 6906	*Esp*I		GC ↓ TNAGC	7
	*Esp*II		n.d.	
Section III				
Microcoleus sp. UTEX 2220	*Mst*I		TGCGCA	8
	*Mst*II	*Sau*I	CC ↓ TNAGG	9
Pseudanabaena sp. PCC 6901	*Psp*I	*Asu*I	GGNCC	10
Pseudanabaena sp. PCC 7409	*Pse*I	*Asu*I	GGNCC	1
Section IV				
Anabaena sp. PCC 6309	*Asu*I		G ↓ GNCC	11
	*Asu*II		TT ↓ CGAA	12
	*Asu*III	*Acy*I	GPuCGPyC	
Anabaena sp. PCC 7108	*Ast*WI	*Acy*I	GPuCGPyC	12, 13
Anabaena sp. PCC 7122	*Acy*I		GPu ↓ CGPyC	14
	*Acy*II		n.d.	
Anabaena sp. PCC 7937	*Avr*I	*Ava*I	C ↓ PyCGPuG	13, 15, 16
	*Avr*II		CCTAGG	16
Anabaena sp. CCAP 1403/1	*Aca*I	*Asu*II	TTCGAA	1
	*Aca*II	*Bam*HI	GGATCC	
	*Aca*III	*Mst*I	TGCGCA	
	*Aca*IV	*Hae*III	GGCC	
Anabaena sp. CCAP 1403/9	*Aoc*I	*Mst*II	CC ↓ TNAGG	1
	*Aoc*II	*Sdu*I	G C GAGCTC T A	
Anabaena sp. CCAP 1403/11	*Aos*I	*Mst*I	TGC ↓ GCA	17
	*Aos*II	*Acy*I	GPu ↓ CGPyC	
	*Aos*III	*Sac*II	CCGCGG	18
Anabaena sp. CCAP 1403/12	*Asp*HI	*Asu*II	TTCGAA	18
	*Asp*HII	*Mst*I	TGCGCA	

(*continued*)

TABLE III (*continued*)

Organism	Endo-nuclease	Isoschizomer	Recognition sequence[a]	Ref.[b]
Anabaena sp. CCAP 1403/13f	*Afl*I	*Ava*II	$G\downarrow G^{A}_{T}CC$	19
	*Afl*II		C ↓ TTAAG	
	*Afl*III		A ↓ CPuPyGT	
Anabaena sp. CCAP 1446/1a	*Ain*I	*Pst*I	CTGCAG	1
	*Ain*II	*Bam*HI	GGATCC	
Anabaena sp. TAi	*Asp*TAiI	*Pst*I	CTGCAG	1
	*Asp*TAiII	*Bam*HI	GGATCC	
	*Asp*TAiIII	*Hae*III	GGCC	
Calothrix sp. PCC 7101	*Tte*I	*Hae*III	GGCC	20
Calothrix sp. PCC 7601	*Fdi*I	*Ava*II	$G\downarrow G^{A}_{T}CC$	21
	*Fdi*II	*Mst*I	TGC ↓ GCA	
Calothrix sp. CCAP 1410/5	*Csc*I	*Sac*II	CCGCGG	22
Cylindrospermum sp. UTEX 1828	*Clc*I	*Pst*I	CTGCAG	1
	*Clc*II	*Mst*I	TGCGCA	
Cylindrospermum sp. UTEX 2014	*Cli*I	*Ava*II	$GG^{A}_{T}CC$	23
	*Cli*II	*Mst*I	TGCGCA	13
	*Cli*III		n.d.	
Nostoc sp. PCC 6302	*Nsp*(6302)I	*Mst*I	TGCGCA	1, 13
Nostoc sp. PCC 6310	*Nsp*(6310)I	*Ava*I	CPyCGPuG	13, 24
	Nsp(6310)II		n.d.	
Nostoc sp. PCC 6411	*Asp*(6411)I	*Ava*I	CPyCGPuG	15
	Asp(6411)II	*Ava*II	$GG^{A}_{T}CC$	
Nostoc sp. PCC 6705	*Nsp*BI	*Asu*II	TTCGAA	15
	*Nsp*BII		$C^{A}_{C}G\downarrow C^{T}_{G}G$	15
Nostoc sp. PCC 6719	*Nsp*(6719)I	*Ava*I	CPyCGPuG	15
	Nsp(6719)II	*Ava*II	$GG^{A}_{T}CC$	
Nostoc sp. PCC 6720	*Acr*I	*Ava*I	CPyCGPuG	1
	*Acr*II	*Bst*EII	GGTNACC	
Nostoc sp. PCC 7118	*Ava*I		C ↓ PyCGPuG	25
	*Ava*II		$G\downarrow G^{A}_{T}CC$	25, 26
	*Ava*III		ATGCAT	27
Nostoc sp. PCC 7119	*Asp*(7119)I	*Ava*I	CPyCGPuG	15
	Asp(7119)II	*Ava*II	$GG^{A}_{T}CC$	
Nostoc sp. PCC 7120	*Asp*(7120)I	*Ava*I	CPyCGPuG	15
	Asp(7120)II	*Ava*II	$GG^{A}_{T}CC$	
Nostoc sp. PCC 7413	*Nsp*HI	*Nsp*CI	PuCATG ↓ Py	15
	*Nsp*HII	*Ava*II	$GG^{A}_{T}CC$	
Nostoc sp. PCC 7524	*Nsp*CI	*Nsp*HI	PuCATG ↓ Py	28
	*Nsp*CII	*Sdu*I	$G\begin{smallmatrix}G\\A\\T\end{smallmatrix}GC\begin{smallmatrix}C\\T\\A\end{smallmatrix}\downarrow C$	

(*continued*)

TABLE III (*continued*)

Organism	Endo-nuclease	Isoschizomer	Recognition sequence[a]	Ref.[b]
	*Nsp*CIII	*Ava*I	CPyCGPuG	
	*Nsp*CIV	*Asu*I	GGNCC	
	*Nsp*CV	*Asu*II	TTCGAA	
Nostoc sp. PCC 8009	*Nsp*MACI	*Bgl*II	A ↓ GATCT	29
Nostoc linckia	*Nli*I	*Ava*I	CPyCGPuG	15
	*Nli*II	*Ava*II	$\mathrm{GG}^{\mathrm{A}}_{\mathrm{T}}\mathrm{CC}$	
Nostoc sp. M-131-G	*Nmu*I	*Ava*I	CPyCGPuG	15
	*Nmu*II	*Ava*II	$\mathrm{GG}^{\mathrm{A}}_{\mathrm{T}}\mathrm{CC}$	
Nostoc sp. 19-6C-C	*Nsp*19-6C-CI	*Ava*II	$\mathrm{GG}^{\mathrm{A}}_{\mathrm{T}}\mathrm{CC}$	13, 23
Nostoc sp. 23-9B	*Nsp*23-9BI	*Ava*II	$\mathrm{GG}^{\mathrm{A}}_{\mathrm{T}}\mathrm{CC}$	1, 13
Nostoc sp. 78-12B	*Nsp*78-12BI	*Asu*I	GGNCC	1, 13
Nostoc sp. UM-3	*Nsp*UM-3I	*Mst*I	TGCGCA	13, 23
	*Nsp*UM-3II	*Asu*I	GGNCC	
	*Nsp*UM-3III		n.d.	
	*Nsp*UM-3IV		n.d.	
Section V				
Fischerella sp. PCC 73103	*Fsp*I	*Mst*I	TGCGCA	30
	*Fsp*II	*Asu*II	TTCGAA	
Fischerella sp. PCC 7414	*Mla*I	*Asu*II	TT ↓ CGAA	31

[a] Arrows indicate the cleavage site in the recognition sequence. n.d., Not determined.

[b] References to Table III:

(1) F. Calléja and A. de Waard, personal communication; (2) R. H. Lau and W. F. Doolittle, *FEBS Lett.* **121,** 200 (1980); (3) P. R. Whitehead and N. L. Brown, *Arch. Microbiol.* **141,** 70 (1985); (4) P. R. Whitehead and N. L. Brown, *FEBS Lett.* **143,** 296 (1982); (5) M. L. Gallagher and W. F. Burke, Jr., *FEMS Microbiol. Lett.* **26,** 317 (1985); (6) F. Calléja, N. Tandeau de Marsac, T. Coursin, H. van Ormondt, and A. de Waard, *Nucleic Acids Res.* **13,** 6745 (1985); (7) F. Calléja, B. M. M. Dekker, T. Coursin, and A. de Waard, *FEBS Lett.* **178,** 69 (1984); (8) T. R. Gingeras, J. P. Milazzo, and R. J. Roberts, *Nucleic Acids Res.* **5,** 4105 (1978); (9) I. Schildkraut, unpublished observations; (10) B. Mulligan and M. Szekeres, unpublished observations; (11) S. G. Hughes, T. Bruce, and K. Murray, *Biochem. J.* **185,** 59 (1980); (12) A. de Waard and M. Duyvesteyn, *Arch. Microbiol.* **128,** 242 (1980); (13) C. P. Wolk and J. Kraus, *Arch. Microbiol.* **131,** 302 (1982); (14) A. de Waard, J. Korsuize, C. P. van Beveren, and J. Maat, *FEBS Lett.* **96,** 106 (1978); (15) M. G. C. Duyvesteyn, J. Korsuize, A. de Waard, A. Vonshak, and C. P. Wolk, *Arch. Microbiol.* **134,** 276 (1983); (16) E. C. Rosenvold and A. Honigman, *Gene* **2,** 273 (1977); (17) A. de Waard, C. P. van Beveren, M. Duyvesteyn, and H. van Ormondt, *FEBS Lett.* **101,** 71 (1979); (18) A. de Waard, personal communication; (19) P. R. Whitehead and N. L. Brown, *J. Gen. Microbiol.* **131,** 951 (1985); (20) B. Siegelman, unpublished observations; (21) C. A. M. J. J. van den Hondel, R. W. van Leen, G. A. van Arkel, M. Duyvesteyn, and A. de Waard, *FEMS Microbiol. Lett.* **16,** 7 (1983); (22) M. G. C. Duyvesteyn, J. Korsuize, and A. de Waard, *Plant Mol. Biol.* **1,** 75 (1981); (23) C. Karreman and A. de Waard, personal communication; (24) E. Flores and C. P. Wolk, personal communication; (25) S. G. Hughes and K. Murray, *Biochem. J.* **185,** 65 (1980); (26) J. G. Sutcliffe and G. M. Church, *Nucleic Acids Res.* **5,** 2313(1978); (27) G. Roizes, P.-C. Nardeux, and R. Monier, *FEBS Lett.* **104,** 39 (1979); (28) J. Reaston, M. G. C. Duyvesteyn, and A. de Waard, *Gene* **20,** 103 (1982); (29) R. H. Lau, L. P. Visentin, S. M. Martin, J. D. Hofman, and W. F. Doolittle, *FEBS Lett.* **179,** 129 (1985); (30) M. Szekeres (New England Biolabs), personal communication (1985); (31) M. Duyvesteyn and A. de Waard, *FEBS Lett.* **111,** 423 (1980).

TABLE IV
CYANOBACTERIAL CLONING VECTORS

Organism	Cyanobacterial plasmid origin		*E. coli* plasmid origin		Shuttle vector or hybrid plasmid				
	Designation	Size (kb)	Designation	Size (kb)	Designation	Size (kb)	Selectable markers	Unique restriction sites[c]	Ref.[e]
Section I									
Synechococcus sp. PCC 6301	pBAI	7.8	pBR322	4.36	pBAS18[a]	12.1	Ap^R	*Eco*RI, *Xho*I	1
Synechococcus sp. PCC 6311	pDF3	8.0	pBR325	5.7	pDF30	14.0	Ap^R, Cm^R	*Xho*I, *Sal*I, *Eco*RI*	2
Synechococcus sp. PCC 7002	pAQ1	4.6	pBR322	4.36	pAQE1 and 2[b]	8.9	Ap^R	*Pvu*II, *Ava*I, *Sal*I, *Bam*HI, *Eco*RI	3
					pAQE5 (*Ava*I site deletion of pAQE2)	8.3	Ap^R	*Pvu*II, *Sal*I, *Bam*HI, *Eco*RI	3
					pAQE12 (pAQE2 deleted of 3 kb)	5.9	Ap^R	*Hin*dIII, *Eco*RI, *Hinc*II*	4
					pAQE15 (pAQE12 derivative containing the polylinker MCS7 from pUC7)	5.9	Ap^R	*Hin*dIII, *Eco*RI, *Bam*HI, *Sal*I, *Acc*I	4
					pAQE17 (pAQE15 derivative containing the Km/Nm^R gene from pRZ102)	7.3	Ap^R, Km/Nm^R	*Hin*dIII, *Eco*RI, *Bam*HI, *Sal*I, *Acc*I, *Ava*I, *Sma*I, *Bgl*II	4
					pAQE17L (pAQE17 derivative containing the *lac* operon from pBRP2)	18.3	Ap^R, Km/Nm^R, Lac^+	*Hin*dIII, *Bam*HI	4
					pAQE17DL (pAQE17L deleted of 3 kb)	15.3	Ap^R, Km/Nm^R, Lac^+	*Hin*dIII, *Bgl*II, *Sal*I	4

	pAQI	4.6	pBR325	5.7	pAQE9 and 10[b]	10.3	Ap^R, Cm^R	*Ava*I, *Sal*I, *Bam*HI	3
					pAQE11 (*Ava*I site deletion of pAQE10)	9.4	Ap^R, Cm^R	*Sal*I, *Bam*HI, *Pvu*II*	3
Synechococcus sp. PCC 7942	pUC1	8.2	pACYC184	4.0	pUC104	12.2	Ap^R, Cm^R	*Eco*RI*, *Xho*I, *Sal*I	5
					pUC105	12.2	Ap^R, Cm^R	*Eco*RI*, *Xho*I, *Sal*I, *Bgl*II, *Bam*HI	5
					pPUC29 (pUC104 + a 1.7 kb *Bgl*II fragment from pHC79 housing the λ *cos* sequence)	14.2	Ap^R, Cm^R	*Eco*RI*, *Xho*I, *Sal*I, *Xba*I, *Bst*EII	6, 7
	pUC13	6.9	pACYC184	4.0	pUC303	11.0	Cm^R, Sm^R	*Eco*RI*, *Xho*I, *Sal*I	8
	pUH24	8.0	pBR322::Cm	—	pLS103	10.1	Ap^R	*Hin*dIII, *Sal*I	9
	pUH24	8.0	pBR328	4.9	pSG111	12.9	Ap^R, Cm^R	*Eco*RI*, *Xho*I, *Sal*I, *Sph*I	10
	pANS (= pUH24)	8.0	pBR325	5.7	pECAN1	13.5	Ap^R, Cm^R	*Eco*RI*, *Sal*I	11
	pANS	8.0	pDPL13	2.5	pPLANB1 and 2[b]	10.2	Ap^R	*Xba*I, *Sma*I, *Nar*I, *Stu*I, *Eco*RI, *Sac*I, *Sal*I	11, 12
					pCB4 (pPLANB2 deleted of a 4 kb *Xho*I fragment)	6.7	Ap^R	*Bam*HI, *Sma*I, *Xba*I, *Nar*I, *Xho*I	11
	pANS (*Bam*HI–*Xho*I fragment)	4.65	pDPL13	2.5	pXB7	7.2	Ap^R	*Bam*HI, *Eco*RI, *Mst*I, *Nae*I, *Nar*I, *Nru*I, *Sac*I, *Sma*I, *Stu*I, *Xba*I	13
	pANS (*Bam*HI–*Xho*I fragment)	4.65	pUC8	2.65	pECAN8	7.3	Ap^R	*Eco*RI, *Sma*I, *Bam*HI, *Sal*I	13
	pANS (*Bam*HI–*Xho*I fragment)	4.65	pDPLK	3.15	pKBX	7.8	Km^R	*Bam*HI, *Eco*RI, *Mst*I, *Nae*I, *Nar*I, *Nru*I, *Sac*I, *Sma*I, *Stu*I, *Xba*I, *Ava*I	13

(*continued*)

TABLE IV (*continued*)

Organism	Cyanobacterial plasmid origin		*E. coli* plasmid origin		Shuttle vector or hybrid plasmid		Selectable markers	Unique restriction sites[c]	Ref.[e]
	Designation	Size (kb)	Designation	Size (kb)	Designation	Size (kb)			
	pUH24	8.0	pBR325	5.7	pSG111M	14.0	Ap^R, Cm^R	*Eco*RI, *Sal*I, *Sph*I, *Xho*I	14
Synechocystis sp. PCC 6803	pUG1 (*Hpa*I fragment)	1.7	pACYC177	3.7	pUF3	5.7	Km^R	*Xho*I*, *Cla*I*, *Sma*I*, *Bam*HI, *Nco*I, *Xba*I, *Pst*I, *Mst*I, *Bgl*I, *Acc*I	15
					pUF311 (result of *in vivo* recombination between pUF3 and pUG1)	8.0	Km^R	*Xho*I*, *Cla*I*, *Sma*I*, *Bam*HI, *Pst*I, *Mst*I, *Bgl*I, *Acc*I	15
					pFCLV7 (pUF311 derivative containing the Cm^R gene from pACYC184)	9.35	Km^R, Cm^R	*Xho*I*, *Cla*I*, *Sma*I*, *Eco*RI*, *Pvu*II*, *Bam*HI, *Pst*I, *Mst*I, *Bgl*I, *Acc*I	15
					pUF12	5.75	Km^R	*Xho*I*, *Cla*I*, *Sma*I*, *Bam*HI, *Nco*I, *Xba*I, *Pst*I, *Mst*I, *Bgl*I, *Acc*I	15
	pSS2 (*Sau*3A fragment)	1.9	pACYC184	4.0	pSE7	5.9	Cm^R	Not mentioned	16
					pSE76 (result of *in vivo* recombination between pSE7 and pSS2)	8.2	Cm^R	*Sal*I*	16

Section III									
Plectonema sp. PCC 6306	pGL3	1.4	pBR328	4.9	pGL5[a]	6.3	Ap^R, Cm^R	*Pst*I*, *Pvu*II*, *Eco*RI*, *Hpa*I, *Hin*dIII, *Bam*HI, *Sph*I, *Sal*I, *Ava*I	17
Section IV									
Calothrix sp. PCC 7601	Endogenous plasmid (fragment)	8.0	pJCF22	4.1	pJCF31	12.1	Cm^R	*Asu*II, *Mst*II, *Nco*I*, *Sma*I*, *Xho*I*	18, 19
	pJCF31 (fragment)	8.0	pJCF22	4.1	pJCF62	12.1	Cm^R, Km/Nm^R	*Mst*II, *Cla*I*, *Nco*I*, *Sma*I*, *Xho*I*	19, 20
Nostoc sp. PCC 6310	pDU1	6.3	pBR322	4.36	pRL5[d]	11.1	Cm^R, Sm^R	*Asu*II, *Cla*I, *Nde*I, *Nar*I$^+$, *Ava*I*, *Sst*I*	14, 21, 22
					pRL6[d]	11.3	Cm^R, Km/Nm^R	*Cla*I, *Nde*I, *Ava*II*, *Bgl*II*, *Bss*HII*, *Pst*I*, *Sph*I*	14, 21, 22
Nostoc sp. PCC 6705	pGL2	3.9	pBR328	4.9	pGL4[a]	8.3	Ap^R, Cm^R	*Pst*I, *Pvu*II, *Eco*RI, *Hpa*I, *Xba*I, *Sph*I, *Sal*I, *Ava*I, *Bam*HI	17
Nostoc sp. PCC 7107	pDU1	6.3	pBR322	4.36	pRL5[d]	11.1	Cm^R, Sm^R	*Asu*II, *Cla*I, *Nde*I, *Nar*I$^+$, *Ava*I*, *Sst*I*	14, 21, 22
					pRL6[d]	11.3	Cm^R, Km/Nm^R	*Cla*I, *Nde*I, *Ava*II*, *Bgl*II*, *Bss*HII*, *Pst*I*, *Sph*I*	14, 21, 22
Nostoc sp. PCC 7118	pDU1	6.3	pBR322	4.36	pRL6[d]	11.3	Cm^R, Km/Nm^R	*Cla*I, *Nde*I, *Ava*II*, *Bgl*II*, *Bss*HII*, *Pst*I*, *Sph*I*	14, 21, 22
Nostoc sp. PCC 7120	pDU1	6.3	pBR322	4.36	pRL6[d]	11.3	Cm^R, Km/Nm^R	*Cla*I, *Nde*I, *Ava*II*, *Bgl*II*, *Bss*HII*, *Pst*I*, *Sph*I*	14, 21, 22

(*continued*)

TABLE IV (*continued*)

Organism	Cyanobacterial plasmid origin		*E. coli* plasmid origin		Shuttle vector or hybrid plasmid				
	Designation	Size (kb)	Designation	Size (kb)	Designation	Size (kb)	Selectable markers	Unique restriction sites[c]	Ref.[e]
					pRL11	10.7	Cm^R, Em^R	*Asu*II, *Cla*I, *Mst*I, *Nae*I, *Nar*I$^+$, *Nde*I, *Nhe*I, *Pss*I, *Bgl*I, *Not*I$^+$, *Nsi*I*, *Sst*I*, *Bal*I*, *Nco*I*	23
					pRL25	9.8	Km/Nm^R	*Asu*II, *Bam*HI, *Cla*I, *Eco*RI, *Nde*I, *Nhe*I$^+$, *Bgl*I, *Sca*I, *Not*I$^+$	24
					pRL25C (cosmid)	10.2	Km/Nm^R	*Asu*II, *Bam*HI, *Eco*RI, *Nde*I, *Nhe*I$^+$, *Ppu*MI, *Bgl*I, *Not*I$^+$, *Sca*I	24
					pRL163	10.3	Ap^R, Km/Nm^R	*Asu*II, *Cla*I, *Eco*RI, *Nde*I, *Nhe*I$^+$, *Bal*I*, *Bss*HII*, *Nco*I*, *Pst*I*, *Rsr*II*, *Sph*I*, *Not*I$^+$	22
					pRL488 (promoter probe)	12.3	Km/Nm^R, *lux*	*Asu*II, *Kpn*I, *Sal*I, *Sma*I, *Sst*I, *Bgl*I	22
Nostoc sp. PCC 73102	pDU1	6.3	pBR322	4.36	pRL6[d]	11.3	Cm^R, Km/Nm^R	*Cla*I, *Nde*I, *Bgl*II*, *Bss*HII*, *Pst*I*, *Ava*II*, *Sph*I*	14, 21, 22

Nostoc sp. PCC 7524	pDU1	6.3	pBR322	4.36	pVW1[a]	10.6	Ap^R	Not mentioned	14
					pVW1C (pVW1 derivative containing the Cm^R gene of pBR328)[a]	11.6	Ap^R, Cm^R	Not mentioned	14
					pRL1 (pVW1C deleted of *Ava*I, *Ava*II, *Avr*II sites)	9.2	Cm^R	*Bal*I*, *Nco*I*, *Nae*I, *Nde*I, *Cla*I, *Bcl*I, *Bgl*I	14
					pRL5 (pRL1 derivative containing the Sm^R gene from R300B)	11.1	Cm^R, Sm^R	*Ava*I*, *Sst*I*, *Nde*I, *Cla*I, *Asu*II, *Nar*I$^+$	14, 22
					pRL6 (pRL1 derivative containing the Km/Nm^R gene from ColE1::Tn5)	11.3	Cm^R, Km/Nm^R	*Ava*II*, *Sph*I*, *Bgl*II*, *Nde*I, *Cla*I, *Pst*I*, *Bss*HII*	14, 22
					pRL8 (pRL1 derivative containing the Em^R gene from pE194)	11.5	Cm^R, Em^R	*Nco*I*, *Bal*I*, *Nae*I, *Ava*II*, *Sst*I*, *Cla*I, *Bgl*I	14
Nostoc sp. CCAP1453/17	pDU1	6.3	pBR322	4.36	pRL6[d]	11.3	Cm^R, Km/Nm^R	*Ava*II*, *Sph*I*, *Bgl*II*, *Pst*I*, *Bss*HII*, *Nde*I, *Cla*I	14, 21, 22
Nostoc sp. M-131	pDU1	6.3	pBR322	4.36	pRL5[d]	11.1	Cm^R, Sm^R	*Asu*II, *Cla*I, *Nde*I, *Nar*I$^+$, *Ava*I*, *Sst*I*	14, 21, 22
					pRL6[d]	11.3	Cm^R, Km/Nm^R	*Ava*II*, *Sph*I*, *Bgl*II*, *Pst*I*, *Bss*HII*, *Nde*I, *Cla*I	14, 21, 22

(*continued*)

TABLE IV (*continued*)

Organism	Cyanobacterial plasmid origin		*E. coli* plasmid origin		Shuttle vector or hybrid plasmid				
	Designation	Size (kb)	Designation	Size (kb)	Designation	Size (kb)	Selectable markers	Unique restriction sites[c]	Ref.[e]
					pRL11	10.7	Cm^R, Em^R	*Asu*II, *Cla*I, *Mst*I, *Nae*I, *Nar*I⁺, *Nde*I, *Nhe*I, *Pss*I, *Nsi*I*, *Sst*I*, *Bal*I*, *Nco*I*, *Bgl*I, *Not*I⁺	23

[a] Hybrid plasmids not shown to behave as stable shuttle vectors.

[b] Plasmids with the insert in different orientations.

[c] (*) Unique restriction sites located within antibiotic resistance markers. (+) Two sites are close together and are still useful for cloning.

[d] See *Nostoc* PCC7524 for more details on the construction.

[e] References to Table IV: (1) K. Shinozaki, N. Tomioka, C. Yamada, and M. Sugiura, *Gene* **19,** 221 (1982); (2) D. Friedberg and J. Seijffers, *Gene* **22,** 267 (1983); (3) J. S. Buzby, R. D. Porter, and S. E. Stevens, Jr., *J. Bacteriol.* **154,** 1446 (1983); (4) J. S. Buzby, R. D. Porter, and S. E. Stevens, Jr., *Science* **230,** 805 (1985); (5) C. J. Kuhlemeier, W. E. Borrias, C. A. M. J. J. van den Hondel, and G. A. van Arkel, *Mol. Gen. Genet.* **184,** 249 (1981); (6) N. Tandeau de Marsac, W. E. Borrias, C. J. Kuhlemeier, A.-M. Castets, G. A. van Arkel, and C. A. M. J. J. van den Hondel, *Gene* **20,** 111 (1982); (7) C. J. Kuhlemeier, V. J. P. Teeuwsen, M. J. T. Janssen, and G. A. van Arkel, *Gene* **31,** 109 (1984); (8) C. J. Kuhlemeier, A. A. M. Thomas, A. van der Ende, R. W. van Leen, W. E. Borrias, C. A. M. J. J. van den Hondel, and G. A. van Arkel, *Plasmid* **10,** 156 (1983); (9) L. A. Sherman and P. van de Putte, *J. Bacteriol.* **150,** 410 (1982); (10) S. S. Golden and L. A. Sherman, *J. Bacteriol.* **155,** 966 (1983); (11) S. Gendel, N. Straus, D. Pulleyblank, and J. Williams, *J. Bacteriol.* **156,** 148 (1983); (12) S. Gendel, N. Straus, D. Pulleyblank, and J. Williams, *FEMS Microbiol. Lett.* **19,** 291 (1983); (13) R. H. Lau and N. A. Straus, *FEMS Lett.* **27,** 253 (1985); (14) C. P. Wolk, A. Vonshak, P. Kehoe, and J. Elhai, *Proc. Natl. Acad. Sci. U.S.A.* **81,** 1561 (1984); (15) F. Chauvat, L. de Vries, A. van der Ende, and G. A. van Arkel, *Mol. Gen. Genet.* **204,** 185 (1986); (16) S. Shestakov, I. Elanskaya, and M. Bibikova, *Abstr. Int. Symp. Photosynth. Prokaryotes, 5th,* p. 109 (1985); (17) G. R. Lambert, J. G. Scott, and N. G. Carr, *FEMS Microbiol. Lett.* **21,** 225(1984); (18) J. Cobley, *Abstr. Int. Symp. Photosynth. Prokaryotes, 5th,* p. 105 (1985); (19) J. Cobley, personal communication; (20) J. Cobley, J. R. Seludo, E. Zerweck, and H. Jaeger, *Abstr. Mol. Biol. Cyanobacteria Workshop, St. Louis* (1987); (21) E. Flores and C. P. Wolk, *J. Bacteriol.* **162,** 1339 (1985); (22) J. Elhai and C. P. Wolk, personal communication; (23) C. P. Wolk, personal communication; (24) C. P. Wolk, Y. Cai, L. Cardemil, E. Flores, B. Hohn, M. Murry, G. Schmetterer, B. Schrautemeier, and R. Wilson, *J. Bacteriol.* **170,** 1239 (1988).

TABLE V
CLONED CYANOBACTERIAL GENES

Organism	Cloned gene[a]	Gene product	Isolation procedure	Sequence	Ref.[b]
Section I					
Synechococcus sp. PCC 6301	*apcA*	Allophycocyanin α subunit	Hybridization with a *Cyanophora paradoxa* DNA probe	Yes	1
	apcB	Allophycocyanin β subunit	Hybridization with a *Cyanophora paradoxa* DNA probe	Yes	1
	apcC	$L_C^{7.8}$ linker polypeptide (phycobilisome core)	Found downstream from *apcB* by sequencing	Yes	1
	atpA	α subunit of ATPase F_1	Hybridization with a *Rhodopseudomonas blastica* DNA probe	Yes	2
	atpB	β subunit of ATPase F_1	Hybridization with a *Rhodopseudomonas blastica* DNA probe	Yes	2
	atpC	γ subunit of ATPase F_1	Found downstream from *atpA* by sequencing	Yes	2
	atpD	δ subunit of ATPase F_1	Found downstream from *atpF* by sequencing	Yes	2
	atpE	ε subunit of ATPase F_1	Found downstream from *atpB* by sequencing	Yes	2
	atpF	*b* subunit of ATPase F_0	Found downstream from *atpG* by sequencing	Yes	2
	atpG	*b′* subunit of ATPase F_0	Found downstream from *atpH* by sequencing	Yes	2
	atpH	*c* subunit of ATPase F_0	Hybridization with a pea DNA probe	Yes	2

(*continued*)

TABLE V (*continued*)

Organism	Cloned gene[a]	Gene product	Isolation procedure	Sequence	Ref.[b]
	atpI	*a* subunit of ATPase F_0	Found upstream from *atpH* by sequencing	Yes	2
	cpcA	Phycocyanin α subunit	Hybridization with synthetic specific oligonucleotides	Yes	3, 4
	cpcB	Phycocyanin β subunit	Hybridization with synthetic specific oligonucleotides	Yes	3, 4
	Gene 1	Unknown function (sequence similarity with the first gene in the *E. coli atp* operon)	Found upstream from *atpI* by sequencing	Yes	2
	petF	Ferredoxin apoprotein	Found downstream from *atpC* by sequencing	Yes	2
	ppc	Phosphoenolpyruvate carboxylase	Complementation of *E. coli ppc* mutations	Yes	5, 6
	psaD, *psaE*	Subunits of photosystem I complex	Hybridization with synthetic oligonucleotides	Yes	7
	rbcL	Ribulose-bisphosphate carboxylase large subunit	Hybridization with a spinach DNA probe	Yes	8, 9
	rbcS	Ribulose-bisphosphate carboxylase small subunit	Hybridization with a pea cDNA probe	Yes	10
	rrnA	5S, 16S, 23S rRNAs, $tRNA^{Ile}$, and $tRNA^{Ala}$	rRNA–DNA hybridization	Yes	11–14
	rrnB	5S, 16S, 23S rRNAs, $tRNA^{Ile}$, and $tRNA^{Ala}$	rRNA–DNA hybridization	Yes	15–17
	URF-1	Unknown function	Found upstream from gene 1 by sequencing	Yes	2
	URF-2	Unknown function	Found downstream from *petF* by sequencing	Yes	2

	URF-3, URF-4	Unknown function	Found upstream from *atpB* by sequencing	Yes	2
	URF-5	Unknown function	Found downstream from *atpE* by sequencing	Yes	2
Synechococcus sp. PCC 7002	*apcA*	Allophycocyanin α subunit	Hybridization with a *Cyanophora paradoxa* DNA probe	Yes	18
	apcB	Allophycocyanin β subunit	Hybridization with a *Cyanophora paradoxa* DNA probe	Yes	18
	apcC	$L_C^{7.8}$ linker polypeptide (phycobilisome core)	Found downstream from *apcB* by sequencing	Yes	18
	apcD	Allophycocyanin B	Hybridization with a *Calothrix* PCC 7601 DNA probe	Yes	18
	argE	Acetylornithine deacetylase	Complementation of *E. coli argE* mutations	No	19
	cpcA	Phycocyanin α subunit	Hybridization with synthetic specific oligonucleotides	Yes	20, 21
	cpcB	Phycocyanin β subunit	Hybridization with synthetic specific oligonucleotides	Yes	20, 21
	cpcC	L_R^{33} linker polypeptide (phycobilisome rod)	Found downstream from *cpcA* by sequencing	Yes	22
	cpcD	$L_R^{9.7}$ linker polypeptide (phycobilisome rod)	Found downstream from *cpcC* by sequencing	Yes	22
	cpcE	Protein required for attachment of phycocyanobilin to phycocyanin α subunit	Found downstream from *cpcD* by sequencing	Yes	18
	cpcF	Protein required for phycocyanin accumulation	Found downstream from *cpcE* by sequencing	Yes	18
	leuB	Isopropylmalate dehydrogenase	Complementation of *E. coli leuB* mutations	No	19
	leuC	α-Isopropylmalate isomerase subunit	Complementation of *E. coli leuC* mutations	No	19

(*continued*)

TABLE V (*continued*)

Organism	Cloned gene[a]	Gene product	Isolation procedure	Sequence	Ref.[b]
	psaA	Chlorophyll *a*-binding apoprotein of photosystem I reaction center	Hybridization with maize DNA probes	Yes	23
	psaB	Chlorophyll *a*-binding apoprotein of photosystem I reaction center	Hybridization with maize DNA probes	Yes	23
	*psbA1**	D1 protein of photosystem II reaction center	Hybridization with a chloroplast DNA probe	Yes	24
	*psbA2**	D1 protein of photosystem II reaction center	Hybridization with a chloroplast DNA probe	No	24
	psbB	CP47 chlorophyll *a*-binding protein of photosystem II	Hybridization with a *Cyanophora paradoxa* DNA probe	Partial	25
	psbC	CP43 chlorophyll *a*-binding protein of photosystem II	Hybridization with a *Cyanophora paradoxa* DNA probe	Partial	25
	*psbD1**, *psbD2**	D2 protein of photosystem II reaction center	Hybridization with a *Cyanophora paradoxa* DNA probe	Yes	25
	psbE, *psbF*	Cytochrome *b*-559 apoproteins	Hybridization with *Cyanophora paradoxa* DNA probes	No	25
	recA	Homologous DNA recombinase	Hybridization with an *E. coli* DNA probe	Yes	26, 27
	rbcL	Ribulose-bisphosphate carboxylase large subunit	Hybridization with a *Cyanophora paradoxa* DNA probe	No	28

Synechococcus sp. PCC 7942	*cpcA*	Phycocyanin α subunit	Hybridization with synthetic oligonucleotides	Yes	29
	cpcB	Phycocyanin β subunit	Hybridization with synthetic oligonucleotides	Yes	29, 30
	*crt**	Carotenoid-binding protein	Antisera against carotenoid-associated thylakoid membrane protein	No	31
	fus	EF-G elongation factor	Hybridization with an *E. coli* DNA probe	No	32
	metF	5,10-Methylenetetrahydrofolate reductase	Use of Tn*901*-induced mutation	Yes	33, 34
	narB	Involved in nitrate reduction	Use of Tn*901*-induced mutation	No	35
	narA	Involved in nitrate reduction	Complementation of *Synechococcus* PCC 7942 *nar* mutations	No	35
	narC	Involved in nitrate reduction	Complementation of *Synechococcus* PCC 7942 *nar* mutations	No	36
	petF1	Ferredoxin I apoprotein	Hybridization with synthetic oligonucleotides	Yes	37
			Hybridization with an *Anabaena* PCC 7937 DNA probe	Yes	38
	*psbA1**, *psbA2**, *psbA3**	D1 protein of photosystem II reaction center	Hybridization with a spinach chloroplast DNA probe	Yes	39
	psbC2	Chlorophyll *a*-binding protein of photosystem II (induced during Fe starvation)	Found upstream from *petG* by sequencing	Yes	40
	*psbD1**, *psbD2**	D2 protein of photosystem II reaction center	Heterologous hybridization	Yes	41
	*petG**	Flavodoxin apoprotein	Hybridization with synthetic oligonucleotides	Yes	42

(*continued*)

TABLE V (*continued*)

Organism	Cloned gene[a]	Gene product	Isolation procedure	Sequence	Ref.[b]
	*petI**	Cytochrome *c*-553 apoprotein	Hybridization with synthetic oligonucleotides	Yes	43
	recA	Homologous DNA recombinase	Hybridization with an *Anabaena* PCC 7937 DNA probe	No	44
	sodB	Iron superoxide dismutase	Hybridization with an *E. coli* DNA probe	Yes	45
	trxM	Thioredoxin *m* apoprotein	Hybridization with synthetic oligonucleotides	Yes	46
	tuf	EF-Tu elongation factor	Hybridization with an *E. coli* DNA probe	No	32
	woxA	33-kDa Mn-stabilizing protein of photosystem II	Screening by antisera	Yes	47
Synechocystis sp. PCC 6701	*cpcA*	Phycocyanin α subunit	Hybridization with a *Calothrix* PCC 7601 DNA probe	Partial	48
	cpcB	Phycocyanin β subunit	Hybridization with a *Calothrix* PCC 7601 DNA probe	Partial	48
Synechocystis sp. PCC 6803	*psbA1**	D1 protein of photosystem II reaction center	Hybridization with an *Amaranthus hybridus* DNA probe	Yes	49
	*psbA2**	D1 protein of photosystem II reaction center	Hybridization with an *Amaranthus hybridus* DNA probe	No	50
	*psbA3**	D1 protein of photosystem II reaction center	Hybridization with a *psbA1* DNA probe	No	51
	psbB	CP47 chlorophyll *a*-binding protein of photosystem II	Hybridization with a *Cyanophora paradoxa* DNA probe	Yes	52
	psbC	CP43 chlorophyll *a*-binding protein of photosystem II	Hybridization with a plastid DNA probe	Yes	50, 53

	*psbD1**, *psbD2**	D2 protein of photosystem II reaction center	Hybridization with a *Chlamydomonas* DNA probe	Yes	54
	psbE, *psbF*	Cytochrome *b*-559 apoproteins	Hybridization with a spinach chloroplast DNA probe	Yes	55
	psbI, *psbJ*	Cytochrome *b*-559 apoproteins	Found downstream from *psbF* by sequencing	Yes	56
	woxA	33-kDa Mn-stabilizing protein of photosystem II	Antisera against spinach 33-kDa polypeptide	Yes	57
Section III					
Pseudanabaena sp. PCC 7409	*apcA*	Allophycocyanin α subunit	Hybridization with a *Cyanophora paradoxa* DNA probe	No	58
	apcB	Allophycocyanin β subunit	Hybridization with a *Cyanophora paradoxa* DNA probe	No	58
	*cpcA1**	Phycocyanin-1 α subunit present in both red and green light	Hybridization with a *Synechococcus* PCC 7002 DNA probe	No	59
	*cpcB1**	Phycocyanin-1 β subunit present in both red and green light	Hybridization with a *Synechococcus* PCC 7002 DNA probe	No	59
	*cpcA2**	Phycocyanin-2 α subunit present in red light only	Hybridization with a *Synechococcus* PCC 7002 DNA probe	No	59
	*cpcB2**	Phycocyanin-2 β subunit present in red light only	Hybridization with a *Synechococcus* PCC 7002 DNA probe	No	59
	cpeA	Phycoerythrin α subunit present in green light only	Hybridization with a *Synechococcus* PCC 7002 DNA probe	Yes	59
	cpeB	Phycoerythrin β subunit present in green light only	Hybridization with a *Synechococcus* PCC 7002 DNA probe	Yes	59
	nifD	Mo–Fe protein (nitrogenase) α subunit	Hybridization with a *Klebsiella pneumoniae* DNA probe	No	60
	nifH	Fe protein (nitrogenase reductase)	Hybridization with a *K. pneumoniae* DNA probe	No	60
	nifK	Mo–Fe protein (nitrogenase) β subunit	Hybridization with a *K. pneumoniae* DNA probe	No	60

(*continued*)

TABLE V (*continued*)

Organism	Cloned gene[a]	Gene product	Isolation procedure	Sequence	Ref.[b]
Spirulina platensis C1	*fus*	EF-G elongation factor	Hybridization with an *E. coli* DNA probe	No	61
	rbcL	Ribulose-bisphosphate carboxylase large subunit	Hybridization with a *Chlamydomonas* DNA probe	No	62
	rbcS	Ribulose bisphosphate carboxylase small subunit	Hybridization with a *Chlamydomonas* DNA probe	No	62
	tuf	EF-Tu elongation factors	Hybridization with an *E. coli* DNA probe	No	61
Section IV					
Anabaena sp. PCC 7937	*petF1*	Ferredoxin I apoprotein	Hybridization with a *Silene pratensis* DNA probe	Yes	63
	recA	Homologous DNA recombinase	Complementation of *E. coli recA* mutations	No	64
Anabaena variabilis	*ppc*	Phosphoenol pyruvate carboxylase	Complementation of *E. coli ppc* mutations	No	65
Calothrix sp. PCC 7601	*apcA1**	Allophycocyanin-1 α subunit	Hybridization with a *Synechococcus* PCC 6301 DNA probe	Yes	66, 67
			Hybridization with a *Cyanophora paradoxa* DNA probe	Partial	68
	apcA2	Allophycocyanin-2 α-type subunit	Hybridization with a *Synechococcus* PCC 6301 DNA probe	Yes	67
	*apcB1**	Allophycocyanin-1 β subunit	Found downstream from *apcA1* by sequencing	Yes	66, 67
			Found downstream from *apcA1* by sequencing	Partial	68
	apcC	$L_C^{7.8}$ linker polypeptide (phycobilisome core)	Found downstream from *apcB1* by sequencing	Yes	67, 68

apcD	Allophycocyanin B	Hybridization with a *Synechococcus* PCC 6301 *apcA* DNA probe	Yes	69
apcE	L_{CM}^{92} "anchor" polypeptide (phycobilisome core)	Found upstream from *apcA1* by sequencing	Partial	67
*cpcA1**	Phycocyanin-1 α subunit present in both red and green light	Found downstream from *cpcB1* by sequencing	Yes	66, 68, 70, 71
*cpcA2**	Phycocyanin-2 α subunit present in red light only	Found downstream from *cpcB2* by sequencing	Yes	66, 67, 68, 71, 72
cpcA3	Phycocyanin-3 α subunit present in both red and green light	Found downstream from *cpcB3* by sequencing	Yes	70
*cpcB1**	Phycocyanin-1 β subunit present in both red and green light	Hybridization with a *Synechococcus* PCC 7002 DNA probe	Yes	66, 70
		Hybridization with a *Cyanophora paradoxa* DNA probe	Yes	68, 71
*cpcB2**	Phycocyanin-2 β subunit present in red light only	Hybridization with a *Synechococcus* PCC 7002 DNA probe	Yes	66, 67, 72
		Hybridization with a *Cyanophora paradoxa* DNA probe	Yes	68, 71
cpcB3	Phycocyanin-3 β subunit present in both red and green light	Hybridization with a *Synechococcus* PCC 7002 DNA probe	Yes	70
*cpcD2**	$L_{R}^{9.7}$ linker polypeptide (phycobilisome rod)	Found downstream from *cpcI2* by sequencing	Yes	73
*cpcE**	Unknown function (sequence similarity with *Synechococcus* PCC 7002 *cpcE*)	Found downstream from *cpcB1* by sequencing	Yes	70

(*continued*)

TABLE V (*continued*)

Organism	Cloned gene[a]	Gene product	Isolation procedure	Sequence	Ref.[b]
	cpcF	Protein required for phycocyanin accumulation	Found downstream from *cpcE* by sequencing	Yes	67
	*cpcH2**	L_R^{39} linker polypeptide (phycobilisome rod) present in red light only	Found upstream from *cpcI2* by sequencing	Yes	73
	*cpcI2**	$L_R^{37.5}$ linker polypeptide (phycobilisome rod) present in red light only	Hybridization with synthetic oligonucleotides	Yes	73
	*cpcH3**	Sequence similarity with rod linker polypeptides L_R^{3g}	Found downstream from *cpcA3* by sequencing	Yes	67
	*cpcI3**	Sequence similarity with rod linker polypeptides $L_R^{37.5}$	Found downstream from *cpcH3* by sequencing	Partial	67
	cpeA	Phycoerythrin α subunit present in green light only	Hybridization with synthetic oligonucleotides and a *Synechococcus* PCC 7002 *cpcB* DNA probe	Yes	74
	cpeB	Phycoerythrin β subunit present in green light only	Found upstream from *cpeA* by sequencing	Yes	74
	*gvpA1**, *gvpA2**	Structural protein of gas vesicles	Hybridization with synthetic oligonucleotides	Yes	75, 76
	gvpC	Unknown function	Found downstream from *gvpA2* by sequencing	Yes	76
	gvpD	Putative structural protein of gas vesicles	Hybridization with *gvpA1*	Yes	67
	IS*701*	Insertion element	Screening of spontaneous pigmentation mutants	Yes	67
	IS*702*	Insertion element	Screening of spontaneous pigmentation mutants	No	34

	orf16g	Unknown function	Found downstream from *thrB* by sequencing	Yes	77
	orfY	Unknown function	Found downstream from *cpeB* by sequencing	Yes	67
	orfZ	Unknown function	Found downstream from *orfY* by sequencing	Yes	67
	*psbA1**	D1 protein of photosystem II reaction centers	Hybridization with a maize plastid DNA probe	Yes	78
	*psbA2**	D1 protein of photosystem II reaction centers	Hybridization with a maize plastid DNA probe	No	78
	thrB	Homoserine kinase	Complementation of *E. coli thrB* mutations	Yes	77
Nostoc sp. PCC 7119	*trx*	Thioredoxin apoprotein	Hybridization with an *E. coli* DNA probe	Yes	79
Nostoc sp. PCC 7120	*als*	Acetolactate synthase	Hybridization with a yeast DNA probe	No	80
	atpA	α subunit of ATPase F_1	Sequence similarity with a *Chlamydomonas* DNA probe	Yes	81
	atpB	β subunit of ATPase F_1	Sequence similarity with a maize DNA probe	Yes	82
	atpC	γ subunit of ATPase F_1	Found downstream from *atpA* by sequencing	Yes	81
	atpD	δ subunit of ATPase F_1	Found upstream from *atpA* by sequencing	Yes	81
	atpE	ε subunit of ATPase F_1	Found downstream from *atpB* by sequencing	Yes	82
	atpF	*b* subunit of ATPase F_0	Found upstream from *atpD* by sequencing	Yes	81
	atpG	*b′* subunit of ATPase F_0	Found upstream from *atpF* by sequencing	Yes	81
	atpH	*c* subunit of ATPase F_0	Found upstream from *atpG* by sequencing	Yes	81

(*continued*)

TABLE V (*continued*)

Organism	Cloned gene[a]	Gene product	Isolation procedure	Sequence	Ref.[b]
	atpI	*a* subunit of ATPase F_0	Found upstream from *atpH* by sequencing	Yes	81
	cpcA	Phycocyanin α subunit	Hybridization with a *Cyanophora paradoxa* DNA probe	Yes	83
	cpcB	Phycocyanin β subunit	Hybridization with a *Cyanophora paradoxa* DNA probe	Yes	83
	cpcC	L_R^{33} linker polypeptide (phycobilisome rod)	Found downstream from *cpcA* by sequencing	Yes	83
	cpcD	$L_R^{9.7}$ linker polypeptide (phycobilisome rod)	Found downstream from *cpcC* by sequencing	Yes	83
	cpcE	Unknown function (sequence similarity with *Calothrix* PCC 7601 and *Synechococcus* PCC 7002 *cpcE*)	Found downstream from *cpcD* by sequencing	Yes	83
	cpcF	Unknown function (sequence similarity with *Calothrix* PCC 7601 and *Synechococcus* PCC 7002 *cpcF*)	Found downstream from *cpcE* by sequencing	Yes	84
	cpcG	L_{CR}^{29} linker polypeptide (phycobilisome core–rod junction)	Found downstream from *cpcF* by sequencing	Yes	84
	Gene 1	Unknown function (sequence similarity with the first gene in the *E. coli atp* operon)	Found upstream from *atpI* by sequencing	Yes	81
	glnA	Glutamine synthetase	Hybridization with an *E. coli* DNA probe	Yes	85, 86

ISA*I*	Putative insertion element	Found linked to *psbA1* by sequencing	Yes	87
ISA*II*	Putative insertion element	Found by sequence similarity with ISA*I*	Yes	87
ISA*III*	Putative insertion element	Found by sequence similarity with ISA*I*	Yes	87
ISA*IV*	Putative insertion element	Found by sequence similarity with ISA*I*	Yes	87
nifB	Involved in FeMoco synthesis (cofactor of the Mo–Fe protein)	Found upstream from *nifS* by sequencing	Yes	88
nifD	Mo–Fe protein (nitrogenase) α subunit	Hybridization with a *K. pneumoniae* DNA probe	Yes	89, 90
*nifH1**	Fe protein (nitrogenase reductase)	Hybridization with a *K. pneumoniae* DNA probe	Yes	89, 91
*nifH2**	Fe protein (nitrogenase reductase)	Hybridization with a *K. pneumoniae* DNA probe	Yes	92
nifK	Mo–Fe protein (nitrogenase) β subunit	Hybridization with a *K. pneumoniae* DNA probe	Yes	89, 93
nifS	Involved in maturation of the Fe protein	Hybridization with a *K. pneumoniae* DNA probe	Yes	89, 94
orf	Unknown function	Found downstream from *nifS* by sequencing	Yes	88
petF1	Ferredoxin I apoprotein	Hybridization with synthetic oligonucleotides	Yes	95
petF	Ferredoxin apoprotein	Found downstream from *nifB* by sequencing	Yes	88
*psbA1**, *psbA2**, *psbA3**, *psbA4**	D1 protein of photosystem II reaction center	Hybridization with a spinach chloroplast DNA probe	Yes	96, 97
rbcL	Ribulose-bisphosphate carboxylase large subunit	Hybridization with maize and *Chlamydomonas* DNA probes	Yes	98

(*continued*)

TABLE V (*continued*)

Organism	Cloned gene[a]	Gene product	Isolation procedure	Sequence	Ref.[b]
	rbcS	Ribulose-bisphosphate carboxylase small subunit	Hybridization with a pea cDNA probe	Yes	99
	trx	Thioredoxin apoprotein	Hybridization with synthetic oligonucleotides	Yes	87
	xisA	Excisase mediating the DNA rearrangement between *nifHD* and *nifK*	Located between *nifD* and *nifK* by expression in *E. coli*	Yes	100
Nostoc sp. PCC 7906	*petA*	Cytochrome *f* apoprotein	Hybridization with a plastid DNA probe	Yes	101
	petB	Cytochrome b_6 apoprotein	Hybridization with a plastid DNA probe	Yes	101
	petC	Rieske Fe–S apoprotein	Hybridization with a spinach cDNA probe	Yes	101
	petD	Subunit IV apoprotein	Hybridization with a plastid DNA probe	Yes	101
Nostoc sp. PCC 8009	*apcA*	Allophycocyanin α subunit	Found downstream from *apcE* by sequencing	Yes	102
	apcB	Allophycocyanin β subunit	Found downstream from *apcA* by sequencing	Partial	102
	apcE	L_{CM}^{95} "anchor" polypeptide (phycobilisome core)	Antisera against *Nostoc* PCC 8009 95-kDa polypeptide	Partial	102
	psaA	Chlorophyll *a*-binding apoprotein of photosystem I reaction center	Hybridization with a *Synechococcus* PCC 7002 DNA probe	No	58
	psaB	Chlorophyll *a*-binding apoprotein of photosystem I reaction center	Hybridization with a *Synechococcus* PCC 7002 DNA probe	No	58

Nostoc sp. 7801	*leuB*	Isopropylmalate dehydrogenase	Complementation of *E. coli leuB* mutations	No	103
Section V					
Chlorogloeopsis sp. PCC 6912	*rbcL*	Ribulose-bisphosphate carboxylase large subunit	Hybridization with a *Synechococcus* PCC 6301 DNA probe	No	104
	rbcS	Ribulose-bisphosphate carboxylase small subunit	Hybridization with a *Synechococcus* PCC 6301 DNA probe	No	104

[a] (*) The published gene designation has been modified or created according to the nomenclature rules proposed in the Introduction.

[b] References to Table V: (1) J. Houmard, D. Mazel, C. Moguet, D. A. Bryant, and N. Tandeau de Marsac, *Mol. Gen. Genet.* **205,** 404 (1986); (2) A. L. Cozens and J. E. Walker, *J. Mol. Biol.* **194,** 359 (1987); (3) L. K. Lind, S. R. Kalla, A. Lönneborg, G. Oquist, and P. Gustafsson, *FEBS Lett.* **188,** 27 (1985); (4) S. R. Kalla, L. Lind, A. Lönneborg, J. Lidholm, G. Oquist, and P. Gustafsson, *Abstr. Int. Symp. Photosynth. Prokaryotes, 5th,* p. 341 (1985); (5) T. Kodaki, F. Katagiri, M. Asano, K. Izui, and H. Katsuki, *J. Biochem.* (*Tokyo*) **97,** 533 (1985); (6) F. Katagiri, T. Kodaki, N. Fujita, K. Izui, and H. Katsuki, *Gene* **38,** 265 (1985); (7) J. Omaha and A. N. Glazer, personal communication; (8) B. Y. Reichelt and S. F. Delaney, *DNA* **2,** 121 (1983); (9) K. Shinozaki, C. Yamada, N. Takahata, and M. Sugiura, *Proc. Natl. Acad. Sci. U.S.A.* **80,** 4050 (1983); (10) K. Shinozaki and M. Sugiura, *Nucleic Acids Res.* **11,** 6957 (1983); (11) N. Tomioka, K. Shinozaki, and M. Sugiura, *Mol. Gen. Genet.* **184,** 359 (1981); (12) N. Tomioka and M. Sugiura, *Mol. Gen. Genet.* **191,** 46 (1983); (13) M. Kumano, N. Tomioka, and M. Sugiura, *Gene* **24,** 219 (1983); (14) N. Tomioka and M. Sugiura, *Mol. Gen. Genet.* **193,** 427 (1984); (15) S. E. Williamson and W. F. Doolittle, *Nucleic Acids Res.* **11,** 225 (1983); (16) S. E. Douglas and W. F. Doolittle, *FEBS Lett.* **166,** 307 (1984); (17) S. E. Douglas and W. F. Doolittle, *Nucleic Acids Res.* **12,** 3373 (1984); (18) D. A. Bryant, *in* "Photosynthetic Light-Harvesting Systems: Structure and Function" (H. Scheer and S. Schneider, eds.), p. 217. de Gruyter, Berlin, 1988; (19) R. D. Porter, J. S. Buzby, A. Pilon, P. I. Fields, J. M. Dubbs, and S. E. Stevens, Jr., *Gene* **41,** 249 (1986); (20) R. de Lorimier, D. A. Bryant, R. D. Porter, W.-Y. Liu, E. Jay, and S. E. Stevens, Jr., *Proc. Natl. Acad. Sci. U.S.A.* **81,** 7946 (1984); (21) T. J. Pilot and J. L. Fox, *Proc. Natl. Acad. Sci. U.S.A.* **81,** 6983 (1984); (22) D. A. Bryant, R. de Lorimier, G. Guglielmi, V. L. Stirewalt, A. Cantrell, and S. E. Stevens, Jr., *Prog. Photosynth. Res.* **4,** 749 (1987); (23) A. Cantrell and D. A. Bryant, *Plant Mol. Biol.* **9,** 453 (1987); (24) J. C. Gingrich, J. S. Buzby, V. L. Stirewalt, and D. A. Bryant, *Photosynth. Res.* (in press); (25) J. C. Gingrich and D. A. Bryant, personal communication; (26) R. C. Murphy, D. A. Bryant, R. D. Porter, and N. Tandeau de Marsac, *J. Bacteriol.* **169,** 2739 (1987); (27) R. C. Murphy and D. A. Bryant, personal communication; (28) S. E. Stevens, Jr., personal communication; (29) R. H. Lau, G. Alvarado-Urbina, and P. C. K. Lau, *Gene,* **52,** 21 (1987); (30) P. C. K. Lau, J. A. Condie, G. Alvarado-Urbina, and R. H. Lau, *Nucleic Acids Res.* **15,** 2394 (1987); (31) K. J. Reddy, K. Masamoto, and L. A. Sherman, *Plant Physiol. Suppl.* **83,** 60 (1987); (32) F. S. Mickel and L. L. Spremulli, *J. Bacteriol.* **166,** 78 (1986); (33) N. Tandeau de Marsac, W. E. Borrias, C. J. Kuhlemeier, A.-M. Castets, G. A. van Arkel, and C. A. M. J. J. van den Hondel, *Gene* **20,** 111 (1982); (34) D. Co, D. Mazel, J. Houmard and N. Tandeau de Marsac, unpublished observations; (35) C. J. Kuhlemeier, T. Logtenberg, W. Stoorvogel, H. A. A. van

(*continued*)

Footnotes to TABLE V (*continued*)

Heugten, W. E. Borrias, and G. A. van Arkel, *J. Bacteriol.* **159,** 36 (1984); (36) C. J. Kuhlemeier, V. J. P. Teeuwsen, M. J. T. Janssen, and G. A. van Arkel, *Gene* **31,** 109 (1984); (37) M. E. Reith, D. E. Laudenbach, and N. A. Straus, *J. Bacteriol.* **168,** 1319 (1986); (38) J. van der Plas, R. P. de Groot, M. R. Woortman, P. J. Weisbeek, and G. A. van Arkel, *Nucleic Acids Res.* **14,** 7804 (1986); (39) S. S. Golden, J. Brusslan, and R. Haselkorn, *EMBO J.* **5,** 2789 (1986); (40) D. E. Laudenbach and N. A. Straus, personal communication; (41) S. S. Golden, personal communication; (42) D. E. Laudenbach, M. E. Reith, and N. A. Straus, *J. Bacteriol.* **170,** 258 (1988); (43) C. McDowell, D. E. Laudenbach, and N. A. Straus, *Abstr. Mol. Biol. Cyanobacteria Workshop, St. Louis* (1987); (44) G. Owttrim and J. R. Coleman, *Abstr. Mol. Biol. Cyanobacteria Workshop, St. Louis* (1987); (45) D. E. Laudenbach, C. J. Trick, and N. A. Straus, personal communication; (46) E. Muller and B. B. Buchanan, *Abstr. Mol. Biol. Cyanobacteria Workshop, St. Louis* (1987); (47) T. Kuwabara, K. J. Reddy, and L. A. Sherman, *Proc. Natl. Acad. Sci. U.S.A.* **84,** 8230 (1987); (48) L. K. Anderson and A. R. Grossman, *Abst. Mol. Biol. Cyanobacteria Workshop, St. Louis* (1987); (49) H. D. Osiewacz and L. McIntosh, *Nucleic Acids Res.* **15,** 10585 (1987); (50) J. G. K. Williams, personal communication; (51) C. Jansson, R. J. Debus, H. D. Osiewacz, M. Gurevitz, and L. McIntosh, *Plant Physiol.* **85,** 1021 (1987); (52) W. F. J. Vermaas, J. G. K. Williams, and C. J. Arntzen, *Plant Mol. Biol.* **8,** 317 (1987); (53) V. A. Dzelzkalns and L. Bogorad, *EMBO J.* **7,** 333 (1988); (54) J. G. K. Williams and D. A. Chisholm, *Prog. Photosynth. Res.* **4,** 809 (1987); (55) H. B. Pakrasi, J. G. K. Williams, and C. J. Arntzen, *EMBO J.* **7,** 325 (1988); (56) H. B. Pakrasi, personal communication; (57) J. B. Philbrick and B. A. Zilinskas, *Abstr. Mol. Biol. Cyanobacteria Workshop, St. Louis* (1987); (58) D. A. Bryant, personal communication; (59) J. M. Dubbs and D. A. Bryant, *Prog. Photosynth. Res.* **4,** 765 (1987); (60) R. K. Singh, S. E. Stevens, Jr., and D. A. Bryant, *FEMS Microbiol. Lett.* **48,** 53 (1987); (61) O. Tiboni, and O. Ciferri, *Abstr. Int. Symp. Photosynth. Prokaryotes, 5th,* p. 361 (1985); (62) O. Tiboni, G. di Pasquale, and O. Ciferri, *Biochim. Biophys. Acta* **783,** 258 (1984); (63) J. van der Plas, R. P. de Groot, P. J. Weisbeek, and G. A. van Arkel, *Nucleic Acids Res.* **14,** 7803 (1986); (64) G. Owttrim and J. R. Coleman, *J. Bacteriol.* **169,** 1824 (1987); (65) T. R. Harrington, B. R. Glick, and N. W. Lem, *Gene* **45,** 113 (1986); (66) N. Tandeau de Marsac and J. Houmard, *in* "The Cyanobacteria" (P. Fay and C. van Baalen, eds.), p. 251. Elsevier,

Amsterdam, 1987; (67) N. Tandeau de Marsac, D. Mazel, T. Damerval, G. Guglielmi, V. Capuano, and J. Houmard, *Photosynth. Res.* (in press) (1988); (68) P. B. Conley, P. G. Lemaux, T. L. Lomax, and A. R. Grossman, *Proc. Natl. Acad. Sci. U.S.A.* **83,** 3924 (1986); (69) J. Houmard, V. Capuano, T. Coursin, and N. Tandeau de Marsac, *Mol. Microbiol.* **2,** 101 (1988); (70) D. Mazel, J. Houmard, and N. Tandeau de Marsac, *Mol. Gen. Genet.* **211,** 296 (1988); (71) P. B. Conley, P. G. Lemaux, and A. Grossman, *J. Mol. Biol.* **199,** 447 (1988); (72) V. Capuano, D. Mazel, N. Tandeau de Marsac, and J. Houmard, *Nucleic Acids Res.* **16,** 1626 (1988); (73) T. L. Lomax, P. B. Conley, J. Schilling, and A. R. Grossman, *J. Bacteriol.* **169,** 2675 (1987); (74) D. Mazel, G. Guglielmi, J. Houmard, W. Sidler, D. A. Bryant, and N. Tandeau de Marsac, *Nucleic Acids Res.* **14,** 8279 (1986); (75) N. Tandeau de Marsac, D. Mazel, D. A. Bryant, and J. Houmard, *Nucleic Acids Res.* **13,** 7223 (1985); (76) T. Damerval, J. Houmard, G. Guglielmi, K. Csiszàr, and N. Tandeau de Marsac, *Gene* **54,** 83 (1987); (77) C. Parsot and D. Mazel, *Mol. Microbiol.* **1,** 45 (1987); (78) B. Mulligan, N. Schultes, L. Chen, and L. Bogorad, *Proc. Natl. Acad. Sci. U.S.A.* **81,** 2693 (1984); (79) C.-J. Lim, F. K. Gleason, and J. A. Fuchs, *J. Bacteriol.* **168,** 1258 (1986); (80) B. J. Mazur and C.-F. Chui, personal communication; (81) D. F. McCarn and S. E. Curtis, personal communication; (82) S. E. Curtis, *J. Bacteriol.* **169,** 80 (1987); (83) W. R. Belknap and R. Haselkorn, *EMBO J.* **6,** 871 (1987); (84) W. R. Belknap and R. Haselkorn, personal communication; (85) R. Fisher, R. Tuli, and R. Haselkorn, *Proc. Natl. Acad. Sci. U.S.A.* **78,** 3393 (1981); (86) N. E. Tumer, S. J. Robinson, and R. Haselkorn, *Nature (London)* **306,** 337 (1983); (87) J. Alam and S. E. Curtis, personal communication; (88) M. E. Mulligan and R. Haselkorn, *Abstr. Mol. Biol. Cyanobacteria Workshop, St. Louis* (1987); (89) D. Rice, B. J. Mazur, and R. Haselkorn, *J. Biol. Chem.* **257,** 13157 (1982); (90) P. J. Lammers and R. Haselkorn, *Proc. Natl. Acad. Sci. U.S.A.* **80,** 4723 (1983); (91) M. Mevarech, D. Rice, and R. Haselkorn, *Proc. Natl. Acad. Sci. U.S.A.* **77,** 6476 (1980); (92) S. J. Robinson and R. Haselkorn, *Abstr. Int. Symp. Photosynth. Prokaryotes, 5th,* p. 346 (1985); (93) B. J. Mazur and C.-F. Chui, *Proc. Natl. Acad. Sci. U.S.A.* **79,** 6782 (1982); (94) R. Haselkorn, *Annu. Rev. Microbiol.* **40,** 525 (1986); (95) J. Alam, R. A. Whitaker, D. W. Krogman, and S. E. Curtis, *J. Bacteriol.* **168,** 1265 (1986); (96) S. E. Curtis and R. Haselkorn, *Plant Mol. Biol.* **3,** 249 (1984); (97) J. Vrba and S. E. Curtis, personal communication; (98) S. E. Curtis and R. Haselkorn, *Proc. Natl. Acad. Sci. U.S.A.* **80,** 1835 (1983); (99) S. A. Nierzwicki-Bauer, S. E. Curtis, and R. Haselkorn, *Proc. Natl. Acad. Sci. U.S.A.* **81,** 5961 (1984); (100) P. J. Lammers, J. W. Golden, and R. Haselkorn, *Cell* **44,** 905 (1986); (101) T. Kallas, S. Spiller, and R. Malkin, *Abstr. Mol. Biol. Cyanobacteria Workshop, St. Louis* (1987); (102) B. A. Zilinskas, K. H. Chen, and D. A. Howell, *Plant Physiol. Suppl.* **83,** 60 (1987); (103) G. A. Cangelosi, C. M. Joseph, J. J. Rosen, and J. C. Meeks, *Arch Microbiol.* **145,** 315 (1986); (104) D. Vakeria, G. A. Codd, A. M. Hawthornthwaite, and W. D. P. Stewart, *Arch. Microbiol.* **145,** 228 (1986).

Addendum

Addendum to Article [22]

By MICHAEL HERDMAN and ROSMARIE RIPPKA

Recent observations[11a] seem to indicate that hormogonium differentiation might not be regulated by light quality per se, but may rather result from a transient metabolic change associated with the shift of light quality: hormogonium differentiation in *Calothrix* PCC 7601 is also observed at high frequency (90–100% of the cellular population)[11a] in white light 5–6 hr after transfer from medium BG-11,[1] containing $NaNO_3$ as nitrogen source, to the same medium devoid of combined nitrogen.[1] In contrast to *Nostoc muscorum* PCC 6719, for which extensive hormogonial development in response to nitrate deprivation was first described by Armstrong *et al.*,[11b] hormogonium differentiation in *Calothrix* PCC 7601 is not provoked by an osmotic change, since replacement of $NaNO_3$ by an equimolar concentration of NaCl does not inhibit the differentiation process.[11a] Experiments performed to determine the commitment time of differentiation revealed that a short period (2 hr) of deprivation of $NaNO_3$ is sufficient to trigger hormogonium formation, following readdition of this nitrogen source.[11a] These results suggest that the metabolic changes leading to hormogonium differentiation in *Calothrix* PCC 7601 are not associated with prolonged (and therefore severe) nitrogen starvation, but may rather be the result of a more immediate (as yet unknown) effect exerted on, for example, the rate of photosynthesis, redox potential, or changes in the activities of enzymes involved in nitrogen metabolism. In support of this theory are some recent transcriptional studies[11c] which seem to indicate that glutamine synthetase might exert a regulatory function in the process of hormogonium differentiation in *Calothrix* PCC 7601 and *Calothrix* PCC 7504: high rates of transcription of the *gln*A gene (encoding glutamine synthetase) can be correlated with repression of hormogonium differentiation in both strains, whereas significantly lower levels of transcription are associated (at least transiently) with efficient hormogonium formation.[11c]

[11a] R. Rippka and M. Herdman, unpublished observations.
[11b] R. E. Armstrong, P. K. Hayes, and A. E. Walsby, *J. Gen. Microbiol.* **128,** 263 (1983).
[11c] T. Damerval, G. Guglielmi, and N. Tandeau de Marsac, unpublished observations.

In view of the complexity of the regulation of hormogonium differentiation, it is advisable to test for this property by one (or all) of the following procedures.

1. For strains which do not undergo chromatic adaptation, the culture is grown to late exponential phase, the cells are harvested, washed thoroughly, and diluted in fresh medium to a density equal to about 10% of that of the parental culture. Incubation conditions, which depend on the strain under examination and are not defined here, are not changed after dilution.
2. The following method[8] permits differentiation of 90–100% of the filaments of *Calothrix* PCC 7601 and may be applicable to other strains which contain phycoerythrin (although light intensities and temperature may need to be modified). The organism is grown at 25° in green light (about 10 $\mu E/m^2/sec$, 450–600 nm, λ_{max} 525 nm; see Ref. 12 for the spectral properties of appropriate filters) in medium BG-11 buffered with 10 m*M* $NaHCO_3$ and gassed with CO_2 (1% in air). When the culture is sufficiently dense (~100–200 Klett units, equivalent to 100–200 μg protein/ml) the cells are centrifuged, washed, resuspended in fresh medium to a Klett of 10–20 units, and placed under red light (550–700 nm,[11d] about 1.5–5 $\mu E/m^2/sec$). Hormogonia are produced relatively synchronously within 18 hr. The same protocol is applicable[8] for cells grown in white light (Osram-L Universal White, 10–50 $\mu E/m^2/sec$), diluted 8–10-fold and transferred to red light (1.5–5 $\mu E/m^2/sec$), the light energy supply per cell thus remaining fairly constant. Similar results are obtained if the cultures are gassed with air and the medium is buffered with 10 m*M* HEPES, pH 7.5, and is supplemented with glucose (0.5%, w/v), a utilizable carbon source for this strain.
3. Since hormogonium differentiation may be triggered by transient nitrogen starvation and/or osmotic changes (see above), the capacity for this property should be examined by the following method[11a] (modified as appropriate depending on the strain to be examined). A culture in mid- to late-exponential phase of growth in the presence of $NaNO_3$ is centrifuged and washed in the same medium without the nitrogen source. Samples of this culture are incubated in the same N-free medium under the same light regime as the parental culture. For *Calothrix* PCC 7601, grown to a Klett of 50–150 units, hormogonium frequency was similar irrespective of the culture density after transfer, dilution not being obligatory under

[11d] N. Tandeau de Marsac, *J. Bacteriol.* **130,** 82 (1977).

these conditions.[11a] To discriminate between the involvement of a transient lack of combined nitrogen or an osmotic effect, appropriate control experiments (replacing $NaNO_3$ by equiosmolar concentrations of, e.g., NaCl, KCl, or glucose[11b]) should be performed.

For all of the above methods it is advisable to carry out the centrifugation and washing at room temperature, since we[11a] have observed for *Calothrix* PCC 7601 that a transient cold shock (4–10°) leads to an increase in hormogonium frequency in white light even in the presence of $NaNO_3$.

Author Index

Numbers in parentheses are footnote reference numbers and indicate that an author's work is referred to although the name is not cited in the text.

A

B

C

D

E

F

G

H

I

J

K

L

M

N

O

P

Q

R

S

T

U

V

W

X

Y

Z

Subject Index

A

B

D

E

G

H

I

K

L

N

O

P

Q

R

S

T

U

V

Z